Main-Group Elements

			13 IIIA	14 IVA	15 VA	16 VIA	17 VIIA	18 VIIIA
								2 **He** 4.002602
			5 **B** 10.811	6 **C** 12.0107	7 **N** 14.00674	8 **O** 15.9994	9 **F** 18.9984032	10 **Ne** 20.1797
10	11 IB	12 IIB	13 **Al** 26.981538	14 **Si** 28.0855	15 **P** 30.973762	16 **S** 32.066	17 **Cl** 35.4527	18 **Ar** 39.948
28 **Ni** 58.6934	29 **Cu** 63.546	30 **Zn** 65.39	31 **Ga** 69.723	32 **Ge** 72.61	33 **As** 74.92160	34 **Se** 78.96	35 **Br** 79.904	36 **Kr** 83.80
46 **Pd** 106.42	47 **Ag** 107.8682	48 **Cd** 112.411	49 **In** 114.818	50 **Sn** 118.710	51 **Sb** 121.760	52 **Te** 127.60	53 **I** 126.90447	54 **Xe** 131.29
78 **Pt** 195.078	79 **Au** 196.96655	80 **Hg** 200.59	81 **Tl** 204.3833	82 **Pb** 207.2	83 **Bi** 208.98038	84 **Po** (209)	85 **At** (210)	86 **Rn** (222)
110 **Uun** (269)	111 **Uuu** (272)	112 **Uub** (277)		114 **Uuq** (289)				

Inner-Transition Metals

63 **Eu** 151.964	64 **Gd** 157.25	65 **Tb** 158.92534	66 **Dy** 162.50	67 **Ho** 164.93032	68 **Er** 167.26	69 **Tm** 168.93421	70 **Yb** 173.04	71 **Lu** 174.967
95 **Am** (243)	96 **Cm** (247)	97 **Bk** (247)	98 **Cf** (251)	99 **Es** (252)	100 **Fm** (257)	101 **Md** (258)	102 **No** (259)	103 **Lr** (262)

Essentials of
General
Chemistry

Essentials of General Chemistry

Darrell D. Ebbing
WAYNE STATE UNIVERSITY

Steven D. Gammon
UNIVERSITY OF IDAHO

Ronald O. Ragsdale
UNIVERSITY OF UTAH

HOUGHTON MIFFLIN COMPANY   BOSTON NEW YORK

Executive Editor: Richard Stratton
Associate Editor: Sara Wise
Senior Project Editor: Maria Morelli
Senior Production/Design Coordinator: Jill Haber
Manufacturing Manager: Florence Cadran
Executive Marketing Manager: Andy Fisher

Warning: This book contains descriptions of chemical reactions and photographs of experiments that are potentially dangerous and harmful if undertaken without proper supervision, equipment, and safety precautions. DO NOT attempt to perform these experiments relying solely on the information presented in this text.

Photo credits: A list of credits follows the Index.

Cover image: Bruce Dale/National Geographic Image Collection

Printed in the U.S.A.

Library of Congress Control Number: 2001133243

ISBN: 0-618-22328-2

23456789-RM-06 05 04 03

Contents in Brief

Basics of Chemistry

Atomic and Molecular Structure

States of Matter and Solutions

Chemical Reactions and Equilibrium

Nuclear Chemistry and Chemistry of the Elements

Contents

1   Chemistry and Measurement   1

2   Atoms, Molecules, and Ions   33

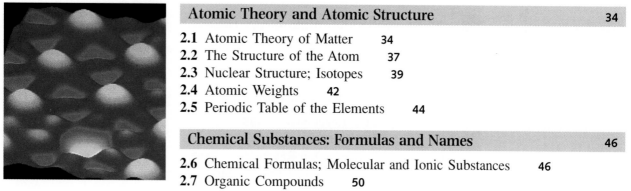

5   The Gaseous State   138

6   Thermochemistry   174

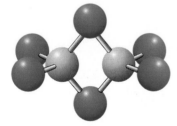

10   Molecular Geometry and Chemical Bonding Theory   292

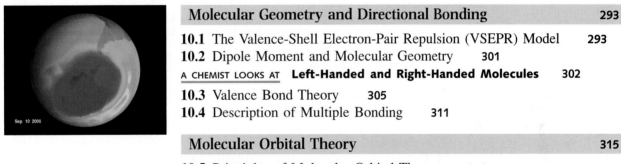

11   States of Matter; Liquids and Solids   328

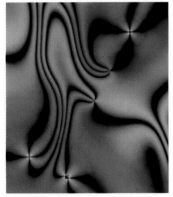

12 Solutions

13 Materials of Technology

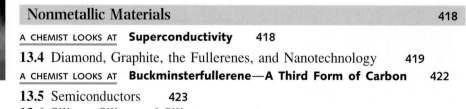

14 Rates of Reaction 435

15 Chemical Equilibrium 480

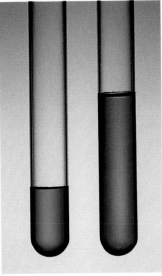

20   Electrochemistry

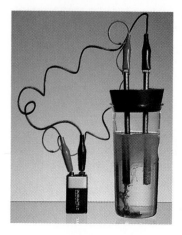
21   Nuclear Chemistry

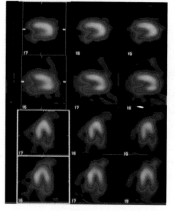

22 Chemistry of the Main-Group Elements

23 The Transition Elements and Coordination Compounds

24   Organic Chemistry   754

Essays

A Chemist Looks At

LIFE SCIENCE

MATERIALS

ENVIRONMENT

FRONTIERS

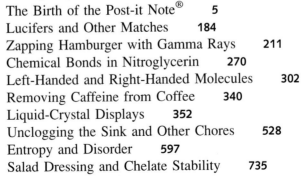

EVERYDAY LIFE

Instrumental Methods

Preface

To the Instructor

The Essentials of General Chemistry embodies the core and essence of Ebbing and Gammon's highly successful *General Chemistry*. There are several reasons for distilling a text as comprehensive as *General Chemistry* into an Essentials text. *The Essentials of General Chemistry* can be completely covered in two semesters without omitting any fundamental areas. It also facilitates the use of a common curriculum at large universities and colleges where there are many instructors and sections. This is a book for the students, and we must not forget that a textbook is the best place for a student to learn the concepts of chemistry. Students can read this shortened version in its entirety, and it is a first-rate book at a lower cost. *The Essentials of General Chemistry* offers the students a friendly, challenging, and stimulating environment for learning and developing the principles of chemistry.

Past users and prospective users will note that we have been able to maintain the spirit and vitality of the parent textbook, even though the text proper has been shortened by 325 pages. We have kept most of the essays that provided excitement and relevancy. The text still includes the lucid explanations that are a hallmark of all seven editions of *General Chemistry*.

What has been cut? The Exercises that usually followed the Examples have been eliminated. (Many of these Exercises have been incorporated into the Lecture Outline for the Essentials book, which is a help for both professors and students.) The Examples are immediately followed by references to similar problems to be found at the end of the chapter. The Media Activities have been moved to the *Technology Guide,* and we have reduced the number of questions and problems at the end of each chapter. For instance, in the chapter on the gaseous state, these questions and problems are reduced from 134 to 80 in number. This will leave most instructors plenty of exercises to use in subsequent years without repeating an assignment. Many times we have eliminated additional explanations and examples of principles that have already been illustrated. The treatment of molecular orbital theory has been largely restricted to simple diatomic molecules. The discussion of colloids has been greatly curtailed, and the section on optical activity in coordination compounds has been omitted. Many professors did not cover Chapter 25 on biological chemistry in *General Chemistry*, so we have essentially dropped that chapter. We have included the section on synthetic polymers in Chapter 24, "Organic Chemistry."

We invite instructors to examine this new Essentials of *General Chemistry* textbook carefully. It is a beautifully illustrated and produced book that presents chemistry in a stimulating, meaningful, and coherent way.

Clear, Lucid Explanations of Chemical Concepts   Just as in *General Chemistry*, we have placed the highest priority on writing clear explanations of chemical concepts. We have made every effort to relate abstract concepts to specific real-world events and have presented topics in a logical, yet flexible, order. By incorporating suggestions from instructors and students, we have refined the writing.

Concise Yet Comprehensive Treatment   This shorter, less expansive text includes all of the important topics expected in a general chemistry text, but with substantial savings in length.

Coherent Problem-Solving Approach We teach problem solving by coupling worked-out Examples with corresponding end-of-chapter Problems. This approach has been one of the cornerstones of the larger *General Chemistry* text. In the present book, we have also included Problem Strategies to underscore how one thinks through the solution to a problem.

We have incorporated two additional study aids that have met with an enthusiastic response in *General Chemistry*: Concept Checks and Conceptual Problems. We have written these problems to encourage students to think about the chemical concepts involved, rather than focusing narrowly on just obtaining a numerical result. Many of these problems include artwork to help students visualize the key concepts.

Chapter Essays Showcasing Chemistry as a Modern Science Our "A Chemist Looks At..." essays highlight applications of chemistry to medicine and health, to frontiers of science, to the environment, and to everyday life. Students immediately see how the material is related to their future careers, their world, and their lives. The essays are designed to engage students' interest, while at the same time showcasing the chemistry involved. Most of the topics concern contemporary research demonstrating how chemistry is a vibrant, constantly changing science that is acutely relevant to our modern world. The essay "Removing Caffeine from Coffee," for example, describes the removal of caffeine from coffee via supercritical carbon dioxide instead of environmentally problematic organic solvents. Among the other essays are "Nitric Oxide and Biological Signaling," which describes the importance of a simple chemical compound in human biology, and "Scanning Tunnel Microscopy," which discusses a tool that makes it possible to view chemistry at the molecular level.

An Enhanced Illustration Program with an Emphasis on Molecular Concepts Most of us are strongly visual in our learning. When we see something, we tend to remember it. Much of the art for this text comes from *General Chemistry*, where it has been constantly refined and improved. In particular, the art focuses on presenting chemistry at the molecular level. We start building the molecular "story" in Chapter 1, and by Chapter 2 we have developed the molecular view and have integrated it into the problem-solving apparatus as well as into the text discussions. We continue in the subsequent chapters to use the molecular view to enhance the presentation of chemical concepts. In Chapter 3, on stoichiometry, for example, a Concept Check asks the student to visualize the concept of limiting reactant in terms of molecular models.

A Chapter on Materials Chapter 13, "Materials of Technology," traces cutting-edge developments in this important area. For example, it includes information on nonmetals, discussing the fullerenes and nanotechnology, ceramics, and composites.

To the Student

You are about to embark on a fascinating intellectual adventure. You are surrounded by chemical materials and substances that play vital roles in your life. How can you gain an understanding and an appreciation of the chemistry that takes place?

Of course, any new subject has concepts and ways of looking at things that at first may seem strange. Chemistry, dealing as it does with all the materials of the universe, from living things to the objects of the heavens, is a complex field. But its central concept is simple: Everything is composed of a selection from about one hundred different kinds of atoms, chemically combined as molecules or other clumps of matter. As abstract as such objects might seem, chemists now have supermicroscopes that can "see" these atoms and molecules.

Having studied and taught chemistry for years, your authors are well aware of the problems that students encounter in their study of chemistry. In writing this book, we have constantly attempted to relate the chemical concepts to specific things in the real world. And we have included an abundance of instructional aids to help you thoroughly understand the ideas presented.

Features of This Book

Each individual learns in a different way. For that reason, we have incorporated a number of different learning aids into the text to help you master the subject. We hope that by becoming familiar with these features, which are listed below, you will be able to tailor-make a study program that meets your particular needs.

Chapter Theme   Each chapter begins with a theme—something specific that reveals the real-world relevance of the chapter topic. For example, Chapter 2 ("Atoms, Molecules, and Ions") opens with a discussion of sodium, chlorine, and sodium chloride. This chapter theme then leads naturally into a series of questions (such as "How do we explain the differences in the properties of different forms of matter?") that we answer later in the chapter.

Vocabulary   Chemists use words in a precise way, and it is important for your "chemistry vocabulary" to include all the terms you will need in order to read and communicate about the subject effectively. When an important new word is introduced in the text, we have flagged it by putting it in *boldface* type. The definition of that word will generally follow, in the same sentence, in *italic* type. All of these key words are collected at the end of the chapter in the list of Important Terms. They also appear, along with a few other words, in the Glossary at the end of the book. In addition, on the student web site, you can use the *Flashcard* program or print them out to practice learning these terms.

Concept Checks and Conceptual Problems   Many of the questions that you encounter daily start out "What do you think will happen if…?" or "Why is this choice the better one?" These questions are not asking for a numerical answer; rather, they are asking you to apply your conceptual knowledge to a problem. To answer this type of question, you need to think critically and to apply a variety of chemical concepts and ideas. Merely knowing a set of formulas and memorizing a series of steps to solve a problem will not help you obtain the answer. In many cases, your conceptual understanding of chemistry will be what you remember and apply later in life.

To help you master the concepts presented in this text, we have placed Concept Checks in the body of every chapter and a section of Conceptual Problems at the end of each chapter. Answers to the Concept Checks and the odd-numbered Conceptual Problems are provided at the end of the book. More detailed discussions of the Concept Checks are included on the student web site.

Problem-Solving Program   Problem solving is an important part of chemistry. By solving problems yourself, you become involved with the subject, and by being involved with the subject, you will learn it. But problem solving is like learning to swim or to play a musical instrument; it becomes easy only with practice. In chemistry, one concept builds upon another, and fact upon fact. The secret of problem solving in chemistry is to know what you learned earlier so well that when you approach a new problem, it is easy to see how to put the pieces together.

Recognizing the importance of problem solving in chemistry, we felt that the burden could be much reduced if we established and followed a consistent problem-solving program. Accordingly, we introduce each problem-solving skill with an Example, in which you are led through the reasoning involved in working out a particular type of problem. Many of these examples include a Problem Strategy that underscores the thinking process involved in solving the problem. At the end of the Example is a list of corresponding end-of-chapter Practice Problems. Solving them will provide immediate reinforcement! Try some of these to gain mastery of each problem-solving skill. Answers to odd-numbered problems appear at the end of the book.

Checklist for Review   When it comes to reviewing, students generally develop their own techniques. We have tried to accommodate these differences by presenting various review possibilities. For example, you may find that the list of Important Terms is useful not only because it is a list of new words but also because, as you look over those words, you see the structure of the chapter. As you mentally note this structure, try to recall the ideas associated with the words. You may also choose to use the *Flashcard* program on the web site to review these terms. Many chapters also introduce one or more mathematical equations to be used in problem solving. In the chapter, these are shaded in color; then, in the Checklist for Review, they are listed as Key Equations. The Summary of Facts and Concepts presents a verbal overview of the chapter. Study this, and as you go over each statement, try to flesh out the main points. Finally, we present a list of Operational Skills. This is a summary of the problem-solving skills introduced or developed in that chapter. Each entry in this extremely useful section tells you what information is needed, and what is to be solved for, in a given type of problem. Each operational skill also cites a numbered Example that illustrates the application of that problem-solving skill.

End-of-Chapter Questions and Problems   This section begins with Review Questions. These have been designed to test your understanding of the ideas introduced in the chapter. Generally, they can be answered by straightforward recall or by simple extension of the chapter material. Following these questions, we have added a section of more in-depth Conceptual Problems. After these problems you will find several sections of problems to help you hone your problem-solving skills. The Practice Problems are keyed to particular topics; the General Problems are not. The Cumulative-Skills Problems give you an opportunity to combine several skills, including skills that you developed in previous chapters.

Complete Instructional Package

This text is complemented by a complete package of print and electronic ancillaries.

For Students and Instructors

Student web site (http://college.hmco.com/, select "chemistry"). Password required. Includes Houghton Mifflin's ACE self-quizzing, interactive molecules, movies, flashcards of key terms and concepts, and links to other useful sites.

Student Technology Package   This package is available on request at an **additional charge** with all new texts. It is also available for sale separately. The package includes

> **Technology Guide**   A handy guide to the technology resources available to accompany *The Essentials of General Chemistry*. This booklet includes Media Activities, which guide students in using the technology resources provided to explore topics and concepts and to solve problems.
>
> *General Chemistry Interactive 5.0* **Student CD.**   A highly interactive CD-ROM that helps students visualize molecular behavior, explore important concepts, and practice working problems. The CD runs on both Macintosh and Windows platforms.
>
> **Passkey.**   A unique 13-digit code that gives students access to web site materials. The passkey is valid for 12 months from sign-up.
>
> **Smarthinking™ online tutoring.**   Passkey required. Web-based tutoring by qualified e-structors at Smarthinking.com.
>
> **Houghton Mifflin's Eduspace™ Online homework system.**   Passkey required. The online homework system provides students with questions that are keyed directly to the text content. Individual student progress is tracked and may be viewed, managed, and downloaded by instructors.

Experiments in General Chemistry,   R. A. D. Wentworth, Indiana University. Forty experiments parallel the material found in the textbook. Each lab exercise has a pre-lab assignment, background information, clear instructions for performing the experiment, and a convenient section for reporting results and observations. An instructor's resource manual is also available. Several experiments in this edition incorporate the use of computers and the Internet.

Student Solutions Manual,   David Bookin, Mount San Jacinto College; Darrell D. Ebbing, Wayne State University; and Steven D. Gammon, University of Idaho. This manual contains detailed solutions to all the odd-numbered problems in the text (in-chapter exercises, end-of-chapter problems, General Problems, and Cumulative-Skills Problems). It also contains answers to the Review Questions, Concept Checks, and Conceptual Problems.

Study Guide for *The Essentials of General Chemistry*,   Larry K. Krannich, University of Alabama at Birmingham. Each chapter of this study guide reinforces the students' understanding of concepts and operational skills presented in the text. It includes the following features for each chapter: a list of key terms and their definitions, a diagnostic test with answers, a summary of major concepts and operational skills, additional practice problems and their solutions, and a chapter post-test with answers.

Qualitative Analysis and Ionic Equilibrium,   George H. Schenk, Wayne State University. This laboratory manual presents a traditional qualitative analysis scheme that can be used with the text or can stand alone. It stresses the chemistry of metal ions and anions. Each exercise has a preliminary report and a final report for pre-lab and post-lab activities.

Workbook Lecture Outline,   Ron Ragsdale, University of Utah. Students can use the outline as a platform to take organized notes on in-class lectures or as a guide to help them process important concepts that they encounter in their own reading. Each outline is set up as a skeleton of the corresponding chapter with enough room to take notes. Students can fill in some of the outline before class so that they can listen more attentively during lecture. Instructors may find the outlines useful in organizing their own lectures as well.

For Instructors

Instructor's Resource Manual with Printed Test Bank, Ron Ragsdale, University of Utah. This manual offers information about chapter essays, suggestions for alternative sequencing of topics, short chapter descriptions, a master list of operational skills, correlation of Cumulative-Skills Problems with text topics, alternative examples for lectures, suggested lecture demonstrations, and a list of overhead transparencies. The printed test bank contains more than 2000 multiple-choice test questions organized by chapter.

HMClassPrep™ Package. This instructor CD-ROM has everything that instructors will need to develop their lectures. It includes PowerPoint slides of text figures, photos, concept checks, and lecture outlines, along with videos, animations, and an online *Instructor's Resource Manual*. The assets can be accessed from the CD-ROM by chapter or asset type. They can be customized to help instructors create lectures and in-class activities in a flexible manner.

HMTesting Computerized Testing System. This flexible test-editing program and comprehensive grade book make it easy to administer tests and to track students' progress. Instructors can administer tests via a network server or the Web. The HMTesting database contains a wealth of algorithmically generated questions and can produce multiple-choice or free-response tests. Instructors can customize questions on the basis of chapter, question format, and/or specific topics.

Transparencies. Numerous two-color and four-color transparencies of figures, tables, and photographs selected from the text are provided.

Instructor web site at college.hmco.com. This web site, created exclusively for instructors, allows access to all student web site resources plus downloadable PowerPoint slides, an on-line version of the *Instructor's Resource Manual,* additional instructor and classroom resources, animations, and relevant links.

Complete Solutions Manual, David Bookin, Mount San Jacinto College; Darrell D. Ebbing, Wayne State University; and Steven D. Gammon, University of Idaho. This complete solutions manual contains worked-out answers to *all* of the problems that appear in *The Essentials of General Chemistry*. This includes detailed, step-by-step solutions for all the Practice Problems, General Problems, and Cumulative-Skills Problems that appear at the end of the chapters. Also provided are answers to all the Review Questions, Concept Checks, and Conceptual Problems. This supplement is intended for the instructor's convenience and for those who want their students to have solutions to all problems.

Instructor's Resource Manual to the Lab Manual, R. A. D. Wentworth, Indiana University. This manual provides instructors with sample results for all pre- and post-lab activities in *Experiments in General Chemistry*.

Content for Course Management Software is available through *WebCT* and *Blackboard.com.* These two distributed learning systems enable you to create a virtual classroom without any knowledge of HTML. Features include assessment tools and a grade book, online file exchange between you and your students, and online syllabi and course descriptions. The customized cartridges for *General Chemistry* feature quizzes, study materials, and exercises related to the text and can be used in conjunction with *The Essentials of General Chemistry*.

Acknowledgments

We especially want to thank Richard Stratton, senior sponsoring editor, and Sara Wise, associate editor, for helping us to achieve our goals for this revision. Besides contributing ideas, insights, and inspiration, they assembled an outstanding group of people to participate in this project. We also want to thank Maria Morelli, senior project editor, and Jill Haber, senior production design coordinator, for their work in melding the final manuscript, line art, and photos into a finished book.

Darrell wishes to thank his wife, Jean, and children, Julie, Linda, and Russell, for their continued support and encouragement over many years of writing. Steve thanks his wife, Jodi, his two children, Katie and Andrew, and his parents, Judy and Dick, for their support and for helping him keep a perspective on the important things in life.

Ron thanks his wife Eileen for all her help, support, and encouragement in his writing endeavors. Eileen was a very positive influence for him as he carried out his role in the generation of *The Essentials of General Chemistry*.

In addition, this project benefited both from the reviews of the seventh edition of *General Chemistry* and from the reviews of the manuscript for *The Essentials of General Chemistry*. We are grateful to the following people for their contribution to both books.

Carey Bissonnette, *University of Waterloo*

Bob Belford, *West Virginia University*

Conrad Bergo, *East Stroudsburg University*

Aaron Brown, *Ventura College*

Tim Champion, *Johnson C. Smith University*

Paul Cohen, *College of New Jersey*

Lee Coombs, *California Polytechnic State University*

Jack Cummins, *Metro State College*

William M. Davis, *The University of Texas at Brownsville*

Earline F. Dikeman, *Kansas State University*

Evelyn S. Erenrich, *Rutgers University*

Greg Ferrence, *Illinois State University*

Renee Gittler, *Penn State Lehigh Valley*

Brian Glaser, *Black Hawk College*

David Grainger, *Colorado State University*

Christopher Grayce, *University of California at Irvine*

John M. Halpin, *New York University*

Carol Handy, *Portland Community College*

Daniel Haworth, *Marquette University*

Gregory Kent Haynes, *Morgan State University*

Robert Henry, *Tarrant County College*

Grant Holder, *Appalachian State University*

Andrew Jorgensen, *University of Toledo*

Kirk Kawagoe, *Fresno City College*

David Kort, *Mississippi State University*

Charles Kosky, *Borough of Manhattan Community College*

Jeffrey Kovac, *University of Tennessee at Knoxville*

Art Landis, *Emporia State University*

Richard Langley, *Stephen F. Austin State University*

Robert Mentore, *Ramapo College*

Joyce Miller, *San Jacinto College (South)*

Bob Morris, *Ball State University*

John Nash, *Purdue University*

Deborah Nycz, *Broward Community College*

Michael A. Quinlan, *University of Southern California*

Joe Rorke, *College of DuPage*

John Schaumloffel, *University of Massachusetts at Dartmouth*

Vernon Thielmann, *Southwest Missouri State University*

Jennifer Travers, *Oregon State University*

Gershon Vincow, *Syracuse University*

Donald Wirz, *University of California at Riverside*

Pete Witt, *Midlands Technical College*

Kim Woodrum, *University of Kentucky*

Michael A. Janusa, *Nicholls State University*

John A. Tossell, *University of Maryland*

Vahe M. Marganian, *Bridgewater State College*

Karl Sohlberg, *Drexel University*

Ernie Grisdale, *Lord Fairfax Community College*

Cindy DeForest Hauser, *Oregon State University*

David Speckhard, *Loras College*

Darrell Ebbing and Steven Gammon would like to acknowledge the key role of Ronald Ragsdale in the realization of a shorter version of *General Chemistry*. Ron has a vast teaching experience with shortened versions. His background includes years of teaching science majors and writing lecture outlines for a number of leading chemical textbooks. His experience is shown in the following roles he has played: chief reader for AP Chemistry, chair of the AP Examination Committee, chief examiner for the International Baccalaureate, and member of the United States Chemistry Olympiad Examination Committee. His expertise is reflected in his having received the National Catalyst Award, the Governor's Medal for Science and Technology, the University of Utah's highest teaching award, the Hatch Teaching Award, and the Utah Award from the Salt Lake Section and the Central Utah Section of the American Chemical Society.

Features of *Essentials of General Chemistry*

Essentials of General Chemistry focuses on the importance of teaching quantitative **problem solving**—a quality complemented by features that focus on key themes of **visualization** and **conceptual understanding**.

VISUALIZATION

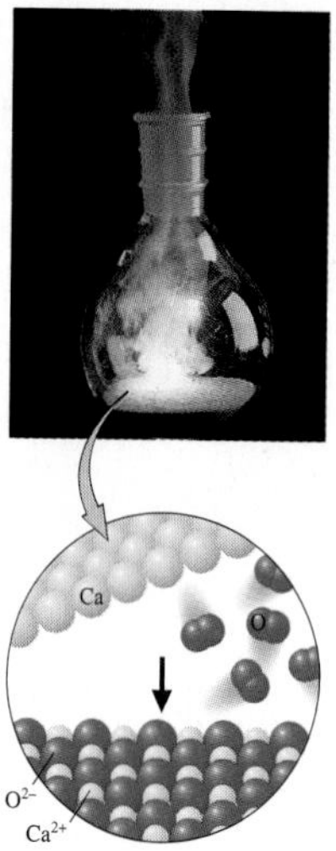

ion gains two electrons to form a copper atom in the metal. The net effect is that two electrons are transferred from each iron atom in the metal to each copper(II) ion.

The concept of *oxidation numbers* was developed as a simple way of keeping track of electrons in a reaction. Using oxidation numbers, you can determine whether or not electrons have been transferred from one atom to another. If electrons have been transferred, an oxidation–reduction reaction has occurred.

Oxidation Numbers

We define the **oxidation number** (or **oxidation state**) of an atom in a substance as *the actual charge of the atom if it exists as a monatomic ion, or a hypothetical charge assigned to the atom in the substance by simple rules.* An oxidation–reduction reaction is one in which one or more atoms change oxidation number, implying that there has been a transfer of electrons.

Consider the combustion of calcium metal in oxygen gas (Figure 4.11).

$$2Ca(s) + O_2(g) \longrightarrow 2CaO(s)$$

This is an oxidation–reduction reaction. To see this, you assign oxidation numbers to the atoms in the equation and then note that the atoms change oxidation number during the reaction.

Since the oxidation number of an atom in an element is always zero, Ca and O in O_2 have oxidation numbers of zero. Another rule follows from the definition of oxidation number: The oxidation number of an atom that exists in a substance as a monatomic ion equals the charge on that ion. So the oxidation number of Ca in CaO

Molecular blowups help students connect the macroscopic to the molecular level.

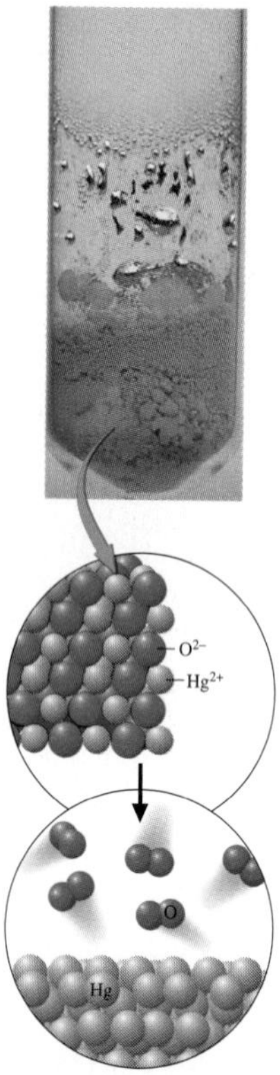

Figure 4.14
Decomposition reaction. The decomposition reaction of mercury(II) oxide into its elements, mercury and oxygen.

Figure 12.1
Immiscible and miscible liquids. *Left:* Water and methylene chloride are immiscible and form two layers. *Right:* Acetone and water are miscible liquids; that is, the two substances dissolve in each other in all proportions.

Liquid Solutions

Most liquid solutions are obtained by dissolving a gas, liquid, or solid in some liquid. Soda water, for example, consists of a solution of carbon dioxide in water. Acetone in water is an example of a liquid–liquid solution. (Immiscible and miscible liquids are shown in Figure 12.1.) Seawater contains both dissolved gases (from air) and solids (mostly sodium chloride).

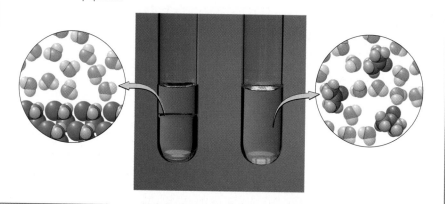

Hydrogen Bonding

CH$_3$F

CH$_3$OH

Figure 11.17
Fluoromethane and methanol. Space-filling molecular models.

It is interesting to compare fluoromethane, CH$_3$F, and methanol, CH$_3$OH (Figure 11.17). They have about the same molecular weight (34 for CH$_3$F and 32 for CH$_3$OH) and about the same dipole moment (1.81 D for CH$_3$F and 1.70 D for CH$_3$OH). You might expect these substances to have about the same intermolecular attractive forces and therefore about the same boiling points. In fact, the boiling points are quite different. Fluoromethane boils at $-78°C$ and is a gas under normal conditions. Methanol boils at 65°C and is normally a liquid. Why?

We have already seen that the properties of water and glycerol cannot be explained in terms of van der Waals forces alone. What water, glycerol, and methanol have in common is one or more —OH groups.

H—O—H

water

glycerol

methanol

> **Many chemical structures are depicted in multiple ways to help students make the leap from symbolic to visual representations.**

$$\Delta H = H_f - H_i = (U_f + PV_f) - (U_i + PV_i)$$

Collecting the internal-energy terms and the pressure-volume terms, you can rewrite this as

$$\Delta H = (U_f - U_i) + P(V_f - V_i) = \Delta U + P\Delta V$$

You write ΔU for $U_f - U_i$ and ΔV for $V_f - V_i$ and rearrange this as follows:

$$\Delta U = \Delta H - P\Delta V$$

The equation says that the internal energy of the system changes in two ways. It changes because energy leaves or enters the system as heat (ΔH), and it changes because the system increases or decreases in volume against the constant pressure of the atmosphere (which requires energy $-P\Delta V$).

Consider a specific reaction. When 2 mol of sodium and 2 mol of water react in a beaker, 1 mol of hydrogen forms and heat evolves.

Because hydrogen gas forms during the reaction, the volume of the system increases. To expand, the system must push back the atmosphere, and this requires energy equal to the pressure-volume work. It may be easier to see this pressure-volume work if you replace the constant pressure of the atmosphere by an equivalent pressure from a piston-and-weight assembly, as in Figure 6.7. When hydrogen is released during the reaction, it pushes upward on the piston and raises the weight. It requires energy to lift a weight upward in a gravitational field. If you calculate this pressure-volume work at 25°C and 1.00 atm pressure, you find that it is $-P\Delta V = -2.5$ kJ.

In the sodium-water reaction, the internal energy changes by -367.5 kJ because heat evolves and changes by -2.5 kJ because pressure-volume work is done. The total change of internal energy is

$$\Delta U = \Delta H - P\Delta V = -367.5 \text{ kJ} - 2.5 \text{ kJ} = -370.0 \text{ kJ}$$

Figure 6.7
Pressure-volume work. In this experiment, we replace the pressure of the atmosphere by a piston-and-weight assembly of equal pressure. As sodium metal reacts with water, the hydrogen gas evolved pushes the piston and weight upward (compare *before* and *after*). It requires work to raise the piston and weight upward in a gravitational field.

As you can see, ΔU does not differ a great deal from ΔH. This is the case in most reactions.

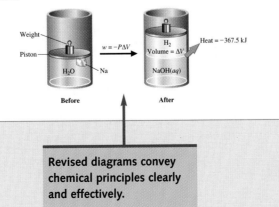

Weight
Piston
H$_2$O — Na
Before

$w = -P\Delta V$

H$_2$
Volume $= \Delta V$
NaOH(aq)
After

Heat $= -367.5$ kJ

> **Revised diagrams convey chemical principles clearly and effectively.**

CONCEPTUAL UNDERSTANDING

Concept Checks throughout the chapters challenge the student to learn the ideas underlying chemistry. An icon identifies conceptually focused material.

CONCEPT CHECK 14.4

Consider the following potential energy curves for two different reactions:

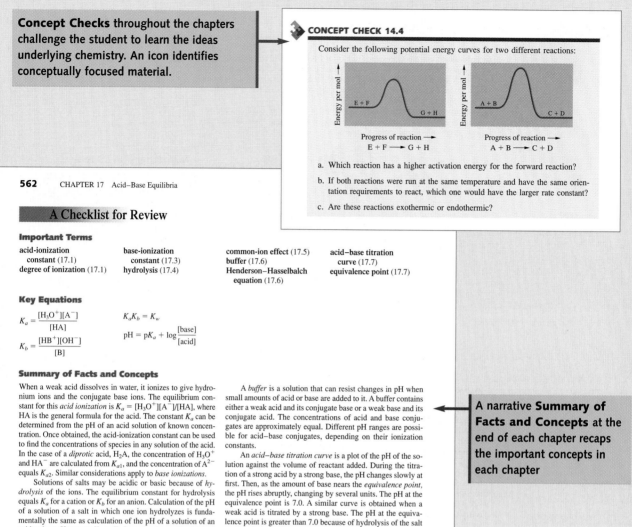

$$E + F \longrightarrow G + H \qquad A + B \longrightarrow C + D$$

a. Which reaction has a higher activation energy for the forward reaction?

b. If both reactions were run at the same temperature and have the same orientation requirements to react, which one would have the larger rate constant?

c. Are these reactions exothermic or endothermic?

---

562 CHAPTER 17 Acid–Base Equilibria

A Checklist for Review

Important Terms

acid-ionization constant (17.1)
degree of ionization (17.1)
base-ionization constant (17.3)
hydrolysis (17.4)
common-ion effect (17.5)
buffer (17.6)
Henderson–Hasselbalch equation (17.6)
acid–base titration curve (17.7)
equivalence point (17.7)

Key Equations

$$K_a = \frac{[H_3O^+][A^-]}{[HA]} \qquad K_a K_b = K_w$$

$$K_b = \frac{[HB^+][OH^-]}{[B]} \qquad pH = pK_a + \log\frac{[base]}{[acid]}$$

Summary of Facts and Concepts

When a weak acid dissolves in water, it ionizes to give hydronium ions and the conjugate base ions. The equilibrium constant for this *acid ionization* is $K_a = [H_3O^+][A^-]/[HA]$, where HA is the general formula for the acid. The constant K_a can be determined from the pH of an acid solution of known concentration. Once obtained, the acid-ionization constant can be used to find the concentrations of species in any solution of the acid. In the case of a *diprotic* acid, H_2A, the concentration of H_3O^+ and HA^- are calculated from K_{a1}, and the concentration of A^{2-} equals K_{a2}. Similar considerations apply to *base ionizations*.

Solutions of salts may be acidic or basic because of *hydrolysis* of the ions. The equilibrium constant for hydrolysis equals K_a for a cation or K_b for an anion. Calculation of the pH of a solution of a salt in which one ion hydrolyzes is fundamentally the same as calculation of the pH of a solution of an acid or base. However, K_a or K_b for an ion is usually obtained from the conjugate base or acid by applying the equation $K_a K_b = K_w$.

A *buffer* is a solution that can resist changes in pH when small amounts of acid or base are added to it. A buffer contains either a weak acid and its conjugate base or a weak base and its conjugate acid. The concentrations of acid and base conjugates are approximately equal. Different pH ranges are possible for acid–base conjugates, depending on their ionization constants.

An *acid–base titration curve* is a plot of the pH of the solution against the volume of reactant added. During the titration of a strong acid by a strong base, the pH changes slowly at first. Then, as the amount of base nears the *equivalence point*, the pH rises abruptly, changing by several units. The pH at the equivalence point is 7.0. A similar curve is obtained when a weak acid is titrated by a strong base. The pH at the equivalence point is greater than 7.0 because of hydrolysis of the salt produced. An indicator must be chosen that changes color within a pH range near the equivalence point.

A narrative **Summary of Facts and Concepts** at the end of each chapter recaps the important concepts in each chapter

---

132 CHAPTER 4 Chemical Reactions

Conceptual Problems

Conceptual Problems at the end of each chapter reinforce principles by asking nonquantitative questions about the material.

4.11 You come across a beaker that contains water, aqueous ammonium acetate, and a precipitate of calcium phosphate.
a. Write the balanced molecular equation for a reaction between two solutions containing ions that could produce this solution.
b. Write the complete ionic equation for the reaction in part a.
c. Write the net ionic equation for the reaction in part a.

4.12 Three acid samples are prepared for titration by 0.01 *M* NaOH:
1. Sample 1 is prepared by dissolving 0.01 mol of HCl in 50 mL of water.
2. Sample 2 is prepared by dissolving 0.01 mol of HCl in 60 mL of water.
3. Sample 3 is prepared by dissolving 0.01 mol of HCl in 70 mL of water.
a. Without performing a formal calculation, compare the concentrations of the three acid samples (rank them from highest to lowest).
b. When performing the titration, which sample, if any, will require the largest volume of the 0.01 *M* NaOH for neutralization?

4.13 Would you expect a precipitation reaction between an ionic compound that is an electrolyte and an ionic compound that is a nonelectrolyte? Justify your answer.

4.14 Equal quantities of the hypothetical strong acid HX, weak acid HA, and weak base BZ are each added to a separate beaker of water, producing the solutions depicted in the drawings. In the drawings, the relative amounts of each substance present in the solution (neglecting the water) are shown. Identify the acid or base that was used to produce each of the solutions (HX, HA, or BZ).

= H_3O^+
= OH^-

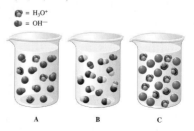

A B C

3.3 Mass Percentages from the Formula

We define the **mass percentage** of A as *the parts of A per hundred parts of the total, by mass*. That is,

$$\text{Mass \% } A = \frac{\text{mass of } A \text{ in the whole}}{\text{mass of the whole}} \times 100\%$$

The next example will provide practice with the concept of mass percentage. In this example we will start with a compound (formaldehyde, CH_2O) whose formula is given and obtain the percentage composition.

EXAMPLE 3.7
Calculating the Percentage Composition from the Formula

Formaldehyde, CH_2O, is a toxic gas with a pungent odor. Large quantities are consumed in the manufacture of plastics, and a water solution of the compound is used to preserve biological specimens. Calculate the mass percentages of the elements in formaldehyde.

◆ PROBLEM STRATEGY

Interpret the formula in molar terms and then convert moles to masses. Thus, 1 mol CH_2O has a mass of 30.0 g and contains 1 mol C (12.0 g), 2 mol H (2 × 1.01 g), and 1 mol O (16.0 g). Divide each mass of element by the molar mass, then multiply by 100.

SOLUTION

$$\% \text{ C} = \frac{12.0 \text{ g}}{30.0 \text{ g}} \times 100\% = \mathbf{40.0\%}$$

$$\% \text{ H} = \frac{2 \times 1.01 \text{ g}}{30.0 \text{ g}} \times 100\% = \mathbf{6.73\%}$$

You can calculate the percentage of O in the same way, but it can also be found by subtracting the percentages of C and H from 100%:

$$\% \text{ O} = 100\% - (40.0\% + 6.73\%) = \mathbf{53.3\%}$$

See Problems 3.35, 3.36, 3.37, and 3.38.

> In-text **Examples** guide students through the logic of solving a type of problem.

> **Problem Strategies** outline the thinking that underlies the numerical solution of the problem. The **Solution** then applies that thinking to a particular problem.

> A **Reference to End-of-Chapter Problems** directs students to other problems of this type.

---

368   CHAPTER 11   States of Matter; Liquids and Solids

Practice Problems

> End-of-chapter **Practice Problems** are keyed to particular topics by a heading. In addition, there are **General Problems** and **Cumulative-Skills Problems**, which require students to combine several skills.

Phase Transitions

11.21 Identify the phase transition occurring in each of the following.
a. The water level in an aquarium tank falls continuously (the tank has no leak).
b. A mixture of scrambled eggs placed in a cold vacuum chamber slowly turns to a powdery solid.
c. Chlorine gas is passed into a very cold test tube where it turns to a yellow liquid.
d. When carbon dioxide gas under pressure exits from a small orifice, it turns to a white "snow."

11.22 Identify the phase transition occurring in each of the following.
a. Mothballs slowly become smaller and eventually disappear.
b. Rubbing alcohol spilled on the palm of the hand feels cool as the volume of liquid decreases.
c. A black deposit of tungsten metal collects on the inside of a lightbulb whose filament is tungsten metal.
d. Raindrops hit a cold metal surface, which becomes covered with ice.

11.23 Use Figure 11.6 to estimate the boiling point of diethyl ether, $(C_2H_5)_2O$, under an external pressure of 350 mmHg.

11.24 Use Figure 11.6 to estimate the boiling point of carbon tetrachloride, CCl_4, under an external pressure of 350 mmHg.

of the water in the flask was raised to 83°C? The heat of vaporization of water at 100°C is 40.7 kJ/mol and the specific heat is 4.18 J/(g · °C).

11.29 Chloroform, $CHCl_3$, a volatile liquid, was once used as an anesthetic but has been replaced by safer compounds. Chloroform boils at 61.7°C and has a heat of vaporization of 31.4 kJ/mol. What is its vapor pressure at 33.0°C?

11.30 Methanol, CH_3OH, a colorless, volatile liquid, was formerly known as wood alcohol. It boils at 65.0°C and has a heat of vaporization of 37.4 kJ/mol. What is its vapor pressure at 22.0°C?

11.31 White phosphorus, P_4, is normally a white, waxy solid melting at 44°C to a colorless liquid. The liquid has a vapor pressure of 400.0 mmHg at 251.0°C and 760.0 mmHg at 280.0°C. What is the heat of vaporization of this substance?

11.32 Carbon disulfide, CS_2, is a volatile, flammable liquid. It has a vapor pressure of 400.0 mmHg at 28.0°C and 760.0 mmHg at 46.5°C. What is the heat of vaporization of this substance?

Phase Diagrams

11.33 Shown here is the phase diagram for compound Z. The triple point of Z is −5.1°C at 3.3 atm and the critical point is 51°C and 99.1 atm.

REAL-LIFE APPLICATIONS

A chapter on **Materials** includes cutting-edge developments and applications of particular value to students in science-related majors.

An STM image of the atomic structure of gallium arsenide, GaAs (Ga, blue; As, red).

Materials of Technology

Each chapter begins with a **chapter theme**, a real-world topic that raises a question that leads into the chapter topic.

In the summer of 2000, scientists at Bell Laboratories and Oxford University reported that they had constructed a motor out of DNA, the chemical material of the genetic code. This molecular-scale motor used DNA as both a structural material and as a fuel to run the motor. Earlier, Harvard University chemists had constructed molecular-scale tweezers out of carbon-atom tubes, showing that they could manipulate clusters of polystyrene molecules. These scientists' efforts represent some of the many ongoing investigations into "nanotechnology," in which one manipulates materials on a molecular scale to create useful devices.

Telecommunications is an example of an area in which the development of new materials has had immense impact, leading to rapid change in the technology. Initially, telecommunications was restricted to voice communication by telephone using copper wires to carry a message in the form of an electrical signal. Today, it is just

407

A CHEMIST LOOKS AT

Removing Caffeine from Coffee

EVERYDAY LIFE

The stimulant in coffee is caffeine, a bitter-tasting white substance with the formula $C_8H_{10}N_4O_2$. (Figure 11.13 shows the structural formula of caffeine.)

For those who like the taste of roasted coffee but don't want the caffeine, decaffeinated coffee is available. A German chemist, Ludwig Roselius, first made "decaf" coffee about 1900 by extracting the caffeine from green coffee beans with the solvent chloroform, $CHCl_3$. Today, though, most of the commercial decaffeinated coffee produced (Figure 11.14) uses *supercritical* carbon dioxide as the extracting fluid.

In a tank of carbon dioxide under pressure (for example, in a CO_2 fire extinguisher), the substance normally exists as the liquid in equilibrium with its gas phase. But we know from the previous text discussion that above 31°C, the two phases, gas and liquid, are replaced by a single fluid phase. So, on a hot summer day, the carbon dioxide in such a tank is above its critical temperature and pressure and exists as the supercritical fluid.

Supercritical carbon dioxide is a near-ideal solvent. Under normal conditions, carbon dioxide is not a very

Figure 11.14
Decaf coffee. Many brands are available.

good solvent for organic substances, but supercritical carbon dioxide readily dissolves many of these substances, including caffeine. It is nontoxic and nonflammable. Carbon dioxide does contribute to the greenhouse effect, but once used the gas can be recirculated for solvent use and not vented to the atmosphere.

Supercritical fluids have gained much attention recently because of the possibility of replacing toxic and environmentally less desirable solvents. For example, at the moment, the usual solvent used to dry-clean clothes is perchloroethylene, CCl_2CCl_2. Although nonflammable and less toxic than carbon tetrachloride, which was the solvent previously used, perchloroethylene is regulated as an air pollutant under the Clean Air Act. Some scientists have shown that you can dry-clean with supercritical carbon dioxide if you use a special detergent.

Substances other than carbon dioxide have also ... properties. For example, ... conditions dissolves ionic ... critical point (374°C, 217 ... solvent for nonpolar sub... nd carbon dioxide promise ... ronmentally unfriendly or-

A CHEMIST LOOKS AT

Taking Your Medicine

LIFE SCIENCE

The expressions "take your medicine" and "it's a bitter pill" are metaphors that refer to doing what's necessary to solve a difficult personal problem. To take your medicine might be unpleasant if it is bitter—and medicines do frequently taste bitter. Why?

A bitter taste appears to be a common feature of a base. It is a fact that many medicinal substances are nitrogen bases, substances that organic chemists call *amines*. Such substances are considered derivatives of ammonia, in which one or more hydrogen atoms have been substituted by carbon-containing groups. Here, for example, are the structures of some amines:

$$H-\overset{\overset{\displaystyle CH_3}{|}}{\underset{..}{N}}-H \qquad H_3C-\overset{\overset{\displaystyle CH_3}{|}}{\underset{..}{N}}-H \qquad H_3C-\overset{\overset{\displaystyle CH_3}{|}}{\underset{..}{N}}-CH_3$$

The nitrogen atom in an amine has a lone pair of electrons that can be donated to form a covalent bond. So, an amine is a Lewis base. But, like ammonia, an amine accepts a hydrogen ion from an acid to form an amine ion, so it is also a Brønsted–Lowry base.

Some of our most important medicinal drugs have originated from plants. Lewis and Clark took Peruvian bark, or cinchona bark, with them as a medicine on their 1804 expedition from the eastern United States to the Pacific Coast and back. The bitter essence of cinchona bark is quinine, an amine drug that has been used to combat malaria. Quinine is responsible for the bitter taste of tonic water, a carbonated beverage (Figure 16.4). Some other amines that come from plants are caffeine (from coffee, a stimulant), atropine (from the deadly nightshade, used to dilate the pupil of the eye for eye exams), and codeine (from the opium poppy, used as a painkiller).

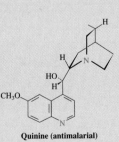

Figure 16.4
Tonic water. *Left:* This carbonated beverage contains quinine, which accounts for its bitter taste. *Right:* Structural formula of quinine.

Quinine (antimalarial)

A suite of robust technology tools is available to augment teaching and learning.

The **Ebbing/Gammon web sites** for students and instructors enrich, enhance, and facilitate teaching and learning by taking the text's three main themes—problem solving, understanding concepts, and visualization—to an interactive level.

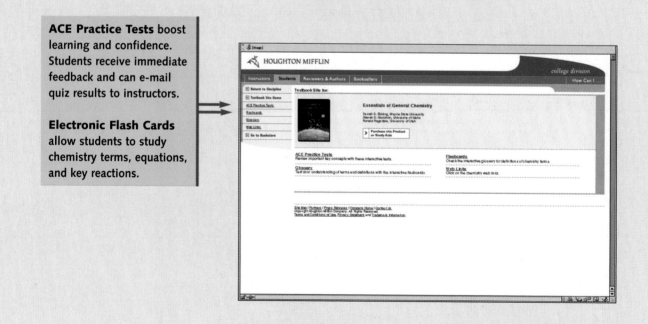

ACE Practice Tests boost learning and confidence. Students receive immediate feedback and can e-mail quiz results to instructors.

Electronic Flash Cards allow students to study chemistry terms, equations, and key reactions.

Optical fibers that use light for data transmission.

Chemistry and Measurement

I n 1964 Barnett Rosenberg and his coworkers at Michigan State University were studying the effects of electricity on bacterial growth. They inserted platinum electrodes into a live bacterial culture and allowed an electric current to pass. After 1 to 2 hours, they noted that cell division in the bacteria stopped. The researchers were very surprised by this result. They were able to show that cell division was inhibited by a substance containing platinum, produced from the platinum electrodes by the electric current. A substance such as this one, the researchers thought, might be useful as an anticancer drug, because cancer involves runaway cell division. Later research confirmed this view, and today the platinum-containing substance *cisplatin* is a leading anticancer drug (Figure 1.1).

This story illustrates three significant reasons to study chemistry. First, chemistry has important practical applications. The development of lifesaving drugs is one, and a complete list would touch upon most areas of modern technology.

Second, chemistry is an intellectual enterprise, a way of explaining our material world. When Rosenberg and his coworkers saw that cell division in the bacteria had

CONTENTS

Figure 1.1
Barnett Rosenberg.
Discoverer of the anticancer
activity of cisplatin.

ceased, they systematically looked for the chemical substance that caused it to cease. They sought a chemical explanation for the occurrence.

Finally, chemistry figures prominently in other fields. Rosenberg's experiment began as a problem in biology; through the application of chemistry it led to an advance in medicine. Whatever your career plans, you will find your knowledge of chemistry is a useful intellectual tool for making important decisions.

An Introduction to Chemistry

All of the objects around you—this book, your pen or pencil, and the things of nature such as rocks, water, and plant and animal substances—constitute the *matter* of the universe. We can define chemistry as the science of the composition and structure of materials and of the changes that materials undergo.

Because chemistry deals with all materials, it is a subject of enormous breadth. It would be difficult to exaggerate the influence of chemistry on modern science and technology or on our ideas about our planet and the universe. In the section that follows, we will take a brief glimpse at modern chemistry and see some of the ways it has influenced technology, science, and modern thought.

1.1   Modern Chemistry: A Brief Glimpse

For thousands of years, human beings have fashioned natural materials into useful products. Modern chemistry certainly has its roots in this endeavor. After the discovery of fire, people began to notice changes in certain rocks and minerals exposed to high temperatures. From these observations came the development of ceramics, glass, and metals, which today are among our most useful materials. Dyes and medicines were other early products obtained from natural substances. For example, the ancient Phoenicians extracted a bright purple dye, known as Tyrian purple, from a species of sea snail. One ounce of Tyrian purple required over 200,000 snails. Because of its brilliant hue and scarcity, the dye became the choice of royalty.

Although chemistry has its roots in early technology, chemistry as a field of study based on scientific principles came into being only in the latter part of the eighteenth century. Chemists began to look at the precise quantities of substances they used in their experiments. From this work came the central principle of modern chemistry: the materials around us are composed of exceedingly small particles called *atoms,* and the precise arrangement of these atoms into *molecules* or more complicated structures accounts for the many different characteristics of materials. Once chemists understood this central principle, they could begin to fashion molecules to order. They could *synthesize* molecules. Tyrian purple, for example, was eventually synthesized from the simpler molecule aniline; see Figure 1.2. Chemists could also correlate molecular structure with the characteristics of materials and so began to fashion materials with special characteristics.

The liquid-crystal displays (LCDs) that you see in watches, calculators, and similar devices (Figure 1.3) are an example of an application that depends on the special characteristics of materials. The liquid crystals used in these displays are a form of matter intermediate in characteristics between those of liquids and those of solid crystals. These liquid crystals are composed of rodlike molecules that tend to align them-

Figure 1.2
Molecular models of Tyrian purple and aniline.
Tyrian purple *(top)* is a dye that was obtained by the early Phoenicians from a species of sea snail. The dye was eventually synthesized from aniline *(bottom)*.

▶ Liquid crystals and liquid-crystal displays are described in the essay at the end of Section 11.6.

selves something like the wood matches in a matchbox. The liquid crystals are held between thin plates that align the molecules in a particular direction, giving the normal light-gray background of the display. These plates are covered with small electrodes, and when any one of them is electrified, the nearby molecules of the liquid crystal are realigned so that they point in a new direction, changing the gray in that area to black. Figure 1.4 shows a model of one of the molecules that forms a liquid crystal; note the rodlike shape of the molecule. Chemists have designed many similar molecules for liquid-crystal applications. ◀

Chemists continue to develop new materials and to discover new properties of old ones. Electronics and communications, for example, have been completely transformed by technological advances in materials. Optical-fiber cables have replaced long-distance telephone cables made of copper wire. Optical fibers are fine threads of extremely pure glass. Because of their purity, these fibers can transmit laser light pulses for miles compared with only a few inches in ordinary glass. Not only is optical-fiber cable cheaper and less bulky than copper cable carrying the same information (Figure 1.5), but by using different colors of light, optical-fiber cable can carry voice, data, and video information at the same time. At the ends of an optical-fiber cable, devices using other new materials convert the light pulses to electrical signals and back, while computer chips constructed from still other materials process the signals (Figure 1.6).

Chemistry has also affected the way we think of the world around us. For example, biochemists and molecular biologists have made a remarkable finding: all forms of life appear to share many of the same molecules and molecular processes. Consider the information of inheritance, the genetic information that is passed on from one generation of an organism to the next. Individual organisms, whether bacteria or human beings, store this information in a particular kind of molecule called deoxyribonucleic acid, or DNA for short (Figure 1.7).

DNA consists of two intertwined molecular chains; each chain consists of links of four different types of molecular pieces, or bases. Just as you record information on a page by stringing together characters (letters, numbers, spaces, and so on), an organism stores the information for reproducing itself in the order of these bases in its DNA. In a multicellular organism, such as a human being, every cell contains the same DNA.

The atomic theory of matter, which forms the basis of modern chemistry, was the

Figure 1.3 *(left)*
A watch that uses a liquid-crystal display.
These liquid-crystal displays are common in watches and laptop computers.

Figure 1.4 *(right)*
Model of a molecule that forms a liquid crystal.
Note that the molecule has a rodlike shape.

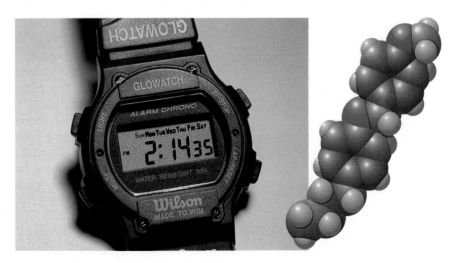

Figure 1.5 *(left)*
Comparison of optical-fiber and copper-wire cables. The fine optical fiber (in the upper center part of the photograph) can carry much more information than copper-wire cables.

Figure 1.6 *(right)*
A gallium-arsenide chip. Using this chip, IBM scientists demonstrated that they could transform light pulses into electronic computer signals at the rate of 3 billion bits of information a second (equivalent to reading 40 encyclopedia volumes a second).

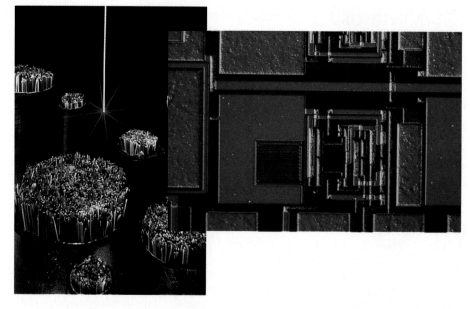

work of the British chemist John Dalton (1766–1844). Throughout his life, Dalton maintained an interest in the science of weather and climate. This interest led Dalton to study the atmosphere and to speculate on its fundamental structure, which eventually led him to his atomic theory. We have some current questions about atmospheric chemistry: Is the earth's climate being changed by human activity? What are the chemical processes involved in the depletion of ozone (a form of oxygen) over Antarctica?

One of our first projects will be to look at this central concept of chemistry, the atomic theory of matter. We will do that in the next chapter, but first we will need some basic vocabulary to talk about science and to describe materials. Then we will need to discuss measurement and units, because measurement is critical for quantitative work.

1.2   Experiment and Explanation

Figure 1.7
A computer-generated model of a fragment of a DNA molecule. DNA contains the hereditary information of an organism that is passed on from one generation to the next.

Experiment and explanation are the heart of chemical research. An **experiment** is *an observation of natural phenomena carried out in a controlled manner so that the results can be duplicated and rational conclusions obtained.* In the chapter opening it was mentioned that Rosenberg studied the effects of electricity on bacterial growth. Temperature and amounts of nutrients in a given volume of bacterial medium are important variables in such experiments.

After a series of experiments, perhaps a researcher sees some relationship or regularity in the results. For instance, Rosenberg noted that in each experiment in which an electric current was passed through a bacterial culture by means of platinum electrodes, the bacteria ceased dividing. If the regularity or relationship is fundamental and we can state it simply, we call it a law. A **law** is *a concise statement or mathematical equation about a fundamental relationship or regularity of nature.* An example is the law of conservation of mass, which says that the mass, or quantity of matter, remains constant during any chemical change.

A **hypothesis** is *a tentative explanation of some regularity of nature.* Having seen

The Birth of the Post-it Note®

Have you ever used a Post-it and wondered where the idea for those little sticky notes came from? You have a chemist to thank for its invention. The story of the Post-it Note is one that illustrates how the creativity and insights of a scientist can result in a product that is as common in the office as the stapler or pen.

In the early 1970s, Art Fry, a 3M scientist, was standing in the choir at his church trying to keep track of all the little bits of paper that marked the music selections for the service. During the service, a number of the markers fell out of the music, making him lose his place. While standing in front of the congregation, he realized that he needed a bookmark that would stick to the book, wouldn't hurt the book, and could be easily detached. To make his plan work, he required an adhesive that would not *permanently* stick things together.

Still thinking about his problem the next day, Fry consulted a colleague, Spencer Silver, who was studying adhesives at the 3M research labs. The study consisted of conducting a series of tests on a range of adhesives to determine the strength of the bond they formed. One of the adhesives that Silver created for the study was an adhesive that always remained sticky. Fry recognized that this adhesive was just what he needed for his bookmark.

As a result he continued to experiment with the bookmark to improve its properties of sticking, detaching, and not hurting the surface to which it was attached. One day, while working on some paperwork, he wrote a question on one of his experimental strips of paper and sent it to his boss stuck to the top of a file folder. His boss then answered the question on the note and returned it attached to some other documents. Fry and his boss then realized that they had invented a new way for people to communicate: the Post-it Note was born (one of the top five office products sold in the United States).

that bacteria ceased to divide when an electric current from platinum electrodes passed through the culture, Rosenberg was eventually able to propose the hypothesis that certain platinum compounds were responsible. He could test his hypothesis by looking for the platinum compound and testing its ability to inhibit cell division.

If a hypothesis successfully passes many tests, it becomes known as a theory. A **theory** is a *tested explanation of basic natural phenomena*. An example is the molecular theory of gases—the theory that all gases are composed of very small particles called molecules. This theory has withstood many tests and has been fruitful in suggesting many experiments. Note that we cannot prove a theory absolutely. It is always possible that further experiments will show the theory to be limited or that someone will develop a better theory. For example, the physics of the motion of objects devised by Isaac Newton withstood experimental tests for more than two centuries, until physicists discovered that the equations do not hold for objects moving near the speed of light. Later physicists showed that very small objects also do not follow Newton's equations. Both discoveries resulted in revolutionary developments in physics. The first led to the theory of relativity; the second to quantum mechanics, which has had an immense impact on chemistry.

The *general* process of advancing scientific knowledge through observation; the framing of laws, hypotheses, or theories; and the conducting of more experiments is called the *scientific method* (Figure 1.8).

1.3 Law of Conservation of Mass

Modern chemistry emerged in the eighteenth century, when chemists began to use the balance systematically as a tool in research (Figure 1.9). **Matter** is *whatever occupies space and can be perceived by our senses*.

Figure 1.8
A representation of the scientific method. The flow diagram shows the general steps in the scientific method. At the right, Rosenberg's work on the development of an anticancer drug illustrates the steps.

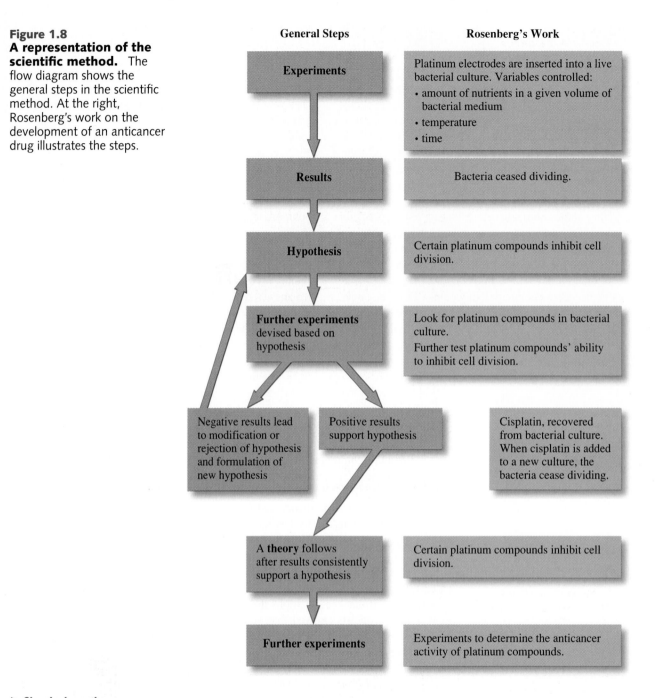

▶ Chemical reactions may involve a gain or loss of heat and other forms of energy. According to Einstein, mass and energy are equivalent. Thus, when energy is lost as heat, mass is also lost, but changes of mass in chemical reactions (billionths of a gram) are too small to detect.

Antoine Lavoisier (1743–1794), a French chemist, was one of the first to insist on the use of the balance in chemical research. By weighing substances before and after chemical change, he demonstrated the **law of conservation of mass,** which states that *the total mass remains constant during a chemical change (chemical reaction).* ◀

In a series of experiments, Lavoisier applied the law of conservation of mass to clarify the phenomenon of burning, or combustion. He showed that when a material burns, a component of air (which he called oxygen) combines chemically with the

Figure 1.9
Laboratory balances.
Left: A simple two-pan balance. Material to be weighed is placed on the left pan and is balanced with weights of known mass placed on the right pan. *Right:* A modern single-pan balance. The mass of the material on the pan appears on the digital readout.

material. For example, when the liquid metal mercury is heated in air, it burns or combines with oxygen to give a red-orange substance, whose modern name is mercury(II) oxide. See Figure 1.10. We can represent the chemical change as follows:

$$\text{Mercury} + \text{oxygen} \longrightarrow \text{mercury(II) oxide}$$

By strongly heating the red-orange substance, Lavoisier was able to decompose it to yield the original mercury and oxygen. The following example illustrates how the law of conservation of mass can be used to study this reaction.

Figure 1.10
Heating mercury in air.
Left: Mercury reacts with oxygen to yield mercury(II) oxide. The color of the oxide varies from red to yellow depending on the particle size. *Right:* When you heat mercury(II) oxide, it decomposes to mercury and oxygen.

EXAMPLE 1.1

Using the Law of Conservation of Mass

A 2.53 gram sample of mercury heated in air produces 2.73 grams of a red-orange residue.

$$\text{Mercury} + \text{oxygen} \longrightarrow \text{red-orange residue}$$

What is the mass of oxygen that reacts? If you strongly heat the red-orange residue, it decomposes to give back the mercury and the oxygen. What is the mass of oxygen released?

▶ PROBLEM STRATEGY

Apply the law of conservation of mass to the reaction. The total mass remains constant during a chemical reaction; that is,

Mass of substances before reaction = mass of substances after reaction

SOLUTION

From the law of conservation of mass,

Mass of mercury + mass of oxygen = mass of red-orange residue

Substituting, you obtain

2.53 grams + mass of oxygen = 2.73 grams

or

Mass of oxygen = (2.73 − 2.53) grams = **0.20 grams**

See Problems 1.19 and 1.20.

Lavoisier set out his views on chemistry in his book, *Traité Élémentaire de Chimie* (Basic Treatise on Chemistry), in 1789. The book was very influential and set the stage for modern chemistry.

You should note the distinction between the terms *mass* and *weight* in precise usage. The weight of an object is the force of gravity exerted on it. The weight is proportional to the mass of the object divided by the square of the distance between the center of mass of the object and that of the earth. ◀ The mass of an object is the same wherever it is measured, but the weight of an object varies.

▶ The force of gravity F between objects whose masses are m_1 and m_2 is Gm_1m_2/r^2, where G is the gravitational constant and r is the distance between the centers of mass of the two objects.

1.4 Matter: Physical State and Chemical Constitution

There are two principal ways of classifying matter: by its physical state as a solid, liquid, or gas and by its chemical constitution as an element, compound, or mixture.

Solids, Liquids, and Gases

Matter exists in different physical forms under different conditions. Water, for example, exists as ice (solid water), as liquid water, and as steam (gaseous water) (Figure

Figure 1.11
Molecular representations of solid, liquid, and gas.
The first beaker contains a solid with a molecular view of the solid; the molecular view depicts the closely packed, immobile atoms that make up the solid structure. The middle beaker contains a liquid with a molecular view of the liquid; the molecular view depicts atoms that are close together but moving freely. The third beaker contains a gas with a molecular view of the gas; the molecular view depicts atoms that are far apart and moving freely.

Solid   Liquid   Gas

1.11). The main identifying characteristic of solids is their rigidity: they tend to maintain their shapes when subjected to outside forces. Liquids and gases, however, are *fluids;* that is, they flow easily and change their shapes in response to slight outside forces.

What distinguishes a gas from a liquid is the characteristic of *compressibility* (and its opposite, *expansibility*). A gas is easily compressible, whereas a liquid is not. In fact, a given quantity of gas can fill a container of almost any size.

These two characteristics, rigidity (or fluidity) and compressibility (or expansibility), can be used to frame definitions of the three states of matter:

solid *the form of matter characterized by rigidity;* a solid is relatively incompressible and has fixed shape and volume.

liquid *the form of matter that is a relatively incompressible fluid;* a liquid has a fixed volume but no fixed shape.

gas *the form of matter that is an easily compressible fluid;* a given quantity of gas will fit into a container of almost any size and shape.

*The three forms of matter—solid, liquid, and gas—*are referred to as the **states of matter.**

Elements, Compounds, and Mixtures

To understand how matter is classified by its chemical constitution, we must first distinguish between physical and chemical changes and between physical and chemical properties. A **physical change** is *a change in the form of matter but not in its chemical identity.* Changes of physical state are examples of physical changes. The process of dissolving one material in another is a further example of a physical change. For instance, you can dissolve sodium chloride (table salt) in water. The water and sodium chloride in this liquid retain their chemical identities and can be separated by distillation. See Figure 1.12. The solution is brought to a boil in the distillation flask and the liquid is obtained by condensing the vapor. The original flask now contains the solid sodium chloride.

A **chemical change,** or **chemical reaction,** is *a change in which one or more kinds of matter are transformed into new matter or several new kinds of matter.* The rusting of iron (iron combines with oxygen) is a chemical change. The original materials (iron and oxygen) combine chemically and cannot be separated by any physical means.

Figure 1.12
Separation by distillation.
You can separate an easily vaporized liquid from another substance by distillation.

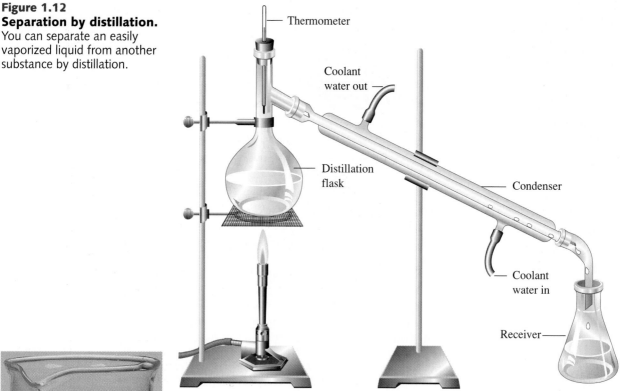

Thermometer

Coolant water out

Distillation flask

Condenser

Coolant water in

Receiver

Figure 1.13
Reaction of sodium with water. Sodium metal flits around the water surface as it reacts briskly, giving off hydrogen gas. The other product is sodium hydroxide, which changes a substance added to the water (phenolphthalein) from colorless to pink.

▶ In Chapter 2, we will redefine an element in terms of atoms.

We characterize a material by its physical and chemical properties. A **physical property** is *a characteristic that can be observed for a material without changing its chemical identity.* Examples are physical state (solid, liquid, or gas), melting point, and color. A **chemical property** is *a characteristic of a material involving its chemical change.* A chemical property of iron is its ability to react with oxygen to produce rust.

Substances   A **substance** is *a kind of matter that cannot be separated into other kinds of matter by any physical process.* It is possible to separate sodium chloride from water by the physical process of distillation. However, sodium chloride is itself a substance and cannot be separated by physical processes into new materials.

No matter what its source, a substance always has the same characteristic properties. Sodium is a solid metal having a melting point of 98°C. The metal also reacts vigorously with water (Figure 1.13). Whether sodium chloride is obtained by burning sodium in chlorine or from seawater, it is a white solid melting at 801°C.

Elements   Millions of substances have been synthesized by chemists, but all are made from a small number of substances known as elements. Lavoisier defined an **element** as *a substance that cannot be decomposed by any chemical reaction into simpler substances.* In 1789 Lavoisier listed 33 substances as elements, of which more than 20 are still so regarded. Today 113 elements are known. Some elements are shown in Figure 1.14. ◀

Compounds   A **compound** is *a substance composed of two or more elements chemically combined.* Joseph Louis Proust (1754–1826) discovered the **law of definite pro-**

Figure 1.14
Some elements.
Center: Sulfur. *From upper right, clockwise:* Arsenic, iodine, magnesium, bismuth, mercury.

portions (also known as the **law of constant composition**): *a pure compound, whatever its source, always contains definite or constant proportions of the elements by mass.* For example, 1.0000 gram of sodium chloride always contains 0.3934 gram of sodium and 0.6066 gram of chlorine, chemically combined. ◄

▶ It is now known that some compounds do not follow the law of definite proportions. These non-stoichiometric compounds, as they are called, are described briefly in Chapter 11.

▶ Chromatography, another example of a physical method used to separate mixtures, is described in the essay at the end of this section.

Mixtures   Most of the materials around us are mixtures. A **mixture** is *a material that can be separated by physical means into two or more substances.* Unlike a pure compound, a mixture has variable composition. When you dissolve sodium chloride in water, you obtain a mixture; its composition depends on the relative amount of sodium chloride dissolved. ◄

A **heterogeneous mixture** is *a mixture that consists of physically distinct parts, each with different properties.* Figure 1.15 shows a heterogeneous mixture of potassium dichromate and iron filings. A **homogeneous mixture** (also known as a **solution**) is *a mixture that is uniform in its properties throughout given samples.* When

Figure 1.15
A heterogeneous mixture.
Left: The mixture on the watch glass consists of potassium dichromate (orange crystals) and iron filings. *Right:* A magnet separates the iron filings from the mixture.

Figure 1.16
Relationships among elements, compounds, and mixtures. Mixtures can be separated by physical processes into substances, and substances can be combined physically into mixtures. Compounds can be separated by chemical reactions into their elements, and elements can be combined chemically to form compounds.

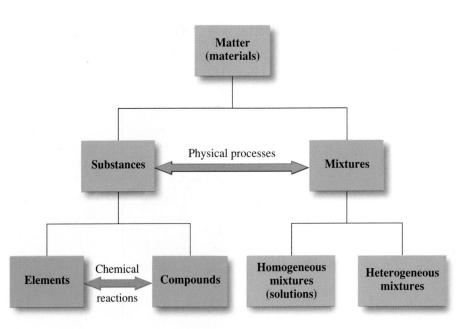

sodium chloride is dissolved in water, you obtain a homogeneous mixture, or solution. Air is a gaseous solution, principally of two elementary substances, nitrogen and oxygen.

Figure 1.16 summarizes the relationships among elements, compounds, and mixtures.

Physical Measurements

Chemists characterize and identify substances by their particular properties. To determine many of these properties requires physical measurements. Cisplatin, the substance featured in the chapter opening, is a yellow substance whose solubility in water is 0.252 gram in 100 grams of water. You can use the solubility, as well as other physical properties, to identify cisplatin. We will now look at the measurement process.

1.5 Measurement and Significant Figures

▶ **2.54 centimeters = 1 inch**

Measurement is the comparison of a physical quantity to be measured with a **unit** of measurement—that is, with *a fixed standard of measurement*. ◀ A steel rod that measures 9.12 times the centimeter unit has a length of 9.12 centimeters.

If you repeat a particular measurement, you usually do not obtain precisely the same result, because each measurement is subject to experimental error. The measured values vary slightly from one another. The term **precision** refers to *the closeness of the set of values obtained from identical measurements of a quantity*. **Accuracy** is a related term; it refers to *the closeness of a single measurement to its true*

Separation of Mixtures by Chromatography

Chromatography is a group of similar separation techniques. Each depends on how fast a substance moves, in a stream of gas or liquid, past a stationary phase to which the substance may be slightly attracted. An example is provided by a simple experiment in *paper chromatography* (see Figure 1.17). In this experiment, a line of ink is

(continued)

Figure 1.17
An illustration of paper chromatography. A line of ink has been drawn along the lower edge of a sheet of paper. The dyes in the ink separate as a solution of methanol and water creeps up the paper.

Figure 1.18
Column chromatography.
(A) A solution containing substances to be separated is poured into the top of a column, which contains powdered chalk. (B) Pure liquid is added to the column, and the substances begin to separate into bands. (C) The substances separate further on the column. Each substance is collected in a separate flask as it comes off the column.

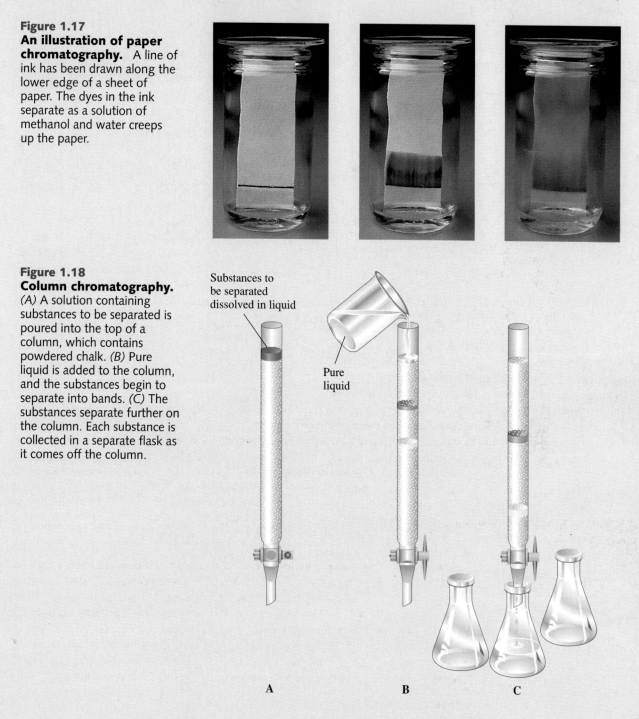

Substances to be separated dissolved in liquid

Pure liquid

A   B   C

drawn near one edge of a sheet of paper, and the paper is placed upright with this edge in a solution of methanol and water. As the solution creeps up the paper, the ink moves upward, separating into a series of different-colored bands that correspond to the different dyes in the ink. Each dye is attracted to the wet paper fibers but with different strengths of attraction. As the solution moves upward, the dyes less strongly attracted to the paper fibers move more rapidly.

The Russian botanist Mikhail Tswett was the first to understand the basis of chromatography and to apply it systematically as a method of separation. In 1906 Tswett separated pigments in plant leaves by *column chromatography*. He first dissolved the pigments from the leaves in petroleum ether, a liquid similar to gasoline. After packing a glass tube or column with powdered chalk, he poured the solution of pigments into the top of the column (see Figure 1.18). When he washed the column by pouring in more petroleum ether, it began to show distinct yellow and green bands. These bands, each containing a pure pigment, became well separated as they moved down the column, so that the pure pigments could be obtained. The name chromatography originates from this early separation of colored substances (the stem *chromato-* means "color"), although the technique is not limited to colored substances.

Gas chromatography (GC) is a more recent separation method. Here the moving stream is a gaseous mixture of vaporized substances plus a gas such as helium, which is called the *carrier*. The stationary material is either a solid or a liquid adhering to a solid, packed in a column. As the gas passes through the column, substances in the mixture are attracted differently to the stationary column packing and thus are separated. Gas chromatography is a rapid, small-scale method of separating mixtures. It is also important in the analysis of mixtures because the time it takes for a substance at a given temperature to travel through the column to a detector (called the *retention time*) is fixed. You can therefore use retention times to help identify substances. Figure 1.19 shows a gas chromatograph and a portion of a computer plot *(chromatogram)*. Each peak on the chromatogram corresponds to a specific substance. The peaks were automatically recorded by the instrument as the different substances in the mixture passed the detector. Chemists have analyzed complicated mixtures by gas chromatography. Analysis of chocolate, for example, shows that it contains over 800 flavor compounds.

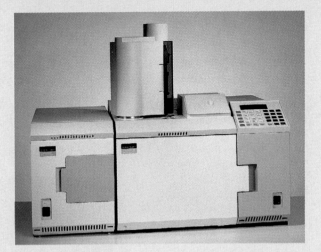

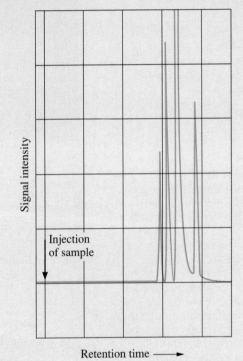

Figure 1.19
Gas chromatography. *Top:* A modern gas chromatograph. *Bottom:* This is a chromatogram of a hexane mixture, showing its separation into four different substances. Such hexane mixtures occur in gasoline; hexane is also used as a solvent to extract the oil from certain vegetable seeds.

Figure 1.20
Figure 1.20
Precision of measurement with a centimeter ruler. The length of the rod is just over 9.1 cm. On successive measurements, we estimate the length by eye at 9.12, 9.11, and 9.13 cm. We record the length as between 9.11 cm and 9.13 cm.

value. In Figure 1.20, a steel rod has been placed near a ruler subdivided into tenths of a centimeter. You can see that the rod measures just over 9.1 cm (cm = centimeter). With care, it is possible to estimate by eye to hundredths of a centimeter. Here you might give the measurement as 9.12 cm. Suppose you measure the length of this rod twice more. You find the values to be 9.11 cm and 9.13 cm. The spread of values indicates the precision with which a measurement can be made by this centimeter ruler. ◄

To indicate the precision of a measured number, we often use the concept of significant figures. **Significant figures** are *those digits in a measured number (or result of a calculation with measured numbers) that include all certain digits plus a final one having some uncertainty.* When you measured the rod, you obtained the values 9.12 cm, 9.11 cm, and 9.13 cm. You could report the result as the average, 9.12 cm. The first two digits (9.1) are certain; the next digit (2) is estimated, so it has some uncertainty.

► Measurements that are of high precision are usually accurate. It is possible, however, to have a systematic error in a measurement. Suppose that, in calibrating a ruler, the first centimeter is made too small by 0.1 cm. Then, although the measurements of length on this ruler are still precise to 0.01 cm, they are accurate to only 0.1 cm.

Number of Significant Figures

Number of significant figures refers to *the number of digits reported for the value of a measured or calculated quantity, indicating the precision of the value.* Thus, there are three significant figures in 9.12 cm, whereas 9.123 cm has four. To count the number of significant figures in a given measured quantity, you observe the following rules:

1. All digits are significant except zeros at the beginning of the number and possibly terminal zeros (one or more zeros at the end of a number). Thus, 9.12 cm, 0.912 cm, and 0.00912 cm all contain three significant figures.

2. Terminal zeros ending at the right of the decimal point are significant. Each of the following has three significant figures: 9.00 cm, 9.10 cm, 90.0 cm.

3. Terminal zeros in a number without an explicit decimal point may or may not be significant. If someone gives a measurement as 900 cm, you do not know whether one, two, or three significant figures are intended. If the person writes 900. cm (note the decimal point), the zeros are significant. More generally, you can remove any uncertainty in such cases by expressing the measurement in scientific notation.

Scientific notation is *the representation of a number in the form $A \times 10^n$, where A is a number with a single nonzero digit to the left of the decimal point and n is an integer, or whole number.* In scientific notation, the measurement 900 cm precise to two significant figures is written 9.0×10^2 cm. If precise to three significant figures, it is written 9.00×10^2 cm. Scientific notation is also convenient for expressing very large or very small quantities. ◄

► See Appendix A for a review of scientific notation.

Significant Figures in Calculations

How do you determine the correct number of significant figures to report for the answer to a calculation? The following are two rules that we use:

> 1. **Multiplication and division.** When multiplying or dividing measured quantities, give as many significant figures in the answer as there are in the measurement with *the least number of significant figures.*
>
> 2. **Addition and subtraction.** When adding or subtracting measured quantities, give the same number of decimal places in the answer as there are in the measurement with *the least number of decimal places.*

Let us see how you apply these rules. Suppose you have a substance believed to be cisplatin, and, in an effort to establish its identity, you measure its solubility (the amount that dissolves in a given quantity of water). You find that 0.0634 gram of the substance dissolves in 25.31 grams of water. The amount dissolving in 100.0 grams is

$$100.0 \text{ grams of water} \times \frac{0.0634 \text{ gram cisplatin}}{25.31 \text{ grams of water}}$$

Performing the arithmetic on a pocket calculator (Figure 1.21), you get 0.2504938 for the numerical part of the answer (100.0 × 0.0634 ÷ 25.31). It would be incorrect to give this number as the final answer. The measurement 0.0634 gram has the least number of significant figures (three). Therefore, you report the answer to three significant figures—that is, 0.250 gram.

Now consider the addition of 184.2 grams and 2.324 grams. On a calculator, you find that 184.2 + 2.324 = 186.524. But because the quantity 184.2 grams has the least number of decimal places—one, whereas 2.324 grams has three—the answer is 186.5 grams.

Exact Numbers

So far we have discussed only numbers that involve uncertainties. However, you will also encounter exact numbers. An **exact number** is *a number that arises when you count items or sometimes when you define a unit.* For example, when you say there are 9 coins in a bottle, you mean exactly 9, not 8.9 or 9.1. Also, when you say there are 12 inches to a foot, you mean exactly 12. Similarly, the inch is defined to be exactly 2.54 centimeters. The conventions of significant figures do not apply to exact numbers. Thus, the 2.54 in the expression "1 inch equals 2.54 centimeters" should not be interpreted as a measured number with three significant figures. In effect, the 2.54 has an infinite number of significant figures. The number of significant figures in a calculation result depends only on the numbers of significant figures in quantities having uncertainties. For example, suppose you want the total mass of 9 coins when each coin has a mass of 3.0 grams. The calculation is

$$3.0 \text{ grams} \times 9 = 27 \text{ grams}$$

You report the answer to two significant figures because 3.0 grams has two significant figures. The number 9 is exact and does not determine the number of significant figures.

Figure 1.21
Significant figures and calculators. Not all of the figures that appear on a calculator display are significant. In performing the calculation 100.0 × 0.0634 ÷ 25.31, the calculator display shows 0.2504938. We would report the answer as 0.250, however, because the factor 0.0634 has the least number of significant figures (three).

Rounding

In reporting the solubility of your substance in 100.0 grams of water as 0.250 gram, you *rounded* the number you read off the calculator (0.2504938). **Rounding** is *the procedure of dropping nonsignificant digits in a calculation result and adjusting the last digit reported*. The general procedure is as follows. Look at the leftmost digit to be dropped.

1. If this digit is 5 or greater, add 1 to the last digit to be retained and drop all digits farther to the right. Thus, rounding 1.2151 to three significant figures gives 1.22.

2. If this digit is less than 5, simply drop it and all digits farther to the right. Rounding 1.2143 to three significant figures gives 1.21.

In doing a calculation of two or more steps, it is desirable to retain nonsignificant digits for intermediate answers. This ensures that accumulated small errors from rounding do not appear in the final result. If you use a calculator, you can simply enter numbers one after the other, performing each arithmetic operation and rounding only the final answer. ◄ To keep track of the correct number of significant figures, you may want to record intermediate answers with a line under the last significant figure, as shown in the solution to part d of the following example.

► The answers for Problems will follow this practice. For most Example Problems where intermediate answers are reported, all answers, intermediate and final, will be rounded.

EXAMPLE 1.2

Using Significant Figures in Calculations

Perform the following calculations and round the answers to the correct number of significant figures (units of measurement have been omitted).

a. $\dfrac{2.568 \times 5.8}{4.186}$

b. $5.41 - 0.398$

c. $3.38 - 3.01$

d. $4.18 - 58.16 \times (3.38 - 3.01)$

⬥ PROBLEM STRATEGY

Parts a, b, and c are relatively straightforward; use the rule for significant figures that applies in each case. In a multistep calculation as in d, proceed step by step, applying the relevant rule to each step. First, do the operations within parentheses. By convention, you perform multiplications and divisions before additions and subtractions. Do not round intermediate answers, but do keep track of the least significant digit, say by underlining it.

SOLUTION

a. The factor 5.8 has the fewest significant figures; therefore, the answer should be reported to two significant figures. Round the answer to **3.6.** b. The number with the least number of decimal places is 5.41. Therefore, round the answer to two decimal places, to **5.01.** c. The answer is **0.37.** Note how you have lost one significant figure in the subtraction. d. You first do the subtraction within parentheses, underlining the least significant digits.

$$4.1\underline{8} - 58.1\underline{6} \times (3.3\underline{8} - 3.0\underline{1}) = 4.1\underline{8} - 58.1\underline{6} \times 0.3\underline{7}$$

Following convention, you do the multiplication before the subtraction.

$$4.1\underline{8} - 58.1\underline{6} \times 0.3\underline{7} = 4.1\underline{8} - 2\underline{1}.5192 = -1\underline{7}.3392$$

The final answer is **−17.**

See Problems 1.35 and 1.36.

CONCEPT CHECK 1.1

a. When you report your weight to someone, how many significant figures do you typically use?

b. What is your weight with two significant figures?

c. Indicate your weight and the number of significant figures you would obtain if you weighed yourself on a truck scale that can measure in 50-kg or 100-lb increments.

1.6 SI Units

▶ In the system of units that Lavoisier used in the eighteenth century, there were 9216 grains to the pound (the *livre*). English chemists of the same period used a system in which there were 7000 grains to the pound—unless they were trained as apothecaries, in which case there were 5760 grains to the pound!

▶ The amount of substance is discussed in Chapter 3, and the electric current (ampere) is introduced in Chapter 20. Luminous intensity will not be used in this book.

The first measurements were probably based on the human body (the length of the foot, for example). In time, fixed standards developed, but these varied from place to place. As science became more quantitative, scientists found that the lack of standard units was a problem. ◀ In 1791 a study committee of the French Academy of Sciences devised the *metric system*. Most nations have since adopted the metric system.

SI Base Units and SI Prefixes

In 1960 the General Conference of Weights and Measures adopted the **International System** of units (or **SI,** after the French *le Système International d'Unités*). This system has seven **SI base units,** *the SI units from which all others can be derived.* Table 1.1 lists these base units and the symbols used to represent them. In this chapter, we will discuss four base quantities: length, mass, time, and temperature. ◀

Table 1.1
SI Base Units

Quantity	Unit	Symbol
Length	meter	m
Mass	kilogram	kg
Time	second	s
Temperature	kelvin	K
Amount of substance	mole	mol
Electric current	ampere	A
Luminous intensity	candela	cd

In SI, a larger or smaller unit for a physical quantity is indicated by an **SI prefix,** which is *a prefix used to indicate a power of ten.* For example, the base unit of length in SI is the meter, and 10^{-2} meter is called a *centi*meter. The SI prefixes are listed in Table 1.2. Only those shown in color will be used in this book.

Length, Mass, and Time

▶ The meter was originally defined in terms of a standard platinum–iridium bar kept at Sèvres, France. In 1983, the meter was defined as the distance traveled by light in a vacuum in 1/299,792,458 seconds.

The **meter (m)** is *the SI base unit of length.* ◀ By combining it with one of the SI prefixes, you can get a unit of appropriate size for any length measurement. For the very small lengths used in chemistry, the nanometer (nm; 1 nanometer $= 10^{-9}$ m) or the picometer (pm; 1 picometer $= 10^{-12}$ m) is an acceptable SI unit.

▶ The present standard of mass is the platinum–iridium kilogram mass kept at the International Bureau of Weights and Measures in Sèvres, France.

The **kilogram (kg)** is *the SI base unit of mass,* equal to about 2.2 pounds. ◀ This is an unusual base unit in that it contains a prefix. In forming other SI mass units, prefixes are added to the word *gram* (g) to give units such as the *milli*gram (mg; 1 mg $= 10^{-3}$ g).

The **second (s)** is *the SI base unit of time.* Combining this unit with prefixes such as *milli-, micro-, nano-,* and *pico-,* you create units appropriate for measuring very rapid events.

Temperature

Temperature is difficult to define precisely, but we all have an intuitive idea of what we mean by it. It is a measure of "hotness." Heat energy passes from a hot object to a cold one, and the quantity of heat passed between the objects depends on the difference in temperature between the two. Therefore, temperature and heat are different, but related, concepts.

A thermometer is a device for measuring temperature. A scale alongside a glass column gives a measure of the temperature. The **Celsius scale** is *the temperature scale in general scientific use.* On this scale, the freezing point of water is 0°C and the boiling point of water at normal barometric pressure is 100°C. However, *the SI base unit of temperature* is the **kelvin (K),** a unit on an *absolute temperature* scale. ◀ On any absolute

▶ Note that the degree sign (°) is not used with the Kelvin scale, and the unit of temperature is simply the kelvin (not capitalized).

Table 1.2
SI Prefixes*

Multiple	Prefix	Symbol
10^6	mega	M
10^3	kilo	k
10^2	hecto	h
10	deka	da
10^{-1}	deci	d
10^{-2}	centi	c
10^{-3}	milli	m
10^{-6}	micro	μ†
10^{-9}	nano	n
10^{-12}	pico	p

*The units highlighted in color are the ones most commonly used.

†Greek letter mu, pronounced "mew."

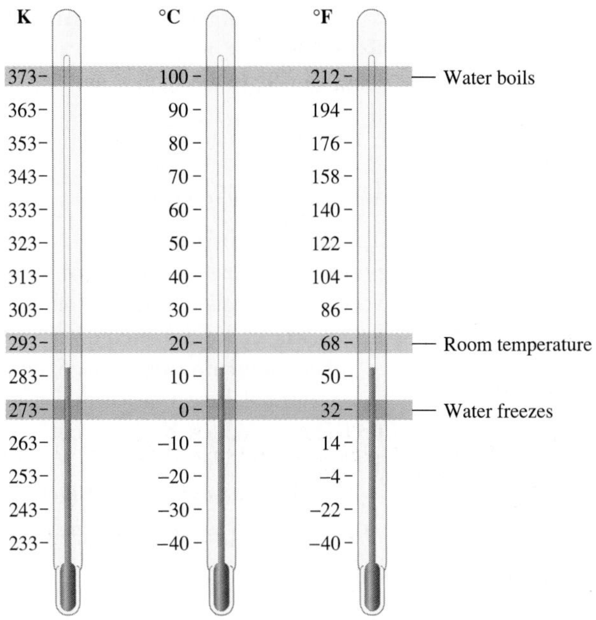

Figure 1.22
Comparison of temperature scales. Room temperature is about 293 K, 20°C, and 68°F. Water freezes at 273.15 K, 0°C, and 32°F. Water boils under normal pressure at 373.15 K, 100°C, and 212°F.

K	°C	°F	
373–	100–	212–	— Water boils
363–	90–	194–	
353–	80–	176–	
343–	70–	158–	
333–	60–	140–	
323–	50–	122–	
313–	40–	104–	
303–	30–	86–	
293–	20–	68–	— Room temperature
283–	10–	50–	
273–	0–	32–	— Water freezes
263–	–10–	14–	
253–	–20–	–4–	
243–	–30–	–22–	
233–	–40–	–40–	

scale, the lowest temperature that can be attained theoretically is zero. The Celsius and the Kelvin scales have equal-size units, but 0°C is equivalent to 273.15 K. Thus, it is easy to convert from one scale to the other, using the formula

$$T_K = \left(t_C \times \frac{1\ K}{1°C}\right) + 273.15\ K$$

where T_K is the temperature in kelvins and t_C is the temperature in degrees Celsius.

Figure 1.22 compares Kelvin, Celsius, and Fahrenheit scales. As the figure shows, 0°C is the same as 32°F (both exact), and 100°C corresponds to 212°F (both exact). Therefore, there are $212 - 32 = 180$ Fahrenheit degrees in the range of 100 Celsius degrees. That is, there are exactly 9 Fahrenheit degrees for every 5 Celsius degrees. Knowing this, and knowing that 0°C equals 32°F, we can derive a formula to convert degrees Celsius to degrees Fahrenheit. ◄

▶ These two temperature conversion formulas are often written K = °C + 273.15 and °F = (1.8 × °C) + 32. Although these yield the correct number, they do not take into account the units.

$$t_F = \left(t_C \times \frac{9°F}{5°C}\right) + 32°F$$

By rearranging this equation, we can obtain a formula for converting degrees Fahrenheit to degrees Celsius:

$$t_C = \frac{5°C}{9°F} \times (t_F - 32°F)$$

EXAMPLE 1.3

Converting from One Temperature Scale to Another

The hottest place on record in North America is Death Valley in California. It reached a temperature of 134°F in 1913. What is this temperature reading in degrees Celsius? in kelvins?

SOLUTION

Substituting, we find that

$$t_C = \frac{5\,°C}{9\,°F} \times (t_F - 32\,°F) = \frac{5\,°C}{9\,°F} \times (134\,°F - 32\,°F) = \mathbf{56.7\,°C}$$

In kelvins,

$$T_K = \left(t_C \times \frac{1\,K}{1\,°C}\right) + 273.15\,K = \left(56.7\,°C \times \frac{1\,K}{1\,°C}\right) + 273.15\,K = \mathbf{329.9\,K}$$

See Problems 1.43, 1.44, 1.45, and 1.46.

 CONCEPT CHECK 1.2

a. Estimate and express the length of your leg in an appropriate metric unit.

b. What would be a reasonable height, in meters, of a three-story building?

c. How would you be feeling if your body temperature was 39°C?

d. Would you be comfortable sitting in a room at 23°C in a short sleeved shirt?

1.7 Derived Units

Once base units have been defined for a system of measurement, you can derive other units from them. Table 1.3 defines a number of derived units. Volume and density are discussed in this section; pressure and energy are discussed later (in Sections 5.1 and 6.1, respectively).

Volume

Volume is defined as length cubed and has the SI unit of cubic meter (m^3). This is too large a unit for normal laboratory work, so we use either cubic decimeters (dm^3) or cubic centimeters (cm^3, also written cc). Traditionally, chemists have used the **liter (L),** which is *a unit of volume equal to a cubic decimeter* (approximately one quart). In fact, most laboratory glassware (Figure 1.23) is calibrated in liters or milliliters

Table 1.3
Derived Units

Quantity	Definition of Quantity	SI Unit
Area	Length squared	m^2
Volume	Length cubed	m^3
Density	Mass per unit volume	kg/m^3
Speed	Distance traveled per unit time	m/s
Acceleration	Speed changed per unit time	m/s^2
Force	Mass times acceleration of object	$kg \cdot m/s^2$ (= newton, N)
Pressure	Force per unit area	$kg/(m \cdot s^2)$ (= pascal, Pa)
Energy	Force times distance traveled	$kg \cdot m^2/s^2$ (= joule, J)

Figure 1.23
Some laboratory glassware. *Left to right:* A 600-mL beaker; a 100-mL graduated cylinder; a 250-mL volumetric flask; a 250-mL Erlenmeyer flask. *Front:* A 5-mL pipet. A graduated cylinder is used to measure approximate volumes of liquids. A pipet is calibrated to deliver a specified volume of liquid when filled to an etched line with a suction bulb.

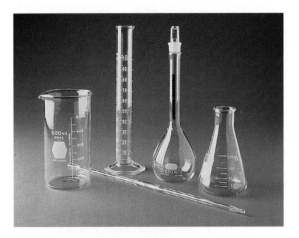

(1000 mL = 1 L). Because 1 dm equals 10 cm, a cubic decimeter, or one liter, equals $(10 \text{ cm})^3 = 1000 \text{ cm}^3$. Therefore, a milliliter equals a cubic centimeter. In summary

$$1 \text{ L} = 1 \text{ dm}^3 \quad \text{and} \quad 1 \text{ mL} = 1 \text{ cm}^3$$

Density

The **density** of an object is its *mass per unit volume*. You can express this as

$$d = \frac{m}{V}$$

where d is the density, m is the mass, and V is the volume.

Density is an important characteristic property of a material. Water, for example, has a density of 1.000 g/cm^3 at 4°C and a density of 0.998 g/cm^3 at 20°C. Lead has a density of 11.3 g/cm^3 at 20°C. (Figures 1.24 and 1.25 dramatically show some relative densities.) ◄ Because the density is characteristic of a substance, it can be helpful in identifying it. Example 1.4 illustrates this point. Density can also be useful in determining whether a substance is pure.

► The density of solid materials on earth ranges from about 1 g/cm^3 to 22.5 g/cm^3 (osmium metal). In the interior of certain stars, the density of matter is truly staggering. Black neutron stars—stars composed of neutrons, or atomic cores compressed by gravity to a superdense state—have densities of about 10^{15} g/cm^3.

EXAMPLE 1.4

Calculating the Density of a Substance

A colorless liquid is believed to be one of the following:

Substance	Density (in g/mL)
n-butyl alcohol	0.810
ethylene glycol	1.114
isopropyl alcohol	0.785
toluene	0.866

To identify the substance, a chemist determined its density. By pouring a sample of the liquid into a graduated cylinder, she found that the volume was 35.1 mL. She also found that the sample weighed 30.5 g. What was the density of the liquid? What was the substance?

Figure 1.24 *(left)*
The relative densities of copper and mercury.
Copper floats on mercury because the density of copper is less than that of mercury.

Figure 1.25 *(right)*
Relative densities of some liquids. Shown are three liquids (dyed so that they will show up clearly): *ortho*-xylene, water, and 1,1,1-trichloroethane. Water (blue layer) is less dense than 1,1,1-trichloroethane (colorless) and floats on it. *Ortho*-xylene (top, golden layer) is less dense than water and floats on it.

SOLUTION

$$d = \frac{m}{V} = \frac{30.5 \text{ g}}{35.1 \text{ mL}} = \textbf{0.869 g/mL}$$

The density of the liquid equals that of **toluene** (within experimental error).

See Problems 1.47 and 1.48.

In addition to characterizing a substance, the density provides a useful relationship between mass and volume. For example, suppose an experiment calls for a certain mass of liquid. Rather than weigh the liquid on a balance, you might instead measure out the corresponding volume. Example 1.5 illustrates this idea.

EXAMPLE 1.5

Using the Density to Relate Mass and Volume

An experiment requires 43.7 g of isopropyl alcohol. Instead of measuring out the sample on a balance, a chemist dispenses the liquid into a graduated cylinder. The density of isopropyl alcohol is 0.785 g/mL. What volume of isopropyl alcohol should he use?

SOLUTION

You rearrange the formula defining the density to obtain the volume.

$$V = \frac{m}{d}$$

Then you substitute into this formula:

$$V = \frac{43.7 \text{ g}}{0.785 \text{ g/mL}} = \textbf{55.7 mL}$$

See Problems 1.49 and 1.50.

 CONCEPT CHECK 1.3

You are working in the office of a precious metal buyer. A miner brings you a nugget of metal that he claims is gold. You suspect that the metal is a form of "fool's gold." What simple experiment could you perform to decide whether or not the miner's nugget is gold?

1.8   Units and Dimensional Analysis (Factor-Label Method)

In performing numerical calculations with physical quantities, enter each quantity as a number with its associated unit. Both the numbers and the units are then carried through the indicated algebraic operations. The advantages of this are twofold:

1. The units for the answer will come out of the calculations.

2. If you make an error in arranging factors in the calculation (for example, if you use the wrong formula), this will become apparent because the final units will be nonsense.

Dimensional analysis (or the **factor-label method**) is *the method of calculation in which one carries along the units for quantities.* As an illustration, suppose you want to find the volume V of a cube, given s, the length of a side of the cube. Because $V = s^3$, if $s = 5.00$ cm, you find that $V = (5.00 \text{ cm})^3 = 5.00^3 \text{ cm}^3$. There is no guesswork about the unit of volume here; it is cubic centimeters (cm^3).

Suppose, however, that you wish to express the volume in liters (L), a metric unit that equals 10^3 cubic centimeters. Write this equality as

$$1 \text{ L} = 10^3 \text{ cm}^3$$

If you divide both sides of the equality by the right-hand quantity, you get

$$\frac{1 \text{ L}}{10^3 \text{ cm}^3} = \frac{10^3 \cancel{\text{ cm}^3}}{10^3 \cancel{\text{ cm}^3}} = 1$$

Observe that you treat units in the same way as algebraic quantities. Note too that the right-hand side now equals 1 and no units are associated with it. Because it is always possible to multiply any quantity by 1 without changing that quantity, you can multiply the previous expression for the volume by the factor $1 \text{ L}/10^3 \text{ cm}^3$ without changing the actual volume. You are changing only the way you express this volume:

$$V = 5.00^3 \, \cancel{cm^3} \times \underbrace{\frac{1 \, L}{10^3 \, \cancel{cm^3}}}_{\substack{\text{converts} \\ \text{cm}^3 \text{ to L}}} = 125 \times 10^{-3} \, L = 0.125 \, L$$

The ratio $1 \, L/10^3 \, cm^3$ is called a **conversion factor** because it is *a factor equal to 1 that converts a quantity expressed in one unit to a quantity expressed in another unit.* ◄ Note that the numbers in this conversion factor are *exact*, because 1 L equals exactly 1000 cm^3.

► It takes more room to explain conversion factors than it does to use them. With practice, you will be able to write the final conversion step without the intermediate algebraic manipulations outlined here.

EXAMPLE 1.6

Converting Units: Metric Unit to Metric Unit

A sample of nitrogen in a glass bulb weighs 243 mg. What is this mass in SI base units of mass (kilograms)?

SOLUTION

Since $1 \, mg = 10^{-3} \, g$, you can write

$$243 \, \cancel{mg} \times \frac{10^{-3} \, g}{1 \, \cancel{mg}} = 2.43 \times 10^{-1} \, g$$

Then, because the prefix *kilo-* means 10^3, you write

$$1 \, kg = 10^3 \, g$$

and

$$2.43 \times 10^{-1} \, \cancel{g} \times \frac{1 \, kg}{10^3 \, \cancel{g}} = \mathbf{2.43 \times 10^{-4} \, kg}$$

Note, however, that you can do the two conversions in one step, as follows:

$$243 \, \cancel{mg} \times \underbrace{\frac{10^{-3} \, \cancel{g}}{1 \, \cancel{mg}}}_{\substack{\text{converts} \\ \text{mg to g}}} \times \underbrace{\frac{1 \, kg}{10^3 \, \cancel{g}}}_{\substack{\text{converts} \\ \text{g to kg}}} = 2.43 \times 10^{-4} \, kg$$

See Problems 1.51 and 1.52.

EXAMPLE 1.7

Converting Units: Metric Volume to Metric Volume

The world's oceans contain approximately $1.35 \times 10^9 \, km^3$ of water. What is this volume in liters?

SOLUTION

$$1.35 \times 10^9 \, \cancel{km^3} \times \underbrace{\left(\frac{10^3 \, \cancel{m}}{1 \, \cancel{km}}\right)^3}_{\substack{\text{converts} \\ \text{km}^3 \text{ to m}^3}} \times \underbrace{\left(\frac{1 \, dm}{10^{-1} \, \cancel{m}}\right)^3}_{\substack{\text{converts} \\ \text{m}^3 \text{ to dm}^3}} = 1.35 \times 10^{21} \, dm^3$$

Because a cubic decimeter is equal to a liter, the volume of the oceans is

$$\mathbf{1.35 \times 10^{21} \, L}$$

The conversion-factor method can be used to convert any unit to another unit, provided a conversion equation exists between the two units. Relationships between certain U.S. units and metric units are given in Table 1.4. You can use these to convert between U.S. and metric units. Suppose you wish to convert 0.547 lb to grams. From Table 1.4, note that 1 lb = 0.4536 kg, or 1 lb = 453.6 g, so the conversion factor from pounds to grams is 453.6 g/1 lb. Therefore,

$$0.547 \text{ lb} \times \frac{453.6 \text{ g}}{1 \text{ lb}} = 248 \text{ g}$$

The next example illustrates a conversion requiring several steps.

EXAMPLE 1.8

Converting Units: Any Unit to Another Unit

How many centimeters are there in 6.51 miles? Use the exact definitions 1 mi = 5280 ft, 1 ft = 12 in, and 1 in = 2.54 cm.

SOLUTION

This problem involves several conversions, which can be done all at once. From the definitions, you obtain the following conversion factors:

$$1 = \frac{5280 \text{ ft}}{1 \text{ mi}} \qquad 1 = \frac{12 \text{ in}}{1 \text{ ft}} \qquad 1 = \frac{2.54 \text{ cm}}{1 \text{ in}}$$

Then,

$$6.51 \text{ mi} \times \underbrace{\frac{5280 \text{ ft}}{1 \text{ mi}}}_{\substack{\text{converts} \\ \text{mi to ft}}} \times \underbrace{\frac{12 \text{ in}}{1 \text{ ft}}}_{\substack{\text{converts} \\ \text{ft to in}}} \times \underbrace{\frac{2.54 \text{ cm}}{1 \text{ in}}}_{\substack{\text{converts} \\ \text{in to cm}}} = \mathbf{1.05 \times 10^6 \text{ cm}}$$

All of the conversion factors are exact, so the number of significant figures in the result is determined by the number of significant figures in 6.51 mi.

See Problems 1.53, 1.54, 1.55, and 1.56.

Table 1.4
Relationships of Some U.S. and Metric Units

Length	Mass	Volume
1 in = 2.54 cm (exact)	1 lb = 0.4536 kg	1 qt = 0.9464 L
1 yd = 0.9144 m (exact)	1 lb = 16 oz (exact)	4 qt = 1 gal (exact)
1 mi = 1.609 km	1 oz = 28.35 g	
1 mi = 5280 ft (exact)		

A Checklist for Review

Important Terms

Note: The number in parentheses denotes the section in which the term is defined.

experiment (1.2)
law (1.2)
hypothesis (1.2)
theory (1.2)
matter (1.3)
law of conservation of mass (1.3)
solid (1.4)
liquid (1.4)
gas (1.4)
states of matter (1.4)
physical change (1.4)

chemical change (chemical reaction) (1.4)
physical property (1.4)
chemical property (1.4)
substance (1.4)
element (1.4)
compound (1.4)
law of definite proportions (law of constant composition) (1.4)
mixture (1.4)
heterogeneous mixture (1.4)

homogeneous mixture (solution) (1.4)
unit (1.5)
precision (1.5)
accuracy (1.5)
significant figures (1.5)
number of significant figures (1.5)
scientific notation (1.5)
exact number (1.5)
rounding (1.5)
International System (SI) (1.6)

SI base units (1.6)
SI prefix (1.6)
meter (m) (1.6)
kilogram (kg) (1.6)
second (s) (1.6)
Celsius scale (1.6)
kelvin (K) (1.6)
liter (L) (1.7)
density (1.7)
dimensional analysis (factor-label method) (1.8)
conversion factor (1.8)

Key Equations

$$T_K = \left(t_C \times \frac{1\ K}{1°C} \right) + 273.15\ K$$

$$t_C = \frac{5°C}{9°F} \times (t_F - 32°F)$$

$$d = m/V$$

Summary of Facts and Concepts

Chemistry is an experimental science in that the facts of chemistry are obtained by *experiment*. These facts are systematized and explained by *theory,* and theory suggests more experiments. The *scientific method* involves this interplay, in which the body of accepted knowledge grows as it is tested by experiment.

Chemistry emerged as a *quantitative science* with the work of the eighteenth-century French chemist Antoine Lavoisier. He made use of the idea that the mass, or quantity of matter, remains constant during a chemical reaction *(law of conservation of mass).*

Matter may be classified by its physical state as a *gas, liquid,* or *solid.* Matter may also be classified by its chemical constitution as an *element, compound,* or *mixture.* Materials are either substances or mixtures of substances. Substances are ei-

ther elements or compounds, which are composed of two or more elements. Mixtures can be separated into substances by physical processes, but compounds can be separated into elements only by chemical reactions.

A quantitative science requires the making of measurements. Any measurement has limited *precision,* which you convey by writing the measured number to a certain number of *significant figures.* The International System (SI) uses a particular selection of metric units. It employs seven *base units* combined with *prefixes* to obtain units of various size. Units for other quantities are derived from these.

Dimensional analysis is a technique of calculating with physical quantities in which units are included and treated in the same way as numbers.

Operational Skills

1. **Using the law of conservation of mass** Given the masses of all substances in a chemical reaction except one, calculate the mass of this one substance. **(EXAMPLE 1.1)**

2. **Using significant figures in calculations** Given an arithmetic setup, report the answer to the correct number of significant figures and round it properly. **(EXAMPLE 1.2)**

3. **Converting from one temperature scale to another** Given a temperature reading on one scale, convert it to another scale—Celsius, Kelvin, or Fahrenheit. (EXAMPLE 1.3)

4. **Calculating the density of a substance** Given the mass and volume of a substance, calculate the density. (EXAMPLE 1.4)

5. **Using the density to relate mass and volume** Given the mass and density of a substance, calculate the volume; or given the volume and density, calculate the mass. (EXAMPLE 1.5)

6. **Converting units** Given an equation relating one unit to another (or a series of such equations), convert a measurement expressed in one unit to a new unit. (EXAMPLES 1.6, 1.7, AND 1.8)

Review Questions

1.1 Discuss some ways in which chemistry has changed technology. Give one or more examples of how chemistry has affected another science.

1.2 Define the terms *experiment* and *theory*. How are theory and experiment related? What is a hypothesis?

1.3 Illustrate the steps in the scientific method using Rosenberg's discovery of the anticancer activity of cisplatin.

1.4 Define the terms *matter* and *mass*. What is the difference between mass and weight?

1.5 State the law of conservation of mass. Describe how you might demonstrate this law.

1.6 A chemical reaction is often accompanied by definite changes in appearance. Figure 1.10 shows the reactions of the metal mercury with oxygen. Describe the changes that occur.

1.7 Give examples of an element, a compound, a heterogeneous mixture, and a homogeneous mixture.

1.8 What is meant by the precision of a measurement? How is it indicated?

1.9 Distinguish between a measured number and an exact number. Give examples of each.

1.10 How does the International System (SI) obtain units of different size from a given unit?

1.11 What is an absolute temperature scale? How are degrees Celsius related to kelvin units?

1.12 Why should units be carried along with numbers in a calculation?

Conceptual Problems

Key: These problems are designed to check your understanding of the concepts associated with some of the main topics presented in the chapter. A strong conceptual understanding of chemistry is the foundation for both applying chemical knowledge and solving chemical problems. These problems vary in level of difficulty and often can be used as a basis for group discussion.

1.13
a. Sodium metal is partially melted. What are the two phases present?
b. A sample of sand is composed of granules of quartz (silicon dioxide) and seashells (calcium carbonate). The sand is mixed with water. What phases are present?

1.14 A material is believed to be a compound. Suppose you have several samples of this material obtained from various places around the world. Comment on what you would expect to find upon observing the melting point and color for each sample. What would you expect to find upon determining the elemental composition for each sample?

1.15 Say you live in a climate where the temperature ranges from −100°F to 20°F and you want to define a new temperature scale, YS (YS is the "Your Scale" temperature scale), which defines this range as 0.0°YS to 100.0°YS.
a. Come up with an equation that would allow you to convert between °F and °YS.
b. Using your equation, what would be the temperature in °F if it were 66°YS?

1.16 You are presented with a piece of metal in a jar. It is your job to determine what the metal is. What are some physical properties that you could measure in order to determine the type of metal? You suspect that the metal might be sodium; what are some chemical properties that you could investigate? (See Section 1.4 for some ideas.)

1.17
a. Which of the following items have a mass of about 1 g?
   a grain of sand   a 5.0-gallon bucket of water
   a paper clip   a brick
   a nickel   a car
b. What is the approximate mass (using SI mass units) of each of the items in part a?

1.18   What is the length of the nail reported to the correct number of significant figures?

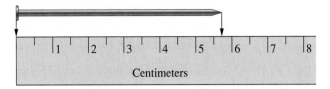

Centimeters

Practice Problems

Key: These problems are for practice in applying problem-solving skills. They are divided by topic. The problems are arranged in matched pairs; the odd-numbered problem of each pair is listed first and its answer is given in the back of the book.

Conservation of Mass

1.19   Zinc metal reacts with yellow crystals of sulfur in a fiery reaction to produce a white powder of zinc sulfide. A chemist determines that 65.4 g of zinc reacts with 32.1 g of sulfur. How many grams of zinc sulfide could be produced from 20.0 g of zinc metal?

1.20   Aluminum metal reacts with bromine, a red-brown liquid with a noxious odor. The reaction is vigorous and produces aluminum bromide, a white crystalline substance. A sample of 27.0 g of aluminum yields 266.7 g of aluminum bromide. How many grams of bromine react with 15.0 g of aluminum?

Solids, Liquids, and Gases

1.21   Give the normal state (solid, liquid, or gas) of each of the following.
a. sodium hydrogen carbonate (baking soda)
b. isopropyl alcohol (rubbing alcohol)
c. copper

1.22   Give the normal state (solid, liquid, or gas) of each of the following.
a. potassium hydrogen tartrate (cream of tartar)
b. carbon (diamond)
c. bromine

Chemical and Physical Changes; Properties of Substances

1.23   The following are properties of substances. Decide whether each is a physical property or a chemical property.
a. Chlorine gas liquefies at −35°C under normal pressure.
b. Hydrogen burns in chlorine gas.
c. Bromine freezes at −7.2°C.
d. Lithium is a soft, silvery-colored metal.

1.24   Decide whether each of the following is a physical property or a chemical property of the substance.
a. Salt substitute, potassium chloride, dissolves in water.
b. Seashells, calcium carbonate, fizz when immersed in vinegar.

c. The gas hydrogen sulfide smells like rotten eggs.
d. Fine steel wool (Fe) can be burned in air.

1.25   Iodine is a solid having somewhat lustrous, blue-black crystals. The crystals vaporize readily to a violet-colored gas. Iodine combines with many metals. For example, aluminum combines with iodine to give aluminum iodide. Identify the physical and the chemical properties of iodine that are cited.

1.26   Mercury(II) oxide is an orange-red solid with a density of 11.1 g/cm^3. It decomposes when heated to give mercury and oxygen. The compound is insoluble in water (does not dissolve in water). Identify the physical and the chemical properties of mercury(II) oxide that are cited.

Elements, Compounds, and Mixtures

1.27   Consider the following separations of materials. State whether a physical process or a chemical reaction is involved in each separation.
a. Sodium chloride is obtained from seawater by evaporation of the water.
b. Mercury is obtained by heating the substance mercury(II) oxide; oxygen is also obtained.
c. Pure water is obtained from ocean water by evaporating the water, then condensing it.
d. Iron is produced from an iron ore that contains the substance iron(III) oxide.

1.28   All of the following processes involve a separation of either a mixture into substances or a compound into elements. For each, decide whether a physical process or a chemical reaction is required.
a. Sodium metal is obtained from the substance sodium chloride.
b. Iron filings are separated from sand by using a magnet.
c. Sugar crystals are separated from a sugar syrup by evaporation of water.
d. Copper is produced when zinc metal is placed in a solution of copper(II) sulfate, a compound.

1.29   Which of the following are pure substances and which are mixtures? For each, list all of the different phases present.
a. bromine liquid and its vapor
b. paint, containing a liquid solution and a dispersed solid pigment
c. baking powder containing sodium hydrogen carbonate and potassium hydrogen tartrate

1.30 Which of the following are pure substances and which are mixtures? For each, list all of the different phases present.
a. a sugar solution with sugar crystals at the bottom
b. ink containing a liquid solution with fine particles of carbon
c. liquid water and steam at 100°C

Significant Figures

1.31 How many significant figures are there in each of the following measurements?
a. 73.0000 g   b. 0.0503 kg
c. 6.300 cm   d. 0.80090 m

1.32 How many significant figures are there in each of the following measurements?
a. 130.0 kg   b. 0.05930 g
c. 0.224800 m   d. 1008 s

1.33 The circumference of the earth at the equator is 40,000 km. This value is precise to two significant figures. Write this in scientific notation to express correctly the number of significant figures.

1.34 The astronomical unit equals the mean distance between the earth and the sun. This distance is 150,000,000 km, which is precise to three significant figures. Express this in scientific notation to the correct number of significant figures.

1.35 Do the indicated arithmetic and give the answer to the correct number of significant figures.
a. $0.71 + 81.8$
b. $934 \times 0.00435 + 107$
c. $(847.89 - 847.73) \times 14673$

1.36 Do the indicated arithmetic and give the answer to the correct number of significant figures.
a. $8.937 - 8.930$
b. $8.937 + 8.930$
c. $0.00015 \times 54.6 + 1.002$

1.37 One sphere has a radius of 5.10 cm; another has a radius of 5.00 cm. What is the difference in volume (in cubic centimeters) between the two spheres? Give the answer to the correct number of significant figures. The volume of a sphere is $(4/3)\pi r^3$, where $\pi = 3.1416$ and r is the radius.

1.38 A solid cylinder of iron of circular cross section with a radius of 1.500 cm has a ruler etched along its length. What is the volume of iron contained between the marks labeled 3.10 cm and 3.50 cm? The volume of a cylinder is $\pi r^2 l$, where $\pi = 3.1416$, r is the radius, and l is the length.

SI Units

1.39 Write the following measurements, without scientific notation, using the appropriate SI prefix.
a. 5.89×10^{-12} s   b. 0.2010 m

1.40 Write the following measurements, without scientific notation, using the appropriate SI prefix.
a. 4.851×10^{-6} g   b. 3.16×10^{-2} m

1.41 Using scientific notation, convert:
a. 6.15 ps to s   b. 3.781 μm to m

1.42 Using scientific notation, convert:
a. 6.20 km to m   b. 1.98 ns to s

Temperature Conversion

1.43 Convert:
a. 68°F to degrees Celsius
b. −70°C to degrees Fahrenheit

1.44 Convert:
a. 121°F to degrees Celsius
b. 58°C to degrees Fahrenheit

1.45 Salt and ice are stirred together to give a mixture to freeze ice cream. The temperature of the mixture is −21.1°C. What is this temperature in degrees Fahrenheit?

1.46 Liquid nitrogen can be used for the quick freezing of foods. The liquid boils at −196°C. What is this temperature in degrees Fahrenheit?

Density

1.47 A liquid with a volume of 8.5 mL has a mass of 6.74 g. The liquid is either octane, ethanol, or benzene, the densities of which are 0.702 g/cm^3, 0.789 g/cm^3, and 0.879 g/cm^3, respectively. What is the identity of the liquid?

1.48 A mineral sample has a mass of 31.5 g and a volume of 3.9 cm^3. The mineral is either sphalerite (density = 4.0 g/cm^3), cassiterite (density = 6.99 g/cm^3), or cinnabar (density = 8.10 g/cm^3). Which is it?

1.49 Platinum has a density of 21.4 g/cm^3. What is the mass of 5.9 cm^3 of this metal?

1.50 What is the mass of a 22.7-mL sample of gasoline, which has a density of 0.70 g/cm^3?

Unit Conversions

1.51 Sodium hydrogen carbonate, known commercially as baking soda, reacts with acidic materials such as vinegar to release carbon dioxide gas. An experiment calls for 0.480 kg of sodium hydrogen carbonate. Express this mass in milligrams.

1.52 The acidic constituent in vinegar is acetic acid. A 10.0-mL sample of a certain vinegar contains 483 mg of acetic acid. What is this mass of acetic acid expressed in micrograms?

1.53 How many grams are there in 3.58 short tons? Note that 1 g = 0.03527 oz (ounces avoirdupois), 1 lb (pound) = 16 oz, and 1 short ton = 2000 lb. (These relations are exact.)

1.54 The calorie, the Btu (British thermal unit), and the joule are units of energy; 1 calorie = 4.184 joules (exact), and 1 Btu = 252.0 calories. Convert 2.17 Btu to joules.

1.55 The first measurement of sea depth was made in 1840 in the central South Atlantic, where a plummet was lowered 2425 fathoms. What is this depth in meters? Note that 1 fathom = 6 ft, 1 ft = 12 in, and 1 in = 2.54×10^{-2} m. (These relations are exact.)

1.56 The estimated amount of recoverable oil from the field at Prudhoe Bay in Alaska is 9.6×10^9 barrels. What is this amount of oil in cubic meters? One barrel = 42 gal (exact), 1 gal = 4 qt (exact), and 1 qt = 9.46×10^{-4} m^3.

General Problems

Key: These problems provide more practice but are not divided by topic or keyed to exercises. Each odd-numbered problem and the even-numbered problem that follows it are similar; answers to odd-numbered problems are given in the back of the book.

1.57 When a mixture of aluminum powder and iron(III) oxide is ignited, it produces molten iron and aluminum oxide. In an experiment, 5.40 g of aluminum was mixed with 18.50 g of iron(III) oxide. At the end of the reaction, the mixture contained 11.17 g of iron, 10.20 g of aluminum oxide, and an undetermined amount of unreacted iron(III) oxide. No aluminum was left. What is the mass of the iron(III) oxide?

1.58 When chlorine gas is bubbled into a solution of sodium bromide, the sodium bromide reacts to give bromine, a red-brown liquid, and sodium chloride (ordinary table salt). A solution was made by dissolving 20.6 g of sodium bromide in 100.0 g of water. After passing chlorine through the solution, investigators analyzed the mixture. It contained 16.0 g of bromine and 11.7 g of sodium chloride. How many grams of chlorine reacted?

1.59 Identify each of the following elements from its properties under normal conditions. (See Table 2.1.)
a. red-brown liquid   b. white, waxy solid
c. soft, yellow metal   d. soft, black solid

1.60 Identify each of the following elements from its properties under normal conditions. (See Table 2.1.)
a. reddish metal
b. brittle, yellow solid
c. greenish-yellow gas
d. silver-colored liquid metal

1.61 Analyses of several samples of a material containing only iron and oxygen gave the following results. Could this material be a compound?

	Mass of Sample	Mass of Iron	Mass of Oxygen
Sample A	1.518 g	1.094 g	0.424 g
Sample B	2.056 g	1.449 g	0.607 g
Sample C	1.873 g	1.335 g	0.538 g

1.62 A red-orange solid contains only mercury and oxygen. Analyses of three different samples gave the following results. Are the data consistent with the hypothesis that the material is a compound?

	Mass of Sample	Mass of Mercury	Mass of Oxygen
Sample A	1.0410 g	0.9641 g	0.0769 g
Sample B	1.5434 g	1.4293 g	0.1141 g
Sample C	1.2183 g	1.1283 g	0.0900 g

1.63 A cubic box measures 39.3 cm on an edge. What is the volume of the box in cubic centimeters? Express the answer to the correct number of significant figures.

1.64 A cylinder with circular cross section has a radius of 2.45 cm and a height of 56.32 cm. What is the volume of the cylinder? Express the answer to the correct number of significant figures.

1.65 An aquarium has a rectangular cross section that is 47.8 in by 12.5 in; it is 19.5 in high. How many U.S. gallons does the aquarium contain? One U.S. gallon equals exactly 231 in^3.

1.66 A spherical tank has a radius of 150.0 in. Calculate the volume of the tank in cubic inches; then convert this to Imperial gallons. The volume of a sphere is $(4/3)\,\pi r^3$, where r is the radius. One Imperial gallon equals 277.4 in^3.

1.67 Perform the following arithmetic setups and express the answers to the correct number of significant figures.
a. $\dfrac{56.1 - 51.1}{6.58}$ b. $\dfrac{56.1 + 51.1}{6.58}$
c. $(9.1 + 8.6) \times 26.91$ d. $0.0065 \times 3.21 + 0.0911$

1.68 Perform the following arithmetic setups and report the answers to the correct number of significant figures.
a. $\dfrac{9.345 - 9.005}{9.811}$ b. $\dfrac{9.345 + 9.005}{9.811}$
c. $(8.12 + 7.53) \times 3.71$ d. $0.71 \times 0.36 + 17.36$

1.69 Write each of the following in terms of the SI base unit (that is, express the prefix as the power of ten).
a. 1.07 ps   b. 5.8 μm
c. 319 nm   d. 15.3 ms

1.70 Write each of the following in terms of the SI base unit (that is, express the prefix as the power of ten).
a. 7.3 mK   b. 275 pm
c. 19.6 ms   d. 45 μm

1.71 Calcium carbonate, a white powder used in toothpastes, antacids, and other preparations, decomposes when heated to about 825°C. What is this temperature in degrees Fahrenheit?

1.72   Sodium hydrogen carbonate (baking soda) starts to decompose to sodium carbonate (soda ash) at about 50°C. What is this temperature in degrees Fahrenheit?

1.73   Zinc metal can be purified by distillation (transforming the liquid metal to vapor, then condensing the vapor back to liquid). The metal boils at normal atmospheric pressure at 1666°F. What is this temperature in degrees Celsius? in kelvin units?

1.74   Iodine is a bluish-black solid. It forms a violet-colored vapor when heated. The solid melts at 236°F. What is this temperature in degrees Celsius? in kelvin units?

1.75   The density of magnesium metal (used in fireworks) is 1.74 g/cm^3. Express this density in SI units (kg/m^3).

1.76   Vanadium metal is added to steel to impart strength. The density of vanadium is 5.96 g/cm^3. Express this in SI units (kg/m^3).

1.77   Some bottles of colorless liquids were being labeled when the technicians accidentally mixed them up and lost track of their contents. A 15.0-mL sample withdrawn from one bottle weighed 22.3 g. The technicians knew that the liquid was either acetone, benzene, chloroform, or carbon tetrachloride (which have densities of 0.792 g/cm^3, 0.899 g/cm^3, 1.489 g/cm^3, and 1.595 g/cm^3, respectively). What was the identity of the liquid?

1.78   A solid will float on any liquid that is more dense than it is. The volume of a piece of calcite weighing 35.6 g is 12.9 cm^3. On which of the following liquids will the calcite float: carbon tetrachloride (density = 1.60 g/cm^3), methylene bromide (density = 2.50 g/cm^3), tetrabromoethane (density = 2.96 g/cm^3), methylene iodide (density = 3.33 g/cm^3)?

1.79   Vinegar contains acetic acid (about 5% by mass). Pure acetic acid has a strong vinegar smell but is corrosive to the skin. What volume of pure acetic acid has a mass of 35.00 g? The density of acetic acid is 1.053 g/mL.

1.80   Ethyl acetate has a characteristic fruity odor and is used as a solvent in paint lacquers and perfumes. An experiment requires 0.021 kg of ethyl acetate. What volume is this (in liters)? The density of ethyl acetate is 0.902 g/mL.

1.81   Convert:
a. 5.91 kg of chrome yellow to milligrams
b. 753 mg of vitamin A to micrograms
c. 90.1 MHz (megahertz), the wavelength of an FM signal, to kilohertz

1.82   Convert:
a. 7.19 μg of cyanocobalamin (vitamin B_{12}) to milligrams
b. 104 pm, the radius of a sulfur atom, to nanometers
c. 0.010 mm, the diameter of a typical blood capillary, to centimeters

1.83   The masses of diamonds and gems are measured in carats. A carat is defined as 200 mg. If a jeweler has 275 carats of diamonds, how many grams does she have?

1.84   Recent world production of gold was 49.6×10^6 troy ounces. One troy ounce equals 31.10 g. What is the world production of gold in metric tons (10^6 g)?

Cumulative-Skills Problems

Note: These problems require two or more operational skills you learned in this chapter. In later chapters, the problems under this heading will combine skills introduced in previous chapters with those given in the current one.

1.85   A steel sphere has a radius of 1.58 in. If this steel has a density of 7.88 g/cm^3, what is the mass of the sphere in grams?

1.86   A weather balloon filled with helium has a diameter of 3.00 ft. What is the mass in grams of the helium in the balloon at 21°C and normal pressure? The density of helium under these conditions is 0.166 g/L.

1.87   The land area of Greenland is 840,000 mi^2, with only 132,000 mi^2 free of perpetual ice. The average thickness of this ice is 5000 ft. Estimate the mass of the ice (assume two significant figures). The density of ice is 0.917 g/cm^3.

1.88   Antarctica, almost completely covered in ice, has an area of 5,500,000 mi^2 with an average height of 7500 ft. Without the ice, the height would be only 1500 ft. Estimate the mass of this ice (two significant figures). The density of ice is 0.917 g/cm^3.

1.89   A sample of an ethanol–water solution has a volume of 54.2 cm^3 and a mass of 49.6 g. What is the percentage of ethanol (by mass) in the solution? (Assume that there is no change in volume when the pure compounds are mixed.) The density of ethanol is 0.789 g/cm^3 and that of water is 0.998 g/cm^3. Alcoholic beverages are rated in *proof,* which is a measure of the relative amount of ethanol in the beverage. Pure ethanol is exactly 200 proof; a solution that is 50% ethanol by volume is exactly 100 proof. What is the proof of the given ethanol–water solution?

1.90   You have a piece of gold jewelry weighing 9.35 g. Its volume is 0.654 cm^3. Assume that the metal is an alloy (mixture) of gold and silver, which have densities of 19.3 g/cm^3 and 10.5 g/cm^3, respectively. Also assume that there is no change in volume when the pure metals are mixed. Calculate the percentage of gold (by mass) in the alloy. The relative amount of gold in an alloy is measured in *karats.* Pure gold is 24 karats; an alloy of 50% gold is 12 karats. State the proportion of gold in the jewelry in karats.

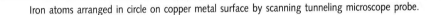

Iron atoms arranged in circle on copper metal surface by scanning tunneling microscope probe.

Atoms, Molecules, and Ions

Sodium is a soft, silvery metal. This metal cannot be handled with bare fingers, because it reacts with any moisture on the skin, causing a burn. Chlorine is a poisonous, greenish-yellow gas with a choking odor. When molten sodium is placed into an atmosphere of chlorine, a dramatic reaction occurs: the sodium bursts into flame and a white, crystalline powder forms (see Figure 2.1). The sodium and chlorine undergo a chemical change—a chemical reaction—to form matter with different chemical and physical properties. How do we explain the differences in properties? And how do we explain chemical reactions such as the burning of sodium in chlorine? This chapter and the next take an introductory look at these basic questions in chemistry.

Figure 2.1
Reaction of sodium and chlorine. *Left:* Sodium metal and chlorine gas. *Right:* A small piece of molten sodium burning in a flask of chlorine. The product is sodium chloride, common table salt.

Atomic Theory and Atomic Structure

All matter is composed of atoms. In this first part of Chapter 2, we explore the atomic theory of matter and the structure of atoms. We also examine the *periodic table,* which is an organizational chart designed to highlight similarities among the elements.

2.1 Atomic Theory of Matter

As we noted in Chapter 1, Lavoisier laid the experimental foundation of modern chemistry. But the British chemist John Dalton (1766–1844) provided the basic theory: all matter—whether element, compound, or mixture—is composed of small particles called atoms. The postulates of Dalton's theory are presented in this section. Note that the terms *element, compound,* and *chemical reaction,* which were defined in Chapter 1 in terms of matter as we normally see it, are redefined here by the postulates of Dalton's theory in terms of atoms.

Postulates of Dalton's Atomic Theory

The main points of Dalton's **atomic theory,** *an explanation of the structure of matter in terms of different combinations of very small particles,* are as follows:

1. All matter is composed of indivisible atoms. An **atom** is *an extremely small particle of matter that retains its identity during chemical reactions.*

2. An **element** is *a type of matter composed of only one kind of atom,* each atom of a given kind having the same properties. Mass is one such property. Thus, the atoms of a given element have a characteristic mass. ◀

▶ As you will see in Section 2.4, it is the *average* mass of an atom that is characteristic of each element on earth.

3. A **compound** is *a type of matter composed of atoms of two or more elements chemically combined in fixed proportions.* Water, for example, consists of hydrogen and oxygen atoms in the ratio of 2 to 1.

4. A **chemical reaction** consists of *the rearrangement of the atoms present in the reacting substances to give new chemical combinations present in the substances formed by the reaction.* Atoms are not created, destroyed, or broken into smaller particles by any chemical reaction.

Today we know that atoms are not truly indivisible; they are themselves made up of particles. Nevertheless, Dalton's postulates are essentially correct.

Atomic Symbols and Models

It is convenient to use symbols for the atoms of the different elements. An **atomic symbol** is *a one- or two-letter notation used to represent an atom corresponding to a particular element.* ◄ Typically, the atomic symbol consists of the first letter, capitalized, from the name of the element, sometimes with an additional letter from the name in lowercase. For example, chlorine has the symbol Cl. Other symbols are derived from a name in another language (usually Latin). Sodium is given the symbol Na from its Latin name, *natrium.* Symbols of selected elements are listed in Table 2.1.

Early in the development of his atomic theory, Dalton built spheres to represent atoms and used combinations of these spheres to represent compounds. Chemists continue to use this idea of representing atoms by three-dimensional models, but because we now know so much more about atoms and molecules than Dalton did, these models have become more refined. Section 2.6 describes these models in more detail.

► The atomic names (temporary) given to newly discovered elements are derived from the atomic number (discussed in Section 2.3) using the numerical roots: *nil* (0), *un* (1), *bi* (2), *tri* (3), *quad* (4), *pent* (5), *hex* (6), *sept* (7), *oct* (8), and *enn* (9) with the ending *-ium.* For example, the element with atomic number 118 is ununoctium; its atomic symbol is Uuo.

Deductions from Dalton's Atomic Theory

Note how atomic theory explains the difference between an element and a compound. Atomic theory also explains two laws we considered earlier. One of these is the law of conservation of mass, which states that the total mass remains constant during a chemical reaction. By postulate 2, every atom has a definite mass. Because a chemical reaction only rearranges the chemical combinations of atoms (postulate 4), the mass must remain constant. The other law explained by atomic theory is the law of definite proportions (constant composition). Postulate 3 defines a compound as a type of matter containing the atoms of two or more elements in definite proportions. Because the atoms have definite mass, compounds must have the elements in definite proportions by mass.

A good theory should not only explain known facts and laws but also predict new ones. The **law of multiple proportions,** deduced by Dalton from his atomic theory, states that *when two elements form more than one compound, the masses of one element in these compounds for a fixed mass of the other element are in ratios of small whole numbers.* To illustrate this law, consider the following. If we take a fixed mass of carbon, 1.000 gram, and react it with oxygen, we end up with two compounds: one that contains 1.3321 grams of oxygen and one that contains 2.6642 grams of oxygen. Note that the ratio of the amounts of oxygen in the two compounds is 2 to 1 (2.6642 ÷ 1.3321). Applying atomic theory, if we assume that the compound that has 1.3321 grams of oxygen to 1.000 gram of carbon is CO (the combination of one carbon atom and one oxygen atom), then the compound that contains twice as much oxygen per 1.000 gram of carbon must be CO_2. The deduction of the law of multiple proportions from atomic theory was important in convincing chemists of the validity of the theory.

Table 2.1
Some Common Elements

Name of Element	Atomic Symbol	Physical Appearance of Element*
Aluminum	Al	Silvery-white metal
Barium	Ba	Silvery-white metal
Bromine	Br	Reddish-brown liquid
Calcium	Ca	Silvery-white metal
Carbon	C	
Graphite		Soft, black solid
Diamond		Hard, colorless crystal
Chlorine	Cl	Greenish-yellow gas
Chromium	Cr	Silvery-white metal
Cobalt	Co	Silvery-white metal
Copper	Cu (from *cuprum*)	Reddish metal
Fluorine	F	Pale yellow gas
Gold	Au (from *aurum*)	Pale yellow metal
Helium	He	Colorless gas
Hydrogen	H	Colorless gas
Iodine	I	Bluish-black solid
Iron	Fe (from *ferrum*)	Silvery-white metal
Lead	Pb (from *plumbum*)	Bluish-white metal
Magnesium	Mg	Silvery-white metal
Manganese	Mn	Gray-white metal
Mercury	Hg (from *hydrargyrum*)	Silvery-white liquid metal
Neon	Ne	Colorless gas
Nickel	Ni	Silvery-white metal
Nitrogen	N	Colorless gas
Oxygen	O	Colorless gas
Phosphorus (white)	P	Yellowish-white, waxy solid
Potassium	K (from *kalium*)	Soft, silvery-white metal
Silicon	Si	Gray, lustrous solid
Silver	Ag (from *argentum*)	Silvery-white metal
Sodium	Na (from *natrium*)	Soft, silvery-white metal
Sulfur	S	Yellow solid
Tin	Sn (from *stannum*)	Silvery-white metal
Zinc	Zn	Bluish-white metal

*Common form of the element under normal conditions.

CONCEPT CHECK 2.1

Like Dalton, chemists continue to model atoms using spheres. Modern models are usually drawn with a computer and use different colors to represent atoms of different elements. Which of the models below represents CO_2?

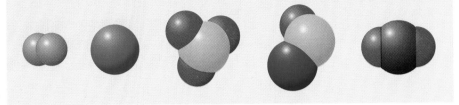

2.2 The Structure of the Atom

Although Dalton postulated that atoms were indivisible particles, experiments conducted around the beginning of the last century showed that atoms themselves consist of particles. These experiments showed that an atom consists of two kinds of particles: a **nucleus,** *the atom's central core, which is positively charged and contains most of the atom's mass,* and one or more electrons. An **electron** is *a very light, negatively charged particle that exists in the region around the atom's positively charged nucleus.*

Discovery of the Electron

In 1897 the British physicist J. J. Thomson (Figure 2.2) conducted a series of experiments that showed that atoms were not indivisible particles. Figure 2.3 shows an experimental apparatus similar to the one used by Thomson. In this apparatus, two electrodes from a high-voltage source are sealed into a glass tube from which the air has been evacuated. The negative electrode is called the *cathode;* the positive one, the *anode.* When the high-voltage current is turned on, the glass tube emits a greenish light. Experiments showed that this greenish light is caused by the interaction of the glass with *cathode rays,* which are rays that originate from the cathode.

Figure 2.2
Joseph John Thomson (1856–1940). J. J. Thomson's scientific ability was recognized early with his appointment as professor of physics in the Cavendish Laboratory at Cambridge University when he was not quite 28 years old. Soon after this appointment, Thomson began research on the discharge of electricity through gases. This work culminated in 1897 with the discovery of the electron. Thomson was awarded the Nobel Prize in physics in 1906.

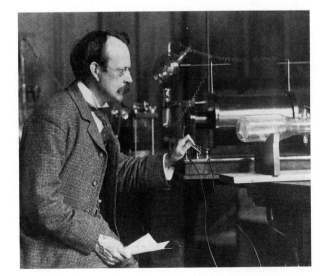

Figure 2.3
Formation of cathode rays. Cathode rays leave the cathode, or negative electrode, and are accelerated toward the anode, or positive electrode. Some of the rays pass through the hole in the anode to form a beam, which then is bent by the electric plates in the tube.

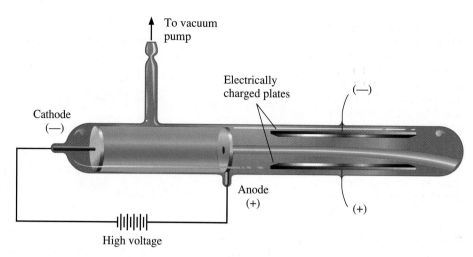

The cathode rays move toward the anode, where some rays pass through a hole to form a beam (Figure 2.3). This beam bends away from the negatively charged plate and toward the positively charged plate. (Cathode rays are not directly visible, but do cause certain materials such as zinc sulfide to glow so you can observe them.) Thomson showed that the characteristics of cathode rays are independent of the material making up the cathode. From such evidence, he concluded that a cathode ray consists of a beam of negatively charged particles (or electrons). ◄

From his experiments, Thomson could also calculate the ratio of the electron's mass, m_e, to its electric charge, e. In 1909, the U.S. physicist Robert Millikan performed a series of ingenious experiments in which he obtained the charge on the electron by observing how a charged drop of oil falls in the presence and in the absence of an electric field (Figure 2.4). From this type of experiment, the charge on the electron is found to be 1.602×10^{-19} coulombs (the *coulomb*, abbreviated C, is a unit of electric charge). If you use this charge with the mass-to-charge ratio of the electron, you obtain an electron mass of 9.109×10^{-31} kg, which is more than 1800 times smaller than the mass of the lightest atom (hydrogen).

► A television tube uses the deflection of cathode rays by magnetic fields. A beam of cathode rays is directed toward a coated screen on the front of the tube, where by varying the magnetism generated by electromagnetic coils, the beam traces a luminescent image.

The Nuclear Model of the Atom

Ernest Rutherford (1871–1937), a British physicist, put forth the idea of the *nuclear model* of the atom in 1911, based on experiments done in his laboratory by Hans Geiger and Ernest Marsden. These scientists observed the effect of bombarding thin gold foil with alpha radiation from radioactive substances such as uranium (Figure 2.5). Geiger and Marsden found that most of the alpha particles passed through a metal foil as though nothing were there, but a few (about 1 in 8000) were scattered at large angles and sometimes almost backward.

According to Rutherford's model, most of the mass of the atom (99.95% or more) is concentrated in a positively charged center, or nucleus, around which the negatively charged electrons move. Although most of the mass of an atom is in its nucleus, the nucleus occupies only a very small portion of the space of the atom. Nuclei have diameters of about 10^{-3} pm, whereas atomic diameters are about 100 pm, a hundred thousand times larger. If you were to use a golf ball to represent the nucleus, the atom would be about 3 miles in diameter!

Figure 2.4
Millikan's oil-drop experiment. An atomizer, or spray bottle, introduces a fine mist of oil drops into the top chamber. Several drops happen to fall through a small hole into the lower chamber, where the experimenter follows the motion of one drop with a microscope. Some of these drops have picked up one or more electrons as a result of friction in the atomizer and have become negatively charged. A negatively charged drop will be attracted upward when the experimenter turns on a current to the electric plates. The drop's upward speed (obtained by timing its rise) is related to its mass-to-charge ratio, from which you can calculate the charge on the electron.

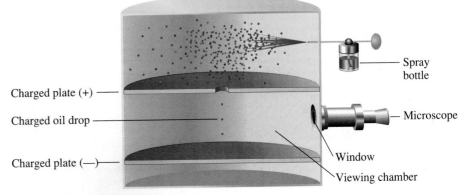

Charged plate (+)
Charged oil drop
Charged plate (—)

Spray bottle
Microscope
Window
Viewing chamber

The nuclear model explains the results of bombarding gold and other metal foils with alpha particles. Alpha particles are helium nuclei that pass through the metal atoms of the foil undeflected by the lightweight electrons. When an alpha particle does happen to hit a nucleus, however, it is scattered at a wide angle because it is deflected by the massive, positively charged nucleus (Figure 2.6).

CONCEPT CHECK 2.2

What would be a feasible model for the atom if Geiger and Marsden had found that 7999 out of 8000 alpha particles were deflected back at the alpha-particle source?

2.3 Nuclear Structure; Isotopes

The nucleus of an atom also has a structure; in chemistry we can consider that the nucleus is composed of two different kinds of particles, protons and neutrons.

An important property of the nucleus is its electric charge. One way to determine the value of the positive charge of a nucleus is by analyzing the distribution of alpha particles scattered from a metal foil. ◄ Each element has a unique nuclear charge that is an integer multiple of the magnitude of the electron charge. This integer, which is characteristic of an element, is called the *atomic number* (Z). A hydrogen atom nucleus, whose magnitude of charge equals that of the electron, has the smallest atomic number, which is one.

▶ The simplest way to obtain the nuclear charge is by analyzing the x rays emitted by the element when irradiated with cathode rays.

Figure 2.5
Alpha-particle scattering from metal foils. Alpha radiation is produced by a radioactive source and formed into a beam by a lead plate with a hole in it. (Lead absorbs the radiation.) Scattered alpha particles are made visible by a zinc sulfide screen, which emits tiny flashes where particles strike it. A movable microscope is used for viewing the flashes.

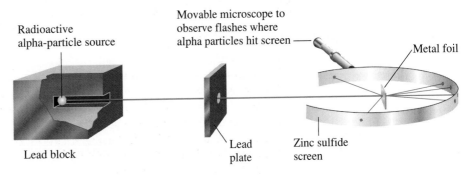

Radioactive alpha-particle source

Movable microscope to observe flashes where alpha particles hit screen

Metal foil

Lead block

Lead plate

Zinc sulfide screen

Figure 2.6
Representation of the scattering of alpha particles by a gold foil. Most of the alpha particles pass through the foil barely deflected. A few, however, collide with gold nuclei and are deflected at large angles. (The relative sizes of nuclei are smaller than can be drawn here.)

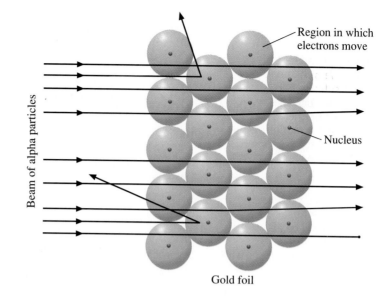

Region in which electrons move

Nucleus

Beam of alpha particles

Gold foil

In 1919, Rutherford discovered that hydrogen nuclei, or what we now call protons, form when alpha particles strike some of the lighter elements, such as nitrogen. A **proton** is *a nuclear particle having a positive charge equal to that of the electron and a mass more than 1800 times that of the electron.*

The **atomic number** (Z) is *the number of protons in the nucleus of an atom.* An **element** is *a substance whose atoms all have the same atomic number.* The inside back cover of the book lists the elements and their atomic numbers (see the Table of Atomic Numbers and Atomic Weights).

The neutron was also discovered by alpha-particle scattering experiments. When beryllium is irradiated with alpha rays, a strongly penetrating radiation is obtained from the metal. In 1932 the British physicist James Chadwick (1891–1974) showed that this penetrating radiation consists of neutral particles. The **neutron** is *a nuclear particle having a mass almost identical to that of the proton but no electric charge.* Table 2.2 compares the masses and charges of the electron and the two nuclear particles, the proton and the neutron.

Now consider the nucleus of the sodium atom. The nucleus contains 11 protons and 12 neutrons. Thus, the charge on the sodium nucleus is simply +11. Similarly, the nucleus of an aluminum atom contains 13 protons and 14 neutrons; the charge on the nucleus is +13.

Table 2.2
Properties of the Electron, Proton, and Neutron

Particle	Mass (kg)	Charge (C)	Mass (amu)*	Charge (e)
Electron	9.10939×10^{-31}	-1.60218×10^{-19}	0.00055	−1
Proton	1.67262×10^{-27}	$+1.60218 \times 10^{-19}$	1.00728	+1
Neutron	1.67493×10^{-27}	0	1.00866	0

*The atomic mass unit (amu) equals 1.66054×10^{-27} kg; it is defined in Section 2.4.

Figure 2.7
A representation of two isotopes of carbon. The drawing shows the basic particles making up the carbon-12 and carbon-13 isotopes. (The relative sizes of the nuclei are much exaggerated here.)

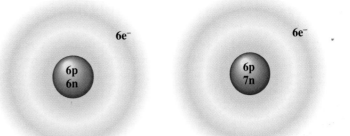

We characterize a nucleus by its atomic number (Z) and its mass number (A). The **mass number** (A) is *the total number of protons and neutrons in a nucleus.* The nucleus of the naturally occurring sodium atom has an atomic number of 11 and a mass number of 23 (11 + 12).

A **nuclide** is *an atom characterized by a definite atomic number and mass number.* The shorthand notation for any nuclide consists of the symbol of the element with the atomic number written as a subscript on the left and the mass number as a superscript on the left. You write the *nuclide symbol* for the naturally occurring sodium nuclide as follows:

$$\text{Mass number} \longrightarrow {}^{23}_{11}\text{Na} \longleftarrow \text{Atomic number}$$

An atom has as many electrons about its nucleus as the nucleus has protons.

All nuclei of the atoms of a particular element have the same atomic number, but the nuclei may have different mass numbers. **Isotopes** are *atoms whose nuclei have the same atomic number but different mass numbers; that is, the nuclei have the same number of protons but different numbers of neutrons.* Naturally occurring oxygen is a mixture of isotopes; it contains 99.759% oxygen-16, 0.037% oxygen-17, and 0.204% oxygen-18. ◄

Figure 2.7 may help you to visualize the relationship among the different subatomic particles in the isotopes of an element. The size of the nucleus as represented in the figure is very much exaggerated in order to show the numbers of protons and neutrons. Electrons, although extremely light particles, move throughout relatively large diffuse regions, or "shells," about the nucleus of an atom.

▶ Isotopes were first suspected in about 1912 when chemically identical elements with different atomic masses were found in radioactive materials. The most convincing evidence, however, came from the mass spectrometer, discussed in Section 2.4.

EXAMPLE 2.1

Writing Nuclide Symbols

What is the nuclide symbol for a nucleus that contains 38 protons and 50 neutrons?

SOLUTION

From the Table of Atomic Numbers and Atomic Weights on the inside back cover of this book, you will note that the element with atomic number 38 is strontium, symbol Sr. The mass number is 38 + 50 = 88. The symbol is ${}^{88}_{38}\text{Sr}$.

2.4 Atomic Weights

A central feature of Dalton's atomic theory was the idea that an atom of an element has a characteristic mass. Now we know that a naturally occurring element may be a mixture of isotopes, each isotope having its own characteristic mass. Thus, what Dalton actually calculated were *average* atomic masses. ◄

Relative Atomic Masses

Dalton found the average mass of one atom relative to the average mass of another. We will refer to these relative atomic masses as *atomic weights*. To see how Dalton obtained his atomic weights, imagine that you burn hydrogen in oxygen. The product of this reaction is water. You find that 1.0000 gram of hydrogen reacts with 7.9367 g of oxygen. To obtain the atomic weight of oxygen (relative to hydrogen), you need to know the relative numbers of hydrogen atoms and oxygen atoms in water. Today we know that water contains two atoms of hydrogen for every atom of oxygen. So the atomic weight of oxygen is $2 \times 7.9367 = 15.873$ times that of the mass of the average hydrogen atom. ◄

Atomic Mass Units

Dalton's hydrogen-based atomic weight scale was eventually replaced by a scale based on oxygen and then, in 1961, by the present carbon-12 mass scale. The relative atomic masses are obtained by an instrument called a *mass spectrometer*. The mass of an atom is compared to the carbon-12 isotope, which is assigned a mass of exactly 12 atomic mass units. One **atomic mass unit (amu)** is, therefore, *a mass unit equal to exactly one-twelfth the mass of a carbon-12 atom.*

On this modern scale, the **atomic weight** of an element is *the average atomic mass for the naturally occurring element, expressed in atomic mass units.* A complete table of atomic weights appears on the inside back cover of this book.

Mass Spectrometry and Atomic Weights

Earlier we noted that you use a mass spectrometer to obtain accurate masses of atoms. Mass spectrometers measure the mass-to-charge ratios of positively charged **ions** (discussed in Section 2.6), which are obtained from atoms by the removal of electrons. Figure 2.8 shows a simplified sketch of a mass spectrometer.

In the mass spectrometer shown in Figure 2.8, neon gas passes through an inlet tube into a chamber where the neon atoms collide with electrons from an electron beam. The force of a collision can knock an electron from a neon atom. The positive neon ions produced this way are drawn toward a negative grid, and some of them pass through slits to form a beam of positive electricity. This beam then travels through a magnetic field. The magnetic field deflects the positively charged ions according to their mass-to-charge ratios. The most massive positively charged ions are deflected the least by the magnetic field. For this reason, each positively charged ion arrives at a definite position at the end of the tube depending on its mass-to-charge ratio. The positively charged neon ions split into three beams corresponding to the three different isotopes of neon. Each ion has a charge of +1 but a mass number of either 20, 21, or 22. ◄

► Some variation of isotopic composition occurs in a number of elements, and this limits the significant figures in their average atomic masses.

► Dalton had assumed the formula for water to be HO. Using this formula and accurate data, he would have obtained 7.9367 for the relative atomic weight of oxygen.

► Neon was the first element to be separated into isotopes. In 1913 J. J. Thomson found the mass of neon to be 20 amu, but he also found a less-abundant mass at 22 amu, which he thought was a contaminant. Later, with improved apparatus, F. W. Aston showed that most elements are mixtures of isotopes.

Figure 2.8
Diagram of a simple mass spectrometer, showing the separation of neon isotopes. Neon gas enters an evacuated chamber, where neon atoms form positive ions when they collide with electrons. Positively charged neon ions, Ne^+, are accelerated from this region by the negative grid and pass between the poles of a magnet. The beam of positively charged ions is split into three beams by the magnetic field according to the mass-to-charge ratios. (The detector is shown here as a photographic plate; in modern spectrometers, the detector is electronic, and the mass positions are recorded on a computer.)

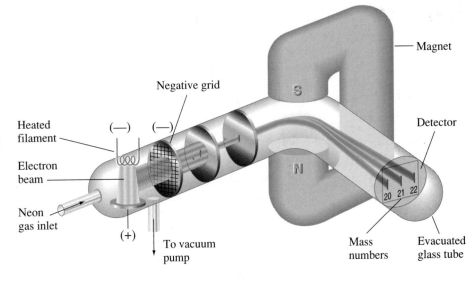

The neon isotopes have the following masses: neon-20, 19.992 amu; neon-21, 20.994 amu; and neon-22, 21.991 amu. You can also obtain the percentage of the total number of neon ions that are neon-20, for example, from the relative number of positively charged neon-20 ions reaching the detector at the end of the tube. Figure 2.9 shows the *mass spectrum* of neon.

The mass spectrum gives us all of the information needed to calculate the atomic weight: the masses of isotopes and their relative numbers, or fractional abundances. The **fractional abundance** of an isotope is *the fraction of the total number of atoms that is composed of a particular isotope.* The fractional abundances of the neon isotopes in naturally occurring neon are neon-20, 0.9051; neon-21, 0.0027; and neon-22, 0.0922.

You calculate the atomic weight of an element by multiplying each isotopic mass by its fractional abundance and summing the values. If you do that for neon using the data given here, you will obtain 20.18 amu. The next example illustrates the calculation in full for chromium.

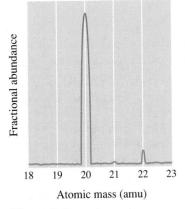

Figure 2.9
The mass spectrum of neon. Neon is separated into its isotopes Ne-20, Ne-21, and Ne-22. The height at each mass peak is proportional to the fraction of that isotope in the element.

EXAMPLE 2.2

Determining Atomic Weight from Isotopic Masses and Fractional Abundances

Chromium, Cr, has the following isotopic masses and fractional abundances:

Mass Number	Isotopic Mass (amu)	Fractional Abundance
50	49.9461	0.0435
52	51.9405	0.8379
53	52.9407	0.0950
54	53.9389	0.0236

What is the atomic weight of chromium?

SOLUTION

Multiply each isotopic mass by its fractional abundance, then sum:

$$49.9461 \text{ amu} \times 0.0435 = 2.17 \text{ amu}$$
$$51.9405 \text{ amu} \times 0.8379 = 43.52 \text{ amu}$$
$$52.9407 \text{ amu} \times 0.0950 = 5.03 \text{ amu}$$
$$53.9389 \text{ amu} \times 0.0236 = \underline{1.27 \text{ amu}}$$
$$51.99 \text{ amu}$$

The atomic weight of chromium is **51.99 amu.**

See Problems 2.29 and 2.30.

2.5 Periodic Table of the Elements

In 1869 the Russian chemist Dmitri Mendeleev (1834–1907; Figure 2.10) and the German chemist J. Lothar Meyer (1830–1895), working independently, made similar discoveries. They found that when they arranged the elements in order of atomic weight, they could place them in horizontal rows, one row under the other, so that the elements in each vertical column have similar properties. *A tabular arrangement of elements in rows and columns, highlighting the regular repetition of properties of the elements,* is called a **periodic table.**

A modern version of the periodic table, with the elements arranged by atomic number rather than atomic weight, is shown in Figure 2.11. Each entry lists the atomic number, atomic symbol, and atomic weight of an element. As we develop the subject matter of chemistry throughout the text, you will see how useful the periodic table is.

Periods and Groups

The basic structure of the periodic table is its division into rows and columns, or periods and groups. A **period** consists of *the elements in any one horizontal row of the periodic table.* A **group** consists of *the elements in any one column of the periodic table.*

The first period of elements consists of only hydrogen (H) and helium (He). The second period has 8 elements, beginning with lithium (Li) and ending with neon (Ne). There is then another period of 8 elements, and this is followed by a period having 18 elements, beginning with potassium (K) and ending with krypton (Kr). The fifth period also has 18 elements. The sixth period actually consists of 32 elements, but in order for the row to fit on a page, part of it appears at the bottom of the table. Otherwise the table would have to be expanded, with the additional elements placed after lanthanum (La, atomic number 57). The seventh period, though not complete, also has some of its elements placed as a row at the bottom of the table.

The numbering frequently seen in North America labels the groups with Roman numerals and A's and B's. In Europe a similar convention has been used, but some columns have the A's and B's interchanged. To eliminate this confusion, the International Union of Pure and Applied Chemistry (IUPAC) suggested a convention in which the columns are numbered 1 to 18. When we refer to an element by its periodic group, we will use the traditional North American convention. The A groups

**Figure 2.10
Dmitri Ivanovich
Mendeleev (1834–1907).**
Mendeleev constructed a periodic table as part of his effort to systematize chemistry. He received many international honors for his work, but his reception in czarist Russia was mixed. He had pushed for political reforms and made many enemies as a result.

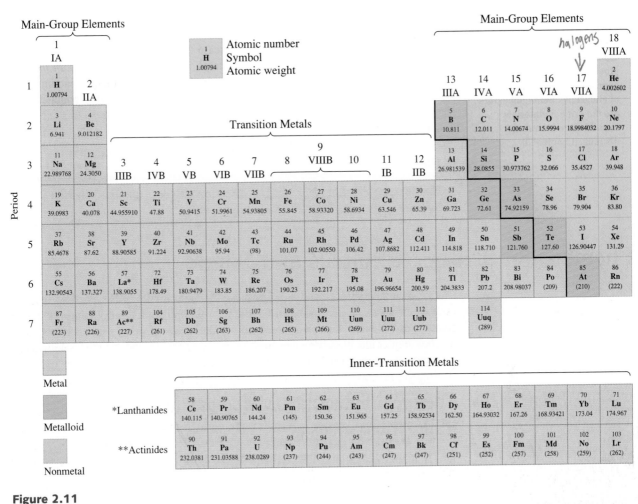

Figure 2.11
A modern form of the periodic table. This table is also given on the inside front cover of the book. The placement in the main body of the table of the two partial rows at the bottom is indicated by an asterisk (*) and a double asterisk (**).

are called *main-group* (or *representative*) *elements;* the B groups are called *transition elements.* The two rows of elements at the bottom of the table are called *inner-transition elements* (the first row is referred to as the *lanthanides;* the second row, as the *actinides*).

As noted earlier, the elements in any one group have similar properties. For example, the elements in Group IA, often known as the *alkali metals,* are soft metals that react easily with water. (Hydrogen, a gas, is an exception and might better be put in a group by itself.) The Group VIIA elements, known as *halogens,* are also reactive elements. We have already noted the vigorous reaction of chlorine and sodium. Bromine, which is a red-brown liquid, also reacts vigorously with sodium.

Metals, Nonmetals, and Metalloids

The elements of the periodic table in Figure 2.11 are divided by a heavy "staircase" line into metals on the left and nonmetals on the right. A **metal** is *a substance or mixture that has a characteristic luster and is generally a good conductor of heat and electricity.* Except for mercury, the metallic elements are solids at room temperature (about 20°C). They are more or less *malleable* (can be hammered into sheets) and *ductile* (can be drawn into wire).

A **nonmetal** is *an element that does not exhibit the characteristics of a metal.*

▶ When these pure semiconductor elements have small amounts of certain other elements added to them (a process called *doping*), they become very good conductors of electricity. Semiconductors are the critical materials in solid-state electronic devices. (Chapter 13)

Many nonmetals are gases (for example, chlorine and oxygen). The solid nonmetals are usually hard, brittle substances. Bromine is the only liquid nonmetal.

Most of the elements bordering the staircase line in the periodic table are metalloids, or semimetals. A **metalloid,** or **semimetal,** is *an element having both metallic and nonmetallic properties.* These elements, such as silicon (Si) and germanium (Ge), are usually good *semiconductors*—elements that, when pure, are poor conductors of electricity at room temperature but become moderately good conductors at higher temperatures. ◀

CONCEPT CHECK 2.3

Consider the elements He, Ne, and Ar. Can you come up with a reason why they are in the same group in the periodic table?

Chemical Substances: Formulas and Names

Atomic theory has developed steadily since Dalton's time and has become the cornerstone of chemistry. It results in an enormous simplification: all of the millions of compounds we know today are composed of the atoms of just a few elements. Now we look more closely at how we describe the composition and structure of chemical substances in terms of atoms.

2.6 Chemical Formulas; Molecular and Ionic Substances

The **chemical formula** of a substance is *a notation that uses atomic symbols with numerical subscripts to convey the relative proportions of atoms of the different elements in the substance.* Consider the formula of aluminum oxide, Al_2O_3. This means that the compound is composed of aluminum atoms and oxygen atoms in the ratio 2 : 3. Consider the formula for sodium chloride, NaCl. When no subscript is written for a symbol, it is assumed to be 1.

Additional information may be conveyed by different kinds of chemical formulas. To understand this, we need to look briefly at two main types of substances: molecular and ionic.

Molecular Substances

▶ Another way to understand the large numbers of molecules involved in relatively small quantities of matter is to consider 1 g of water (about one-fifth teaspoon). It contains 3.3×10^{22} water molecules. If you had a penny for every molecule in this quantity of water, the height of your stack of pennies would be about 300 million times the distance from the earth to the sun.

A **molecule** is *a definite group of atoms that are chemically bonded together—that is, tightly connected by attractive forces.* The nature of these strong forces is discussed in Chapters 9 and 10. A *molecular substance* is a substance that is composed of molecules all of which are alike. The molecules in such a substance are so small that even extremely minute samples contain tremendous numbers of them. One billionth (10^{-9}) of a drop of water, for example, contains about 2 trillion (2×10^{12}) water molecules. ◀

A **molecular formula** *gives the exact number of different atoms of an element in a molecule.* The hydrogen peroxide molecule contains two hydrogen atoms and two oxygen atoms chemically bonded. Therefore, its molecular formula is H_2O_2. Other simple molecular substances are water, H_2O; ammonia, NH_3; carbon dioxide, CO_2;

Figure 2.12
Examples of molecular and structural formulas and molecular models.
Three common molecules—water, ammonia, and ethanol—are shown.

	Water	Ammonia	Ethanol
Molecular formula	H_2O	NH_3	C_2H_6O

Structural formula	H—O—H	H—N—H
		     |
		     H
			H H
			   | |
			 H—C—C—O—H
			   | |
			   H H

Molecular model (ball-and-stick type)	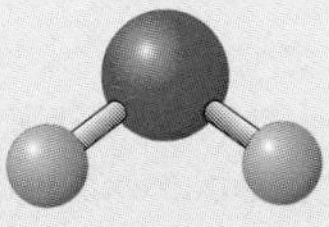	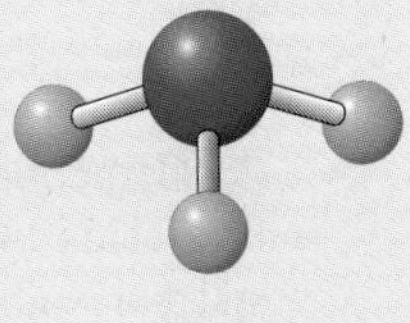	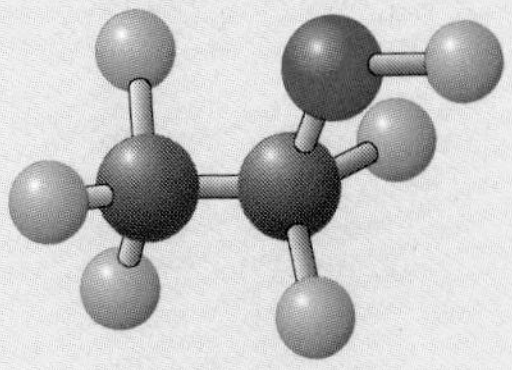
Molecular model (space-filling type)	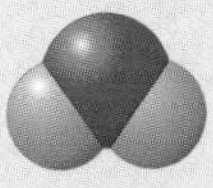	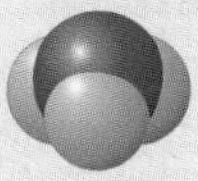	

and ethanol (ethyl alcohol), C_2H_6O. In Chapter 3, you will see how we determine such formulas.

The atoms in a molecule are chemically bonded in a definite way. A *structural formula* is a chemical formula that shows how the atoms are bonded to one another. For example, it is known that each of the hydrogen atoms in the water molecule is bonded to the oxygen atom. Thus, the structural formula of water is H—O—H. A line joining two atomic symbols in such a formula represents the chemical bond connecting the atoms. Figure 2.12 shows some structural formulas. Structural formulas are sometimes condensed in writing. For example, the structural formula of ethanol may be written CH_3CH_2OH or C_2H_5OH, depending on the detail you want to convey.

The atoms in a molecule are not only connected in definite ways but exhibit definite spatial arrangements as well. Chemists often construct molecular models as an aid in visualizing the shapes and sizes of molecules. Figure 2.12 shows molecular models for several compounds. While the ball-and-stick type of model shows the bonds and bond angles clearly, the space-filling type gives a more realistic feeling of the space occupied by the atoms.

Some elements are molecular substances and are represented by molecular formulas. Chlorine, for example, is a molecular substance and has the formula Cl_2, each molecule being composed of two chlorine atoms bonded together. Sulfur consists of molecules composed of eight atoms; its molecular formula is S_8. Helium and neon are composed of isolated atoms; their formulas are He and Ne, respectively. Other elements, such as carbon (in the form of graphite or diamond), do not have a simple molecular structure but consist of a very large, indefinite number of atoms bonded together. These elements are represented simply by their atomic symbols. Models of some elementary substances are shown in Figure 2.13.

An important class of molecular substances is the polymers. **Polymers** are *very large molecules that are made up of a number of smaller molecules repeatedly linked together.* **Monomers** are *the small molecules that are linked together to form the polymer.*

Figure 2.13
Molecular models of some elementary substances. *Left to right:* Chlorine, Cl_2, white phosphorus, P_4, and sulfur, S_8.

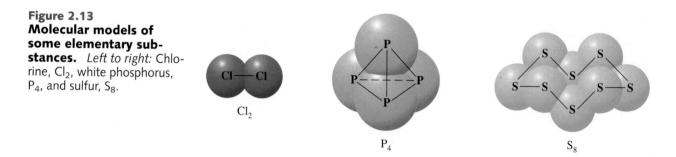

Polymers are both natural and synthetic. Wool and silk are natural polymers of amino acids linked by peptide bonds. Nylon® for fabrics, Kevlar® for bulletproof vests, and Nomex® for flame-retardant clothing all contain peptide bonds. Plastics and rubbers are polymers that are made from carbon- and hydrogen-containing monomers. Even the Teflon® coating on cookware is a polymer that is the result of linking CF_2CF_2 monomers (Figure 2.14). From these examples, it is obvious that polymers are very important molecular materials that we use every day in a wide variety of applications.

Ionic Substances

Although many substances are molecular, others are composed of ions. An **ion** is *an electrically charged particle obtained from an atom or molecule by adding or removing electrons.*

During the formation of certain compounds atoms can become ions. Metal atoms tend to lose electrons, whereas nonmetals tend to gain electrons. When a metal atom such as sodium and a nonmetal atom such as chlorine approach one another, an electron can transfer from the metal atom to the nonmetal atom to produce ions.

An atom that picks up an extra electron becomes *a negatively charged ion,* called an **anion.** An atom that loses an electron becomes *a positively charged ion,* called a **cation.** A sodium atom, for example, can lose an electron to form a sodium cation (denoted Na^+). A chlorine atom can gain an electron to form a chloride anion (denoted Cl^-).

Some ions consist of two or more atoms chemically bonded but having an excess or deficiency of electrons so that the unit has an electric charge. An example is the

Figure 2.14
$F_2C{=}CF_2$ monomer and Teflon®. *Left:* Model of the monomer used to make Teflon®, CF_2CF_2. *Center:* Model showing linkage of CF_2CF_2 monomers that make Teflon®. *Right:* Pan with Teflon® coating.

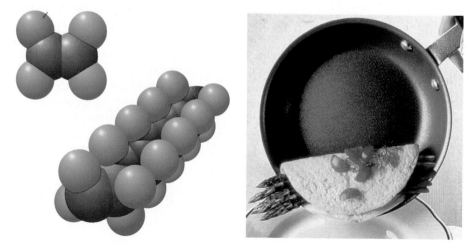

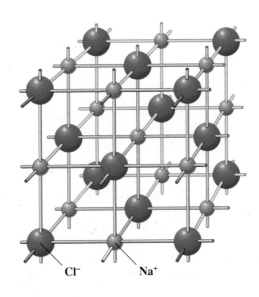

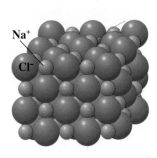

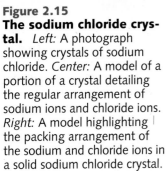

Figure 2.15
The sodium chloride crystal. *Left:* A photograph showing crystals of sodium chloride. *Center:* A model of a portion of a crystal detailing the regular arrangement of sodium ions and chloride ions. *Right:* A model highlighting the packing arrangement of the sodium and chloride ions in a solid sodium chloride crystal.

sulfate ion, SO_4^{2-}. The superscript $2-$ indicates an excess of two electrons on the group of atoms.

An **ionic compound** is *a compound composed of cations and anions*. The strong attraction between positive and negative charges holds the ions together in a regular arrangement in space. For example, in sodium chloride, each Na^+ ion is surrounded by six Cl^- ions, and each Cl^- ion is surrounded by six Na^+ ions. The result is a *crystal*, a solid having a regular three-dimensional arrangement of ions. Figure 2.15 shows sodium chloride crystals and two types of models used to depict the arrangement of the ions in the crystal.

The formula of an ionic compound is written by giving the smallest possible integer number of different ions in the substance, except that the charges on the ions are omitted so that the formulas merely indicate the atoms involved. For example, sodium chloride is written NaCl (not Na^+Cl^-). Iron(III) sulfate is a compound consisting of iron(III) ions, Fe^{3+}, and sulfate ions, SO_4^{2-}, in the ratio 2 : 3. The formula is written $Fe_2(SO_4)_3$, in which parentheses enclose the formula of an ion composed of more than one atom (again, omitting ion charges); parentheses are used only when there are two or more such ions.

The **formula unit** of a substance is *the group of atoms or ions explicitly symbolized in the formula*. For example, the formula unit of water is the H_2O molecule. The formula unit of iron(III) sulfate, $Fe_2(SO_4)_3$, consists of two Fe^{3+} ions and three SO_4^{2-} ions.

All substances, including ionic compounds, are electrically neutral. You can use this fact to obtain the formula of an ionic compound, given the formulas of the ions. This is illustrated in the following example.

EXAMPLE 2.3

Writing an Ionic Formula, Given the Ions

a. Chromium(III) oxide is used as a green paint pigment (Figure 2.16). It is a compound composed of Cr^{3+} and O^{2-} ions. What is the formula of chromium(III) oxide?

Figure 2.16
Chromium(III) oxide. The compound is used as a green paint pigment.

b. Strontium oxide is a compound composed of Sr^{2+} and O^{2-} ions. Write the formula of this compound.

▶ PROBLEM STRATEGY

Because a compound is neutral, the sum of positive and negative charges equals zero. Consider the ionic compound $CaCl_2$, which consists of one Ca^{2+} ion and two Cl^- ions. The sum of the charges is:

$$\underbrace{1 \times (+2)}_{\text{Positive charge}} + \underbrace{2 \times (-1)}_{\text{Negative charge}} = 0$$

Note that the number of calcium ions in $CaCl_2$ equals the magnitude of charge on the chloride ion (1), whereas the number of chloride ions in $CaCl_2$ equals the magnitude of charge on the calcium ion (2).

SOLUTION

a. Two Cr^{3+} ions have a total charge of $6+$, and three O^{2-} ions have a total charge of $6-$, giving the combination a net charge of zero. The simplest ratio of Cr^{3+} to O^{2-} is $2:3$, and the formula is $\mathbf{Cr_2O_3}$. Note that the charge (without its sign) on one ion becomes the subscript of the other ion.

$$Cr^{③+} \diagdown O^{②-} \qquad \text{or} \qquad Cr_2O_3$$

b. You see that equal numbers of Sr^{2+} and O^{2-} ions will give a neutral compound. Thus, the formula is SrO. If you use the units of charge to find the subscripts, you get

$$Sr^{②+} \diagdown O^{②-} \qquad \text{or} \qquad Sr_2O_2$$

The final formula is **SrO,** because this gives the simplest ratio of ions.

See Problems 2.43 and 2.44.

2.7 Organic Compounds

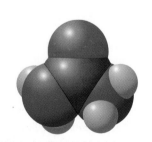

Figure 2.17
Molecular model of urea
(CH_4N_2O). Urea was the first organic molecule deliberately synthesized by a chemist.

An important class of *molecular substances that contain carbon combined with other elements, such as hydrogen, oxygen, and nitrogen,* is **organic compounds.** Historically, organic compounds were restricted to those that could be produced only from living entities. In 1828 a German chemist, Friedrick Wöhler, synthesized urea (a molecular compound in human urine, CH_4N_2O, Figure 2.17) from the molecular compounds ammonia (NH_3) and cyanic acid (HNCO). His work clearly demonstrated that a given compound is exactly the same whether it comes from a living entity or is synthesized.

Organic compounds make up the majority of all known compounds. Since 1957, more than 13 million (60%) of the recorded substances in an international materials registry have been listed as organic. The proteins, amino acids, enzymes, and DNA that make up your body are all either individual organic molecules or contain organic molecules. Table sugar, peanut oil, and antibiotic medicines are all examples of organic molecules as well.

The simplest organic compounds are hydrocarbons. **Hydrocarbons** are *those compounds containing only hydrogen and carbon.* Common examples include methane (CH_4), ethane (C_2H_6), propane (C_3H_8), acetylene (C_2H_2), and benzene

CH$_4$
Methane

C$_2$H$_6$
Ethane

C$_3$H$_8$
Propane

C$_2$H$_2$
Acetylene

C$_6$H$_6$
Benzene

CH$_3$OH
Methanol

CH$_3$CH$_2$OCH$_2$CH$_3$
Diethyl ether

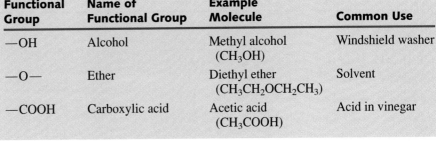

Table 2.3
Examples of Organic Functional Groups

Functional Group	Name of Functional Group	Example Molecule	Common Use
—OH	Alcohol	Methyl alcohol (CH$_3$OH)	Windshield washer
—O—	Ether	Diethyl ether (CH$_3$CH$_2$OCH$_2$CH$_3$)	Solvent
—COOH	Carboxylic acid	Acetic acid (CH$_3$COOH)	Acid in vinegar

(C$_6$H$_6$). Hydrocarbons are used extensively as sources of energy for heating our homes, for powering internal combustion engines, and for generating electricity. They also are the starting materials for most plastics.

The chemistry of organic molecules is often determined by groups of atoms in the molecule that have characteristic chemical properties. A **functional group** is a *reactive portion of a molecule that undergoes predictable reactions.* An *alcohol* is a molecule that contains an —OH functional group. The term *ether* indicates that an organic molecule contains an oxygen atom between two carbon atoms as in diethyl ether (CH$_3$CH$_2$OCH$_2$CH$_3$). Table 2.3 contains a few examples of organic functional groups along with example compounds.

CONCEPT CHECK 2.4

Identify the following compounds as being a hydrocarbon, an alcohol, an ether, or a carboxylic acid.

(a)   (b)   (c)   (d)

2.8 Naming Simple Compounds

Before the structural basis of chemical substances became established, compounds were named after people, places, or particular characteristics. Examples are Glauber's salt (sodium sulfate, discovered by J. R. Glauber), sal ammoniac (ammonium chloride, named after the ancient Egyptian deity Ammon from the temple near which the substance was made), and washing soda (sodium carbonate, used for softening wash water). Today several million compounds are known and a system is required for naming compounds. **Chemical nomenclature** is *the systematic naming of chemical compounds.*

Table 2.4
Common Monatomic Ions of the Main-Group Elements*

	IA	IIA	IIIA	IVA	VA	VIA	VIIA
Period 1							H^-
Period 2	Li^+	Be^{2+}	B	C	N^{3-}	O^{2-}	F^-
Period 3	Na^+	Mg^{2+}	Al^{3+}	Si	P	S^{2-}	Cl^-
Period 4	K^+	Ca^{2+}	Ga^{3+}	Ge	As	Se^{2-}	Br^-
Period 5	Rb^+	Sr^{2+}	In^{3+}	Sn^{2+}	Sb	Te^{2-}	I^-
Period 6	Cs^+	Ba^{2+}	Tl^+, Tl^{3+}	Pb^{2+}	Bi^{3+}		

*Elements shown in color do not normally form compounds having monatomic ions.

Inorganic compounds are *composed of elements other than carbon.* A few notable exceptions to this classification scheme include carbon monoxide, carbon dioxide, carbonates, and the cyanides; all contain carbon and yet are generally considered to be inorganic.

We first look at the naming of ionic compounds. Then, we look at the naming of some simple molecular compounds, including binary molecular compounds and acids. Finally, we look at hydrates of ionic compounds.

Ionic Compounds

Ionic compounds, as we saw in the previous section, are substances composed of ions. Most ionic compounds contain metal and nonmetal atoms; for example, NaCl. (The ammonium salts, such as NH_4Cl, are a prominent exception.) You name an ionic compound by giving the name of the cation followed by the name of the anion. Before you can name ionic compounds, you need to be able to write and name ions.

A **monatomic ion** is *an ion formed from a single atom.* Table 2.4 lists common monatomic ions of the main-group elements.

Rules for Predicting the Charges on Monatomic Ions

1. Most of the main-group metallic elements have one monatomic cation with a charge equal to the group number in the periodic table (the Roman numeral).

2. Some metallic elements have more than one cation. These elements have common cations with a charge equal to the group number minus 2, in addition to having a cation with a charge equal to the group number. Example: The common ion of lead is Pb^{2+}. (The group number is 4; the charge is 4 − 2.)

3. Most transition elements form more than one cation. Most of these elements have one ion with a charge of +2. Example: Iron has common cations Fe^{2+} and Fe^{3+}. Copper has common cations Cu^+ and Cu^{2+}.

4. The charge on a monatomic anion for a nonmetallic main-group element equals the group number minus 8. Example: Oxygen has the monatomic anion O^{2-}. (The group number is 6; the charge is 6 − 8.)

Rules for Naming Monatomic Ions

1. Monatomic cations are named after the element if there is only one such ion.

2. If there is more than one monatomic cation of an element, Rule 1 is not sufficient. The *Stock system* of nomenclature names the cations after the element, but follows this by a Roman numeral in parentheses denoting the charge on the ion. Example: Fe^{2+} is the iron(II) ion and Fe^{3+} is the iron(III) ion. ◄

   In an older system of nomenclature, such ions are named by adding the suffixes *-ous* and *-ic* to a stem name of the element (which may be from the Latin) to indicate the ions of lower and higher charge, respectively. Example: Fe^{2+} is called ferrous ion; Fe^{3+}, ferric ion. Cu^{+} is called cuprous ion; Cu^{2+}, cupric ion.

   Table 2.5 lists some common cations of the transition elements. Most of these elements have more than one ion, so require the Stock nomenclature system or the older suffix system. A few, such as zinc, have only a single ion that is normally encountered, and you usually name them by just the metal name. You would not be wrong, however, if, for example, you named Zn^{2+} as zinc(II) ion.

3. The names of the monatomic anions are obtained from a stem name of the element followed by the suffix *-ide*. Example: Br^{-} is called bromide ion, from the stem name *brom-* for bromine and the suffix *-ide*.

   A **polyatomic ion** is *an ion consisting of two or more atoms chemically bonded together and carrying a net electric charge*. Table 2.6 lists some common polyatomic ions. The first two are cations (Hg_2^{2+} and NH_4^{+}). There are no simple rules for writing the formulas of such ions.

   Most of the ions in Table 2.6 are *oxoanions* (also called *oxyanions*), which consist of oxygen with another element. Sulfur, for example, forms the oxoanions sulfate ion, SO_4^{2-}, and sulfite ion, SO_3^{2-}.

   Note that the names of the oxoanions have a stem name from the central element, plus a suffix *-ate* or *-ite*. These suffixes denote the relative number of oxygen atoms in the oxoanions of a given central element. The name of the oxoanion with the greater number of oxygen atoms has the suffix *-ate;* the name of the oxoanion with the lesser number of oxygen atoms has the suffix *-ite*. The sulfite ion, SO_3^{2-}, and sulfate ion, SO_4^{2-}, are examples. Unfortunately, the suffixes do not tell you the actual number of oxygen atoms in the oxoanion, only the relative number.

   In some cases, the two suffixes *-ite* and *-ate* are not enough when there are more than two oxoanions of a given characteristic element. Table 2.6 lists four oxoanions

► The Roman numeral actually denotes the *oxidation state,* or *oxidation number,* of the atom in the compound. For a monatomic ion, the oxidation state equals the charge. Otherwise, the oxidation state is a hypothetical charge assigned in accordance with certain rules; see Section 4.5.

Table 2.5
Common Cations of the Transition Elements

Ion	Ion Name	Ion	Ion Name	Ion	Ion Name
Cr^{3+}	Chromium(III) or chromic	Co^{2+}	Cobalt(II) or cobaltous	Zn^{2+}	Zinc
Mn^{2+}	Manganese(II) or manganous	Ni^{2+}	Nickel(II) or nickel	Ag^{+}	Silver
Fe^{2+}	Iron(II) or ferrous	Cu^{+}	Copper(I) or cuprous	Cd^{2+}	Cadmium
Fe^{3+}	Iron(III) or ferric	Cu^{2+}	Copper(II) or cupric	Hg^{2+}	Mercury(II) or mercuric

Table 2.6
Some Common Polyatomic Ions

Name	Formula	Name	Formula
Mercury(I) or mercurous	Hg_2^{2+}	Nitrite	NO_2^-
Ammonium	NH_4^+	Nitrate	NO_3^-
Cyanide	CN^-	Hydroxide	OH^-
Carbonate	CO_3^{2-}	Peroxide	O_2^{2-}
Hydrogen carbonate (or bicarbonate)	HCO_3^-	Phosphate	PO_4^{3-}
		Monohydrogen phosphate	HPO_4^{2-}
Acetate	$C_2H_3O_2^-$	Dihydrogen phosphate	$H_2PO_4^-$
Oxalate	$C_2O_4^{2-}$	Sulfite	SO_3^{2-}
Hypochlorite	ClO^-	Sulfate	SO_4^{2-}
Chlorite	ClO_2^-	Hydrogen sulfite (or bisulfite)	HSO_3^-
Chlorate	ClO_3^-		
Perchlorate	ClO_4^-	Hydrogen sulfate (or bisulfate)	HSO_4^-
Chromate	CrO_4^{2-}		
Dichromate	$Cr_2O_7^{2-}$	Thiosulfate	$S_2O_3^{2-}$
Permanganate	MnO_4^-		

of chlorine: ClO^-, ClO_2^-, ClO_3^-, and ClO_4^-. In such cases, the two prefixes *hypo-* and *per-* are used in addition to the two suffixes. The two oxoanions with the least number of oxygen atoms (ClO^- and ClO_2^-) are named using the suffix *-ite,* and the prefix *hypo-* is used with the ClO^- ion. The two oxoanions with the greatest number of oxygen atoms (ClO_3^- and ClO_4^-) are named using the suffix *-ate,* and the prefix *per-* is used with the ClO_4^- ion.

Some of the polyatomic ions in Table 2.6 are oxoanions bonded to one or more hydrogen ions (H^+). They are sometimes referred to as *acid anions,* because acids are substances that provide H^+ ions. As an example, monohydrogen phosphate ion, HPO_4^{2-}, is essentially a phosphate ion (PO_4^{3-}) to which a hydrogen ion (H^+) has bonded. In an older terminology, ions such as hydrogen carbonate and hydrogen sulfate were called bicarbonate and bisulfate, respectively.

The last anion in the table is thiosulfate ion, $S_2O_3^{2-}$. The prefix *thio-* means that an oxygen atom in the root ion name (sulfate, SO_4^{2-}) has been replaced by a sulfur atom.

There are only a few common polyatomic cations. Mercury(I) ion is one of the few common metal ions that is not monatomic; its formula is Hg_2^{2+}. The ion charge indicated in parentheses is the charge per metal atom. The ammonium ion, NH_4^+, is one of the few common cations composed of only nonmetal atoms.

Now that we have discussed the naming of ions, let us look at the naming of ionic compounds. Example 2.4 illustrates how you name an ionic compound given its formula.

EXAMPLE 2.4

Naming an Ionic Compound from Its Formula

Name the following: a. Mg_3N_2, b. $CrSO_4$.

PROBLEM STRATEGY

Note that the compounds in a and b contain both metal and nonmetal atoms, so we expect them to be ionic compounds. The first thing to do is write the formulas of the cations and anions in the compound, then name the ions. Look first for any monatomic cations and anions whose charges are predictable. For those metals having more than one cation, you will need to deduce the charge on the metal ion from the charge on the anion. You will need to memorize or have available the formulas and names of the polyatomic anions.

SOLUTION

a. Magnesium, a Group IIA metal, is expected to form only a 2+ ion (Mg^{2+}, the magnesium ion). Nitrogen (Group VA) is expected to form an anion of charge equal to the group number minus 8 (N^{3-}, the nitride ion). The name of the compound is **magnesium nitride.**

b. Chromium is a transition element and has more than one monatomic ion. You can find the charge on the Cr ion if you know the formula of the anion. From Table 2.6, you see that the SO$_4$ in $CrSO_4$ refers to the anion SO_4^{2-} (the sulfate ion). Therefore, the Cr cation must be Cr^{2+} to give electrical neutrality. So, the compound is **chromium(II) sulfate.**

See Problems 2.45 and 2.46.

EXAMPLE 2.5

Writing the Formula from the Name of an Ionic Compound

Write formulas for the following compounds: a. iron(II) phosphate b. titanium(IV) oxide.

SOLUTION

a. Iron(II) phosphate contains the iron(II) ion, Fe^{2+}, and the phosphate ion, PO_4^{3-}. Now use the method of Example 2.3 to obtain the formula. The formula is $Fe_3(PO_4)_2$.

b. Titanium(IV) oxide is composed of titanium(IV) ions, Ti^{4+}, and oxide ions, O^{2-}. The formula of titanium(IV) oxide is TiO_2.

See Problems 2.47 and 2.48.

Binary Molecular Compounds

A **binary compound** is *a compound composed of only two elements*. Binary compounds composed of a metal and a nonmetal are usually ionic and are named as ionic compounds. Binary compounds composed of two nonmetals or metalloids are usually molecular and are named using a prefix system. In this system, you usually name the two elements using the order given by the formula of the compound.

Table 2.8
Some Oxoanions and Their Corresponding Oxoacids

Oxoanion		Oxoacid	
CO_3^{2-}	Carbon*ate* ion	H_2CO_3	Carbon*ic* acid
NO_2^-	Nit*rite* ion	HNO_2	Nit*rous* acid
NO_3^-	Nit*rate* ion	HNO_3	Nit*ric* acid
PO_4^{3-}	Phosph*ate* ion	H_3PO_4	Phosphor*ic* acid
SO_3^{2-}	Sulf*ite* ion	H_2SO_3	Sulfur*ous* acid
SO_4^{2-}	Sulf*ate* ion	H_2SO_4	Sulfur*ic* acid
ClO^-	*Hypo*chlor*ite* ion	$HClO$	*Hypo*chlor*ous* acid
ClO_2^-	Chlor*ite* ion	$HClO_2$	Chlor*ous* acid
ClO_3^-	Chlor*ate* ion	$HClO_3$	Chlor*ic* acid
ClO_4^-	*Per*chlor*ate* ion	$HClO_4$	*Per*chlor*ic* acid

For example, the acid name corresponding to sulfate ion is sulfuric acid. Table 2.8 lists some oxoanions and their corresponding oxoacids.

Some binary compounds of hydrogen and nonmetals yield acidic solutions when dissolved in water. These *solutions* are named like compounds by using the prefix *hydro-* and the suffix *-ic* with the stem name of the nonmetal, followed by the word *acid*. We denote the solution by the formula of the binary compound followed by (*aq*) for aqueous (water) solution. The corresponding binary compound can be distinguished from the solution by appending the state of the compound to the formula. Here are some examples:

Binary Compound	Acid Solution
$HCl(g)$, hydrogen chloride	*hydro*chloric *acid*, $HCl(aq)$
$HF(g)$, hydrogen fluoride	*hydro*fluoric *acid*, $HF(aq)$

EXAMPLE 2.9

Writing the Name and Formula of an Anion from the Acid

Selenium has an oxoacid, H_2SeO_4, called selenic acid. What is the formula and name of the corresponding anion?

SOLUTION

When you remove two H^+ ions from H_2SeO_4, you obtain the SeO_4^{2-} ion. You name the ion from the acid by replacing *-ic* with *-ate*. The anion is called the **selenate ion.**

See Problems 2.51 and 2.52.

Figure 2.19
Copper(II) sulfate. The hydrate $CuSO_4 \cdot 5H_2O$ is blue; the anhydrous compound, $CuSO_4$, is white.

Hydrates

A **hydrate** is *a compound that contains water molecules weakly bound in its crystals.* When an aqueous solution of copper(II) sulfate is evaporated, blue crystals form in which each $CuSO_4$ is associated with five molecules of water. The formula of the hydrate is written $CuSO_4 \cdot 5H_2O$. When the blue crystals of the hydrate are heated, the water is driven off, leaving behind white crystals of *anhydrous* copper(II) sulfate (see Figure 2.19). Hydrates are named from the *anhydrous* compound, followed by the word *hydrate* with a prefix to indicate the number of water molecules per formula unit of the compound.

EXAMPLE 2.10
Naming a Hydrate from Its Formula

Epsom salts has the formula $MgSO_4 \cdot 7H_2O$. What is the chemical name of the substance?

SOLUTION

$MgSO_4$ is magnesium sulfate. $MgSO_4 \cdot 7H_2O$ is **magnesium sulfate heptahydrate.**

See Problems 2.53 and 2.54.

 CONCEPT CHECK 2.5

You take a job with the U.S. Environmental Protection Agency (EPA) inspecting college chemistry laboratories. On the first day of the job, while inspecting Generic University, you encounter a bottle with the formula Al_2Q_3 that was used as an unknown compound in an experiment. Before you send the compound off to the EPA lab for analysis you want to narrow down the possibilities for element Q. What are the likely (real element) candidates for element Q?

Chemical Reactions: Equations

An important feature of atomic theory is its explanation of a chemical reaction as a rearrangement of the atoms of substances. The reaction of sodium with chlorine described in the chapter opening involves the rearrangement of the atoms of sodium and chlorine to give sodium chloride. This reaction is conveniently represented by a chemical equation.

2.9 Writing Chemical Equations

A **chemical equation** is *the symbolic representation of a chemical reaction in terms of chemical formulas.* For example, the burning of sodium in chlorine to produce sodium chloride is written

$$2Na + Cl_2 \longrightarrow 2NaCl$$

The formulas on the left side of an equation (before the arrow) represent the reactants. The arrow means "react to form" or "yield." The formulas on the right side represent the products. Coefficients in front of the formula give the relative number of formula units involved in the reaction. Coefficients of 1 are usually understood, not written.

In many cases, it is useful to indicate the states or phases of the substances in an equation. Use the following phase labels:

$$(g) = \text{gas}, \ (l) = \text{liquid}, \ (s) = \text{solid}, \ (aq) = \text{aqueous solution}$$

When you use these labels, the previous equation becomes

$$2Na(s) + Cl_2(g) \longrightarrow 2NaCl(s)$$

If the reactants are heated to make the reaction go, you can indicate this by putting the symbol Δ (capital Greek delta) over the arrow. For example,

$$2NaNO_3(s) \xrightarrow{\Delta} 2NaNO_2(s) + O_2(g)$$

When an aqueous solution of hydrogen peroxide, H_2O_2, comes in contact with platinum, the hydrogen peroxide decomposes into water and oxygen. The platinum acts as a *catalyst*, a substance that speeds up a reaction without undergoing any net change itself. The equation for this reaction is as follows: ◄

▶ Platinum is a silvery-white metal used for jewelry. It is also valuable as a catalyst for many reactions, including those that occur in the catalytic converters of automobiles.

$$2H_2O_2(aq) \xrightarrow{Pt} 2H_2O(l) + O_2(g)$$

2.10 Balancing Chemical Equations

When the coefficients in a chemical equation are correctly given, the numbers of atoms of each element are equal on both sides of the arrow. The equation is then said to be *balanced*. A chemical reaction involves simply a recombination of the atoms; none are destroyed and none are created. Consider the burning of natural gas, which is composed mostly of methane, CH_4. You describe this as

$$\underset{\substack{\text{one molecule} \\ \text{of methane}}}{CH_4} + \underset{\substack{\text{two molecules} \\ \text{of oxygen}}}{2O_2} \underset{\substack{\text{react to} \\ \text{form}}}{\longrightarrow} \underset{\substack{\text{one molecule} \\ \text{of carbon dioxide}}}{CO_2} + \underset{\substack{\text{two molecules} \\ \text{of water}}}{2H_2O}$$

Figure 2.20 represents the reaction in terms of molecular models.

Consider the burning of propane gas (Figure 2.21). By experiment, you determine that propane reacts with oxygen to give carbon dioxide and water. The formulas of propane, oxygen, carbon dioxide, and water are C_3H_8, O_2, CO_2, and H_2O, respectively. Then you can write

$$C_3H_8 + O_2 \longrightarrow CO_2 + H_2O$$

Figure 2.20
Representation of the reaction of methane with oxygen. Molecular models represent the reaction of CH_4 with O_2 to give CO_2 and H_2O.

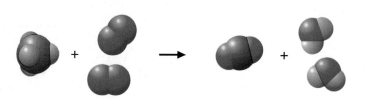

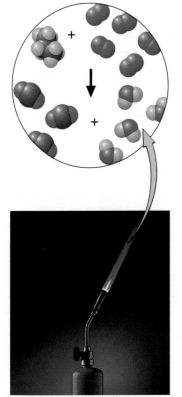

Figure 2.21
The burning of propane gas. Propane gas, C_3H_8, from the tank burns by reacting with oxygen, O_2, in air to give carbon dioxide, CO_2, and water, H_2O.

▶ Balancing equations this way is relatively quick for all but the most complicated equations. A special method for such equations is described in Chapters 4 and 20.

To balance equations, you select coefficients that will make the numbers of atoms of each element equal on both sides of the equation.

Because there are three carbon atoms on the left side of the equation (C_3H_8), you must have three carbon atoms on the right. This is achieved by writing 3 for the coefficient of CO_2.

$$\underline{1}C_3H_8 + O_2 \longrightarrow \underline{3}CO_2 + H_2O$$

Similarly, there are eight hydrogen atoms on the left (in C_3H_8), so you write 4 for the coefficient of H_2O.

$$\underline{1}C_3H_8 + O_2 \longrightarrow \underline{3}CO_2 + \underline{4}H_2O$$

The coefficients on the right are now determined, and there are ten oxygen atoms on this side of the equation: 6 from the three CO_2 molecules and 4 from the four H_2O molecules. You now write 5 for the coefficient of O_2. We drop the coefficient 1 from C_3H_8 and have the balanced equation.

$$C_3H_8 + 5O_2 \longrightarrow 3CO_2 + 4H_2O$$

The combustion of propane can be equally well represented by

$$2C_3H_8 + 10O_2 \longrightarrow 6CO_2 + 8H_2O$$

in which the coefficients of the previous equation have been doubled. Usually, however, it is preferable to *write the coefficients so that they are the smallest whole numbers possible.*

The method we have outlined—called *balancing by inspection*—is essentially a trial-and-error method. ◀ It can be made easier, however, by observing the following rule:

> Balance first the atoms for elements that occur in only one substance on each side of the equation.

Remember that you cannot change a subscript in any formula; only the coefficients can be altered to balance an equation.

EXAMPLE 2.11

Balancing Simple Equations

Balance the following equations.

a. $H_3PO_3 \longrightarrow H_3PO_4 + PH_3$ b. $Ca + H_2O \longrightarrow Ca(OH)_2 + H_2$

c. $Fe_2(SO_4)_3 + NH_3 + H_2O \longrightarrow Fe(OH)_3 + (NH_4)_2SO_4$

PROBLEM STRATEGY

Look at the chemical equation for atoms of elements that occur in only one substance on each side of the equation. Begin by balancing the equation in one of these atoms. Use the number of atoms on the left side of the arrow as the coefficient of the substance containing that element on the right side, and vice versa. After balancing one element in an equation, the rest become easier.

SOLUTION

a. Oxygen occurs in just one of the products (H_3PO_4). It is therefore easiest to balance O atoms first. To do this, note that H_3PO_3 has three O atoms; use 3 as the coefficient of H_3PO_4 on the right side. There are four O atoms in H_3PO_4 on the right side; use 4 as the coefficient of H_3PO_3 on the left side.

$$\underline{4}H_3PO_3 \longrightarrow \underline{3}H_3PO_4 + PH_3$$

This equation is now also balanced in P and H atoms. Thus, the balanced equation is

$$\mathbf{4H_3PO_3 \longrightarrow 3H_3PO_4 + PH_3}$$

b. The equation is balanced in Ca atoms as it stands.

$$\underline{1}Ca + H_2O \longrightarrow \underline{1}Ca(OH)_2 + H_2$$

O atoms occur in only one reactant and in only one product, so they are balanced next.

$$\underline{1}Ca + \underline{2}H_2O \longrightarrow \underline{1}Ca(OH)_2 + H_2$$

The equation is now also balanced in H atoms. Thus, the answer is

$$\mathbf{Ca + 2H_2O \longrightarrow Ca(OH)_2 + H_2}$$

c. To balance the equation in Fe atoms, you write

$$\underline{1}Fe_2(SO_4)_3 + NH_3 + H_2O \longrightarrow \underline{2}Fe(OH)_3 + (NH_4)_2SO_4$$

You balance the S atoms by placing the coefficient 3 for $(NH_4)_2SO_4$.

$$\underline{1}Fe_2(SO_4)_3 + NH_3 + H_2O \longrightarrow \underline{2}Fe(OH)_3 + \underline{3}(NH_4)_2SO_4$$

Now you balance the N atoms.

$$\underline{1}Fe_2(SO_4)_3 + \underline{6}NH_3 + H_2O \longrightarrow \underline{2}Fe(OH)_3 + \underline{3}(NH_4)_2SO_4$$

To balance the O atoms, you first count the number of O atoms on the right (18). Then you count the number of O atoms in substances on the left with known coefficients. There are 12 O's in $Fe_2(SO_4)_3$; hence, the number of remaining O's (in H_2O) must be $18 - 12 = 6$.

$$\underline{1}Fe_2(SO_4)_3 + \underline{6}NH_3 + \underline{6}H_2O \longrightarrow \underline{2}Fe(OH)_3 + \underline{3}(NH_4)_2SO_4$$

Finally, note that the equation is now balanced in H atoms. The answer is

$$\mathbf{Fe_2(SO_4)_3 + 6NH_3 + 6H_2O \longrightarrow 2Fe(OH)_3 + 3(NH_4)_2SO_4}$$

Check each final equation by counting the number of atoms of each element on both sides of the equations.

See Problems 2.59 and 2.60.

A Checklist for Review

Important Terms

atomic theory (2.1)
atom (2.1)
element (2.1, 2.3)
compound (2.1)
chemical reaction (2.1)
atomic symbol (2.1)
law of multiple proportions (2.1)
nucleus (2.2)
electron (2.2)
proton (2.3)
atomic number (Z) (2.3)
neutron (2.3)

mass number (A) (2.3)
nuclide (2.3)
isotope (2.3)
atomic mass unit (amu) (2.4)
atomic weight (2.4)
fractional abundance (2.4)
periodic table (2.5)
period (of periodic table) (2.5)
group (of periodic table) (2.5)
metal (2.5)

nonmetal (2.5)
metalloid (semimetal) (2.5)
chemical formula (2.6)
molecule (2.6)
molecular formula (2.6)
polymer (2.6)
monomer (2.6)
ion (2.6)
anion (2.6)
cation (2.6)
ionic compound (2.6)
formula unit (2.6)

organic compound (2.7)
hydrocarbon (2.7)
functional group (2.7)
chemical nomenclature (2.7)
inorganic compound (2.8)
monatomic ion (2.8)
polyatomic ion (2.8)
binary compound (2.8)
oxoacid (2.8)
hydrate (2.8)
chemical equation (2.9)

Summary of Facts and Concepts

Atomic theory is central to chemistry. According to this theory, all matter is composed of small particles, or *atoms*. Each *element* is composed of the same kind of atom, and a *compound* is composed of two or more elements chemically combined in fixed proportions. A *chemical reaction* consists of the rearrangement of the atoms present in the reacting substances to give new chemical combinations present in the substances formed by the reaction. Although Dalton considered atoms to be the ultimate particles of matter, we now know that atoms themselves have structure. An atom has a *nucleus* and *electrons*.

Thomson established that cathode rays consist of negatively charged particles, or electrons, that are constituents of all atoms. Thomson measured the mass-to-charge ratio of the electron, and later Millikan measured its charge. From these measurements, the electron was found to be more than 1800 times lighter than the lightest atom.

Rutherford proposed the nuclear model of the atom to account for the results of experiments in which alpha particles were scattered from metal foils. According to this model, the atom consists of a nucleus, around which the electrons exist. The nucleus has most of the mass of the atom and consists of *protons* and *neutrons*. Each atom has a nucleus with a specific number of protons (*atomic number*), and around the nucleus are an equal number of electrons. The number of protons plus neutrons in a nucleus equals the *mass number*. Atoms whose nuclei have the same number of protons but different numbers of neutrons are called *isotopes*.

The *atomic weight* can be calculated from the isotopic masses and *fractional abundances* of the isotopes in a naturally occurring element. These data can be determined by the use of a mass spectrometer.

The elements can be arranged in rows and columns by atomic number to form the *periodic table*. Elements in a given *group* have similar properties. (A *period* is a row in the periodic table.) Elements on the left and at the center of the table are *metals;* those on the right are *nonmetals*.

A *chemical formula* is a notation used to convey the relative proportions of the atoms of the different elements in a substance. If the substance is molecular, the formula gives the precise number of each kind of atom in the molecule. If the substance is ionic, the formula gives the relative number of different ions in the compound.

Chemical nomenclature is the systematic naming of compounds based on their formulas or structures.

We represent a reaction by a *chemical equation*, writing a chemical formula for each reactant and product. The coefficients in the equation indicate the relative numbers of reactant and product molecules or formula units. We determine the coefficients by *balancing* the numbers of each kind of atom on both sides of the equation.

Operational Skills

1. **Writing nuclide symbols** Given the number of protons and neutrons in a nucleus, write its nuclide symbol. **(EXAMPLE 2.1)**
2. **Determining atomic weight from isotopic masses and fractional abundances** Given the isotopic masses (in atomic mass units) and fractional isotopic abundances for a naturally occurring element, calculate its atomic weight. **(EXAMPLE 2.2)**
3. **Writing an ionic formula, given the ions** Given the formulas of a cation and an anion, write the formula of the ionic compound of these ions. **(EXAMPLE 2.3)**
4. **Writing the name of a compound from its formula, or vice versa** Given the formula of a simple compound (ionic, binary molecular, acid, or hydrate), write the name

(EXAMPLES 2.4, 2.6, 2.10), or vice versa **(EXAMPLES 2.5, 2.7)**.

5. **Writing the name of a binary molecular compound from its molecular model** Given the molecular model of a binary compound, write the name. **(EXAMPLE 2.8)**
6. **Writing the name and formula of an anion from the acid** Given the name and formula of an oxoacid, write the name and formula of the oxoanion; or from the name and formula of the oxoanion, write the formula and name of the oxoacid. **(EXAMPLE 2.9)**
7. **Balancing simple equations** Given the formulas of the reactants and products in a chemical reaction, obtain the coefficients of the balanced equation. **(EXAMPLE 2.11)**

Review Questions

2.1 Describe atomic theory and discuss how it explains the great variety of different substances. How does it explain chemical reactions?

2.2 Two compounds of iron and chlorine, A and B, contain 1.270 g and 1.904 g of chlorine, respectively, for each gram of iron. Show that these amounts are in the ratio 2 : 3. Is this consistent with the law of multiple proportions? Explain.

2.3 Explain the operation of a cathode-ray tube. Describe the deflection of cathode rays by electrically charged plates placed within the cathode-ray tube. What does this imply about cathode rays?

2.4 Explain Millikan's oil-drop experiment.

2.5 Describe the nuclear model of the atom. How does this model explain the results of alpha-particle scattering from metal foils?

2.6 Describe how protons and neutrons were discovered to be constituents of nuclei.

2.7 Briefly explain how a mass spectrometer works. What kinds of information does one obtain from the instrument?

2.8 Define the term *atomic weight*. Why might the values of atomic weights on a planet elsewhere in the universe be different from those on earth?

2.9 What is the name of the element in Group IVA and Period 5?

2.10 Cite some properties that are characteristic of a metal.

2.11 Ethane consists of molecules with two atoms of carbon and six atoms of hydrogen. Write the molecular formula for ethane.

2.12 What is the fundamental difference between an organic substance and an inorganic substance? Write chemical formulas of three inorganic molecules that contain carbon.

2.13 Give an example of a binary compound that is ionic. Give an example of a binary compound that is molecular.

2.14 Which of the following models represent a(n):

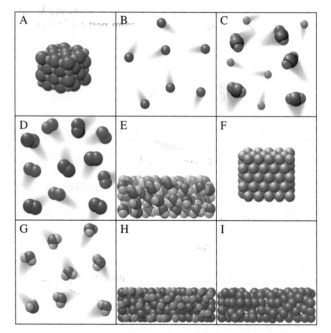

a. element
b. compound
c. mixture
d. ionic solid
e. gas made up of an element and a compound
f. mixture of elements
g. solid element
h. solid
i. liquid

2.15 The compounds CuCl and $CuCl_2$ were formerly called cuprous chloride and cupric chloride, respectively. What are their names using the Stock system of nomenclature? What are the advantages of the Stock system of nomenclature over the former one?

2.16 Explain what is meant by the term *balanced chemical equation.*

Conceptual Problems

2.17 A friend is trying to balance the following equation:

$$N_2 + H_2 \longrightarrow NH_3$$

He presents you with his version of the "balanced" equation:

$$N + H_3 \longrightarrow NH_3$$

You immediately recognize that he has committed a serious error; however, he argues that there is nothing wrong, since the equation is balanced. What reason can you give to convince him that his "method" of balancing the equation is flawed?

2.18 You have the mythical metal element "X" that can exist as X^+, X^{2+}, and X^{5+} ions.
a. What would be the chemical formula for compounds formed from the combination of each of the X ions and SO_4^{2-}?
b. If the name of the element X is exy, what would be the names of each of the compounds from part a of this problem?

2.19 Match the molecular model with the correct chemical formula: CH_3OH, NH_3, KCl, H_2O.

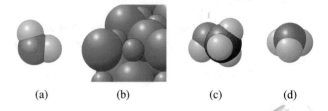

(a)   (b)   (c)   (d)

2.20 Currently, the atomic mass unit (amu) is based on being exactly one-twelfth the mass of a carbon-12 atom and is equal to 1.66×10^{-27} kg.
a. If the amu were based on sodium-23 with a mass equal to exactly 1/23 of the mass of a sodium-23 atom, would the mass of the amu be different?
b. If the new mass of the amu based on sodium-23 is 1.67×10^{-27} kg, how would the mass of a hydrogen atom, in amu, compare with the current mass of a hydrogen atom in amu?

Practice Problems

2.21 What is the name of the element represented by each of the following atomic symbols?
a. Ne   b. Zn   c. Ag   d. Mg

2.22 Give the atomic symbol for each of the following elements.
a. cobalt   b. aluminum   c. neon   d. titanium

Electrons, Protons, and Neutrons

2.23 The following table gives the number of protons and neutrons in the nuclei of various atoms. Which atom is the isotope of atom A? Which atom has the same mass number as atom A?

	Protons	Neutrons
Atom A	17	19
Atom B	16	19
Atom C	18	18

2.24 The following table gives the number of protons and neutrons in the nuclei of various atoms. Which atom is the isotope of atom A? Which atom has the same mass number as atom A?

	Protons	Neutrons
Atom A	31	39
Atom B	32	38
Atom C	38	50

2.25 Naturally occurring chlorine is a mixture of the isotopes Cl-35 and Cl-37. How many protons and how many neutrons are there in each isotope? How many electrons are there in the neutral atoms?

2.26 Naturally occurring lithium is a mixture of ^6_3Li and ^7_3Li. Give the number of protons, neutrons, and electrons in the neutral atom of each isotope.

Atomic Weights

2.27 Ammonia is a gas with a characteristic pungent odor. It is sold as a water solution for use in household cleaning. The gas is a compound of nitrogen and hydrogen in the atomic ratio

1 : 3. A sample of ammonia contains 7.933 g N and 1.712 g H. What is the atomic mass of N relative to H?

2.28 Hydrogen sulfide is a gas with the odor of rotten eggs. The gas can sometimes be detected in automobile exhaust. It is a compound of hydrogen and sulfur in the atomic ratio 2 : 1. A sample of hydrogen sulfide contains 0.587 g H and 9.330 g S. What is the atomic mass of S relative to H?

2.29 An element has three naturally occurring isotopes with the following masses and abundances:

Isotopic Mass (amu)	Fractional Abundance
38.964	0.9326
39.964	1.000×10^{-4}
40.962	0.0673

Calculate the atomic weight of this element. What is the identity of the element?

2.30 An element has three naturally occurring isotopes with the following masses and abundances:

Isotopic Mass (amu)	Fractional Abundance
27.977	0.9221
28.976	0.0470
29.974	0.0309

Calculate the atomic weight of this element. What is the identity of the element?

2.31 While traveling to a distant universe, you discover the hypothetical element, "X." You obtain a representative sample of the element and discover that it is made up of two isotopes, X-23 and X-25. To help your science team calculate the atomic mass of the substance, you send the following drawing of your sample with your report.

In the report, you also inform the science team that the gold atoms are X-23, which have an isotopic mass of 23.02 amu, and the green atoms are X-25, which have an isotopic mass of 25.147 amu. What is the atomic mass of element X?

2.32 While roaming a parallel universe, you discover the hypothetical element, "Z." You obtain a representative sample of the element and discover that it is made up of two isotopes, Z-47 and Z-51. To help your science team calculate the atomic mass of the substance, you send the following drawing of your sample with your report.

In the report, you also inform the science team that the blue atoms are Z-47, which have an isotopic mass of 47.510 amu, and the orange atoms are Z-51, which have an isotopic mass of 51.126 amu. What is the atomic mass of element Z?

Periodic Table

2.33 Identify the group and period for each of the following. Refer to a periodic table. Label each as a metal, nonmetal, or metalloid.
a. C   b. Po   c. Cr   d. Mg   e. B

2.34 Refer to a periodic table and obtain the group and period for each of the following elements. Also determine whether the element is a metal, nonmetal, or metalloid.
a. Si   b. F   c. Ca   d. Co   e. Xe

2.35 Give one example (atomic symbol and name) for each of the following.
a. a main-group (representative) element in the second period
b. an alkali metal
c. a transition element in the fourth period
d. a lanthanide element

2.36 Give one example (atomic symbol and name) for each of the following.
a. a transition element in the sixth period
b. a halogen
c. a main-group (representative) element in the third period
d. an actinide element

Molecular and Ionic Substances

2.37 A 1.50-g sample of nitrous oxide (an anesthetic, sometimes called laughing gas) contains 2.05×10^{22} N_2O molecules. How many nitrogen atoms are in this sample? How many nitrogen atoms are in 1.00 g of nitrous oxide?

2.38 Nitric acid is composed of HNO_3 molecules. A sample weighing 3.50 g contains 3.34×10^{22} HNO_3 molecules. How many nitrogen atoms are in this sample? How many oxygen atoms are in 4.50 g of nitric acid?

2.39 Write the molecular formula for each of the following compounds represented by molecular models.

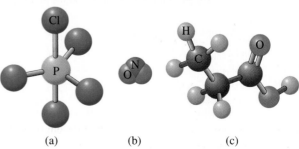

(a)   (b)   (c)

2.40 Write the molecular formula for each of the following compounds represented by molecular models.

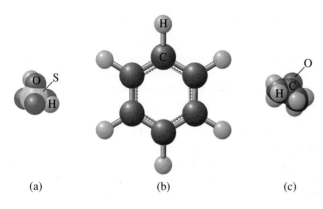

(a)   (b)   (c)

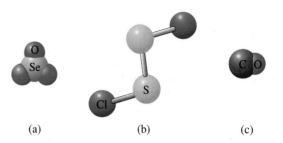

(a)   (b)   (c)

2.50 Write the systematic name for each of the following molecules represented by a molecular model.

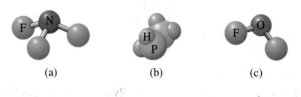

(a)   (b)   (c)

2.41 Iron(III) nitrate has the formula $Fe(NO_3)_3$. What is the ratio of iron atoms to oxygen atoms in this compound?

2.42 Ammonium phosphate, $(NH_4)_3PO_4$, has how many hydrogen atoms for each oxygen atom?

2.43 Write the formula for the compound of each of the following pairs of ions.
a. Fe^{3+} and CN^- b. K^+ and SO_4^{2-}
c. Li^+ and N^{3-} d. Ca^{2+} and P^{3-}

2.44 For each of the following pairs of ions, write the formula of the corresponding compound.
a. Co^{2+} and Br^- b. NH_4^+ and SO_4^{2-}
c. Na^+ and PO_4^{3-} d. Fe^{3+} and NO_3^-

Chemical Nomenclature

2.45 Name the following compounds.
a. Na_2SO_4 b. CaO
c. $CuCl$ d. Cr_2O_3

2.46 Name the following compounds.
a. Na_2O b. Mn_2O_3
c. NH_4HCO_3 d. $Cu(NO_3)_2$

2.47 Write the formulas of:
a. lead(II) dichromate
b. barium hydrogen carbonate
c. cesium sulfide
d. iron(II) acetate

2.48 Write the formulas of:
a. sodium thiosulfate
b. copper(I) hydroxide
c. calcium hydrogen carbonate
d. nickel(II) nitride

2.49 Write the systematic name for each of the following compounds represented by a molecular model.

2.51 Give the name and formula of the acid corresponding to each of the following oxoanions.
a. bromate ion, BrO_3^-
b. hyponitrite ion, $N_2O_2^{2-}$
c. disulfite ion, $S_2O_5^{2-}$
d. arsenate ion, AsO_4^{3-}

2.52 Give the name and formula of the acid corresponding to each of the following oxoanions.
a. selenite ion, SeO_3^{2-}
b. sulfite ion, SO_3^{2-}
c. hypoiodite ion, IO^-
d. diphosphate ion, $P_2O_7^{4-}$

2.53 Glauber's salt has the formula $Na_2SO_4 \cdot 10H_2O$. What is the chemical name of this substance?

2.54 Emerald-green crystals of the substance $NiSO_4 \cdot 6H_2O$ are used in nickel plating. What is the chemical name of this compound?

2.55 Iron(II) sulfate heptahydrate is a blue-green, crystalline compound used to prepare other iron compounds. What is the formula of iron(II) sulfate heptahydrate?

2.56 Cobalt(II) chloride hexahydrate has a pink color. It loses water on heating and changes to a blue-colored compound. What is the formula of cobalt(II) chloride hexahydrate?

Chemical Equations

2.57 For the balanced chemical equation $Pb(NO_3)_2 + K_2CO_3 \longrightarrow PbCO_3 + 2KNO_3$, how many oxygen atoms are on the left side?

2.58 In the equation $2PbS + O_2 \longrightarrow 2PbO + 2SO_2$, how many oxygen atoms are there on the right side? Is the equation balanced as written?

2.59 Balance the following equations.
a. $Sn + NaOH \longrightarrow Na_2SnO_2 + H_2$
b. $Al + Fe_3O_4 \longrightarrow Al_2O_3 + Fe$
c. $CH_3OH + O_2 \longrightarrow CO_2 + H_2O$
d. $P_4O_{10} + H_2O \longrightarrow H_3PO_4$
e. $PCl_5 + H_2O \longrightarrow H_3PO_4 + HCl$

2.60 Balance the following equations.
a. $Cl_2O_7 + H_2O \longrightarrow HClO_4$
b. $MnO_2 + HCl \longrightarrow MnCl_2 + Cl_2 + H_2O$
c. $Na_2S_2O_3 + I_2 \longrightarrow NaI + Na_2S_4O_6$
d. $Al_4C_3 + H_2O \longrightarrow Al(OH)_3 + CH_4$
e. $NO_2 + H_2O \longrightarrow HNO_3 + NO$

2.61 Solid calcium phosphate and aqueous sulfuric acid solution react to give calcium sulfate, which comes out of the solution as a solid. The other product is phosphoric acid, which remains in solution. Write a balanced equation for the reaction using complete formulas for the compounds with phase labels.

2.62 Solid potassium metal reacts with water, giving a solution of potassium hydroxide and releasing hydrogen gas. Write a balanced equation for the reaction using complete formulas for the compounds with phase labels.

General Problems

2.63 Two samples of different compounds of nitrogen and oxygen have the following compositions. Show that the compounds follow the law of multiple proportions. What is the ratio of oxygen in the two compounds for a fixed amount of nitrogen?

	Amount N	Amount O
Compound A	1.206 g	2.755 g
Compound B	1.651 g	4.714 g

2.64 Two samples of different compounds of sulfur and oxygen have the following compositions. Show that the compounds follow the law of multiple proportions. What is the ratio of oxygen in the two compounds for a fixed amount of sulfur?

	Amount S	Amount O
Compound A	1.210 g	1.811 g
Compound B	1.783 g	1.779 g

2.65 Compounds of europium, Eu, are used to make color television screens. The europium nucleus has a charge of +63. How many electrons are there in the neutral atom? in the Eu^{3+} ion?

2.66 Cesium, Cs, is used in photoelectric cells ("electric eyes"). The cesium nucleus has a charge of +55. What is the number of electrons in the neutral atom? in the Cs^+ ion?

2.67 A nucleus of mass number 80 contains 45 neutrons. An atomic ion of this element has 36 electrons in it. Write the symbol for this atomic ion (give the symbol for the nucleus and give the ionic charge as a right superscript).

2.68 One isotope of a metallic element has mass number 119 and has 69 neutrons in the nucleus. An atomic ion has 48 electrons. Write the symbol for this ion (give the symbol for the nucleus and give the ionic charge as a right superscript).

2.69 Silver has two naturally occurring isotopes, one of mass 106.91 amu and the other of mass 108.90 amu. Find the fractional abundances for these two isotopes. The atomic weight is 107.87 amu.

2.70 Obtain the fractional abundances for the two naturally occurring isotopes of copper. The masses of the isotopes are $^{63}_{29}Cu$, 62.9298 amu; $^{65}_{29}Cu$, 64.9278 amu. The atomic weight is 63.546 amu.

2.71 Identify the following elements, giving their names and atomic symbols.
a. a nonmetal that is normally a liquid
b. a normally gaseous element in Group IA
c. a transition element in Group VB, Period 5
d. the halogen in Period 2

2.72 Identify the following elements, giving their names and atomic symbols.
a. a normally gaseous element in Group VIA
b. a metal that is normally a liquid
c. a main-group element in Group IIIA, Period 2
d. the alkali metal in Period 4

2.73 Write formulas for all the ionic compounds that can be formed by these ions: Na^+, Ni^{2+}, SO_4^{2-}, and Cl^-.

2.74 Write formulas for all the ionic compounds that can be formed by these ions: Ca^{2+}, Cr^{3+}, O^{2-}, and NO_3^-.

2.75 Give the formulas for the following compounds.
a. mercury(I) sulfide   b. cobalt(III) sulfite
c. ammonium dichromate   d. aluminum sulfide

2.76 Give the formulas for the following compounds.
a. hydrogen peroxide   b. aluminum phosphate
c. lead(IV) nitride   d. boron trifluoride

2.77 Name the following molecular compounds.
a. $AsBr_3$ b. H_2Se c. P_2O_5 d. SiO_2

2.78 Name the following molecular compounds.
a. ClF_4 b. CS_2 c. NF_3 d. SF_6

2.79 Balance the following equations.
a. $C_2H_6 + O_2 \longrightarrow CO_2 + H_2O$
b. $P_4O_6 + H_2O \longrightarrow H_3PO_3$
c. $KClO_3 \longrightarrow KCl + KClO_4$
d. $(NH_4)_2SO_4 + NaOH \longrightarrow NH_3 + H_2O + Na_2SO_4$

2.80 Balance the following equations.
a. $NaOH + H_3PO_4 \longrightarrow Na_3PO_4 + H_2O$
b. $SiCl_4 + H_2O \longrightarrow SiO_2 + HCl$

c. $Ca_3(PO_4)_2 + C \longrightarrow Ca_3P_2 + CO$
d. $H_2S + O_2 \longrightarrow SO_2 + H_2O$

2.81 A monatomic ion has a charge of +2. The nucleus of the ion has a mass number of 62. The number of neutrons in the nucleus is 1.21 times that of the number of protons. How many electrons are in the ion? What is the name of the element?

2.82 A monatomic ion has a charge of +3. The nucleus of the ion has a mass number of 45. The number of neutrons in the nucleus is 1.14 times that of the number of protons. How many electrons are in the ion? What is the name of the element?

Cumulative-Skills Problems

2.83 A sample of green crystals of nickel(II) sulfate heptahydrate was heated carefully to produce the bluish-green nickel(II) sulfate hexahydrate. What are the formulas of the hydrates? If 8.753 g of the heptahydrate produces 8.192 g of the hexahydrate, how many grams of anhydrous nickel(II) sulfate could be obtained?

2.84 Cobalt(II) sulfate heptahydrate has pink-colored crystals. When heated carefully, it produces cobalt(II) sulfate monohydrate, which has red crystals. What are the formulas of these hydrates? If 3.548 g of the heptahydrate yields 2.184 g of the monohydrate, how many grams of the anhydrous cobalt(II) sulfate could be obtained?

2.85 A sample of metallic element X, weighing 3.177 g, combines with 0.6015 L of O_2 gas (at normal pressure and 20.0°C) to form the metal oxide with the formula XO. If the density of O_2 gas under these conditions is 1.330 g/L, what is the mass of this oxygen? The atomic weight of oxygen is 15.9994 amu. What is the atomic weight of X? What is the identity of X?

2.86 A sample of metallic element X, weighing 4.315 g, combines with 0.4810 L of Cl_2 gas (at normal pressure and 20.0°C) to form the metal chloride with the formula XCl. If the density of Cl_2 gas under these conditions is 2.948 g/L, what is the mass of the chlorine? The atomic weight of chlorine is 35.453 amu. What is the atomic weight of X? What is the identity of X?

Reaction of zinc and iodine to produce zinc iodide.

Calculations with Chemical Formulas and Equations

Acetic acid (ah-see'-tik acid) is a colorless liquid with a sharp, vinegary odor. In fact, vinegar contains acetic acid, which accounts for vinegar's odor and sour taste. Vinegar results from the fermentation of wine or cider by certain bacteria. These bacteria require oxygen, and the overall chemical change is the reaction of ethanol (alcohol) in wine with oxygen to give acetic acid.

Laboratory preparation of acetic acid may also start from ethanol, which reacts with oxygen in two steps. First, ethanol reacts with oxygen to yield acetaldehyde, in addition to water. In the second step, the acetaldehyde reacts with more oxygen to produce acetic acid. (The human body also produces acetaldehyde and then acetic acid from alcohol, as it attempts to eliminate alcohol from the system.)

This chapter focuses on two basic questions, which we can illustrate using these compounds: How do you determine the chemical formula of a substance such as acetic acid? How much acetic acid can you prepare from a given quantity of ethanol? You must know the formulas of all the substances involved in a reaction before you can write the chemical equation, and you need the balanced chemical equation to determine the quantitative relationships among the different substances in the reaction. We begin by discussing how you relate number of atoms or molecules to grams of substance, because this is the key to answering both questions.

Mass and Moles of Substance

You buy a quantity of groceries in several ways. You often purchase items such as oranges and lemons by counting out a particular number. Some things, such as eggs, can be purchased in a "package" that represents a known quantity; for example, a dozen. Bulk foods, such as peanuts or candy, are usually purchased by mass, because it is too tedious to count them out, and because we know that a given mass yields a certain quantity of the item. All three of these methods are used by chemists to determine the quantity of matter: counting, using a package that represents a quantity, and measuring the mass. For a chemist, it is a relatively easy matter to weigh a substance to obtain the mass. However, the number of atoms or molecules in even a seemingly minute amount is much too large to count. Nevertheless, chemists are interested in knowing such numbers. How many atoms of carbon are there in one molecule of acetic acid? How many molecules of acetic acid can be obtained from one molecule of ethanol? To count the numbers of atoms, molecules, and ions, we must look at the concept of molecular weight and formula weight and introduce the package chemists call the mole.

3.1 Molecular Weight and Formula Weight

We can easily extend the concept of atomic weight to include molecular weight. The **molecular weight** (MW) of a substance is *the sum of the atomic weights of all the atoms in a molecule of the substance.* For example, the molecular weight of water, H_2O, is 18.0 amu (2 × 1.0 amu from two H atoms plus 16.0 amu from one O atom).

The **formula weight** (FW) of a substance is *the sum of the atomic weights of all atoms in a formula unit of the compound,* whether molecular or not. Sodium chloride, with the formula unit NaCl, has a formula weight of 58.44 amu (22.99 amu from Na plus 35.45 amu from Cl). NaCl is ionic, so strictly speaking the expression "molecular weight of NaCl" has no meaning. ◀

▶ Some chemists use the term *molecular weight* in a less strict sense for ionic as well as molecular compounds.

EXAMPLE 3.1

Calculating the Formula Weight from a Formula

Calculate the formula weight of each of the following to three significant figures, using a table of atomic weights (AW): a. chloroform, $CHCl_3$; b. iron(III) sulfate, $Fe_2(SO_4)_3$.

◆ PROBLEM STRATEGY

Obtain the sum of the atomic weights for the atoms in each formula. Remember in part b that the number of atoms of each element within parentheses in the formula should be multiplied by the subscript to the parentheses.

SOLUTION

a. The calculation is

$$
\begin{array}{lll}
1 \times \text{AW of C} = & & 12.0 \text{ amu} \\
1 \times \text{AW of H} = & & 1.0 \text{ amu} \\
3 \times \text{AW of Cl} = 3 \times 35.45 \text{ amu} = & 106.4 \text{ amu} \\
\hline
\text{FW of CHCl}_3 = & & 119.4 \text{ amu}
\end{array}
$$

The answer rounded to three significant figures is **119 amu.**

b. The calculation is

$$
\begin{array}{lll}
2 \times \text{AW of Fe} = & 2 \times 55.8 \text{ amu} = & 111.6 \text{ amu} \\
3 \times \text{AW of S} = & 3 \times 32.1 \text{ amu} = & 96.3 \text{ amu} \\
3 \times 4 \times \text{AW of O} = & 12 \times 16.00 \text{ amu} = & 192.0 \text{ amu} \\
\hline
\text{FW of Fe}_2(\text{SO}_4)_3 = & & 399.9 \text{ amu}
\end{array}
$$

The answer rounded to three significant figures is **4.00×10^2 amu.**

See Problems 3.15 and 3.16.

EXAMPLE 3.2

Calculating the Formula Weight from Molecular Models

For the following two compounds, write the molecular formula and calculate the formula weight to four significant figures:

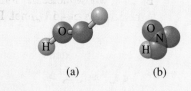

(a)   (b)

SOLUTION

a. This molecular model is of a molecule that is composed of two O and two H atoms. For inorganic compounds, the elements in a chemical formula are written in order such that the most metalliclike element is listed first. Hence, the chemical formula is **H_2O_2.** Using the same approach as Example 3.1, calculating the formula weight yields **34.02 amu.**

b. This molecular model represents a molecule made up of one N atom, three O atoms, and one H atom. The chemical formula is then **HNO_3.** The formula weight is **63.01 amu.**

See Problems 3.17 and 3.18.

3.2 The Mole Concept

A 10.0 g sample of ethanol (less than 3 teaspoonsful) contains 1.31×10^{23} molecules. Imagine a device that counts molecules at the rate of one million per second. It would take more than four billion years—nearly the age of the earth—for this device to count that many molecules! Chemists have adopted the *mole concept* as a convenient way to deal with the enormous numbers of molecules or ions in the samples they work with.

Definition of Mole and Molar Mass

A **mole** (symbol **mol**) is defined as *the quantity of a given substance that contains as many molecules or formula units as the number of atoms in exactly 12 g of carbon-12.*

The number of atoms in a 12-g sample of carbon-12 is called **Avogadro's number** (to which we give the symbol N_A). Recent measurements of this number give the value 6.0221367×10^{23}, which to three significant figures is 6.02×10^{23}.

A mole of a substance contains Avogadro's number of molecules (or formula units). The term *mole,* like a dozen or a gross, thus refers to a particular number of things. A dozen eggs equals 12 eggs, a gross of pencils equals 144 pencils, and a mole of ethanol equals 6.02×10^{23} ethanol molecules.

In using the term *mole* for ionic substances, we mean the number of formula units of the substance. For example, a mole of sodium carbonate, Na_2CO_3, is a quantity containing $6.02\ E23\ Na_2CO_3$ units which contain $2 \times 6.02 \times 10^{23}\ Na^+$ ions and $1 \times 6.02 \times 10^{23}\ CO_3^{2-}$ ions. ◄

When using the term *mole,* it is important to specify the formula of the unit to avoid any misunderstanding. For example, a mole of oxygen atoms (with the formula O) contains 6.02×10^{23} O atoms. A mole of oxygen molecules (formula O_2) contains $6.02 \times 10^{23}\ O_2$ molecules—that is, $2 \times 6.02 \times 10^{23}$ O atoms.

The **molar mass** of a substance is *the mass of one mole of the substance.* Carbon-12 has a molar mass of exactly 12 g/mol, by definition.

> For all substances, the molar mass in grams per mole is numerically equal to the formula weight in atomic mass units.

Ethanol, whose molecular formula is C_2H_5OH, has a molecular weight of 46.1 amu and a molar mass of 46.1 g/mol. Figure 3.1 shows molar amounts of different substances.

► Sodium carbonate, Na_2CO_3, is a white, crystalline solid known commercially as soda ash. Large amounts of soda ash are used in the manufacture of glass. The hydrated compound, $Na_2CO_3 \cdot 10H_2O$, is known as washing soda.

Ethanol
C_2H_5OH

Figure 3.1
One mole each of various substances. *Clockwise from top left:* 1-octanol ($C_8H_{17}OH$); mercury(II) iodide (HgI_2); methanol (CH_3OH); sulfur (S_8).

EXAMPLE 3.3

Calculating the Mass of an Atom or Molecule

a. What is the mass in grams of a chlorine atom, Cl?

b. What is the mass in grams of a hydrogen chloride molecule, HCl?

◆ PROBLEM STRATEGY

Because each mole of substance contains Avogadro's number of units of that substance, dividing the mass of a substance in one mole by Avogadro's number gives the mass of that unit of substance.

SOLUTION

a. The atomic weight of Cl is 35.5 amu, so the molar mass of Cl is 35.5 g/mol. Dividing 35.5 g by 6.02×10^{23} gives the mass of one atom.

$$\text{Mass of a Cl atom} = \frac{35.5\ \text{g}}{6.02 \times 10^{23}} = \mathbf{5.90 \times 10^{-23}\ g}$$

b. The molecular weight of HCl equals 36.5 amu. Therefore, 1 mol HCl contains 36.5 g HCl and

$$\text{Mass of an HCl molecule} = \frac{36.5 \text{ g}}{6.02 \times 10^{23}} = \textbf{6.06} \times \textbf{10}^{-23} \textbf{ g}$$

See Problems 3.19 and 3.20.

Mole Calculations

Now that you know how to find the mass of one mole of substance, there are two important questions to ask. First, how much does a given number of moles of a substance weigh? Second, how many moles of a given formula unit does a given mass of substance contain? Both questions are easily answered using dimensional analysis, or the conversion-factor method. ◄

▶ **Alternatively, because the molar mass is the mass per mole, you can relate mass and moles by means of the formula**

Molar mass = mass/moles

To illustrate, consider the conversion of grams of ethanol, C_2H_5OH, to moles of ethanol. The molar mass of ethanol is 46.1 g/mol, so we write

$$1 \text{ mol } C_2H_5OH = 46.1 \text{ g } C_2H_5OH$$

Thus, the factor converting grams of ethanol to moles of ethanol is 1 mol C_2H_5OH/46.1 g C_2H_5OH. To convert moles of ethanol to grams of ethanol, we simply invert the conversion factor (46.1 g C_2H_5OH/1 mol C_2H_5OH). Note that the unit you are converting *from* is on the bottom of the conversion factor; the unit you are converting *to* is on the top.

Again, suppose you are going to prepare acetic acid from 10.0 g of ethanol, C_2H_5OH. Convert 10.0 g C_2H_5OH to moles C_2H_5OH by multiplying by the appropriate conversion factor.

$$10.0 \text{ g } C_2H_5OH \times \frac{1 \text{ mol } C_2H_5OH}{46.1 \text{ g } C_2H_5OH} = 0.217 \text{ mol } C_2H_5OH$$

The following examples further illustrate this conversion-factor technique.

Figure 3.2
Reaction of zinc and iodine. Heat from the reaction of the elements causes some iodine to vaporize (violet vapor).

EXAMPLE 3.4

Converting Moles of Substance to Grams

Zinc iodide, ZnI_2, can be prepared by the direct combination of elements (Figure 3.2). A chemist determines from the amounts of elements that 0.0654 mol ZnI_2 can form. How many grams of zinc iodide is this?

PROBLEM STRATEGY

Use the formula weight to write the factor that converts from mol ZnI_2 to g ZnI_2.

SOLUTION

The molar mass of ZnI_2 is 319 g/mol. Therefore,

$$0.0654 \text{ mol } ZnI_2 \times \frac{319 \text{ g } ZnI_2}{1 \text{ mol } ZnI_2} = \textbf{20.9 g } \textbf{ZnI}_2$$

See Problems 3.21 and 3.22.

Figure 3.3
Preparation of lead(II) chromate. When lead(II) nitrate solution (colorless) is added to potassium chromate solution (clear yellow), bright yellow solid lead(II) chromate forms (giving a cloudlike formation of fine crystals).

EXAMPLE 3.5

Converting Grams of Substance to Moles

Lead(II) chromate, $PbCrO_4$, is a yellow paint pigment (called chrome yellow) prepared by a precipitation reaction (Figure 3.3). In a preparation, 45.6 g of lead(II) chromate is obtained as a precipitate. How many moles of $PbCrO_4$ is this?

SOLUTION

The molar mass of $PbCrO_4$ is 323 g/mol. That is,

$$1 \text{ mol } PbCrO_4 = 323 \text{ g } PbCrO_4$$

Therefore,

$$45.6 \text{ g } PbCrO_4 \times \frac{1 \text{ mol } PbCrO_4}{323 \text{ g } PbCrO_4} = \textbf{0.141 mol } PbCrO_4$$

See Problems 3.23 and 3.24.

EXAMPLE 3.6

Calculating the Number of Molecules in a Given Mass

How many molecules are there in a 3.46-g sample of hydrogen chloride, HCl?

PROBLEM STRATEGY

The number of molecules in a sample is related to moles of compound (1 mol HCl = 6.02×10^{23} HCl molecules). Therefore, if you first convert grams HCl to moles, then you can convert moles to number of molecules.

SOLUTION

$$3.46 \text{ g } HCl \times \frac{1 \text{ mol } HCl}{36.5 \text{ g } HCl} \times \frac{6.02 \times 10^{23} \text{ HCl molecules}}{1 \text{ mol } HCl}$$

$$= \textbf{5.71} \times \textbf{10}^{22} \textbf{ HCl molecules}$$

See Problems 3.27, 3.28, 3.29, and 3.30.

CONCEPT CHECK 3.1

You have 1.5 moles of tricycles.

a. How many moles of seats do you have?

b. How many moles of tires do you have?

c. How could you use parts a and b as an analogy to teach a friend about the number of moles of OH^- ions in 1.5 moles of $Mg(OH)_2$?

Determining Chemical Formulas

When a chemist has discovered a new compound, the first question to answer is, What is the formula? To answer, you determine the amounts of the elements for a given amount of compound. This is conveniently expressed as **percentage composition.** You then determine the formula from this percentage composition. If the compound is a molecular substance, it is also necessary to know the molecular weight of the compound.

3.3 Mass Percentages from the Formula

We define the **mass percentage** of A as *the parts of A per hundred parts of the total, by mass.* That is,

$$\text{Mass \% A} = \frac{\text{mass of } A \text{ in the whole}}{\text{mass of the whole}} \times 100\%$$

The next example will provide practice with the concept of mass percentage. In this example we will start with a compound (formaldehyde, CH_2O) whose formula is given and obtain the percentage composition.

EXAMPLE 3.7

Calculating the Percentage Composition from the Formula

Formaldehyde, CH_2O, is a toxic gas with a pungent odor. Large quantities are consumed in the manufacture of plastics, and a water solution of the compound is used to preserve biological specimens. Calculate the mass percentages of the elements in formaldehyde.

PROBLEM STRATEGY

Interpret the formula in molar terms and then convert moles to masses. Thus, 1 mol CH_2O has a mass of 30.0 g and contains 1 mol C (12.0 g), 2 mol H (2 × 1.01 g), and 1 mol O (16.0 g). Divide each mass of element by the molar mass, then multiply by 100.

SOLUTION

$$\% \text{ C} = \frac{12.0 \text{ g}}{30.0 \text{ g}} \times 100\% = \textbf{40.0\%}$$

$$\% \text{ H} = \frac{2 \times 1.01 \text{ g}}{30.0 \text{ g}} \times 100\% = \textbf{6.73\%}$$

You can calculate the percentage of O in the same way, but it can also be found by subtracting the percentages of C and H from 100%:

$$\% \text{ O} = 100\% - (40.0\% + 6.73\%) = \textbf{53.3\%}$$

See Problems 3.35, 3.36, 3.37, and 3.38.

Formaldehyde
CH_2O

EXAMPLE 3.8

Calculating the Mass of an Element in a Given Mass of Compound

How many grams of carbon are there in 83.5 g of formaldehyde, CH_2O? Use the percentage composition obtained in the previous example (40.0% C, 6.73% H, 53.3% O).

SOLUTION

CH_2O is 40.0% C, so the mass of carbon in 83.5 g CH_2O is

$$83.5 \text{ g} \times 0.400 = \textbf{33.4 g}$$

3.4 Elemental Analysis: Percentages of Carbon, Hydrogen, and Oxygen

▶ Sodium hydroxide reacts with carbon dioxide according to the following equations:

$$NaOH + CO_2 \longrightarrow NaHCO_3$$
$$2NaOH + CO_2 \longrightarrow$$
$$Na_2CO_3 + H_2O$$

The first step in the determination of the formula of a new compound is to obtain its percentage composition. Consider the determination of the percentages of carbon, hydrogen, and oxygen in compounds containing only these three elements. Burn a sample of the compound of known mass and get CO_2 and H_2O. Next relate the masses of CO_2 and H_2O to the masses of carbon and hydrogen. Then calculate the mass percentages of C and H. Determine the mass percentage of O by difference.

Figure 3.4 shows an apparatus used to find the amount of carbon and hydrogen in a compound. The compound is burned in a stream of oxygen. As a result of the combustion, every mole of carbon (C) in the compound ends up as a mole of carbon dioxide (CO_2), and every mole of hydrogen (H) ends up as one-half mole of water (H_2O). The water is collected by a drying agent. The carbon dioxide is collected by chemical reaction with sodium hydroxide, NaOH. ◀ By weighing the U-tubes containing the drying agent and the sodium hydroxide before and after combustion, it is possible to determine the masses of water and carbon dioxide produced. From these data, you can calculate the percentage composition of the compound.

The next example shows how to determine the percentage composition of acetic acid from combustion data.

Figure 3.4
Combustion method for determining the percentages of carbon and hydrogen in a compound. The compound is placed in the sample dish and is heated by the furnace. Vapor of the compound burns in O_2 in the presence of CuO pellets, giving CO_2 and H_2O. The water vapor is collected by a drying agent, and CO_2 combines with the sodium hydroxide.

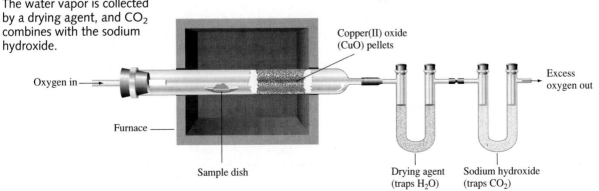

Oxygen in →

Copper(II) oxide
(CuO) pellets

Excess
oxygen out

Furnace

Sample dish

Drying agent
(traps H_2O)

Sodium hydroxide
(traps CO_2)

Mass Spectrometry and Molecular Formula

Sophisticated instruments have become indispensable in modern chemical analysis and research. One such instrument is the mass spectrometer. The mass spectrum of a compound is used in combination with other information to identify a substance or to obtain the molecular formula and structure of a newly prepared compound.

Positive ions of molecules, like those of atoms, can be generated by bombarding the gas, or vapor of the substance, with electrons. The mass spectra of molecules, however, are usually much more complicated than those of atoms. One reason is that the molecular ions produced often break into fragments, giving several different kinds of positive ions. Consider the CH_2Cl_2 molecule (methylene chloride). When this molecule is struck by a high-energy electron, a positive ion $CH_2Cl_2^+$ may form.

$$CH_2Cl_2 + e^- \longrightarrow CH_2Cl_2^+ + 2e^-$$

The $CH_2Cl_2^+$ ion gains a great deal of energy from the collision of CH_2Cl_2 with the electron, and the ion frequently loses this energy by breaking into smaller pieces. One way is

$$CH_2Cl_2^+ \longrightarrow CH_2Cl^+ + Cl$$

Thus, the original molecule, even one as simple as CH_2Cl_2, can give rise to a number of ions.

The second reason for the complexity of the mass spectrum of a molecular substance is that many of the atoms in any ion can occur with different isotopic mass, so each ion often has many peaks. The mass spectrum of methylene chloride, CH_2Cl_2, is shown in Figure 3.5. Fourteen peaks are clearly visible. A larger molecule can give an even more complicated spectrum.

Because of the complexity of the mass spectrum, it can be used as a "fingerprint" in identifying a compound. Only methylene chloride has exactly the spectrum shown in Figure 3.5. Thus, by comparing the mass spectrum of an unknown substance with those in a catalog of mass spectra of known compounds, you can determine its identity.

The mass spectrum itself contains a wealth of information about molecular structure. The most intense peaks at the greatest mass often correspond to the ion from the original molecule and give you the molecular weight. Thus, you would expect the peaks at mass 84 and mass 86 to be from the original molecular ion, so the molecular weight is approximately 84 to 86.

An elemental analysis is also possible. The two most intense peaks in Figure 3.5 are at 84 amu and 49 amu. They differ by 35 amu. Perhaps the original molecule that gives the peak at 84 amu contains a chlorine-35

EXAMPLE 3.9

Calculating the Percentages of C and H by Combustion

Acetic acid contains only C, H, and O. A 4.24-mg sample of acetic acid is completely burned. It gives 6.21 mg of carbon dioxide and 2.54 mg of water. What is the mass percentage of each element in acetic acid?

PROBLEM STRATEGY

Convert the mass of CO_2 to moles of CO_2. Then convert this to moles of C, noting that 1 mol C produces 1 mol CO_2. Finally, convert to mass of C. Similarly, convert the mass of H_2O to mol H_2O, then to mol H, and finally to mass of H. Once you have the masses of C and H, you can calculate the mass percentages. Subtract from 100% to get %O.

SOLUTION

Following is the calculation of grams C in the original acetic acid:

$$6.21 \times 10^{-3} \text{ g } CO_2 \times \frac{1 \text{ mol } CO_2}{44.0 \text{ g } CO_2} \times \frac{1 \text{ mol C}}{1 \text{ mol } CO_2} \times \frac{12.0 \text{ g C}}{1 \text{ mol C}}$$

$$= 1.69 \times 10^{-3} \text{ g C (or 1.69 mg C)}$$

Figure 3.5

Mass spectrum of methylene chloride, CH_2Cl_2. The lines at higher mass correspond to ions of the original molecule. Several lines occur because of the presence of different isotopes in the ion.

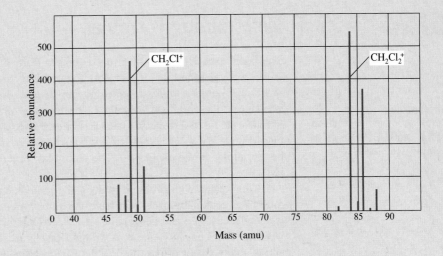

atom, which is lost to give the peak at 49 amu. If this is true, you should expect a weaker peak at mass 86, corresponding to the original molecular ion with chlorine-37 in place of a chlorine-35 atom.

The relative heights of the peaks, which depend on the natural abundances of atoms, are also important, because they give you additional information about the elements present. The relative heights can also tell you how many atoms of a given element are in the original molecule. Naturally occurring chlorine is 75.8% chlorine-35

and 24.2% chlorine-37. If the original molecule contained only one Cl atom, the peaks at 84 amu and 86 amu would be in the ratio 0.758 : 0.242. That is, the peak at 86 amu would be about one-third the height of the one at 84 amu. In fact, the relative height is twice this value. This means that the molecular ion contains two chlorine atoms, because the chance that any such ion contains one chlorine-37 atom is then twice as great. Thus, simply by comparing relative peak heights, you can both confirm the presence of particular elements and obtain the molecular formula.

For hydrogen, you note that 1 mol H_2O yields 2 mol H, so you write

$$2.54 \times 10^{-3} \text{ g } H_2O \times \frac{1 \text{ mol } H_2O}{18.0 \text{ g } H_2O} \times \frac{2 \text{ mol } H}{1 \text{ mol } H_2O} \times \frac{1.01 \text{ g } H}{1 \text{ mol } H}$$
$$= 2.85 \times 10^{-4} \text{ g H (or 0.285 mg H)}$$

Now calculate the mass percentages of C and H in acetic acid.

$$\text{Mass \% C} = \frac{1.69 \text{ mg}}{4.24 \text{ mg}} \times 100\% = 39.9\%$$

$$\text{Mass \% H} = \frac{0.285 \text{ mg}}{4.24 \text{ mg}} \times 100\% = 6.72\%$$

You find the mass percentage of oxygen by subtracting the sum of these percentages from 100%:

$$\text{Mass \% O} = 100\% - (39.9\% + 6.72\%) = 53.4\%$$

See Problems 3.39 and 3.40.

3.5 Determining Formulas

The percentage composition of a compound leads directly to its empirical formula. An **empirical formula** (or **simplest formula**) for a compound is *the formula of a substance written with the smallest integer (whole number) subscripts*. For most ionic substances, the empirical formula is the formula of the compound. ◄ This is often not the case for molecular substances. For example, hydrogen peroxide has the molecular formula H_2O_2. The empirical formula, however, merely tells you the ratio of numbers of atoms in the compound. The empirical formula of hydrogen peroxide (H_2O_2) is HO.

▶ The formula of sodium peroxide, an ionic compound of Na^+ and O_2^{2-}, is Na_2O_2. Its empirical formula is NaO.

Compounds with different molecular formulas can have the same empirical formula, and such substances will have the *same percentage composition*. An example is acetylene, C_2H_2, and benzene, C_6H_6. Acetylene is a gas used as a fuel and also in welding. Benzene, in contrast, is a liquid that is used in the manufacture of plastics and is a component of gasoline. Table 3.1 illustrates how these two compounds, with the same empirical formula, but different molecular formulas, also have different chemical structures.

To obtain the molecular formula of a substance, you need two pieces of information: (1) the percentage composition, from which the empirical formula can be determined; and (2) the molecular weight.

Empirical Formula from the Composition

The empirical formula of a compound shows the ratios of numbers of atoms in the compound. You can find this formula from the composition of the compound by converting masses of the elements to moles.

Table 3.1
Molecular Models of Two Compounds That Have the Empirical Formula CH.
Although benzene and acetylene have the same empirical formula, they do not have the same molecular formula or structure.

Compound	Empirical Formula	Molecular Formula	Molecular Model
Acetylene	CH	C_2H_2	
Benzene	CH	C_6H_6	

Figure 3.6
Chromium compounds of different colors.
Clockwise from top: Potassium chromate, K_2CrO_4; chromium(VI) oxide, CrO_3; chromium(III) sulfate, $Cr_2(SO_4)_3$; chromium metal; potassium dichromate, $K_2Cr_2O_7$; chromium(III) oxide, Cr_2O_3.

EXAMPLE 3.10
Determining the Empirical Formula from Percentage Composition (General)

Chromium forms compounds of various colors; see Figure 3.6. Sodium dichromate is a bright orange, crystalline substance. An analysis of sodium dichromate gives the following mass percentages: 17.5% Na, 39.7% Cr, and 42.8% O. What is its empirical formula?

PROBLEM STRATEGY

Assume for the purposes of this calculation that you have 100.0 g of substance. Then the mass of each element in the sample equals the numerical value of the percentage. Now convert the masses to moles and divide each mole number by the smallest. In this example, you do not obtain a series of integers, or whole numbers, from this division. You will need to find a whole-number factor to multiply these results by to obtain integers. Normally this factor will be 2 or 3, though it might be larger.

SOLUTION

Of the 100.0 g of sodium dichromate, 17.5 g is Na, 39.7 g is Cr, and 42.8 g is O. You convert these amounts to moles.

$$17.5 \text{ g Na} \times \frac{1 \text{ mol Na}}{23.0 \text{ g Na}} = 0.761 \text{ mol Na}$$

$$39.7 \text{ g Cr} \times \frac{1 \text{ mol Cr}}{52.0 \text{ g Cr}} = 0.763 \text{ mol Cr}$$

$$42.8 \text{ g O} \times \frac{1 \text{ mol O}}{16.0 \text{ g O}} = 2.68 \text{ mol O}$$

Divide all the mole numbers by the smallest one.

$$\text{For Na: } \frac{0.761 \text{ mol}}{0.761 \text{ mol}} = 1.00$$

$$\text{For Cr: } \frac{0.763 \text{ mol}}{0.761 \text{ mol}} = 1.00$$

$$\text{For O: } \frac{2.68 \text{ mol}}{0.761 \text{ mol}} = 3.52$$

If you round off the last digit, which is subject to experimental error, you get $Na_{1.0}Cr_{1.0}O_{3.5}$. In this case, the subscripts are not all integers. However, they can be made into integers by multiplying each one by 2; you get $Na_{2.0}Cr_{2.0}O_{7.0}$. Thus, the empirical formula is $\mathbf{Na_2Cr_2O_7}$.

See Problems 3.43 and 3.44.

Molecular Formula from Empirical Formula

The molecular formula of a compound is a multiple of its empirical formula. For example, the molecular formula of acetylene, C_2H_2, is equivalent to $(CH)_2$, and the

Figure 3.7
Molecular model of acetic acid. The hydrogen atom (blue atom at the top right) attached to the oxygen atom (red atom) is easily lost, so it is acidic. The three hydrogen atoms attached to the carbon atom (black atom at the left) are not acidic.

molecular formula of benzene, C_6H_6, is equivalent to $(CH)_6$. Therefore, the molecular weight is some multiple of the empirical formula weight. For any molecular compound,

$$\text{Molecular weight} = n \times \text{empirical formula weight}$$

where n is the number of empirical formula units in the molecule. The molecular formula is obtained by multiplying the subscripts of the empirical formula by n, where

$$n = \frac{\text{molecular weight}}{\text{empirical formula weight}}$$

After determining the empirical formula for a compound, calculate its empirical formula weight. From an experimental determination of its molecular weight, you can calculate n and then the molecular formula. The next example illustrates how you use percentage composition and molecular weight to determine the molecular formula of acetic acid.

EXAMPLE 3.11

Determining the Molecular Formula from Percentage Composition and Molecular Weight

In Example 3.9, we found the percentage composition of acetic acid to be 39.9% C, 6.7% H, and 53.4% O. Determine the empirical formula. The molecular weight of acetic acid was determined by experiment to be 60.0 amu. What is its molecular formula?

SOLUTION

A sample of 100.0 g of acetic acid contains 39.9 g C, 6.7 g H, and 53.4 g O. Converting these masses to moles gives 3.33 mol C, 6.6 mol H, and 3.34 mol O. Dividing the mole numbers by the smallest one gives 1.00 for C, 2.0 for H, and 1.00 for O. **The empirical formula of acetic acid is CH_2O.** (Note that the percentage composition of acetic acid is the same as that of formaldehyde—see Example 3.7—so they must have the same empirical formula.) The empirical formula weight is 30.0 amu. So

$$n = \frac{\text{molecular weight}}{\text{empirical formula weight}} = \frac{60.0\ \text{amu}}{30.0\ \text{amu}} = 2.00$$

The molecular formula of acetic acid is $(CH_2O)_2$, or $C_2H_4O_2$.

See Problems 3.45, 3.46, 3.47, and 3.48.

▶ The condensed structural formula for acetaldehyde is CH_3CHO. Its structural formula is

Determining this structural formula requires additional information.

The formula of acetic acid is often written $HC_2H_3O_2$ to indicate that one of the hydrogen atoms is acidic while the other three are not (Figure 3.7). The formula of acetaldehyde is C_2H_4O. The equations for the preparation of acetic acid are described in the chapter opener. ◀

$$2C_2H_5OH + O_2 \longrightarrow 2C_2H_4O + 2H_2O$$
$$\text{ethanol} \qquad\qquad\qquad \text{acetaldehyde}$$

In practice, the reaction is carried out with the reactants in the gas phase at about 400°C using silver as a catalyst.

The second step consists of reacting acetaldehyde with oxygen to obtain acetic acid. Acetaldehyde liquid is mixed with a catalyst—manganese(II) acetate—and air is bubbled through it. The balanced equation is

$$2C_2H_4O + O_2 \longrightarrow 2HC_2H_3O_2$$
$$\text{acetaldehyde} \qquad\qquad \text{acetic acid}$$

With the balanced equation, we can answer quantitative questions such as, How much acetic acid can you obtain from a 10.0-g sample of acetaldehyde?

 CONCEPT CHECK 3.2

A friend has some questions about empirical formulas and molecular formulas.

a. For a problem that asked him to determine the empirical formula, he came up with the answer $C_2H_8O_2$. Is this a possible answer to the problem?

b. For another problem he came up with the answer $C_{1.5}H_4$ as the empirical formula. Is this answer correct? Explain.

c. Your friend completed a problem of the same type as Example 3.11. His answers indicate that the compound had an empirical formula of C_3H_8O and the molecular formula C_3H_8O. Is this result possible?

Stoichiometry: Quantitative Relations in Chemical Reactions

In Chapter 2, we described a chemical equation as a representation of what occurs when molecules react. We will now study chemical equations more closely to answer questions about **stoichiometry** (pronounced "stoy-key-om′-e-tree"), *the calculation of the quantities of reactants and products involved in a chemical reaction.* Stoichiometry is based on the chemical equation and on the relationship between mass and moles. In the next sections, we will use the industrial Haber process for the production of ammonia to illustrate stoichiometric calculations.

3.6 Molar Interpretation of a Chemical Equation

In the Haber process for producing ammonia, NH_3, nitrogen (from the atmosphere) reacts with hydrogen at high temperature and pressure.

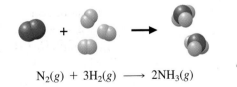

$$N_2(g) + 3H_2(g) \longrightarrow 2NH_3(g)$$

Hydrogen is usually obtained from natural gas or petroleum and so is relatively expensive. Thus, an important question to answer is, How much hydrogen is required to give a particular quantity of ammonia?

To answer quantitative questions, you must first look at the balanced chemical equation: one N_2 molecule and three H_2 molecules react to produce two NH_3 molecules (see models above) or one mole of N_2 reacts with three moles of H_2 to give two moles of NH_3.

Because moles can be converted to mass, you can also give a mass interpretation of a chemical equation. The molar masses of N_2, H_2, and NH_3 are 28.0, 2.02, and 17.0 g/mol, respectively. Therefore 28.0 g of N_2 reacts with 3×2.02 g of H_2 to yield 2×17.0 g of NH_3.

We summarize these three interpretations as follows:

$$
\begin{array}{ccccc}
N_2 & + & 3H_2 & \longrightarrow & 2NH_3 \\
1 \text{ molecule } N_2 & + & 3 \text{ molecules } H_2 & \longrightarrow & 2 \text{ molecules } NH_3 \quad \text{(molecular interpretation)} \\
1 \text{ mol } N_2 & + & 3 \text{ mol } H_2 & \longrightarrow & 2 \text{ mol } NH_3 \quad \text{(molar interpretation)} \\
28.0 \text{ g } N_2 & + & 3 \times 2.02 \text{ g } H_2 & \longrightarrow & 2 \times 17.0 \text{ g } NH_3 \quad \text{(mass interpretation)}
\end{array}
$$

Suppose you ask how many grams of nitrogen will react with 6.06 g (3×2.02 g) of hydrogen. You see from the last equation that the answer is 28.0 g N_2. We formulated this question for one mole of nitrogen. You may ask how much hydrogen is needed to yield 907 kg of ammonia. The solution to this problem depends on the fact that *the number of moles involved in a reaction is proportional to the coefficients in the balanced chemical equation.*

3.7 Amounts of Substances in a Chemical Reaction

A balanced chemical equation relates the amounts of substances in a reaction. The coefficients in the equation can be given a molar interpretation, and can be used to calculate the moles of product obtained from any given moles of reactant. This type of calculation can also be used to answer questions about masses of reactants and products.

Suppose you have a mixture of H_2 and N_2, and 4.8 mol H_2 in this mixture reacts with N_2 to produce NH_3. How many moles of NH_3 can you produce from this quantity of H_2?

The balanced chemical equation tells you that 3 mol H_2 produce 2 mol NH_3. You can express this as a conversion factor: ◀

▶ This conversion factor simply expresses the fact that the mole ratio of NH_3 to H_2 in the reaction is 2 to 3.

$$
\underbrace{\frac{2 \text{ mol } NH_3}{3 \text{ mol } H_2}}_{\substack{\text{Converts from} \\ \text{mol } H_2 \text{ to mol } NH_3}}
$$

Multiplying any quantity of H_2 by this conversion factor mathematically converts that quantity of H_2 to the quantity of NH_3 as specified by the balanced chemical equation.

To calculate the quantity of NH_3 produced from 4.8 mol H_2, you write 4.8 mol H_2 and multiply this by the preceding conversion factor:

$$
4.8 \text{ mol } H_2 \times \frac{2 \text{ mol } NH_3}{3 \text{ mol } H_2} = 3.2 \text{ mol } NH_3
$$

Note how the unit mol H_2 cancels to give the answer in mol NH_3.

It is just as easy to calculate the moles of reactant needed to obtain the specified moles of product. To set up the conversion factor, you refer to the balanced chemical equation and place the quantity you are converting from on the bottom and the quantity you are converting to on the top.

$$
\underbrace{\frac{3 \text{ mol } H_2}{2 \text{ mol } NH_3}}_{\substack{\text{Converts from} \\ \text{mol } NH_3 \text{ to mol } H_2}}
$$

Now consider the problem: How much hydrogen is needed to yield 907 kg of ammonia? The balanced chemical equation directly relates moles of substances, not masses. Therefore,

$$\text{Mass NH}_3 \longrightarrow \text{mol NH}_3 \longrightarrow \text{mol H}_2 \longrightarrow \text{mass H}_2$$

The calculation to convert 907 kg NH_3, or 9.07×10^5 g NH_3, to mol NH_3 is as follows:

$$9.07 \times 10^5 \text{ g NH}_3 \times \underbrace{\frac{1 \text{ mol NH}_3}{17.0 \text{ g NH}_3}}_{\substack{\text{Converts from} \\ \text{g NH}_3 \text{ to mol NH}_3}} = 5.34 \times 10^4 \text{ mol NH}_3$$

Now you convert from moles NH_3 to moles H_2.

$$5.34 \times 10^4 \text{ mol NH}_3 \times \underbrace{\frac{3 \text{ mol H}_2}{2 \text{ mol NH}_3}}_{\substack{\text{Converts from} \\ \text{mol NH}_3 \text{ to mol H}_2}} = 8.01 \times 10^4 \text{ mol H}_2$$

Finally, you convert moles H_2 to grams H_2.

$$8.01 \times 10^4 \text{ mol H}_2 \times \underbrace{\frac{2.02 \text{ g H}_2}{1 \text{ mol H}_2}}_{\substack{\text{Converts from} \\ \text{mol H}_2 \text{ to g H}_2}} = 1.62 \times 10^5 \text{ g H}_2$$

Once you feel comfortable with the individual conversions, you can do this type of calculation in a single step by multiplying successively by conversion factors, as follows:

$$9.07 \times 10^5 \text{ g NH}_3 \times \frac{1 \text{ mol NH}_3}{17.0 \text{ g NH}_3} \times \frac{3 \text{ mol H}_2}{2 \text{ mol NH}_3} \times \frac{2.02 \text{ g H}_2}{1 \text{ mol H}_2} =$$
$$1.62 \times 10^5 \text{ g H}_2 \text{ (or 162 kg H}_2)$$

Figure 3.8 illustrates this calculation diagrammatically.

The following example illustrates this type of calculation.

Figure 3.8
Steps in a stoichiometric calculation. You convert the mass of substance A in a reaction to moles of substance A, then to moles of another substance B, and finally to mass of substance B.

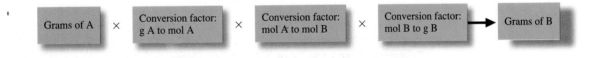

Grams of A × Conversion factor: g A to mol A × Conversion factor: mol A to mol B × Conversion factor: mol B to g B → Grams of B

Figure 3.9
Hematite. The name of this iron mineral stems from the Greek word for blood, which alludes to the color of certain forms of the mineral.

EXAMPLE 3.12

Relating the Quantity of Reactant to Quantity of Product

Hematite, Fe_2O_3, is an important ore of iron; see Figure 3.9. (An ore is a natural substance from which the metal can be profitably obtained.) The free metal is obtained by reacting hematite with carbon monoxide, CO, in a blast furnace. Carbon monoxide is formed in the furnace by partial combustion of carbon. The reaction is

$$Fe_2O_3(s) + 3CO(g) \longrightarrow 2Fe(s) + 3CO_2(g)$$

How many grams of iron can be produced from 1.00×10^3 g Fe_2O_3?

PROBLEM STRATEGY

This calculation involves the conversion of a quantity of Fe_2O_3 to a quantity of Fe. In solving this problem, we assume that there is enough CO for the complete reaction of the Fe_2O_3. An essential feature of this type of calculation is that you must use the information from the balanced chemical equation to convert from the moles of a given substance to the moles of another substance. Therefore, you first convert the mass of Fe_2O_3 to moles of Fe_2O_3. Then, using the relationship from the balanced chemical equation, you convert moles of Fe_2O_3 to moles of Fe. Finally, you convert the moles of Fe to grams of Fe.

SOLUTION

The calculation is as follows:

$$1.00 \times 10^3 \text{ g } Fe_2O_3 \times \frac{1 \text{ mol } Fe_2O_3}{160 \text{ g } Fe_2O_3} \times \frac{2 \text{ mol } Fe}{1 \text{ mol } Fe_2O_3} \times \frac{55.8 \text{ g } Fe}{1 \text{ mol } Fe} = \textbf{698 g Fe}$$

See Problems 3.53 and 3.54.

CONCEPT CHECK 3.3

The main reaction of a charcoal grill is $C(s) + O_2(g) \longrightarrow CO_2(g)$. Which of the statements below are incorrect? Why?

a. 1 atom of carbon reacts with 1 molecule of oxygen to produce 1 molecule of CO_2.

b. 1 g of C reacts with 1 g of O_2 to produce 2 grams of CO_2.

c. 12 g of C reacts with 32 g of O_2 to produce 44 g of CO_2.

d. 1 mol of C reacts with 0.5 mol of O_2 to produce 1 mol of CO_2.

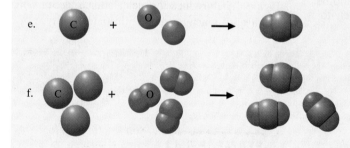

3.8 Limiting Reactant; Theoretical and Percentage Yields

Often reactants are added to a reaction vessel in amounts different from the molar proportions given by the chemical equation. In such cases, only one of the reactants may be completely consumed at the end of the reaction, whereas some amounts of other reactants will remain unreacted. The **limiting reactant** (or **limiting reagent**) is *the reactant that is entirely consumed when a reaction goes to completion.* A reactant that is not completely consumed is often referred to as an *excess reactant*. Once one of the reactants is used up, the reaction stops.

An analogy may help you understand the limiting reactant problem. Suppose you are supervising the assembly of automobiles. Your plant has in stock 300 steering wheels and 900 tires, plus an excess of every other needed component. How many autos can you assemble from this stock? Here is the "balanced equation" for the auto assembly:

$$1 \text{ steering wheel} + 4 \text{ tires} + \text{other components} \longrightarrow 1 \text{ auto}$$

One way to solve this problem is to calculate the number of autos that you could assemble from each component. Looking at the equation, you can determine that from 300 steering wheels, you could assemble 300 autos; from 900 tires you could assemble $900 \div 4 = 225$ autos.

How many autos can you produce, 225 or 300? Note that by the time you have assembled 225 autos, you will have exhausted your stock of tires, so no more autos can be assembled. Tires are the "limiting reactant."

First, calculate the numbers of autos you could assemble from the number of steering wheels and from the number of tires available. From the balanced equation, you can see that one steering wheel is equivalent to one auto, and four tires are equivalent to one auto. Therefore:

$$300 \text{ steering wheels} \times \frac{1 \text{ auto}}{1 \text{ steering wheel}} = 300 \text{ autos}$$

$$900 \text{ tires} \times \frac{1 \text{ auto}}{4 \text{ tires}} = 225 \text{ autos}$$

Comparing the numbers of autos produced (225 autos versus 300 autos), you conclude that tires are the limiting component.

Now consider a chemical reaction, the burning of hydrogen in oxygen.

$$2H_2(g) + O_2(g) \longrightarrow 2H_2O(g)$$

Suppose you put 1 mol H_2 and 1 mol O_2 into a reaction vessel. How many moles of H_2O will be produced? Calculate the moles of H_2O that you could produce from each quantity of reactant.

$$1 \text{ mol } H_2 \times \frac{2 \text{ mol } H_2O}{2 \text{ mol } H_2} = 1 \text{ mol } H_2O$$

$$1 \text{ mol } O_2 \times \frac{2 \text{ mol } H_2O}{1 \text{ mol } O_2} = 2 \text{ mol } H_2O$$

Comparing these results, you see that hydrogen, H_2, yields the least amount of product, so it must be the limiting reactant. Oxygen, O_2, is the excess reactant.

We can summarize the limiting-reactant problem as follows. Suppose you are given the amounts of reactants added to a vessel, and you wish to calculate the amount of product obtained when the reaction is complete. The problem is twofold: (1) you must first identify the limiting reactant; (2) you then calculate the amount of product from the amount of limiting reactant. The next examples illustrate the steps.

EXAMPLE 3.13

Calculating with a Limiting Reactant (Involving Moles)

Zinc metal reacts with hydrochloric acid by the following reaction:

$$Zn(s) + 2HCl(aq) \longrightarrow ZnCl_2(aq) + H_2(g)$$

If 0.30 mol Zn is added to hydrochloric acid containing 0.52 mol HCl, how many moles of H_2 are produced?

PROBLEM STRATEGY

STEP 1   Which is the limiting reactant? To answer this, using the relationship from the balanced chemical equation, you take each reactant in turn and ask how much product (H_2) would be obtained if each were totally consumed. The reactant that gives the smaller amount of product is the limiting reactant.

STEP 2   Calculate the amount of product from the limiting reactant.

SOLUTION

STEP 1

$$0.30 \text{ mol Zn} \times \frac{1 \text{ mol } H_2}{1 \text{ mol Zn}} = 0.30 \text{ mol } H_2$$

$$0.52 \text{ mol HCl} \times \frac{1 \text{ mol } H_2}{2 \text{ mol HCl}} = 0.26 \text{ mol } H_2$$

You see that hydrochloric acid must be the limiting reactant and that some zinc must be left unconsumed. (Zinc is the excess reactant.)

STEP 2   Since HCl is the limiting reactant, the amount of H_2 produced must be **0.26 mol.**

See Problems 3.57 and 3.58.

EXAMPLE 3.14

Calculating with a Limiting Reactant (Involving Masses)

In a process for producing acetic acid, oxygen is bubbled into acetaldehyde, CH_3CHO, containing manganese(II) acetate (catalyst) under pressure at 60°C.

$$2CH_3CHO(l) + O_2(g) \longrightarrow 2HC_2H_3O_2(l)$$

In a laboratory test of this reaction, 20.0 g CH_3CHO and 10.0 g O_2 were put into a reaction vessel. a. How many grams of acetic acid can be produced by this reaction from these amounts of reactants? b. How many grams of the excess reactant remain after the reaction is complete?

SOLUTION

a. How much acetic acid is produced?

STEP 1   To determine which reactant is limiting, you convert grams of each reactant (20.0 g CH_3CHO and 10.0 g O_2) to moles of product, $HC_2H_3O_2$. Acetaldehyde has a molar mass of 44.1 g/mol, and oxygen has a molar mass of 32.0 g/mol.

$$20.0 \text{ g } CH_3CHO \times \frac{1 \text{ mol } CH_3CHO}{44.1 \text{ g } CH_3CHO} \times \frac{2 \text{ mol } HC_2H_3O_2}{2 \text{ mol } CH_3CHO} = 0.454 \text{ mol } HC_2H_3O_2$$

$$10.0 \text{ g } O_2 \times \frac{1 \text{ mol } O_2}{32.0 \text{ g } O_2} \times \frac{2 \text{ mol } HC_2H_3O_2}{1 \text{ mol } O_2} = 0.625 \text{ mol } HC_2H_3O_2$$

Thus, acetaldehyde, CH_3CHO, is the limiting reactant, so 0.454 mol $HC_2H_3O_2$ was produced.

STEP 2 Convert 0.454 mol $HC_2H_3O_2$ to grams of $HC_2H_3O_2$.

$$0.454 \text{ mol } HC_2H_3O_2 \times \frac{60.1 \text{ g } HC_2H_3O_2}{1 \text{ mol } HC_2H_3O_2} = \textbf{27.3 g } HC_2H_3O_2$$

b. How much of the excess reactant (oxygen) was left over? Convert the moles of acetic acid to grams of oxygen (the quantity of oxygen needed to produce this amount of acetic acid).

$$0.454 \text{ mol } HC_2H_3O_2 \times \frac{1 \text{ mol } O_2}{2 \text{ mol } HC_2H_3O_2} \times \frac{32.0 \text{ g } O_2}{1 \text{ mol } O_2} = 7.26 \text{ g } O_2$$

You started with 10.0 g O_2, so the quantity remaining is

$$(10.0 - 7.26) \text{ g } O_2 = \textbf{2.7 g } O_2 \qquad \text{(mass remaining)}$$

See Problems 3.59 and 3.60.

The **theoretical yield** of product is *the maximum amount of product that can be obtained by a reaction from given amounts of reactants.* It is the amount calculated from the stoichiometry based on the limiting reactant. In Example 3.14, the theoretical yield of acetic acid is 27.3 g. In practice, the *actual yield* of a product may be much less for several possible reasons. First, some product may be lost during the process of separating it from the final reaction mixture. Second, there may be other, competing reactions that occur simultaneously with the reactant on which the theoretical yield is based. Finally, many reactions appear to stop before they reach completion; they give mixtures of reactants and products. ◄

▶ Such reactions reach chemical equilibrium. We will discuss equilibrium quantitatively in Chapter 15.

It is important to know the actual yield from a reaction in order to make economic decisions about a preparation method. The reactants for a given method may not be too costly per kilogram, but if the actual yield is very low, the final cost can be very high. The **percentage yield** of product is *the actual yield (experimentally determined) expressed as a percentage of the theoretical yield (calculated).*

$$\text{Percentage yield} = \frac{\text{actual yield}}{\text{theoretical yield}} \times 100\%$$

To illustrate the calculation of percentage yield, recall that the theoretical yield of acetic acid calculated in Example 3.14 was 27.3 g. If the actual yield of acetic acid obtained in an experiment, using the amounts of reactants given in Example 3.14, is 23.8 g, then

$$\text{Percentage yield of } HC_2H_3O_2 = \frac{23.8 \text{ g}}{27.3 \text{ g}} \times 100\% = 87.2\%$$

CONCEPT CHECK 3.4

You perform the hypothetical reaction of an element, $X_2(g)$, with another element, $Y(g)$, to produce $XY(g)$.

a. Write the balanced chemical equation for the reaction.

b. If X_2 and Y were mixed in the quantities shown in the container below and allowed to react, which of the following is the correct representation of the contents of the container after the reaction has occurred?

c. Using the information presented in part b, identify the limiting reactant.

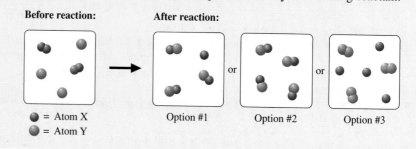

Before reaction:

After reaction:

Option #1 or Option #2 or Option #3

● = Atom X
● = Atom Y

A Checklist for Review

Important Terms

molecular weight (3.1)
formula weight (3.1)
mole (mol) (3.2)
Avogadro's number
  (N_A) (3.2)

molar mass (3.2)
percentage composition
  (3.3)
mass percentage (3.3)

empirical (simplest)
  formula (3.5)
stoichiometry (3.6)

limiting reactant
  (reagent) (3.8)
theoretical yield (3.8)
percentage yield (3.8)

Key Equations

$$\text{Mass \% } A = \frac{\text{mass of } A \text{ in the whole}}{\text{mass of the whole}} \times 100\%$$

$$\text{Percentage yield} = \frac{\text{actual yield}}{\text{theoretical yield}} \times 100\%$$

$$n = \frac{\text{molecular weight}}{\text{empirical formula weight}}$$

Summary of Facts and Concepts

A *formula weight* equals the sum of the atomic weights of the atoms in the formula of a compound. If the formula corresponds to that of a molecule, this sum of atomic weights equals the *molecular weight* of the compound. The mass of *Avogadro's number* (6.02×10^{23}) of formula units—that is, the mass of one *mole* of substance—equals the mass in grams that corresponds to the numerical value of the formula weight in amu. This mass is called the *molar mass*.

The *empirical formula (simplest formula)* of a compound is obtained from the *percentage composition* of the substance.

To calculate the empirical formula, you convert mass percentages to ratios of moles, which, when expressed in smallest whole numbers, give the subscripts in the formula. A molecular formula is a multiple of the empirical formula; this multiple is determined from the experimental value of the molecular weight.

A chemical equation may be interpreted in terms of moles of reactants and products, as well as in terms of molecules. The maximum amount of product from a reaction is determined by the *limiting reactant*.

Operational Skills

Note: A table of atomic weights is necessary for most of these skills.

1. **Calculating the formula weight from a formula or molecular model** Given the formula of a compound

and a table of atomic weights, calculate the formula weight. **(EXAMPLES 3.1, 3.2)**

2. **Calculating the mass of an atom or molecule** Using the molar mass and Avogadro's number, calculate the

mass of an atom or molecule in grams.   **(EXAMPLE 3.3)**

3. **Converting moles of substance to grams, and vice versa**   Given the moles of a compound with a known formula, calculate the mass **(EXAMPLE 3.4)**. Or, given the mass of a compound with a known formula, calculate the moles.   **(EXAMPLE 3.5)**

4. **Calculating the number of molecules in a given mass**   Given the mass of a sample of a molecular substance and its formula, calculate the number of molecules in the sample.   **(EXAMPLE 3.6)**

5. **Calculating the percentage composition from the formula**   Given the formula of a compound, calculate the mass percentages of the elements in it.   **(EXAMPLE 3.7)**

6. **Calculating the mass of an element in a given mass of compound**   Given the mass percentages of elements in a given mass of a compound, calculate the mass of any element.   **(EXAMPLE 3.8)**

7. **Calculating the percentages of C and H by combustion**   Given the masses of CO_2 and H_2O obtained from the combustion of a known mass of a compound of C, H, and O, compute the mass percentage of each element.   **(EXAMPLE 3.9)**

8. **Determining the empirical formula from percentage composition**   Given the masses of elements in a known mass of compound, or given its percentage composition, obtain the empirical formula.   **(EXAMPLE 3.10)**

9. **Determining the molecular formula from percentage composition and molecular weight**   Given the empirical formula and molecular weight of a substance, obtain its molecular formula.   **(EXAMPLE 3.11)**

10. **Relating quantities in a chemical equation**   Given a chemical equation and the amount of one substance, calculate the amount of another substance involved in the reaction.   **(EXAMPLE 3.12)**

11. **Calculating with a limiting reactant**   Given the amounts of reactants and the chemical equation, find the limiting reactant; then calculate the amount of a product.   **(EXAMPLES 3.13, 3.14)**

Review Questions

3.1   What is the difference between a formula weight and a molecular weight? Could a given substance have both a formula weight and a molecular weight?

3.2   Describe in words how to obtain the formula weight of a compound from the formula.

3.3   One mole of N_2 contains how many N_2 molecules? How many N atoms are there in one mole of N_2? One mole of iron(III) sulfate, $Fe_2(SO_4)_3$, contains how many moles of SO_4^{2-} ions? how many moles of O atoms?

3.4   Explain what is involved in determining the composition of a compound of C, H, and O by combustion.

3.5   Explain what is involved in obtaining the empirical formula from the percentage composition.

3.6   A substance has the molecular formula $C_6H_{12}O_2$. What is its empirical formula?

3.7   Describe in words the meaning of the equation

$$CH_4 + 2O_2 \longrightarrow CO_2 + 2H_2O$$

using a molecular, a molar, and then a mass interpretation.

3.8   Explain how a chemical equation can be used to relate the masses of different substances involved in a reaction.

3.9   What is a limiting reactant in a reaction mixture? Explain how it determines the amount of product.

3.10   Come up with some examples of limiting reactants that use the concept but don't involve chemical reactions.

Conceptual Problems

3.11   You react nitrogen and hydrogen in a container to produce ammonia, $NH_3(g)$. The following figure depicts the contents of the container after the reaction is complete.

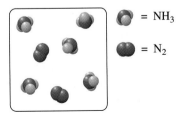

    ● = NH_3

    ● = N_2

a. Write a balanced chemical equation for the reaction.

b. What is the limiting reactant?

c. How many molecules of the limiting reactant would you need to add to the container in order to have a complete reaction (convert all reactants to products)?

3.12   Propane, C_3H_8, is the fuel of choice in a gas barbecue. When burning, the balanced equation is

$$C_3H_8 + 5O_2 \longrightarrow 3CO_2 + 4H_2O$$

a. What is the limiting reactant when cooking with a gas grill?

b. When using a gas grill you can sometimes turn the gas up to the point at which the flame becomes yellow and smokey. In terms of the chemical reaction, what is happening?

3.13   A critical point to master in becoming proficient at solving problems is evaluating whether or not your answer is reasonable. A friend asks you to look over her homework to see if she has done the calculations correctly. Shown below are descriptions of some of her answers. Without using your calculator or doing calculations on paper, see if you can judge the answers below as being reasonable or ones that will require her to go back and work the problems again.
a. 0.33 mole of an element has a mass of 1.0×10^{-3} g.
b. The mass of one molecule of water is 1.80×10^{-10} g.
c. There are 3.01×10^{23} atoms of Na in 0.500 mol of Na.

3.14   An exciting chemical demonstration involves the reaction of hydrogen and oxygen to produce water vapor:

$$2H_2(g) + O_2(g) \longrightarrow 2H_2O(g)$$

The reaction is carried out in soap bubbles or balloons that are filled with the reactants. We get the reaction to proceed by igniting the bubbles or balloons. The more H_2O that is formed during the reaction, the bigger the bang. Explain the following observations.
a. When a bubble containing equal amounts of H_2 and O_2 is ignited, a sizable bang results.
b. When a bubble containing a ratio of 2 to 1 in the amounts of H_2 and O_2 is ignited, the loudest bang results.
c. When a bubble containing just O_2 is ignited, virtually no sound is made.

Practice Problems

Formula Weights and Mole Calculations

3.15   Find the formula weights of the following substances to three significant figures.
a. methanol, CH_3OH
b. potassium carbonate, K_2CO_3

3.16   Find the formula weights of the following substances to three significant figures.
a. acetic acid, $HC_2H_3O_2$
b. calcium hydroxide, $Ca(OH)_2$

3.17   Calculate the formula weights of the following molecules to three significant figures.

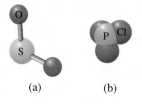

(a)   (b)

3.18   Calculate the formula weights of the following molecules to three significant figures.

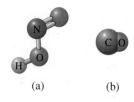

(a)   (b)

3.19   Calculate the mass (in grams) of each of the following species.
a. Na atom   b. CH_3Cl molecule

3.20   Calculate the mass (in grams) of each of the following species.
a. Fe atom   b. N_2O molecule

3.21   Calculate the mass in grams of the following.
a. 0.15 mol Na   b. 2.78 mol CH_2Cl_2

3.22   Calculate the mass in grams of the following.
a. 0.205 mol Fe   b. 5.8 mol CO_2

3.23   Obtain the moles of substance in the following.
a. 2.86 g C   b. 76 g C_4H_{10}

3.24   Obtain the moles of substance in the following.
a. 2.57 g As   b. 41.4 g N_2H_4

3.25   Calcium sulfate, $CaSO_4$, is a white, crystalline powder. Gypsum is a mineral, or natural substance, that is a hydrate of calcium sulfate. A 1.000-g sample of gypsum contains 0.791 g $CaSO_4$. How many moles of $CaSO_4$ are there in this sample? Assuming that the rest of the sample is water, how many moles of H_2O are there in the sample? Show that the result is consistent with the formula $CaSO_4 \cdot 2H_2O$.

3.26   A 1.547-g sample of blue copper(II) sulfate pentahydrate, $CuSO_4 \cdot 5H_2O$, is heated carefully to drive off the water. The white crystals of $CuSO_4$ that are left behind have a mass of 0.989 g. How many moles of H_2O were in the original sample? Show that the relative molar amounts of $CuSO_4$ and H_2O agree with the formula of the hydrate.

3.27   Calculate the following.
a. number of atoms in 32.0 g Br_2
b. number of molecules in 62 g NH_3
c. number of $SO_4{}^{2-}$ ions in 14.3 g $Cr_2(SO_4)_3$

3.28   Calculate the following.
a. number of atoms in 8.71 g I_2
b. number of molecules in 14.9 g N_2O_5
c. number of Ca^{2+} ions in 3.91 g $Ca_3(PO_4)_2$

3.29 Carbon tetrachloride is a colorless liquid used in the manufacture of fluorocarbons and as an industrial solvent. How many molecules are there in 7.58 mg of carbon tetrachloride?

3.30 Chlorine trifluoride is a colorless, reactive gas used in nuclear fuel reprocessing. How many molecules are there in a 5.88-mg sample of chlorine trifluoride?

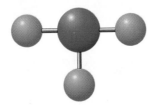

Mass Percentage

3.31 Phosphorus oxychloride is the starting compound for preparing substances used as flame retardants for plastics. An 8.53-mg sample of phosphorus oxychloride contains 1.72 mg of phosphorus. What is the mass percentage of phosphorus in the compound?

3.32 Ethyl mercaptan is an odorous substance added to natural gas to make leaks easily detectable. A sample of ethyl mercaptan weighing 3.17 mg contains 1.64 mg of sulfur. What is the mass percentage of sulfur in the substance?

3.33 A fertilizer is advertised as containing 15.8% nitrogen (by mass). How much nitrogen is there in 4.15 kg of fertilizer?

3.34 Seawater contains 0.0065% (by mass) of bromine. How many grams of bromine are there in 1.00 L of seawater? The density of seawater is 1.025 g/cm^3.

Chemical Formulas

3.35 Calculate the percentage composition for each of the following compounds (three significant figures).
a. CO   b. NaH_2PO_4

3.36 Calculate the percentage composition for each of the following compounds (three significant figures).
a. NO   b. $KClO_4$

3.37 Calculate the mass percentage of each element in toluene, represented by the following molecular model.

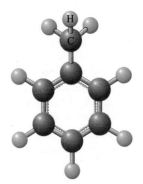

3.38 Calculate the mass percentage of each element in 2-propanol, represented by the following molecular model.

3.39 Ethylene glycol is used as an automobile antifreeze and in the manufacture of polyester fibers. The name glycol stems from the sweet taste of this poisonous compound. Combustion of 6.38 mg of ethylene glycol gives 9.06 mg CO_2 and 5.58 mg H_2O. The compound contains only C, H, and O. What are the mass percentages of the elements in ethylene glycol?

3.40 Phenol, commonly known as carbolic acid, was used by Joseph Lister as an antiseptic for surgery in 1865. Its principal use today is in the manufacture of phenolic resins and plastics. Combustion of 5.23 mg of phenol yields 14.67 mg CO_2 and 3.01 mg H_2O. Phenol contains only C, H, and O. What is the percentage of each element in this substance?

3.41 An oxide of osmium (symbol Os) is a pale yellow solid. If 2.89 g of the compound contains 2.16 g of osmium, what is its empirical formula?

3.42 An oxide of tungsten (symbol W) is a bright yellow solid. If 5.34 g of the compound contains 4.23 g of tungsten, what is its empirical formula?

3.43 Two compounds have the same composition: 92.25% C and 7.75% H.
a. Obtain the empirical formula corresponding to this composition.
b. One of the compounds has a molecular weight of 52.03 amu; the other, of 78.05 amu. Obtain the molecular formulas of both compounds.

3.44 Two compounds have the same composition: 85.62% C and 14.38% H.

a. Obtain the empirical formula corresponding to this composition.

b. One of the compounds has a molecular weight of 28.03 amu; the other, of 56.06 amu. Obtain the molecular formulas of both compounds.

3.45 Putrescine, a substance produced by decaying animals, has the empirical formula C_2H_6N. Several determinations of molecular weight give values in the range of 87 to 90 amu. Find the molecular formula of putrescine.

3.46 Compounds of boron with hydrogen are called boranes. One of these boranes has the empirical formula BH_3 and a molecular weight of 28 amu. What is its molecular formula?

3.47 Oxalic acid is a toxic substance used by laundries to remove rust stains. Its composition is 26.7% C, 2.2% H, and 71.1% O (by mass), and its molecular weight is 90 amu. What is its molecular formula?

3.48 Adipic acid is used in the manufacture of nylon. The composition of the acid is 49.3% C, 6.9% H, and 43.8% O (by mass), and the molecular weight is 146 amu. What is the molecular formula?

Stoichiometry: Quantitative Relations in Reactions

3.49 Ethylene, C_2H_4, burns in oxygen to give carbon dioxide, CO_2, and water. Write the equation for the reaction, giving molecular, molar, and mass interpretations below the equation.

3.50 Hydrogen sulfide gas, H_2S, burns in oxygen to give sulfur dioxide, SO_2, and water. Write the equation for the reaction, giving molecular, molar, and mass interpretations below the equation.

3.51 Iron in the form of fine wire burns in oxygen to form iron(III) oxide.

$$4Fe(s) + 3O_2(g) \longrightarrow 2Fe_2O_3(s)$$

How many moles of O_2 are needed to produce 5.21 mol Fe_2O_3?

3.52 Nickel(II) chloride reacts with sodium phosphate to precipitate nickel(II) phosphate.

$$3NiCl_2(aq) + 2Na_3PO_4(aq) \longrightarrow Ni_3(PO_4)_2(s) + 6NaCl(aq)$$

How many moles of nickel(II) chloride are needed to produce 0.479 mol nickel(II) phosphate?

3.53 Tungsten metal, W, is used to make incandescent bulb filaments. The metal is produced from the yellow tungsten(VI) oxide, WO_3, by reaction with hydrogen.

$$WO_3(s) + 3H_2(g) \longrightarrow W(s) + 3H_2O(g)$$

How many grams of tungsten can be obtained from 4.81 kg of hydrogen with excess tungsten(VI) oxide?

3.54 Acrylonitrile, C_3H_3N, is the starting material for the production of a kind of synthetic fiber (acrylics). It can be made from propylene, C_3H_6, by reaction with nitric oxide, NO.

$$4C_3H_6(g) + 6NO(g) \longrightarrow 4C_3H_3N(g) + 6H_2O(g) + N_2(g)$$

How many grams of acrylonitrile are obtained from 651 kg of propylene and excess NO?

3.55 The following reaction, depicted using molecular models, is used to make carbon tetrachloride, CCl_4, a solvent and starting material for the manufacture of fluorocarbon refrigerants and aerosol propellants.

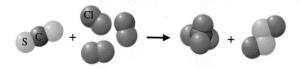

Calculate the number of grams of carbon disulfide, CS_2, needed for a laboratory-scale reaction with 62.7 g of chlorine, Cl_2.

3.56 Using the following reaction (depicted using molecular models), large quantities of ammonia are burned in the presence of a platinum catalyst to give nitric oxide, as the first step in the preparation of nitric acid.

Suppose a vessel contains 12.2 g of NH_3, how many grams of O_2 are needed for a complete reaction?

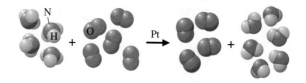

Limiting Reactant; Theoretical and Percentage Yields

3.57 Potassium superoxide, KO_2, is used in rebreathing gas masks to generate oxygen.

$$4KO_2(s) + 2H_2O(l) \longrightarrow 4KOH(s) + 3O_2(g)$$

If a reaction vessel contains 0.25 mol KO_2 and 0.15 mol H_2O, what is the limiting reactant? How many moles of oxygen can be produced?

3.58 Solutions of sodium hypochlorite, NaClO, are sold as a bleach (such as Clorox). They are prepared by the reaction of chlorine with sodium hydroxide.

$$2NaOH(aq) + Cl_2(g) \longrightarrow NaCl(aq) + NaClO(aq) + H_2O(l)$$

If you have 1.23 mol of NaOH in solution and 1.47 mol of Cl_2 gas available to react, which is the limiting reactant? How many moles of NO(g) could be obtained?

3.59 Methanol, CH_3OH, is prepared industrially from the gas-phase catalytic balanced reaction that has been depicted here using molecular models.

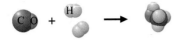

In a laboratory test, a reaction vessel was filled with 35.4 g CO and 10.2 g H_2. How many grams of methanol would be produced in a complete reaction? Which reactant remains unconsumed at the end of the reaction? How many grams of it remain?

3.60 Carbon disulfide, CS_2, burns in oxygen. Complete combustion gives the balanced reaction that has been depicted here using molecular models.

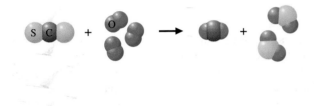

Calculate the grams of sulfur dioxide, SO_2, produced when a mixture of 30.0 g of carbon disulfide and 35.0 g of oxygen reacts. Which reactant remains unconsumed at the end of the combustion? How many grams remain?

3.61 Aspirin (acetylsalicylic acid) is prepared by heating salicylic acid, $C_7H_6O_3$, with acetic anhydride, $C_4H_6O_3$. The other product is acetic acid, $C_2H_4O_2$.

$$C_7H_6O_3 + C_4H_6O_3 \longrightarrow C_9H_8O_4 + C_2H_4O_2$$

What is the theoretical yield (in grams) of aspirin, $C_9H_8O_4$, when 2.00 g of salicylic acid is heated with 4.00 g of acetic anhydride? If the actual yield of aspirin is 1.98 g, what is the percentage yield?

3.62 Methyl salicylate (oil of wintergreen) is prepared by heating salicylic acid, $C_7H_6O_3$, with methanol, CH_3OH.

$$C_7H_6O_3 + CH_3OH \longrightarrow C_8H_8O_3 + H_2O$$

In an experiment, 1.50 g of salicylic acid is reacted with 11.20 g of methanol. The yield of methyl salicylate, $C_8H_8O_3$, is 1.31 g. What is the percentage yield?

General Problems

3.63 Caffeine, the stimulant in coffee and tea, has the molecular formula $C_8H_{10}N_4O_2$. Calculate the mass percentage of each element in the substance. Give the answers to three significant figures.

3.64 Morphine, a narcotic substance obtained from opium, has the molecular formula $C_{17}H_{19}NO_3$. What is the mass percentage of each element in morphine (to three significant figures)?

3.65 A moth repellent, *para*-dichlorobenzene, has the composition 49.1% C, 2.7% H, and 48.2% Cl. Its molecular weight is 147 amu. What is its molecular formula?

3.66 Sorbic acid is added to food as a mold inhibitor. Its composition is 64.3% C, 7.2% H, and 28.5% O, and its molecular weight is 112 amu. What is its molecular formula?

3.67 Thiophene is a liquid compound of the elements C, H, and S. A sample of thiophene weighing 7.96 mg was burned in oxygen, giving 16.65 mg CO_2. Another sample was subjected to a series of reactions that transformed all of the sulfur in the compound to barium sulfate. If 4.31 mg of thiophene gave 11.96 mg of barium sulfate, what is the empirical formula of thiophene? Its molecular weight is 84 amu. What is its molecular formula?

3.68 Aniline, a starting compound for urethane plastic foams, consists of C, H, and N. Combustion of such compounds yields CO_2, H_2O, and N_2 as products. If the combustion of 9.71 mg of aniline yields 6.63 mg H_2O and 1.46 mg N_2, what is its empirical formula? The molecular weight of aniline is 93 amu. What is its molecular formula?

3.69 A sample of limestone (containing calcium carbonate, $CaCO_3$) weighing 438 mg is treated with oxalic acid, $H_2C_2O_4$, to give calcium oxalate, CaC_2O_4.

$$CaCO_3(s) + H_2C_2O_4(aq) \longrightarrow$$
$$CaC_2O_4(s) + H_2O(l) + CO_2(g)$$

The mass of the calcium oxalate is 472 mg. What is the mass percentage of calcium carbonate in this limestone?

3.70 A titanium ore contains rutile (TiO_2) plus some iron oxide and silica. When it is heated with carbon in the presence of chlorine, titanium tetrachloride, $TiCl_4$, is formed.

$$TiO_2(s) + C(s) + 2Cl_2(g) \longrightarrow TiCl_4(g) + CO_2(g)$$

Titanium tetrachloride, a liquid, can be distilled from the mixture. If 35.4 g of titanium tetrachloride is recovered from 17.4 g of crude ore, what is the mass percentage of TiO_2 in the ore (assuming all TiO_2 reacts)?

3.71 Ethylene oxide, C_2H_4O, is made by the oxidation of ethylene, C_2H_4.

$$2C_2H_4(g) + O_2(g) \longrightarrow 2C_2H_4O(g)$$

Ethylene oxide is used to make ethylene glycol for automobile antifreeze. In a pilot study, 10.6 g of ethylene gave 9.91 g of ethylene oxide. What is the percentage yield of ethylene oxide?

3.72 Nitrobenzene, $C_6H_5NO_2$, an important raw material for the dye industry, is prepared from benzene, C_6H_6, and nitric acid, HNO_3.

$$C_6H_6(l) + HNO_3(l) \longrightarrow C_6H_5NO_2(l) + H_2O(l)$$

When 22.4 g of benzene and an excess of HNO_3 are used, what is the theoretical yield of nitrobenzene? If 28.7 g of nitrobenzene is recovered, what is the percentage yield?

3.73 Zinc metal can be obtained from zinc oxide, ZnO, by reaction at high temperature with carbon monoxide, CO.

$$ZnO(s) + CO(g) \longrightarrow Zn(s) + CO_2(g)$$

The carbon monoxide is obtained from carbon.

$$2C(s) + O_2(g) \longrightarrow 2CO(g)$$

What is the maximum amount of zinc that can be obtained from 75.0 g of zinc oxide and 50.0 g of carbon?

3.74 Hydrogen cyanide, HCN, can be made by a two-step process. First, ammonia is reacted with O_2 to give nitric oxide, NO.

$$4NH_3(g) + 5O_2(g) \longrightarrow 4NO(g) + 6H_2O(g)$$

Then nitric oxide is reacted with methane, CH_4.

$$2NO(g) + 2CH_4(g) \longrightarrow 2HCN(g) + 2H_2O(g) + H_2(g)$$

When 24.2 g of ammonia and 25.1 g of methane are used, how many grams of hydrogen cyanide can be produced?

Cumulative-Skills Problems

3.75 A 0.500-g mixture of Cu_2O and CuO contains 0.425 g Cu. What is the mass of CuO in the mixture?

3.76 A mixture of Fe_2O_3 and FeO was found to contain 72.00% Fe by mass. What is the mass of Fe_2O_3 in 0.500 g of this mixture?

3.77 Hemoglobin is the oxygen-carrying molecule of red blood cells, consisting of a protein and a nonprotein substance. The nonprotein substance is called heme. A sample of heme weighing 35.2 mg contains 3.19 mg of iron. If a heme molecule contains one atom of iron, what is the molecular weight of heme?

3.78 Penicillin V was treated chemically to convert sulfur to barium sulfate, $BaSO_4$. An 8.19-mg sample of penicillin V gave 5.46 mg $BaSO_4$. What is the percentage of sulfur in penicillin V? If there is one sulfur atom in the molecule, what is the molecular weight?

3.79 A 3.41-g sample of a metallic element, M, reacts completely with 0.0158 mol of a gas, X_2, to form 4.52 g MX. What are the identities of M and X?

3.80 1.92 g M^+ ion reacts with 0.158 mol X^- ion to produce a compound, MX_2, which is 86.8% X by mass. What are the identities of M^+ and X^-?

Beakers with precipitates of lead(II) iodide and mercury(II) iodide.

Chemical Reactions

Chemical reactions are the heart of chemistry. Some reactions, such as those accompanying a forest fire or the explosion of dynamite, are quite dramatic. Others are much less obvious, although all chemical reactions must involve detectable change. A chemical reaction involves a change from reactant substances to product substances, and the product substances will have physical and chemical properties different from those of the reactants.

Figure 4.1 shows an experimenter adding a colorless solution of potassium iodide, KI, to a colorless solution of lead(II) nitrate, $Pb(NO_3)_2$. What you see is the formation of a cloud of bright yellow crystals where the two solutions have come into contact, clear evidence of a chemical reaction. The bright yellow crystals are lead(II) iodide, PbI_2, one of the reaction products. We call a solid that forms during a chemical reaction in solution a *precipitate;* the reaction is a precipitation reaction.

In this chapter, we will discuss the major types of chemical reactions, including precipitation reactions. Some of the most important reactions we will describe involve ions in aqueous (water) solution. Therefore, we will first look at these ions and see how we represent by chemical equations the reactions involving ions in aqueous solution.

Some questions we will answer are: What is the evidence for ions in solution? How do we write chemical equations for reactions involving ions? How can we classify and describe the many reactions we observe so that we can begin to understand them? What is the quantitative description of solutions and reactions in solution?

CONTENTS

Figure 4.1
Reaction of potassium iodide solution and lead(II) nitrate solution. The reactant solutions are colorless, but one of the products, lead(II) iodide, forms as a yellow precipitate.

Ions in Aqueous Solution

You probably have heard that you should not operate electrical equipment while standing in water. And you may have read a murder mystery in which the victim was electrocuted when an electrical appliance "accidentally" fell into his or her bath water. Actually, if the water were truly pure, the person would be safe from electrocution, because pure water is a nonconductor of electricity. Bath water or water as it flows from the faucet, however, is a *solution* of water with small amounts of dissolved substances in it, and these dissolved substances make the solution an electrical conductor. This allows an electric current to flow from an electrical appliance to the human body. Let us look at the nature of such solutions.

4.1 Ionic Theory of Solutions and Solubility Rules

Chemists began studying the electrical behavior of substances in the early nineteenth century, and they knew that you could make pure water electrically conducting by dissolving certain substances in it. In 1884, the young Swedish chemist Svante Arrhenius proposed the *ionic theory of solutions* to account for this conductivity. He said that certain substances produce freely moving ions when they dissolve in water, and these ions conduct an electric current in an aqueous solution. ◀

▶ Arrhenius submitted his ionic theory as part of his doctoral dissertation to the faculty at Uppsala, Sweden, in 1884. It was not well received and he barely passed. In 1903, however, he was awarded the Nobel Prize in chemistry for this theory.

Suppose you dissolve sodium chloride, NaCl, in water. From our discussion in Section 2.6, you may remember that sodium chloride is an ionic solid consisting of sodium ions, Na^+, and chloride ions, Cl^-, held in a regular, fixed array. When you dissolve solid sodium chloride in water, the Na^+ and Cl^- ions go into solution as freely moving ions. Now suppose you dip electric wires that are connected to the poles of a battery into a solution of sodium chloride. The wire that connects to the positive pole of the battery attracts the negatively charged chloride ions in solution, because of their opposite charges. Similarly, the wire connected to the negative pole of the battery attracts the positively charged sodium ions in solution (Figure 4.2). Thus, the ions in the solution begin to move, and these moving charges form the elec-

Figure 4.2
Motion of ions in solution. Ions are in fixed positions in a crystal. During the solution process, however, ions leave the crystal and become freely moving. Note that Na^+ ions (small gray spheres) are attracted to the negative wire, whereas Cl^- ions (large green spheres) are attracted to the positive wire.

tric current in the solution. (In a wire, it is moving electrons that constitute the electric current.)

Now consider pure water. Water consists of molecules, each of which is electrically neutral. Since each molecule carries no net electric charge, it carries no overall electric charge when it moves. Thus, pure water is a nonconductor of electricity.

In summary, although water is itself nonconducting, it has the ability to dissolve various substances, some of which go into solution as freely moving ions. An aqueous solution of ions is electrically conducting.

Electrolytes and Nonelectrolytes

We can divide the substances that dissolve in water into two broad classes, electrolytes and nonelectrolytes. An **electrolyte** is *a substance that dissolves in water to give an electrically conducting solution.* Sodium chloride, table salt, is an example of an electrolyte. When most ionic substances dissolve in water, ions that were in fixed sites in the crystalline solid go into the surrounding aqueous solution, where they are free to move about. The resulting solution is conducting because the moving ions form an electric current. Thus, in general, *ionic solids that dissolve in water are electrolytes.*

Not all electrolytes are ionic substances. Certain molecular substances dissolve in water to form ions. The resulting solution is electrically conducting, and so we say that the molecular substance is an electrolyte. An example is hydrogen chloride gas, $HCl(g)$, which is a molecular substance. Hydrogen chloride gas dissolves in water, giving $HCl(aq)$, which in turn produces hydrogen ions, H^+, and chloride ions, Cl^-, in aqueous solution. (The solution of H^+ and Cl^- ions is called hydrochloric acid.)

$$HCl(aq) \xrightarrow{\text{H}_2\text{O}} H^+(aq) + Cl^-(aq)$$

We will look more closely at molecular electrolytes, such as HCl, at the end of this section.

A **nonelectrolyte** is *a substance that dissolves in water to give a nonconducting or very poorly conducting solution.* A common example is sucrose, $C_{12}H_{22}O_{11}$, which is ordinary table sugar. Another example is methanol, CH_3OH, a compound used in car window washer solution. Both of these are molecular substances. The solution process occurs because molecules of the substance mix with molecules of water. Molecules are electrically neutral and cannot carry an electric current, so the solution is electrically nonconducting.

Observing the Electrical Conductivity of a Solution

Figure 4.3 shows a simple apparatus that allows you to observe the ability of a solution to conduct an electric current. The apparatus has two electrodes, here they are flat metal plates, dipping into the solution in a beaker. One electrode connects directly to a battery through a wire. The other electrode connects by a wire to a light bulb that connects with another wire to the other side of the battery. For an electrical current to flow from the battery, there must be a complete circuit, which allows the current to flow from the positive pole of the battery through the circuit to the negative pole of the battery. To have a complete circuit, the solution in the beaker must conduct electricity, as the wires do. If the solution is conducting, the circuit is complete and the bulb lights. If the solution is nonconducting, the circuit is incomplete and the bulb does not light.

Figure 4.3
Testing the electrical conductivity of a solution.
Left: Pure water does not conduct; therefore the bulb does not light. *Right:* A solution of sodium chloride allows the current to pass through it, and the bulb lights.

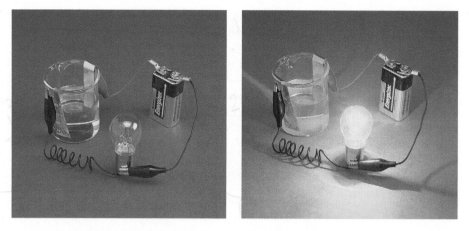

The beaker shown on the left side of Figure 4.3 contains pure water. Because the bulb is not lit, we conclude that pure water is a nonconductor. The beaker shown on the right side of Figure 4.3 contains a solution of sodium chloride in water. In this case, the bulb burns brightly, showing that the solution is a very good conductor of electricity.

How brightly the bulb lights tells you whether the solution is a very good conductor (contains a "strong" electrolyte) or only a moderately good conductor (contains a "weak" electrolyte).

Strong and Weak Electrolytes

When electrolytes dissolve in water they produce ions, but to varying extents. A **strong electrolyte** is *an electrolyte that exists in solution almost entirely as ions.* Most ionic solids that dissolve in water do so by going into the solution almost completely as ions, so they are strong electrolytes. An example is sodium chloride.

$$NaCl(s) \xrightarrow{\text{H}_2\text{O}} Na^+(aq) + Cl^-(aq)$$

A **weak electrolyte** is *an electrolyte that dissolves in water to give a relatively small percentage of ions.* These are generally molecular substances. Ammonia, NH_3, is an example. Pure ammonia is a gas that readily dissolves in water and goes into solution as ammonia molecules, $NH_3(aq)$. Ammonia molecules react with water to form ammonium ions, NH_4^+, and hydroxide ions, OH^-.

$$NH_3(aq) + H_2O(l) \longrightarrow NH_4^+(aq) + OH^-(aq)$$

However, these ions, $NH_4^+ + OH^-$, react with each other to give back ammonia molecules and water molecules.

$$NH_4^+(aq) + OH^-(aq) \longrightarrow NH_3(aq) + H_2O(l)$$

Both reactions, the original one and its reverse, occur constantly and simultaneously. We denote this situation by writing a single equation with a double arrow:

$$NH_3(aq) + H_2O(l) \rightleftharpoons NH_4^+(aq) + OH^-(aq)$$

As a result, just a small percentage of the NH_3 molecules have reacted at any given moment to form ions. Thus, ammonia is a weak electrolyte. See Figure 4.4.

Figure 4.4
Comparing strong and weak electrolytes. The apparatus is similar to that in Figure 4.3, but this time strong and weak electrolytes are compared. The solution on the left is of HCl (a strong electrolyte) and that on the right is of NH_3 (a weak electrolyte). Note how much more brightly the bulb on the left burns compared with that on the right.

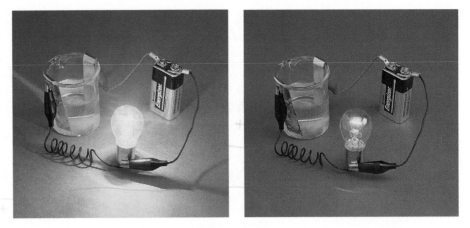

Most soluble molecular substances are either nonelectrolytes or weak electrolytes. An exception is hydrogen chloride gas, HCl(*g*), which is a strong electrolyte. We represent its reaction with H_2O by an equation with a single arrow:

$$HCl(aq) \longrightarrow H^+(aq) + Cl^-(aq)$$

Solubility Rules

Substances vary widely in their *solubility,* or ability to dissolve, in water. Some compounds, such as sodium chloride and ethyl alcohol (CH_3CH_2OH), dissolve readily and are said to be *soluble*. Others, such as calcium carbonate (which occurs naturally as limestone and marble) and benzene (C_6H_6), have quite limited solubilities and are thus said to be *insoluble*.

Soluble ionic compounds form solutions that are strong electrolytes. To predict the solubility of ionic compounds, chemists have developed solubility rules. Table 4.1 lists eight solubility rules for ionic compounds. Example 4.1 illustrates how to use the rules.

EXAMPLE 4.1

Using the Solubility Rules

Determine whether the following compounds are soluble or insoluble in water.

a. Hg_2Cl_2 b. KI   c. lead(II) nitrate

SOLUTION

a. According to Rule 3 in Table 4.1, most compounds that contain chloride are soluble. However, Hg_2Cl_2 is listed as one of the exceptions to this rule. Therefore, **Hg_2Cl_2 is not soluble in water.**

b. According to Rule 1, Group IA compounds are soluble, and according to Rule 3, most iodides are soluble. Therefore, **KI is soluble in water.**

c. According to Rule 2, compounds containing nitrates, NO_3^-, are soluble. Therefore, **lead(II) nitrate, $Pb(NO_3)_2$, is soluble in water.**

Table 4.1
Solubility Rules for Ionic Compounds

Rule	Applies to	Statement	Exceptions
1	Li^+, Na^+, K^+, NH_4^+	Group IA and ammonium compounds are soluble.	—
2	$C_2H_3O_2^-, NO_3^-$	Acetates and nitrates are soluble.	—
3	Cl^-, Br^-, I^-	Most chlorides, bromides, and iodides are soluble.	$AgCl, Hg_2Cl_2, PbCl_2, AgBr, HgBr_2, Hg_2Br_2,$ $PbBr_2, AgI, HgI_2, Hg_2I_2, PbI_2$
4	SO_4^{2-}	Most sulfates are soluble.	$CaSO_4, SrSO_4, BaSO_4, Ag_2SO_4, Hg_2SO_4,$ $PbSO_4$
5	CO_3^{2-}	Most carbonates are insoluble.	Group IA carbonates, $(NH_4)_2CO_3$
6	PO_4^{3-}	Most phosphates are insoluble.	Group IA phosphates, $(NH_4)_3PO_4$
7	S^{2-}	Most sulfides are insoluble.	Group IA sulfides, $(NH_4)_2S$
8	OH^-	Most hydroxides are insoluble.	Group IA hydroxides, $Ca(OH)_2, Sr(OH)_2,$ $Ba(OH)_2$

precipitate

Let us summarize the main points in this section. Soluble substances are either electrolytes or nonelectrolytes. Nonelectrolytes form nonconducting aqueous solutions because they dissolve completely as molecules. Electrolytes form electrically conducting solutions in water because they dissolve to give ions in solution. Electrolytes can be strong or weak. Almost all soluble ionic substances are strong electrolytes. Soluble molecular substances usually are nonelectrolytes or weak electrolytes; the latter solution consists primarily of molecules, but has a small percentage of ions. The solubility rules can be used to predict the solubility of ionic compounds in water.

4.2 Molecular and Ionic Equations

For a reaction involving ions, we have a choice of chemical equations, depending on the kind of information we want to convey. We can represent such a reaction by a *molecular equation,* a *complete ionic equation,* or a *net ionic equation.*

To illustrate these different kinds of equations, consider the preparation of calcium carbonate. This white, fine powdery compound is used as a paper filler to brighten and retain ink, as an antacid, and as a mild abrasive in toothpastes. One way to prepare this compound is to react calcium hydroxide with sodium carbonate. Let us look at the different ways to write the equation for this reaction.

Molecular Equations

You could write the equation for this reaction as follows:

$$Ca(OH)_2(aq) + Na_2CO_3(aq) \longrightarrow CaCO_3(s) + 2NaOH(aq)$$

formula

We call this a **molecular equation,** which is *a chemical equation in which the reactants and products are written as if they were molecular substances, even though they may actually exist in solution as ions.* The molecular equation is useful because it is explicit about what the reactant solutions are and what products you obtain. The equation says that you add aqueous solutions of calcium hydroxide and sodium carbonate to the reaction vessel. The insoluble, white calcium carbonate precipitates. After you remove the precipitate, you are left with a solution of sodium hydroxide. The molecular equation closely describes what you actually do.

Complete Ionic Equations

(Full).

Although a molecular equation is useful in describing the actual reactant and product substances, it does not give you an ionic-theory interpretation of the reaction. Because this kind of information is useful, you often need to rewrite the molecular equation as an ionic equation.

Again, consider the reaction of $Ca(OH)_2$ and Na_2CO_3. Both are soluble ionic substances and therefore strong electrolytes when they dissolve in water. If you want to emphasize that the solution contains freely moving ions, it would be better to write $Ca^{2+}(aq) + 2OH^-(aq)$ in place of $Ca(OH)_2(aq)$. Also write $2Na^+(aq) + CO_3^{2-}(aq)$ in place of $Na_2CO_3(aq)$. The reactant side of the equation becomes

$$Ca^{2+}(aq) + 2OH^-(aq) + 2Na^+(aq) + CO_3^{2-}(aq) \longrightarrow$$

Thus, the reaction mixture begins as a solution of four different kinds of ions.

Now let us look at the product side of the equation. One product is the precipitate $CaCO_3(s)$. According to the solubility rules, this is an insoluble ionic compound so it will exist in water as a solid. On the other hand, NaOH is a soluble ionic substance and therefore a strong electrolyte, which we denote by writing $Na^+(aq) + OH^-(aq)$. The complete equation is

$$Ca^{2+}(aq) + 2OH^-(aq) + 2Na^+(aq) + CO_3^{2-}(aq) \longrightarrow$$
$$CaCO_3(s) + 2Na^+(aq) + 2OH^-(aq)$$

The purpose of such a *complete ionic equation* is to represent each substance by its predominant form in the reaction mixture. For example, if the substance is a soluble ionic compound, it probably dissolves as individual ions. In a complete ionic equation, you represent the compound as separate ions. If the substance is a weak electrolyte, it is present in solution primarily as molecules, so you represent it by its molecular formula. If the substance is an insoluble ionic compound, you represent it by the formula of the compound.

Thus, a **complete ionic equation** is *a chemical equation in which strong electrolytes (such as soluble ionic compounds) are written as separate ions in the solution.* You represent other reactants and products by the formulas of the compounds, indicating any soluble substance by (*aq*) after its formula and any insoluble solid substance by (*s*) after its formula.

Net Ionic Equations

In the complete ionic equation representing the reaction of calcium hydroxide and sodium carbonate, some ions (OH^- and Na^+) appear on both sides of the equation. This means that nothing happens to these ions as the reaction occurs. They are called spectator ions. A **spectator ion** is *an ion in an ionic equation that does not take part in the reaction.* You can cancel such ions from both sides to express the essential reaction that occurs.

Figure 4.5
Limestone formations. It is believed that most limestone formed as a precipitate of calcium carbonate (and other carbonates) from seawater. The photograph shows limestone formations at Bryce Point, Bryce Canyon National Park, Utah. More than sixty million years ago, this area was covered by seawater.

$$Ca^{2+}(aq) + 2\cancel{OH^-(aq)} + 2\cancel{Na^+(aq)} + CO_3^{2-}(aq) \longrightarrow$$
$$CaCO_3(s) + 2\cancel{Na^+(aq)} + 2\cancel{OH^-(aq)}$$

The resulting equation is

$$Ca^{2+}(aq) + CO_3^{2-}(aq) \longrightarrow CaCO_3(s)$$

This is the **net ionic equation,** *an ionic equation from which spectator ions have been canceled.* It shows that the reaction that actually occurs at the ionic level is between calcium ions and carbonate ions to form solid calcium carbonate.

From the net ionic equation, you can see that mixing any solution of calcium ion with any solution of carbonate ion will give you this same reaction. For example, the strong electrolyte calcium nitrate, $Ca(NO_3)$, dissolves readily in water to provide a source of calcium ions. Similarly, the strong electrolyte potassium carbonate, K_2CO_3, dissolves readily in water to provide a source of carbonate ions. When you mix solutions of these two compounds, you obtain a solution of calcium ions and carbonate ions, which react to form the insoluble calcium carbonate. The other product is potassium nitrate, a soluble ionic compound. The molecular equation representing the reaction is

$$Ca(NO_3)_2(aq) + K_2CO_3(aq) \longrightarrow CaCO_3(s) + 2KNO_3(aq)$$

You obtain the complete ionic equation from this molecular equation by rewriting each of the soluble ionic compounds as ions, but retaining the formula for the precipitate $CaCO_3(s)$:

$$Ca^{2+}(aq) + 2\cancel{NO_3^-(aq)} + 2\cancel{K^+(aq)} + CO_3^{2-}(aq) \longrightarrow$$
$$CaCO_3(s) + 2\cancel{K^+(aq)} + 2\cancel{NO_3^-(aq)}$$

The net ionic equation is

$$Ca^{2+}(aq) + CO_3^{2-}(aq) \longrightarrow CaCO_3(s)$$

Note that the net ionic equation is identical to the one obtained from the reaction of $Ca(OH)_2$ and Na_2CO_3.

The value of the net ionic equation is its generality. For example, seawater contains Ca^{2+} and CO_3^{2-} ions from various sources. Whatever the sources of these ions, you expect them to react to form a precipitate of calcium carbonate (Figure 4.5).

EXAMPLE 4.2

Writing Net Ionic Equations

Write a net ionic equation for each of the following molecular equations.

a. $2HClO_4(aq) + Ca(OH)_2(aq) \longrightarrow Ca(ClO_4)_2(aq) + 2H_2O(l)$
Perchloric acid, $HClO_4$, is a strong electrolyte, forming H^+ and ClO_4^- ions in solution. $Ca(ClO_4)_2$ is a soluble ionic compound.

b. $HC_2H_3O_2(aq) + NaOH(aq) \longrightarrow NaC_2H_3O_2(aq) + H_2O(l)$
Acetic acid, $HC_2H_3O_2$, is a molecular substance and a weak electrolyte.

SOLUTION

a. According to the solubility rules presented in Table 4.1 and the problem statement, $Ca(OH)_2$ and $Ca(ClO_4)_2$ are soluble ionic compounds so they are strong electrolytes. The problem statement notes that $HClO_4$ is also a strong electrolyte. You write each strong electrolyte in the form of separate ions. Water, H_2O, is a nonelectrolyte (or very weak electrolyte), so you retain its molecular formula. The complete ionic equation is

$$2H^+(aq) + 2ClO_4^-(aq) + Ca^{2+}(aq) + 2OH^-(aq) \longrightarrow$$
$$Ca^{2+}(aq) + 2ClO_4^-(aq) + 2H_2O(l)$$

After canceling spectator ions and dividing by 2, you get the following net ionic equation:

$$H^+(aq) + OH^-(aq) \longrightarrow H_2O(l)$$

b. According to the solubility rules, NaOH and $NaC_2H_3O_2$ are soluble ionic compounds, so they are strong electrolytes. The problem statement notes that $HC_2H_3O_2$ is a weak electrolyte, which you write by its molecular formula. Water, H_2O, is a nonelectrolyte, so you retain its molecular formula also. The complete ionic equation is

$$HC_2H_3O_2(aq) + Na^+(aq) + OH^-(aq) \longrightarrow Na^+(aq) + C_2H_3O_2^-(aq) + H_2O(l)$$

and the net ionic equation is

$$HC_2H_3O_2(aq) + OH^-(aq) \longrightarrow C_2H_3O_2^-(aq) + H_2O(l)$$

See Problems 4.17 and 4.18.

Types of Chemical Reactions

Among the several million known substances, many millions of chemical reactions are possible. Beginning students are often bewildered by the possibilities. How can I know when two substances will react when they are mixed? How can I predict the products? Although it is not possible to give completely general answers to these questions, it is possible to make sense of chemical reactions. Most of the reactions we will study belong to one of three types:

1. Precipitation reactions. In these reactions, you mix solutions of two ionic substances and a solid ionic substance (a precipitate) forms.

2. Acid–base reactions. An acid substance reacts with a substance called a base. Such reactions involve the transfer of a proton between reactants.

3. Oxidation–reduction reactions. These involve the transfer of electrons between reactants.

We will look at each of these types of reactions.

4.3 Precipitation Reactions

A precipitation reaction occurs in aqueous solution because one product is insoluble. To predict whether a precipitate will form when you mix two solutions of ionic compounds, you need to know whether any of the potential products that might form are insoluble or not.

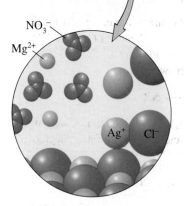

Figure 4.6
Reaction of magnesium chloride and silver nitrate. Magnesium chloride solution is added to a beaker of silver nitrate solution. A white precipitate of silver chloride forms.

▶ The reactants $MgCl_2$ and $AgNO_3$ must be added in correct amounts; otherwise, the excess reactant will remain along with the product $Mg(NO_3)_2$.

Predicting Precipitation Reactions

Suppose you mix together solutions of magnesium chloride, $MgCl_2$, and silver nitrate, $AgNO_3$. You can write the potential reactants as follows:

$$MgCl_2 + AgNO_3 \longrightarrow$$

This equation has the form of an exchange reaction. An **exchange (or metathesis) reaction** is *a reaction between compounds that, when written as a molecular equation, appears to involve the exchange of parts between the two reactants.*

If you exchange the anions between magnesium chloride and silver nitrate, you get silver chloride and magnesium nitrate. The formulas are $AgCl$ and $Mg(NO_3)_2$. The balanced equation is

$$MgCl_2 + 2AgNO_3 \longrightarrow 2AgCl + Mg(NO_3)_2$$

The potential products are silver chloride and magnesium nitrate. According to Rule 3, silver chloride is one of the exceptions to the general solubility of chlorides. Therefore, we predict that the silver chloride is insoluble. Magnesium nitrate is soluble according to Rule 2.

Now we can append the appropriate phase labels to the compounds in the preceding equation.

$$MgCl_2(aq) + 2AgNO_3(aq) \longrightarrow 2AgCl(s) + Mg(NO_3)_2(aq)$$

We predict that reaction occurs because silver chloride is insoluble and precipitates from the reaction mixture. Figure 4.6 shows the formation of the white silver chloride from this reaction. If you separate the precipitate from the solution by pouring it through filter paper, the solution that passes through (the filtrate) contains magnesium nitrate which you could obtain by evaporating the water. ◀

To see the reaction that occurs on an ionic level, you need to rewrite the molecular equation as a net ionic equation. You first write the soluble ionic compounds in the form of ions, leaving the formula of the precipitate unchanged.

$$Mg^{2+}(aq) + 2Cl^-(aq) + 2Ag^+(aq) + 2NO_3^-(aq) \longrightarrow$$
$$2AgCl(s) + Mg^{2+}(aq) + 2NO_3^-(aq)$$

After canceling spectator ions and reducing the coefficients to the smallest whole numbers, you obtain the net ionic equation:

$$Ag^+(aq) + Cl^-(aq) \longrightarrow AgCl(s)$$

This equation represents the essential reaction that occurs: Ag^+ ions and Cl^- ions in aqueous solution react to form solid silver chloride.

If silver chloride were soluble, a reaction would not have occurred. Because Ag^+ and Cl^- react to give the precipitate $AgCl$, the effect is to remove these ions from the reaction mixture as an insoluble compound and leave behind a solution of $Mg(NO_3)_2$.

EXAMPLE 4.3

Deciding Whether Precipitation Occurs

For each of the following, decide whether a precipitation reaction occurs. If it does, write the balanced molecular equation, and then the net ionic equation. If no reaction occurs, write the compounds followed by an arrow and then *NR* (no reaction).

a. Aqueous solutions of sodium chloride and iron(II) nitrate are mixed.

b. Aqueous solutions of aluminum sulfate and sodium hydroxide are mixed.

SOLUTION

a. The formulas of the compounds are $NaCl$ and $Fe(NO_3)_2$. Exchanging anions, you get sodium nitrate, $NaNO_3$, and iron(II) chloride, $FeCl_2$. The equation for the exchange reaction is

$$NaCl + Fe(NO_3)_2 \longrightarrow NaNO_3 + FeCl_2 \quad \text{(not balanced)}$$

Referring to Table 4.1, note that $NaCl$ and $NaNO_3$ are soluble. Also, iron(II) nitrate is soluble and iron(II) chloride is soluble. Since there is no precipitate, no reaction occurs. You obtain simply an aqueous solution of the four different ions (Na^+, Cl^-, Fe^{2+}, and NO_3^-). For the answer, we write

$$\textbf{NaCl}(aq) + \textbf{Fe(NO}_3\textbf{)}_2(aq) \longrightarrow \textbf{NR}$$

b. The formulas of the compounds are $Al_2(SO_4)_3$ and $NaOH$. Exchanging anions, you get aluminum hydroxide, $Al(OH)_3$, and sodium sulfate, Na_2SO_4. The equation for the exchange reaction is

$$Al_2(SO_4)_3 + NaOH \longrightarrow Al(OH)_3 + Na_2SO_4 \quad \text{(not balanced)}$$

From Table 4.1, you see that $Al_2(SO_4)_3$ is soluble, $NaOH$ and Na_2SO_4 are soluble, and $Al(OH)_3$ is insoluble. Thus, aluminum hydroxide precipitates. The balanced molecular equation with phase labels is

$$\textbf{Al}_2\textbf{(SO}_4\textbf{)}_3(aq) + \textbf{6NaOH}(aq) \longrightarrow \textbf{2Al(OH)}_3(s) + \textbf{3Na}_2\textbf{SO}_4(aq)$$

To get the net ionic equation, you write the soluble ionic compounds as ions and cancel spectator ions.

$$2Al^{3+}(aq) + 3SO_4^{2-}\cancel{(aq)} + 6\cancel{Na^+(aq)} + 6OH^-(aq) \longrightarrow$$
$$2Al(OH)_3(s) + 6\cancel{Na^+(aq)} + 3SO_4^{2-}\cancel{(aq)}$$

The net ionic equation is

$$\textbf{Al}^{3+}(aq) + \textbf{3OH}^-(aq) \longrightarrow \textbf{Al(OH)}_3(s)$$

Thus, aluminum ion reacts with hydroxide ion to precipitate aluminum hydroxide.

See Problems 4.19, 4.20, 4.21, and 4.22.

CONCEPT CHECK 4.1

Your lab partner tells you that she mixed two solutions that contain ions. You analyze the solution and find that it contains the ions and precipitate shown in the beaker.

a. Write the molecular equation for the reaction.

b. Write the complete ionic equation for the reaction.

c. Write the net ionic equation for the reaction.

$Na^+(aq)$

$C_2H_3O_2^-(aq)$

$SrSO_4(s)$

4.4 Acid–Base Reactions

Figure 4.7
**Household acids and
bases.** Shown are a variety
of household products that are
either acids or bases.

Acids and bases are some of the most important electrolytes. You can recognize acids and bases by some simple properties. Acids have a sour taste. Solutions of bases, on the other hand, have a bitter taste and a soapy feel. (Of course, you should never taste laboratory chemicals.) Some examples of acids are acetic acid, present in vinegar; citric acid, a constituent of lemon juice; and hydrochloric acid, found in the digestive fluid of the stomach. An example of a base is aqueous ammonia, often used as a household cleaner. Table 4.2 lists further examples; see also Figure 4.7.

Another simple property of acids and bases is their ability to cause color changes in certain dyes. An **acid–base indicator** is *a dye used to distinguish between acidic and basic solutions by means of the color changes it undergoes in these solutions.* Such dyes are common in natural materials. The amber color of tea, for example, is lightened by the addition of lemon juice (citric acid). Red cabbage juice changes to green then yellow when a base is added (Figure 4.8). The green and yellow colors change back to red when an acid is added. Litmus is a common laboratory acid–base indicator. This dye, produced from certain species of lichens, turns red in acidic solution and blue in basic solution. Phenolphthalein (fee′ nol thay′ leen), another laboratory acid–base indicator, is colorless in acidic solution and pink in basic solution.

Table 4.2
Common Acids and Bases

Name	Formula	Remarks
Acids:		
Acetic acid	$HC_2H_3O_2$	Found in vinegar
Acetylsalicylic acid	$HC_9H_7O_4$	Aspirin
Ascorbic acid	$H_2C_6H_6O_6$	Vitamin C
Citric acid	$H_3C_6H_5O_7$	Found in lemon juice (citrus fruits)
Hydrochloric acid	HCl	Found in gastric juice (digestive fluid in stomach)
Sulfuric acid	H_2SO_4	Battery acid
Bases:		
Ammonia	NH_3	Aqueous solution used as a household cleaner
Calcium hydroxide	$Ca(OH)_2$	Slaked lime (used in mortar for building construction)
Magnesium hydroxide	$Mg(OH)_2$	Milk of magnesia (antacid and laxative)
Sodium hydroxide	$NaOH$	Drain cleaners, oven cleaners

Figure 4.8
Red cabbage juice as an acid–base indicator.
Left: Preparation of red cabbage juice. The beaker (green solution) contains red cabbage juice and sodium bicarbonate (baking soda). *Right:* Red cabbage juice has been added from a pipet to solutions in the beakers. These solutions vary in acidity from highly acidic on the left to highly basic on the right.

An important chemical characteristic of acids and bases is the way they react with one another. To understand these acid–base reactions, we need to have precise definitions of the terms *acid* and *base*.

Definitions of Acid and Base

When Arrhenius developed his ionic theory of solutions, he also gave the classic definitions of acids and bases. According to Arrhenius, an **acid** is *a substance that produces hydrogen ions, H$^+$, when it dissolves in water.* An example is nitric acid, HNO$_3$, a molecular substance that dissolves in water to give H$^+$ and NO$_3^-$.

$$HNO_3(aq) \xrightarrow{\text{H}_2\text{O}} H^+(aq) + NO_3^-(aq)$$

An Arrhenius **base** is *a substance that produces hydroxide ions, OH$^-$, when it dissolves in water.* For example, sodium hydroxide is a base.

$$NaOH(s) \xrightarrow{\text{H}_2\text{O}} Na^+(aq) + OH^-(aq)$$

The molecular substance ammonia, NH$_3$, is also a base because it yields hydroxide ions when it reacts with water. ◄

▶ Solutions of ammonia are sometimes called ammonium hydroxide and given the formula NH$_4$OH(aq), based on analogy with NaOH. No NH$_4$OH molecule or compound has ever been found, however.

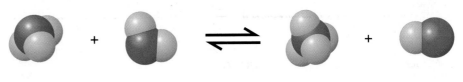

$$NH_3(aq) + H_2O(l) \rightleftharpoons NH_4^+(aq) + OH^-(aq)$$

Although the Arrhenius concept of acids and bases is useful, it is somewhat limited. For example, it tends to single out the OH$^-$ ion as the source of base character, when other ions or molecules can play a similar role. In 1923, Johannes N. Brønsted and Thomas M. Lowry independently noted that many reactions involve nothing more than the transfer of a proton (H$^+$) between reactants. In this view, acid–base reactions are *proton-transfer reactions.*

Brønsted and Lowry defined an **acid** as *the species (molecule or ion) that donates a proton to another species in a proton-transfer reaction.* They defined a **base** as *the species (molecule or ion) that accepts a proton in a proton-transfer reaction.* In the reaction of ammonia with water, the H_2O molecule is the acid, because it donates a proton. The NH_3 molecule is a base, because it accepts a proton.

$$\overset{\text{H}^+}{\overbrace{NH_3(aq) + H_2O(l)}} \rightleftharpoons NH_4^+(aq) + OH^-(aq)$$
$$\underset{\text{base}}{} \quad \underset{\text{acid}}{}$$

The dissolution of nitric acid, HNO_3, in water is actually a proton-transfer reaction, although the following equation, which we used as an illustration of an Arrhenius acid, does not spell that out.

$$HNO_3(aq) \xrightarrow{\text{H}_2\text{O}} H^+(aq) + NO_3^-(aq)$$

To see that this is a proton-transfer reaction, we need to clarify the structure of the hydrogen ion, $H^+(aq)$. This ion consists of a proton (H^+) in association with water molecules, which is what (aq) means. This is not a weak association, however, because the proton (or hydrogen nucleus) would be expected to attract electrons strongly to itself. In fact, the $H^+(aq)$ ion might be better thought of as a proton chemically bonded to a water molecule to give the H_3O^+ ion, with other water molecules less strongly associated with this ion, which we represent by the phase-labeled formula $H_3O^+(aq)$. Written in this form, we usually call this the *hydronium ion*. The hydrogen ion, $H^+(aq)$, and the hydronium ion, $H_3O^+(aq)$, represent precisely the same physical ion. For simplicity, we often write the formula for this ion as $H^+(aq)$. ◄

Now let us rewrite the preceding equation by replacing $H^+(aq)$ by $H_3O^+(aq)$. To maintain a balanced equation, we will also need to add $H_2O(l)$ to the left side.

► The $H^+(aq)$ ion has variable structure in solution. In solids, there is evidence for $[H(H_2O)_n]^+$, where n is 1, 2, 3, 4, or 6.

$$HNO_3(aq) + H_2O(l) \longrightarrow H_3O^+(aq) + NO_3^-(aq)$$

The HNO_3 molecule is an acid (proton donor) and H_2O is a base (proton acceptor). Note that H_2O may function as an acid or a base, depending on the other reactant.

The Arrhenius definitions and those of Brønsted and Lowry are essentially equivalent for aqueous solutions, although their points of view are different. For instance, sodium hydroxide and ammonia are bases in the Arrhenius view because they increase the percentage of OH^- ion in the aqueous solution. They are bases in the Brønsted–Lowry view because they provide species that can accept protons. ◄

► We will discuss the Brønsted–Lowry concept of acids and bases more thoroughly in Chapter 16.

Strong and Weak Acids and Bases

A **strong acid** is *an acid that ionizes completely in water; it is a strong electrolyte.* Hydrochloric acid, $HCl(aq)$, is another example of a strong acid.

$$HCl(aq) + H_2O(l) \longrightarrow H_3O^+(aq) + Cl^-(aq)$$

Table 4.3 lists six common strong acids. Most of the other acids we will discuss are weak acids.

Table 4.3
Common Strong Acids and Bases

Strong Acids	Strong Bases
$HClO_4$	LiOH
H_2SO_4	NaOH
HI	KOH
HBr	$Ca(OH)_2$
HCl	$Sr(OH)_2$
HNO_3	$Ba(OH)_2$

A **weak acid** is *an acid that only partly ionizes in water; it is a weak electrolyte.* An example of a weak acid is hydrocyanic acid, HCN(*aq*). The hydrogen cyanide molecule, HCN, reacts with water to produce a small percentage of ions in solution.

$$HCN(aq) + H_2O(l) \rightleftharpoons H_3O^+(aq) + CN^-(aq)$$

A **strong base** is *a base that is present in aqueous solution entirely as ions, one of which is OH$^-$; it is a strong electrolyte.* The ionic compound sodium hydroxide, NaOH, is an example of a strong base. It dissolves in water as Na$^+$ and OH$^-$.

$$NaOH(s) \xrightarrow{H_2O} Na^+(aq) + OH^-(aq)$$

The hydroxides of Groups IA and IIA elements, except for beryllium hydroxide, are strong bases (see Table 4.3).

A **weak base** is *a base that is only partly ionized in water; it is a weak electrolyte.* Ammonia, NH_3, is an example.

$$NH_3(aq) + H_2O(l) \rightleftharpoons NH_4^+(aq) + OH^-(aq)$$

You will find it important to be able to identify an acid or base as strong or weak. When you write an ionic equation, you represent strong acids and bases by the ions they form and weak acids and bases by the formulas of the compounds. The next example will give you some practice identifying acids and bases as strong or weak.

EXAMPLE 4.4

Classifying Acids and Bases as Strong or Weak

Identify each of the following compounds as a strong or weak acid or base:

a. LiOH b. $HC_2H_3O_2$ c. HBr d. HNO_2

SOLUTION

Refer to Table 4.3 for the common strong acids and bases. You can assume that other acids and bases are weak.

a. As noted in Table 4.3, **LiOH is a strong base.**

b. Acetic acid, $HC_2H_3O_2$, is not one of the strong acids listed in Table 4.3; therefore, we assume **$HC_2H_3O_2$ is a weak acid.**

c. As noted in Table 4.3, **HBr is a strong acid.**

d. Nitrous acid, HNO_2, is not one of the strong acids listed in Table 4.3; therefore, we assume **HNO_2 is a weak acid.**

See Problems 4.23 and 4.24.

Neutralization Reactions

One of the chemical properties of acids and bases is that they neutralize one another. A **neutralization reaction** is *a reaction of an acid and a base that results in an ionic compound and possibly water. The ionic compound that is a product of a neutralization reaction* is called a **salt.** Salts can be obtained from neutralization reactions such as the following:

$$2HCl(aq) + Ca(OH)_2(aq) \longrightarrow CaCl_2(aq) + 2H_2O(l)$$

      acid   base   salt

$$HCN(aq) + KOH(aq) \longrightarrow KCN(aq) + H_2O(l)$$

    acid   base   salt

The salt formed in a neutralization reaction consists of cations obtained from the base and anions obtained from the acid.

We wrote these reactions as molecular equations to make explicit the reactant compounds and the salts produced. However, to discuss the essential reactions that occur, you need to write the net ionic equations.

The first reaction involves a strong acid, $HCl(aq)$, and a strong base, $Ca(OH)_2(aq)$. Writing these strong electrolytes in the form of ions gives the following complete ionic equation:

$$2H^+(aq) + 2\cancel{Cl^-(aq)} + \cancel{Ca^{2+}(aq)} + 2OH^-(aq) \longrightarrow \cancel{Ca^{2+}(aq)} + 2\cancel{Cl^-(aq)} + 2H_2O(l)$$

Canceling spectator ions and dividing by 2 gives the net ionic equation:

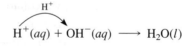

$$H^+(aq) + OH^-(aq) \longrightarrow H_2O(l)$$

Note the transfer of a proton from the hydrogen ion (or hydronium ion, H_3O^+) to the hydroxide ion.

The second reaction involves $HCN(aq)$, a weak acid, and $KOH(aq)$, a strong base; the product is KCN, a strong electrolyte. The net ionic equation is:

$$HCN(aq) + OH^-(aq) \longrightarrow CN^-(aq) + H_2O(l)$$

In each of these examples, hydroxide ions latch strongly onto protons to form water. Because water is a very stable substance, it effectively provides the driving force of the reaction.

EXAMPLE 4.5

Writing an Equation for a Neutralization

Write the molecular equation and then the net ionic equation for the neutralization of nitrous acid, HNO_2, by sodium hydroxide, NaOH, both in aqueous solution. Use an arrow with H^+ over it to show the proton transfer.

Table 4.5
Rules for Assigning Oxidation Numbers

Rule	Applies to	Statement
1	Elements	The oxidation number of an atom in an element is zero.
2	Monatomic ions	The oxidation number of an atom in a monatomic ion equals the charge on the ion.
3	Oxygen	The oxidation number of oxygen is -2 in most of its compounds. (An exception is O in H_2O_2 and other peroxides, where the oxidation number is -1.)
4	Hydrogen	The oxidation number of hydrogen is $+1$ in most of its compounds. (The oxidation number of hydrogen is -1 in binary compounds with a metal, such as CaH_2.)
5	Halogens	The oxidation number of fluorine is -1 in all of its compounds. Each of the other halogens (Cl, Br, I) has an oxidation number of -1 in binary compounds, except when the other element is another halogen above it in the periodic table or the other element is oxygen.
6	Compounds and ions	The sum of the oxidation numbers of the atoms in a compound is zero. The sum of the oxidation numbers of the atoms in a polyatomic ion equals the charge on the ion.

Oxidation-Number Rules

So far, we have used two rules for obtaining oxidation numbers: (1) the oxidation number of an atom in an element is zero, and (2) the oxidation number of an atom in a monatomic ion equals the charge on the ion. These and several other rules for assigning oxidation numbers are given in Table 4.5.

In molecular substances, we use these rules to give the *approximate* charges on the atoms. Consider the molecule SO_2. An oxygen atom in SO_2 takes on a negative charge relative to the sulfur atom. The magnitude of the charge on an oxygen atom in a molecule is not a full -2 charge as in the O^{2-} ion. However, it is convenient to assign an oxidation number of -2 to oxygen in SO_2 to help us express the approximate charge distribution in the molecule. Rule 3 in Table 4.5 says that an oxygen atom has an oxidation number of -2 in most of its compounds.

Rules 4 and 5 are similar in that they tell you what to expect for the oxidation number of certain elements in their compounds.

Rule 6 states that the sum of the oxidation numbers of the atoms in a compound is zero. Because any compound is electrically neutral, the sum of the charges on its atoms must be zero. This rule is easily extended to ions: the sum of the oxidation numbers (hypothetical charges) of the atoms in a polyatomic ion equals the charge on the ion.

You can use Rule 6 to obtain the oxidation number of one atom in a compound or ion, if you know the oxidation numbers of the other atoms in the compound or ion. Consider the SO_2 molecule. According to Rule 6,

$$(\text{Oxidation number of S}) + 2 \times (\text{oxidation number of O}) = 0$$

or

$$(\text{Oxidation number of S}) + 2 \times (-2) = 0.$$

Therefore,

$$\text{Oxidation number of S (in } SO_2) = -2 \times (-2) = +4$$

The next example illustrates the assignment of oxidation numbers.

EXAMPLE 4.7

Assigning Oxidation Numbers

Use the rules from Table 4.5 to obtain the oxidation number of the chlorine atom in each of the following: (a) $HClO_4$ (perchloric acid), (b) ClO_3^- (chlorate ion).

SOLUTION

a. For perchloric acid, Rule 6 gives the equation

$$(\text{Oxidation number of H}) + (\text{oxidation number of Cl}) + 4 \times$$
$$(\text{oxidation number of O}) = 0$$

Using Rules 3 and 4, you obtain

$$(+1) + (\text{oxidation number of Cl}) + 4 \times (-2) = 0$$

Therefore,

$$\text{Oxidation number of Cl (in } HClO_4) = -(+1) - 4 \times (-2) = \mathbf{+7}$$

b. For the chlorate ion, Rule 6 gives the equation

$$(\text{Oxidation number of Cl}) + 3 \times (\text{oxidation number of O}) = -1$$

Using Rule 3, you obtain

$$(\text{Oxidation number of Cl}) + 3 \times (-2) = -1$$

Therefore,

$$\text{Oxidation number of Cl (in } ClO_3^-) = -1 - 3 \times (-2) = \mathbf{+5}$$

See Problems 4.31, 4.32, 4.33, and 4.34.

Describing Oxidation–Reduction Reactions

A net ionic equation can be written in terms of two half-reactions.

$$\overset{0}{Fe}(s) + \overset{+2}{Cu^{2+}}(aq) \longrightarrow \overset{+2}{Fe^{2+}}(aq) + \overset{0}{Cu}(s)$$

A **half-reaction** is *one of two parts of an oxidation–reduction reaction, one part of which involves a loss of electrons and the other a gain of electrons.* The half-reactions are

$$\overset{0}{Fe}(s) \longrightarrow \overset{+2}{Fe^{2+}}(aq) + 2e^- \qquad \text{(electrons lost by Fe)}$$

$$\overset{+2}{Cu^{2+}}(aq) + 2e^- \longrightarrow \overset{0}{Cu}(s) \qquad \text{(electrons gained by } Cu^{2+})$$

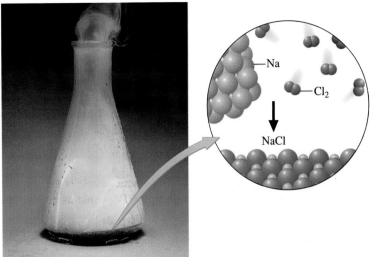

Figure 4.13
Combination reaction.
Left: Sodium metal and chlorine gas. *Right:* The spectacular combination reaction of sodium and chlorine.

Oxidation is *the half-reaction in which there is a loss of electrons by a species (or an increase of oxidation number of an atom).* **Reduction** is *the half-reaction in which there is a gain of electrons by a species (or a decrease in the oxidation number of an atom).*

An **oxidizing agent** is *a species that oxidizes another species; it is itself reduced.* Similarly, a **reducing agent** is *a species that reduces another species; it is itself oxidized.* In our example reaction, the copper(II) ion is the oxidizing agent, whereas iron metal is the reducing agent.

The relationships among these terms are shown in the following diagram for the reaction of iron with copper(II) ion.

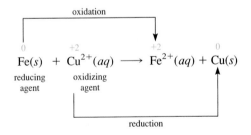

Some Common Oxidation–Reduction Reactions

Many of the oxidation–reduction reactions can be described as combination, decomposition, displacement, or combustion reactions.

Combination Reactions   A **combination reaction** is *a reaction in which two substances combine to form a third substance.* However, not all combination reactions are oxidation–reduction reactions. The reaction of sodium and chlorine (Figure 4.13)

$$2Na(s) + Cl_2(g) \longrightarrow 2NaCl(s)$$

is a redox reaction. Antimony and chlorine also combine in a fiery reaction.

$$2Sb + 3Cl_2 \longrightarrow 2SbCl_3$$

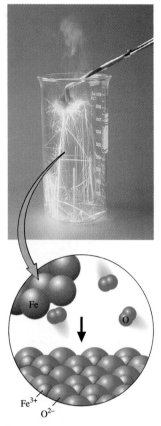

Figure 4.16
Combustion reaction. The combustion of iron wool. Iron reacts with oxygen to produce iron(III) oxide, Fe_2O_3. The reaction is similar to the rusting of iron but is much faster.

At first glance, the equation representing the reaction of zinc with silver(I) ions in solution might appear to be balanced.

$$Zn(s) + Ag^+(aq) \longrightarrow Zn^{2+}(aq) + Ag(s)$$

However, because a balanced chemical equation must have a charge balance as well as a mass balance, this equation is not balanced. Let us apply the half-reaction method for balancing this equation.

Half-Reaction Method Applied to Simple Oxidation–Reduction Equations

The *half-reaction method* consists of first separating the equation into two half-reactions, one for oxidation, the other for reduction. You balance each half-reaction, then combine them to obtain a balanced oxidation–reduction reaction. Here is an illustration of the process.

$$\overset{0}{Zn}(s) + \overset{+1}{Ag^+}(aq) \longrightarrow \overset{+2}{Zn^{2+}}(aq) + \overset{0}{Ag}(s)$$

Write the half-reactions in an unbalanced form.

$$Zn \longrightarrow Zn^{2+} \qquad \text{(oxidation)}$$
$$Ag^+ \longrightarrow Ag \qquad \text{(reduction)}$$

Next, balance the charge in each equation by adding electrons to the more positive side to create balanced half-reactions. Following this procedure, the balanced half-reactions are:

$$Zn \longrightarrow Zn^{2+} + 2e^- \qquad \text{(oxidation half-reaction)}$$
$$Ag^+ + e^- \longrightarrow Ag \qquad \text{(reduction half-reaction)}$$

Since each Ag^+ is capable of gaining only one electron, we need to double the amount of Ag^+ in order for it to accept all of the electrons produced by Zn during oxidation. Therefore, we multiply each half-reaction by a factor (integer) so that when we add them together, the electrons cancel. We multiply the first equation by 1 and multiply the second equation by 2.

$$1 \times (Zn \longrightarrow Zn^{2+} + 2e^-)$$
$$\underline{2 \times (Ag^+ + e^- \longrightarrow Ag)}$$
$$Zn + 2Ag^+ + 2e^- \longrightarrow Zn^{2+} + 2Ag + 2e^-$$

The electrons cancel, which finally yields the balanced oxidation–reduction equation:

$$Zn(s) + 2Ag^+(aq) \longrightarrow Zn^{2+}(aq) + 2Ag(s)$$

Example 4.8 further illustrates this technique.

EXAMPLE 4.8

Balancing Simple Oxidation–Reduction Reactions by the Half-Reaction Method

Consider a more difficult problem, the combination (oxidation–reduction) reaction of magnesium metal and nitrogen gas:

$$Mg(s) + N_2(g) \longrightarrow Mg_3N_2(s)$$

Apply the half-reaction method to balance this equation.

SOLUTION

Identify the oxidation states of the elements:

$$\overset{0}{Mg}(s) + \overset{0}{N_2}(g) \longrightarrow \overset{+2\ -3}{Mg_3N_2}(s)$$

In this problem, a molecular compound, nitrogen (N_2), is undergoing reduction. When a species undergoing reduction or oxidation is a molecule, write the formula of the molecule in the half-reaction. Also, make sure that both the mass and the charge are balanced.

$$Mg \longrightarrow Mg^{2+} + 2e^- \quad \text{(balanced oxidation half-reaction)}$$
$$N_2 + 6e^- \longrightarrow 2N^{3-} \quad \text{(balanced reduction half-reaction)}$$

We now need to multiply each half-reaction by a factor that will cancel the electrons.

$$3 \times (Mg \longrightarrow Mg^{2+} + 2e^-)$$
$$1 \times (N_2 + 6e^- \longrightarrow 2N^{3-})$$
$$\overline{3Mg + N_2 + \cancel{6e^-} \longrightarrow 3Mg^{2+} + 2N^{3-} + \cancel{6e^-}}$$

From inspecting the coefficients in this reaction, looking at the original equation, and knowing that Mg^{2+} and N^{3-} will combine to form an ionic compound (Mg_3N_2), we can rewrite the equation in the following form:

$$3Mg(s) + N_2(g) \longrightarrow Mg_3N_2(s)$$

See Problems 4.37 and 4.38.

Working with Solutions

The majority of the chemical reactions discussed in this chapter take place in solution. In a solid, the molecules or ions in a crystal tend to occupy approximately fixed positions, so that the chance of two molecules or ions coming together to react is small. In liquid solutions, reactant molecules are free to move throughout the liquid. When you run reactions in liquid solutions, it is convenient to dispense the amounts of reactants by measuring out volumes of reactant solutions. In the next two sections, we will discuss calculations involved in making up solutions, and in Section 4.10 we will describe stoichiometric calculations involving such solutions.

4.7 Molar Concentration

When we dissolve a substance in a liquid, we call the substance the *solute* and the liquid the *solvent*. The general term *concentration* refers to the quantity of solute in a standard quantity of solution. Qualitatively, we say that a solution is *dilute* when the solute concentration is low and *concentrated* when the solute concentration is high. For commercially available solutions, the term *concentrated* refers to the maximum, or near maximum, concentration available. For example, concentrated aqueous ammonia contains about 28% NH_3 by mass.

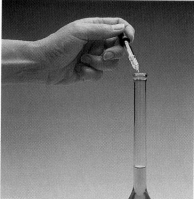

Figure 4.17
**Preparing a 0.200 *M*
CuSO₄ solution.** *Left:*
0.0500 mol CuSO₄ · 5H₂O
(12.48 g) is weighed on a
platform balance. *Center:* The
copper(II) sulfate pentahydrate
is transferred carefully to the
volumetric flask. *Right:* Water
is added to bring the solution
level to the mark on the neck
of the 250-mL volumetric
flask. The molarity is 0.0500
mol/0.250 L = 0.200 *M*.

In this example, we expressed the concentration quantitatively by giving the mass percentage of solute—that is, the mass of solute in 100 g of solution. However, we need a unit of concentration that is convenient for dispensing reactants in solution, such as one that specifies moles of solute per solution volume.

Molar concentration, or **molarity (*M*),** is defined as *the moles of solute dissolved in one liter (cubic decimeter) of solution.*

$$\text{Molarity } (M) = \frac{\text{moles of solute}}{\text{liters of solution}}$$

If you want to prepare a solution that is 0.200 *M* CuSO₄, you place 0.200 mol CuSO₄ in a 1.000-L volumetric flask, or a proportional amount in a flask of a different size (Figure 4.17). You add a small quantity of water to dissolve the CuSO₄. Then you fill the flask with additional water to the mark on the neck and mix the solution.

EXAMPLE 4.9

Calculating Molarity from Mass and Volume

A sample of NaNO₃ weighing 0.38 g is placed in a 50.0-mL volumetric flask. The flask is then filled with water to the mark on the neck, dissolving the solid. What is the molarity of the resulting solution?

SOLUTION

You find that 0.38 g NaNO₃ is $4.\underline{4}7 \times 10^{-3}$ mol NaNO₃; the last significant figure is underlined. The volume of solution is 50.0×10^{-3} L, so the molarity is

$$\text{Molarity} = \frac{4.\underline{4}7 \times 10^{-3} \text{ mol NaNO}_3}{50.0 \times 10^{-3} \text{ L soln}} = \textbf{0.089 } \textit{\textbf{M}} \textbf{ NaNO}_3$$

See Problems 4.39 and 4.40.

The advantage of molarity as a concentration unit is that the amount of solute is related to the volume of solution. For a specified mass of substance, you can measure out a definite volume of solution. As the following example illustrates, molarity can be used as a factor for converting from moles of solute to liters of solution, and vice versa.

EXAMPLE 4.10

Using Molarity as a Conversion Factor

An experiment calls for the addition to a reaction vessel of 0.184 g of sodium hydroxide, NaOH, in aqueous solution. How many milliliters of 0.150 M NaOH should be added?

SOLUTION

Here is the calculation. (The molar mass of NaOH is 40.0 g/mol.)

$$0.184 \text{ g NaOH} \times \frac{1 \text{ mol NaOH}}{40.0 \text{ g NaOH}} \times \frac{1 \text{ L soln}}{0.150 \text{ mol NaOH}}$$

$$= 3.07 \times 10^{-2} \text{ L soln (or 30.7 mL)}$$

You need to add **30.7 mL** of 0.150 M NaOH solution to the reaction vessel.

See Problems 4.41, 4.42, 4.43, 4.44, 4.45, and 4.46.

4.8 Diluting Solutions

Commercially available aqueous ammonia (28.0% NH_3) is 14.8 M NH_3. Suppose, however, that you want a solution that is 1.00 M NH_3. You need to dilute the concentrated solution with a definite quantity of water. For this purpose, you must know the relationship between the molarity of the solution before dilution (the *initial molarity*) and that after dilution (the *final molarity*).

To obtain this relationship, first recall the equation defining molarity:

$$\text{Molarity} = \frac{\text{moles of solute}}{\text{liters of solution}}$$

You can rearrange this to give

$$\text{Moles of solute} = \text{molarity} \times \text{liters of solution} = M \times V$$

Because the moles of solute does not change during the dilution (Figure 4.18),

$$M_i \times V_i = M_f \times V_f$$

where i and f represent the initial and final conditions, respectively. (Note: You can use any volume units, but both V_i and V_f must be in the same units.)

EXAMPLE 4.11

Diluting a Solution

You are given a solution of 14.8 M NH_3. How many milliliters of this solution do you require to give 100.0 mL of 1.00 M NH_3 when diluted?

↓ Add water
(solvent)

SOLUTION

You know the final volume (100.0 mL), final concentration (1.00 M), and initial concentration (14.8 M). You write the dilution formula and rearrange it to give the initial volume.

$$M_i V_i = M_f V_f$$

$$V_i = \frac{M_f V_f}{M_i}$$

Now you substitute the known values into the right side of the equation.

$$V_i = \frac{1.00\,M \times 100.0\text{ mL}}{14.8\,M} = \textbf{6.76 mL}$$

See Problems 4.47 and 4.48.

Figure 4.18
Molecular view of the dilution process. *Top:* A molecular view of a solution of Cl_2 dissolved in water. *Bottom:* The solution after performing a dilution by adding water. Note how the number of moles of Cl_2 in the container does not change when performing the dilution, only the concentration changes. In this particular case, the concentration of Cl_2 drops to half of the starting concentration because the volume was doubled.

CONCEPT CHECK 4.3

Consider the following beakers. Each contains a solution of the hypothetical atom X.

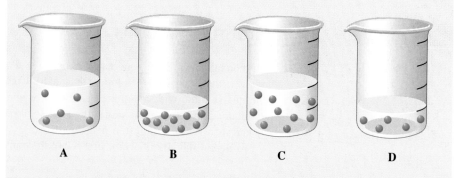

A   B   C   D

a. Arrange the beakers in order of increasing concentration of X.

b. Without adding or removing X, what specific things could you do to make the concentrations of X equal in each beaker? (*Hint:* Think about dilutions.)

Quantitative Analysis

Analytical chemistry deals with the determination of composition of materials. The materials that one might analyze include air, water, food, hair, body fluids, pharmaceutical preparations, and so forth. The analysis of materials is divided into qualitative and quantitative analysis. *Qualitative analysis* involves the identification of substances or species present in a material. For instance, you might determine that a sample of water contains lead(II) ion. **Quantitative analysis** involves *the determination of the amount of a substance or species present in a material.*

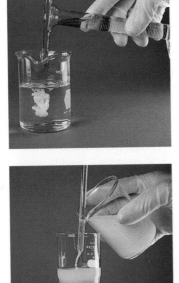

4.9 Gravimetric Analysis

Gravimetric analysis is *a type of quantitative analysis in which the amount of a species in a material is determined by converting the species to a product that can be isolated completely and weighed.* Precipitation reactions are frequently used in gravimetric analyses. The precipitate, or solid formed in the reaction, is filtered from the solution, dried, and weighed.

As an example of a gravimetric analysis, consider the problem of determining the amount of lead (Pb^{2+}) in a sample of drinking water. Lead(II) sulfate is a very insoluble compound of lead(II) ion. When sodium sulfate, Na_2SO_4, is added to a solution containing Pb^{2+}, lead(II) sulfate precipitates as $PbSO_4$. If you assume that the lead is present in solution as lead(II) nitrate, you can write the following equation for the reaction:

$$Na_2SO_4(aq) + Pb(NO_3)_2(aq) \longrightarrow 2NaNO_3(aq) + PbSO_4(s)$$

You can separate the white precipitate of lead(II) sulfate from the solution by filtration. Figure 4.19 shows a similar analysis.

Figure 4.19
Gravimetric analysis for barium ion.   *Top:* A solution of potassium chromate (yellow) is poured down a stirring rod into a solution containing an unknown amount of barium ion, Ba^{2+}. The yellow precipitate that forms is barium chromate, $BaCrO_4$. *Bottom:* The solution is filtered by pouring it into a crucible containing a porous glass partition. Afterward, the crucible is heated to dry the barium chromate. By weighing the crucible before and afterward, you can determine the mass of precipitate.

EXAMPLE 4.12

Determining the Amount of a Species by Gravimetric Analysis

A 1.000-L sample of polluted water was analyzed for lead(II) ion, Pb^{2+}, by adding an excess of sodium sulfate to it. The mass of lead(II) sulfate that precipitated was 229.8 mg. What is the mass of lead in a liter of the water? Give the answer as milligrams of lead per liter of solution.

▶ PROBLEM STRATEGY

All of the lead in the water solution is precipitated as lead(II) sulfate, $PbSO_4$. If you determine the percentage of lead in $PbSO_4$, you can calculate the quantity of lead in the water sample.

SOLUTION

Following Example 3.7, you obtain the mass percentage of Pb in $PbSO_4$ by dividing the molar mass of Pb by the molar mass of $PbSO_4$, then multiplying by 100%:

$$\%Pb = \frac{207.2 \text{ g/mol}}{303.3 \text{ g/mol}} \times 100\% = 68.32\%$$

Therefore, the 1.000-L sample of water contains

$$\text{Amount Pb in sample} = 229.8 \text{ mg } PbSO_4 \times 0.6832 = 157.0 \text{ mg Pb}$$

The water sample contains **157.0 mg Pb per liter.**

See Problems 4.49 and 4.50.

4.10 Volumetric Analysis

As you saw earlier, you can use molarity as a conversion factor, and in this way you can calculate the volume of solution that is equivalent to a given mass of solute (see Example 4.10).

EXAMPLE 4.13

Calculating the Volume of Reactant Solution Needed

Consider the reaction of sulfuric acid, H_2SO_4, with sodium hydroxide, NaOH.

$$H_2SO_4(aq) + 2NaOH(aq) \longrightarrow 2H_2O(l) + Na_2SO_4(aq)$$

Suppose a beaker contains 35.0 mL of 0.175 M H_2SO_4. How many milliliters of 0.250 M NaOH must be added to react completely with the sulfuric acid?

PROBLEM STRATEGY

You convert from 35.0 mL H_2SO_4 solution to moles H_2SO_4 (using the molarity of H_2SO_4), then to moles NaOH (from the chemical equation). Finally, you convert this to volume of NaOH solution (using the molarity of NaOH).

SOLUTION

The calculation is as follows:

$$35.0 \times 10^{-3} \text{ L } H_2SO_4 \text{ soln} \times \frac{0.175 \text{ mol } H_2SO_4}{1 \text{ L } H_2SO_4 \text{ soln}} \times \frac{2 \text{ mol NaOH}}{1 \text{ mol } H_2SO_4} \times$$

$$\frac{1 \text{ L NaOH soln}}{0.250 \text{ mol NaOH}} = 4.90 \times 10^{-2} \text{ L NaOH soln}$$

Thus, 35.0 mL of 0.175 M sulfuric acid solution reacts exactly with **49.0 mL** of 0.250 M sodium hydroxide solution.

See Problems 4.53 and 4.54.

An important method for determining the amount of a particular substance is based on measuring the volume of reactant solution. **Titration** is *a procedure for determining the amount of substance* A *by adding a carefully measured volume of a solution with known concentration of* B *until the reaction of* A *and* B *is just complete.* **Volumetric analysis** is *a method of analysis based on titration.*

Figure 4.20 shows a flask containing hydrochloric acid with an unknown amount of HCl being titrated with sodium hydroxide solution, NaOH, of known molarity. The reaction is

$$NaOH(aq) + HCl(aq) \longrightarrow NaCl(aq) + H_2O(l)$$

▶ An indicator is a substance that undergoes a color change when a reaction approaches completion. See Section 4.4.

To the HCl solution are added a few drops of phenolphthalein indicator. ◀ Phenolphthalein is colorless in the hydrochloric acid but turns pink at the completion of the reaction of NaOH with HCl. Sodium hydroxide with a concentration of 0.207 M is contained in a buret. The solution in the buret is added to the HCl in the flask until the phenolphthalein just changes from colorless to pink. At this point, the reaction is complete and the volume of NaOH that reacts with the HCl is read from the buret.

Figure 4.20
Titration of an unknown amount of HCl with NaOH. *Left:* The flask contains HCl and a few drops of phenolphthalein indicator; the buret contains 0.207 *M* NaOH (the buret reading is 44.97 mL). *Center:* NaOH was added to the solution in the flask until a persistent faint pink color was reached, marking the endpoint of the titration (the buret reading is 49.44 mL). The amount of HCl can be determined from the volume of NaOH used (4.47 mL); see Example 4.14. *Right:* The addition of several drops of NaOH solution beyond the endpoint gives a deep pink color.

EXAMPLE 4.14

Calculating the Quantity of Substance in a Titrated Solution

A flask contains a solution with an unknown amount of HCl. This solution is titrated with 0.207 *M* NaOH. It takes 4.47 mL NaOH to complete the reaction. What is the mass of the HCl?

PROBLEM STRATEGY

You convert the volume of NaOH (4.47×10^{-3} L NaOH solution) to moles NaOH (from the molarity of NaOH). Then you convert moles NaOH to moles HCl (from the chemical equation). Finally, you convert moles HCl to grams HCl.

SOLUTION

The calculation is as follows:

$$4.47 \times 10^{-3} \text{ L NaOH soln} \times \frac{0.207 \text{ mol NaOH}}{1 \text{ L NaOH soln}} \times \frac{1 \text{ mol HCl}}{1 \text{ mol NaOH}} \times \frac{36.5 \text{ g HCl}}{1 \text{ mol HCl}}$$

$$= \mathbf{0.0338 \text{ g HCl}}$$

CONCEPT CHECK 4.4

Consider three flasks, each containing 0.10 mol of acid. You need to learn something about the acids in each of the flasks, so you perform titration using an NaOH solution. Here are the results of the experiment:

Flask A   10 mL of NaOH required for neutralization

Flask B   20 mL of NaOH required for neutralization

Flask C   30 mL of NaOH required for neutralization

a. What have you learned about each of these acids from performing the experiment?

b. Could you use the results of this experiment to determine the concentration of the NaOH? If not, what assumption about the molecular formulas of the acids would allow you to make the concentration determination?

A Checklist for Review

Important Terms

electrolyte (4.1)
nonelectrolyte (4.1)
strong electrolyte (4.1)
weak electrolyte (4.1)
molecular equation (4.2)
complete ionic equation (4.2)
spectator ion (4.2)
net ionic equation (4.2)
exchange (metathesis) reaction (4.3)
acid–base indicator (4.4)

acid (Arrhenius) (4.4)
base (Arrhenius) (4.4)
acid (Brønsted–Lowry) (4.4)
base (Brønsted–Lowry) (4.4)
strong acid (4.4)
weak acid (4.4)
strong base (4.4)
weak base (4.4)
neutralization reaction (4.4)
salt (4.4)
polyprotic acid (4.4)

oxidation number (oxidation state) (4.5)
oxidation–reduction reaction (redox reaction) (4.5)
half-reaction (4.5)
oxidation (4.5)
reduction (4.5)
oxidizing agent (4.5)
reducing agent (4.5)
combination reaction (4.5)

decomposition reaction (4.5)
displacement reaction (single-replacement reaction) (4.5)
combustion reaction (4.5)
molar concentration (molarity) (*M*) (4.7)
quantitative analysis (4.9)
gravimetric analysis (4.9)
titration (4.10)
volumetric analysis (4.10)

Key Equations

$$\text{Molarity } (M) = \frac{\text{moles of solute}}{\text{liters of solution}}$$

$$M_i \times V_i = M_f \times V_f$$

Summary of Facts and Concepts

Reactions often involve ions in aqueous solution. Many of the compounds in such reactions are *electrolytes,* which are substances that dissolve in water to give ions. Electrolytes that exist in solution almost entirely as ions are called *strong electrolytes.* Electrolytes that dissolve in water to give a relatively small percentage of ions are called *weak electrolytes.* The *solubility rules* can be used to predict the extent to which an ionic compound will dissolve in water. Most soluble ionic compounds are strong electrolytes.

We can represent a reaction involving ions in one of three different ways, depending on what information we want to convey. A *molecular equation* is one in which substances are written as if they were molecular, even though they are ionic. This equation closely describes what you actually do in the laboratory. However, this equation does not describe what is happening at the level of ions and molecules. For that purpose, we rewrite the molecular equation as a *complete ionic equation* by replacing the formulas for strong electrolytes by their ion formulas. If you cancel *spectator ions* from the complete ionic equation, you obtain the *net ionic equation.*

Most of the important reactions we will consider in this course can be divided into three major classes: (1) precipitation reactions, (2) acid–base reactions, and (3) oxidation–reduction reactions. A *precipitation reaction* occurs in aqueous solution because one product is insoluble.

Acids are substances that yield hydrogen ions in aqueous solution or donate protons. *Bases* are substances that yield hydroxide ions in aqueous solution or accept protons. These *acid–base reactions* are proton-transfer reactions. In this chapter, we covered neutralization reactions (reactions of acids and bases to yield salts) and reactions of certain salts with acids to yield a gas.

Oxidation–reduction reactions are reactions involving a transfer of electrons from one species to another. The concept of *oxidation numbers* helps us describe this type of reaction. The atom that increases in oxidation number is said to undergo *oxidation;* the atom that decreases in oxidation number is said to undergo *reduction.* Oxidation and reduction must occur together in a reaction. Many oxidation–reduction reactions fall into the following categories: combination reactions, decomposition reactions, displacement reactions, and combustion reactions.

Molar concentration, or *molarity,* is the moles of solute in a liter of solution. Knowing the molarity allows you to calculate the amount of solute in any volume of solution. Because the moles of solute are constant during the *dilution of a solution,* you can determine to what volume to dilute a concentrated solution to give one of desired molarity.

Quantitative analysis involves the determination of the amount of a species in a material. *Titration* is a method of chemical analysis in which you measure the volume of solution of known molarity that reacts with a compound of unknown amount.

Operational Skills

1. **Using the solubility rules**   Given the formula of an ionic compound, predict its solubility in water. **(EXAMPLE 4.1)**
2. **Writing net ionic equations**   Given a molecular equation, write the corresponding net ionic equation. **(EXAMPLE 4.2)**
3. **Deciding whether precipitation occurs**   Using solubility rules, decide whether two soluble ionic compounds react to form a precipitate. If they do, write the net ionic equation. **(EXAMPLE 4.3)**
4. **Classifying acids and bases as strong or weak**   Given the formula of an acid or a base, classify it as strong or weak. **(EXAMPLE 4.4)**
5. **Writing an equation for a neutralization**   Given an acid and a base, write the molecular equation and then the net ionic equation for the neutralization reaction. **(EXAMPLE 4.5)**
6. **Writing an equation for a reaction with gas formation**   Given the reaction between a carbonate, sulfite, or sulfide and an acid, write the molecular and the net ionic equations. **(EXAMPLE 4.6)**
7. **Assigning oxidation numbers**   Given the formula of a simple compound or ion, obtain the oxidation numbers of the atoms, using the rules for assigning oxidation numbers. **(EXAMPLE 4.7)**
8. **Balancing simple oxidation–reduction reactions by the half-reaction method**   Given an oxidation–reduction reaction, balance it. **(EXAMPLE 4.8)**
9. **Calculating molarity from mass and volume**   Given the mass of the solute and the volume of the solution, calculate the molarity. **(EXAMPLE 4.9)**
10. **Using molarity as a conversion factor**   Given the volume and molarity of a solution, calculate the amount of solute. Or, given the amount of solute and the molarity of a solution, calculate the volume. **(EXAMPLE 4.10)**
11. **Diluting a solution**   Calculate the volume of solution of known molarity required to make a specified volume of solution with different molarity. **(EXAMPLE 4.11)**
12. **Determining the amount of a species by gravimetric analysis**   Given the amount of a precipitate in a gravimetric analysis, calculate the amount of a related species. **(EXAMPLE 4.12)**
13. **Calculating the volume of reactant solution needed**   Given the chemical equation, calculate the volume of solution of known molarity of one substance that just reacts with a given volume of solution of another substance. **(EXAMPLE 4.13)**
14. **Calculating the quantity of substance in a titrated solution**   Calculate the mass of one substance that reacts with a given volume of solution of known molarity of another substance. **(EXAMPLE 4.14)**

Review Questions

4.1   Explain why some electrolyte solutions are strongly conducting, whereas others are weakly conducting.

4.2   Define the terms *strong electrolyte* and *weak electrolyte.* Give an example of each.

4.3   What is a *spectator ion?* Illustrate with a complete ionic reaction.

4.4   What is a net ionic equation? What is the value in using a net ionic equation? Give an example.

4.5   What are the major types of chemical reactions? Give a brief description and an example of each.

4.6   Describe in words how you would prepare pure crystalline $AgCl$ and $NaNO_3$ from solid $AgNO_3$ and solid $NaCl$.

4.7   Give an example of a neutralization reaction. Label the acid, base, and salt.

4.8   Give an example of a polyprotic acid and write equations for the successive neutralizations of the acidic hydrogen atoms of the acid molecule to produce a series of salts.

4.9   Why must oxidation and reduction occur together in a reaction?

4.10   Give an example of a displacement reaction. What is the oxidizing agent? What is the reducing agent?

Conceptual Problems

4.11 You come across a beaker that contains water, aqueous ammonium acetate, and a precipitate of calcium phosphate.

a. Write the balanced molecular equation for a reaction between two solutions containing ions that could produce this solution.

b. Write the complete ionic equation for the reaction in part a.

c. Write the net ionic equation for the reaction in part a.

4.12 Three acid samples are prepared for titration by 0.01 *M* NaOH:

1. Sample 1 is prepared by dissolving 0.01 mol of HCl in 50 mL of water.

2. Sample 2 is prepared by dissolving 0.01 mol of HCl in 60 mL of water.

3. Sample 3 is prepared by dissolving 0.01 mol of HCl in 70 mL of water.

a. Without performing a formal calculation, compare the concentrations of the three acid samples (rank them from highest to lowest).

b. When performing the titration, which sample, if any, will require the largest volume of the 0.01 *M* NaOH for neutralization?

4.13 Would you expect a precipitation reaction between an ionic compound that is an electrolyte and an ionic compound that is a nonelectrolyte? Justify your answer.

4.14 Equal quantities of the hypothetical strong acid HX, weak acid HA, and weak base BZ are each added to a separate beaker of water, producing the solutions depicted in the drawings. In the drawings, the relative amounts of each substance present in the solution (neglecting the water) are shown. Identify the acid or base that was used to produce each of the solutions (HX, HA, or BZ).

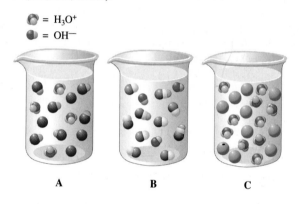

A   B   C

Practice Problems

Solubility Rules

4.15 Using solubility rules, decide whether the following ionic solids are soluble or insoluble in water. If they are soluble, indicate what ions you would expect to be present in solution.

a. AgBr

b. Li_2SO_4

c. $Ca_3(PO_4)_2$

d. Na_2CO_3

4.16 Using solubility rules, decide whether the following ionic solids are soluble or insoluble in water. If they are soluble, indicate what ions you would expect to be present in solution.

a. $(NH_4)_2SO_4$

b. $BaCO_3$

c. $PbSO_4$

d. $Ca(NO_3)_2$

Ionic Equations

4.17 Write net ionic equations for the following molecular equations. HBr is a strong electrolyte.

a. $HBr(aq) + KOH(aq) \longrightarrow KBr(aq) + H_2O(l)$

b. $AgNO_3(aq) + NaBr(aq) \longrightarrow AgBr(s) + NaNO_3(aq)$

c. $CaS(aq) + 2HBr(aq) \longrightarrow CaBr_2(aq) + H_2S(g)$

d. $NaOH(aq) + NH_4Br(aq) \longrightarrow$
$$NaBr(aq) + NH_3(g) + H_2O(l)$$

4.18 Write net ionic equations for the following molecular equations. HBr is a strong electrolyte.

a. $HBr(aq) + NH_3(aq) \longrightarrow NH_4Br(aq)$

b. $2HBr(aq) + Ba(OH)_2(aq) \longrightarrow 2H_2O(l) + BaBr_2(aq)$

c. $Pb(NO_3)_2(aq) + 2NaBr(aq) \longrightarrow PbBr_2(s) + 2NaNO_3(aq)$

d. $MgCO_3(s) + H_2SO_4(aq) \longrightarrow$
$$MgSO_4(aq) + H_2O(l) + CO_2(g)$$

Precipitation

4.19 Write the molecular equation and the net ionic equation for each of the following aqueous reactions. If no reaction occurs, write *NR* after the arrow.

a. $FeSO_4 + NaCl \longrightarrow$

b. $Na_2CO_3 + MgBr_2 \longrightarrow$

c. $MgSO_4 + NaOH \longrightarrow$

d. $NiCl_2 + NaBr \longrightarrow$

4.20 Write the molecular equation and the net ionic equation for each of the following aqueous reactions. If no reaction occurs, write *NR* after the arrow.

a. $AgNO_3 + NaI \longrightarrow$
b. $Ba(NO_3)_2 + K_2SO_4 \longrightarrow$
c. $Mg(NO_3)_2 + K_2SO_4 \longrightarrow$
d. $CaCl_2 + Al(NO_3)_3 \longrightarrow$

4.21 For each of the following, write molecular and net ionic equations for any precipitation reaction that occurs. If no reaction occurs, indicate this.
a. Solutions of barium nitrate and lithium sulfate are mixed.
b. Solutions of sodium bromide and calcium nitrate are mixed.
c. Solutions of aluminum sulfate and sodium hydroxide are mixed.
d. Solutions of calcium bromide and sodium phosphate are mixed.

4.22 For each of the following, write molecular and net ionic equations for any precipitation reaction that occurs. If no reaction occurs, indicate this.
a. Zinc chloride and sodium sulfide are dissolved in water.
b. Sodium sulfide and calcium chloride are dissolved in water.
c. Magnesium sulfate and potassium iodide are dissolved in water.
d. Magnesium sulfate and potassium carbonate are dissolved in water.

Strong and Weak Acids and Bases

4.23 Classify each of the following as a strong or weak acid or base.
a. HF
b. KOH
c. $HClO_4$
d. HIO

4.24 Classify each of the following as a strong or weak acid or base.
a. HBrO
b. HCNO
c. $Sr(OH)_2$
d. HI

Neutralization Reactions

4.25 Complete and balance each of the following molecular equations (in aqueous solution); include phase labels. Then, for each, write the net ionic equation.
a. $NaOH + HNO_3 \longrightarrow$
b. $HCl + Ba(OH)_2 \longrightarrow$
c. $HC_2H_3O_2 + Ca(OH)_2 \longrightarrow$
d. $NH_3 + HNO_3 \longrightarrow$

4.26 Complete and balance each of the following molecular equations (in aqueous solution); include phase labels. Then, for each, write the net ionic equation.
a. $Al(OH)_3 + HCl \longrightarrow$
b. $HBr + Sr(OH)_2 \longrightarrow$
c. $Ba(OH)_2 + HC_2H_3O_2 \longrightarrow$
d. $HNO_3 + KOH \longrightarrow$

4.27 For each of the following, write the molecular equation, including phase labels. Then write the net ionic equation. Note that the salts formed in these reactions are soluble.
a. the neutralization of hydrobromic acid with calcium hydroxide solution
b. the reaction of solid aluminum hydroxide with nitric acid
c. the reaction of aqueous hydrogen cyanide with calcium hydroxide solution
d. the neutralization of lithium hydroxide solution by aqueous hydrogen cyanide

4.28 For each of the following, write the molecular equation, including phase labels. Then write the net ionic equation. Note that the salts formed in these reactions are soluble.
a. the neutralization of lithium hydroxide solution by aqueous perchloric acid
b. the reaction of barium hydroxide solution and aqueous nitrous acid
c. the reaction of sodium hydroxide solution and aqueous nitrous acid
d. the neutralization of aqueous hydrogen cyanide by aqueous strontium hydroxide

Reactions Evolving a Gas

4.29 The following reactions occur in aqueous solution. Complete and balance the molecular equations using phase labels. Then write the net ionic equations.
a. $CaS + HBr \longrightarrow$
b. $MgCO_3 + HNO_3 \longrightarrow$
c. $K_2SO_3 + H_2SO_4 \longrightarrow$

4.30 The following reactions occur in aqueous solution. Complete and balance the molecular equations using phase labels. Then write the net ionic equations.
a. $BaCO_3 + HNO_3 \longrightarrow$
b. $K_2S + HCl \longrightarrow$
c. $CaSO_3(s) + HI \longrightarrow$

Oxidation Numbers

4.31 Obtain the oxidation number for the element noted in each of the following.
a. Ga in Ga_2O_3
b. Nb in NbO_2
c. Br in $KBrO_4$
d. Mn in K_2MnO_4

4.32 Obtain the oxidation number for the element noted in each of the following.
a. Cr in CrO_3
b. Hg in Hg_2Cl_2
c. Ga in $Ga(OH)_3$
d. P in Na_3PO_4

4.33 Obtain the oxidation number for the element noted in each of the following.
a. N in NH_2^-
b. I in IO_3^-
c. Al in $Al(OH)_4^-$
d. P in $H_2PO_4^-$

4.34 Obtain the oxidation number for the element noted in each of the following.
a. N in NO_2^-
b. Cr in CrO_4^{2-}
c. Zn in $Zn(OH)_4^{2-}$
d. As in $H_2AsO_3^-$

Describing Oxidation–Reduction Reactions

4.35 In the following reactions, label the oxidizing agent and the reducing agent.
a. $P_4(s) + 5O_2(g) \longrightarrow P_4O_{10}(s)$
b. $Co(s) + Cl_2(g) \longrightarrow CoCl_2(s)$

4.36 In the following reactions, label the oxidizing agent and the reducing agent.
a. $ZnO(s) + C(s) \longrightarrow Zn(g) + CO(g)$
b. $8Fe(s) + S_8(s) \longrightarrow 8FeS(s)$

Balancing Oxidation–Reduction Reactions

4.37 Balance the following oxidation–reduction reactions by the half-reaction method.
a. $CuCl_2(aq) + Al(s) \longrightarrow AlCl_3(aq) + Cu(s)$
b. $Cr^{3+}(aq) + Zn(s) \longrightarrow Cr(s) + Zn^{2+}(aq)$

4.38 Balance the following oxidation–reduction reactions by the half-reaction method.
a. $FeI_3(aq) + Mg(s) \longrightarrow Fe(s) + MgI_2(aq)$
b. $H_2(g) + Ag^+(aq) \longrightarrow Ag(s) + H^+(aq)$

Molarity

4.39 A sample of 0.0256 mol of iron(III) chloride, $FeCl_3$, was dissolved in water to give 25.0 mL of solution. What is the molarity of the solution?

4.40 A 50.0-mL volume of $AgNO_3$ solution contains 0.0285 mol $AgNO_3$ (silver nitrate). What is the molarity of the solution?

4.41 What volume of 0.120 M $CuSO_4$ is required to give 0.150 mol of copper(II) sulfate, $CuSO_4$?

4.42 How many milliliters of 0.126 M $HClO_4$ (perchloric acid) are required to give 0.00752 mol $HClO_4$?

4.43 Heme, obtained from red blood cells, binds oxygen, O_2. How many moles of heme are there in 75 mL of 0.0019 M heme solution?

4.44 Insulin is a hormone that controls the use of glucose in the body. How many moles of insulin are required to make up 28 mL of 0.0048 M insulin solution?

4.45 How many grams of sodium dichromate, $Na_2Cr_2O_7$, should be added to a 50.0-mL volumetric flask to prepare 0.025 M $Na_2Cr_2O_7$ when the flask is filled to the mark with water?

4.46 Describe how you would prepare 2.50×10^2 mL of 0.20 M Na_2SO_4. What mass (in grams) of sodium sulfate, Na_2SO_4, is needed?

4.47 You wish to prepare 0.12 M HNO_3 from a stock solution of nitric acid that is 15.8 M. How many milliliters of the stock solution do you require to make up 1.00 L of 0.12 M HNO_3?

4.48 A chemist wants to prepare 0.25 M HCl. Commercial hydrochloric acid is 12.4 M. How many milliliters of the commercial acid does the chemist require to make up 1.50 L of the dilute acid?

Gravimetric Analysis

4.49 A chemist added an excess of sodium sulfate to a solution of a soluble barium compound to precipitate all of the barium ion as barium sulfate, $BaSO_4$. How many grams of barium ion are in a 458-mg sample of the barium compound if a solution of the sample gave 513 mg $BaSO_4$ precipitate? What is the mass percentage of barium in the compound?

4.50 A soluble iodide was dissolved in water. Then an excess of silver nitrate, $AgNO_3$, was added to precipitate all of the iodide ion as silver iodide, AgI. If 1.545 g of the soluble iodide gave 2.185 g of silver iodide, how many grams of iodine are in the sample of soluble iodide? What is the mass percentage of iodine, I, in the compound?

4.51 A compound of iron and chlorine is soluble in water. An excess of silver nitrate was added to precipitate the chloride ion as silver chloride. If a 134.8-mg sample of the compound gave 304.8 mg AgCl, what is the formula of the compound?

4.52 A 1.345-g sample of a compound of barium and oxygen was dissolved in hydrochloric acid to give a solution of barium ion, which was then precipitated with an excess of potassium chromate to give 2.012 g of barium chromate, $BaCrO_4$. What is the formula of the compound?

Volumetric Analysis

4.53 What volume of 0.250 M HNO_3 (nitric acid) reacts with 42.4 mL of 0.150 M Na_2CO_3 (sodium carbonate) in the following reaction?

$$2HNO_3(aq) + Na_2CO_3(aq) \longrightarrow$$
$$2NaNO_3(aq) + H_2O(l) + CO_2(g)$$

4.54 A flask contains 49.8 mL of 0.150 M $Ca(OH)_2$ (calcium hydroxide). How many milliliters of 0.350 M Na_2CO_3 (sodium carbonate) are required to react completely with the calcium hydroxide in the following reaction?

$$Na_2CO_3(aq) + Ca(OH)_2(aq) \longrightarrow CaCO_3(s) + 2NaOH(aq)$$

4.55 A solution of hydrogen peroxide, H_2O_2, is titrated with a solution of potassium permanganate, $KMnO_4$. The reaction is

$$5H_2O_2(aq) + 2KMnO_4(aq) + 3H_2SO_4(aq) \longrightarrow$$
$$5O_2(g) + 2MnSO_4(aq) + K_2SO_4(aq) + 8H_2O(l)$$

It requires 51.7 mL of 0.145 M $KMnO_4$ to titrate 20.0 g of the solution of hydrogen peroxide. What is the mass percentage of H_2O_2 in the solution?

4.56 A 3.33-g sample of iron ore is transformed to a solution of iron(II) sulfate, $FeSO_4$, and this solution is titrated with 0.150 M $K_2Cr_2O_7$ (potassium dichromate). If it requires 41.4 mL of potassium dichromate solution to titrate the iron(II) sulfate solution, what is the percentage of iron in the ore? The reaction is

$$6FeSO_4(aq) + K_2Cr_2O_7(aq) + 7H_2SO_4(aq) \longrightarrow$$
$$3Fe_2(SO_4)_3(aq) + Cr_2(SO_4)_3(aq) + 7H_2O(l) + K_2SO_4(aq)$$

General Problems

4.57 Magnesium metal reacts with hydrobromic acid to produce hydrogen gas and a solution of magnesium bromide. Write the molecular equation for this reaction. Then write the corresponding net ionic equation.

4.58 Aluminum metal reacts with perchloric acid to produce hydrogen gas and a solution of aluminum perchlorate. Write the molecular equation for this reaction. Then write the corresponding net ionic equation.

4.59 Nickel(II) sulfate solution reacts with lithium hydroxide solution to produce a precipitate of nickel(II) hydroxide and a solution of lithium sulfate. Write the molecular equation for this reaction. Then write the corresponding net ionic equation.

4.60 Potassium sulfate solution reacts with barium bromide solution to produce a precipitate of barium sulfate and a solution of potassium bromide. Write the molecular equation for this reaction. Then write the corresponding net ionic equation.

4.61 Complete and balance each of the following molecular equations, including phase labels, if a reaction occurs. Then write the net ionic equation. If no reaction occurs, write *NR* after the arrow.
a. $Sr(OH)_2 + HC_2H_3O_2 \longrightarrow$
b. $NH_4I + CsCl \longrightarrow$
c. $NaNO_3 + CsCl \longrightarrow$
d. $NH_4I + AgNO_3 \longrightarrow$

4.62 Complete and balance each of the following molecular equations, including phase labels, if a reaction occurs. Then write the net ionic equation. If no reaction occurs, write *NR* after the arrow.
a. $HClO_4 + BaCO_3 \longrightarrow$
b. $H_2CO_3 + Sr(OH)_2 \longrightarrow$
c. $K_3PO_4 + MgCl_2 \longrightarrow$
d. $FeSO_4 + MgCl_2 \longrightarrow$

4.63 Describe in words how you would do each of the following preparations. Then give the molecular equation for each preparation.
a. $CuCl_2(s)$ from $CuSO_4(s)$
b. $Ca(C_2H_3O_2)_2(s)$ from $CaCO_3(s)$
c. $NaNO_3(s)$ from $Na_2SO_3(s)$
d. $MgCl_2(s)$ from $Mg(OH)_2(s)$

4.64 Describe in words how you would do each of the following preparations. Then give the molecular equation for each preparation.
a. $MgCl_2(s)$ from $MgCO_3(s)$
b. $NaNO_3(s)$ from $NaCl(s)$
c. $Al(OH)_3(s)$ from $Al(NO_3)_3(s)$
d. $HCl(aq)$ from $H_2SO_4(aq)$

4.65 Classify each of the following reactions as a combination reaction, decomposition reaction, displacement reaction, or combustion reaction.
a. When they are heated, ammonium dichromate crystals, $(NH_4)_2Cr_2O_7$, decompose to give nitrogen, water vapor, and solid chromium(III) oxide, Cr_2O_3.
b. When aqueous ammonium nitrite, NH_4NO_2, is heated, it gives nitrogen and water vapor.
c. When gaseous ammonia, NH_3, reacts with hydrogen chloride gas, HCl, fine crystals of ammonium chloride, NH_4Cl, are formed.
d. Aluminum added to an aqueous solution of sulfuric acid, H_2SO_4, forms a solution of aluminum sulfate, $Al_2(SO_4)_3$. Hydrogen gas is released.

4.66 Classify each of the following reactions as a combination reaction, decomposition reaction, displacement reaction, or combustion reaction.
a. When solid calcium oxide, CaO, is exposed to gaseous sulfur trioxide, SO_3, solid calcium sulfate, $CaSO_4$, is formed.
b. Calcium metal (solid) reacts with water to produce a solution of calcium hydroxide, $Ca(OH)_2$, and hydrogen gas.

c. When solid sodium hydrogen sulfite, $NaHSO_3$, is heated, solid sodium sulfite, Na_2SO_3, sulfur dioxide gas, SO_2, and water vapor are formed.

d. Magnesium reacts with bromine to give magnesium bromide, $MgBr_2$.

4.67   An aqueous solution contains 4.50 g of calcium chloride, $CaCl_2$, per liter. What is the molarity of $CaCl_2$? When calcium chloride dissolves in water, the calcium ions, Ca^{2+}, and chloride ions, Cl^-, in the crystal go into the solution. What is the molarity of each ion in the solution?

4.68   An aqueous solution contains 3.45 g of iron(III) sulfate, $Fe_2(SO_4)_3$, per liter. What is the molarity of $Fe_2(SO_4)_3$? When the compound dissolves in water, the Fe^{3+} ions and SO_4^{2-} ions in the crystal go into the solution. What is the molar concentration of each ion in the solution?

4.69   A stock solution of potassium dichromate, $K_2Cr_2O_7$, is made by dissolving 89.3 g of the compound in 1.00 L of solution. How many milliliters of this solution are required to prepare 1.00 L of 0.100 M $K_2Cr_2O_7$?

4.70   A 71.2-g sample of oxalic acid, $H_2C_2O_4$, was dissolved in 1.00 L of solution. How would you prepare 1.00 L of 0.150 M $H_2C_2O_4$ from this solution?

4.71   A solution contains 6.00% (by mass) NaBr (sodium bromide). The density of the solution is 1.046 g/cm³. What is the molarity of NaBr?

4.72   An aqueous solution contains 4.00% NH_3 (ammonia) by mass. The density of the aqueous ammonia is 0.979 g/mL. What is the molarity of NH_3 in the solution?

4.73   A barium mineral was dissolved in hydrochloric acid to give a solution of barium ion. An excess of potassium sulfate was added to 50.0 mL of the solution, and 1.128 g of barium sulfate precipitate formed. Assume that the original solution was barium chloride. What was the molarity of $BaCl_2$ in this solution?

4.74   Bone was dissolved in hydrochloric acid, giving 50.0 mL of solution containing calcium chloride, $CaCl_2$. To precipitate the calcium ion from the resulting solution, an excess of potassium oxalate was added. The precipitate of calcium ox-

alate, CaC_2O_4, weighed 1.437 g. What was the molarity of $CaCl_2$ in the solution?

4.75   You have a sample of a rat poison whose active ingredient is thallium(I) sulfate. You analyze this sample for the mass percentage of active ingredient by adding potassium iodide to precipitate yellow thallium(I) iodide. If the sample of rat poison weighed 759.0 mg and you obtained 212.2 mg of the dry precipitate, what is the mass percentage of the thallium(I) sulfate in the rat poison?

4.76   An antacid tablet has calcium carbonate as the active ingredient; other ingredients include a starch binder. You dissolve the tablet in hydrochloric acid and filter off insoluble material. You add potassium oxalate to the filtrate (containing calcium ion) to precipitate calcium oxalate. If a tablet weighing 0.680 g gave 0.6332 g of calcium oxalate, what is the mass percentage of active ingredient in the tablet?

4.77   A 0.608-g sample of fertilizer contained nitrogen as ammonium sulfate, $(NH_4)_2SO_4$. It was analyzed for nitrogen by heating with sodium hydroxide.

$$(NH_4)_2SO_4(s) + 2NaOH(aq) \longrightarrow$$
$$Na_2SO_4(aq) + 2H_2O(l) + 2NH_3(g)$$

The ammonia was collected in 46.3 mL of 0.213 M HCl (hydrochloric acid), with which it reacted.

$$NH_3(g) + HCl(aq) \longrightarrow NH_4Cl(aq)$$

This solution was titrated for excess hydrochloric acid with 44.3 mL of 0.128 M NaOH.

$$NaOH(aq) + HCl(aq) \longrightarrow NaCl(aq) + H_2O(l)$$

What is the percentage of nitrogen in the fertilizer?

4.78   An antacid tablet contains sodium hydrogen carbonate, $NaHCO_3$, and inert ingredients. A 0.500-g sample of powdered tablet was mixed with 50.0 mL of 0.190 M HCl (hydrochloric acid). The mixture was allowed to stand until it reacted.

$$NaHCO_3(s) + HCl(aq) \longrightarrow NaCl(aq) + H_2O(l) + CO_2(g)$$

The excess hydrochloric acid was titrated with 47.1 mL of 0.128 M NaOH (sodium hydroxide).

$$HCl(aq) + NaOH(aq) \longrightarrow NaCl(aq) + H_2O(l)$$

What is the percentage of sodium hydrogen carbonate in the antacid?

Cumulative-Skills Problems

4.79 Lead(II) nitrate reacts with cesium sulfate in an aqueous precipitation reaction. What are the formulas of lead(II) nitrate and cesium sulfate? Write the molecular equation and net ionic equation for the reaction. What are the names of the products? Give the molecular equation for another reaction that produces the same precipitate.

4.80 Silver nitrate reacts with strontium chloride in an aqueous precipitation reaction. What are the formulas of silver nitrate and strontium chloride? Write the molecular equation and net ionic equation for the reaction. What are the names of the products? Give the molecular equation for another reaction that produces the same precipitate.

4.81 What volume of a solution of ethanol, C_2H_6O, that is 94.0% ethanol by mass contains 0.200 mol C_2H_6O? The density of the solution is 0.807 g/mL.

4.82 What volume of a solution of ethylene glycol, $C_2H_6O_2$, that is 56.0% ethylene glycol by mass contains 0.350 mol $C_2H_6O_2$? The density of the solution is 1.072 g/mL.

4.83 A 10.0-mL sample of potassium iodide solution was analyzed by adding an excess of silver nitrate solution to produce silver iodide crystals, which were filtered from the solution.

$$KI(aq) + AgNO_3(aq) \longrightarrow KNO_3(aq) + AgI(s)$$

If 2.183 g of silver iodide was obtained, what was the molarity of the original KI solution?

4.84 A 25.0-mL sample of sodium sulfate solution was analyzed by adding an excess of barium chloride solution to produce barium sulfate crystals, which were filtered from the solution.

$$Na_2SO_4(aq) + BaCl_2(aq) \longrightarrow 2NaCl(aq) + BaSO_4(s)$$

If 5.719 g of barium sulfate was obtained, what was the molarity of the original Na_2SO_4 solution?

4.85 A metal, M, was converted to the sulfate, $M_2(SO_4)_3$. Then a solution of the sulfate was treated with barium chloride to give barium sulfate crystals, which were filtered off.

$$M_2(SO_4)_3(aq) + 3BaCl_2(aq) \longrightarrow 2MCl_3(aq) + 3BaSO_4(s)$$

If 1.200 g of the metal gave 6.026 g of barium sulfate, what is the atomic weight of the metal? What is the metal?

4.86 A metal, M, was converted to the chloride, MCl_2. Then a solution of the chloride was treated with silver nitrate to give silver chloride crystals, which were filtered from the solution.

$$MCl_2(aq) + 2AgNO_3(aq) \longrightarrow M(NO_3)_2(aq) + 2AgCl(s)$$

If 2.434 g of the metal gave 7.964 g of silver chloride, what is the atomic weight of the metal? What is the metal?

4.87 Phosphoric acid is prepared by dissolving phosphorus(V) oxide, P_4O_{10}, in water. What is the balanced equation for this reaction? How many grams of P_4O_{10} are required to make 1.50 L of aqueous solution containing 5.00% phosphoric acid by mass? The density of the solution is 1.025 g/mL.

4.88 Iron(III) chloride can be prepared by reacting iron metal with chlorine. What is the balanced equation for this reaction? How many grams of iron are required to make 2.50 L of aqueous solution containing 9.00% iron(III) chloride by mass? The density of the solution is 1.067 g/mL.

4.89 Determine the volume of sulfuric acid solution needed to prepare 37.4 g of aluminum sulfate, $Al_2(SO_4)_3$, by the reaction

$$2Al(s) + 3H_2SO_4(aq) \longrightarrow Al_2(SO_4)_3(aq) + 3H_2(g)$$

The sulfuric acid solution, whose density is 1.104 g/mL, contains 15.0% H_2SO_4 by mass.

4.90 Determine the volume of sodium hydroxide solution needed to prepare 26.2 g sodium phosphate, Na_3PO_4, by the reaction

$$3NaOH(aq) + H_3PO_4(aq) \longrightarrow Na_3PO_4(aq) + 3H_2O(l)$$

The sodium hydroxide solution, whose density is 1.133 g/mL, contains 12.0% NaOH by mass.

4.91 The active ingredients of an antacid tablet contained only magnesium hydroxide and aluminum hydroxide. Complete neutralization of a sample of the active ingredients required 48.5 mL of 0.187 M hydrochloric acid. The chloride salts from this neutralization were obtained by evaporation of the filtrate from the titration; they weighed 0.4200 g. What was the percentage by mass of magnesium hydroxide in the active ingredients of the antacid tablet?

4.92 The active ingredients in an antacid tablet contained only calcium carbonate and magnesium carbonate. Complete reaction of a sample of the active ingredients required 41.33 mL of 0.08750 M hydrochloric acid. The chloride salts from the reaction were obtained by evaporation of the filtrate from this titration; they weighed 0.1900 g. What was the percentage by mass of the calcium carbonate in the active ingredients of the antacid tablet?

5

A chemical reaction producing gas bubbles in solution.

The Gaseous State

Gases are composed of small molecules and they have several characteristics that distinguish them from liquids and solids. You can compress gases into smaller and smaller volumes. Anyone who has transported gases such as oxygen compressed in steel cylinders can appreciate this (Figure 5.1). Also, for gases, unlike for liquids and solids, you can relate pressure, volume, temperature, and molar amount of substance by the ideal gas law.

Kinetic-molecular theory describes a gas as composed of molecules in constant motion. This theory helps explain the simple relationship that exists among the pressure, volume, temperature, and amount of a gas.

This chapter introduces the empirical gas laws and the kinetic-molecular theory that explains these laws. We will first consider the concept of pressure.

Gas Laws

5.1 Gas Pressure and Its Measurement

Figure 5.1
Cylinders of gas. Gases such as oxygen and nitrogen can be transported as compressed gases in steel cylinders. Large volumes of gas at normal pressures can be compressed into a small volume. Note the pressure gauges.

▶ Atmospheric, or barometric, pressure depends not only on altitude but also on weather conditions. The "highs" and "lows" of weather reports refer to high- and low-pressure air masses. A high is associated with fair weather; a low brings unsettled weather, or storms.

Pressure is defined as *the force exerted per unit area of surface*. A coin resting on a table exerts a force, and therefore a pressure, downward on the table due to gravity. The air above the table exerts an additional pressure on the table.

Let us calculate the pressure on a table from a perfectly flat coin with a radius and mass equal to that of a new penny (9.3 mm in radius and 2.5 g). The force equals the mass of the coin times the constant acceleration of gravity. The constant acceleration of gravity is 9.81 m/s^2, and the force on the coin due to gravity is

$$\text{Force} = \text{mass} \times \text{constant acceleration of gravity}$$
$$= (2.5 \times 10^{-3} \text{ kg}) \times (9.81 \text{ m/s}^2) = 2.5 \times 10^{-2} \text{ kg} \cdot \text{m/s}^2$$

The cross-sectional area of the coin is $\pi \times (\text{radius})^2 = 3.14 \times (9.3 \times 10^{-3} \text{ m})^2 = 2.7 \times 10^{-4} \text{ m}^2$. Therefore,

$$\text{Pressure} = \frac{\text{force}}{\text{area}} = \frac{2.5 \times 10^{-2} \text{ kg} \cdot \text{m/s}^2}{2.7 \times 10^{-4} \text{ m}^2} = 93 \text{ kg/(m} \cdot \text{s}^2)$$

The SI unit of pressure, $kg/(m \cdot s^2)$, is given the name **pascal (Pa),** after the French physicist Blaise Pascal (1623–1662), who studied fluid pressure. Note that the pressure exerted by the coin is approximately 100 Pa. The pressure exerted by the atmosphere is about 1000 times larger, or about 100,000 Pa. Thus, the pascal is an extremely small unit.

Chemists have traditionally used units of pressure based on the mercury barometer. A **barometer** is *a device for measuring the pressure of the atmosphere*. The mercury barometer consists of a glass tube about one meter long, filled with mercury and inverted in a dish of mercury (Figure 5.2). The height of the mercury is a direct measure of the atmospheric pressure. ◀ A mercury column placed in a sealed flask, as in Figure 5.3, measures the gas pressure in the flask. It acts as a **manometer,** *a device that measures the pressure of a gas or liquid in a vessel*.

The unit **millimeters of mercury (mmHg),** also called the **torr** (after Evangelista Torricelli, who invented the mercury barometer in 1643), is *a unit of pressure equal to that exerted by a column of mercury 1 mm high at 0.00°C*. The **atmosphere (atm)** is a related *unit of pressure equal to exactly 760 mmHg*.

The general relationship between the pressure P and the height h of a liquid column in a barometer or manometer is

$$P = gdh$$

Here g is the constant acceleration of gravity (9.81 m/s^2) and d is the density of the liquid in the manometer. If g, d, and h are in SI units, the pressure is given in pascals. Table 5.1 summarizes the relationships among the various units of pressure.

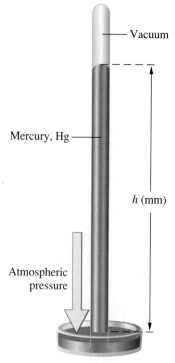

Vacuum

Mercury, Hg

h (mm)

Atmospheric
pressure

Figure 5.2
A mercury barometer.
The height h is proportional to
the barometric pressure. For
that reason, the pressure is
often given as the height of
the mercury column, in units
of millimeters of mercury,
mmHg.

Table 5.1
Important Units of Pressure

Unit	Relationship or Definition
Pascal (Pa)	$kg/(m \cdot s^2)$
Atmosphere (atm)	$1\ atm = 1.01325 \times 10^5\ Pa \simeq 100\ kPa$
mmHg, or torr	$760\ mmHg = 1\ atm$

EXAMPLE 5.1

Converting Units of Pressure

The pressure of a gas in a flask is measured to be 797.7 mmHg using a
mercury-filled manometer as in Figure 5.3. What is this pressure in pascals and
atmospheres?

 PROBLEM STRATEGY

To convert to the appropriate units, use the conversion factors presented in Table
5.1 and the techniques presented in Chapter 1. The conversion 760 mmHg = 1
atm is exact.

SOLUTION

Conversion to pascals:

$$797.7\ mmHg \times \frac{1.01325 \times 10^5\ Pa}{760\ mmHg} = \mathbf{1.064 \times 10^5\ Pa}$$

Conversion to atmospheres:

$$797.7\ mmHg \times \frac{1\ atm}{760\ mmHg} = \mathbf{1.050\ atm}$$

See Problems 5.19 and 5.20.

CONCEPT CHECK 5.1

Suppose that you set up two barometers like the one shown in Figure 5.2. In
one of the barometers you use mercury, and in the other you use water. Which
of the barometers would have a higher column of liquid, the one with Hg or
H_2O? Explain your answer.

5.2 Empirical Gas Laws

All gases under moderate conditions behave quite simply with respect to pressure,
temperature, volume, and molar amount.

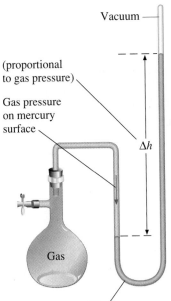

Vacuum

(proportional to gas pressure)

Gas pressure on mercury surface

Δh

Gas

Mercury

Figure 5.3
A flask equipped with a closed-tube manometer.
The gas pressure in the flask is proportional to the difference in heights between the liquid levels in the manometer, Δh.

Boyle's Law: Relating Volume and Pressure

One characteristic property of a gas is its *compressibility*. By comparison, liquids and solids are relatively incompressible. The compressibility of gases was first studied quantitatively by Robert Boyle in 1661. When he poured mercury into the open end of a J-shaped tube, the volume of the enclosed gas decreased (Figure 5.4). From such experiments, he formulated the law now known by his name. According to **Boyle's law,** *the volume of a sample of gas at a given temperature varies inversely with the applied pressure.* That is, $V \propto 1/P$, where V is the volume, P is the pressure, and $\propto$ means "is proportional to." Thus, if the pressure is doubled, the volume is halved.

Boyle's law can also be expressed in the form of an equation.

$$PV = \text{constant} \quad \text{(for a given amount of gas at fixed temperature)}$$

Table 5.2 gives some pressure and volume data for 1.000 g O_2 at 0°C. Note that the product of the pressure and volume is nearly constant. By plotting the volume of the oxygen at different pressures (as shown in Figure 5.5), you obtain a graph showing the inverse relationship of P and V.

You can use Boyle's law to calculate the volume occupied by a gas when the pressure changes. You can write P_i and V_i for the initial pressure and initial volume, and P_f and V_f for the final pressure and final volume. Because the temperature does not change, the product of the pressure and volume remains constant. Thus, you can write

$$P_f V_f = P_i V_i$$

Dividing both sides of the equation by P_f gives

$$V_f = V_i \times \frac{P_i}{P_f}$$

Figure 5.4
Boyle's experiment.
The volume of the gas at normal atmospheric pressure (760 mmHg) is 100 mL. When the pressure is doubled by adding 760 mm of mercury, the volume is halved (to 50 mL). Tripling the pressure decreases the volume to one-third of the original (to 33 mL).

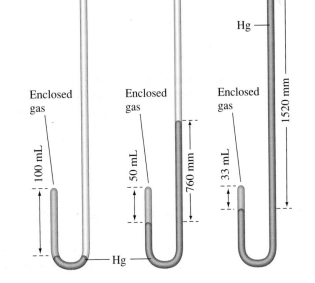

Enclosed gas

100 mL

Enclosed gas

50 mL

760 mm

Enclosed gas

33 mL

Hg

1520 mm

Hg

Table 5.2
Pressure–Volume Data for 1.000 g O_2 at 0°C

P (atm)	V (L)	PV
0.2500	2.801	0.7002
0.5000	1.400	0.7000
0.7500	0.9333	0.7000
1.000	0.6998	0.6998
2.000	0.3495	0.6990
3.000	0.2328	0.6984
4.000	0.1744	0.6976
5.000	0.1394	0.6970

(Increasing pressure ↓) (Increasing volume ↑)

Figure 5.5
Gas pressure–volume relationship. *(A)* Plot of volume vs. pressure for a sample of oxygen. The volume (of 1.000 g O_2 at 0°C) decreases with increasing pressure. When the pressure is doubled (from 0.50 atm to 1.00 atm), the volume is halved (from 1.40 L to 0.70 L). *(B)* Plot of $1/V$ vs. pressure (at constant temperature) for the same sample. The straight line indicates that volume varies inversely with pressure.

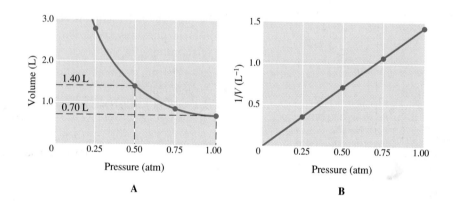

A

B

EXAMPLE 5.2

Using Boyle's Law

A volume of air occupying 12.0 dm³ at 98.9 kPa is compressed to a pressure of 119.0 kPa. The temperature remains constant. What is the new volume?

PROBLEM STRATEGY

Putting the data for the problem in tabular form, you see what data you have and what you must find. This will suggest the method of solution.

$$V_i = 12.0 \text{ dm}^3 \qquad P_i = 98.9 \text{ kPa}$$
$$V_f = \quad ? \qquad P_f = 119.0 \text{ kPa}$$

Temperature and moles remain constant.

Because P and V vary but temperature and moles are constant, you use Boyle's law.

SOLUTION

Application of Boyle's law gives

$$V_f = V_i \times \frac{P_i}{P_f} = 12.0 \text{ dm}^3 \times \frac{98.9 \text{ kPa}}{119.0 \text{ kPa}} = \mathbf{9.97 \text{ dm}^3}$$

Figure 5.6
Effect of temperature on volume of a gas.
A balloon immersed in liquid nitrogen shrinks because the air inside contracts in volume. When the balloon is removed from the liquid nitrogen, the air inside warms up and the balloon expands to its original size.

▶ The first ascent of a hot-air balloon carrying people was made on November 21, 1783. A few days later, Jacques Alexandre Charles made an ascent in a hydrogen-filled balloon. On landing, the balloon was attacked and torn to shreds by terrified peasants armed with pitchforks.

▶ The Kelvin temperature scale and a more formal equation for temperature conversion were discussed in Section 1.6.

Note that the pressure on the gas increases, so the gas is compressed (gives a smaller volume), and the ratio of pressures is less than 1. Thus, you could get the above result by simply multiplying the volume by the ratio of pressures, choosing the ratio so it is less than 1. (If you had expected the gas volume to increase, you would have chosen the ratio to be greater than 1.)

See Problems 5.21 and 5.22.

Before leaving the subject of Boyle's law, we should note that the pressure–volume product for a gas is not precisely constant. You can see this from the *PV* data given in Table 5.2 for oxygen. In fact, all gases follow Boyle's law at low to moderate pressures but deviate from this law at high pressures. The extent of deviation depends on the gas. We will return to this point at the end of the chapter.

Charles's Law: Relating Volume and Temperature

Temperature also affects gas volume. When you immerse an air-filled balloon in liquid nitrogen ($-196°C$), the balloon shrinks (Figure 5.6). After the balloon is removed from the liquid nitrogen, it returns to its original size. A gas contracts when cooled and expands when heated.

One of the first quantitative observations of gases at different temperatures was made by Jacques Alexandre Charles in 1787. Charles was a French physicist and a pioneer in hot-air and hydrogen-filled balloons. ◀ Later, John Dalton (in 1801) and Joseph Louis Gay-Lussac (in 1802) continued these kinds of experiments, which showed that a sample of gas at a fixed pressure increases in volume *linearly* with temperature (Figure 5.7).

When you *extrapolate* the straight lines in Figure 5.7 to zero volume, you find that they all intersect at a temperature of $-273.15°C$. These extrapolations show that we can express the volume variation of a gas with temperature more simply by choosing a different thermometer scale. We can use a temperature scale equal to degrees Celsius plus 273.15, which is the *Kelvin scale* (an absolute scale). ◀

$$K = °C + 273.15$$

If we write T for the temperature on the Kelvin scale, we obtain

$$V = bT$$

This is **Charles's law,** which we can state as follows: *the volume occupied by any sample of gas at a constant pressure is directly proportional to the absolute temperature.* Thus, doubling the absolute temperature of a gas doubles its volume.

Charles's law can be rearranged into a form that is very useful for computation.

$$\frac{V}{T} = \text{constant} \qquad \text{(for a given amount of gas at a fixed pressure)}$$

Consider a sample of gas at a fixed pressure, and suppose the temperature changes from its initial value T_i to a final value T_f. How does the volume change? Because the volume divided by absolute temperature is constant, you can write

$$\frac{V_f}{T_f} = \frac{V_i}{T_i}$$

Figure 5.7
Linear relationship of gas volume and temperature at constant pressure.
The graph shows gas volume versus temperature for a given mass of gas at 1.00 atm pressure. This linear relationship is independent of amount or kind of gas. Note that all lines extrapolate to −273°C at zero volume.

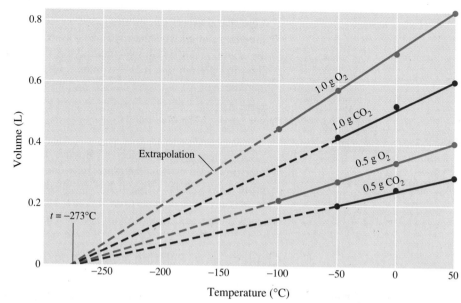

Or, rearranging slightly,

$$V_f = V_i \times \frac{T_f}{T_i}$$

Note that to obtain the final volume, the initial volume is multiplied by a ratio of absolute temperatures.

EXAMPLE 5.3

Using Charles's Law

Earlier we found that the total volume of oxygen that can be obtained from a particular tank at 1.00 atm and 21°C is 785 L (including the volume remaining in the tank). What would be this volume of oxygen if the temperature had been 28°C?

SOLUTION

$$T_i = (21 + 273)\text{K} = 294 \text{ K}$$
$$T_f = (28 + 273)\text{K} = 301 \text{ K}$$

Following is the data table:

$$V_i = 785 \text{ L} \quad P_i = 1.00 \text{ atm} \quad T_i = 294 \text{ K}$$
$$V_f = ? \quad P_f = 1.00 \text{ atm} \quad T_f = 301 \text{ K}$$

Note that T varies and P remains constant, so V must change.

$$V_f = V_i \times \frac{T_f}{T_i} = 785 \text{ L} \times \frac{301 \text{ K}}{294 \text{ K}} = \mathbf{804 \text{ L}}$$

Note that the temperature increases, so you expect the volume to increase. This means that the ratio of absolute temperatures is greater than 1. You could get the foregoing result by simply multiplying the volume by the ratio of absolute temperatures, choosing the ratio so that it is greater than 1.

See Problems 5.25 and 5.26.

Combined Gas Law: Relating Volume, Temperature, and Pressure

Boyle's law ($V \propto 1/P$) and Charles's law ($V \propto T$) can be combined and expressed in a single equation:

$$\frac{P_f V_f}{T_f} = \frac{P_i V_i}{T_i}$$

which rearranges to

$$V_f = V_i \times \frac{P_i}{P_f} \times \frac{T_f}{T_i}$$

Thus, the final volume is obtained by multiplying the initial volume by ratios of pressures and absolute temperatures.

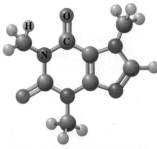

Figure 5.8
Oxidation of an organic compound with copper(II) oxide. *Top:* Molecular model of caffeine, an organic compound. *Center:* A chemist heats caffeine with copper(II) oxide, CuO. CuO oxidizes C and H in the compound to CO_2 and H_2O, and converts N to N_2. *Bottom:* Copper metal produced by the reaction.

EXAMPLE 5.4

Using the Combined Gas Law

Modern determination of %N in an organic compound is an automated version of one developed by the French chemist Jean-Baptiste Dumas in 1830. The Dumas method uses hot copper(II) oxide to oxidize C and H in the compound to CO_2 and H_2O (both are trapped chemically), and to convert N in the compound to N_2 gas (Figure 5.8). Using the Dumas method, 39.8 mg of caffeine gives 10.1 cm³ of nitrogen gas at 23°C and 746 mmHg. What is the volume of nitrogen at 0°C and 760 mmHg?

SOLUTION

You first express the temperatures in kelvins:

$$T_i = (23 + 273) \text{ K} = 296 \text{ K}$$
$$T_f = (0 + 273) \text{ K} = 273 \text{ K}$$

Now you put the data for the problem in tabular form:

$V_i = 10.1 \text{ cm}^3$	$P_i = 746 \text{ mmHg}$	$T_i = 296 \text{ K}$
$V_f = $?	$P_f = 760 \text{ mmHg}$	$T_f = 273 \text{ K}$

From the data, you see that you need to use Boyle's and Charles's laws combined to find how V varies as P and T change.

$$V_f = V_i \times \frac{P_i}{P_f} \times \frac{T_f}{T_i} = 10.1 \text{ cm}^3 \times \frac{746 \text{ mmHg}}{760 \text{ mmHg}} \times \frac{273 \text{ K}}{296 \text{ K}} = \textbf{9.14 cm}^3$$

See Problems 5.27 and 5.28.

Avogadro's Law: Relating Volume and Amount

In 1808 the French chemist Joseph Louis Gay-Lussac (1778–1850) concluded from experiments on gas reactions that the volumes of reactant gases at the same pressure and temperature are in ratios of small whole numbers. For example, two volumes of hydrogen react with one volume of oxygen.

$$2H_2(g) + O_2(g) \longrightarrow 2H_2O(g)$$

2 volumes   1 volume

Three years later, the Italian chemist Amedeo Avogadro (1776–1856) interpreted the law of *combining volumes* in terms of what we now call **Avogadro's law:** *equal volumes of any two gases at the same temperature and pressure contain the same number of molecules.* Thus, two volumes of hydrogen contain twice the number of molecules as in one volume of oxygen, in agreement with the chemical equation for the reaction.

One mole of any gas contains the same number of molecules (Avogadro's number = 6.02×10^{23}). The *volume of one mole of gas* is called the **molar gas volume,** V_m. Volumes of gases are often compared at **standard temperature and pressure** **(STP),** *the reference conditions for gases chosen by convention to be 0°C and 1 atm pressure.* At STP, the molar gas volume is found to be 22.4 L/mol.

We can re-express Avogadro's law as follows: the molar gas volume at a given temperature and pressure is a specific constant independent of the nature of the gas.

$$V_m = \text{specific constant } (= 22.4 \text{ L/mol at STP})$$
(depending on T and P but independent of the gas)

5.3 The Ideal Gas Law

In the previous section, we discussed the empirical gas laws. Here we will show that these laws can be combined into one equation, called the *ideal gas equation.* We can combine Boyle's law and Charles's law into the equation

$$V = \text{constant} \times \frac{T}{P} \qquad \text{(for a given amount of gas)}$$

This "constant" is independent of the temperature and pressure but does depend on the amount of gas. For one mole, the constant will have a specific value, which we will denote as R. The molar volume, V_m, is

$$V_m = R \times \frac{T}{P}$$

According to Avogadro's law, the molar volume at a specific value of T and P is a constant independent of the nature of the gas, and this implies that R is a constant independent of the gas. The **molar gas constant, R,** is *the constant of proportionality relating the molar volume of a gas to T/P.* Values of R in various units are given in Table 5.3.

Table 5.3
Molar Gas Constant in Various Units

Value of R
0.082058 L • atm/(K • mol)
8.3145 J/(K • mol)*
8.3145 kg • m²/(s² • K • mol)
8.3145 kPa • dm³/(K • mol)
1.9872 cal/(K • mol)*

*The units of pressure times volume are the units of energy—for example, joules (J) or calories (cal).

The preceding equation can be written for n moles of gas by multiplying both sides by n.

$$\underbrace{nV_m}_{V} = \frac{nRT}{P} \qquad \text{or} \qquad PV = nRT$$

Because V_m is the volume per mole, nV_m is the total volume V. *The equation $PV = nRT$, which combines all of the gas laws,* is called the **ideal gas law.**

$$PV = nRT$$

The ideal gas law includes all the information contained in Boyle's, Charles's, and Avogadro's laws. In fact, starting with the ideal gas law, you can derive any of the other gas laws.

The limitations that apply to Boyle's, Charles's, and Avogadro's laws also apply to the ideal gas law. That is, the ideal gas law is most accurate for low to moderate pressures and for temperatures that are not too low. ◄

▶ No gas is "ideal." But the ideal gas law is very useful even though it is only an approximation. The behavior of real gases is described at the end of this chapter.

Calculations Using the Ideal Gas Law

The type of problem to which Boyle's and Charles's laws are applied involves a change in conditions (P, V, or T) of a gas. The ideal gas law allows us to solve another type of problem: given any three of the quantities P, V, n, and T, calculate the unknown quantity. Example 5.5 illustrates such a problem.

EXAMPLE 5.5

Using the Ideal Gas Law

How many grams of oxygen, O_2, are there in a 50.0-L gas cylinder at 21°C when the oxygen pressure is 15.7 atm?

PROBLEM STRATEGY

In asking for the mass of oxygen, we are in effect asking for moles of gas, n. The problem gives P, V, and T, so you can use the ideal gas law to solve for n. The proper value to use for R depends on the units of P and V.

SOLUTION

The data given in the problem are

Variable	Value
P	15.7 atm
V	50.0 L
T	(21 + 273) K = 294 K
n	?

Solving the ideal gas law for n gives

$$n = \frac{PV}{RT}$$

A CHEMIST LOOKS AT

Nitric Oxide Gas and Biological Signaling

LIFE SCIENCE

In 1998, the Nobel committee awarded its prize in physiology or medicine to three scientists for the astounding discovery that nitric oxide gas, NO, functions as the signaling agent between biological cells in a wide variety of chemical processes. Until this discovery, biochemists had thought that the major chemical reactions in a cell always involved very large molecules. Now they discovered that a simple gas, NO, could have a central role in cell chemistry.

Figure 5.9
Nitroglycerin patch. Nitroglycerin has been used for more than a hundred years to treat and prevent angina attacks. Here a patient wears a patch that dispenses nitroglycerin steadily over a period of time.

Prizewinners Robert Furchgott and Louis Ignarro, independently, unraveled the role of nitric oxide in blood-pressure regulation. Cells in the lining of arteries detect increased blood pressure and respond by producing nitric oxide. NO rapidly diffuses through the artery wall to cells in the surrounding muscle tissue. In response, the muscle tissue relaxes, the blood vessel expands, and the blood pressure drops.

In a related discovery, another prizewinner, Ferid Murad, explained how nitroglycerin works to alleviate the intense chest pain of an angina attack, which results from reduced blood flow to the heart muscle as a result of partial blockage of arteries by plaque. Physicians have prescribed nitroglycerin for angina for more than a century, knowing only that it works. Murad found that nitroglycerin breaks down in the body to form nitric oxide, which relaxes the arteries, allowing greater blood flow to the heart. Alfred Nobel, were he alive today, would no doubt be stunned by this news. Nobel, who established the prizes bearing his name, made his fortune in the nineteenth century from his invention of dynamite, a mixture of nitroglycerin with clay that tamed the otherwise hazardous explosive. When Nobel had heart trouble, his physician recommended that he eat a small quantity of nitroglycerin; he refused. In a letter, he wrote "It is ironical that I am now ordered by my physician to eat nitroglycerin." Today, a patient may take either nitroglycerin pills (containing tenths of a milligram of compound in a stabilized mixture) for occasional use or can use a chest patch to dispense nitroglycerin continuously to the skin, where it is absorbed (Figure 5.9).

Substituting the data gives

$$n = \frac{15.7 \text{ atm} \times 50.0 \text{ L}}{0.0821 \text{ L} \cdot \text{atm}/(\text{K} \cdot \text{mol}) \times 294 \text{ K}} = 32.5 \text{ mol}$$

and converting moles to mass of oxygen yields

$$32.5 \text{ mol O}_2 \times \frac{32.0 \text{ g O}_2}{1 \text{ mol O}_2} = 1.04 \times 10^3 \text{ g O}_2$$

See Problems 5.33 and 5.34.

Figure 5.10
Hot-air ballooning.
A propane gas burner on board a balloon heats the air. The heated air expands, occupying a larger volume; it therefore has a lower density than the surrounding air. The hot air and balloon rise.

Gas Density; Molecular-Weight Determination

The density of a substance is its mass divided by volume, and because the volume of a gas varies with temperature and pressure, the density of a gas also varies with temperature and pressure (Figure 5.10). In the next example, we use the ideal gas law to calculate the density of a gas at any temperature and pressure.

EXAMPLE 5.6

Calculating Gas Density

What is the density of oxygen, O_2, in grams per liter at 25°C and 0.850 atm?

PROBLEM STRATEGY

The density of a gas is often expressed in g/L. Using the ideal gas law you can calculate the moles of O_2 in 1 L of O_2. Next, you can convert the moles of O_2 to a mass of O_2, keeping in mind that this mass is the amount of O_2 per liter of O_2 (g/1 L), which is the density.

SOLUTION

The data given are

Variable	Value
P	0.850 atm
V	1 L (exact value)
T	(25 + 273) K = 298 K
n	?

Therefore,

$$n = \frac{PV}{RT} = \frac{0.850 \text{ atm} \times 1 \text{ L}}{0.0821 \text{ L} \cdot \text{atm}/(\text{K} \cdot \text{mol}) \times 298 \text{ K}} = 0.0347 \text{ mol}$$

Now convert mol O_2 to grams.

$$0.0347 \text{ mol } O_2 \times \frac{32.0 \text{ g } O_2}{1 \text{ mol } O_2} = 1.11 \text{ g } O_2$$

Therefore, the density of O_2 at 25°C and 0.850 atm is **1.11 g/L.**

See Problems 5.35 and 5.36.

The density of a gas is directly proportional to its molecular weight. Bromine, whose molecular weight is five times that of oxygen, is thus more than five times as dense as air. Figure 5.11 shows bromine gas being poured into a beaker. It displaces the air and fills the beaker.

The relation between density and molecular weight of a gas suggests that you could use a measurement of gas density to determine its molecular weight. As an illustration, consider the determination of the molecular weight of halothane, an inhalation anesthetic. The density of halothane vapor at 71°C (344 K) and 768 mmHg (1.01 atm) is 7.05 g/L. To obtain the molecular weight, you calculate the moles of vapor in a given volume. The molar mass equals mass divided by the moles in the

Figure 5.11
A gas whose density is greater than that of air.
The reddish-brown gas being poured from the flask is bromine.

same volume. From the density, you see that one liter of vapor has a mass of 7.05 g. The moles in this volume are obtained from the ideal gas law.

$$n = \frac{PV}{RT} = \frac{1.01 \text{ atm} \times 1 \text{ L}}{0.0821 \text{ L} \cdot \text{atm}/(\text{K} \cdot \text{mol}) \times 344 \text{ K}} = 0.0358 \text{ mol}$$

Therefore, the molar mass, M_m, is

$$M_m = \frac{m}{n} = \frac{7.05 \text{ g}}{0.0358 \text{ mol}} = 197 \text{ g/mol}$$

Thus, the molecular weight is 197 amu.

From the ideal gas law, you can obtain an explicit relationship between the molecular weight and density of a gas. Recall that the molar mass ($M_m = m/n$) when expressed in grams per mole is numerically equal to the molecular weight. If you substitute $n = m/M_m$ into the ideal gas law, $PV = nRT$, you obtain

$$PV = \frac{m}{M_m} RT \qquad \text{or} \qquad PM_m = \frac{m}{V} RT$$

But m/V equals the density, d. Substituting this gives

$$PM_m = dRT$$

We can illustrate the use of this equation by solving Example 5.6 again. We asked, "What is the density of oxygen, O_2, in grams per liter at 25°C (298 K) and 0.850 atm?" You rearrange the previous equation to give an explicit formula for the density and then substitute into this formula.

$$d = \frac{PM_m}{RT} = \frac{0.850 \text{ atm} \times 32.0 \text{ g/mol}}{0.0821 \text{ L} \cdot \text{atm}/(\text{K} \cdot \text{mol}) \times 298 \text{ K}} = 1.11 \text{ g/L}$$

Note that by giving M_m in g/mol and R in L · atm/(K · mol), you obtain the density in g/L.

To obtain the molecular weight directly, you rearrange the equation $PM_m = dRT$ to give an explicit formula for M_m, then substitute into it.

$$M_m = \frac{dRT}{P}$$

 CONCEPT CHECK 5.2

Three 3.0-L flasks, each at a pressure of 878 mmHg, are in a room. The flasks contain He, Ar, and Xe, respectively.

a. Which of the flasks contains the most atoms of gas?

b. Which of the flasks has the greatest density of gas?

c. If the He flask was heated and the Ar flask cooled, which of the three flasks would be at the highest pressure?

5.4 Stoichiometry Problems Involving Gas Volumes

In Chapter 3 you learned how to find the mass of one substance in a chemical reaction from the mass of another substance in the reaction. Now that you know how to use the ideal gas law, we can extend these types of problems to include gas volumes. You solve such a problem by breaking it into two problems, one involving stoichiometry and the other involving the ideal gas law. The next example illustrates this method.

Figure 5.12
Automobile air bag.
Automobiles are being equipped with air bags that inflate during a front-end collision. Most air bags are inflated with nitrogen obtained from the rapid reaction of sodium azide, NaN_3 (the azide ion is N_3^-).

EXAMPLE 5.7

Solving Stoichiometry Problems Involving Gas Volumes

Automobiles are being equipped with air bags that inflate on collision to protect the occupants from injury (Figure 5.12). Many such air bags are inflated with nitrogen, N_2, using the rapid reaction of sodium azide, NaN_3, and iron(III) oxide, Fe_2O_3, which is initiated by a spark. The overall reaction is

$$6NaN_3(s) + Fe_2O_3(s) \longrightarrow 3Na_2O(s) + 2Fe(s) + 9N_2(g)$$

How many grams of sodium azide would be required to provide 75.0 L of nitrogen gas at 25°C and 748 mmHg?

PROBLEM STRATEGY

The chemical equation relates the moles of NaN_3 to moles of N_2. Use the ideal gas law to relate the volume of N_2 to moles of N_2 and hence moles of NaN_3.

SOLUTION

Here are the available data:

Variable	Value
P	$748 \text{ mmHg} \times \dfrac{1 \text{ atm}}{760 \text{ mmHg}} = 0.984 \text{ atm}$
V	75.0 L
T	$(25 + 273) \text{ K} = 298 \text{ K}$
n	?

Rearrange the ideal gas law to obtain n.

$$n = \frac{PV}{RT}$$

Substituting from the data gives you the moles of N_2.

$$n = \frac{0.984 \text{ atm} \times 75.0 \text{ L}}{0.0821 \text{ L} \cdot \text{atm}/(\text{K} \cdot \text{mol}) \times 298 \text{ K}} = 3.02 \text{ mol}$$

From the moles of N_2, you use the chemical equation to obtain moles of NaN_3.

$$3.02 \text{ mol } N_2 \times \frac{6 \text{ mol } NaN_3}{9 \text{ mol } N_2} = 2.01 \text{ mol } NaN_3$$

Calculate the grams of NaN_3 using the molecular weight of NaN_3.

$$2.01 \text{ mol } NaN_3 \times \frac{65.01 \text{ g } NaN_3}{1 \text{ mol } NaN_3} = \textbf{131 g } NaN_3$$

See Problems 5.41, 5.42, 5.43, and 5.44.

5.5 Gas Mixtures; Law of Partial Pressures

While studying the composition of air, John Dalton concluded in 1801 that each gas in a mixture of unreactive gases acts, as far as its pressure is concerned, as though it were the only gas in the mixture. To illustrate, consider two 1-L flasks, one filled with helium to a pressure of 152 mmHg at a given temperature and the other filled with hydrogen to a pressure of 608 mmHg at the same temperature. Suppose all of the helium in the one flask is put in with the hydrogen in the other flask (see Figure 5.13). After the gases are mixed in one flask, each gas occupies a volume of one liter, just as before, and has the same temperature. Thus, the pressure exerted by helium in the mixture is 152 mmHg, and the pressure exerted by hydrogen in the mixture is 608 mmHg. The total pressure exerted by the gases in the mixture is 152 mmHg + 608 mmHg = 760 mmHg.

Partial Pressures and Mole Fractions

Figure 5.13
An illustration of Dalton's law of partial pressures. The valve connecting the flasks and the valve at the funnel are opened so that flask A fills with mineral oil. The helium flows into flask B (having the same volume as flask A), where it mixes with hydrogen. Each gas exerts the pressure it would exert if the other were not there.

The pressure exerted by a particular gas in a mixture is the **partial pressure** of that gas. According to **Dalton's law of partial pressures,** *the sum of the partial pressures of all the different gases in a mixture is equal to the total pressure of the mixture.*

If you let P be the total pressure and P_A, P_B, P_C, . . . be the partial pressures of the component gases in a mixture, the law of partial pressures can be written as

$$P = P_A + P_B + P_C + \cdots$$

The individual partial pressures follow the ideal gas law. For component A,

$$P_A V = n_A RT$$

where n_A is the number of moles of component A.

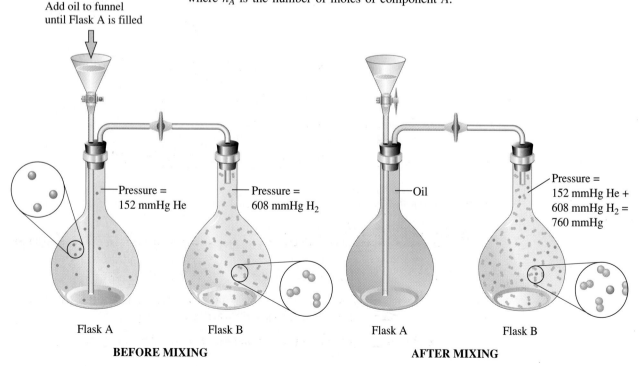

Add oil to funnel until Flask A is filled

Pressure = 152 mmHg He

Pressure = 608 mmHg H$_2$

Oil

Pressure = 152 mmHg He + 608 mmHg H$_2$ = 760 mmHg

Flask A   Flask B

BEFORE MIXING

Flask A   Flask B

AFTER MIXING

The composition of a gas mixture is often described in terms of the mole fractions of component gases. The **mole fraction** of a component gas is *the fraction of moles of that component in the total moles of gas mixture*. Because the pressure of a gas is proportional to moles, the mole fraction also equals the partial pressure divided by total pressure.

$$\text{Mole fraction of } A = \frac{n_A}{n} = \frac{P_A}{P}$$

Mole percent equals mole fraction $\times$ 100. Mole percent is equivalent to the percentage of the molecules that are component molecules.

EXAMPLE 5.8

Calculating Partial Pressures and Mole Fractions of a Gas in a Mixture

A 1.00-L sample of dry air at 25°C and 786 mmHg contains 0.925 g N_2, plus other gases including oxygen, argon, and carbon dioxide. a. What is the partial pressure (in mmHg) of N_2 in the air sample? b. What is the mole fraction and mole percent of N_2 in the mixture?

SOLUTION

a. Each gas in a mixture follows the ideal gas law. To calculate the partial pressure of N_2, you convert 0.923 g N_2 to moles N_2.

$$0.923 \text{ g } N_2 \times \frac{1 \text{ mol } N_2}{28.0 \text{ g } N_2} = 0.0330 \text{ mol } N_2$$

You substitute into the ideal gas law (noting that 25°C is 298 K).

$$P_{N_2} = \frac{n_{N_2} RT}{V}$$

$$= \frac{0.0330 \text{ mol} \times 0.0821 \text{ L} \cdot \text{atm}/(K \cdot \text{mol}) \times 298 \text{ K}}{1.00 \text{ L}}$$

$$= 0.807 \text{ atm } (= \textbf{613 mmHg})$$

b. The mole fraction of N_2 in air is

$$\text{Mole fraction of } N_2 = \frac{P_{N_2}}{P} = \frac{613 \text{ mmHg}}{786 \text{ mmHg}} = \textbf{0.780}$$

Air contains **78.0 mole percent** of N_2.

▶ **CONCEPT CHECK 5.3**

A flask equipped with a valve contains 3.0 mol of H_2 gas. You introduce 3.0 mol of Ar gas into the flask via the valve and then seal the flask.

a. What happens to the pressure of just the H_2 gas in the flask after the introduction of the Ar? If it changes, by what factor does it do so?

b. How do the pressures of the Ar and the H_2 in the flask compare?

c. How does the total pressure in the flask relate to the pressures of the two gases?

Collecting Gases over Water

A useful application of the law of partial pressures arises when you collect gases over water (a method used for gases that do not dissolve appreciably in water). Figure 5.14 shows how a gas, produced by chemical reaction in the flask, is collected by leading it to an inverted tube, where it displaces water. As gas bubbles through the water, the gas picks up molecules of water vapor that mix with it. The partial pressure of water vapor in the gas mixture in the collection tube depends only on the temperature. This partial pressure of water vapor is called the *vapor pressure* of water. ◄ Values of the vapor pressure of water at various temperatures are listed in Table 5.4. The following example shows how to find the partial pressure and then the mass of the collected gas.

▶ **Vapor pressure is the maximum partial pressure of the vapor in the presence of the liquid. It is defined more precisely in Chapter 11.**

EXAMPLE 5.9

Calculating the Amount of Gas Collected over Water

Hydrogen gas is produced by the reaction of hydrochloric acid, HCl, on zinc metal.

$$2HCl(aq) + Zn(s) \longrightarrow ZnCl_2(aq) + H_2(g)$$

The gas is collected over water. If 156 mL of gas is collected at 19°C and 769 mmHg total pressure, what is the mass of hydrogen collected?

Figure 5.14
Collection of gas over water. Hydrogen, prepared by the reaction of zinc with hydrochloric acid, is led to an inverted tube initially filled with water. When the gas-collection tube is adjusted so that the water level in the tube is at the same height as the level in the beaker, the gas pressure in the tube equals the barometric pressure (769 mmHg). The total gas pressure equals the sum of the partial pressure of the hydrogen (752 mmHg) and the vapor pressure of water (17 mmHg). For clarity, hydrochloric acid is shown in light brown and hydrogen gas in light blue.

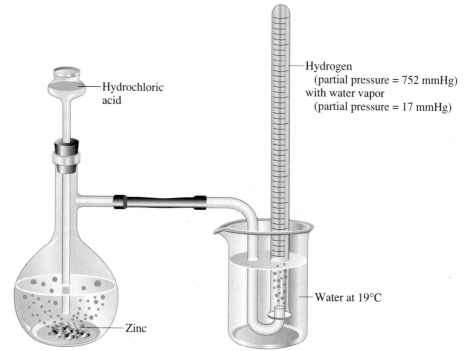

Hydrochloric acid

Hydrogen (partial pressure = 752 mmHg) with water vapor (partial pressure = 17 mmHg)

Water at 19°C

Zinc

Table 5.4
Vapor Pressure of Water at Various Temperatures

Temperature (°C)	Pressure (mmHg)
0	4.6
10	9.2
15	12.8
17	14.5
19	16.5
21	18.7
23	21.1
25	23.8
27	26.7
30	31.8
40	55.3
60	149.4
80	355.1
100	760.0

PROBLEM STRATEGY

The gas collected is hydrogen mixed with water vapor. To obtain the amount of hydrogen, you must use Dalton's law (Step 1). Then you can calculate the moles of hydrogen from the ideal gas law (Step 2).

SOLUTION

STEP 1 The vapor pressure of water at 19°C is 16.5 mmHg. From Dalton's law of partial pressures, you know that the total gas pressure equals the partial pressure of hydrogen, P_{H_2}, plus the partial pressure of water, P_{H_2O}.

$$P = P_{H_2} + P_{H_2O}$$

Substituting and solving for the partial pressure of hydrogen, you get

$$P_{H_2} = P - P_{H_2O} = (769 - 16.5) \text{ mmHg} = 752 \text{ mmHg}$$

STEP 2 Now you can use the ideal gas law to find the moles of hydrogen collected. The data are

Variable	Value
P	$752 \text{ mmHg} \times \dfrac{1 \text{ atm}}{760 \text{ mmHg}} = 0.989 \text{ atm}$
V	$156 \text{ mL} = 0.156 \text{ L}$
T	$(20 + 273) \text{ K} = 293 \text{ K}$
n	?

From the ideal gas law, $PV = nRT$, you have

$$n = \frac{PV}{RT} = \frac{0.989 \text{ atm} \times 0.156 \text{ L}}{0.0821 \text{ L} \cdot \text{atm}/(\text{K} \cdot \text{mol}) \times 293 \text{ K}} = 0.00641 \text{ mol}$$

Then you convert moles of H_2 to grams of H_2.

$$0.00641 \text{ mol } H_2 \times \frac{2.02 \text{ g } H_2}{1 \text{ mol } H_2} = \textbf{0.0129 g } H_2$$

See Problems 5.49 and 5.50.

Kinetic-Molecular Theory

In the following sections, you will see how the interpretation of a gas in terms of the **kinetic-molecular theory** (or simply **kinetic theory**) leads to the ideal gas law. *According to this theory, a gas consists of molecules in constant random motion.* Kinetic energy, E_k, is the energy associated with the motion of an object of mass m. From physics,

$$E_k = \tfrac{1}{2}m \times (\text{speed})^2$$

5.6 Kinetic Theory of an Ideal Gas

The kinetic explanation of gas pressure is that it results from the continual bombardment of the container walls by constantly moving molecules (Figure 5.15). The kinetic

Figure 5.15
Kinetic-theory model of gas pressure. According to kinetic theory, gas pressure is the result of the bombardment of the container walls by constantly moving molecules.

▶ The statements that there are no intermolecular forces and that the volume of molecules is negligible are simplifications that lead to the ideal gas law. But intermolecular forces are needed to explain how we get the liquid state from the gaseous state; it is intermolecular forces that hold molecules together in the liquid state.

theory of gases was developed by a number of influential physicists, including James Clerk Maxwell (1859) and Ludwig Boltzmann (in the 1870s). Throughout the last half of the nineteenth century, research continued on the kinetic theory, making it a cornerstone of our present view of molecular substances.

Postulates of Kinetic Theory

Physical theories are often given in terms of *postulates:* the basic statements from which all conclusions or predictions of a theory are deduced. The postulates are accepted as long as the predictions from the theory agree with experiment.

The kinetic theory of an *ideal gas* (a gas that follows the ideal gas law) is based on five postulates.

POSTULATE 1 *Gases are composed of molecules whose size is negligible compared with the average distance between them.* Most of the volume occupied by a gas is empty space. This means that you can usually ignore the volume occupied by the molecules.

POSTULATE 2 *Molecules move randomly in straight lines in all directions and at various speeds.* This means that properties of a gas that depend on the motion of molecules, such as pressure, will be the same in all directions.

POSTULATE 3 *The forces of attraction or repulsion between two molecules (intermolecular forces) in a gas are very weak or negligible, except when they collide.* ◀ This means that a molecule will continue moving in a straight line with undiminished speed until it collides with another gas molecule or with the walls of the container.

POSTULATE 4 *When molecules collide with one another, the collisions are elastic.* In an elastic collision, the total kinetic energy remains constant; no kinetic energy is lost. The collision of steel spheres is nearly elastic (that is, the spheres bounce off each other and continue moving) (Figure 5.16).

POSTULATE 5 *The average kinetic energy of a molecule is proportional to the absolute temperature.* This postulate establishes what we mean by temperature from a molecular point of view: the higher the temperature, the greater the molecular kinetic energy.

Qualitative Interpretation of the Gas Laws

A qualitative description of the kinetic theory as applied to Avogadro's law is illustrated in Figure 5.17. The explanation is dependent upon the force and frequency of the collisions with the container's wall.

In Figure 5.18, similar concepts are applicable to Charles's law. In order for the pressure in container B to be equal to that in A, the molecules must have a higher speed, which results from an increase in temperature.

The Ideal Gas Law from Kinetic Theory

One of the most important features of kinetic theory is its explanation of the ideal gas law. To show how you can get the ideal gas law from kinetic theory, we will first find an expression for the pressure of a gas.

According to kinetic theory, the pressure of a gas, P, will be proportional to the

Figure 5.16
Elastic collision of steel balls. *Left:* The steel ball at the left is lifted to give it energy. *Right:* When the steel ball is released, the energy is transmitted through elastic collisions to the ball on the right. Note that the ball at the right rises as high as the original ball was lifted.

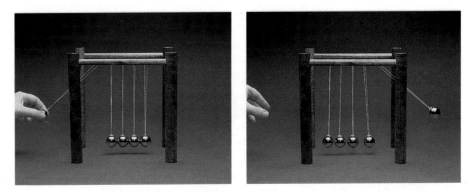

frequency of molecular collisions with a surface and to the average force exerted by a molecule in collision.

$$P \propto \text{frequency of collisions} \times \text{average force}$$

The average force exerted by a molecule during a collision depends on its mass m and its average speed u. The frequency of collisions is also proportional to the average speed u. Frequency of collisions is inversely proportional to the gas volume V, because the larger the volume, the less often a given molecule strikes the container walls. Finally, the frequency of collisions is proportional to the number of molecules N in the gas volume. Putting these factors together gives

$$P \propto \left(u \times \frac{1}{V} \times N\right) \times mu$$

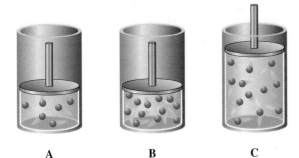

A   B   C

Figure 5.17
Molecular description of Avogadro's law. *(A)* A container where the gas molecules are at atmospheric pressure ($P_{inside} = P_{atm}$). The pressure in the container is due to the force and frequency of molecular collisions with the container walls. *(B)* The container after increasing the number of gas molecules while not allowing the piston to move. Due to the greater concentration of gas (more moles of gas in the same volume), the frequency of collisions of the gas molecules with the walls of the container has increased, causing the pressure to increase ($P_{inside} > P_{atm}$). *(C)* Container after the molecules are allowed to move the piston and increase the volume. The concentration of gas molecules and frequency of collisions with the container walls have decreased, and the pressure of the gas molecules inside the container is again equal to atmospheric pressure ($P_{inside} = P_{atm}$).

A   B

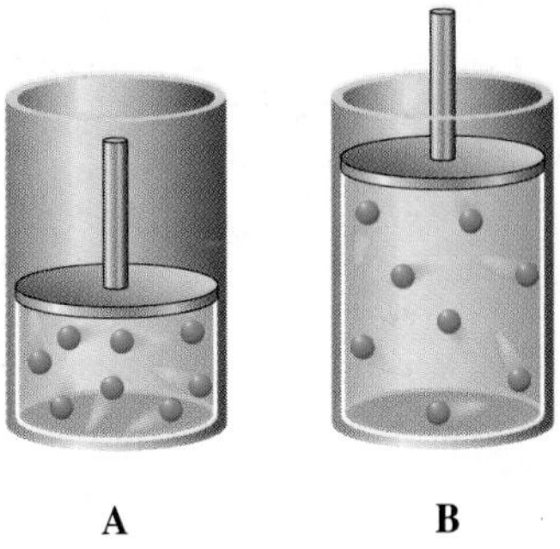

Figure 5.18
Molecular description of Charles's law. *(A)* A fixed number of moles of gas in a container at room temperature. The pressure in the container is due to the force and frequency of molecular collisions with the container walls. *(B)* Container after the volume has been doubled, resulting in a decrease in the concentration (number of molecules per unit of volume), which leads to a reduction in the frequency of molecular collisions with the container walls, thereby reducing the pressure.

Bringing the volume to the left side, you get

$$PV \propto Nmu^2$$

> ▶ Recall that kinetic energy is defined as $\frac{1}{2} m$ multiplied by (speed)^2.

Because the average kinetic energy of a molecule of mass m and average speed u is $\frac{1}{2} mu^2$, PV is proportional to the average kinetic energy of a molecule. ◀ Moreover, the average kinetic energy is proportional to the absolute temperature (Postulate 5). Noting that the number of molecules, N, is proportional to the moles of molecules, n, you have

$$PV \propto nT$$

You can write this as an equation by inserting a constant of proportionality, R, which you can identify as the molar gas constant.

$$PV = nRT$$

The next two sections give additional deductions from kinetic theory.

5.7 Molecular Speeds; Diffusion and Effusion

The principal tenet of kinetic theory is that molecules are in constant random motion. Now we will look at the speeds of molecules and at some conclusions of kinetic theory regarding molecular speeds.

Molecular Speeds

The British physicist James Clerk Maxwell (1831–1879) showed theoretically—and it has since been demonstrated experimentally—that molecular speeds are distributed as shown in Figure 5.19. This distribution of speeds depends on the temperature. At any temperature, the molecular speeds vary widely, but most are close to the average speed.

The **root-mean-square (rms) molecular speed,** u, is *a type of average molecular speed, equal to the speed of a molecule having the average molecular kinetic energy.*

Figure 5.19
Maxwell's distribution of molecular speeds. The distributions of speeds of H_2 molecules are shown for 0°C and 500°C. Note that the speed corresponding to the maximum in the curve (the most probable speed) increases with temperature.

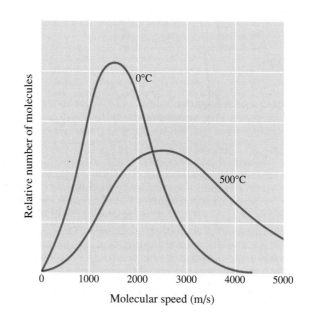

It is given by the following formula:

$$u = \sqrt{\frac{3RT}{M_m}}$$

where R is the molar gas constant, T is the absolute temperature, and M_m is the molar mass for the gas. This result follows from Postulate 5 of kinetic theory. ◄

▶ According to kinetic theory, the total kinetic energy of a mole of any gas equals $\frac{3}{2}RT$. The average kinetic energy of a molecule, which by definition is $\frac{1}{2}mu^2$, is obtained by dividing $\frac{3}{2}RT$ by Avogadro's number, N_A. Therefore, the kinetic energy is $\frac{1}{2}mu^2 = 3RT/(2N_A)$. Hence $u^2 = 3RT/(N_A m)$. Or, noting that $N_A m$ equals the molar mass M_m, you get $u^2 = 3RT/M_m$, from which you get the text equation.

In applying this equation, *care must be taken to use consistent units.* If SI units are used for R (= 8.31 kg · m²/(s² · K · mol)), T (K), and M_m (kg/mol), as in the following example, the rms speed will be in meters per second.

Values of the rms speed calculated from this formula indicate that molecular speeds are astonishingly high. For example, the rms speed of H_2 molecules at 20°C is 1.90×10^3 m/s (over 4000 mi/hr).

EXAMPLE 5.10

Calculating the rms Speed of Gas Molecules

Calculate the rms speed of O_2 molecules in a tank at 21°C and 15.7 atm.

SOLUTION

The rms molecular speed is independent of pressure but does depend on the absolute temperature, which is (21 + 273) K = 294 K. The molar mass of O_2 is 32.0×10^{-3} kg/mol, and $R = 8.31$ kg · m²/(s² · K · mol). Hence,

$$u = \left(\frac{3 \times 8.31 \text{ kg} \cdot \text{m}^2/(\text{s}^2 \cdot \text{K} \cdot \text{mol}) \times 294 \text{ K}}{32.0 \times 10^{-3} \text{ kg/mol}}\right)^{1/2} = \textbf{479 m/s}$$

See Problems 5.51, 5.52, 5.53, and 5.54.

Diffusion and Effusion

The pleasant odor of apple pie baking in the oven quickly draws people to the kitchen. The spread of an odor is easily explained by kinetic theory. All molecules are in constant, chaotic motion, and eventually a cluster of molecules of a particular substance will spread out to occupy a larger and larger space. Gaseous **diffusion** is *the process whereby a gas spreads out through another gas to occupy the space uniformly.* After sufficient time, the chaotic motion of molecules results in a complete mixing of the gases. Figure 5.20 demonstrates the diffusion of ammonia gas through air.

When you think about the kinetic-theory calculations of molecular speed, you might ask why diffusion is not even much faster than it is. Why does it take minutes for the gas to diffuse throughout a room when the molecules are moving at perhaps a thousand miles per hour? This was, in fact, one of the first criticisms of kinetic theory. The answer is simply that a molecule never travels very far in one direction before it collides with another molecule and moves off in another direction. If you could trace out the path of an individual molecule, it would be a zigzagging trail.

Although the rate of diffusion certainly depends in part on the average molecular speed, the effect of molecular collisions makes the theoretical picture a bit complicated. *Effusion,* like diffusion, is a process involving the flow of a gas but is theoretically much simpler.

Figure 5.20
Gaseous diffusion.
Left: Concentrated aqueous ammonia in the beaker evolves ammonia gas into the glass tube, which contains a strip of wet indicator paper. *Right:* The indicator changes color as the ammonia gas diffuses upward through the air in the tube.

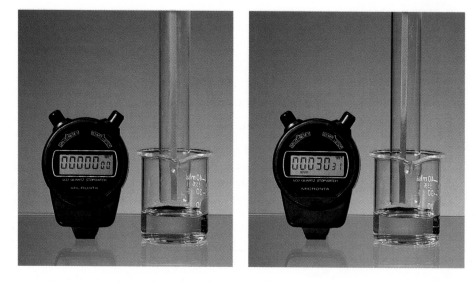

If you place a container of gas in a vacuum and then make a very small hole in the container, the gas molecules escape through the hole at the same speed they had in the container (Figure 5.21). *The process in which a gas flows through a small hole in a container* is called **effusion.** It was first studied by Thomas Graham, who discovered in 1846 that the rate of effusion of a gas is inversely proportional to the square root of its density. Today we usually state **Graham's law of effusion** in terms of molecular weight: *the rate of effusion of gas molecules from a particular hole is inversely proportional to the square root of the molecular weight of the gas at constant temperature and pressure.*

Let us consider a kinetic theory analysis of an effusion experiment. Suppose the hole in the container is made small enough so that the gas molecules continue to move randomly. When a molecule happens to encounter the hole, it leaves the container. All you have to consider for effusion is the rate at which molecules encounter the hole in the container.

Figure 5.21
Model of gaseous effusion.   According to kinetic theory, gas molecules are in constant random motion. When a molecule on the left side of the box happens to hit the pinhole in the partition between the two parts of the box, it passes (or effuses) to the right side. The rate of effusion depends on the speed of the molecules— the faster the molecules move, the more likely they are to encounter the pinhole and pass from the left side of the box to the right side.

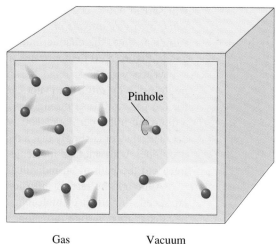

Gas   Vacuum

Figure 5.22
The hydrogen fountain.
The demonstrator places a beaker containing hydrogen gas over the porous clay container. Hydrogen effuses into the porous container faster than air effuses out. As a result, the pressure inside the porous container and the flask connected to it increases, forcing colored water out the side tube as a stream.

The rate of effusion of molecules from a container depends on three factors: the cross-sectional area of the hole; the number of molecules per unit volume; and the average molecular speed.

If you compare the effusion of different gases from the same container at the same temperature and pressure, the first two factors will be the same. The average molecular speeds will be different, however.

Because the average molecular speed essentially equals $\sqrt{3RT/M_m}$, the rate of effusion is inversely proportional to the square root of the molar mass. The derivation of Graham's law from kinetic theory was considered a triumph of the theory and greatly strengthened confidence in its validity.

$$\text{Rate of effusion of molecules} \propto \frac{1}{\sqrt{M_m}} \quad \text{(for the same container at constant } T \text{ and } P\text{)}$$

The hydrogen fountain, shown in Figure 5.22, is dependent on the differences in rates of effusion of gases. (Can you explain why a helium-filled balloon loses pressure after a few hours?)

CO_2
$M_m = 44.0 \text{ g/mol}$

SO_2
$M_m = 64.1 \text{ g/mol}$

EXAMPLE 5.11

Calculating the Ratio of Effusion Rates of Gases

Calculate the ratio of effusion rates of molecules of carbon dioxide, CO_2, and sulfur dioxide, SO_2, from the same container and at the same temperature and pressure.

SOLUTION

The two rates of effusion are inversely proportional to the square roots of their molar masses, so you can write

$$\frac{\text{Rate of effusion of } CO_2}{\text{Rate of effusion of } SO_2} = \sqrt{\frac{M_m(SO_2)}{M_m(CO_2)}}$$

where $M_m(SO_2)$ is the molar mass of SO_2 (64.1 g/mol) and $M_m(CO_2)$ is the molar mass of CO_2 (44.0 g/mol). Substituting these molar masses into the formula gives

$$\frac{\text{Rate of effusion of CO}_2}{\text{Rate of effusion of SO}_2} = \sqrt{\frac{64.1 \text{ g/mol}}{44.0 \text{ g/mol}}} = \mathbf{1.21}$$

In other words, carbon dioxide effuses 1.21 times faster than sulfur dioxide.

See Problems 5.55 and 5.56.

Graham's law has practical application in the preparation of fuel rods for nuclear fission reactors. Such reactors depend on the fact that the uranium-235 nucleus undergoes fission (splits) when bombarded with neutrons. When the nucleus splits, several neutrons are emitted and a large amount of energy is liberated. These neutrons bombard more uranium-235 nuclei, and the process continues with the evolution of more energy. However, natural uranium consists of 99.27% uranium-238 (which does not undergo fission) and only 0.72% uranium-235 (which does undergo fission). A uranium fuel rod must contain about 3% uranium-235 to sustain the nuclear reaction.

To increase the percentage of uranium-235 in a sample of uranium (a process called *enrichment*), one first prepares uranium hexafluoride, UF_6, a white, crystalline solid that is easily vaporized. Uranium hexafluoride vapor is allowed to pass through a series of porous membranes. Each membrane has many small holes through which the vapor can effuse. Because the UF_6 molecules with the lighter isotope of uranium travel about 0.4% faster than the UF_6 molecules with the heavier isotope, the gas that passes through first is somewhat richer in uranium-235. When this vapor passes through another membrane, the uranium-235 vapor becomes further concentrated. It takes many effusion stages to reach the necessary enrichment.

5.8 Real Gases

In Table 5.2, we found that the pressure–volume product for O_2 at 0°C was not quite constant, particularly at high pressures. Experiments show that the ideal gas law describes the behavior of real gases quite well at low pressures and moderate temperatures, but not at high pressures and low temperatures.

Figure 5.23 shows the behavior of the pressure–volume product at various pressures for several gases. The gases deviate noticeably from ideal gas behavior at high pressures. Also, the deviations differ for each kind of gas. We can explain why this is so by examining the postulates of kinetic theory, from which the ideal gas law can be derived.

Postulate 1 of kinetic theory says that the volume of space occupied by molecules is negligible compared with the total gas volume. At low pressures, where the volume of individual molecules is negligible compared with the total volume available (Figure 5.24A), the ideal gas law is a good approximation. At higher pressures, where the volume of individual molecules becomes important (Figure 5.24B), the space through which a molecule can move is significantly different from V, the entire gas volume.

Postulate 3 says that the forces of attraction between molecules (intermolecular forces) in a gas are very weak or negligible. This is a good approximation at low

Figure 5.23
Pressure–volume product of gases at different pressures.   The bottom graph shows the pressure–volume product of one mole of various gases at 0°C and at different pressures. The top graph shows values at low pressure. For an ideal gas, the pressure–volume product is constant.

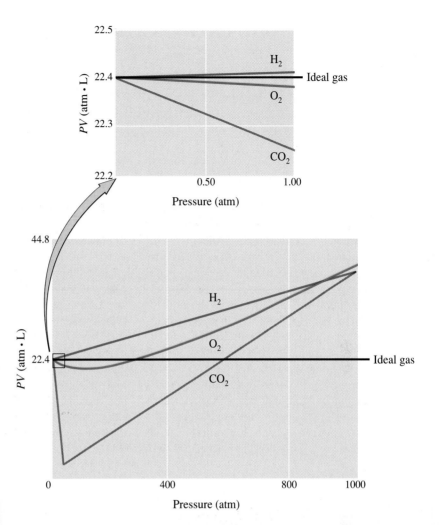

A   B

Figure 5.24
Effect of molecular volume at high pressure.
(A) At low pressures, the volume of molecules is a small fraction of the total volume and can be neglected, as in the ideal gas law. *(B)* At high pressures, the volume of molecules is a significant fraction of the total volume and cannot be neglected. The ideal gas law is no longer a good approximation.

pressures because these forces diminish rapidly as the distance between molecules increases. Intermolecular forces become significant at higher pressures. Because of these intermolecular forces, the actual pressure of a gas is less than that predicted by ideal gas behavior.

The Dutch physicist J. D. van der Waals (1837–1923) was the first to account for these deviations of a real gas from ideal gas behavior. The **van der Waals equation** *accounts for deviations from ideal behavior.*

$$\left(P + \frac{n^2 a}{V^2}\right)(V - nb) = nRT$$

The constants a and b are chosen to fit experiment as closely as possible. Table 5.5 gives values of van der Waals constants for several gases. The volume available for a molecule to move in equals the gas volume, V, minus the volume occupied by molecules. So you replace V in the ideal gas law with $V - nb$, where nb represents the volume occupied by n moles of molecules. The force per unit wall area, or pressure, is reduced from that assumed in the ideal gas law by a factor proportional to n^2/V^2. So, you replace P in the ideal gas law by $P + n^2 a/V^2$.

Carbon Dioxide Gas and the Greenhouse Effect

ENVIRONMENT

Yes, I know, I shouldn't have left that chocolate bar on the seat of my car, which I parked on a sunny street near the university. You and I know that the interior of such a car can become quite hot. Have you ever wondered why? If the sun's heat energy can get into the car through its window glass, why can't it leave equally well?

Parked cars are similar to greenhouses, which people have used for centuries to protect plants from cold weather. The Swedish chemist Svante Arrhenius, who you may remember from the previous chapter (for his theory of ionic solutions), realized that carbon dioxide in the atmosphere acts like the glass in a greenhouse. His calculations, published in 1898, were the first to show how sensitive the temperature of the earth might be to the percentage of carbon dioxide in the atmosphere. Arrhenius's reference to the earth's atmospheric envelope, with its carbon dioxide, as a "hothouse" has evolved into our present-day term *greenhouse effect*.

Here is the explanation of the greenhouse effect. The principal gases in the atmosphere are oxygen, O_2, and nitrogen, N_2. These gases are transparent to visible light from the sun, and when this light reaches the surface of the earth, it is absorbed and converted to heat. This heat causes atoms in the earth's surface to vibrate, which then radiate the heat energy as infrared radiation, or heat rays (Figure 5.25). Neither oxygen nor nitrogen absorb infrared radiation, so if the earth's atmosphere contained only these gases, the infrared radiation, or heat rays, would simply escape into outer space; there would be no greenhouse effect. However, other gases in the atmosphere, especially carbon dioxide, do absorb infrared radiation, and it is this absorption that warms the atmosphere and eventually also the earth's surface, giving us a greenhouse effect. The greater the percentage of carbon dioxide in the atmosphere, the warmer the earth should be. (Glass, acting like carbon dioxide, allows visible light to pass into a greenhouse, but absorbs infrared radiation, effectively trapping the heat.)

The greenhouse effect has become associated with the concept of *global warming*. Records kept since the nineteenth century clearly show that the amount of carbon dioxide in the atmosphere has increased dramatically, from about 284 ppm (parts per million) in 1830 to about 370 ppm in 2000 (Figure 5.26). (Parts per million is equivalent to mole percent $\times$ 10^{-4}.) Scientists

Figure 5.25
Greenhouse effect of certain gases in the atmosphere.
The atmosphere is transparent to sunlight, except for radiation in the far ultraviolet. When sunlight reaches the earth's surface, it is absorbed. The heated surface then radiates the energy back as infrared radiation. The major components of the atmosphere, O_2 and N_2, are transparent to this radiation, but gases such as CO_2, H_2O, CH_4, and chlorofluorocarbons, do absorb infrared rays. These substances in the atmosphere then reradiate the infrared rays, with the result that a significant fraction of the radiation returns to earth. In effect, these gases trap the radiation, acting like the glass on a greenhouse.

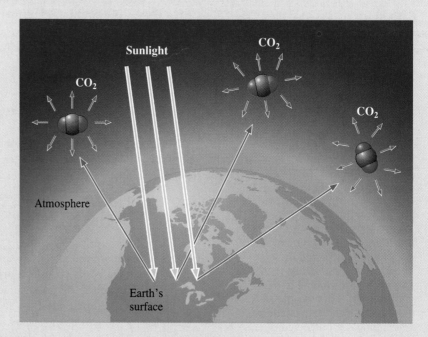

believe that our burning of fossil fuels (coal, oil, and natural gas) is responsible for this change. Moreover, temperature measurements made at the surface of the earth during the past 100 years indicate an average temperature increase of between 0.4°C and 0.8°C. Thus, many climate scientists believe that human activities are at least partially responsible for global warming and, hence, they predict drastic changes in climate. But to complicate matters, satellite measurements of upper atmosphere temperatures made during the last 20 years do not show an increase in temperature. So, a number of climate scientists question the idea of human-induced global warming, claiming that what we are seeing is a natural climate variation.

In January 2000, the U.S. National Research Council (NRC) released a report of a panel that was trying to reconcile these views. They concluded that the increase in surface temperature is real. But John M. Wallace, professor of atmospheric sciences at the University of Washington and chair of the NRC panel, said, "The rapid increase in the Earth's surface temperature over the past 20 years is not necessarily representative of how the atmosphere is responding to long-term, human-induced changes. . . ." He cautioned against politicizing a scientific investigation and noted that "We still don't have all the answers." Climate is enormously complicated, depending not only on the concentration of carbon dioxide in the atmosphere (greenhouse effect), but also on dust particles (which reflect sunlight), natural variation in radiation from the sun, and many other variables. Despite the complications of climate research, the majority of scientists do believe that human activities are affecting our climate.

Figure 5.26
Concentrations of carbon dioxide in the atmosphere (average annual values).
Concentrations of carbon dioxide in the atmosphere have been increasing steadily since about 1830. The blue curve is derived from ice core samples (containing air bubbles) obtained from Antarctica. The purple curve is derived from direct measurements of atmospheric carbon dioxide collected since 1958 at Mauna Loa, Hawaii. (*Source of blue curve:* D.M. Etheridge, L.P. Steele, R.L. Langenfelds, R.J. Francey, J.M. Barnola and V.I. Morgan. 1998. Historical CO_2 records from the Law Dome DE08, DE08-2, and DSS ice cores. In Trends: A Compendium of Data on Global Change. Carbon Dioxide Information Analysis Center, Oak Ridge National Laboratory, U.S. Department of Energy, Oak Ridge, Tenn., U.S.A. *Source of purple curve:* C.D. Keeling and T.P. Whorf. 1999. Atmospheric CO_2 records from sites in the SIO air sampling network. In Trends: A Compendium of Data on Global Change. Carbon Dioxide Information Analysis Center, Oak Ridge National Laboratory, U.S. Department of Energy, Oak Ridge, Tenn., U.S.A.)

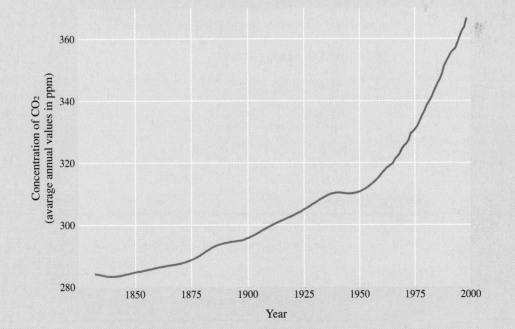

Table 5.5
van der Waals Constants for Some Gases

Gas	a $L^2 \cdot atm/mol^2$	b L/mol
Carbon dioxide, CO_2	3.658	0.04286
Ethane, C_2H_6	5.570	0.06499
Ethanol, C_2H_5OH	12.56	0.08710
Helium, He	0.0346	0.0238
Hydrogen, H_2	0.2453	0.02651
Oxygen, O_2	1.382	0.03186
Sulfur dioxide, SO_2	6.865	0.05679
Water, H_2O	5.537	0.03049

(From Lide, David R., Ed., in *CRC Handbook of Chemistry and Physics*, 74th ed., pp. 6–48. Copyright CRC Press Inc., Boca Raton, Florida, 1993. With the permission of CRC Press, Inc.)

The next example illustrates the use of the van der Waals equation to calculate the pressure exerted by a gas under given conditions.

EXAMPLE 5.12

Using the van der Waals Equation

If sulfur dioxide were an ideal gas, the pressure at 0.0°C exerted by 1.000 mole occupying 22.41 L would be 1.000 atm (22.41 L is the molar volume of an ideal gas at STP). Use the van der Waals equation to estimate the pressure of this volume of 1.000 mol SO_2 at 0.0°C. See Table 5.5 for values of a and b.

PROBLEM STRATEGY

You first rearrange the van der Waals equation to give P in terms of the other variables.

$$P = \frac{nRT}{V - nb} - \frac{n^2 a}{V^2}$$

Now you substitute $R = 0.08206$ L $\cdot$ atm/(K $\cdot$ mol), $T = 273.2$ K, $V = 22.41$ L, $a = 6.865$ L^2 $\cdot$ atm/mol^2, and $b = 0.05679$ L/mol.

SOLUTION

$$P = \frac{1.000 \text{ mol} \times 0.08206 \text{ L} \cdot atm/(K \cdot mol) \times 273.2 \text{ K}}{22.41 \text{ L} - (1.000 \text{ mol} \times 0.05679 \text{ L/mol})} -$$

$$\frac{(1.000 \text{ mol})^2 \times 6.865 \text{ L}^2 \cdot atm/mol^2}{(22.41 \text{ L})^2}$$

$$= 1.003 \text{ atm} - 0.014 \text{ atm} = \textbf{0.989 atm}$$

See Problems 5.57 and 5.58.

A Checklist for Review

Important Terms

pressure (5.1)
pascal (Pa) (5.1)
barometer (5.1)
manometer (5.1)
millimeters of mercury
  (mmHg or torr) (5.1)
atmosphere (atm) (5.1)
Boyle's law (5.2)

Charles's law (5.2)
Avogadro's law (5.2)
molar gas volume (V_m) (5.2)
standard temperature and
  pressure (STP) (5.2)
molar gas constant (R) (5.3)
ideal gas law (5.3)
partial pressure (5.5)

Dalton's law of partial
  pressures (5.5)
mole fraction (5.5)
kinetic-molecular theory
  of gases (kinetic theory)
  (5.6)
root-mean-square (rms)
  molecular speed (5.7)

diffusion (5.7)
effusion (5.7)
Graham's law of effusion
  (5.7)
van der Waals equation
  (5.8)

Key Equations

$PV = \text{constant}$ (constant n, T)

$\dfrac{V}{T} = \text{constant}$ (constant n, P)

$V_m = \text{specific constant}$

(depending on T, P; independent of gas)

$PV = nRT$

$PM_m = dRT$

$P = P_A + P_B + P_C + \ldots$

$$\text{Mole fraction of A} = \frac{n_A}{n} = \frac{P_A}{P}$$

$$u = \sqrt{\frac{3RT}{M_m}}$$

$$\text{Rate of effusion} \propto \frac{1}{\sqrt{M_m}}$$

(same container at constant T, P)

$$\left(P + \frac{n^2 a}{V^2}\right)(V - nb) = nRT$$

Summary of Facts and Concepts

The *pressure* of a gas equals the force exerted per unit area of surface. It is measured by a *manometer* in units of *pascals, millimeters of mercury,* or *atmospheres.*

    Gases at low to moderate pressures and moderate temperatures follow the same simple relationships, or gas laws. Thus, for a given amount of gas at constant temperature, the volume varies inversely with pressure *(Boyle's law).* Also, for a given amount of gas at constant pressure, the volume is directly proportional to the absolute temperature *(Charles's law).* These two laws, together with *Avogadro's law* (equal volumes of any two gases at the same temperature and pressure contain the same number of molecules), can be formulated as one equation, $PV = nRT$ *(ideal gas law).* This equation gives the relationship among P, V, n, and T for a gas. It also relates these quantities for each component in a gas mixture. The total gas pressure, P, equals the sum of the partial pressures of each component *(law of partial pressures).*

    The ideal gas law can be explained by the *kinetic-molecular theory.* We define an ideal gas as consisting of molecules with negligible volume that are in constant random motion. In the ideal gas model, there are no intermolecular forces between molecules, and the average kinetic energy of molecules is proportional to the absolute temperature. From kinetic theory, one can show that the *rms molecular speed* equals $\sqrt{3RT/M_m}$. Given this result, one can derive *Graham's law of effusion:* the rate of effusion of gas molecules under identical conditions is inversely proportional to the square root of the molecular weight.

    Real gases deviate from the ideal gas law at high pressure and low temperature. From kinetic theory, we expect these deviations, because real molecules do have volume and intermolecular forces do exist. These two factors are partially accounted for in the *van der Waals equation.*

Operational Skills

1. **Converting units of pressure**   Given the pressure of a gas in one unit of pressure (Pa, mmHg, or atm), convert to another unit of pressure. **(EXAMPLE 5.1)**

2. **Using the empirical gas laws**   Given an initial volume occupied by a gas, calculate the final volume when the pressure changes at fixed temperature **(EXAMPLE 5.2)**;

when the temperature changes at fixed pressure (EXAMPLE 5.3); and when both pressure and temperature change (EXAMPLE 5.4).

3. **Using the ideal gas law**   Given any three of the variables P, V, T, and n for a gas, calculate the fourth from the ideal gas law.   (EXAMPLE 5.5)

4. **Relating gas density and molecular weight**   Given the molecular weight, calculate the density of a gas for a particular temperature and pressure (EXAMPLE 5.6); or, given the gas density, calculate the molecular weight.

5. **Solving stoichiometry problems involving gas volumes**   Given the volume (or mass) of one substance in a reaction, calculate the mass (or volume) of another produced or used up.   (EXAMPLE 5.7)

6. **Calculating partial pressures and mole fractions of a gas in a mixture**   Given the masses of gases in a mixture, calculate the partial pressures and mole fractions.   (EXAMPLE 5.8)

7. **Calculating the amount of gas collected over water**   Given the volume, total pressure, and temperature of gas collected over water, calculate the mass of the dry gas.   (EXAMPLE 5.9)

8. **Calculating the rms speed of gas molecules**   Given the molecular weight and temperature of a gas, calculate the rms molecular speed.   (EXAMPLE 5.10)

9. **Calculating the ratio of effusion rates of gases**   Given the molecular weights of two gases, calculate the ratio of rates of effusion (EXAMPLE 5.11); or, given the relative effusion rates of a known and an unknown gas, obtain the molecular weight of the unknown gas.

10. **Using the van der Waals equation**   Given n, T, V, and the van der Waals constants a and b for a gas, calculate the pressure from the van der Waals equation.   (EXAMPLE 5.12)

Review Questions

5.1   Define *pressure*. From the definition, obtain the SI unit of pressure in terms of SI base units.

5.2   What variables determine the height of the liquid in a manometer?

5.3   Explain how you would set up an absolute temperature scale based on the Fahrenheit scale, knowing that absolute zero is −273.15°C.

5.4   State Avogadro's law in words. How does this law explain the law of combining volumes? Use the gas reaction

$$N_2 + 3H_2 \longrightarrow 2NH_3$$

as an example in your explanation.

5.5   What does the term *molar gas volume* mean? What is the molar gas volume (in liters) at STP for an ideal gas?

5.6   Starting from Boyle's, Charles's, and Avogadro's laws, obtain the ideal gas law, $PV = nRT$.

5.7   What is the value of R in units of L · mmHg/(K · mol)?

5.8   Give the postulates of kinetic theory and state any evidence that supports them.

5.9   What is the origin of gas pressure, according to kinetic theory?

5.10   How does the rms molecular speed depend on absolute temperature? on molar volume?

5.11   Explain why a gas appears to diffuse more slowly than average molecular speeds might suggest.

5.12   What is effusion? Why does a gas whose molecules have smaller mass effuse faster than one whose molecules have larger mass?

5.13   Under what conditions does the behavior of a real gas begin to differ significantly from the ideal gas law?

5.14   What is the physical meaning of the a and b constants in the van der Waals equation?

Conceptual Problems

5.15   Using the concepts developed in this chapter, explain the following observations.
a. Automobile tires are flatter on cold days.
b. You are not supposed to dispose of aerosol cans in a fire.
c. The lid of a water bottle pops off when the bottle sits in the sun.

5.16   Consider the following gas container equipped with a movable piston.

a. By what factor (increase by 1, decrease by 1.5, etc.) would you change the pressure if you wanted the volume to change from volume C to volume D?

b. If the piston were moved from volume C to volume A, by what factor would the pressure change?

c. By what factor would you change the temperature in order to change from volume C to volume B?

d. If you increased the number of moles of gas in the container by a factor of 2, by what factors would the pressure and the volume change?

5.17 Consider the following setup, which shows identical containers connected by a tube with a valve that is presently closed. The container on the left has 1.0 mol of H_2 gas; the container on the right has 1.0 mol of O_2.

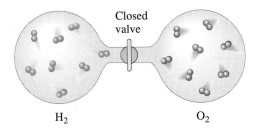

Note: Acceptable answers to some of these questions might be "both" or "neither one."

a. Which container has the greatest density of gas?

b. Which container has molecules that are moving at a faster average molecular speed?

c. Which container has more molecules?

d. If the valve is opened, will the pressure in each of the containers change? If it does, how will it change (increase, decrease, or no change)?

e. 2.0 mol of Ar is added to the system with the valve open. What fraction of the total pressure will be due to the H_2?

5.18 Two identical He-filled balloons, each with a volume of 20 L, are allowed to rise into the atmosphere. One rises to an altitude of 3000 m while the other rises to 6000 m.

a. Assuming that the balloons are at the same temperature, which balloon has the greater volume?

b. What information would you need in order to calculate the volume of each of the balloons at their respective heights?

Practice Problems

Note: In these problems, the final zeros given in temperatures and pressures (for example, 20°C, 760 mmHg) are significant figures.

Units of Pressure

5.19 A gas in a closed-tube manometer (Figure 5.3) has a measured pressure of 0.047 atm. Calculate the pressure in mmHg.

5.20 The barometric pressure measured outside an airplane at 9 km (30,000 ft) was 259 mmHg. Calculate the pressure in kPa.

Empirical Gas Laws

5.21 Suppose you had a 3.15-L sample of neon gas at 21°C and a pressure of 0.951 atm. What would be the volume of this gas if the pressure were increased to 1.292 atm while the temperature remained constant?

5.22 You fill a balloon with helium gas to a volume of 2.68 L at 23°C and 789 mmHg. Now you release the balloon. What would be the volume of helium if its pressure changes to 556 mmHg but the temperature is unchanged?

5.23 A McLeod gauge measures low gas pressures by compressing a known volume of the gas at constant temperature. If 315 cm^3 of gas is compressed to a volume of 0.0457 cm^3 under a pressure of 2.51 kPa, what was the original gas pressure?

5.24 If 456 dm^3 of krypton at 101 kPa and 21°C is compressed into a 27.0-dm^3 tank at the same temperature, what is the pressure of krypton in the tank?

5.25 A sample of nitrogen gas at 18°C and 760 mmHg has a volume of 2.67 mL. What is the volume at 0°C and 1 atm of pressure?

5.26 A mole of gas at 0°C and 760 mmHg occupies 22.41 L. What is the volume at 20°C and 760 mmHg?

5.27 A bacterial culture isolated from sewage produced 35.5 mL of methane, CH_4, at 31°C and 753 mmHg. What is the volume of this methane at standard temperature and pressure (0°C, 760 mmHg)?

5.28 Pantothenic acid is a B vitamin. Using the Dumas method, you find that a sample weighing 71.6 mg gives 3.84 mL of nitrogen gas at 23°C and 785 mmHg. What is the volume of nitrogen at STP?

5.29 In the presence of a platinum catalyst, ammonia, NH_3, burns in oxygen, O_2, to give nitric oxide, NO, and water vapor. How many volumes of nitric oxide are obtained from one volume of ammonia, assuming each gas is at the same temperature and pressure?

5.30 Methanol, CH_3OH, can be produced in industrial plants by reacting carbon dioxide with hydrogen in the presence of a catalyst. Water is the other product. How many volumes of hydrogen are required for each volume of carbon dioxide when each gas is at the same temperature and pressure?

Ideal Gas Law

5.31 Starting from the ideal gas law, prove that the volume of a mole of gas is inversely proportional to the pressure at constant temperature (Boyle's law).

5.32 Starting from the ideal gas law, prove that the volume of a mole of gas is directly proportional to the absolute temperature at constant pressure (Charles's law).

5.33 An experiment calls for 3.50 mol of chlorine, Cl_2. What volume would this be if the gas volume is measured at 34°C and 3.00 atm?

5.34 According to your calculations, a reaction should yield 5.67 g of oxygen, O_2. What do you expect the volume to be at 23°C and 0.985 atm?

5.35 What is the density of ammonia gas, NH_3, at 31°C and 751 mmHg? Obtain the density in grams per liter.

5.36 Calculate the density of hydrogen sulfide gas, H_2S, at 56°C and 967 mmHg. Obtain the density in grams per liter.

5.37 A chemist vaporized a liquid compound and determined its density. If the density of the vapor at 90°C and 753 mmHg is 1.585 g/L, what is the molecular weight of the compound?

5.38 You vaporize a liquid substance at 100°C and 755 mmHg. The volume of 0.548 g of vapor is 237 mL. What is the molecular weight of the substance?

5.39 Ammonium chloride, NH_4Cl, is a white solid. When heated to 325°C, it gives a vapor that is a mixture of ammonia and hydrogen chloride.

$$NH_4Cl(s) \longrightarrow NH_3(g) + HCl(g)$$

Suppose someone contends that the vapor consists of NH_4Cl molecules rather than a mixture of NH_3 and HCl. Could you decide between these alternative views on the basis of gas-density measurements? Explain.

5.40 Phosphorus pentachloride, PCl_5, is a white solid that sublimes (vaporizes without melting) at about 100°C. At higher temperatures, the PCl_5 vapor decomposes to give phosphorus trichloride, PCl_3, and chlorine, Cl_2.

$$PCl_5(g) \longrightarrow PCl_3(g) + Cl_2(g)$$

How could gas-density measurements help to establish that PCl_5 vapor is decomposing?

Stoichiometry with Gas Volumes

5.41 Calcium carbide reacts with water to produce acetylene gas, C_2H_2.

$$CaC_2(s) + 2H_2O(l) \longrightarrow Ca(OH)_2(aq) + C_2H_2(g)$$

Calculate the volume (in liters) of acetylene produced at 26°C and 684 mmHg from 0.050 mol CaC_2 and excess H_2O.

5.42 Magnesium metal reacts with hydrochloric acid to produce hydrogen gas, H_2.

$$Mg(s) + 2HCl(aq) \longrightarrow MgCl_2(aq) + H_2(g)$$

Calculate the volume (in liters) of hydrogen produced at 28°C and 665 mmHg from 0.0840 mol Mg and excess HCl.

5.43 Lithium hydroxide, LiOH, is used in spacecraft to recondition the air by absorbing the carbon dioxide exhaled by astronauts. The reaction is

$$2LiOH(s) + CO_2(g) \longrightarrow Li_2CO_3(s) + H_2O(l)$$

What volume of carbon dioxide gas at 21°C and 781 mmHg could be absorbed by 327 g of lithium hydroxide?

5.44 Magnesium burns in air to produce magnesium oxide, MgO, and magnesium nitride, Mg_3N_2. Magnesium nitride reacts with water to give ammonia.

$$Mg_3N_2(s) + 6H_2O(l) \longrightarrow 3Mg(OH)_2(s) + 2NH_3(g)$$

What volume of ammonia gas at 24°C and 753 mmHg will be produced from 4.56 g of magnesium nitride?

5.45 Ammonium sulfate is used as a nitrogen and sulfur fertilizer. It is produced by reacting ammonia with sulfuric acid. Write the balanced equation for the reaction of gaseous ammonia with sulfuric acid solution. What volume (in liters) of ammonia at 15°C and 1.15 atm is required to produce 150.0 g of ammonium sulfate?

5.46 Sodium hydrogen carbonate is also known as baking soda. When this compound is heated, it decomposes to sodium carbonate, carbon dioxide, and water vapor. Write the balanced equation for this reaction. What volume (in liters) of carbon dioxide gas at 75°C and 756 mmHg will be produced from 26.8 g of sodium hydrogen carbonate?

Gas Mixtures

5.47 The gas from a certain volcano had the following composition in mole percent (that is, mole fraction × 100): 65.0% CO_2, 25.0% H_2, 5.4% HCl, 2.8% HF, 1.7% SO_2, and 0.1% H_2S. What would be the partial pressure of each of these gases if the total pressure of volcanic gas were 760 mmHg?

5.48 In a series of experiments, the U.S. Navy developed an undersea habitat. In one experiment, the mole percent composition of the atmosphere in the undersea habitat was 79.0% He,

17.0% N_2, and 4.0% O_2. What will the partial pressure of each gas be when the habitat is 58.8 m below sea level, where the pressure is 6.91 atm?

5.49 Formic acid, $HCHO_2$, is a convenient source of small quantities of carbon monoxide. When warmed with sulfuric acid, formic acid decomposes to give CO gas.

$$HCHO_2(l) \longrightarrow H_2O(l) + CO(g)$$

If 3.85 L of carbon monoxide was collected over water at 25°C and 689 mmHg, how many grams of formic acid were consumed?

5.50 An aqueous solution of ammonium nitrite, NH_4NO_2, decomposes when heated to give off nitrogen, N_2.

$$NH_4NO_2(s) \longrightarrow 2H_2O(g) + N_2(g)$$

This reaction may be used to prepare pure nitrogen. How many grams of ammonium nitrite must have reacted if 3.75 dm^3 of nitrogen gas was collected over water at 19°C and 97.8 kPa?

Molecular Speeds; Effusion

5.51 Uranium hexafluoride, UF_6, is a white solid that sublimes (vaporizes without melting) at 57°C under normal atmospheric pressure. The compound is used to separate uranium isotopes by effusion. What is the rms speed (in m/s) of a uranium hexafluoride molecule at 57°C?

5.52 For a spacecraft or a molecule to leave the moon, it must reach the escape velocity (speed) of the moon, which is 2.37 km/s. The average daytime temperature of the moon's surface is 365 K. What is the rms speed (in m/s) of a hydrogen molecule at this temperature? How does this compare with the escape velocity?

5.53 At what temperature would CO_2 molecules have an rms speed equal to that of H_2 molecules at 25°C?

5.54 At what temperature does the rms speed of O_2 molecules equal 400. m/s?

5.55 If 4.83 mL of an unknown gas effuses through a hole in a plate in the same time it takes 9.23 mL of argon, Ar, to effuse through the same hole under the same conditions, what is the molecular weight of the unknown gas?

5.56 A given volume of nitrogen, N_2, required 68.3 s to effuse from a hole in a chamber. Under the same conditions, another gas required 85.6 s for the same volume to effuse. What is the molecular weight of this gas?

van der Waals Equation

5.57 Calculate the pressure of ethanol vapor, $C_2H_5OH(g)$, at 82.0°C if 1.000 mol $C_2H_5OH(g)$ occupies 30.00 L. Use the van der Waals equation (see Table 5.5 for data). Compare with the result from the ideal gas law.

5.58 Calculate the pressure of water vapor at 120.0°C if 1.000 mol of water vapor occupies 32.50 L. Use the van der Waals equation (see Table 5.5 for data). Compare with the result from the ideal gas law.

General Problems

5.59 A glass tumbler containing 243 cm^3 of air at 1.00×10^2 kPa (the barometric pressure) and 20°C is turned upside down and immersed in a body of water to a depth of 20.5 m. The air in the glass is compressed by the weight of water above it. Calculate the volume of air in the glass, assuming the temperature and barometric pressure have not changed.

5.60 The density of air at 20°C and 1.00 atm is 1.205 g/L. If this air were compressed at the same temperature to equal the pressure at 30.0 m below sea level, what would be its density? Assume the barometric pressure is constant at 1.00 atm. The density of seawater is 1.025 g/cm^3.

5.61 A balloon containing 5.0 dm^3 of gas at 14°C and 100.0 kPa rises to an altitude of 2000. m, where the temperature is 20°C. The pressure of gas in the balloon is now 79.0 kPa. What is the volume of gas in the balloon?

5.62 A volume of air is taken from the earth's surface, at 15°C and 1.00 atm, to the stratosphere, where the temperature is −20°C and the pressure is 1.00×10^{-3} atm. By what factor is the volume increased?

5.63 A radioactive metal atom decays (goes to another kind of atom) by emitting an alpha particle (He^{2+} ion). The alpha particles are collected as helium gas. A sample of helium with a volume of 12.05 mL was obtained at 765 mmHg and 23°C. How many atoms decayed during the period of the experiment?

5.64 The combustion method used to analyze for carbon and hydrogen can be adapted to give percentage N by collecting the nitrogen from combustion of the compound as N_2. A sample of a compound weighing 8.75 mg gave 1.77 mL N_2 at 25°C and 749 mmHg. What is the percentage N in the compound?

5.65 A person exhales about 5.8×10^2 L of carbon dioxide per day (at STP). The carbon dioxide exhaled by an astronaut is absorbed from the air of a space capsule by reaction with lithium hydroxide, LiOH.

$$2LiOH(s) + CO_2(g) \longrightarrow Li_2CO_3(s) + H_2O(l)$$

How many grams of lithium hydroxide are required per astronaut per day?

5.66   Pyruvic acid, $HC_3H_3O_3$, is involved in cell metabolism. It can be assayed for (that is, the amount of it determined) by using a yeast enzyme. The enzyme makes the following reaction go to completion:

$$HC_3H_3O_3(aq) \longrightarrow C_2H_4O(aq) + CO_2(g)$$

If a sample containing pyruvic acid gives 20.3 mL of carbon dioxide gas, CO_2, at 349 mmHg and 30°C, how many grams of pyruvic acid are there in the sample?

5.67   Liquid oxygen was first prepared by heating potassium chlorate, $KClO_3$, in a closed vessel to obtain oxygen at high pressure. The oxygen was cooled until it liquefied.

$$2KClO_3(s) \longrightarrow 2KCl(s) + 3O_2(g)$$

If 170. g of potassium chlorate reacts in a 2.50-L vessel, which was initially evacuated, what pressure of oxygen will be attained when the temperature is finally cooled to 25°C? Use the preceding chemical equation and ignore the volume of solid product.

5.68   Raoul Pictet, the Swiss physicist who first liquefied oxygen, attempted to liquefy hydrogen. He heated potassium formate, $KCHO_2$, with KOH in a closed 2.50-L vessel.

$$KCHO_2(s) + KOH(s) \longrightarrow K_2CO_3(s) + H_2(g)$$

If 75.0 g of potassium formate reacts in a 2.50-L vessel, which was initially evacuated, what pressure of hydrogen will be attained when the temperature is finally cooled to 21°C? Use the preceding chemical equation and ignore the volume of solid product.

5.69   A 41.41-mL sample of a 0.1250 M acid reacts with an excess of Na_2CO_3 to form 150.0 mL CO_2 at 646 mmHg and 27°C. If the acid is either HCl or H_2SO_4, which is it?

5.70   A 48.90-mL sample of a 0.2040 M acid reacts with an excess of Na_2CO_3 to form 125.0 mL CO_2 at 722 mmHg and 17°C. If the acid is either HCl or H_2SO_4, which is it?

5.71   If the rms speed of NH_3 molecules is found to be 0.510 km/s, what is the temperature (in degrees Celsius)?

5.72   If the rms speed of He atoms in the exosphere (highest region of the atmosphere) is 3.53×10^3 m/s, what is the temperature (in kelvins)?

5.73   Carbon monoxide, CO, and oxygen, O_2, react according to

$$2CO(g) + O_2(g) \longrightarrow 2CO_2(g)$$

Assuming that the reaction takes place and goes to completion, determine what substances remain and what their partial pressures are after the valve is opened in the apparatus represented in the accompanying figure. Also assume that the temperature is fixed at 300 K.

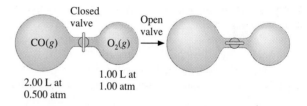

5.74   Suppose the apparatus shown in the figure accompanying Problem 5.73 contains H_2 at 0.500 atm in the left vessel separated from O_2 at 1.00 atm in the other vessel. The valve is then opened. If H_2 and O_2 react to give H_2O when the temperature is fixed at 533 K, what substances remain and what are their partial pressures after reaction?

Cumulative-Skills Problems

5.75   A sample of natural gas is 85.2% methane, CH_4, and 14.8% ethane, C_2H_6, by mass. What is the density of this mixture at 18°C and 771 mmHg?

5.76   A sample of a breathing mixture for divers contained 34.3% helium, He; 51.7% nitrogen, N_2; and 14.0% oxygen, O_2 (by mass). What is the density of this mixture at 22°C and 755 mmHg?

5.77   A sample of sodium peroxide, Na_2O_2, was reacted with an excess of water.

$$2Na_2O_2(s) + 2H_2O(l) \longrightarrow 4NaOH(aq) + O_2(g)$$

All of the sodium peroxide reacted, and the oxygen was collected over water at 21°C. The barometric pressure was 771 mmHg. The apparatus was similar to that shown in Figure 5.14. However, the level of water inside the tube was 25.0 cm above the level of water outside the tube. If the volume of gas in the tube is 31.0 mL, how many grams of sodium peroxide were in the sample?

5.78   A sample of zinc metal was reacted with an excess of hydrochloric acid.

$$Zn(s) + 2HCl(aq) \longrightarrow ZnCl_2(aq) + H_2(g)$$

All of the zinc reacted, and the hydrogen gas was collected over water at 18°C; the barometric pressure was 751 mmHg.

The apparatus was similar to that shown in Figure 5.14, but the level of water inside the tube was 31.0 cm above the level outside the tube. If the volume of gas in the tube is 22.1 mL, how many grams of zinc were there in the sample?

5.79   A mixture contained calcium carbonate, $CaCO_3$, and magnesium carbonate, $MgCO_3$. A sample of this mixture weighing 7.85 g was reacted with excess hydrochloric acid. The reactions are

$$CaCO_3(g) + 2HCl(aq) \longrightarrow CaCl_2(aq) + H_2O(l) + CO_2(g)$$
$$MgCO_3(s) + 2HCl(aq) \longrightarrow MgCl_2(aq) + H_2O(l) + CO_2(g)$$

If the sample reacted completely and produced 1.94 L of carbon dioxide, CO_2, at 25°C and 785 mmHg, what were the percentages of $CaCO_3$ and $MgCO_3$ in the mixture?

5.80   A mixture contained zinc sulfide, ZnS, and lead sulfide, PbS. A sample of the mixture weighing 6.12 g was reacted with an excess of hydrochloric acid. The reactions are

$$ZnS(s) + 2HCl(aq) \longrightarrow ZnCl_2(aq) + H_2S(g)$$
$$PbS(s) + 2HCl(aq) \longrightarrow PbCl_2(aq) + H_2S(g)$$

If the sample reacted completely and produced 1.049 L of hydrogen sulfide, H_2S, at 23°C and 745 mmHg, what were the percentages of ZnS and PbS in the mixture?

6

Ammonium dichromate decomposes with the release of heat in a fiery reaction.

Thermochemistry

Nearly all chemical reactions involve either the release or the absorption of heat, a form of energy. The burning of coal and gasoline are dramatic examples of chemical reactions in which a great deal of heat is released. Chemical reactions that absorb heat are usually less dramatic. If crystals of barium hydroxide octahydrate, $Ba(OH)_2 \cdot 8H_2O$, are mixed with crystals of ammonium nitrate, NH_4NO_3, in a flask, the solids form first a slush, then a liquid. Because the reaction mixture absorbs heat from the surroundings, the flask feels cool. It soon becomes so cold that if it is set in a puddle of water on a board, the water freezes (Figure 6.1).

In this chapter, we will be concerned with the quantity of heat released or absorbed in a chemical reaction. How do you measure the quantity of heat released or absorbed by a chemical reaction? To what extent can you relate the quantity of heat involved in a given reaction to the quantities of heat in other reactions?

Figure 6.1
A reaction that absorbs heat. Two crystalline substances, barium hydroxide octahydrate and an ammonium salt, are mixed thoroughly in a flask. Then the flask is set in a puddle of water on a board. In a couple of minutes, the flask and board are frozen solidly together. The board can then be inverted with the flask frozen to it.

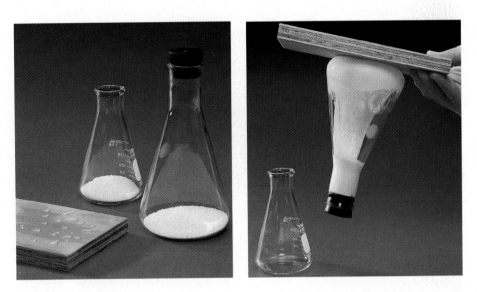

Understanding Heats of Reaction

Thermodynamics is the science of the relationships between heat and other forms of energy. *Thermochemistry* is one area of thermodynamics. It concerns the study of the quantity of heat absorbed or evolved by chemical reactions. Heat measurements provide data needed to determine whether a particular chemical reaction occurs and, if so, to what extent. ◄

▶ These questions concern chemical equilibrium and will be discussed in Chapter 19.

6.1   Energy and Its Units

We can define **energy** briefly as *the potential or capacity to move matter.* Energy exists in different forms that can be interconverted. You can see the relationship of a given form of energy to the motion of matter by following its interconversions into different forms.

Consider the interconversions of energy in a steam-driven electrical generator. A fuel is burned to heat water and generate steam. The steam expands against a piston (or turbine), which is connected to a drive shaft that turns an electrical coil in a magnetic field. Electricity is generated in the coil. The fuel contains chemical energy, which is converted to heat. Part of the heat is then converted to motion of the drive shaft, and this motion is converted to electrical energy. You could send the electricity into a lightbulb, converting electrical energy to heat energy and light energy. Photovoltaic cells can convert light back to electricity, which could be used to run a motor that can move matter (Figure 6.2).

Kinetic Energy; Units of Energy

▶ In the previous chapter, we used the symbol *u* for *average* molecular speed. Here *v* is the speed of an individual object or particle.

Kinetic energy is *the energy associated with an object by virtue of its motion.* An object of mass m and speed or velocity v has kinetic energy E_k equal to

$$E_k = \tfrac{1}{2}mv^2 \quad ◄$$

Figure 6.2
Conversion of light energy to kinetic energy. *Left:* The winning solar-powered vehicle from the 1999 Sunrayce, a race featuring vehicles designed and built by North American universities. The panel of photovoltaic cells is at the rear of the vehicle. *Right:* Solar-powered lawn mower. The panel of photovoltaic cells is on the handlebar.

Consider the kinetic energy of a person whose mass is 59.0 kg and whose speed is 26.8 m/s (60 miles per hour). So

$$E_k = \tfrac{1}{2} \times (59.0 \text{ kg}) \times (26.8 \text{ m/s})^2 = 2.12 \times 10^4 \text{ kg} \cdot \text{m}^2/\text{s}^2$$

Because you substituted SI units of mass and speed, you obtain the SI unit of energy, $kg \cdot m^2/s^2$, which is given the name **joule (J)** after the English physicist James Prescott Joule (1818–1889). You see that a person weighing 130 lb and traveling 60 miles per hour has a kinetic energy equal to 2.12×10^4 J, or 21.2 kJ (21.2 kilojoules).

The joule is an extremely small unit. To appreciate its size, note that the *watt* is a measure of the quantity of energy used per unit time and equals 1 joule per second. A kilowatt-hour, the unit by which electric energy is sold, equals 3600 kilowatt-seconds, or 3.6 million joules. A household might use something like 1000 kilowatt-hours (3.6 billion joules) of electricity in a month.

The **calorie (cal)** is *a non-SI unit of energy commonly used by chemists, origi-nally defined as the amount of energy required to raise the temperature of one gram of water by one degree Celsius.* In 1925 the calorie was defined in terms of the joule:

$$1 \text{ cal} = 4.184 \text{ J} \qquad \text{(exact definition)}$$

EXAMPLE 6.1

Calculating Kinetic Energy

A good pitcher can throw a baseball so that it travels between 60 and 80 miles per hour. A regulation baseball weighing 143 g (0.143 kg) travels 75 miles per hour (33.5 m/s). What is the kinetic energy of this baseball in joules? in calories?

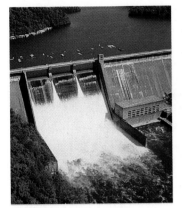

Figure 6.3
Potential energy and kinetic energy. Water at the top of the dam has potential energy. As the water falls over the dam, this potential energy is converted to kinetic energy.

SOLUTION

You substitute into the formula $E_k = \frac{1}{2}mv^2$ using SI units.

$$E_k = \frac{1}{2} \times 0.143 \text{ kg} \times (33.5 \text{ m/s})^2 = \textbf{80.2 J}$$

Using the conversion factor 1 cal/4.184 J we obtain

$$80.2 \text{ J} \times \frac{1 \text{ cal}}{4.184 \text{ J}} = \textbf{19.2 cal}$$

See Problems 6.23, 6.24, 6.25, and 6.26.

Potential Energy

Potential energy is *the energy an object has by virtue of its position in a field of force.* For example, water at the top of a dam has potential energy (in addition to whatever kinetic energy it may possess). You can calculate this potential energy of the water from the formula $E_p = mgh$. Here E_p is the potential energy of a quantity of water at the top of the dam, m is the mass of the water, g is the constant acceleration of gravity, and h is the height of the water measured from some standard level. The choice of this standard level is arbitrary, because only *differences* of potential energy are important in any physical situation. It is convenient to choose the standard level to be the surface of the earth.

The potential energy of the water at the top of the dam is converted to kinetic energy when the water falls to a lower level. As the water falls, it moves more quickly. The potential energy decreases and the kinetic energy increases (Figure 6.3).

Internal Energy

Consider the total energy of a quantity of water as it moves over the dam. This water as a whole has kinetic energy and potential energy. However, we know that water is made up of molecules, which are made up of smaller particles, electrons and nuclei. Each of these particles also has kinetic energy and potential energy. *The sum of the kinetic and potential energies of the particles making up a substance* is referred to as the **internal energy,** U, of the substance. Therefore, the total energy, E_{tot}, of a quantity of water equals the sum of its kinetic and potential energies as a whole ($E_k + E_p$) plus its internal energy.

$$E_{tot} = E_k + E_p + U$$

Normally when you study a substance in the laboratory, the substance is at rest in a vessel. Its kinetic energy as a whole is zero. Moreover, its potential energy as a whole is constant, and you can take it to be zero. In this case, the total energy of the substance equals its internal energy, U.

Law of Conservation of Energy

When water falls over a dam, potential energy is converted into kinetic energy. Some of the kinetic energy of the water may also be converted into random molecular motion—that is, into internal energy of the water. The total energy, E_{tot}, of the water, however, remains constant, equal to the sum of the kinetic energy, E_k, the potential energy, E_p, and the internal energy, U, of the water.

This result can be stated more generally as the **law of conservation of energy:** *energy may be converted from one form to another, but the total quantity of energy remains constant.*

CONCEPT CHECK 6.1

A solar-powered water pump has photovoltaic cells that protrude from top panels. These cells collect energy from sunlight, storing it momentarily in a battery, which later runs an electric motor that pumps water up to a storage tank on a hill. What energy conversions are involved in using sunlight to pump water into the storage tank?

6.2 Heat of Reaction

In the chapter opening, we mentioned chemical reactions (such as the burning of coal) that evolve, or release, heat. We also described a reaction that absorbs heat. Both types of reaction involve a *heat of reaction.* To understand this concept, you need to know what is meant by a thermodynamic *system* and its *surroundings* and to have a precise definition of the term *heat.*

Suppose that you are interested in studying the change of a thermodynamic property (such as internal energy) during a physical or chemical change. *The substance or mixture of substances under study in which a change occurs* is called the **thermodynamic system** (or simply **system**). The **surroundings** are *everything in the vicinity of the thermodynamic system* (Figure 6.4).

Definition of Heat

Heat is defined as *the energy that flows into or out of a system because of a difference in temperature between the thermodynamic system and its surroundings.* As long as a system and its surroundings are in thermal contact (that is, they are not thermally insulated from one another), energy (heat) flows between them to establish temperature equality, or *thermal equilibrium.* Heat flows from a region of higher temperature to one of lower temperature; once the temperatures become equal, heat flow stops. Note that once heat flows into a system, it appears in the system as an increase in its internal energy.

We can explain this flow of energy between two regions of different temperatures in terms of kinetic-molecular theory. Imagine two vessels in contact, each containing a gas, one gas hotter than the other. The average speed of molecules in the hotter gas is greater than that of molecules in the colder gas. But as the molecules in their random motions collide with the vessel walls, they lose energy to or gain energy from the walls. The faster molecules tend to slow down, while the slower molecules tend to speed up. Eventually, the average speeds of the molecules in the two vessels (and therefore the temperatures of the two gases) become equal. The net result is that energy has transferred through the vessel walls from the hot gas to the cold gas.

Heat is denoted by the symbol q. The algebraic sign of q is chosen to be positive if heat is absorbed by the system and negative if heat is evolved. The sign of q can be remembered this way: when heat is absorbed by a system, energy is *added* to it; q is assigned a positive quantity. On the other hand, when heat is evolved by a system, energy is *subtracted* from it; q is assigned a negative number.

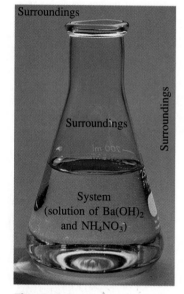

Figure 6.4
Illustration of a thermodynamic system. The *system* consists of the portion of the universe that we choose to study; in this case, it is a solution of $Ba(OH)_2$ and NH_4NO_3. Everything else, including the reaction vessel, constitutes the *surroundings.*

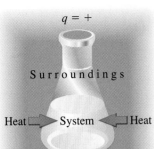

$q = +$

Surroundings

Heat ⟶ System ⟵ Heat

Endothermic

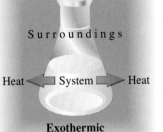

$q = -$

Surroundings

Heat ⟵ System ⟶ Heat

Exothermic

Heat of Reaction

The **heat of reaction** (at a given temperature) is *the value of q required to return a system to the given temperature at the completion of the reaction.*

Chemical reactions or physical changes are classified as exothermic or endothermic. An **exothermic process** is *a chemical reaction or a physical change in which heat is evolved.* An **endothermic process** is *a chemical reaction or a physical change in which heat is absorbed.* Experimentally, you note that in the exothermic reaction, the reaction flask initially warms; in the endothermic reaction, the reaction flask initially cools. We can summarize as follows:

Type of Reaction	Experimental Effect Noted	Result on System	Sign of q
Endothermic	Reaction vessel cools (heat is absorbed)	Energy added	$+$
Exothermic	Reaction vessel warms (heat is evolved)	Energy subtracted	$-$

Suppose that in an experiment 1 mol of methane burns in oxygen and evolves 890 kJ of heat: $CH_4(g) + 2O_2(g) \longrightarrow CO_2(g) + 2H_2O(l)$. The reaction is exothermic. Therefore the heat of reaction, q, is -890 kJ.

Or consider the reaction described in the chapter opening, in which crystals of barium hydroxide octahydrate, $Ba(OH)_2 \cdot 8H_2O$, react with crystals of ammonium nitrate, NH_4NO_3.

$$Ba(OH)_2 \cdot 8H_2O(s) + 2NH_4NO_3(s) \longrightarrow 2NH_3(g) + 10H_2O(l) + Ba(NO_3)_2(aq)$$

When 1 mol $Ba(OH)_2 \cdot 8H_2O$ reacts with 2 mol NH_4NO_3, the reaction mixture absorbs 170.4 kJ of heat. The reaction is endothermic. Therefore the heat of reaction, q, is $+170.4$ kJ.

6.3 Enthalpy and Enthalpy Change

The heat absorbed or evolved by a reaction depends on the conditions under which the reaction occurs. Usually, a reaction takes place in a vessel open to the atmosphere, and therefore at the constant pressure of the atmosphere. We will assume that this is the case and write the heat of reaction as q_p, the subscript p indicating that the process occurs at constant pressure.

Enthalpy

There is a property of substances called enthalpy (en′-thal-py) that is related to the heat of reaction q_p. **Enthalpy** (denoted H) is *an extensive property of a substance that can be used to obtain the heat absorbed or evolved in a chemical reaction.* (An extensive property is a property that depends on the amount of substance. Other examples of extensive properties are mass and volume.)

Enthalpy is a state function. A **state function** is *a property of a system that depends only on its present state, which is determined by variables such as temperature and pressure, and is independent of any previous history of the system.* This means that a change in enthalpy does not depend on how the change was made, but only on the initial state and final state of the system. The camping-hiking analogy in

Figure 6.5
An analogy to illustrate a state function. The two campsites differ by 3600 ft in altitude. This difference in altitude is independent of the path taken between the campsites. The distance traveled and the things encountered along the way, however, do depend on the path taken. Altitude here is analogous to a thermodynamic state function.

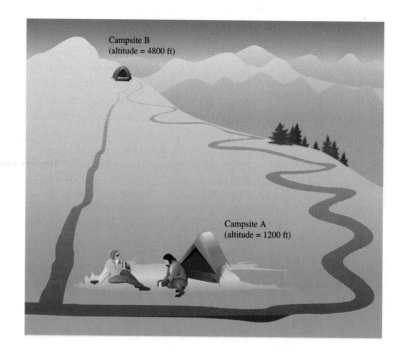

Campsite B (altitude = 4800 ft)

Campsite A (altitude = 1200 ft)

PE related to
Enthalpy
same thing

Figure 6.5 may clarify the point. The difference in altitude of the two campsites is independent of the route you take from one to the other. Altitude is analogous to a state function, whereas distance traveled is not.

Enthalpy of Reaction

The change in enthalpy for a reaction at a given temperature and pressure (called the **enthalpy of reaction**) is obtained by subtracting the enthalpy of the reactants from the enthalpy of the products. We will use the symbol Δ (meaning "change in") and write the change in enthalpy as ΔH. Thus $\Delta H = H_{final} - H_{initial}$. Since you start from reactants and end with products, the enthalpy of reaction is

$$\Delta H = H(\text{products}) - H(\text{reactants})$$

Because H is a state function, it depends only on the initial state (the reactants) and the final state (the products).

The key relation in this chapter is that between enthalpy change and heat of reaction:

$$\Delta H = q_p$$

The enthalpy of reaction equals the heat of reaction at constant pressure.

Consider the reaction at 25°C of sodium and water, carried out in a beaker open to the atmosphere at 1.00 atm pressure.

$$2Na(s) + 2H_2O(l) \longrightarrow 2NaOH(aq) + H_2(g)$$

The metal and water react vigorously and heat evolves. Experiment shows that 2 mol of sodium reacts with 2 mol of water to evolve 367.5 kJ of heat and you write $q_p = -367.5$ kJ. Therefore, the change of enthalpy for the reaction is $\Delta H = -367.5$ kJ. Figure 6.6 shows an *enthalpy diagram* for this reaction.

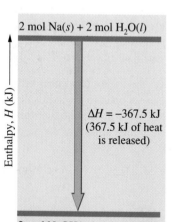

2 mol Na(s) + 2 mol H₂O(l)

Enthalpy, H (kJ)

$\Delta H = -367.5$ kJ
(367.5 kJ of heat is released)

2 mol NaOH(aq) + 1 mol H₂(g)

Figure 6.6
An enthalpy diagram.
When 2 mol Na(s) and 2 mol H₂O(l) react to give 2 mol NaOH(aq) and 1 mol H₂(g), 367.5 kJ of heat is released, and the enthalpy of the system decreases by 367.5 kJ.

Enthalpy and Internal Energy

In the preceding discussion, we noted that the enthalpy change equals the heat of reaction at constant pressure. It is useful at this point to note briefly the relationship of enthalpy to internal energy.

The enthalpy, H, is defined precisely as the internal energy, U, plus pressure, P, times volume, V.

$$H = U + PV$$

We will label the initial quantities (those for reactants) with a subscript i and the final quantities (those for products) with a subscript f. Then

$$\Delta H = H_f - H_i = (U_f + PV_f) - (U_i + PV_i)$$

Collecting the internal-energy terms and the pressure-volume terms, you can rewrite this as

$$\Delta H = (U_f - U_i) + P(V_f - V_i) = \Delta U + P\Delta V$$

You write ΔU for $U_f - U_i$ and ΔV for $V_f - V_i$ and rearrange this as follows:

$$\Delta U = \Delta H - P\Delta V$$

The equation says that the internal energy of the system changes in two ways. It changes because energy leaves or enters the system as heat (ΔH), and it changes because the system increases or decreases in volume against the constant pressure of the atmosphere (which requires energy $-P\Delta V$).

Consider a specific reaction. When 2 mol of sodium and 2 mol of water react in a beaker, 1 mol of hydrogen forms and heat evolves.

Because hydrogen gas forms during the reaction, the volume of the system increases. To expand, the system must push back the atmosphere, and this requires energy equal to the pressure–volume work. It may be easier to see this pressure–volume work if you replace the constant pressure of the atmosphere by an equivalent pressure from a piston-and-weight assembly, as in Figure 6.7. When hydrogen is released during the reaction, it pushes upward on the piston and raises the weight. It requires energy to lift a weight upward in a gravitational field. If you calculate this pressure–volume work at 25°C and 1.00 atm pressure, you find that it is $-P\Delta V = -2.5$ kJ.

In the sodium–water reaction, the internal energy changes by -367.5 kJ because heat evolves and changes by -2.5 kJ because pressure–volume work is done. The total change of internal energy is

$$\Delta U = \Delta H - P\Delta V = -367.5 \text{ kJ} - 2.5 \text{ kJ} = -370.0 \text{ kJ}$$

As you can see, ΔU does not differ a great deal from ΔH. This is the case in most reactions.

Figure 6.7
Pressure–volume work.
In this experiment, we replace the pressure of the atmosphere by a piston-and-weight assembly of equal pressure. As sodium metal reacts with water, the hydrogen gas evolved pushes the piston and weight upward (compare *before* and *after*). It requires work to raise the piston and weight upward in a gravitational field.

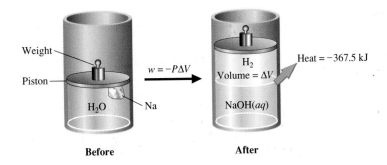

6.4 Thermochemical Equations

A **thermochemical equation** is *the chemical equation for a reaction (including phase labels) in which the equation is given a molar interpretation, and the enthalpy of reaction for these molar amounts is written directly after the equation.* For the reaction of sodium and water, you would write

$$2Na(s) + 2H_2O(l) \longrightarrow 2NaOH(aq) + H_2(g); \Delta H = -367.5 \text{ kJ}$$

This equation says that 2 mol of sodium reacts with 2 mol of water to produce 2 mol of sodium hydroxide and 1 mol of hydrogen gas, and 367.5 kJ of heat evolves.

Note that the thermochemical equation includes phase labels because ΔH depends on the phase of the substances. Consider the reaction of hydrogen and oxygen to produce water. If the product is water vapor, the heat released is 483.7 kJ.

$$2H_2(g) + O_2(g) \longrightarrow 2H_2O(g); \Delta H = -483.7 \text{ kJ}$$

On the other hand, if the product is liquid water, the heat released is 571.7 kJ.

$$2H_2(g) + O_2(g) \longrightarrow 2H_2O(l); \Delta H = -571.7 \text{ kJ}$$

► It takes 44.0 kJ of heat to vaporize 1 mol of liquid water at 25°C.

In this case, additional heat is released when water vapor condenses to liquid. ◄

EXAMPLE 6.2

Writing Thermochemical Equations

Aqueous sodium hydrogen carbonate solution (baking soda solution) reacts with hydrochloric acid to produce aqueous sodium chloride, water, and carbon dioxide gas. The reaction absorbs 11.8 kJ of heat at constant pressure for each mole of sodium hydrogen carbonate. Write the thermochemical equation for the reaction.

SOLUTION

You first write the balanced chemical equation.

$$NaHCO_3(aq) + HCl(aq) \longrightarrow NaCl(aq) + H_2O(l) + CO_2(g)$$

Because the reaction absorbs heat, the enthalpy of reaction for molar amounts of this equation is +11.8 kJ. The thermochemical equation is

$$\textbf{NaHCO}_3\textbf{(aq) + HCl(aq)} \longrightarrow \textbf{NaCl(aq) + H}_2\textbf{O(l) + CO}_2\textbf{(g); } \Delta H = +11.8 \text{ kJ}$$

See Problems 6.29 and 6.30.

Two important rules for manipulating thermochemical equations are:

1. When a thermochemical equation is multiplied by any factor, the value of ΔH for the new equation is obtained by multiplying the value of ΔH in the original equation by that same factor.

2. When a chemical equation is reversed, the value of ΔH is reversed in sign.

Consider the thermochemical equation for the synthesis of ammonia.

$$N_2(g) + 3H_2(g) \longrightarrow 2NH_3(g); \Delta H = -91.8 \text{ kJ}$$

Suppose you want the thermochemical equation to show what happens when twice as many moles of nitrogen and hydrogen react to produce ammonia. Doubling the previous equation, you obtain

$$2N_2(g) + 6H_2(g) \longrightarrow 4NH_3(g); \Delta H = -184 \text{ kJ}$$

Suppose you reverse the first equation we wrote for the synthesis of ammonia. Then the thermochemical equation is

$$2NH_3(g) \longrightarrow N_2(g) + 3H_2(g); \Delta H = +91.8 \text{ kJ}$$

If you want to express this in terms of 1 mol of ammonia, you simply multiply this equation by a factor of $\frac{1}{2}$. ◀

▶ If you use the molar interpretation of a chemical equation, there is nothing unreasonable about using such coefficients as $\frac{1}{2}$ and $\frac{3}{2}$.

EXAMPLE 6.3

Manipulating Thermochemical Equations

When 2 mol $H_2(g)$ and 1 mol $O_2(g)$ react to give liquid water, 572 kJ of heat evolves.

$$2H_2(g) + O_2(g) \longrightarrow 2H_2O(l); \Delta H = -572 \text{ kJ}$$

Write this equation for 1 mol of liquid water. Give the reverse equation, in which 1 mol of liquid water dissociates into hydrogen and oxygen.

SOLUTION

You multiply the coefficients and ΔH by $\frac{1}{2}$:

$$H_2(g) + \tfrac{1}{2}O_2(g) \longrightarrow H_2O(l); \Delta H = -286 \text{ kJ}$$

Reversing the equation, you get

$$H_2O(l) \longrightarrow H_2(g) + \tfrac{1}{2}O_2(g); \Delta H = +286 \text{ kJ}$$

See Problems 6.31 and 6.32.

CONCEPT CHECK 6.2

Natural gas consists primarily of methane, CH_4. It is used in a process called *steam reforming* to prepare a gaseous mixture of carbon monoxide and hydrogen for industrial use.

$$CH_4(g) + H_2O(g) \longrightarrow CO(g) + 3H_2(g); \Delta H = 206 \text{ kJ}$$

The reverse reaction, the reaction of carbon monoxide and hydrogen, has been explored as a way to prepare methane (synthetic natural gas). Which of the following are exothermic? Of these, which one is the most exothermic?

a. $CH_4(g) + H_2O(g) \longrightarrow CO(g) + 3H_2(g)$

b. $2CH_4(g) + 2H_2O(g) \longrightarrow 2CO(g) + 6H_2(g)$

c. $CO(g) + 3H_2(g) \longrightarrow CH_4(g) + H_2O(g)$

d. $2CO(g) + 6H_2(g) \longrightarrow 2CH_4(g) + 2H_2O(g)$

Lucifers and Other Matches

Samuel Jones, an Englishman, patented one of the first kinds of matches in 1828. It consisted of a glass bead containing sulfuric acid surrounded by a coating of sugar with an oxidizing agent. You ignited the match by breaking the bead using a pair of pliers or, if you were more daring, your teeth. This action released the acid, which ignited an exothermic reaction in the surrounding combustible materials.

Later Jones began to market a friction match discovered, but not patented, by John Walker. Walker, who had been experimenting with explosives, discovered this match one day when he tried to remove a small glob of a dried mixture from a stick. He rubbed the stick on the floor and was surprised when it burst into flame. Jones called his matches "Lucifers." They were well named; when lighted, they gave off a shower of sparks and smoky fumes with the acrid odor of sulfur dioxide. Jones had every box inscribed with the warning "Persons whose lungs are delicate should by no means use Lucifers."

A few years later, a Frenchman, Charles Sauria, invented the white phosphorus match, which became an immediate success. When rubbed on a rough surface, the match lighted easily, without hazardous sparks, and smelled better than Lucifers. The match head contained white phosphorus, an oxidizing agent, and glue. White

phosphorus is a yellowish-white, waxy substance, often sold in the form of sticks looking something like fat crayons. Unlike crayons, though, white phosphorus ignites spontaneously in air (so it is generally stored under water). The glue in the match mixture had two purposes: it protected the white phosphorus from air, and it held the match mixture firmly together. The white phosphorus match had one serious drawback. White phosphorus is quite poisonous. Workers in match factories often began to show the agonizing symptoms of "phossy jaw," from white phosphorus poisoning, in which the jawbone disintegrates. The manufacture of white phosphorus matches was outlawed in the early 1900s.

The head of the "strike-anywhere" match, which you can buy today at any grocery store, contains the relatively nontoxic tetraphosphorus trisulfide, P_4S_3, and potassium chlorate, $KClO_3$ (Figure 6.8). By rubbing the match head against a surface, you create enough heat by friction to ignite the match material. Tetraphosphorus trisulfide then burns in air in a very exothermic reaction.

$$P_4S_3(s) + 8O_2(g) \longrightarrow P_4O_{10}(s) + 3SO_2(g);$$
$$\Delta H = -3677 \text{ kJ}$$

Safety matches have a head containing mostly an oxidizing agent and require a striking surface containing nonpoisonous red phosphorus.

Figure 6.8
Strike-anywhere matches. *Left:* The head of the match contains tetraphosphorus trisulfide and potassium chlorate. *Right:* The substances in the match head react when frictional heat ignites the mixture.

6.5   Applying Stoichiometry to Heats of Reaction

Consider the reaction of methane, CH_4, burning in oxygen at constant pressure. How much heat could you obtain from 10.0 g of methane? The thermochemical equation is

$$CH_4(g) + 2O_2(g) \longrightarrow CO_2(g) + 2H_2O(l); \Delta H = -890.3 \text{ kJ}$$

The calculation involves the following conversions:

Grams of CH_4 $\longrightarrow$ moles of CH_4 $\longrightarrow$ kilojoules of heat

$$10.0 \text{ g } CH_4 \times \frac{1 \text{ mol } CH_4}{16.0 \text{ g } CH_4} \times \frac{-890.3 \text{ kJ}}{1 \text{ mol } CH_4} = -556 \text{ kJ} \blacktriangleleft$$

▶ If you wanted the result in kilocalories, you could convert this answer as follows:

$$556 \text{ kJ} \times \frac{1 \text{ kcal}}{4.184 \text{ kJ}} = 133 \text{ kcal}$$

EXAMPLE 6.4

Calculating the Heat of Reaction from the Stoichiometry

How much heat is evolved when 9.07×10^5 g of ammonia is produced from its elements? (Assume that the reaction occurs at constant pressure.)

$$N_2(g) + 3H_2(g) \longrightarrow 2NH_3(g); \Delta H = -91.8 \text{ kJ}$$

PROBLEM STRATEGY

The calculation involves converting grams of NH_3 to moles of NH_3 and then to kilojoules of heat.

Grams of NH_3 $\longrightarrow$ moles of NH_3 $\longrightarrow$ kilojoules of heat

SOLUTION

$$9.07 \times 10^5 \text{ g } NH_3 \times \frac{1 \text{ mol } NH_3}{17.0 \text{ g } NH_3} \times \frac{-91.8 \text{ kJ}}{2 \text{ mol } NH_3} = -2.45 \times 10^6 \text{ kJ}$$

See Problems 6.33 and 6.34.

6.6   Measuring Heats of Reaction

Now that you have a firm idea of what heats of reaction are, how would you measure them?

Heat Capacity and Specific Heat

The **heat capacity** (C) of a sample of substance is *the quantity of heat needed to raise the temperature of the sample of substance one degree Celsius (or one kelvin).* Changing the temperature of the sample from an initial temperature t_i to a final temperature t_f requires heat equal to

▶ The heat capacity will depend on whether the process is constant-pressure or constant-volume. We will assume a constant-pressure process unless otherwise stated.

$$q = C\Delta t$$

where Δt is the change of temperature and equals $t_f - t_i$. ◀

Table 6.1
Specific Heats and Molar Heat Capacities of Some Substances*

Substance	Specific Heat J/(g · °C)	Molar Heat Capacity J/(mol · °C)
Aluminum, Al	0.901	24.3
Copper, Cu	0.384	24.4
Ethanol, C_2H_5OH	2.43	112.2
Iron, Fe	0.449	25.1
Water, H_2O	4.18	75.3

*Values are for 25°C.

Suppose a piece of iron requires 6.70 J of heat to raise the temperature by one degree Celsius. Its heat capacity is therefore 6.70 J/°C. The quantity of heat required to raise the temperature of the piece of iron from 25.0°C to 35.0°C is

$$q = C\Delta t = (6.70 \text{ J/°C}) \times (35.0°C - 25.0°C) = 67.0 \text{ J}$$

Heat capacity is directly proportional to the amount of substance. The *molar heat capacity* of a substance is its heat capacity for one mole of substance.

Heat capacities are also compared for one-gram amounts of substances. The **specific heat capacity** (or simply **specific heat**) is *the quantity of heat required to raise the temperature of one gram of a substance by one degree Celsius (or one kelvin) at constant pressure*. To find the heat q required to raise the temperature of a sample, you multiply the specific heat of the substance, s, by the mass in grams, m, and the temperature change, Δt.

$$q = s \times m \times \Delta t$$

The specific heats and molar heat capacities of a few substances are listed in Table 6.1. Example 6.5 illustrates the use of the preceding equation.

EXAMPLE 6.5
Relating Heat and Specific Heat

Calculate the heat absorbed by 15.0 g of water to raise its temperature from 20.0°C to 50.0°C (at constant pressure). The specific heat of water is 4.18 J/(g · °C).

SOLUTION
You substitute into the equation

$$q = s \times m \times \Delta t$$

The temperature change is

$$\Delta t = t_f - t_i = 50.0°C - 20.0°C = +30.0°C$$

Therefore,

$$q = 4.18 \text{ J/(g} \cdot \text{°C)} \times 15.0 \text{ g} \times (+30.0\text{°C}) = \mathbf{1.88 \times 10^3 \text{ J}}$$

See Problems 6.37 and 6.38.

Measurement of Heat of Reaction

You measure the heat of reaction in a **calorimeter,** *a device used to measure the heat absorbed or evolved during a physical or chemical change.* The device can be as simple as the apparatus sketched in Figure 6.9, which consists of an insulated container (for example, a pair of polystyrene coffee cups) with a thermometer. The coffee-cup calorimeter is a constant-pressure calorimeter where the heat can be directly related to the enthalpy change, ΔH.

For reactions involving gases, a *bomb calorimeter* is generally used (Figure 6.10). To measure the heat released when graphite burns in oxygen, a sample of graphite is placed in a small cup in the calorimeter. The graphite is surrounded by oxygen, and the graphite and oxygen are sealed in a steel vessel. An electrical circuit is activated to start the burning of the graphite. The bomb is surrounded by water in an insulated container, and the heat of reaction is calculated from the temperature change of the calorimeter.

Because the reaction occurs in a bomb calorimeter, the volume remains constant, and under these conditions the heat of reaction does not in general equal ΔH; a small correction is usually needed. This correction is negligible when the reaction does not involve gases or when the number of moles of reactant gas equals the number of moles of product gas. ◀

▶ At the end of Section 6.3, we noted that $\Delta H = \Delta U + P\Delta V$. The heat at constant volume equals ΔU. To obtain ΔH, you must add a correction, $P\Delta V$. The change in volume is significant only if there are changes in gas volumes. From the ideal gas law, the molar volume of a gas is RT/P. If the moles of gas that have reacted is n_i and the moles of gas produced is n_f, then $P\Delta V$ equals essentially $RT \times (n_f - n_i)$, or $RT\Delta n$, where Δn refers to the change of moles of gas. To obtain the correction in joules, we use $R = 8.31 \text{ J/(K} \cdot \text{mol)}$. Thus, RT at 25°C (298 K) is $8.31 \text{ J/(K} \cdot \text{mol)} \times 298 \text{ K} = 2.48 \times 10^3 \text{ J/mol}$. The final result is $\Delta H = \Delta U + (2.48 \text{ kJ}) \times \Delta n$.

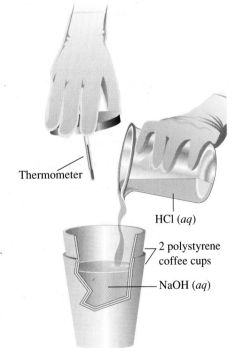

Figure 6.9
A simple coffee-cup calorimeter. This calorimeter is made of two nested polystyrene coffee cups. The outer cup helps to insulate the reaction mixture from its surroundings. After the reactants are added to the inner cup, the calorimeter is covered to reduce heat loss by evaporation and convection. The heat of reaction is determined by noting the temperature rise or fall.

Thermometer

HCl (*aq*)

2 polystyrene coffee cups

NaOH (*aq*)

Figure 6.10
A bomb calorimeter. The heat of a reaction involving a gas (here the heat of combustion of graphite) is conveniently determined in a sealed vessel called a bomb. The reaction is started by an ignition coil running through the graphite sample.

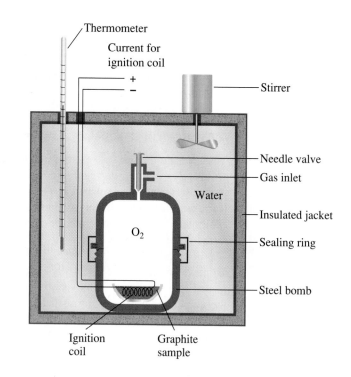

EXAMPLE 6.6

Calculating ΔH from Calorimetric Data

Suppose 0.562 g of graphite is placed in a calorimeter with an excess of oxygen at 25.00°C and 1 atm pressure (Figure 6.10). The equation is

$$C(graphite) + O_2(g) \longrightarrow CO_2(g)$$

The calorimeter temperature rises from 25.00°C to 25.89°C. The heat capacity of the calorimeter and its contents was determined in a separate experiment to be 20.7 kJ/°C. What is the heat of reaction at 25.00°C and 1 atm pressure?

PROBLEM STRATEGY

The heat released by the reaction is absorbed by the calorimeter and its contents. Let q_{rxn} be the quantity of heat from the reaction mixture, and let C_{cal} be the heat capacity of the calorimeter and contents. The quantity of heat absorbed by the calorimeter is $C_{cal}\Delta t$. This will have the same magnitude as q_{rxn}, but the opposite sign: $q_{rxn} = -C_{cal}\Delta t$.

SOLUTION

The heat from the graphite sample is

$$q_{rxn} = -C_{cal}\Delta t = -20.7 \text{ kJ/°C} \times (25.89°C - 25.00°C)$$
$$= -20.7 \text{ kJ/°C} \times 0.89°C = -18.4 \text{ kJ}$$

The factor to convert grams C to kJ heat is -18.4 kJ/0.562 g C.

The conversion of 1 mol C to kJ heat for 1 mol (ΔH) is

$$1 \ \text{mol C} \times \frac{12 \ \text{g C}}{1 \ \text{mol C}} \times \frac{-18.4 \ \text{kJ}}{0.562 \ \text{g C}} = -3.9 \times 10^2 \ \text{kJ}$$

(The final answer has been rounded to two significant figures.) When 1 mol of carbon burns, 3.9×10^2 kJ of heat is released. You can summarize the results by the thermochemical equation:

$$\text{C(graphite)} + \text{O}_2(g) \longrightarrow \text{CO}_2(g); \Delta H = -3.9 \times 10^2 \ \text{kJ}$$

See Problems 6.39, 6.40, 6.41, and 6.42.

Using Heats of Reaction

Now we want to find how heats of reaction can be used. We will see that the ΔH for one reaction can be obtained from the ΔH's of other reactions.

6.7   Hess's Law

The enthalpy change for a chemical reaction is independent of the path by which the products are obtained. In 1840, the Russian chemist Germain Henri Hess, a professor at the University of St. Petersburg, discovered this result by experiment. **Hess's law of heat summation** states that *for a chemical equation that can be written as the sum of two or more steps, the enthalpy change for the overall equation equals the sum of the enthalpy changes for the individual steps.*

Suppose you would like to find the enthalpy change for the combustion of graphite (carbon) to carbon monoxide. The direct determination is very difficult, because once carbon monoxide forms it reacts further with oxygen to yield carbon dioxide. If you do the experiment in a limited quantity of oxygen, you obtain a mixture of carbon monoxide and carbon dioxide. How can you obtain the enthalpy change for the preparation of pure carbon monoxide from graphite and oxygen?

The answer is to apply Hess's law. To do this, imagine that the combustion of graphite to carbon monoxide takes place in two separate steps:

$$2\text{C(graphite)} + 2\text{O}_2(g) \longrightarrow 2\text{CO}_2(g) \qquad \text{(first step)}$$
$$2\text{CO}_2(g) \longrightarrow 2\text{CO}(g) + \text{O}_2(g) \qquad \text{(second step)}$$

The net result is the combustion of 2 mol of graphite in 1 mol of oxygen to give 2 mol of carbon monoxide.

$$2\text{C(graphite)} + \cancel{2}\text{O}_2(g) \longrightarrow \cancel{2\text{CO}_2(g)}$$
$$\underline{\cancel{2\text{CO}_2(g)} \longrightarrow 2\text{CO}(g) + \cancel{\text{O}_2(g)}}$$
$$2\text{C(graphite)} + \text{O}_2(g) \longrightarrow 2\text{CO}(g)$$

According to Hess's law, the enthalpy change for the overall equation (which is the equation you want) equals the sum of the enthalpy changes for the two steps. Now you need to determine the enthalpy changes for the separate steps.

You can determine the enthalpy change for the first step by simply burning graphite in an excess of oxygen, as described in Example 6.6. The result is $\Delta H = -393.5$ kJ per mole of CO_2 formed. For 2 mol CO_2, you multiply by 2.

$$2\text{C(graphite)} + 2\text{O}_2(g) \longrightarrow 2\text{CO}_2(g); \Delta H = (-393.5 \ \text{kJ}) \times (2)$$

Figure 6.11
Enthalpy diagram illustrating Hess's law.
The diagram shows two different ways to go from graphite and oxygen (reactants) to carbon monoxide (products). Going by way of reactions 1 and 2 is equivalent to the direct reaction 3.

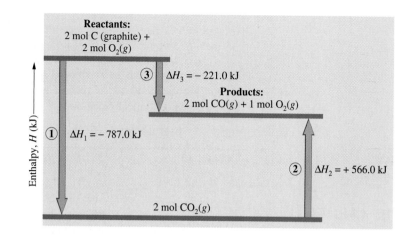

The second step, the decomposition of carbon dioxide, is not an easy experiment. However, the reverse of this decomposition is simply the combustion of carbon monoxide. You could determine the ΔH for that combustion by burning carbon monoxide in an excess of oxygen. The experiment is similar to the one for the combustion of graphite to carbon dioxide.

$$2CO(g) + O_2(g) \longrightarrow 2CO_2(g); \Delta H = -566.0 \text{ kJ}$$

From the properties of thermochemical equations (Section 6.4), you know that the enthalpy change for the reverse reaction is simply (-1) times the original reaction.

$$2CO_2(g) \longrightarrow 2CO(g) + O_2(g); \Delta H = (-566.0 \text{ kJ}) \times (-1)$$

If you now add these two steps and add their enthalpy changes, you obtain the chemical equation and the enthalpy change for the combustion of carbon monoxide, which is what you wanted.

$2C(\text{graphite}) + 2O_2(g) \longrightarrow 2CO_2(g)$	$\Delta H_1 = (-393.5 \text{ kJ}) \times (2)$
$2CO_2(g) \qquad\qquad \longrightarrow 2CO(g) + O_2(g)$	$\Delta H_2 = (-566.0 \text{ kJ}) \times (-1)$
$2C(\text{graphite}) + O_2(g) \longrightarrow 2CO(g)$	$\Delta H_3 = -221.0 \text{ kJ}$

You see that the combustion of 2 mol of graphite to give 2 mol of carbon monoxide has an enthalpy change of -221.0 kJ. Figure 6.11 gives an enthalpy diagram showing the relationship among the enthalpy changes for this calculation.

The next example gives another illustration of how Hess's law can be used to calculate the enthalpy change for a reaction from the enthalpy values for other reactions.

EXAMPLE 6.7

Applying Hess's Law

What is the enthalpy of reaction, ΔH, for the formation of tungsten carbide, WC, from the elements? (Tungsten carbide is very hard and is used to make cutting tools and rock drills.)

$$W(s) + C(\text{graphite}) \longrightarrow WC(s)$$

The enthalpy change for this reaction is difficult to measure directly, because the reaction occurs at 1400°C. However, the heats of combustion of the elements and of tungsten carbide can be measured easily:

$$2W(s) + 3O_2(g) \longrightarrow 2WO_3(s); \Delta H = -1680.6 \text{ kJ} \qquad \textbf{(1)}$$

$$C(\text{graphite}) + O_2(g) \longrightarrow CO_2(g); \Delta H = -393.5 \text{ kJ} \qquad \textbf{(2)}$$

$$2WC(s) + 5O_2(g) \longrightarrow 2WO_3(s) + 2CO_2(g); \Delta H = -2391.6 \text{ kJ} \qquad \textbf{(3)}$$

PROBLEM STRATEGY

You need to multiply Equations 1, 2, and 3 by factors so that when you add the three equations you obtain the desired equation for the formation of WC(s). To obtain these factors, compare Equations 1, 2, and 3 in turn with the desired equation. For instance, note that Equation 1 has 2W(s) on the left side, whereas the desired equation has W(s). Therefore, you multiply Equation 1 (and its ΔH) by $\frac{1}{2}$.

SOLUTION

Multiplying Equation 1 by $\frac{1}{2}$, you obtain

$$W(s) + \tfrac{3}{2}O_2(g) \longrightarrow WO_3(s); \Delta H = \tfrac{1}{2} \times (-1680.6 \text{ kJ}) = -840.3 \text{ kJ}$$

Compare Equation 2 with the desired equation. Both have C(graphite) on the left side; therefore, you leave Equation 2 as it is. Now, compare Equation 3 with the desired equation. Equation 3 has 2WC(s) on the left side, whereas the desired equation has WC(s) on the right side. Hence, you reverse Equation 3 and multiply it (and its ΔH) by $\frac{1}{2}$.

$$WO_3(s) + CO_2(g) \longrightarrow WC(s) + \tfrac{5}{2}O_2(g); \Delta H = -\tfrac{1}{2} \times (-2391.6 \text{ kJ}) = 1195.8 \text{ kJ}$$

Note that the ΔH is obtained by multiplying the value for Equation 3 by $-\frac{1}{2}$. Now these three equations and the corresponding ΔH's are added together.

$W(s) + \tfrac{3}{2}O_2(g) \longrightarrow WO_3(s)$	$\Delta H = -840.3 \text{ kJ}$
$C(\text{graphite}) + O_2(g) \longrightarrow CO_2(g)$	$\Delta H = -393.5 \text{ kJ}$
$WO_3(s) + CO_2(g) \longrightarrow WC(s) + \tfrac{5}{2}O_2(g)$	$\Delta H = 1195.8 \text{ kJ}$
$W(s) + C(\text{graphite}) \longrightarrow WC(s)$	$\Delta H = -38.0 \text{ kJ}$

See Problems 6.43 and 6.44.

CONCEPT CHECK 6.3

The heat of fusion (also called heat of melting), ΔH_{fus}, of ice is the enthalpy change for

$$H_2O(s) \longrightarrow H_2O(l); \Delta H_{fus}$$

Similarly, the heat of vaporization, ΔH_{vap}, of liquid water is the enthalpy change for

$$H_2O(l) \longrightarrow H_2O(g); \Delta H_{vap}$$

How is the heat of sublimation, ΔH_{sub}, the enthalpy change for the reaction

$$H_2O(s) \longrightarrow H_2O(g); \Delta H_{sub}$$

related to ΔH_{fus} and ΔH_{vap}?

6.8   Standard Enthalpies of Formation

The term **standard state** refers to *the standard thermodynamic conditions chosen for substances when listing or comparing thermodynamic data: 1 atm pressure and the specified temperature (usually 25°C).* ◄ These standard conditions are indicated by a superscript degree sign (°). The enthalpy change for a reaction in which reactants in their standard states yield products in their standard states is denoted $\Delta H°$.

As we will show, it is sufficient to tabulate just the enthalpy changes for formation reactions—that is, for reactions in which compounds are formed from their elements. To specify the formation reaction precisely, however, we must specify the exact form of each element.

Some elements exist in the same physical state in two or more distinct forms. For example, oxygen occurs both as dioxygen and as ozone with O_3 molecules. Dioxygen gas is odorless; ozone gas has a characteristic pungent odor. Solid carbon has two principal crystalline forms: graphite and diamond. Graphite is a soft, black, crystalline substance; diamond is a hard, usually colorless crystal. The elements oxygen and carbon are said to exist in different allotropic forms. An **allotrope** is *one of two or more distinct forms of an element in the same physical state* (Figure 6.12).

The **reference form** of an element for the purpose of specifying the formation reaction is usually *the stablest form (physical state and allotrope) of the element under standard thermodynamic conditions.* ◄

Table 6.2 lists standard enthalpies of formation of substances (a longer table is given in Appendix B). The **standard enthalpy of formation** (also called the **standard heat of formation**) of a substance, denoted $\Delta H_f°$, is *the enthalpy change for the formation of one mole of the substance in its standard state from its elements in their reference form and in their standard states.*

Consider the standard enthalpy of formation of liquid water. The standard enthalpy change is −285.8 kJ per mole of H_2O. Therefore, the thermochemical equation is

$$H_2(g) + \tfrac{1}{2}O_2(g) \longrightarrow H_2O(l); \ \Delta H_f° = -285.8 \text{ kJ}$$

The values of standard enthalpies of formation listed in Table 6.2 and in other tables are determined by direct measurement in some cases and by applying Hess's law in others.

Note that the standard enthalpy of formation of an element will depend on the form of the element. For example, the $\Delta H_f°$ for diamond equals the enthalpy change

► The International Union of Pure and Applied Chemistry (IUPAC) recommends that the standard pressure be 1 bar (1×10^5 Pa). Thermodynamic tables are becoming available for 1 bar pressure, and in the future such tables will probably replace those for 1 atm.

► Although the reference form is usually the stablest allotrope of an element, the choice is essentially arbitrary as long as one is consistent.

Figure 6.12
Allotropes of sulfur.
Left: An evaporating dish contains rhombic sulfur, the stable form of the element at room temperature. *Right:* When this sulfur is melted, then cooled, it forms long needles of monoclinic sulfur, another allotrope. At room temperature, monoclinic sulfur will slowly change back to rhombic sulfur. Both forms contain the molecule S_8, depicted by the model.

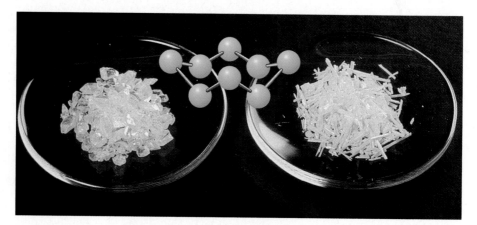

Table 6.2
Standard Enthalpies of Formation (at 25°C)*

Formula	ΔH_f° (kJ/mol)	Formula	ΔH_f° (kJ/mol)
$e^-(g)$	0	*Silicon*	
Hydrogen		$Si(s)$	0
$H^+(aq)$	0	$SiO_2(s)$	−910.9
$H(g)$	218.0	$SiF_4(g)$	−1548
$H_2(g)$	0	*Lead*	
Sodium		$Pb(s)$	0
$Na^+(g)$	609.8	$PbO(s)$	−219
$Na^+(aq)$	−239.7	$PbS(s)$	−98.3
$Na(g)$	107.8	*Nitrogen*	
$Na(s)$	0	$N(g)$	473
$NaCl(s)$	−411.1	$N_2(g)$	0
$NaHCO_3(s)$	−947.7	$NH_3(g)$	−45.9
$Na_2CO_3(s)$	−1130.8	$NH_4^+(aq)$	−132.8
Calcium		$NO(g)$	90.3
$Ca^{2+}(aq)$	−543.0	$NO_2(g)$	33.2
$Ca(s)$	0	$HNO_3(aq)$	−206.6
$CaO(s)$	−635.1	*Oxygen*	
$CaCO_3(s)$ (calcite)	−1206.9	$O(g)$	249.2
Carbon		$O_2(g)$	0
$C(g)$	715.0	$O_3(g)$	143
$C(graphite)$	0	$OH^-(aq)$	−229.9
$C(diamond)$	1.9	$H_2O(g)$	−241.8
$CO(g)$	−110.5	$H_2O(l)$	−285.8
$CO_2(g)$	−393.5	*Sulfur*	
$HCO_3^-(aq)$	−691.1	$S(g)$	279
$CH_4(g)$	−74.9	$S_2(g)$	129
$C_2H_4(g)$	52.5	$S_8(rhombic)$	0
$C_2H_6(g)$	−84.7	$S_8(monoclinic)$	2
$C_6H_6(l)$	49.0	$SO_2(g)$	−296.8
$HCHO(g)$	−116	$H_2S(g)$	−20
$CH_3OH(l)$	−238.6	*Fluorine*	
$CS_2(g)$	117	$F^-(g)$	−255.6
$CS_2(l)$	87.9	$F^-(aq)$	−329.1
$HCN(g)$	135	$F_2(g)$	0
$HCN(l)$	105	$HF(g)$	−273
$CCl_4(g)$	−96.0	*Chlorine*	
$CCl_4(l)$	−139	$Cl^-(aq)$	−167.5
$CH_3CHO(g)$	−166	$Cl(g)$	121.0
$C_2H_5OH(l)$	−277.6		*(continued)*

*See Appendix B for additional values.

Table 6.2   (Continued)

Formula	ΔH_f° (kJ/mol)	Formula	ΔH_f° (kJ/mol)
$Cl_2(g)$	0	$I_2(s)$	0
$HCl(g)$	−92.3	**Silver**	
Bromine		$Ag^+(g)$	1026.4
$Br^-(g)$	−218.9	$Ag^+(aq)$	105.9
$Br^-(aq)$	−120.9	$Ag(s)$	0
$Br_2(l)$	0	$AgF(s)$	−203
Iodine		$AgCl(s)$	−127.0
$I^-(g)$	−194.7	$AgBr(s)$	−99.5
$I^-(aq)$	−55.9	$AgI(s)$	−62.4

*See Appendix B for additional values.

from the stablest form of carbon (graphite) to diamond. The thermochemical equation is

$$C(\text{graphite}) \longrightarrow C(\text{diamond}); \Delta H_f^\circ = 1.9 \text{ kJ}$$

On the other hand, the ΔH_f° for graphite equals zero. Note the values of ΔH_f° for the elements listed in Table 6.2; the reference forms have zero values.

Now let us see how to use standard enthalpies of formation to find the standard enthalpy change for a reaction.

$$CH_4(g) + 4Cl_2(g) \longrightarrow CCl_4(l) + 4HCl(g); \Delta H^\circ = ?$$

From Table 6.2 you find the enthalpies of formation for $CH_4(g)$, $CCl_4(l)$, and $HCl(g)$. You can then write the following thermochemical equations:

$$C(\text{graphite}) + 2H_2(g) \longrightarrow CH_4(g); \Delta H_f^\circ = -74.9 \text{ kJ} \quad\quad \textbf{(1)}$$

$$C(\text{graphite}) + 2Cl_2(g) \longrightarrow CCl_4(l); \Delta H_f^\circ = -139 \text{ kJ} \quad\quad \textbf{(2)}$$

$$\tfrac{1}{2}H_2(g) + \tfrac{1}{2}Cl_2(g) \longrightarrow HCl(g); \Delta H_f^\circ = -92.3 \text{ kJ} \quad\quad \textbf{(3)}$$

You now apply Hess's law. Since you want CH_4 to appear on the left, and CCl_4 and 4HCl on the right, you reverse Equation 1 and add Equation 2 and 4 × Equation 3.

$CH_4(g) \longrightarrow$ C̶(̶g̶r̶a̶p̶h̶i̶t̶e̶)̶ $+$ 2̶H̶₂̶(̶g̶)̶	$(-74.9 \text{ kJ}) \times (-1)$	
C̶(̶g̶r̶a̶p̶h̶i̶t̶e̶)̶ $+ 2Cl_2(g) \longrightarrow CCl_4(l)$	$(-139 \text{ kJ}) \times (1)$	
2̶H̶₂̶(̶g̶)̶ $+ 2Cl_2(g) \longrightarrow 4HCl(g)$	$(-92.3 \text{ kJ}) \times (4)$	
$CH_4(g) + 4Cl_2(g) \longrightarrow CCl_4(l) + 4HCl(g)$	$\Delta H^\circ = -433 \text{ kJ}$	

The setup of this calculation can be greatly simplified once you closely examine what you are doing. Note that the ΔH_f° for each compound is multiplied by its coefficient in the chemical equation whose ΔH° you are calculating. Moreover, the ΔH_f° for each reactant is multiplied by a negative sign. You can symbolize the enthalpy of formation of a substance by writing the formula in parentheses following ΔH_f°. Then

$$\Delta H^\circ = [\Delta H_f^\circ(CCl_4) + 4\,\Delta H_f^\circ(HCl)] - [\Delta H_f^\circ(CH_4) + 4\,\Delta H_f^\circ(Cl_2)]$$
$$= [(-139) + 4(-92.3)] \text{ kJ} - [(-74.9) + 4(0)] \text{ kJ}$$
$$= -433 \text{ kJ}$$

In general, you can calculate the $\Delta H°$ for a reaction by the equation

$$\Delta H° = \Sigma\, n\, \Delta H_f°(\text{products}) - \Sigma\, m\, \Delta H_f°(\text{reactants})$$

Here Σ is the mathematical symbol meaning "the sum of," and m and n are the coefficients of the substances in the chemical equation.

EXAMPLE 6.8

Calculating the Heat of Phase Transition from Standard Enthalpies of Formation

Use values of $\Delta H_f°$ to calculate the heat of vaporization, $\Delta H_{vap}°$, of carbon disulfide at 25°C. The vaporization process is

$$CS_2(l) \longrightarrow CS_2(g)$$

CS_2

SOLUTION

Here is the equation for the vaporization, with values of $\Delta H_f°$ multiplied by coefficients (here, all 1's).

$$CS_2(l) \longrightarrow CS_2(g)$$
$$1(88) \qquad\quad 1(117) \quad (kJ)$$

The calculation is

$$\Delta H_{vap}° = \Sigma\, n\, \Delta H_f°(\text{products}) - \Sigma\, m\, \Delta H_f°(\text{reactants})$$
$$= \Delta H_f°[CS_2(g)] - \Delta H_f°[CS_2(l)]$$
$$= (117 - 88)\ kJ = \mathbf{29\ kJ}$$

See Problems 6.47 and 6.48.

EXAMPLE 6.9

Calculating the Enthalpy of Reaction from Standard Enthalpies of Formation

Large quantities of ammonia are used to prepare nitric acid. The first step consists of the catalytic oxidation of ammonia to nitric oxide, NO.

$$4NH_3(g) + 5O_2(g) \xrightarrow{\text{Pt}} 4NO(g) + 6H_2O(g)$$

What is the standard enthalpy change for this reaction?

SOLUTION

Here is the equation with the $\Delta H_f°$'s recorded beneath it:

$$4NH_3(g) + 5O_2(g) \longrightarrow 4NO(g) + 6H_2O(g)$$
$$4(-45.9) \quad\ 5(0) \qquad\quad\ 4(90.3) \quad\ 6(-241.8) \quad (kJ)$$

$$\Delta H° = \Sigma\, n\, \Delta H_f°(\text{products}) - \Sigma\, m\, \Delta H_f°(\text{reactants})$$
$$= [4\,\Delta H_f°(NO) + 6\,\Delta H_f°(H_2O)] - [4\,\Delta H_f°(NH_3) + 5\,\Delta H_f°(O_2)]$$
$$= [4(90.3) + 6(-241.8)]\ kJ - [4(-45.9) + 5(0)]\ kJ$$
$$= \mathbf{-906\ kJ}$$

Be very careful of arithmetical signs—they are a likely source of mistakes. Here you must use the ΔH_f° for $H_2O(g)$, not for $H_2O(l)$.

See Problems 6.49 and 6.50.

Enthalpies of formation can also be defined for ions. It is not possible to make thermal measurements on individual ions, so we must arbitrarily define the standard enthalpy of formation of one ion as zero. Then values for all other ions can be deduced from calorimetric data. The standard enthalpy of formation of $H^+(aq)$ is taken as zero. Values of ΔH_f° for some ions are given in Table 6.2.

6.9 Fuels—Foods, Commercial Fuels, and Rocket Fuels

The earliest use of fuels for heat came with the control of fire. This major advance allowed the human species to migrate from tropical savannas and eventually to inhabit most of the earth. During the mid-eighteenth century, the discovery of the steam engine, which converts the chemical energy latent in fuels to mechanical energy, ushered in the Industrial Revolution. Today fuels are absolutely necessary for every facet of modern technology. In this section we will look at foods as fuels; at fossil fuels (which include gas, oil, and coal); at coal gasification and liquefaction; and at rocket fuels.

Foods as Fuels

Foods fill three needs of the body: they supply substances for the growth and repair of tissue, they supply substances for the synthesis of compounds used in the regulation of body processes, and they supply energy. About 80% of the energy we need is for heat. The rest is used for muscular action, chemical processes, and other body processes. ◄

▶ The human body requires about as much energy in a day as does a 100-watt lightbulb.

The body generates energy from food by combustion, so the overall enthalpy change is the same as the heat of combustion. You can get some idea of the energy available from carbohydrate foods by looking at glucose ($C_6H_{12}O_6$). The thermochemical equation is

$$C_6H_{12}O_6(s) + 6O_2(g) \longrightarrow 6CO_2(g) + 6H_2O(l); \; \Delta H^\circ = -2803 \text{ kJ}$$

One gram of glucose yields 15.6 kJ (3.73 kcal) of heat when burned.

A representative fat is glyceryl trimyristate, $C_{45}H_{86}O_6$. The equation for its combustion is

$$C_{45}H_{86}O_6(s) + \tfrac{127}{2}O_2(g) \longrightarrow 45CO_2(g) + 43H_2O(l); \; \Delta H^\circ = -27{,}820 \text{ kJ}$$

▶ In the popular literature of nutrition, the kilocalorie is referred to as the Calorie (capital C). Thus, these values are given as 4.0 Calories and 9.0 Calories.

One gram of this fat yields 38.5 kJ (9.20 kcal) of heat when burned. The average values quoted for carbohydrates and fats are 4.0 kcal/g and 9.0 kcal/g, respectively. ◄ Note that fats contain more than twice the fuel value per gram as do carbohydrates.

Fossil Fuels

All of the fossil fuels in existence today were created millions of years ago when aquatic plants and animals were buried and compressed by layers of sediment at the bottoms of swamps and seas. This organic matter was converted by bacterial decay and pressure to petroleum (oil), gas, and coal. Figure 6.13 gives the percentages of the total energy consumed in the United States from various sources. The fossil fuels account for nearly 90% of the total.

Anthracite, or hard coal, the oldest variety of coal, was laid down as long as 250 million years ago and may contain over 80% carbon. Bituminous coal, a younger variety of coal, has between 45% and 65% carbon. Fuel values of coals are rated in Btu's (British thermal units) per pound, which are essentially heats of combustion per pound of coal. A typical value is 13,200 Btu/lb. A Btu equals 1054 J, so 1 Btu/lb equals 2.32 J/g. Therefore, the combustion of coal in oxygen yields about 30.6 kJ/g. You can compare this value with the heat of combustion of pure carbon (graphite).

$$C(graphite) + O_2(g) \longrightarrow CO_2(g); \Delta H° = -393.5 \, kJ$$

The value given in the equation is for 1 mol (12.0 g) of carbon. Per gram, you get 32.8 kJ/g, which is comparable with the values obtained for coal.

Natural gas and petroleum are very convenient fluid fuels, being easily transportable and having no ash. Purified natural gas is primarily methane, CH_4, but it also contains small amounts of ethane, C_2H_6; propane, C_3H_8; and butane, C_4H_{10}. The fuel values of natural gas are close to that for the heat of combustion of methane:

$$CH_4(g) + 2O_2(g) \longrightarrow CO_2(g) + 2H_2O(g); \Delta H° = -802 \, kJ$$

This value of $\Delta H°$ is equivalent to 50.1 kJ per gram of fuel.

Petroleum is a very complicated mixture of compounds. Gasoline, which is obtained from petroleum by chemical and physical processes, contains many different hydrocarbons (compounds of carbon and hydrogen). One such hydrocarbon is octane, C_8H_{18}. The combustion of octane evolves 5074 kJ of heat per mole.

$$C_8H_{18}(l) + \tfrac{25}{2}O_2(g) \longrightarrow 8CO_2(g) + 9H_2O(g); \Delta H° = -5074 \, kJ$$

This value of $\Delta H°$ is equivalent to 44.4 kJ/g. These combustion values indicate another reason why the fluid fossil fuels are popular: they release more heat per gram than coal does.

Coal Gasification and Liquefaction

The major problem with petroleum and natural gas as fuels is their relative short supply. It has been estimated that petroleum supplies will be 80% depleted by about the year 2030. Natural-gas supplies may be depleted even sooner.

Coal supplies, on the other hand, are sufficient to last several more centuries. This abundance has spurred much research into developing commercial methods for converting coal to the more easily handled liquid and gaseous fuels. Most of these methods begin with converting coal to carbon monoxide, CO.

$$C(s) + H_2O(g) \longrightarrow CO(g) + H_2(g)$$

In this reaction, steam is passed over hot coal. Once a mixture of carbon monoxide and hydrogen is obtained, it can be transformed by various reactions into useful products. For example, this mixture is reacted over a catalyst to give methane.

$$CO(g) + 3H_2(g) \longrightarrow CH_4(g) + H_2O(g)$$

An added advantage of coal gasification and coal liquefaction is that sulfur, normally present in fossil fuels, can be removed during the process. The burning of sulfur-containing coal is a major source of air pollution and acid rain.

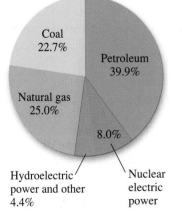

Figure 6.13
Sources of energy consumed in the United States (1996). Data are from *Monthly Energy Review On-line*, http://www.eia.doe. gov.

Figure 6.14
The launching of the _Columbia_ space shuttle.
The solid fuel for the booster rockets is a mixture of aluminum metal powder and other materials with ammonium perchlorate as the oxidizer. A cloud of aluminum oxide forms as the rockets burn.

Rocket Fuels

Rockets are self-contained missiles propelled by the ejection of gases from an orifice. Rockets are believed to have originated with the Chinese—perhaps before the thirteenth century. However, it was not until the twentieth century that rocket propulsion began to be studied seriously, and since World War II rockets have become major weapons. Space exploration with satellites propelled by rocket engines began in 1957 with the Russian satellite _Sputnik I._ Today weather and communications satellites are regularly put into orbit about the earth using rocket engines.

One of the factors determining which fuel and oxidizer to use is the mass of the fuel and oxidizer required. Hydrogen is the element of lowest density, and at the same time it reacts exothermically with oxygen to give water. The thermochemical equation for the combustion of hydrogen is

$$H_2(g) + \tfrac{1}{2}O_2(g) \longrightarrow H_2O(g); \ \Delta H° = -242 \text{ kJ}$$

This value of $\Delta H°$ is equivalent to 120 kJ/g of fuel (H_2) compared with 50 kJ/g of methane. The second and third stages of the _Saturn V_ launch vehicle that sent a three-man Apollo crew to the moon used a hydrogen/oxygen system.

The first stage of liftoff used kerosene and oxygen, and an unbelievable 550 metric tons of kerosene were burned in 2.5 minutes! It is interesting to calculate the average rate of energy production in this 2.5-minute interval. Kerosene is approximately $C_{12}H_{26}$. The thermochemical equation is

$$C_{12}H_{26}(l) + \tfrac{37}{2}O_2(g) \longrightarrow 12CO_2(g) + 13H_2O(g); \ \Delta H° = -7513 \text{ kJ}$$

This value of $\Delta H°$ is equivalent to 44.1 kJ/g. Each second, the average energy produced was 1.62×10^{11} J. This is equivalent to 1.62×10^{11} watts, or 217 million horsepower.

The landing module for the Apollo mission used a fuel made of hydrazine, N_2H_4, and a derivative of hydrazine. The oxidizer was dinitrogen tetroxide, N_2O_4. The reaction of the oxidizer with hydrazine is

$$2N_2H_4(l) + N_2O_4(l) \longrightarrow 3N_2(g) + 4H_2O(g); \ \Delta H° = -1049 \text{ kJ}$$

Solid propellants are also used as rocket fuels. The mixture used in the booster rockets of the _Columbia_ space shuttle (Figure 6.14) was a fuel containing aluminum metal powder. An oxidizer of ammonium perchlorate, NH_4ClO_4, was mixed with the fuel.

A Checklist for Review

Important Terms

energy (6.1)
kinetic energy (6.1)
joule (J) (6.1)
calorie (cal) (6.1)
potential energy (6.1)
internal energy (6.1)
law of conservation of
   energy (6.1)

thermodynamic system (or
   system) (6.2)
surroundings (6.2)
heat (6.2)
heat of reaction (6.2)
exothermic process (6.2)
endothermic process (6.2)
enthalpy (6.3)

state function (6.3)
enthalpy of reaction (6.3)
thermochemical equation
   (6.4)
heat capacity (6.6)
specific heat capacity
   (specific heat) (6.6)
calorimeter (6.6)

Hess's law of heat standard
   summation (6.7)
standard state (6.8)
allotrope (6.8)
reference form (6.8)
standard enthalpy of
   formation (standard heat
   of formation) (6.8)

Key Equations

$E_k = \frac{1}{2}mv^2$ $q = C \, \Delta t$

$1 \text{ cal} = 4.184 \text{ J}$ $q = s \times m \times \Delta t$

$\Delta H = q_p$ $\Delta H° = \Sigma \, n \, \Delta H_f°(\text{products}) - \Sigma \, m \, \Delta H_f°(\text{reactants})$

Summary of Facts and Concepts

Energy exists in various forms, including *kinetic energy* and *potential energy*. The SI unit of energy is the *joule* (1 calorie = 4.184 joules). The *internal energy* of a substance is the sum of the kinetic energies and potential energies of the particles making up the substance. According to the *law of conservation of energy,* the total quantity of energy remains constant.

Reactions absorb or evolve definite quantities of heat under given conditions. At constant pressure, this *heat of reaction* is the *enthalpy of reaction,* ΔH. The chemical equation plus ΔH for molar amounts of reactants is referred to as the *thermochemical equation.* With it, you can calculate the heat for any

amount of substance by *applying stoichiometry to heats of reaction.* One measures the heat of reaction in a *calorimeter.* Direct calorimetric determination of the heat of reaction requires a reaction that goes to completion without other reactions occurring at the same time. Otherwise, the heat or enthalpy of reaction is determined indirectly from other enthalpies of reaction by using *Hess's law of heat summation.* Thermochemical data are conveniently tabulated as *enthalpies of formation.* If you know the values for each substance in an equation, you can easily compute the enthalpy of reaction. As an application of thermochemistry, the last section of the chapter discusses fuels.

Operational Skills

1. **Calculating kinetic energy**   Given the mass and speed of an object, calculate the kinetic energy.   **(EXAMPLE 6.1)**
2. **Writing thermochemical equations**   Given a chemical equation, states of substances, and the quantity of heat absorbed or evolved for molar amounts, write the thermochemical equation.   **(EXAMPLE 6.2)**
3. **Manipulating thermochemical equations**   Given a thermochemical equation, write the thermochemical equation for different multiples of the coefficients or for the reverse reaction.   **(EXAMPLE 6.3)**
4. **Calculating the heat of reaction from the stoichiometry**   Given the value of ΔH for a chemical equation, calculate the heat of reaction for a given mass of reactant or product.   **(EXAMPLE 6.4)**
5. **Relating heat and specific heat**   Given any three of the quantities q, s, m, and Δt, calculate the fourth one.   **(EXAMPLE 6.5)**

6. **Calculating ΔH from calorimetric data**   Given the amounts of reactants and the temperature change of a calorimeter of specified heat capacity, calculate the heat of reaction.   **(EXAMPLE 6.6)**
7. **Applying Hess's law**   Given a set of reactions with enthalpy changes, calculate ΔH for a reaction obtained from these other reactions by using Hess's law.   **(EXAMPLE 6.7)**
8. **Calculating the heat of phase transition from standard enthalpies of formation**   Given a table of standard enthalpies of formation, calculate the heat of phase transition.   **(EXAMPLE 6.8)**
9. **Calculating the enthalpy of reaction from standard enthalpies of formation**   Given a table of standard enthalpies of formation, calculate the enthalpy of reaction.   **(EXAMPLE 6.9)**

Review Questions

6.1   Define *energy, kinetic energy, potential energy,* and *internal energy.*

6.2   Define the joule in terms of SI base units.

6.3   What is the original definition of the calorie? What is the present definition?

6.4   Describe the interconversions of potential and kinetic energy in a moving pendulum. A moving pendulum eventually comes to rest. Has the energy been lost? If not, what has happened to it?

6.5   Suppose heat flows into a vessel containing a gas. As the heat flows into the gas, what happens to the gas molecules? What happens to the internal energy of the gas?

6.6   Define an *exothermic* reaction and an *endothermic* reaction. Give an example of each.

6.7   The internal energy of a substance is a state function. What does this mean?

6.8   Why is it important to give the states of the reactants and products when giving an equation for ΔH?

6.9 Consider the reaction of methane, CH_4, with oxygen, O_2, discussed in Section 6.5. How would you set up the calculation if the problem had been to compute the heat if 10.0 g H_2O were produced (instead of 10.0 g CH_4 reacted)?

6.10 Define the heat capacity of a substance. Define the specific heat of a substance.

6.11 What property of enthalpy provides the basis of Hess's law? Explain.

6.12 What is meant by the thermodynamic standard state?

6.13 What is meant by the reference form of an element? What is the standard enthalpy of formation of an element in its reference form?

6.14 What is meant by the standard enthalpy of formation of a substance?

6.15 What is a fuel? What are the fossil fuels?

6.16 Give chemical equations for the conversion of carbon in coal to methane, CH_4.

Conceptual Problems

6.17 A small car is traveling at twice the speed of a larger car, which has twice the mass of the smaller car. Which car has the greater kinetic energy? (Or do they both have the same kinetic energy?)

6.18 The equation for the combustion of butane, C_4H_{10}, is

$$C_4H_{10}(g) + \tfrac{13}{2}O_2(g) \longrightarrow 4CO_2(g) + 5H_2O(g)$$

Which one of the following generates the least heat? Why?
a. burning one mole of butane
b. reacting one mole of oxygen with excess butane
c. producing one mole of carbon dioxide by burning butane
d. producing one mole of water by burning butane

6.19 You have two samples of different metals, metal A and metal B, each having the same mass. You heat both metals to 95°C and then place each one into separate beakers containing the same quantity of water at 25°C.
a. You measure the temperatures of the water in the two beakers when each metal has cooled by 10°C and find that the temperature of the water with metal A is higher than the temperature of the water with metal B. Which metal has the greater specific heat? Explain.
b. After waiting a period of time, the temperature of the water in each beaker rises to a maximum value. In which beaker does the water rise to the higher value, the one with metal A or the one with metal B? Explain.

6.20 A block of aluminum and a block of iron, both having the same mass, are removed from a freezer and placed outside on a warm day. When the same quantity of heat has flowed into each block, which block will be warmer? Assume that neither block has yet reached the outside temperature. (See Table 6.1 for the specific heats of the metals.)

6.21 Tetraphosphorus trisulfide, P_4S_3, burns in excess oxygen to give tetraphosphorus decoxide, P_4O_{10}, and sulfur dioxide, SO_2. Suppose you have measured the enthalpy change for this reaction. How could you use it to obtain the enthalpy of formation of P_4S_3? What other data do you need?

6.22 A soluble salt, MX_2, is added to water in a beaker. The equation for the dissolving of the salt is:

$$MX_2(s) \longrightarrow M^{2+}(aq) + 2X^-(aq); \quad \Delta H > 0$$

a. Immediately after the salt dissolves, is the solution warmer or colder?
b. Indicate the direction of heat flow, in or out of the beaker, while the salt dissolves.
c. After the salt dissolves and the water returns to room temperature, what is the value of q for the system?

Practice Problems

Energy and Its Units

6.23 A car whose mass is 4.85×10^3 lb is traveling at a speed of 57 miles per hour. What is the kinetic energy of the car in joules? in calories? See Table 1.4 for conversion factors.

6.24 A bullet weighing 235 grains is moving at a speed of 2.52×10^3 ft/s. Calculate the kinetic energy of the bullet in joules and in calories. One grain equals 0.0648 g.

6.25 Chlorine dioxide, ClO_2, is a reddish-yellow gas used in bleaching paper pulp. The average speed of a ClO_2 molecule at 25°C is 306 m/s. What is the kinetic energy (in joules) of a ClO_2 molecule moving at this speed?

6.26 Nitrous oxide, N_2O, has been used as a dental anesthetic. The average speed of an N_2O molecule at 25°C is 379 m/s. Calculate the kinetic energy (in joules) of an N_2O molecule traveling at this speed.

Heat of Reaction

6.27 Nitric acid, a source of many nitrogen compounds, is produced from nitrogen dioxide. An old process for making nitrogen dioxide employed nitrogen and oxygen.

$$N_2(g) + 2O_2(g) \longrightarrow 2NO_2(g)$$

The reaction absorbs 66.4 kJ of heat per 2 mol NO_2 produced. Is the reaction endothermic or exothermic? What is the value of q?

6.28 Hydrogen cyanide is used in the manufacture of clear plastics such as Lucite and Plexiglas. It is prepared from ammonia and natural gas (CH_4).

$$2NH_3(g) + 3O_2(g) + 2CH_4(g) \longrightarrow 2HCN(g) + 6H_2O(g)$$

The reaction evolves 939 kJ of heat per 2 mol HCN formed. Is the reaction endothermic or exothermic? What is the value of q?

Thermochemical Equations

6.29 When 1 mol of iron metal reacts with hydrochloric acid at constant temperature and pressure to produce hydrogen gas and aqueous iron(II) chloride, 87.9 kJ of heat evolves. Write a thermochemical equation for this reaction.

6.30 When 2 mol of potassium chlorate crystals decompose to potassium chloride crystals and oxygen gas at constant temperature and pressure, 44.7 kJ of heat is given off. Write a thermochemical equation for this reaction.

6.31 When white phosphorus burns in air, it produces phosphorus(V) oxide.

$$P_4(s) + 5O_2(g) \longrightarrow P_4O_{10}(s); \Delta H = -2940 \text{ kJ}$$

What is ΔH for the following equation?

$$P_4O_{10}(s) \longrightarrow P_4(s) + 5O_2(g)$$

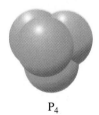

P_4

6.32 Carbon disulfide burns in air, producing carbon dioxide and sulfur dioxide.

$$CS_2(l) + 3O_2(g) \longrightarrow CO_2(g) + 2SO_2(g); \Delta H = -1075 \text{ kJ}$$

What is ΔH for the following equation?

$$\tfrac{1}{2}CS_2(l) + \tfrac{3}{2}O_2(g) \longrightarrow \tfrac{1}{2}CO_2(g) + SO_2(g)$$

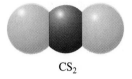

CS_2

Stoichiometry of Reaction Heats

6.33 Colorless nitric oxide, NO, combines with oxygen to form nitrogen dioxide, NO_2, a brown gas.

$$2NO(g) + O_2(g) \longrightarrow 2NO_2(g); \Delta H = -114 \text{ kJ}$$

What is the enthalpy change per gram of nitric oxide?

6.34 Hydrogen, H_2, is used as a rocket fuel. The hydrogen is burned in oxygen to produce water vapor.

$$2H_2(g) + O_2(g) \longrightarrow 2H_2O(g); \Delta H = -484 \text{ kJ}$$

What is the enthalpy change per gram of hydrogen?

6.35 Propane, C_3H_8, is a common fuel gas. Use the following to calculate the grams of propane you would need to provide 369 kJ of heat.

$$C_3H_8(g) + 5O_2(g) \longrightarrow 3CO_2(g) + 4H_2O(g);$$
$$\Delta H = -2044 \text{ kJ}$$

6.36 Ethanol, C_2H_5OH, is mixed with gasoline and sold as gasohol. Use the following to calculate the grams of ethanol needed to provide 293 kJ of heat.

$$C_2H_5OH(l) + 3O_2(g) \longrightarrow 2CO_2(g) + 3H_2O(g);$$
$$\Delta H = -1235 \text{ kJ}$$

Heat Capacity and Calorimetry

6.37 You wish to heat water to make coffee. How much heat (in joules) must be used to raise the temperature of 0.180 kg of tap water (enough for one cup of coffee) from 19°C to 96°C (near the ideal brewing temperature)? Assume the specific heat is that of pure water, 4.18 J/(g · °C).

6.38 An iron skillet weighing 1.63 kg is heated on a stove to 178°C. Suppose the skillet is cooled to room temperature, 21°C. How much heat energy (in joules) must be removed to effect this cooling? The specific heat of iron is 0.449 J/(g · °C).

6.39 When 15.3 g of sodium nitrate, $NaNO_3$, was dissolved in water in a calorimeter, the temperature fell from 25.00°C to 21.56°C. If the heat capacity of the solution and the calorimeter is 1071 J/°C, what is the enthalpy change when 1 mol of sodium nitrate dissolves in water? The solution process is

$$NaNO_3(s) \longrightarrow Na^+(aq) + NO_3^-(aq); \Delta H = ?$$

6.40 When 23.6 g of calcium chloride, $CaCl_2$, was dissolved in water in a calorimeter, the temperature rose from 25.0°C to 38.7°C. If the heat capacity of the solution and the calorimeter is 1258 J/°C, what is the enthalpy change when 1 mol of calcium chloride dissolves in water? The solution process is

$$CaCl_2(s) \longrightarrow Ca^{2+}(aq) + 2Cl^-(aq); \Delta H = ?$$

6.41 A sample of ethanol, C_2H_5OH, weighing 2.84 g was burned in an excess of oxygen in a bomb calorimeter. The temperature of the calorimeter rose from 25.00°C to 33.73°C. If the heat capacity of the calorimeter and contents was 9.63 kJ/°C, what is the value of q for burning 1 mol of ethanol at constant volume and 25.00°C? The reaction is

$$C_2H_5OH(l) + 3O_2(g) \longrightarrow 2CO_2(g) + 3H_2O(l)$$

6.42 A sample of benzene, C_6H_6, weighing 3.51 g was burned in an excess of oxygen in a bomb calorimeter. The temperature of the calorimeter rose from 25.00°C to 37.18°C. If the heat capacity of the calorimeter and contents was 12.05 kJ/°C, what is the value of q for burning 1 mol of benzene at constant volume and 25.00°C? The reaction is

$$C_6H_6(l) + \tfrac{15}{2}O_2(g) \longrightarrow 6CO_2(g) + 3H_2O(l)$$

Hess's Law

6.43 Hydrazine, N_2H_4, is a colorless liquid used as a rocket fuel. What is the enthalpy change for the process in which hydrazine is formed from its elements?

$$N_2(g) + 2H_2(g) \longrightarrow N_2H_4(l)$$

Use the following reactions and enthalpy changes:

$$N_2H_4(l) + O_2(g) \longrightarrow N_2(g) + 2H_2O(l); \Delta H = -622.2 \text{ kJ}$$
$$H_2(g) + \tfrac{1}{2}O_2(g) \longrightarrow H_2O(l); \Delta H = -285.8 \text{ kJ}$$

6.44 Hydrogen peroxide, H_2O_2, is a colorless liquid whose solutions are used as a bleach and an antiseptic. H_2O_2 can be prepared in a process whose overall change is

$$H_2(g) + O_2(g) \longrightarrow H_2O_2(l)$$

Calculate the enthalpy change using the following data:

$$H_2O_2(l) \longrightarrow H_2O(l) + \tfrac{1}{2}O_2(g); \Delta H = -98.0 \text{ kJ}$$
$$2H_2(g) + O_2(g) \longrightarrow 2H_2O(l); \Delta H = -571.6 \text{ kJ}$$

6.45 Compounds with carbon–carbon double bonds, such as ethylene, C_2H_4, add hydrogen in a reaction called hydrogenation.

$$C_2H_4(g) + H_2(g) \longrightarrow C_2H_6(g)$$

Calculate the enthalpy change for this reaction, using the following combustion data:

$$C_2H_4(g) + 3O_2(g) \longrightarrow 2CO_2(g) + 2H_2O(l);$$
$$\Delta H = -1401 \text{ kJ}$$
$$C_2H_6(g) + \tfrac{7}{2}O_2(g) \longrightarrow 2CO_2(g) + 3H_2O(l);$$
$$\Delta H = -1550 \text{ kJ}$$
$$H_2(g) + \tfrac{1}{2}O_2(g) \longrightarrow H_2O(l); \Delta H = -286 \text{ kJ}$$

6.46 Acetic acid, CH_3COOH, is contained in vinegar. Suppose acetic acid was formed from its elements, according to the following equation:

$$2C(\text{graphite}) + 2H_2(g) + O_2(g) \longrightarrow CH_3COOH(l)$$

Find the enthalpy change, ΔH, for this reaction, using the following data:

$$CH_3COOH(l) + 2O_2(g) \longrightarrow 2CO_2(g) + 2H_2O(l);$$
$$\Delta H = -871 \text{ kJ}$$
$$C(\text{graphite}) + O_2(g) \longrightarrow CO_2(g); \Delta H = -394 \text{ kJ}$$
$$H_2(g) + \tfrac{1}{2}O_2(g) \longrightarrow H_2O(l); \Delta H = -286 \text{ kJ}$$

Standard Enthalpies of Formation

6.47 The cooling effect of alcohol on the skin is due to its evaporation. Calculate the heat of vaporization of ethanol (ethyl alcohol), C_2H_5OH.

$$C_2H_5OH(l) \longrightarrow C_2H_5OH(g); \Delta H° = ?$$

The standard enthalpy of formation of $C_2H_5OH(l)$ is -277.6 kJ/mol and that of $C_2H_5OH(g)$ is -235.4 kJ/mol.

6.48 Carbon tetrachloride, CCl_4, is a liquid used as an industrial solvent and in the preparation of fluorocarbons. What is the heat of vaporization of carbon tetrachloride?

$$CCl_4(l) \longrightarrow CCl_4(g); \Delta H° = ?$$

Use standard enthalpies of formation (Table 6.2).

6.49 Iron is obtained from iron ore by reduction with carbon monoxide. The overall reaction is

$$Fe_2O_3(s) + 3CO(g) \longrightarrow 2Fe(s) + 3CO_2(g)$$

Calculate the standard enthalpy change for this equation. See Appendix B for data.

6.50 The first step in the preparation of lead from its ore (galena, PbS) consists of roasting the ore.

$$2PbS(s) + 3O_2(g) \longrightarrow 2SO_2(g) + 2PbO(s)$$

Calculate the standard enthalpy change for this reaction, using enthalpies of formation (see Appendix B).

6.51 The Group IIA carbonates decompose when heated. For example,

$$MgCO_3(s) \longrightarrow MgO(s) + CO_2(g)$$

General Problems

6.53 Liquid hydrogen peroxide has been used as a propellant for rockets. Hydrogen peroxide decomposes into oxygen and water, giving off heat energy equal to 686 Btu per pound of propellant. What is this energy in joules per gram of hydrogen peroxide? (1 Btu = 252 cal; see also Table 1.4.)

6.54 Hydrogen is an ideal fuel in many respects; for example, the product of its combustion, water, is nonpolluting. The heat given off in burning hydrogen to gaseous water is 5.16×10^4 Btu per pound. What is this heat energy in joules per gram? (1 Btu = 252 cal; see also Table 1.4.)

6.55 When calcium carbonate, $CaCO_3$ (the major constituent of limestone and seashells), is heated, it decomposes to calcium oxide (quicklime).

$$CaCO_3(s) \longrightarrow CaO(s) + CO_2(g); \Delta H = 178.3 \text{ kJ}$$

How much heat is required to decompose 21.3 g of calcium carbonate?

6.56 Calcium oxide (quicklime) reacts with water to produce calcium hydroxide (slaked lime).

$$CaO(s) + H_2O(l) \longrightarrow Ca(OH)_2(s); \Delta H = -65.2 \text{ kJ}$$

The heat released by this reaction is sufficient to ignite paper. How much heat is released when 24.5 g of calcium oxide reacts?

6.57 Formic acid, $HCHO_2$, was first discovered in ants (*formica* is Latin for "ant"). In an experiment, 5.48 g of formic acid was burned at constant pressure.

$$2HCHO_2(l) + O_2(g) \longrightarrow 2CO_2(g) + 2H_2O(l)$$

If 30.3 kJ of heat evolved, what is ΔH per mole of formic acid?

6.58 Acetic acid, $HC_2H_3O_2$, is the sour constituent of vinegar (*acetum* is Latin for "vinegar"). In an experiment, 3.58 g of acetic acid was burned.

$$HC_2H_3O_2(l) + 2O_2(g) \longrightarrow 2CO_2(g) + 2H_2O(l)$$

Use enthalpies of formation (Appendix B) and calculate the heat required to decompose 10.0 g of magnesium carbonate.

6.52 The Group IIA carbonates decompose when heated. For example,

$$BaCO_3(s) \longrightarrow BaO(s) + CO_2(g)$$

Use enthalpies of formation (Appendix B) and calculate the heat required to decompose 10.0 g of barium carbonate.

If 52.0 kJ of heat evolved, what is ΔH per mole of acetic acid?

6.59 A piece of lead of mass 121.6 g was heated by an electrical coil. From the resistance of the coil, the current, and the time the current flowed, it was calculated that 235 J of heat was added to the lead. The temperature of the lead rose from 20.4°C to 35.5°C. What is the specific heat of the lead?

6.60 The specific heat of copper metal was determined by putting a piece of the metal weighing 35.4 g in hot water. The quantity of heat absorbed by the metal was calculated to be 47.0 J from the temperature drop of the water. What was the specific heat of the metal if the temperature of the metal rose 3.45°C?

6.61 In a calorimetric experiment, 6.48 g of lithium hydroxide, LiOH, was dissolved in water. The temperature of the calorimeter rose from 25.00°C to 36.66°C. What is ΔH for the solution process?

$$LiOH(s) \longrightarrow Li^+(aq) + OH^-(aq)$$

The heat capacity of the calorimeter and its contents is 547 J/°C.

6.62 When 21.45 g of potassium nitrate, KNO_3, was dissolved in water in a calorimeter, the temperature fell from 25.00°C to 14.14°C. What is ΔH for the solution process?

$$KNO_3(s) \longrightarrow K^+(aq) + NO_3^-(aq)$$

The heat capacity of the calorimeter and its contents is 682 J/°C.

6.63 A 10.00-g sample of acetic acid, $HC_2H_3O_2$, was burned in a bomb calorimeter in an excess of oxygen.

$$HC_2H_3O_2(l) + 2O_2(g) \longrightarrow 2CO_2(g) + 2H_2O(l)$$

The temperature of the calorimeter rose from 25.00°C to 35.81°C. If the heat capacity of the calorimeter and its contents is 13.43 kJ/°C, what is the enthalpy change for the reaction?

6.64 The sugar arabinose, $C_5H_{10}O_5$, is burned completely in oxygen in a calorimeter.

$$C_5H_{10}O_5(s) + 5O_2(g) \longrightarrow 5CO_2(g) + 5H_2O(l)$$

Burning a 0.548-g sample caused the temperature to rise from 20.00°C to 20.54°C. The heat capacity of the calorimeter and its contents is 15.8 kJ/°C. Calculate ΔH for the combustion reaction per mole of arabinose.

6.65 Hydrogen sulfide, H_2S, is a poisonous gas with the odor of rotten eggs. The reaction for the formation of H_2S from the elements is

$$H_2(g) + \tfrac{1}{8}S_8(\text{rhombic}) \longrightarrow H_2S(g)$$

Use Hess's law to obtain the enthalpy change for this reaction from the following enthalpy changes:

$$H_2S(g) + \tfrac{3}{2}O_2(g) \longrightarrow H_2O(g) + SO_2(g); \Delta H = -519 \text{ kJ}$$
$$H_2(g) + \tfrac{1}{2}O_2(g) \longrightarrow H_2O(g); \Delta H = -242 \text{ kJ}$$
$$\tfrac{1}{8}S_8(\text{rhombic}) + O_2(g) \longrightarrow SO_2(g); \Delta H = -297 \text{ kJ}$$

6.66 Ethylene glycol, $HOCH_2CH_2OH$, is used as antifreeze. It is produced from ethylene oxide, C_2H_4O, by the reaction

$$C_2H_4O(g) + H_2O(l) \longrightarrow HOCH_2CH_2OH(l)$$

Use Hess's law to obtain the enthalpy change for this reaction from the following enthalpy changes:

$$2C_2H_4O(g) + 5O_2(g) \longrightarrow 4CO_2(g) + 4H_2O(l);$$
$$\Delta H = -2612.2 \text{ kJ}$$
$$HOCH_2CH_2OH(l) + \tfrac{5}{2}O_2(g) \longrightarrow 2CO_2(g) + 3H_2O(l);$$
$$\Delta H = -1189.8 \text{ kJ}$$

6.67 Calcium oxide, CaO, is prepared by heating calcium carbonate (from limestone and seashells).

$$CaCO_3(s) \longrightarrow CaO(s) + CO_2(g)$$

Calculate the standard enthalpy of reaction, using enthalpies of formation. The ΔH_f° of CaO(s) is -635 kJ/mol. Other values are given in Table 6.2.

6.68 Sodium carbonate, Na_2CO_3, is used to manufacture glass. It is obtained from sodium hydrogen carbonate, $NaHCO_3$, by heating.

$$2NaHCO_3(s) \longrightarrow Na_2CO_3(s) + H_2O(g) + CO_2(g)$$

Calculate the standard enthalpy of reaction, using enthalpies of formation (Table 6.2).

6.69 Calculate the heat released when 2.000 L O_2 with a density of 1.11 g/L at 25°C reacts with an excess of hydrogen to form liquid water at 25°C.

6.70 Calculate the heat released when 4.000 L Cl_2 with a density of 2.46 g/L at 25°C reacts with an excess of sodium metal to form solid sodium chloride at 25°C.

6.71 Ammonium nitrate is an oxidizing agent and can give rise to explosive mixtures. A mixture of 2.00 mol of powdered aluminum and 3.00 mol of ammonium nitrate crystals reacts exothermically, yielding nitrogen gas, water vapor, and aluminum oxide. How many grams of the mixture are required to provide 245 kJ of heat? See Appendix B for data.

6.72 The thermite reaction is a very exothermic reaction; it has been used to produce liquid iron for welding. A mixture of 2 mol of powdered aluminum metal and 1 mol of iron(III) oxide yields liquid iron and solid aluminum oxide. How many grams of the mixture are needed to produce 348 kJ of heat? See Appendix B for data.

Cumulative-Skills Problems

6.73 What will be the final temperature of a mixture made from 25.0 g of water at 15.0°C, from 45.0 g of water at 50.0°C, and from 15.0 g of water at 37.0°C?

6.74 What will be the final temperature of a mixture made from equal masses of the following: water at 25.0°C, ethanol at 35.5°C, and iron at 95°C?

6.75 How much heat is released when a mixture containing 10.0 g NH_3 and 20.0 g O_2 reacts by the equation

$$4NH_3(g) + 5O_2(g) \longrightarrow 4NO(g) + 6H_2O(g);$$
$$\Delta H^\circ = -906 \text{ kJ}$$

6.76 How much heat is released when a mixture containing 10.0 g CS_2 and 10.0 g Cl_2 reacts by the equation

$$CS_2(g) + 3Cl_2(g) \longrightarrow S_2Cl_2(g) + CCl_4(g);$$
$$\Delta H^\circ = -232 \text{ kJ}$$

6.77 Consider the Haber process:

$$N_2(g) + 3H_2(g) \longrightarrow 2NH_3(g); \Delta H^\circ = -91.8 \text{ kJ}$$

The density of ammonia at 25°C and 1.00 atm is 0.696 g/L. The density of nitrogen, N_2, is 1.145 g/L, and the molar heat capacity is 29.12 J/(mol · °C). (a) How much heat is evolved in the production of 1.00 L of ammonia at 25°C and 1.00 atm? (b) What percentage of this heat is required to heat the nitrogen required for this reaction (0.500 L) from 25°C to 400°C, the temperature at which the Haber process is run?

6.78 An industrial process for manufacturing sulfuric acid, H_2SO_4, uses hydrogen sulfide, H_2S, from the purification of natural gas. In the first step of this process, the hydrogen sulfide is burned to obtain sulfur dioxide, SO_2.

$$2H_2S(g) + 3O_2(g) \longrightarrow 2H_2O(l) + 2SO_2(g);$$
$$\Delta H° = -1125 \text{ kJ}$$

The density of sulfur dioxide at 25°C and 1.00 atm is 2.62 g/L, and the molar heat capacity is 30.2 J/(mol · °C). (a) How much heat would be evolved in producing 1.00 L of SO_2 at 25°C and 1.00 atm? (b) Suppose heat from this reaction is used to heat 1.00 L of SO_2 from 25°C and 1.00 atm to 500°C for its use in the next step of the process. What percentage of the heat evolved is required for this?

6.79 The carbon dioxide exhaled in the breath of astronauts is often removed from the spacecraft by reaction with lithium hydroxide.

$$2LiOH(s) + CO_2(g) \longrightarrow Li_2CO_3(s) + H_2O(l)$$

Estimate the grams of lithium hydroxide required per astronaut per day. Assume that each astronaut requires 2.50×10^3 kcal of energy per day. Further assume that this energy can be equated to the heat of combustion of a quantity of glucose, $C_6H_{12}O_6$, to $CO_2(g)$ and $H_2O(l)$. From the amount of glucose required to give 2.50×10^3 kcal of heat, calculate the amount of CO_2 produced and hence the amount of LiOH required. The $\Delta H_f°$ for glucose(s) is -1273 kJ/mol.

6.80 A rebreathing gas mask contains potassium superoxide, KO_2, which reacts with moisture in the breath to give oxygen.

$$4KO_2(s) + 2H_2O(l) \longrightarrow 4KOH(s) + 3O_2(g)$$

Estimate the grams of potassium superoxide required to supply a person's oxygen needs for one hour. Assume a person requires 1.00×10^2 kcal of energy for this time period. Further assume that this energy can be equated to the heat of combustion of a quantity of glucose, $C_6H_{12}O_6$, to $CO_2(g)$ and $H_2O(l)$. From the amount of glucose required to give 1.00×10^2 kcal of heat, calculate the amount of oxygen consumed and hence the amount of KO_2 required. The $\Delta H_f°$ for glucose(s) is -1273 kJ/mol.

The colored flames from several metal compounds.

Quantum Theory of the Atom

According to Rutherford's model (Section 2.2), an atom consists of a nucleus many times smaller than the atom itself, with electrons occupying the remaining space. How are the electrons distributed in this space? Or, we might ask, what are the electrons doing in the atom?

The answer was to come from an unexpected area: the study of colored flames. When metal compounds burn in a flame, they emit bright colors. The spectacular colors of fireworks are due to the burning of metal compounds. Lithium and strontium compounds give a deep red color; barium compounds, a green color; and copper compounds, a bluish-green color.

Although the red flames of lithium and strontium appear similar, the light from each can be resolved (separated) by means of a prism into distinctly different colors. This resolution easily distinguishes the two elements. A prism disperses the colors of

EMISSION (LINE) SPECTRA

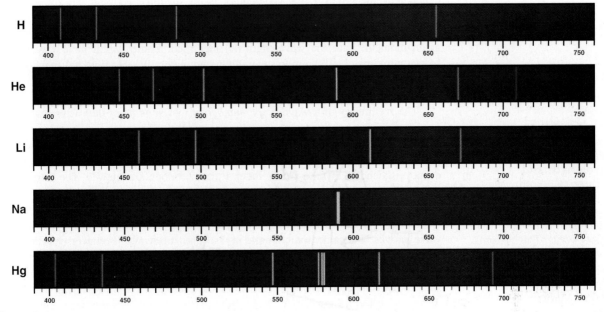

Figure 7.1
Emission (line) spectra of some elements. The lines correspond to visible light emitted by atoms. (Wavelengths of lines are given in nanometers.)

white light just as small raindrops spread the colors of sunlight into a rainbow or spectrum. But the light from a flame, when passed through a prism, reveals something other than a rainbow. Instead of the continuous range of color from red to yellow to violet, the spectrum of lithium shows a red line, a yellow line, and two blue lines against a black background. (See Figure 7.1.)

Each element, in fact, has a characteristic line spectrum because of the emission of light from atoms in the hot gas. The spectra can be used to identify elements. How is it that each atom emits particular colors of light? What does a line spectrum tell us about the structure of an atom? If you know something about the structures of atoms, can you explain the formation of ions and molecules? We will answer these questions in this and the next few chapters.

Light Waves, Photons, and the Bohr Theory

The present theory of the electronic structure of atoms started with an explanation of the colored light produced in hot gases and flames. Before we discuss this, we need to describe the nature of light.

7.1 The Wave Nature of Light

If you drop a stone into one end of a quiet pond, the impact of the stone with the water starts an up-and-down motion of the water surface. This up-and-down motion travels outward from where the stone hit; it is a familiar example of a wave. A *wave* is a continuously repeating change. Light is also a wave. It consists of oscillations in electric and magnetic fields that can travel through space. Visible light, x rays, and radio waves are all forms of *electromagnetic radiation*.

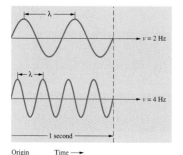

Origin   Time ——▶

Figure 7.2
Relation between wavelength and frequency. Both waves are traveling at the same speed. The top wave has a wavelength twice that of the bottom wave. The bottom wave, however, has twice the frequency of the top wave.

You characterize a wave by its wavelength and frequency. The **wavelength,** denoted by the Greek letter λ (lambda), is *the distance between any two adjacent identical points of a wave.* Thus, the wavelength is the distance between two adjacent peaks or troughs of a wave. Figure 7.2 shows a cross section of a wave at a given moment, with the wavelength (λ) identified. Radio waves have wavelengths from approximately 100 mm to several hundred meters. Visible light has much shorter wavelengths, about 10^{-6} m. Wavelengths of visible light are often given in nanometers (1 nm = 10^{-9} m).

The **frequency** of a wave is *the number of wavelengths of that wave that pass a fixed point in one unit of time* (usually one second). For example, imagine you are anchored in a small boat on a pond when a stone is dropped into the water. The number of wavelengths that pass you in one second is the frequency of that wave. Frequency is denoted by the Greek letter ν (nu, pronounced "new"). The unit of frequency is /s, or s^{-1}, also called the *hertz* (Hz).

The wavelength and frequency of a wave are related to each other. Figure 7.2 shows two waves, each traveling from left to right at the same speed; that is, each wave moves the same total length in 1s. The top wave, however, has a wavelength twice that of the bottom wave. In 1s, two complete wavelengths of the top wave move left to right from the origin. It has a frequency of 2/s, or 2 Hz. In the same time, four complete wavelengths of the bottom wave move left to right from the origin. It has a frequency of 4 Hz. Note that for two waves traveling with a given speed, wavelength and frequency are inversely related: the greater the wavelength, the lower the frequency, and vice versa. The product $\nu\lambda$ is the total length of the wave that has passed the point in 1s. This length of wave per second is the speed of the wave. For light of speed c

$$c = \nu\lambda$$

The speed of light waves in a vacuum is a constant and is independent of wavelength or frequency. This speed is 3.00×10^8 m/s, which is the value for c that we use in the following examples. ◀

▶ To understand how fast the speed of light is, it might help to realize that it takes only 2.5 s for radar waves (which travel at the speed of light) to leave earth, bounce off the moon, and return—a total distance of 478,000 miles.

EXAMPLE 7.1

Obtaining the Wavelength of Light from Its Frequency

What is the wavelength of the yellow sodium emission, which has a frequency of 5.09×10^{14}/s?

SOLUTION

The frequency and wavelength are related by the formula $c = \nu\lambda$. You rearrange this formula to give

$$\lambda = \frac{c}{\nu}$$

in which c is the speed of light (3.00×10^8 m/s). Substituting,

$$\lambda = \frac{3.00 \times 10^8 \text{ m/s}}{5.09 \times 10^{14}\text{/s}} = 5.89 \times 10^{-7} \text{ m, or } \textbf{589 nm}$$

See Problems 7.21 and 7.22.

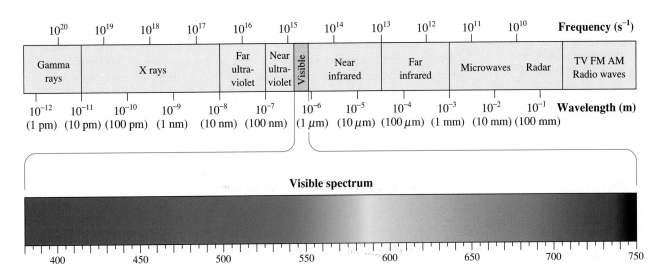

Figure 7.3
The electromagnetic spectrum. Divisions between regions are not defined precisely.

EXAMPLE 7.2

Obtaining the Frequency of Light from Its Wavelength

What is the frequency of violet light with a wavelength of 408 nm?

SOLUTION

You rearrange the equation relating frequency and wavelength to give

$$\nu = \frac{c}{\lambda}$$

Substituting for λ (408 nm = 408×10^{-9} m) gives

$$\nu = \frac{3.00 \times 10^8 \text{ m/s}}{408 \times 10^{-9} \text{ m}} = \mathbf{7.35 \times 10^{14}/s} \blacktriangleleft$$

See Problems 7.23 and 7.24.

▶ In 1854, Kirchhoff found that each element has a unique spectrum. Later, Bunsen and Kirchhoff developed the prism spectroscope and used it to confirm their discovery of two new elements, cesium (in 1860) and rubidium (in 1861).

The range of frequencies or wavelengths of electromagnetic radiation is called the **electromagnetic spectrum,** shown in Figure 7.3. Visible light extends from the violet end of the spectrum, which has a wavelength of about 400 nm, to the red end, with a wavelength of less than 800 nm. Infrared radiation has wavelengths greater than 800 nm, and ultraviolet radiation has wavelengths less than 400 nm.

7.2   Quantum Effects and Photons

Isaac Newton, who studied the properties of light in the seventeenth century, believed that light consisted of a beam of particles. In 1801, however, British physicist Thomas Young showed that light, like waves, could be diffracted. *Diffraction* is a property of waves in which the waves spread out when they encounter an obstruction or small hole about the size of the wavelength.

▶ **Max Planck was professor of physics at the University of Berlin when he did this research. He received the Nobel Prize in physics for it in 1918.**

▶ **Radiation emitted from the human body and warm objects is mostly infrared, which is detected by burglar alarms, military night-vision scopes, and similar equipment.**

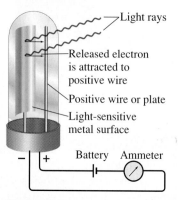

Figure 7.4
The photoelectric effect.
Light shines on a metal surface, knocking out electrons. The metal surface is contained in an evacuated tube, which allows the ejected electrons to be accelerated to a positively charged plate. As long as light of sufficient frequency shines on the metal, free electrons are produced and a current flows through the tube. When the light is turned off, the current stops flowing.

▶ **Albert Einstein obtained his Ph.D. in 1905. In the same year he published four groundbreaking research papers: one on the photoelectric effect (for which he received the Nobel Prize in physics in 1921); two on special relativity; and one on Brownian motion (which led to experiments that tested kinetic-molecular theory and ended remaining doubts of the existence of atoms and molecules).**

By the early part of the twentieth century, the wave theory of light appeared to be well entrenched. But in 1905 the German physicist Albert Einstein (1879–1955) discovered that he could explain a phenomenon known as the *photoelectric effect* by postulating that light had both wave and particle properties. Einstein based this idea on the work of the German physicist Max Planck (1858–1947).

Planck's Quantization of Energy

In 1900 Max Planck found a theoretical formula that exactly describes the intensity of light of various frequencies emitted by a hot solid at different temperatures. ◀ Earlier, others had shown experimentally that the light of maximum intensity from a hot solid varies in a definite way with temperature. A solid glows red at 750°C, then white as the temperature increases to 1200°C. As the temperature increases, more yellow and blue light become mixed with the red, giving white light. ◀

According to Planck, the atoms of the solid oscillate, or vibrate, with a definite frequency ν, depending on the solid. But in order to reproduce the results of experiments on glowing solids, he found it necessary to accept a strange idea. An atom could have only certain energies of vibration, E—those allowed by the formula

$$E = nh\nu, \qquad n = 1, 2, 3, \ldots$$

where h is a constant, now called **Planck's constant,** *with the value 6.63×10^{-34} $J \cdot s$.* The value of n must be 1 or 2 or some other whole number. Thus, the only energies a vibrating atom can have are $h\nu$, $2h\nu$, $3h\nu$, and so forth.

The numbers symbolized by n are called *quantum numbers.* The vibrational energies of the atoms are said to be *quantized;* that is, the possible energies are limited to certain values. Quantization of energy seems contradicted by everyday experience. Imagine the same concept being applied to the energy of a moving automobile. Quantization of a car's energy would mean that only certain speeds were possible. A car could travel at, say, 10, 20, or 30 miles per hour (mi/hr) but not at any intermediate speeds! For car energies, this is unreasonable.

Photoelectric Effect

Albert Einstein boldly extended Planck's work to include the structure of light itself. Einstein reasoned that if a vibrating atom changed energy, say from $3h\nu$ to $2h\nu$, it would decrease in energy by $h\nu$, and this energy would be emitted as a bit (or quantum) of light energy. He therefore postulated that light consists of quanta (now called **photons**), or *particles of electromagnetic energy, with energy E proportional to the observed frequency of the light:* ◀

$$E = h\nu$$

The **photoelectric effect** is *the ejection of electrons from the surface of a metal or from another material when light shines on it* (see Figure 7.4). Electrons are ejected, however, only when the frequency of light exceeds a certain *threshold value* characteristic of the particular metal. For example, although violet light will cause potassium metal to eject electrons, no amount of red light has any effect.

To explain this dependence of the photoelectric effect on the frequency, Einstein assumed that an electron is ejected from a metal when it is struck by a single photon. Therefore, this photon must have at least enough energy to remove the electron

Zapping Hamburger with Gamma Rays

EVERYDAY LIFE

In 1993 in Seattle, Washington, four children died and hundreds of people became sick from food poisoning when they ate undercooked hamburgers containing a dangerous strain of a normally harmless bacteria, *Escherichia coli*. A similar bout of food poisoning containing the dangerous strain of *E. coli* occurred in the summer of 1997; 17 people became ill. This time the tainted hamburger was traced to a meat processor in Nebraska, which immediately recalled 25 million pounds of ground beef. Within months, the U.S. Food and Drug Administration approved the irradiation of red meat by high-energy gamma rays to kill harmful bacteria.

The idea of using high-energy radiation to disinfect foods is not new (Figure 7.5). It has been known for about 50 years that high-energy radiation kills bacteria and molds in foods, prolonging their shelf life. Gamma rays kill organisms by breaking up the DNA molecules within their cells.

Currently, the gamma rays used in food irradiation come from the radioactive decay of cobalt-60. It is perhaps this association with radioactivity that has slowed the acceptance of irradiation as a food disinfectant. However, food is not made radioactive by irradiating it

Figure 7.5
Gamma irradiation of foods. Here, fresh produce has been irradiated to reduce spoilage and to ensure that the produce does not contain harmful bacteria.

with gamma rays. Irradiation of foods with gamma rays may soon join pasteurization as a way to protect our food supply from harmful bacteria.

from the attractive forces of the metal. A photon of red light has insufficient energy to remove an electron from potassium. But a photon corresponding to the threshold frequency has just enough energy, and at higher frequencies it has more than enough energy. When the photon hits the metal, its energy $h\nu$ is taken up by the electron.

The wave and particle pictures of light should be regarded as complementary views of the same physical entity. This is called the *wave–particle duality* of light. The equation $E = h\nu$ displays this duality; E is the energy of a light particle or photon, and ν is the frequency of the associated wave. Neither the wave nor the particle view alone is a complete description of light.

EXAMPLE 7.3

Calculating the Energy of a Photon

The red spectral line of lithium occurs at 671 nm (6.71×10^{-7} m). Calculate the energy of one photon of this light.

SOLUTION

The frequency of this light is

$$\nu = \frac{c}{\lambda} = \frac{3.00 \times 10^8 \text{ m/s}}{6.71 \times 10^{-7} \text{ m}} = 4.47 \times 10^{14}/\text{s}$$

Hence, the energy of one photon is

$$E = h\nu = 6.63 \times 10^{-34} \text{ J} \cdot s \times 4.47 \times 10^{14}/s = \mathbf{2.96 \times 10^{-19} \text{ J}}$$

See Problems 7.27, 7.28, 7.29, and 7.30.

7.3 The Bohr Theory of the Hydrogen Atom

According to Rutherford's nuclear model, the atom consists of a nucleus with most of the mass of the atom and a positive charge, around which move enough electrons to make the atom electrically neutral. But this model, offered in 1911, posed a dilemma. Using the then-current theory, one could show that an electrically charged particle (such as an electron) that revolves around a center would continuously lose energy as electromagnetic radiation. As an electron in an atom lost energy, it would spiral into the nucleus. The stability of the atom could not be explained.

A solution to this theoretical dilemma was found in 1913 by Niels Bohr (1885–1962), a Danish physicist (see Figure 7.6). Using the work of Planck and Einstein, Bohr applied a new theory to hydrogen. Before we look at Bohr's theory, we need to consider the line spectra of atoms.

Atomic Line Spectra

With a prism we can spread out the light from a bulb to give a **continuous spectrum**—that is, *a spectrum containing light of all wavelengths* (see Figure 7.7). The light emitted by a heated gas, however, yields different results. Rather than a continuous spectrum, with all colors of the rainbow, we obtain a **line spectrum**—*a spectrum showing only certain colors or specific wavelengths of light*. When the light from a hydrogen gas discharge tube is separated into its components by a prism, it gives a spectrum of lines, each line corresponding to a given wavelength. The light produced in the discharge tube is emitted by hydrogen atoms. Figure 7.1 shows the line spectrum of the hydrogen atom, as well as line spectra of other atoms.

The line spectrum of the hydrogen atom in the visible region consists of only four lines (a red, a blue-green, a blue, and a violet), although others appear in the infrared and ultraviolet regions. In 1885 J. J. Balmer showed that the wavelengths λ in the visible spectrum of hydrogen could be reproduced by a simple formula:

$$\frac{1}{\lambda} = 1.097 \times 10^7/\text{m}\left(\frac{1}{2^2} - \frac{1}{n^2}\right)$$

Here n is some whole number (integer) greater than 2. By substituting $n = 3$, for example, and calculating $1/\lambda$ and then λ, one finds $\lambda = 6.56 \times 10^{-7}$ m, or 656 nm, a wavelength corresponding to red light. The wavelengths of the other lines are obtained by successively substituting $n = 4$, $n = 5$, and $n = 6$.

Bohr's Postulates

Bohr set down the following postulates to account for (1) the stability of the hydrogen atom and (2) the line spectrum of the atom.

Figure 7.6
Niels Bohr (1885–1962).
After Bohr developed his quantum theory of the hydrogen atom, he used his ideas to explain the periodic behavior of the elements. He received the Nobel Prize in physics in 1922.

Figure 7.7
Dispersion of white light by a prism. White light, entering at the left, strikes a prism, which disperses the light into a continuous spectrum of wavelengths.

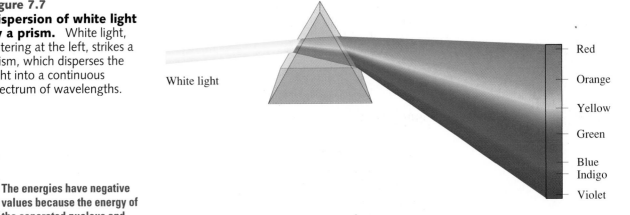

White light

Red
Orange
Yellow
Green
Blue
Indigo
Violet

▶ The energies have negative values because the energy of the separated nucleus and electron is taken to be zero. As the nucleus and electron come together to form a stable state of the atom, energy is released and the energy becomes less than zero, or negative.

1. Energy-level Postulate An electron can have only *specific energy values in an atom*, which are called its **energy levels.** Therefore, the atom itself can have only specific total energy values.

Bohr devised a rule for quantization that could be applied to the motion of an electron in an atom. From this he derived the following formula for the energy levels of the electron in the hydrogen atom:

$$E = -\frac{R_H}{n^2} \qquad n = 1, 2, 3, \ldots \infty \qquad \text{(for H atom)}$$

where R_H is a constant (expressed in energy units) with the value 2.179×10^{-18} J. Different values of the possible energies of the electron are obtained by putting in different values of n. Here n is called the *principal quantum number.* The diagram in Figure 7.8 shows the energy levels of the electron in the H atom. ◀

2. Transitions Between Energy Levels An electron in an atom can change energy only by going from one energy level to another energy level.

We explain the emission of light by atoms to give a line spectrum as follows: An electron in a higher energy level undergoes a transition to a lower energy level (see Figure 7.8). In this process, the electron loses energy, which is emitted as a photon. In other words, the final energy of the electron plus the energy of the photon equals the initial energy of the electron:

$$E_f + h\nu = E_i$$

When you rearrange this, you obtain

$$\text{Energy of emitted photon} = h\nu = E_i - E_f$$

By substituting values of the energy levels of the electron in the H atom, which Bohr had derived, he was able to reproduce Balmer's formula.

To show how Bohr obtained Balmer's formula, write n_i for the principal quantum number of the initial energy level, and n_f for the principal quantum number of the final energy level. Then, from Postulate 1,

$$E_i = -\frac{R_H}{n_i^2} \qquad \text{and} \qquad E_f = -\frac{R_H}{n_f^2}$$

Energy (E) ⟶

0    $n = \infty$
$-R_H/36$ $n = 6$
$-R_H/25$ $n = 5$
$-R_H/16$ $n = 4$
$-R_H/9$ $n = 3$

$-R_H/4$ $n = 2$

$-R_H$ $n = 1$

Figure 7.8
Energy-level diagram for the electron in the hydrogen atom. Energy is plotted on the vertical axis (in fractional multiples of R_H). The arrow represents an electron transition (discussed in Postulate 2) from level $n = 4$ to level $n = 2$. Light of wavelength 486 nm (blue-green) is emitted. (See Example 7.4 for the calculation of this wavelength.)

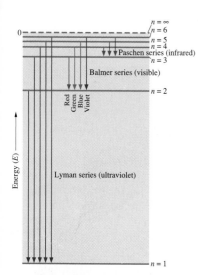

Figure 7.9
Transitions of the electron in the hydrogen atom.
The diagram shows the Lyman, Balmer, and Paschen series of transitions that occur for n_f = 1, 2, and 3, respectively.

You substitute these into the previous equation (the one shaded green), which states that the energy of the emitted photon, $h\nu$, equals $E_i - E_f$.

$$h\nu = E_i - E_f = \left(-\frac{R_H}{n_i^2}\right) - \left(-\frac{R_H}{n_f^2}\right)$$

That is,

$$h\nu = R_H\left(\frac{1}{n_f^2} - \frac{1}{n_i^2}\right)$$

Now recalling that $\nu = c/\lambda$, you can rewrite this as

$$\frac{1}{\lambda} = \frac{R_H}{hc}\left(\frac{1}{n_f^2} - \frac{1}{n_i^2}\right)$$

By substituting $R_H = 2.179 \times 10^{-18}$ J, $h = 6.626 \times 10^{-34}$ J · s, and $c = 2.998 \times 10^8$ m/s, you find that $R_H/hc = 1.097 \times 10^7$/m, which is the constant given in the Balmer formula. In Balmer's formula, the quantum number n_f is 2. If you change n_f to other integers, you obtain different series of lines for the spectrum of the H atom (see Figure 7.9).

EXAMPLE 7.4

Determining the Wavelength or Frequency of a Hydrogen Atom Transition

What is the wavelength of light emitted when the electron in a hydrogen atom undergoes a transition from energy level $n = 4$ to level $n = 2$?

PROBLEM STRATEGY

You start with the formula for the energy levels of the H atom, $E = -R_H/n^2$, and obtain the energy change for the transition. This energy difference will equal the energy of the photon, from which you can calculate frequency, then wavelength, of the emitted light. (Although you could do this problem by "plugging into" the previously derived equation for $1/\lambda$, the method followed here requires you to remember only key formulas, shown in color in the text.)

SOLUTION

From the formula for the energy levels, you know that

$$E_i = \frac{-R_H}{4^2} = \frac{-R_H}{16} \qquad \text{and} \qquad E_f = \frac{-R_H}{2^2} = \frac{-R_H}{4}$$

You subtract the lower value from the higher value, to get a positive result. Because this result equals the energy of the photon, you equate it to $h\nu$:

$$\left(\frac{-R_H}{16}\right) - \left(\frac{-R_H}{4}\right) = \frac{-4R_H + 16R_H}{64} = \frac{-R_H + 4R_H}{16} = \frac{3R_H}{16} = h\nu$$

The frequency of the light emitted is

$$\nu = \frac{3R_H}{16\,h} = \frac{3}{16} \times \frac{2.179 \times 10^{-18}\,\cancel{J}}{6.63 \times 10^{-34}\,\cancel{J}\cdot\text{s}} = 6.17 \times 10^{14}/\text{s}$$

Since $\lambda = c/\nu$,

$$\lambda = \frac{3.00 \times 10^8 \text{ m/s}}{6.17 \times 10^{14}/\text{s}} = 4.86 \times 10^{-7} \text{ m, or } \textbf{486 nm}$$

The color is blue-green (see Figure 7.8).

See Problems 7.31, 7.32, 7.35, and 7.36.

According to Bohr's theory, the *emission* of light from an atom occurs when an electron undergoes a transition from an upper energy level to a lower one. Normally, the electron in a hydrogen atom exists in its lowest, or $n = 1$, level. To get into a higher energy level, the electron must gain energy, or be *excited*. The excitation of atoms and the subsequent emission of light are most likely to occur in a hot gas, where atoms have large kinetic energies.

Bohr's theory explains not only the emission but also the *absorption* of light. When an electron in the hydrogen atom undergoes a transition from $n = 3$ to $n = 2$, a photon of red light (wavelength 656 nm) is emitted. When red light of wavelength 656 nm shines on a hydrogen atom in the $n = 2$ level, a photon can be absorbed. If the photon is absorbed, the energy is gained by the electron, which undergoes a transition to the $n = 3$ level. Materials that have a color, such as dyed textiles and painted walls, appear colored because of the absorption of light. For example, when white light falls on a substance that absorbs red light, the color components that are not absorbed, the yellow and blue light, are reflected. The substance appears blue-green.

 CONCEPT CHECK 7.1

An atom has a line spectrum consisting of a red line and a blue line. Assume that each line corresponds to a transition between two adjacent energy levels. Sketch an energy-level diagram with three energy levels that might explain this line spectrum, indicating the transitions on this diagram. Consider the transition from the highest energy level on this diagram to the lowest energy level. How would you describe the color or region of the spectrum corresponding to this transition?

Quantum Mechanics and Quantum Numbers

Bohr's theory firmly established the concept of atomic energy levels. It was unsuccessful, however, in accounting for the details of atomic structure and in predicting energy levels for atoms other than hydrogen. Further understanding of atomic structure required other theoretical developments.

7.4 Quantum Mechanics

Current ideas about atomic structure depend on the principles of *quantum mechanics,* a theory that applies to submicroscopic particles such as electrons. The development of this theory was stimulated by the discovery of the de Broglie relation.

Lasers and Compact Disc Players

MATERIALS

Lasers are sources of intense, highly directed beams of *monochromatic* light—light of very narrow wavelength range. Figure 7.10 shows a laser being used in an industrial research project. The word *laser* is an acronym meaning *l*ight *a*mplification by *s*timulated *e*mission of *r*adiation. Many different kinds of lasers now exist, but the general principle of a laser can be understood by looking at the ruby laser, the first type constructed (in 1960).

Ruby is aluminum oxide containing a small concentration of chromium(III) ions, Cr^{3+}, in place of some aluminum ions. The electron transitions in a ruby laser are those of Cr^{3+} ions in solid Al_2O_3. An energy-level diagram of Cr^{3+} in ruby is shown in Figure 7.11. Most of the Cr^{3+} ions are initially in the lowest energy level (level 1). If you shine light of wavelength 545 nm on a ruby crystal, the light is absorbed and Cr^{3+} ions undergo transitions from level 1 to level 3. A few of these ions in level 3 emit photons and return to level 1, but most of them undergo *radiationless transitions* to level 2. In these transitions, the ions lose energy as heat to the ruby crystal, rather than emit photons.

The Cr^{3+} ion in level 2, like any species in a level other than the lowest, is unstable and in time will undergo a transition to level 1. Thus, you expect the Cr^{3+} ions in level 2 to undergo transitions to level 1 with the

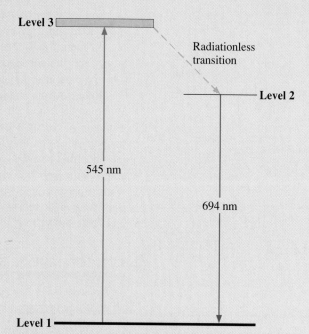

Figure 7.11
Energy levels of the chromium(III) ion in ruby. The energy levels have been numbered by increasing energy from 1 to 3. (Level 3 is broadened in the solid state.) When chromium(III) ions absorb the 545-nm portion of the light from a flash lamp, electrons in the ions first undergo transitions to level 3 and then radiationless transitions to level 2. (The energy is lost as heat.) Electrons accumulate in level 2 until more exist in level 2 than in level 1. Then a photon from a spontaneous emission from level 2 to level 1 can stimulate the emission from other atoms in level 2. These photons move back and forth between reflective surfaces, stimulating additional emissions and forming a laser pulse at 694 nm.

Figure 7.10
Lasers in industrial research. A laser beam is being used in a research project to design a metal-cutting tool.

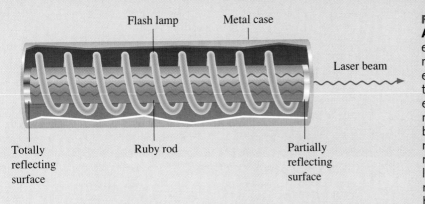

Figure 7.12
A ruby laser. A flash lamp encircles a ruby rod. Light of 545 nm (green) from the flash "pumps" electrons from level 1 to level 3, then level 2, from which stimulated emission forms a laser pulse at 694 nm (red). The stimulated emission bounces back and forth between reflective surfaces at the ends of the ruby rod, building up a coherent laser beam. One end has a partially reflective surface to allow the laser beam to exit from the ruby.

spontaneous emission of photons. (The wavelength of this emission is 694 nm.) However, this spontaneous emission of Cr^{3+} is relatively slow. If you flash a ruby rod with a bright light at 545 nm, most of the Cr^{3+} ions end up in level 2 for perhaps a fraction of a millisecond (which is quite a long time, compared with that of most excited states of atoms—they exist only about 10^{-8} s). This buildup of many excited species is crucial to the operation of a laser. If these excited ions can be triggered to emit simultaneously, or nearly simultaneously, an intense emission will be obtained.

The process of *stimulated emission* is ideal for this triggering. When a photon corresponding to 694 nm encounters a Cr^{3+} ion in level 2, it stimulates the ion to undergo the transition from level 2 to level 1. The ion emits a photon corresponding to exactly the same wavelength as the original photon. In place of just one photon, there are now two photons, the original one and the one obtained by stimulated emission. The net effect is to increase the intensity of the light at this wavelength. Thus, a weak light at 694 nm can be amplified by stimulated emission of the excited ruby.

We can now explain the operation of a ruby laser. A sketch of the laser is shown in Figure 7.12. It consists of a ruby rod a centimeter or less in diameter and 2 cm to 10 cm long. The ruby rod is silvered at one end to provide a completely reflective surface; the other end has a partially reflective surface. A coiled flash-lamp tube is placed around the ruby rod. When the flash lamp is discharged, a bright light is emitted, and light at 545 nm (green) is absorbed by the ruby. Most of the Cr^{3+} ions end up in level 2. A few of these ions spontaneously emit photons corre-

sponding to 694 nm (red light), and these photons stimulate other ions to emit. As the photons are reflected back and forth between the reflective surfaces, more and more ions are stimulated to emit, and the light quickly builds up in intensity until it passes through the partially reflective surface as a pulse of laser light at 694 nm.

Laser light is *coherent*. This means that the waves forming the beam are all in phase; that is, the waves have their maxima and minima at the same points in space and time. The property of coherence of a laser beam is used in compact disc (CD) audio players (Figure 7.13). Music is encoded on the disc in the form of pits, or indentations, on a spiral track. When the disc is played, a small laser beam scans the track and is reflected back to a detector. Light reflected from an indentation is out of phase with light from the laser and interferes with it. Because of this interference, the reflected beam is diminished in intensity and gives a diminished detector signal. Fluctuations in the signal are then converted to sound.

Other properties of laser light are used in different applications. The ability to focus intense light on a spot is used in the surgical correction of a detached retina in the eye. In effect, the laser beam is used to do "spot welding" of the retina. Laser printers also employ the intensity of a laser beam. These printers follow the principle of photocopiers but use a computer to direct the laser light in a pattern of dots to form an image. In chemical research, laser beams provide intense monochromatic light to locate energy levels in molecules, to study the products of very fast chemical reactions, and to analyze samples for small amounts of substances.

(continued)

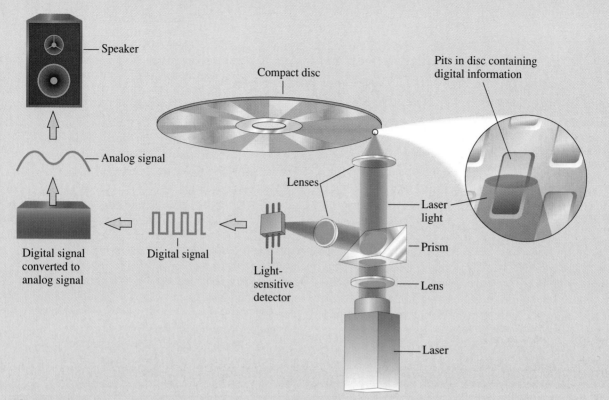

Figure 7.13

How a compact disc player works. Music or other information is encoded in the form of pits on a compact disc. Lenses direct laser light to the disc, which reflects it back. Light reflected from a pit is out of phase with light from the laser, and these two waves interfere, reducing the intensity of the wave. As the disc moves around, light is reflected from either the main surface at full intensity (on) or from a pit at reduced intensity (off). A light-sensitive detector converts this reflected light to a digital signal (a series of on and off pulses), and a digital-to-analog converter changes this signal to an analog signal that the speakers can accept.

de Broglie Relation

According to Einstein, light has not only wave properties, which we characterize by frequency and wavelength, but also particle properties. For example, a particle of light, the photon, has a definite energy $E = h\nu$. One can also show that the photon has momentum. (The momentum of a particle is the product of its mass and speed.) This momentum, mc, is related to the wavelength of the light: $mc = h/\lambda$ or $\lambda = h/mc$. ◄

In 1923 the French physicist Louis de Broglie reasoned that if light (considered as a wave) exhibits particle aspects, then perhaps particles of matter show characteristics of waves under the proper circumstances. He therefore postulated that a particle of matter of mass m and speed v has an associated wavelength, by analogy with light:

► A photon has a rest mass of zero but a relativistic mass m as a result of its motion. Einstein's equation $E = mc^2$ relates this relativistic mass to the energy of the photon, which also equals $h\nu$. Therefore, $mc^2 = h\nu$, or $mc = h\nu/c = h/\lambda$.

$$\lambda = \frac{h}{mv}$$

The equation $\lambda = h/mv$ is called the **de Broglie relation.**

If matter has wave properties, why are they not commonly observed? Calculation using the de Broglie relation shows that a baseball (0.145 kg) moving at about

Figure 7.14
Scanning electron microscope image. This image is of a wasp's head. Color has been added by computer for contrast in discerning different parts of the image.

▶ In 1986, Thomas H. Newman collected a $1000 prize offered 26 years earlier by Richard Feynman, physics Nobel Prize winner. Newman, a student of R. Fabian Pease at Stanford University, satisfied Feynman's challenge to reduce a normal printed page to 1/25,000th of its linear dimensions. Newman used a specially constructed electron-beam tool to etch the first page of *A Tale of Two Cities* onto an area 5.6 μm by 5.6 μm. The image could be read with an electron microscope, as Feynman required.

▶ Schrödinger received the Nobel Prize in physics in 1933 for his wave formulation of quantum mechanics. Werner Heisenberg won the Nobel Prize the previous year for his matrix-algebra formulation of quantum mechanics. The two formulations yield identical results.

60 mi/hr (27 m/s) has a wavelength of about 10^{-34} m, a value so incredibly small that such waves cannot be detected. On the other hand, electrons moving at near the speed of light have wavelengths the size of a few hundred picometers (1 pm = 10^{-12} m).

The wave property of electrons was first demonstrated in 1927 by C. Davisson and L. H. Germer in the United States and by George Paget Thomson in Britain. They showed that a beam of electrons, just like x rays, can be diffracted by a crystal. The German physicist Ernst Ruska used this wave property to construct the first *electron microscope* in 1933; he shared the 1986 Nobel Prize in physics for this work. The *resolving power,* or ability to distinguish detail, of a microscope that uses waves depends on their wavelength. To resolve detail the size of several hundred picometers, we need a wavelength on that order. Electrons are readily focused with electric and magnetic fields. Figure 7.14 shows a photograph taken with an electron microscope, displaying the detail that is possible with this instrument. ◀

EXAMPLE 7.5

Applying the de Broglie Relation

What is the wavelength (in picometers) associated with an electron, whose mass is 9.11×10^{-31} kg, traveling at a speed of 4.19×10^6 m/s? (This speed can be attained by an electron accelerated between two charged plates differing by 50.0 volts; voltages in the kilovolt range are used in electron microscopes.)

SOLUTION

$$\lambda = \frac{h}{mv} = \frac{6.63 \times 10^{-34} \, \text{J} \cdot \text{s}}{9.11 \times 10^{-31} \, \text{kg} \times 4.19 \times 10^6 \, \text{m/s}} = 1.74 \times 10^{-10} \, \text{m} = \mathbf{174 \ pm}$$

See Problems 7.37 and 7.38.

Wave Functions

In 1926 Erwin Schrödinger, guided by de Broglie's work, devised a theory that could be used to find the wave properties of electrons in atoms and molecules. *The branch of physics that mathematically describes the wave properties of submicroscopic particles* is called **quantum mechanics** or **wave mechanics.** ◀

We will discuss some of the most important conclusions of the theory. Our usual concept of motion comes from what we see in the everyday world. We might, for instance, visually follow a ball that has been thrown. The path of the ball is given by its position and velocity at various times. We are therefore conditioned to think in terms of a continuous path for moving objects. In Bohr's theory, the electron was thought of as moving about, or orbiting, the nucleus in the way the earth orbits the sun. Quantum mechanics vastly changes this view of motion.

In 1927 Werner Heisenberg showed from quantum mechanics that it is impossible to know simultaneously, with absolute precision, both the position and the momentum of a particle such as an electron. Heisenberg's **uncertainty principle** is *a relation that states that the product of the uncertainty in position and the uncertainty in momentum of a particle can be no smaller than Planck's constant divided by* 4π. Thus, letting Δx be the uncertainty in the x coordinate of the particle and letting Δp_x be the uncertainty in the momentum in the x direction, we have

$$(\Delta x)(\Delta p_x) \geq \frac{h}{4\pi}$$

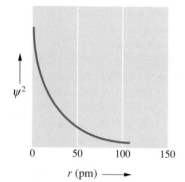

Figure 7.15
Plot of ψ^2 for the lowest energy level of the hydrogen atom. The square of the wave function is plotted versus the distance, r, from the nucleus.

The uncertainty principle says that the more precisely you know the position (the smaller Δx, Δy, and Δz), the less well you know the momentum of the particle (the larger Δp_x, Δp_y, and Δp_z).

The uncertainty principle is only significant for particles of very small mass such as electrons. You can see this by noting that the momentum equals mass times velocity, so $p_x = mv_x$. The preceding relation becomes

$$(\Delta x)(\Delta v_x) \geq \frac{h}{4\pi m}$$

For dust particles and baseballs, where m is relatively large, the term on the right becomes nearly zero, and the uncertainties of position and velocity are quite small. For electrons the uncertainties in position and momentum are normally quite large.

Although quantum mechanics does not allow us to describe the electron in the hydrogen atom as moving in an orbit, it does allow us to make *statistical* statements about where we would find the electron if we were to look for it. For example, we can obtain the *probability* of finding an electron at a certain point in a hydrogen atom. Although we cannot say that an electron will definitely be at a particular position at a given time, we can say that the electron is likely (or not likely) to be at this position.

Information about a particle in a given energy level (such as an electron in an atom) is contained in a mathematical expression called a *wave function*, denoted by the Greek letter psi, ψ. The wave function is obtained by solving an equation of quantum mechanics (Schrödinger's equation). Its square, ψ^2, gives the probability of finding the particle within a region of space.

The wave function and its square, ψ^2, have values for all locations about a nucleus. Figure 7.15 shows values of ψ^2 for the electron in the lowest energy level of the hydrogen atom along a line starting from the nucleus. Note that ψ^2 is large near the nucleus ($r = 0$), indicating that the electron is most likely to be found in this region. The value of ψ^2 decreases rapidly as the distance from the nucleus increases, but ψ^2 never goes to exactly zero, although the probability does become extremely small at large distances from the nucleus.

Figure 7.16 shows another view of this electron probability. The graph plots the probability of finding the electron in different spherical shells at particular distances from the nucleus, rather than the probability at a point. Even though the probability of finding the electron at a point near the nucleus is high, the volume of any shell there is small. Therefore, the probability of finding the electron *within a shell* is greatest at some distance from the nucleus. This distance just happens to equal the radius that Bohr calculated for an electron orbit in his model.

Figure 7.16
Probability of finding an electron in a spherical shell about the nucleus.
(A) The diagram shows the probability density for an electron in a hydrogen atom. The region is marked off in shells about the nucleus. *(B)* The graph shows the probability of finding the electron within shells at various distances from the nucleus (radial probability). The curve exhibits a maximum, which means that the radial probability is greatest for a given distance from the nucleus.

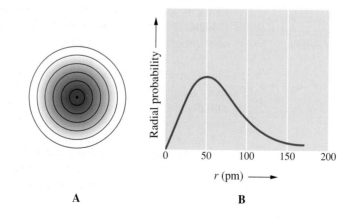

A   B

7.5 Quantum Numbers and Atomic Orbitals

According to quantum mechanics, each electron in an atom is described by four different quantum numbers, three of which (n, l, and m_l) specify the wave function that gives the probability of finding the electron at various points in space. ◄ *A wave function for an electron in an atom* is called an **atomic orbital.** An atomic orbital is pictured qualitatively by describing the region of space where there is high probability of finding the electrons. A fourth quantum number (m_s) refers to a magnetic property of electrons called *spin.*

► Three different quantum numbers are needed because there are three dimensions to space.

Quantum Numbers

The allowed values and general meaning of each of the four quantum numbers of an electron in an atom are as follows:

1. Principal Quantum Number (n) *This quantum number is the one on which the energy of an electron in an atom principally depends; it can have any positive value: 1, 2, 3, and so on.* The energy of an electron in an atom depends *principally* on n. The smaller n is, the lower the energy.

The *size* of an orbital also depends on n. The larger the value of n is, the larger the orbital. Orbitals of the same quantum state n are said to belong to the same *shell.*

2. Angular Momentum Quantum Number (l) (Also Called *Azimuthal Quantum Number*) *This quantum number distinguishes orbitals of given n having different shapes; it can have any integer value from 0 to n − 1.* Within each shell of quantum number n, there are n different kinds of orbitals, each with a distinctive shape denoted by an l quantum number. For example, if an electron has a principal quantum number of 3, the possible values for l are 0, 1, and 2.

Although the energy of an orbital is principally determined by the n quantum number, the energy also depends somewhat on the l quantum number (except for the H atom). For a given n, the energy of an orbital increases with l.

Orbitals of the same n but different l are said to belong to different *subshells* of a given shell. The different subshells are usually denoted by letters as follows:

Letter	*s*	*p*	*d*	*f*	*g* . . .
l	0	1	2	3	4 . . .

To denote a subshell within a particular shell, we write the value of the n quantum number for the shell, followed by the letter designation for the subshell. For example, $2p$ denotes a subshell with quantum numbers $n = 2$ and $l = 1$. ◄

► The rather odd choice of letter symbols for *l* quantum numbers survives from old spectroscopic terminology (describing the lines in a spectrum as *s*harp, *p*rincipal, *d*iffuse, and *f*undamental).

3. Magnetic Quantum Number (m_l) *This quantum number distinguishes orbitals of given n and l—that is, of given energy and shape but having a different orientation in space; the allowed values are the integers from −l to +l.* For $l = 0$ (s subshell), the allowed m_l quantum number is 0 only; there is only one orbital in the s subshell. For $l = 1$ (p subshell), $m_l = -1$, 0, and $+1$; there are three different orbitals in the p subshell. The orbitals have the same shape but different orientations in space. In addition, each orbital of a given subshell has the same energy. Note that there are $2l + 1$ orbitals in each subshell of quantum number l.

Scanning Tunneling Microscopy

The 1986 Nobel Prize in physics went to three physicists for their work in developing microscopes for viewing extremely small objects. Half of the prize went to Ernst Ruska for the development of the electron microscope, described earlier. The other half was awarded to Gerd Binnig and Heinrich Rohrer, at IBM's research laboratory in Zurich, Switzerland, for their invention of the *scanning tunneling microscope* in 1981. This instrument makes possible the viewing of atoms and molecules on a solid surface (Figure 7.17).

The scanning tunneling microscope consists of a tungsten metal needle with an extremely fine point (the probe) placed close to the sample to be viewed (Figure 7.18). If the probe is close enough to the sample, electrons can tunnel from the probe to the sample. The probability for this can be increased by having a small voltage applied between the probe and sample. Electrons tunneling from the probe to the sample give rise to a measurable electric current. The magnitude of the current depends on the distance between the probe and the sample (as well as on the wave function of the atom in the sample). By adjusting this distance, the current can be maintained at a fixed value. As the probe scans the sample, it moves toward or away from the sample to maintain a fixed current; in effect, the probe follows the contours of the sample.

Researchers routinely use the scanning tunneling microscope to study the arrangement of atoms and molecules on surfaces. Figure 7.19 shows 48 iron atoms that were arranged in a circle on a copper surface. Each of the iron atoms looks like a sharp mountain peak rising from a plain. Inside this "quantum corral" (as it has come to be called), you can see the wavelike distribution of electrons trapped within. Here is graphic proof of the wavelike nature of electrons!

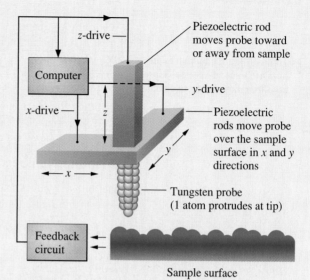

Figure 7.18
The scanning tunneling microscope. A tunneling current flows between the probe and the sample when there is a small voltage between them. A feedback circuit, which provides this voltage, senses the current and varies the voltage on a piezoelectric rod (z-drive) in order to keep the distance constant between the probe and sample. A computer has been programmed to provide voltages to the x-drive and the y-drive to move the probe over the surface of the sample.

Figure 7.19
Quantum corral. IBM scientists used a scanning tunneling microscope probe to arrange 48 iron atoms in a circle on a copper metal surface. The iron atoms appear as the tall peaks in the diagram. Note the wavelike ripple of electrons trapped within the circle of atoms (called *a quantum corral*). [IBM Research Division, Almaden Research Center; research done by Donald M. Eigler and coworkers.]

Figure 7.17
Scanning tunneling microscope image of benzene molecules on a metal surface.
Benzene molecules, C_6H_6, are arranged in a regular array on a rhodium metal surface.

Table 7.1
Permissible Values of Quantum Numbers for Atomic Orbitals

n	l	m_l*	Subshell Notation	Number of Orbitals in the Subshell
1	0	0	$1s$	1
2	0	0	$2s$	1
2	1	$-1, 0, +1$	$2p$	3
3	0	0	$3s$	1
3	1	$-1, 0, +1$	$3p$	3
3	2	$-2, -1, 0, +1, +2$	$3d$	5
4	0	0	$4s$	1
4	1	$-1, 0, +1$	$4p$	3
4	2	$-2, -1, 0, +1, +2$	$4d$	5
4	3	$-3, -2, -1, 0, +1, +2, +3$	$4f$	7

*Any one of the m_l quantum numbers may be associated with the n and l quantum numbers on the same line.

4. Spin Quantum Number (m_s)   *This quantum number refers to the two possible orientations of the spin axis of an electron; possible values are $+\frac{1}{2}$ and $-\frac{1}{2}$.* An electron acts as though it were spinning on its axis like the earth. Such an electron spin would give rise to a circulating electric charge that would generate a magnetic field. In this way, an electron behaves like a small bar magnet, with a north and a south pole. ◄

▶ Electron spin will be discussed further in Section 8.1.

Table 7.1 lists the permissible quantum numbers for all orbitals through the $n = 4$ shell. These values follow from the rules just given. Energies for the orbitals are shown in Figure 7.20 for the hydrogen atom. Note that all orbitals with the same principal quantum number n have the same energy. For atoms with more than one electron, however, only orbitals in the same subshell (denoted by a given n and l) have the same energy. We will have more to say about orbital energies in Chapter 8.

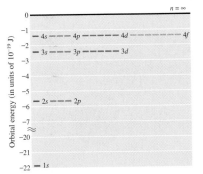

Figure 7.20
Orbital energies of the hydrogen atom.   The lines for each subshell indicate the number of different orbitals of that subshell. (Note break in the energy scale.)

EXAMPLE 7.6

Using the Rules for Quantum Numbers

State whether each of the following sets of quantum numbers is permissible for an electron in an atom. If a set is not permissible, explain why.

a. $n = 1$, $l = 1$, $m_l = 0$, $m_s = +\frac{1}{2}$

b. $n = 3$, $l = 1$, $m_l = -2$, $m_s = -\frac{1}{2}$

c. $n = 2$, $l = 1$, $m_l = 0$, $m_s = +\frac{1}{2}$

d. $n = 2$, $l = 0$, $m_l = 0$, $m_s = 1$

SOLUTION

a. **Not permissible.** The l quantum number is equal to n; it must be less than n.

b. **Not permissible.** The magnitude of the m_l quantum number (that is, the m_l value, ignoring its sign) must not be greater than l.

99% contour

1s orbital

99% contour

2s orbital

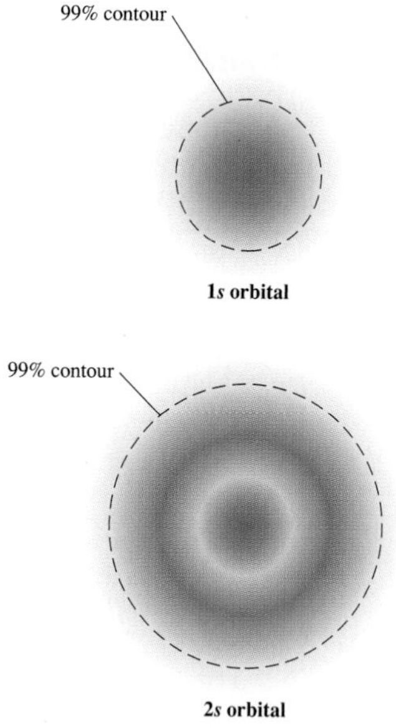

Figure 7.21
Cross-sectional representations of the probability distributions of *s* orbitals. In a 1*s* orbital, the probability distribution is largest near the nucleus. In a 2*s* orbital, it is greatest in a spherical shell about the nucleus. Note the relative "size" of the orbitals, indicated by the 99% contours.

c. **Permissible.**

d. **Not permissible.** The m_s quantum number can be only $+\frac{1}{2}$ or $-\frac{1}{2}$.

See Problems 7.45 and 7.46.

Atomic Orbital Shapes

An *s* orbital has a spherical shape, though specific details of the probability distribution depend on the value of *n*. Figure 7.21 shows cross-sectional representations of the probability distributions of a 1*s* and a 2*s* orbital. The color shading is darker where the electron is more likely to be found. In the case of a 1*s* orbital, the electron is most likely to be found near the nucleus. The shading becomes lighter as the distance from the nucleus increases, indicating that the electron is less likely to be found there.

The orbital does not abruptly end at some particular distance from the nucleus. We can gauge the "size" of the orbital by means of the *99% contour*. The electron has a 99% probability of being found within the space of the 99% contour (the sphere indicated by the dashed line in the diagram).

A 2*s* orbital differs in detail from a 1*s* orbital. The electron in a 2*s* orbital is likely to be found in two regions, one near the nucleus and the other in a spherical shell about the nucleus. (The electron is most likely to be here.) The 99% contour shows that the 2*s* orbital is larger than the 1*s* orbital.

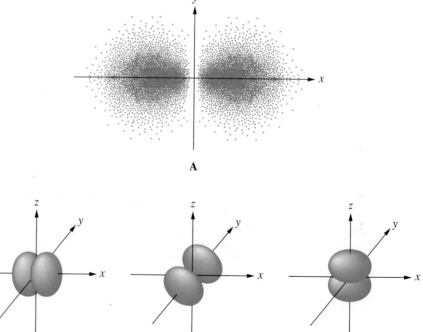

Figure 7.22
The 2*p* orbitals. *(A)* Electron distribution in the 2p_x orbital. Note that it consists of two lobes oriented along the *x*-axis. *(B)* Orientations of the three 2*p* orbitals. The drawings depict the general shape and orientation of the orbitals, but not the detailed electron distribution as in *(A)*.

2p_x orbital

2p_y orbital

2p_z orbital

B

Figure 7.23
The five 3*d* orbitals. These are labeled by subscripts, as in d_{xy}, that describe their mathematical characteristics.

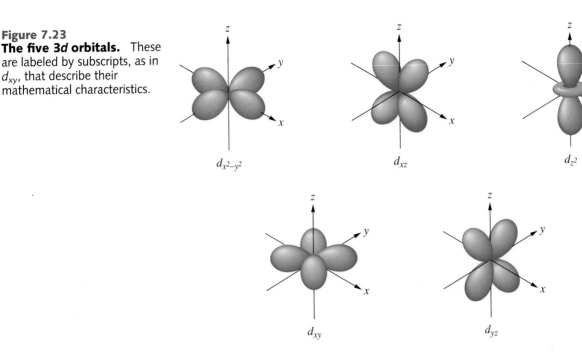

There are three *p* orbitals in each *p* subshell. All *p* orbitals have the same basic shape (two lobes arranged along a straight line with the nucleus between the lobes) but differ in their orientations in space. Because the three orbitals are set at right angles to each other, we can show each one as oriented along a different coordinate axis (Figure 7.22). We denote these orbitals as $2p_x$, $2p_y$, and $2p_z$. A $2p_x$ orbital has its greatest electron probability along the *x*-axis, a $2p_y$ orbital along the *y*-axis, and a $2p_z$ orbital along the *z*-axis.

There are five *d* orbitals, which have more complicated shapes than do *s* and *p* orbitals. These are represented in Figure 7.23.

A Checklist for Review

Important Terms

wavelength (λ) (7.1)
frequency (ν) (7.1)
electromagnetic
  spectrum (7.1)
Planck's constant (7.2)
photons (7.2)

photoelectric effect (7.2)
continuous spectrum (7.3)
line spectrum (7.3)
energy levels (7.3)
de Broglie relation (7.4)

quantum (wave)
  mechanics (7.4)
uncertainty principle (7.4)
atomic orbital (7.5)
principal quantum number
  (*n*) (7.5)

angular momentum
  quantum number (*l*) (7.5)
magnetic quantum number
  (m_l) (7.5)
spin quantum number
  (m_s) (7.5)

Key Equations

$c = \nu\lambda$

$E = h\nu$

$E = -\dfrac{R_H}{n^2} \qquad n = 1, 2, 3, \ldots \infty \qquad$ (for H atom)

Energy of emitted photon $= h\nu = E_i - E_f$

$\lambda = \dfrac{h}{mv}$

Summary of Facts and Concepts

One way to study the electronic structure of the atom is to analyze the electromagnetic radiation that is emitted from an atom. Electromagnetic radiation is characterized by its *wavelength λ* and frequency *ν*, and these quantities are related to the speed of light c ($c = \nu\lambda$).

Einstein showed that light consists of particles *(photons)*, each of energy $E = h\nu$, where h is Planck's constant. According to Bohr, electrons in an atom have *energy levels,* and when an electron in a higher energy level drops (or undergoes a transition) to a lower energy level, a photon is emitted. The energy of the photon equals the difference in energy between the two levels.

Electrons and other particles of matter have both particle and wave properties. For a particle of mass m and speed v, the wavelength is related to momentum mv by the *de Broglie relation:* $\lambda = h/mv$. The wave properties of a particle are described by a *wave function,* from which we can get the probability of finding the particle in different regions of space.

Each electron in an atom is characterized by four different quantum numbers. The distribution of an electron in space—its *atomic orbital*—is characterized by three of these quantum numbers: the principal quantum number, the angular momentum quantum number, and the magnetic quantum number. The fourth quantum number (spin quantum number) describes the magnetism of the electron.

Operational Skills

1. **Relating wavelength and frequency of light**   Given the frequency of light, calculate the wavelength, or vice versa.   **(EXAMPLES 7.1, 7.2)**
2. **Calculating the energy of a photon**   Given the frequency or wavelength of light, calculate the energy associated with one photon.   **(EXAMPLE 7.3)**
3. **Determining the wavelength or frequency of a hydrogen atom transition**   Given the initial and final principal quantum numbers for an electron transition in the hydro-

gen atom, calculate the frequency or wavelength of light emitted **(EXAMPLE 7.4)**. You need the value of R_H.
4. **Applying the de Broglie relation**   Given the mass and speed of a particle, calculate the wavelength of the associated wave.   **(EXAMPLE 7.5)**
5. **Using the rules for quantum numbers**   Given a set of quantum numbers n, l, m_l, and m_s, state whether that set is permissible for an electron.   **(EXAMPLE 7.6)**

Review Questions

7.1   Give a brief wave description of light. What are two characteristics of light waves?

7.2   What is the mathematical relationship among the different characteristics of light waves? State the meaning of each of the terms in the equation.

7.3   Planck originated the idea that energies can be quantized. What does the term *quantized* mean? What was Planck trying to explain when he was led to the concept of quantization of energy? Give the formula he arrived at and explain each of the terms in the formula.

7.4   In your own words, explain the photoelectric effect. How does the photon concept explain this effect?

7.5   Describe the wave–particle picture of light.

7.6   Give the equation that relates particle properties of light. Explain the meaning of each symbol in the equation.

7.7   Physical theory at the time Rutherford proposed his nuclear model of the atom was not able to explain how this model could give a stable atom. Explain the nature of this difficulty.

7.8   Explain the main features of Bohr's theory. Do these features solve the difficulty alluded to in Question 7.7?

7.9   Explain the process of emission of light by an atom.

7.10   Explain the process of absorption of light by an atom.

7.11   What is the evidence for electron waves? Give a practical application.

7.12   Give the possible values of (a) the principal quantum number, (b) the angular momentum quantum number, (c) the magnetic quantum number, and (d) the spin quantum number.

7.13   What is the notation for the subshell in which $n = 4$ and $l = 3$? How many orbitals are in this subshell?

7.14   What is the general shape of an s orbital? of a p orbital?

Conceptual Problems

7.15 Consider two beams of the same yellow light. Imagine that one beam has its wavelength doubled; the other has its frequency doubled. Which of these two beams is then in the ultraviolet region?

7.16 Some infrared radiation has a wavelength that is 1000 times larger than that of a certain visible light. This visible light has a frequency that is 1000 times smaller than that of some X radiation. How many times more energy is there in a photon of this X radiation than there is in a photon of the infrared radiation?

7.17 Three emission lines involving three energy levels in an atom occur at wavelengths x, $1.5x$, and $3.0x$ nanometers. Which wavelength corresponds to the transition from the highest to the lowest of the three energy levels?

7.18 An atom emits yellow light when an electron makes the transition from the $n = 5$ to the $n = 1$ level. In separate experiments, suppose you bombarded the $n = 1$ level of this atom with red light, yellow light (obtained from the previous emission), and blue light. In which experiment or experiments would the electron be promoted to the $n = 5$ level?

7.19 Which of the following particles has the longest wavelength?
a. an electron traveling at x meters per second
b. a proton traveling at x meters per second
c. a proton traveling at $2x$ meters per second

7.20 Imagine a world in which the rule for the l quantum number is that values start with 1 and go up to n. The rules for the n and m_l quantum numbers are unchanged from those of our world. Write the quantum numbers for the first two shells (i.e., $n = 1$ and $n = 2$).

Practice Problems

Electromagnetic Waves

7.21 Radio waves in the AM region have frequencies in the range 550 to 1600 kilocycles per second (550 to 1600 kHz). Calculate the wavelength corresponding to a radio wave of frequency 1.365×10^6/s (that is, 1365 kHz).

7.22 Microwaves have frequencies in the range 10^9 to 10^{12}/s (cycles per second), equivalent to between 1 gigahertz and 1 terahertz. What is the wavelength of microwave radiation whose frequency is 1.258×10^{10}/s?

7.23 Light with a wavelength of 478 nm lies in the blue region of the visible spectrum. Calculate the frequency of this light.

7.24 Calculate the frequency associated with light of wavelength 656 nm. (This corresponds to one of the wavelengths of light emitted by the hydrogen atom.)

7.25 The meter was defined in 1963 as the length equal to 1,650,763.73 wavelengths of the orange-red radiation emitted by the krypton-86 atom (the meter has since been redefined). What is the wavelength of this transition? What is the frequency?

7.26 The second is defined as the time it takes for 9,192,631,770 wavelengths of a certain transition of the cesium-133 atom to pass a fixed point. What is the frequency of this electromagnetic radiation? What is the wavelength?

Photons

7.27 What is the energy of a photon corresponding to radio waves of frequency 1.365×10^6/s?

7.28 What is the energy of a photon corresponding to microwave radiation of frequency 1.258×10^{10}/s?

7.29 The green line in the atomic spectrum of thallium has a wavelength of 535 nm. Calculate the energy of a photon of this light.

7.30 Indium compounds give a blue-violet flame test. The atomic emission responsible for this blue-violet color has a wavelength of 451 nm. Obtain the energy of a single photon of this wavelength.

Bohr Theory

7.31 An electron in a hydrogen atom in the level $n = 5$ undergoes a transition to level $n = 3$. What is the frequency of the emitted radiation?

7.32 Calculate the frequency of electromagnetic radiation emitted by the hydrogen atom in the electron transition from $n = 4$ to $n = 3$.

7.33 Calculate the shortest wavelength of the electromagnetic radiation emitted by the hydrogen atom in undergoing a transition from the $n = 6$ level.

7.34   Calculate the longest wavelength of the electromagnetic radiation emitted by the hydrogen atom in undergoing a transition from the $n = 6$ level.

7.35   What is the difference in energy between the two levels responsible for the violet emission line of the calcium atom at 422.7 nm?

7.36   What is the difference in energy between the two levels responsible for the ultraviolet emission line of the magnesium atom at 285.2 nm?

de Broglie Waves

Note: Masses of the electron, proton, and neutron are listed on the inside back cover of this book.

7.37   What is the wavelength of a neutron traveling at a speed of 4.15 km/s? (Neutrons of these speeds are obtained from a nuclear pile.)

7.38   What is the wavelength of a proton traveling at a speed of 6.21 km/s? What would be the region of the spectrum for electromagnetic radiation of this wavelength?

7.39   At what speed must an electron travel to have a wavelength of 10.0 pm?

7.40   At what speed must a neutron travel to have a wavelength of 10.0 pm?

Atomic Orbitals

7.41   If the n quantum number of an atomic orbital is 4, what are the possible values of l? If the l quantum number is 3, what are the possible values of m_l?

7.42   The n quantum number of an atomic orbital is 5. What are the possible values of l? What are the possible values of m_l if the l quantum number is 4?

7.43   Give the notation (using letter designations for l) for the subshells denoted by the following quantum numbers.
a. $n = 6, l = 2$
b. $n = 5, l = 4$
c. $n = 4, l = 3$
d. $n = 6, l = 1$

7.44   Give the notation (using letter designations for l) for the subshells denoted by the following quantum numbers.
a. $n = 3, l = 1$
b. $n = 4, l = 2$
c. $n = 4, l = 0$
d. $n = 5, l = 3$

7.45   Explain why each of the following sets of quantum numbers would not be permissible for an electron, according to the rules for quantum numbers.
a. $n = 1, l = 0, m_l = 0, m_s = +1$
b. $n = 1, l = 3, m_l = +3, m_s = +\frac{1}{2}$
c. $n = 3, l = 2, m_l = +3, m_s = -\frac{1}{2}$
d. $n = 0, l = 1, m_l = 0, m_s = +\frac{1}{2}$
e. $n = 2, l = 1, m_l = -1, m_s = +\frac{3}{2}$

7.46   State which of the following sets of quantum numbers would be possible and which impossible for an electron in an atom.
a. $n = 0, l = 0, m_l = 0, m_s = +\frac{1}{2}$
b. $n = 1, l = 1, m_l = 0, m_s = +\frac{1}{2}$
c. $n = 1, l = 0, m_l = 0, m_s = -\frac{1}{2}$
d. $n = 2, l = 1, m_l = -2, m_s = +\frac{1}{2}$
e. $n = 2, l = 1, m_l = -1, m_s = +\frac{1}{2}$

General Problems

7.47   The blue line of the strontium atom emission has a wavelength of 461 nm. What is the frequency of this light? What is the energy of a photon of this light?

7.48   The barium atom has an emission with wavelength 554 nm (green). Calculate the frequency of this light and the energy of a photon of this light.

7.49   The energy of a photon is 4.10×10^{-19} J. What is the wavelength of the corresponding light? What is the color of this light?

7.50   The energy of a photon is 3.34×10^{-19} J. What is the wavelength of the corresponding light? What is the color of this light?

7.51   The photoelectric work function of a metal is the minimum energy needed to eject an electron by irradiating the metal with light. For calcium, this work function equals 4.34×10^{-19} J. What is the minimum frequency of light for the photoelectric effect in calcium?

7.52 The photoelectric work function for magnesium is 5.90×10^{-19} J. (The work function is the minimum energy needed to eject an electron from the metal by irradiating it with light.) Calculate the minimum frequency of light required to eject electrons from magnesium.

7.53 Calculate the wavelength of the Balmer line of the hydrogen spectrum in which the initial n quantum number is 5 and the final n quantum number is 2.

7.54 Calculate the wavelength of the Balmer line of the hydrogen spectrum in which the initial n quantum number is 3 and the final n quantum number is 2.

7.55 A hydrogen-like ion has a nucleus of charge $+Ze$ and a single electron outside this nucleus. The energy levels of these ions are $-Z^2 R_H/n^2$ (where $Z =$ atomic number). Calculate the wavelength of the transition from $n = 3$ to $n = 2$ for He^+, a hydrogen-like ion. In what region of the spectrum does this emission occur?

7.56 What is the wavelength of the transition from $n = 4$ to $n = 3$ for Li^{2+}? In what region of the spectrum does this emis-

sion occur? Li^{2+} is a hydrogen-like ion. Such an ion has a nucleus of charge $+Ze$ and a single electron outside this nucleus. The energy levels of the ion are $-Z^2 R_H/n^2$, where Z is the atomic number.

7.57 What is the number of different orbitals in each of the following subshells?
a. $3d$
b. $4f$
c. $4p$
d. $5s$

7.58 What is the number of different orbitals in each of the following subshells?
a. $5f$
b. $5g$
c. $6s$
d. $5p$

7.59 List the possible subshells for the $n = 6$ shell.

7.60 List the possible subshells for the $n = 7$ shell.

Cumulative-Skills Problems

7.61 The energy required to dissociate the Cl_2 molecule to Cl atoms is 239 kJ/mol Cl_2. If the dissociation of a Cl_2 molecule were accomplished by the absorption of a single photon whose energy was exactly the quantity required, what would be its wavelength (in meters)?

7.62 The energy required to dissociate the H_2 molecule to H atoms is 432 kJ/mol H_2. If the dissociation of an H_2 molecule were accomplished by the absorption of a single photon whose energy was exactly the quantity required, what would be its wavelength (in meters)?

7.63 A microwave oven heats by radiating food with microwave radiation, which is absorbed by the food and converted to heat. Suppose an oven's radiation wavelength is 12.5 cm. A container with 0.250 L of water was placed in the oven, and the temperature of the water rose from $20.0°C$ to $100.0°C$. How many photons of this microwave radiation were required? Assume that all the energy from the radiation was used to raise the temperature of the water.

7.64 Warm objects emit electromagnetic radiation in the infrared region. Heat lamps employ this principle to generate infrared radiation. Water absorbs infrared radiation with wavelengths near 2.80 μm. Suppose this radiation is absorbed by the water and converted to heat. A 1.00-L sample of water absorbs infrared radiation, and its temperature increases from $20.0°C$ to $30.0°C$. How many photons of this radiation are used to heat the water?

7.65 Light with a wavelength of 425 nm fell on a potassium surface, and electrons were ejected at a speed of 4.88×10^5 m/s. What energy was expended in removing an electron from the metal? Express the answer in joules (per electron) and in kilojoules per mole (of electrons).

7.66 Light with a wavelength of 405 nm fell on a strontium surface, and electrons were ejected. If the speed of an ejected electron is 3.36×10^5 m/s, what energy was expended in removing the electron from the metal? Express the answer in joules (per electron) and in kilojoules per mole (of electrons).

Sodium metal reacts rapidly with water, producing hydrogen gas, which catches fire.

Electron Configurations and Periodicity

Marie Curie, a Polish-born French chemist, and her husband, Pierre, announced the discovery of radium in 1898 (Figure 8.1). They had separated a very radioactive mixture from pitchblende, an ore of uranium. This mixture was primarily a compound of barium. When the mixture was heated in a flame, however, it gave a new atomic line spectrum, in addition to the spectrum for barium. The Curies based their discovery of a new element on this finding. It took them four more years to obtain a pure compound of radium. Radium, in most of its chemical and physical properties, is similar to the nonradioactive element barium.

Chemists had long known that groups of elements have similar properties. In 1869 Dmitri Mendeleev found that when the elements were arranged in a particular way, they fell into columns, with elements in the same column displaying similar properties. Thus, Mendeleev placed beryllium, calcium, strontium, and barium in one column. Now, with the Curies' discovery, radium was added to this column.

Figure 8.1
**Marie Sklodowska Curie
(1867–1934).** Marie Sklodowska
Curie, born in Warsaw, Poland,
began her doctoral work with Henri
Becquerel soon after he discovered
the spontaneous radiation emitted by
uranium salts. She found this radiation
to be an atomic property and coined
the word *radioactivity* for it. In
1903 the Curies and Becquerel were
awarded the Nobel Prize in physics for
their discovery of radioactivity. Three
years later, Pierre Curie was killed
in a carriage accident. Marie Curie
continued their work on radium and
in 1911 was awarded the Nobel Prize
in chemistry for the discovery of
polonium and radium and the isolation
of pure radium metal. This was the first
time a scientist had received two Nobel
awards. (Since then two others have
been so honored.)

Mendeleev's arrangement of the elements, the *periodic table,* was originally based
on the observed chemical and physical properties of the elements and their compounds. We now explain this arrangement in terms of the electronic structure of atoms.

Electronic Structure of Atoms

In Chapter 7 we found that an electron in an atom has four quantum numbers—n, l,
m_l, and m_s—associated with it. The spin quantum number, m_s, describes the spin orientation of an electron. We will look further at electron spin; then we will discuss
how electrons are distributed among the possible orbitals.

8.1   Electron Spin and the Pauli Exclusion Principle

Otto Stern and Walther Gerlach first observed electron spin magnetism in 1921. They
directed a beam of silver atoms into the field of a specially designed magnet. The
same experiment can be done with hydrogen atoms. The beam of hydrogen atoms is
split into two by the magnetic field; half of the atoms are bent in one direction and
half in the other (see Figure 8.2). The fact that the atoms are affected by the laboratory magnet shows that they themselves act as magnets.

The beam of hydrogen atoms is split into two because the electron in each atom
behaves as a tiny magnet with only two possible orientations. In effect, the electron
acts as though it were a ball of spinning charge (Figure 8.3), and like a circulating
electric charge the electron would create a magnetic field. The resulting directions of
spin magnetism correspond to spin quantum numbers $m_s = +\frac{1}{2}$ and $m_s = -\frac{1}{2}$. ◄

▶ Protons and many nuclei also
have spin.

Figure 8.2
The Stern–Gerlach experiment. The diagram shows the experiment using hydrogen atoms (simpler to interpret theoretically), although the original experiment employed silver atoms. A beam of hydrogen atoms (shown in blue) is split into two by a nonuniform magnetic field. One beam consists of atoms each with an electron having $m_s = +\frac{1}{2}$; the other beam consists of atoms each having an electron with $m_s = -\frac{1}{2}$.

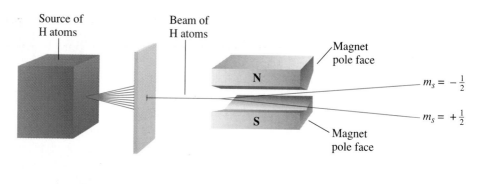

Electron Configurations and Orbital Diagrams

An **electron configuration** of an atom is *a particular distribution of electrons among available subshells.* The notation for a configuration lists the subshell symbols, one after the other, with a superscript giving the number of electrons in that subshell. For example, a configuration of the lithium atom (atomic number 3) with two electrons in the 1s subshell and one electron in the 2s subshell is written $1s^2 2s^1$.

The notation for a configuration gives the number of electrons in each subshell, but we use *a diagram to show how the orbitals of a subshell are occupied by electrons.* It is called an **orbital diagram.** An orbital is represented by a circle. Each group of orbitals in a subshell is labeled by its subshell notation. An electron in an orbital is shown by an arrow; the arrow points up when $m_s = +\frac{1}{2}$ and down when $m_s = -\frac{1}{2}$. The orbital diagram

$$1s \quad 2s \quad 2p$$

shows the electronic structure of an atom in which there are two electrons in the 1s orbital, two electrons in the 2s subshell, and one electron in the 2p subshell ($m_s = +\frac{1}{2}$). The electron configuration is $1s^2 2s^2 2p^1$.

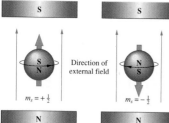

Figure 8.3
A representation of electron spin. The two possible spin orientations are indicated by the models. By convention, the spin direction is given as shown by the large arrow on the spin axis. Electrons behave as tiny bar magnets, as shown in the figure.

Pauli Exclusion Principle

Not all of the conceivable arrangements of electrons among the orbitals of an atom are physically possible. The **Pauli exclusion principle,** which summarizes experimental observations, states that *no two electrons in an atom can have the same four quantum numbers.* If one electron in an atom has the quantum numbers $n = 1$, $l = 0$, $m_l = 0$, and $m_s = +\frac{1}{2}$, no other electron can have these same quantum numbers.

Because there are only two possible values of m_s, an orbital can hold no more than two electrons. In an orbital diagram, an orbital with two electrons must be written with arrows pointing in opposite directions. We can restate the Pauli exclusion principle:

Pauli exclusion principle: An orbital can hold at most two electrons, and then only if the electrons have opposite spins.

Each subshell holds a maximum of twice as many electrons as the number of orbitals in the subshell. The maximum number of electrons in various subshells is given in the following table.

Subshell	Number of Orbitals	Maximum Number of Electrons
$s\ (l = 0)$	1	2
$p\ (l = 1)$	3	6
$d\ (l = 2)$	5	10
$f\ (l = 3)$	7	14

EXAMPLE 8.1

Applying the Pauli Exclusion Principle

Which of the following orbital diagrams or electron configurations are possible and which are impossible, according to the Pauli exclusion principle? Explain.

a. 
    1s 2s   2p

b. 
    1s 2s   2p

c. 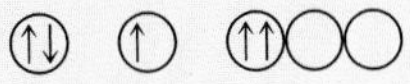
    1s 2s   2p

d. $1s^3 2s^1$

e. $1s^2 2s^1 2p^7$

f. $1s^2 2s^2 2p^6 3s^2 3p^6 3d^8 4s^2$

SOLUTION

a. **Possible** orbital diagram. b. **Impossible** orbital diagram; there are three electrons in the $2s$ orbital. c. **Impossible** orbital diagram; there are two electrons in a $2p$ orbital with the same spin. d. **Impossible** electron configuration; there are three electrons in the $1s$ subshell (one orbital). e. **Impossible** electron configuration; there are seven electrons in the $2p$ subshell (which can hold only six electrons). f. **Possible**. Note that the $3d$ subshell can hold as many as ten electrons.

See Problems 8.27, 8.28, 8.29, and 8.30.

8.2 Building-Up Principle and the Periodic Table

The configuration associated with the lowest energy level of the atom corresponds to a quantum-mechanical state called the *ground state*. Other configurations correspond to *excited states*, associated with energy levels other than the lowest. For example, the ground state of the sodium atom has the electron configuration $1s^2 2s^2 2p^6 3s^1$. Other electron configurations represent excited states. ◄

The chemical properties of an atom are related primarily to the electron configuration of its ground state. Table 8.1 lists the experimentally determined ground-state electron configurations of atoms $Z = 1$ to $Z = 36$.

▶ The transition of the sodium atom from the excited state $1s^2 2s^2 2p^6 3p^1$ to the ground state $1s^2 2s^2 2p^6 3s^1$ is accompanied by the emission of yellow light at 589 nm. Excited states of an atom are needed to describe its spectrum.

Table 8.1
Ground-State Electron Configurations of Atoms Z = 1 to 36

Z	Element	Configuration	Z	Element	Configuration
1	H	$1s^1$	19	K	$1s^2 2s^2 2p^6 3s^2 3p^6 4s^1$
2	He	$1s^2$	20	Ca	$1s^2 2s^2 2p^6 3s^2 3p^6 4s^2$
3	Li	$1s^2 2s^1$	21	Sc	$1s^2 2s^2 2p^6 3s^2 3p^6 3d^1 4s^2$
4	Be	$1s^2 2s^2$	22	Ti	$1s^2 2s^2 2p^6 3s^2 3p^6 3d^2 4s^2$
5	B	$1s^2 2s^2 2p^1$	23	V	$1s^2 2s^2 2p^6 3s^2 3p^6 3d^3 4s^2$
6	C	$1s^2 2s^2 2p^2$	24	Cr	$1s^2 2s^2 2p^6 3s^2 3p^6 3d^5 4s^1$
7	N	$1s^2 2s^2 2p^3$	25	Mn	$1s^2 2s^2 2p^6 3s^2 3p^6 3d^5 4s^2$
8	O	$1s^2 2s^2 2p^4$	26	Fe	$1s^2 2s^2 2p^6 3s^2 3p^6 3d^6 4s^2$
9	F	$1s^2 2s^2 2p^5$	27	Co	$1s^2 2s^2 2p^6 3s^2 3p^6 3d^7 4s^2$
10	Ne	$1s^2 2s^2 2p^6$	28	Ni	$1s^2 2s^2 2p^6 3s^2 3p^6 3d^8 4s^2$
11	Na	$1s^2 2s^2 2p^6 3s^1$	29	Cu	$1s^2 2s^2 2p^6 3s^2 3p^6 3d^{10} 4s^1$
12	Mg	$1s^2 2s^2 2p^6 3s^2$	30	Zn	$1s^2 2s^2 2p^6 3s^2 3p^6 3d^{10} 4s^2$
13	Al	$1s^2 2s^2 2p^6 3s^2 3p^1$	31	Ga	$1s^2 2s^2 2p^6 3s^2 3p^6 3d^{10} 4s^2 4p^1$
14	Si	$1s^2 2s^2 2p^6 3s^2 3p^2$	32	Ge	$1s^2 2s^2 2p^6 3s^2 3p^6 3d^{10} 4s^2 4p^2$
15	P	$1s^2 2s^2 2p^6 3s^2 3p^3$	33	As	$1s^2 2s^2 2p^6 3s^2 3p^6 3d^{10} 4s^2 4p^3$
16	S	$1s^2 2s^2 2p^6 3s^2 3p^4$	34	Se	$1s^2 2s^2 2p^6 3s^2 3p^6 3d^{10} 4s^2 4p^4$
17	Cl	$1s^2 2s^2 2p^6 3s^2 3p^5$	35	Br	$1s^2 2s^2 2p^6 3s^2 3p^6 3d^{10} 4s^2 4p^5$
18	Ar	$1s^2 2s^2 2p^6 3s^2 3p^6$	36	Kr	$1s^2 2s^2 2p^6 3s^2 3p^6 3d^{10} 4s^2 4p^6$

Building-Up Principle (Aufbau Principle)

Most of the configurations in Table 8.1 can be explained in terms of the **building-up principle** (or **Aufbau principle**), *a scheme used to reproduce the electron configurations of atoms by successively filling subshells with electrons.* Following this principle, you obtain the electron configuration of an atom by successively filling subshells in the following order: 1s, 2s, 2p, 3s, 3p, 4s, 3d, 4p, 5s, 4d, 5p, 6s, 4f, 5d, 6p, 7s, 5f. This order reproduces the experimentally determined electron configurations (with some exceptions, which we will discuss later). You need not memorize this order, as you can very easily obtain it from the periodic table.

By filling orbitals of lowest energy first, you usually get the lowest total energy (ground state) of the atom. Recall that the energy of an orbital depends only on the quantum numbers n and l. ◄ Orbitals with the same n and l but different m_l—that is, different orbitals of the same subshell—have the same energy. The energy depends primarily on n, increasing with its value. Except for the H atom, the energies of orbitals with the same n increase with the l quantum number. A 3p orbital has slightly greater energy than a 3s orbital.

When subshells have nearly the same energy, however, the building-up order is not strictly determined by the order of their energies. The ground-state configurations, which we are trying to predict by the building-up order, are determined by the *total* energies of the atoms. The total energy of an atom depends not only on the energies

► The quantum numbers and characteristics of orbitals were discussed in Section 7.5.

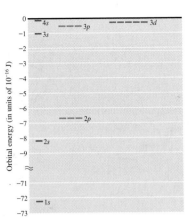

Figure 8.4
Orbital energies for the scandium atom (Z = 21).
Note that in the scandium atom, unlike the hydrogen atom, the subshells for each n are spread apart in energy. Thus, the $2p$ energy is above the $2s$. Similarly, the $n = 3$ subshells are spread to give the order $3s < 3p < 3d$. The $3d$ subshell energy is now just below the $4s$. (Values for this figure were calculated from theory by Charlotte F. Fischer, Vanderbilt University.)

of the subshells but also on the energies of interaction among the different subshells. It so happens that for all elements with $Z = 21$ or greater, the energy of the $3d$ subshell is lower than the energy of the $4s$ subshell (Figure 8.4), which is opposite to the building-up order.

Now you can see how to reproduce the electron configurations of Table 8.1 using the building-up principle. In the case of the simplest atom, hydrogen, you obtain the ground state by placing the single electron into the $1s$ orbital, giving the configuration $1s^1$. With helium ($Z = 2$), the first electron goes into the $1s$ orbital, as in hydrogen, followed by the second electron. The configuration is $1s^2$.

You continue this way through the elements, each time increasing Z by 1 and adding another electron. In lithium ($Z = 3$), the first two electrons give the configuration $1s^2$, like helium, but the third electron goes into the next higher orbital, because the $1s$ orbital is now filled. This gives the configuration $1s^2 2s^1$. In beryllium ($Z = 4$), the fourth electron fills the $2s$ orbital, giving the configuration $1s^2 2s^2$.

Using the abbreviation [He] for $1s^2$, the configurations are

$$Z = 3 \quad \text{lithium} \quad 1s^2 2s^1 \quad \text{or} \quad [\text{He}]2s^1$$
$$Z = 4 \quad \text{beryllium} \quad 1s^2 2s^2 \quad \text{or} \quad [\text{He}]2s^2$$

With boron ($Z = 5$), the electrons begin filling the $2p$ subshell. You get

$$Z = 5 \quad \text{boron} \quad 1s^2 2s^2 2p^1 \quad \text{or} \quad [\text{He}]2s^2 2p^1$$
$$Z = 6 \quad \text{carbon} \quad 1s^2 2s^2 2p^2 \quad \text{or} \quad [\text{He}]2s^2 2p^2$$
$$\vdots$$
$$Z = 10 \quad \text{neon} \quad 1s^2 2s^2 2p^6 \quad \text{or} \quad [\text{He}]2s^2 2p^6$$

With sodium ($Z = 11$), the $3s$ orbital begins to fill. Using the abbreviation [Ne] for $1s^2 2s^2 2p^6$, you have

$$Z = 11 \quad \text{sodium} \quad 1s^2 2s^2 2p^6 3s^1 \quad \text{or} \quad [\text{Ne}]3s^1$$
$$Z = 12 \quad \text{magnesium} \quad 1s^2 2s^2 2p^6 3s^2 \quad \text{or} \quad [\text{Ne}]3s^2$$

Then the $3p$ subshell begins to fill.

$$Z = 13 \quad \text{aluminum} \quad 1s^2 2s^2 2p^6 3s^2 3p^1 \quad \text{or} \quad [\text{Ne}]3s^2 3p^1$$
$$\vdots$$
$$Z = 18 \quad \text{argon} \quad 1s^2 2s^2 2p^6 3s^2 3p^6 \quad \text{or} \quad [\text{Ne}]3s^2 3p^6$$

With the $3p$ subshell filled, a stable configuration has been attained; argon is an unreactive element.

Now the $4s$ orbital begins to fill. You get [Ar]$4s^1$ for potassium ($Z = 19$) and [Ar]$4s^2$ for calcium. At this point the $3d$ subshell begins to fill. You get [Ar]$3d^1 4s^2$ for scandium ($Z = 21$), [Ar]$3d^2 4s^2$ for titanium ($Z = 22$), and [Ar]$3d^3 4s^2$ for vanadium ($Z = 23$). Note that we have written the configurations with subshells arranged in order by shells. This puts the subshells involved in chemical reactions at the far right.

Let us skip to zinc ($Z = 30$). The $3d$ subshell has filled; the configuration is [Ar]$3d^{10} 4s^2$. Now the $4p$ subshell begins to fill, starting with gallium ($Z = 31$), configuration [Ar]$3d^{10} 4s^2 4p^1$, and ending with krypton ($Z = 36$), configuration [Ar]$3d^{10} 4s^2 4p^6$.

Electron Configurations and the Periodic Table

The pattern for the ground-state electron configurations of the atoms explains the periodic table. Note that neon, argon, and krypton have configurations in which a p subshell has just filled. (Helium has a filled $1s$ subshell.) These elements are the first members of the group called *noble gases* because of their relative unreactivity.

Look now at the configurations of beryllium, magnesium, and calcium, members of the group of *alkaline earth metals* (Group IIA), which are similar, moderately reactive elements.

beryllium	$1s^2 2s^2$	or	$[He]2s^2$
magnesium	$1s^2 2s^2 2p^6 3s^2$	or	$[Ne]3s^2$
calcium	$1s^2 2s^2 2p^6 3s^2 3p^6 4s^2$	or	$[Ar]4s^2$

Each of these configurations consists of a **noble-gas core,** that is, *an inner-shell configuration corresponding to one of the noble gases,* plus two outer electrons with an ns^2 configuration.

The elements boron, aluminum, and gallium (Group IIIA) also have similarities. Their configurations are

boron	$1s^2 2s^2 2p^1$	or	$[He]2s^2 2p^1$
aluminum	$1s^2 2s^2 2p^6 3s^2 3p^1$	or	$[Ne]3s^2 3p^1$
gallium	$1s^2 2s^2 2p^6 3s^2 3p^6 3d^{10} 4s^2 4p^1$	or	$[Ar]3d^{10} 4s^2 4p^1$

Boron and aluminum have noble-gas cores plus three electrons with the configuration $ns^2 np^1$. Gallium has an additional filled $3d$ subshell. *The noble-gas core together with $(n-1)d^{10}$ electrons* is often referred to as a **pseudo-noble-gas core,** because these electrons usually are not involved in chemical reactions.

An electron in an atom outside the noble-gas or pseudo-noble-gas core is called a **valence electron.** Such electrons are primarily involved in chemical reactions, and similarities among the configurations of valence electrons (the *valence-shell configurations*) account for similarities of the chemical properties among groups of elements. Figure 8.5 shows a periodic table with the valence-shell configurations included. Note the similarity in electron configuration within any group (column) of elements.

The *main-group* (or *representative*) *elements* all have valence-shell configurations $ns^a np^b$ as the outer s or p subshell is being filled. Similarly, in the *d-block transition elements* (often called simply *transition elements* or *transition metals*), a d subshell is being filled. In the *f-block transition elements* (or *inner-transition elements*), an f subshell is being filled. (See Figure 8.5 for the configurations of these elements.)

Exceptions to the Building-Up Principle

The building-up principle reproduces most of the ground-state configurations correctly. For chromium the building-up principle predicts the configuration $[Ar]3d^4 4s^2$. It is found experimentally to be $[Ar]3d^5 4s^1$. These two configurations are actually very close in total energy because of the closeness in energies of the $3d$ and $4s$ orbitals (note Figure 8.4). For copper we predict the configuration $[Ar]3d^9 4s^2$, but experiment shows the ground-state configuration to be $[Ar]3d^{10} 4s^1$.

Remember that the configuration predicted by the building-up principle is very close in energy to the ground-state configuration (if it is not the ground state). Most of the qualitative conclusions regarding the chemistry of an element are not materially affected by arguing from the configuration given by the building-up principle. ◄

▶ **More exceptions occur among the heavier transition elements, where the outer subshells are very close together. We must concede that simplicity was not the uppermost concern in the construction of the universe!**

Main–Group Elements
s subshell fills

Main–Group Elements
p subshell fills

	Atomic number
1	
H	Symbol
$1s^1$	Valence-shell configuration

Transition Metals
d subshell fills

Inner–Transition Metals
f subshell fills

Period	IA	IIA	IIIB	IVB	VB	VIB	VIIB	VIIIB	VIIIB	VIIIB	IB	IIB	IIIA	IVA	VA	VIA	VIIA	VIIIA
1	1 **H** $1s^1$																	2 **He** $1s^2$
2	3 **Li** $2s^1$	4 **Be** $2s^2$											5 **B** $2s^2 2p^1$	6 **C** $2s^2 2p^2$	7 **N** $2s^2 2p^3$	8 **O** $2s^2 2p^4$	9 **F** $2s^2 2p^5$	10 **Ne** $2s^2 2p^6$
3	11 **Na** $3s^1$	12 **Mg** $3s^2$											13 **Al** $3s^2 3p^1$	14 **Si** $3s^2 3p^2$	15 **P** $3s^2 3p^3$	16 **S** $3s^2 3p^4$	17 **Cl** $3s^2 3p^5$	18 **Ar** $3s^2 3p^6$
4	19 **K** $4s^1$	20 **Ca** $4s^2$	21 **Sc** $3d^1 4s^2$	22 **Ti** $3d^2 4s^2$	23 **V** $3d^3 4s^2$	24 **Cr** $3d^5 4s^1$	25 **Mn** $3d^5 4s^2$	26 **Fe** $3d^6 4s^2$	27 **Co** $3d^7 4s^2$	28 **Ni** $3d^8 4s^2$	29 **Cu** $3d^{10} 4s^1$	30 **Zn** $3d^{10} 4s^2$	31 **Ga** $4s^2 4p^1$	32 **Ge** $4s^2 4p^2$	33 **As** $4s^2 4p^3$	34 **Se** $4s^2 4p^4$	35 **Br** $4s^2 4p^5$	36 **Kr** $4s^2 4p^6$
5	37 **Rb** $5s^1$	38 **Sr** $5s^2$	39 **Y** $4d^1 5s^2$	40 **Zr** $4d^2 5s^2$	41 **Nb** $4d^4 5s^1$	42 **Mo** $4d^5 5s^1$	43 **Tc** $4d^5 5s^2$	44 **Ru** $4d^7 5s^1$	45 **Rh** $4d^8 5s^1$	46 **Pd** $4d^{10}$	47 **Ag** $4d^{10} 5s^1$	48 **Cd** $4d^{10} 5s^2$	49 **In** $5s^2 5p^1$	50 **Sn** $5s^2 5p^2$	51 **Sb** $5s^2 5p^3$	52 **Te** $5s^2 5p^4$	53 **I** $5s^2 5p^5$	54 **Xe** $5s^2 5p^6$
6	55 **Cs** $6s^1$	56 **Ba** $6s^2$	57 **La*** $5d^1 6s^2$	72 **Hf** $5d^2 6s^2$	73 **Ta** $5d^3 6s^2$	74 **W** $5d^4 6s^2$	75 **Re** $5d^5 6s^2$	76 **Os** $5d^6 6s^2$	77 **Ir** $5d^7 6s^2$	78 **Pt** $5d^9 6s^1$	79 **Au** $5d^{10} 6s^1$	80 **Hg** $5d^{10} 6s^2$	81 **Tl** $6s^2 6p^1$	82 **Pb** $6s^2 6p^2$	83 **Bi** $6s^2 6p^3$	84 **Po** $6s^2 6p^4$	85 **At** $6s^2 6p^5$	86 **Rn** $6s^2 6p^6$
7	87 **Fr** $7s^1$	88 **Ra** $7s^2$	89 **Ac**** $6d^1 7s^2$	104 **Rf** $6d^2 7s^2$	105 **Db** $6d^3 7s^2$	106 **Sg** $6d^4 7s^2$	107 **Bh** $6d^5 7s^2$	108 **Hs** $6d^6 7s^2$	109 **Mt** $6d^7 7s^2$	110 **Uun** $6d^8 7s^2$	111 **Uuu** $6d^9 7s^2$	112 **Uub** $6d^{10} 7s^2$	114 **Uuq** $7s^2 7p^2$					

*Lanthanides

58 **Ce** $4f^1 5d^1 6s^2$	59 **Pr** $4f^3 6s^2$	60 **Nd** $4f^4 6s^2$	61 **Pm** $4f^5 6s^2$	62 **Sm** $4f^6 6s^2$	63 **Eu** $4f^7 6s^2$	64 **Gd** $4f^7 5d^1 6s^2$	65 **Tb** $4f^9 6s^2$	66 **Dy** $4f^{10} 6s^2$	67 **Ho** $4f^{11} 6s^2$	68 **Er** $4f^{12} 6s^2$	69 **Tm** $4f^{13} 6s^2$	70 **Yb** $4f^{14} 6s^2$	71 **Lu** $4f^{14} 5d^1 6s^2$

**Actinides

90 **Th** $6d^2 7s^2$	91 **Pa** $5f^2 6d^1 7s^2$	92 **U** $5f^3 6d^1 7s^2$	93 **Np** $5f^4 6d^1 7s^2$	94 **Pu** $5f^6 7s^2$	95 **Am** $5f^7 7s^2$	96 **Cm** $5f^7 6d^1 7s^2$	97 **Bk** $5f^9 7s^2$	98 **Cf** $5f^{10} 7s^2$	99 **Es** $5f^{11} 7s^2$	100 **Fm** $5f^{12} 7s^2$	101 **Md** $5f^{13} 7s^2$	102 **No** $5f^{14} 7s^2$	103 **Lr** $5f^{14} 6d^1 7s^2$

Figure 8.5
A periodic table. This table shows the valence-shell configurations of the elements.

8.3 Writing Electron Configurations Using the Periodic Table

To discuss bonding and the chemistry of the elements coherently, you must be able to reproduce the atomic configurations with ease.

One approach is to recall the structure of the periodic table. Because that structure is basic, it offers a sound way to remember the building-up order. Figure 8.6 shows a periodic table stressing this pattern. The value of *n* is obtained from the period (row) number. In the red area, an $(n - 1)d$ subshell is being filled.

You read the building-up order by starting with the first period, in which the 1*s* subshell is being filled. In the second period, you have 2*s* (violet area); then, staying in the same period but jumping across, you have 2*p* (blue area). In the third period, you have 3*s* and 3*p*; in the fourth period, 4*s* (violet area), 3*d* (red area), and 4*p* (blue area). This pattern should become clear enough to visualize with a periodic table that

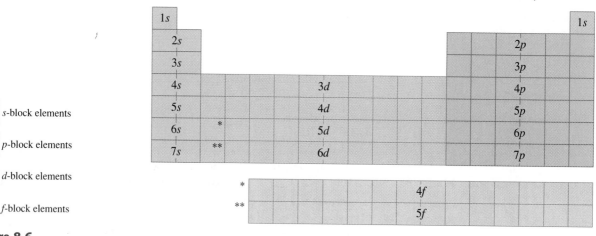

s-block elements

p-block elements

d-block elements

f-block elements

Figure 8.6
A periodic table illustrating the building-up order. The colored areas of elements show the different subshells that are filling with those elements.

is not labeled with the subshells, such as the one on the inside front cover of this book. The detailed method is illustrated in the next example.

EXAMPLE 8.2

Determining the Configuration of an Atom Using the Building-Up Principle

Use the building-up principle to obtain the configuration for the ground state of the gallium atom ($Z = 31$). Give the configuration in complete form (do not abbreviate for the core). What is the valence-shell configuration?

SOLUTION

From a periodic table, you get the following building-up order:

	$1s$	$2s$ $2p$	$3s$ $3p$	$4s$ $3d$ $4p$
Period:	first	second	third	fourth

Now you fill the subshells with electrons, remembering that you have a total of 31 electrons to distribute. You get

$$1s^2 2s^2 2p^6 3s^2 3p^6 4s^2 3d^{10} 4p^1$$

Or, if you rearrange the subshells by shells, you write

$$\mathbf{1s^2 2s^2 2p^6 3s^2 3p^6 3d^{10} 4s^2 4p^1}$$

The valence-shell configuration is $\mathbf{4s^2 4p^1}$.

In many cases, you need only the configuration of the outer electrons. You can determine this from the position of the element in the periodic table. Recall that the valence-shell configuration of a main-group element is $ns^a np^b$, where n, the principal quantum number of the outer shell, also equals the period number for the element. The total number of valence electrons, which equals $a + b$, can be obtained from the group number. For example, gallium is in Period 4, so $n = 4$. It is in Group IIIA, so the number of valence electrons is 3. This gives the valence-shell configuration $4s^2 4p^1$. The configuration of outer shells of a transition element is obtained in a similar fashion. The next example gives the details.

EXAMPLE 8.3

Determining the Configuration of an Atom Using the Period and Group Numbers

What are the configurations for the outer electrons of a. tellurium, $Z = 52$, and b. nickel, $Z = 28$?

SOLUTION

a. You locate tellurium in a periodic table and find it to be in Period 5, Group VIA. Thus, it is a main-group element, and the outer subshells are $5s$ and $5p$. These subshells contain six electrons, because the group is VIA. The valence-shell configuration is $\mathbf{5s^2 5p^4}$. b. Nickel is a Period 4 transition element, in which the general form of the outer-shell configuration is $3d^{a-2}4s^2$. To determine a, you note that it equals the Roman numeral group number up to iron (8). After that you count Co as 9 and Ni as 10. Hence, the outer-shell configuration is $\mathbf{3d^8 4s^2}$.

See Problems 8.35 and 8.36.

 CONCEPT CHECK 8.1

Two elements in Period 3 are adjacent to one another in the periodic table. The ground-state atom of one element has only s electrons in its valence shell; the other one has at least one p electron in its valence shell. Identify the elements.

8.4   Orbital Diagrams of Atoms; Hund's Rule

We have not yet described how the electrons are arranged within each subshell. There may be several different ways of arranging electrons in a particular configuration. Consider the carbon atom ($Z = 6$) with the ground-state configuration $1s^2 2s^2 2p^2$. Three possible arrangements are given in the following orbital diagrams.

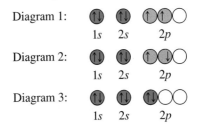

These orbital diagrams show different states of the carbon atom. Each state has a different energy and, as you will see, different magnetic characteristics.

Hund's Rule

In about 1927, Friedrich Hund discovered an empirical rule determining the lowest-energy arrangement of electrons in a subshell. **Hund's rule** states that *the lowest-energy arrangement of electrons in a subshell is obtained by putting electrons into separate orbitals of the subshell with the same spin before pairing electrons.* Let

Table 8.2
Orbital Diagrams for the Ground States of Atoms from $Z = 1$ to $Z = 10$

Atom	Z	Configuration	Orbital Diagram 1s	2s	2p
Hydrogen	1	$1s^1$	↑	○	○○○
Helium	2	$1s^2$	↑↓	○	○○○
Lithium	3	$1s^2 2s^1$	↑↓	↑	○○○
Beryllium	4	$1s^2 2s^2$	↑↓	↑↓	○○○
Boron	5	$1s^2 2s^2 2p^1$	↑↓	↑↓	↑○○
Carbon	6	$1s^2 2s^2 2p^2$	↑↓	↑↓	↑↑○
Nitrogen	7	$1s^2 2s^2 2p^3$	↑↓	↑↓	↑↑↑
Oxygen	8	$1s^2 2s^2 2p^4$	↑↓	↑↓	↑↓↑↑
Fluorine	9	$1s^2 2s^2 2p^5$	↑↓	↑↓	↑↓↑↓↑
Neon	10	$1s^2 2s^2 2p^6$	↑↓	↑↓	↑↓↑↓↑↓

us see how this would apply to the carbon atom, whose ground-state configuration is $1s^2 2s^2 2p^2$. The first four electrons go into the $1s$ and $2s$ orbitals. The next two electrons go into separate $2p$ orbitals, with both electrons having the same spin, following Hund's rule.

To apply Hund's rule to the oxygen atom, whose ground-state configuration is $1s^2 2s^2 2p^4$, we place the first seven electrons as follows:

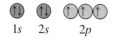

The last electron is paired with one of the $2p$ electrons to give a doubly occupied orbital. The orbital diagram for the ground state of the oxygen atom is

Table 8.2 gives orbital diagrams for the ground states of the first ten elements.

EXAMPLE 8.4

Applying Hund's Rule

Write an orbital diagram for the ground state of the iron atom.

PROBLEM STRATEGY

First obtain the electron configuration. Then draw circles for the orbitals of each subshell. A filled subshell should have doubly occupied orbitals. For a partially filled subshell, apply Hund's rule, putting electrons into separate orbitals with the same spin before pairing electrons.

SOLUTION

The electron configuration of the iron atom is $1s^2 2s^2 2p^6 3s^2 3p^6 3d^6 4s^2$. All the subshells except the $3d$ are filled. In placing the six electrons in the $3d$ subshell, you note that the first five go into separate $3d$ orbitals with their spin arrows in the same direction. The sixth electron must doubly occupy a $3d$ orbital. The orbital diagram is

You can write this diagram in abbreviated form using [Ar] for the argon-like core of the iron atom:

$$[\text{Ar}] \quad \boxed{⇅}\boxed{↑}\boxed{↑}\boxed{↑}\boxed{↑} \quad \boxed{⇅}$$
$$\qquad\qquad \textbf{3d} \qquad\qquad \textbf{4s}$$

See Problems 8.37 and 8.38.

Magnetic Properties of Atoms

An electron in an atom behaves like a small magnet, but the magnetic attractions from two electrons that are opposite in spin cancel each other. As a result, an atom that has only doubly occupied orbitals has no net spin magnetism. However, an atom with *unpaired* electrons does exhibit a net magnetism.

The magnetic properties of an atom can be observed. The most direct way is to determine whether the atomic substance is attracted to the field of a strong magnet. A **paramagnetic substance** is *a substance that is weakly attracted by a magnetic field, and this attraction is generally the result of unpaired electrons.* ◀ For example, sodium vapor has been found experimentally to be paramagnetic. Sodium atoms contain unpaired electrons. (The configuration is [Ne]$3s^1$.) A **diamagnetic substance** is *a substance that is not attracted by a magnetic field or is very slightly repelled by such a field. This property generally means that the substance has only paired electrons.* Mercury vapor is found experimentally to be diamagnetic. Mercury vapor consists of mercury atoms, with the electron configuration [Xe]$4f^{14}5d^{10}6s^2$, which has only paired electrons.

We expect the different orbital diagrams presented at the beginning of this section for the carbon atom to have different magnetic properties. Diagram 1, predicted by Hund's rule to be the ground state, would give a magnetic atom, whereas the other diagrams would not. If we could prepare a vapor of carbon atoms, it should be attracted to a magnet. (It should be paramagnetic.) It is difficult to prepare a vapor of free carbon atoms in sufficient concentration to observe a result. However, the visible–ultraviolet spectrum of carbon atoms can be obtained from dilute vapor. From an analysis of this spectrum, it is possible to conclude that the ground-state atom is magnetic, which is consistent with the prediction of Hund's rule.

▶ The strong, permanent magnetism seen in iron objects is called *ferromagnetism* and is due to the cooperative alignment of electron spins in many iron atoms. Paramagnetism is a much weaker effect. Nevertheless, paramagnetic substances can be attracted to a strong magnet. Liquid oxygen is composed of paramagnetic O_2 molecules. When poured over a magnet, the liquid clings to the poles. (See Figure 10.25.)

Periodicity of the Elements

You have seen that the periodic table that Mendeleev discovered in 1869 can be explained by the periodicity of the ground-state electron configurations of the atoms. Now we will look at various aspects of the periodicity of the elements.

8.5 Mendeleev's Predictions from the Periodic Table

One of Mendeleev's periodic tables is reproduced in Figure 8.7. In this early form of the periodic table, within each column some elements were placed toward the left side and some toward the right. With some exceptions, the elements on a given side have similar properties. Mendeleev left spaces in his periodic table for what he felt were undiscovered elements. There are blank spaces in his row 5, for example, one directly under aluminum and another under silicon (looking at just the elements on the right side of the column). By writing the known elements in this row with their atomic weights, he could determine approximate values (between the known ones) for the missing elements (values in parentheses).

Cu	Zn	—	—	As	Se	Br
63 amu	65 amu	(68 amu)	(72 amu)	75 amu	78 amu	80 amu

The Group III element directly under aluminum Mendeleev called eka-aluminum, with the symbol Ea. (*Eka* is the Sanskrit word meaning "first"; thus eka-aluminum is the first element under aluminum.) The known Group III elements have oxides of the form R_2O_3, so Mendeleev predicted that eka-aluminum would have an oxide with the formula Ea_2O_3.

The physical properties of this undiscovered element could be predicted by comparing values for the neighboring known elements. For eka-aluminum Mendeleev predicted a density of 5.9 g/cm^3, a low *melting point*, and a high *boiling point*.

In 1874 the French chemist Paul-Émile Lecoq de Boisbaudran found two previously unidentified lines in the atomic spectrum of a sample of sphalerite (a zinc sulfide mineral). Realizing he was on the verge of a discovery, Lecoq de Boisbaudran quickly prepared a large batch of the zinc mineral, from which he isolated a gram of

Figure 8.7
Mendeleev's periodic table. This one was published in 1872.

Reihen	Gruppe I. — R^2O	Gruppe II. — RO	Gruppe III. — R^2O^3	Gruppe IV. RH^4 RO^2	Gruppe V. RH^3 R^2O^5	Gruppe VI. RH^2 RO^3	Gruppe VII. RH R^2O^7	Gruppe VIII. — RO^4
1	H = 1							
2	Li = 7	Be = 9,4	B = 11	C = 12	N = 14	O = 16	F = 19	
3	Na = 23	Mg = 24	Al = 27,3	Si = 28	P = 31	S = 32	Cl = 35,5	
4	K = 39	Ca = 40	— = 44	Ti = 48	V = 51	Cr = 52	Mn = 55	Fe = 56, Co = 59, Ni = 59, Cu = 63.
5	(Cu = 63)	Zn = 65	— = 68	— = 72	As = 75	Se = 78	Br = 80	
6	Rb = 85	Sr = 87	?Yt = 88	Zr = 90	Nb = 94	Mo = 96	— = 100	Ru = 104, Rh =104, Pd = 106, Ag = 108.
7	(Ag = 108)	Cd = 112	In = 113	Su = 118	Sb = 122	Te = 125	J = 127	
8	Cs = 133	Ba = 137	?Di = 138	?Ce = 140	—	—	—	— — —
9	(—)	—	—		—	—	—	
10	—	—	?Er = 178	?La = 180	Ta = 182	W = 184	—	Os = 195, Ir = 197, Pt = 198, Au = 199.
11	(Au = 199)	Hg = 200	Tl =204	Pb = 207	Bi = 208	—	—	
12	—	—	—	Th = 231	—	U = 240	—	— — —

a new element. He called this new element gallium. The properties of gallium were remarkably close to those Mendeleev predicted for eka-aluminum.

Property	Predicted for Eka-Aluminum	Found for Gallium
Atomic weight	68 amu	69.7 amu
Formula of oxide	Ea_2O_3	Ga_2O_3
Density of the element	5.9 g/cm^3	5.91 g/cm^3
Melting point of the element	Low	30.1°C
Boiling point of the element	High	1983°C

The predictive power of Mendeleev's periodic table was demonstrated again when scandium (eka-boron) was discovered in 1879 and germanium (eka-silicon) in 1886. Both elements had properties remarkably like those predicted by Mendeleev. These early successes won acceptance for the organizational and predictive power of the periodic table.

8.6 Some Periodic Properties

The electron configurations of the atoms display a periodic variation with increasing atomic number. The **periodic law** states that *when the elements are arranged by atomic number, their physical and chemical properties vary periodically.* We will look at three physical properties of an atom: atomic radius, ionization energy, and electron affinity. These three quantities, especially ionization energy and electron affinity, are important in discussions of chemical bonding (Chapter 9).

Atomic Radius

An atom does not have a definite size, because the statistical distribution of electrons does not abruptly end but merely decreases to very small values as the distance from the nucleus increases. This can be seen in Figure 8.8. Consequently, atomic size must

Figure 8.8
Electron distribution for the argon atom.   This is a radial distribution, showing the probability of finding an electron at a given distance from the nucleus. The distribution shows three maxima, for the $n=1$, $n=2$, and $n=3$ shells. The outermost maximum occurs at 66 pm; then the distribution falls steadily, becoming negligibly small after several hundred picometers.

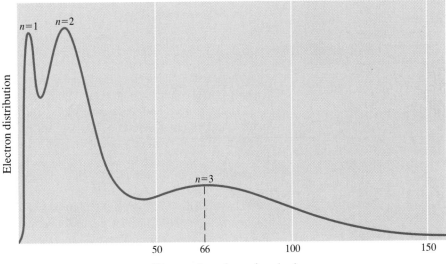

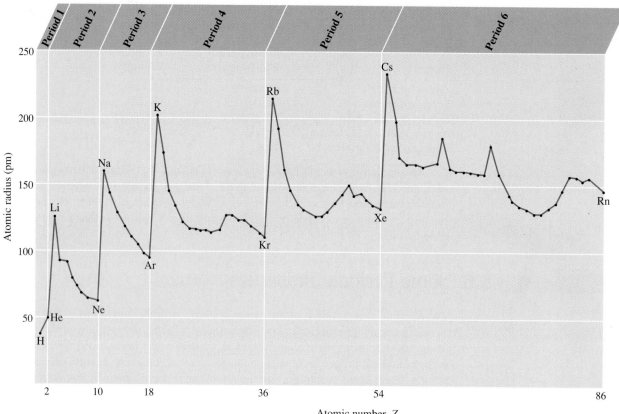

Figure 8.9
Atomic radius (covalent radius) versus atomic number. Note that the curve is periodic (tends to repeat). Each period of elements begins with the Group IA atom, and the atomic radius tends to decrease until the Group VIIIA atom. (Values for He, Ne, and Ar are estimated because there are no known compounds.)

be defined in a somewhat arbitrary manner, so various measures of atomic size exist. The atomic radii plotted in Figure 8.9 and also represented in Figure 8.10 are *covalent radii*, which are obtained from measurements of distances between the nuclei of atoms in the chemical bonds of molecular substances.

Figures 8.9 and 8.10 show the following general trends in size of atomic radii:

1. Within each period, the atomic radius tends to decrease with increasing atomic number (nuclear charge).

2. Within each group (vertical column), the atomic radius tends to increase with the period number.

The atomic radius increases greatly going from any noble-gas atom to the following Group IA atom, giving the curve in Figure 8.9 a saw-tooth appearance.

These general trends in atomic radius can be explained if you look at the two factors that primarily determine the size of the outermost orbital. One factor is the principal quantum number n of the orbital. The other factor is the effective nuclear charge acting on an electron in the orbital; increasing the effective nuclear charge reduces the size of the orbital by pulling the electrons inward. The **effective nuclear charge** is *the positive charge that an electron experiences from the nucleus, equal to the nuclear charge but reduced by any shielding or screening from any intervening electron distribution.* Consider the effective nuclear charge on the $2s$ electron in the lithium atom. The nuclear charge is $3e$, but the effect of this charge on the $2s$ electron is reduced by the distribution of the two $1s$ electrons lying between the nucleus and the $2s$ electron (roughly, each core electron reduces the nuclear charge by $1e$).

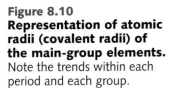

Figure 8.10
Representation of atomic radii (covalent radii) of the main-group elements.
Note the trends within each period and each group.

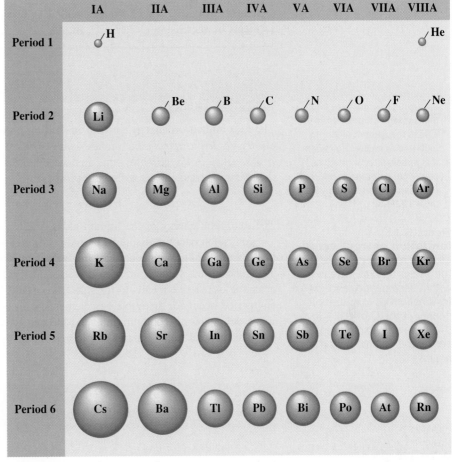

Consider a given period of elements. The principal quantum number of the outer orbitals remains constant. However, the effective nuclear charge increases, because the nuclear charge increases and the number of core electrons remains constant. Consequently, the size of the outermost orbital and, therefore, the radius of the atom decrease with increasing Z in any period.

Now consider a given column of elements. The effective nuclear charge remains nearly constant (approximately equal to e times the number of valence electrons), but n gets larger. You observe that the atomic radius increases.

EXAMPLE 8.5

Determining Relative Atomic Sizes from Periodic Trends

Refer to a periodic table and use the trends noted for size of atomic radii to arrange the following in order of increasing atomic radius: Al, C, Si.

SOLUTION

Note that C is above Si in Group IVA. Therefore, the radius of C is smaller than that of Si (the atomic radius increases going down a group of elements). Note that Al and Si are in the same period. Therefore the radius of Si is smaller than that of Al (radius decreases with Z in a period). Hence, the order of elements by increasing radius is **C, Si, Al.**

See Problems 8.41 and 8.42.

▶ Ionization energies are often measured in electron volts (eV). This is the amount of energy imparted to an electron when it is accelerated through an electrical potential of one volt. One electron volt is equivalent to 96.5 kJ/mol.

Ionization Energy

The **first ionization energy** (or **first ionization potential**) of an atom is *the minimum energy needed to remove the highest-energy (that is, the outermost) electron from the neutral atom in the gaseous state.* For the lithium atom, the first ionization energy is the energy needed for the following process:

$$\text{Li}(1s^2 2s^1) \longrightarrow \text{Li}^+(1s^2) + \text{e}^-$$

The ionization energy of the lithium atom is 520 kJ/mol. ◀

Ionization energies display a periodic variation when plotted against atomic number (Figure 8.11). Within any period, values tend to increase with atomic number. Thus, the lowest values in a period are found for the Group IA elements. It is characteristic of reactive metals to lose electrons easily. The largest ionization energies in any period occur for the noble-gas elements. This is partly responsible for the stability of the noble-gas configurations and the unreactivity of the noble gases.

This general trend—increasing ionization energy with atomic number in a given

Figure 8.11
Ionization energy versus atomic number. Note that the values tend to increase within each period, except for small drops in ionization energy at the Group IIIA and VIA elements. Large drops occur when a new period begins.

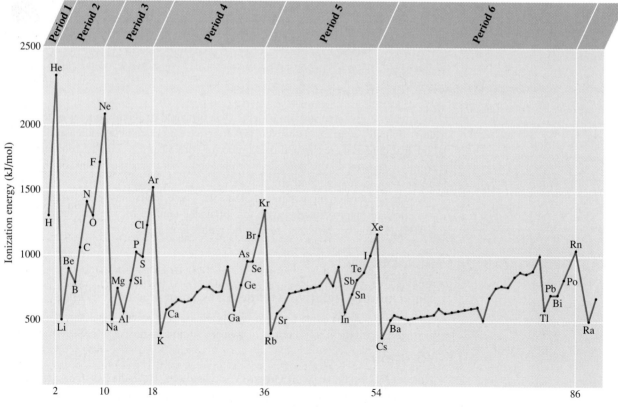

period—can be explained as follows: The energy needed to remove an electron from the outer shell is proportional to the effective nuclear charge divided by the average distance between electron and nucleus. (This distance is inversely proportional to the effective nuclear charge.) Hence, the ionization energy is proportional to the square of the effective nuclear charge and increases going across a period.

Small deviations from this general trend occur. A IIIA element (ns^2np^1) has smaller ionization energy than the preceding IIA element (ns^2). Apparently the np electron of the IIIA element is more easily removed than one of the ns electrons of the preceding IIA element. Also note that a VIA element (ns^2np^4) has smaller ionization energy than the preceding VA element. As a result of electron repulsion, it is easier to remove an electron from the doubly occupied np orbital of the VIA element than from a singly occupied orbital of the preceding VA element.

Ionization energies tend to decrease going down any column of main-group elements. This is because atomic size increases going down the column.

EXAMPLE 8.6

Determining Relative Ionization Energies from Periodic Trends

Using a periodic table only, arrange the following elements in order of increasing ionization energy: Ar, Se, S.

SOLUTION

Note that Se is below S in Group VIA. Therefore, the ionization energy of Se should be less than that of S. Also, S and Ar are in the same period, with Z increasing from S to Ar. Therefore, the ionization energy of S should be less than that of Ar. Hence the order is **Se, S, Ar.**

See Problems 8.43 and 8.44.

The electrons of an atom can be removed successively. The energies required at each step are known as the *first* ionization energy, the *second* ionization energy, and so forth. Table 8.3 lists the successive ionization energies of the first ten elements. Note that the ionization energies for a given element increase as more electrons are removed. The first and second ionization energies of beryllium (electron configuration $1s^22s^2$) are 899 kJ/mol and 1757 kJ/mol, respectively. The second ionization energy is the energy needed to remove a $2s$ electron from the positive ion Be^+. Its value is greater than that of the first ionization energy because the electron is being removed from a positive ion.

Note that there is a large jump in value from the second ionization energy of Be (1757 kJ/mol) to the third ionization energy (14,848 kJ/mol). The third ionization energy corresponds to removing an electron from the core of the atom (from the $1s$ orbital)—that is, from a noble-gas configuration ($1s^2$). A vertical line in Table 8.3 separates the energies needed to remove valence electrons from those needed to remove core electrons. For each element, a large increase in ionization energy occurs when this line is crossed. The large increase results from the fact that once the valence electrons are removed, further ionizations become much more difficult. We will see in Chapter 9 that metal atoms often form compounds by losing valence electrons; core electrons are not significantly involved in the formation of compounds.

Table 8.3
Successive Ionization Energies of the First Ten Elements (kJ/mol)*

Element	First	Second	Third	Fourth	Fifth	Sixth	Seventh
H	1312						
He	2372	5250					
Li	520	7298	11,815				
Be	899	1757	14,848	21,006			
B	801	2427	3660	25,025	32,826		
C	1086	2353	4620	6222	37,829	47,276	
N	1402	2857	4578	7475	9445	53,265	64,358
O	1314	3388	5300	7469	10,989	13,326	71,333
F	1681	3374	6020	8407	11,022	15,164	17,867
Ne	2081	3952	6122	9370	12,177	15,238	19,998

*Ionization energies to the right of a vertical line correspond to removal of electrons from the core of the atom.

Electron Affinity

When a neutral atom in the gaseous state picks up an electron to form a stable negative ion, energy is released. For example, a chlorine atom can pick up an electron to give a chloride ion, Cl^-, and 349 kJ/mol of energy is released. You write the process as follows:

$$Cl([Ne]3s^23p^5) + e^- \longrightarrow Cl^-([Ne]3s^23p^6)$$

The **electron affinity** is *the energy change for the process of adding an electron to a neutral atom in the gaseous state to form a negative ion.* If the negative ion is stable, the energy change for its formation is a negative number. Thus, the electron affinity of Cl is −349 kJ/mol. Large negative numbers such as this indicate that a very stable negative ion is formed. Small negative numbers indicate that a less stable ion is formed. Table 8.4 gives the electron affinities of the main-group elements. ◀

▶ Electron affinities are also defined with positive values for the formation of stable ions.

Electron affinities, *E.A.,* have a periodic variation, just as atomic radii and ionization energies do, though somewhat more complicated. To see this variation, consider each of the main groups of elements beginning with Group IA. All of the Group IA elements have small negative electron affinities. When you add an electron to a lithium atom, for example, it goes into the 2s orbital to form a moderately stable negative ion, releasing energy.

$$Li(1s^22s^1) + e^- \longrightarrow Li^-(1s^22s^2); \ E.A. = -60 \text{ kJ/mol}$$

None of the Group IIA elements form stable negative ions; that is, the electron affinities of these elements are positive. Each of the atoms has a filled *ns* subshell, so that if you were to add an electron, it would have to go into the next higher energy subshell (*np*). (For a similar reason, the Group VIIIA elements also do not form stable negative ions.)

From Group IIIA to Group VIIA, the added electron goes into the *np* subshell of the valence shell. With the exception of the Group VA elements, the electron affinities tend toward more negative values as you progress to the right through these ele-

Table 8.4
Electron Affinities of the Main-Group Elements (kJ/mol)*

Period	IA	IIIA	IVA	VA	VIA	VIIA
1	H −73					
2	Li −60	B −27	C −122	N 0	O −141	F −328
3	Na −53	Al −44	Si −134	P −72	S −200	Cl −349
4	K −48	Ga −30	Ge −120	As −77	Se −195	Br −325
5	Rb −47	In −30	Sn −121	Sb −101	Te −190	I −295
6	Cs −45	Tl −30	Pb −110	Bi −110	Po −180	At −270

*Atoms of the alkaline earth metals (Group IIA) and the noble gases (Group VIIIA) do not form stable negative ions.

ments in any period (more energy is released and stabler negative ions are formed). The electron affinity of a Group VA element is generally less negative than the preceding Group IVA element.

Broadly speaking, the general trend is toward more negative electron affinities from left to right in any period. Note especially that the Group VIA and Group VIIA elements have the largest negative electron affinities of any of the main-group elements. Therefore, these elements have compounds containing monatomic anions (such as F^- and O^{2-}).

CONCEPT CHECK 8.2

Given the following information for element E, identify the element's group in the periodic table: The electron affinity of E is positive (that is, it does not form a stable negative ion). The first ionization energy of E is less than the second ionization energy, which in turn is very much less than its third ionization energy.

8.7 Periodicity in the Main-Group Elements

The chemical and physical properties of the main-group elements clearly display periodic character. For instance, the metallic elements lie to the left of the "staircase" line in the periodic table (see inside front cover); nonmetals lie to the right. So, as you move left to right in any row of the periodic table, the metallic character of the elements decreases. As you progress down a column, however, the elements tend to increase in metallic character.

These variations of metallic–nonmetallic character can be attributed in part to the ionization energies of the corresponding atoms. Elements with low ionization energy tend to be metallic, whereas those with high ionization energy tend to be nonmetallic. So, it is not surprising that the metallic–nonmetallic character of an element is periodic.

The basic–acidic behavior of the oxides of the elements is a good indicator of the metallic–nonmetallic character of the elements. Oxides are classified as basic or acidic depending on their reactions with acids and bases. A **basic oxide** is *an oxide that reacts with acids*. Most metal oxides are basic. An **acidic oxide** is *an oxide that reacts with bases*. Most nonmetal oxides are acidic oxides. An **amphoteric oxide** is *an oxide that has both basic and acidic properties*.

In the following brief descriptions of the main-group elements, we will note the metallic–nonmetallic behavior of the elements, as well as the basic–acidic character of the oxides. Although elements in a given group are expected to be similar, the degree of similarity does vary among the groups. The alkali metals (Group IA) show marked similarities, as do the halogens (Group VIIA). On the other hand, the Group IVA elements range from a nonmetal (carbon) at the top of the column to a metal (lead) at the bottom.

Hydrogen ($1s^1$)

Although the electron configuration of hydrogen would seem to place the element in Group IA, its properties are quite different, and it seems best to consider the element as belonging in a group by itself. The element is a colorless gas composed of H_2 molecules. ◄

▶ At very high pressures, however, hydrogen is believed to have metallic properties.

Group IA Elements, the Alkali Metals (ns^1)

The *alkali metals* are soft and reactive, with the reactivities increasing as you move down the column of elements. All of the metals react with water to produce hydrogen.

$$2Li(s) + 2H_2O(l) \longrightarrow 2LiOH(aq) + H_2(g)$$

The vigor of the reaction increases from lithium (moderate) to rubidium (violent). All of the alkali metals form basic oxides with the general formula R_2O.

Group IIA Elements, the Alkaline Earth Metals (ns^2)

The alkaline earth metals are also chemically reactive but much less so than the alkali metals. Reactivities increase going down the group. The alkaline earth metals form basic oxides with the general formula RO.

Group IIIA Elements (ns^2np^1)

Groups IA and IIA exhibit only slight increases in metallic character down a column, but with Group IIIA we see a significant increase. The first Group IIIA element, boron, is a metalloid. Other elements in this group—aluminum, gallium, indium, and thallium—are metals. (Gallium is a curious metal; it melts easily in the palm of the hand—see Figure 8.12.)

The oxides in this group have the general formula R_2O_3. Boron oxide, B_2O_3, is an acidic oxide; aluminum oxide, Al_2O_3, and gallium oxide, Ga_2O_3, are amphoteric

Figure 8.12
Gallium. This metal melts from the heat of a hand.

oxides. The change in the oxides from acidic to amphoteric to basic is indicative of an increase in metallic character of the elements.

Group IVA Elements (ns^2np^2)

This group shows the most distinct change in metallic character. It begins with the non-metal carbon, C, followed by the metalloids silicon, Si, and germanium, Ge, and then the metals tin, Sn, and lead, Pb. Both tin and lead were known to the ancients. ◄

All the elements in this group form oxides with the general formula RO_2, which progress from acidic to amphoteric. Carbon dioxide, CO_2, an acidic oxide, is a gas. (Carbon also forms the monoxide, CO.) Silicon dioxide, SiO_2, an acidic oxide, exists as quartz and white sand (particles of quartz). Germanium dioxide, GeO_2, is acidic, though less so than silicon dioxide. Tin dioxide, SnO_2, an amphoteric oxide, is found as the mineral cassiterite, the principal ore of tin. Lead dioxide, PbO_2, is amphoteric. Lead has a more stable monoxide, PbO. Figure 8.13 shows oxides of some Group IVA elements.

Group VA Elements (ns^2np^3)

The Group VA elements also show the distinct transition from nonmetal (nitrogen, N, and phosphorus, P) to metalloid (arsenic, As, and antimony, Sb) to metal (bismuth, Bi). Nitrogen occurs as a colorless, odorless, relatively unreactive gas with N_2 molecules; white phosphorus is a white, waxy solid with P_4 molecules. Gray arsenic is a brittle solid with metallic luster; antimony is a brittle solid with a silvery, metallic luster. Bismuth is a hard, lustrous metal with a pinkish tinge.

The Group VA elements form oxides with empirical formulas R_2O_3 and R_2O_5. In some cases, the molecular formulas are twice these formulas—that is, R_4O_6 and R_4O_{10}. Nitrogen has the acidic oxides N_2O_3 and N_2O_5, although it also has other, better known oxides, such as NO. Phosphorus has the acidic oxides P_4O_6 and P_4O_{10}. Arsenic has the acidic oxides As_2O_3 and As_2O_5; antimony has the amphoteric oxides Sb_2O_3 and Sb_2O_5; and bismuth has the basic oxide Bi_2O_3.

Group VIA Elements, the Chalcogens (ns^2np^4)

These elements, the *chalcogens* (pronounced kal'-ke-jens), show the transition from nonmetal (oxygen, O, sulfur, S, and selenium, Se) to metalloid (tellurium, Te) to metal (polonium, Po). Oxygen occurs as a colorless, odorless gas with O_2 molecules. It also has an allotrope, ozone, with molecular formula O_3. Sulfur is a brittle, yellow solid with molecular formula S_8. Tellurium is a shiny gray, brittle solid; polonium is a silvery metal.

Sulfur, selenium, and tellurium form oxides with the formulas RO_2 and RO_3. (Sulfur burns in air to form sulfur dioxide; see Figure 8.14.) These oxides, except for TeO_2, are acidic; TeO_2 is amphoteric. Polonium has an oxide PoO_2, which is amphoteric, though more basic in character than TeO_2.

Group VIIA Elements, the Halogens (ns^2np^5)

The *halogens* are reactive nonmetals with the general molecular formula X_2, where X symbolizes a halogen. Fluorine, F_2, is a pale yellow gas; chlorine, Cl_2, a pale greenish-yellow gas; bromine, Br_2, a reddish-brown liquid; and iodine, I_2, a bluish-

► Bronze, one of the first *alloys* (metallic mixtures) used by humans, contains about 90% copper and 10% tin. Bronze melts at a lower temperature than copper but is much harder.

Figure 8.13
Oxides of some Group IVA elements. Powdered lead monoxide (yellow), lead dioxide (dark brown), tin dioxide (white), and crystalline silicon dioxide (clear quartz).

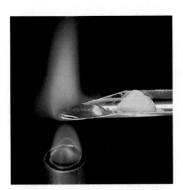

Figure 8.14
Burning sulfur.
The combustion of sulfur in air gives a blue flame. The product is primarily sulfur dioxide, detectable by its acrid odor.

black solid that has a violet vapor (see Figure 8.15). Little is known about the chemistry of astatine, At, because all isotopes are radioactive with very short half-lives. (The half-life of a radioactive isotope is the time it takes for half of the isotope to decay, or break down, to another element.) It might be expected to be a metalloid.

Each halogen forms several compounds with oxygen; these are generally unstable, acidic oxides.

Group VIIIA Elements, the Noble Gases (ns^2np^6)

The Group VIIIA elements exist as gases consisting of uncombined atoms. For a long time these elements were thought to be chemically inert, because no compounds were known. Then, in the early 1960s, several compounds of xenon were prepared. Now compounds are also known for krypton and radon. These elements are known as the *noble gases* because of their relative unreactivity.

CONCEPT CHECK 8.3

An element is a metalloid that forms an acidic oxide with the formula R_2O_5. Identify the element.

Figure 8.15
The halogens.   From left to right, the flasks contain chlorine, bromine, and iodine. (Bromine and iodine were warmed to produce the vapors; bromine is normally a reddish-brown liquid and iodine a bluish-black solid.)

A Checklist for Review

Important Terms

electron configuration (8.1)
orbital diagram (8.1)
Pauli exclusion principle (8.1)
building-up (Aufbau) principle (8.2)

noble-gas core (8.2)
pseudo-noble-gas core (8.2)
valence electron (8.2)
Hund's rule (8.4)
paramagnetic substance (8.4)

diamagnetic substance (8.4)
periodic law (8.6)
effective nuclear charge (8.6)
first ionization energy (first ionization potential) (8.6)

electron affinity (8.6)
basic oxide (8.7)
acidic oxide (8.7)
amphoteric oxide (8.7)

Summary of Facts and Concepts

To understand the similarities that exist among the members of a group of elements, it is necessary to know the *electron configurations* for the ground states of atoms. The ground-state configuration of an atom represents the electron arrangement that has the lowest total energy. This arrangement can be reproduced by the *building-up principle*, where electrons fill the subshells in particular order (the building-up order) consistent with the Pauli exclusion principle. The arrangement of electrons in partially filled subshells is governed by *Hund's rule*.

Elements in the same group of the periodic table have similar valence-shell configurations. Chemical and physical properties of the elements show periodic behavior. *Atomic radii*, for example, tend to decrease across any period (left to right) and increase down any group. *First ionization energies* tend to increase across a period and decrease down a group. *Electron affinities* of the Group VIA and Group VIIA elements have large negative values.

Operational Skills

1. **Applying the Pauli exclusion principle** Given an orbital diagram or electron configuration, decide whether it is possible or not, according to the Pauli exclusion principle. **(EXAMPLE 8.1)**
2. **Determining the configuration of an atom using the building-up principle** Given the atomic number of an atom, write the complete electron configuration for the ground state, according to the building-up principle. **(EXAMPLE 8.2)**
3. **Determining the configuration of an atom using the period and group numbers** Given the period and group for an element, write the configuration of the outer electrons. **(EXAMPLE 8.3)**
4. **Applying Hund's rule** Given the electron configuration for the ground state of an atom, write the orbital diagram. **(EXAMPLE 8.4)**
5. **Applying periodic trends** Using the known trends and referring to a periodic table, arrange a series of elements in order by atomic radius **(EXAMPLE 8.5)** or ionization energy **(EXAMPLE 8.6)**.

Review Questions

8.1 Describe the experiment of Stern and Gerlach. How are the results for the hydrogen atom explained?

8.2 Describe the model of electron spin given in the text. What are the restrictions on electron spin?

8.3 What is the maximum number of electrons that can occupy a g subshell ($l = 4$)?

8.4 Define each of the following: noble-gas core, pseudo-noble-gas core, valence electron.

8.5 Give two different possible orbital diagrams for the $1s^2 2s^2 2p^4$ configuration of the oxygen atom, one of which should correspond to the ground state. Label the diagram for the ground state.

8.6 Define the terms *diamagnetic substance* and *paramagnetic substance*. Does the ground-state oxygen atom give a diamagnetic or a paramagnetic substance? Explain.

8.7 What kind of subshell is being filled in Groups IA and IIA? in Groups IIIA to VIIIA? in the transition elements? in the lanthanides and actinides?

8.8 How was Mendeleev able to predict the properties of gallium before it was discovered?

8.9 Describe the major trends that emerge when atomic radii are plotted against atomic number. Describe the trends observed when first ionization energies are plotted against atomic number.

8.10 What main group in the periodic table has elements with the most negative electron affinities for each period? What electron configurations of neutral atoms have only unstable negative ions?

8.11 The ions Na^+ and Mg^{2+} occur in chemical compounds, but the ions Na^{2+} and Mg^{3+} do not. Explain.

8.12 Describe the major trends in metallic character observed in the periodic table of the elements.

8.13 Distinguish between an acidic and a basic oxide. Give examples of each.

8.14 What is the name of the alkali metal atom with valence-shell configuration $5s^1$?

8.15 What would you predict for the atomic number of the halogen under astatine in the periodic table?

8.16 List the elements in Groups IIIA to VIA in the same order as in the periodic table. Label each element as a metal, a metalloid, or a nonmetal. Does each column of elements display the expected trend of increasing metallic characteristics?

8.17 For the list of elements you made for Question 8.16, note whether the oxides of each element are acidic, basic, or amphoteric.

8.18 Write an equation for the reaction of potassium metal with water.

8.19 From what is said in Section 8.7 about Group IIA elements, list some properties of barium.

8.20 Match each description in the left column with the appropriate element in the right column.

a. A waxy, white solid, normally stored under water
b. A yellow solid that burns in air
c. A reddish-brown liquid
d. A soft, light metal that reacts vigorously with water

Sulfur
Sodium
White phosphorus
Bromine

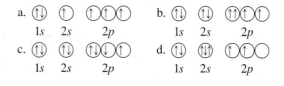

Conceptual Problems

8.21 Two elements in Period 5 are adjacent to one another in the periodic table. The ground-state atom of one element has only s electrons in its valence shell; the other has at least one d electron in an unfilled shell. Identify the elements.

8.22 Two elements are in the same column of the periodic table, one above the other. The ground-state atom of one element has two s electrons in its outer shell, and no d electrons anywhere in its configuration. The other element has d electrons in its configuration. Identify the elements.

8.23 You travel to an alternate universe where the atomic orbitals are different from those on earth, but all other aspects of the atoms are the same. In this universe, you find that the first (lowest energy) orbital is filled with three electrons and the second orbital can hold a maximum of nine electrons. You discover an element Z that has five electrons in its atom. Would you expect Z to be more likely to form a cation or an anion? Indicate a possible charge on this ion.

8.24 Would you expect to find an element having both a very large (positive) first ionization energy and an electron affinity that is much less than zero (large but negative)? Explain.

8.25 Two elements are in the same group, one following the other. One is a metalloid, the other is a metal. Both form oxides of the formula RO_2; the first is acidic, the next is amphoteric. Identify the two elements.

8.26 A metalloid has an acidic oxide of the formula R_2O_3. The element has no oxide of the formula R_2O_5. What is the name of the element?

Practice Problems

Pauli Exclusion Principle

8.27 Which of the following orbital diagrams are allowed by the Pauli exclusion principle? Explain how you arrived at this decision. Give the electron configuration for the allowed ones.

8.28 Which of the following orbital diagrams are allowed and which are not allowed by the Pauli exclusion principle? Explain. For those that are allowed, write the electron configuration.

8.29 Which of the following electron configurations are possible? Explain why the others are not.
a. $1s^22s^12p^6$ b. $1s^22s^22p^63s^13d^6$
c. $1s^22s^22p^8$ d. $1s^22s^22p^63s^23d^{12}$

8.30 Choose the electron configurations that are possible from among the following. Explain why the others are impossible.
a. $1s^22s^32p^6$ b. $1s^22s^22p^4$
c. $1s^22s^22p^83s^23p^63d^7$ d. $1s^22s^22p^63s^13d^9$

Building-up Principle and Hund's Rule

8.31 Give the electron configuration of the ground state of chlorine, using the building-up principle.

8.32 Use the building-up principle to obtain the ground-state configuration of phosphorus.

8.33 Use the building-up principle to obtain the electron configuration of the ground state of vanadium.

8.34 Give the electron configuration of the ground state of cobalt, using the building-up principle.

8.35 Cadmium is a Group IIB element in Period 5. What would you expect for the valence-shell configuration of cadmium?

8.36 Titanium is a Group IVB element in Period 4. What would you expect for the configuration of outer electrons of titanium?

8.37 Write the orbital diagram for the ground state of nickel. The electron configuration is $[Ar]3d^84s^2$.

8.38 Write the orbital diagram for the ground state of terbium. The electron configuration is $[Xe]4f^96s^2$.

8.39 Write an orbital diagram for the ground state of the potassium atom. Is the atomic substance diamagnetic or paramagnetic?

8.40 Write an orbital diagram for the ground state of the zinc atom. Is the atomic substance diamagnetic or paramagnetic?

Periodic Trends

8.41 Order the following elements by increasing atomic radius according to what you expect from periodic trends: Se, S, As.

8.42 Using periodic trends, arrange the following elements in order of increasing atomic radius: F, S, Cl.

8.43 Using periodic trends, arrange the following elements by increasing ionization energy: Ar, Na, Cl, Al.

8.44 Arrange the following elements in order of increasing ionization energy: Mg, Ca, S. Do not look at Figure 8.11.

8.45 From what you know in a general way about electron affinities, state which member of each of the following pairs has the greater negative value: (a) As, Br (b) F, Li.

8.46 From what you know in a general way about electron affinities, state which member of each of the following pairs has the greater negative value: (a) Cl, S (b) Se, K.

General Problems

8.47 Obtain the valence-shell configuration of the polonium atom, Po, using the position of this atom in the periodic table.

8.48 Obtain the valence-shell configuration of the thallium atom, Tl, using the position of this atom in the periodic table.

8.49 Write the orbital diagram for the ground state of the arsenic atom. Give all orbitals.

8.50 Write the orbital diagram for the ground state of the germanium atom. Give all orbitals.

8.51 For eka-lead, predict the electron configuration, whether the element is a metal or nonmetal, and the formula of an oxide.

8.52 For eka-bismuth, predict the electron configuration, whether the element is a metal or nonmetal, and the formula of an oxide.

8.53 Write the orbital diagram corresponding to the ground state of Nb, whose configuration is $[Kr]4d^45s^1$.

8.54 Write the orbital diagram for the ground state of ruthenium. The configuration is $[Kr]4d^75s^1$.

8.55 Match each set of characteristics on the left with an element in the column at the right.
a. A reactive nonmetal; the atom has a large negative electron affinity
b. A soft metal; the atom has low ionization energy
c. A metalloid that forms an oxide of formula R_2O_3
d. A chemically unreactive gas

Sodium (Na)
Antimony (Sb)
Argon (Ar)
Chlorine (Cl_2)

8.56   Match each element on the right with a set of characteristics on the left.

a. A reactive, pale yellow gas; the atom has a large negative electron affinity

Oxygen (O_2)
Gallium (Ga)
Barium (Ba)
Fluorine (F_2)

b. A soft metal that reacts with water to produce hydrogen

c. A metal that forms an oxide of formula R_2O_3

d. A colorless gas; the atom has a moderately large negative electron affinity

8.57   Find the electron configuration of the element with $Z = 23$. From this, give its group and period in the periodic table. Classify the element as a main-group, a d-block transition, or an f-block transition element.

8.58   Find the electron configuration of the element with $Z = 33$. From this, give its group and period in the periodic table. Is this a main-group, a d-block transition, or an f-block transition element?

Cumulative-Skills Problems

8.59   A 2.50-g sample of barium reacted completely with water. What is the equation for the reaction? How many milliliters of dry H_2 evolved at 21°C and 748 mmHg?

8.60   A sample of cesium metal reacted completely with water, evolving 48.1 mL of dry H_2 at 19°C and 768 mmHg. What is the equation for the reaction? What was the mass of cesium in the sample?

8.61   What is the formula of radium oxide? What is the percentage of radium in this oxide?

8.62   What is the formula of hydrogen telluride? What is the percentage of tellurium in this compound?

8.63   How much energy would be required to ionize 5.00 mg of Na(g) atoms to $Na^+(g)$ ions? The first ionization energy of Na atoms is 496 kJ/mol.

8.64   How much energy is evolved when 2.65 mg of Cl(g) atoms adds electrons to give $Cl^-(g)$ ions?

8.65   The lattice energy of an ionic solid such as NaCl is the enthalpy change ΔH for the process in which the solid changes to ions. For example,

$$NaCl(s) \longrightarrow Na^+(g) + Cl^-(g) \qquad \Delta H = 786 \text{ kJ/mol}$$

Assume that the ionization energy and electron affinity are ΔH values for the processes defined by those terms. The ionization energy of Na is 496 kJ/mol. Use this, the electron affinity from Table 8.4, and the lattice energy of NaCl to calculate ΔH for the following process:

$$Na(g) + Cl(g) \longrightarrow NaCl(s)$$

8.66   Calculate ΔH for the following process:

$$K(g) + Br(g) \longrightarrow KBr(s)$$

The lattice energy of KBr is 689 kJ/mol, and the ionization energy of K is 419 kJ/mol. The electron affinity of K is given in Table 8.4. See Problem 8.65.

The shape of snowflakes results from bonding (and intermolecular) forces.

Ionic and Covalent Bonding

The properties of a substance, such as sodium chloride (Figure 9.1), are determined in part by the chemical bonds that hold the atoms together. A *chemical bond* is a strong attractive force that exists between certain atoms in a substance. In Chapter 2 we described how sodium (a silvery metal) reacts with chlorine (a pale greenish-yellow gas) to produce sodium chloride (table salt, a white solid). The substances in this reaction are quite different, as are their chemical bonds. Sodium chloride, NaCl, consists of Na^+ and Cl^- ions held in a regular arrangement, or crystal, by *ionic bonds*. Ionic bonding results from the attractive force of oppositely charged ions.

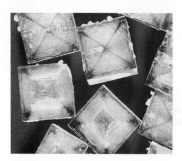

Figure 9.1
Sodium chloride crystals.
Natural crystals of sodium chloride mineral (halite).

A second kind of chemical bond is a *covalent bond*. In a covalent bond, two atoms share valence electrons, which are attracted to the positively charged cores of both atoms. For example, chlorine consists of Cl_2 molecules. A covalent bond holds the two atoms together. This is consistent with the equal sharing of electrons that you would expect between identical atoms.

Metallic bonding, seen in sodium and other metals, represents another important type of bonding. A crystal of sodium metal consists of a regular arrangement of sodium atoms. The valence electrons of these atoms move throughout the crystal, attracted to the positive cores of all Na^+ ions. This attraction holds the crystal together.

What determines the type of bonding in each substance? How do you describe the bonding in various substances? In this chapter we will look at some simple, useful concepts of bonding that can help us answer these questions.

Ionic Bonds

The first explanation of chemical bonding was suggested by the properties of *salts*. Salts are generally crystalline solids that melt at high temperatures. Sodium chloride, for example, melts at 801°C. A molten salt (the liquid after melting) conducts an electric current. A salt dissolved in water gives a solution that also conducts an electric current. This suggests the possibility that ions exist in solids, held together by the attraction of opposite charges.

9.1 Describing Ionic Bonds

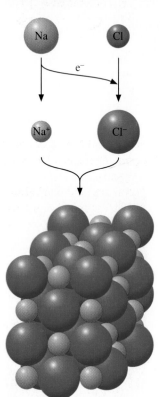

An **ionic bond** is *a chemical bond formed by the electrostatic attraction between positive and negative ions*. The bond forms between two atoms when one or more electrons are transferred from the valence shell of one atom to the valence shell of the other. The atom that loses electrons becomes a *cation* (positive ion), and the atom that gains electrons becomes an *anion* (negative ion). Any given ion tends to attract as many neighboring ions of opposite charge as possible. When large numbers of ions gather together, they form an ionic solid.

To understand why ionic bonding occurs, consider the transfer of a valence electron from a sodium atom to a chlorine atom. You can represent the electron transfer by the following equation:

$$Na([Ne]3s^1) + Cl([Ne]3s^23p^5) \longrightarrow Na^+([Ne]) + Cl^-([Ne]3s^23p^6)$$

The sodium atom has lost its $3s$ electron and has taken on the neon configuration, [Ne]. The chlorine atom has accepted the electron into its $3p$ subshell and has taken on the argon configuration. Once a cation or anion forms, it attracts ions of opposite charge. Within the sodium chloride crystal, NaCl, every Na^+ ion is surrounded by six Cl^- ions, and every Cl^- ion by six Na^+ ions (Figure 2.15).

Lewis Electron-Dot Symbols

You can simplify the preceding equation for the electron transfer between Na and Cl by writing Lewis electron-dot symbols for the atoms and monatomic ions. A **Lewis electron-dot symbol** is *a symbol in which the electrons in the valence shell of an atom or ion are represented by dots placed around the letter symbol of the element.*

Table 9.1 lists Lewis symbols and corresponding valence-shell electron configurations for the atoms of the second and third periods. Note that dots are placed one to each side of a letter symbol until all four sides are occupied. Then the dots are written two to a side until all valence electrons are accounted for. [This pairing of dots does not always correspond to the pairing of electrons in the ground state. Thus, you write $\cdot \dot{\text{B}} \cdot$ for boron, rather than $:\text{B}\cdot$, which more closely corresponds to the ground-state configuration $[\text{He}]2s^2 2p^1$. The first symbol better reflects boron's chemistry, in which each single electron (single dot) tends to be involved in bond formation.]

The equation representing the transfer of an electron from the sodium atom to the chlorine atom is

$$\text{Na}\cdot \, + \, \cdot \ddot{\underset{..}{\text{Cl}}}: \; \longrightarrow \; \text{Na}^+ \, + \, [:\ddot{\underset{..}{\text{Cl}}}:]^-$$

The noble-gas configurations of the ions are apparent from the symbols. No dots are shown for the cation. (All valence electrons have been removed, leaving the noble-gas core.) Eight dots are shown in brackets for the anion (noble-gas configuration $ns^2 np^6$).

EXAMPLE 9.1

Using Lewis Symbols to Represent Ionic Bond Formation

Use Lewis electron-dot symbols to represent the transfer of electrons from magnesium to fluorine atoms to form ions with noble-gas configurations.

SOLUTION

The Lewis symbols for the atoms are $:\ddot{\text{F}}\cdot$ and $\cdot \text{Mg}\cdot$ (see Table 9.1). The magnesium atom loses two electrons to assume a noble-gas configuration. But because a fluorine atom can accept only one electron to fill its valence shell, two fluorine atoms must take part in the electron transfer. We can represent this electron transfer as follows:

$$:\ddot{\underset{..}{\text{F}}}\cdot \, + \, \cdot \text{Mg}\cdot \, + \, \cdot \ddot{\underset{..}{\text{F}}}: \; \longrightarrow \; [:\ddot{\underset{..}{\text{F}}}:]^- \, + \, \text{Mg}^{2+} \, + \, [:\ddot{\underset{..}{\text{F}}}:]^-$$

Energy Involved in Ionic Bonding

It is instructive to look at the energy changes involved in ionic bond formation. From this analysis, you can gain further understanding of why certain atoms bond ionically and others do not.

If atoms bond, there should be a net decrease in energy, because the bonded state is more stable and at a lower energy level. Consider again the formation of an ionic bond between a sodium atom and a chlorine atom. You can think of this as occurring

---

Table 9.1
Lewis Electron-Dot Symbols for Atoms of the Second and Third Periods

Period	IA ns^1	IIA ns^2	IIIA $ns^2 np^1$	IVA $ns^2 np^2$	VA $ns^2 np^3$	VIA $ns^2 np^4$	VIIA $ns^2 np^5$	VIIIA $ns^2 np^6$
Second	Li$\cdot$	$\cdot$Be$\cdot$	$\cdot$B$\cdot$	$\cdot$C$\cdot$	$:$N$\cdot$	$:$O$\cdot$	$:$F$\cdot$	$:$Ne$:$
Third	Na$\cdot$	$\cdot$Mg$\cdot$	$\cdot$Al$\cdot$	$\cdot$Si$\cdot$	$:$P$\cdot$	$:$S$\cdot$	$:$Cl$\cdot$	$:$Ar$:$

Figure 9.2
Energetics of ionic bonding. The transfer of an electron from an Na atom to a Cl atom is not in itself energetically favorable; it requires 147 kJ/mol of energy (Step 1). However, 493 kJ of energy is released when these oppositely charged ions come together to form ion pairs (Step 2a). And additional energy (293 kJ) is released when these ion pairs form the solid NaCl crystal (Step 2b). The lattice energy released when one mole each of Na^+ and Cl^- ions react to produce NaCl(s) is 786 kJ/mol, and the overall process of NaCl formation is energetically favorable, releasing 639 kJ/mol if one starts with gaseous Na and Cl atoms.

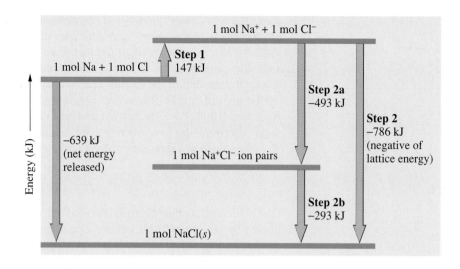

▶ Ionization energies and electron affinities of atoms were discussed in Section 8.6.

in two steps: (1) An electron is transferred between the two separate atoms to give ions. (2) The ions then attract one another to form an ionic bond. In reality, the transfer of the electron and the formation of an ionic bond occur simultaneously, rather than in discrete steps, as the atoms approach one another.

The first step, removing the electron from the sodium atom, requires 496 kJ/mol. Adding the electron to the chlorine atom releases −349 kJ/mol. ◀ It requires more energy to remove an electron from the sodium atom than is gained when the electron is added to the chlorine atom [(496 − 349) kJ/mol, or 147 kJ/mol, to form ions (Figure 9.2, Step 1)].

When positive and negative ions bond, however, more than enough energy is released to supply the 147 kJ/mol. What principally determines the energy released when ions bond is the attraction of oppositely charged ions. You can estimate this energy from *Coulomb's law.*

$$E = \frac{kQ_1Q_2}{r}$$

The ions have electric charges of Q_1 and Q_2 and r is the separation distance of 2.82 × 10^{-10} m. Here k is a physical constant, equal to 8.99 × 10^9 J · m/C² (C is the symbol for coulomb); the charge on Na^+ is $+e$ and that on Cl^- is $-e$, where e equals 1.602 × 10^{-19} C. Therefore,

$$E = \frac{-(8.99 \times 10^9 \text{ J} \cdot \text{m/C}^2) \times (1.602 \times 10^{-19} \text{ C})^2}{2.82 \times 10^{-10} \text{ m}} = -8.18 \times 10^{-19} \text{ J}$$

For the energy of one mole of NaCl, we multiply by Avogadro's number, 6.02 × 10^{23}. We obtain −493 kJ/mol when one mole of Na^+ and one mole of Cl^- come together to form NaCl ion pairs (Figure 9.2, Step 2a). ◀

▶ The energy value −493 kJ/mol is approximate because of the simplifying assumption we made.

The attraction of oppositely charged ions does not stop with the bonding of pairs of ions. The maximum attraction of ions of opposite charge with the minimum repulsion of ions of the same charge is obtained with the formation of the crystalline solid. Then additional energy is released (Figure 9.2, Step 2b).

The energy released when a crystalline solid forms from ions is related to the lattice energy of the solid. The **lattice energy** is *the change in energy that occurs when*

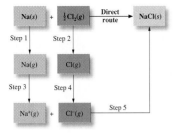

Figure 9.3
Born–Haber cycle for NaCl. The formation of NaCl(s) from the elements is accomplished by two different routes. The direct route is the formation reaction (shown in boldface), and the enthalpy change is ΔH_f°. The indirect route occurs in five steps.

an ionic solid is separated into isolated ions in the gas phase. For sodium chloride, the process is

$$NaCl(s) \longrightarrow Na^+(g) + Cl^-(g)$$

The energy required for this process is the lattice energy of NaCl. You can calculate this energy from Coulomb's law, or obtain an experimental value from thermodynamic data. The value, $+786$ kJ/mol, is for just the opposite process of the one we were considering earlier, in which ions come together to bond, forming the solid. The net energy obtained when gaseous Na and Cl atoms form solid NaCl is $(-786 + 147)$ kJ/mol $= -639$ kJ/mol.

Two elements bond ionically if the ionization energy of one is sufficiently small and the electron affinity of the other is sufficiently large and negative. This situation exists between a reactive metal and a reactive nonmetal. In general, bonding between a metal and a nonmetal is ionic.

Lattice Energies from the Born–Haber Cycle

Direct experimental determination of the lattice energy of an ionic solid is difficult. However, this quantity can be indirectly determined from experiment by means of a thermochemical "cycle" originated by Max Born and Fritz Haber in 1919 and now called the *Born–Haber cycle*. The reasoning is based on Hess's law.

To obtain the lattice energy of NaCl, you think of solid sodium chloride being formed from the elements by two different routes, as shown in Figure 9.3. In one route, NaCl(s) is formed directly from the elements, Na(s) and $\frac{1}{2}Cl_2(g)$. The enthalpy change for this is ΔH_f°, which is given in Table 6.2 as -411 kJ per mole of NaCl. The second route consists of five steps:

1. *Sublimation of sodium.* Sodium is vaporized to a gas of sodium atoms. The enthalpy change for this process is 108 kJ per mole of sodium.

2. *Dissociation of chlorine.* Chlorine molecules are dissociated to atoms. The enthalpy change for this equals the Cl—Cl bond dissociation energy, which is 240 kJ per mole of bonds, or 120 kJ per mole of Cl atoms.

3. *Ionization of sodium.* Sodium atoms are ionized to Na^+ ions. The enthalpy change equals 496 kJ per mole of Na.

4. *Formation of chloride ion.* The enthalpy change for this is the electron affinity of atomic chlorine, which equals -349 kJ per mole of Cl atoms.

5. *Formation of NaCl(s) from ions.* The ions Na^+ and Cl^- formed in Steps 3 and 4 combine to give solid sodium chloride, just the reverse of the lattice energy (breaking the solid into ions). If we let U be the lattice energy, the enthalpy change for Step 5 is $-U$.

Let us write these five steps and add them. We also add the corresponding enthalpy changes, following Hess's law.

$$
\begin{array}{llll}
Na(s) & \longrightarrow Na(g) & \Delta H_1 = & 108 \text{ kJ} \\
\frac{1}{2}Cl_2(g) & \longrightarrow Cl(g) & \Delta H_2 = & 120 \text{ kJ} \\
Na(g) & \longrightarrow Na^+(g) + e^-(g) & \Delta H_3 = & 496 \text{ kJ} \\
Cl(g) + e^-(g) & \longrightarrow Cl^-(g) & \Delta H_4 = & -349 \text{ kJ} \\
Na^+(g) + Cl^-(g) & \longrightarrow NaCl(s) & \Delta H_5 = & -U \\
\hline
Na(s) + \frac{1}{2}Cl_2(g) & \longrightarrow NaCl(s) & \Delta H_f^\circ = & 375 \text{ kJ} - U
\end{array}
$$

The final equation is simply the formation reaction for $NaCl(s)$. The enthalpy change for this formation reaction is $375 \text{ kJ} - U$. But the enthalpy of formation has been determined calorimetrically and equals -411 kJ. Equating these two values, we get

$$375 \text{ kJ} - U = -411 \text{ kJ}$$

Solving for U yields the lattice energy of NaCl, which is 786 kJ.

Properties of Ionic Substances

Typically, ionic substances are high-melting solids. Sodium chloride, NaCl, ordinary salt, melts at 801°C, and magnesium oxide, MgO, a ceramic, melts at 2800°C. The explanation for the high melting points of these substances is simple. The temperature at which melting occurs depends on the strength of the interaction between the atoms or ions. Typical ionic solids require high temperatures for this process to occur because of the strong interactions between the ions.

Coulomb's law also explains why magnesium oxide, which is composed of ions having double charges (Mg^{2+} and O^{2-}), has such a high melting point compared with sodium chloride (Na^+ and Cl^-). Coulomb interactions depend on the product of the ion charges. For NaCl, this is 1×1, whereas for MgO, this is 2×2, or 4 times greater. (Coulomb's law also depends on the distance between charges, or ions, which is about the same in NaCl and MgO.) The much larger Coulomb interaction in MgO requires a much greater temperature to initiate melting.

The liquid melt from an ionic solid consists of ions, and so the liquid conducts an electrical current. If the ionic solid dissolves in a molecular liquid, such as water, the resulting solution consists of ions dispersed among molecules; the solution also conducts an electrical current.

Recently, chemists have prepared ionic substances that have very low melting points. This property is the result of large, nonspherical cations that lead to especially weak ionic bonding. Some of these salts are liquids at room temperature.

9.2 Electron Configurations of Ions

Ions of the Main-Group Elements

In Chapter 2, in discussing the naming of ionic compounds, we listed the common monatomic ions of the main-group elements (Table 2.4). Most of the cations are obtained by removing all the valence electrons from the atoms of metallic elements. Table 9.2 lists the first through the fourth ionization energies of Na, Mg, and Al. The energy needed to remove the first electron from the Na atom is only 496 kJ/mol (first ionization energy). But the energy required to remove another electron is nearly ten times greater (4562 kJ/mol). The electron in this case must be taken from Na^+, which has a neon configuration. Magnesium and aluminum atoms are similar. Their valence electrons are easily removed, but the energy needed to take an electron from either of the ions that result (Mg^{2+} and Al^{3+}) is extremely high. That is why no compounds are found with ions having charges greater than the group number.

The loss of successive electrons from an atom requires increasingly more energy. Consequently, Group IIIA elements show less tendency to form ionic compounds. Boron, in fact, forms no compounds with B^{3+} ions. However, the tendency to form ions becomes greater going down any column of the periodic table because of decreasing ionization energy. The remaining elements of Group IIIA do form compounds containing 3+ ions.

Table 9.2
Ionization Energies of Na, Mg, and Al (in kJ/mol)*

| Element | Successive Ionization Energies | | | |
	First	Second	Third	Fourth
Na	496	4,562	6,912	9,543
Mg	738	1,451	7,733	10,540
Al	578	1,817	2,745	11,577

*Energies for the ionization of valence electrons lie to the left of the colored line.

There is also a tendency for Group IIIA to VA elements of higher periods, particularly Period 6, to form compounds with ions having a positive charge of two less than the group number. Thallium in IIIA, Period 6, has compounds with 1+ ions and compounds with 3+ ions. Ions with charge equal to the group number minus two are obtained when the np electrons of an atom are lost but the ns^2 electrons are retained.

Few compounds of 4+ ions are known because the energy required to form ions is so great. The first three elements of Group IVA—C, Si, and Ge—are nonmetals (or metalloids) and usually form covalent rather than ionic bonds. Tin and lead, however, commonly form compounds with 2+ ions. For example, tin forms tin(II) chloride, $SnCl_2$, which is an ionic compound. It also forms tin(IV) chloride, $SnCl_4$, but this is a covalent compound. Bismuth, in Group VA, is a metallic element that forms compounds of Bi^{3+}, where only the $6p$ electrons have been lost.

Group VIA and Group VIIA elements, whose atoms have the largest negative electron affinities, would be expected to form monatomic ions by gaining electrons to give noble-gas configurations. An atom of a Group VIIA element picks up one electron to give a 1− anion; examples are F^- and Cl^-. (Hydrogen also forms compounds of the 1− ion, H^-. The hydride ion, H^-, has a $1s^2$ configuration, like the noble-gas atom helium.) An atom of a Group VIA element picks up two electrons to give a 2− anion; examples are O^{2-} and S^{2-}. Although the electron affinity of nitrogen is zero, the N^{3-} ion is stable in the presence of certain positive ions, including Li^+ and those of the alkaline earth elements.

To summarize, the common monatomic ions found in compounds of the main-group elements fall into three categories (see Table 2.4):

1. Cations of Groups IA to IIIA having noble-gas or pseudo-noble-gas configurations. The ion charges equal the group numbers.

2. Cations of Groups IIIA to VA having ns^2 electron configurations. The ion charges equal the group numbers minus two. Examples are Tl^+, Sn^{2+}, Pb^{2+}, and Bi^{3+}.

3. Anions of Groups VA to VIIA having noble-gas configurations. The ion charges equal the group numbers minus eight.

EXAMPLE 9.2

Writing the Electron Configuration and Lewis Symbol for a Main-Group Ion

Write the electron configuration and the Lewis symbol for N^{3-}.

Ionic Liquids and Green Chemistry

FRONTIERS

Modern chemists work to expand the boundaries of known materials. They have discovered plastics that conduct electricity like metals and materials that look solid, but are so porous they are almost as light as air. Now chemists have produced room-temperature ionic liquids (Figure 9.4). Most of these are clear, well-behaved substances that look and pour much like water, and like water their strength is as solvents (liquids that dissolve other substances). In fact, they promise to be "super solvents." Given a material—an organic substance, a plastic, or even a rock—researchers believe you will be able to find an ionic liquid that is capable of dissolving it!

Copyright 2000 American Chemical Society

Figure 9.4
Room-temperature ionic liquid. Hélène Olivier of the French Petroleum Institute displays a tube containing an ionic liquid (including a blue nickel compound) with a hydrocarbon liquid above it. Typical liquids, such as a hydrocarbon, are molecular substances, whereas ionic liquids are composed of ions.

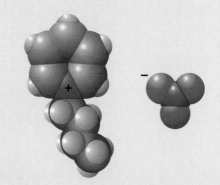

Figure 9.5
The ions composing an ionic liquid. Shown here are space-filling models of the ions composing N-butylpyridinium nitrate. Note the bulky cation and small nitrate anion.

The high melting point of sodium chloride is easily explained. It consists of small, spherical ions that pack closely together. Thus, the ions interact strongly, giving a solid with a high melting point. Room-temperature ionic liquids, by contrast, consist of large, nonspherical cations with various anions (Figure 9.5). It is the large, bulky cations that keep the ions from packing closely; the large distances between ions result in weak interactions, yielding a substance whose melting point is often well below room temperature.

The demand for *green chemistry,* the commercial production of chemicals using environmentally sound methods, has spurred much of the research into ionic liquids. Many chemical processes use volatile organic solvents. These solvents are liquids that evaporate easily into the surrounding air, where they can contribute to air pollution. Organic solvents are often flammable, too. Ionic liquids are neither volatile nor flammable. In addition to these environmental rewards, however, ionic liquids appear to offer another bonus: The proper choice of ionic liquid may improve the yield and lower the costs of a chemical process.

SOLUTION

The electron configuration of the N atom is $[He]2s^2 2p^3$. By gaining three electrons, the atom assumes a $3-$ charge and the neon configuration $\mathbf{[He]2s^2 2p^6}$. The Lewis symbol is

$$[\:\ddot{\underset{..}{N}}\:]^{3-}$$

See Problems 9.23, 9.24, 9.25, and 9.26.

Transition-Metal Ions

Most transition elements form several cations of different charges. For example, iron has the cations Fe^{2+} and Fe^{3+}. Neither has a noble-gas configuration.

In forming ions in compounds, the atoms of transition elements generally lose the ns electrons first; then they may lose one or more $(n-1)d$ electrons. The $2+$ ions are common for the transition elements. Many transition elements also form $3+$ ions by losing one electron in addition to the two ns electrons. Table 2.5 lists some common transition-metal ions. Many compounds of transition-metal ions are colored because of transitions involving d electrons, whereas the compounds of the main-group elements are usually colorless (Figure 9.6).

EXAMPLE 9.3

Writing Electron Configurations of Transition-Metal Ions

Write the electron configurations of Fe^{2+} and Fe^{3+}.

SOLUTION

The electron configuration of the Fe atom $(Z = 26)$ is $[Ar]3d^6 4s^2$. To obtain the configuration of Fe^{2+}, remove the $4s^2$ electrons. To obtain the configuration of Fe^{3+}, also remove a $3d$ electron. **The configuration of Fe^{2+} is $[Ar]3d^6$, and that of Fe^{3+} is $[Ar]3d^5$.**

See Problems 9.27 and 9.28.

Figure 9.6
Common transition-metal cations in aqueous solution. *Left to right:* Cr^{3+} (red-violet), Mn^{2+} (pale pink), Fe^{2+} (pale green), Fe^{3+} (pale yellow), Co^{2+} (pink), Ni^{2+} (green), Cu^{2+} (blue), Zn^{2+} (colorless).

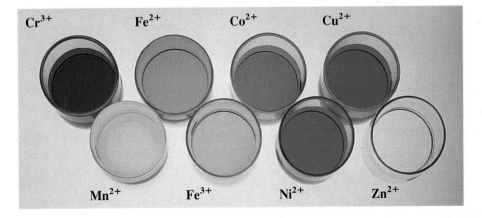

CONCEPT CHECK 9.1

The following are electron configurations for some ions. Which ones would you expect to see in chemical compounds? State the concept or rule you used to decide for or against any ion.

a. Fe^{2+} [Ar]$3d^4 4s^2$ b. N^{2-} [He]$2s^2 2p^5$ c. Zn^{2+} [Ar]$3d^{10}$

d. Na^{2+} [He]$2s^2 2p^5$ e. Ca^{2+} [Ne]$3s^2 3p^6$

9.3 Ionic Radii

The **ionic radius** is *a measure of the size of the spherical region around the nucleus of an ion within which the electrons are most likely to be found.* Defining an ionic radius is somewhat arbitrary because an electron distribution never abruptly ends. However, if we imagine ions to be spheres of definite size, we can obtain their radii from known distances between nuclei in crystals.

To understand how you compute ionic radii, consider the determination of the radius of an I^- ion in lithium iodide. Figure 9.7 depicts a layer of ions in LiI. The distance between adjacent iodine nuclei equals twice the I^- radius. From x-ray diffraction experiments, the iodine–iodine distance is found to be 426 pm. Therefore, the I^- radius in LiI is $\frac{1}{2} \times 426$ pm = 213 pm. Other crystals give approximately the same radius for the I^- ion. Table 9.3, which lists average values of ionic radii obtained from many compounds of the main-group elements, gives 216 pm for the I^- radius.

The values of ionic radii compare with atomic radii in ways that you might expect. For example, you expect a cation to be smaller and an anion to be larger than the corresponding atom (see Figure 9.8).

A cation formed when an atom loses all its valence electrons is smaller than the atom because it has one less shell of electrons. Because an anion has more electrons than the atom, the electron–electron repulsion is greater, so the valence orbitals expand. Thus, the anion radius is larger than the atomic radius.

The ionic radii of the main-group elements shown in Table 9.3 follow a regular pattern, just as atomic radii do. *Ionic radii increase down any column because of the addition of electron shells.*

The pattern across a period becomes clear if you look first at the cations, then at the anions. For example, in the third period we have

Cation	Na^+	Mg^{2+}	Al^{3+}
Radius (pm)	95	65	50

All of these ions have the neon configuration $1s^2 2s^2 2p^6$, but different nuclear charges; they are isoelectronic. **Isoelectronic** *refers to different species having the same number of electrons.* To understand the decrease in radius from Na^+ to Al^{3+}, imagine the nuclear charge increases. With each increase of charge, the orbitals contract due to the greater attractive force of the nucleus. Thus, in any isoelectronic sequence of atomic ions, the ionic radius decreases with increasing atomic number.

The anions in the third-period elements are much larger than the cations in the same period. This abrupt increase in ionic radius is due to the fact that the anions S^{2-} and Cl^- have configurations with one more shell of electrons than the cations. And

A

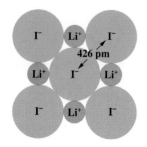

B

Figure 9.7
Determining the iodide ion radius in the lithium iodide (LiI) crystal. *(A)* A three-dimensional view of the crystal. *(B)* Cross section through a layer of ions. Iodide ions are assumed to be spheres in contact with one another. The distance between iodine nuclei (426 pm) is determined experimentally. One-half of this distance (213 pm) equals the iodide ion radius.

Na
[He] $2s^2 2p^6 3s^1$

Na$^+$
[He] $2s^2 2p^6$

Cl
[Ne] $3s^2 3p^5$

Cl$^-$
[Ne] $3s^2 3p^6$

Figure 9.8
Comparison of atomic and ionic radii. Note that the sodium atom loses its outer shell in forming the Na$^+$ ion. Thus, the cation is smaller than the atom. The Cl$^-$ ion is larger than the Cl atom, because the same nuclear charge holds a greater number of electrons less strongly.

Table 9.3
Ionic Radii (in pm) of Some Main-Group Elements

Period	IA	IIA	IIIA	VIA	VIIA
2	Li$^+$	Be^{2+}		O^{2-}	F$^-$
	60	31		140	136
3	Na$^+$	Mg^{2+}	Al^{3+}	S^{2-}	Cl$^-$
	95	65	50	184	181
4	K$^+$	Ca^{2+}	Ga^{3+}	Se^{2-}	Br$^-$
	133	99	62	198	195
5	Rb$^+$	Sr^{2+}	In^{3+}	Te^{2-}	I$^-$
	148	113	81	221	216
6	Cs$^+$	Ba^{2+}	Tl^{3+}		
	169	135	95		

because these anions also constitute an isoelectronic sequence, the ionic radius decreases with atomic number (as with the atoms):

Anion	S^{2-}	Cl$^-$
Radius (pm)	184	181

Thus, *in general, across a period the cations decrease in radius. When you reach the anions, there is an abrupt increase in radius, and then the radius again decreases.*

EXAMPLE 9.4

Using Periodic Trends to Obtain Relative Ionic Radii

Without looking at Table 9.3, arrange the following ions in order of decreasing ionic radius: F$^-$, Mg^{2+}, O^{2-}. (You may use a periodic table.)

SOLUTION

Note that F$^-$, Mg^{2+}, and O^{2-} are isoelectronic. If you arrange them by increasing nuclear charge, they will be in order of decreasing ionic radius. **The order is O^{2-}, F$^-$, Mg^{2+}.**

See Problems 9.31 and 9.32.

Covalent Bonds

▶ G. N. Lewis (1875–1946) was professor of chemistry at the University of California at Berkeley. Well known for work on chemical bonding, he was also noted for his research in molecular spectroscopy and thermodynamics.

In the preceding sections we looked at ionic substances, which are typically high-melting solids. Many substances, however, are molecular—gases, liquids, or low-melting solids consisting of molecules (Figure 9.9). The forces that hold atoms together in a molecular substance cannot be understood on the basis of an ionic model. In 1916 Gilbert Newton Lewis proposed that the strong attractive force between two atoms in a molecule results from a **covalent bond,** *a chemical bond formed by the sharing of a pair of electrons between atoms.* ◀ We will discuss the descriptive aspects of covalent bonding in the following sections.

9.4 Describing Covalent Bonds

Consider the formation of a covalent bond between two H atoms to give the H_2 molecule. As the atoms approach one another, their $1s$ orbitals begin to overlap. Each electron can then occupy the space around both atoms. In other words, the two electrons can be shared by the atoms (Figure 9.10). The electrons are attracted simultaneously by the positive charges of the two hydrogen nuclei. This attraction that bonds the electrons to both nuclei is the force holding the atoms together. Although ions do not exist in H_2, the force that holds the atoms together can still be regarded as arising from the attraction of oppositely charged particles: nuclei and electrons.

It is interesting to see how the potential energy of the atoms changes as they approach and then bond. Figure 9.11 shows the potential energy of the atoms for various distances between nuclei. The potential energy of the atoms when they are some distance apart is indicated by a position on the potential-energy curve at the far right. As the atoms approach (moving from right to left on the potential-energy curve), the potential energy gets lower and lower. The decrease in energy is a reflection of the bonding of the atoms. Eventually the potential energy reaches a minimum value. The distance between nuclei at this minimum energy is called the *bond length* of H_2. It is the normal distance between nuclei in the molecule.

Now imagine the reverse process. You start with the H_2 molecule, the atoms at their normal bond length apart. To separate the atoms in the molecule, energy must be added. The energy that must be added is called the *bond dissociation energy*. The larger the bond dissociation energy, the stronger is the bond.

Lewis Formulas

You can represent the formation of the covalent bond in H_2 from atoms as follows:

$$\text{H}\cdot + \cdot\text{H} \longrightarrow \text{H:H}$$

This uses the Lewis electron-dot symbol for the hydrogen atoms and represents the covalent bond by a pair of dots. Recall that the two electrons from the covalent bond spend part of the time in the region of each atom. In this sense, each atom in H_2 has a helium configuration. We can draw a circle about each atom to emphasize this.

The formation of a bond between H and Cl to give an HCl molecule can be represented in a similar way.

$$\text{H}\cdot + \cdot\ddot{\underset{\cdot\cdot}{\text{Cl}}}: \longrightarrow \left(\text{H}\!:\!\ddot{\underset{\cdot\cdot}{\text{Cl}}}:\right)$$

Unpaired electrons on each atom pair up to form a covalent bond. The pair of electrons is shared by the two atoms. Each atom then acquires a noble-gas configuration of electrons, the H atom having two electrons about it (as in He), and the Cl atom having eight valence electrons about it (as in Ar).

A formula using dots to represent valence electrons is called a **Lewis electron-dot formula.** An electron pair represented by a pair of dots in such a formula is either a **bonding pair** *(an electron pair shared between two atoms)* or a **lone,** or **nonbonding, pair** *(an electron pair that remains on one atom and is not shared).* For example,

Figure 9.9
Two molecular substances.
Top: Iodoform, CHI_3, is a low-melting, yellow solid (m.p. 120°C). *Bottom:* Carbon tetrachloride, CCl_4, is a colorless liquid.

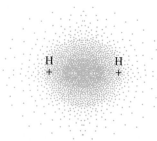

Figure 9.10
The electron probability distribution for the H_2 molecule. The electrons occupy the space around both atoms.

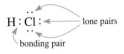

Frequently, the number of covalent bonds formed by an atom equals the number of unpaired electrons shown in its Lewis symbol. Consider the formation of NH_3.

$$3H\cdot \; + \; \cdot \overset{\cdot}{N}: \; \longrightarrow \; H:\overset{\cdot\cdot}{\underset{H}{N}}:$$

In many instances, the number of bonds formed by an atom in Groups IVA to VIIA equals the number of unpaired electrons, which is eight minus the group number. For example, a nitrogen atom (Group VA) forms $8 - 5 = 3$ covalent bonds. ◄

▶ The numbers of unpaired electrons in the Lewis symbols for the atoms of elements in Groups IA and IIIA equal the group numbers. But, except for the first elements of these groups, the atoms usually form ionic bonds.

Coordinate Covalent Bonds

When bonds form between atoms that both donate an electron, you have

$$A\cdot \; + \; \cdot B \; \longrightarrow \; A:B$$

However, it is possible for both electrons to come from the same atom. A **coordinate covalent bond** is *a bond formed when both electrons of the bond are donated by one atom:*

$$A \; + \; :B \; \longrightarrow \; A:B$$

An example is the formation of the ammonium ion, in which an electron pair on the N atom of NH_3 forms a bond with H^+.

$$H^+ \; + \; :NH_3 \; \longrightarrow \; \left[H:\overset{\cdot\cdot}{\underset{H}{N}}:H \right]^+$$

The new N—H bond is clearly identical to the other N—H bonds.

Figure 9.11
Potential-energy curve for H_2. The stable molecule occurs at the bond distance corresponding to the minimum in the potential-energy curve.

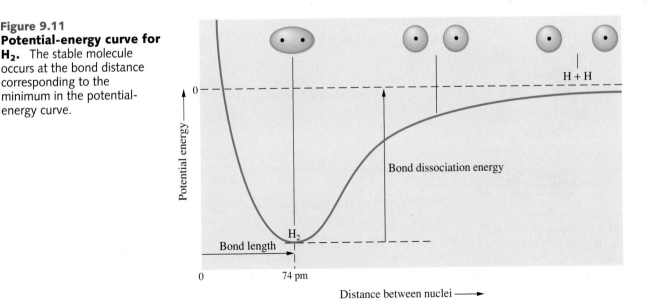

Chemical Bonds in Nitroglycerin

EVERYDAY LIFE

Nitroglycerin gained a nasty reputation soon after its discovery in 1846. Unless kept cold, the pale yellow, oily liquid detonates from even the mildest vibration.

Here is the structure of nitroglycerin (also see the molecular model in Figure 9.12):

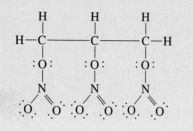

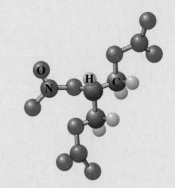

Figure 9.12
Molecular model of nitroglycerin. In this ball-and-stick model, carbon atoms are black, hydrogen atoms are light blue, oxygen atoms are red, and nitrogen atoms are dark blue.

Octet Rule

In each of the molecules we have described so far, the atoms have obtained noble-gas configurations through the sharing of electrons. *The tendency of atoms in molecules to have eight electrons in their valence shells (two for hydrogen atoms)* is known as the **octet rule.** Many of the molecules we will discuss follow the octet rule.

Multiple Bonds

A **single bond** is *a covalent bond in which a single pair of electrons is shared by two atoms*. But it is possible for atoms to share additional electron pairs. A **double bond** is *one in which two pairs of electrons are shared by two atoms*. A **triple bond** is *one in which three pairs of electrons are shared by two atoms*. As examples, consider ethylene, C_2H_4, and acetylene, C_2H_2. Their Lewis formulas are

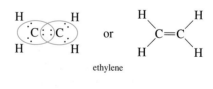

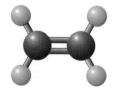

Ethylene

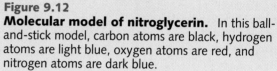

Acetylene

Note the octet of electrons on each C atom. Double bonds form primarily with C, N, O, and S atoms. Triple bonds form mostly to C and N atoms.

With just a little jostling, nitroglycerin, $C_3H_5(ONO_2)_3$, can rearrange its atoms to give stable products:

$$4C_3H_5(ONO_2)_3(l) \longrightarrow$$
$$6N_2(g) + 12CO_2(g) + 10H_2O(g) + O_2(g)$$

The stability of the products results from their strong bonds, which are much stronger than those in nitroglycerin. The explosive force of the reaction results from both the rapid reaction and from the large volume increase on forming gaseous products.

In 1867, the Swedish chemist Alfred Nobel discovered that nitroglycerin behaved better when absorbed on diatomaceous earth, a crumbly rock, giving an explosive mixture that Nobel called *dynamite*. Part of the wealth he made from his explosive factories was left in trust to establish the Nobel Prizes. Explosives have many peacetime uses, including road building, mining, and demolition (Figure 9.13).

Source: Adapted from an essay in *Introductory Chemistry,* second edition, by Ebbing and Wentworth (Houghton Mifflin, 1998), p. 308.

Figure 9.13
Demolition of a building. Explosives are placed at predetermined positions so that when they are detonated, the building collapses on itself.

9.5 Polar Covalent Bonds; Electronegativity

Figure 9.14
The HCl molecule. A molecular model.

A covalent bond involves the sharing of at least one pair of electrons between two atoms. When the atoms are alike, as in the case of the H—H bond of H_2, the bonding electrons are shared equally. But when the two atoms are of different elements, the bonding electrons need not be shared equally. A **polar covalent bond** is *a covalent bond in which the bonding electrons spend more time near one atom than the other.* For example, in the case of the HCl molecule (Figure 9.14), the bonding electrons spend more time near the chlorine atom than the hydrogen atom.

You can consider the polar bond as intermediate between a *nonpolar* bond, as in H_2, and an ionic bond, as in NaCl. From this point of view, an ionic bond is simply an extreme example of a polar covalent bond. To illustrate, we can represent the bonding in H_2, HCl, and NaCl with electron-dot formulas as follows:

$$\text{H : H} \qquad \text{H : } \overset{..}{\underset{..}{\text{Cl}}} \text{:} \qquad \text{Na}^+ \; \text{:} \overset{..}{\underset{..}{\text{Cl}}} \text{:}^-$$

Nonpolar covalent   Polar covalent   Ionic

Thus, it is possible to arrange different bonds to form a gradual transition from nonpolar covalent to ionic.

Electronegativity is *a measure of the ability of an atom in a molecule to draw bonding electrons to itself.* Several electronegativity scales have been proposed. A widely used scale was derived by Linus Pauling from bond energies. Pauling's electronegativity values are given in Figure 9.15. Because electronegativities depend somewhat on the bonds formed, these values are actually average ones. ◀

Fluorine, the most electronegative element, was assigned a value of 4.0 on Pauling's scale. Lithium, at the left end of the same period, has a value of 1.0. *In general, electronegativity increases from left to right and decreases from top to bottom in the periodic table.* Metals are the least electronegative elements (they are *electropositive*) and nonmetals the most electronegative.

▶ Linus Pauling received the Nobel Prize in chemistry in 1954 for his work on the nature of the chemical bond. In 1962 he received the Nobel Peace Prize.

| | H 2.1 | | | | | | | | | | | | | | | | |

IA	IIA												IIIA	IVA	VA	VIA	VIIA
Li 1.0	Be 1.5												B 2.0	C 2.5	N 3.0	O 3.5	F 4.0
Na 0.9	Mg 1.2	IIIB	IVB	VB	VIB	VIIB		VIIIB		IB	IIB		Al 1.5	Si 1.8	P 2.1	S 2.5	Cl 3.0
K 0.8	Ca 1.0	Sc 1.3	Ti 1.5	V 1.6	Cr 1.6	Mn 1.5	Fe 1.8	Co 1.8	Ni 1.8	Cu 1.9	Zn 1.6	Ga 1.6	Ge 1.8	As 2.0	Se 2.4	Br 2.8
Rb 0.8	Sr 1.0	Y 1.2	Zr 1.4	Nb 1.6	Mo 1.8	Tc 1.9	Ru 2.2	Rh 2.2	Pd 2.2	Ag 1.9	Cd 1.7	In 1.7	Sn 1.8	Sb 1.9	Te 2.1	I 2.5
Cs 0.7	Ba 0.9	La-Lu 1.1-1.2	Hf 1.3	Ta 1.5	W 1.7	Re 1.9	Os 2.2	Ir 2.2	Pt 2.2	Au 2.4	Hg 1.9	Tl 1.8	Pb 1.8	Bi 1.9	Po 2.0	At 2.2
Fr 0.7	Ra 0.9	Ac-No 1.1-1.7														

Figure 9.15
Electronegativities of the elements. The values given are those of Pauling.

The absolute value of the difference in electronegativity of two bonded atoms gives a rough measure of the polarity to be expected in a bond. When this difference is small, the bond is nonpolar. When it is large, the bond is polar (or, if the difference is very large, perhaps ionic). The electronegativity differences for the bonds H—H, H—Cl, and Na—Cl are 0.0, 0.9, and 2.1, respectively. Differences in electronegativity explain why ionic bonds usually form between a metal atom and a nonmetal atom. On the other hand, covalent bonds form primarily between two nonmetals because the electronegativity differences are small.

EXAMPLE 9.5

Using Electronegativities to Obtain Relative Bond Polarities

Use electronegativity values (Figure 9.15) to arrange the following bonds in order of increasing polarity: P—H, H—O, C—Cl.

SOLUTION

The absolute values of the electronegativity differences are P—H, 0.0; H—O, 1.4; C—Cl, 0.5. Hence, the order is **P—H, C—Cl, H—O.**

See Problems 9.37 and 9.38.

You can use an electronegativity scale to predict the direction in which the electrons shift during bond formation; the electrons are pulled toward the more electronegative atom. For example, consider the H—Cl bond. Because the Cl atom ($X = 3.0$) is more electronegative than the H atom ($X = 2.1$), the bonding electrons in H—Cl are pulled toward Cl. Because the bonding electrons spend most of their time around the Cl atom, that end of the bond acquires a partial negative charge (indicated $\delta-$). The H-atom end of the bond has a partial positive charge ($\delta+$).

$$\overset{\delta+ \quad \delta-}{\text{H—Cl}}$$

The HCl molecule is a *polar molecule*.

9.6 Writing Lewis Electron-Dot Formulas

The Lewis electron-dot formula of a molecule shows how atoms are bonded. Bonding electron pairs are indicated either by two dots or by a dash. In addition to the bonding electrons, however, an electron-dot formula shows the positions of lone pairs of electrons. In this section we will discuss the steps for writing the electron-dot formula for a molecule made from atoms of the main-group elements.

Before you can write the Lewis formula of a molecule (or a polyatomic ion), you must know the *skeleton structure* of the molecule. The skeleton structure tells you which atoms are bonded to one another.

Normally the skeleton structure must be found by experiment. For simple molecules, you can often predict the skeleton structure. For instance, many small molecules or polyatomic ions consist of a central atom around which are bonded atoms of greater electronegativity, such as F, Cl, and O. Note the structural formula (and molecular model) of phosphorus oxychloride, $POCl_3$ (Figure 9.16). The phosphorus atom is surrounded by more electronegative atoms. (In some cases, H atoms surround a more electronegative atom as in H_2O and NH_3, but H cannot be a central atom because it normally forms only one bond.) Similarly, *oxoacids* are substances in which O atoms (and possibly other electronegative atoms) are bonded to a central atom, with one or more H atoms usually bonded to O atoms. An example is chlorosulfonic acid, HSO_3Cl (Figure 9.17).

Another useful idea for predicting skeleton structures is that molecules or polyatomic ions often have symmetrical structures. For example, disulfur dichloride, S_2Cl_2, is symmetrical, with the more electronegative Cl atoms around the S atoms: Cl—S—S—Cl.

Once you know the skeleton structure of a molecule, you can find the Lewis formula by applying the following four steps:

STEP 1: *Calculate the total number of valence electrons* for the molecule by summing the number of valence electrons for each atom. If you are writing the Lewis formula of a polyatomic anion, you *add* the number of negative charges to this total. For a polyatomic cation, you *subtract* the number of positive charges from the total.

STEP 2: *Write the skeleton structure of the molecule or ion,* connecting every bonded pair of atoms by a pair of dots (or a dash).

STEP 3: *Distribute electrons to the atoms surrounding the central atom (or atoms)* to satisfy the octet rule for these surrounding atoms.

STEP 4: *Distribute the remaining electrons as pairs to the central atom (or atoms),* after subtracting the number of electrons already distributed from the total found in Step 1. If there are fewer than eight electrons on the central atom, this suggests that a multiple bond is present. To obtain a multiple bond, move one or two electron pairs (depending on whether the bond is to be double or triple) from a surrounding atom to the bond connecting the central atom. Atoms that often form multiple bonds are C, N, O, and S.

The next several examples illustrate how to write the Lewis electron-dot formula for a small molecule, given the molecular formula.

Cl
|
Cl—P—O
|
Cl

Figure 9.16
The phosphorus oxychloride molecule.

O
‖
O—S—O—H
|
Cl

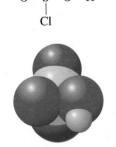

Figure 9.17
The chlorosulfonic acid molecule.

EXAMPLE 9.6

Writing Lewis Formulas (Single Bonds Only)

Sulfur dichloride, SCl_2, is a red, fuming liquid used in the manufacture of insecticides. Write the Lewis formula for the molecule.

SOLUTION

The number of valence electrons from an atom equals the group number: 6 for S, 7 for each Cl, for a total of 20 electrons. You expect the skeleton structure to have S as the central atom, with the more electronegative Cl atoms bonded to it. After connecting atoms by electron pairs and distributing electrons to the outer atoms, you have

$$:\ddot{C}l : S : \ddot{C}l : \quad \text{or} \quad :\ddot{C}l-S-\ddot{C}l :$$

This accounts for 8 electron pairs, or 16 electrons. Subtracting this from the total number of electrons (20) gives 4 electrons, or 2 electron pairs. You place these on the central atom (S). The final Lewis formula is

$$:\ddot{C}l : \ddot{S} : \ddot{C}l : \quad \textbf{or} \quad :\ddot{C}l-\ddot{S}-\ddot{C}l :$$

See Problems 9.39 and 9.40.

SCl_2

EXAMPLE 9.7

Writing Lewis Formulas (Including Multiple Bonds)

Carbonyl chloride, or phosgene, $COCl_2$, is a highly toxic gas used as a starting material for the preparation of polyurethane plastics. What is the electron-dot formula of $COCl_2$?

SOLUTION

The total number of valence electrons is $4 + 6 + (2 \times 7) = 24$. You expect the more electropositive atom, C, to be central, with the O and Cl atoms bonded to it. After distributing electron pairs to these surrounding atoms, you have

$$:\ddot{C}l : C : \ddot{O} : \quad \text{or} \quad :\ddot{C}l-C-\ddot{O} :$$
$$\quad\quad :\ddot{C}l : \quad\quad\quad\quad\quad :\ddot{C}l :$$

This accounts for all 24 valence electrons, leaving only 6 electrons on C. This is two fewer than an octet, so you move a pair of electrons on the O atom to give a carbon–oxygen double bond. The electron-dot formula of $COCl_2$ is

$$:\ddot{C}l : C :: \ddot{O} \quad \textbf{or} \quad :\ddot{C}l-C=\ddot{O}$$
$$\quad\quad :\ddot{C}l : \quad\quad\quad\quad :\ddot{C}l :$$

Check to see that each atom has an octet.

$COCl_2$

EXAMPLE 9.8

Writing Lewis Formulas (Ionic Species)

Obtain the electron-dot formula (Lewis formula) of the BF_4^- ion.

SOLUTION

The total valence electrons provided by the boron and four fluorine atoms is $3 + (4 \times 7) = 31$. Because the anion has a charge of $1-$, it has one more electron than is provided by the neutral atoms. Thus, the total number of valence electrons is 32 (or 16 electron pairs). You assume that the skeleton structure has boron as the central atom, with the more electronegative F atoms bonded to it. After connecting the B and F atoms by bonds and placing electron pairs around the F atoms to satisfy the octet rule, you obtain

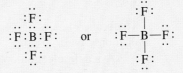

This uses up all 32 electrons. The charge on the entire ion is indicated by the minus sign written as a superscript to square brackets enclosing the electron-dot formula.

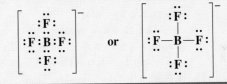

See Problems 9.41 and 9.42.

BF_4^-

▶ CONCEPT CHECK 9.2

Each of the following may seem, at first glance, to be plausible electron-dot formulas for the molecule N_2F_2. Most, however, are incorrect for some reason. What concepts or rules apply to each, either to cast it aside or to keep it as the correct formula?

9.7 Delocalized Bonding: Resonance

We have assumed up to now that the bonding electrons are localized in the region between two atoms. In some cases, however, this assumption does not fit the experimental data. Suppose, for example, that you try to write an electron-dot formula for ozone, O_3. You find that you can write two formulas:

$$
\begin{array}{cc}
\overset{\cdots}{O} & \overset{\cdots}{O} \\
\overset{\cdots}{O} \!=\!\! \overset{\cdots}{O}\overset{\cdots}{O} \!: & :\overset{\cdots}{O}\! \overset{\cdots}{O}\!=\!\!\overset{\cdots}{O} \\
A & B
\end{array}
$$

In formula A, the oxygen–oxygen bond on the left is a double bond and the oxygen–oxygen bond on the right is a single bond. In formula B, the situation is just

► **The lengths of the two oxygen–oxygen bonds (that is, the distances between the atomic nuclei) are each 128 pm.**

the opposite. Experiment shows, however, that both bonds in O_3 are identical. Therefore, neither formula can be correct. ◄

According to theory, one of the bonding pairs in ozone is spread over the region of all three atoms rather than associated with a particular oxygen–oxygen bond. This is called **delocalized bonding,** *a type of bonding in which a bonding pair of electrons is spread over a number of atoms rather than localized between two.* We might symbolically describe the delocalized bonding in ozone as follows:

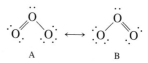

The broken line indicates a bonding pair of electrons that spans three nuclei rather than only two.

A single electron-dot formula cannot properly describe delocalized bonding. Instead, a resonance description is often used. According to the **resonance description,** *you describe the electron structure of a molecule having delocalized bonding by writing all possible electron-dot formulas.* These are called the *resonance formulas* of the molecule. The actual electron distribution of the molecule is a composite of these resonance formulas.

The electron structure of ozone can be described in terms of the two resonance formulas presented at the start of this section. By convention, we usually write all of the resonance formulas and connect them by double-headed arrows. For ozone we would write

Unfortunately, this notation can be misinterpreted. It does not mean that the ozone molecule flips back and forth between two forms. There is only one ozone molecule. The double-headed arrow means that you should form a mental picture of the molecule by fusing the various resonance formulas.

Attempting to write electron-dot formulas leads you to recognize that delocalized bonding exists in many molecules. Whenever you can write several plausible electron-dot formulas—which often differ merely in the allocation of single and double bonds to the same kinds of atoms (as in ozone)—you can expect delocalized bonding.

EXAMPLE 9.9
Writing Resonance Formulas

Describe the electron structure of the carbonate ion, CO_3^{2-}, in terms of electron-dot formulas.

SOLUTION

One possible electron-dot formula for the carbonate ion is

Because you expect all carbon–oxygen bonds to be equivalent, you must describe the electron structure in resonance terms.

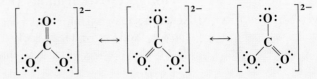

You expect one electron pair to be delocalized over the region of all three carbon–oxygen bonds.

See Problems 9.43 and 9.44.

CO_3^{2-}

Metals are extreme examples of delocalized bonding. A sodium metal crystal, for example, can be regarded as an array of Na^+ ions surrounded by a "sea" of electrons (Figure 9.18). The valence, or bonding, electrons are delocalized over the entire metal crystal. The freedom of these electrons to move throughout the crystal is responsible for the electrical conductivity of a metal.

9.8   Exceptions to the Octet Rule

Many molecules composed of atoms of the main-group elements have electronic structures that satisfy the octet rule, but a number of them do not. A few molecules, such as NO, have an odd number of electrons and so cannot satisfy the octet rule. Other exceptions to the octet rule fall into two groups—one group of molecules with an atom having fewer than eight valence electrons around it and the other group of molecules with an atom having more than eight valence electrons around it.

Phosphorus pentafluoride is an exception in which the central atom has more than eight valence electrons around it. Each molecule consists of a phosphorus atom surrounded by fluorine atoms (Figure 9.19). The phosphorus atom has ten valence electrons around it:

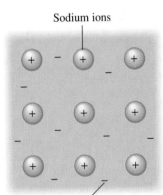

Sodium ions

Valence electrons move throughout metal

Figure 9.18
Delocalized bonding in sodium metal.   The metal consists of positive sodium ions in a "sea" of valence electrons. Valence (bonding) electrons are free to move throughout the metal crystal (beige area).

The octet rule stems from the fact that the main-group elements in most cases employ only an ns and three np valence-shell orbitals in bonding, and these orbitals hold eight electrons. Elements of the second period are restricted to these orbitals, but from the third period on, the elements also have unfilled nd orbitals, which may be used in bonding. For example, the valence-shell configuration of phosphorus is $3s^2 3p^3$. Using just these $3s$ and $3p$ orbitals, the phosphorus atom can accept only three additional electrons, forming three covalent bonds (as in PF$_3$). However, more bonds can be formed if the empty $3d$ orbitals of the atom are used. If each of the five electrons of the phosphorus atom is paired with unpaired electrons of fluorine atoms, PF$_5$ can be formed. In this way, phosphorus forms both the trifluoride and the pentafluoride. By contrast, nitrogen (which has no available d orbitals in its valence shell) forms only the trifluoride, NF_3.

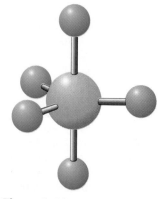

Figure 9.19
Phosphorus pentafluoride, PF₅. A molecular model.

EXAMPLE 9.10

Writing Lewis Formulas (Exceptions to the Octet Rule)

Xenon, a noble gas, forms a number of compounds. One of these is xenon tetra-fluoride, XeF_4, a white, crystalline solid first prepared in 1962. What is the electron-dot formula of the XeF_4 molecule?

SOLUTION

There are 8 valence electrons from the Xe atom and 7 from each F atom, for a total of 36 valence electrons. For the skeleton structure, you draw the Xe atom surrounded by the electronegative F atoms. After placing electron pairs on the F atoms to satisfy the octet rule for them, you have

This accounts for 16 pairs, or 32 electrons. A total of 36 electrons is available, so you put an additional $36 - 32 = 4$ electrons (2 pairs) on the Xe atom. The Lewis formula is

---

See Problems 9.45 and 9.46.

The other group of exceptions to the octet rule consists mostly of molecules containing Group IIA or IIIA atoms. Consider boron trifluoride, BF_3. The molecule consists of boron surrounded by the much more electronegative fluorine atoms. The total number of valence electrons is $3 + (3 \times 7) = 24$. If you connect boron and fluorine atoms by electron pairs and fill out the fluorine atoms with octets of electrons, you obtain:

It is possible to write resonance formulas in which there is multiple bonding between fluorine and boron. Although the chemistry of BF_3 appears to support an electron structure where boron has six electrons around it, it combines with the lone pair of electrons from a Lewis base. For example, the reaction with ammonia can be depicted as

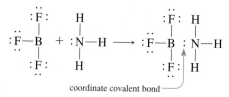

coordinate covalent bond

In this reaction, a coordinate covalent bond forms between the boron and nitrogen atoms, and the boron achieves an octet of electrons.

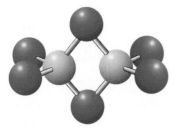

Figure 9.20
The Al$_2$Cl$_6$ molecule. A molecular model.

Other examples of molecules with Group IIIA atoms (such as Al) or Group IIA atoms (such as Be) display electron structures similar to that of boron trifluoride. For example, the BeF$_2$ molecule (found in the vapor over the heated solid BeF$_2$) has the Lewis formula

$$:\ddot{F}:Be:\ddot{F}:$$

Aluminum chloride, AlCl$_3$, offers an interesting study in bonding. At room temperature, the substance is a white, crystalline solid and an ionic compound, as might be expected for a binary compound of a metal and a nonmetal. However, the substance has a relatively low melting point (192°C) for an ionic compound. Apparently this is due to the fact that instead of melting to a liquid of ions, as happens with most ionic solids, the compound forms Al$_2$Cl$_6$ molecules (Figure 9.20), with Lewis formula

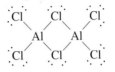

Each atom has an octet of electrons around it. Note that two of the Cl atoms are in *bridge* positions, with each Cl atom having two covalent bonds. When this liquid is heated, it vaporizes as Al$_2$Cl$_6$ molecules.

9.9 Formal Charge and Lewis Formulas

In this section, we will describe how you can use the concept of formal charge to determine the best Lewis formula. Formal charge is also a help in writing the skeleton structure of a molecule.

The **formal charge** of an atom in a Lewis formula is *the hypothetical charge you obtain by assuming that bonding electrons are equally shared between bonded atoms and that the electrons of each lone pair belong completely to one atom.* Use the following *rules for formal charge* to assign the valence electrons to individual atoms:

1. Half of the electrons of a bond are assigned to each atom in the bond (counting each dash as two electrons).

2. Both electrons of a lone pair are assigned to the atom to which the lone pair belongs.

You calculate the formal charge on an atom by taking the number of valence electrons on the free atom (equal to the group number) and subtracting the number of electrons you assigned the atom by the rules of formal charge; that is,

Formal charge = valence electrons on free atom $-\frac{1}{2}$(number of electrons in a bond)
$-$ (number of lone-pair electrons)

As a check on your work, you should note that the sum of the formal charges equals the charge on the molecular species (zero for a neutral molecule).

Let us apply these rules for determining the formal charges for two different skeleton structures for H$_2$O$_2$.

$$H-\ddot{O}-\ddot{O}-H \qquad H-\ddot{O}-\ddot{O}:$$
$$\qquad\qquad\qquad\qquad\qquad | $$
$$\qquad\qquad\qquad\qquad\qquad H$$

In both structures the H atoms have zero formal charges because each hydrogen atom has one valence electron and one electron from the H—O bond. So the formal charge is 1 valence electron minus 1 bonding electron, which sums to zero. In the first structure the oxygen atoms also have zero formal charges, as each oxygen has 6 valence electrons and 6 electrons assigned to it by the rules of formal charge (4 bonding electrons $\times \frac{1}{2}$ + 2 lone pairs $\times$ 2). In the second structure the central oxygen atom has a +1 formal charge ($6 - [\frac{1}{2}(6) + 1 \times 2] = +1$). Therefore, the outer oxygen atom has a formal charge of -1, as the formal charges must be equal to zero for a molecule. Let's confirm the -1 formal charge on the second oxygen atom by calculating its value: Formal charge $= 6 - [\frac{1}{2}(2) + 3 \times 2] = -1$.

We indicate the formal charges in a Lewis formula by inserting circled numbers near the atoms (writing + for +1 and − for −1). For the formulas given earlier for H_2O_2, the formal charges are shown as follows:

$$\ddot{\text{H}}-\overset{..}{\underset{..}{\text{O}}}-\overset{..}{\underset{..}{\text{O}}}-\text{H} \qquad \text{H}-\overset{\oplus}{\underset{}{\overset{..}{\text{O}}}}-\overset{\ominus}{\underset{..}{\overset{..}{\text{O}}}}:$$
$$\text{H}$$

Which structure should we select for H_2O_2 (hydrogen peroxide)? The following two rules are useful in making the selection.

RULE A Whenever you can write several Lewis formulas for a molecule, choose the one having the lowest magnitudes of formal charges.

RULE B When two proposed Lewis formulas for a molecule have the same magnitudes of formal charges, choose the one having the negative formal charge on the more electronegative atom.

The first formula has zero formal charges for all atoms. So by Rule A, this formula most closely approximates the actual electron distribution.

EXAMPLE 9.11

Using Formal Charges to Determine the Best Lewis Formula

Write the Lewis formula that best describes the charge distribution in the sulfuric acid molecule, H_2SO_4, according to the rules of formal charge.

SOLUTION

Assume a skeleton structure in which the S atom is surrounded by the more electronegative O atoms; the H atoms are then attached to two of the O atoms. The following is a possible Lewis formula having single bonds:

$$:\overset{..}{\text{O}}:$$
$$\text{H}-\overset{..}{\underset{..}{\text{O}}}-\overset{|}{\underset{|}{\text{S}}}-\overset{..}{\underset{..}{\text{O}}}-\text{H}$$
$$:\underset{..}{\text{O}}:$$

You calculate the formal charge of each O atom bonded to an H atom as follows: You assign the O atom 2 electrons from the two single bonds and 4 electrons from the two lone pairs, for a total of 6 electrons. Because the number of valence electrons on the O atom is 6, the formal charge is $6 - 6 = 0$. For each of the other two O atoms, you assign 1 electron from the one single bond and

6 electrons from the three lone pairs, for a total of 7. The formal charge is $6 - 7 = -1$. For the S atom, you assign 4 electrons from the four single bonds, for a total of 4. Because the number of valence electrons is 6, the formal charge is $6 - 4 = +2$. The formal charge of each of the H atoms is $1 - 1 = 0$. The Lewis formula with formal charges is

$$
\begin{array}{c}
:\overset{\displaystyle ..}{O}:^{\ominus} \\[2pt]
\quad\ |\,\textcircled{+2}.. \\[-2pt]
H-\overset{..}{\underset{..}{O}}-S-\overset{..}{O}-H \\[-2pt]
\quad\ | \\[2pt]
:\overset{\displaystyle ..}{O}:^{\ominus}
\end{array}
$$

Note that the sum of the formal charges equals 0, as it should for a neutral molecule.

You can also write a Lewis formula that has zero formal charges for atoms, if you form sulfur–oxygen double bonds:

$$
\begin{array}{c}
:O: \\[-2pt]
\ \|\ \\[-2pt]
H-\overset{..}{\underset{..}{O}}-S-\overset{..}{\underset{..}{O}}-H \\[-2pt]
\ \|\ \\[-2pt]
:O:
\end{array}
$$

To the top O atom, you assign 2 electrons from the double bond and 4 electrons from the two lone pairs, for a total of 6 electrons. The formal charge is $6 - 6 = 0$. To the S atom, you assign a total of 6 electrons from bonds. The formal charge is $6 - 6 = 0$. The formal charges of the other atoms are also 0. Thus, this Lewis formula should be a better representation of the electron distribution in the H_2SO_4 molecule.

See Problems 9.47 and 9.48.

CONCEPT CHECK 9.3

Which of the models shown below most accurately represents the hydrogen cyanide molecule, HCN? Write the electron-dot formula that most closely agrees with this model. State any concept or rule you used in arriving at your answer.

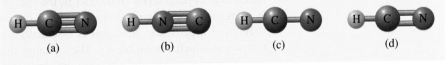

| (a) | (b) | (c) | (d) |

9.10 Bond Length and Bond Order

Bond length (or **bond distance**) is *the distance between the nuclei in a bond*. Bond lengths are determined experimentally using x-ray diffraction or the analysis of molecular spectra. Knowing the bond length in a particular molecule can sometimes provide a clue to the type of bonding present.

In many cases bond lengths for covalent single bonds in compounds can be predicted from covalent radii. **Covalent radii** are *values assigned to atoms in such a way that the sum of covalent radii of atoms A and B predicts an approximate A—B bond length*. Thus, the radius of the Cl atom might be taken to be half the

Resonance

9.43 Write resonance descriptions for the following:
a. NO_2^- b. HNO_3

9.44 Write resonance descriptions for the following:
a. FNO_2 b. SO_3

Exceptions to the Octet Rule

9.45 Write Lewis formulas for the following:
a. XeF_2 b. SeF_4 c. TeF_6 d. XeF_5^+

9.46 Write Lewis formulas for the following:
a. I_3^- b. ClF_3 c. IF_4^- d. BrF_5

Formal Charge and Lewis Formulas

9.47 Write a Lewis formula for each of the following, assuming that the octet rule holds for the atoms. Then obtain the formal charges of the atoms.
a. O_3 b. CO   c. HNO_3

9.48 Write a Lewis formula for each of the following, assuming that the octet rule holds for the atoms. Then obtain the formal charges of the atoms.
a. ClNO   b. $POCl_3$ c. N_2O (NNO)

Bond Length, Bond Order, and Bond Energy

9.49 Calculate the bond length for each of the following single bonds, using covalent radii (Table 9.4):
a. C—H   b. S—Cl   c. Br—Cl   d. Si—O

9.50 Calculate the C—H and C—Cl bond lengths in chloroform, $CHCl_3$, using values for the covalent radii from Table 9.4. How do these values compare with the experimental values: C—H, 107 pm; C—Cl, 177 pm?

9.51 Use bond energies (Table 9.5) to estimate ΔH for the following gas-phase reaction.

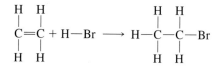

This is called an "addition" reaction, because a compound (HBr) is added across the double bond.

9.52 A commercial process for preparing ethanol (ethyl alcohol), C_2H_5OH, consists of passing ethylene gas, C_2H_4, and steam over an acid catalyst (to speed up the reaction). The gas-phase reaction is

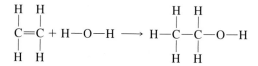

Estimate ΔH for this reaction, using bond energies (Table 9.5).

General Problems

9.53 For each of the following pairs of elements, state whether the binary compound formed is likely to be ionic or covalent. Give the formula and name of the compound.
a. Sr, O   b. C, Br   c. Ga, F   d. N, Br

9.54 For each of the following pairs of elements, state whether the binary compound formed is likely to be ionic or covalent. Give the formula and name of the compound.
a. Na, S   b. Al, F   c. Ca, Cl   d. Si, Br

9.55 Give the Lewis formula for the arsenate ion, AsO_4^{3-}. Write the formula of lead(II) arsenate.

9.56 Give the Lewis formula for the selenite ion, SeO_3^{2-}. Write the formula of aluminum selenite.

9.57 Iodic acid, HIO_3, is a colorless, crystalline compound. What is the electron-dot formula of iodic acid?

9.58 Selenic acid, H_2SeO_4, is a crystalline substance and a strong acid. Write the electron-dot formula of selenic acid.

9.59 Sodium amide, known commercially as sodamide, is used in preparing indigo, the dye used to color blue jeans. It is an ionic compound with the formula $NaNH_2$. What is the electron-dot formula of the amide anion, NH_2^-?

9.60 Lithium aluminum hydride, $LiAlH_4$, is an important reducing agent (an element or compound that generally has a strong tendency to give up electrons in its chemical reactions). Write the electron-dot formula of the AlH_4^- ion.

9.61 Write electron-dot formulas for the following:
a. $SeOCl_2$ b. CSe_2 c. $GaCl_4^-$ d. C_2^{2-}

9.62 Write electron-dot formulas for the following:
a. $POBr_3$ b. Si_2H_6 c. IF_2^+ d. NO^+

9.63 Give resonance descriptions for the following:
a. SeO_2 b. N_2O_4

9.64 Give resonance descriptions for the following:
a. O_3 b. $C_2O_4^{2-}$

6 electrons from the three lone pairs, for a total of 7. The formal charge is 6 − 7 = −1. For the S atom, you assign 4 electrons from the four single bonds, for a total of 4. Because the number of valence electrons is 6, the formal charge is 6 − 4 = +2. The formal charge of each of the H atoms is 1 − 1 = 0. The Lewis formula with formal charges is

$$
\begin{array}{c}
: \overset{..}{O} : ^{\ominus} \\
| \overset{(+2)}{} \\
H - \overset{..}{\underset{..}{O}} - S - \overset{..}{\underset{..}{O}} - H \\
| \\
: \overset{..}{O} : ^{\ominus}
\end{array}
$$

Note that the sum of the formal charges equals 0, as it should for a neutral molecule.

You can also write a Lewis formula that has zero formal charges for atoms, if you form sulfur–oxygen double bonds:

$$
\begin{array}{c}
: \mathbf{O} : \\
\| \\
\mathbf{H} - \overset{..}{\underset{..}{\mathbf{O}}} - \mathbf{S} - \overset{..}{\underset{..}{\mathbf{O}}} - \mathbf{H} \\
\| \\
: \mathbf{O} :
\end{array}
$$

To the top O atom, you assign 2 electrons from the double bond and 4 electrons from the two lone pairs, for a total of 6 electrons. The formal charge is 6 − 6 = 0. To the S atom, you assign a total of 6 electrons from bonds. The formal charge is 6 − 6 = 0. The formal charges of the other atoms are also 0. Thus, this Lewis formula should be a better representation of the electron distribution in the H_2SO_4 molecule.

See Problems 9.47 and 9.48.

CONCEPT CHECK 9.3

Which of the models shown below most accurately represents the hydrogen cyanide molecule, HCN? Write the electron-dot formula that most closely agrees with this model. State any concept or rule you used in arriving at your answer.

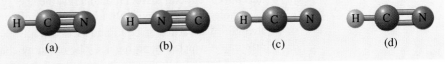

(a) (b) (c) (d)

9.10 Bond Length and Bond Order

Bond length (or **bond distance**) is *the distance between the nuclei in a bond*. Bond lengths are determined experimentally using x-ray diffraction or the analysis of molecular spectra. Knowing the bond length in a particular molecule can sometimes provide a clue to the type of bonding present.

In many cases bond lengths for covalent single bonds in compounds can be predicted from covalent radii. **Covalent radii** are *values assigned to atoms in such a way that the sum of covalent radii of atoms A and B predicts an approximate A—B bond length*. Thus, the radius of the Cl atom might be taken to be half the

Table 9.4
Single-Bond Covalent Radii for Nonmetallic Elements (in pm)

				H
				37
B	C	N	O	F
88	77	70	66	64
	Si	P	S	Cl
	117	110	104	99
		As	Se	Br
		121	117	114
			Te	I
			137	133

▶ Although a double bond is stronger than a single bond, it is not necessarily less reactive. Ethylene, $CH_2{=}CH_2$, for example, is more reactive than ethane, $CH_3{-}CH_3$, where carbon atoms are linked through a single bond. The reactivity of ethylene results from the simultaneous formation of a number of strong, single bonds. See Section 9.11.

Cl—Cl bond length (198 pm). The covalent radius of Cl would be $\frac{1}{2} \times 198$ pm = 99 pm. Table 9.4 lists single-bond covalent radii for nonmetallic elements. To predict the bond length of C—Cl, you add the covalent radii of the two atoms, C and Cl. You get (77 + 99) pm = 176 pm, which compares favorably with the experimental value of 178 pm found in most compounds.

The **bond order,** defined in terms of the Lewis formula, is *the number of pairs of electrons in a bond.* For example, in C : C the bond order is 1 (single bond); in C : : C the bond order is 2 (double bond). Bond length depends on bond order. As the bond order increases, the bond strength increases and the nuclei are pulled inward, decreasing the bond length. Look at carbon–carbon bonds. The average C—C bond length is 154 pm, whereas C=C is 134 pm long and C≡C is 120 pm long. ◀

EXAMPLE 9.12
Relating Bond Order and Bond Length

Consider the molecules N_2H_4, N_2, and N_2F_2. Which molecule has the shortest nitrogen–nitrogen bond? Which has the longest nitrogen–nitrogen bond?

SOLUTION

First write the Lewis formulas:

$$\text{H}-\overset{\displaystyle \text{H}}{\underset{\displaystyle \text{H}}{\text{N}}}-\overset{\displaystyle \text{H}}{\underset{\displaystyle \text{H}}{\text{N}}}-\text{H} \qquad :\text{N}{\equiv}\text{N}: \qquad :\ddot{\text{F}}-\ddot{\text{N}}{=}\ddot{\text{N}}-\ddot{\text{F}}:$$

$$N_2H_4 \qquad\qquad N_2 \qquad\qquad N_2F_2$$

The **nitrogen–nitrogen bond should be shortest in N_2, where it is a triple bond, and longest in N_2H_4, where it is a single bond.** (Experimental values for the nitrogen–nitrogen bond lengths are 109 pm for N_2, 122 pm for N_2F_2, and 147 pm for N_2H_4.)

9.11 Bond Energy

In Section 9.4, when we described the formation of a covalent bond, we introduced the concept of bond dissociation energy, the energy required to break a particular bond in a molecule (see Figure 9.11). The bond dissociation energy is a measure of the strength of a particular bond and is essentially the *enthalpy change* for a gas-phase reaction in which a bond breaks. The enthalpy change, ΔH, is the heat absorbed in a reaction carried out at constant pressure (heat of reaction).

Consider the experimentally determined enthalpy changes for the breaking, or dissociation, of a C—H bond in methane, CH_4, and in ethane, C_2H_6, in the gas phase:

$$\text{H}-\overset{\displaystyle \text{H}}{\underset{\displaystyle \text{H}}{\text{C}}}-\text{H}(g) \longrightarrow \text{H}-\overset{\displaystyle \text{H}}{\underset{\displaystyle \text{H}}{\text{C}}}(g) + \text{H}(g) \qquad \Delta H = 435 \text{ kJ}$$

$$\text{H}-\overset{\displaystyle \text{H}}{\underset{\displaystyle \text{H}}{\text{C}}}-\overset{\displaystyle \text{H}}{\underset{\displaystyle \text{H}}{\text{C}}}-\text{H}(g) \longrightarrow \text{H}-\overset{\displaystyle \text{H}}{\underset{\displaystyle \text{H}}{\text{C}}}-\overset{\displaystyle \text{H}}{\underset{\displaystyle \text{H}}{\text{C}}}(g) + \text{H}(g) \qquad \Delta H = 410 \text{ kJ}$$

Note that the ΔH values are approximately the same in the two cases. Comparisons of this sort lead to the conclusion that we can obtain approximate values of energies of various bonds.

We define the A—B **bond energy** (denoted *BE*) as *the average enthalpy change for the breaking of an A—B bond in a molecule in the gas phase.* For example, to calculate a value for the C—H bond energy, or *BE*(C—H), you might look at the experimentally determined enthalpy change for the breaking of all the C—H bonds in methane:

$$CH_4(g) \longrightarrow C(g) + 4H(g); \Delta H = 1662 \text{ kJ}$$

Because four C—H bonds are broken, you obtain an average value for breaking one C—H bond by dividing the enthalpy change for the reaction by 4:

$$BE(\text{C—H}) = \tfrac{1}{4} \times 1662 \text{ kJ} = 416 \text{ kJ}$$

Table 9.5 lists values of some bond energies. Because it takes energy to break a bond, bond energies are always positive numbers. When a bond is formed, the energy is equal to the negative of the bond energy (energy is released).

Bond energy is a measure of the strength of a bond: the larger the bond energy, the stronger the chemical bond. Note from Table 9.5 that the bonds C—C, C=C, and C≡C have energies of 346, 602, and 835 kJ/mol, respectively.

You can use this table of bond energies to estimate heats of reaction, or enthalpy changes, ΔH, for gaseous reactions. To illustrate this, find ΔH for the following reaction (Figure 9.21):

$$CH_4(g) + Cl_2(g) \longrightarrow CH_3Cl(g) + HCl(g)$$

Table 9.5
Bond Energies (in kJ/mol)*

	Single Bonds								
	H	**C**	**N**	**O**	**S**	**F**	**Cl**	**Br**	**I**
H	432								
C	411	346							
N	386	305	167						
O	459	358	201	142					
S	363	272	—	—	226				
F	565	485	283	190	284	155			
Cl	428	327	313	218	255	249	240		
Br	362	285	—	201	217	249	216	190	
I	295	213	—	201	—	278	208	175	149

Multiple Bonds					
C=C	602	C=N	615	C=O	745 (799 in CO_2)
C≡C	835	C≡N	887	C≡O	1072
N=N	418	N=O	607	S=O (in SO_2)	532
N≡N	942	O=O	494	S=O (in SO_3)	469

*Data are taken from J. E. Huheey, Keiter, and Keiter, *Inorganic Chemistry,* 4th ed. (New York: HarperCollins, 1993), pp. A21–A34.

Figure 9.21
Reaction of methane with chlorine. Molecular models illustrate the reaction
$CH_4 + Cl_2 \longrightarrow CH_3Cl + HCl$.

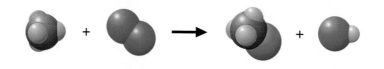

You can *imagine* that the reaction takes place in steps involving the breaking and forming of bonds. Starting with the reactants, you suppose that one C—H bond and the Cl—Cl bond break.

$$
\begin{array}{c}
\text{H} \\
| \\
\text{H—C—H} + \text{Cl—Cl} \longrightarrow \text{H—C} + \text{H} + \text{Cl} + \text{Cl} \\
| \\
\text{H}
\end{array}
$$

The enthalpy change is BE(C—H) + BE(Cl—Cl). Now you reassemble the fragments to give the products.

$$
\begin{array}{c}
\text{H} \\
| \\
\text{H—C} + \text{H} + \text{Cl} + \text{Cl} \longrightarrow \text{H—C—Cl} + \text{H—Cl} \\
| \\
\text{H}
\end{array}
$$

In this case, C—Cl and H—Cl bonds are formed, and the enthalpy change equals the negative of the bond energies $-BE$(C—Cl) − BE(H—Cl). Substituting bond-energy values from Table 9.5, you get the enthalpy of reaction.

$$\Delta H \simeq BE(\text{C—H}) + BE(\text{Cl—Cl}) - BE(\text{C—Cl}) - BE(\text{H—Cl})$$
$$= (411 + 240 - 327 - 428) \text{ kJ}$$
$$= -104 \text{ kJ}$$

The negative sign means that heat is released by the reaction.

> In general, the enthalpy of reaction is (approximately) equal to the sum of the bond energies for bonds broken minus the sum of the bond energies for bonds formed.

Rather than calculating a heat of reaction from bond energies, you usually obtain the heat of reaction from thermochemical data. If the thermochemical data are not known, however, bond energies can give you a reasonable estimate.

Bond energies are perhaps of greatest value when you try to explain heats of reaction or understand the relative stabilities of compounds. In general, a reaction is exothermic if weak bonds are replaced by strong bonds (see Figure 9.22).

Figure 9.22
Explosion of nitrogen triiodide–ammonia complex.   The red-brown
complex of nitrogen triiodide and ammonia is so sensitive to explosion that it can
be detonated with a feather. Nitrogen–iodine single bonds are replaced by very
stable nitrogen–nitrogen triple bonds (N₂) and iodine–iodine single bonds (I₂).

EXAMPLE 9.13

Estimating ΔH from Bond Energies

Polyethylene is formed by linking many ethylene molecules into long chains.
Estimate the enthalpy change per mole of ethylene for this reaction (shown
below), using bond energies.

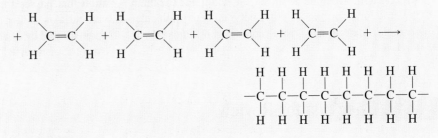

SOLUTION

Imagine the reaction to involve the breaking of the carbon–carbon double bonds
and the formation of carbon–carbon single bonds. For a very long chain, the
net result is that for every C=C bond broken, two C—C bonds are formed.

$$\Delta H \simeq [602 - (2 \times 346)] \text{ kJ} = \mathbf{-90 \text{ kJ}}$$

See Problems 9.51 and 9.52.

Infrared Spectroscopy and Vibrations of Chemical Bonds

A chemical bond acts like a stiff spring connecting nuclei. As a result, the nuclei in a molecule vibrate, rather than maintaining fixed positions relative to each other. Nuclear vibration is depicted in Figure 9.23, which shows a spring model of HCl.

This vibration of molecules is revealed in their absorption of infrared radiation. The frequency of radiation absorbed equals the frequencies of nuclear vibrations. For example, the H—Cl bond vibrates at a frequency of 8.652×10^{13} vibrations per second. If radiation of this frequency falls on the molecule, it absorbs the radiation, which is in the infrared region, and begins vibrating more strongly.

The infrared absorption spectrum of a molecule of even moderate size can have a rather complicated appearance. Figure 9.24 shows the infrared (IR) spectrum of ethyl butyrate, a compound present in pineapple flavor. The complicated appearance of the IR spectrum is actually an advantage. Two different compounds are unlikely to have exactly the same IR spectrum. Therefore, the IR spectrum can act as a compound's "fingerprint."

The IR spectrum of a compound can also yield structural information. Suppose you would like to obtain the structural formula of ethyl butyrate. The molecular formula, determined from combustion analysis, is $C_6H_{12}O_2$.

Figure 9.23
Vibration of the HCl molecule. The vibrating molecule is represented here by a spring model. The atoms in the molecule vibrate; that is, they move constantly back and forth.

Instead of plotting an IR spectrum in frequency units (since the frequencies are very large), you usually would give the frequencies in *wavenumbers*, which are proportional to frequency. To get the wavenumber, you divide the frequency by the speed of light expressed in centimeters per second. For example, HCl absorbs at $(8.652 \times 10^{13}\ s^{-1})/(2.998 \times 10^{10}\ cm/s) = 2886\ cm^{-1}$ (wavenumbers). Wavenumber, or sometimes wavelength, is plotted along the horizontal axis.

Percent transmittance—that is, the percent of radiation that passes through a sample—is plotted on the vertical axis. When a molecule absorbs radiation of a given frequency or wavenumber, this is seen in the spectrum as an inverted spike (peak) at that wavenumber.

A Checklist for Review

Important Terms

ionic bond (9.1)
Lewis electron-dot
    symbol (9.1)
lattice energy (9.1)
ionic radius (9.3)
isoelectronic (9.3)
covalent bond (9.4)

Lewis electron-dot
    formula (9.4)
bonding pair (9.4)
lone (nonbonding)
    pair (9.4)
coordinate covalent
    bond (9.4)

octet rule (9.4)
single bond (9.4)
double bond (9.4)
triple bond (9.4)
polar covalent bond (9.5)
electronegativity (9.5)
delocalized bonding (9.7)

resonance description (9.7)
formal charge (9.9)
bond length (bond
    distance) (9.10)
covalent radii (9.10)
bond order (9.10)
bond energy (9.11)

Summary of Facts and Concepts

An *ionic bond* is a strong attractive force holding ions together. An ionic bond can form between two atoms by the transfer of electrons from the valence shell of one atom to the valence shell of the other. Many similar ions attract one another to form a crystalline solid, in which positive ions are surrounded by negative ions and negative ions are surrounded by positive ions. As a result, ionic solids are typically high-melting solids. Monatomic cations of the main-group elements have charges

equal to the group number (or in some cases, the group number minus two). Monatomic anions of the main-group elements have charges equal to the group number minus eight.

A *covalent bond* is a strong attractive force that holds two atoms together by their sharing of electrons. These bonding electrons are attracted simultaneously to both atomic nuclei. If an electron pair is not equally shared, the bond is *polar*. This polarity results from the difference in *electronegativities* of the atoms.

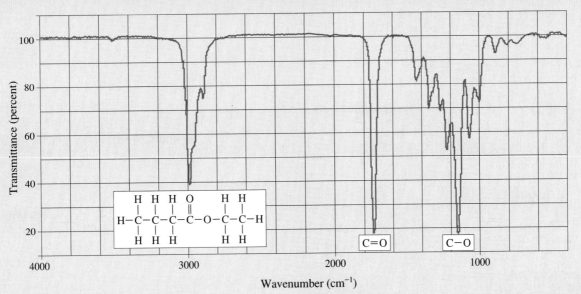

Figure 9.24
Infrared spectrum of ethyl butyrate. Note the peaks corresponding to vibrations of C=O and C—O bonds. The molecular structure is shown at the bottom left. [From NIST Mass Spec Data Center, S.E. Stein, director, "IR and Mass Spectra" in *NIST Chemistry WebBook, NIST Standard Reference Database Number 69*, Eds. W.G. Mallard and P.J. Linstrom, February 2000, National Institute of Standards and Technology, Gaithersburg, MD, 20899 (http//webbook. nist.gov). © 1991, 1994, 1996, 1997, 1998, 1999, 2000 copyright by the U.S. Secretary of Commerce on behalf of the United States of America. All rights reserved.]

Certain structural features of molecules appear as absorption peaks in definite regions of the infrared spectrum. For example, the absorption peak at 1730 cm^{-1} is characteristic of the C=O bond. With some knowledge of where various bonds absorb, one can identify other peaks, including that of C—O at 1180 cm^{-1}. (Generally, the IR peak for an A—B bond occurs at lower wavenumber than for an A=B bond.) The IR spectrum does not reveal the complete structure, but it provides important clues. Data from other instruments, such as the mass spectrometer, give additional clues.

Lewis electron-dot formulas are simple representations of the valence-shell electrons of atoms in molecules and ions. You can apply simple rules to draw these formulas. In molecules with *delocalized bonding*, it is not possible to describe accurately the electron distribution with a single Lewis formula. For these molecules, you must use *resonance*. Although the atoms in Lewis formulas often satisfy the octet rule, *exceptions to the octet rule* are not uncommon. You can obtain the Lewis formulas for these exceptions by following the rules for writing Lewis formulas. The concept of *formal charge* will often help you decide which of several Lewis formulas gives the best description of a molecule or ion.

Bond lengths can be estimated from the *covalent radii* of atoms. Bond length depends on *bond order;* as the bond order increases, the bond length decreases. The A—B *bond energy* is the average enthalpy change when an A—B bond is broken. You can use bond energies to estimate ΔH for gas-phase reactions.

Operational Skills

Note: A periodic table is needed for skills 1, 2, 3, 5, and 6.

1. **Using Lewis symbols to represent ionic bond formation** Given a metallic and a nonmetallic main-group element, use Lewis symbols to represent the transfer of electrons to form ions of noble-gas configurations. **(EXAMPLE 9.1)**

2. **Writing electron configurations of ions** Given an ion, write the electron configuration. For an ion of a main-group element, give the Lewis symbol. **(EXAMPLES 9.2, 9.3)**

3. **Using periodic trends to obtain relative ionic radii** Given a series of ions, arrange them in order of increasing ionic radius. **(EXAMPLE 9.4)**

4. **Using electronegativities to obtain relative bond polarities**   Given the electronegativities of the atoms, arrange a series of bonds in order by polarity.   **(EXAMPLE 9.5)**

5. **Writing Lewis formulas**   Given the molecular formula of a simple compound or ion, write the Lewis electron-dot formula.   **(EXAMPLES 9.6, 9.7, 9.8, 9.10)**

6. **Writing resonance formulas**   Given a simple molecule with delocalized bonding, write the resonance description.   **(EXAMPLE 9.9)**

7. **Using formal charges to determine the best Lewis formula**   Given two or more Lewis formulas, use formal charges to determine which formula best describes the electron distribution or gives the most plausible molecular structure.   **(EXAMPLE 9.11)**

8. **Relating bond order and bond length**   Know the relationship between bond order and bond length.   **(EXAMPLE 9.12)**

9. **Estimating ΔH from bond energies**   Given a table of bond energies, estimate the heat of reaction.   **(EXAMPLE 9.13)**

Review Questions

9.1   Describe the formation of a sodium chloride crystal from atoms.

9.2   Explain what energy terms are involved in the formation of an ionic solid from atoms. In what way should these terms change (become larger or smaller) to give the lowest energy possible for the solid?

9.3   Define lattice energy for potassium bromide.

9.4   Why do most monatomic cations of the main-group elements have a charge equal to the group number? Why do most monatomic anions of these elements have a charge equal to the group number minus eight?

9.5   The 2+ ions of transition elements are common. Explain why this might be expected.

9.6   Describe the trends shown by the radii of the monatomic ions for the main-group elements both across a period and down a column.

9.7   Give an example of a molecule that has a coordinate covalent bond.

9.8   The octet rule correctly predicts the Lewis formula of many molecules involving main-group elements. Explain why this is so.

9.9   Describe the general trends in electronegativities of the elements in the periodic table both across a period and down a column.

9.10   What is the qualitative relationship between bond polarity and electronegativity difference?

9.11   What is a resonance description of a molecule? Why is this concept required if we wish to retain Lewis formulas as a description of the electron structure of molecules?

9.12   Describe the kinds of exceptions to the octet rule that we encounter in compounds of the main-group elements. Give examples.

9.13   What is the relationship between bond order and bond length? Use an example to illustrate it.

9.14   Define bond energy. Explain how one can use bond energies to estimate the heat of reaction.

Conceptual Problems

9.15   Examine each of the following electron-dot formulas and decide whether the formula is correct, or whether you could write a formula that better approximates the electron structure of the molecule. State which concepts or rules you use in each case to arrive at your conclusion.

a. $:\!N\!:\!N\!:$ b. $:\!\overset{..}{\underset{..}{Cl}}\!:\!C\!:\!:\!:\!N\!:$

c. $H\!:\!C\!:\!:\!\overset{..}{\underset{..}{F}}$ d. $H\!:\!\overset{..}{\underset{..}{O}}\!:\!:\!C\!:\!:\!\overset{..}{\underset{..}{O}}\!:\!H$
       $:\!\overset{..}{\underset{..}{O}}\!:$

9.16   Which of the following represent configurations of thallium ions in compounds? Explain your decision in each case.

a. Tl^{2+} $[Xe]4f^{14}5d^{10}6p^{1}$ b. Tl^{3+} $[Xe]4f^{14}5d^{10}$

c. Tl^{4+} $[Xe]4f^{14}5d^{9}$ d. Tl^{+} $[Xe]4f^{14}5d^{10}6s^{2}$

9.17   For each of the following molecular formulas, draw the most reasonable skeleton structure.

a. CH_2Cl_2 b. HNO_2 c. NOF   d. N_2O_4

What rule or concept did you use to obtain each structure?

9.18 For each of the following molecular models, write an appropriate Lewis formula.

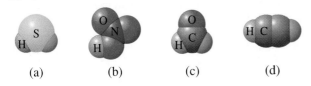

(a)   (b)   (c)   (d)

9.19 Sodium, Na, reacts with element X to form an ionic compound with the formula Na_3X.

a. What is the formula of the compound you expect to form when calcium, Ca, reacts with element X?
b. Would you expect this compound to be ionic or molecular?

9.20 Below are a series of resonance formulas for N_2O (nitrous oxide). Rank these in terms of how closely you think each one represents the true electron structure of the molecule. State the rules and concepts you use to do this ranking.

a. $:\!\ddot{N}\!—\!N\!\equiv\!O\!:$ b. $:\!N\!\equiv\!N\!—\!\ddot{O}\!:$ c. $:\!\ddot{N}\!=\!N\!=\!\ddot{O}$

Practice Problems

Ionic Bonding

9.21 Write Lewis symbols for the following:
a. In   b. In^{3+} c. P   d. P^{3-}

9.22 Write Lewis symbols for the following:
a. Ba   b. Ba^{2+} c. I   d. I^-

9.23 For each of the following, write the electron configuration and Lewis symbol:
a. Ca^{2+} b. Se   c. Se^{2-} d. Br^-

9.24 For each of the following, write the electron configuration and Lewis symbol:
a. Rb^+ b. I   c. I^- d. Te^{2-}

9.25 Write the electron configurations of Bi and Bi^{3+}.

9.26 Write the electron configurations of Sn and Sn^{2+}.

9.27 Give the electron configurations of Ni^{2+} and Ni^{3+}.

9.28 Give the electron configurations of Cu^+ and Cu^{2+}.

Ionic Radii

9.29 Arrange the members of each of the following pairs in order of increasing radius and explain the order:
a. I^-, I b. Ca^{2+}, Ca

9.30 Arrange the members of each of the following pairs in order of increasing radius and explain the order:
a. Rb, Rb^+ b. Se, Se^{2-}

9.31 Arrange the following in order of increasing ionic radius: F^-, Na^+, and N^{3-}. Explain this order. (You may use a periodic table.)

9.32 Arrange the following in order of increasing ionic radius: I^-, Cs^+, and Te^{2-}. Explain this order. (You may use a periodic table.)

Covalent Bonding

9.33 Use Lewis symbols to show the reaction of atoms to form phosphine, PH_3. Indicate bonding pairs and lone pairs in the electron-dot formula of this compound.

9.34 Use Lewis symbols to show the reaction of atoms to form hydrogen sulfide. Indicate which electron pairs in the Lewis formula of H_2S are bonding and which are lone pairs.

Polar Covalent Bonds; Electronegativity

9.35 Using a periodic table (not Figure 9.15), arrange the following in order of increasing electronegativity:
a. P, O, N   b. Na, Al, Mg   c. C, Al, Si

9.36 With the aid of a periodic table (not Figure 9.15), arrange the following in order of increasing electronegativity:
a. Sr, Cs, Ba   b. Ca, Ge, Ga   c. P, As, S

9.37 Arrange the following bonds in order of increasing polarity using electronegativities of atoms: Si—O, C—Br, As—Br.

9.38 Decide which of the following bonds is least polar on the basis of electronegativities of atoms: H—Se, P—Cl, N—Cl.

Writing Lewis Formulas

9.39 Write Lewis formulas for the following molecules:
a. Br_2 b. H_2Se c. NF_3

9.40 Write Lewis formulas for the following molecules:
a. BrF   b. PBr_3 c. NOF

9.41 Write Lewis formulas for the following ions:
a. ClO^- b. $SnCl_3^-$ c. S_2^{2-}

9.42 Write Lewis formulas for the following ions:
a. IBr_2^+ b. ClF_2^+ c. CN^-

Resonance

9.43 Write resonance descriptions for the following:
a. NO_2^- b. HNO_3

9.44 Write resonance descriptions for the following:
a. FNO_2 b. SO_3

Exceptions to the Octet Rule

9.45 Write Lewis formulas for the following:
a. XeF_2 b. SeF_4 c. TeF_6 d. XeF_5^+

9.46 Write Lewis formulas for the following:
a. I_3^- b. ClF_3 c. IF_4^- d. BrF_5

Formal Charge and Lewis Formulas

9.47 Write a Lewis formula for each of the following, assuming that the octet rule holds for the atoms. Then obtain the formal charges of the atoms.
a. O_3 b. CO   c. HNO_3

9.48 Write a Lewis formula for each of the following, assuming that the octet rule holds for the atoms. Then obtain the formal charges of the atoms.
a. ClNO   b. $POCl_3$ c. N_2O (NNO)

Bond Length, Bond Order, and Bond Energy

9.49 Calculate the bond length for each of the following single bonds, using covalent radii (Table 9.4):
a. C—H   b. S—Cl   c. Br—Cl   d. Si—O

9.50 Calculate the C—H and C—Cl bond lengths in chloroform, $CHCl_3$, using values for the covalent radii from Table 9.4. How do these values compare with the experimental values: C—H, 107 pm; C—Cl, 177 pm?

9.51 Use bond energies (Table 9.5) to estimate ΔH for the following gas-phase reaction.

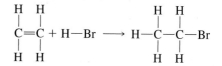

This is called an "addition" reaction, because a compound (HBr) is added across the double bond.

9.52 A commercial process for preparing ethanol (ethyl alcohol), C_2H_5OH, consists of passing ethylene gas, C_2H_4, and steam over an acid catalyst (to speed up the reaction). The gas-phase reaction is

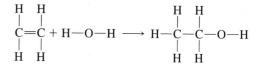

Estimate ΔH for this reaction, using bond energies (Table 9.5).

General Problems

9.53 For each of the following pairs of elements, state whether the binary compound formed is likely to be ionic or covalent. Give the formula and name of the compound.
a. Sr, O   b. C, Br   c. Ga, F   d. N, Br

9.54 For each of the following pairs of elements, state whether the binary compound formed is likely to be ionic or covalent. Give the formula and name of the compound.
a. Na, S   b. Al, F   c. Ca, Cl   d. Si, Br

9.55 Give the Lewis formula for the arsenate ion, AsO_4^{3-}. Write the formula of lead(II) arsenate.

9.56 Give the Lewis formula for the selenite ion, SeO_3^{2-}. Write the formula of aluminum selenite.

9.57 Iodic acid, HIO_3, is a colorless, crystalline compound. What is the electron-dot formula of iodic acid?

9.58 Selenic acid, H_2SeO_4, is a crystalline substance and a strong acid. Write the electron-dot formula of selenic acid.

9.59 Sodium amide, known commercially as sodamide, is used in preparing indigo, the dye used to color blue jeans. It is an ionic compound with the formula $NaNH_2$. What is the electron-dot formula of the amide anion, NH_2^-?

9.60 Lithium aluminum hydride, $LiAlH_4$, is an important reducing agent (an element or compound that generally has a strong tendency to give up electrons in its chemical reactions). Write the electron-dot formula of the AlH_4^- ion.

9.61 Write electron-dot formulas for the following:
a. $SeOCl_2$ b. CSe_2 c. $GaCl_4^-$ d. C_2^{2-}

9.62 Write electron-dot formulas for the following:
a. $POBr_3$ b. Si_2H_6 c. IF_2^+ d. NO^+

9.63 Give resonance descriptions for the following:
a. SeO_2 b. N_2O_4

9.64 Give resonance descriptions for the following:
a. O_3 b. $C_2O_4^{2-}$

9.65 The atoms in N_2O_5 are connected as follows:

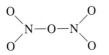

No attempt has been made here to indicate whether a bond is single or double or whether there is resonance. Obtain the Lewis formula (or formulas). The N—O bond lengths are 118 pm and 136 pm. Indicate the lengths of the bonds in the compound.

9.66 Methyl nitrite has the structure

No attempt has been made here to indicate whether a bond is single or double or whether there is resonance. Obtain the Lewis formula (or formulas). The N—O bond lengths are 122 pm and 137 pm. Indicate the lengths of the N—O bonds in the compound.

9.67 Use bond energies to estimate ΔH for the reaction

$$N_2F_2(g) + F_2(g) \longrightarrow N_2F_4(g)$$

9.68 Use bond energies to estimate ΔH for the reaction

$$HCN(g) + 2H_2(g) \longrightarrow CH_3NH_2(g)$$

Cumulative-Skills Problems

9.69 Phosphorous acid, H_3PO_3, has the structure $(HO)_2PHO$, in which one H atom is bonded to the P atom, and two H atoms are bonded to O atoms. For each bond to an H atom, decide whether it is polar or nonpolar. Assume that only polar-bonded H atoms are acidic. Write the balanced equation for the complete neutralization of phosphorous acid with sodium hydroxide. A 200.0-mL sample of H_3PO_3 requires 22.50 mL of $0.1250\ M$ NaOH for complete neutralization. What is the molarity of the H_3PO_3 solution?

9.70 Hypophosphorous acid, H_3PO_2, has the structure $(HO)PH_2O$, in which two H atoms are bonded to the P atom, and one H atom is bonded to an O atom. For each bond to an H atom, decide whether it is polar or nonpolar. Assume that only polar-bonded H atoms are acidic. Write the balanced equation for the complete neutralization of hypophosphorous acid with sodium hydroxide. A 200.0-mL sample of H_3PO_2 requires 22.50 mL of $0.1250\ M$ NaOH for complete neutralization. What is the molarity of the H_3PO_2 solution?

9.71 An ionic compound has the following composition (by mass): Mg, 10.9%; Cl, 31.8%; O, 57.3%. What are the formula and name of the compound? Write the Lewis formulas for the ions.

9.72 An ionic compound has the following composition (by mass): Ca, 30.3%; N, 21.2%; O, 48.5%. What are the formula and name of the compound? Write the Lewis formulas for the ions.

9.73 A gaseous compound has the following composition by mass: C, 25.0%; H, 2.1%; F, 39.6%; O, 33.3%. Its molecular weight is 48.0 amu. Write the Lewis formula for the molecule.

9.74 A liquid compound used in dry cleaning contains 14.5% C and 85.5% Cl by mass and has a molecular weight of 166 amu. Write the Lewis formula for the molecule.

9.75 Calculate the enthalpy of reaction for

$$HCN(g) \longrightarrow H(g) + C(g) + N(g)$$

from enthalpies of formation (see Appendix C). Given that the C—H bond energy is 411 kJ/mol, obtain a value for the C≡N bond energy. Compare your result with the value given in Table 9.5.

9.76 Assume the values of the C—H and C—C bond energies given in Table 9.5. Then, using data given in Appendix C, calculate the C=O bond energy in acetaldehyde,

Compare your result with the value given in Table 9.5.

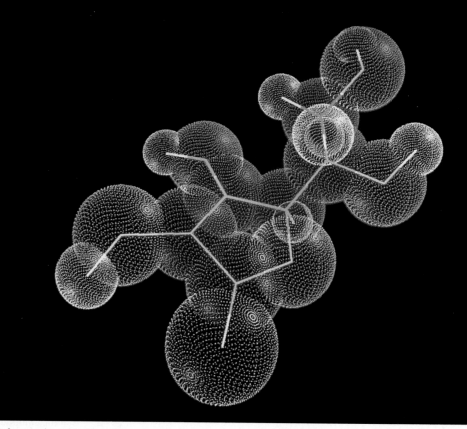

A framework model of the vitamin C molecule showing its three-dimensional shape.

Molecular Geometry and Chemical Bonding Theory

W e know from various experiments that molecules have three-dimensional structures. Consider two molecules: boron trifluoride, BF_3, and phosphorus trifluoride, PF_3. Both have the same general formula, AX_3, but their molecular structures are very different (Figure 10.1). Boron trifluoride is a planar molecule. The angle between any two B—F bonds is 120°. The geometry or general shape of the BF_3 molecule is said to be *trigonal planar*. Phosphorus trifluoride is nonplanar, and the angle between any two P—F bonds is 96°. The geometry of the PF_3 molecule is said to be *trigonal pyramidal*. How do you explain such different molecular geometries?

Subtle differences in structure are also possible. Consider the following molecular structures, which differ only in the arrangement of the atoms about the C=C bond; such molecules are referred to as *isomers*.

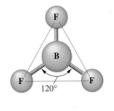

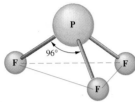

Figure 10.1
Molecular models of BF₃ and PF₃. Although both molecules have the general formula AX₃, boron trifluoride is planar (flat), whereas phosphorus trifluoride is pyramidal (pyramid shaped).

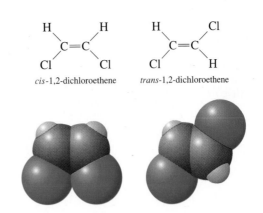

cis-1,2-dichloroethene   *trans*-1,2-dichloroethene

In the *cis* compound, both H atoms are on the same side of the C=C bond; in the *trans* compound, they are on opposite sides. These structural formulas represent entirely different compounds, as their boiling points clearly demonstrate. *Cis*-1,2-dichloroethene boils at 60°C; *trans*-1,2-dichloroethene boils at 48°C. How do you explain the existence of *cis* and *trans* compounds?

In this chapter we discuss how you explain the geometries of molecules in terms of their electronic structures. We also explore two theories of chemical bonding: valence bond theory and molecular orbital theory.

Molecular Geometry and Directional Bonding

Molecular geometry is *the general shape of a molecule, as determined by the relative positions of the atomic nuclei.* There is a simple model that allows you to predict molecular geometries, or shapes, from Lewis formulas. This valence-shell electron-pair model usually predicts the correct general shape of a molecule. It does not explain chemical bonding, however. For this you must look at a theory, such as valence bond theory, that is based on quantum mechanics.

10.1 The Valence-Shell Electron-Pair Repulsion (VSEPR) Model

▶ The acronym VSEPR is pronounced "vesper."

The **valence-shell electron-pair repulsion (VSEPR) model** *predicts the shapes of molecules and ions by assuming that the valence-shell electron pairs are arranged about each atom so that electron pairs are kept as far away from one another as possible, thus minimizing electron-pair repulsions.* ◀ For example, if there are only two electron pairs in the valence shell of an atom, these pairs tend to be at opposite sides of the nucleus, so that repulsion is minimized. This gives a *linear* arrangement of electron pairs (Figure 10.2).

If three electron pairs are in the valence shell of an atom, they tend to be arranged in a plane directed toward the corners of a triangle of equal sides. This arrangement is *trigonal planar* (Figure 10.2).

Four electron pairs in the valence shell of an atom tend to have a *tetrahedral* arrangement. That is, if you imagine the atom at the center of a regular tetrahedron, each region of space in which an electron pair mainly lies extends toward a corner,

Figure 10.2
Arrangement of electron pairs about an atom.
Black lines give the directions of electron pairs about atom A. The blue lines merely help depict the geometric arrangement of electron pairs.

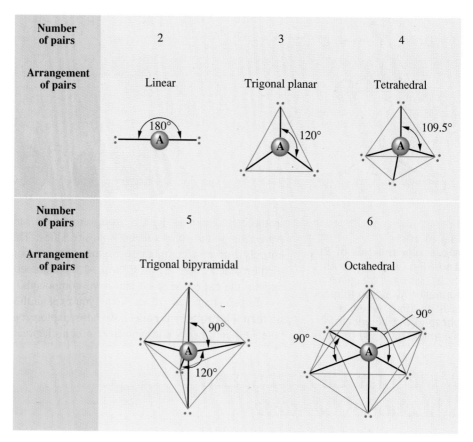

Number of pairs	2	3	4
Arrangement of pairs	Linear	Trigonal planar	Tetrahedral
	180°	120°	109.5°

Number of pairs	5	6
Arrangement of pairs	Trigonal bipyramidal	Octahedral
	90° 120°	90° 90°

or vertex (Figure 10.2). (A regular tetrahedron is a geometrical shape with four faces, each an equilateral triangle.) The regions of space mainly occupied by electron pairs are directed at approximately 109.5° angles to one another.

When you determine the geometry of a molecule experimentally, you locate the positions of the atoms, not the electron pairs. To predict the relative positions of atoms around a given atom using the VSEPR model, you first note the arrangement of valence-shell electron pairs around that central atom. Some of these electron pairs are bonding pairs and some are lone pairs. *The direction in space of the bonding pairs gives you the molecular geometry.* To consider examples of the method, we will first look at molecules AX_n in which the central atom A is surrounded by two, three, or four valence-shell electron pairs.

Central Atom with Two, Three, or Four Valence-Shell Electron Pairs

Examples of geometries in which two, three, or four valence-shell electron pairs surround a central atom are shown in Figure 10.3.

Two Electron Pairs (Linear Arrangement) To find the geometry of the molecule AX_n, you first determine the number of valence-shell electron pairs around atom A. You can get this information from the electron-dot formula. For example, following the rules given in Section 9.6, you can find the electron-dot formula for the BeF_2 molecule. ◄

▶ Beryllium fluoride, BeF_2, is normally a solid. The BeF_2 molecule exists in the vapor phase at high temperature.

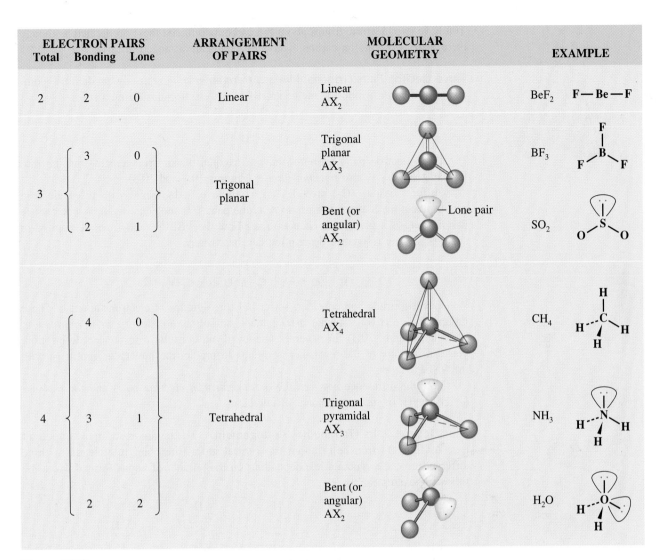

ELECTRON PAIRS			ARRANGEMENT OF PAIRS	MOLECULAR GEOMETRY		EXAMPLE	
Total	Bonding	Lone					
2	2	0	Linear	Linear AX$_2$		BeF$_2$	F—Be—F
3	3	0	Trigonal planar	Trigonal planar AX$_3$		BF$_3$	
	2	1		Bent (or angular) AX$_2$	—Lone pair	SO$_2$	
4	4	0	Tetrahedral	Tetrahedral AX$_4$		CH$_4$	
	3	1		Trigonal pyramidal AX$_3$		NH$_3$	
	2	2		Bent (or angular) AX$_2$		H$_2$O	

Figure 10.3
Molecular geometries.   Central atom with two, three, or four valence-shell electron pairs. The notation in the column at the right provides a convenient way to represent the shape of a molecule drawn on a flat surface. A straight line represents a bond or electron pair in the plane of the page. A dashed line represents a bond or electron pair extending behind the page. A wedge represents a bond or electron pair extending above (in front of) the page.

BeF$_2$

$$:\!\overset{..}{F}\!:Be:\!\overset{..}{F}\!:$$

There are two electron pairs in the valence shell for beryllium, and the VSEPR model predicts that they will have a linear arrangement (see Figure 10.2). Fluorine atoms are bonded in the same direction as the electron pairs.

The VSEPR model can also be applied to molecules with multiple bonds. In this case, each multiple bond is treated as though it were a single electron pair. To predict the geometry of carbon dioxide, first write its electron-dot formula.

$$:\!\overset{..}{O}\!::C::\!\overset{..}{O}\!:$$

BF$_3$

You have two electron groups about the carbon atom, and these are treated as though there were two pairs on carbon. Thus, the bonds are arranged linearly.

Three Electron Pairs (Trigonal Planar Arrangement) Let us now predict the geometry of the boron trifluoride molecule, BF$_3$. The electron-dot formula is

$$
\begin{array}{c}
:\ddot{F}: \\
:\ddot{F}:\ddot{B}:\ddot{F}:
\end{array}
$$

The three electron pairs on boron have a trigonal planar arrangement, and the molecular geometry is also *trigonal planar* (Figures 10.1 and 10.3).

Sulfur dioxide, SO$_2$, provides an example of a molecule with three electron groups about the S atom; one group is a lone pair. This molecule requires a resonance description, so it gives us a chance to see how the VSEPR model can be applied in such cases. The resonance description can be written

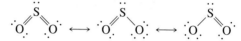

Whichever formula you consider, sulfur has three groups of electrons about it. These groups have a trigonal planar arrangement, so sulfur dioxide is a *bent*, or *angular*, molecule (Figure 10.3). In general, whenever you consider a resonance description, you can use any of the resonance formulas to predict the molecular geometry with the VSEPR model.

Note the difference between the *arrangement of electron pairs* and the *molecular geometry*, or the arrangement of nuclei.

CH$_4$

NH$_3$

H$_2$O

Four Electron Pairs (Tetrahedral Arrangement) The common and most important case of four electron pairs about the central atom (the octet rule) leads to three different molecular geometries, depending on the number of bonds formed. Note the following examples.

$$
\begin{array}{ccc}
H & H & H \\
\ddots & \ddots & \\
H:\ddot{C}:H & :\ddot{N}:H & :\ddot{O}:H \\
\ddots & \ddots & \ddots \\
H & H &
\end{array}
$$

	CH$_4$	NH$_3$	H$_2$O
Molecular geometry:	tetrahedral	trigonal pyramidal	bent

In each of these examples, the electron pairs are arranged tetrahedrally, and two or more atoms are bonded in these tetrahedral directions to give the different geometries (Figure 10.3).

When all electron pairs are bonding, as in methane, CH$_4$, the molecular geometry is *tetrahedral*. When three of the pairs are bonding and one pair is nonbonding, as in ammonia, NH$_3$, the molecular geometry is *trigonal pyramidal*. Note that the nitrogen atom is at the apex of the pyramid, and the three hydrogen atoms extend downward to form the triangular base of the pyramid.

Now you can answer the question: Why is the BF$_3$ molecule trigonal planar and the PF$_3$ molecule trigonal pyramidal? The difference occurs because in BF$_3$ there are three pairs of electrons around boron, and in PF$_3$ there are four pairs of electrons around phosphorus. As in NH$_3$, there are three bonding pairs and one lone pair in PF$_3$.

Steps in the Prediction of Geometry by the VSEPR Model Let us summarize the steps to follow in order to predict the geometry of an AX$_n$ molecule or ion by the VSEPR method. (All X atoms of AX$_n$ need not be identical.)

1. Determine from the electron-dot formula how many electron pairs are around the central atom. Count a multiple bond as one pair.

2. Arrange the electron pairs as shown in Figure 10.2.

3. Obtain the molecular geometry as shown in Figure 10.3.

CONCEPT CHECK 10.1

An atom in a molecule is surrounded by four pairs of electrons: one lone pair and three bonding pairs. Describe how the four electron pairs are arranged about the atom. How are any three of these pairs arranged in space? What is the geometry about this central atom, taking into account just the bonded atoms?

EXAMPLE 10.1

Predicting Molecular Geometries (Two, Three, or Four Electron Pairs)

Predict the geometry of the following molecules or ions, using the VSEPR method: a. $BeCl_2$; b. NO_2^-; c. $SiCl_4$.

SOLUTION

a. Write Lewis formulas. Distribute the valence electrons to the skeleton structure of $BeCl_2$ as follows:

$$:\overset{..}{Cl}:Be:\overset{..}{Cl}:$$

The two pairs on Be have a linear arrangement, indicating a **linear** molecular geometry for $BeCl_2$. b. The nitrite ion, NO_2^-, has the following resonance description:

$$\left[:\overset{..}{O}\diagdown\overset{\overset{..}{N}}{}\diagup\overset{..}{O}:\right]^- \longleftrightarrow \left[:\overset{..}{O}\diagdown\overset{\overset{..}{N}}{}\diagup\overset{..}{O}:\right]^-$$

The N atom has three electron groups (one single bond, one double bond, and one lone pair) about it. Therefore the molecular geometry of the NO_2^- ion is **bent.** c. The electron-dot formula of $SiCl_4$ is

$$\begin{array}{c} :\overset{..}{Cl}: \\ :\overset{..}{Cl}:Si:\overset{..}{Cl}: \\ :\overset{..}{Cl}: \end{array}$$

The molecular geometry is **tetrahedral.**

See Problems 10.17, 10.18, 10.19, and 10.20.

Bond Angles and the Effect of Lone Pairs

The VSEPR model allows you to predict the approximate angles between bonds in molecules. For example, it tells you that CH_4 should have a tetrahedral geometry and that the H—C—H bond angles should be 109.5° (see Figure 10.2). Because all of the valence-shell electron pairs about the carbon atom are bonding and all of the bonds are alike, you expect the CH_4 molecule to have an exact tetrahedral geometry. However, if one or more of the electron pairs is nonbonding or if there are dissimilar

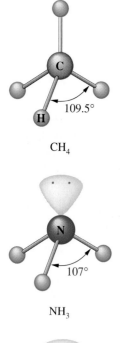

CH₄

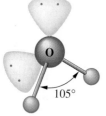

NH₃

H₂O

Figure 10.4
**H—A—H bond angles
in some molecules.**
Experimentally determined
bond angles are shown for
CH₄, NH₃, and H₂O,
represented here by models.

bonds, all four valence-shell electron pairs will not be alike. Then you expect the bond angles to deviate from 109.5°. Experimentally determined H—A—H bond angles (A is the central atom) in CH₄, NH₃, and H₂O are shown in Figure 10.4.

The increase or decrease of bond angles from the ideal values is often predictable. *A lone pair tends to require more space than a corresponding bonding pair.* You can explain this as follows: A lone pair of electrons is attracted to only one atomic core, whereas a bonding pair is attracted to two. As a result, the lone pair is spatially more diffuse, while the bonding pair is drawn more tightly to the nuclei. Consider the trigonal pyramidal molecule NH₃. The lone pair on the nitrogen atom requires more space than the bonding pairs. Therefore, the N—H bonds are effectively pushed away from the lone pair, and the H—N—H bond angles become smaller than the tetrahedral value of 109.5°. The experimental value of an H—N—H bond angle is a few degrees smaller (107°). But the trigonal pyramidal molecule PF₃ has an F—P—F bond angle of 96°, which is significantly smaller than the exact tetrahedral value.

Multiple bonds require more space than single bonds because of the greater number of electrons. The C=O bond in the formaldehyde molecule, CH₂O, requires more space than the C—H bonds. You predict that the H—C—H bond angle will be smaller than 120°. Similarly, you expect the H—C—H bond angles in the ethylene molecule, CH₂CH₂, to be smaller than the trigonal planar value. Experimental values are shown in Figure 10.5.

Central Atom with Five or Six Valence-Shell Electron Pairs

Five electron pairs tend to have a *trigonal bipyramidal arrangement.* The electron pairs tend to be directed to the corners of a trigonal bipyramid, a figure formed by placing the face of one tetrahedron onto the face of another tetrahedron (see Figure 10.2).

Six electron pairs tend to have an *octahedral arrangement.* The electron pairs tend to be directed to the corners of a regular octahedron, a figure that has eight triangular faces and six vertexes, or corners (see Figure 10.2).

Five Electron Pairs (Trigonal Bipyramidal Arrangement) Large atoms like phosphorus can accommodate more than eight valence electrons. The phosphorus atom in phosphorus pentachloride, PCl₅, has five electron pairs in its valence shell. With five electron pairs around phosphorus, all bonding, PCl₅ should have a *trigonal bipyramidal geometry* (Figure 10.6). Note, however, that the vertexes of the trigonal bipyramid are not all equivalent. Thus, the directions to which the electron pairs point are not equivalent. Two of the directions, called *axial directions*, form an axis through the central atom. They are 180° apart. The other three directions are called *equatorial directions*. These point toward the corners of the equilateral triangle that lies on

Figure 10.5
**H—C—H bond angles in
molecules with carbon
double bond.** Bond angles
are shown for formaldehyde,
CH₂O, and ethylene, CH₂CH₂,
represented here by models.

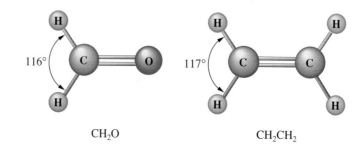

CH₂O   CH₂CH₂

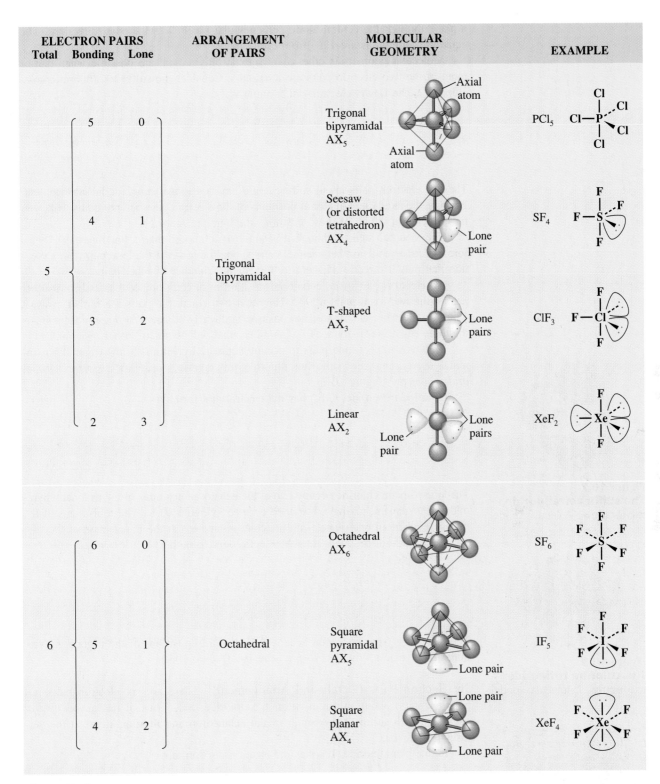

ELECTRON PAIRS			ARRANGEMENT OF PAIRS	MOLECULAR GEOMETRY	EXAMPLE
Total	Bonding	Lone			
5	5	0	Trigonal bipyramidal	Trigonal bipyramidal AX$_5$	PCl$_5$
	4	1		Seesaw (or distorted tetrahedron) AX$_4$	SF$_4$
	3	2		T-shaped AX$_3$	ClF$_3$
	2	3		Linear AX$_2$	XeF$_2$
6	6	0	Octahedral	Octahedral AX$_6$	SF$_6$
	5	1		Square pyramidal AX$_5$	IF$_5$
	4	2		Square planar AX$_4$	XeF$_4$

Figure 10.6
Molecular geometries.   Central atom with five or six valence-shell electron pairs.

a plane through the central atom, perpendicular (at 90°) to the axial directions. The equatorial directions are 120° from each other.

Molecular geometries other than trigonal bipyramidal are possible when one or more of the five electron pairs are lone pairs. Consider the sulfur tetrafluoride molecule, SF_4. The Lewis electron-dot formula is

The five electron pairs about sulfur should have a trigonal bipyramidal arrangement. Because the axial and equatorial positions of the electron pairs are not equivalent, you must decide in which of these positions the lone pair appears.

The lone pair acts as though it is somewhat "larger" than a bonding pair. Therefore, the total repulsion between all pairs should be lower if the lone pair is in a position that puts it directly adjacent to the smallest number of other pairs.

An electron pair in an equatorial position is directly adjacent to only *two* other pairs—the two axial pairs at 90°. The other two equatorial pairs are farther away at 120°. An electron pair in an axial position is directly adjacent to *three* other pairs—the three equatorial pairs at 90°. The other axial pair is much farther away at 180°.

Thus you expect lone pairs to occupy equatorial positions in the trigonal bipyramidal arrangement. For sulfur tetrafluoride, this gives a *seesaw* (or *distorted tetrahedral*) *geometry* (Figure 10.7).

Chlorine trifluoride, ClF_3, has the electron-dot formula

The lone pairs on chlorine occupy two of the equatorial positions of the trigonal bipyramidal arrangement, giving a *T-shaped geometry* (Figure 10.8). The four atoms of the molecule all lie in one plane, with the chlorine nucleus at the intersection of the "T."

Xenon difluoride, XeF_2, has the electron-dot formula

The three lone pairs on xenon occupy the equatorial positions of the trigonal bipyramidal arrangement, giving a *linear geometry* (Figure 10.9).

Six Electron Pairs (Octahedral Arrangement) There are six electron pairs (all bonding pairs) about sulfur in sulfur hexafluoride, SF_6. Thus, it has an *octahedral geometry,* with sulfur at the center of the octahedron and fluorine atoms at the vertexes (Figure 10.6).

Iodine pentafluoride, IF_5, has the electron-dot formula

Figure 10.7
The sulfur tetrafluoride molecule. Molecular model.

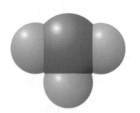

Figure 10.8
The chlorine trifluoride molecule. Molecular model.

Figure 10.9
The xenon difluoride molecule. Molecular model.

The lone pair on iodine occupies one of the six equivalent positions in the octahedral arrangement, giving a *square pyramidal geometry* (Figure 10.10). The name derives from the shape formed by drawing lines between atoms.

Xenon tetrafluoride, XeF_4, has the electron-dot formula

The two lone pairs on xenon occupy opposing positions in the octahedral arrangement to minimize their repulsion. The result is a *square planar geometry* (Figure 10.11).

Figure 10.10
The iodine pentafluoride molecule. Molecular model.

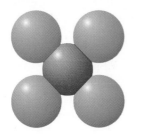

Figure 10.11
The xenon tetrafluoride molecule. Molecular model.

EXAMPLE 10.2

Predicting Molecular Geometries (Five or Six Electron Pairs)

What do you expect for the geometry of tellurium tetrachloride, $TeCl_4$?

▶ PROBLEM STRATEGY

Draw the Lewis formula for a molecule and determine the number of electron groups about the central atom.

SOLUTION

First you distribute the valence electrons to the Cl atoms to satisfy the octet rule. Then you allocate the remaining valence electrons to the central atom, Te. The electron-dot formula is

There are five electron pairs in the valence shell of Te in $TeCl_4$. Of these, four are bonding pairs and one is a lone pair. The arrangement of electron pairs is trigonal bipyramidal. You expect the lone pair to occupy an equatorial position, so $TeCl_4$ has a **seesaw** molecular geometry. (See Figure 10.6; 4 bonding pairs, 1 lone pair.)

See Problems 10.21 and 10.22.

10.2 Dipole Moment and Molecular Geometry

The VSEPR model provides a simple procedure for predicting the geometry of a molecule. Information about the geometry of a molecule can sometimes be obtained from an experimental quantity called the dipole moment. The **dipole moment** is *a quantitative measure of the degree of charge separation in a molecule*. The polarity of a bond, such as that in HCl, is characterized by a separation of electric charge. We can represent this in HCl by indicating partial charges, δ^+ and δ^-, on the atoms.

$$\overset{\delta^+}{H}—\overset{\delta^-}{Cl}$$

Left-Handed and Right-Handed Molecules

EVERYDAY LIFE

This morning, half asleep, I (D. E.) tried to put my left glove on my right hand. Silly mistake, of course. A person's two hands are similar but not identical. If I look at my hands, say, with both palms toward me, the thumbs point in opposite directions. But, if I now hold the palm of my right hand toward a mirror and compare that image with my actual left hand, I see that the thumbs point in the same direction. A person's left hand looks like the mirror image of his or her right hand, and vice versa. You can mentally superimpose one hand onto the mirror image of the other hand. Yet the two hands are themselves nonsuperimposable. Some molecules have this "handedness" property too, and this possibility gives rise to a subtle kind of isomerism.

The presence of handedness in molecules depends on the fact that the atoms in the molecules occupy specific places in three-dimensional space. In most organic molecules, this property depends on the tetrahedral bonding of the carbon atom. Consider a carbon atom bonded to four different kinds of groups, say, —H, —OH, —CH₃, and —COOH. (The resulting molecule is called lactic acid.) Two isomers are possible in which one isomer is the mirror image of the other, but the two isomers are themselves not identical (they are nonsuperimposable) (see Figure 10.12). However, all four groups bonded to this carbon atom must be different to have this isomerism. If you replace the —OH group of lactic acid by, say, an H atom, the molecule and its mirror image become identical. There are no isomers; there is just one kind of molecule.

The two lactic acid isomers shown in Figure 10.12 are labeled D-lactic acid and L-lactic acid (D is for *dextro,* meaning right; L is for *levo,* meaning left). They might be expected to have quite similar properties, and they do. They have identical melting points, boiling points, and

solubilities, for instance. And if we try to prepare lactic acid in the laboratory, we almost always get an equal molar mixture of the D and L isomers. But the biological origins of the molecules are different. Only D-lactic acid occurs in sour milk, whereas only L-lactic acid forms in muscle tissue after exercising. (If you overexercise, lactic acid accumulates in muscle tissue, causing soreness.)

Biological molecules frequently have the handedness character. The chemical substance responsible for the flavor of spearmint is L-carvone. The substance responsible for the flavor of caraway seeds, which you see in some rye bread, is D-carvone (see Figure 10.13). Here are two substances whose molecules differ only in being mirror images of one another. Yet, they have strikingly different flavors: one is minty, the other is pungent aromatic.

Figure 10.13
Spearmint and caraway. Spearmint leaves contain L-carvone; caraway seeds contain D-carvone.

Figure 10.12
Isomers of lactic acid.
Note that the two molecules labeled D-lactic acid and L-lactic acid are mirror images and cannot be superimposed on one another. See Figure 10.3 for an explanation of the notation used to represent the tetrahedral geometry.

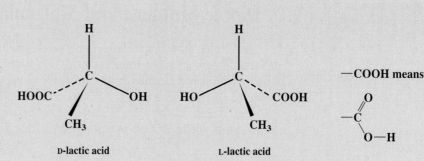

D-lactic acid

L-lactic acid

—COOH means

Figure 10.14
Alignment of polar molecules by an electric field. *(A)* A polar substance is placed between metal plates. *(B)* When the plates are connected to an electric voltage (whose direction is shown by the + and − signs), the polar molecules align so that their negative ends point toward the positive plate.

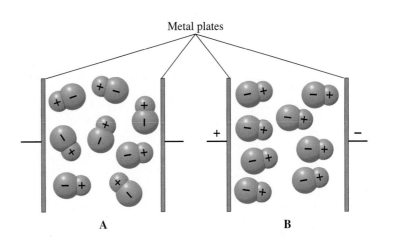

Metal plates

A   B

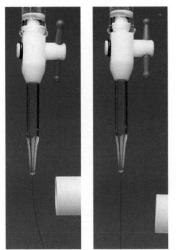

Figure 10.15
Attraction of a polar liquid to an electrified rod. *Left:* Water is a polar liquid; it is attracted to the electrically charged rod. The capacitance of the polar liquid (due to the alignment of polar molecules) stabilizes a charge separation induced in the column of liquid by the charged rod. Charges of sign opposite to those on the rod form on the liquid surface near the rod, resulting in a net attraction. *Right:* Carbon tetrachloride, CCl_4, is a nonpolar liquid; it is not attracted to the glass rod.

Any molecule that has a net separation of charge, as in HCl, has a dipole moment. A molecule in which the distribution of electric charge is equivalent to charges $+\delta$ and $-\delta$ separated by a distance d has a dipole moment equal to δd. Dipole moments are usually measured in units of *debyes* (D). In SI units, dipole moments are measured in coulomb-meters (C · m), and $1\ D = 3.34 \times 10^{-30}\ C \cdot m$.

Measurements of dipole moments are based on the fact that polar molecules can be oriented by an electric field. Figure 10.14 shows an electric field generated by charged plates. Note that the polar molecules tend to align themselves and that this orientation of the molecules affects the *capacitance* of the charged plates. Consequently, measurements of the capacitance of plates separated by different substances can be used to obtain the dipole moments. Figure 10.15 shows a simple demonstration that distinguishes between a polar liquid and a nonpolar liquid. The polar liquid, but not the nonpolar one, is attracted to a charged rod.

You can sometimes relate the presence or absence of a dipole moment in a molecule to its molecular geometry. For example, consider the carbon dioxide molecule. Each carbon–oxygen bond has a polarity in which the more electronegative oxygen atom has a partial negative charge.

$$\overset{\delta^-}{O} = \overset{2\delta^+}{C} = \overset{\delta^-}{O}$$

We will denote the dipole-moment contribution from each bond (the bond dipole) by an arrow with a positive sign at one end (↦). The dipole-moment arrow points from the positive partial charge toward the negative partial charge. Thus, you can rewrite the formula for carbon dioxide as

$$\overset{\leftharpoonup\ \ \rightharpoonup}{O = C = O}$$

Each bond dipole, like a force, is a *vector* quantity; that is, it has both magnitude and direction. Like forces, two bond dipoles of equal magnitude but opposite direction cancel each other. (Think of two groups of people in a tug of war. As long as each group pulls on the rope with the same force but in the opposite direction, there is no movement—the net force is zero.) Because the two carbon–oxygen bonds in CO_2 are equal but point in opposite directions, they give a net dipole moment of zero for the molecule.

For comparison, consider the water molecule. The bond dipoles point from the hydrogen atoms toward the more electronegative oxygen.

Table 10.1
Relationship Between Molecular Geometry and Dipole Moment

Formula	Molecular Geometry	Dipole Moment*
AX	Linear	Can be nonzero
AX_2	Linear	Zero
	Bent	Can be nonzero
AX_3	Trigonal planar	Zero
	Trigonal pyramidal	Can be nonzero
	T-shaped	Can be nonzero
AX_4	Tetrahedral	Zero
	Square planar	Zero
	Seesaw	Can be nonzero
AX_5	Trigonal bipyramidal	Zero
	Square pyramidal	Can be nonzero
AX_6	Octahedral	Zero

*All X atoms are assumed to be identical.

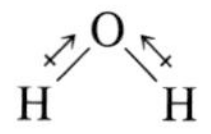

Here, the two bond dipoles do not point directly toward or away from each other. As a result, they add together to give a nonzero dipole moment for the molecule. The dipole moment of H_2O has been observed to be 1.94 D. If the H_2O molecule were linear, the dipole moment would be zero.

The analysis we have just made for two different geometries of AX_2 molecules can be extended to other AX_n molecules (in which all X atoms are identical). Table 10.1 summarizes the relationship between molecular geometry and dipole moment. Those geometries in which A—X bonds are directed symmetrically about the central atom (for example, linear, trigonal planar, and tetrahedral) give molecules of zero dipole moment; that is, the molecules are *nonpolar*. Those geometries in which the X atoms tend to be on one side of the molecule (for example, bent and trigonal pyramidal) can have nonzero dipole moments; that is, they can give *polar* molecules.

EXAMPLE 10.3

Relating Dipole Moment and Molecular Geometry

Each of the following molecules has a nonzero dipole moment. Select the molecular geometry that is consistent with this information. Explain your reasoning.

a. SO_2 linear, bent

b. PH_3 trigonal planar, trigonal pyramidal

SOLUTION

a. In the linear geometry, the S—O bond contributions to the dipole moment would cancel, giving a zero dipole moment. That would not happen in the **bent** geometry; hence, this must be the geometry for the SO_2 molecule. b. In the trigonal planar geometry, the bond contributions to the dipole moment would cancel, giving a zero dipole moment. That would not occur in the **trigonal pyramidal** geometry; hence, this is a possible molecular geometry for PH_3.

See Problems 10.23 and 10.24.

 CONCEPT CHECK 10.2

Two molecules, each with the general formula AX_3, have different dipole moments. Molecule Y has a dipole moment of zero, whereas molecule Z has a nonzero dipole moment. From this information, what can you say about the geometries of Y and Z?

10.3 Valence Bond Theory

The VSEPR model is usually a satisfactory method for predicting molecular geometries. To understand bonding and electronic structure, however, you must look to quantum mechanics. We will look in a qualitative way at the basic ideas involved in **valence bond theory,** *an approximate theory to explain the electron pair or covalent bond.*

Basic Theory

According to valence bond theory, a bond forms between two atoms when the following conditions are met:

1. An orbital on one atom comes to occupy a portion of the same region of space as an orbital on the other atom. The two orbitals are said to *overlap*.

2. The total number of electrons in both orbitals is no more than two.

As the orbital of one atom overlaps the orbital of another, the electrons in the orbitals begin to move about both atoms. Because the electrons are attracted to both nuclei at once, they pull the atoms together. Strength of bonding depends on the amount of overlap; the greater the overlap, the greater the bond strength. The two orbitals cannot contain more than two electrons because a given region of space can hold only two electrons of opposite spin.

For example, consider the formation of the H_2 molecule from atoms. Each atom has the electron configuration $1s^1$. As the H atoms approach each other, their $1s$ orbitals begin to overlap and a covalent bond forms (Figure 10.16). Valence bond theory also explains why two He atoms (each $1s^2$) do not bond. Suppose two He atoms approach one another and their $1s$ orbitals begin to overlap. As the orbitals begin to overlap, each electron pair strongly repels the other. The atoms come together, then fly apart.

Figure 10.16
Formation of H_2.
The H—H bond forms when the $1s$ orbitals, one from each atom, overlap.

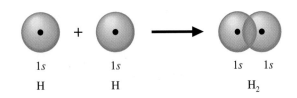

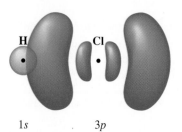

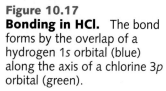

1s   3p

Figure 10.17
Bonding in HCl. The bond forms by the overlap of a hydrogen 1s orbital (blue) along the axis of a chlorine 3p orbital (green).

Because the strength of bonding depends on orbital overlap, orbitals other than s orbitals bond only in given directions. *Orbitals bond in the directions in which they protrude or point, to obtain maximum overlap.* Consider the bonding between a hydrogen atom and a chlorine atom to give the HCl molecule. A chlorine atom has the electron configuration $[Ne]3s^23p^5$. The bonding of the hydrogen atom has to occur with the singly occupied 3p orbital of chlorine. For the strongest bonding, the 1s orbital of hydrogen must overlap along the axis of the singly occupied 3p orbital of chlorine (see Figure 10.17).

Hybrid Orbitals

From what has been said, you might expect the number of bonds formed by a given atom to equal the number of unpaired electrons in its valence shell. Chlorine, whose orbital diagram is

1s   2s   2p   3s   3p

has one unpaired electron and forms one bond. Oxygen, whose orbital diagram is

1s   2s   2p

has two unpaired electrons and forms two bonds, as in H_2O.

However, consider the carbon atom, whose orbital diagram is

1s   2s   2p

You might expect this atom to bond to two hydrogen atoms to form the CH_2 molecule. Although this molecule is known to be present momentarily during some reactions, it is very reactive and cannot be isolated. But methane, CH_4, in which the carbon atom bonds to four hydrogen atoms, is well known. In fact, a carbon atom usually forms four bonds.

You might explain this as follows: Four unpaired electrons are formed when an electron from the 2s orbital of the carbon atom is *promoted* (excited) to the vacant 2p orbital.

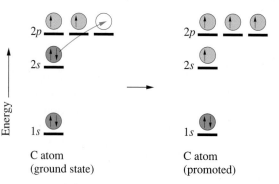

It would require energy to promote the carbon atom this way, but more than enough energy would be obtained from the formation of two additional covalent bonds. One bond would form from the overlap of the carbon 2s orbital with a hydrogen 1s orbital. Each of the other three bonds would form from a carbon 2p orbital and a hydrogen 1s orbital.

▶ **Infrared spectroscopy shows that CH₄ has four equivalent C—H bonds.**

Experiment shows, however, that the four C—H bonds in methane are identical. ◀ This implies that the carbon orbitals involved in bonding are also equivalent. For this reason, valence bond theory assumes that the four valence orbitals of the carbon atom combine during the bonding process to form four new, but equivalent, hybrid orbitals. **Hybrid orbitals** are *orbitals used to describe bonding that are obtained by taking combinations of atomic orbitals of the isolated atoms.* In this case, a set of hybrid orbitals is constructed from one *s* orbital and three *p* orbitals, so they are called *sp³* hybrid orbitals. Calculations from theory show that each *sp³* hybrid orbital has a large lobe pointing in one direction and a small lobe pointing in the opposite direction. The four *sp³* hybrid orbitals point in tetrahedral directions. Figure 10.18*A* shows the shape of a single *sp³* orbital; *B* shows a stylized set of four orbitals pointing in tetrahedral directions.

The C—H bonds in methane, CH₄, are described by valence bond theory as the overlapping of each *sp³* hybrid orbital of the carbon atom with 1*s* orbitals of hydrogen atoms (see Figure 10.18*C*). Thus, the bonds are arranged tetrahedrally, which is predicted by the VSEPR model. You can represent the hybridization of carbon and the bonding of hydrogen to the carbon atom in methane as follows:

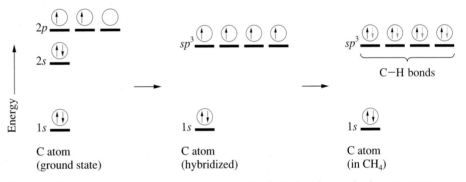

Here the blue arrows represent electrons originally belonging to hydrogen atoms.

Hybrid orbitals can be formed from various numbers of atomic orbitals. *The number of hybrid orbitals formed always equals the number of atomic orbitals used.* For example, if you combine an *s* orbital and two *p* orbitals to get a set of equivalent orbitals, you get three hybrid orbitals (called *sp²* hybrid orbitals). A set of hybrid orbitals always has definite directional characteristics. Here, all three *sp²* hybrid

Figure 10.18
Spatial arrangement of *sp³* hybrid orbitals.
(A) The shape of a single *sp³* hybrid orbital. Each orbital consists of two lobes. One lobe is small, but dense, and concentrated near the nucleus. The other lobe is large, but diffuse. Bonding occurs with the large lobe, since it extends farther from the nucleus. (B) The four hybrid orbitals are arranged tetrahedrally in space. (Small lobes are omitted here for clarity, and large lobes are stylized and greatly narrowed for ease in depicting the directional bonding.) (C) Bonding in CH₄. Each C—H bond is formed by the overlap of a 1*s* orbital from a hydrogen atom and an *sp³* hybrid orbital of the carbon atom.

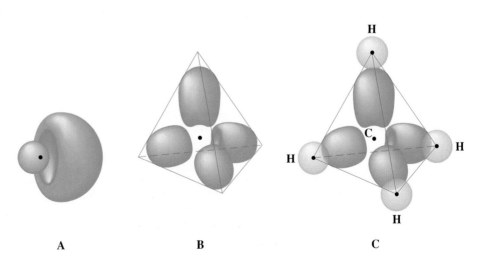

A   B   C

Table 10.2
Kinds of Hybrid Orbitals

Hybrid Orbitals	Geometric Arrangement	Number of Orbitals	Example
sp	Linear	2	Be in BeF_2
sp^2	Trigonal planar	3	B in BF_3
sp^3	Tetrahedral	4	C in CH_4
sp^3d	Trigonal bipyramidal	5	P in PCl_5
sp^3d^2	Octahedral	6	S in SF_6

orbitals lie in a plane and are directed at 120° angles to one another; that is, they have a trigonal planar arrangement. Some possible hybrid orbitals and their geometric arrangements are listed in Table 10.2 and shown in Figure 10.19. Note that the geometric arrangements of hybrid orbitals are the same as those for electron pairs in the VSEPR model.

Only two of the three p orbitals are used to form sp^2 hybrid orbitals. The unhybridized p orbital is perpendicular to the plane of the sp^2 hybrid orbitals. Similarly, only one of the three p orbitals is used to form sp hybrid orbitals. The two unhybridized p orbitals are perpendicular to the axis of the sp hybrid orbitals and perpendicular to each other. We will use these facts when we discuss multiple bonding in the next section.

Now that you know something about hybrid orbitals, let us develop a general scheme for describing the bonding about the central atom. First, notice from Table

Figure 10.19
Diagrams of hybrid orbitals showing their spatial arrangements.
Each lobe shown is one hybrid orbital (small lobes omitted for clarity).

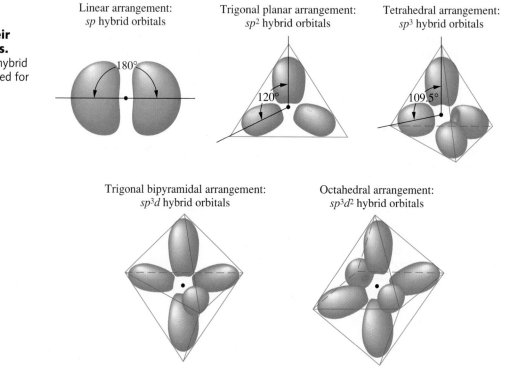

Linear arrangement:
sp hybrid orbitals

Trigonal planar arrangement:
sp^2 hybrid orbitals

Tetrahedral arrangement:
sp^3 hybrid orbitals

Trigonal bipyramidal arrangement:
sp^3d hybrid orbitals

Octahedral arrangement:
sp^3d^2 hybrid orbitals

10.2 that there is a relationship between the type of hybrid orbitals on an atom and the geometric arrangement of those orbitals. Thus, if you know the geometric arrangement, you know what hybrid orbitals to use in the bond description of the central atom. Your first task is to obtain the geometric arrangement about the central atom. In lieu of information about the geometry, you can use the VSEPR model. To obtain the bonding description about any atom in a molecule, you proceed as follows:

Step 1: Write the Lewis electron-dot formula of the molecule.

Step 2: From the Lewis formula, use the VSEPR model to obtain the arrangement of electron pairs about this atom.

Step 3: From the geometric arrangement of the electron pairs, deduce the type of hybrid orbitals on this atom required for the bonding description (see Table 10.2).

Step 4: Assign valence electrons to the hybrid orbitals of this atom one at a time, pairing them only when necessary.

Step 5: Form bonds to this atom by overlapping singly occupied orbitals of other atoms with the singly occupied hybrid orbitals of this atom.

As an application of this scheme, let us look at the BF_3 molecule and obtain the bond description of the boron atom. Following Step 1, you write the Lewis formula of BF_3.

Now you apply the VSEPR model to the boron atom (Step 2). There are three electron pairs about the boron atom, so they are expected to have a planar trigonal arrangement. Looking at Table 10.2 (Step 3), you note that the three sp^2 hybrid orbitals have a trigonal planar arrangement. In Step 4, you assign the valence electrons of the boron atom to the hybrid orbitals. Finally, in Step 5, you imagine three fluorine atoms approaching the boron atom. The singly occupied $2p$ orbital on a fluorine atom overlaps one of the sp^2 hybrid orbitals on boron, forming a covalent bond. Three such B—F bonds form. Note that one of the $2p$ orbitals of boron remains unhybridized and is unoccupied by electrons. It is oriented perpendicular to the molecular plane. You can summarize these steps using orbital diagrams as follows:

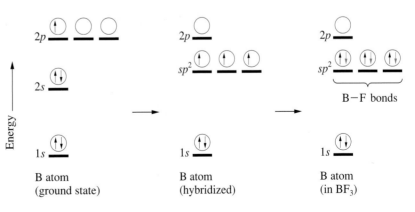

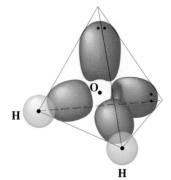

Figure 10.20
Bonding in H₂O. Orbitals on oxygen are *sp³* hybridized; bonding is tetrahedral (see Example 10.4).

EXAMPLE 10.4

Applying Valence Bond Theory (Two, Three, or Four Electron Pairs)

Describe the bonding in H_2O according to valence bond theory. Assume that the molecular geometry is the same as given by the VSEPR model.

SOLUTION

The Lewis formula for H_2O is

$$H\!-\!\overset{\cdot\cdot}{\underset{|}{O}}:$$
$$H$$

Note that there are four pairs of electrons about the oxygen atom. According to the VSEPR model, these are directed tetrahedrally, and from Table 10.2 you see that you should use *sp³* hybrid orbitals. Each O—H bond is formed by the overlap of a 1*s* orbital of a hydrogen atom with one of the singly occupied *sp³* hybrid orbitals of the oxygen atom. You can represent the bonding to the oxygen atom in H_2O as follows (see also Figure 10.20):

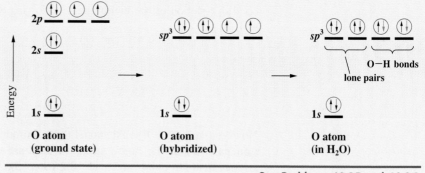

See Problems 10.25 and 10.26.

The actual angle between O—H bonds in H_2O has been experimentally determined to be 104.5°. You should note that the hybrid orbitals for bonds and lone pairs are not exactly equivalent. The lone pairs are somewhat larger than bonding pairs. Because they take up more space, the lone pairs push the bonding pairs closer together than they are in the exact tetrahedral case.

EXAMPLE 10.5

Applying Valence Bond Theory (Five or Six Electron Pairs)

Describe the bonding in XeF_4 using hybrid orbitals.

SOLUTION

The Lewis formula of XeF_4 is

The xenon atom will require six orbitals to describe the bonding. This suggests that you use sp^3d^2 hybrid orbitals on xenon (according to Table 10.2). Each fluorine atom has one singly occupied orbital, so you assume that this orbital is used in bonding. Each Xe—F bond is formed by the overlap of a xenon sp^3d^2 hybrid orbital with a singly occupied fluorine $2p$ orbital. You can summarize this as follows:

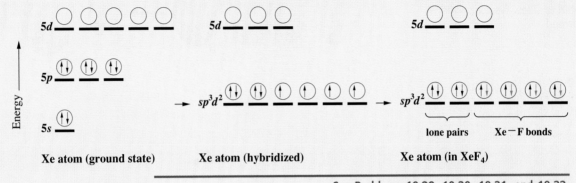

See Problems 10.29, 10.30, 10.31, and 10.32.

10.4 Description of Multiple Bonding

Now we consider the possibility that *more than one* orbital from each bonding atom might overlap, resulting in a multiple bond.

As an example, consider the ethylene molecule.

One hybrid orbital is needed for each bond (whether a single or a multiple bond) and for each lone pair. Because each carbon atom is bonded to three other atoms and there are no lone pairs, three hybrid orbitals are needed. This suggests the use of sp^2 hybrid orbitals on each carbon atom. Thus, during bonding, the $2s$ orbital and two of the $2p$ orbitals of each carbon atom form three hybrid orbitals having trigonal planar orientation. A third $2p$ orbital on each carbon atom remains unhybridized and is perpendicular to the plane of the three sp^2 hybrid orbitals.

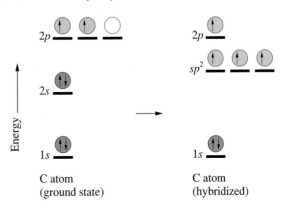

Figure 10.21
Sigma and pi bonds.
(A) The formation of a σ bond by the overlap of two *s* orbitals. *(B)* A σ bond can also be formed by the overlap of two *p* orbitals along their axes. *(C)* When two *p* orbitals overlap sideways, a π bond is formed.

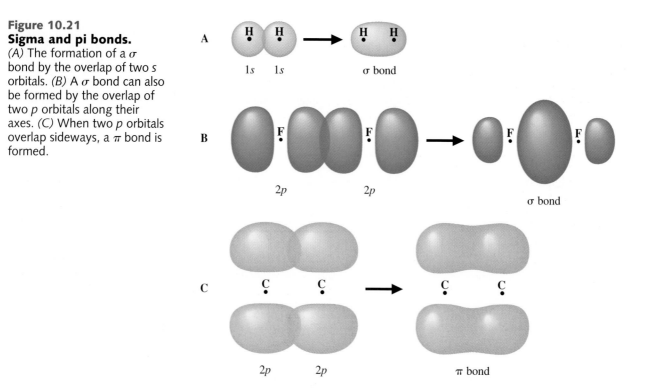

To describe the multiple bonding in ethylene, we must distinguish between two kinds of bonds. A σ **(sigma) bond** *has a cylindrical shape about the bond axis.* It is formed either when two *s* orbitals overlap, as in H_2 (Figure 10.21*A*), or when an orbital with directional character, such as a *p* orbital or a hybrid orbital, overlaps another orbital *along their axes* (Figure 10.21*B*). The bonds we discussed in the previous section are σ bonds.

A π **(pi) bond** *has an electron distribution above and below the bond axis.* It is formed by the *sideways* overlap of two parallel *p* orbitals (see Figure 10.21*C*). A sideways overlap will not give as strong a bond as an along-the-axis overlap of two *p* orbitals. A π bond occurs when two parallel orbitals are still available after strong σ bonds have formed.

Now imagine that the separate atoms of ethylene move into their normal molecular positions. Each sp^2 hybrid carbon orbital overlaps a 1*s* orbital of a hydrogen atom or an sp^2 hybrid orbital of another carbon atom to form a σ bond (Figure 10.22*A*). Together, the σ bonds give the molecular framework of ethylene.

As you see from the orbital diagram for the hybridized C atom, a single 2*p* orbital still remains on each carbon atom. These orbitals are perpendicular to the plane of the hybrid orbitals; that is, they are perpendicular to the —CH_2 plane. Note that the two —CH_2 planes can rotate about the carbon–carbon axis without affecting the overlap of the hybrid orbitals. As these planes rotate, the 2*p* orbitals also rotate. When the —CH_2 planes rotate so that the 2*p* orbitals become parallel, the orbitals overlap to give a π bond (Figure 10.22*B*).

You therefore describe the carbon–carbon double bond as one σ bond and one π bond. Note that when the two 2*p* orbitals are parallel, the two —CH_2 ends of the molecule lie in the same plane. Thus, the formation of a π bond "locks" the two ends into a flat, rigid molecule.

Figure 10.22
Bonding in ethylene.
(A) The σ-bond framework in ethylene, formed by the overlap of sp^2 hybrid orbitals on C atoms and $1s$ orbitals on H atoms. (B) The formation of the π bonds in ethylene. When the $2p$ orbitals are perpendicular to one another, there is no overlap and no bond formation. When the two —CH$_2$ groups rotate so that the $2p$ orbitals are parallel, a π bond forms.

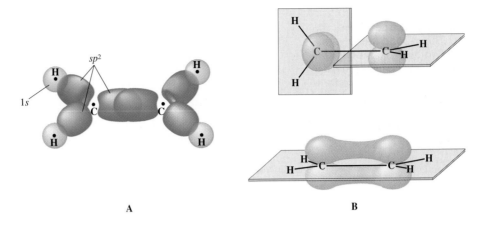

A   B

You can describe the triple bonding in acetylene, H—C≡C—H, in similar fashion. Because each carbon atom is bonded to two other atoms and there are no lone pairs, two hybrid orbitals are needed. This suggests sp hybridization (see Table 10.2).

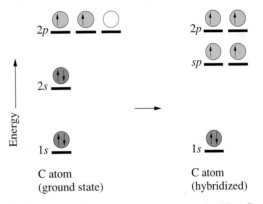

These sp hybrid orbitals have a linear arrangement, so the H—C—C—H geometry is linear. Bonds formed by the overlap of these hybrid orbitals are σ bonds. The two $2p$ orbitals not used to construct hybrid orbitals are perpendicular to the bond axis and to each other. They are used to form two π bonds. Thus, the carbon–carbon triple bond consists of one σ bond and two π bonds (see Figure 10.23).

EXAMPLE 10.6

Applying Valence Bond Theory (Multiple Bonding)

Describe the bonding on a given N atom in dinitrogen difluoride, N_2F_2, using valence bond theory.

SOLUTION

The electron-dot formula of N_2F_2 is

Figure 10.23
Bonding in acetylene.
(A) The σ-bond framework.
(B) Two 2p orbitals on each carbon atom begin to overlap (symbolized by lines) to form two π bonds.

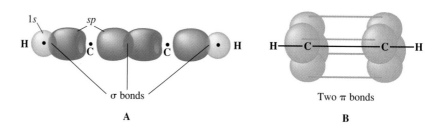

σ bonds

A

Two π bonds

B

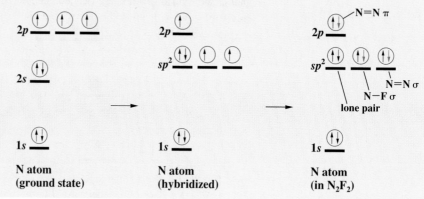

Note that a double bond is described as a π bond plus a σ bond. Hybrid orbitals are needed to describe each σ bond and each lone pair (a total of three hybrid orbitals for each N atom). This suggests sp^2 hybridization. According to this description, one of the sp^2 hybrid orbitals is used to form the N—F bond, another to form the σ bond of N=N, and the third to hold the lone pair on the N atom. The 2p orbitals on each N atom overlap to form the π bond of N=N. Hybridization and bonding of the N atoms are shown as follows:

A

B

Figure 10.24
Lack of geometric isomers in 1,2-dichloroethane.
Because of rotation about the carbon–carbon bond in 1,2-dichloroethane, geometric isomers are not possible. Note that the molecule pictured in (A) can be twisted easily to give the molecule pictured in (B).

The π-bond description of the double bond agrees well with experiment. The *geometric, or cis–trans, isomers* of the compound 1,2-dichloroethene (described in the chapter opening) illustrate this. *Isomers* are compounds of the same molecular formula but with different arrangements of the atoms. (The numbers in the name 1,2-dichloroethene refer to the positions of the chlorine atoms; one chlorine atom is attached to carbon atom 1 and the other to carbon atom 2.) The structures of these isomers of 1,2-dichloroethene are

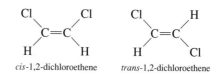

cis-1,2-dichloroethene   *trans*-1,2-dichloroethene

To transform one isomer into the other, one end of the molecule must be rotated as the other remains fixed. For this to happen, the π bond must be broken. Breaking the π bond requires considerable energy, so the *cis* and *trans* compounds are not easily interconverted. Contrast this with 1,2-dichloroethane (Figure 10.24), in which the two ends of the molecule can rotate without breaking any bonds. Here isomers corresponding to different spatial orientations of the two chlorine atoms cannot be prepared, because the two ends rotate freely with respect to one another. There is only one compound.

The *cis* and *trans* isomers have different properties. The *cis* isomer of 1,2-dichloroethene boils at 60°C; the *trans* compound, at 48°C. The *trans* compound has no dipole moment because it is symmetrical (the polar C—Cl bonds point in opposite directions and so cancel). However, the *cis* compound has a dipole moment of 1.85 D.

CONCEPT CHECK 10.3

An atom in a molecule has one single bond and one triple bond to other atoms. What hybrid orbitals do you expect for this atom? Describe how you arrive at your answer.

Molecular Orbital Theory

Although simple valence bond theory satisfactorily describes most of the molecules we encounter, it does not apply to all molecules. For example, according to this theory any molecule with an even number of electrons should be diamagnetic because we assume the electrons to be paired and to have opposite spins. In fact, a few molecules with an even number of electrons are paramagnetic (attracted to a magnet), indicating that some of the electrons are not paired. The best-known example of such a paramagnetic molecule is O_2 (Figure 10.25). Molecular orbital theory provides a straightforward explanation of the paramagnetism of O_2.

Molecular orbital theory is *a theory of the electronic structure of molecules in terms of molecular orbitals, which may spread over several atoms or the entire molecule.* This theory views the electronic structure of molecules to be much like the electronic structure of atoms. Each molecular orbital has a definite energy. To obtain the ground state of a molecule, electrons are put into orbitals of lowest energy, consistent with the Pauli exclusion principle, just as in atoms.

10.5 Principles of Molecular Orbital Theory

You can think of a molecular orbital as being formed from a combination of atomic orbitals. As atoms approach each other and their atomic orbitals overlap, molecular orbitals are formed.

Bonding and Antibonding Orbitals

Consider the H_2 molecule. As the atoms approach to form the molecule, their $1s$ orbitals overlap. One molecular orbital is obtained by adding the two $1s$ orbitals (see Figure 10.26). Note that where the atomic orbitals overlap, their values sum to give a larger result. This means that in this molecular orbital, electrons are often found in the region between the two nuclei where the electrons can hold the nuclei together. *Molecular orbitals that are concentrated in regions between nuclei* are called **bonding orbitals.** The bonding orbital in H_2, which we have just described, is denoted σ_{1s}. The σ (sigma) means that the molecular orbital has a cylindrical shape about the bond axis.

Another molecular orbital is obtained by subtracting the $1s$ orbital on one atom from the $1s$ orbital on the other (Figure 10.26). When the orbitals are subtracted, the

Figure 10.25
Paramagnetism of oxygen, O_2. Liquid oxygen is poured between the poles of a strong magnet. Oxygen adheres to the poles, showing that it is paramagnetic.

Figure 10.26
Formation of bonding and antibonding orbitals from 1s orbitals of hydrogen atoms. When the two 1s orbitals overlap, they can either add to give a bonding molecular orbital or subtract to give an antibonding molecular orbital.

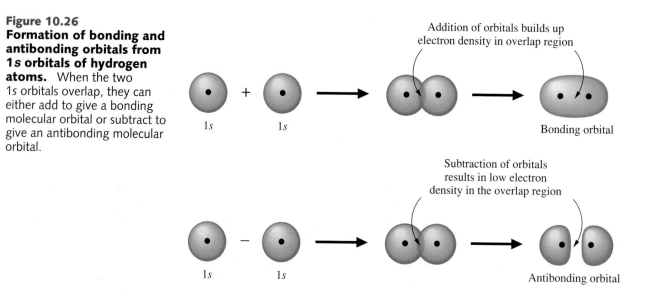

resulting values in the region of the overlap are close to zero. This means that in this molecular orbital, the electrons spend little time between the nuclei. *Molecular orbitals that are concentrated in regions other than between two nuclei* are called **antibonding orbitals.** The antibonding orbital in H_2 is denoted σ^*_{1s}.

Figure 10.27 shows the energies of the molecular orbitals σ_{1s} and σ^*_{1s} relative to the atomic orbitals. Energies of the separate atomic orbitals are represented by heavy black lines at the far left and right. The energies of the molecular orbitals are shown by heavy black lines in the center. Note that the energy of a bonding orbital is less than that of the separate atomic orbitals, whereas the energy of an antibonding orbital is higher.

You obtain the electron configuration for the ground state of H_2 by placing the two electrons (one from each atom) into the lower-energy orbital (see Figure 10.27). The orbital diagram is

⇅○
σ_{1s} σ^*_{1s}

and the electron configuration is $(\sigma_{1s})^2$. Because the energy of the two electrons is lower than their energies in the isolated atoms, the H_2 molecule is stable.

You obtain a similar set of orbitals when you consider the approach of two helium atoms. To obtain the ground state of He_2, two electrons from each atom fill the molecular orbitals. Two electrons go into the σ_{1s} orbital and two go into the σ^*_{1s} orbital.

Figure 10.27
Relative energies of the 1s orbital of the H atom and the σ_{1s} and σ^*_{1s} molecular orbitals of H_2.
Arrows denote occupation of the σ_{1s} orbital by electrons in the ground state of H_2.

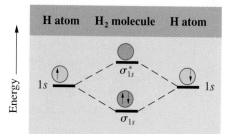

The orbital diagram is

$$\sigma_{1s} \quad \sigma_{1s}^*$$

and the configuration is $(\sigma_{1s})^2(\sigma_{1s}^*)^2$. The energy decrease from the bonding electrons is offset by the energy increase from the antibonding electrons. Hence, He_2 is not a stable molecule.

Bond Order

The term *bond order* refers to the number of bonds that exist between two atoms. In molecular orbital theory, the bond order of a diatomic molecule is defined as one-half the difference between the number of electrons in bonding orbitals, n_b, and the number of electrons in antibonding orbitals, n_a. ◄

▶ For a Lewis formula, the bond order equals the number of electron pairs shared between two atoms (see Section 9.10).

$$\text{Bond order} = \tfrac{1}{2}(n_b - n_a)$$

For H_2, which has two bonding electrons,

$$\text{Bond order} = \tfrac{1}{2}(2 - 0) = 1$$

That is, H_2 has a single bond. For He_2, which has two bonding and two antibonding electrons, the bond order is zero.

Bond orders need not be whole numbers. For example, the H_2^+ molecular ion has the configuration $(\sigma_{1s})^1$ and a bond order of $\tfrac{1}{2}(1 - 0) = \tfrac{1}{2}$.

Factors That Determine Orbital Interaction

Now let us consider the Li_2 molecule. Each Li atom has $1s$ and $2s$ orbitals. Which orbitals interact to form molecular orbitals? To find out, you need to understand the factors that determine orbital interaction.

The strength of the interaction between two atomic orbitals to form molecular orbitals is determined by two factors: (1) the energy difference between the interacting orbitals, and (2) the magnitude of their overlap. *For the interaction to be strong, the energies of the two orbitals must be approximately equal and the overlap must be large.*

From this last statement, you see that when two Li atoms approach one another to form Li_2, only like orbitals on the two atoms interact appreciably. The $2s$ orbital of one lithium atom interacts with the $2s$ orbital of the other atom. Because the $2s$ orbitals are outer orbitals, they are able to overlap and interact strongly when the atoms approach. As in H_2, these atomic orbitals interact to give a bonding orbital (denoted σ_{2s}) and an antibonding orbital (denoted σ_{2s}^*). Figure 10.28 gives the relative energies of the orbitals.

You obtain the ground-state configuration of Li_2 by putting six electrons (three from each atom) into the molecular orbitals of lowest energy. The configuration of the diatomic molecule Li_2 is

$$Li_2 \qquad (\sigma_{1s})^2(\sigma_{1s}^*)^2(\sigma_{2s})^2$$

The $(\sigma_{1s})^2(\sigma_{1s}^*)^2$ part of the configuration is often abbreviated KK (which denotes the K shells, or inner shells, of the two atoms). These electrons do not have a significant effect on bonding.

$$Li_2 \qquad KK(\sigma_{2s})^2$$

In calculating bond order, we can ignore KK. Thus, the bond order is 1.

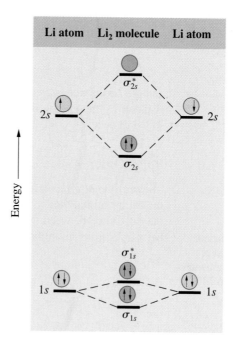

In Be$_2$, the energy diagram is similar to that in Figure 10.28. You have eight electrons to distribute (four from each Be atom). The ground-state configuration of Be$_2$ is

$$Be_2 \qquad KK(\sigma_{2s})^2(\sigma_{2s}^*)^2$$

Note that the configuration has two bonding and two antibonding electrons outside the K shells. Thus, the bond order is $\frac{1}{2}(2 - 2) = 0$. No bond is formed.

10.6 Electron Configurations of Diatomic Molecules of the Second-Period Elements

The previous section looked at the electron configurations of some simple molecules: H$_2$, He$_2$, Li$_2$, and Be$_2$. These are **homonuclear diatomic molecules**—that is, *molecules composed of two like nuclei.* (**Heteronuclear diatomic molecules** are *molecules composed of two different nuclei*—for example, CO and NO.) To find the electron configurations of other homonuclear diatomic molecules, we need to have additional molecular orbitals.

Now we need to consider the formation of molecular orbitals from *p* atomic orbitals. There are two different ways in which 2*p* atomic orbitals can interact. One set of 2*p* orbitals can overlap along their axes to give one bonding and one antibonding σ orbital (σ_{2p} and σ_{2p}^*). The other two sets of 2*p* orbitals then overlap sideways to give two bonding and two antibonding π orbitals (π_{2p} and π_{2p}^*). See Figure 10.29.

Figure 10.30 shows the relative energies of the molecular orbitals obtained from 2*s* and 2*p* atomic orbitals. This order of molecular orbitals reproduces the known elec-

Figure 10.29
The different ways in which 2p orbitals can interact. When the 2p orbitals overlap along their axes, they form σ_{2p} and σ_{2p}^* molecular orbitals. When they overlap sideways, they form π_{2p} and π_{2p}^* molecular orbitals.

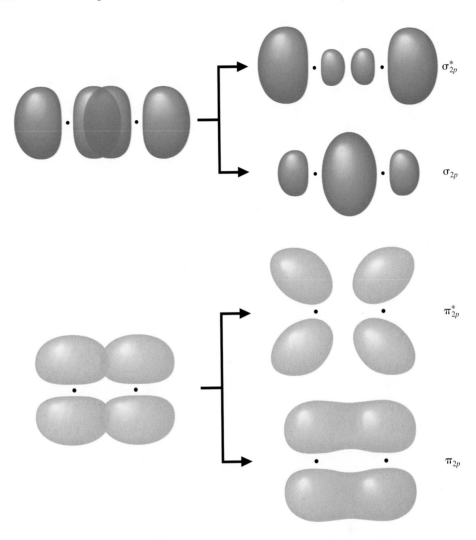

σ_{2p}^*

σ_{2p}

π_{2p}^*

π_{2p}

▶ **This order gives the correct number of electrons in subshells, even though in O$_2$ and F$_2$ the energy of σ_{2p} is below π_{2p}.**

tron configurations of homonuclear diatomic molecules composed of elements in the second row of the periodic table. ◀ The order of filling is

$$\sigma_{2s} \quad \sigma_{2s}^* \quad \pi_{2p} \quad \sigma_{2p} \quad \pi_{2p}^* \quad \sigma_{2p}^*$$

Note that there are two orbitals in the π subshell and two orbitals in the π^* subshell. Because each orbital can hold two electrons, a π or π^* subshell can hold four electrons.

EXAMPLE 10.7

Describing Molecular Orbital Configurations (Homonuclear Diatomic Molecules)

Give the orbital diagram of the O$_2$ molecule. Is the molecular substance diamagnetic or paramagnetic? What is the electron configuration? What is the bond order of O$_2$?

Figure 10.30
Relative energies of molecular orbitals of homonuclear diatomic molecules (excluding K shells). The arrows show the occupation of molecular orbitals by the valence electrons of N_2.

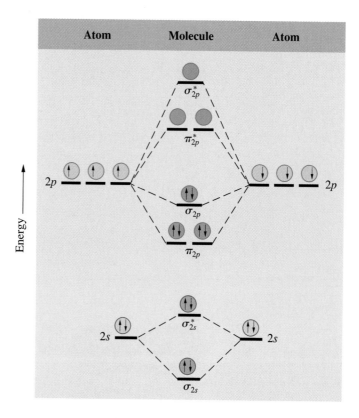

SOLUTION

There are 12 valence electrons in O_2 (six from each atom), which occupy the molecular orbitals as shown in the following orbital diagram:

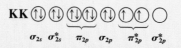

Note that the two electrons in the π^*_{2p} subshell must go into different orbitals with their spins in the same direction (Hund's rule). Because there are two unpaired electrons, the molecular substance is **paramagnetic.** The electron configuration is

$$KK(\sigma_{2s})^2(\sigma^*_{2s})^2(\pi_{2p})^4(\sigma_{2p})^2(\pi^*_{2p})^2$$

There are eight bonding electrons and four antibonding electrons. Therefore,

$$\text{Bond order} = \tfrac{1}{2}(8 - 4) = \mathbf{2}$$

See Problems 10.35 and 10.36.

Table 10.3 compares experimentally determined bond lengths, bond dissociation energies, and magnetic character of the second-period homonuclear diatomic molecules. Note that as the bond order increases, bond length tends to decrease and bond dissociation energy tends to increase.

Table 10.3
Theoretical Bond Orders and Experimental Data for the Second-Period Homonuclear Diatomic Molecules

Molecule	Bond Order	Bond Length (pm)	Bond Dissociation Energy (kJ/mol)	Magnetic Character
Li_2	1	267	110	Diamagnetic
Be_2	0	*	*	*
B_2	1	159	290	Paramagnetic
C_2	2	124	602	Diamagnetic
N_2	3	110	942	Diamagnetic
O_2	2	121	494	Paramagnetic
F_2	1	142	155	Diamagnetic
Ne_2	0	*	*	*

The symbol * means that no stable molecule has been observed.

When the atoms in a heteronuclear diatomic molecule are close to one another in a row of the periodic table, the molecular orbitals have the same relative order of energies as those for homonuclear diatomic molecules.

EXAMPLE 10.8
Describing Molecular Orbital Configurations (Heteronuclear Diatomic Molecules)

Write the orbital diagram for nitric oxide, NO. What is the bond order of NO?

SOLUTION

You assume that the order of filling of orbitals is the same as for homonuclear diatomic molecules. There are 11 valence electrons in NO. Thus, the orbital diagram is

$$KK \; \sigma_{2s} \; \sigma_{2s}^* \; \pi_{2p} \; \sigma_{2p} \; \pi_{2p}^* \; \sigma_{2p}^*$$

Because there are eight bonding and three antibonding electrons,

$$\text{Bond order} = \tfrac{1}{2}(8 - 3) = \tfrac{5}{2}$$

A CHEMIST LOOKS AT

Stratospheric Ozone (An Absorber of Ultraviolet Rays)

ENVIRONMENT

Ozone is a reactive form of oxygen having a bent molecular geometry. Ozone forms in the atmosphere whenever O_2 is irradiated by ultraviolet light or subjected to electrical discharges.

Although ozone occurs at ground level and is an ingredient of *photochemical smog*, it is an essential component of the *stratosphere*, which occurs at about 10 to 15 km above ground level. Ozone in the stratosphere absorbs ultraviolet radiation between 200 and 300 nm. Radiation from the sun contains ultraviolet rays that are harmful to the DNA of biological organisms. Oxygen, O_2, absorbs the most energetic of these ultraviolet rays in the earth's upper atmosphere, but only ozone in the strat-

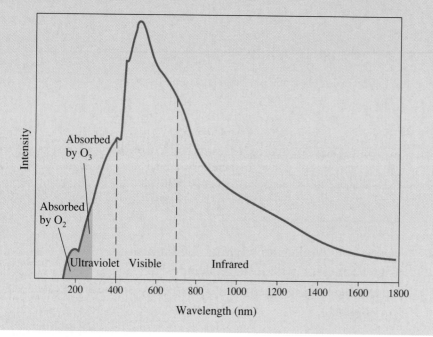

Figure 10.31
The solar spectrum.
Radiation below 200 nm is absorbed in the upper atmosphere by O_2. Ultraviolet radiation between 200 and 300 nm is absorbed by O_3 in the stratosphere.

A Checklist for Review

Important Terms

molecular geometry (10.1)
valence-shell electron-pair
 repulsion (VSEPR)
 model (10.1)
dipole moment (10.2)

valence bond theory (10.3)
hybrid orbitals (10.3)
σ (sigma) bond (10.4)
π (pi) bond (10.4)

molecular orbital
 theory (10.5)
bonding orbitals (10.5)
antibonding orbitals (10.5)

homonuclear diatomic
 molecules (10.6)
heteronuclear diatomic
 molecules (10.6)

osphere absorbs the remaining ultraviolet radiation that is destructive to life on earth (Figure 10.31).

In 1974, Mario J. Molina and F. Sherwood Rowland expressed concern that chlorofluorocarbons (CFCs), such as $CClF_3$ and CCl_2F_2, would be a source of chlorine atoms that could catalyze the decomposition of ozone in the stratosphere, so that the ozone would be destroyed faster than it could be produced. The chlorofluorocarbons are relatively inert compounds used as refrigerants, spray-can propellants, and blowing agents (substances used to produce plastic foams). As a result of their inertness, however, CFCs concentrate in the atmosphere, where they steadily rise into the stratosphere. Once in the stratosphere, ultraviolet light decomposes them to form chlorine atoms, which react with ozone to form ClO and O_2. The ClO molecules react with oxygen atoms in the stratosphere to regenerate Cl atoms.

$$Cl(g) + O_3(g) \longrightarrow ClO(g) + O_2(g)$$
$$ClO(g) + O(g) \longrightarrow Cl(g) + O_2(g)$$
$$\overline{O_3(g) + O(g) \longrightarrow 2O_2(g)}$$

The net result is the decomposition of ozone to dioxygen. Chlorine atoms are consumed in the first step but regenerated in the second. Thus, chlorine atoms are not used up, so they function as a catalyst.

In 1985, British researchers reported that the ozone over the Antarctic had been declining precipitously each spring for several years, although it returned the following winter (Figure 10.32). Researchers found that wherever ozone is depleted in the stratosphere, chlorine monoxide (ClO) appears, as expected if chlorine atoms are catalyzing the depletion (see the previous reactions). The chlorofluorocarbons are the suspected source of these chlorine atoms. Many nations, including the

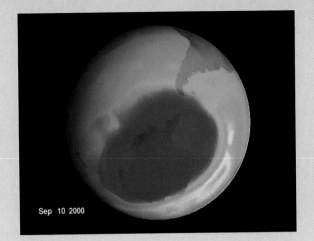

Sep 10 2000

Figure 10.32
Computer map of the South Polar region showing total stratospheric ozone. Data were obtained by the Total Ozone Mapping Spectrometer (TOMS) aboard the *Nimbus-7* satellite in September 2000. The ozone hole is shown in dark blue, in the region near the South Pole (Antarctica).

United States, have pledged to phase out the usage of CFCs, and manufacturers are hurrying to substitute safer compounds. Auto manufacturers, for instance, now use hydrofluorocarbons (HFCs), such as CF_3CH_2F, which do not contain Cl atoms, as refrigerants in air conditioners.

In October 1995, Sherwood Rowland (University of California, Irvine), Mario Molina (Massachusetts Institute of Technology), and Paul Crutzen (Max Planck Institute, Mainz, Germany) were awarded the Nobel Prize in chemistry for their work on stratospheric ozone depletion.

Summary of Facts and Concepts

Molecular geometry refers to the spatial arrangement of atoms in a molecule. The *valence-shell electron-pair repulsion (VSEPR) model* is a simple model for predicting molecular geometries. It is based on the idea that the valence-shell electron pairs are arranged symmetrically about an atom to minimize electron-pair repulsion. The geometry about an atom is then determined by the directions of the bonding pairs. Information about the geometry of a molecule can sometimes be obtained from the experimentally determined presence or absence of a *dipole moment*.

The bonding and geometry in a molecule can be described in terms of *valence bond theory.* In this theory, a bond is formed by the overlap of orbitals from two atoms. *Hybrid orbitals,* a set of equivalent orbitals formed by combining atomic orbitals, are often needed to describe this bond. Multiple bonds occur via the overlap of atomic orbitals to give *σ bonds* and *π bonds*. *Cis–trans* isomers result from the molecular rigidity imposed by a *π* bond.

Molecular orbital theory can also be used to explain bonding in molecules. According to this theory, electrons in a molecule occupy orbitals that may spread over the entire molecule. You can think of these molecular orbitals as constructed from atomic orbitals. Thus, when two atoms approach to form a diatomic molecule, the atomic orbitals interact to form *bonding* and *antibonding* molecular orbitals. The electron configuration of a diatomic molecule such as O_2 can be predicted from the order of filling of the molecular orbitals. From this configuration, you can predict the bond order and whether the molecular substance is diamagnetic or paramagnetic.

Operational Skills

1. **Predicting molecular geometries** Given the formula of a simple molecule, predict its geometry, using the VSEPR model. **(EXAMPLES 10.1, 10.2)**

2. **Relating dipole moment and molecular geometry** State what geometries of a molecule AX$_n$ are consistent with the information that the molecule has a nonzero dipole moment. **(EXAMPLE 10.3)**

3. **Applying valence bond theory** Given the formula of a

simple molecule, describe its bonding, using valence bond theory. **(EXAMPLES 10.4, 10.5, 10.6)**

4. **Describing molecular orbital configurations** Given the formula of a diatomic molecule obtained from first- or second-period elements, deduce the molecular orbital configuration, the bond order, and whether the molecular substance is diamagnetic or paramagnetic. **(EXAMPLES 10.7, 10.8)**

Review Questions

10.1 Describe the main features of the VSEPR model.

10.2 According to the VSEPR model, what are the arrangements of two, three, four, five, and six valence-shell electron pairs about an atom?

10.3 Why is a lone pair expected to occupy an equatorial position instead of an axial position in the trigonal bipyramidal arrangement?

10.4 Why is it possible for a molecule to have polar bonds, yet have a dipole moment of zero?

10.5 Explain in terms of valence bond theory why the orbitals used on an atom give rise to a particular geometry about that atom.

10.6 What is the difference between a sigma bond and a pi bond?

10.7 Describe the bonding in ethylene, C_2H_4, in terms of valence bond theory.

10.8 How does the valence bond description of a carbon–carbon double bond account for *cis–trans* isomers?

10.9 What are the differences between a bonding and an antibonding molecular orbital of a diatomic molecule?

10.10 What factors determine the strength of interaction between two atomic orbitals to form a molecular orbital?

10.11 Describe the formation of bonding and antibonding molecular orbitals resulting from the interaction of two 2s orbitals.

10.12 Describe the formation of molecular orbitals resulting from the interaction of two 2p orbitals.

Conceptual Problems

10.13 Match the following molecular substances with one of the molecular models (i) to (iv) that correctly depicts the geometry of the corresponding molecule.
a. SeO_2 b. $BeCl_2$ c. PBr_3 d. BCl_3

(i)

(ii)

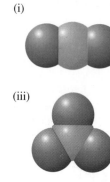

(iii)

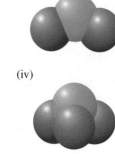

(iv)

10.14 An atom in a molecule has two bonds to other atoms and one lone pair. What kind of hybrid orbitals do you expect for this atom? Describe how you arrived at your answer.

10.15 Two compounds have the same molecular formula, $C_2H_2Br_2$. One has a dipole moment; the other does not. Both compounds react with bromine, Br_2, to produce the same compound. This reaction is a generally accepted test for double bonds, and each bromine atom of Br_2 adds to a different atom of the double bond. What is the identity of the original compounds? Describe the argument you use.

10.16 A neutral molecule is identified as a tetrafluoride, XF_4, where X is an unknown atom. If the molecule has a dipole moment of 0.63 D, can you give some possibilities for the identity of X?

Practice Problems

The VSEPR Model

10.17 Predict the shape or geometry of the following molecules, using the VSEPR model.
a. SCl_2 b. CF_4 c. $COCl_2$ d. NF_3

10.18 Use the electron-pair repulsion model to predict the geometry of the following molecules:
a. $GeCl_2$ b. $AsCl_3$ c. SO_3 d. XeO_4

10.19 Predict the geometry of the following ions, using the electron-pair repulsion model.
a. NO_3^- b. ClO_4^- c. SCN^- d. SO_4^{2-}

10.20 Use the VSEPR model to predict the geometry of the following ions:
a. N_3^- b. BH_4^- c. SO_3^{2-} d. NO_2^-

10.21 What geometry is expected for the following molecules, according to the VSEPR model?
a. PF_5 b. BrF_3 c. BrF_5 d. SCl_4

10.22 From the electron-pair repulsion model, predict the geometry of the following molecules:
a. ClF_5 b. SbF_5 c. SeF_4 d. TeF_6

Dipole Moment and Molecular Geometry

10.23 (a) The molecule AsF_3 has a dipole moment of 2.59 D. Which of the following geometries are possible: trigonal planar, trigonal pyramidal, or T-shaped? (b) The molecule H_2S has a dipole moment of 0.97 D. Is the geometry linear or bent?

10.24 (a) The molecule BrF_3 has a dipole moment of 1.19 D. Which of the following geometries are possible: trigonal planar, trigonal pyramidal, or T-shaped? (b) The molecule $TeCl_4$ has a dipole moment of 2.54 D. Is the geometry tetrahedral, seesaw, or square planar?

Valence Bond Theory

10.25 What hybrid orbitals would be expected for the central atom in each of the following molecules or ions?

(a)

(b)

10.26 What hybrid orbitals would be expected for the central atom in each of the following molecules or ions?

(a)   (b)

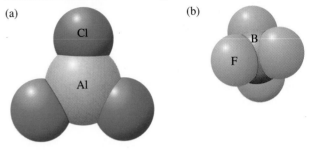

10.27 (a) Mercury(II) chloride dissolves in water to give poorly conducting solutions, indicating that the compound is largely nonionized in solution—it dissolves as $HgCl_2$ molecules. Describe the bonding of the $HgCl_2$ molecule, using valence bond theory. (b) Phosphorus trichloride, PCl_3, is a colorless liquid with a highly irritating vapor. Describe the bonding in the PCl_3 molecule, using valence bond theory. Use hybrid orbitals.

10.28 (a) Nitrogen trifluoride, NF_3, is a relatively unreactive, colorless gas. How would you describe the bonding in the NF_3 molecule in terms of valence bond theory? Use hybrid orbitals. (b) Silicon tetrafluoride, SiF_4, is a colorless gas formed when hydrofluoric acid attacks silica (SiO_2) or glass. Describe the bonding in the SiF_4 molecule, using valence bond theory.

10.29 What hybrid orbitals would be expected for the central atom in each of the following?
a. BrF_5 b. BrF_3 c. $AsCl_5$ d. ClF_4^+

10.30 What hybrid orbitals would be expected for the central atom in each of the following?
a. PCl_5 b. SeF_6 c TeF_4 d. IF_4^-

10.31 Phosphorus pentachloride is normally a white solid. It exists in this state as the ionic compound $[PCl_4^+][PCl_6^-]$. Describe the electron structure of the PCl_6^- ion in terms of valence bond theory.

10.32 Iodine, I_2, dissolves in an aqueous solution of iodide ion, I^-, to give the triiodide ion, I_3^-. This ion consists of one iodine atom bonded to two others. Describe the bonding of I_3^- in terms of valence bond theory.

10.33 The hyponitrite ion, $^-O—N{=}N—O^-$, exists in solid compounds as the *trans* isomer. Using valence bond theory, explain why *cis–trans* isomers might be expected for this ion. Draw structural formulas of the *cis–trans* isomers.

10.34 Fumaric acid, $C_4H_4O_4$, occurs in the metabolism of glucose in the cells of plants and animals. It is used commercially in beverages. The structural formula of fumaric acid is

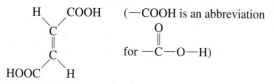

(—COOH is an abbreviation

for —$\overset{\overset{\textstyle O}{\|}}{C}$—O—H)

Maleic acid is the *cis* isomer of fumaric acid. Using valence bond theory, explain why these isomers are possible.

General Problems

10.37 Predict the molecular geometry of the following:
a. $SnCl_2$ b. COF_2 c. ICl_2^- d. PCl_6^-

10.38 Predict the molecular geometry of the following:
a. HOF   b. NH_4^+ c. PF_5 d. ClF_4^+

10.39 Describe the hybrid orbitals used by each carbon atom in the following molecules:

a.
$$\underset{H}{\overset{H}{}}C=\underset{\underset{H}{|}}{\overset{\overset{H}{|}}{C}}-\underset{\underset{H}{|}}{\overset{\overset{H}{|}}{C}}-OH \qquad b.\ N\equiv C-C\equiv N$$

10.40 Describe the hybrid orbitals used by each nitrogen atom in the following molecules:

a. H—$\underset{\underset{H}{|}}{\overset{\overset{H}{|}}{N}}$—$\underset{\underset{H}{|}}{\overset{\overset{H}{|}}{N}}$—H   b. $N\equiv C-C\equiv N$

10.41 Explain how the dipole moment could be used to distinguish between the *cis* and *trans* isomers of 1,2-dibromoethene:

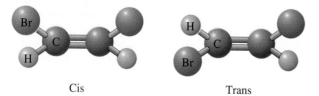

Cis   Trans

Molecular Orbital Theory

10.35 Describe the electronic structure of each of the following, using molecular orbital theory. Calculate the bond order of each and decide whether it should be stable. For each, state whether the substance is diamagnetic or paramagnetic.
a. B_2 b. B_2^+ c. O_2^-

10.36 Use molecular orbital theory to describe the bonding in the following. For each one, find the bond order and decide whether it is stable. Is the substance diamagnetic or paramagnetic?
a. C_2^+ b. Ne_2 c. C_2^-

10.42 Two compounds have the formula $Pt(NH_3)_2Cl_2$. (Compound B is *cisplatin*, mentioned in the opening to Chapter 1.) They have square planar structures. One is expected to have a dipole moment; the other is not. Which one would have a dipole moment?

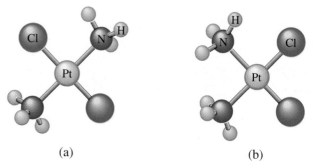

(a)   (b)

10.43 What is the molecular orbital configuration of HeH^+? Do you expect the ion to be stable?

10.44 What is the molecular orbital configuration of He_2^+? Do you expect the ion to be stable?

10.45 Calcium carbide, CaC_2, consists of Ca^{2+} and C_2^{2-} (acetylide) ions. Write the molecular orbital configuration and bond order of the acetylide ion, C_2^{2-}.

10.46 Sodium peroxide, Na_2O_2, consists of Na^+ and O_2^{2-} (peroxide) ions. Write the molecular orbital configuration and bond order of the peroxide ion, O_2^{2-}.

10.47 The oxygen–oxygen bond in O_2^+ is 112 pm and in O_2 is 121 pm. Explain why the bond length in O_2^+ is shorter than in O_2. Would you expect the bond length in O_2^- to be longer or shorter than that in O_2? Why?

10.48 The nitrogen–nitrogen bond distance in N_2 is 109 pm. On the basis of bond orders, would you expect the bond distance in N_2^+ to be less than or greater than 109 pm? Answer the same question for N_2^-.

Cumulative-Skills Problems

10.49 A molecular compound is composed of 60.4% Xe, 22.1% O, and 17.5% F, by mass. If the molecular weight is 217.3 amu, what is the molecular formula? What is the Lewis formula? Predict the molecular geometry using the VSEPR model. Describe the bonding, using valence bond theory.

10.50 A molecular compound is composed of 58.8% Xe, 7.2% O, and 34.0% F, by mass. If the molecular weight is 223 amu, what is the molecular formula? What is the Lewis formula? Predict the molecular geometry using the VSEPR model. Describe the bonding, using valence bond theory.

10.51 Each of the following compounds has a nitrogen–nitrogen bond: N_2, N_2H_4, N_2F_2. Match each compound with one of the following bond lengths: 110 pm, 122 pm, 145 pm. Describe the geometry about one of the N atoms in each compound. What hybrid orbitals are needed to describe the bonding in valence bond theory?

10.52 The bond length in C_2 is 131 pm. Compare this with the bond lengths in C_2H_2 (120 pm), C_2H_4 (134 pm), and C_2H_6 (153 pm). What bond order would you predict for C_2 from its bond length? Does this agree with the molecular orbital configuration you would predict for C_2?

10.53 Draw resonance formulas of the nitric acid molecule, HNO_3. What is the geometry about the N atom? What is the hybridization on N? Use bond energies and one Lewis formula for HNO_3 to estimate ΔH_f° for $HNO_3(g)$. The actual value of ΔH_f° for $HNO_3(g)$ is -135 kJ/mol, which is lower than the estimated value because of stabilization of HNO_3 by resonance. The *resonance energy* is defined as ΔH_f°(estimated) $- \Delta H_f^\circ$ (actual). What is the resonance energy of HNO_3?

10.54 One resonance formula of benzene, C_6H_6, is

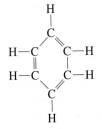

What is the other resonance formula? What is the geometry about a carbon atom? What hybridization would be used in valence bond theory to describe the bonding? The ΔH_f° for $C_6H_6(g)$ is -83 kJ/mol; ΔH_f° for $C(g)$ is 715 kJ/mol. Obtain the resonance energy of benzene. (See Problem 10.53.)

Changes of State

In the chapter opener, we discussed the change of solid carbon dioxide directly to a gas. Such *a change of a substance from one state to another* is called a **change of state** or **phase transition.** A *phase* is a homogeneous portion of a system.

11.2 Phase Transitions

In general, each of the three states of a substance can change into either of the other states as depicted in the diagram below.

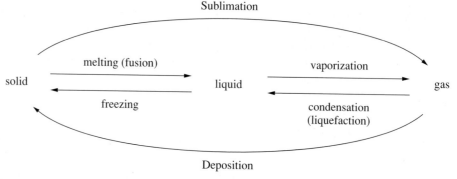

The sublimation of iodine is shown in Figure 11.3.

Figure 11.3
Sublimation of iodine.
Top: The beaker contains iodine crystals, I_2; a dish of ice rests on top of the beaker. *Bottom:* Iodine has an appreciable vapor pressure even below its melting point (114°C); thus, when heated carefully, the solid sublimes without melting. The vapor deposits as crystals on the cool underside of the dish.

Vapor Pressure

Liquids, and even some solids, are continuously vaporizing. If a liquid is in a closed vessel with space above it, a partial pressure of the vapor state builds up in this space. The **vapor pressure** of a liquid is *the partial pressure of the vapor over the liquid, measured at equilibrium at a given temperature.* To understand what we mean by equilibrium, let us look at a simple method for measuring vapor pressure.

You introduce a few drops of water from a medicine dropper into the mercury column of a barometer (Figure 11.4). Being less dense than mercury, the water rises in the tube to the top of the mercury, where it vaporizes.

To understand the process of vaporization, it is necessary to realize that the molecules in a liquid have a distribution of kinetic energies (Figure 11.5). Those molecules moving away from the surface and toward the vapor phase will escape only if their kinetic energies are greater than a certain minimum value equal to the potential energy from the attraction of molecules in the body of the liquid.

As molecules at the surface gain sufficient kinetic energy (through collisions with neighboring molecules), they escape the liquid and go into the space at the top of the barometer column. More and more water molecules begin to fill this space. As vaporization of water proceeds, the mercury column moves downward. As the number of molecules in the vapor state increases, more and more gaseous molecules join the liquid water.

The rate of condensation steadily increases as the number of molecules in the volume of vapor increases, until the rate at which molecules are condensing on the liquid equals the rate at which molecules are vaporizing.

Table 11.1
Melting Points and Boiling Points (at 1 atm) of Several Substances

Name	Type of Solid*	Melting Point, °C	Boiling Point, °C
Neon, Ne	Molecular	−249	−246
Hydrogen sulfide, H_2S	Molecular	−86	−61
Chloroform, $CHCl_3$	Molecular	−64	62
Water, H_2O	Molecular	0	100
Acetic acid, $HC_2H_3O_2$	Molecular	17	118
Mercury, Hg	Metallic	−39	357
Sodium, Na	Metallic	98	883
Tungsten, W	Metallic	3410	5660
Cesium chloride, CsCl	Ionic	645	1290
Sodium chloride, NaCl	Ionic	801	1413
Magnesium oxide, MgO	Ionic	2800	3600
Quartz, SiO_2	Covalent network	1610	2230
Diamond, C	Covalent network	3550	4827

*Types of solids are discussed in Section 11.5.

The heat needed for the vaporization of a liquid is called the **heat of vaporization** and is denoted ΔH_{vap}. At 100°C, the heat of vaporization of water is 40.7 kJ per mole. ◄

▶ The fact that heat is required for vaporization can be demonstrated by evaporating a dish of water inside a vessel attached to a vacuum pump. With the pressure kept low, water evaporates quickly enough to freeze the water remaining in the dish.

$$H_2O(l) \longrightarrow H_2O(g); \; \Delta H_{vap} = 40.7 \text{ kJ/mol}$$

Note that much more heat is required for vaporization than for melting. Melting needs only enough energy for the molecules to escape from their sites in the solid. For vaporization, enough energy must be supplied to break most of the intermolecular attractions.

Figure 11.8
Heating curve for water.
Heat is being added at a constant rate to a system containing water. Note the flat regions of the curve. When heat is added during a phase transition the temperature does not change.

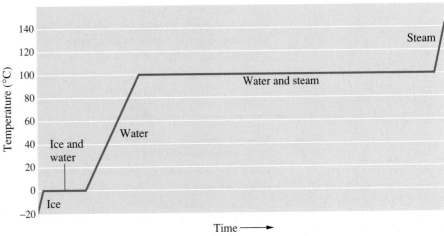

CCl_2F_2

EXAMPLE 11.1

Calculating the Heat Required for a Phase Change of a Given Mass of Substance

A particular refrigerator cools by evaporating liquefied dichlorodifluoromethane, CCl_2F_2. How many kilograms of this liquid must be evaporated to freeze a tray of water at 0°C to ice at 0°C? The mass of the water is 525 g, the heat of fusion of ice is 6.01 kJ/mol, and the heat of vaporization of dichlorodifluoromethane is 17.4 kJ/mol.

SOLUTION

The heat that must be removed to freeze 525 g of water at 0°C is

$$525 \text{ g } H_2O \times \frac{1 \text{ mol } H_2O}{18.0 \text{ g } H_2O} \times \frac{-6.01 \text{ kJ}}{1 \text{ mol } H_2O} = -175 \text{ kJ}$$

The minus sign indicates that heat energy is taken away from the water. Consequently, the vaporization of dichlorodifluoromethane absorbs 175 kJ of heat. The mass of CCl_2F_2 that must be vaporized to absorb this quantity of heat is

$$175 \text{ kJ} \times \frac{1 \text{ mol } CCl_2F_2}{17.4 \text{ kJ}} \times \frac{121 \text{ g } CCl_2F_2}{1 \text{ mol } CCl_2F_2} = 1.22 \times 10^3 \text{ g } CCl_2F_2$$

1.22 kg of dichlorodifluoromethane must be evaporated.

See Problems 11.25 and 11.26.

Clausius–Clapeyron Equation: Relating Vapor Pressure and Liquid Temperature

We noted earlier that the vapor pressure of a substance depends on temperature. The variation of vapor pressure with temperature of some liquids was given in Figure 11.6. It can be shown that the logarithm of the vapor pressure of a liquid or solid varies with the absolute temperature according to the following approximate relation:

$$\ln P = -\frac{A}{T} + B$$

Here $\ln P$ is the natural logarithm of the vapor pressure, and A and B are positive constants. You can confirm this relation for the liquids shown in Figure 11.6 by replotting the data. If you put $y = \ln P$ and $x = 1/T$, the previous relation becomes

$$y = -Ax + B$$

This means that if you plot $\ln P$ versus $1/T$, you should get a straight line with slope $-A$. The data of Figure 11.6 are replotted this way in Figure 11.9. ◄

► **See Appendix A for the graphing of a straight line.**

The previous equation has been derived from thermodynamics by assuming the vapor behaves like an ideal gas. The result, known as the *Clausius–Clapeyron equation,* shows that the constant A is proportional to the heat of vaporization of the liquid, ΔH_{vap}.

$$\ln P = \frac{-\Delta H_{vap}}{RT} + B$$

Using this equation and taking two different temperatures, we can generate two equations. Taking the difference between these two equations will give us the two-point form of the Clausius–Clapeyron equation.

$$\ln \frac{P_2}{P_1} = \frac{\Delta H_{vap}}{R} \left(\frac{1}{T_1} - \frac{1}{T_2} \right)$$

The next example illustrates the use of this two-point form of the Clausius–Clapeyron equation.

EXAMPLE 11.2

Calculating the Vapor Pressure at a Given Temperature

Estimate the vapor pressure of water at 85°C. Note that the normal boiling point of water is 100°C and that its heat of vaporization is 40.7 kJ/mol.

SOLUTION

The two-point form of the Clausius–Clapeyron equation relates five quantities: P_1, T_1, P_2, T_2, and ΔH_{vap}. The normal boiling point is the temperature at which the vapor pressure equals 760 mmHg. Thus, you can let P_1 equal 760 mmHg and T_1 equal 373 K (100°C). Then P_2 would be the vapor pressure at T_2, or 358 K (85°C). The heat of vaporization, ΔH_{vap}, is 40.7×10^3 J/mol. Note that we have expressed this quantity in joules per mole to agree with the units of the molar gas constant, R, which equals 8.31 J/(K · mol). You can now substitute into the two-point equation (the equation shaded in color just before this example).

$$\ln \frac{P_2}{760 \text{ mmHg}} = \frac{40.7 \times 10^3 \text{ J/mol}}{8.31 \text{ J/(K} \cdot \text{mol)}} \left(\frac{1}{373 \text{ K}} - \frac{1}{358 \text{ K}} \right)$$

$$= (4898 \text{ K}) \times (-1.123 \times 10^{-4}/\text{K}) = -0.550$$

Solve this equation for P_2, which is the vapor pressure of water at 85°C. To do so, you take the antilogarithm of both sides of the equation. This gives

$$\frac{P_2}{760 \text{ mmHg}} = \text{antilog}(-0.550)$$

The antilog(-0.550) is $e^{-0.550} = 0.577$. (Appendix A describes how to obtain the antilogarithm.) Thus,

$$\frac{P_2}{760 \text{ mmHg}} = 0.577$$

Therefore,

$$P_2 = 0.577 \times 760 \text{ mmHg} = \textbf{439 mmHg}$$

This value may be compared with the experimental value of 434 mmHg given in Table 5.4. Differences from the experimental value stem from the fact that the Clausius–Clapeyron equation is an approximate relation.

See Problems 11.29, 11.30, 11.31, and 11.32.

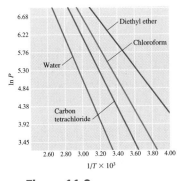

Figure 11.9
Plot of the logarithm of vapor pressure versus 1/T. The liquids shown in Figure 11.6 are replotted in this graph. Note the straight-line relationship.

The Clausius–Clapeyron equation can also be used to solve for ΔH_{vap}, given values for the pressures and temperatures.

11.3 Phase Diagrams

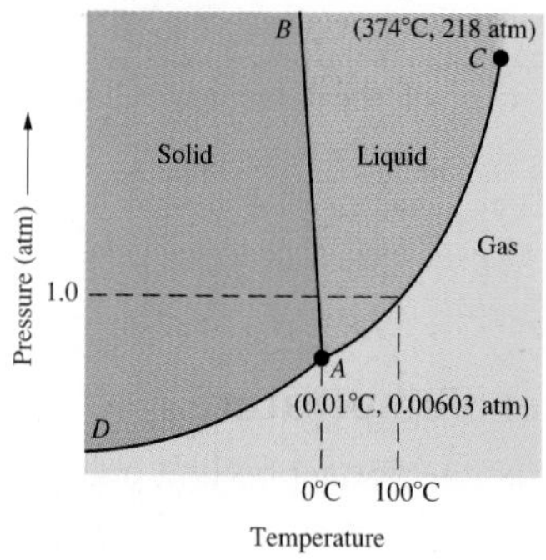

A **phase diagram** is *a graphical way to summarize the conditions under which the different states of a substance are stable.*

Melting-Point Curve

Figure 11.10
Phase diagram for water (not to scale). The curves *AB*, *AC*, and *AD* divide the diagram into regions that give combinations of temperature and pressure for which only one state is stable. Along any curve, the two states from the adjoining regions are in equilibrium.

Figure 11.10 is a phase diagram for water. It consists of three curves that divide the diagram into regions labeled "solid," "liquid," and "gas." In each region, the indicated state is stable. Every point on each of the curves indicates experimentally determined temperatures and pressures at which two states are in equilibrium. Thus, the curve *AB*, dividing the solid region from the liquid region, represents the conditions under which the solid and liquid are in equilibrium. This curve gives the melting points of the solid at various pressures.

Usually, the melting point is only slightly affected by pressure. For this reason, the melting-point curve *AB* is nearly vertical. If a liquid is more dense than the solid, as is the case for water, the melting point decreases with pressure. The melting-point curve in such cases leans slightly to the left. In the case of ice, the decrease is indeed slight—only 1°C for a pressure increase of 133 atm. For most substances, the liquid state is less dense than the solid. In that case, the melting-point curve leans slightly to the right.

Vapor-Pressure Curves for the Liquid and the Solid

The curve *AC* that divides the liquid region from the gaseous region gives the vapor pressures of the liquid at various temperatures. It also gives the boiling points of the liquid for various pressures. The boiling point of water at 1 atm is shown on the phase diagram (Figure 11.10).

The curve *AD* that divides the solid region from the gaseous region gives the vapor pressures of the solid at various temperatures. This curve intersects the other curves at point *A*, the **triple point,** which is *the point on a phase diagram representing the temperature and pressure at which three phases of a substance coexist in equilibrium.* ◄

▶ Because the triple point for water occurs at a definite temperature, it is used to define the Kelvin temperature scale. The temperature of water at its triple point is defined to be 273.16 K (0.01°C).

Suppose a solid is warmed at a pressure below the pressure at the triple point. In a phase diagram, this corresponds to moving along a horizontal line below the triple point. You can see from Figure 11.10 that such a line will intersect curve *AD*, the vapor-pressure curve for the solid. Thus, the solid will pass directly into the gas; that is, the solid will sublime. Freeze-drying of a food (or brewed coffee) is accomplished by placing the frozen food in a vacuum (below 0.00603 atm) so that the ice in it sublimes. Because the food can be dried at a lower temperature than if heat-dried, it retains more flavor and can often be reconstituted by simply adding water.

The triple point of carbon dioxide is at −57°C and 5.1 atm (Figure 11.11). Therefore, the solid sublimes if warmed at any pressure below 5.1 atm. This is why solid carbon dioxide sublimes at normal atmospheric pressure (1 atm). Above 5.1 atm, however, the solid melts if warmed.

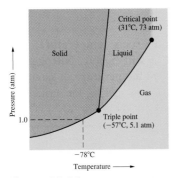

Figure 11.11
Phase diagram for carbon dioxide (not to scale). At normal atmospheric pressure (1 atm), the solid sublimes when warmed.

Critical Temperature and Pressure

Imagine an experiment in which liquid and gaseous carbon dioxide are sealed in a thick-walled glass vessel at 20°C. At this temperature, the liquid is in equilibrium with its vapor at a pressure of 57 atm. You observe that the liquid and vapor are separated by a well-defined boundary, or meniscus. Now suppose the temperature is raised. The vapor pressure increases, and at 30°C it is 71 atm. Then, as the temperature approaches 31°C, a curious thing happens. The meniscus becomes fuzzy and less well defined. At 31°C, the meniscus disappears altogether. Above this temperature there is only one fluid state, called a *supercritical fluid* (see Figure 11.12).

The temperature above which the liquid state of a substance no longer exists regardless of the pressure is called the **critical temperature.** *The vapor pressure at the critical temperature* is called the **critical pressure.** It is the minimum pressure that must be applied to a gas at the critical temperature to liquefy it.

On a phase diagram, the preceding experiment corresponds to following the vapor-pressure curve where the liquid and vapor are in equilibrium. Note that this curve in Figure 11.11 ends at a point at which the temperature and pressure have their critical values. This is the *critical point.* If you look at the phase diagram for water, you will see that the vapor-pressure curve for the liquid similarly ends, at point *C*, which is the critical point for water. In this case, the critical temperature is 374°C and the critical pressure is 218 atm.

Many important gases cannot be liquefied at room temperature. Nitrogen, for example, has a critical temperature of -147°C. This means the gas cannot be liquefied until the temperature is below -147°C.

EXAMPLE 11.3

Relating the Conditions for the Liquefaction of Gases to the Critical Temperature

The critical temperatures of ammonia and nitrogen are 132°C and -147°C, respectively. Explain why ammonia can be liquefied at room temperature by merely compressing the gas to a high enough pressure, whereas the compression of nitrogen requires a low temperature as well.

Figure 11.12
Observing the critical phenomenon. *(A)* Carbon dioxide liquid in equilibrium with its vapor at 20°C. *(B)* The same two states in equilibrium at 30°C (just below the critical point); the densities of the liquid and vapor are becoming equal. *(C)* Carbon dioxide at 31°C (the critical temperature); the liquid and vapor now have the same densities—in fact, the distinction between liquid and vapor has disappeared, resulting in what is called a supercritical fluid.

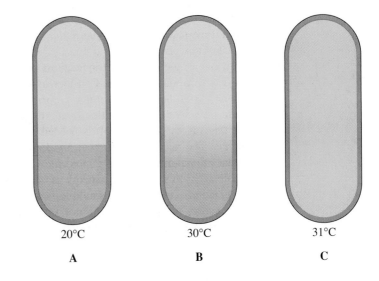

SOLUTION

The critical temperature of ammonia is well above room temperature. If ammonia gas at room temperature is compressed, it will liquefy. **Nitrogen, however, has a critical temperature well below room temperature.** It cannot be liquefied by compression unless the temperature is below the critical temperature.

See Problems 11.35 and 11.36.

 CONCEPT CHECK 11.1

When camping at high altitude, you need to pay particular attention to changes in cooking times for foods that are boiled in water. If you like eggs that are boiled for 10 minutes near sea level, would you have to cook them for a longer or a shorter time at 3200 m to get the egg you like? Be sure to explain your answer.

Liquid State

The experimental values of boiling points, vapor pressures, critical temperatures, and heats of vaporization can be explained in terms of intermolecular forces. You have seen that molecules tend to escape the liquid state and form a vapor. The vapor pressure is the equilibrium partial pressure of this vapor over the liquid; it increases with temperature. The boiling point is the temperature at which the vapor pressure equals the pressure applied to the liquid. Both vapor pressure and boiling point are important properties of a liquid. Values of the vapor pressure for some liquids at 20°C are listed in Table 11.2. As will be noted in the next section, intermolecular forces are related to molecular structure.

11.4 Intermolecular Forces; Explaining Liquid Properties

Many of the physical properties of liquids can be explained in terms of **intermolecular forces,** *the forces of interaction between molecules.* These forces are normally weakly attractive.

One of the most direct indications of the attraction between molecules is the heat of vaporization of liquids. Consider a substance like neon, which consists of single atoms. Neon liquefies when the temperature is lowered to −246°C at 1.00 atm pressure. The heat of vaporization of the liquid at this temperature is 1.77 kJ/mol. Some of this energy (0.23 kJ/mol) is needed to push back the atmosphere when the vapor forms. The remaining energy (1.54 kJ/mol) must be supplied to overcome intermolecular attractions. Because a molecule in a liquid is surrounded by several neighbor molecules, this remaining energy is some multiple of a single molecule–molecule interaction. Typically, this multiple is about 5, so we expect the neon–neon interaction energy to be about 0.3 kJ/mol. Other experiments yield a similar value. By comparison, the energy of attraction between two hydrogen atoms in the hydrogen molecule is 432 kJ/mol. Thus, the energy of attraction between neon atoms is about a thousand times smaller than that between atoms in a chemical bond.

Table 11.2
Properties of Some Liquids at 20°C

Substance	Molecular Weight (amu)	Vapor Pressure (mmHg)
Water, H_2O	18	1.8×10^1
Carbon dioxide, CO_2	44	4.3×10^4
Pentane, C_5H_{12}	72	4.4×10^2
Glycerol, $C_3H_8O_3$	92	1.6×10^{-4}
Chloroform, $CHCl_3$	119	1.7×10^2
Carbon tetrachloride, CCl_4	154	8.7×10^1
Bromoform, $CHBr_3$	253	3.9×10^0

Attractive intermolecular forces can be larger than those in neon. For example, chlorine, Cl_2, and bromine, Br_2, have intermolecular attractive energies of 3.0 kJ/mol and 4.3 kJ/mol, respectively.

Three types of attractive forces are known to exist between neutral molecules: dipole–dipole forces, London (or dispersion) forces, and hydrogen bonding forces. The term **van der Waals forces** is *a general term for those intermolecular forces that include dipole–dipole and London forces.* ◀ Van der Waals forces are the weak attractive forces in a large number of substances, including Ne, Cl_2, and Br_2, which we just discussed. Hydrogen bonding occurs in substances containing hydrogen atoms bonded to certain very electronegative atoms. Approximate energies of intermolecular attractions are compared with those of chemical bonds in Table 11.3.

▶ The Dutch physicist Johannes van der Waals (1837–1923) was the first to suggest the importance of intermolecular forces and used the concept to derive his equation for gases. (See van der Waals equation, Section 5.8.)

Dipole–Dipole Forces

Polar molecules can attract one another through dipole–dipole forces. The **dipole–dipole force** is *an attractive intermolecular force resulting from the tendency of polar molecules to align themselves such that the positive end of one molecule is near the negative end of another.* A polar molecule has a dipole moment as a result of the electronic structure of the molecule. ◀ For example, hydrogen chloride, HCl, is a polar

▶ Dipole moments were discussed in Section 10.2.

Table 11.3
Types of Intermolecular and Chemical Bonding Interactions

Type of Interaction	Approximate Energy (kJ/mol)
Intermolecular	
Van der Waals (dipole–dipole, London)	0.1 to 10
Hydrogen bonding	10 to 40
Chemical bonding	
Ionic	100 to 1000
Covalent	100 to 1000

Removing Caffeine from Coffee

EVERYDAY LIFE

The stimulant in coffee is caffeine, a bitter-tasting white substance with the formula $C_8H_{10}N_4O_2$. (Figure 11.13 shows the structural formula of caffeine.)

For those who like the taste of roasted coffee but don't want the caffeine, decaffeinated coffee is available. A German chemist, Ludwig Roselius, first made "decaf" coffee about 1900 by extracting the caffeine from green coffee beans with the solvent chloroform, $CHCl_3$. Today, though, most of the commercial decaffeinated coffee produced (Figure 11.14) uses *supercritical* carbon dioxide as the extracting fluid.

In a tank of carbon dioxide under pressure (for example, in a CO_2 fire extinguisher), the substance normally exists as the liquid in equilibrium with its gas phase. But we know from the previous text discussion that above 31°C, the two phases, gas and liquid, are replaced by a single fluid phase. So, on a hot summer day, the carbon dioxide in such a tank is above its critical temperature and pressure and exists as the supercritical fluid.

Supercritical carbon dioxide is a near-ideal solvent. Under normal conditions, carbon dioxide is not a very

Figure 11.14
Decaf coffee. Many brands are available.

good solvent for organic substances, but supercritical carbon dioxide readily dissolves many of these substances, including caffeine. It is nontoxic and nonflammable. Carbon dioxide does contribute to the greenhouse effect, but once used the gas can be recirculated for solvent use and not vented to the atmosphere.

Supercritical fluids have gained much attention recently because of the possibility of replacing toxic and environmentally less desirable solvents. For example, at the moment, the usual solvent used to dry-clean clothes is perchloroethylene, CCl_2CCl_2. Although nonflammable and less toxic than carbon tetrachloride, which was the solvent previously used, perchloroethylene is regulated as an air pollutant under the Clean Air Act. Some scientists have shown that you can dry-clean with supercritical carbon dioxide if you use a special detergent.

Substances other than carbon dioxide have also shown intriguing solvent properties. For example, whereas water under normal conditions dissolves ionic and polar substances, above its critical point (374°C, 217 atm) it becomes an excellent solvent for nonpolar substances. Supercritical water and carbon dioxide promise to replace many toxic or environmentally unfriendly organic solvents.

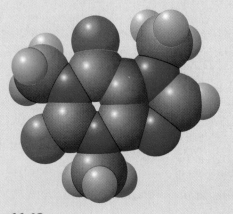

Figure 11.13
The caffeine molecule. Ball-and-stick model of caffeine, $C_8H_{10}N_4O_2$.

molecule because of the difference in electronegativities of the H and Cl atoms. Figure 11.15 shows the alignment of the polar HCl molecules. Note that this alignment creates a net attraction between molecules. This attractive force is partly responsible for the fact that hydrogen chloride becomes a liquid when cooled to −85°C. At this temperature, HCl molecules have a low enough kinetic energy for the intermolecular forces to hold the molecules in the liquid state.

London (Dispersion) Forces

In nonpolar molecules, there can be no dipole–dipole force. Yet such substances liquefy, so there must be another type of attractive intermolecular force. Even in hydrogen chloride, calculations show that the dipole–dipole force accounts for less than 20% of the total energy of attraction. What is the nature of the additional attractive force?

In 1930 Fritz London found he could account for a weak attraction between any two molecules. Let us use neon to illustrate his argument. The electrons of this atom move about the nucleus so that over a period of time they have a spherical distribution. Experimentally, you find that the atom has no dipole moment. However, at any instant, more electrons may be on one side of the nucleus than on the other. The atom becomes a small, instantaneous dipole, with one side having a partial negative charge, δ^-, and the other side having a partial positive charge, δ^+ (see Figure 11.16A).

Now imagine that another neon atom is close to this first atom—say, next to the partial negative charge that appeared at that instant. This partial negative charge repels the electrons in the second atom, so at this instant it too becomes a small dipole (see Figure 11.16B). Note that these two dipoles are oriented with the partial negative charge of one atom next to the partial positive charge of the other atom. Therefore, an attractive force exists between the two atoms.

Electrons are in constant motion, but the motion of electrons on one atom affects the motion of electrons on the other atom. As a result, the *instantaneous dipoles* of the atoms tend to change together, maintaining a net attractive force (compare *B* and *C* of Figure 11.16). Such instantaneous changes of electron distributions occur in all molecules. For this reason, an attractive force always exists between any two molecules.

London forces (also called **dispersion forces**) are *the weak attractive forces between molecules resulting from the small, instantaneous dipoles that occur because of the varying positions of the electrons during their motion about nuclei.*

London forces tend to increase with molecular weight because the forces increase in strength with the number of electrons. Also, larger molecular weight often means larger atoms, which are more *polarizable* (more easily distorted to give instantaneous dipoles because the electrons are farther from the nuclei). The relationship of London

Figure 11.15
Alignment of polar molecules of HCl.
Molecules tend to line up in the solid so that positive ends point to negative ends.

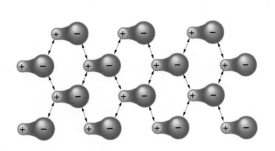

Solid

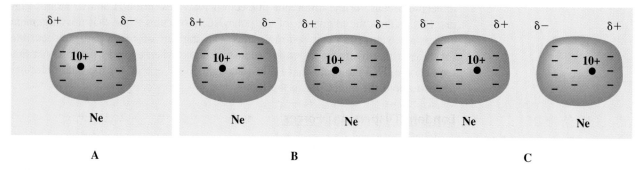

Figure 11.16
Origin of the London force. *(A)* At some instant, there are more electrons on one side of a neon atom than on the other. *(B)* If this atom is near another neon atom, the electrons on that atom are repelled. The result is two instantaneous dipoles, which give an attractive force.

forces to molecular weight is only approximate, however. For molecules of about the same molecular weight, the more compact one is probably less polarizable, so the London forces are smaller. Consider the three isomers of pentane, C_5H_{12}:

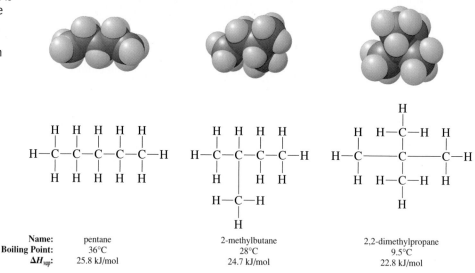

Name:	pentane	2-methylbutane	2,2-dimethylpropane
Boiling Point:	36°C	28°C	9.5°C
ΔH_{vap}:	25.8 kJ/mol	24.7 kJ/mol	22.8 kJ/mol

These isomers differ in the arrangement of the atoms. Pentane is a long chain of carbon atoms. The other two molecules, however, have increasingly more compact arrangements of atoms. As a result, you expect London forces to decrease from pentane to 2-methylbutane and from 2-methylbutane to 2,2-dimethylpropane. This is reflected in the physical properties listed above.

Van der Waals Forces and the Properties of Liquids

The vapor pressure of a liquid depends on intermolecular forces, because the ease or difficulty with which a molecule leaves the liquid depends on the strength of its attraction to other molecules. When the intermolecular forces in the liquid are strong, you expect the vapor pressure to be low.

Let us look again at Table 11.2, which lists vapor pressures at 20°C. The intermolecular attractions in these liquids (except for water and glycerol) are due entirely to van der Waals forces. Of the van der Waals forces, the London force is always present and usually dominant. The dipole–dipole force is usually appreciable only in small polar molecules or in large molecules having very large dipole moments.

The liquids in Table 11.2 are listed in order of increasing molecular weight. Therefore, London forces would be expected to increase for these liquids from the top to the bottom of the table. If London forces are the dominant attractive forces in these liquids, the vapor pressures should decrease from top to bottom in the table. And this is what you see, except for water and glycerol. (Their vapor pressures are relatively low, and an additional force, hydrogen bonding, is needed to explain them.)

The normal boiling point of a liquid must depend on intermolecular forces, because it is related to the vapor pressure. Thus, it is lowest for liquids with the weakest intermolecular forces.

Hydrogen Bonding

It is interesting to compare fluoromethane, CH_3F, and methanol, CH_3OH (Figure 11.17). They have about the same molecular weight (34 for CH_3F and 32 for CH_3OH) and about the same dipole moment (1.81 D for CH_3F and 1.70 D for CH_3OH). You might expect these substances to have about the same intermolecular attractive forces and therefore about the same boiling points. In fact, the boiling points are quite different. Fluoromethane boils at $-78°C$ and is a gas under normal conditions. Methanol boils at $65°C$ and is normally a liquid. Why?

We have already seen that the properties of water and glycerol cannot be explained in terms of van der Waals forces alone. What water, glycerol, and methanol have in common is one or more —OH groups.

CH_3F

CH_3OH

Figure 11.17
Fluoromethane and methanol. Space-filling molecular models.

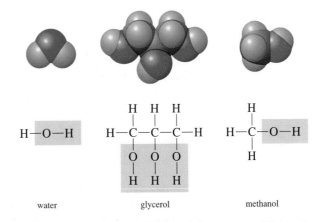

Molecules that have the —OH group are subject to an additional attractive force called hydrogen bonding. **Hydrogen bonding** is *a weak to moderate attractive force that exists between a hydrogen atom covalently bonded to a very electronegative atom, X, and a lone pair of electrons on another small, electronegative atom, Y.* Hydrogen bonding is represented in structural drawings by a series of dots.

$$—X—H\cdot\cdot Y—$$

Usually, hydrogen bonding is seen in cases where X and Y are the atoms F, O, or N.

Figure 11.18A shows a plot of boiling point versus molecular weight for the hydrides (binary compounds with hydrogen) of the Group VIA elements. If London forces were the only intermolecular forces present in this series of compounds, the boiling points should increase regularly with molecular weight. You do see such a regular increase in boiling point for H_2S, H_2Se, and H_2Te. On the other hand, H_2O has a much higher boiling point than you would expect if only London forces were

Figure 11.18
Boiling point versus molecular weight for hydrides. (A) Plot for the Group VIA hydrides. Note that the boiling points of H₂S, H₂Se, and H₂Te increase fairly smoothly, as you would expect if London forces are dominant. H₂O has a much higher boiling point, indicating an additional intermolecular attraction. (B) Plot for the hydrides of Groups VIIA, VA, and IVA.

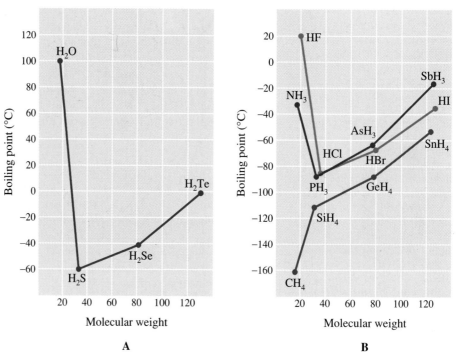

A

B

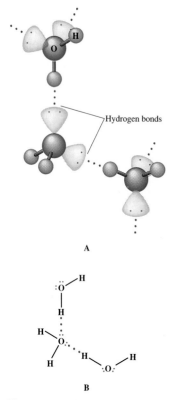

A

B

Figure 11.19
Hydrogen bonding in water. (A) The electrons in the O—H bonds of H₂O molecules are attracted to the oxygen atoms, leaving the positively charged protons partially exposed. A proton on one water molecule is attracted to a lone pair on an oxygen atom in another water molecule. (B) Lewis structures of water with hydrogen bonding between molecules.

present. This is consistent with the view that hydrogen bonding exists in H₂O but is essentially absent in H₂S, H₂Se, and H₂Te.

The hydrogen atom of one water molecule is attracted to the electron pair of another water molecule (Figure 11.19). Many H₂O molecules can be linked this way to form clusters of molecules. Similar hydrogen bonding is seen in other molecules containing —O—H groups.

Figure 11.18B shows plots of boiling point versus molecular weight for Group VIIA, VA, and IVA hydrides. Hydrogen fluoride, HF, and ammonia, NH₃, have particularly high boiling points compared with other hydrides of their periodic group.

The hydrogen bonding attraction between two water molecules may be explained in part on the basis of the dipole moment of the —O—H bond. The partial positive charge on the hydrogen atom of one water molecule is attracted to the partial negative charge of a lone pair on another water molecule. In addition, however, a hydrogen atom covalently bonded to an electronegative atom appears to be special. The electrons in the —O—H bond are drawn to the O atom, leaving the concentrated positive charge of the proton partially exposed. This concentrated positive charge is strongly attracted to a lone pair of electrons on another O atom. Figure 11.20 shows the hydrogen bonding between two biologically important molecules.

EXAMPLE 11.4

Identifying Intermolecular Forces

What kinds of intermolecular forces (London, dipole–dipole, hydrogen bonding) are expected in the following substances?

Figure 11.20
Hydrogen bonding between two biologically important molecules.
This computer-drawn figure shows the electron densities of two molecules (guanine and cytosine); the electron densities increase as the color changes from blue to green to yellow to red. Note the three H bonds between the two molecules. Each H bond is between an H atom with low electron density (exposing the positive charge on the atom) and the lone electron pair (high density) of either an O or N atom. (The molecules guanine and cytosine are two of the four bases used as codes in DNA.) (Electron density figure taken from G.D. Shusterman, A.J. Shusterman, *J. Chem. Ed.*, 74, 771, 1997.)

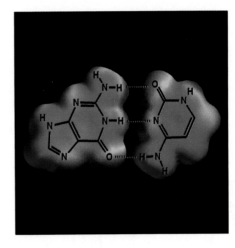

a. methane, CH_4
b. (see model at right)
c. butanol (butyl alcohol), $CH_3CH_2CH_2CH_2OH$

SOLUTION

a. Methane is a nonpolar molecule. Hence, the only intermolecular attractions are **London forces.** b. Trichloromethane is an unsymmetrical molecule with polar bonds. Thus, we expect **dipole–dipole forces,** in addition to **London forces.** c. Butanol has a hydrogen atom attached to an oxygen atom. Therefore, you expect **hydrogen bonding.** Because the molecule is polar (from the O—H bond), you also expect **dipole–dipole forces. London forces** exist too, because such forces exist between all molecules.

See Problems 11.37 and 11.38.

EXAMPLE 11.5

Determining Relative Vapor Pressure on the Basis of Intermolecular Attraction

For each of the following pairs, choose the substance you expect to have the lower vapor pressure at a given temperature: a. carbon dioxide (CO_2) or sulfur dioxide (SO_2); b. dimethyl ether (CH_3OCH_3) or ethanol (CH_3CH_2OH).

SOLUTION

a. London forces increase roughly with molecular weight. The molecular weights for SO_2 and CO_2 are 64 amu and 44 amu, respectively. Therefore, the London forces between SO_2 molecules should be greater than between CO_2 molecules. Moreover, because SO_2 is polar but CO_2 is not, there are dipole–dipole forces between SO_2 molecules but not between CO_2 molecules. **You conclude that sulfur dioxide has the lower vapor pressure.** Experimental values of the vapor pressures at 20°C are: CO_2, 56.3 atm; SO_2, 3.3 atm.

b. Dimethyl ether and ethanol have the same molecular formulas but different structural formulas. The structural formulas are

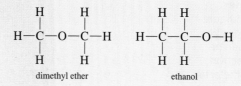

The molecular weights are equal, and therefore the London forces are approximately the same. However, there will be strong hydrogen bonding in ethanol but not in dimethyl ether. **You expect ethanol to have the lower vapor pressure.** Experimental values of vapor pressures at 20°C are: CH_3OCH_3, 4.88 atm; CH_3CH_2OH, 0.056 atm.

See Problems 11.39, 11.40, 11.41, and 11.42.

CONCEPT CHECK 11.2

A common misconception is that the following chemical reaction occurs when boiling water:

$$2H_2O(l) \longrightarrow 2H_2(g) + O_2(g)$$

instead of

$$H_2O(l) \longrightarrow H_2O(g)$$

a. What evidence do you have that the second reaction is correct?

b. How would the enthalpy of the wrong reaction compare with that of the correct reaction?

c. How could you calculate the enthalpy change for the wrong reaction (Chapter 6)?

Solid State

A solid is a nearly incompressible state of matter with a well-defined shape, because the units (atoms, molecules, or ions) making up the solid are in close contact and in fixed positions or sites. In the next section, we will look at the kinds of forces holding the units together in different types of solids.

11.5 Classification of Solids by Type of Attraction of Units

A solid consists of structural units—atoms, molecules, or ions—that are attracted to one another strongly enough to give a rigid substance. One way to classify solids is by the type of force holding the structural units together. In some cases, these forces are intermolecular and hold molecules together. In other cases, these forces are chemical bonds (metallic, ionic, or covalent) and hold atoms together. From this point of view, then, there are four different types of solids: molecular, metallic, ionic, and covalent. It should be noted, however, that a given ionic bond may have considerable covalent character or vice versa, so the distinction between ionic and covalent solids is not a precise one.

Types of Solids

A **molecular solid** is *a solid that consists of atoms or molecules held together by intermolecular forces*. Many solids are of this type. Examples are solid neon, solid water (ice), and solid carbon dioxide (dry ice).

A **metallic solid** is *a solid that consists of positive cores of atoms held together by a surrounding "sea" of electrons (metallic bonding)*. In this kind of bonding, positively charged atomic cores are surrounded by delocalized electrons. Examples include iron, copper, and silver.

An **ionic solid** is *a solid that consists of cations and anions held together by the electrical attraction of opposite charges (ionic bonds)*. Examples are cesium chloride, sodium chloride, and zinc sulfide (but ZnS has considerable covalent character).

A **covalent network solid** is *a solid that consists of atoms held together in large networks or chains by covalent bonds*. Diamond is an example of a three-dimensional network solid. Every carbon atom in diamond is covalently bonded to four others, so an entire crystal might be considered an immense molecule. It is possible to have similar two-dimensional "sheet" and one-dimensional "chain" molecules, with atoms held together by covalent bonds. Two examples are graphite (sheets) and asbestos (chains). Figure 11.21 shows the structures of diamond and graphite.

Table 11.4 summarizes these four types of solids.

Figure 11.21
Structures of diamond and graphite.   Diamond is a three-dimensional network solid; every carbon atom is covalently bonded to four others. Graphite consists of sheets of carbon atoms covalently bonded to form a hexagonal pattern of atoms; the sheets are held together by van der Waals forces.

EXAMPLE 11.6
Identifying Types of Solids

Which of the four basic types of solids would you expect the following substances to be? a. solid ammonia, NH_3; b. cesium, Cs; c. cesium iodide, CsI; d. silicon, Si.

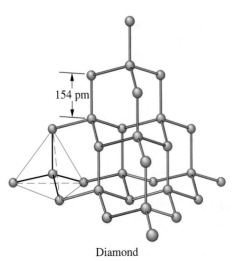

154 pm

Diamond

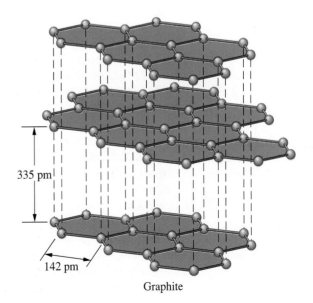

335 pm

142 pm

Graphite

Table 11.4
Types of Solids

Type of Solid	Structural Units	Attractive Forces Between Structural Units	Examples
Molecular	Atoms or molecules	Intermolecular forces	Ne, H_2O, CO_2
Metallic	Atoms (positive cores surrounded by electron "sea")	Metallic bonding (extreme delocalized bond)	Fe, Cu, Ag
Ionic	Ions	Ionic bonding	CsCl, NaCl, ZnS
Covalent network	Atoms	Covalent bonding	Diamond, graphite, asbestos

SOLUTION

a. Ammonia has a molecular structure; therefore, it freezes as a **molecular solid.**
b. Cesium is a metal; it is a **metallic solid.** c. Cesium iodide is an ionic substance; it exists as an **ionic solid.** d. Silicon atoms might be expected to form covalent bonds to other silicon atoms, as carbon does in diamond. A **covalent network solid** would result.

See Problems 11.45 and 11.46.

Physical Properties

Many physical properties of a solid can be directly related to its structure. Let us look at several of these properties.

Melting Point and Structure Table 11.1 lists the melting points for various solids and also gives the type of solid. For a solid to melt, the forces holding the structural units in their sites must be overcome, at least partially. In a molecular solid, these forces are weak intermolecular attractions. Thus, molecular solids tend to have low melting points (usually below 300°C). At room temperature, many molecular substances are either liquids (such as water and ethanol) or gases (such as carbon dioxide and ammonia). Note also that the melting points of molecular solids reflect the strengths of intermolecular attractions, just as boiling points do.

▶ Both molecular and covalent network solids have covalent bonds, but none of the covalent bonds are broken during the melting of a molecular solid.

For ionic solids and covalent network solids to melt, chemical bonds must be broken. For that reason, the melting points of these types of solids are relatively high. The melting point of the ionic solid sodium chloride is 801°C; that of magnesium oxide is 2800°C. Melting points of covalent network solids are generally quite high. Quartz, for example, melts at 1610°C; diamond, at 3550°C. See Table 11.1. ◀

▶ Lattice energy is the energy needed to separate a crystal into isolated ions in the gaseous state. It represents the strength of attraction of ions in the solid.

The difference between the melting points of sodium chloride and magnesium oxide can be explained in terms of the charges on the ions. You expect the melting point to rise with lattice energy. ◀ Lattice energies, however, depend on the product of the ionic charges (and inversely on the distance between charges). For sodium chloride (Na^+Cl^-) this ion-charge product is $(+1) \times (-1) = -1$, whereas for magnesium

Figure 11.22
Behavior of crystals when struck. Lead is malleable; rock salt cracks along crystal planes when struck.

oxide ($Mg^{2+}O^{2-}$) it is $(+2) \times (-2) = -4$. Thus, the lattice energy is greater for magnesium oxide, and this is reflected in its much higher melting point.

Metals often have high melting points, but there is considerable variability. Mercury, a liquid at room temperature, melts at $-39°C$. Tungsten melts at $3410°C$, the highest melting point of any metallic element. In general, melting points are low for the Group IA and IIA elements but increase as you move right in the periodic table to the transition metals. The elements in the center of the transition series (such as tungsten) have high melting points. Then, as you move farther to the right, the melting points decrease and are again low for the Group IIB elements.

EXAMPLE 11.7

Determining Relative Melting Points Based on Types of Solids

Arrange the following elements in order of increasing melting point: silicon, hydrogen, lithium. Explain your reasoning.

SOLUTION

Hydrogen is a molecular substance (H_2) with a very low molecular weight (2 amu). You expect it to have a very low melting point. Lithium is a Group IA element and is expected to be a relatively low-melting metal. Except for mercury, however, all metals are solids below 25°C. So the melting point of lithium is well above that of hydrogen. Silicon might be expected to have a covalent network structure like carbon and to have a high melting point. Therefore, the list of these elements in order of increasing melting point is **hydrogen, lithium, silicon.**

See Problems 11.47 and 11.48.

Hardness and Structure   Hardness depends on how easily the structural units of a solid can be moved relative to one another and therefore on the strength of attractive forces between the units. Thus, molecular crystals, with weak intermolecular forces, are rather soft compared with ionic crystals, in which the attractive forces are much stronger. A three-dimensional covalent network solid is usually quite hard because of the rigidity given to the structure by strong covalent bonds throughout it. Thus diamond and silicon carbide (SiC) are among the hardest substances known.

We should add that molecular and ionic crystals are generally brittle because they fracture easily along crystal planes. Metallic crystals, by contrast, are *malleable;* that is, they can be shaped by hammering (Figure 11.22).

Electrical Conductivity and Structure   One of the characteristic properties of metals is good electrical conductivity. The delocalized valence electrons are easily moved by an electrical field and are responsible for carrying the electric current. By contrast, most covalent and ionic solids are nonconductors, because the electrons are localized to particular atoms or bonds. Ionic substances do become conducting in the liquid state, however, because the ions can move.

11.6 Crystalline Solids; Crystal Lattices and Unit Cells

Solids can be crystalline or amorphous. A **crystalline solid** *is composed of one or more crystals; each crystal has a well-defined ordered structure in three dimensions.* Sodium chloride and sucrose (table sugar) are examples of crystalline substances. Metals are usually compact masses of crystals. An **amorphous solid** *has a disordered structure; it lacks the well-defined arrangement of basic units (atoms, molecules, or ions) found in a crystal.* A glass is an amorphous solid obtained by cooling a liquid rapidly enough that its basic units are "frozen" in random positions before they can assume an ordered crystalline arrangement. ◀

Crystal Lattices

A crystal is a three-dimensional ordered arrangement of basic units. (The basic unit is either an atom, molecule, or ion, depending on the type of crystal.) The ordered structure of a crystal is conveniently described in terms of a *crystal lattice,* a repeating pattern.

We define a **crystal lattice** as *the geometric arrangement of lattice points of a crystal, in which we choose one lattice point at the same location within each of the basic units of the crystal.* Figure 11.23 shows the crystal structure of copper metal. If we place a point at the same location in each copper atom (say at the center of the atom), we obtain the crystal lattice for copper. The crystal lattice and crystal structure are not the same. We can imagine building the crystal structure for copper from its crystal lattice by placing a copper atom at each lattice point. Many different crystals have the same crystal lattice. For instance, the crystal lattice for solid methane, CH_4, is the same (except for the distances between points) as that for copper. We obtain the methane crystal structure by placing CH_4 molecules at the lattice points. The crystal lattice shows only the arrangement of basic units of the crystal.

You can imagine dividing a crystal into many equivalent cells, or *unit cells,* so that these cells become the "bricks" from which you mentally construct the crystal. The **unit cell** of a crystal is *the smallest boxlike unit from which you can imagine constructing a crystal by stacking the units in three dimensions.*

There are 14 different kinds of unit cells. Each unit-cell shape can be characterized by the angles between the edges of the unit cell and the relative lengths of those edges. A crystal belonging to the cubic system, for example, has 90° angles between the edges of the unit cell, and all edges are of equal length. We will describe the cubic crystal system in some detail.

▶ When quartz crystals (SiO_2) are melted and then cooled rapidly, they form a glass called silica glass. Though ideal for some purposes, silica glass is difficult to work with because of its high temperature of softening. This temperature can be lowered by adding various oxides, such as Na_2O and CaO. In practice, ordinary glass is formed by melting sand (quartz crystals) with sodium carbonate and calcium carbonate. The carbonates decompose to the oxides plus carbon dioxide.

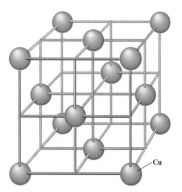

Figure 11.23
Crystal structure and crystal lattice of copper. The copper atoms have been shrunk so that the crystal structure is more visible. The crystal lattice is the geometric arrangement of lattice points, which you can take to be the centers of the atoms. Lines have been drawn to emphasize the geometry of the lattice.

Cubic Unit Cells

The cubic crystal system has three possible cubic unit cells: simple (or primitive) cubic, body-centered cubic, and face-centered cubic. These cells are illustrated in Figure 11.24. A **simple cubic unit cell** is *a cubic unit cell in which lattice points are situated only at the corners.* A **body-centered cubic unit cell** is *a cubic unit cell in which there is a lattice point at the center of the cubic cell in addition to those at the corners.* A **face-centered cubic unit cell** is *a cubic unit cell in which there are lattice points at the centers of each face of the unit cell in addition to those at the corners.*

The simplest crystal structures are those in which there is only a single atom at each lattice point. Most metals are examples. Copper metal has a face-centered cubic unit cell with one copper atom at each lattice point (see Figure 11.23). The unit cells

Figure 11.24
Cubic unit cells.   The simple cubic unit cell has lattice points only at the corners; the body-centered cubic unit cell also has a lattice point at the center of the cell; and the face-centered cubic unit cell has lattice points at the centers of each face in addition to those at the corners.

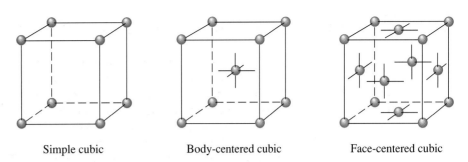

Simple cubic   Body-centered cubic   Face-centered cubic

for this and other cubic cells of simple atomic crystals are shown in Figure 11.25. Note that only the portions of the atoms actually within the unit cell are shown. For certain applications, you will need to know the number of atoms in a unit cell of such an atomic crystal. The next example shows how you count atoms in a unit cell.

EXAMPLE 11.8

Determining the Number of Atoms per Unit Cell

How many atoms are there in the face-centered cubic unit cell of an atomic crystal having one atom at each lattice point? Remember that atoms at the corners and faces of the unit cell are shared with adjoining unit cells (see Figure 11.25).

SOLUTION

Each atom at the center of a face is shared with one other unit cell. Hence, only one-half of the atom belongs to a particular unit cell. But there are six faces on a cube. This gives three whole atoms from the faces. Each corner atom is shared among eight unit cells. Hence, only one-eighth of a corner atom belongs to a particular unit cell. But there are eight corners on a cube. Therefore, the corners contribute one whole atom. Thus, there is a total of four atoms in a face-centered cubic unit cell. To summarize,

$$6 \text{ faces} \times \frac{\frac{1}{2} \text{ atom}}{\text{face}} = 3 \text{ atoms}$$

$$8 \text{ corners} \times \frac{\frac{1}{8} \text{ atom}}{\text{corner}} = 1 \text{ atom}$$

Total   **4 atoms**

See Problems 11.51 and 11.52.

Figure 11.25
Space-filling representation of cubic unit cells.   Only that portion of each atom belonging to a unit cell is shown. Note that a corner atom is shared with eight unit cells and a face atom is shared with two.

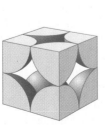 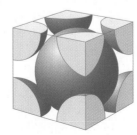 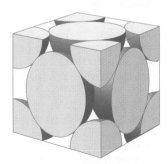

Simple cubic   Body-centered cubic   Face-centered cubic

Liquid-Crystal Displays

EVERYDAY LIFE

Liquid-crystal displays (LCDs) are common in calculators and wristwatches (see Figure 1.3). Liquid crystals, which are the basis of these displays, generally consist of rodlike molecules in a phase that is intermediate in order between a liquid and a crystalline solid. (Figure 1.4 shows a typical liquid-crystal molecule.) Several different types of liquid-crystal phases exist, differing in the type of molecular order. The *nematic phase* is shown in Figure 11.26. The molecules in this liquid-crystal phase are aligned in one direction, much like the matches in a matchbox. Otherwise, the molecules have random positions, as in a liquid.

Liquid-crystal displays use nematic liquid crystals. The molecular order in a nematic liquid crystal, which results from weak intermolecular forces, is easily disrupted. For this reason, the liquid crystals flow like a liquid. Because of the weakness of the intermolecular forces, the molecules in a nematic phase are easily realigned along new directions.

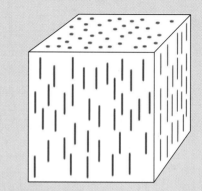

Figure 11.26
Nematic liquid crystal. The rodlike molecules are aligned in the same direction, although the positions of the molecules are random.

A liquid-crystal display uses this ease of molecular reorientation to change areas of the display from light to dark, resulting in the patterns you see on the display. The display consists of liquid crystals contained be-

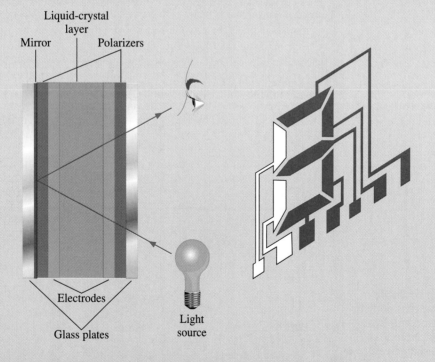

Liquid-crystal layer
Mirror   Polarizers

Electrodes
Glass plates
Light source

Figure 11.27
A liquid-crystal display (LCD). *Left:* The display consists of a layer of liquid crystals between glass plates. Sheets of polarizers and transparent electrodes are also included in the "sandwich." *Right:* Portion of one set of electrodes. The electrodes are shaped to form numbers when certain of them have a voltage applied to them. Here, those that combine to form a 3 are shown in gray.

tween glass plates whose interior surfaces are treated to align the molecules in a given direction. The space between the glass plates also contains polarizer sheets (similar to the Polaroid lenses in sunglasses) and transparent electrodes to reorient the molecules (Figure 11.27). When a voltage to a set of electrodes in some area of the display is turned on, the molecules of the liquid crystal in that area reorient along a new direction. When the voltage is turned off, the molecules return to their original orientation.

The purpose of the polarizer sheets is to make the change of molecular orientation visible by using the ability of the liquid-crystal layer to change the plane of polarized light. Light generally consists of waves that are vibrating in various planes. A polarizer filters out all waves except those vibrating in a given plane, resulting in what is called *polarized light*. If you were to hold a second polarizer up to this polarized light, the light would pass or not, depending on the relative orientations of the two polarizers. (You can observe this by taking the lenses from an old pair of Polaroid sunglasses and holding them up so you can look through both lenses at once. When you rotate the second polarizer with respect to the first, you see that more or less light passes through, depending on the relative orientations of the polarizers.)

Figure 11.28 shows how the liquid-crystal display works. Light from a source outside the display passes through the first polarizer, resulting in polarized light, which then passes through the liquid-crystal layer. When the voltage to a set of electrodes in an area of the display is off, the liquid-crystal layer in that area rotates the plane of the polarized light so that it can pass through the second polarizer to the mirror and be reflected back. The display area appears bright. When the voltage is turned on, however, the plane of the polarized light is not rotated. In that case, the light cannot pass through the second polarizer to the mirror, and the display area appears dark. The pattern on the display is created by turning sets of electrodes on and off.

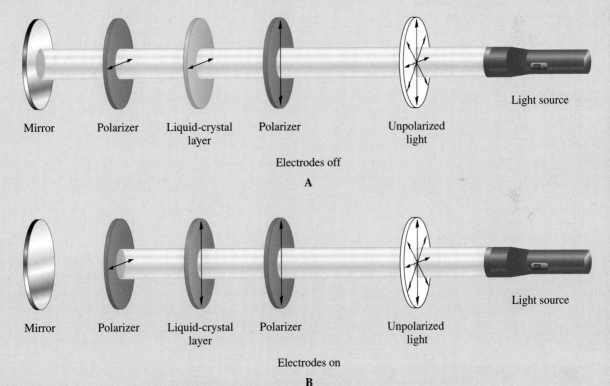

Figure 11.28
How the liquid-crystal display works. *(A)* Unpolarized light from an external light source (at the right) is filtered by the first polarizer, giving light polarized along the plane indicated by the double-headed arrow. The plane of polarization of the light is rotated by the liquid-crystal layer, so it can pass through the second polarizer to the mirror, where the light is reflected back. This area of the display appears bright. *(B)* The molecules of the liquid-crystal layer are realigned when the electrodes are turned on. This time the liquid crystal layer does not rotate the plane of the polarized light, so it cannot pass through to the mirror. This area of the display appears dark.

Crystal Defects

So far we have assumed that crystals have perfect order. In fact, real crystals have various defects or imperfections. These are principally of two kinds: chemical impurities and defects in the formation of the lattice.

Ruby is an example of a crystal with a chemical impurity. The crystal is mainly colorless aluminum oxide, Al_2O_3, but occasional aluminum ions, Al^{3+}, are replaced by chromium(III) ions, Cr^{3+}, which give a red color.

Various lattice defects occur during crystallization. Crystal planes may be misaligned, or sites in the crystal lattice may remain vacant. For example, there might be an equal number of sodium ion vacancies and chloride ion vacancies in a sodium chloride crystal. It is also possible to have an unequal number of cation and anion vacancies in an ionic crystal. For example, iron(II) oxide, FeO, usually crystallizes with some iron(II) sites left unoccupied. Enough of the remaining iron atoms have +3 charges to give an electrically balanced crystal. As a result, there are more oxygen atoms than iron atoms in the crystal. Moreover, the exact composition of the crystal can vary, so the formula FeO is only approximate. Such a compound, whose composition varies slightly from its idealized formula, is said to be *nonstoichiometric*. Other examples of nonstoichiometric compounds are Cu_2O and Cu_2S. Each usually has less copper than expected from the formula. The recently discovered ceramic materials having superconducting properties at relatively high temperatures are nonstoichiometric compounds. An example is yttrium barium copper oxide, $YBa_2Cu_3O_{7-x}$, where x is approximately 0.1. It is a compound with oxygen atom vacancies.

11.7 Structures of Some Crystalline Solids

We have described the structure of crystals in a general way. Now we want to look in detail at the structure of several crystalline solids that represent the different types: molecular, metallic, ionic, and covalent network.

Molecular Solids; Closest Packing

The simplest molecular solids are the frozen noble gases—for example, solid neon. In this case, the molecules are single atoms and the intermolecular interactions are London forces. The maximal attraction is obtained when each atom is surrounded by the largest possible number of other atoms. The problem, then, is simply to find how identical spheres can be packed as tightly as possible into a given space.

To form a layer of close-packed spheres, you place each row of spheres in the crevices of the adjoining rows. This is illustrated in Figure 11.29A. Place the next layer of spheres in the hollows in the first layer. Only half of these hollows can be filled with spheres. Once you have placed a sphere in a given hollow, it partially covers three neighboring hollows (Figure 11.29B) and completely determines the pattern of the layer (Figure 11.29C).

When you come to fill the third layer, you find that you have a choice of sites for spheres, labeled x and y in Figure 11.29C. Note that each of the x sites has a sphere in the first layer directly beneath it, but the y sites do not. When the spheres in the third layer are placed in the x sites, so that the third layer repeats the first layer, you label this stacking *ABA*. When successive layers are placed so that the spheres of each layer are directly over a layer that is one layer away, you get a stacking that

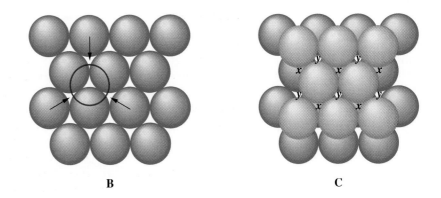

A   B   C

Figure 11.29
Closest packing of spheres. *(A)* Each layer is formed by placing spheres in the crevices of the adjoining row. *(B)* When a sphere is placed in a hollow on top of a layer, three other hollows are partially covered. *(C)* There are two types of hollows on top of the second layer, labeled *x* and *y*. The *x* sites are directly above the spheres in the first layer; the *y* sites are not. Spheres of a third layer may occupy either all of the *x* sites or all of the *y* sites.

▶ If you pack into a crate as many perfectly round oranges of the same size as possible, they will be close packed. However, they need not be either hexagonal or cubic close packed, because the *A*-, *B*-, and *C*-type layers can be placed at random (except that similar layers cannot be adjacent). A possible arrangement might be *ABCABABABC* · · ·. It is the forces between structural units in a close-packed crystal that determine whether one of the ordered structures, either cubic or hexagonal, will occur.

you label *ABABABA* · · · . (This notation indicates that the *A* layers are directly over other *A* layers, whereas the *B* layers are directly over other *B* layers.) The result is a **hexagonal close-packed structure** (hcp), *a crystal structure composed of close-packed atoms (or other units) with the stacking* ABABABA · · · ; *the structure has a hexagonal unit cell.*

When the spheres in the third layer are placed in *y* sites, so that the third layer is over neither the first nor the second layer, you get a stacking that you label *ABC*. The fourth layer must be over either the *A* or the *B* layer. If subsequent layers are stacked so that they are over the layer two layers below, you get the stacking *ABCAB-CABCA* · · · , which results in a **cubic close-packed structure** (ccp), *a crystal structure composed of close-packed atoms (or other units) with the stacking ABCAB-CABCA* · · · . ◀

The cubic close-packed lattice is identical to the lattice having a face-centered cubic unit cell. To see this, you take portions of four layers from the cubic close-packed array (Figure 11.30, left). When these are placed together, they form a cube, as shown in Figure 11.30, right.

In any close-packed arrangement, each interior atom is surrounded by 12 nearest-neighbor atoms. *The number of nearest-neighbor atoms of an atom* is called its **coordination number.** Thinking of atoms as hard spheres, one can calculate that the spheres occupy 74% of the space of the crystal. There is no way of packing identical spheres so that an atom has a coordination number greater than 12 or the spheres occupy more than 74% of the space of the crystal. All the noble-gas solids have cubic close-packed crystals except helium, which has a hexagonal close-packed crystal.

Metallic Solids

If you were to assume that metallic bonding is completely nondirectional, you would expect metals to crystallize in one of the close-packed structures, as the noble-gas elements do. Copper and silver, for example, are cubic close packed (that is, face-centered cubic with one atom at each lattice point). However, a body-centered cubic arrangement of spheres (having one sphere at each lattice point of a body-centered cubic lattice) occupies almost as great a percentage of space as the close-packed ones. In a body-centered arrangement, 68% of the space is occupied by spheres, compared with the maximal value of 74% in the close-packed arrangement. Each atom in a body-centered lattice has a coordination number of 8, rather than the 12 of each atom in a close-packed lattice.

Figure 11.31 lists the structures of the metallic elements. With few exceptions, they have either a close-packed or a body-centered structure.

Figure 11.30
The cubic close-packed structure has a face-centered cubic lattice.
Left: An "exploded" view of portions of four layers of the cubic close-packed structure. *Right:* These layers form a face-centered cubic unit cell.

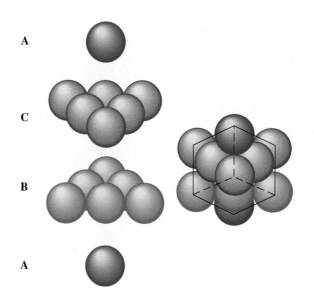

Ionic Solids

The description of ionic crystals is complicated by the fact that you must give the positions in the crystal structure of both the cations and the anions. We will look at the structures of three different cubic crystals with the general formula MX (where M is the metal and X is the nonmetal): cesium chloride (CsCl), sodium chloride (NaCl), and cubic zinc sulfide or zinc blende (ZnS). Many ionic substances of the general formula MX have crystal structures that are one of these types. Once you have learned these structures, you can relate the structures of many other compounds to them.

Cesium chloride consists of positive ions and negative ions of about equal size arranged in a cubic array. Figure 11.32 shows the unit cell of cesium chloride. It shows both a space-filling model and a model in which the ions are shrunk in size in order to display more clearly their relative positions. Note that there are chloride ions at each corner of the cube and a cesium ion at the center. The unit cell contains one Cs^+ ion and one Cl^- ion. (Figure 11.32 has $\frac{1}{8}$ Cl^- ion at each of eight corners plus one Cs^+ ion at the center of the unit cell.) Ammonium chloride, NH_4Cl, and thallium chloride, TlCl, are examples of other ionic compounds having this "cesium chloride" structure.

Figure 11.31
Crystal structures of metals. Most metals have one of the close-packed structures (hcp = hexagonal close-packed; ccp = cubic close-packed) or a body-centered cubic structure (bcc; manganese has a complicated bcc lattice, with several atoms at each lattice point). Other structures are body-centered tetragonal (bct), orthorhombic (or), and rhombohedral (rh).

IA	IIA										IB	IIB	IIIA	IVA	VA
Li bcc	**Be** hcp														
Na bcc	**Mg** hcp	IIIB	IVB	VB	VIB	VIIB		VIIIB			IB	IIB	**Al** ccp		
K bcc	**Ca** ccp	**Sc** hcp	**Ti** hcp	**V** bcc	**Cr** bcc	**Mn** bcc	**Fe** bcc	**Co** hcp	**Ni** ccp	**Cu** ccp	**Zn** hcp	**Ga** or		
Rb bcc	**Sr** ccp	**Y** hcp	**Zr** hcp	**Nb** bcc	**Mo** bcc	**Tc** hcp	**Ru** hcp	**Rh** ccp	**Pd** ccp	**Ag** ccp	**Cd** hcp	**In** bct	**Sn** bct	
Cs bcc	**Ba** bcc	**La** hcp	**Hf** hcp	**Ta** bcc	**W** bcc	**Re** hcp	**Os** hcp	**Ir** ccp	**Pt** ccp	**Au** ccp	**Hg** rh	**Tl** hcp	**Pb** ccp	**Bi** rh

Figure 11.32
Cesium chloride unit cell.
Cesium chloride consists of interpenetrating simple cubic lattices of Cs^+ and Cl^- ions. The figure shows a unit cell with Cl^- ions at the corners of the unit cell and a Cs^+ ion at the center. (A space-filling model is on the left; a model with ions shrunk in size to emphasize the structure is on the right.) An alternative unit cell would have Cs^+ ions at the corners with a Cl^- ion at the center.

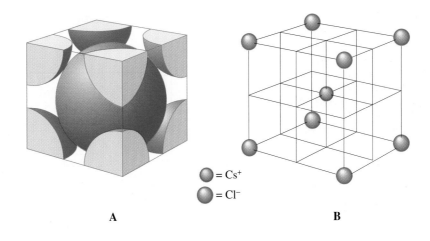

= Cs^+
= Cl^-

A   B

Figure 11.33 shows the unit cell of sodium chloride. It has Cl^- ions at the corners as well as at the center of each cube face. Because the chloride ions are so much larger than the sodium ions, the Cl^- ions are nearly touching and form an approximate cubic close-packed structure of Cl^- ions. The Na^+ ions are arranged in cavities of this close-packed structure. Some other compounds that crystallize in this structure are potassium chloride, KCl; calcium oxide, CaO; and silver chloride, AgCl.

Zinc sulfide, ZnS, crystallizes in either of two structures. The zinc sulfide mineral having cubic structure is called zinc blende or sphalerite. Zinc sulfide also exists as hexagonal crystals. (The mineral is called wurtzite.) A solid substance that can occur in more than one crystal structure is said to be *polymorphic*.

Figure 11.34 shows the unit cell of zinc blende. Sulfide ions are at the corners and at the center of each face of the unit cell. The positions of the zinc ions can be described if you imagine the unit cell divided into eight cubic parts, or subcubes. Zinc ions are at the centers of alternate subcubes. The number of S^{2-} ions per unit cell is four (8 corner ions $\times \frac{1}{8}$ + 6 face ions $\times \frac{1}{2}$), and the number of Zn^{2+} ions is four (all are completely within the unit cell). Thus, there are equal numbers of Zn^{2+} and S^{2-} ions, as expected by the formula ZnS.

The unit cell of zinc blende can also be taken to have zinc ions at the corners and face centers, with sulfide ions in alternate subcubes. Zinc oxide, ZnO, and beryllium oxide, BeO, also have the "zinc blende" structure.

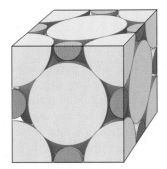

= Cl^-
= Na^+

Figure 11.33
Sodium chloride unit cell.
Sodium chloride has a face-centered cubic lattice. The unit cell shown here (as a space-filling model) has Cl^- ions at the corners and at the centers of the faces of the cube. The Cl^- ions are much larger than Na^+ ions, so the Cl^- ions are nearly touching and have approximately a cubic close-packed structure; the Na^+ ions are in cavities in this structure.

Covalent Network Solids

The structures of covalent network solids are determined primarily by the directions of covalent bonds. Diamond is a simple example. It is one allotropic form of the element carbon, in which each carbon atom is covalently bonded to four other carbon atoms in tetrahedral directions to give a three-dimensional covalent network (see Figure 11.21). The unit cell of diamond is similar to that of zinc blende (Figure 11.34); the Zn^{2+} and S^{2-} ions are replaced by carbon atoms, as in Figure 11.35. The crystal lattice, which is face-centered cubic, can be obtained by placing lattice points at the centers of alternate carbon atoms (either those shown in dark gray or those shown in light gray). Silicon, germanium, and gray tin (a nonmetallic allotrope) have "diamond" structures.

Graphite is another allotrope of carbon. It consists of large, flat sheets of carbon atoms covalently bonded to form hexagonal arrays, and these sheets are stacked one

Figure 11.34

Zinc blende (cubic ZnS) unit cell. Zinc sulfide, ZnS, crystallizes in two different forms, or polymorphs: a hexagonal form (wurtzite) and a face-centered cubic form (zinc blende, or sphalerite), which is shown here. Sulfide ions, S^{2-}, are at the corners and at the centers of the faces of the unit cell; Zn^{2+} ions are in alternate subcubes of the unit cell. The unit cell can also be described as having Zn^{2+} ions at each corner and at the center of each face, with S^{2-} ions in alternate subcubes.

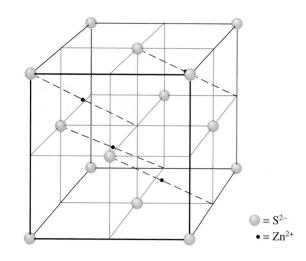

$\bigcirc = S^{2-}$
$\bullet = Zn^{2+}$

on top of the other (see Figure 11.21). You can describe each sheet in terms of resonance formulas such as the following:

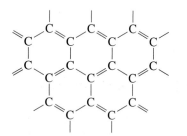

(one resonance formula)

This is just one of many possible resonance formulas, because others can be drawn by moving the double bonds around. This means that the π electrons are delocalized over a plane of carbon atoms. You can also describe this delocalization in molecular orbital terms. You imagine that each carbon atom is sp^2 hybridized, so each atom has three sp^2 hybrid orbitals in the molecular plane and an additional $2p$ orbital perpendicular to this plane. A σ bond forms between two carbon atoms when an sp^2 hybrid orbital on one atom overlaps an sp^2 hybrid orbital on another. The result is a planar σ-bond framework. The $2p$ orbitals of all the atoms overlap to form π orbitals encompassing the entire plane of atoms.

Each carbon–carbon distance within a graphite sheet is 142 pm, which is almost midway between the length of a C—C bond (154 pm) and a C=C bond (134 pm) and which indicates an intermediate bond order (see Figure 11.21). The two-dimensional sheets, or layers, are stacked one on top of the other, and the attraction between sheets results from London forces. The distance between layers is 335 pm because London forces are weaker than covalent bonding forces.

You can explain several properties of graphite on the basis of its structure. The layer structure, which you see in electron micrographs (Figure 11.36), tends to separate easily, so that sheets of graphite slide over one another. The "lead" in a pencil is graphite. The use of graphite to lubricate locks and other closely fitting metal surfaces depends on the ability of graphite particles to move over one another as the layers slide. Another property of graphite is its electrical conductivity. Electrical conductivity is unusual in covalent network solids, because normally the electrons in such bonds

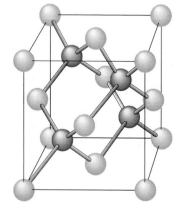

Figure 11.35

Diamond unit cell. The unit cell of diamond can be obtained from the unit cell of zinc blende by replacing Zn^{2+} ions by carbon atoms (shown here as dark spheres) and S^{2-} ions by carbon atoms (shown here as light spheres). Note that each carbon atom is bonded to four others.

are localized. Graphite is a moderately good conductor, however, because delocalization leads to mobile electrons within each carbon atom layer.

> **CONCEPT CHECK 11.3**
>
> Shown here is a representation of a unit cell for a crystal. The orange balls are atom A, and the black balls are atom B.
>
>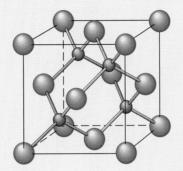
>
> a. What is the chemical formula of the compound that has this unit cell ($A_?B_?$)?
>
> b. Consider the configuration of the A atoms. Is this a cubic unit cell? If so, which type?

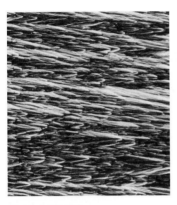

Figure 11.36
Electron micrograph of graphite. Note the layer structure.

11.8 Calculations Involving Unit-Cell Dimensions

You can determine the structure and dimensions of a unit cell by diffraction methods, which we will describe briefly in the next section. Once you know the unit-cell dimensions and the structure of a crystal, however, some interesting calculations are possible. For instance, suppose you find that a metallic solid has a face-centered cubic unit cell with all atoms at lattice points, and you determine the edge length of the unit cell. From this unit-cell dimension, you can calculate the volume of the unit cell. Then, knowing the density of the metal, you can calculate the mass of the atoms in the unit cell. Because you know that the unit cell is face-centered cubic with all atoms at lattice points and that such a unit cell has four atoms, you can obtain the mass of an individual atom. This determination of the mass of a single atom gave one of the first accurate determinations of Avogadro's number. The calculations are shown in the next example.

> **EXAMPLE 11.9**
> ### Calculating Atomic Mass from Unit-Cell Dimension and Density
>
> X-ray diffraction from crystals provides one of the most accurate ways of determining Avogadro's number. Silver crystallizes in a face-centered cubic lattice with all atoms at the lattice points. The length of an edge of the unit cell was determined by x-ray diffraction to be 408.6 pm. The density of silver is 10.50 g/cm^3. Calculate the mass of a silver atom. Then, using the known value of the atomic weight, calculate Avogadro's number.

PROBLEM STRATEGY

Knowing the edge length of a unit cell, you can calculate the unit-cell volume. Then, from the density, you can find the mass of the unit cell and hence the mass of a silver atom.

SOLUTION

You obtain the volume, V, of the unit cell by cubing the length of an edge.

$$V = (4.086 \times 10^{-10}\,\text{m})^3 = 6.822 \times 10^{-29}\,\text{m}^3$$

The density, d, of silver in grams per cubic meter is

$$d = 10.50\,\frac{\text{g}}{\text{cm}^3} \times \left(\frac{1\,\text{cm}}{10^{-2}\,\text{m}}\right)^3 = 1.050 \times 10^7\,\text{g/m}^3$$

Density is mass per volume; hence, the mass of a unit cell equals the density times the volume of the unit cell.

$$
\begin{aligned}
m &= dV \\
&= 1.050 \times 10^7\,\text{g/m}^3 \times 6.822 \times 10^{-29}\,\text{m}^3 \\
&= 7.163 \times 10^{-22}\,\text{g}
\end{aligned}
$$

Because there are four atoms in a face-centered unit cell having one atom at each lattice point (see Example 11.8), the mass of a silver atom is

$$\text{Mass of 1 Ag atom} = \tfrac{1}{4} \times 7.163 \times 10^{-22}\,\text{g} = \mathbf{1.791 \times 10^{-22}\,g}$$

The known atomic weight of silver is 107.87 amu. Thus, the molar mass (Avogadro's number of atoms) is 107.87 g/mol, and Avogadro's number, N_A, is

$$N_A = \frac{107.87\,\text{g/mol}}{1.791 \times 10^{-22}\,\text{g}} = \mathbf{6.023 \times 10^{23}/mol}$$

If you know or assume the structure of an atomic crystal, you can calculate the length of the unit-cell edge from the density of the substance. This is illustrated in the next example. Agreement of this value with that obtained from x-ray diffraction confirms that your view of the structure of the crystal is correct.

EXAMPLE 11.10

Calculating Unit-Cell Dimension from Unit-Cell Type and Density

Platinum crystallizes in a face-centered cubic lattice with all atoms at the lattice points. It has a density of 21.45 g/cm³ and an atomic weight of 195.08 amu. From these data, calculate the length of a unit-cell edge. Compare this with the value of 392.4 pm obtained from x-ray diffraction.

PROBLEM STRATEGY

The mass of an atom, and hence the mass of a unit cell, can be calculated from the atomic weight. Knowing the density and the mass of the unit cell, we can calculate the volume and the edge length of a unit cell.

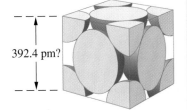

392.4 pm?

Platinum
$d = 21.45$ g/cm^3
atomic weight = 195.08 amu

SOLUTION

You can use Avogadro's number (6.022×10^{23}/mol) to convert the molar mass of platinum (195.08 g/mol) to the mass per atom.

$$\frac{195.08 \text{ g Pt}}{1 \text{ mol Pt}} \times \frac{1 \text{ mol Pt}}{6.022 \times 10^{23} \text{ Pt atoms}} = \frac{3.239 \times 10^{-22} \text{ g Pt}}{1 \text{ Pt atom}}$$

Because there are four atoms per unit cell, the mass per unit cell can be calculated as follows:

$$\frac{3.239 \times 10^{-22} \text{ g Pt}}{1 \text{ Pt atom}} \times \frac{4 \text{ Pt atoms}}{1 \text{ unit cell}} = \frac{1.296 \times 10^{-21} \text{ g}}{1 \text{ unit cell}}$$

The volume of the unit cell is

$$V = \frac{m}{d} = \frac{1.296 \times 10^{-21} \text{ g}}{21.45 \text{ g/cm}^3} = 6.042 \times 10^{-23} \text{ cm}^3$$

If the edge length of the unit cell is denoted as l, the volume is $V = l^3$. Hence, the edge length is

$$l = \sqrt[3]{V}$$
$$= \sqrt[3]{6.042 \times 10^{-23} \text{ cm}^3}$$
$$= 3.924 \times 10^{-8} \text{ cm (or } 3.924 \times 10^{-10} \text{ m)}$$

Thus, the edge length is 392.4 pm, which is in excellent agreement with the x-ray diffraction value.

See Problems 11.53 and 11.54.

11.9 Determining Crystal Structure by X-Ray Diffraction

▶ X-ray diffraction has been used to obtain the structure of proteins, which are large molecules essential to life. John Kendrew and Max Perutz received the Nobel Prize in chemistry in 1962 for their x-ray work in determining the structure of myoglobin (MW = 17,600 amu) and hemoglobin (MW = 66,000 amu). Hemoglobin is the oxygen-carrying protein of the red blood cells, and myoglobin is the oxygen-carrying protein in muscle tissue.

The determination of crystal structure by x-ray diffraction is one of the most important ways of determining the structures of molecules. Because of its ordered structure, a crystal consists of repeating planes of the same kind of atom. These planes can act as reflecting surfaces for x rays. When x rays are reflected from these planes, they show a *diffraction pattern,* which can be recorded on a photographic plate as a series of spots (see Figure 11.37). By analyzing the diffraction pattern, you can determine the positions of the atoms in the unit cell of the crystal. Once you have determined the positions of each atom in the unit cell of a molecular solid, you have also found the positions of the atoms in the molecule. ◀

Consider the scattering, or reflection, of x rays from a crystal. Only at certain angles of reflection do the x rays remain in phase. At some angles the rays constructively interfere (giving dark areas on a negative, corresponding to large x-ray intensity). At other angles the rays destructively interfere (giving light areas on a photographic plate, corresponding to low x-ray intensity). Note the dark and light areas on the photographic negative of the diffraction pattern shown in Figure 11.37.

INSTRUMENTAL METHODS

Automated X-Ray Diffractometry

Max von Laue, a German physicist, was the first to suggest the use of x rays for the determination of crystal structure. Soon afterward, in 1913, the British physicists William Bragg and his son Lawrence developed the method on which modern crystal-structure determination is based. They realized that the atoms in a crystal form reflecting planes for x rays, and from this idea they derived the fundamental equation of crystal-structure determination.

$$n\lambda = 2d \sin \theta, n = 1, 2, 3, \ldots$$

The *Bragg equation* relates the wavelength of x rays, λ, to the distance between atomic planes, d, and the angle of reflection, θ. Note that reflections occur at several angles, corresponding to different integer values of n.

A molecular crystal has many different atomic planes, so that it reflects an x-ray beam in many different directions. By analyzing the intensities and angular directions of the reflected beams, you can determine the exact positions of all the atoms in the unit cell of the crystal and therefore obtain the structure of the molecule. The problem of obtaining the x-ray data (intensities and angular directions of the reflections) and then analyzing them, however, is not trivial. Originally, the reflected x rays were recorded on photographic plates. After taking many pictures, the scientist would pore over the negatives, measuring the densities of the spots and their positions on the plates. Then he or she would work through lengthy and laborious calculations to analyze the data. Even with early computers, the determination of a molecular structure required a year or more.

With the development of electronic x-ray detectors and minicomputers, x-ray diffraction has become automated, so that the time and effort of determining the structure of a molecule have been substantially reduced. Now frequently the most difficult task is preparing a suitable crystal. The crystal should be several tenths of a millimeter in each dimension and without significant defects. Such crystals of protein molecules, for example, can be especially difficult to prepare.

Once a suitable crystal has been obtained, the structure of a molecule of moderate size can often be determined in a day or so. The crystal is mounted on a glass fiber (or in a glass capillary containing an inert gas, if the substance reacts with air) and placed on a pin or spindle within the circular assembly of the x-ray diffractometer (Figure 11.38). The crystal and x-ray detector (placed on the opposite side of the crystal from the x-ray tube) rotate under computer control, while the computer records the intensities and angles of thousands of x-ray reflection spots. After computer analysis of the data, the molecular structure is printed out.

You can relate the resulting pattern produced by the diffraction to the structure of the crystal. You can determine the type of unit cell and its size; if the solid is composed of molecules, you can determine the position of each atom in the molecule.

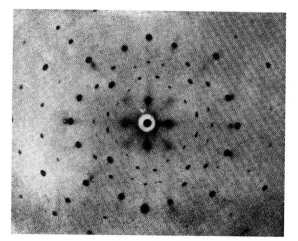

Figure 11.37
A crystal diffraction pattern. This diffraction pattern was obtained by diffracting x rays from a sodium chloride crystal and collecting the image on photographic film. The resulting negative shows dark spots on a lighter background.

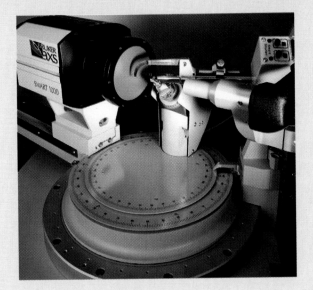

Figure 11.38
Automated x-ray diffractometer. *(Top)* The single-crystal specimen is mounted on a glass fiber, which is placed on a spindle within the circular assembly of the diffractometer. A new data collection system (left of specimen) reduces data collection time from several days to several hours. *(Bottom)* The schematic diagram shows the diffracted rays being detected by a photograph. In a modern diffractometer, the final data collection is done with a fixed electronic detector, and the crystal is rotated. The data are collected and analyzed by a computer accompanying the diffractometer.

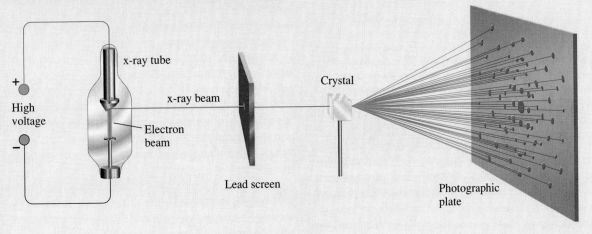

High voltage

x-ray tube

x-ray beam

Electron beam

Lead screen

Crystal

Photographic plate

A CHEMIST LOOKS AT

Water (A Special Substance for Planet Earth)

ENVIRONMENT

Water is the only liquid substance (other than petroleum) to be found on earth in significant amounts. This liquid is also readily convertible under conditions on earth to the solid and gaseous forms. Water has several unusual properties that set it apart from other substances. For example, its solid phase, ice, is less dense than liquid water, whereas for most substances the solid phase is more dense than the liquid. In addition, water has an unusually large heat capacity. These are some of the properties that are important in determining conditions favorable to life. In fact, it is difficult to think of life without water.

The unusual properties of water are largely linked to its ability to form hydrogen bonds. For example, ice is less dense than liquid water because ice has an open, hydrogen-bonded structure (see Figure 11.39). Each oxygen atom in the structure of ice is surrounded

(continued)

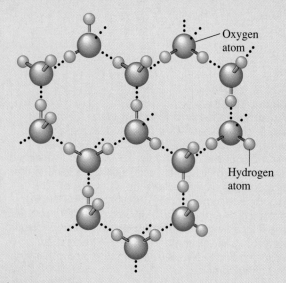

Figure 11.39
The structure of ice. Oxygen atoms are represented by large red spheres; hydrogen atoms, by small spheres. Each oxygen atom is tetrahedrally surrounded by four hydrogen atoms. Two are close, giving the H_2O molecule. Two are farther away, held by hydrogen bonding (represented by three dots). The distribution of hydrogen atoms in these two types of positions is random.

tetrahedrally by four hydrogen atoms: two that are close and covalently bonded to give the H_2O molecule, and two others that are farther away and held by hydrogen bonds. The tetrahedral angles give rise to a three-dimensional structure that contains open space. When ice melts, hydrogen bonds break, the ice structure partially disintegrates, and the molecules become more compactly arranged, leading to a more dense liquid.

Even liquid water has significant hydrogen bonding. Only about 15% of the hydrogen bonds are broken when ice melts. You might view the liquid as composed of ice-like clusters in which hydrogen bonds are continually breaking and forming, so clusters disappear and new ones appear. This is sometimes referred to as the "flickering cluster" model of liquid water.

The unusually large heat of vaporization of water has an important effect on the earth's weather. Evaporation of the surface waters absorbs over 30% of the solar energy reaching the earth's surface. This energy is released when the water vapor condenses, and thunderstorms and hurricanes may result. In the process, the waters of the earth are circulated and the freshwater sources are replenished. This natural cycle of water from the oceans to freshwater sources and its return to the ocean is called the *hydrologic cycle* (Figure 11.40).

Although evaporation and condensation of water play a dominant role in our weather, other properties are important as well. The exceptionally large heat capacity of bodies of water has an important moderating effect on the surrounding temperature by warming cold air masses in winter and cooling warm air masses in summer. Worldwide, the oceans are most important, but even inland lakes have a pronounced effect. For example, the Great Lakes give Detroit a more moderate winter than cities somewhat farther south but with no lakes nearby.

The fact that ice is less dense than water means that it forms on top of the liquid when freezing occurs. This has far-reaching effects, both for weather and for aquatic animals. When ice forms on a body of water, it insulates the underlying water from the cold air and limits further freezing. Fish depend on this for winter survival. Consider what would happen to a lake if ice were more dense than water. The ice would freeze from the bottom of the lake upward. Without the insulating effect at the surface, the lake could well freeze solid, killing the fish. Spring thaw would be prolonged, because the insulating effect of the surface water would make it take much longer for the ice at the bottom of a lake to melt.

The solvent properties of water are also unusual. Water is both a polar substance and a hydrogen-bonding molecule. As a result, water dissolves many substances, including ionic and polar compounds. These properties make water the premier solvent, biologically and industrially.

As a result of the solvent properties of water, the naturally occurring liquid always contains dissolved materials, particularly ionic substances. *Hard water* contains certain metal ions, such as Ca^{2+} and Mg^{2+}. These ions react with soaps, which are sodium salts of stearic acid and similar organic acids, to produce a curdy precipitate of calcium and magnesium salts. This precipitate adheres to clothing and bathtubs (as bathtub ring). Removing Ca^{2+} and Mg^{2+} ions from hard water is referred to as *water softening*.

Water is often softened by *ion exchange*. Ion exchange is a process whereby a water solution is passed through a column of a material that replaces one kind of ion in solution with another kind. Home and commercial water softeners contain cation-exchange resins, which are insoluble macromolecular substances (substances consisting of giant molecules) to which negatively charged groups are chemically bonded. The negative charges are counterbalanced by ions such as Na^+. When hard water containing the Ca^{2+} ion passes through a column of this resin, the Na^+ ion in the resin is replaced by the Ca^{2+} ion.

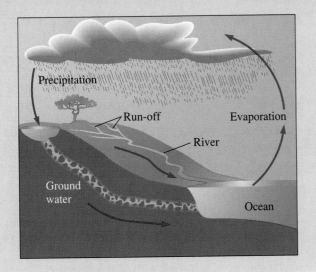

$$2NaR(s) + Ca^{2+}(aq) \longrightarrow CaR_2(s) + 2Na^+(aq)$$
$$(R^- = \text{anion of exchange resin})$$

The water passing through the column now contains Na^+ in place of Ca^{2+} and has been softened. Once the resin has been completely converted to the calcium salt, it can be regenerated by flushing the column with a concentrated solution of NaCl to reverse the previous reaction.

Figure 11.40
The hydrologic cycle. Ocean water evaporates to form clouds. Then, precipitation (rain and snow) from these clouds replenishes freshwater sources. Water from these sources eventually returns to the oceans via rivers, run-off, or groundwater.

A Checklist for Review

Important Terms

change of state (phase transition) (11.1)
vapor pressure (11.2)
boiling point (11.2)
freezing point (11.2)
melting point (11.2)
heat of fusion (11.2)
heat of vaporization (11.2)
phase diagram (11.3)

triple point (11.3)
critical temperature (11.3)
critical pressure (11.3)
intermolecular forces (11.4)
van der Waals forces (11.4)
dipole–dipole force (11.4)
London (dispersion) forces (11.4)
hydrogen bonding (11.4)

molecular solid (11.5)
metallic solid (11.5)
ionic solid (11.5)
covalent network solid (11.5)
crystalline solid (11.6)
amorphous solid (11.6)
crystal lattice (11.6)
unit cell (11.6)
simple cubic unit cell (11.6)

body-centered cubic unit cell (11.6)
face-centered cubic unit cell (11.6)
hexagonal close-packed structure (11.7)
cubic close-packed structure (11.7)
coordination number (11.7)

Key Equation

$$\ln \frac{P_2}{P_1} = \frac{\Delta H_{vap}}{R} \left(\frac{1}{T_1} - \frac{1}{T_2} \right)$$

Summary of Facts and Concepts

Gases are composed of molecules in constant random motion throughout mostly empty space. This explains why gases are compressible fluids. Liquids are also composed of molecules in constant random motion, but the molecules are more tightly packed. Thus, liquids are incompressible fluids. Solids are composed of atoms, molecules, or ions that are in close contact and oscillate about fixed sites. Thus, solids are incompressible and rigid rather than fluid.

Melting and freezing are examples of changes of state, or *phase transitions*. Vapor pressure, boiling point, and melting point are properties of substances that involve phase transitions. Vapor pressure increases with temperature; according to the *Clausius–Clapeyron equation*, the logarithm of the vapor

pressure varies inversely with the absolute temperature. The conditions under which a given state of a substance exists are graphically summarized in a *phase diagram*.

The three kinds of attractive intermolecular forces are *dipole–dipole forces, London forces,* and *hydrogen bonding.* London forces are weak attractive forces present in all molecules; London forces tend to increase with increasing molecular weight. Vapor pressure tends to decrease with increasing molecular weight, whereas the boiling point tends to increase (unless hydrogen bonding is present). Hydrogen bonding occurs in substances containing H atoms bonded to F, O, or N atoms. When hydrogen bonding is present, vapor pressure tends to be lower than otherwise expected, whereas the boiling point tends to be higher.

Solids can be classified in terms of the type of force between the structural units; there are molecular, metallic, ionic, and covalent network solids. Melting point, hardness, and electrical conductivity are properties that can be related to the structure of the solid.

Solids can be crystalline or amorphous. A crystalline solid has an ordered arrangement of structural units placed at *crystal lattice* points. We may think of a crystal as constructed from *unit cells.* Cubic unit cells are of three kinds: simple cubic, body-centered cubic, and face-centered cubic. One of the most important ways of determining the structure of a crystalline solid is by *x-ray diffraction.*

Operational Skills

1. **Calculating the heat required for a phase change of a given mass of substance** Given the heat of fusion (or vaporization) of a substance, calculate the amount of heat required to melt (or vaporize) a given quantity of that substance. **(EXAMPLE 11.1)**
2. **Calculating vapor pressures and heats of vaporization** Given the vapor pressure of a liquid at one temperature and its heat of vaporization, calculate the vapor pressure at another temperature **(EXAMPLE 11.2)**
3. **Relating the conditions for the liquefaction of gases to the critical temperature** Given the critical temperature and pressure of a substance, describe the conditions necessary for liquefying the gaseous substance. **(EXAMPLE 11.3)**
4. **Identifying intermolecular forces** Given the molecular structure, state the kinds of intermolecular forces expected for a substance. **(EXAMPLE 11.4)**
5. **Determining relative vapor pressure on the basis of intermolecular attraction** Given two liquids, decide on the basis of intermolecular forces which has the higher

vapor pressure at a given temperature or which has the lower boiling point. **(EXAMPLE 11.5)**
6. **Identifying types of solids** From what you know about the bonding in a solid, classify it as a molecular, metallic, ionic, or covalent network. **(EXAMPLE 11.6)**
7. **Determining the relative melting points based on types of solids** Given a list of substances, arrange them in order of increasing melting point from what you know of their structures. **(EXAMPLE 11.7)**
8. **Determining the number of atoms per unit cell** Given the description of a unit cell, find the number of atoms per cell. **(EXAMPLE 11.8)**
9. **Calculating atomic mass from unit-cell dimension and density** Given the edge length of the unit cell, the crystal structure, and the density of a metal, calculate the mass of a metal atom. **(EXAMPLE 11.9)**
10. **Calculating unit-cell dimension from unit-cell type and density** Given the unit-cell structure, the density, and the atomic weight for an element, calculate the edge length of the unit cell. **(EXAMPLE 11.10)**

Review Questions

11.1 List the different phase transitions that are possible and give examples of each.

11.2 Describe vapor pressure in molecular terms. What do we mean by saying it involves a dynamic equilibrium?

11.3 Explain why 15 g of steam at 100°C melts more ice than 15 g of liquid water at 100°C.

11.4 Why is the heat of fusion of a substance smaller than its heat of vaporization?

11.5 Explain why evaporation leads to cooling of the liquid.

11.6 Describe the behavior of a liquid and its vapor in a closed vessel as the temperature increases.

11.7 Gases that cannot be liquefied at room temperature merely by compression are called "permanent" gases. How could you liquefy such a gas?

11.8 The pressure in a cylinder of nitrogen continuously decreases as gas is released from it. On the other hand, a cylinder of propane maintains a constant pressure as propane is released. Explain this difference in behavior.

11.9 Explain what is meant by hydrogen bonding. Describe the hydrogen bonding between two H_2O molecules.

11.10 Why do molecular substances have relatively low melting points?

11.11 Describe the face-centered cubic unit cell.

11.12 Describe the structure of thallium(I) iodide, which has the same structure as cesium chloride.

11.13 What is the coordination number of Cs^+ in CsCl? of Na^+ in NaCl? of Zn^{2+} in ZnS?

11.14 Explain in words how Avogadro's number could be obtained from the unit-cell edge length of a cubic crystal. What other data are required?

Conceptual Problems

11.15 Shown here is a curve of the distribution of kinetic energies of the molecules in a liquid at an arbitrary temperature *T*.

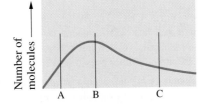

The lines marked A, B, and C represent the point where each of the molecules for three different liquids (liquid A, liquid B, and liquid C) has enough kinetic energy to escape into the gas phase (see Figure 11.5 for more information). Write a brief explanation for each of your answers to the following questions.

a. Which of the molecules—A, B, or C—would have the majority of the molecules in the gas phase at temperature *T*?

b. Which of the molecules—A, B, or C—has the strongest intermolecular attractions?

c. Which of the molecules would have the lowest vapor pressure at temperature *T*?

11.16 Consider a substance X with a $\Delta H_{vap} = 20.3$ kJ/mol and $\Delta H_{fus} = 9.0$ kJ/mol. The melting point, freezing point, and heat capacities of both the solid and liquid X are identical to those of water.

a. If you place one beaker containing 50 g of X at $-10°C$ and another beaker with 50 g of H_2O at $-10°C$ on a hot plate and start heating them, which material will reach the boiling point first?

b. Which of the materials from part a, X or H_2O, would completely boil away first?

c. On a piece of graph paper, draw the heating curves for H_2O and X. How do the heating curves reflect your answers from parts a and b?

11.17 Using the information presented in this chapter, explain why farmers spray water above and on their fruit trees on still nights when they know the temperature is going to drop below 0°C. (*Hint:* Totally frozen fruit is what the farmers are trying to avoid.)

11.18 You are presented with three bottles, each containing a different liquid: bottle A, bottle B, and bottle C. Bottle A's label states that it is an ionic compound with a boiling point of 35°C. Bottle B's label states that it is a molecular compound with a boiling point of 29.2°C. Bottle C's label states that it is a molecular compound with a boiling point of 67.1°C.

a. Which of the compounds is most likely to be incorrectly identified?

b. If Bottle A were a molecular compound, which of the compounds has the strongest intermolecular attractions?

c. If Bottle A were a molecular compound, which of the compounds would have the highest vapor pressure?

11.19 Shown here is a representation of a unit cell for a crystal. The orange balls are atom A, and the black balls are atom B.

a. What is the chemical formula of the compound that has this unit cell ($A_?B_?$)?

b. Consider the configuration of the A atoms. Is this a cubic unit cell? If so, which type?

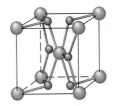

11.20 Assuming normal winter conditions ($-1.5°C$ and 1.0 atm pressure), consult the phase diagram for water (Figure 11.10) and come up with a reason why ice skates and sleds slide so well on solid water. Keep in mind that sleds and ice skates do not typically slide well on other solid surfaces (concrete and metal, for example).

Practice Problems

Phase Transitions

11.21 Identify the phase transition occurring in each of the following.
a. The water level in an aquarium tank falls continuously (the tank has no leak).
b. A mixture of scrambled eggs placed in a cold vacuum chamber slowly turns to a powdery solid.
c. Chlorine gas is passed into a very cold test tube where it turns to a yellow liquid.
d. When carbon dioxide gas under pressure exits from a small orifice, it turns to a white "snow."

11.22 Identify the phase transition occurring in each of the following.
a. Mothballs slowly become smaller and eventually disappear.
b. Rubbing alcohol spilled on the palm of the hand feels cool as the volume of liquid decreases.
c. A black deposit of tungsten metal collects on the inside of a lightbulb whose filament is tungsten metal.
d. Raindrops hit a cold metal surface, which becomes covered with ice.

11.23 Use Figure 11.6 to estimate the boiling point of diethyl ether, $(C_2H_5)_2O$, under an external pressure of 350 mmHg.

11.24 Use Figure 11.6 to estimate the boiling point of carbon tetrachloride, CCl_4, under an external pressure of 350 mmHg.

11.25 Isopropyl alcohol, $CH_3CHOHCH_3$, is used in rubbing alcohol mixtures. Alcohol on the skin cools by evaporation. How much heat is absorbed by the alcohol if 5.00 g evaporates? The heat of vaporization of isopropyl alcohol is 42.1 kJ/mol.

11.26 Liquid butane, C_4H_{10}, is stored in cylinders to be used as a fuel. Suppose 39.3 g of butane gas is removed from a cylinder. How much heat must be provided to vaporize this much gas? The heat of vaporization of butane is 21.3 kJ/mol.

11.27 A quantity of ice at 0°C is added to 64.3 g of water in a glass at 55°C. After the ice melted, the temperature of the water in the glass was 15°C. How much ice was added? The heat of fusion of water is 6.01 kJ/mol and the specific heat is 4.18 J/(g · °C).

11.28 Steam at 100°C was passed into a flask containing 275 g of water at 21°C, where the steam condensed. How many grams of steam must have condensed if the temperature

of the water in the flask was raised to 83°C? The heat of vaporization of water at 100°C is 40.7 kJ/mol and the specific heat is 4.18 J/(g · °C).

11.29 Chloroform, $CHCl_3$, a volatile liquid, was once used as an anesthetic but has been replaced by safer compounds. Chloroform boils at 61.7°C and has a heat of vaporization of 31.4 kJ/mol. What is its vapor pressure at 33.0°C?

11.30 Methanol, CH_3OH, a colorless, volatile liquid, was formerly known as wood alcohol. It boils at 65.0°C and has a heat of vaporization of 37.4 kJ/mol. What is its vapor pressure at 22.0°C?

11.31 White phosphorus, P_4, is normally a white, waxy solid melting at 44°C to a colorless liquid. The liquid has a vapor pressure of 400.0 mmHg at 251.0°C and 760.0 mmHg at 280.0°C. What is the heat of vaporization of this substance?

11.32 Carbon disulfide, CS_2, is a volatile, flammable liquid. It has a vapor pressure of 400.0 mmHg at 28.0°C and 760.0 mmHg at 46.5°C. What is the heat of vaporization of this substance?

Phase Diagrams

11.33 Shown here is the phase diagram for compound Z. The triple point of Z is −5.1°C at 3.3 atm and the critical point is 51°C and 99.1 atm.

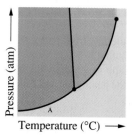

a. What is the state of Z at position A?
b. If we increase the temperature from the compound at position A to 60°C while holding the pressure constant, what is the state of Z?
c. If we take the compound starting under the conditions of part b and reduce the temperature to 20°C and increase the pressure to 65 atm, what is the state of Z?
d. Would it be possible to make the compound starting under the conditions of part c a solid by increasing just the pressure?

11.34 Shown here is the phase diagram for compound X. The triple point of X is −25.1°C at 0.50 atm and the critical point is 22°C and 21.3 atm.

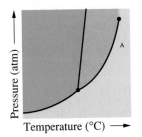

a. What is the state of X at position A?
b. If we decrease the temperature from the compound at position A to −28.2°C while holding the pressure constant, what is the state of X?
c. If we take the compound starting under the conditions of part b and increase the temperature to 15.3°C and decrease the pressure to 0.002 atm, what is the state of X?
d. Would it be possible to make the compound starting under the conditions of part c a solid by increasing just the pressure?

11.35 Which of the following substances can be liquefied by applying pressure at 25°C? For those that cannot, describe the conditions under which they *can* be liquefied.

Substance	Critical Temperature	Critical Pressure
Sulfur dioxide, SO_2	158°C	78 atm
Acetylene, C_2H_2	36°C	62 atm
Methane, CH_4	−82°C	46 atm
Carbon monoxide, CO	−140°C	35 atm

11.36 A tank of gas at 21°C has a pressure of 1.0 atm. Using the data in the table, answer the following questions. Explain your answers.
a. If the tank contains carbon tetrafluoride, CF_4, is the liquid state also present?
b. If the tank contains butane, C_4H_{10}, is the liquid state also present?

Substance	Boiling Point at 1 atm	Critical Temperature	Critical Pressure
CF_4	−128°C	−46°C	41 atm
C_4H_{10}	−0.5°C	152°C	38 atm

Intermolecular Forces and Properties of Liquids

11.37 For each of the following substances, list the kinds of intermolecular forces expected.

a.

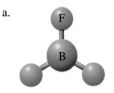

b. isopropyl alcohol, $CH_3CHOHCH_3$
c. hydrogen iodide, HI

11.38 Which of the following compounds would you expect to exhibit *only* London forces?
a. carbon tetrachloride, CCl_4

b.

c. phosphorus trichloride, PCl_3

11.39 Methane, CH_4, reacts with chlorine, Cl_2, to produce a series of chlorinated hydrocarbons: methyl chloride (CH_3Cl), methylene chloride (CH_2Cl_2), chloroform ($CHCl_3$), and carbon tetrachloride (CCl_4). Which compound has the lowest vapor pressure at room temperature? Explain.

11.40 The halogens form a series of compounds with each other, which are called interhalogens. Examples are bromine chloride (BrCl), iodine bromide (IBr), bromine fluoride (BrF), and chlorine fluoride (ClF). Which compound is expected to have the lowest boiling point at any given pressure? Explain.

11.41 Predict the order of increasing vapor pressure at a given temperature for the following compounds:
a. FCH_2CH_2F
b. $HOCH_2CH_2OH$
c. FCH_2CH_2OH
Explain why you chose this order.

11.42 Predict the order of increasing vapor pressure at a given temperature for the following compounds:
a. $CH_3CH_2CH_2CH_2OH$
b. $HOCH_2CH_2CH_2OH$
c. $CH_3CH_2OCH_2CH_3$
Explain why you chose this order.

11.43 List the following substances in order of increasing boiling point:

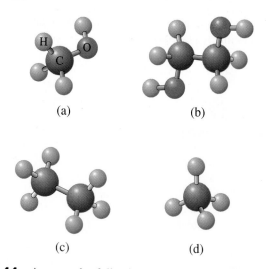

(a) (b)

(c) (d)

11.44 Arrange the following compounds in order of increasing boiling point.

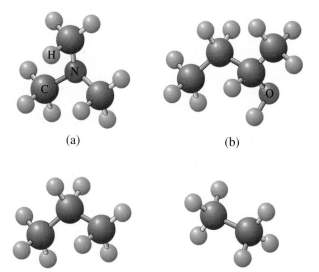

(a) (b)

(c) (d)

Types of Solids

11.45 Classify each of the following solid elements as molecular, metallic, ionic, or covalent network.
a. tin, Sn
b. germanium, Ge
c. sulfur, S_8

11.46 Which of the following do you expect to be molecular solids?
a. sodium hydroxide, NaOH
b. solid ethane, C_2H_6
c. nickel, Ni

11.47 Arrange the following compounds in order of increasing melting point.

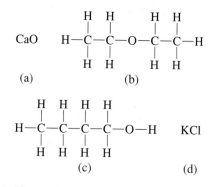

CaO

(a) (b)

(c) (d)

11.48 Arrange the following substances in order of increasing melting point.

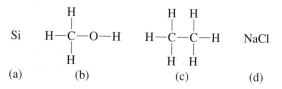

Si

(a) (b) (c) (d)

11.49 Associate each of the solids Co, LiCl, SiC, and CHI_3 with one of the following sets of properties.
a. A white solid melting at 613°C; the liquid is electrically conducting, although the solid is not.
b. A very hard, blackish solid subliming at 2700°C.
c. A yellow solid with a characteristic odor having a melting point of 120°C.
d. A gray, lustrous solid melting at 1495°C; both the solid and liquid are electrical conductors.

11.50 Associate each of the solids BN, P_4S_3, Pb, and $CaCl_2$ with one of the following sets of properties.
a. A bluish-white, lustrous solid melting at 327°C; the solid is soft and malleable.
b. A white solid melting at 772°C; the solid is an electrical nonconductor but dissolves in water to give a conducting solution.
c. A yellowish-green solid melting at 172°C.
d. A very hard, colorless substance melting at about 3000°C.

Crystal Structure

11.51 How many atoms are there in a simple cubic unit cell of an atomic crystal in which all atoms are at lattice points?

11.52 How many atoms are there in a body-centered cubic unit cell of an atomic crystal in which all atoms are at lattice points?

11.53 Copper metal has a face-centered cubic structure with all atoms at lattice points and a density of 8.93 g/cm³. Its atomic weight is 63.5 amu. Calculate the edge length of the unit cell.

11.54 Barium metal has a body-centered cubic lattice with all atoms at lattice points; its density is 3.51 g/cm³. From these data and the atomic weight, calculate the edge length of a unit cell.

11.55 Gold has cubic crystals whose unit cell has an edge length of 407.9 pm. The density of the metal is 19.3 g/cm³. From these data and the atomic weight, calculate the number of gold atoms in a unit cell, assuming all atoms are at lattice points. What type of cubic lattice does gold have?

11.56 Chromium forms cubic crystals whose unit cell has an edge length of 288.5 pm. The density of the metal is 7.20 g/cm³. Use these data and the atomic weight to calculate the number of atoms in a unit cell, assuming all atoms are at lattice points. What type of cubic lattice does chromium have?

11.57 Tungsten has a body-centered cubic lattice with all atoms at the lattice points. The edge length of the unit cell is 316.5 pm. The atomic weight of tungsten is 183.8 amu. Calculate its density.

11.58 Lead has a face-centered cubic lattice with all atoms at lattice points and a unit-cell edge length of 495.0 pm. Its atomic weight is 207.2 amu. What is the density of lead?

General Problems

11.59 The percent relative humidity of a sample of air is found as follows: (partial pressure of water vapor/vapor pressure of water) × 100. A sample of air at 21°C was cooled to 15°C, where moisture began to condense as dew. What was the relative humidity of the air at 21°C?

11.60 A sample of air at 21°C has a relative humidity of 58%. At what temperature will water begin to condense as dew? (See Problem 11.59.)

11.61 The vapor pressure of benzene is 100.0 mmHg at 26.1°C and 400.0 mmHg at 60.6°C. What is the boiling point of benzene at 760.0 mmHg?

11.62 The vapor pressure of water is 17.5 mmHg at 20.0°C and 355.1 mmHg at 80.0°C. Calculate the boiling point of water at 760.0 mmHg.

11.63 Describe the behavior of carbon dioxide gas when compressed at the following temperatures:
a. 20°C   b. −70°C   c. 40°C
The triple point of carbon dioxide is −57°C and 5.1 atm, and the critical point is 31°C and 73 atm.

11.64 Describe the behavior of iodine vapor when cooled at the following pressures:
a. 120 atm   b. 1 atm   c. 50 mmHg
The triple point of iodine is 114°C and 90.1 mmHg, and the critical point is 512°C and 116 atm.

11.65 Describe the formation of hydrogen bonds in propanol, $CH_3CH_2CH_2OH$. Represent possible hydrogen bonding structures in propanol by using structural formulas and the conventional notation for a hydrogen bond.

11.66 Describe the formation of hydrogen bonds in hydrogen peroxide, H_2O_2. Represent possible hydrogen bonding structures in hydrogen peroxide by using structural formulas and the conventional notation for a hydrogen bond.

11.67 Consider the elements Al, Si, P, and S from the third row of the periodic table. In each case, identify the type of solid the element would form.

11.68 The elements in Problem 11.67 form the fluorides AlF_3, SiF_4, PF_3, and SF_4. In each case, identify the type of solid formed by the fluoride.

11.69 Decide which substance in each of the following pairs has the lower melting point. Explain how you made each choice.
a. potassium chloride, KCl; or calcium oxide, CaO

b. carbon tetrachloride,

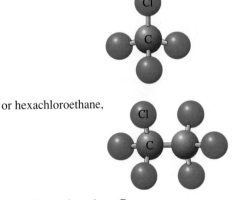

or hexachloroethane,

c. zinc, Zn; or chromium, Cr
d. acetic acid, CH_3COOH; or ethyl chloride, C_2H_5Cl

11.70   Decide which substance in each of the following pairs has the lower melting point. Explain how you made each choice.
a. magnesium oxide, MgO; or hexane, C_6H_{14}
b. 1-propanol,

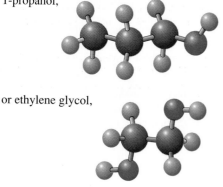

or ethylene glycol,

c. silicon, Si; or sodium, Na
d. methane, CH_4; or silane, SiH_4

11.71   Iridium metal, Ir, crystallizes in a face-centered cubic (close-packed) structure. The edge length of the unit cell was found by x-ray diffraction to be 383.9 pm. The density of iridium is 22.42 g/cm^3. Calculate the mass of an iridium atom. Use Avogadro's number to calculate the atomic weight of iridium.

11.72   The edge length of the unit cell of tantalum metal, Ta, is 330.6 pm; the unit cell is body-centered cubic (one atom at each lattice point). Tantalum has a density of 16.69 g/cm^3. What is the mass of a tantalum atom? Use Avogadro's number to calculate the atomic weight of tantalum.

11.73   Use your answer to Problem 11.53 to calculate the radius of the copper atom. Assume that copper atoms are spheres. Then note that the spheres on any face of a unit cell touch along the diagonal.

11.74   Rubidium metal has a body-centered cubic structure (with one atom at each lattice point). The density of the metal is 1.532 g/cm^3. From this information and the atomic weight, calculate the edge length of the unit cell. Now assume that rubidium atoms are spheres. Each corner sphere of the unit cell touches the body-centered sphere. Calculate the radius of a rubidium atom.

11.75   Account for the following observations:
a. Both diamond and silicon carbide are very hard, whereas graphite is both soft and slippery.
b. Carbon dioxide is a gas, whereas silicon dioxide is a high-melting solid.

11.76   Greater variation exists between the properties of the first and second members of a family in the periodic table than between other members. Discuss this observation for the oxygen family using the following data.

Element	Boiling Point, °C	Compound	Boiling Point, °C
O_2	-183	H_2O	100
S_8	445	H_2S	-61
Se_8	685	H_2Se	-42

11.77   Use chemical principles to discuss the following observations:
a. CO_2 sublimes at $-78°C$, whereas SiO_2 boils at 2200°C.
b. HF boils at 19°C, whereas HCl boils at $-85°C$.
c. CF_4 boils at $-128°C$, whereas SiF_4 boils at $-86°C$.

11.78   a. Draw Lewis structures of each of the following compounds: LiH, NH_3, CH_4, CO_2.
b. Which of these has the highest boiling point? Why?
c. Which of these has the lowest boiling point? Why?
d. Which of these has the next-to-highest boiling point? Why?

Cumulative-Skills Problems

11.79   How much heat is needed to vaporize 10.0 mL of liquid hydrogen cyanide, HCN, at 25.0°C? The density of the liquid is 0.687 g/mL. Use standard heats of formation, which are given in Appendix C.

11.80   How much heat is needed to vaporize 20.0 mL of liquid methanol, CH_3OH, at 25.0°C? The density of the liquid is 0.787 g/mL. Use standard heats of formation, which are given in Appendix C.

11.81   How much heat must be added to 12.5 g of solid white phosphorus, P_4, at 25.0°C to give the liquid at its melting point, 44.1°C? The heat capacity of solid white phosphorus is 95.4 J/(K · mol); its heat of fusion is 2.63 kJ/mol.

11.82 How much heat must be added to 25.0 g of solid sodium, Na, at 25.0°C to give the liquid at its melting point, 97.8°C? The heat capacity of solid sodium is 28.2 J/(K · mol), and its heat of fusion is 2.60 kJ/mol.

11.83 Acetic acid, CH_3COOH, forms stable pairs of molecules held together by two hydrogen bonds.

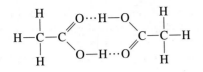

Such molecules—themselves formed by the association of two simpler molecules—are called dimers. The vapor over liquid acetic acid consists of a mixture of monomers (single acetic acid molecules) and dimers. At 100.6°C the total pressure of vapor over liquid acetic acid is 436 mmHg. If the vapor consists of 0.630 mole fraction of the dimer, what are the masses of monomer and dimer in 1.000 L of the vapor? What is the density of the vapor?

11.84 The total pressure of vapor over liquid acetic acid at 71.3°C is 146 mmHg. If the density of the vapor is 0.702 g/L, what is the mole fraction of dimer in the vapor? See Problem 11.83.

Sugar cube dissolving in water.

Solutions

Most chemical reactions are run in solution. Solutions have particular properties that are useful. When gold is used for jewelry, it is alloyed with a small amount of silver. Gold–silver alloys are not only harder than pure gold, but they also melt at lower temperatures and are therefore easier to cast.

Solubility, the amount of one substance that dissolves in another, varies with temperature and, in some cases, with pressure. The variation of solubility with pressure can be a useful property. Acetylene gas, C_2H_2, for example, is used as a fuel in welding torches. It can be transported safely under pressure in cylinders as a solution in acetone, CH_3COCH_3. Acetone is a liquid, and at 1 atm pressure 1 L of the liquid dissolves 27 g of acetylene. But at 12 atm, the pressure in a full cylinder, the same quantity of acetone dissolves 320 g of acetylene, so that more can be transported.

Another useful property of solutions is their lower melting or freezing points compared with those of the major component. We have already mentioned gold–silver alloys. The use of ethylene glycol, CH_2OHCH_2OH, as an automobile antifreeze depends on the same property. Water containing ethylene glycol freezes at temperatures below the freezing point of pure water.

What determines how much the freezing point of a solution is lowered? How does the solubility of a substance change when conditions such as pressure and temperature change? These are some of the questions we will address in this chapter.

Solution Formation

A *solution* is a homogeneous mixture of two or more substances, consisting of ions or molecules. From the examples of solutions mentioned in the chapter opening, you see that they may be quite varied in their characteristics. To begin our discussion, let us look at the different types of solutions we might encounter.

12.1 Types of Solutions

Solutions may exist in any of the three states of matter; that is, they may be gases, liquids, or solids. The terms *solute* and *solvent* refer to the components of a solution. The **solute,** *in the case of a solution of a gas or solid dissolved in a liquid, is the gas or solid; in other cases, the solute is the component in smaller amount.* The **solvent,** *in a solution of a gas or solid dissolved in a liquid, is the liquid; in other cases, the solvent is the component in greater amount.*

Gaseous Solutions

In general, nonreactive gases or vapors can mix in all proportions to give a gaseous mixture. *Fluids that mix with or dissolve in each other in all proportions* are said to be **miscible fluids.** Gases are thus miscible. (If two fluids do not mix but form two layers, they are said to be *immiscible.*) Air, which is a mixture of oxygen, nitrogen, and smaller amounts of other gases, is an example of a gaseous solution.

Figure 12.1
Immiscible and miscible liquids. *Left:* Water and methylene chloride are immiscible and form two layers. *Right:* Acetone and water are miscible liquids; that is, the two substances dissolve in each other in all proportions.

Liquid Solutions

Most liquid solutions are obtained by dissolving a gas, liquid, or solid in some liquid. Soda water, for example, consists of a solution of carbon dioxide in water. Acetone in water is an example of a liquid–liquid solution. (Immiscible and miscible liquids are shown in Figure 12.1.) Seawater contains both dissolved gases (from air) and solids (mostly sodium chloride).

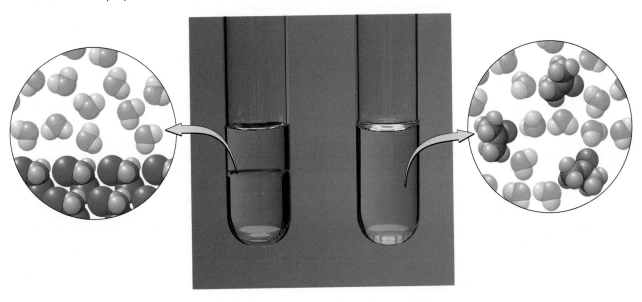

▶ Sodium melts at 98°C and potassium melts at 63°C. When the two metals are mixed to give a solution that is 20% sodium, for example, the melting point is lowered to −10°C. Potassium−sodium alloy is used as a heat-transfer medium in nuclear reactors.

It is also possible to make a liquid solution by mixing two solids together. Consider a potassium−sodium alloy. Both potassium and sodium are solids at room temperature, but a liquid solution results when the mixture contains 10% to 50% sodium. ◀

Solid Solutions

Solid solutions such as gold−silver alloys are possible. Dental-filling alloy is a solution of mercury (a liquid) in silver (a solid), with small amounts of other metals.

12.2 Solubility and the Solution Process

The amount of substance that will dissolve in a solvent depends on both the substance and the solvent. We describe the amount that dissolves in terms of solubility.

Solubility; Saturated Solutions

To understand the concept of solubility, consider the process of dissolving sodium chloride in water. Sodium chloride dissolves in water as Na^+ and Cl^- ions. Suppose you stir 40.0 g of sodium chloride crystals into 100 mL of water at 20°C. Sodium ions and chloride ions leave the surface of the crystals and enter the solution. The ions move about at random in the solution and may by chance collide with a crystal and stick, thus returning to the crystalline state. As the sodium chloride continues to dissolve, more ions enter the solution, and the rate at which they return to the crystalline state increases. Eventually, a *dynamic equilibrium* is reached in which the rate at which ions leave the crystals equals the rate at which ions return to the crystals. You write the dynamic equilibrium this way:

$$NaCl(s) \xrightleftharpoons{H_2O} Na^+(aq) + Cl^-(aq)$$

At equilibrium, no more sodium chloride appears to dissolve; 36.0 g has gone into solution, leaving 4.0 g of crystals at the bottom of the vessel. You have a **saturated solution**—that is, *a solution that is in equilibrium with respect to a given dissolved substance* (Figure 12.2). The **solubility** of sodium chloride in water (*the amount that dissolves in a given quantity of water at a given temperature to give a saturated solution*) is 36.0 g/100 mL at 20°C. An **unsaturated solution** is *a solution not in equilibrium with respect to a given dissolved substance and in which more of the substance can dissolve.*

Sometimes it is possible to obtain a **supersaturated solution,** *a solution that contains more dissolved substance than a saturated solution does.* For example, the sol-

Figure 12.2
A saturated solution.
When 40.0 g NaCl is stirred into 100 mL H₂O, 36.0 g dissolves at 20°C, leaving 4.0 g of the crystalline solid on the bottom of the beaker. This solution is saturated.

40.0 g NaCl

+

100 mL H₂O

=

Saturated solution containing 100 mL H₂O and 36.0 g NaCl

The additional 4.0 g NaCl remains undissolved

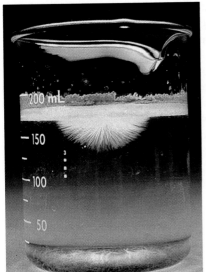

Figure 12.3
Crystallization from a supersaturated solution of sodium acetate. *Left:* Crystallization begins to occur when a small crystal of sodium acetate is added. *Center, right:* Within seconds, crystal growth spreads from the original crystal throughout the solution.

ubility of sodium thiosulfate, $Na_2S_2O_3$, in water at 100°C is 231 g/100 mL. But at room temperature, the solubility is much less—about 50 g/100 mL. Suppose you prepare a solution saturated with sodium thiosulfate at 100°C. You might expect that as the water solution is cooled, sodium thiosulfate would crystallize out. In fact, if the solution is slowly cooled to room temperature, this does not occur. Instead the result is a solution in which 231 g of sodium thiosulfate is dissolved in 100 mL of cold water, compared with the 50 g you would normally expect to find dissolved.

Supersaturated solutions are not in equilibrium with the solid substance. If a small crystal of sodium thiosulfate is added to a supersaturated solution, the excess immediately crystallizes out (see Figure 12.3).

Factors in Explaining Solubility

The solubilities of substances in one another vary widely. You might find a substance miscible in one solvent but nearly insoluble in another. As a general rule, "like dissolves like." That is, similar substances dissolve one another. Oil is miscible in gasoline. On the other hand, oil does not mix with water. Water is a polar substance, whereas hydrocarbons are not. What factors are involved in solubility?

The solubility of one substance in another can be explained in terms of two factors. One is the natural tendency of substances to mix. This is sometimes also referred to as the natural tendency toward disorder. ◀ Figure 12.4A shows a vessel divided

▶ *Entropy* is a measure of disorder. Thus, the tendency toward disorder can be expressed as a tendency toward increasing entropy. The mixing of substances increases entropy. This increase is one of the primary factors leading to the spontaneous formation of a solution.

Figure 12.4
The mixing of gas molecules. (A) A vessel is divided by a removable partition with oxygen on the left and nitrogen on the right. (B) When the partition is removed, molecules of the two gases mix through their random motions.

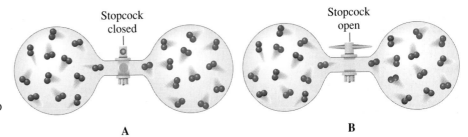

Stopcock closed

Stopcock open

A   **B**

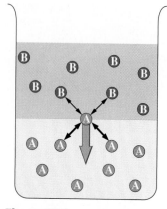

Figure 12.5
The immiscibility of liquids. Suppose an A molecule moves from liquid A into liquid B. If the intermolecular attraction between two A molecules is much stronger than the intermolecular attraction between an A molecule and a B molecule, the net force of attraction tends to pull the A molecule back into liquid A. Thus, liquid A will be immiscible with liquid B.

into two parts, with oxygen on the left and nitrogen on the right. If you remove the partition, the molecules of the two gases begin to mix. Ultimately, the molecules become thoroughly mixed through their random motions (Figure 12.4B).

If the process of dissolving one substance in another involved nothing more than simple mixing, you would expect substances to be completely soluble in one another. This is only sometimes the case. Usually, substances have limited solubility in one another. A factor that can limit solubility is the relative forces of attraction between species. Suppose there are strong attractions between solute species and strong attractions between solvent species, but weak attractions between solute and solvent species. In that case, the strongest attractions are maintained so long as the solute and solvent species do not mix. The lowest energy of the solute–solvent system is obtained then also.

The solubility of a solute in a solvent depends on a balance between the natural tendency for the solute and solvent species to mix and the tendency for a system to have the lowest energy possible.

Molecular Solutions

The simplest example of a molecular solution is one gas dissolved in another gas. Air is an example. The intermolecular forces in gases are weak. The only solubility factor of importance is the natural tendency for molecules to mix. Gases are therefore miscible.

Substances may be miscible even when the intermolecular forces are not negligible. Consider the solution of the two similar liquid hydrocarbons heptane, C_7H_{16}, and octane, C_8H_{18}, which are components of gasoline. The intermolecular attractions are due to London forces, and those between heptane and octane molecules are nearly equal to those between octane and octane molecules and heptane and heptane molecules. There are no favored attractions, so octane and heptane molecules tend to move freely through one another.

As a counterexample, consider the mixing of octane with water. There are strong hydrogen bonding forces between water molecules. For octane to mix with water, hydrogen bonds must be broken and replaced by much weaker London forces between water and octane. In this case, the maximum forces of attraction among molecules (and therefore the lower energy) result if the octane and water remain unmixed. Therefore, octane and water are nearly immiscible. (See Figure 12.5.)

The statement "like dissolves like" succinctly expresses these observations. That is, substances with similar intermolecular attractions are usually soluble in one another. The two similar hydrocarbons heptane and octane are completely miscible, whereas octane and water (with dissimilar intermolecular attractions) are immiscible.

For the series of alcohols listed in Table 12.1, the solubility in water decreases from miscible to slightly soluble. Water and alcohols are alike in having —OH groups through which strong hydrogen bonding attractions arise (Figure 12.6). The attraction between a methanol molecule, CH_3OH, and a water molecule is nearly as strong as that between two methanol molecules or between two water molecules. Methanol and water are miscible, as are ethanol and water and 1-propanol and water. However, as the hydrocarbon end, R—, of the alcohol becomes the more prominent portion of the molecule, the alcohol becomes less like water. Therefore, the solubilities of alcohols decrease with increasing length of R.

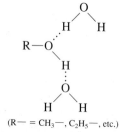

(R— = CH₃—, C₂H₅—, etc.)

Figure 12.6
Hydrogen bonding of water molecules and alcohol molecules.

Ionic Solutions

Ionic substances differ markedly in their solubilities in water. For example, sodium chloride, NaCl, has a solubility of 36 g per 100 mL of water at room temperature,

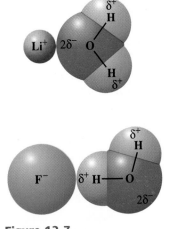

Figure 12.7
Attraction of water molecules to ions because of the ion–dipole force.
The O end of H_2O orients toward the cation, whereas an H atom of H_2O orients toward the anion.

▶ Hydration of ions also occurs in crystalline solids. For example, copper(II) sulfate pentahydrate, $CuSO_4 \cdot 5H_2O$, contains five water molecules for each $CuSO_4$ formula unit in the crystal. Substances like this, called hydrates, were discussed in Chapter 2.

▶ According to Coulomb's law, the energy of attraction of two ions is proportional to the product of the ion charges and inversely proportional to the distance between the centers of the ions.

Table 12.1
Solubilities of Alcohols in Water

Name	Formula	Solubility in H_2O (g/100 g H_2O at 20°C)
Methanol	CH_3OH	Miscible
Ethanol	CH_3CH_2OH	Miscible
1-Propanol	$CH_3CH_2CH_2OH$	Miscible
1-Butanol	$CH_3CH_2CH_2CH_2OH$	7.9
1-Pentanol	$CH_3CH_2CH_2CH_2CH_2OH$	2.7
1-Hexanol	$CH_3CH_2CH_2CH_2CH_2CH_2OH$	0.6

whereas calcium phosphate, $Ca_3(PO_4)_2$, has a solubility of only 0.002 g per 100 mL of water. In most cases, these differences in solubility can be explained in terms of the different energies of attraction between ions in the crystal and between ions and water.

The energy of attraction between an ion and a water molecule is due to an *ion–dipole force.* In the case of a positive ion (Li^+, for example), water molecules orient with their oxygen atoms toward the ion. In the case of a negative ion (for instance, F^-), water molecules orient with their hydrogen atoms toward the ion (see Figure 12.7).

The attraction of ions for water molecules is called **hydration.** ◀ Hydration of ions favors the dissolving of an ionic solid in water. Figure 12.8 illustrates the hydration of ions in the dissolving of a lithium fluoride crystal. Ions on the surface become hydrated and then move into the body of the solution as hydrated ions.

If the hydration of ions were the only factor in the solution process, you would expect all ionic solids to be soluble in water. The ions in a crystal, however, are very strongly attracted to one another. Therefore, the solubility of an ionic solid depends not only on the energy of hydration of ions but also on *lattice energy,* the energy holding ions together in the crystal lattice. Lattice energy works against the solution process.

Lattice energies depend on the charges on the ions (as well as on the distance between the centers of neighboring positive and negative ions). The greater the magnitude of ion charge, the greater is the lattice energy. ◀ For this reason, you might expect substances with singly-charged ions to be comparatively soluble and those with multiply-charged ions to be less soluble. This is borne out by the fact that compounds of the alkali metal ions (such as Na^+ and K^+) and ammonium ions (NH_4^+) are generally soluble, whereas those of phosphate ions (PO_4^{3-}), for example, are generally insoluble.

Lattice energy is also inversely proportional to the distance between neighboring ions, and this distance depends on the sum of the radii of the ions. In the series of alkaline earth hydroxides—$Mg(OH)_2$, $Ca(OH)_2$, $Sr(OH)_2$, $Ba(OH)_2$—the lattice energy decreases as the radius of the alkaline earth ion increases (from Mg^{2+} to Ba^{2+}). If the lattice energy alone determines the trend in solubilities, you should expect the solubility to increase from magnesium hydroxide to barium hydroxide. Magnesium hydroxide is insoluble in water, and barium hydroxide is soluble. But this is not the whole story. The energy of hydration also depends on ionic radius. A small

Figure 12.8
The dissolving of lithium fluoride in water. Ions on the surface of the crystal can hydrate, that is, associate with water molecules; ions at the corners are especially easy to remove because they are held by fewer lattice forces. Ions are completely hydrated in the aqueous phase and move off into the body of the liquid.

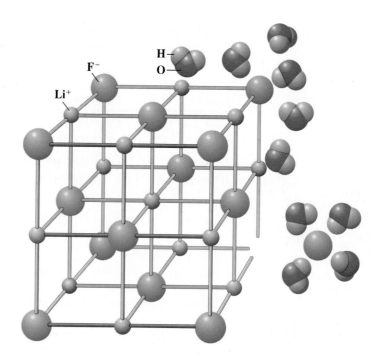

ion has a concentrated electric charge and a strong electric field that attracts water molecules. Therefore, the energy of hydration is greatest for a small ion such as Mg^{2+} and least for a large ion such as Ba^{2+}. If the energy of hydration of ions alone determined the trend in solubilities, you would expect the solubilities to decrease from magnesium hydroxide to barium hydroxide, rather than to increase.

The explanation for the observed solubility trend in the alkaline earth hydroxides is that the lattice energy decreases more rapidly in the series $Mg(OH)_2$, $Ca(OH)_2$, $Sr(OH)_2$, and $Ba(OH)_2$ than does the energy of hydration in the series of ions Mg^{2+}, Ca^{2+}, Sr^{2+}, and Ba^{2+}. For this reason, the lattice-energy factor dominates this solubility trend.

You see the opposite solubility trend when the energy of hydration decreases more rapidly, so that it dominates the trend. Consider the alkaline earth sulfates. Here the lattice energy depends on the sum of the radius of the cation and the radius of the sulfate ion. Because the sulfate ion, SO_4^{2-}, is much larger than the hydroxide ion, OH^-, the percent change in lattice energy in going through the series of sulfates from $MgSO_4$ to $BaSO_4$ is smaller than in the hydroxides. The lattice energy changes less, and the energy of hydration of the cation decreases by a greater amount. Now the energy of hydration dominates the solubility trend, and the solubility decreases from magnesium sulfate to barium sulfate. Magnesium sulfate is soluble in water, and barium sulfate is insoluble.

12.3 Effects of Temperature and Pressure on Solubility

In general, the solubility of a substance depends on temperature. For example, the solubility of ammonium nitrate in 100 mL of water is 118 g at 0°C and 811 g at 100°C.

Hemoglobin Solubility and Sickle-Cell Anemia

LIFE SCIENCE

Sickle-cell anemia was the first inherited disease shown to have a specific molecular basis. In people with the disease, the red blood cells tend to become elongated (sickle-shaped) when the concentration of oxygen (O_2) is low (see Figure 12.9). Once the red blood cells have sickled, they can no longer function in their normal capacity as oxygen carriers, and they often break apart. Moreover, the sickled cells clog capillaries, interfering with the blood supply to vital organs.

In 1949 Linus Pauling showed that people with sickle-cell anemia have abnormal hemoglobin. Hemoglobin is normally present in solution within the red blood cells. But in people with sickle-cell anemia, the unoxygenated hemoglobin readily comes out of solution. It produces a fibrous precipitate that deforms the cell, giving it the characteristic sickle shape.

Hemoglobins are large molecules (molecular weight about 64,000 amu) consisting of four protein chains. Normal and sickle-cell hemoglobins are almost exactly alike, except that the chain of the hemoglobin responsible for sickle-cell anemia differs from normal hemoglobin in one place. In this place, the normal hemoglobin has the group

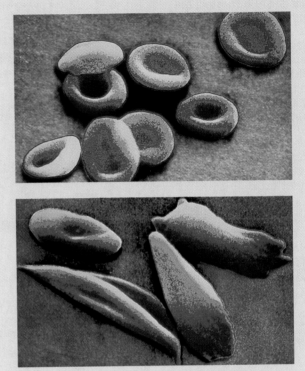

Figure 12.9
Red blood cells. Scanning electron micrographs of normal (top) and sickled (bottom) red blood cells. The color is added by computer.

$$
\begin{array}{c}
| \\
CH_2 \\
| \\
CH_2 \\
| \\
HO-C=O
\end{array}
$$

which helps confer water solubility on the molecule because of the polarity of the group and its ability to form hydrogen bonds. The abnormal hemoglobin has the following hydrocarbon group:

$$
\begin{array}{c}
| \\
H-C-CH_3 \\
| \\
CH_3
\end{array}
$$

Hydrocarbon groups are nonpolar. This small change makes the molecule less water-soluble.

Figure 12.10
Instant hot and cold compress packs. The inner bag of the instant hot compress contains calcium chloride, $CaCl_2$. When the inner bag is broken, the $CaCl_2$ dissolves exothermically in water. The instant cold compress contains an inner bag of ammonium nitrate, NH_4NO_3, which when broken apart allows the NH_4NO_3 to dissolve in the water in the outer bag. The solution process is endothermic, so the bag feels cold.

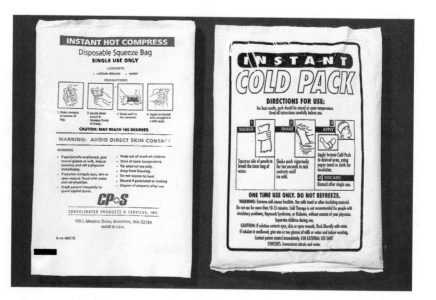

Temperature Change

Most gases become less soluble in water at higher temperatures. The first bubbles that appear when tap water is heated are bubbles of air released as the increasing temperature reduces the solubility of air in water. In contrast, most ionic solids become more soluble in water with rising temperature.

For example, potassium nitrate, KNO_3, changes solubility dramatically from 14 g/100 g H_2O at 0°C to 245 g/100 g H_2O at 100°C. Copper sulfate, $CuSO_4$, shows a moderate increase in solubility over this temperature interval. Sodium chloride, NaCl, increases only slightly in solubility with temperature.

A number of ionic compounds decrease in solubility with increasing temperature. Calcium sulfate, $CaSO_4$, and calcium hydroxide, $Ca(OH)_2$, are common examples. They are slightly soluble compounds that become even less soluble at higher temperatures.

Heat can be released or absorbed when ionic substances are dissolved in water. In some cases, this *heat of solution* is quite noticeable. When sodium hydroxide is dissolved in water, the solution becomes hot. On the other hand, when ammonium nitrate is dissolved in water, the solution becomes very cold. This cooling effect from the dissolving of ammonium nitrate in water is used in instant cold packs. An instant cold pack consists of a bag of NH_4NO_3 crystals inside a bag of water (Figure 12.10). When the inner bag is broken, NH_4NO_3 dissolves in the water. Heat is absorbed, so the bag feels cold. Similarly, hot packs are available, containing either $CaCl_2$ or $MgSO_4$, which dissolve in water with the evolution of heat.

Pressure Change

In general, pressure change has little effect on the solubility of a liquid or solid in water, but the solubility of a gas is very much affected by pressure. The qualitative effect of a change in pressure on the solubility of a gas can be predicted from Le Chatelier's principle. **Le Chatelier's principle** *states that when a system in equilibrium is disturbed by a change of temperature, pressure, or concentration variable,*

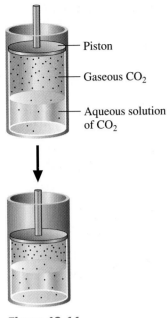

Figure 12.11
Effect of pressure on gas solubility. When the piston is pushed down, increasing the partial pressure of carbon dioxide, more gas dissolves (which tends to reduce the carbon dioxide partial pressure).

the system shifts in equilibrium composition in a way that tends to counteract this change of variable. Let us see how Le Chatelier's principle can predict the effect of a change in pressure on gas solubility.

Imagine a cylindrical vessel that is fitted with a movable piston and contains carbon dioxide over its saturated water solution (Figure 12.11). The equilibrium is

$$CO_2(g) \rightleftharpoons CO_2(aq)$$

Suppose you increase the partial pressure of CO_2 by pushing the piston down. This change of partial pressure, according to Le Chatelier's principle, shifts the equilibrium composition in a way that tends to counteract the pressure increase. From the preceding equation, you see that the partial pressure of CO_2 decreases if more CO_2 dissolves. The system comes to a new equilibrium, in which more CO_2 has dissolved. Conversely, when the partial pressure of carbon dioxide is reduced, its solubility is decreased. A bottle of carbonated beverage fizzes when the cap is removed: as the partial pressure of carbon dioxide is reduced, gas comes out of solution (Figure 12.12).

Henry's Law: Relating Pressure to the Solubility of a Gas in a Liquid

The effect of pressure on the solubility of a gas in a liquid can be predicted quantitatively. According to **Henry's law,** *the solubility of a gas is directly proportional to the partial pressure of the gas above the solution.* Expressed mathematically, the law is

$$S = k_H P$$

where S is the solubility of the gas (expressed as mass of solute per unit volume of solvent), k_H is Henry's law constant for the gas for a particular liquid at a given temperature, and P is the partial pressure of the gas. The next example shows how this formula is used.

EXAMPLE 12.1

Applying Henry's Law

In the chapter opening, we noted that 27 g of acetylene, C_2H_2, dissolves in 1 L of acetone at 1.0 atm pressure. If the partial pressure of acetylene is increased to 12 atm, what is its solubility in acetone?

▶ PROBLEM STRATEGY

Let S_1 be the solubility of the gas at partial pressure P_1, and let S_2 be the solubility at partial pressure P_2. Then you can write Henry's law for both pressures.

$$S_1 = k_H P_1$$
$$S_2 = k_H P_2$$

Dividing the second equation by the first, you get

$$\frac{S_2}{S_1} = \frac{k_H P_2}{k_H P_1} \quad \text{or} \quad \frac{S_2}{S_1} = \frac{P_2}{P_1}$$

Figure 12.12
Sudden release of pressure from a carbonated beverage.
A carbonated beverage is produced by dissolving carbon dioxide in beverage solution under pressure. More carbon dioxide dissolves at the higher pressure than otherwise. When this pressure is suddenly released, carbon dioxide is less soluble, and the excess bubbles from the solution.

You can use this relation to find the solubility at one pressure given the solubility at another.

SOLUTION

At 1.0 atm partial pressure of acetylene (P_1), the solubility S_1 is 27 g C_2H_2 per liter of acetone. For a partial pressure of 12 atm (P_2),

$$\frac{S_2}{27 \text{ g } C_2H_2/\text{L acetone}} = \frac{12 \text{ atm}}{1.0 \text{ atm}}$$

Then

$$S_2 = \frac{27 \text{ g } C_2H_2}{\text{L acetone}} \times \frac{12}{1.0} = \frac{3.2 \times 10^2 \text{ g } C_2H_2}{\text{L acetone}}$$

At 12 atm partial pressure of acetylene, 1 L of acetone dissolves **3.2×10^2 g of acetylene.**

See Problems 12.27 and 12.28.

CONCEPT CHECK 12.1

Most fish have a very difficult time surviving at elevations much above 3500 m. How could Henry's law be used to account for this fact?

Colligative Properties

The addition of ethylene glycol, CH_2OHCH_2OH, to water lowers the freezing point of water below 0°C. For example, if 0.010 mol of ethylene glycol is added to 1 kg of water, the freezing point is lowered to −0.019°C. The magnitude of freezing-point lowering is directly proportional to the number of ethylene glycol molecules added to a quantity of water. The same lowering is observed with the addition of other non-electrolyte substances.

Freezing-point lowering is a colligative property. **Colligative properties** of solutions are *properties that depend on the concentration of solute molecules or ions in solution but not on the chemical identity of the solute.* We will look into ways of expressing the concentration of a solution.

12.4 Ways of Expressing Concentration

The *concentration* of a solute is the amount of solute dissolved in a given quantity of solvent or solution. The quantity of solvent or solution can be expressed in terms of volume or in terms of mass or molar amount. Thus, there are several ways of expressing the concentration of a solution.

Recall that the *molarity* of a solution is the moles of solute in a liter of solution. ◄

$$\text{Molarity} = \frac{\text{moles of solute}}{\text{liters of solution}}$$

► Molarity was discussed in
Section 4.7.

Some other concentration units are mass percentage of solute, molality, and mole fraction.

Mass Percentage of Solute

Solution concentration is sometimes expressed in terms of the **mass percentage of solute**—that is, *the percentage by mass of solute contained in a solution.*

$$\text{Mass percentage of solute} = \frac{\text{mass of solute}}{\text{mass of solution}} \times 100\%$$

For example, an aqueous solution that is 3.5% sodium chloride by mass contains 3.5 g of NaCl in 100.0 g of solution. It could be prepared by dissolving 3.5 g of NaCl in 96.5 g of water (100.0 − 3.5 = 96.5).

EXAMPLE 12.2

Calculating with Mass Percentage of Solute

How would you prepare 425 g of an aqueous solution containing 2.40% by mass of sodium acetate, $NaC_2H_3O_2$?

SOLUTION

The mass of sodium acetate in 425 g of solution is

$$\text{Mass of } NaC_2H_3O_2 = 425 \text{ g} \times 0.0240 = 10.2 \text{ g}$$

The quantity of water in the solution is

$$\text{Mass of } H_2O = \text{mass of solution} - \text{mass of } NaC_2H_3O_2 = 425 \text{ g} - 10.2 \text{ g} = 415 \text{ g}$$

You would prepare the solution by dissolving 10.2 g of sodium acetate in 415 g of water.

See Problems 12.29 and 12.30.

Molality

The **molality** of a solution is *the moles of solute per kilogram of solvent.*

$$\text{Molality} = \frac{\text{moles of solute}}{\text{kilograms of solvent}}$$

Note that molality (denoted *m*) is defined in terms of *mass of solvent*, and molarity is defined in terms of *volume of solution*.

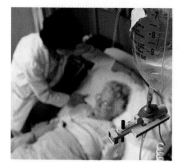

Figure 12.13
Intravenous feeding of glucose. Glucose, $C_6H_{12}O_6$, is the principal energy substance of the body and is transported to various cells by the blood. Here glucose is being fed intravenously to a patient.

EXAMPLE 12.3

Calculating the Molality of Solute

Glucose, $C_6H_{12}O_6$, is a sugar that occurs in fruits. It is also known as "blood sugar" because it is found in blood and is the body's main source of energy (see Figure 12.13). What is the molality of a solution containing 5.67 g of glucose dissolved in 25.2 g of water?

SOLUTION

The moles of glucose (MW = 180.2 amu) in 5.67 g are found as follows:

$$5.67 \text{ g } C_6H_{12}O_6 \times \frac{1 \text{ mol } C_6H_{12}O_6}{180.2 \text{ g } C_6H_{12}O_6} = 0.0315 \text{ mol } C_6H_{12}O_6$$

The mass of water is 25.2 g, or 25.2×10^{-3} kg.

$$\text{Molality} = \frac{0.0315 \text{ mol } C_6H_{12}O_6}{25.2 \times 10^{-3} \text{ kg solvent}} = \textbf{1.25 } \textit{m } \textbf{C}_6\textbf{H}_{12}\textbf{O}_6$$

See Problems 12.31, 12.32, 12.33, and 12.34.

Mole Fraction

The **mole fraction** of a component substance A (X_A) in a solution is defined as *the moles of component substance divided by the total moles of solution* (that is, moles of solute plus solvent).

$$X_A = \frac{\text{moles of substance } A}{\text{total moles of solution}}$$

The sum of the mole fractions of all the components of a solution equals 1. Multiplying mole fractions by 100 gives *mole percent*. For a two-component solution that is 0.15 mole fraction solute, you could say that 15% of the molecules are solute molecules and 85% are solvent molecules.

EXAMPLE 12.4

Calculating the Mole Fractions of Components

What are the mole fractions of glucose and water in a solution containing 5.67 g of glucose, $C_6H_{12}O_6$, dissolved in 25.2 g of water?

SOLUTION

This is the glucose solution described in Example 12.3. There you found that 5.67 g of glucose equals 0.0315 mol of glucose. The moles of water in the solution are

$$25.2 \text{ g } H_2O \times \frac{1 \text{ mol } H_2O}{18.0 \text{ g } H_2O} = 1.40 \text{ mol } H_2O$$

Hence, the total moles of solution are

$$1.40 \text{ mol} + 0.0315 \text{ mol} = 1.432 \text{ mol}$$

(You retain an extra figure in the answer for further computation.) Finally, you get

$$\textbf{Mole fraction glucose} = \frac{0.0315 \text{ mol}}{1.432 \text{ mol}} = \textbf{0.0220}$$

$$\textbf{Mole fraction water} = \frac{1.40 \text{ mol}}{1.432 \text{ mol}} = \textbf{0.978}$$

The sum of the mole fractions is 1.000.

Conversion of Concentration Units

It is relatively easy to interconvert concentration units when they are expressed in terms of mass or moles of solute and solvent, as the following example shows.

EXAMPLE 12.5

Converting Molality to Mole Fractions

An aqueous solution is 0.120 m glucose, $C_6H_{12}O_6$. What are the mole fractions of each component in the solution?

PROBLEM STRATEGY

A 0.120 m glucose solution contains 0.120 mol of glucose in 1.00 kg of water. After converting 1.00 kg H_2O to moles, you can calculate the mole fraction using its definition.

SOLUTION

The moles of H_2O in 1.00 kg of water is

$$1.00 \times 10^3 \text{ g } H_2O \times \frac{1 \text{ mol } H_2O}{18.0 \text{ g } H_2O} = 55.6 \text{ mol } H_2O$$

Hence,

$$\textbf{Mole fraction glucose} = \frac{0.120 \text{ mol}}{(0.120 + 55.6) \text{ mol}} = \textbf{0.00215}$$

$$\textbf{Mole fraction water} = \frac{55.6 \text{ mol}}{(0.120 + 55.6) \text{ mol}} = \textbf{0.998}$$

See Problems 12.35 and 12.36.

Glucose

To convert molality to molarity and vice versa, you must know the density of the solution. As the next example shows, molality and molarity are approximately equal in dilute aqueous solutions. Why?

EXAMPLE 12.6

Converting Molarity to Molality

An aqueous solution is 0.907 M $Pb(NO_3)_2$. What is the molality of lead nitrate, $Pb(NO_3)_2$, in this solution? The density of the solution is 1.252 g/mL.

PROBLEM STRATEGY

There is 0.907 mol $Pb(NO_3)_2$ per liter of solution. Consider 1 L ($= 1.000 \times 10^3$ mL) of solution and calculate its mass, using the density. You can then calculate the mass of lead nitrate and find the mass of water by difference. Now you can calculate the molality.

SOLUTION

$$\text{Mass of solution} = \text{density} \times \text{volume}$$
$$= 1.252 \text{ g/mL} \times 1.000 \times 10^3 \text{ mL}$$
$$= 1.252 \times 10^3 \text{ g}$$

The mass of lead nitrate is

$$0.907 \text{ mol } Pb(NO_3)_2 \times \frac{331.2 \text{ g } Pb(NO_3)_2}{1 \text{ mol } Pb(NO_3)_2} = 3.00 \times 10^2 \text{ g } Pb(NO_3)_2$$

The mass of water is

$$\text{Mass of } H_2O = \text{mass of solution} - \text{mass of } Pb(NO_3)_2$$
$$= 1.252 \times 10^3 \text{ g} - 3.00 \times 10^2 \text{ g}$$
$$= 9.52 \times 10^2 \text{ g} (= 0.952 \text{ kg})$$

Hence, the molality of lead nitrate in this solution is

$$\frac{0.907 \text{ mol } Pb(NO_3)_2}{0.952 \text{ kg solvent}} = \textbf{0.953 } \boldsymbol{m} \textbf{ } Pb(NO_3)_2$$

See Problems 12.37 and 12.38.

12.5 Vapor Pressure of a Solution

During the nineteenth century, chemists observed that the vapor pressure of a volatile solvent was lowered by addition of a nonvolatile solute. **Vapor-pressure lowering** of a solvent is *a colligative property equal to the vapor pressure of the pure solvent minus the vapor pressure of the solution.* For example, water at 20°C has a vapor pressure of 17.54 mmHg. Ethylene glycol, CH_2OHCH_2OH, is a liquid whose vapor pressure at 20°C is relatively low; it can be considered to be nonvolatile compared with water. An aqueous solution containing 0.0100 mole fraction of ethylene glycol has a vapor pressure of 17.36 mmHg. Thus, the vapor-pressure lowering, ΔP, of water is

$$\Delta P = 17.54 \text{ mmHg} - 17.36 \text{ mmHg} = 0.18 \text{ mmHg}$$

Figure 12.14 shows a demonstration of vapor-pressure lowering.

In about 1886, the French chemist François Marie Raoult observed that the partial vapor pressure of solvent over a solution of a nonelectrolyte solute depends on the mole fraction of solvent in the solution. Consider a solution of volatile solvent,

Figure 12.14
Demonstration of vapor-pressure lowering. *(A)* Two beakers, each containing water solutions, are placed under a bell jar. The solution in the left beaker is less concentrated than that in the right beaker, so its vapor pressure is greater. As a result, vapor leaves the solution on the left (which becomes more concentrated) and condenses on the solution on the right (which becomes less concentrated). *(B)* After some time, the two solutions become equal in concentration and in vapor pressure.

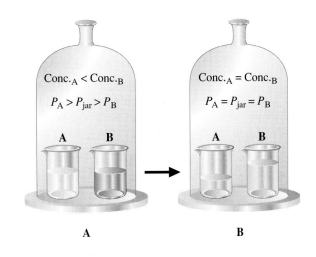

A, and nonelectrolyte solute, *B*, which may be volatile or nonvolatile. According to **Raoult's law,** *the partial pressure of solvent, P_A, over a solution equals the vapor pressure of the pure solvent, P_A°, times the mole fraction of solvent, X_A, in the solution.*

$$P_A = P_A^\circ X_A$$

If the solute is nonvolatile, P_A is the total vapor pressure of the solution. Because the mole fraction of solvent in a solution is always less than 1, the vapor pressure of the solution of a nonvolatile solute is less than that for the pure solvent. In general, Raoult's law is observed to hold for dilute solutions. If the solvent and solute are chemically similar, Raoult's law may hold for all mole fractions.

You can obtain an explicit expression for the vapor-pressure lowering of a solvent in a solution assuming Raoult's law holds and that the solute is a nonvolatile nonelectrolyte. The vapor-pressure lowering, ΔP, is

$$\Delta P = P_A^\circ - P_A$$

Substituting Raoult's law gives

$$\Delta P = P_A^\circ - P_A^\circ X_A = P_A^\circ(1 - X_A)$$

But the sum of the mole fractions of the components of a solution must equal 1; that is, $X_A + X_B = 1$. So $X_B = 1 - X_A$. Therefore,

$$\Delta P = P_A^\circ X_B$$

From this equation, you can see that the vapor-pressure lowering is a colligative property—one that depends on the concentration, but not on the nature, of the solute. Thus, if the mole fraction of ethylene glycol, X_B, in an aqueous solution is doubled, the vapor-pressure lowering is doubled. Also, because the previous equation does not depend on the characteristics of the solute (other than its being nonvolatile and a nonelectrolyte), a solution that is 0.010 mole fraction urea, $(NH_2)_2CO$, has the same vapor-pressure lowering as one that is 0.010 mole fraction ethylene glycol. The next example illustrates the use of the previous equation.

EXAMPLE 12.7

Calculating Vapor-Pressure Lowering

Calculate the vapor-pressure lowering of water when 5.67 g of glucose, $C_6H_{12}O_6$, is dissolved in 25.2 g of water at 25°C. The vapor pressure of water at 25°C is 23.8 mmHg. What is the vapor pressure of the solution?

SOLUTION

This is the glucose solution described in Example 12.4. According to the calculations performed there, the solution is 0.0220 mole fraction glucose. Therefore, the vapor-pressure lowering is

$$\Delta P = P_A^{\circ}X_B = 23.8 \text{ mmHg} \times 0.0220 = \textbf{0.524 mmHg}$$

The vapor pressure of the solution is

$$P_A = P_A^{\circ} - \Delta P = (23.8 - 0.524) \text{ mmHg} = \textbf{23.3 mmHg}$$

See Problems 12.39 and 12.40.

 CONCEPT CHECK 12.2

You need to boil a water-based solution at a temperature lower than 100°C. What kind of liquid could you add to the water to make this happen?

12.6 Boiling-Point Elevation and Freezing-Point Depression

The *normal* boiling point of a liquid is the temperature at which its vapor pressure equals 1 atm. Because the addition of a nonvolatile solute to a liquid reduces its vapor pressure, the temperature must be increased to a value greater than the normal boiling point to achieve a vapor pressure of 1 atm. Figure 12.15 shows the vapor-pressure curve of a solution. The curve is below the vapor-pressure curve of the pure liquid solvent. The **boiling-point elevation,** ΔT_b, is *a colligative property of a solution equal to the boiling point of the solution minus the boiling point of the pure solvent.*

The boiling-point elevation, ΔT_b, is found to be proportional to the molal concentration, c_m, of the solution (for dilute solutions).

$$\Delta T_b = K_b c_m$$

The constant of proportionality, K_b (called the *boiling-point-elevation constant*), depends only on the solvent. Table 12.2 lists values of K_b, as well as boiling points, for some solvents. Benzene, for example, has a boiling-point-elevation constant of 2.61°C/m.

Figure 12.15 also shows the effect of a dissolved solute on the freezing point of a solution. Usually it is the pure solvent that freezes out of solution. Sea ice, for example, is almost pure water. For that reason, the vapor-pressure curve for the solid is unchanged. Therefore, the freezing point shifts to a lower temperature. The **freezing-point depression,** ΔT_f, is *a colligative property of a solution equal to the freezing point of the pure solvent minus the freezing point of the solution.* (ΔT_f is shown in Figure 12.15.)

Figure 12.15
Phase diagram showing the effect of a nonvolatile solute on freezing point and boiling point. Note that the freezing point (T_f) is lowered and the boiling point (T_b) is elevated.

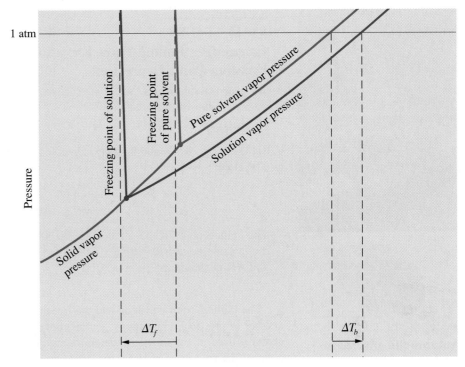

Freezing-point depression, ΔT_f, like boiling-point elevation, is proportional to the molal concentration, c_m (for dilute solutions).

$$\Delta T_f = K_f c_m$$

Here K_f is the *freezing-point-depression constant* and depends only on the solvent. Table 12.2 gives values of K_f for some solvents. The freezing-point-depression constant for benzene is 5.07°C/m.

Table 12.2
Boiling-Point-Elevation Constants (K_b) and Freezing-Point-Depression Constants (K_f)

Solvent	Formula	Boiling Point (°C)	Freezing Point (°C)	K_b (°C/m)	K_f (°C/m)
Acetic acid	$HC_2H_3O_2$	118.5	16.60	3.08	3.59
Benzene	C_6H_6	80.2	5.455	2.61	5.065
Camphor	$C_{10}H_{16}O$	—	179.5	—	40
Carbon disulfide	CS_2	46.3	—	2.40	—
Cyclohexane	C_6H_{12}	80.74	6.55	2.79	20.0
Ethanol	C_2H_5OH	78.3	—	1.07	—
Water	H_2O	100.000	0.000	0.512	1.858

[Data are taken from Landolt-Börnstein, 6th ed., *Zahlenwerte und Functionen aus Physik, Chemie, Astronomie, Geophysik, und Technik*, Vol. II, Part IIa (Heidelberg: © Springer-Verlag, 1960), pp. 844–849 and pp. 918–919.]

Figure 12.16
Automobile antifreeze mixtures. The main ingredient of automobile antifreeze mixtures is ethylene glycol, CH_2OHCH_2OH. This substance has low vapor pressure, so it does not easily vaporize away. Ethylene glycol is also used to produce polyester fibers and plastics. The substance has a sweet taste (the name glycol derives from the Greek word *glukus,* meaning "sweet") but is poisonous.

EXAMPLE 12.8

Calculating Boiling-Point Elevation and Freezing-Point Depression

An aqueous solution is 0.0222 *m* glucose. What are the boiling point and the freezing point of this solution?

SOLUTION

Table 12.2 gives K_b and K_f for water as 0.512°C/*m* and 1.86°C/*m*, respectively. Therefore,

$$\Delta T_b = K_b c_m = 0.512°C/m \times 0.0222\ m = 0.0114°C$$
$$\Delta T_f = K_f c_m = 1.86°C/m \times 0.0222\ m = 0.0413°C$$

The boiling point of the solution is 100.000°C + 0.0114°C = **100.011°C,** and the freezing point is 0.000°C − 0.0413°C = **−0.041°C.** Note that ΔT_b is added and ΔT_f is subtracted.

The boiling-point elevation and the freezing-point depression of solutions have a number of practical applications. We mentioned in the chapter opening that ethylene glycol is used in automobile radiators as an antifreeze because it lowers the freezing point of the coolant (Figure 12.16). The same substance also helps prevent the radiator coolant from boiling away by elevating the boiling point. Sodium chloride and calcium chloride are spread on icy roads in the winter to lower the melting point of ice and snow below the temperature of the surrounding air. Salt–ice mixtures are used as freezing mixtures in home ice-cream makers.

Colligative properties are also used to obtain molecular weights. Although the mass spectrometer is now often used for routine determinations of the molecular weight of pure substances, colligative properties are still employed to obtain information about the species in solution. Freezing-point depression is often used because it is simple to determine a melting point or a freezing point (Figure 12.17). From the freezing-point lowering, you can calculate the molal concentration, and from the molality, you can obtain the molecular weight. The next example illustrates this calculation.

EXAMPLE 12.9

Calculating the Molecular Weight of a Solute from Molality

A solution is prepared by dissolving 0.131 g of a substance in 25.4 g of water. The molality of the solution is determined by freezing-point depression to be 0.056 *m*. What is the molecular weight of the substance?

PROBLEM STRATEGY

Molality equals moles of solute (moles of substance) divided by kilograms of solvent (H_2O). By substituting values for molality and kilograms H_2O, you can solve for moles of substance. The molar mass of the substance equals mass of substance (0.131 g) divided by moles of substance.

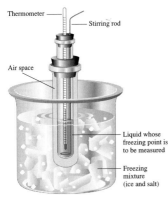

Figure 12.17
Determination of the freezing point of a liquid.
The liquid is cooled by means of a freezing mixture. To control the rate of temperature decrease, the liquid is separated from the freezing mixture by an air space.

SOLUTION

The molality of an aqueous solution is

$$\text{Molality} = \frac{\text{moles of substance}}{\text{kg } H_2O}$$

Hence, for the given solution,

$$0.056 \frac{\text{mol}}{\text{kg } H_2O} = \frac{\text{moles of substance}}{25.4 \times 10^{-3} \text{ kg } H_2O}$$

Rearranging this equation, you get

$$\text{Moles of substance} = 0.056 \frac{\text{mol}}{\text{kg } H_2O} \times 25.4 \times 10^{-3} \text{ kg } H_2O = 1.42 \times 10^{-3} \text{ mol}$$

(An extra digit is retained for further computation.) The molar mass of substance equals

$$\text{Molar mass} = \frac{0.131 \text{ g}}{1.42 \times 10^{-3} \text{ mol}} = 92.3 \text{ g/mol}$$

You round the answer to two significant figures. The molecular weight is **92 amu.**

See Problems 12.43 and 12.44.

12.7 Osmosis

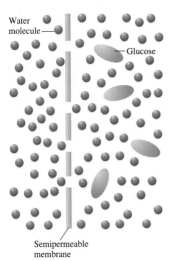

Figure 12.18
A semipermeable membrane separating water and an aqueous solution of glucose. Here the membrane is depicted as being similar to a sieve—the pores are too small to allow the glucose molecules to pass.

Certain membranes allow solvent molecules to pass through them but not solute molecules, particularly not those of large molecular weight. Such a membrane is called *semipermeable* and might be an animal bladder, a vegetable tissue, or a piece of cellophane. Figure 12.18 depicts the operation of a semipermeable membrane. **Osmosis** is *the phenomenon of solvent flow through a semipermeable membrane to equalize the solute concentrations on both sides of the membrane.* When two solutions of the same solvent are separated by a semipermeable membrane, solvent molecules migrate through the membrane in both directions. However, the rate of solvent migration is faster from the solution of low concentration to the solution of high concentration.

Figure 12.19 shows an experiment that demonstrates osmosis. A concentrated glucose solution is placed in an inverted funnel whose mouth is sealed with a semipermeable membrane. The funnel containing the glucose solution is then placed in a beaker of pure water. As water flows from the beaker through the membrane into the funnel, the liquid level rises in the stem of the funnel.

The glucose solution continues to rise up the funnel stem until the downward pressure exerted by the solution above the membrane eventually stops the upward flow of solvent (water). In general, **osmotic pressure** is *a colligative property of a solution equal to the pressure that, when applied to the solution, just stops osmosis.*

The osmotic pressure π of a solution is related to the molar concentration of solute, M:

$$\pi = MRT$$

Here R is the gas constant and T is the absolute temperature.

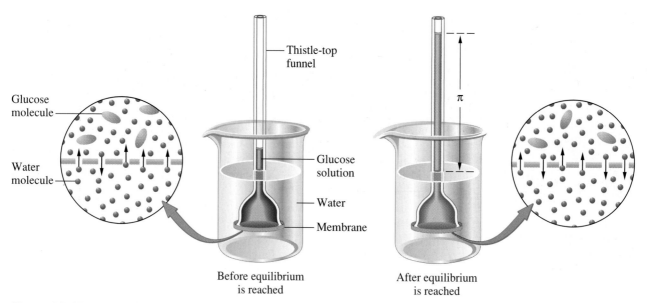

Thistle-top funnel

π

Glucose molecule

Water molecule

Glucose solution

Water

Membrane

Before equilibrium is reached

After equilibrium is reached

Figure 12.19
An experiment in osmosis.
Left: Before reaching equilibrium, water molecules pass through the membrane into the glucose solution in the inverted funnel at a greater rate than they return (osmosis). This causes the liquid level in the funnel to rise. *Right:* The column of liquid in the funnel eventually reaches sufficient height to exert a downward pressure, π (the osmotic pressure), causing the rates of water molecules entering and leaving the glucose solution to become equal. Once the osmotic pressure, π, has been reached, osmosis stops.

Often osmosis is used to determine the molecular weights of polymeric substances. Polymers are very large molecules generally made up from a simple, repeating unit. Although polymer solutions can be fairly concentrated in terms of grams per liter, on a mole-per-liter basis they may be quite dilute. The freezing-point depression is usually too small to measure, though the osmotic pressure may be appreciable.

EXAMPLE 12.10

Calculating Osmotic Pressure

The formula for low-molecular-weight starch is $(C_6H_{10}O_5)_n$, where n averages 2.00×10^2. When 0.798 g of starch is dissolved in 100.0 mL of water solution, what is the osmotic pressure at 25°C?

▶ PROBLEM STRATEGY

Obtain the molecular weight of $(C_6H_{10}O_5)_{200}$, and use it to obtain the molarity of the starch solution. Substitute into the formula for the osmotic pressure.

SOLUTION

The molecular weight of $(C_6H_{10}O_5)_{200}$ is 32,400 amu. The number of moles in 0.798 g of starch is

$$0.798 \text{ g starch} \times \frac{1 \text{ mol starch}}{32,400 \text{ g starch}} = 2.46 \times 10^{-5} \text{ mol starch}$$

The molarity of the solution is

$$\frac{2.46 \times 10^{-5} \text{ mol}}{0.1000 \text{ L solution}} = 2.46 \times 10^{-4} \text{ mol/L solution}$$

and the osmotic pressure at 25°C is

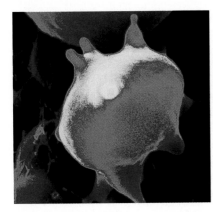

(A)

(B)

(C)

Figure 12.20
The importance of osmotic pressure to cells. Shown here are color-enhanced electron micrographs of red blood cells in solutions of various osmotic pressures. *(A)* The osmotic pressure of the solution is greater than that of the cell; the cell has collapsed. *(B)* The osmotic pressure of the solution is equal to that of the cell; the cell has its normal round shape with depressed center. *(C)* The osmotic pressure of the solution is less than that of the cell; the cell has a bloated shape.

▶ Osmotic pressure appears to be important for the rising of sap in a tree. During the day, water evaporates from the leaves of the tree, so the aqueous solution in the leaves becomes more concentrated and the osmotic pressure increases. Sap flows upward to dilute the water solution in the leaves.

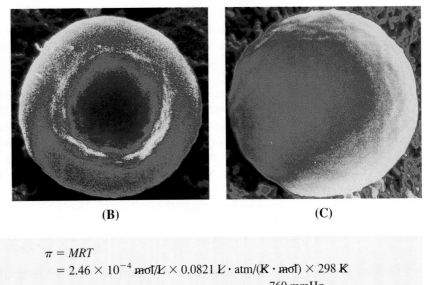

$$\pi = MRT$$
$$= 2.46 \times 10^{-4} \text{ mol/L} \times 0.0821 \text{ L} \cdot \text{atm}/(K \cdot \text{mol}) \times 298 \text{ K}$$
$$= 6.02 \times 10^{-3} \text{ atm} = 6.02 \times 10^{-3} \text{ atm} \times \frac{760 \text{ mmHg}}{1 \text{ atm}}$$
$$= \textbf{4.58 mmHg}$$

See Problems 12.45 and 12.46.

Osmosis is important in many biological processes. ◀ A cell might be thought of as an aqueous solution enclosed by a semipermeable membrane. The solution surrounding the cell must have an osmotic pressure equal to that within the cell. Otherwise, water would either leave the cell, dehydrating it, or enter the cell and possibly burst the membrane. For intravenous feeding (adding a nutrient solution to the venous blood supply of a patient), it is necessary that the nutrient solution have exactly the osmotic pressure of blood plasma. If it does not, the blood cells may collapse or burst as a result of osmosis (Figure 12.20).

The process of *reverse osmosis* has been applied to the problem of purifying water. In particular, the method has been used to *desalinate* ocean water. In normal osmosis, the solvent flows from a dilute solution through a membrane to a more concentrated solution. By applying a pressure equal to the osmotic pressure to the more concentrated solution, the process of osmosis can be stopped. By applying an even greater pressure, the osmotic process can be reversed. Then, solvent flows from the concentrated solution (which could be ocean water) through a membrane to the more dilute solution (which could be more or less pure water).

CONCEPT CHECK 12.3

Explain why pickles are stored in a brine (salt) solution. What would the pickles look like if they were stored in water?

12.8 Colligative Properties of Ionic Solutions

To explain the colligative properties of ionic solutions, you must realize that it is the total concentration of ions, rather than the concentration of ionic substance, that is important. For example, the freezing-point depression of 0.100 m sodium chloride solution is nearly twice that of 0.100 m glucose solution. Each formula unit of NaCl gives two particles. You can write the freezing-point lowering more generally as

$$\Delta T_f = iK_f c_m$$

where i is the number of ions resulting from each formula unit and c_m is the molality computed on the basis of the formula of the ionic compound. The equations for the other colligative properties must be similarly modified by the factor i.

Actually, the freezing points of ionic solutions agree with the previous equation only when the solutions are quite dilute. The i values calculated from the freezing-point depression are usually smaller than the number of ions obtained from the formula unit. For example, a 0.029 m aqueous solution of potassium sulfate, K_2SO_4, has a freezing point of $-0.14°C$. Hence,

$$i = \frac{\Delta T_f}{K_f c_m} = \frac{0.14°C}{1.86°C/m \times 0.029 \, m} = 2.6$$

▶ The value of i equal to $\Delta T_f/K_f c_m$ is often called the *van't Hoff factor*. Thus, the van't Hoff factor for 0.029 m K_2SO_4 is 2.6.

You might have expected a value of 3, on the basis of the fact that K_2SO_4 ionizes to give three ions. ◀ In 1923 Peter Debye and Erich Hückel were able to show that the colligative properties of salt solutions could be explained by assuming that the salt is completely ionized in solution but that the *activities,* or effective concentrations, of the ions are less than their actual concentrations as a result of the electrical interactions of the ions in solution. The *Debye–Hückel theory* allows us to calculate these activities. When this is done, excellent agreement is obtained for dilute solutions.

EXAMPLE 12.11

Determining Colligative Properties of Ionic Solutions

Estimate the freezing point of a 0.010 m aqueous solution of aluminum sulfate, $Al_2(SO_4)_3$. Assume the value of i based on the formula of the compound.

SOLUTION

When aluminum sulfate, $Al_2(SO_4)_3$, dissolves in water, it dissociates into five ions.

$$Al_2(SO_4)_3(s) \xrightarrow{H_2O} 2Al^{3+}(aq) + 3SO_4{}^{2-}(aq)$$

Therefore, you assume $i = 5$. The freezing-point depression is

$$\Delta T_f = iK_f c_m = 5 \times 1.86°C/m \times 0.010 \, m = 0.093°C$$

The estimated freezing point of the solution is $0.000°C - 0.093°C = $ **$-0.093°C$.**

See Problems 12.47 and 12.48.

CONCEPT CHECK 12.4

Each of the following substances is dissolved in a separate 10.0-L container of water: 1.5 mol NaCl, 1.3 mol Na_2SO_4, 2.0 mol $MgCl_2$, and 2.0 mol KBr. Without doing extensive calculations, rank the boiling points of each of the solutions from highest to lowest.

Colloid Formation

Colloids appear homogeneous like solutions, but they consist of comparatively large particles of one substance dispersed throughout another substance.

12.9 Colloids

A **colloid** is *a dispersion of particles of one substance (the dispersed phase) throughout another substance or solution (the continuous phase).* Fog is an example of a colloid: it consists of very small water droplets (dispersed phase) in air (continuous phase). A colloid differs from a true solution in that the dispersed particles are larger than normal molecules, though they are too small to be seen with a microscope. The particles range from about 1×10^3 pm to about 2×10^5 pm in size.

Tyndall Effect

▶ Although all gases and liquids scatter light, the scattering from a pure substance or true solution is quite small and usually not detectable. However, because of the considerable depth of the atmosphere, the scattering of light by air molecules can be seen. The blue color of the sky is due to the fact that blue light is scattered more easily than red light.

Although a colloid appears to be homogeneous because the dispersed particles are quite small, it can be distinguished from a true solution by its ability to scatter light. *The scattering of light by colloidal-size particles* is known as the **Tyndall effect.** ◀ For example, the atmosphere appears to be a clear gas, but a ray of sunshine against a dark background shows up many fine dust particles by light scattering. Similarly, when a beam of light is directed through clear gelatin (a colloid, not a true solution), the beam becomes visible by the scattering of light from colloidal gelatin particles. The beam appears as a ray passing through the solution (Figure 12.21). When the same experiment is performed with a true solution, such as an aqueous solution of sodium chloride, the beam of light is not visible.

Types of Colloids

Colloids are characterized according to the state (solid, liquid, or gas) of the dispersed phase and of the continuous phase. Table 12.3 lists various types of colloids and some examples of each. Fog and smoke are **aerosols,** which are *liquid droplets or solid particles dispersed throughout a gas.* An **emulsion** consists of *liquid droplets dispersed throughout another liquid* (as particles of butterfat are dispersed through homogenized milk). A **sol** consists of *solid particles dispersed in a liquid.*

Figure 12.21
A demonstration of the Tyndall effect by a colloid.
A light beam is visible perpendicular to its path only if light is scattered toward the viewer. The vessel on the left contains a colloid, which scatters light. The vessel on the right contains a true solution, which scatters negligible light.

Table 12.3
Types of Colloids

Continuous Phase	Dispersed Phase	Name	Example
Gas	Liquid	Aerosol	Fog, mist
Gas	Solid	Aerosol	Smoke
Liquid	Gas	Foam	Whipped cream
Liquid	Liquid	Emulsion	Mayonnaise (oil dispersed in water)
Liquid	Solid	Sol	$AgCl(s)$ dispersed in H_2O
Solid	Gas	Foam	Pumice, plastic foams
Solid	Liquid	Gel	Jelly, opal (mineral with liquid inclusions)
Solid	Solid	Solid sol	Ruby glass (glass with dispersed metal)

Hydrophilic and Hydrophobic Colloids

Colloids in which the continuous phase is water are divided into two major classes: hydrophilic colloids and hydrophobic colloids. A **hydrophilic colloid** is *a colloid in which there is a strong attraction between the dispersed phase and the continuous phase (water).* Many such colloids consist of macromolecules (very large molecules) dispersed in water. Except for the large size of the dispersed molecules, these colloids are like normal solutions. Protein solutions, such as gelatin in water, are hydrophilic colloids. Gelatin molecules are attracted to water molecules by London forces and hydrogen bonding.

A **hydrophobic colloid** is *a colloid in which there is a lack of attraction between the dispersed phase and the continuous phase (water).* Hydrophobic colloids are basically unstable. Given sufficient time, the dispersed phase aggregates into larger particles. In this behavior, they are quite unlike true solutions and hydrophilic colloids. The time taken to separate may be extremely long, however. A colloid of gold particles in water prepared by Michael Faraday in 1857 is still preserved in the British Museum in London. This colloid is hydrophobic as well as a sol (solid particles dispersed in water).

Association Colloids

When molecules or ions that have both a hydrophobic and a hydrophilic end are dispersed in water, they associate, or aggregate, to form colloidal-sized particles, or micelles. A **micelle** is *a colloidal-sized particle formed in water by the association of molecules or ions that each have a hydrophobic end and a hydrophilic end.* The hydrophobic ends point inward toward one another, and the hydrophilic ends are on the outside of the micelle facing the water. *A colloid in which the dispersed phase consists of micelles* is called an **association colloid.**

Ordinary soap in water provides an example of an association colloid. Soap consists of compounds such as sodium stearate, $C_{17}H_{35}COONa$. The stearate ion has a long hydrocarbon end that is hydrophobic (because it is nonpolar) and a carboxyl group (COO^-) at the other end that is hydrophilic (because it is ionic).

Figure 12.22
A stearate micelle in a water solution.   Stearate ions associate in groups (micelles), with their hydrocarbon ends pointing inward. The ionic ends, on the outside of the micelle, point into the water solution.

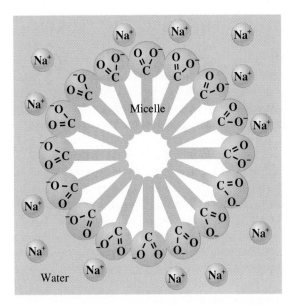

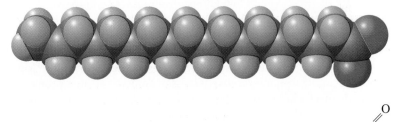

$$CH_3CH_2CH_2CH_2CH_2CH_2CH_2CH_2CH_2CH_2CH_2CH_2CH_2CH_2CH_2CH_2CH_2C{\overset{O}{\underset{O^-}{\diagup}}}$$

hydrophobic end   hydrophilic end

stearate ion

In water solution, the stearate ions associate into micelles in which the hydrocarbon ends point inward toward one another and away from the water, and ionic carboxyl groups are on the outside of the micelle facing the water (Figure 12.22).

The cleansing action of soap occurs because oil and grease can be absorbed into the hydrophobic centers of soap micelles and washed away (Figure 12.23). Synthetic detergents also form association colloids. Sodium lauryl sulfate is a synthetic detergent present in toothpastes and shampoos.

Figure 12.23
Formation of an association colloid with soap.   *Left:* Vegetable oil floating on water (dyed green). *Right:* When the mixture is shaken with soap, an emulsion forms as the oil droplets are absorbed into soap micelles.

$$CH_3CH_2CH_2CH_2CH_2CH_2CH_2CH_2CH_2CH_2CH_2CH_2OSO_3^-\ Na^+$$
sodium lauryl sulfate

It has a hydrophilic sulfate group ($-OSO_3^-$) and a hydrophobic dodecyl group ($C_{12}H_{25}-$, the hydrocarbon end).

A CHEMIST LOOKS AT

The World's Smallest Test Tubes

FRONTIERS

Phospholipids are the main constituents of the membranes of living cells. These membranes enclose the chemical substances of a cell and allow these substances to carry on the reactions needed for life processes unimpeded by random events external to the membrane. Recently, chemists have prepared cell-like structures—*lipid vesicles*—from phospholipids and have used them to carry out chemical reactions by combining the contents of two lipid vesicles, with diameters from about 50 nanometers to several micrometers. Richard Zare, at Stanford University, who was one of those who described these experiments, has called these lipid vesicles "the world's smallest test tubes."

Like the soap ions, phospholipids have the intriguing property of self-assembly into groups. Soap ions in water naturally group themselves into micelles. Phospholipids, however, have a chemical structure that, while similar to that of soap ions, precludes their assembly into micelles. Rather, they form layer structures. And under the right circumstances, a layer may fold back on itself like the skin of a basketball to enclose a space, forming a vesicle.

Lecithins are typical phospholipids. Lipids are biological substances like fats and oils that are soluble in or-

ganic solvents; phospholipids contain phosphate groups. Here is the general structure of a lecithin:

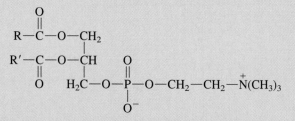

R and R′ denote long hydrocarbon groups, with perhaps 14 to 18 carbon atoms.

Note that, like a soap ion, a phospholipid has a hydrophobic end (the two hydrocarbon groups) and a hydrophilic end (an ionic portion containing a phosphate group, $-PO_4^--$). Like soap ions, phospholipid molecules in water tend to associate so that their hydrophilic (or ionic) "heads" dip into the water, with their hydrophobic (or hydrocarbon) "tails" pointing away. However, the hydrocarbon tails of a phospholipid molecule are too bulky to associate into micelles and instead form a *bilayer,* a layer two molecules thick. Figure 12.24 shows a model of a cell membrane that displays this bilayer structure—two phospholipid layers, their hydrocarbon tails pointing inward toward each other. The hy-

A Checklist for Review

Important Terms

solute (12.1)
solvent (12.1)
miscible fluids (12.1)
saturated solution (12.2)
solubility (12.2)
unsaturated solution (12.2)
supersaturated
  solution (12.2)
hydration (of ions) (12.2)

Le Chatelier's
  principle (12.3)
Henry's law (12.3)
colligative properties
  (12.4)
mass percentage of
  solute (12.4)
molality (12.4)
mole fraction (12.4)

vapor-pressure
  lowering (12.5)
Raoult's law (12.5)
boiling-point elevation (12.6)
freezing-point
  depression (12.6)
osmosis (12.7)
osmotic pressure (12.7)
colloid (12.9)

Tyndall effect (12.9)
aerosols (12.9)
emulsion (12.9)
sol (12.9)
hydrophilic colloid (12.9)
hydrophobic colloid (12.9)
micelle (12.9)
association colloid (12.9)

drophilic heads are rendered in blue; the hydrophobic tails are in green. Note that the interior of a membrane (shown in green) is "oily" from the hydrocarbon groups, and therefore repels molecules or ions dissolved in the water solutions outside and inside a cell. (Protein molecules embedded in the cell membrane form channels or active pumps to transport specific molecules and ions across the cell membrane.)

In one of their experiments, Zare and his colleagues brought two phospholipid vesicles, each about a micrometer in diameter and containing different reactant substances, together under a special microscope. When the vesicles just touched, the researchers delivered an electrical pulse to them that opened a small pore in each vesicle. The two vesicles coalesced into one, and their contents reacted. The scientists detected the product molecules from the orange fluorescence they gave off when radiated with ultraviolet light. The techniques these researchers are developing should allow scientists to mimic the conditions in biological cells. And lipid vesicles might one day be designed to deliver drugs to specific cells, say cancer cells.

Specific proteins associated with the membrane surface

Specific proteins embedded in phospholipid bilayer

Figure 12.24 A model of a cell membrane. The membrane consists of a phospholipid bilayer with protein molecules inserted in it.

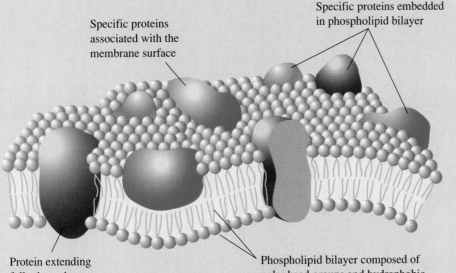

Protein extending fully through the membrane

Phospholipid bilayer composed of polar head groups and hydrophobic tails of phospholipid molecules

Key Equations

$$S = k_H P$$

$$\text{Mass percentage of solute} = \frac{\text{mass of solute}}{\text{mass of solution}} \times 100\%$$

$$\text{Molality} = \frac{\text{moles of solute}}{\text{kilograms of solvent}}$$

$$X_A = \frac{\text{moles of substance } A}{\text{total moles of solution}}$$

$$P_A = P_A^\circ X_A$$

$$\Delta P = P_A^\circ X_B$$

$$\Delta T_b = K_b c_m$$

$$\Delta T_f = K_f c_m$$

$$\pi = MRT$$

Summary of Facts and Concepts

Solutions are homogeneous mixtures and can be gases, liquids, or solids. Often one substance will dissolve in another only to a limited extent. The maximum amount that dissolves at equilibrium is the *solubility* of that substance. Solubility is explained in terms of the natural tendency toward disorder and by the tendency to result in the strongest forces of attraction between species. The dissolving of one molecular substance in another is limited when intermolecular forces strongly favor the unmixed substances. When two substances have similar types of intermolecular forces, they tend to be soluble ("like dissolves like"). Ionic substances differ greatly in their solubilities in water, because solubility depends on the relative values of *lattice energy* and *hydration energy*.

Solubilities usually vary with temperature. At higher temperatures, most gases will become less soluble in water, whereas most ionic solids will become more soluble. Pressure has a significant effect on the solubility only of a gas. A gas is more soluble in a liquid if the partial pressure of the gas is increased, which is in agreement with *Le Chatelier's principle*. According to *Henry's law,* the solubility of a gas is directly proportional to the partial pressure of the gas.

Colligative properties of solutions depend only on the concentration of solute particles. The concentration may be defined by the *molarity, mass percentage of solute, molality,* or *mole fraction*. One example of a colligative property is the *vapor-pressure lowering* of a volatile solvent by the addition of a nonvolatile solute. According to *Raoult's law,* the vapor pressure of this solution depends on the mole fraction of solvent. Because adding a nonvolatile solute lowers the vapor pressure, the boiling point must be raised to bring the vapor pressure back to 1 atm *(boiling-point elevation)*. Such a solution also exhibits a *freezing-point depression*. Boiling-point elevation and freezing-point depression are colligative properties. *Osmosis* is another colligative property. In osmosis, there is a flow of solvent through a semipermeable membrane in order to equalize the concentrations of solutions on the two sides of the membrane. Colligative properties may be used to measure solute concentration. In this way, one can determine the molecular weight of the solute. The colligative properties of an ionic solution depend on the concentration of ions.

A *colloid* is a dispersion of particles of one substance (about 1×10^3 pm to 2×10^5 pm in size) throughout another. Colloids can be detected by the *Tyndall effect* (the scattering of light by colloidal-size particles). Colloids are characterized by the state of the dispersed phase and the state of the continuous phase. An aerosol, for example, consists of liquid droplets or solid particles dispersed in a gas. A *hydrophilic colloid* consists of a dispersed phase that is attracted to the water. Many of these colloids have macromolecules dissolved in water. In *hydrophobic colloids,* there is little attraction between the dispersed phase and the water. An *association colloid* consists of molecules with a hydrophobic end and a hydrophilic end dispersed in water. These molecules associate into colloidal-size groups, or *micelles*.

Operational Skills

1. **Applying Henry's law**   Given the solubility of a gas at one pressure, find its solubility at another pressure. **(EXAMPLE 12.1)**
2. **Calculating solution concentration**   Given the mass percent of solute, state how to prepare a given mass of solution **(EXAMPLE 12.2)**. Given the masses of solute and solvent, find the molality **(EXAMPLE 12.3)** and mole fractions **(EXAMPLE 12.4)**.
3. **Converting concentration units**   Given the molality of a solution, calculate the mole fractions of solute and solvent; and given the mole fractions, calculate the molality **(EXAMPLE 12.5)**. Given the density, find the molality from the molarity **(EXAMPLE 12.6)**.
4. **Calculating vapor-pressure lowering**   Given the mole fraction of solute in a solution of nonvolatile, undissociated solute and the vapor pressure of pure solvent, calculate the vapor-pressure lowering and vapor pressure of the solution. **(EXAMPLE 12.7)**

5. **Calculating boiling-point elevation and freezing-point depression**   Given the molality of a solution of nonvolatile, undissociated solute, calculate the boiling-point elevation and freezing-point depression. **(EXAMPLE 12.8)**
6. **Calculating molecular weights**   Given the masses of solvent and solute and the molality of the solution, find the molecular weight of the solute **(EXAMPLE 12.9)**.
7. **Calculating osmotic pressure**   Given the molarity and the temperature of a solution, calculate its osmotic pressure.   **(EXAMPLE 12.10)**
8. **Determining colligative properties of ionic solutions**   Given the concentration of ionic compound in a solution, calculate the magnitude of a colligative property; if *i* is not given, assume the value based on the formula of the ionic compound.   **(EXAMPLE 12.11)**

Review Questions

12.1   Give one example of each: a gaseous solution, a liquid solution, a solid solution.

12.2   What are the two factors needed to explain the differences in solubilities of substances?

12.3   Using the concept of hydration, describe the process of dissolving a sodium chloride crystal in water.

12.4   What is the usual solubility behavior of an ionic compound in water when the temperature is raised? Give an example of an exception to this behavior.

12.5   Give one example of each: a salt whose heat of solution is exothermic and a salt whose heat of solution is endothermic.

12.6   What do you expect to happen to a concentration of dissolved gas in a solution as the solution is heated?

12.7   Pressure has an effect on the solubility of oxygen in water but a negligible effect on the solubility of sugar in water. Why?

12.8 Four ways were discussed to express the concentration of a solute in solution. Identify them and define each concentration unit.

12.9 Explain why the boiling point of a solution containing a nonvolatile solute is higher than the boiling point of a pure solvent.

12.10 List two applications of freezing-point depression.

12.11 Explain the process of reverse osmosis to produce drinkable water from ocean water.

12.12 Give an example of an aerosol, a foam, an emulsion, a sol, and a gel.

12.13 Explain on the basis that "like dissolves like" why glycerol, $CH_2OHCHOHCH_2OH$, is miscible in water but benzene, C_6H_6, has very limited solubility in water.

12.14 Explain how soap removes oil from a fabric.

Conceptual Problems

12.15 Even though the oxygen demands of trout and bass are different, they can exist in the same body of water. However, if the temperature of the water in the summer gets above about 23°C, the trout begin to die, but not the bass. Why is this the case?

12.16 You want to purchase a salt to melt snow and ice on your sidewalk. Which one of the following salts would best accomplish your task using the least amount: KCl, $CaCl_2$, PbS_2, $MgSO_4$, or AgCl?

12.17 Ten grams of the hypothetical ionic compounds XZ and YZ are each placed in a separate 2.0-L beaker of water. XZ completely dissolves, whereas YZ is insoluble. The energy of hydration of the Y^+ ion is greater than that of the X^+ ion. Explain this difference in solubility.

12.18 Small amounts of a nonvolatile, nonelectrolyte solute and a volatile solute are each dissolved in separate beakers containing 1 kg of water. If the number of moles of each solute is equal:
a. Which solution will have the higher vapor pressure?
b. Which solution will boil at a higher temperature?

12.19 A green leafy salad wilts if left too long in a salad dressing containing vinegar and salt. Explain what happens.

12.20 People have proposed towing icebergs to arid parts of the earth as a way to deliver freshwater. Explain why icebergs do not contain salts although they are formed by the freezing of ocean water (i.e., saltwater).

Practice Problems

Types of Solutions

12.21 Give an example of a liquid solution prepared by dissolving a gas in a liquid.

12.22 Give an example of a solid solution prepared from two solids.

Solubility

12.23 Arrange the following substances in order of increasing solubility in hexane, C_6H_{14}: CH_2OHCH_2OH, $C_{10}H_{22}$, H_2O.

12.24 Indicate which of the following is more soluble in ethanol, C_2H_5OH: acetic acid, CH_3COOH, or stearic acid, $C_{17}H_{35}COOH$.

12.25 Which of the following ions would be expected to have the greater energy of hydration, Mg^{2+} or Al^{3+}?

12.26 Which of the following ions would be expected to have the greater energy of hydration, F^- or Cl^-?

12.27 The solubility of carbon dioxide in water is 0.161 g CO_2 in 100 mL of water at 20°C and 1.00 atm. A soft drink is carbonated with carbon dioxide gas at 5.50 atm pressure. What is the solubility of carbon dioxide in water at this pressure?

12.28 Nitrogen, N_2, is soluble in blood and can cause intoxication at sufficient concentration. For this reason, the U.S. Navy advises divers using compressed air not to go below 125 feet. The total pressure at this depth is 4.79 atm. If the solubility of nitrogen at 1.00 atm is 1.75×10^{-3} g/100 mL of water, and the mole percent of nitrogen in air is 78.1, what is the solubility of nitrogen in water from air at 4.79 atm?

Solution Concentration

12.29 How would you prepare 72.5 g of an aqueous solution that is 2.50% potassium iodide, KI, by mass?

12.30 How would you prepare 455 g of an aqueous solution that is 6.50% sodium sulfate, Na_2SO_4, by mass?

12.31 Vanillin, $C_8H_8O_3$, occurs naturally in vanilla extract and is used as a flavoring agent. A 39.1-mg sample of vanillin was dissolved in 168.5 mg of diphenyl ether, $(C_6H_5)_2O$. What is the molality of vanillin in the solution?

12.32 Lauryl alcohol, $C_{12}H_{25}OH$, is prepared from coconut oil; it is used to make sodium lauryl sulfate, a synthetic detergent. What is the molality of lauryl alcohol in a solution of 17.1 g lauryl alcohol dissolved in 165 g ethanol, C_2H_5OH?

12.33 Fructose, $C_6H_{12}O_6$, is a sugar occurring in honey and fruits. The sweetest sugar, it is nearly twice as sweet as sucrose (cane or beet sugar). How much water should be added to 3.50 g of fructose to give a 0.125 m solution?

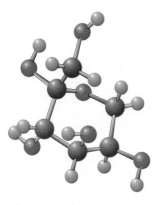

Fructose

12.34 Caffeine, $C_8H_{10}N_4O_2$, is a stimulant found in tea and coffee. A sample of the substance was dissolved in 45.0 g of chloroform, $CHCl_3$, to give a 0.0946 m solution. How many grams of caffeine were in the sample?

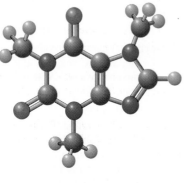

Caffeine

12.35 A bleaching solution contains sodium hypochlorite, NaClO, dissolved in water. The solution is 0.650 m NaClO. What is the mole fraction of sodium hypochlorite?

12.36 An antiseptic solution contains hydrogen peroxide, H_2O_2, in water. The solution is 0.655 m H_2O_2. What is the mole fraction of hydrogen peroxide?

12.37 A solution of vinegar is 0.763 M acetic acid, $HC_2H_3O_2$. The density of the vinegar is 1.004 g/mL. What is the molal concentration of acetic acid?

12.38 A beverage contains tartaric acid, $H_2C_4H_4O_6$, a substance obtained from grapes during wine making. If the beverage is 0.271 M tartaric acid, what is the molal concentration? The density of the solution is 1.016 g/mL.

Colligative Properties

12.39 Calculate the vapor pressure at 35°C of a solution made by dissolving 20.2 g of sucrose, $C_{12}H_{22}O_{11}$, in 70.1 g of water. The vapor pressure of pure water at 35°C is 42.2 mmHg. What is the vapor-pressure lowering of the solution? (Sucrose is nonvolatile.)

12.40 What is the vapor pressure at 23°C of a solution of 1.20 g of naphthalene, $C_{10}H_8$, in 25.6 g of benzene, C_6H_6? The vapor pressure of pure benzene at 23°C is 86.0 mmHg; the vapor pressure of naphthalene can be neglected. Calculate the vapor-pressure lowering of the solution.

12.41 An aqueous solution of a molecular compound freezes at −0.086°C. What is the molality of the solution?

12.42 Urea, $(NH_2)_2CO$, is dissolved in 100.0 g of water. The solution freezes at −0.085°C. How many grams of urea were dissolved to make this solution?

12.43 Safrole is contained in oil of sassafras and was once used to flavor root beer. A 2.39-mg sample of safrole was dissolved in 103.0 mg of diphenyl ether. The solution had a melting point of 25.70°C. Calculate the molecular weight of safrole. The freezing point of pure diphenyl ether is 26.84°C, and the freezing-point-depression constant, K_f, is 8.00°C/m.

12.44 Butylated hydroxytoluene (BHT) is used as an antioxidant in processed foods. (It prevents fats and oils from becoming rancid.) A solution of 2.500 g of BHT in 100.0 g of benzene had a freezing point of 4.880°C. What is the molecular weight of BHT?

12.45 Dextran is a polymeric carbohydrate produced by certain bacteria. It is used as a blood plasma substitute. An aqueous solution contains 0.582 g of dextran in 106 mL of solution at 21°C. It has an osmotic pressure of 1.47 mmHg. What is the average molecular weight of the dextran?

12.46 Arginine vasopressin is a pituitary hormone. It helps regulate the amount of water in the blood by reducing the flow of urine from the kidneys. An aqueous solution containing 21.6 mg of vasopressin in 100.0 mL of solution has an osmotic pressure at 25°C of 3.70 mmHg. What is the molecular weight of the hormone?

12.47 A 0.0140-g sample of an ionic compound with the formula $Cr(NH_3)_5Cl_3$ was dissolved in water to give 25.0 mL of solution at 25°C. The osmotic pressure was determined to be 119 mmHg. How many ions are obtained from each formula unit when the compound is dissolved in water?

12.48 In a mountainous location, the boiling point of pure water is found to be 95°C. How many grams of sodium chloride must be added to 1 kg of water to bring the boiling point back to 100°C? Assume that $i = 2$.

Colloids

12.49 Give the type of colloid (aerosol, foam, emulsion, sol, or gel) that each of the following represents.
a. rain cloud   b. milk of magnesia
c. soapsuds   d. silt in water

12.50 Give the type of colloid (aerosol, foam, emulsion, sol, or gel) that each of the following represents.
a. ocean spray   b. beaten egg white
c. dust cloud   d. salad dressing

General Problems

12.51 An aqueous solution is 8.50% ammonium chloride, NH_4Cl, by mass. The density of the solution is 1.024 g/mL. What are the molality, mole fraction, and molarity of NH_4Cl in the solution?

12.52 An aqueous solution is 27.0% lithium chloride, LiCl, by mass. The density of the solution is 1.127 g/mL. What are the molality, mole fraction, and molarity of LiCl in the solution?

12.53 A 75-g sample of a gaseous fuel mixture contains 0.51 mole fraction propane, C_3H_8; the remainder of the mixture is butane, C_4H_{10}. What are the masses of propane and butane in the sample?

12.54 The diving atmosphere used by the U.S. Navy in its undersea Sea-Lab experiments consisted of 0.036 mole fraction O_2 and 0.056 mole fraction N_2, with helium (He) making up the remainder. What are the masses of nitrogen, oxygen, and helium in a 7.84-g sample of this atmosphere?

12.55 A sample of potassium aluminum sulfate 12-hydrate, $KAl(SO_4)_2 \cdot 12H_2O$, containing 118.6 mg is dissolved in 1.000 L of solution. Calculate the following for the solution:
a. The molarity of $KAl(SO_4)_2$.
b. The molarity of SO_4^{2-}.
c. The molality of $KAl(SO_4)_2$, assuming that the density of the solution is 1.00 g/mL.

12.56 A sample of aluminum sulfate 18-hydrate, $Al_2(SO_4)_3 \cdot 18H_2O$, containing 159.3 mg is dissolved in 1.000 L of solution. Calculate the following for the solution:
a. The molarity of $Al_2(SO_4)_3$.
b. The molarity of SO_4^{2-}.
c. The molality of $Al_2(SO_4)_3$, assuming that the density of the solution is 1.00 g/mL.

12.57 Urea, $(NH_2)_2CO$, has been used to melt ice from sidewalks, because the use of salt is harmful to plants. If the saturated aqueous solution contains 44% urea by mass, what is the freezing point? (The answer will be approximate, because the equation in the text applies accurately only to dilute solutions.)

12.58 Calcium chloride, $CaCl_2$, has been used to melt ice from roadways. Given that the saturated solution is 32% $CaCl_2$ by mass, estimate the freezing point.

12.59 The osmotic pressure of blood at 37°C is 7.7 atm. A solution that is given intravenously must have the same osmotic pressure as the blood. What should be the molarity of a glucose solution to give an osmotic pressure of 7.7 atm at 37°C?

12.60 Maltose, $C_{12}H_{22}O_{11}$, is a sugar produced by malting (sprouting) grain. A solution of maltose at 25°C has an osmotic pressure of 5.50 atm. What is the molar concentration of maltose?

12.61 Which aqueous solution has the lower freezing point, 0.10 m $CaCl_2$ or 0.10 m glucose?

12.62 Which aqueous solution has the lower boiling point, 0.10 m KCl or 0.10 m $CaCl_2$?

12.63 Commercially, sulfuric acid is obtained as a 98% solution. If this solution is 18 M, what is its density? What is its molality?

12.64 Phosphoric acid is usually obtained as an 85% phosphoric acid solution. If it is 15 M, what is the density of this solution? What is its molality?

12.65 A compound of manganese, carbon, and oxygen contains 28.17% Mn and 30.80% C. When 0.125 g of this compound is dissolved in 5.38 g of cyclohexane, the solution freezes at 5.28°C. What is the molecular formula of this compound?

12.66 A compound of cobalt, carbon, and oxygen contains 28.10% C and 34.47% Co. When 0.147 g of this compound is dissolved in 6.72 g of cyclohexane, the solution freezes at 5.23°C. What is the molecular formula of this compound?

Cumulative-Skills Problems

12.67 The lattice enthalpy of sodium chloride, $\Delta H°$ for

$$NaCl(s) \longrightarrow Na^+(g) + Cl^-(g)$$

is 787 kJ/mol; the heat of solution in making up 1 M NaCl(aq) is +4.0 kJ/mol. From these data, obtain the sum of the heats of hydration of Na^+ and Cl^-. That is, obtain the sum of $\Delta H°$ values for

$$Na^+(g) \longrightarrow Na^+(aq)$$
$$Cl^-(g) \longrightarrow Cl^-(aq)$$

If the heat of hydration of Cl^- is −338 kJ/mol, what is the heat of hydration of Na^+?

12.68 The lattice enthalpy of potassium chloride is 717 kJ/mol; the heat of solution in making up 1 M KCl(aq) is +18.0 kJ/mol. Using the value for the heat of hydration of Cl^- given in Problem 12.67, obtain the heat of hydration of K^+. Compare this with the value you obtained for Na^+ in Problem 12.67. Explain the relative values of Na^+ and K^+.

12.69 A solution is made up by dissolving 15.0 g MgSO$_4$ · 7H$_2$O in 100.0 g of water. What is the molality of MgSO$_4$ in this solution?

12.70 A solution is made up by dissolving 15.0 g Na$_2$CO$_3$ · 10H$_2$O in 100.0 g of water. What is the molality of Na$_2$CO$_3$ in this solution?

12.71 An aqueous solution is 15.0% by mass of copper(II) sulfate pentahydrate, $CuSO_4 \cdot 5H_2O$. What is the molarity of $CuSO_4$ in this solution at 20°C? The density of this solution at 20°C is 1.167 g/mL.

12.72 An aqueous solution is 20.0% by mass of sodium thiosulfate pentahydrate, $Na_2S_2O_3 \cdot 5H_2O$. What is the molarity of $Na_2S_2O_3$ in this solution at 20°C? The density of this solution at 20°C is 1.174 g/mL.

12.73 A compound of carbon, hydrogen, and oxygen was burned in oxygen, and 1.000 g of the compound produced 1.434 g CO_2 and 0.783 g H_2O. In another experiment, 0.1107 g of the compound was dissolved in 25.0 g of water. This solution had a freezing point of −0.0894°C. What is the molecular formula of the compound?

12.74 A compound of carbon, hydrogen, and oxygen was burned in oxygen, and 1.000 g of the compound produced 1.418 g CO_2 and 0.871 g H_2O. In another experiment, 0.1103 g of the compound was dissolved in 45.0 g of water. This solution had a freezing point of −0.0734°C. What is the molecular formula of the compound?

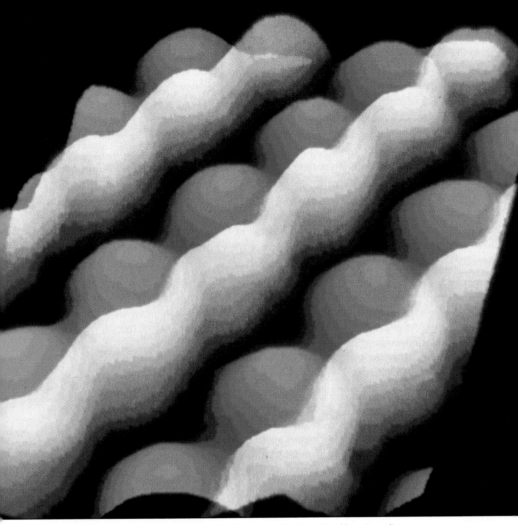

An STM image of the atomic structure of gallium arsenide, GaAs (Ga, blue; As, red).

Materials of Technology

In the summer of 2000, scientists at Bell Laboratories and Oxford University reported that they had constructed a motor out of DNA, the chemical material of the genetic code. This molecular-scale motor used DNA as both a structural material and as a fuel to run the motor. Earlier, Harvard University chemists had constructed molecular-scale tweezers out of carbon-atom tubes, showing that they could manipulate clusters of polystyrene molecules. These scientists' efforts represent some of the many ongoing investigations into "nanotechnology," in which one manipulates materials on a molecular scale to create useful devices.

Telecommunications is an example of an area in which the development of new materials has had immense impact, leading to rapid change in the technology. Initially, telecommunications was restricted to voice communication by telephone using copper wires to carry a message in the form of an electrical signal. Today, it is just

407

Figure 13.1
Racing bicycle. The frame is made of a special high-chromium steel, which is strong but lightweight.

as likely that fibers of pure glass, or fiber-optic cables, carry the message in the form of light pulses. An advantage of fiber-optic cable is that it can carry much more information than a copper-wire cable of similar size. Fiber-optic cables can easily carry television and Internet signals as well as voice messages. At present, fiber-optic communications requires that electrical signals be converted to light pulses then back to electrical signals for processing by, say, a computer. The question scientists are now asking is this: Can the communications processing be done entirely with light, possibly resulting in faster communications and more compact equipment? This, of course, would require new materials with special properties.

In this chapter, we begin by discussing metals and metallurgy. Metals are central to modern technology, and scientists continue to develop new metals with special properties. (See Figure 13.1.) We will then look at some nonmetallic materials, including the fullerenes, which are new materials made entirely of carbon atoms. We end the chapter by looking at *composites,* materials made of two or more different kinds of materials. Bone is a natural composite of calcium phosphate crystals and a protein called collagen. Calcium phosphate lends structural strength to bone, but by itself is brittle. Collagen binds the calcium phosphate crystals into a stable, somewhat elastic whole. Modern human-made composites try to mimic and extend the properties of such natural composites.

Metals and Metallurgy

Metals are electrical conductors. Over 4 million metric tons (4×10^9 kg) of copper are used annually worldwide for electrical purposes alone (Figure 13.2). Metals are also generally good conductors of heat. Metals are more or less malleable and ductile. By *malleable,* we mean you can pound the material into flat sheets; by *ductile,* we mean you can draw it into a wire. Thus we can define a **metal** as *a material that is lustrous (shiny), has high electrical and heat conductivities, and is malleable and ductile.*

A metal need not be a pure element; it can be a compound or mixture. An **alloy** is *a material with metallic properties that is either a compound or a mixture.* If the alloy is a mixture, it may be homogeneous (a solution) or heterogeneous. Most commercial metals are alloys consisting usually of one metal with small quantities of some

Figure 13.2
Metals as electrical conductors. *Left:* A power transmission line; the wire is aluminum. *Right:* Aluminum and copper wire.

Table 13.1
Ten Most Abundant Elements in the Earth's Crust

Element	Mass Percentage	Element	Mass Percentage
1. Oxygen	46.4	6. Sodium	2.8
2. Silicon	27.7	7. Potassium	2.6
3. Aluminum	8.1	8. Magnesium	2.1
4. Iron	5.0	9. Titanium	0.4
5. Calcium	3.6	10. Hydrogen	0.1

other metals. Gold jewelry, for example, is made from an alloy that is a solid solution of gold containing some silver. Pure gold is too soft for jewelry; the alloy is much harder.

13.1 Natural Sources of the Metallic Elements

A major source of metals and their compounds is the earth's crust. Mine shafts go to depths of less than about 4 km (compared with the earth's radius of 6371 km); therefore, we access only the thinnest outer layer of the earth. Table 13.1 lists the ten most abundant elements in the earth's crust and the mass percentage of each element in this part of the earth.

The two most abundant elements in the earth's crust are oxygen and silicon. These elements, along with various metals, are the ingredients of *silicates,* which are the chief substances of rocks and clay (Figure 13.3). Clay and many rocks are *aluminosilicates,* silicates in which aluminum is an important constituent. These aluminosilicates account for the vast quantity of aluminum in the earth's crust. ◄

The commercial sources of most metals are not silicates, however, because generally metals are not easily obtained from them. Despite the abundance and availability of clay, an economical method of producing aluminum from clay has yet to be discovered. The usual compounds from which metals are obtained are oxides (and hydroxides), sulfides, and carbonates. Aluminum is produced from bauxite.

Bauxite is technically a *rock;* geologists define a rock as a naturally occurring solid material composed of one or more minerals. A **mineral** is *a naturally occurring inorganic solid substance or solid solution with a definite crystalline structure.* One of the constituents of bauxite is gibbsite, which is a mineral form of aluminum hydroxide, $Al(OH)_3$. Corundum is an oxide mineral of aluminum; its formula is Al_2O_3. Some of the aluminum atoms in corundum may be replaced by chromium, forming a mineral that is a red solid solution of Al_2O_3 and Cr_2O_3. When this mineral is of gemstone quality, it is known as ruby (Figure 13.4).

An **ore** is *a rock or mineral from which a metal or nonmetal can be economically produced.* Bauxite is the principal ore of aluminum. Iron, the fourth most abundant element in the earth's crust, has several ores, the most important of which is the mineral hematite. Hematite, which is iron(III) oxide, Fe_2O_3, varies in form and color. The earthy, red-colored form is known as red ocher; in addition to its use as an ore, it is used as a paint pigment (Figure 13.5). Metal ores are not evenly distributed over the earth. Table 13.2 lists some of the ores of important metals, along with their country of origin.

▶ The structure of silicates and aluminosilicates is discussed in Section 13.6.

Figure 13.3
Pictured Rocks. The Pictured Rocks along the shore of Lake Superior in Michigan are sandstone cliffs composed principally of silicon and oxygen. The beautiful colors of the cliffs are due to streaks of hematite, or iron(III) oxide.

13.2 Metallurgy

No doubt nuggets of gleaming metal in sand and rock initially fascinated early humans simply because of their appearance. Eventually, however, someone discovered that metals could be fashioned into tools. The first metals used were probably those that occurred naturally in the free state, such as gold. Later, it was found that heating certain rocks in a hot charcoal fire yielded metals. We realize now that these rocks contain metal compounds that can be reduced to the metallic state by reaction with charcoal (carbon) or the carbon monoxide obtained from the partial burning of charcoal. Copper was probably the first metal produced this way. Later, it was discovered that rocks containing copper and tin compounds yielded bronze, the first manufactured alloy. It was from these beginnings that metallurgy arose. **Metallurgy** is *the scientific study of the production of metals from their ores and the making of alloys having various useful properties.* In this section, we will look at the basic steps in the production of a metal from its ore:

1. *Preliminary treatment.* Usually, an ore must first be treated in some way to concentrate its metal-containing portion. During the preliminary treatment, the metal-containing mineral is separated from less desirable parts of the ore.

2. *Reduction.* Unless the metal occurs free, the metal compound obtained from preliminary treatment has to be reduced. Electrolysis or chemical reduction may be used, depending on the metal.

3. *Refining.* Once the free metal is obtained, it may have to be purified (refined).

Preliminary Treatment

A metal ore contains varying quantities of economically worthless material along with the mineral containing the metallic element. *The worthless portion of an ore* is called **gangue** (pronounced "gang"). To separate the useful mineral from the gangue, both physical and chemical methods are used, depending on the ore.

Figure 13.4
Ruby. Ruby is aluminum oxide, Al_2O_3, containing a small percentage of chromium(III) oxide, Cr_2O_3.

Figure 13.5
Yellow and red ocher. Yellow and red ocher are natural, earthy forms of iron(III) oxide, Fe_2O_3. The colors depend on the crystal form, which can be varied by strong heating.

Table 13.2
Principal Ores of Some Important Metals

Element	Ore	Where Found
Aluminum, Al	Bauxite, $AlO(OH)$ and $Al(OH)_3$	Surinam, Jamaica, Guyana
Chromium, Cr	Chromite, $FeCr_2O_4$	Russia, South Africa
Copper, Cu	Chalcopyrite, $CuFeS_2$	USA, Chile, Canada, Kazakhstan
Iron, Fe	Hematite, Fe_2O_3	Australia, Ukraine, USA
Lead, Pb	Galena, PbS	USA, Australia, Canada
Lithium, Li	Spodumene, $LiAlSi_2O_6$	USA, Canada, Brazil
Mercury, Hg	Cinnabar, HgS	Spain, Algeria
Nickel, Ni	Pentlandite, NiS–FeS mixture	Canada
Tin, Sn	Cassiterite, SnO_2	Malaysia, Bolivia
Zinc, Zn	Sphalerite, ZnS	Canada, USA, Australia

Flotation is *a physical method of separating a mineral from the gangue that depends on differences in their wettabilities by a liquid solution* (Figure 13.6). *Flotation agents* are used to coat the metal-bearing mineral particles selectively so they are wet less easily by water than is the gangue. Particles of metal-bearing mineral become coated with the flotation agent. A molecule of this substance has a polar end, which is attracted to the mineral, and a water-repellent (hydrophobic) end. ◀ Mineral particles become coated with molecules of the flotation agent with their hydrophobic ends pointing outward. The coated mineral particles are more strongly attracted to air bubbles than to the water solution and so float to the surface, where they are trapped by the froth. The froth is removed at the top of the flotation vessel and the gangue is removed at the bottom. Sulfide ores, such as those of copper, lead, and zinc, are concentrated in this way.

▶ The action of a flotation agent has some similarity to detergent action; see Section 12.9.

The **Bayer process** is *a chemical procedure in which purified aluminum oxide, Al_2O_3, is separated from the aluminum ore bauxite.* It depends on the fact that aluminum hydroxide is amphoteric, meaning that it behaves both as a base and as an acid. Bauxite, as we noted earlier, contains aluminum hydroxide, $Al(OH)_3$, and aluminum oxide hydroxide, $AlO(OH)$, as well as relatively worthless constituents such as silicate minerals with some iron oxides. When bauxite is mixed with hot, aqueous sodium hydroxide solution, the amphoteric aluminum minerals, along with some silicates, dissolve in the strong base. The aluminum minerals give the aluminate ion, $Al(OH)_4^-$.

$$Al(OH)_3(s) + OH^-(aq) \longrightarrow Al(OH)_4^-(aq)$$
$$AlO(OH)(s) + OH^-(aq) + H_2O(l) \longrightarrow Al(OH)_4^-(aq)$$

Other substances, including the iron oxides, remain undissolved and are filtered off. As the hot solution of sodium aluminate cools, aluminum hydroxide precipitates, leaving the soluble silicates in solution.

$$Al(OH)_4^-(aq) \longrightarrow Al(OH)_3(s) + OH^-(aq)$$

The aluminum hydroxide is finally *calcined* (heated strongly in a furnace) to produce aluminum oxide.

$$2Al(OH)_3(s) \xrightarrow{\Delta} Al_2O_3(s) + 3H_2O(g)$$

Bauxite is also a source of gallium, the element under aluminum in the periodic table. Gallium is important in preparing gallium arsenide for solid-state devices, including lasers for compact disc players and light-emitting diodes, or LEDs (Figure 13.7). Sodium gallate, $Na[Ga(OH)_4]$, is more soluble than the corresponding aluminum compound, and it remains in the filtrate from the Bayer process after aluminum hydroxide has been filtered off.

Once an ore is concentrated, it may be necessary to convert the mineral to a compound more suitable for reduction. **Roasting** is *the process of heating a mineral in air to obtain the oxide.* Sulfide minerals, such as zinc ore (containing the mineral sphalerite, ZnS), are usually roasted before reducing them to the metal.

$$2ZnS(s) + 3O_2(g) \longrightarrow 2ZnO(s) + 2SO_2(g); \Delta H° = -884 \text{ kJ}$$

The roasting process is exothermic, because it is essentially the burning of the sulfide ore. Once the ore has been heated to initiate roasting, additional heating is unnecessary.

One of the products of the roasting of a sulfide ore is sulfur dioxide. Sulfur dioxide oxidizes further to yield sulfur trioxide, which dissolves in moisture in the air to

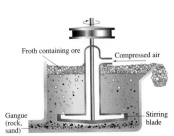

Froth containing ore

Compressed air

Gangue (rock, sand)

Stirring blade

Figure 13.6
Flotation process for concentrating certain ores. The ore attaches to bubbles of air and is carried off in the froth. The gangue settles to the bottom of the tank, where it is withdrawn.

yield *acid rain*. Considerable damage to the environment has occurred as a result. Modern ore-roasting facilities capture most of the sulfur dioxide and produce sulfuric acid. ◄

► Acid rain is discussed in the essay at the end of Section 17.2.

Reduction

A metal, if it is not already free, must be obtained from one of its compounds by reduction, using either electrolysis or chemical reduction.

Electrolysis The reactive main-group metals, such as lithium and sodium, are obtained by *electrolysis*. Electrolysis uses an electric current to reduce a metal compound to the metal. Lithium, for example, is obtained commercially by electrolysis of molten lithium chloride, LiCl.

Magnesium is another reactive metal obtained by electrolysis of the chloride. Magnesium ion, Mg^{2+}, is the third most abundant dissolved ion in the oceans, after Cl^- and Na^+. The oceans, therefore, are an essentially inexhaustible supply of magnesium ion, from which the metal can be obtained. The **Dow process,** *a commercial method for isolating magnesium from seawater,* depends on the fact that magnesium ion can be precipitated from aqueous solution by adding a base. Figure 13.8 shows a flowchart for the process. Seashells provide the source of the base (calcium hydroxide). Seashells are principally calcium carbonate, $CaCO_3$, which when heated decomposes to calcium oxide, CaO, and carbon dioxide, CO_2. Calcium oxide reacts with water to produce calcium hydroxide. The base, calcium hydroxide, provides hydroxide ions that react with magnesium ions in seawater to precipitate magnesium hydroxide:

$$Mg^{2+}(aq) + 2OH^-(aq) \longrightarrow Mg(OH)_2(s)$$

The magnesium hydroxide is allowed to settle out in ponds; here it gives a suspension, or slurry, of $Mg(OH)_2$. The slurry is filtered to recover the magnesium hydroxide precipitate, which is then treated with hydrochloric acid to yield magnesium chloride. The dry magnesium chloride salt is melted and electrolyzed at 700°C to yield magnesium metal and chlorine gas.

$$MgCl_2(l) \xrightarrow{\text{electrolysis}} Mg(l) + Cl_2(g)$$

Aluminum is manufactured from the aluminum oxide obtained in the Bayer process using electrolysis. The method for doing this was discovered in 1886 by Charles Martin Hall (then a student at Oberlin College) in the United States and, independently, by Paul Héroult in France. (Both men were 22 years old at the time.) Before this discovery, aluminum was a precious metal, and in the 1850s it was so rare and exotic that the French court used aluminum tableware at royal functions in place of the usual silver and gold utensils.

The **Hall–Héroult process** is *the commercial method for producing aluminum by the electrolysis of a molten mixture of aluminum oxide in cryolite,* Na_3AlF_6. Originally, the process used natural cryolite. Today, the Hall–Héroult process uses synthetic cryolite, or sodium hexafluoroaluminate, produced by reacting aluminum hydroxide from the Bayer process with sodium hydroxide and hydrofluoric acid.

$$3NaOH(aq) + Al(OH)_3(s) + 6HF(aq) \longrightarrow Na_3AlF_6(aq) + 6H_2O(l)$$

A Hall–Héroult electrolytic cell consists of a rectangular steel shell lined first with an insulating material, then with carbon (from baked petroleum coke) to form the negative electrode (Figure 13.9). The other (positive) electrode is also made from

Figure 13.7
Preparation of gallium arsenide, GaAs, semiconductor material.
The bars in the tube are gallium; the small pieces in the tube near the furnace are arsenic. Gallium arsenide is used to make semiconductor diode lasers for use in fiber optic communications, compact disc players, and laser printers. Also, computer chips are being developed using gallium arsenide, because it conducts an electrical signal five times faster than presently used silicon.

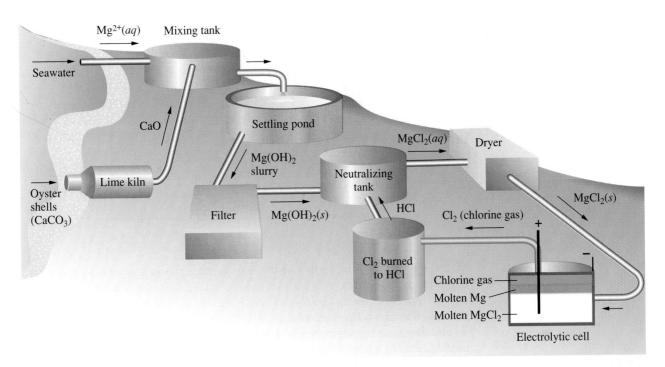

Figure 13.8
Dow process for producing magnesium from seawater. Oyster shells or other seashells are calcined in a kiln to produce calcium oxide, which when added to seawater precipitates magnesium ion as magnesium hydroxide. This is neutralized with hydrochloric acid to give magnesium chloride. Electrolysis of molten $MgCl_2$ yields magnesium metal and chlorine. Hydrochloric acid can be recovered by burning the chlorine with natural gas.

carbon. The electrolyte consists of molten cryolite, at about 1000°C, into which some aluminum oxide is dissolved. The overall electrolysis reaction is

$$2Al_2O_3(s) + 3C(s) \xrightarrow{\text{electrolysis}} 4Al(l) + 3CO_2(g)$$

The positive carbon electrodes are consumed in the electrolysis and must be replaced periodically; aluminum oxide is continually added to the electrolyte bath.

The energy required to produce aluminum by the Hall–Héroult process is considerable—about 14,000 kilowatt-hours per ton of aluminum. (A watt equals one joule per second, so a kilowatt-hour equals 3.6 million joules.) To put this in perspective, note that a typical household might use about 1000 kilowatt-hours per month. Therefore, the energy used to produce one ton of aluminum is sufficient to supply electricity to about 14 homes for one month. Recycling of aluminum would save much of this energy. Only about 7% of the energy required to produce aluminum from aluminum oxide is needed to obtain aluminum metal from used cans and scrap.

Chemical Reduction The cheapest chemical reducing agent is some form of carbon, such as hard coal or coke. Coke is the solid residue left from coal after its volatile constituents have been driven off by heating in the absence of air. Some important metals, including iron and zinc, are obtained by reduction of their compounds with carbon (or carbon monoxide, which is a product of partial oxidation of carbon).

Iron is produced by reduction in a blast furnace. A mixture of iron ore (containing an oxide such as Fe_2O_3), coke, and limestone (a rock that is mainly calcium carbonate) is added at the top of the furnace (Figure 13.10), and a blast of heated air enters at the bottom. The reactions are somewhat complicated, but essentially coke burns to produce carbon monoxide, which reacts with iron oxides to reduce them to iron and carbon dioxide. For example, the reduction of Fe_2O_3 by CO is:

$$Fe_2O_3(s) + 3CO(g) \longrightarrow 2Fe(l) + 3CO_2(g)$$

Figure 13.9
Hall–Héroult cell for the production of aluminum. Aluminum oxide is electrolyzed in molten cryolite (the electrolyte). Molten aluminum forms at the negative electrode (tank lining), where it is periodically withdrawn.

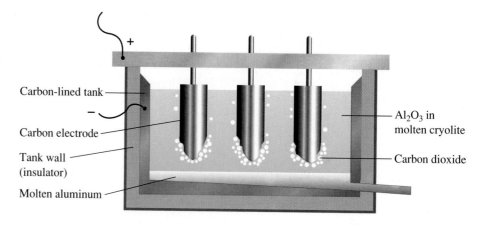

Carbon-lined tank

Carbon electrode

Tank wall (insulator)

Molten aluminum

Al_2O_3 in molten cryolite

Carbon dioxide

Molten iron flows to the bottom of the blast furnace, where it is drawn off. Impurities in the iron ore react with calcium oxide obtained from the limestone to produce a glassy material called *slag*. Molten slag collects in a layer floating on the molten iron and is drawn off periodically.

Hydrogen and reactive metals, such as sodium, magnesium, and aluminum, are used as reducing agents when carbon is unsuitable. Tungsten, for example, reacts with carbon to produce tungsten carbide, WC, a relatively unreactive and extremely hard substance. ◄ The metal is prepared from tungsten(VI) oxide, a canary-yellow compound obtained from the processing of tungsten ore. The oxide is reduced by heating it in a stream of hydrogen gas.

► **Tungsten carbide, WC, is used for cutting tools because of its hardness even at high temperatures.**

$$WO_3(s) + 3H_2(g) \longrightarrow W(s) + 3H_2O(g)$$

Figure 13.10
Blast furnace for the reduction of iron ore to iron metal. Iron ore, coke, and limestone are added at the top of the blast furnace, and heated air enters at the bottom, where it burns the coke. The carbon monoxide from the burning reduces the iron ore to metallic iron.

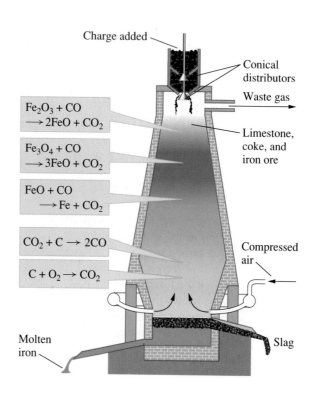

Charge added

Conical distributors

Waste gas

$Fe_2O_3 + CO$
$\longrightarrow 2FeO + CO_2$

$Fe_3O_4 + CO$
$\longrightarrow 3FeO + CO_2$

$FeO + CO$
$\longrightarrow Fe + CO_2$

$CO_2 + C \rightarrow 2CO$

$C + O_2 \rightarrow CO_2$

Limestone, coke, and iron ore

Compressed air

Molten iron

Slag

One of the most important uses of tungsten metal is in the production of filaments for incandescent bulbs (Figure 13.11). This usage depends on the fact that tungsten has the highest melting point (3410°C) and highest boiling point (5900°C) of any metal. To be useful, the incandescent filament in a lightbulb must not melt and should not vaporize excessively. The tungsten metal filament does slowly vaporize, and you can often see where metal has condensed as a black coating on the inside surface of a burned-out bulb.

Refining

Often the metal obtained from reduction contains impurities and must be purified, or refined, before it can be used. Zinc, for example, contains lead, cadmium, and iron as impurities. Zinc has a relatively low boiling point (908°C) and is refined by distillation.

Copper metal used for electrical wire must be quite pure; otherwise its electrical resistance is too large. More than 90% of the copper produced by reduction of copper ores is further refined, or purified, by electrolysis. The electrolysis cell consists of two electrodes, one of impure copper connected to the positive pole of a battery and the other of pure copper connected to the negative pole. The electrodes dip into a solution of copper ions. Copper(II) ions from the solution combine with electrons from the negative pole of the battery, leaving copper atoms that plate out onto the pure copper electrode.

$$Cu^{2+}(aq) + 2e^- \longrightarrow Cu(s)$$

Electrons from the impure copper flow to the positive pole of the battery, leaving copper(II) ions behind that enter the solution. These copper(II) ions replace the ones that plate out onto the pure copper electrode:

$$Cu(s) \longrightarrow Cu^{2+}(aq) + 2e^-$$

The overall result is that copper leaves the impure electrode and plates out onto the pure copper electrode, enlarging it at the expense of the other electrode. ◄

Figure 13.11
Tungsten filament of an incandescent lightbulb.
A coiled wire of tungsten becomes white-hot when an electric current flows through it. The wire is enclosed in a glass bulb containing gases that do not react with the tungsten, such as nitrogen and argon. The gases carry the heat away from the wire, which would otherwise overheat and boil away.

▶ This method of refining copper is described further at the end of Section 20.10.

CONCEPT CHECK 13.1

Why must a mineral containing a metal usually be reduced to obtain the free metal?

13.3   Bonding in Metals

The special properties of a metal result from its delocalized bonding, in which bonding electrons are spread over a number of atoms. In this section, we will look first at the "electron-sea" model of a metal and then at the molecular orbital theory of bonding in metals.

Electron-Sea Model of Metals

A very simple picture of a metal depicts an array of positive ions surrounded by a "sea" of valence electrons free to move over the entire metal crystal. When the metal is connected to a source of electric current, the electrons easily move away from the negative side of the electric source and toward the positive side, forming an electric current in the metal. In other words, the metal is a conductor of electric current because of the mobility of the valence electrons. A metal is also a good heat conductor because the mobile electrons can carry additional kinetic energy across the metal. ◄

► The electron-sea model of metallic bonding is described briefly in Section 9.7; see Figure 9.18.

► Molecular orbital theory is discussed in Sections 10.5 and 10.6.

Molecular Orbital Theory of Metals

The electron-sea model of metals is a simplified view that accounts in only a qualitative way for properties of a metal such as electrical conductivity. Molecular orbital theory gives a more detailed picture of the bonding in a metal. ◄

Recall that molecular orbitals form between two atoms when atomic orbitals on the atoms overlap. In some cases, the atomic orbitals on three or more atoms overlap to form molecular orbitals that encompass all of the atoms. These molecular orbitals are said to be *delocalized*. The number of molecular orbitals that form by the overlap of atomic orbitals always equals the number of atomic orbitals. In a metal, the outer orbitals of an enormous number of metal atoms overlap to form an enormous number of molecular orbitals that are delocalized over the metal. As a result, a large number of energy levels are crowded together into "bands." Because of these energy bands, the molecular orbital theory of metals is often referred to as *band theory*.

Sodium provides a simple illustration of band theory. Imagine that you build a crystal of sodium by bringing sodium atoms together one at a time, and during this process you follow the formation of molecular orbitals and associated energy levels. Each isolated sodium atom has the electron configuration $[Ne]3s^1$. When two sodium atoms approach each other, their $3s$ orbitals overlap to form two molecular orbitals (a bonding molecular orbital and an antibonding molecular orbital). The inner-core electrons, represented as $[Ne]$ in the electron configuration, remain essentially non-bonding. Now imagine that a third atom is brought up to this diatomic molecule. The three $3s$ orbitals overlap to form three molecular orbitals, each orbital encompassing the entire Na_3 molecule. When a large number N (on the order of Avogadro's number) of sodium atoms have been brought together to form a crystal, the atoms will have formed N molecular orbitals encompassing the entire crystal.

Figure 13.12 shows that at each stage the number of energy levels grows and the energy levels spread until the molecular orbital levels merge into a *band* of essentially continuous energies. We call this the $3s$ band of the sodium metal. A $3s$ band

Figure 13.12
Formation of an energy band in sodium metal.
Note that the number of energy levels grows until the levels merge into a continuous band of energies. The lower half of the band is occupied by electrons.

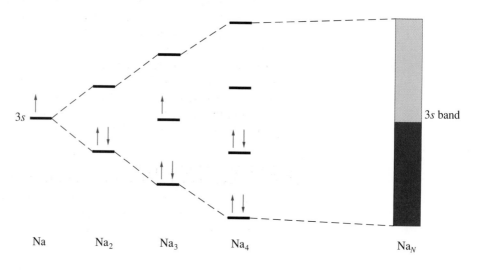

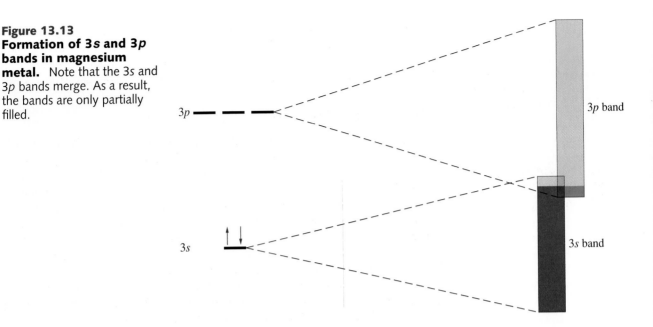

Figure 13.13
Formation of 3s and 3p bands in magnesium metal. Note that the 3s and 3p bands merge. As a result, the bands are only partially filled.

formed from N atoms will have N orbitals that hold a maximum of $2N$ electrons. Because each sodium atom has one valence electron, N atoms will supply N $3s$ electrons and will therefore half-fill the $3s$ band.

Electrons become free to move throughout a crystal when they are excited to unoccupied orbitals of a band. In sodium metal, this requires very little energy; because of the half-filled $3s$ band, unoccupied orbitals lie just above the occupied orbitals of highest energy. When a voltage is applied to a metal crystal, electrons are excited to the unoccupied orbitals and move toward the positive pole of the voltage source.

The explanation of the electrical conductivity of magnesium metal is somewhat more complicated. A magnesium atom has the configuration $[Ne]3s^2$. As in sodium, the $3s$ orbitals of magnesium metal overlap to form a $3s$ band. If this were the whole story, you would expect the $2N$ valence electrons of the atoms to completely fill the $3s$ band. Therefore, if a small voltage were applied, the electrons would have no place to go. You would expect magnesium to be an insulator, but in fact it is a conductor.

Magnesium has unoccupied $3p$ bands formed from unoccupied $3p$ orbitals of the atoms. As the orbitals of the individual atoms interact in forming the metal, the energy levels spread so that the bottom of the $3p$ band merges with the top of the $3s$ band (Figure 13.13). Imagine electrons filling the $3s$ band. When the electrons reach the energy where the two bands have merged, electrons begin to fill orbitals in both bands. As a result, the $3s$ and $3p$ bands of magnesium metal are only partially filled by the time you have accounted for all $2N$ valence electrons. Therefore, when a voltage is applied to the metal, the highest-energy electrons are easily excited into the unoccupied orbitals, giving an electrical conductor.

CONCEPT CHECK 13.2

A semiconductor is a material with a separation between the filled band and an unfilled band called an energy gap. Would it take more or less energy to make a semiconductor carry current than it would a metal?

Superconductivity

MATERIALS

A *superconductor* is a material that abruptly loses its resistance to an electric current when cooled to a definite characteristic temperature. This means that an electric current will flow in a superconductor without heat loss, unlike the current in a typical conductor. Once a current has been started in a superconducting circuit, it continues to flow indefinitely. Another intriguing property of a superconductor is its perfect diamagnetism. This means that the superconductor completely repels magnetic field lines. Figure 13.14 shows a magnet suspended in midair over one of the newly discovered ceramic superconductors. The magnet seems to levitate.

Superconductivity was discovered in 1911 by the Dutch physicist Heike Kamerlingh Onnes soon after he found a way to liquefy helium. Kamerlingh Onnes found that mercury metal suddenly loses all resistance to an electric current when cooled to 4 K. Superconductors first became useful when a niobium metal alloy was found to become superconducting at about 23 K and to remain superconducting even when large currents flow through it. It became possible to construct superconducting magnets with high magnetic fields by starting an electric current in a superconducting circuit. Such magnets are being used in medical magnetic resonance imaging.

In 1986, Johannes Georg Bednorz and Karl Alexander Müller of IBM discovered that certain copper oxide ceramic materials became superconducting at 30 K, and within months researchers had found similar materials that become superconducting at 125 K. (They won the 1987

Figure 13.14
Levitation of a magnet by a superconductor.
The magnet (samarium–cobalt alloy) is supported above the ceramic superconductor (approximate formula $YBa_2Cu_3O_7$). The ceramic becomes superconducting when it is cooled by liquid nitrogen.

Nobel Prize in physics for their discovery.) Perhaps materials can be found that are superconducting even at room temperature. What remains is to determine how to fabricate such superconducting materials into wires and similar objects with the ability to carry large currents.

Nonmetallic Materials

About 20,000 years ago, humans found that they could harden clay by cooking it in a fire; but once they discovered how to increase the temperature of a fire through the use of furnaces, the kinds of materials available to them multiplied quickly. By firing wet clay in furnaces or kilns, they could produce pottery vessels. At even higher temperatures, they could produce glass.

These materials, fired clay products and glass, are still vitally important today, but to these more traditional materials, we can now add a wealth of sophisticated materials with special properties. The new superconductors mentioned in the essay on superconductivity are just one example of these new materials.

In the following sections, we look at several nonmetallic materials with applications in modern technology. We begin with a discussion of the different allotropes of carbon: diamond, graphite, and the fullerenes. The fullerenes are recently discovered molecular forms of the element carbon, in which the carbon atoms form hollow balls and tubes that may make them important as catalysts or possibly as drug-delivery materials. Diamond shows promise as a material that might supersede silicon in its role as a material for solid-state electronics.

13.4 Diamond, Graphite, the Fullerenes, and Nanotechnology

Two allotropic forms of carbon, diamond and graphite, are covalent network solids. Their structures were discussed in Section 11.7. Diamond is a three-dimensional network structure in which every carbon atom is covalently bonded to four other carbon atoms. To move one plane of atoms in the diamond crystal relative to another requires the breaking of many strong carbon–carbon bonds. Because of this, diamond is the hardest material known, and its principal industrial use is as an abrasive (grinding material).

Graphite has an entirely different structure. It has a layer structure, with layers being attracted to one another by London forces. Within a layer, each carbon atom is covalently bonded to three other atoms, giving a flat layer of carbon-atom hexagons. The bonding is sp^2 with delocalized π bonds. As a result of the delocalized bonding, graphite is an electrical conductor, unlike diamond, which is an insulator.

In 1985, Richard Smalley and Robert Curl of the United States and Harold Kroto of the United Kingdom discovered the first of a series of molecular forms of the element carbon: C_{60}, which they called *buckminsterfullerene*. (See the essay at the end of this section.) Since then, scientists have discovered a series of related carbon-atom molecules, called *fullerenes*.

Diamond

Early chemists, intrigued by diamond's rarity as a natural material, hoped to produce it from graphite. The higher density of diamond compared with that of graphite suggests (by applying Le Chatelier's principle) that high pressure would facilitate the transformation of graphite to diamond. ◄ In 1897, the French chemist Henri Moissan tried to achieve high pressures by quickly solidifying carbon in molten iron. He did find transparent crystals in the solid mixture, but these were later shown not to be diamond. It was not until 1955 that scientists at General Electric succeeded in synthesizing diamond. Using a transition-metal catalyst, the process required temperatures of about 2000°C and pressure of about 100,000 atm. The synthetic diamonds produced by this process are generally very small and most useful as an abrasive, or grinding, material. Gem-quality diamonds have been made, but they cannot compete in price with natural diamonds. Figure 13.15 shows some synthetic diamonds.

More recent research into producing diamond from graphite involves the process of *chemical vapor deposition*. In general, a chemical reaction in a gas or vapor produces a product that can be deposited onto a solid material, or substrate. One procedure for producing diamond films consists of passing a mixture of methane, CH_4, and hydrogen, H_2, over a hot filament to produce reactive species that deposit diamond onto a solid substrate (Figure 13.16). The methane provides the carbon atoms for the diamond film. Hydrogen, it is believed, forms hydrogen atoms on the hot filament, and these hydrogen atoms then play a pivotal role in the formation of diamond. The

▶ Le Chatelier's principle was discussed in Section 12.3.

Figure 13.15
Synthetic diamonds.
The diamonds *(right)* were made by heating graphite *(left)* under high pressure with a catalyst. Pencil "lead" is graphite mixed with clay.

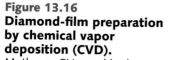

Figure 13.16
Diamond-film preparation by chemical vapor deposition (CVD).
Methane, CH_4, and hydrogen, H_2, react in the region of a hot filament to produce reactive species that deposit diamond on the substrate.

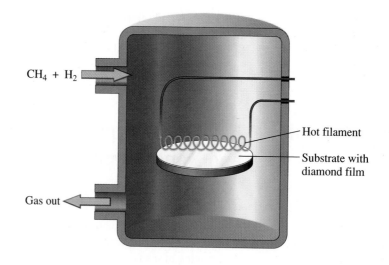

hydrogen atoms are thought to abstract hydrogen atoms from methane, forming reactive methyl radicals, CH_3, and hydrogen, H_2. The methyl radicals diffuse to the substrate surface where they form carbon–carbon bonds, ultimately yielding tiny crystals of diamond. Any graphite that happens to form reacts quickly with hydrogen, and so is etched away, leaving the diamond crystals as a film.

One application of diamond films is as a coating on cutting tools. Another application of diamond films depends on the high thermal conductivity of diamond, which can conduct heat away four times faster than copper metal. Diamond films are now being marketed as a base for microelectronic parts that generate heat.

Although diamond is an electrical insulator, like silicon and other materials, it becomes a conductor when small quantities of other substances, such as boron, are added to it. We say that the diamond has been "doped" and behaves as a *semiconductor*. In principle, diamond could supplant silicon as the material for constructing microelectronics devices.

Graphite

The "lead" in a lead pencil is actually graphite, which occurs naturally as a mineral. It is the layer structure of graphite, with the weak attractive forces between those layers and the delocalized bonding electrons within each layer, that is responsible for the substance's use as the lead in pencils. The delocalized electrons, spread over a layer of graphite, give rise to many close energy levels, similar to those in a metal. These energy levels absorb all the wavelengths of visible light, rendering the material black. ◄

▶ See the end of Section 7.3 for a discussion of the relationship between absorption of light and the color observed for a material.

Today most graphite is used for electrodes, for example, to produce aluminum. Although natural graphite is still mined as a material for lead pencils, most of the graphite of modern industry is manufactured by heating coke in an electric furnace to about 3000°C.

The development of lightweight rechargeable batteries as sources of portable energy has become a high priority, and graphite has become one of the most studied materials for use as the negative electrode in these batteries. One of the most successful of these is the lithium-ion battery (Figure 13.17). The battery consists of a positive electrode of lithium cobalt oxide, $LiCoO_2$, or similar compound, with graphite as the other electrode. During charging of the battery, lithium ions (Li^+) are pulled

Figure 13.17
Lithium-ion batteries.
Left: Discharge of a lithium-ion battery. Initially, the battery was charged by connecting the two electrodes to an external voltage to draw lithium ions from the $LiCoO_2$ electrode to the graphite electrode, where lithium ions reside in the space between the graphite layers. During discharge, the lithium ions move from the graphite electrode back to the $LiCoO_2$ electrode, which the ions find energetically more favorable. Electrons move in the external circuit in the same direction as the lithium ions to maintain electrical balance in the battery. *Right:* Two commercial lithium-ion batteries.

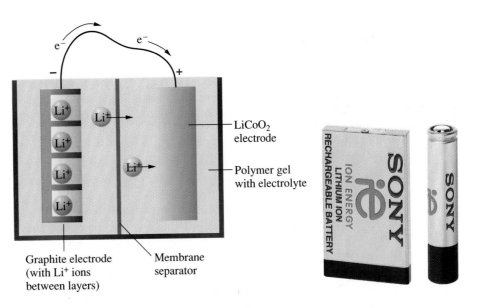

by the application of an external voltage from the lithium compound toward the graphite, where the ions penetrate the layer structure. When the battery is discharged, the lithium ions flow from the graphite toward the lithium compound electrode. Electric-charge balance is maintained in the battery during discharge as electrons flow through an external circuit (giving a flow of electricity) from the graphite electrode to the lithium compound electrode. Lithium-ion batteries are lightweight and can go through many charge–discharge cycles. They are being used in cell phones and laptop computers, and they are being studied for use in electric vehicles.

Fullerenes and Nanotechnology

The *fullerenes* comprise a family of molecules each consisting of a closed cage of carbon atoms arranged in pentagons as well as hexagons. The most symmetrical fullerene is buckminsterfullerene, which consists of 12 pentagons and 20 hexagons of carbon atoms. (See the essay at the end of this section.) Other fullerenes consist of graphite-like sheets of carbon atoms rolled into tubes and capped at each end by half of a buckminsterfullerene molecule. These fullerenes can be extremely long tubes of carbon (consisting of a million or more carbon atoms) and are referred to as carbon *nanotubes*.

The unique characteristics of buckminsterfullerene have sparked interest in possible applications of the substance. A novel feature of the molecule is that the carbon atoms of the cage structure contain just enough empty space to enclose an atom. Atoms such as potassium have been encapsulated within the buckminsterfullerene molecule. Possibly, this ability of buckminsterfullerene could be used as a special storage device. Other fullerenes also have spaces within their cage structures, and encapsulation of substances within these materials is being investigated. Buckminsterfullerene also shows the ability to pick up atoms of metals within the spaces between molecules of the crystal. Buckminsterfullerene with potassium atoms in these spaces has shown the properties of a superconductor, in which the electrical resistance of the material drops to zero below about 18 K. The fullerenes also display catalytic behavior. For example, the fullerenes have been shown to act as catalysts in converting methane to ethane and higher hydrocarbons. Also, it is possible to attach transition-

Buckminsterfullerene—A Third Form of Carbon

MATERIALS

Until recently, carbon was thought to occur in only two principal forms: diamond and graphite. In 1985, Harold W. Kroto (from the University of Sussex in Brighton, England) approached Richard E. Smalley and Robert F. Curl (at Rice University in Houston, Texas) to do some experiments to simulate the conditions in certain stars to see what sorts of carbon-containing molecules might be produced. The research group at Rice had previously constructed an instrument in which they used an intense laser beam to vaporize solids. The hot vapor produced in this way could then be directed as a molecular beam into a mass spectrometer, where the molecular weights of the species in the vapor could be measured.

The experiments on the vaporization of graphite produced surprising results. Molecular clusters of 2 to 30 carbon atoms were found, as expected; but in addition, the mass spectrum of the vapor consistently showed the presence of a particularly abundant molecule, C_{60}. Why was this molecule so stable? Kroto, Smalley, and Curl wrestled with this problem and eventually came to the conclusion that the molecule must be like a piece of graphite sheet that somehow closed back on itself to form a closed-dome structure. Smalley set about trying to construct a model of C_{60} by gluing paper polygons together. He discovered that he could obtain a very stable, closed polygon with 60 vertices by starting with a pentagon and attaching hexagons to each of its five edges. To this bowl-shaped structure he attached more pentagons and hexagons, producing a paper soccer ball (Figure 13.18). Kroto and Smalley named the molecule buckminsterfullerene, after R. Buckminster Fuller, who studied closed-dome architectural structures constructed from polygons.

In 1990, buckminsterfullerene was prepared in gram quantities. Once the reddish-brown substance was available in sufficient quantity, researchers were able to verify the soccer-ball structure of the C_{60} molecule.

In 1996, the Nobel Prize in chemistry was awarded to professors Curl, Kroto, and Smalley for their discovery.

Figure 13.18

Structure of buckminsterfullerene.
Left: A frame model of C_{60}. Carbon atoms are at the corners of each polygon. Each bond is intermediate between a single and a double bond, similar to the bonding in graphite. *Right:* The buckminsterfullerene molecule is often called a "buckyball," because of its soccer-ball appearance.

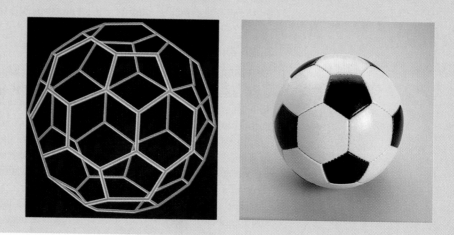

metal atoms (such as osmium) to buckminsterfullerene, and these derivatives might yield unique catalysts.

Carbon nanotubes have become an especially active area of research, especially in **nanotechnology** *(an area of technology in which one manipulates materials on a molecular scale to create useful devices)*. These nanotubes are especially strong and, depending on the structure, can be as electrically conducting as metals. Chemists at Harvard University have shown that they can make carbon nanotubes bend by applying a voltage to them. They have used this property to construct molecular-scale tweezers, as we mentioned in the chapter opening.

13.5 Semiconductors

Semiconductor materials are the basis of solid-state electronics devices, such as transistors, light-emitting diodes (LEDs), and laser diodes. A CD player, for example, uses transistors for its electronic circuits, as well as a laser diode to "read" a compact disc. (See the essay *Lasers and Compact Disc Players* at the end of Section 7.3.) The basic component of most of these electronic devices is silicon, a metalloid. A metalloid is a **semiconducting element,** which is *an element that has low electrical conductivity at room temperature, but increases in conductivity as the temperature increases.* (This contrasts with the behavior of metals, which are generally good electrical conductors, although their conductivity decreases with increasing temperature.) A striking property of a semiconducting element is that you can very much increase the electrical conductivity of the pure element by adding minute quantities of certain other elements to it, a process that is called *doping.*

To understand what happens when you dope a semiconducting element, let's consider the doping of silicon with phosphorus. You first start with very pure silicon. Phosphorus has five valence electrons, whereas silicon has only four. By doping silicon with phosphorus, you replace a few silicon atoms in the solid with phosphorus atoms. After each phosphorus atom bonds to four silicon atoms, an electron is left over (Figure 13.19A). The extra electrons in phosphorus-doped silicon are free to conduct an electric current, so the doped silicon becomes a good conductor. It is called an n-*type semiconductor,* because the current is carried by negative charges (electrons).

Another type of doped semiconductor is a p-*type semiconductor.* You make these by doping the semiconductor element with another element that contains fewer valence electrons. For example, suppose you dope silicon with boron, which has only three valence electrons. By doping, you replace a few of the silicon atoms in the solid with boron atoms. After a boron atom forms bonds to four silicon atoms, one of the boron–silicon bonds has only one electron in it. In other words, the orbital has a vacancy for an electron. You can think of this vacancy as a positively charged "hole" in the bonding orbital (Figure 13.19B). An electron from a neighboring atom can move to occupy this hole. Now, a hole exists on the neighboring atom, and an electron from an atom next to it can move into that hole. This process can continue, and the result is the movement of positively charged holes in the semiconductor. The boron-doped silicon becomes a conductor, with the electrical charge being carried by positively charged holes, hence the name *p*-type semiconductor.

Figure 13.19
Effect of doping silicon.
(A) Silicon doped with phosphorus. The extra electrons (denoted e^-) from phosphorus atoms are free to conduct a current. (B) Silicon doped with boron. A bond with a missing electron is equivalent to a positively charged hole, indicated here as a plus sign in a circle.

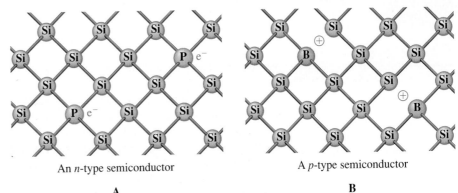

An *n*-type semiconductor

A *p*-type semiconductor

A

B

Figure 13.20
A _p–n_ junction as a rectifier. _Top:_ When a source of electricity is connected as shown, an electric current flows. Note the direction of flow of charge carriers (positive holes and electrons). _Bottom:_ When the wire connections are reversed, no current flows. Note the depletion of charge carriers near the junction of the two semiconductors.

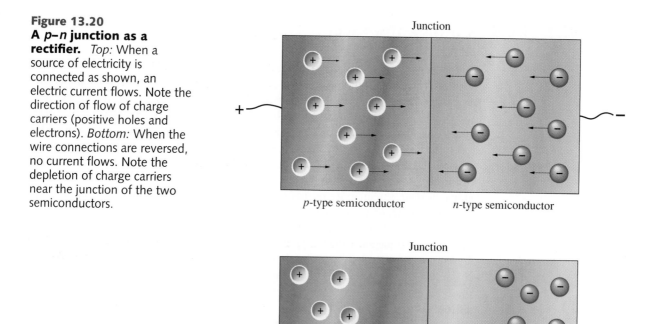

Junction

_p_-type semiconductor   _n_-type semiconductor

Junction

_p_-type semiconductor   _n_-type semiconductor

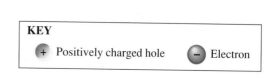

KEY
(+) Positively charged hole   (–) Electron

Figure 13.21
An integrated circuit chip. The chip is just above the date on the penny. Its base material is silicon; miniature circuits with electronic components are constructed on the chip.

Often a solid-state electronic device depends on the electrical property of a p–n _junction_. A p–n junction consists of a _p_-type semiconductor joined to an _n_-type semiconductor. It can function as a _rectifier,_ a device that allows current to flow in one direction but not the other. Imagine a p–n junction with wires connected to it from a source of direct current as shown in Figure 13.20, _top._ Positive holes are continuously generated in the _p_-type semiconductor by the positive wire, and electrons are released into the _n_-type semiconductor from the negative wire. The two charge carriers (positive holes and electrons) move toward the junction, where they combine. The result is a continuous flow of current through the p–n junction. Now suppose that the wire connections are reversed, as in Figure 13.20, _bottom._ Positive holes are attracted to the negative wire, and electrons are attracted to the positive wire. The region of the junction quickly becomes depleted in charge carriers, so the p–n junction becomes nonconducting, and no current can flow.

By combining _p_-type and _n_-type semiconductors, you can construct various kinds of _transistors,_ devices for controlling electric signals. Some of the latest computer chips have microscopic electrical circuits integrated with as many as a million transistors per square centimeter of surface area (Figure 13.21).

13.6 Silicon, Silica, and Silicates

If we were to list the materials of technology, silicon would certainly be prominent on that list. As we mentioned in the previous section, silicon is the basic material in the semiconductor devices that make up CD players, computers, and other electronics gear. But oxygen compounds of silicon also are important to technology. Quartz crystals, a form of silica (also called silicon dioxide, SiO_2), for example, are used to control frequencies in radio transmitters and watches. And many practical materials such as cement and bricks are silicate materials. Silicates are compounds of silicon and oxygen with one or more metallic elements.

Silicon

About 95% of the earth's crust is silica and silicate rocks and minerals. Elemental silicon is obtained by reducing quartz sand with coke in an electric furnace at 3000°C.

$$SiO_2(l) + 2C(s) \longrightarrow Si(l) + 2CO(g)$$

Silicon has a diamond-like structure. It is a hard, lustrous gray solid and is used to make alloys and solid-state electronic devices.

For the manufacture of solid-state devices, it is necessary to start with extremely pure silicon (no more than $10^{-8}\%$ impurities). You first convert the impure element to silicon tetrachloride, $SiCl_4$, which is a low-boiling liquid (b.p. 58°C) that can be purified by distillation.

$$Si(s) + 2Cl_2(g) \longrightarrow SiCl_4(g)$$

You then reduce the purified silicon tetrachloride by passing the vapor with hydrogen through a hot tube, where pure silicon crystallizes on the surface of a pure silicon rod.

$$SiCl_4(g) + 2H_2(g) \longrightarrow Si(s) + 4HCl(g)$$

The silicon prepared in this process is polycrystalline. In a final step, this polycrystalline silicon is fashioned into a rod that consists of a single crystal, from which high-purity silicon wafers can be cut (Figure 13.22).

Silica (Silicon Dioxide)

Silica, silicon dioxide, has several different crystalline forms. The most important of these is quartz, which is a constituent of many rocks. It is the weathering of these rocks that releases quartz particles, a major component of most kinds of sands. Amethyst is a gem form of quartz; it contains a small quantity of Fe_2O_3, which is responsible for its purple color. **Silica,** SiO_2, is *a covalent network solid in which each silicon atom is covalently bonded in tetrahedral directions to four oxygen atoms; each oxygen atom is in turn bonded to another silicon atom* (Figure 13.23).

Quartz crystals have a very interesting and useful property: they exhibit the *piezoelectric effect.* In a piezoelectric crystal, compression of the crystal in a particular direction causes an electric voltage to develop across it. Such crystals are used in phonograph pickups and microphones to convert sound vibrations to alternating electric currents. The opposite effect is also possible: an alternating electric current applied to a piezoelectric crystal can make it vibrate. When cut to precise dimensions, the

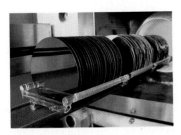

Figure 13.22
High-purity silicon rod and wafers cut from it. Silicon wafers form the base material for integrated circuit chips used in solid-state electronic devices.

Figure 13.23
Structure of silica (SiO₂).
Left: A silicon atom is bonded tetrahedrally to four oxygen atoms, giving an SiO₄ tetrahedron. Each of the oxygen atoms on this tetrahedron bonds to silicon atoms on other tetrahedra, giving a three-dimensional structure. *Right:* Shown is a fragment of silica structure with three SiO₄ tetrahedra bonded together.

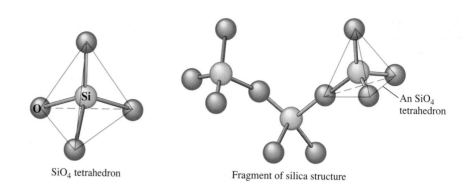

SiO₄ tetrahedron   Fragment of silica structure   An SiO₄ tetrahedron

crystal responds most strongly to a certain vibrational frequency. Such crystals are used to control the frequency of an alternating electric current. When the alternating current frequency deviates from the natural frequency of the crystal, a feedback mechanism adjusts the alternating current frequency. Quartz crystals are used to control radio and television frequencies, as well as clocks.

Silicates

A **silicate** is *a compound of silicon and oxygen (with one or more metals) that may be formally regarded as a derivative of silicic acid, H_4SiO_4 or $Si(OH)_4$*. Silicic acid has never been isolated, although solutions containing anions of the acid are well known. When silica is melted with sodium carbonate, it forms a soluble material referred to commercially as *water glass*. These solutions contain various silicate ions, such as

$$H-O-\underset{\underset{O_-}{|}}{\overset{\overset{O^-}{|}}{Si}}-O-H$$

The solution also contains ions with two or more silicon atoms, which form by condensation reactions. A **condensation reaction** is *a reaction in which two molecules or ions are chemically joined by the elimination of a small molecule such as H_2O*. For example, the silicic acid anion can react in a condensation reaction with another such anion to form a disilicate anion (anion of two silicon atoms):

$$H-O-\underset{\underset{O_-}{|}}{\overset{\overset{O^-}{|}}{Si}}-O-H + H-O-\underset{\underset{O_-}{|}}{\overset{\overset{O^-}{|}}{Si}}-O-H \longrightarrow H-O-\underset{\underset{O_-}{|}}{\overset{\overset{O^-}{|}}{Si}}-O-\underset{\underset{O_-}{|}}{\overset{\overset{O^-}{|}}{Si}}-O-H + H_2O$$

Silicate anions containing long Si—O—S— . . . chains can result from such condensation reactions. Materials exhibiting chains and networks of this sort of silicon–oxygen bonding are common. The silicate minerals all have this type of structure.

An enormous variety of silicate minerals exists, but all of their structures have SiO₄ tetrahedra as basic units. A few minerals contain SiO_4^{4-} as discrete ions; they are referred to as *orthosilicates*. Zircon, $ZrSiO_4$, is an example of such a silicate mineral. Zircon is the principal ore of zirconium metal. Because of the brilliance of zircon (a result of its high refractive index), colorless crystals of this mineral are used as a diamond-like gem.

Figure 13.24
Linking of two SiO₄ tetrahedra. The SiO₄ tetrahedra are linked through a common oxygen atom. Note the negative charges on the O atoms that are not shared by Si atoms.

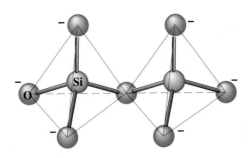

Most silicate minerals have more complicated structures in which SiO_4 tetrahedra are linked to other SiO_4 tetrahedra through common oxygen atoms (Figure 13.24). In such structures, the O atoms attached to only one Si atom have a negative charge, giving silicate anions. These negative charges are balanced by metal ions in silicate minerals. Beryl, for example, is a beryllium mineral with the formula $Be_3Al_2(SiO_3)_6$. The silicate anion in beryl consists of six SiO_4 tetrahedra formed into a cycle, as shown in Figure 13.25 (cyclic silicate anion). Emerald is gem-quality beryl having a deep green color that results from a small quantity of chromium in the crystal. In other silicate minerals, the SiO_4 tetrahedra are linked to form anions that are long chains or sheets (Figure 13.25, chain and double-chain silicate anions).

In the aluminosilicate minerals, aluminum atoms substitute for some of the silicon atoms. An **aluminosilicate mineral** is *a mineral consisting of silicate sheets or three-dimensional networks in which some of the SiO_4 tetrahedra of a silicate structure have been replaced by AlO_4 tetrahedra*. If you imagine a three-dimensional structure such as quartz (SiO_2) with some of the SiO_4 tetrahedra replaced by AlO_4 tetrahedra, you obtain a three-dimensional network aluminosilicate. Because the nucleus of the aluminum atom has one less positive charge than that of the silicon atom, such minerals require another metal atom for charge balance. Orthoclase, which occurs in granite, is an example of a three-dimensional network aluminosilicate. The empirical formula of the aluminosilicate anion is $AlSi_3O_8^-$; the formula of orthoclase is $KAlSi_3O_8$.

Figure 13.26 shows several silicate minerals. The one on the left is an example of asbestos, a class of fibrous minerals. In some cases, the fibrous structure results from long-chain silicate anions in the mineral. However, the common asbestos mineral

Figure 13.25
Structures of some silicate minerals. A number of silicate minerals consist of finite silicate anions, such as SiO_4^{4-} and $Si_2O_7^{6-}$. Beryl contains the cyclic anion $Si_6O_{18}^{12-}$. Other silicate minerals have anions with very long chains or double chains.

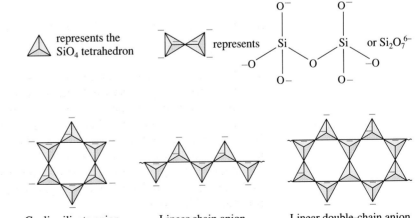

Cyclic silicate anion   Linear chain anion   Linear double-chain anion

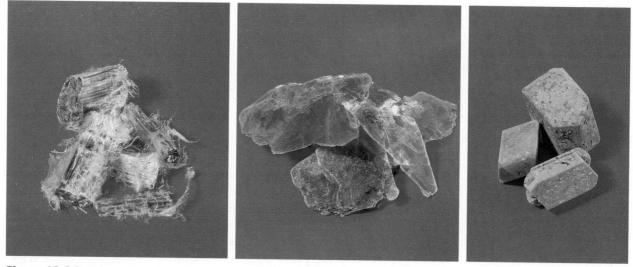

Figure 13.26
Some silicate minerals.
Left: Asbestos; note the fibers on its surface. *Center:* Mica; this mineral cleaves easily into sheets. *Right:* Orthoclase, a feldspar mineral; this is a three-dimensional network silicate.

chrysotile is actually a sheet silicate mineral in which the sheets tend to break up into long pieces that roll up into fine tubes. Mica is a sheet silicate mineral that tends to break easily into visible pieces of mineral sheets (Figure 13.26, center). Figure 13.26 (right) shows the three-dimensional aluminosilicate orthoclase, a very hard mineral.

CONCEPT CHECK 13.3

For certain chemical reactions, quartz containers are used instead of glass. What would the meniscus of water look like in a quartz test tube? Give an explanation for your answer.

13.7 Ceramics

The word *ceramics* comes from a Greek word for pottery, which are objects made by firing clay. Fired clay is a major medium for producing objects of art and other objects including tiles, bricks, and dinnerware. However, modern research has increased in number the kinds of materials we can make that are chemically similar to these traditional ceramics. Consequently, a broader definition has emerged: **ceramics** are *nonmetallic, inorganic solids that are hard and brittle and usually produced at elevated temperature*. This definition would include glass.

Traditional Ceramics and Glass

Clay is a natural, earthy mixture of very small crystals of certain silicate sheet minerals. These minerals form by the weathering of granite, the rock that composes the backbone of mountain ranges. Clay minerals easily adsorb water, and wet clay is moldable. The wet platelike crystals adhere to one another to give a *plastic* mass. It is this plasticity that allows the potter to form useful and artistic shapes.

Once the potter has shaped the wet clay into a vase or other object, he or she allows the clay to dry and then fires it. During the elevated temperatures of the firing, complex irreversible changes occur that result first in a hard, porous material. As the temperature rises, some of the more easily fusible clay minerals begin to melt, forming a glass. The surface tension of the melt draws the liquid and crystals together so that, on cooling, the solid ceramic is more dense than clay fired at lower temperatures and it is no longer porous. It consists of particles suspended in a glass matrix.

Most pottery is *glazed*. After firing, the ceramic object is coated with a water suspension of silica and other oxides; then it is fired again. The oxides melt and react to form a glassy coating, or glaze. In the case of earthenware, the glaze gives the ceramic a nonporous covering. For most pots, glazing simply provides a decorative shell.

Figure 13.27 shows a porcelain vase in which the glaze displays beautiful needlelike crystal formations. The artist carefully formulated a glaze mixture of silica and zinc oxide, in addition to other oxides. He then fired the vase with its glaze, carefully controlling the temperature, until only a few microscopic crystals of zinc oxide remained. These microscopic crystals formed the nuclei on which zinc orthosilicate crystallized.

Pottery is an ancient art, going back thousands of years. Glassmaking is a newer art, dating from about 2500 B.C. Glass is an amorphous, or noncrystalline, solid. When a pure liquid substance freezes, it usually forms a crystalline solid. It is often possible, however, to cool a liquid below its equilibrium freezing point to give a *supercooled* liquid. The term *glass,* in the broadest sense, refers to any supercooled liquid whose viscosity is so high that it has the properties of a solid. Ordinary glass is a silicate produced by fusing silica (SiO_2) with other oxides (or substances such as carbonates that yield oxides when heated). Common glass (called soda-lime glass) contains sodium and calcium oxides, in addition to silica.

Some Recent Ceramic Materials

Research in ceramics has resulted in many new kinds of materials with diverse applications. For example, ceramics and glass have been joined in the production of glass–ceramics. These are made by adding materials to a glass to introduce many crystal nuclei into it. The glass–ceramic stovetop pots shown in Figure 13.28 were made from a lithium aluminosilicate glass to which titanium and zirconium oxides were added. Small crystals of zirconium titanate precipitate in the molten glass and form nuclei on which lithium aluminosilicate crystallizes. The resulting glass–ceramic has a very low thermal expansion, so it does not break easily when subjected to large temperature changes. It remains transparent because the crystals are extremely small and their refractive index is close to that of the glass.

Silicon carbide, SiC, is a ceramic material widely used as an abrasive in grinding wheels and cutting tools. Silicon nitride, Si_3N_4, is a newer ceramic. It is available in several crystalline forms, and recently researchers have discovered a high-pressure form that is especially hard. Because of its hardness, silicon nitride has been used in special cutting tools. It also shows some promise as a material for high-temperature engine components. Ceramics have recently been used to make medical implants and bone replacement materials. High-purity alumina, Al_2O_3, is nearly inert in the human body, so its high strength and wear resistance makes it useful for hip replacement devices (Figure 13.29). Although porcelain ceramics have traditionally been used as electrical insulators, some of the newer ceramics are electrical conductors. They are used, for example, in electrodes and heating elements. The ceramic superconductors mentioned in the essay on page 418 are extreme examples of electrical conductors. At low enough temperatures, these ceramic materials lose all electrical resistance.

Figure 13.27
A vase with a crystalline glaze. The crystals are zinc orthosilicate. Artist Stuart Gray has carefully experimented with glaze compositions and firing temperatures to obtain the effect shown.

Figure 13.28
Glass–ceramic cookware. This transparent glass–ceramic cookware is made by Corning, Inc.

13.8 Composites

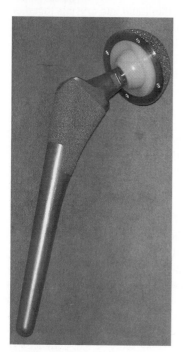

Figure 13.29
Hip Replacement Device.
Shown here is a hip replacement device with an alumina ceramic joint head.

A **composite,** as we mentioned in the chapter opening, is a *material constructed of two or more different kinds of materials.* We mentioned bone as a natural composite of calcium phosphate crystals and a protein collagen. Bone, as a composite, is much stronger than either of its components and less subject to fracture than the calcium phosphate structure would be by itself. An example of a commercial composite is epoxy plastic reinforced with carbon fibers; it is used in aircraft parts, golf clubs, fishing rods, and so forth. The carbon fibers are made from textile fibers by pyrolysis (heating the textile until it decomposes). At high temperatures, the carbon fibers become graphitized, or transformed to a more or less crystalline graphite structure. These carbon fibers have great strength (Figure 13.30).

Composites frequently consist of a material with embedded fibers, which in the example just given is epoxy plastic with embedded carbon fibers. The epoxy plastic is strong but subject to fracture. The embedded fibers impede a fracture that might start, thus enhancing the stability of the structure. Another example of a composite is silicon carbide ceramic with embedded silicon carbide fibers. These fibers are made by pyrolysis of a polymer of dimethylsilane, $(CH_3)_2SiH_2$. By itself, silicon carbide ceramic is brittle, like most ceramics. The silicon carbide fibers, however, break up any fracture lines that may form, making the composite much less brittle.

Figure 13.30
Carbon fibers. Carbon fibers are made by pyrolyzing textile fibers. They are used to make composites.

A Checklist for Review

Important Terms

metal (13.1)
alloy (13.1)
mineral (13.1)
ore (13.1)
metallurgy (13.2)
gangue (13.2)

flotation (13.2)
Bayer process (13.2)
roasting (13.2)
Dow process (13.2)
Hall–Héroult process (13.2)

nanotechnology (13.4)
semiconducting
    element (13.5)
silica (13.6)
silicate (13.6)

condensation reaction (13.6)
aluminosilicate
    mineral (13.6)
ceramics (13.7)
composite (13.8)

Summary of Facts and Concepts

Metals may be pure elements, or they may be *alloys,* which can be either compounds or mixtures. Metallic elements and their compounds are obtained principally from the earth's crust, most of which is composed of metal silicates. The chief sources of metals, however, are not silicates but oxide, carbonate, and sulfide *minerals,* which exist in *ore* deposits widely scattered over the earth.

An important part of *metallurgy* is the production of metals from their ores. This involves three basic steps: *preliminary treatment, reduction,* and *refining.* In preliminary treatment, the ore is concentrated in its metal-containing mineral. The concentrated ore may also require a process such as *roasting,* in which the metal compound is transformed to one that is more easily reduced. In reduction, the metal compound is reduced to the metal, by either electrolysis or chemical reduction. In refining, the metal is purified, or freed of contaminants.

The *electron-sea model* is a simple depiction of a metal as an array of positive ions surrounded by delocalized valence electrons. Molecular orbital theory gives a more detailed picture of the bonding in metals. Because the energy levels in a metal crowd together into bands, this picture of metal bonding is often referred to as *band theory.* According to band theory, the electrons in a crystal become free to move when they are excited to the unoccupied orbitals of a band. In a metal, this requires very little energy, because unoccupied orbitals lie just above the occupied orbitals of highest energy.

Carbon is an important nonmetallic material. It has several allotropes, including diamond, graphite, and the fullerenes. *Diamond* has a three-dimensional network structure in which each carbon atom bonds to four others with sp^3 orbitals. Scientists first synthesized diamond from graphite in 1955 using high temperature and pressure. The process produces diamond grit, which is used for drill bits and cutting wheels. More recently, researchers have made diamond films from graphite using *chemical vapor deposition. Graphite* has a layer structure, in which the carbon atoms in each layer bond to three other atoms with sp^2 orbitals. Most commercial graphite is produced by heating coke in an electric furnace; it is used to make electrodes. Graphite forms one of the electrodes in the new lithium-ion battery. The *fullerenes* comprise a family of molecules with a closed cage of carbon atoms arranged in pentagons and hexagons. The most symmetrical member is *buckminster-fullerene,* C_{60}. Some of the fullerenes consist of tubes of carbon atoms, each end of a tube being capped by half of a buckminsterfullerene molecule; these fullerene tube molecules are referred to as carbon *nanotubes.*

Semiconducting elements form the basis of solid-state electronics devices. A striking property of these elements is that they experience large increases in conductivity when *doped* with small quantities of certain other elements. When silicon is doped with phosphorus, it becomes an n-*type semiconductor,* in which an electric current is carried by electrons. However, when silicon is doped with boron, it becomes a p-*type semiconductor,* in which an electric current is carried by positively charged holes. Joining a *p*-type semiconductor to an *n*-type semiconductor gives a *p–n* junction, which can function as a *rectifier.*

Silicon is a prominent material of technology. The element is the basic material used in semiconductor devices for solid-state electronics. These devices require silicon of extreme purity. The impure element is converted to silicon tetrachloride, which is purified by distillation, after which the silicon tetrachloride is reduced to pure silicon. *Silica* is chemically known as silicon dioxide; quartz is a common form of silica. It exhibits the *piezoelectric effect,* which is used to control radio, television, and clock frequencies. *Silicates* are formed by *condensation reactions;* an enormous variety of silicate minerals exists. The *aluminosilicate* minerals are silicates in which some SiO_4 tetrahedra are replaced by AlO_4 tetrahedra.

The word *ceramics* comes from a Greek word for pottery. Traditional ceramics production begins by forming wet clay, then firing it to make a ceramic object such as a pot or sculpture. Glass is an amorphous, or noncrystalline, solid produced by fusing silica with other oxides. Ceramics have been extended to include materials other than fired clay and glass. Examples are the glass–ceramics, used in making stovetop pots, silicon carbide (used as an abrasive), silicon nitride (used in making cutting tools and engine components), and alumina (for replacement of hip joints).

Composites are constructed of two or more materials. An example is the composite made of epoxy plastic with embedded carbon fibers. The fibers impede any fracture that might start, thus enhancing the stability of the material.

Operational Skills

Note: The problem-solving skills used in this chapter are discussed in previous chapters.

Review Questions

13.1   What is an alloy? Give an example.

13.2   Give four characteristics of a metal.

13.3   Define *mineral, rock,* and *ore.* Is bauxite a mineral or a rock? Explain.

13.4   Ores are usually metal compounds. What kinds of metal compounds most commonly occur in metal ores?

13.5   What are the basic steps in the production of a pure metal from a natural source? Illustrate each step with the preparation of aluminum.

13.6   How does flotation separate an ore mineral from impurities?

13.7   Carbon is often used as a cheap reducing agent, but it is sometimes unsuitable. Give an example in which another reducing agent is preferred.

13.8   Write equations for the preparation of magnesium metal from seawater, limestone, and hydrochloric acid.

13.9   Draw a flowchart for the preparation of aluminum from its ore. Label each step.

13.10   Explain metallic conduction in terms of molecular orbital theory (band theory).

13.11   Describe the formation of an energy band in a metal. Assume that the metal is formed by bringing metal atoms together one at a time.

13.12   Briefly describe three allotropes of carbon.

13.13   Describe some commercial uses of graphite.

13.14   Describe some possible applications of buckminsterfullerene or of the other fullerenes.

13.15   Explain the difference between an *n*-type and a *p*-type semiconductor.

13.16   Describe the steps in preparing ultrapure silicon from quartz sand.

13.17   Explain how a quartz crystal is used to regulate an alternating current frequency.

13.18   Use structural formulas to illustrate the hypothetical condensation reaction of two silicic acid molecules, $Si(OH)_4$.

13.19   What is an aluminosilicate mineral? Give the name and formula of such a mineral.

13.20   Define the term *ceramics.* Give examples of some commercial applications of clay ceramics.

13.21   Give some uses or potential uses of the newer (nonclay) ceramics.

13.22   Describe the production of glass.

13.23   What is a glass–ceramic? Describe an example and give an application.

13.24   What is a *composite?* Give an example of a natural composite and an example of a human-made composite.

Conceptual Problems

13.25   Unless zinc is purified, cadmium is normally an impurity in the metal. Why might you expect this to be the case?

13.26   Electrolysis is used to obtain some metals from their compounds. List some metals obtained this way.

13.27   Aluminum is the third most abundant element (first most abundant metal) in the earth's crust. Does this mean that aluminum ores are widespread and plentiful? Explain.

13.28   The text says that the higher density of diamond compared with graphite suggests that the application of high pressure would facilitate the transformation of graphite to diamond. Explain the reasoning behind this statement.

13.29   Diamond is an insulator, but when small amounts of boron are added it becomes a conductor. What is the explanation for this change in conduction? What are the electric current carriers?

13.30   Is a quartz crystal (SiO_2) a mineral or a rock?

13.31   Orthoclase feldspar has the chemical formula $KAlSi_3O_8$. What other cations could replace those in this chemical formula? Write the resulting formula for two such minerals.

13.32   Cutting wheels have been made from alumina containing fine fibers of silicon carbide. How do the silicon carbide fibers help alumina in this application?

Practice Problems

Metallurgy

13.33 Lead metal is purified by electrolysis in a manner similar to that for copper; the electrolyte is lead(II) hexafluorosilicate, $PbSiF_6$. Describe the process in more detail, giving the half-reactions.

13.34 Nickel metal can be purified by electrolysis in a manner similar to that for copper; the electrolyte is nickel(II) chloride or sulfate. Describe the process in more detail, giving the half-reactions.

13.35 How many kilograms of iron can be produced from 2.00 kg of hydrogen, H_2, when you reduce iron(III) oxide?

13.36 How many kilograms of manganese can be produced from 1.00 kg of aluminum, Al, when you reduce manganese(IV) oxide?

13.37 Using thermodynamic data given in Appendix B, obtain $\Delta H°$ for the roasting of galena, PbS:

$$PbS(s) + \tfrac{3}{2}O_2(g) \longrightarrow PbO(s) + SO_2(g)$$

Is the reaction endothermic or exothermic?

13.38 Using thermodynamic data given in Appendix B, obtain $\Delta H°$ for the reduction of pyrolusite, MnO_2, by aluminum. Is this reaction endothermic or exothermic?

Bonding in Metals

13.39 Sketch a diagram showing the formation of energy levels from the valence orbitals for K, K_2, K_3, and K_n. On the diagram, place arrows indicating how the electrons fill these energy levels.

13.40 Sketch a diagram showing the formation of energy levels from the valence orbitals for Ca, Ca_2, Ca_3, and Ca_n. On the diagram, place arrows indicating how the electrons fill these energy levels.

13.41 How many energy levels are there in the valence band of a single crystal of sodium with a mass of 1.00 mg?

13.42 How many energy levels are there in the $3p$ band of a single crystal of magnesium with a mass of 1.00 mg?

Diamond, Graphite, the Fullerenes, and Nanotechnology

13.43 Look up the densities of graphite and diamond in a handbook, and compare the values for these allotropes of carbon. According to Le Chatelier's principle, which of these allotropes should be the stabler form at high pressure?

13.44 Using data for enthalpies of formation, calculate the standard change of enthalpy when graphite reacts with O_2 to form CO_2. Do the same calculation but for the reaction of diamond with O_2. Which of these values corresponds to the standard enthalpy of formation of CO_2?

13.45 A mixture of methane, CH_4, and hydrogen, H_2, is passed over a hot filament to provide a gas containing carbon atoms. If all of the carbon atoms in the methane are converted to diamond film, what volume of methane at STP is needed to produce 1.00 mg of diamond film?

13.46 A mixture of ethane, C_2H_6, and hydrogen, H_2, is passed over a hot filament to provide a gas containing carbon atoms. If all of the carbon atoms in the ethane are converted to diamond film, what volume of methane at STP is needed to produce 1.00 mg of diamond film?

13.47 Draw a portion of the structure of the buckminsterfullerene molecule, say, a hexagon and a pentagon of carbon atoms, using one resonance form. Draw another resonance form. Describe the bonding at the carbon atoms in terms of hybrid orbitals.

13.48 Draw a portion of the structure of a fullerene nanotube molecule, say, several hexagons, using one resonance form. Draw another resonance form. Describe the bonding at the carbon atoms in terms of hybrid orbitals.

Semiconductors

13.49 Which of the following are n-type semiconductors?
a. C(diamond) doped with B
b. Ge doped with As
c. Si doped with Al
d. Ge doped with P

13.50 Which of the following are p-type semiconductors?
a. C(diamond) doped with B
b. Ge doped with As
c. Si doped with Al
d. Ge doped with P

Silicon, Silica, and Silicates

13.51 Silicon has a diamond-like structure, with each Si atom bonded to four other Si atoms. Describe the bonding in silicon in terms of hybrid orbitals.

13.52 Silicon carbide, SiC, has a structure in which each Si atom is bonded to four C atoms, and each C atom is bonded to four Si atoms. Describe the bonding in terms of hybrid orbitals.

A reaction whose rate has been extensively studied under various conditions is the decomposition of dinitrogen pentoxide, N_2O_5. When this substance is heated in the gas phase, it decomposes to nitrogen dioxide and oxygen:

$$2N_2O_5(g) \longrightarrow 4NO_2(g) + O_2(g)$$

We will look at this reaction in some detail. The questions we will pose include: How is the rate of a reaction like this measured? What conditions affect the rate of a reaction? How do you express the relationship of rate of a reaction to the variables that affect rate? What happens at the molecular level when N_2O_5 decomposes to NO_2 and O_2?

Reaction Rates

Figure 14.1
Catalytic decomposition of hydrogen peroxide.
The hydrogen peroxide decomposes rapidly when hydrobromic acid is added to an aqueous solution. One of the products is oxygen gas, which bubbles vigorously from the solution. In addition, some HBr is oxidized to Br_2, as can be seen from the red color of the liquid and vapor.

What variables affect reaction rates? Some reactions are fast and others are slow, but the rate of any given reaction may be affected by the following factors:

1. *Concentrations of reactants.* Often the rate of reaction increases when the concentration of a reactant is increased. A piece of steel wool burns with some difficulty in air (20% O_2) but bursts into a dazzling white flame in pure oxygen. In some reactions, however, the rate is unaffected by the concentration of a particular reactant, as long as it is present at some concentration.

2. *Concentration of catalyst.* A **catalyst** is *a substance that increases the rate of reaction without being consumed in the overall reaction.* The catalyst does not appear in the balanced chemical equation. A solution of pure hydrogen peroxide, H_2O_2, is stable, but when hydrobromic acid, HBr(aq), is added, H_2O_2 decomposes rapidly into H_2O and O_2 (Figure 14.1).

$$2H_2O_2(aq) \xrightarrow{\text{HBr}(aq)} 2H_2O(l) + O_2(g)$$

3. *Temperature at which the reaction occurs.* Usually reactions speed up when the temperature increases. Reactions during cooking go faster at higher temperatures.

4. *Surface area of a solid reactant or catalyst.* If a reaction involves a solid with a gas or liquid, the surface area of the solid affects the reaction rate. The greater the surface area per unit volume, the faster is the reaction (Figure 14.2).

14.1 Definition of Reaction Rate

The rate of a reaction is the amount of product formed or the amount of reactant used up per unit of time. So that a rate calculation does not depend on the total quantity of reaction mixture used, you express the rate for a unit volume of the mixture. Therefore, the **reaction rate** is *the increase in molar concentration of product of a reaction per unit time or the decrease in molar concentration of reactant per unit time.* The usual unit of reaction rate is moles per liter per second, mol/(L · s).

Consider the gas-phase reaction discussed in the chapter opening:

$$2N_2O_5(g) \longrightarrow 4NO_2(g) + O_2(g)$$

The rate for this reaction could be found by observing the increase in molar concentration of O_2 produced. You denote the molar concentration of a substance by enclosing its formula in square brackets. Thus, $[O_2]$ is the molar concentration of O_2. In a given time interval Δt, the molar concentration of oxygen, $[O_2]$, in the reaction vessel increases by the amount $\Delta[O_2]$. The rate of the reaction is given by

$$\text{Rate of formation of oxygen} = \frac{\Delta[O_2]}{\Delta t}$$

This equation gives the *average* rate over the time interval Δt. If the time interval is very short, the equation gives the *instantaneous* rate—that is, the rate at a particular instant of time. The instantaneous rate is also the value of $\Delta[O_2]/\Delta t$ for the tangent at a given instant (the straight line that just touches the curve of concentration versus time at a given point). See Figure 14.3. ◀ See the tangent at 2400 s.

Figure 14.3 shows the increase in concentration of O_2 during the decomposition of N_2O_5. It shows the calculation of average rates at two positions on the curve. For example, when the time changes from 600 s to 1200 s ($\Delta t = 600$ s), the O_2 concentration increases by 0.0015 mol/L ($= \Delta[O_2]$). Therefore, the average rate $= \Delta[O_2]/\Delta t =$ (0.0015 mol/L)/600 s $= 2.5 \times 10^{-6}$ mol/(L · s). Later, during the time interval from 4200 s to 4800 s, the average rate is 5×10^{-7} mol/(L · s). Note that the rate decreases as the reaction proceeds.

Because the amounts of products and reactants are related by stoichiometry, any substance in the reaction can be used to express the rate of reaction. In the case of the decomposition of N_2O_5 to NO_2 and O_2, we gave the rate in terms of the rate of formation of oxygen, $\Delta[O_2]/\Delta t$. However, you can also express it in terms of the rate of decomposition of N_2O_5.

$$\text{Rate of decomposition of } N_2O_5 = -\frac{\Delta[N_2O_5]}{\Delta t}$$

Note the negative sign. It always occurs in a rate expression for a reactant in order to indicate a decrease in concentration and to give a positive value for the rate. Thus, because $[N_2O_5]$ decreases, $\Delta[N_2O_5]$ is negative and $-\Delta[N_2O_5]/\Delta t$ is positive.

The rate of decomposition of N_2O_5 and the rate of formation of oxygen are easily related. Two moles of N_2O_5 decompose for each mole of oxygen formed, so the rate of decomposition of N_2O_5 is twice the rate of formation of oxygen. To equate the rates, you must divide the rate of decomposition of N_2O_5 by 2 (its coefficient in the balanced chemical equation).

$$\frac{\Delta[O_2]}{\Delta t} = -\frac{1}{2}\frac{\Delta[N_2O_5]}{\Delta t}$$

▶ In calculus, the rate at a given moment (the instantaneous rate) is given by the derivative $d[O_2]/dt$.

Figure 14.2
Effect of large surface area on the rate of reaction. Lycopodium powder (from the tiny spores of a club moss) ignites easily to produce a yellow flame. The powder has a large surface area per volume and burns rapidly in air.

EXAMPLE 14.1

Relating the Different Ways of Expressing Reaction Rates

Consider the reaction of nitrogen dioxide with fluorine to give nitryl fluoride, NO_2F.

$$2NO_2(g) + F_2(g) \longrightarrow 2NO_2F(g)$$

How is the rate of formation of NO_2F related to the rate of reaction of fluorine?

SOLUTION

You write

$$\text{Rate of formation of } NO_2F = \frac{\Delta[NO_2F]}{\Delta t}$$

Figure 14.3
The instantaneous rate and the calculation of the average rate. The average rate of formation of O_2 during the decomposition of N_2O_5 was calculated during two different time intervals. When the time changes from 600 s to 1200 s, the average rate is 2.5×10^{-6} mol/(L · s). Later, when the time changes from 4200 s to 4800 s, the average rate has slowed to 5×10^{-7} mol/(L · s). Thus, the rate of the reaction decreases as the reaction proceeds.

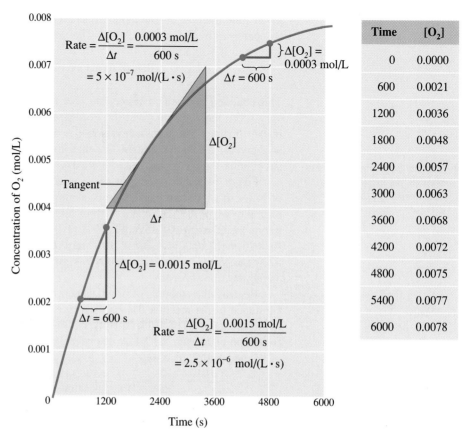

Time	$[O_2]$
0	0.0000
600	0.0021
1200	0.0036
1800	0.0048
2400	0.0057
3000	0.0063
3600	0.0068
4200	0.0072
4800	0.0075
5400	0.0077
6000	0.0078

and

$$\text{Rate of reaction of } F_2 = -\frac{\Delta[F_2]}{\Delta t}$$

You divide each rate by the corresponding coefficient and then equate them:

$$\frac{1}{2}\frac{\Delta[NO_2F]}{\Delta t} = -\frac{\Delta[F_2]}{\Delta t}$$

See Problems 14.23 and 14.24.

EXAMPLE 14.2

Calculating the Average Reaction Rate

Calculate the average rate of decomposition of N_2O_5, $-\Delta[N_2O_5]/\Delta t$, by the reaction

$$2N_2O_5(g) \longrightarrow 4NO_2(g) + O_2(g)$$

during the time interval from $t = 600$ s to $t = 1200$ s (regard all time figures as significant). Use the following data:

Time	[N₂O₅]
600 s	$1.24 \times 10^{-2}\, M$
1200 s	$0.93 \times 10^{-2}\, M$

SOLUTION

Average rate of decomposition of $N_2O_5 = -\dfrac{\Delta[N_2O_5]}{\Delta t} =$

$$-\dfrac{(0.93 - 1.24) \times 10^{-2}\, M}{(1200 - 600)\ s} = -\dfrac{-0.31 \times 10^{-2}\, M}{600\ s} = \mathbf{5.2 \times 10^{-6}\, M/s}$$

Note that this rate is twice the rate of formation of O_2 in the same time interval (within the experimental error of the value given in the preceding text discussion).

See Problems 14.25 and 14.26.

 CONCEPT CHECK 14.1

Shown here is a plot of the concentration of a reactant D versus time.

a. How do the instantaneous rates at points A and B compare?

b. Is the rate for this reaction constant at all points in time?

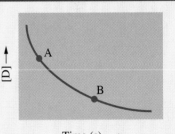

14.2 Experimental Determination of Rate

To obtain the rate of a reaction, you must determine the concentration of a reactant or product during the course of the reaction. One way to do this for a slow reaction is to withdraw samples from the reaction vessel at various times and analyze them. The rate of the reaction of ethyl acetate with water in acidic solution was one of the first to be determined this way. ◀

▶ Ethyl acetate is a liquid with a fruity odor that belongs to a class of organic (carbon-containing) compounds called esters.

$$\underset{\text{ethyl acetate}}{CH_3CH_2O\overset{O}{\overset{\|}{C}}CH_3} + H_2O \xrightarrow{\text{H}^+} \underset{\text{ethanol}}{CH_3CH_2OH} + \underset{\text{acetic acid}}{HO\overset{O}{\overset{\|}{C}}CH_3}$$

This reaction is slow, so the amount of acetic acid produced is easily obtained by titration before any significant further reaction occurs.

More convenient are techniques that can continuously follow the progress of a reaction by observing the change in some physical property of the system. For example, if a gas reaction involves a change in the number of molecules, the pressure of the system changes. By following the pressure change as the reaction proceeds, you can obtain the reaction rate. The decomposition of dinitrogen pentoxide in the gas

Figure 14.4
An experiment to follow the concentration of N_2O_5 as the decomposition proceeds. The total pressure is measured during the reaction at 45°C. Pressure values can be related to the concentrations of N_2O_5, NO_2, and O_2 in the flask.

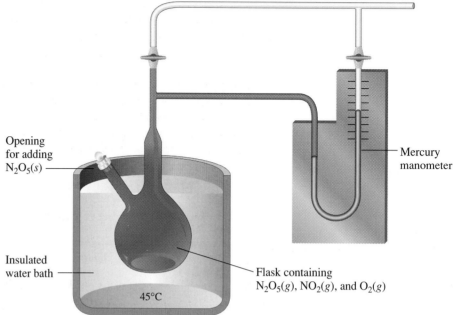

Opening for adding $N_2O_5(s)$

Mercury manometer

Insulated water bath

45°C

Flask containing $N_2O_5(g)$, $NO_2(g)$, and $O_2(g)$

phase has been studied this way. Dinitrogen pentoxide crystals are sealed in a vessel equipped with a manometer (see Figure 14.4). The vessel is then plunged into a water bath at 45°C, at which temperature the solid vaporizes and the gas decomposes.

$$2N_2O_5(g) \longrightarrow 4NO_2(g) + O_2(g)$$

Manometer readings are taken at regular time intervals, and the pressure values are converted to partial pressures or concentrations of N_2O_5. The rates of reaction during various time intervals can be calculated as described in the previous section.

Another physical property used to follow the progress of a reaction is color, or the absorption of light by some species. Consider the reaction

$$ClO^-(aq) + I^-(aq) \longrightarrow IO^-(aq) + Cl^-(aq)$$

The hypoiodite ion, IO^-, absorbs at the blue end of the spectrum near 400 nm. The intensity of this absorption is proportional to $[IO^-]$, and you can use the absorption to determine the reaction rate. You can also follow the decomposition of N_2O_5 from the intensity of the red-brown color of the product NO_2.

14.3 Dependence of Rate on Concentration

Experimentally, it has been found that a reaction rate depends on the concentrations of certain reactants as well as the concentration of catalyst, if there is one. Consider the reaction of nitrogen dioxide with fluorine to give nitryl fluoride, NO_2F.

$$2NO_2(g) + F_2(g) \longrightarrow 2NO_2F(g)$$

The rate of this reaction is observed to be proportional to the concentration of nitrogen dioxide. When the concentration of nitrogen dioxide is doubled, the rate doubles. The rate is also proportional to the concentration of fluorine; doubling the concentration of fluorine also doubles the rate.

A **rate law** is *an equation that relates the rate of a reaction to the concentrations of reactants (and catalyst) raised to various powers.* The following equation is the rate law for the foregoing reaction:

$$\text{Rate} = k[NO_2][F_2]$$

Note that in this rate law both reactant concentrations have an exponent of 1. Here k, called the **rate constant,** is *a proportionality constant in the relationship between rate and concentrations.* It has a fixed value at any given temperature, but it varies with temperature. Whereas the units of rate are usually given as mol/(L · s), the units of k depend on the form of the rate law. For the previous rate law,

$$k = \frac{\text{rate}}{[NO_2][F_2]}$$

from which you get the following unit for k:

$$\frac{\text{mol}/(\text{L} \cdot \text{s})}{(\text{mol/L})^2} = \text{L}/(\text{mol} \cdot \text{s})$$

As a more general example, consider the reaction of substances A and B to give D and E, according to the balanced equation

$$a\text{A} + b\text{B} \xrightarrow{\text{C}} d\text{D} + e\text{E} \qquad \text{C} = \text{catalyst}$$

You could write the rate law in the form

$$\text{Rate} = k[\text{A}]^m[\text{B}]^n[\text{C}]^p$$

The exponents m, n, and p are frequently, but not always, integers. *They must be determined experimentally* and they cannot be obtained simply by looking at the balanced equation. For example, note that the exponents in the equation Rate = $k[NO_2][F_2]$ have no relationship to the coefficients in the balanced equation $2NO_2 + F_2 \longrightarrow 2NO_2F$.

Once you know the rate law for a reaction and have found the value of the rate constant, you can calculate the rate of a reaction for any values of reactant concentrations. As you will see later, knowledge of the rate law is also useful in understanding how the reaction occurs at the molecular level.

Reaction Order

You can classify a reaction by its orders. The **reaction order** with respect to a given reactant species equals *the exponent of the concentration of that species in the rate law, as determined experimentally.* For the reaction of NO_2 with F_2 to give NO_2F, the reaction is first order with respect to the NO_2 because the exponent of $[NO_2]$ in the rate law is 1. Similarly, the reaction is first order with respect to F_2.

The *overall order* of a reaction equals the sum of the orders of the reactant species in the rate law. In this example, the overall order is 2.

Reactions display a variety of reaction orders. Two examples follow:

1. Nitric oxide, NO, reacts with hydrogen according to the equation

$$2NO(g) + 2H_2(g) \longrightarrow N_2(g) + 2H_2O(g)$$

The experimentally determined rate law is

$$\text{Rate} = k[NO]^2[H_2]$$

Thus, the reaction is second order in NO, first order in H_2, and third order overall.

2. Acetone, CH_3COCH_3, reacts with iodine in acidic solution.

$$CH_3COCH_3(aq) + I_2(aq) \xrightarrow{H^+} CH_3COCH_2I(aq) + HI(aq)$$

The experimentally determined rate law is

$$\text{Rate} = k[CH_3COCH_3][H^+]$$

The reaction is first order in acetone. It is zero order in iodine; that is, the rate law contains the factor $[I_2]^0 = 1$. Therefore, the rate does not depend on the concentration of I_2, as long as some concentration of I_2 is present. Note that the reaction is first order in the catalyst, H^+. Thus, the overall order is 2.

Although reaction orders frequently have whole-number values (particularly 1 or 2), they can be fractional. Zero and negative orders are also possible.

EXAMPLE 14.3

Determining the Order of Reaction from the Rate Law

Bromide ion is oxidized by bromate ion in acidic solution.

$$5Br^-(aq) + BrO_3^-(aq) + 6H^+(aq) \longrightarrow 3Br_2(aq) + 3H_2O(l)$$

The experimentally determined rate law is

$$\text{Rate} = k[Br^-][BrO_3^-][H^+]^2$$

What is the order of reaction with respect to each reactant species? What is the overall order of reaction?

SOLUTION

The reaction is **first order with respect to Br⁻** and **first order with respect to BrO₃⁻**; it is **second order with respect to H⁺** (an order with respect to a species equals the exponent of its concentration). **The reaction is fourth order overall** (= 1 + 1 + 2).

See Problems 14.29 and 14.30.

Determining the Rate Law

The experimental determination of the rate law for a reaction requires that you find the order of the reaction with respect to each reactant and any catalyst. The *initial-rate method* is a simple way to obtain reaction orders. It consists of doing a series of experiments in which the initial, or starting, concentrations of reactants are varied. Then the initial rates are compared, from which the reaction orders can be deduced.

To see how this method works, again consider the reaction

$$2N_2O_5(g) \longrightarrow 4NO_2(g) + O_2(g)$$

You observe this reaction in two experiments. In Experiment 2, the initial concentration of N_2O_5 is twice that in Experiment 1. You then note the initial rate of disappearance of N_2O_5 in each case. The initial concentrations and corresponding initial rates for the two experiments are given in the following table: ◄

▶ In Figure 14.3, the slope of the tangent to the curve at $t = 0$ equals the initial rate of appearance of O_2, which equals one-half the initial rate of disappearance of N_2O_5.

	Initial N_2O_5 Concentration	Initial Rate of Disappearance of N_2O_5
Experiment 1	1.0×10^{-2} mol/L	4.8×10^{-6} mol/(L · s)
Experiment 2	2.0×10^{-2} mol/L	9.6×10^{-6} mol/(L · s)

The rate law for this reaction will have the concentration of reactant raised to a power m.

$$\text{Rate} = k[N_2O_5]^m$$

The value of m (the reaction order) must be determined from the experimental data. Note that when the N_2O_5 concentration is doubled, you get a new rate, Rate′, given by the following equation:

$$\text{Rate}' = k(2[N_2O_5])^m = 2^m k[N_2O_5]^m$$

This rate is 2^m times the original rate.

You can now see how the rate is affected when the concentration is doubled for various choices of m. Suppose $m = 2$. You get $2^m = 2^2 = 4$. That is, when the initial concentration is doubled, the rate is multiplied by 4. We summarize the results for various choices of m in Table 14.1.

Suppose you divide the initial rate of reaction of N_2O_5 from Experiment 2 by the initial rate from Experiment 1.

$$\frac{9.6 \times 10^{-6} \text{ mol/(L · s)}}{4.8 \times 10^{-6} \text{ mol/(L · s)}} = 2.0$$

You see that when the N_2O_5 concentration is doubled, the rate is doubled. This corresponds to the case $m = 1$. The rate law must have the form

$$\text{Rate} = k[N_2O_5]$$

You can determine the value of the rate constant k by substituting values of the rate and N_2O_5 concentrations from any of the experiments into the rate law. Using values from Experiment 2, you get

$$9.6 \times 10^{-6} \text{ mol/(L · s)} = k \times 2.0 \times 10^{-2} \text{ mol/L}$$

Hence,

$$k = \frac{9.6 \times 10^{-6}/s}{2.0 \times 10^{-2}} = 4.8 \times 10^{-4}/s$$

Table 14.1
Effect on Rate of Doubling the Initial Concentration of Reactant

m	Rate Is Multiplied by:
-1	$\frac{1}{2}$
0	1
1	2
2	4

EXAMPLE 14.4

Determining the Rate Law from Initial Rates

Iodide ion is oxidized in acidic solution to triiodide ion, I_3^-, by hydrogen peroxide.

$$H_2O_2(aq) + 3I^-(aq) + 2H^+(aq) \longrightarrow I_3^-(aq) + 2H_2O(l)$$

A series of four experiments was run at different concentrations, and the initial rates of I_3^- formation were determined (see table). a. From these data, obtain the reaction orders with respect to H_2O_2, I^-, and H^+. b. Then find the rate constant.

	Initial Concentrations (mol/L)			Initial Rate [mol/(L · s)]
	H_2O_2	**I^-**	**H^+**	
Exp. 1	0.010	0.010	0.00050	1.15×10^{-6}
Exp. 2	0.020	0.010	0.00050	2.30×10^{-6}
Exp. 3	0.010	0.020	0.00050	2.30×10^{-6}
Exp. 4	0.010	0.010	0.00100	1.15×10^{-6}

SOLUTION

a. Comparing Experiment 1 and Experiment 2, you see that when the H_2O_2 concentration is doubled (with other concentrations constant), the rate is doubled. From Table 14.1, you see that $m = 1$. The reaction is first order in H_2O_2.

To solve a problem such as this in a general way, you write the rate law for two experiments.

$$(\text{Rate})_1 = k[H_2O_2]_1^m [I^-]_1^n [H^+]_1^p$$

$$(\text{Rate})_2 = k[H_2O_2]_2^m [I^-]_2^n [H^+]_2^p$$

Now you divide the second equation by the first.

$$\frac{(\text{Rate})_2}{(\text{Rate})_1} = \frac{k[H_2O_2]_2^m [I^-]_2^n [H^+]_2^p}{k[H_2O_2]_1^m [I^-]_1^n [H^+]_1^p}$$

The rate constant cancels. Grouping the terms, you obtain

$$\frac{(\text{Rate})_2}{(\text{Rate})_1} = \left(\frac{[H_2O_2]_2}{[H_2O_2]_1}\right)^m \left(\frac{[I^-]_2}{[I^-]_1}\right)^n \left(\frac{[H^+]_2}{[H^+]_1}\right)^p$$

Now you substitute values from Experiment 1 and Experiment 2. (In a ratio the units cancel and can be omitted.)

$$\frac{2.30 \times 10^{-6}}{1.15 \times 10^{-6}} = \left[\frac{0.020}{0.010}\right]^m \left[\frac{0.010}{0.010}\right]^n \left[\frac{0.00050}{0.00050}\right]^p$$

This gives $2 = 2^m$, from which you obtain $m = 1$.

The exponents n and p are obtained in the same way. Comparing Experiment 1 and Experiment 3, you see that doubling the I^- concentration (with the other concentrations constant) doubles the rate. Therefore, $n = 1$ (the reaction is first order in I^-). Finally, comparing Experiment 1 and Experiment 4, you see that doubling the H^+ concentration (holding other concentrations constant) has no effect on the rate. Therefore, $p = 0$ (the reaction is zero order in H^+). Because $[H^+]^0 = 1$, the rate law is

$$\textbf{Rate} = k[\textbf{H}_2\textbf{O}_2][\textbf{I}^-]$$

The reaction orders with respect to H_2O_2, I^-, and H^+ are **1, 1, and 0,** respectively.

b. You calculate the rate constant by substituting values from any of the experiments into the rate law. Using Experiment 1, you obtain

$$1.15 \times 10^{-6} \frac{\text{mol}}{\text{L} \cdot \text{s}} = k \times 0.010 \frac{\text{mol}}{\text{L}} \times 0.010 \frac{\text{mol}}{\text{L}}$$

$$k = \frac{1.15 \times 10^{-6}/\text{s}}{0.010 \times 0.010 \times \text{mol/L}} = \textbf{1.2} \times \textbf{10}^{-2} \textbf{ L/(mol} \cdot \textbf{s)}$$

See Problems 14.31, 14.32, 14.33, and 14.34.

CONCEPT CHECK 14.2

Rate laws are not restricted to chemical systems; they are used to help describe many "everyday" events. For example, a rate law for tree growth might look something like this:

Rate of growth = (soil type)^w(temperature)^x(light)^y(fertilizer)^z

In this equation, like chemical rate equations, the exponents need to be determined by experiment. (Can you think of some other factors?)

a. Say you are a famous physician trying to determine the factors that influence the rate of aging in humans. Develop a rate law, like the one above, that would take into account at least four factors that affect the rate of aging.

b. Explain what you would need to do in order to determine the exponents in your rate law.

c. Consider smoking to be one of the factors in your rate law. You conduct an experiment and find that a person smoking two packs of cigarettes a day quadruples (4×) the rate of aging over that of a one-pack-a-day smoker. Assuming that you could hold all other factors in your rate law constant, what will be the exponent of the smoking term in your rate law?

14.4 Change of Concentration with Time

A rate law tells you how the rate of a reaction depends on reactant concentrations at a particular moment. But often you would like to have a mathematical relationship showing how a reactant concentration changes over a period of time. Using it, you could answer questions such as: How long does it take for this reaction to be 50% complete? to be 90% complete?

Moreover, as you will see, knowing exactly how the concentrations change with time for different rate laws suggests ways of plotting the experimental data on a graph. Graphical plotting provides an alternative to the initial-rate method for determining the rate law.

Concentration–Time Equations

First-Order Rate Law Let us first look at first-order rate laws. The decomposition of nitrogen pentoxide,

$$2N_2O_5(g) \longrightarrow 4NO_2(g) + O_2(g)$$

has the following rate law:

$$\text{Rate} = -\frac{\Delta[N_2O_5]}{\Delta t} = k[N_2O_5]$$

Using calculus, one can show that such a first-order rate law leads to the following relationship between N_2O_5 concentration and time: ◄

$$\ln\frac{[N_2O_5]_t}{[N_2O_5]_0} = -kt$$

The symbol 0 represents the initial concentration at time $t = 0$.

This equation enables you to calculate the concentration of N_2O_5 at any time, once you are given the initial concentration and the rate constant. Also, you can find the time it takes for the N_2O_5 concentration to decrease to a particular value.

More generally, let A be a substance that reacts to give products according to the equation

$$a\text{A} \longrightarrow \text{products}$$

Suppose that this reaction has a first-order rate law

$$\text{Rate} = -\frac{\Delta[A]}{\Delta t} = k[A]$$

Using calculus, you get the following equation:

$$\ln\frac{[A]_t}{[A]_0} = -kt$$

Here $[A]_t$ is the concentration of reactant A at time t, and $[A]_0$ is the initial concentration. The ratio $[A]_t/[A]_0$ is the fraction of reactant remaining at time t. The next example illustrates how you work with this equation.

► The general derivation using calculus is as follows. Substituting [A] for $[N_2O_5]$, the rate law becomes

$$\frac{-d[A]}{dt} = k[A]$$

You rearrange this to give

$$\frac{-d[A]}{[A]} = k\, dt$$

Integrating from time = 0 to time = t,

$$-\int_{[A]_0}^{[A]_t} \frac{d[A]}{[A]} = k\int_0^t dt$$

gives

$$-\{\ln[A]_t - \ln[A]_0\} = k(t - 0)$$

This can be rearranged to give the equation in the text.

EXAMPLE 14.5

Using the Concentration–Time Equation for a First-Order Reaction

The decomposition of N_2O_5 to NO_2 and O_2 is first order, with a rate constant of 4.80×10^{-4}/s at 45°C. a. If the initial concentration is 1.65×10^{-2} mol/L, what is the concentration after 825 s? b. How long would it take for the concentration of N_2O_5 to decrease to 1.00×10^{-2} mol/L from its initial value, given in a?

SOLUTION

a. You need to use the equation relating concentration to time (the shaded equation given just before this example). You get

$$\ln\frac{[N_2O_5]_t}{1.65 \times 10^{-2}\,\text{mol/L}} = -4.80 \times 10^{-4}\text{/s} \times 825\,\text{s} = -0.396$$

To solve for $[N_2O_5]_t$, you take the antilogarithm of both sides. This removes the ln from the left and gives antilog(-0.396), or $e^{-0.396}$, on the right, which equals 0.673.

$$\frac{[N_2O_5]_t}{1.65 \times 10^{-2} \text{ mol/L}} = 0.673$$

Hence,

$$[N_2O_5]_t = 1.65 \times 10^{-2} \text{ mol/L} \times 0.673 = \textbf{0.0111 mol/L}$$

b. You substitute into the equation relating concentration to time.

$$\ln \frac{1.00 \times 10^{-2} \text{ mol/L}}{1.65 \times 10^{-2} \text{ mol/L}} = -4.80 \times 10^{-4}/\text{s} \times t$$

The left side equals -5.01; the right side equals $-4.80 \times 10^{-4}/\text{s} \times t$. Hence,

$$5.01 = 4.80 \times 10^{-4}/\text{s} \times t$$

Or,

$$t = \frac{5.01}{4.80 \times 10^{-4}/\text{s}} = \textbf{1.04} \times \textbf{10}^{\textbf{3}} \textbf{ s} \qquad \textbf{(17.4 min)}$$

See Problems 14.35 and 14.36.

Second-Order Rate Law   Consider the reaction

$$a\text{A} \longrightarrow \text{products}$$

and suppose it has the second-order rate law

$$\text{Rate} = -\frac{\Delta[\text{A}]}{\Delta t} = k[\text{A}]^2$$

An example is the decomposition of nitrogen dioxide at moderately high temperatures (300°C to 400°C).

$$2NO_2(g) \longrightarrow 2NO(g) + O_2(g)$$

Using calculus, you can obtain the following relationship between the concentration of A and the time.

$$\frac{1}{[\text{A}]_t} = kt + \frac{1}{[\text{A}]_0}$$

Using this equation, you can calculate the concentration of NO_2 at any time during its decomposition if you know the rate constant and the initial concentration. At 330°C, the rate constant for the decomposition of NO_2 is 0.775 L/(mol · s). Suppose the initial concentration is 0.0030 mol/L. What is the concentration of NO_2 after 645 s? By substituting into the previous equation, you get

$$\frac{1}{[NO_2]_t} = 0.775 \text{ L/(mol · s)} \times 645 \text{ s} + \frac{1}{0.0030 \text{ mol/L}} = 8.3 \times 10^2 \text{ L/mol}$$

If you invert both sides of the equation, you find that $[NO_2]_t = 0.0012$ mol/L. Thus, after 645 s, the concentration of NO_2 decreased from 0.0030 mol/L to 0.0012 mol/L.

Half-Life of a Reaction

▶ The concept of half-life is also used to characterize a radioactive nucleus, whose radioactive decay is a first-order process. This is discussed in Chapter 21.

As a reaction proceeds, the concentration of a reactant decreases, because it is being consumed. The **half-life,** $t_{1/2}$, of a reaction is *the time it takes for the reactant concentration to decrease to one-half of its initial value.* ◀

For a first-order reaction, such as the decomposition of dinitrogen pentoxide, the half-life is independent of the initial concentration. To see this, substitute into the equation

$$\ln \frac{[N_2O_5]_t}{[N_2O_5]_0} = -kt$$

In one half-life, the N_2O_5 concentration decreases by one-half, from its initial value, $[N_2O_5]_0$, to the value $[N_2O_5]_t = \frac{1}{2}[N_2O_5]_0$. After substituting, the equation becomes

$$\ln \frac{\frac{1}{2}[N_2O_5]_0}{[N_2O_5]_0} = -kt_{1/2}$$

The expression on the left equals $\ln \frac{1}{2} = -0.693$. Hence,

$$0.693 = kt_{1/2}$$

Solving for the half-life, $t_{1/2}$, you get

$$t_{1/2} = \frac{0.693}{k}$$

Because the rate constant for the decomposition of N_2O_5 at 45°C is 4.80×10^{-4}/s, the half-life is

$$t_{1/2} = \frac{0.693}{4.80 \times 10^{-4}/\text{s}} = 1.44 \times 10^3 \text{ s}$$

Thus, the half-life is 1.44×10^3 s, or 24.0 min.

You see that the half-life of N_2O_5 does not depend on the initial concentration of N_2O_5. This means that the half-life is the same at any time during the reaction. If the initial concentration is 0.0165 mol/L, after one half-life (24.0 min) the concentration decreases to $\frac{1}{2} \times 0.0165$ mol/L $= 0.0083$ mol/L. After another half-life (another 24.0 min), the N_2O_5 concentration decreases to $\frac{1}{2} \times 0.0083$ mol/L $= 0.0042$ mol/L. Every time one half-life passes, the N_2O_5 concentration decreases by one-half again (see Figure 14.5).

The foregoing result for the half-life for the first-order decomposition of N_2O_5 is perfectly general. That is, for the general first-order rate law the half-life is related to the rate constant, but is independent of concentration.

$$t_{1/2} = \frac{0.693}{k}$$

Figure 14.5
A graph illustrating that the half-life of a first-order reaction is independent of initial concentration. In one half-life (1440 s, or 24.0 min), the concentration decreases by one-half, from 0.0165 mol/L to $\frac{1}{2} \times 0.0165$ mol/L = 0.0083 mol/L. After each succeeding half-life (24.0 min), the concentration decreases by one-half again, from 0.0083 mol/L to $\frac{1}{2} \times 0.0083$ mol/L = 0.0042 mol/L, then to $\frac{1}{2} \times 0.0042$ mol/L = 0.0021 mol/L, and so forth.

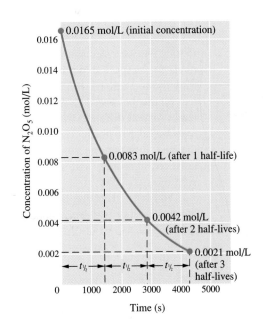

EXAMPLE 14.6

Relating the Half-Life of a Reaction to the Rate Constant

Sulfuryl chloride, SO_2Cl_2, is a colorless, corrosive liquid whose vapor decomposes in a first-order reaction to sulfur dioxide and chlorine.

At 320°C, the rate constant is 2.20×10^{-5}/s. a. What is the half-life of SO_2Cl_2 vapor at this temperature? b. How long would it take for 50.0% of the SO_2Cl_2 to decompose? How long would it take for 75.0% of the SO_2Cl_2 to decompose?

SOLUTION

a.
$$t_{1/2} = \frac{0.693}{k} = \frac{0.693}{2.20 \times 10^{-5}/s} = \textbf{3.15} \times \textbf{10}^4 \textbf{ s}$$

b. The half-life is the time required for one-half (50.0%) of the SO_2Cl_2 to decompose. This is 3.15×10^4 s, or **8.75 hr.** After another half-life, one-half of the remaining SO_2Cl_2 decomposes. The total decomposed is $\frac{1}{2} + (\frac{1}{2} \times \frac{1}{2}) = \frac{3}{4}$, or 75.0%. The time required is two half-lives, or 2×8.75 hr = **17.5 hr.**

See Problems 14.37 and 14.38.

It can be shown by reasoning similar to that given previously that the half-life of a second-order rate law, Rate = $k[A]^2$, is $1/(k[A]_0)$. In this case, the half-life depends on initial concentration and becomes larger as time goes on. Consider the decomposition of NO_2 at 330°C. It takes 430 s for the concentration to decrease by one-half from 0.0030 mol/L to 0.0015 mol/L. However, it takes 860 s (twice as long) for the concentration to decrease by one-half again. The fact that the half-life changes with time is evidence that the reaction is not first order.

Graphing of Kinetic Data

Earlier you saw that the order of a reaction can be determined by comparing initial rates for several experiments in which different initial concentrations are used (initial-rate method). It is also possible to determine the order of a reaction by graphical plotting of the data for a particular experiment. The experimental data are plotted in several different ways, first assuming a first-order reaction, then a second-order reaction, and so forth. The order of the reaction is determined by which graph gives the best fit to the experimental data. To illustrate, we will look at how the plotting is done for first-order and second-order reactions and then compare graphs for a specific reaction.

You have seen that the first-order rate law, $-\Delta[A]/\Delta t = k[A]$, gives the following relationship between concentration of A and time:

$$\ln\frac{[A]_t}{[A]_0} = -kt$$

This equation can be rewritten in a slightly different form, which you can identify with the equation of a straight line. Using the property of logarithms that $\log(A/B) = \log A - \log B$, you get ◄

▶ $\ln x = 2.303 \log x$

$$\ln[A]_t = -kt + \ln[A]_0$$

A straight line has the mathematical form $y = mx + b$, when y is plotted on the vertical axis against x on the horizontal axis. ◄ Let us now make the following identifications:

▶ See Appendix A for a discussion of the mathematics of a straight line.

$$\underbrace{\ln[A]_t}_{y} = \underbrace{-k}_{m}\ \underbrace{t}_{x} + \underbrace{\ln[A]_0}_{b}$$

This means that if you plot $\ln[A]_t$ on the vertical axis against the time t on the horizontal axis, you will get a straight line for a first-order reaction. Figure 14.6 shows a plot of $\ln[N_2O_5]$ at various times during the decomposition reaction. The fact that the points lie on a straight line is confirmation that the rate law is first order.

Figure 14.6
Plot of $\ln[N_2O_5]$ versus time.   A straight line can be drawn through the experimental points (colored dots). The fact that the straight line fits the experimental data so well confirms that the rate law is first order.

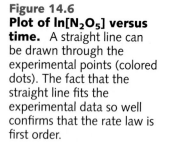

Time	[N₂O₅]	ln[N₂O₅]
0	0.0165	−4.104
600	0.0124	−4.390
1200	0.0093	−4.678
1800	0.0071	−4.948
2400	0.0053	−5.240
3000	0.0039	−5.547
3600	0.0029	−5.843

Time (s)

You can obtain the rate constant for the reaction from the slope, m, of the straight line.

$$-k = m \quad \text{or} \quad k = -m$$

You calculate the slope of this curve in the same way you obtained the average rate of reaction from kinetic data (Example 14.2). You select two points far enough apart that when you subtract to obtain Δx and Δy for the slope, you do not lose significant figures. Using the first and last points in Figure 14.6, you get

$$m = \frac{\Delta y}{\Delta x} = \frac{(-5.843) - (-4.104)}{(3600 - 0) \text{ s}} = \frac{-1.739}{3600 \text{ s}} = -4.83 \times 10^{-4}/\text{s}$$

Therefore, $k = 4.84 \times 10^{-4}/\text{s}$. (Two points were selected directly from the experimental data. In precise work, you would first draw the straight line that best fits the experimental data points and then calculate the slope of this line.)

The second-order rate law, $-\Delta[A]/\Delta t = k[A]^2$, gives the following relationship between concentration of A and time:

$$\underbrace{\frac{1}{[A]_t}}_{} = \underbrace{kt}_{} + \underbrace{\frac{1}{[A]_0}}_{}$$

$$y \quad = mx + \quad b$$

In this case, you get a straight line if you plot $1/[A]_t$ on the vertical axis against the time t on the horizontal axis for a second-order reaction.

As an illustration of the determination of reaction order by graphical plotting, consider the decomposition of NO_2 at 330°C.

$$2NO_2(g) \longrightarrow 2NO(g) + O_2(g)$$

The concentrations of NO_2 for various times are given in the table of data in Figure 14.7. In Figure 14.7A, we have plotted $\ln[NO_2]$ against t, and in Figure 14.7B we have plotted $1/[NO_2]$ against t. Only in (B) do the points closely follow a straight line, indicating that the rate law is second order. That is,

$$\text{Rate of disappearance of } NO_2 = -\frac{\Delta[NO_2]}{\Delta t} = k[NO_2]^2$$

You can obtain the rate constant, k, for a second-order reaction from the slope of the line, similar to the way you did for a first-order reaction. In this case, however, the slope equals k, as you can see from the preceding equation. Choosing the first and last points in Figure 14.7B, you get

$$k = \frac{\Delta y}{\Delta x} = \frac{(379 - 100) \text{ L/mol}}{(360 - 0) \text{ s}} = 0.775 \text{ L/(mol} \cdot \text{s)}$$

Table 14.2 summarizes the relationships discussed in this section for first-order and second-order reactions.

CONCEPT CHECK 14.3

A reaction believed to be either first or second order has a half-life of 20 s at the beginning of the reaction but a half-life of 40 s some time later. What is the order of the reaction?

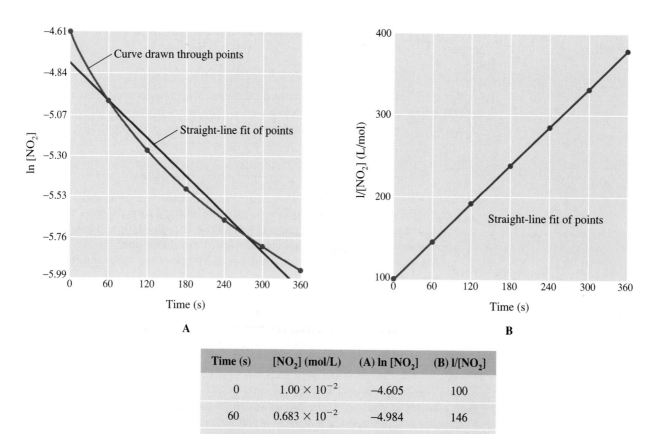

Time (s)	[NO$_2$] (mol/L)	(A) ln [NO$_2$]	(B) l/[NO$_2$]
0	1.00×10^{-2}	−4.605	100
60	0.683×10^{-2}	−4.984	146
120	0.518×10^{-2}	−5.263	193
180	0.418×10^{-2}	−5.476	239
240	0.350×10^{-2}	−5.656	286
300	0.301×10^{-2}	−5.805	332
360	0.264×10^{-2}	−5.938	379

Figure 14.7
Plotting the data for the decomposition of nitrogen dioxide at 330°C.
(A) Plot of ln[NO$_2$] against time. Note that a straight line does not fit the points well.
(B) Plot of 1/[NO$_2$] against time. Note how closely the points follow the straight line,
indicating that the decomposition is second order.

Table 14.2
Relationships for First-Order and Second-Order Reactions

Order	Rate Law	Concentration–Time Equation	Half-Life	Graphical Plot
1	Rate $= k[A]$	$\ln \dfrac{[A]_t}{[A]_0} = -kt$	$0.693/k$	ln [A] vs t
2	Rate $= k[A]^2$	$\dfrac{1}{[A]_t} = kt + \dfrac{1}{[A]_0}$	$1/(k[A]_0)$	$\dfrac{1}{[A]}$ vs t

14.5 Temperature and Rate; Collision and Transition-State Theories

As we noted earlier, the rate of reaction depends on temperature. This shows up in the rate law through the rate constant, which is found to vary with temperature (Figure 14.8). Consider the reaction of nitric oxide, NO, with chlorine, Cl_2, to give nitrosyl chloride, NOCl, and chlorine atoms.

$$NO(g) + Cl_2(g) \longrightarrow NOCl(g) + Cl(g)$$

The rate constant k for this reaction is 4.9×10^{-6} L/(mol · s) at 25°C and 1.5×10^{-5} L/(mol · s) at 35°C. In this case the rate constant and therefore the rate are more than tripled for a 10°C rise in temperature. ◀ How do you explain this strong dependence of reaction rate on temperature?

▶ The change in rate constant with temperature varies considerably from one reaction to another. In many cases, the rate of reaction approximately doubles for a 10°C rise, and this is often given as an approximate rule.

Collision Theory

Why the rate constant depends on temperature can be explained by collision theory. **Collision theory** of reaction rates is *a theory that assumes that, for reaction to occur, reactant molecules must collide with an energy greater than some minimum value and with the proper orientation. The minimum energy of collision required for two molecules to react* is called the **activation energy, E_a.** The value of E_a depends on the particular reaction.

In collision theory, the rate constant for a reaction is given as a product of three factors: (1) Z, the collision frequency; (2) f, the fraction of collisions having energy greater than the activation energy; and (3) p, the fraction of collisions that occur with the reactant molecules properly oriented. Thus,

$$k = Zfp$$

To have a specific reaction to relate the concepts to, consider the gas-phase reaction of NO with Cl_2, mentioned previously. An NO molecule collides with a Cl_2 molecule. If the collision has sufficient energy and if the molecules are properly oriented, reaction occurs.

Collision frequency Z, the frequency with which the reactant molecules collide, depends on temperature. This dependence of collision frequency on temperature does not explain why reaction rates usually change greatly with small temperature increases. As the temperature rises, the gas molecules move faster and therefore collide more frequently. ◀ From kinetic theory, one can show that at 25°C, a 10°C rise in temperature increases the collision frequency by about 2%. If you were to assume that each collision of reactant molecules resulted in reaction, you would conclude that the rate would increase with temperature at the same rate as the collision frequency— that is, by 2% for a 10°C rise in temperature. This clearly does not explain the tripling of the rate (a 300% increase) that you see in the reaction of NO with Cl_2.

▶ According to the kinetic theory of gases, the rms molecular speed equals $\sqrt{3RT/M_m}$ (see Section 5.7).

You see that the collision frequency varies only slowly with temperature. However, f, the fraction of molecular collisions having energy greater than the activation energy, changes rapidly in most reactions with even small temperature changes. It can be shown that f is related to the activation energy, E_a, this way:

$$f = e^{-E_a/RT}$$

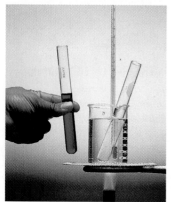

Figure 14.8
Effect of temperature on reaction rate. *Top:* Each test tube contains potassium permanganate, $KMnO_4$, and oxalic acid, $H_2C_2O_4$, at the same concentrations. Permanganate ion oxidizes oxalic acid to CO_2 and H_2O. One test tube was placed in a beaker of warm water (40°C); the other was kept at room temperature (20°C). *Bottom:* After 10 minutes, the test tube at 40°C showed noticeable reaction, whereas the other did not.

Here $e = 2.718\ldots$, and R is the gas constant, which equals 8.31 J/(mol · K). For the reaction of NO with Cl_2, the activation energy is 8.5×10^4 J/mol. At 25°C (298 K), the fraction of collisions with sufficient energy for reaction is 1.2×10^{-15}. Thus, the number of collisions of reactant molecules that actually result in reaction is extremely small. But the frequency of collisions is very large, so the reaction rate, which depends on the product of these quantities, is not small. If the temperature is raised by 10°C to 35°C, the fraction of collisions of NO and Cl_2 molecules with sufficient energy for reaction is 3.8×10^{-15}, over three times larger than the value at 25°C! The tripling of the reaction rate is explained by the temperature dependence of f.

From the previous equation relating f to E_a, you see that f decreases with increasing values of E_a. Because the rate constant depends on f, this means that reactions with large activation energies have small rate constants and that reactions with small activation energies have large rate constants.

The reaction rate also depends on p, the proper orientation of the reactant molecules when they collide. This factor is independent of temperature changes. You can see why it is important that the reactant molecules be properly oriented by looking in some detail at the reaction of NO with Cl_2. Figure 14.9 shows two possible molecular collisions. In Figure 14.9A, the NO and Cl_2 molecules collide properly oriented for reaction. The NO molecule approaches with its N atom toward the Cl_2 molecule. In addition, the angle of approach is about that expected for the formation of bonds in the product molecule NOCl. In Figure 14.9B, an NO molecule approaches with its O atom toward the Cl_2 molecule. Because this orientation does not allow the formation of a bond between the N atom and a Cl atom, it is ineffective for reaction. The NO and Cl_2 molecules come together and then fly apart.

Transition-State Theory

Collision theory explains some important features of a reaction, but it is limited in that it does not explain the role of activation energy. **Transition-state theory** *explains the reaction resulting from the collision of two molecules in terms of an activated complex.* An **activated complex** (transition state) is *an unstable grouping of atoms that can break up to form products.* We can represent the formation of the activated complex this way:

$$O{=}N + Cl{-}Cl \longrightarrow [O{=}N\text{----}Cl\text{----}Cl]$$

When the molecules come together with proper orientation, an N—Cl bond begins to form. At the same time, the kinetic energy of the collision is absorbed by the activated complex as a vibrational motion of the atoms. This energy becomes concentrated in the bonds denoted by the dashed lines and can flow between them. If, at some moment, sufficient energy becomes concentrated in one of the bonds of the activated complex, that bond breaks or falls apart. Depending on whether the N----Cl or Cl----Cl bond breaks, the activated complex either reverts to the reactants or yields the products.

$$\underset{\text{reactants}}{O{=}N + Cl_2} \longleftarrow \underset{\text{activated complex}}{[O{=}N\text{----}Cl\text{----}Cl]} \longrightarrow \underset{\text{products}}{O{=}N{-}Cl + Cl}$$

Potential-Energy Diagrams for Reactions

It is instructive to consider a potential-energy diagram for the reaction of NO with Cl_2. We can represent this reaction by the equation

$$NO + Cl_2 \longrightarrow NOCl_2^{\ddagger} \longrightarrow NOCl + Cl$$

Figure 14.9
Importance of molecular orientation in the reaction of NO and Cl₂. *(A)* NO approaches with its N atom toward Cl₂, and an N—Cl bond forms. Also, the angle of approach is close to that in the product NOCl. *(B)* NO approaches with its O atom toward Cl₂. No N—Cl bond can form, so NO and Cl₂ collide and then fly apart.

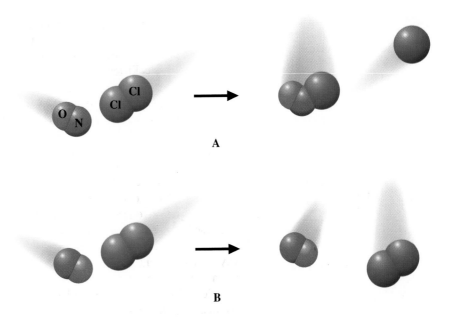

A

B

Here $NOCl_2^{\ddagger}$ denotes the activated complex. Figure 14.10 shows the change in potential energy (indicated by the solid curve) that occurs during the progress of the reaction. The potential-energy curve starts at the left with the potential energy of the reactants, $NO + Cl_2$. Moving along the curve toward the right, the potential energy increases to a maximum, corresponding to the activated complex. Farther to the right, the potential energy decreases to that of the products, $NOCl + Cl$.

At the start, the NO and Cl₂ molecules have a certain quantity of kinetic energy. The total energy of these molecules equals their potential energy plus their kinetic

Figure 14.10
Potential-energy curve (not to scale) for the endothermic reaction NO + Cl₂ ⟶ NOCl + Cl. For NO and Cl₂ to react, at least 85 kJ/mol of energy must be supplied by the collision of reactant molecules. Once the activated complex forms, it may break up to products, releasing 2 kJ/mol of energy. The difference, (85 − 2) kJ/mol = 83 kJ/mol, is the heat energy absorbed, Δ*H*.

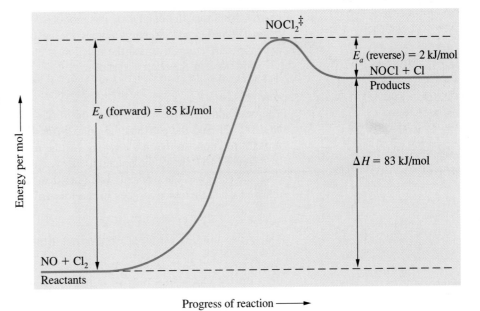

Figure 14.11
Potential-energy curve for an exothermic reaction.
The energy of the reactants is higher than that of the products, so heat energy is released when the reaction goes in the forward direction.

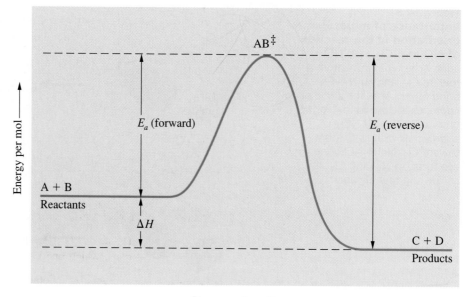

energy. As the reaction progresses the potential energy increases because the outer electrons of the two molecules repel. The kinetic energy decreases, and the molecules slow down. Only if the reactant molecules have sufficient kinetic energy is it possible for the potential energy to increase to the value for the activated complex. This kinetic energy must be equal to or greater than the difference in energy between the activated complex and the reactant molecules (85 kJ/mol). The energy difference is the activation energy for the forward reaction.

At the maximum in the potential-energy curve, the reactant molecules have come together as the activated complex. When the activated complex breaks up into products, the products go to lower potential energy (by 2 kJ/mol) and gain in kinetic energy. Note that the energy of the products is higher than the energy of the reactants. The difference in energy equals the heat of reaction, ΔH. Because the energy increases, ΔH is positive and the reaction is endothermic.

Now look at the reverse reaction:

$$NOCl + Cl \longrightarrow NO + Cl_2$$

The activation energy is 2 kJ/mol, the difference in energy of the initial species, NOCl + Cl, and the activated complex. This is a smaller quantity than that for the forward reaction, so the rate constant for the reverse reaction is larger.

Figure 14.11 shows the potential-energy curve for an exothermic reaction. In this case, the energy of the reactants is higher than that of the products, so heat energy is released when the reaction goes in the forward direction.

CONCEPT CHECK 14.4

Consider the following potential energy curves for two different reactions:

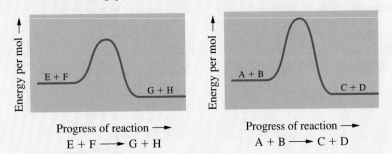

a. Which reaction has a higher activation energy for the forward reaction?

b. If both reactions were run at the same temperature and have the same orientation requirements to react, which one would have the larger rate constant?

c. Are these reactions exothermic or endothermic?

14.6 Arrhenius Equation

Rate constants for most chemical reactions closely follow an equation of the form

$$k = Ae^{-E_a/RT}$$

▶ Arrhenius published this equation in 1889 and suggested that molecules must be given enough energy to become "activated" before they could react. Collision and transition-state theories, which enlarged on this concept, were developed later (1920s and 1930s, respectively).

The mathematical equation $k = Ae^{-E_a/RT}$, which expresses the dependence of the rate constant on temperature, is called the **Arrhenius equation,** after its formulator, the Swedish chemist Svante Arrhenius. ◀ Here e is the base of natural logarithms, 2.718 . . . ; E_a is the activation energy; R is the gas constant, 8.31 J/(K · mol); and T is the absolute temperature. *The symbol A in the Arrhenius equation, which is assumed to be a constant,* is called the **frequency factor.** The frequency factor is related to the frequency of collisions with proper orientation (pZ). (The frequency factor does have a slight dependence on temperature, as you see from collision theory, but usually it can be ignored.)

It is useful to recast Arrhenius's equation in logarithmic form. Taking the natural logarithm of both sides of the Arrhenius equation gives

$$\ln k = \ln A - \frac{E_a}{RT}$$

Let us make the following identification of symbols:

$$\underbrace{\ln k}_{y} = \underbrace{\ln A}_{b} + \underbrace{\left(\frac{-E_a}{R}\right)}_{m}\underbrace{\left(\frac{1}{T}\right)}_{x}$$

This shows that if you plot $\ln k$ against $1/T$, you should get a straight line. The slope of this line is $-E_a/R$, from which you can obtain the activation energy E_a. The intercept is $\ln A$. Figure 14.12 shows a plot of $\ln k$ versus $1/T$ for the data given in Table 14.3. It demonstrates that the points do lie on a straight line.

Table 14.3
Rate Constant for the Decomposition of N_2O_5 at Various Temperatures

Temperature (°C)	k (/s)
45.0	4.8×10^{-4}
50.0	8.8×10^{-4}
55.0	1.6×10^{-3}
60.0	2.8×10^{-3}

Figure 14.12
Plot of ln k versus 1/T.
The logarithm of the
rate constant for the
decomposition of N_2O_5 (data
from Table 14.3) is plotted
versus 1/T. A straight line is
then fitted to the points; the
slope equals $-E_a/(2.303\,R)$.

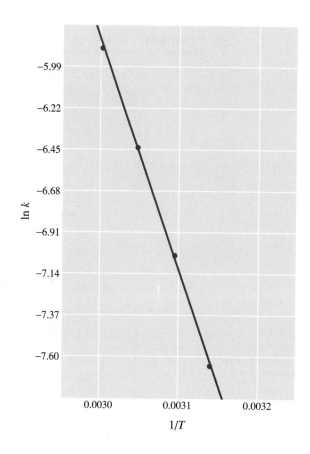

The Arrhenius equation can be put into the following two-point equation that is useful for computation:

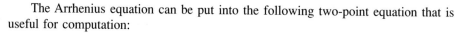

$$\ln \frac{k_2}{k_1} = \frac{E_a}{R}\left(\frac{1}{T_1} - \frac{1}{T_2}\right)$$

EXAMPLE 14.7

Using the Arrhenius Equation

The rate constant for the formation of hydrogen iodide from the elements

$$H_2(g) + I_2(g) \longrightarrow 2HI(g)$$

is 2.7×10^{-4} L/(mol · s) at 600 K and 3.5×10^{-3} L/(mol · s) at 650 K.
a. Find the activation energy E_a. b. Then calculate the rate constant at 700 K.

SOLUTION

a.

$$\ln \frac{3.5 \times 10^{-3}}{2.7 \times 10^{-4}} = \frac{E_a}{8.31 \text{ J/(mol} \cdot \text{K)}}\left(\frac{1}{600 \text{ K}} - \frac{1}{650 \text{ K}}\right)$$

$$\ln 1.30 \times 10^1 = 1.11 = \frac{E_a}{8.31 \text{ J/mol}} \times (1.28 \times 10^{-4})$$

Hence,

$$E_a = \frac{2.56 \times 8.31 \text{ J/mol}}{1.28 \times 10^{-4}} = \mathbf{1.66 \times 10^5 \text{ J/mol}}$$

b. Substitute $E_a = 1.66 \times 10^5$ J/mol and

$$k_1 = 2.7 \times 10^{-4} \text{ L/(mol} \cdot \text{s)} \quad (T_1 = 600 \text{ K})$$
$$k_2 = \text{unknown as yet} \quad (T_2 = 700 \text{ K})$$

You get

$$\ln \frac{k_2}{2.7 \times 10^{-4} \text{ L/(mol} \cdot \text{s)}} = \frac{1.66 \times 10^5 \text{ J/mol}}{8.31 \text{ J/(mol} \cdot \text{K)}} \times \left(\frac{1}{600 \text{ K}} - \frac{1}{700 \text{ K}} \right) = 4.77$$

Taking antilogarithms,

$$\frac{k_2}{2.7 \times 10^{-4} \text{ L/(mol} \cdot \text{s)}} = e^{4.77} = 1.2 \times 10^2$$

Hence,

$$k_2 = (1.2 \times 10^2) \times (2.7 \times 10^{-4}) \text{ L/(mol} \cdot \text{s)} = \mathbf{3.2 \times 10^{-2} \text{ L/(mol} \cdot \text{s)}}$$

See Problems 14.45 and 14.46.

Reaction Mechanisms

A balanced chemical equation is a description of the overall result of a chemical reaction. However, what actually happens at the molecular level may be more involved than is represented by this single equation. The reaction may take place in several steps. In the next sections, we will examine some reactions and see how the rate law can give us information about these steps, or *elementary reactions*.

14.7   Elementary Reactions

Consider the reaction of nitrogen dioxide with carbon monoxide.

$$NO_2(g) + CO(g) \longrightarrow NO(g) + CO_2(g) \quad \text{(net chemical equation)}$$

At temperatures below 500 K, this gas-phase reaction is believed to take place in two steps.

$$NO_2 + NO_2 \longrightarrow NO_3 + NO \quad \text{(elementary reaction)}$$
$$NO_3 + CO \longrightarrow NO_2 + CO_2 \quad \text{(elementary reaction)}$$

Each step, called an **elementary reaction,** is *a single molecular event, such as a collision of molecules, resulting in a reaction. The set of elementary reactions whose overall effect is given by the net chemical equation* is called the **reaction mechanism.**

According to the reaction mechanism just given, two NO_2 molecules collide and react to give the product molecule NO and the reaction intermediate NO_3. A **reaction intermediate** is *a species produced during a reaction that does not appear in the net equation because it reacts in a subsequent step in the mechanism.* Often the reaction intermediate has a fleeting existence and cannot be isolated from the reaction mixture.

The NO_3 molecule reacts quickly with CO to give the product molecules NO_2 and CO_2.

The overall chemical equation is obtained by adding the steps and canceling species that occur on both sides.

$$NO_2 + NO_2 \longrightarrow NO_3 + NO \qquad \text{(elementary reaction)}$$
$$NO_3 + CO \longrightarrow NO_2 + CO_2 \qquad \text{(elementary reaction)}$$

$$NO_2 + \cancel{NO_2} + \cancel{NO_3} + CO \longrightarrow \cancel{NO_3} + NO + \cancel{NO_2} + CO_2 \qquad \text{(overall equation)}$$

EXAMPLE 14.8

Writing the Overall Chemical Equation from a Mechanism

Carbon tetrachloride is obtained by chlorinating chloroform. The mechanism for the gas-phase chlorination of $CHCl_3$ is

$$Cl_2 \rightleftharpoons 2Cl \qquad \text{(elementary reaction)}$$
$$Cl + CHCl_3 \longrightarrow HCl + CCl_3 \qquad \text{(elementary reaction)}$$
$$Cl + CCl_3 \longrightarrow CCl_4 \qquad \text{(elementary reaction)}$$

Obtain the net, or overall, chemical equation from this mechanism.

SOLUTION

The first step produces two Cl atoms (a reaction intermediate). One Cl atom is used in the second step and another is used in the third step. Thus, all Cl atoms cancel. Similarly, the intermediate CCl_3, produced in the second step, is used up in the third step. You can cancel all Cl and CCl_3 species.

$$Cl_2 \rightleftharpoons \cancel{2Cl}$$
$$\cancel{Cl} + CHCl_3 \longrightarrow HCl + \cancel{CCl_3}$$
$$\cancel{Cl} + \cancel{CCl_3} \longrightarrow CCl_4$$

$$Cl_2 + CHCl_3 \longrightarrow HCl + CCl_4 \qquad \textbf{(overall equation)}$$

See Problems 14.49 and 14.50.

Molecularity

Elementary reactions are classified according to their molecularity. The **molecularity** is *the number of molecules on the reactant side of an elementary reaction.* A **unimolecular reaction** is *an elementary reaction that involves one reactant molecule;* a **bimolecular reaction** is *an elementary reaction that involves two reactant molecules.* Bimolecular reactions are the most common. Unimolecular reactions are best illustrated by decomposition of some previously excited species. Some gas-phase reactions are thought to occur in a **termolecular reaction,** *an elementary reaction that involves three reactant molecules.* Higher molecularities are not encountered.

As an example of a unimolecular reaction, consider the elementary process in which an energetically excited ozone molecule (symbolized by O_3^*) spontaneously decomposes.

$$O_3^* \longrightarrow O_2 + O$$

A molecule may be excited to a higher level if it collides with another molecule or absorbs a photon. The energy of the excited molecule is distributed among its three nuclei as vibrational energy. In the redistribution of this energy the excited ozone molecule can decompose into an oxygen molecule and an oxygen atom.

All of the steps in the mechanism of the reaction of NO_2 with CO, given earlier in this section, are bimolecular. Consider the first step, which involves the reaction of two NO_2 molecules.

$$NO_2 + NO_2 \longrightarrow NO_3 + NO$$

When these two NO_2 molecules come together, they form an activated complex of the six atoms, $(NO_2)_2$, which immediately breaks into two new molecules, NO_3 and NO.

The overall reaction of two atoms—say, two Br atoms—to form a diatomic molecule (Br_2) is normally a termolecular process. When two bromine atoms collide, they form an excited bromine molecule, Br_2^*. This excited molecule immediately flies apart, re-forming the atoms, unless another atom or molecule is present just at the moment of molecule formation to take away the excess energy. Suppose an argon atom and the two bromine atoms all collide at the same moment.

$$Br + Br + Ar \longrightarrow Br_2 + Ar^*$$

Energy that would have been left with the bromine molecule is now picked up by the argon atom (giving the energized atom Ar^*). The bromine molecule is stabilized by being left in a lower energy level.

EXAMPLE 14.9

Determining the Molecularity of an Elementary Reaction

What is the molecularity of each step in the mechanism described in Example 14.8?

$$Cl_2 \rightleftharpoons 2Cl$$
$$Cl + CHCl_3 \longrightarrow HCl + CCl_3$$
$$Cl + CCl_3 \longrightarrow CCl_4$$

SOLUTION

The molecularity of any elementary reaction equals the number of reactant molecules. Thus, the forward part of the first step is **unimolecular;** the reverse of the first step, the second step, and the third step are each **bimolecular.**

Rate Equation for an Elementary Reaction

There is no necessarily simple relationship between the overall reaction and the rate law that you observe for it. However, when you are dealing with an elementary reaction, the rate does have a simple, predictable form.

> For an elementary reaction, the rate is proportional to the product of the concentrations of each reactant molecule.

To understand this, let us look at the different possibilities. Consider the unimolecular elementary reaction

$$A \longrightarrow B + C$$

For each A molecule there is a definite probability, or chance, that it will decompose into B and C molecules. The more A molecules there are in a given volume, the more

A molecules that can decompose in that volume per unit time. In other words, the rate of reaction is proportional to the concentration of A.

$$\text{Rate} = k[A]$$

Now consider a bimolecular elementary reaction, such as

$$A + B \longrightarrow C + D$$

For the reaction to occur, the reactant molecules A and B must collide. The rate of formation of product is proportional to the frequency of molecular collisions, because a definite fraction of the collisions produces reaction. The frequency of collisions is proportional to the number of A molecules times the number of B molecules. Therefore, the rate of this elementary reaction is proportional to [A][B].

$$\text{Rate} = k[A][B]$$

A termolecular elementary reaction has a rate equation that is obtained by similar reasoning. For the elementary reaction

$$A + B + C \longrightarrow D + E$$

the rate is proportional to the concentrations of A, B, and C.

$$\text{Rate} = k[A][B][C]$$

Any reaction you observe is likely to consist of several elementary steps, and the rate law that you find is the combined result of these steps. This is why you cannot predict the rate law by looking at the overall equation.

EXAMPLE 14.10

Writing the Rate Equation for an Elementary Reaction

Write rate equations for each of the following elementary reactions.

a. Ozone is converted to O_2 by NO in a single step.

$$O_3 + NO \longrightarrow O_2 + NO_2$$

b. The recombination of iodine atoms occurs as follows:

$$I + I + M \longrightarrow I_2 + M^*$$

where M is some atom or molecule that absorbs energy from the reaction.

c. An H_2O molecule absorbs energy; some time later enough of this energy flows into one O—H bond to break it.

$$H_2O \longrightarrow H + O\text{—}H$$

SOLUTION

The rate equation can be written directly from the elementary reaction (but *only* for an elementary reaction).

a. **Rate** $= k[O_3][NO]$

b. **Rate** $= k[I]^2[M]$

c. **Rate** $= k[H_2O]$

14.8 The Rate Law and the Mechanism

The mechanism of a reaction cannot be observed directly. A mechanism is devised to explain the experimental observations. It is like the explanation provided by a detective to explain a crime in terms of the clues found. Other explanations may be possible, and further clues may make one of these other explanations seem more plausible than the currently accepted one. So it is with reaction mechanisms. They are accepted provisionally, with the understanding that further experiments may lead you to accept another mechanism as the more probable explanation. ◄

An important clue in understanding the mechanism of a reaction is the rate law. The reason for its importance is that once you assume a mechanism, you can predict the rate law. If this prediction does not agree with the experimental rate law, the assumed mechanism must be wrong. Take, for example, the overall equation

$$2NO_2(g) + F_2(g) \longrightarrow 2NO_2F(g)$$

If you follow the rate of disappearance of F_2, you observe that it is directly proportional to the concentration of NO_2 and F_2.

$$\text{Rate} = k[NO_2][F_2] \qquad \text{(experimental rate law)}$$

This rate law is a summary of the experimental data. Assume that the reaction occurs in a single elementary reaction.

$$NO_2 + NO_2 + F_2 \longrightarrow NO_2F + NO_2F \quad \text{(elementary reaction)}$$

This, then, is your assumed mechanism. Because this is an elementary reaction, you can immediately write the rate law predicted by it.

$$\text{Rate} = k[NO_2]^2[F_2] \qquad \text{(predicted rate law)}$$

However, this does not agree with experiment, and your assumed mechanism must be discarded. You conclude that the reaction occurs in more than one step.

> ▶ You see the scientific method in operation here, which recalls the discussion in Chapter 1. Experiments have been made from which you determine the rate law. Then a mechanism is devised to explain the rate law. This mechanism in turn suggests more experiments. These may confirm the explanation or may disagree with it. If they disagree, a new mechanism must be devised that explains all of the experimental evidence.

Rate-Determining Step

The reaction of NO_2 with F_2 is believed to occur in the following steps:

$$NO_2 + F_2 \xrightarrow{k_1} NO_2F + F \qquad \text{(slow step)}$$
$$F + NO_2 \xrightarrow{k_2} NO_2F \qquad \text{(fast step)}$$

The net result of the mechanism must be equivalent to the net result of the overall equation. By adding the two steps together, you can see that this is the case.

$$\begin{array}{rcl} NO_2 + F_2 & \longrightarrow & NO_2F + F \\ F + NO_2 & \longrightarrow & NO_2F \\ \hline 2NO_2 + F_2 & \longrightarrow & 2NO_2F \end{array}$$

The F atom is a reaction intermediate.

The mechanism must also be in agreement with the experimental rate law. Note that the second step is assumed to be much faster than the first, so that as soon as NO_2 and F_2 react, the F atom that is formed reacts with an NO_2 molecule to give another NO_2F molecule. Therefore, the rate law is determined completely by the rate-determining step. The **rate-determining step** is *the slowest step in the reaction mechanism.*

To understand better the significance of the rate-determining step, suppose you and a friend want to start a study group by sending invitation cards to some students in your class. You write a lengthy note on each card explaining the study group, taking an average of 2.0 minutes per card. Your friend puts the card in an envelope, affixes a computer-printed address label, seals the envelope, and stamps it, taking 0.5 minute per card. How long does it take to do 100 cards? What is the average time taken per card (*rate* of completing the cards)?

Because you take longer than your friend to do each card, your friend can complete the task for one card while you are working on another. It takes you a total of (100 cards × 2.0 min/card) to do your step in the task. When you have finished the last card, your friend still has to place the card in the envelope, and so forth. Therefore, you have to add an additional 0.5 min to the total time.

$$\text{Total time for 100 cards} = (100 \text{ cards} \times 2.\underline{0} \text{ min/card}) + 0.\underline{5} \text{ min}$$
$$= 2\underline{0}0.5 \text{ min}$$

We have underlined the last significant figure.

The rate of completing the invitation cards is

$$\text{Rate} = 2\underline{0}0.5 \text{ min}/100 \text{ cards} = 2.0 \text{ min/card}$$

The rate is essentially the time it takes you to compose a note. The time for your friend to finish the task is insignificant compared with the total time. The rate for the task equals the rate for the slower step (the rate-determining step).

The rate-determining step in the reaction of NO_2 and F_2 is the first step in the mechanism, in which NO_2 reacts with F_2 to produce NO_2F and an F atom. The rate equation for this rate-determining step of the mechanism is

$$\text{Rate} = k_1[NO_2][F_2]$$

This should equal the experimental rate law, which it does if you equate k_1 to k (the experimental rate constant). This agreement is not absolute evidence that the mechanism is correct.

EXAMPLE 14.11

Determining the Rate Law from a Mechanism with an Initial Slow Step

Ozone reacts with nitrogen dioxide to produce oxygen and dinitrogen pentoxide.

$$O_3(g) + 2NO_2(g) \longrightarrow O_2(g) + N_2O_5(g)$$

The proposed mechanism is

$$O_3 + NO_2 \longrightarrow NO_3 + O_2 \qquad \text{(slow)}$$
$$NO_3 + NO_2 \longrightarrow N_2O_5 \qquad \text{(fast)}$$

What is the rate law predicted by this mechanism?

SOLUTION

The first step in the mechanism is rate determining, because its rate is much slower than the second step. We can write the rate law directly from this step.

$$\textbf{Rate} = k[O_3][NO_2]$$

See Problems 14.51 and 14.52.

Mechanisms with an Initial Fast Step

A somewhat more complicated situation occurs when the rate-determining step follows an initial fast, equilibrium step. The decomposition of dinitrogen pentoxide,

$$2N_2O_5(g) \longrightarrow 4NO_2(g) + O_2(g) \qquad \text{(overall equation)}$$

is believed to follow this type of mechanism.

$$N_2O_5 \underset{k_{-1}}{\overset{k_1}{\rightleftharpoons}} NO_2 + NO_3 \qquad \text{(fast, equilibrium)}$$

$$NO_2 + NO_3 \xrightarrow{k_2} NO + NO_2 + O_2 \qquad \text{(slow)}$$

$$NO_3 + NO \xrightarrow{k_3} 2NO_2 \qquad \text{(fast)}$$

To give the overall stoichiometry, you need to multiply the first step by two. Note that there are two reaction intermediates, NO_3 and NO. Let us show that the above mechanism is consistent with the experimentally determined rate law,

$$\text{Rate} = k[N_2O_5]$$

The second step in the mechanism is assumed to be much slower than the other steps. Hence, the rate law predicted from this mechanism is

$$\text{Rate} = k_2[NO_2][NO_3]$$

However, this equation cannot be compared directly with experiment because it is written in terms of the reaction intermediate, NO_3. The experimental rate law will be written in terms of substances that occur in the chemical equation, not of reaction intermediates. Re-express the rate equation, eliminating $[NO_3]$ by looking at the first step in the mechanism.

This step is fast and reversible. The rate of the forward reaction is

$$\text{Forward rate} = k_1[N_2O_5]$$

and the rate of the reverse reaction is

$$\text{Reverse rate} = k_{-1}[NO_2][NO_3]$$

When the reaction first begins, there are no NO_2 or NO_3 molecules, and the reverse rate is zero. But as N_2O_5 dissociates, the concentration of N_2O_5 decreases and the concentrations of NO_2 and NO_3 increase. Soon the two rates become equal, such that N_2O_5 molecules form as often as other N_2O_5 molecules dissociate. The first step has reached *dynamic equilibrium.* Because these elementary reactions are much faster than the second step, this equilibrium is reached before any significant reaction by the second step occurs. Moreover, this equilibrium is maintained throughout the reaction. ◄

► Chemical equilibrium is discussed in detail in the next chapter.

At equilibrium, the forward and reverse rates are equal, so you can write

$$k_1[N_2O_5] = k_{-1}[NO_2][NO_3]$$

$$[NO_3] = \frac{k_1}{k_{-1}} \frac{[N_2O_5]}{[NO_2]}$$

Substituting into the rate law, you get

$$\text{Rate} = k_2[NO_2][NO_3] = k_2[\cancel{NO_2}] \times \frac{k_1}{k_{-1}} \frac{[N_2O_5]}{[\cancel{NO_2}]}$$

or

$$\text{Rate} = k_2 \frac{k_1}{k_{-1}} [N_2O_5]$$

Thus, if you identify $k_1 k_2 / k_{-1}$ as k, you reproduce the experimental rate law.

The mechanism must also be in agreement with the overall equation for the reaction. Although the first step is essentially in equilibrium, the products of this step (NO_2 and NO_3) are being continuously used in the subsequent steps. For these steps to proceed, the first step must effectively produce two molecules of NO_3. Thus, the net result of the mechanism is as follows:

$$
\begin{array}{r}
2(N_2O_5 \longrightarrow NO_2 + \cancel{NO_3}) \\
\cancel{NO_2} + \cancel{NO_3} \longrightarrow \cancel{NO} + \cancel{NO_2} + O_2 \\
\cancel{NO_3} + \cancel{NO} \longrightarrow 2NO_2 \\
\hline
2N_2O_5 \longrightarrow 4NO_2 + O_2
\end{array}
$$

The net result of the mechanism is equivalent to the overall equation for the reaction.

EXAMPLE 14.12

Determining the Rate Law from a Mechanism with an Initial Fast, Equilibrium Step

Nitric oxide can be reduced with hydrogen gas to give nitrogen and water vapor.

$$2NO(g) + 2H_2(g) \longrightarrow N_2(g) + 2H_2O(g) \qquad \text{(overall equation)}$$

A proposed mechanism is

$$2NO \underset{k_{-1}}{\overset{k_1}{\rightleftharpoons}} N_2O_2 \qquad \text{(fast, equilibrium)}$$

$$N_2O_2 + H_2 \xrightarrow{k_2} N_2O + H_2O \qquad \text{(slow)}$$

$$N_2O + H_2 \xrightarrow{k_3} N_2 + H_2O \qquad \text{(fast)}$$

What rate law is predicted by this mechanism?

SOLUTION

According to the rate-determining step,

$$\text{Rate} = k_2[N_2O_2][H_2]$$

You try to eliminate N_2O_2 from the rate law by looking at the first step, which is fast and reaches equilibrium. At equilibrium, the forward rate and the reverse rate are equal.

$$k_1[NO]^2 = k_{-1}[N_2O_2]$$

Therefore, $[N_2O_2] = (k_1/k_{-1})[NO]^2$, so

$$\text{Rate} = \frac{k_2 k_1}{k_{-1}} [NO]^2[H_2]$$

Experimentally, you should observe the rate law

$$\textbf{Rate} = k[NO]^2[H_2]$$

where we have replaced the constants $k_2 k_1 / k_{-1}$ with k.

See Problems 14.53 and 14.54.

14.9 Catalysis

A catalyst is a substance that has the power of speeding up a reaction without being consumed by it. In theory, you could add a catalyst to a reaction mixture and, after the reaction, separate that catalyst and use it over and over again.

Catalysts are of enormous importance to the chemical industry, because they allow a reaction to occur with a reasonable rate at a much lower temperature. Moreover, catalysts are often quite specific—they increase the rate of certain reactions, but not others. For instance, an industrial chemist can start with a mixture of carbon monoxide and hydrogen and produce methane gas using one catalyst or produce gasoline using another catalyst. *Enzymes* are the marvelously selective catalysts employed by biological organisms. A biological cell contains thousands of different enzymes that direct the chemical processes that occur in the cell.

How can we explain how a catalyst can influence a reaction without being consumed by it? Briefly, the catalyst must participate in at least one step of a reaction and be regenerated in a later step. Consider the commercial preparation of sulfuric acid, H_2SO_4, from sulfur dioxide, SO_2. The first step involves the reaction of SO_2 with O_2 to produce sulfur trioxide, SO_3. For this reaction to occur at an economical rate, it requires a catalyst. An early industrial process employed nitric oxide, NO, as the catalyst.

$$2SO_2(g) + O_2(g) \xrightarrow{\text{NO}} 2SO_3(g)$$

Here is a proposed mechanism:

$$2NO + O_2 \longrightarrow 2NO_2$$
$$NO_2 + SO_2 \longrightarrow NO + SO_3$$

Two molecules of NO are used up in the first step and are regenerated in the second step.

Note that the catalyst is an active participant in the reaction. But how does this participation explain the increase in speed of the catalyzed reaction over the uncatalyzed reaction? The Arrhenius equation (given at the beginning of Section 14.6) provides an answer. The catalyzed reaction mechanism makes available a reaction path having an increased overall rate of reaction. It increases this rate by decreasing the activation energy E_a.

The depletion of ozone in the stratosphere by Cl atoms provides an example of the lowering of activation energy by a catalyst. Ozone provides protection against biologically destructive, short-wavelength ultraviolet radiation from the sun. Some recent ozone depletion in the stratosphere is believed to result from the Cl-catalyzed decomposition of O_3. Cl atoms in the stratosphere originate from the decomposition of chlorofluorocarbons (CFCs), which are compounds manufactured as refrigerants, aerosol propellants, and so forth. These Cl atoms react with ozone to form ClO and O_2, and the ClO reacts with O atoms to produce Cl and O_2.

$$\begin{aligned} Cl(g) + O_3(g) &\longrightarrow ClO(g) + O_2(g) \\ ClO(g) + O(g) &\longrightarrow Cl(g) + O_2(g) \\ \hline O_3(g) + O(g) &\longrightarrow 2O_2(g) \end{aligned}$$

▶ Ozone depletion in the stratosphere was discussed in an essay at the end of Chapter 10.

The net result is the decomposition of ozone with O atoms to produce O_2. Figure 14.13 shows the potential-energy curves for the uncatalyzed and the catalyzed reactions. The uncatalyzed reaction has such a large activation energy that its rate is extremely low. ◀

Figure 14.13
Comparison of activation energies in the uncatalyzed and catalyzed decompositions of ozone.
The uncatalyzed reaction is $O_3 + O \longrightarrow O_2 + O_2$. Catalysis by Cl atoms provides an alternate pathway with lower activation energy, and therefore a faster reaction.

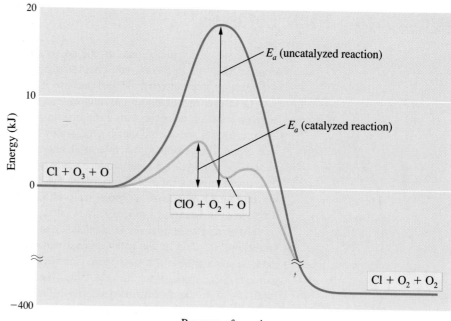

Homogeneous Catalysis

The oxidation of sulfur dioxide using nitric oxide as a catalyst is an example of **homogeneous catalysis,** which is *the use of a catalyst in the same phase as the reacting species.* The catalyst NO and the reacting species SO_2 and O_2 are all in the gas phase. Another example occurs in the oxidation of thallium(I) to thallium(III) by cerium(IV) in aqueous solution.

$$2Ce^{4+}(aq) + Tl^+(aq) \longrightarrow 2Ce^{3+}(aq) + Tl^{3+}(aq)$$

The uncatalyzed reaction is very slow; presumably it involves the collision of three positive ions. The reaction can be catalyzed by manganese(II) ion, with each step bimolecular. The mechanism is thought to be

$$
\begin{aligned}
Ce^{4+} + Mn^{2+} &\longrightarrow Ce^{3+} + Mn^{3+} \\
Ce^{4+} + Mn^{3+} &\longrightarrow Ce^{3+} + Mn^{4+} \\
Mn^{4+} + Tl^+ &\longrightarrow Tl^{3+} + Mn^{2+}
\end{aligned}
$$

Figure 14.14
Heterogeneous catalysts.
The catalyst (for example, platinum or palladium) is absorbed on support materials, such as carbon and alumina.

▶ Understanding the chemical processes occurring at surfaces is being advanced by new techniques, including x-ray photoelectron spectroscopy and scanning tunneling microscopy.

Heterogeneous Catalysis

Some of the most important industrial reactions involve **heterogeneous catalysis—** that is, *the use of a catalyst that exists in a different phase from the reacting species, usually a solid catalyst in contact with a gaseous or liquid solution of reactants* (Figure 14.14). Such surface, or heterogeneous, catalysis is thought to occur by chemical adsorption of the reactants onto the surface of the catalyst. *Adsorption* is the attraction of molecules to a surface. In *physical adsorption,* the attraction is provided by weak intermolecular forces. **Chemisorption,** by contrast, is *the binding of a species to a surface by chemical bonding forces.* It may happen that bonds in the species are broken during chemisorption, and this may provide the basis of catalytic action in certain cases. ◀

An example of heterogeneous catalysis involving chemisorption is provided by *catalytic hydrogenation*. This is the addition of H_2 to a compound, such as one with a carbon–carbon double bond, using a catalyst of platinum or nickel metal. Vegetable oils, which contain carbon–carbon double bonds, are changed to solid fats (shortening) when the bonds are catalytically hydrogenated. In the case of ethylene, C_2H_4, the equation is

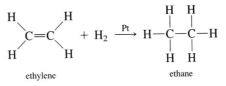

ethylene   ethane

A mechanism for this reaction is represented by the four steps shown in Figure 14.15. In *(A)*, ethylene and hydrogen molecules diffuse to the catalyst surface, where, as shown in *(B)*, they undergo chemisorption. The pi electrons of ethylene form bonds to the metal, and hydrogen molecules break into H atoms that bond to the metal. In *(C)*, two H atoms migrate to an ethylene molecule bonded to the catalyst, where they react to form ethane. Then, in *(D)*, because ethane does not bond to the metal, it diffuses away. Note that the catalyst surface that was used in *(B)* is regenerated in *(D)*.

At the beginning of this section, we described the catalytic oxidation of SO_2 to SO_3, the anhydride of sulfuric acid. In an early process, the homogeneous catalyst NO was used. Today, in the *contact process*, a heterogeneous catalyst, Pt or V_2O_5, is used. Surface catalysts are used in the catalytic converters of automobiles to convert substances that would be atmospheric pollutants, such as CO and NO, into harmless substances, such as CO_2 and N_2.

Enzyme Catalysis

Almost all enzymes are protein molecules with molecular weights ranging to over a million amu. An enzyme has enormous catalytic activity. Enzymes are also highly specific, each enzyme acting only on a specific substance, catalyzing it to undergo a particular reaction. For example, the enzyme sucrase, which is present in the digestive fluid of the small intestine, catalyzes the reaction of sucrose with water to form

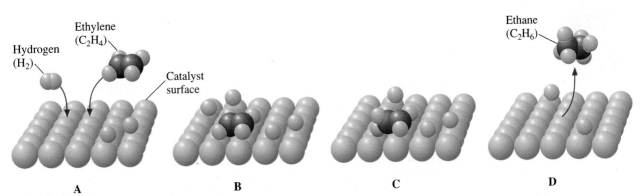

A   B   C   D

Figure 14.15
Proposed mechanism of catalytic hydrogenation of C_2H_4. *(A)* C_2H_4 and H_2 molecules diffuse to the catalyst. *(B)* The molecules form bonds to the catalyst surface. (The H_2 molecules dissociate to atoms in the process.) *(C)* H atoms migrate to the C_2H_4 molecule, where they react to form C_2H_6. *(D)* C_2H_6 diffuses away from the catalyst.

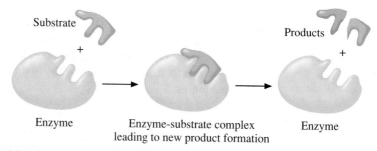

Figure 14.16
Enzyme action (lock-and-key model). The enzyme has an active site to which the substrate binds to form an enzyme–substrate complex. The active site of the enzyme acts like a lock into which the substrate (key) fits. While bound to the enzyme, the substrate may have bonds weakened or new bonds formed to yield the products, which leave the enzyme.

the simpler sugars glucose and fructose. *The substance whose reaction the enzyme catalyzes* is called the **substrate.** Thus, the enzyme sucrase acts on the substrate sucrose.

Figure 14.16 shows schematically how an enzyme acts. The enzyme molecule is a protein chain that tends to fold into a roughly spherical form with an *active site* at which the substrate molecule binds and the catalysis takes place. The substrate molecule, S, fits into the active site on the enzyme molecule, E, somewhat in the way a key fits into a lock, forming an enzyme–substrate complex, ES. (The lock-and-key model is only a rough approximation, because the active site on an enzyme deforms somewhat to fit the substrate molecule.) In effect, the active site "recognizes" the substrate and gives the enzyme its specificity. On binding to the enzyme, the substrate may have bonds weaken or new bonds form that help yield the products, P.

$$E + S \longrightarrow ES \longrightarrow E + P$$

The formation of the enzyme–substrate complex provides a new pathway to products with a lower activation energy.

A Checklist for Review

Important Terms

catalyst (14.1)
reaction rate (14.1)
rate law (14.3)
rate constant (14.3)
reaction order (14.3)
half-life (14.4)
collision theory (14.5)

activation energy (14.5)
transition-state
  theory (14.5)
activated complex (14.5)
Arrhenius equation (14.6)
frequency factor (14.6)
elementary reaction (14.7)

reaction mechanism (14.7)
reaction intermediate (14.7)
molecularity (14.7)
unimolecular
  reaction (14.7)
bimolecular reaction (14.7)
termolecular reaction (14.7)

rate-determining step (14.8)
homogeneous
  catalysis (14.9)
heterogeneous
  catalysis (14.9)
chemisorption (14.9)
substrate (14.9)

Key Equations

$$\ln\frac{[A]_t}{[A]_0} = -kt \qquad\qquad t_{1/2} = \frac{0.693}{k} \qquad\qquad \ln\frac{k_2}{k_1} = \frac{E_a}{R}\left(\frac{1}{T_1} - \frac{1}{T_2}\right)$$

Summary of Facts and Concepts

The *reaction rate* is defined as the increase in moles of product per liter per second (or as the decrease in moles of reactant per liter per second). Rates of reaction are determined by following the change of concentration of a reactant or product, either by chemical analysis or by observing a physical property. It is found that reaction rates are proportional to concentrations of reactants raised to various powers (usually 1 or 2, but they can be fractional or negative). The *rate law* mathematically expresses the relationship between rates and concentrations for a chemical reaction. Although the rate law tells you how the rate of a reaction depends on concentrations at a given moment, it is possible to transform a rate law to show how concentrations change with time. The *half-life* of a reaction is the time it takes for the reaction concentration to decrease to one-half of its original concentration.

Reaction rates can often double or triple with a 10°C rise in temperature. The effect of temperature on the rate can be explained by *collision theory*. According to this theory, two molecules react after colliding only when the energy of collision is greater than the *activation energy* and when the molecules are properly oriented. It is the rapid increase in the fraction of collisions having energy greater than the activation energy that explains the large temperature dependence of reaction rates. *Transition-state theory* explains reaction rates in terms of the formation of an *activated complex* of the colliding molecules. The *Arrhenius equation* is a mathematical relationship showing the dependence of a rate constant on temperature.

A chemical equation describes the overall result of a chemical reaction that may take place in one or more steps. The steps are called *elementary reactions,* and the set of elementary reactions that describes what is happening at the molecular level in an overall reaction is called the *reaction mechanism.* In some cases, the overall reaction involves a *reaction intermediate*—a species produced in one step but used up in a subsequent one. The rate of the overall reaction depends on the rate of the slowest step (the *rate-determining step*). This rate is proportional to the product of the concentrations of each reactant molecule in that step. If this step involves a reaction intermediate, its concentration can be eliminated by using the relationship of the concentrations in the preceding fast, *equilibrium* step.

A *catalyst* is a substance that speeds up a chemical reaction without being consumed by it. The catalyst is used up in one step of the reaction mechanism but regenerated in a later step. Catalytic activity operates by providing a reaction mechanism that has lower activation energy. Catalysis is classified as *homogeneous catalysis* if the substances react within one phase and as *heterogeneous catalysis* if the substances in a gas or liquid phase react at the surface of a solid catalyst. Many industrial reactions involve heterogeneous catalysis. Enzymes are biological catalysts (usually proteins). *Enzyme catalysis* is highly specific, so each enzyme acts on a particular kind of *substrate.*

Operational Skills

1. **Relating the different ways of expressing reaction rates** Given the balanced equation for a reaction, relate the different possible ways of defining the rate of the reaction. **(EXAMPLE 14.1)**
2. **Calculating the average reaction rate** Given the concentration of reactant or product at two different times, calculate the average rate of reaction over that time interval. **(EXAMPLE 14.2)**
3. **Determining the order of reaction from the rate law** Given an empirical rate law, obtain the orders with respect to each reactant (and catalyst, if any) and the overall order. **(EXAMPLE 14.3)**
4. **Determining the rate law from initial rates** Given initial concentrations and initial-rate data (in which the concentrations of all species are changed one at a time, holding the others constant), find the rate law for the reaction. **(EXAMPLE 14.4)**
5. **Using the concentration-time equation for a first-order reaction** Given the rate constant and initial reactant concentration for a first-order reaction, calculate the reactant concentration after a definite time, or calculate the time it takes for the concentration to decrease to a prescribed value. **(EXAMPLE 14.5)**

6. **Relating the half-life of a reaction to the rate constant** Given the rate constant for a reaction, calculate the half-life. **(EXAMPLE 14.6)**
7. **Using the Arrhenius equation** Given the values of the rate constant for two temperatures, find the activation energy and calculate the rate constant at a third temperature. **(EXAMPLE 14.7)**
8. **Writing the overall chemical equation from a mechanism** Given a mechanism for a reaction, obtain the overall chemical equation. **(EXAMPLE 14.8)**
9. **Determining the molecularity of an elementary reaction** Given an elementary reaction, state the molecularity. **(EXAMPLE 14.9)**
10. **Writing the rate equation for an elementary reaction** Given an elementary reaction, write the rate equation. **(EXAMPLE 14.10)**
11. **Determining the rate law from a mechanism** Given a mechanism with an initial slow step, obtain the rate law **(EXAMPLE 14.11)**. Given a mechanism with an initial fast, equilibrium step, obtain the rate law **(EXAMPLE 14.12)**.

Review Questions

14.1 List the four variables or factors that can affect the rate of reaction.

14.2 Define the rate of reaction of HBr in the following reaction. How is this related to the rate of formation of Br_2?

$$4HBr(g) + O_2(g) \longrightarrow 2Br_2(g) + 2H_2O(g)$$

14.3 Give at least two physical properties that might be used to determine the rate of a reaction.

14.4 The reaction

$$3I^-(aq) + H_3AsO_4(aq) + 2H^+(aq) \longrightarrow$$
$$I_3^-(aq) + H_3AsO_3(aq) + H_2O(l)$$

is found to be first order with respect to each of the reactants. Write the rate law. What is the overall order?

14.5 The rate of a reaction is quadrupled when the concentration of one reactant is doubled. What is the order of the reaction with respect to this reactant?

14.6 A rate law is one-half order with respect to a reactant. What is the effect on the rate when the concentration of this reactant is doubled?

14.7 The reaction $A(g) \longrightarrow B(g) + C(g)$ is known to be first order in $A(g)$. It takes 25 s for the concentration of $A(g)$ to decrease by one-half of its initial value. How long does it take for the concentration of $A(g)$ to decrease to one-fourth of its initial value? to one-eighth of its initial value?

14.8 What two factors determine whether a collision between two reactant molecules will result in reaction?

14.9 Sketch a potential-energy diagram for the exothermic, elementary reaction

$$A + B \longrightarrow C + D$$

and on it denote the activation energies for the forward and reverse reactions. Also indicate the reactants, products, and activated complex.

14.10 Rate constants for reactions often follow the Arrhenius equation. Write this equation and then identify each term in it with the corresponding factor or factors from collision theory. Give a physical interpretation of each of those factors.

14.11 By means of an example, explain what is meant by the term *reaction intermediate*.

14.12 Why is it generally impossible to predict the rate law for a reaction on the basis of the chemical equation only?

14.13 The dissociation of N_2O_4 into NO_2,

$$N_2O_4(g) \rightleftharpoons 2NO_2(g)$$

is believed to occur in one step. Obtain the concentration of N_2O_4 in terms of the concentration of NO_2 and the rate constants for the forward and reverse reactions, when the reactions have come to equilibrium.

14.14 How does a catalyst speed up a reaction? How can a catalyst be involved in a reaction without being consumed by it?

14.15 Compare physical adsorption and chemisorption (chemical adsorption).

14.16 Describe the steps in the catalytic hydrogenation of ethylene.

Conceptual Problems

14.17 Consider the reaction $3A \longrightarrow 2B + C$.
a. One rate expression for the reaction is

$$\text{Rate of formation of C} = +\frac{\Delta[C]}{\Delta t}$$

Write two other rate expressions for this reaction in this form.
b. Using your two rate expressions, if you calculated the average rate of the reaction over the same time interval, would the rates be equal?
c. If your answer to part b was no, write two rate expressions that would give an equal rate when calculated over the same time interval.

14.18 Given the reaction $2A + B \longrightarrow C + 3D$, can you write the rate law for this reaction? If so, write the rate law; if not, why?

14.19 You perform some experiments for the reaction $A \longrightarrow B + C$ and determine the rate law has the form

$$\text{Rate} = k[A]^x$$

Calculate the value of exponent x for each of the following.
a. [A] is tripled and you observe no rate change.
b. [A] is doubled and the rate doubles.
c. [A] is tripled and the rate goes up by a factor of 27.

14.20 A friend of yours runs a reaction and generates the following plot. She explains that in following the reaction, she measured the concentration of a compound that she calls "E."

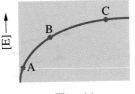

a. Your friend tells you that E is either a reactant or product. Which is it and why?

b. Is the average rate faster between points A and B or B and C? Why?

14.21 Given the hypothetical plot shown here for the concentration of compound Y versus time, answer the following questions.

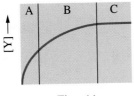

a. In which region of the curve does the rate have a constant value (A, B, or C)?

b. In which region of the curve is the rate the fastest (A, B, or C)?

Practice Problems

Reaction Rates

14.23 Relate the rate of decomposition of NO_2 to the rate of formation of O_2 for the following reaction:

$$2NO_2(g) \longrightarrow 2NO(g) + O_2(g)$$

14.24 For the reaction of hydrogen with iodine

$$H_2(g) + I_2(g) \longrightarrow 2HI(g)$$

relate the rate of disappearance of hydrogen gas to the rate of formation of hydrogen iodide.

14.25 Ammonium nitrite, NH_4NO_2, decomposes in solution, as shown here.

$$NH_4NO_2(aq) \longrightarrow N_2(g) + 2H_2O(l)$$

The concentration of NH_4^+ ion at the beginning of an experiment was $0.500\ M$. After 3.00 hours, it was $0.432\ M$. What is the average rate of decomposition of NH_4NO_2 in this time interval?

14.26 Iron(III) chloride is reduced by tin(II) chloride.

$$2FeCl_3(aq) + SnCl_2(aq) \longrightarrow 2FeCl_2(aq) + SnCl_4(aq)$$

14.22 You carry out the following reaction by introducing N_2O_4 into an evacuated flask and observing the concentration change of the product over time.

$$N_2O_4(g) \longrightarrow 2NO_2(g)$$

Which one of the curves shown here reflects the data collected for this reaction?

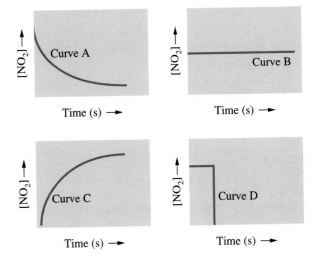

The concentration of Fe^{3+} ion at the beginning of an experiment was $0.03586\ M$. After 4.00 min, it was $0.02638\ M$. What is the average rate of reaction of $FeCl_3$ in this time interval?

14.27 Azomethane, CH_3NNCH_3, decomposes according to the following equation:

$$CH_3NNCH_3(g) \longrightarrow C_2H_6(g) + N_2(g)$$

The initial concentration of azomethane was 1.50×10^{-2} mol/L. After 9.00 min, this concentration decreased to 1.01×10^{-2} mol/L. Obtain the average rate of reaction during this time interval. Express the answer in units of $mol/(L \cdot s)$.

14.28 Nitrogen dioxide, NO_2, decomposes upon heating to form nitric oxide and oxygen according to the following equation:

$$2NO_2(g) \longrightarrow 2NO(g) + O_2(g)$$

At the beginning of an experiment, the concentration of nitrogen dioxide in a reaction vessel was 0.1103 mol/L. After 60.0 s, the concentration decreased to 0.1076 mol/L. What is the average rate of decomposition of NO_2 during this time interval, in $mol/(L \cdot s)$?

Rate Laws

14.29 Oxalic acid, $H_2C_2O_4$, is oxidized by permanganate ion to CO_2 and H_2O.

$$2MnO_4^-(aq) + 5H_2C_2O_4(aq) + 6H^+(aq) \longrightarrow$$
$$2Mn^{2+}(aq) + 10CO_2(g) + 8H_2O(l)$$

The rate law is

$$\text{Rate} = k[MnO_4^-][H_2C_2O_4]$$

What is the order with respect to each reactant? What is the overall order?

14.30 Iron(II) ion is oxidized by hydrogen peroxide in acidic solution.

$$H_2O_2(aq) + 2Fe^{2+}(aq) + 2H^+(aq) \longrightarrow$$
$$2Fe^{3+}(aq) + 2H_2O(l)$$

The rate law is

$$\text{Rate} = k[H_2O_2][Fe^{2+}]$$

What is the order with respect to each reactant? What is the overall order?

14.31 In experiments on the decomposition of azomethane,

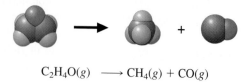

$$CH_3NNCH_3(g) \longrightarrow C_2H_6(g) + N_2(g)$$

the following data were obtained:

	Initial Concentration of Azomethane	Initial Rate
Exp. 1	$1.13 \times 10^{-2}\ M$	$2.8 \times 10^{-6}\ M/s$
Exp. 2	$2.26 \times 10^{-2}\ M$	$5.6 \times 10^{-6}\ M/s$

What is the rate law? What is the value of the rate constant?

14.32 Ethylene oxide, C_2H_4O, decomposes when heated to give methane and carbon monoxide.

$$C_2H_4O(g) \longrightarrow CH_4(g) + CO(g)$$

The following kinetic data were observed for the reaction at 688 K:

	Initial Concentration of Ethylene Oxide	Initial Rate
Exp. 1	$0.00272\ M$	$5.57 \times 10^{-7}\ M/s$
Exp. 2	$0.00544\ M$	$1.11 \times 10^{-6}\ M/s$

Find the rate law and the value of the rate constant for this reaction.

14.33 Nitric oxide, NO, reacts with hydrogen to give nitrous oxide, N_2O, and water.

$$2NO(g) + H_2(g) \longrightarrow N_2O(g) + H_2O(g)$$

In a series of experiments, the following initial rates of disappearance of NO were obtained:

	Initial Concentrations		Initial Rate of Reaction of NO
	NO	H_2	
Exp. 1	$6.4 \times 10^{-3}\ M$	$2.2 \times 10^{-3}\ M$	$2.6 \times 10^{-5}\ M/s$
Exp. 2	$12.8 \times 10^{-3}\ M$	$2.2 \times 10^{-3}\ M$	$1.0 \times 10^{-4}\ M/s$
Exp. 3	$6.4 \times 10^{-3}\ M$	$4.5 \times 10^{-3}\ M$	$5.1 \times 10^{-5}\ M/s$

Find the rate law and the value of the rate constant for the reaction of NO.

14.34 In a kinetic study of the reaction

$$2NO(g) + O_2(g) \longrightarrow 2NO_2(g)$$

the following data were obtained for the initial rates of disappearance of NO:

	Initial Concentrations		Initial Rate of Reaction of NO
	NO	O_2	
Exp. 1	$0.0125\ M$	$0.0253\ M$	$0.0281\ M/s$
Exp. 2	$0.0250\ M$	$0.0253\ M$	$0.112\ M/s$
Exp. 3	$0.0125\ M$	$0.0506\ M$	$0.0561\ M/s$

Obtain the rate law. What is the value of the rate constant?

Change of Concentration with Time; Half-Life

14.35 Ethyl chloride, CH_3CH_2Cl, used to produce tetraethyllead gasoline additive, decomposes when heated to give ethylene and hydrogen chloride.

$$CH_3CH_2Cl(g) \longrightarrow C_2H_4(g) + HCl(g)$$

The reaction is first order. In an experiment, the initial concentration of ethyl chloride was 0.00100 M. After heating at 500°C for 155 s, this was reduced to 0.00067 M. What was the concentration of ethyl chloride after a total of 256 s?

14.36 Cyclobutane, C_4H_8, consisting of molecules in which four carbon atoms form a ring, decomposes when heated to give ethylene.

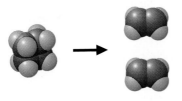

$$C_4H_8(g) \longrightarrow 2C_2H_4(g)$$

The reaction is first order. In an experiment, the initial concentration of cyclobutane was 0.00150 M. After heating at 450°C for 455 s, this was reduced to 0.00119 M. What was the concentration of cyclobutane after a total of 750 s?

14.37 Methyl isocyanide, CH_3NC, isomerizes when heated to give acetonitrile (methyl cyanide), CH_3CN.

$$CH_3NC(g) \longrightarrow CH_3CN(g)$$

The reaction is first order. At 230°C, the rate constant for the isomerization is 6.3×10^{-4}/s. What is the half-life? How long would it take for the concentration of CH_3NC to decrease to 50.0% of its initial value? to 25.0% of its initial value?

14.38 Dinitrogen pentoxide, N_2O_5, decomposes when heated in carbon tetrachloride solvent.

$$N_2O_5 \longrightarrow 2NO_2 + \tfrac{1}{2}O_2(g)$$

If the rate constant for the decomposition of N_2O_5 is 6.2×10^{-4}/min, what is the half-life? (The rate law is first order in N_2O_5.) How long would it take for the concentration of N_2O_5 to decrease to 25% of its initial value? to 12.5% of its initial value?

14.39 In the presence of excess thiocyanate ion, SCN^-, the following reaction is first order in chromium(III) ion, Cr^{3+}; the rate constant is 2.0×10^{-6}/s.

$$Cr^{3+}(aq) + SCN^-(aq) \longrightarrow Cr(SCN)^{2+}(aq)$$

What is the half-life in hours? How many hours would be required for the initial concentration of Cr^{3+} to decrease to each of the following values: 25.0% left, 12.5% left, 6.25% left, 3.125% left?

14.40 In the presence of excess thiocyanate ion, SCN^-, the following reaction is first order in iron(III) ion, Fe^{3+}; the rate constant is 1.27/s.

$$Fe^{3+}(aq) + SCN^-(aq) \longrightarrow Fe(SCN)^{2+}(aq)$$

What is the half-life in seconds? How many seconds would be required for the initial concentration of Fe^{3+} to decrease to each of the following values: 25.0% left, 12.5% left, 6.25% left, 3.125% left? What is the relationship between these times and the half-life?

14.41 Chlorine dioxide oxidizes iodide ion in aqueous solution to iodine; chlorine dioxide is reduced to chlorite ion.

$$2ClO_2(aq) + 2I^-(aq) \longrightarrow 2ClO_2^-(aq) + I_2(aq)$$

The order of the reaction with respect to ClO_2 was determined by starting with a large excess of I^-, so that its concentration was essentially constant. Then

$$\text{Rate} = k[ClO_2]^m[I^-]^n = k'[ClO_2]^m$$

where $k' = k[I^-]^n$. Determine the order with respect to ClO_2 and the rate constant k' by plotting the following data assuming first- and then second-order kinetics. [Data from H. Fukutomi and G. Gordon, *J. Am. Chem. Soc.*, *89*, 1362 (1967).]

Time (s)	$[ClO_2]$ (mol/L)
0.00	4.77×10^{-4}
1.00	4.31×10^{-4}
2.00	3.91×10^{-4}
3.00	3.53×10^{-4}

14.42 Methyl acetate, CH_3COOCH_3, reacts in basic solution to give acetate ion, CH_3COO^-, and methanol, CH_3OH.

$$CH_3COOCH_3(aq) + OH^-(aq) \longrightarrow$$
$$CH_3COO^-(aq) + CH_3OH(aq)$$

The overall order of the reaction was determined by starting with methyl acetate and hydroxide ion at the same concentrations, so $[CH_3COOCH_3] = [OH^-] = x$. Then

$$\text{Rate} = k[CH_3COOCH_3]^m[OH^-]^n = kx^{m+n}$$

Determine the overall order and the value of the rate constant by plotting the following data assuming first- and then second-order kinetics.

Time (min)	$[CH_3COOCH_3]$ (mol/L)
0.00	0.01000
3.00	0.00740
4.00	0.00683
5.00	0.00634

Rate and Temperature

14.43 Sketch a potential-energy diagram for the reaction of nitric oxide with ozone.

$$NO(g) + O_3(g) \longrightarrow NO_2(g) + O_2(g)$$

The activation energy for the forward reaction is 10 kJ; the $\Delta H°$ is -200 kJ. What is the activation energy for the reverse reaction? Label your diagram appropriately.

14.44   Sketch a potential-energy diagram for the decomposition of nitrous oxide.

$$N_2O(g) \longrightarrow N_2(g) + O(g)$$

The activation energy for the forward reaction is 251 kJ; the $\Delta H°$ is +167 kJ. What is the activation energy for the reverse reaction? Label your diagram appropriately.

14.45   In a series of experiments on the decomposition of dinitrogen pentoxide, N_2O_5, rate constants were determined at two different temperatures. At 35°C, the rate constant was 1.4×10^{-4}/s; at 45°C, the rate constant was 5.0×10^{-4}/s. What is the activation energy for this reaction? What is the value of the rate constant at 55°C?

14.46   The reaction

$$2NOCl(g) \longrightarrow 2NO(g) + Cl_2(g)$$

has rate-constant values for the reaction of NOCl of 9.3×10^{-6}/s at 350 K and 6.9×10^{-4}/s at 400 K. Calculate the activation energy for the reaction. What is the rate constant at 425 K?

14.47   The following values of the rate constant were obtained for the decomposition of nitrogen dioxide at various temperatures. Plot the logarithm of k versus $1/T$ and from the graph obtain the energy of activation.

Temperature (°C)	k (L/mol · s)
320	0.527
330	0.776
340	1.121
350	1.607

14.48   The following values of the rate constant were obtained for the decomposition of hydrogen iodide at various temperatures. Plot the logarithm of k versus $1/T$ and from the graph obtain the energy of activation.

Temperature (°C)	k (L/mol · s)
440	2.69×10^{-3}
460	6.21×10^{-3}
480	1.40×10^{-2}
500	3.93×10^{-2}

Reaction Mechanisms; Catalysis

14.49   Nitric oxide, NO, is believed to react with chlorine according to the following mechanism:

$$NO + Cl_2 \rightleftharpoons NOCl_2 \quad \text{(elementary reaction)}$$
$$NOCl_2 + NO \longrightarrow 2NOCl \quad \text{(elementary reaction)}$$

Identify any reaction intermediate. What is the overall equation?

14.50   The decomposition of ozone is believed to occur in two steps.

$$O_3 \rightleftharpoons O_2 + O \quad \text{(elementary reaction)}$$
$$O_3 + O \longrightarrow 2O_2 \quad \text{(elementary reaction)}$$

Identify any reaction intermediate. What is the overall reaction?

14.51   The isomerization of cyclopropane, C_3H_6, is believed to occur by the mechanism shown in the following equations:

$$C_3H_6 + C_3H_6 \xrightarrow{k_1} C_3H_6 + C_3H_6^* \quad \text{(Step 1)}$$
$$C_3H_6^* \xrightarrow{k_2} CH_2{=}CHCH_3 \quad \text{(Step 2)}$$

Here $C_3H_6^*$ is an excited cyclopropane molecule. At low pressure, Step 1 is much slower than Step 2. Derive the rate law for this mechanism at low pressure. Explain.

14.52   The thermal decomposition of nitryl chloride, NO_2Cl,

$$2NO_2Cl(g) \longrightarrow 2NO_2(g) + Cl_2(g)$$

is thought to occur by the mechanism shown in the following equations:

$$NO_2Cl \xrightarrow{k_1} NO_2 + Cl \quad \text{(slow step)}$$
$$NO_2Cl + Cl \xrightarrow{k_2} NO_2 + Cl_2 \quad \text{(fast step)}$$

What rate law is predicted by this mechanism?

14.53   The reaction

$$H_2(g) + I_2(g) \longrightarrow 2HI(g)$$

may occur by the following mechanism:

$$I_2 \underset{k_{-1}}{\overset{k_1}{\rightleftharpoons}} 2I \quad \text{(fast, equilibrium)}$$
$$I + I + H_2 \xrightarrow{k_2} 2HI \quad \text{(slow)}$$

What rate law is predicted by the mechanism?

14.54   Ozone decomposes to oxygen gas.

$$2O_3(g) \longrightarrow 3O_2(g)$$

A proposed mechanism for this decomposition is

$$O_3 \underset{k_{-1}}{\overset{k_1}{\rightleftharpoons}} O_2 + O \quad \text{(fast, equilibrium)}$$
$$O_3 + O \xrightarrow{k_2} 2O_2 \quad \text{(slow)}$$

What is the rate law derived from this mechanism?

14.55 The following is a possible mechanism for a reaction involving hydrogen peroxide in aqueous solution; only a small amount of sodium bromide was added to the reaction mixture:

$$H_2O_2 + Br^- \longrightarrow BrO^- + H_2O$$
$$H_2O_2 + BrO^- \longrightarrow Br^- + H_2O + O_2$$

What is the overall reaction? What species is acting as a catalyst? Are there any reaction intermediates?

General Problems

14.57 A study of the decomposition of azomethane,

$$CH_3NNCH_3(g) \longrightarrow C_2H_6(g) + N_2(g)$$

gave the following concentrations of azomethane at various times:

Time	$[CH_3NNCH_3]$
0 min	$1.50 \times 10^{-2}\ M$
10 min	$1.29 \times 10^{-2}\ M$
20 min	$1.10 \times 10^{-2}\ M$
30 min	$0.95 \times 10^{-2}\ M$

Obtain the average rate of decomposition in units of M/s for each time interval.

14.58 Nitrogen dioxide decomposes when heated.

$$2NO_2(g) \longrightarrow 2NO(g) + O_2(g)$$

During an experiment, the concentration of NO_2 varied with time in the following way:

Time	$[NO_2]$
0.0 min	0.1103 M
1.0 min	0.1076 M
2.0 min	0.1050 M
3.0 min	0.1026 M

Obtain the average rate of decomposition of NO_2 in units of M/s for each time interval.

14.59 You can write the rate law for the decomposition of azomethane as

$$\text{Rate} = k[CH_3NNCH_3]^n \quad \text{or} \quad k = \frac{\text{rate}}{[CH_3NNCH_3]^n}$$

when n is the order of the reaction. Note that when you divide the rates at various times by the concentrations raised to the correct power n, you should get the same number (the rate constant k). Verify that the decomposition of azomethane is first order by dividing each average rate in a time interval (obtained in Problem 14.57) by the average concentration in that interval. Note that each calculation gives nearly the same value. Take the average of these values to obtain the rate constant.

14.56 Consider the following mechanism for a reaction in aqueous solution and indicate the species acting as a catalyst:

$$NH_2NO_2 + OH^- \rightleftharpoons H_2O + NHNO_2^-$$
$$NHNO_2^- \longrightarrow N_2O(g) + OH^-$$

Explain why you believe this species is a catalyst. What is the overall reaction? What substance might be added to the reaction mixture to give the catalytic activity?

14.60 Use the technique described in Problem 14.59 to verify that the decomposition of nitrogen dioxide is second order. That is, divide each average rate in a time interval (obtained in Problem 14.58) by the square of the average concentration in that interval. Note that each calculation gives nearly the same value. Take the average of these calculated values to obtain the rate constant.

14.61 A compound decomposes by a first-order reaction. If the concentration of the compound is 0.0250 M after 65 s when the initial concentration was 0.0350 M, what is the concentration of the compound after 88 s?

14.62 A compound decomposes by a first-order reaction. The concentration of compound decreases from 0.1180 M to 0.0950 M in 5.2 min. What fraction of the compound remains after 7.1 min?

14.63 Plot the data given in Problem 14.57 to verify that the decomposition of azomethane is first order. Determine the rate constant from the slope of the straight-line plot of log $[CH_3NNCH_3]$ versus time.

14.64 The decomposition of aqueous hydrogen peroxide in a given concentration of catalyst yielded the following data:

Time	0.0 min	5.0 min	10.0 min	15.0 min
$[H_2O_2]$	0.1000 M	0.0804 M	0.0648 M	0.0519 M

Verify that the reaction is first order. Determine the rate constant for the decomposition of H_2O_2 (in units of /s) from the slope of the straight-line plot of ln $[H_2O_2]$ versus time.

14.65 The decomposition of nitrogen dioxide,

$$2NO_2(g) \longrightarrow 2NO(g) + O_2(g)$$

has a rate constant of 0.498 M/s at 319°C and a rate constant of 1.81 M/s at 354°C. What are the values of the activation energy and the frequency factor for this reaction? What is the rate constant at 420°C?

14.66 A second-order reaction has a rate constant of $8.7 \times 10^{-4}/(M \cdot s)$ at 30°C. At 40°C, the rate constant is $1.8 \times 10^{-3}/(M \cdot s)$. What are the activation energy and frequency factor for this reaction? Predict the value of the rate constant at 45°C.

14.67 Nitryl bromide, NO_2Br, decomposes into nitrogen dioxide and bromine.

$$2NO_2Br(g) \longrightarrow 2NO_2(g) + Br_2(g)$$

A proposed mechanism is

$$NO_2Br \longrightarrow NO_2 + Br \quad \text{(slow)}$$
$$NO_2Br + Br \longrightarrow NO_2 + Br_2 \quad \text{(fast)}$$

Write the rate law predicted by this mechanism.

14.68 Tertiary butyl chloride reacts in basic solution according to the equation

$$(CH_3)_3CCl + OH^- \longrightarrow (CH_3)_3COH + Cl^-$$

The accepted mechanism for this reaction is

$$(CH_3)_3CCl \longrightarrow (CH_3)_3C^+ + Cl^- \quad \text{(slow)}$$
$$(CH_3)_3C^+ + OH^- \longrightarrow (CH_3)_3COH \quad \text{(fast)}$$

What should be the rate law for this reaction?

14.69 Urea, $(NH_2)_2CO$, can be prepared by heating ammonium cyanate, NH_4OCN.

$$NH_4OCN \longrightarrow (NH_2)_2CO$$

This reaction may occur by the following mechanism:

$$NH_4^+ + OCN^- \underset{k_{-1}}{\overset{k_1}{\rightleftharpoons}} NH_3 + HOCN \quad \text{(fast, equilibrium)}$$

$$NH_3 + HOCN \overset{k_2}{\longrightarrow} (NH_2)_2CO \quad \text{(slow)}$$

What is the rate law predicted by this mechanism?

14.70 Acetone reacts with iodine in acidic aqueous solution to give monoiodoacetone.

$$CH_3COCH_3 + I_2 \overset{H^+}{\longrightarrow} CH_3COCH_2I + HI$$

acetone   monoiodoacetone

A possible mechanism for this reaction is

$$CH_3\overset{O}{\overset{\|}{C}}CH_3 + H_3O^+ \underset{k_{-1}}{\overset{k_1}{\rightleftharpoons}} CH_3\overset{\overset{H}{|}\overset{O^+}{\|}}{C}CH_3 + H_2O$$
(fast, equilibrium)

$$CH_3\overset{\overset{H}{|}\overset{O^+}{\|}}{C}CH_3 + H_2O \overset{k_2}{\longrightarrow} CH_3\overset{\overset{H}{|}\overset{O}{|}}{C}=CH_2 + H_3O^+ \quad \text{(slow)}$$

$$CH_3\overset{\overset{H}{|}\overset{O}{|}}{C}=CH_2 + I_2 \overset{k_3}{\longrightarrow} CH_3\overset{O}{\overset{\|}{C}}CH_2I + HI \quad \text{(fast)}$$

Write the rate law that you derive from this mechanism.

14.71 A study of the gas-phase oxidation of nitrogen monoxide at 25°C and 1.00 atm pressure gave the following results:

$$2NO(g) + O_2(g) \longrightarrow NO_2(g)$$

	Conc. NO, mol/L	Conc. O_2, mol/L	Initial Rate
Exp. 1	4.5×10^{-2}	2.2×10^{-2}	0.80×10^{-2} mol/(L · s)
Exp. 2	4.5×10^{-2}	4.5×10^{-2}	1.60×10^{-2} mol/(L · s)
Exp. 3	9.0×10^{-2}	9.0×10^{-2}	1.28×10^{-1} mol/(L · s)
Exp. 4	3.8×10^{-1}	4.6×10^{-3}	?

a. What is the experimental rate law for the reaction above?
b. What is the initial rate of the reaction in Experiment 4?

14.72 The reaction of water with CH_3Cl in acetone as a solvent is represented by the equation

$$CH_3Cl + H_2O \longrightarrow CH_3OH + HCl$$

The rate of the reaction doubles when the concentration of CH_3Cl is doubled and it quadruples when the concentration of H_2O is doubled.
a. What is the unit for k?
b. Calculate k if CH_3OH is formed at a rate of 1.50 M/s when $[CH_3Cl] = [H_2O] = 0.40$ M.

14.73 Draw a potential-energy diagram for an uncatalyzed exothermic reaction. On the same diagram, indicate the change that results upon the addition of a catalyst. Discuss the role of a catalyst in changing the rate of reaction.

14.74 Draw and label the potential-energy curve for the reaction

$$N_2O_4(g) \rightleftharpoons 2NO_2(g); \Delta H = 57 \text{ kJ}$$

The activation energy for the reverse reaction is 23 kJ. Note ΔH and E_a on the diagram. What is the activation energy for the forward reaction? For which reaction (forward or reverse) will the reaction rate be most sensitive to a temperature increase? Explain.

14.75 What is meant by the term *rate of a chemical reaction?* Why does the rate of a reaction normally change with time? When does the rate of a chemical reaction equal the rate constant?

14.76 Briefly discuss the factors that affect the rates of chemical reactions. Which of the factors affect the magnitude of the rate constant? Which factor(s) do not affect the magnitude of the rate constant? Why?

Cumulative-Skills Problems

14.77 Hydrogen peroxide in aqueous solution decomposes by a first-order reaction to water and oxygen. The rate constant for this decomposition is 7.40×10^{-4}/s. What quantity of heat energy is initially liberated per second from 2.00 L of solution that is 1.50 M H_2O_2? See Appendix B for data.

14.78 Nitrogen dioxide reacts with carbon monoxide by the overall equation

$$NO_2(g) + CO(g) \longrightarrow NO(g) + CO_2(g)$$

At a particular temperature, the reaction is second order in NO_2 and zero order in CO. The rate constant is 0.515 L/(mol · s). How much heat energy evolves per second initially from 3.50 L of reaction mixture containing 0.0250 M NO_2? See Appendix B for data. Assume the enthalpy change is constant with temperature.

14.79 Nitric oxide reacts with oxygen to give nitrogen dioxide.

$$2NO(g) + O_2(g) \longrightarrow 2NO_2(g)$$

The rate law is $-\Delta[NO]/\Delta t = k[NO]^2[O_2]$, where the rate constant is 1.16×10^{-5} L^2/(mol^2 · s) at 339°C. A vessel contains NO and O_2 at 339°C. The initial partial pressures of NO and O_2 are 155 mmHg and 345 mmHg, respectively. What is the rate of decrease of partial pressure of NO (in mmHg per second)? (*Hint:* From the ideal gas law, obtain an expression for the molar concentration of a particular gas in terms of its partial pressure.)

14.80 Nitric oxide reacts with hydrogen as follows:

$$2NO(g) + H_2(g) \longrightarrow N_2O(g) + H_2O(g)$$

The rate law is $-\Delta[H_2]/\Delta t = k[NO]^2[H_2]$, where k is 1.10×10^{-7} L^2/(mol^2 · s) at 826°C. A vessel contains NO and H_2 at 826°C. The partial pressures of NO and H_2 are 144 mmHg and 324 mmHg, respectively. What is the rate of decrease of partial pressure of NO? See Problem 14.79.

15

Oscillating pattern formed by a reaction far from equilibrium.

480

Chemical Equilibrium

Chemical reactions often seem to stop before they are complete. Such reactions are *reversible*. That is, the original reactants form products, but then the products react with themselves to give back the original reactants. Actually two reactions are occurring, and the eventual result is a mixture of reactants and products, rather than simply a mixture of products.

Consider the gaseous reaction in which carbon monoxide and hydrogen react to produce methane and steam:

$$CO(g) + 3H_2(g) \longrightarrow CH_4(g) + H_2O(g)$$
$$\text{methane}$$

This reaction, which requires a catalyst to occur at a reasonable rate, is called *catalytic methanation*. It is a potentially useful reaction because it can be used to produce methane from coal or organic wastes from which carbon monoxide can be obtained (Figure 15.1).

Catalytic methanation is a reversible reaction, and depending on the reaction conditions, the final reaction mixture will have varying amounts of the products methane and steam, as well as the starting substances carbon monoxide and hydrogen. It is

Figure 15.1
Coal gasification plant.
Gaseous fuels can be made from coal in a number of ways. One way (described in the text) is to react coal with steam to produce a mixture of CO and H_2, which then reacts by catalytic methanation to yield CH_4, the major component of natural gas.

also possible to start with methane and steam and, under the right conditions, form a mixture that is predominantly carbon monoxide and hydrogen. The process is called *steam reforming*.

$$CH_4(g) + H_2O(g) \longrightarrow CO(g) + 3H_2(g)$$

The processes of catalytic methanation and steam reforming illustrate the reversibility of chemical reactions. Starting with CO and H_2 and using the right conditions, you can form predominantly CH_4 and H_2O. Starting with CH_4 and H_2O and using different conditions, you can obtain a reaction mixture that is predominantly CO and H_2. An important question is: What conditions favor the production of CH_4 and H_2O, and what conditions favor the production of CO and H_2?

Certain reactions appear to stop before they are complete. The reaction mixture consists of both reactants and products in definite concentrations. Such a reaction mixture is said to have reached *chemical equilibrium*. In earlier chapters, we discussed other types of equilibria, including the equilibrium between a liquid and its vapor and the equilibrium between a solid and its saturated solution. We will now see how to determine the composition of a reaction mixture at equilibrium and how to alter this composition by changing the conditions for the reaction.

Describing Chemical Equilibrium

Many chemical reactions are like the catalytic methanation reaction. Such reactions can be made to go predominantly in one direction or the other depending on the conditions. Let us look more closely at this reversibility and see how to characterize it quantitatively.

15.1 Chemical Equilibrium—A Dynamic Equilibrium

▶ The concept of dynamic equilibrium in chemical reactions was briefly mentioned in Section 14.8. Discussions on vapor pressure (Section 11.2) and solubility (Section 12.2) give detailed explanations of the role of dynamic equilibria.

When substances react, they eventually form a mixture of reactants and products in *dynamic equilibrium*. ◀ This dynamic equilibrium consists of a forward reaction, in which substances react to give products, and a reverse reaction, in which products react to give the original reactants.

Consider the catalytic methanation reaction. It consists of forward and reverse reactions, as represented by the chemical equation

$$CO(g) + 3H_2(g) \Longleftrightarrow CH_4(g) + H_2O(g)$$

Suppose you put 1.000 mol CO and 3.000 mol H_2 into a 10.00-L vessel at 1200 K. At first these concentrations are large, but as the substances react their concentrations decrease (Figure 15.2). The rate of the forward reaction is large at first but steadily decreases. On the other hand, the concentrations of CH_4 and H_2O, which are zero at first, increase with time. The rate of the reverse reaction starts at zero and steadily increases. The forward rate decreases and the reverse rate increases until eventually the rates become equal. The concentrations of reactants and products no longer change, and the reaction mixture has reached *equilibrium*.

Chemical equilibrium is *the state reached by a reaction mixture when the rates of forward and reverse reactions have become equal.* If you observe the reaction mixture, you see no net change, although the forward and reverse reactions are continuing. Thus the equilibrium is a *dynamic* process.

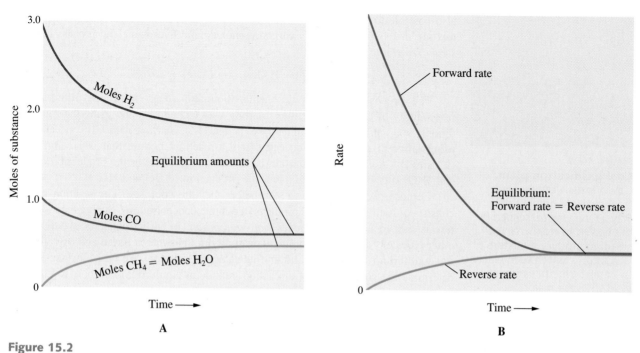

Figure 15.2
Catalytic methanation reaction approaches equilibrium. *(A)* The experiment begins with 1.000 mol CO and 3.000 mol H_2 in a 10.00-L vessel. Note that the amounts of substances become constant at equilibrium. *(B)* The forward rate is large at first but steadily decreases, whereas the reverse rate starts at zero and steadily increases. Eventually both rates become equal (at equilibrium).

Suppose you place known amounts of reactants in a vessel and let the mixture come to equilibrium. To obtain the composition of the equilibrium mixture, you need to determine the amount of only one of the substances, as shown in the following example.

EXAMPLE 15.1

Applying Stoichiometry to an Equilibrium Mixture

Carbon monoxide and hydrogen react according to the following equation:

$$CO(g) + 3H_2(g) \rightleftharpoons CH_4(g) + H_2O(g)$$

When 1.000 mol CO and 3.000 mol H_2 are placed in a 10.00-L vessel at 1200 K and allowed to come to equilibrium, the mixture is found to contain 0.387 mol H_2O. How many moles of each substance are present?

▶ PROBLEM STRATEGY

The problem is essentially one of stoichiometry. It involves initial, or *starting*, amounts of reactants. These amounts *change* as reaction occurs. Later, the reaction comes to *equilibrium*, and you analyze the reaction mixture for the amount of one of the reactants or products. It is convenient to solve this problem by first setting up a table in which you write the starting, change, and equilibrium values of each substance under the balanced equation. This way you can easily see what you have to calculate.

Amount (mol)	$CO(g)$ +	$3H_2(g)$ $\rightleftharpoons$	$CH_4(g)$ +	$H_2O(g)$
Starting				
Change				
Equilibrium				

Fill in the starting amounts. You are not given explicit values for the changes that occur, so you let x be the molar change. That is, each product increases by x moles multiplied by the coefficients of the substances in the balanced equation. Reactants decrease by x moles multiplied by the corresponding coefficients. Equilibrium values are equal to starting values plus the changes.

SOLUTION

Using the information given in the problem, you set up the following table:

Amount (mol)	$CO(g)$ +	$3H_2(g)$ $\rightleftharpoons$	$CH_4(g)$ +	$H_2O(g)$
Starting	1.000	3.000	0	0
Change	$-x$	$-3x$	$+x$	$+x$
Equilibrium	$1.000 - x$	$3.000 - 3x$	x	$x = 0.387$

The problem statement gives the equilibrium amount for H_2O. This tells you that $x = 0.387$ mol. You calculate equilibrium amounts for other substances from the expressions given in the table, using this value of x.

$$\text{Equilibrium amount CO} = (1.000 - x) \text{ mol}$$
$$= (1.000 - 0.387) \text{ mol}$$
$$= 0.613 \text{ mol}$$

$$\text{Equilibrium amount } H_2 = (3.000 - 3x) \text{ mol}$$
$$= (3.000 - 3 \times 0.387) \text{ mol}$$
$$= 1.839 \text{ mol}$$

$$\text{Equilibrium amount } CH_4 = x \text{ mol} = 0.387 \text{ mol}$$

Therefore, the amounts of substances in the equilibrium mixture are **0.613 mol CO, 1.839 mol H_2, 0.387 mol CH_4, and 0.387 mol H_2O.**

See Problems 15.15, 15.16, 15.17, and 15.18.

CONCEPT CHECK 15.1

Two substances A and B react to produce substance C. When reactant A decreases by molar amount x, product C increases by molar amount x. When reactant B decreases by molar amount x, product C increases by molar amount $2x$. Write the chemical equation for the reaction.

15.2 The Equilibrium Constant

In the preceding example, we found that when 1.000 mol CO and 3.000 mol H_2 react in a 10.00-L vessel by catalytic methanation at 1200 K, they give an equilibrium mixture containing 0.613 mol CO, 1.839 mol H_2, 0.387 mol CH_4, and 0.387 mol H_2O. Let us call this Experiment 1. Now consider a similar experiment, Experiment 2, in which you start with an additional mole of carbon monoxide. That is, you place 2.000 mol CO and 3.000 mol H_2 in a 10.00-L vessel at 1200 K. At equilibrium, you find that the vessel contains 1.522 mol CO, 1.566 mol H_2, 0.478 mol CH_4, and 0.478 mol H_2O. The equilibrium composition depends on the amounts of starting substances. Nevertheless, you will see that all of the equilibrium compositions for a reaction at a given temperature are related by a quantity called the *equilibrium constant*.

Definition of the Equilibrium Constant K_c

The **equilibrium-constant expression** for a reaction is *an expression obtained by multiplying the concentrations of products, dividing by the concentrations of reactants, and raising each concentration term to a power equal to the coefficient in the chemical equation.* The **equilibrium constant** K_c is *the value obtained for the equilibrium-constant expression when equilibrium concentrations are substituted.* ◀ For the reaction

$$aA + bB \rightleftharpoons cC + dD$$

▶ In Section 15.3, you will see that you ignore concentrations of pure liquids and solids when you write the equilibrium-constant expression. The concentrations of these substances are constant and are incorporated into the value of K_c.

you have

$$K_c = \frac{[C]^c[D]^d}{[A]^a[B]^b}$$

Here you denote the molar concentration of a substance by writing its formula in square brackets. The subscript c on the equilibrium constant means that it is defined in terms of molar concentrations. The **law of mass action** is *a relation that states that the values of the equilibrium-constant expression K_c are constant for a particular reaction at a given temperature, whatever equilibrium concentrations are substituted.* ◀

▶ "Active mass" was an early term for concentration; hence the term *mass action.*

As the following example illustrates, the equilibrium-constant expression is defined in terms of the balanced chemical equation.

EXAMPLE 15.2

Writing Equilibrium-Constant Expressions

a. Write the equilibrium-constant expression K_c for catalytic methanation.

$$CO(g) + 3H_2(g) \rightleftharpoons CH_4(g) + H_2O(g)$$

b. Write the equilibrium-constant expression K_c for the reverse of the previous reaction.

$$CH_4(g) + H_2O(g) \rightleftharpoons CO(g) + 3H_2(g)$$

c. Write the equilibrium-constant expression K_c for the synthesis of ammonia.

$$N_2(g) + 3H_2(g) \rightleftharpoons 2NH_3(g)$$

d. Write the equilibrium-constant expression K_c when the equation for the previous reaction is written

$$\tfrac{1}{2}N_2(g) + \tfrac{3}{2}H_2(g) \rightleftharpoons NH_3(g)$$

SOLUTION

a. The expression for the equilibrium constant is

$$K_c = \frac{[\mathbf{CH_4}][\mathbf{H_2O}]}{[\mathbf{CO}][\mathbf{H_2}]^3}$$

b. When the equation is written in reverse order, the expression for K_c is inverted:

$$K_c = \frac{[\mathbf{CO}][\mathbf{H_2}]^3}{[\mathbf{CH_4}][\mathbf{H_2O}]}$$

c. The equilibrium constant for $N_2 + 3H_2 \rightleftharpoons 2NH_3$ is

$$K_c = \frac{[NH_3]^2}{[N_2][H_2]^3}$$

d. When the coefficients in the equation in part c are multiplied by $\frac{1}{2}$ to give

$$\tfrac{1}{2}N_2 + \tfrac{3}{2}H_2 \rightleftharpoons NH_3$$

the equilibrium-constant expression becomes

$$K_c = \frac{[NH_3]}{[N_2]^{1/2}[H_2]^{3/2}}$$

which is the square root of the previous expression.

See Problems 15.19 and 15.20.

Equilibrium: A Kinetics Argument

The law of mass action was first stated by the Norwegian chemists Cato Guldberg and Peter Waage in 1867. They were led to this law by a kinetics argument. To understand the argument, consider the decomposition of dinitrogen tetroxide.

Dinitrogen tetroxide is a colorless substance that decomposes, when warmed, to give nitrogen dioxide, NO_2, a brown substance. Suppose you start with dinitrogen tetroxide gas and heat it above room temperature. At this higher temperature, N_2O_4 decomposes to NO_2, which can be noted by the increasing intensity of the brown color of the reaction mixture.

$$N_2O_4(g) \longrightarrow 2NO_2(g)$$

Once some NO_2 is produced, it can react to re-form N_2O_4. Call the decomposition of N_2O_4 the forward reaction and the formation of N_2O_4 the reverse reaction. These are elementary reactions, and you can immediately write the rate equations. The rate of the forward reaction is $k_f[N_2O_4]$, where k_f is the rate constant for the forward reaction. The rate of the reverse reaction is $k_r[NO_2]^2$, where k_r is the rate constant for the reverse reaction.

At first, the rate of the forward reaction is relatively large, and the rate of the reverse reaction is relatively small; as the concentration of N_2O_4 decreases, so the forward rate decreases. At the same time, the concentration of NO_2 increases, so the reverse rate increases. Eventually the two rates become equal and the reaction has reached a dynamic equilibrium.

You can write: Rate of forward reaction = rate of reverse reaction, or

$$k_f[N_2O_4] = k_r[NO_2]^2$$

Rearranging gives

$$\frac{k_f}{k_r} = \frac{[NO_2]^2}{[N_2O_4]}$$

The right side is the equilibrium-constant expression for the decomposition of dinitrogen tetroxide. The left side is the ratio of rate constants. Thus, the equilibrium-constant expression is constant for a given temperature, and you can identify the equilibrium constant K_c as the ratio of rate constants for the forward and reverse reactions.

$$K_c = \frac{k_f}{k_r}$$

Obtaining Equilibrium Constants for Reactions

Earlier, we gave data from the results of two experiments, Experiments 1 and 2, involving catalytic methanation. By substituting the molar concentrations from these two experiments into the equilibrium-constant expression for the reaction, you can show that you get the same value for both experiments. This value equals K_c for methanation at 1200 K. ◀

▶ It is also possible to determine equilibrium constants from thermochemical data, as described in Section 19.6.

Experiment 1 The equilibrium composition is 0.613 mol CO, 1.839 mol H_2, 0.387 mol CH_4, and 0.387 mol H_2O. The volume of the reaction vessel is 10.00 L, so the equilibrium concentration of CO is

$$[CO] = \frac{0.613 \text{ mol}}{10.00 \text{ L}} = 0.0613 \text{ } M$$

Similarly, the other equilibrium concentrations are $[H_2] = 0.1839 \text{ } M$, $[CH_4] = 0.0387 \text{ } M$, and $[H_2O] = 0.0387 \text{ } M$.

Substitute these values into the equilibrium expression for catalytic methanation. It is the usual practice to write equilibrium constants without units. We will follow that practice here. ◀ Substitution of concentrations into the equilibrium-constant expression gives

▶ In thermodynamics, the equilibrium constant is defined in terms of *activities*, rather than concentrations. For an ideal mixture, the activity of a substance is the ratio of its concentration (or partial pressure if a gas) to a standard concentration of 1 *M* (or partial pressure of 1 atm), so that units cancel. Thus, activities have numerical values but no units.

$$K_c = \frac{[CH_4][H_2O]}{[CO][H_2]^3} = \frac{(0.0387)(0.0387)}{(0.0613)(0.1839)^3} = 3.93$$

Experiment 2 The equilibrium composition is 1.522 mol CO, 1.566 mol H_2, 0.478 mol CH_4, and 0.478 mol H_2O and the concentrations are $[CO] = 0.1522 \text{ } M$, $[H_2] = 0.1566 \text{ } M$, $[CH_4] = 0.0478 \text{ } M$, and $[H_2O] = 0.0478 \text{ } M$. Substituting gives

$$K_c = \frac{[CH_4][H_2O]}{[CO][H_2]^3} = \frac{(0.0478)(0.0478)}{(0.1522)(0.1566)^3} = 3.91$$

Within the precision of the data, these values (3.93 and 3.91) are the same. Moreover, experiment shows that when you start with CH_4 and H_2O, instead of CO and H_2, an equilibrium mixture is again reached that yields the same value of K_c (see Experiment 3, Figure 15.3).

The following example provides practice in evaluating an equilibrium constant from equilibrium compositions.

EXAMPLE 15.3

Obtaining an Equilibrium Constant from Reaction Composition

Hydrogen iodide decomposes according to the equation

$$2HI(g) \rightleftharpoons H_2(g) + I_2(g)$$

The amount of I_2 in the reaction mixture can be determined from the intensity of the violet color of I_2. When 4.00 mol HI was placed in a 5.00-L vessel at 458°C, the equilibrium mixture was found to contain 0.442 mol I_2. What is the value of K_c for the decomposition of HI at this temperature?

Figure 15.3
Some equilibrium compositions for the methanation reaction.
Different starting concentrations have been used in each experiment (all at 1200 K). Experiments 1 and 2 start with different concentrations of reactants CO and H_2. Experiment 3 starts with the products CH_4 and H_2O. All three experiments yield essentially the same value of K_c.

	Starting concentrations	Equilibrium concentrations	Calculated value of K_c
Experiment 1	0.1000 M CO 0.3000 M H_2	0.0613 M CO 0.1839 M H_2 0.0387 M CH_4 0.0387 M H_2O	$K_c = 3.93$
Experiment 2	0.2000 M CO 0.3000 M H_2	0.1522 M CO 0.1566 M H_2 0.0478 M CH_4 0.0478 M H_2O	$K_c = 3.91$
Experiment 3	0.1000 M CH_4 0.1000 M H_2O	0.0613 M CO 0.1839 M H_2 0.0387 M CH_4 0.0387 M H_2O	$K_c = 3.93$

SOLUTION

To obtain the concentrations of HI and I_2, you divide the molar amounts by the volume of the reaction vessel (5.00 L).

$$\text{Starting concentration of HI} = \frac{4.00 \text{ mol}}{5.00 \text{ L}} = 0.800 \ M$$

$$\text{Equilibrium concentration of } I_2 = \frac{0.442 \text{ mol}}{5.00 \text{ L}} = 0.0884 \ M$$

From these values, you set up the following table:

Concentration (M)	$2HI(g)$	$\rightleftharpoons$	$H_2(g)$	+	$I_2(g)$
Starting	0.800		0		0
Change	$-2x$		x		x
Equilibrium	$0.800 - 2x$		x		$x = 0.0884$

The equilibrium concentrations of substances can be evaluated from the expressions given in the last line of this table. You know that the equilibrium concentration of I_2 is 0.0884 M and that this equals x. Therefore,

$$[HI] = (0.800 - 2x) \ M = (0.800 - 2 \times 0.0884) \ M = 0.623 \ M$$

$$[H_2] = x = 0.0884 \ M$$

Now you substitute into the equilibrium-constant expression for the reaction. From the chemical equation, you write

$$K_c = \frac{[H_2][I_2]}{[HI]^2}$$

Substituting,

$$K_c = \frac{(0.0884)(0.0884)}{(0.623)^2} = \textbf{0.0201}$$

See Problems 15.25, 15.26, 15.27, and 15.28.

The Equilibrium Constant K_p

In discussing gas-phase equilibria, it is often convenient to write the equilibrium constant in terms of partial pressures of gases rather than concentrations. Note that the concentration of a gas is proportional to its partial pressure at a fixed temperature ($P = (n/V)RT$ where n/V is the concentration).

When you express an equilibrium constant for a gaseous reaction in terms of partial pressures, you call it K_p. For catalytic methanation,

$$CO(g) + 3H_2(g) \rightleftharpoons CH_4(g) + H_2O(g)$$

the equilibrium expression in terms of partial pressures becomes

$$K_p = \frac{P_{CH_4}P_{H_2O}}{P_{CO}P_{H_2}{}^3}$$

In general, the value of K_p is different from that of K_c. From the relationship for molar concentration $n/V = P/RT$, one can show that

$$K_p = K_c(RT)^{\Delta n}$$

where Δn is the sum of the coefficients of *gaseous* products in the chemical equation minus the sum of the coefficients of *gaseous* reactants. For the methanation reaction ($K_c = 3.92$), in which 2 mol of gaseous products ($CH_4 + H_2O$) are obtained from 4 mol of gaseous reactants ($CO + 3H_2$), Δn equals $2 - 4 = -2$. The usual unit of partial pressures in K_p is atmospheres; therefore, the value of R is 0.0821 L · atm/(K · mol). Hence,

$$K_p = 3.92 \times (0.0821 \times 1200)^{-2} = 4.04 \times 10^{-4}$$

Equilibrium Constant for the Sum of Reactions

The equilibrium constants for various chemical reactions can be used to obtain the equilibrium constants of other reactions. The rule is: *if a given chemical equation can be obtained by taking the sum of other equations, the equilibrium constant for the given equation equals the product of the equilibrium constants of the other equations.*

As an application of this rule, consider the following reactions at 1200 K:

$$CO(g) + 3H_2(g) \rightleftharpoons CH_4(g) + H_2O(g); K_1 = 3.92 \qquad \text{(Reaction 1)}$$

$$CH_4(g) + 2H_2S(g) \rightleftharpoons CS_2(g) + 4H_2(g); K_2 = 3.3 \times 10^4 \quad \text{(Reaction 2)}$$

When you take the sum of these two equations, you get

$$CO(g) + 2H_2S(g) \rightleftharpoons CS_2(g) + H_2O(g) + H_2(g) \qquad \text{(Reaction 3)}$$

According to the rule, the equilibrium constant for this reaction, K_3, is $K_3 = K_1K_2$. This result is easy to verify. Substituting the expressions for K_1 and K_2 in the product K_1K_2 gives

$$K_1K_2 = \frac{[CH_4][H_2O]}{[CO][H_2]^3} \times \frac{[CS_2][H_2]^4}{[CH_4][H_2S]^2} = \frac{[CS_2][H_2O][H_2]}{[CO][H_2S]^2}$$

▶ You may also use equilibrium expressions in terms of pressures in this calculation.

which is the equilibrium-constant expression for Reaction 3. The value of the equilibrium constant is $K_1K_2 = 3.92 \times (3.3 \times 10^4) = 1.3 \times 10^5$. ◀

15.3 Heterogeneous Equilibria; Solvents in Homogeneous Equilibria

A **homogeneous equilibrium** is *an equilibrium that involves reactants and products in a single phase.* Catalytic methanation is an example of a homogeneous equilibrium. A **heterogeneous equilibrium** is *an equilibrium involving reactants and products in more than one phase.* For example, the reaction of iron metal filings with steam to produce iron oxide, Fe_3O_4, and hydrogen involves two phases.

$$3Fe(s) + 4H_2O(g) \rightleftharpoons Fe_3O_4(s) + 4H_2(g)$$

In writing the equilibrium-constant expression for a heterogeneous equilibrium, you omit concentration terms for pure solids and liquids. For the reaction of iron with steam, you would write

$$K_c = \frac{[H_2]^4}{[H_2O]^4}$$

Concentrations of Fe and Fe_3O_4 are omitted, because whereas the concentration of a gas can have various values, the concentration of a pure solid or a pure liquid is a constant at a given temperature.

The fact that the concentrations of both Fe and Fe_3O_4 do not appear in the equilibrium-constant expression means that the equilibrium is not affected by the amounts of these substances, *as long as some of each is present.* ◄

▶ If you use the thermodynamic equilibrium constant, which is defined in terms of activities, you find that the activity of a pure solid or liquid is 1. Therefore, when you write the equilibrium constant, the activities of pure solids and liquids need not be given explicitly.

EXAMPLE 15.4

Writing K_c for a Reaction with Pure Solids or Liquids

a. Quicklime (calcium oxide, CaO) is prepared by heating a source of calcium carbonate, $CaCO_3$, such as limestone or seashells.

$$CaCO_3(s) \rightleftharpoons CaO(s) + CO_2(g)$$

Write the expression for K_c.

b. You can write the equilibrium-constant expression for a physical equilibrium, such as vaporization, as well as for a chemical equilibrium. Write the expression for K_c for the vaporization of water.

$$H_2O(l) \rightleftharpoons H_2O(g)$$

SOLUTION

In writing the equilibrium expressions, you ignore pure liquid and solid phases.
a. $K_c = [CO_2]$; b. $K_c = [H_2O(g)]$.

See Problems 15.33 and 15.34.

Using the Equilibrium Constant

We have described how a chemical reaction reaches equilibrium and how you can characterize this equilibrium by the equilibrium constant. Now we want to see the ways in which an equilibrium constant can be used. Consider the following uses:

1. *Qualitatively interpreting the equilibrium constant.* Does a particular equilibrium favor products or reactants?

2. *Predicting the direction of reaction.* In which direction will a reaction proceed with the introduction of both reactants and products?

3. *Calculating equilibrium concentrations.* Can you determine the composition at equilibrium for any set of starting concentrations by using the value of K_c?

15.4 Qualitatively Interpreting the Equilibrium Constant

If the value of the equilibrium constant is large, you know immediately that the products are favored at equilibrium. Consider the synthesis of ammonia from its elements.

$$N_2(g) + 3H_2(g) \rightleftharpoons 2NH_3(g)$$

At 25°C the equilibrium constant K_c equals 4.1×10^8. This means that the numerator (product concentrations) is 4.1×10^8 times larger than the denominator (reactant concentrations). In other words, at this temperature the reaction favors the formation of ammonia at equilibrium.

If the value of the equilibrium constant is small, the reactants are favored at equilibrium. As an example, consider the reaction of nitrogen and oxygen to give nitric oxide, NO:

▶ The equilibrium constant does become large enough at higher temperatures (about 2000°C) to give appreciable amounts of nitric oxide. This is why air ($N_2 + O_2$) in flames and in auto engines becomes a source of the pollutant NO.

$$N_2(g) + O_2(g) \rightleftharpoons 2NO(g)$$

The equilibrium constant K_c equals 4.6×10^{-31} at 25°C. ◀

When the equilibrium constant is neither large nor small (around 1), neither reactants nor products are strongly favored. For example, in the case of the methanation reaction, the equilibrium constant K_c equals 3.92 at 1200 K. You found that if you start with 1.000 mol CO and 3.000 mol H_2 in a 10.00-L vessel, the equilibrium composition is 0.613 mol CO, 1.839 mol H_2, 0.387 mol CH_4, and 0.387 mol H_2O.

15.5 Predicting the Direction of Reaction

Suppose a gaseous mixture from an industrial plant has the following composition at 1200 K: 0.0200 *M* CO, 0.0200 *M* H_2, 0.00100 *M* CH_4, and 0.00100 *M* H_2O. If the mixture is passed over a catalyst at 1200 K, would the reaction

$$CO(g) + 3H_2(g) \rightleftharpoons CH_4(g) + H_2O(g)$$

go toward the right or the left?

To answer this question, you substitute the concentrations of substances into the *reaction quotient* and compare its value to K_c. The **reaction quotient, Q_c,** is *an expression that has the same form as the equilibrium-constant expression but whose concentration values are not necessarily those at equilibrium.* So

$$Q_c = \frac{[CH_4]_i[H_2O]_i}{[CO]_i[H_2]_i^3}$$

where the subscript *i* indicates concentrations at a particular instant *i*. Therefore

$$Q_c = \frac{(0.00100)(0.00100)}{(0.0200)(0.0200)^3} = 6.25$$

For the reaction mixture to go to equilibrium, the value of Q_c must decrease from 6.25 to 3.92 (the equilibrium constant). This will happen if the numerator of Q_c decreases and the denominator increases. Thus, the gaseous mixture will give more CO and H_2.

Consider the problem more generally. You are given a reaction mixture that is not yet at equilibrium. You would like to know in what direction the reaction will go as it approaches equilibrium. To answer this, you substitute the concentrations of substances from the mixture into the reaction quotient Q_c. Then, you compare Q_c to the equilibrium constant K_c. Then

> If $Q_c > K_c$, the reaction will go to the left.
> If $Q_c < K_c$, the reaction will go to the right.
> If $Q_c = K_c$, the reaction mixture is at equilibrium.

EXAMPLE 15.5

Using the Reaction Quotient

A 50.0-L reaction vessel contains 1.00 mol N_2, 3.00 mol H_2, and 0.500 mol NH_3. Will more ammonia, NH_3, be formed or will it dissociate when the mixture goes to equilibrium at 400°C? The equation is

$$N_2(g) + 3H_2(g) \rightleftharpoons 2NH_3(g)$$

K_c is 0.500 at 400°C.

SOLUTION

The composition of the gas has been given in terms of moles. You convert these to molar concentrations by dividing by the volume (50.0 L). This gives 0.0200 M N_2, 0.0600 M H_2, and 0.0100 M NH_3. Substituting these concentrations into the reaction quotient gives

$$Q_c = \frac{[NH_3]_i^2}{[N_2]_i[H_2]_i^3} = \frac{(0.0100)^2}{(0.0200)(0.0600)^3} = 23.1$$

Because $Q_c = 23.1$ is greater than $K_c = 0.500$, the reaction will go to the left as it approaches equilibrium. Therefore, **ammonia will dissociate.**

See Problems 15.37 and 15.38.

CONCEPT CHECK 15.2

Carbon monoxide and hydrogen react in the presence of a catalyst to form methanol, CH_3OH:

$$CO(g) + 2H_2(g) \rightleftharpoons CH_3OH(g)$$

An equilibrium mixture of these three substances is suddenly compressed so that the concentrations of all substances initially double. In what direction does the reaction go as a new equilibrium is attained?

15.6 Calculating Equilibrium Concentrations

Once you have determined the equilibrium constant for a reaction, you can use it to calculate the concentrations of substances in an equilibrium mixture. The next example illustrates a simple type of equilibrium problem.

EXAMPLE 15.6

Obtaining One Equilibrium Concentration Given the Others

A gaseous mixture contains 0.30 M CO, 0.10 M H_2, and 0.020 M H_2O, plus an unknown amount of CH_4. This mixture is at equilibrium at 1200 K.

$$CO(g) + 3H_2(g) \rightleftharpoons CH_4(g) + H_2O(g)$$

What is the concentration of CH_4? The equilibrium constant K_c equals 3.92.

SOLUTION

The equilibrium-constant equation is

$$K_c = \frac{[CH_4][H_2O]}{[CO][H_2]^3}$$

Substituting the known concentrations and the value of K_c gives

$$3.92 = \frac{[CH_4](0.020)}{(0.30)(0.10)^3}$$

You can now solve for $[CH_4]$.

$$[CH_4] = \frac{(3.92)(0.30)(0.10)^3}{(0.020)} = 0.059$$

The concentration of CH_4 in the mixture is **0.059 mol/L.**

Usually you begin a reaction with known starting quantities of substances and want to calculate the quantities at equilibrium. The next example illustrates the steps used to solve this type of problem.

EXAMPLE 15.7

Solving an Equilibrium Problem

The reaction

$$CO(g) + H_2O(g) \rightleftharpoons CO_2(g) + H_2(g)$$

is used to increase the ratio of hydrogen in synthesis gas (mixtures of CO and H_2). Suppose you start with 1.00 mol each of carbon monoxide and water in a 50.0-L vessel. How many moles of each substance are in the equilibrium mixture at 1000°C? The equilibrium constant K_c at this temperature is 0.58.

PROBLEM STRATEGY

The solution of an equilibrium problem involves three steps. In Step 1, you use the given information to set up a table of starting, change, and equilibrium concentrations. Use the initial amounts to obtain the starting concentrations. Express the change concentrations in terms of x. In Step 2, you substitute the equilibrium concentrations in the table into the equilibrium-constant expression and equate this to the value of the equilibrium constant. In Step 3, you solve this equilibrium-constant equation.

SOLUTION

STEP 1   The *starting concentrations* of CO and H_2O are

$$[CO] = [H_2O] = \frac{1.00 \text{ mol}}{50.0 \text{ L}} = 0.0200 \text{ mol/L}$$

The starting concentrations of the products, CO_2 and H_2, are 0. Let x be the moles of CO_2 and H_2 formed per liter. Similarly, x moles each of CO and H_2O are consumed. You write the changes for CO and H_2O as $-x$. You obtain the *equilibrium concentrations* by adding the change in concentrations to the starting concentrations, as shown in the following table:

Concentrations (M)	$CO(g)$	+	$H_2O(g)$	$\rightleftharpoons$	$CO_2(g)$	+	$H_2(g)$
Starting	0.0200		0.0200		0		0
Change	$-x$		$-x$		$+x$		$+x$
Equilibrium	$0.0200 - x$		$0.0200 - x$		x		x

STEP 2   Now substitute into the equilibrium-constant equation,

$$K_c = \frac{[CO_2][H_2]}{[CO][H_2O]}$$

the equilibrium concentrations and you get

$$0.58 = \frac{(x)(x)}{(0.0200 - x)(0.0200 - x)}$$

STEP 3   You now solve this equilibrium equation for the value of x. Note that the right-hand side is a perfect square. If you take the square root of both sides, you get

$$0.76 = \frac{x}{0.0200 - x}$$

Rearranging the equation gives

$$x = \frac{0.0200 \times 0.76}{1.76} = 0.0086$$

If you substitute for x in the last line of the table, the equilibrium concentrations are $0.0114\ M$ CO, $0.0114\ M$ H_2O, $0.0086\ M$ CO_2, and $0.0086\ M$ H_2. To find the moles of each substance in the 50.0-L vessel, you multiply the concentrations by the volume of the vessel. For example, the amount of CO is

$$0.0114 \text{ mol/L} \times 50.0 \text{ L} = 0.570 \text{ mol}$$

The equilibrium composition of the reaction mixture is **0.570 mol CO, 0.570 mol H_2O, 0.43 mol CO_2, and 0.43 mol H_2.**

See Problems 15.41 and 15.42.

In the previous example, if you had not started with the same number of moles of reactants, you would not have gotten an equation with a perfect square. In that case you would have had to solve a quadratic equation, as illustrated below. ◄

► A quadratic equation of the form

$$ax^2 + bx + c = 0$$

has the solutions

$$x = \frac{-b \pm \sqrt{b^2 - 4ac}}{2a}$$

This equation for x is called the *quadratic formula*. (See Appendix A.)

EXAMPLE 15.8

Solving an Equilibrium Problem (Involving a Quadratic Equation in x)

Hydrogen and iodine react according to the equation

$$H_2(g) + I_2(g) \rightleftharpoons 2HI(g)$$

Suppose 1.00 mol H_2 and 2.00 mol I_2 are placed in a 1.00-L vessel. How many moles of substances are in the gaseous mixture when it comes to equilibrium at 458°C? The equilibrium constant K_c at this temperature is 49.7.

SOLUTION

STEP 1 The concentrations of substances are as follows:

Concentrations (M)	$H_2(g)$	+	$I_2(g)$	$\rightleftharpoons$	$2HI(g)$
Starting	1.00		2.00		0
Change	$-x$		$-x$		$2x$
Equilibrium	$1.00 - x$		$2.00 - x$		$2x$

Note that the changes in concentrations equal x multiplied by the coefficient of that substance in the balanced chemical equation.

STEP 2 Substituting into the equilibrium-constant equation,

$$K_c = \frac{[HI]^2}{[H_2][I_2]}$$

you get

$$49.7 = \frac{(2x)^2}{(1.00 - x)(2.00 - x)}$$

STEP 3 Use the quadratic formula to solve for x. The equation rearranges to give

$$(1.00 - x)(2.00 - x) = (2x)^2/49.7 = 0.0805x^2$$

or

$$0.920x^2 - 3.00x + 2.00 = 0$$

Hence,

$$x = \frac{3.00 \pm \sqrt{9.00 - 7.36}}{1.84} = 1.63 \pm 0.70$$

A quadratic equation has two mathematical solutions. You obtain one solution by taking the positive sign and the other by taking the negative sign. Thus

$$x = 2.33 \quad \text{and} \quad x = 0.93$$

However, $x = 2.33$ gives a negative value to $1.00 - x$, which is physically impossible. Substitute the value of $x = 0.93$ into the last line of the table in

Step 1 to get the equilibrium concentrations, and then multiply these by the volume of the vessel (1.00 L) to get the amounts of substances. The last line is rewritten

Concentrations (M)	$H_2(g)$	+	$I_2(g)$	$\rightleftharpoons$	$2HI(g)$
Equilibrium	$1.00 - x = 0.07$		$2.00 - x = 1.07$		$2x = 1.86$

The equilibrium composition is **0.07 mol H_2, 1.07 mol I_2, and 1.86 mol HI.**

See Problems 15.43 and 15.44.

Changing the Reaction Conditions; Le Chatelier's Principle

Obtaining the maximum amount of product from a reaction depends on the proper selection of reaction conditions. By changing these conditions, you can increase or decrease the yield of product. There are three ways to alter the equilibrium composition of a gaseous reaction mixture and possibly increase the yield of product:

1. Changing the concentrations by removing products or adding reactants.

2. Changing the partial pressure of gaseous reactants and products by changing the volume.

3. Changing the temperature.

15.7   Removing Products or Adding Reactants

One way to increase the yield of a desired product is to change concentrations in a reaction mixture by removing a product or adding a reactant. Consider the methanation reaction,

$$CO(g) + 3H_2(g) \rightleftharpoons CH_4(g) + H_2O(g)$$

If you place 1.000 mol CO and 3.000 mol H_2 in a 10.00-L reaction vessel, the equilibrium composition at 1200 K is 0.613 mol CO, 1.839 mol H_2, 0.387 mol CH_4, and 0.387 mol H_2O. Can you alter this composition?

To answer this question, you can apply **Le Chatelier's principle,** which states that *when a system in chemical equilibrium is disturbed by a change of temperature, pressure, or a concentration, the system shifts in equilibrium composition in a way that tends to counteract this change of variable.* ◀ Suppose you remove or add a substance to the equilibrium mixture in order to alter its concentration. Chemical reaction then occurs to partially restore the initial concentration of the removed or added substance. (However, if the concentration of substance cannot be changed, as in the case of a pure solid or liquid reactant or product, changes in amount will have no effect on the equilibrium composition.)

For example, suppose that water vapor is removed from the reaction vessel containing the equilibrium mixture for methanation. Le Chatelier's principle predicts that net chemical change will occur to partially reinstate the original concentration of water

▶ Le Chatelier's principle was introduced in Section 12.3, where it was used to determine the effect of pressure on solubility of gases in liquids.

Table 15.1
Effect of Removing Water Vapor from a Methanation Mixture (in a 10.00-L Vessel)

Stage of Process	Mol CO	Mol H_2	Mol CH_4	Mol H_2O
Original reaction mixture	0.613	1.839	0.387	0.387
After removing water (before equilibrium)	0.613	1.839	0.387	0
When equilibrium is re-established	0.491	1.473	0.509	0.122

vapor. This means that the methanation reaction momentarily goes in the forward direction,

$$CO(g) + 3H_2(g) \longrightarrow CH_4(g) + H_2O(g)$$

until equilibrium is re-established. Going in the forward direction, the concentrations of both water vapor and methane increase.

Cool the reaction mixture quickly to condense the water. Liquid water could be removed and the gases reheated until equilibrium is again established. The concentration of water vapor would build up as the concentration of methane increases. Table 15.1 lists the amounts of each substance at each stage of this process. Note how the yield of methane has been improved.

It is often useful to add an excess of a cheap reactant in order to force the reaction toward more products. Figure 15.4 illustrates the effect of adding or removing a reactant on chemical equilibrium.

You can look at these situations in terms of the reaction quotient, Q_c. Consider the methanation reaction, in which

$$Q_c = \frac{[CH_4]_i[H_2O]_i}{[CO]_i[H_2]_i^3}$$

Figure 15.4
Effect on a chemical equilibrium of changing the concentration of a substance. The concentration is changed by adding or removing a substance. (A) The center beaker contains an orange precipitate of HgI_2 in equilibrium with I^- (colorless) and HgI_4^{2-} (pale yellow): $HgI_2(s) + 2I^-(aq) \rightleftharpoons HgI_4^{2-}(aq)$. (B) The solution of NaI (right beaker) is added to the equilibrium mixture in the center beaker. By adding I^-, the reaction is forced to the right: $HgI_2(s) + 2I^-(aq) \longrightarrow HgI_4^{2-}(aq)$. The orange precipitate of HgI_2 disappears. (C) A solution of NaClO (left beaker) is now added to the reaction mixture; the NaClO removes I^- by oxidizing it to IO^-. By removing I^-, the reaction shifts to the left: $HgI_2(s) + 2I^-(aq) \longleftarrow HgI_4^{2-}(aq)$. An orange precipitate of HgI_2 appears.

If the reaction mixture is at equilibrium, $Q_c = K_c$. Suppose you remove some H_2O from the equilibrium mixture. Now $Q_c < K_c$, and from what we said in Section 15.5, you know the reaction will proceed in the forward direction to restore equilibrium.

EXAMPLE 15.9

Applying Le Chatelier's Principle When a Concentration Is Altered

Predict the direction of reaction when H_2 is removed from a mixture (lowering its concentration) in which the following equilibrium has been established:

$$H_2(g) + I_2(g) \rightleftharpoons 2HI(g)$$

SOLUTION

When H_2 is removed from the reaction mixture, lowering its concentration, the reaction goes in the **reverse direction** (more HI dissociates to H_2 and I_2) to partially restore the H_2 that was removed.

$$H_2(g) + I_2(g) \longleftarrow 2HI(g)$$

See Problems 15.45 and 15.46.

15.8 Changing the Pressure and Temperature

The optimum conditions for catalytic methanation involve moderately elevated temperatures and normal to moderately high pressures.

$$CO(g) + 3H_2(g) \xrightarrow[\text{1 atm–100 atm}]{\text{230°C–450°C}} CH_4(g) + H_2O(g)$$

Let us see whether we can gain insight into why these might be the optimum conditions for the reaction.

Effect of Pressure Change

A pressure change *obtained by changing the volume* can affect the yield of product in a gaseous reaction if the reaction involves a change in total moles of gas. The methanation reaction, $CO + 3H_2 \rightleftharpoons CH_4 + H_2O$, is an example of a change in moles of gas.

To see the effect of such a pressure change, consider what happens when an equilibrium mixture from the methanation reaction is compressed to one-half of its original volume at a fixed temperature (see Figure 15.5). The total pressure is doubled. Because the partial pressures and therefore the concentrations of reactants and products have changed, the mixture is no longer at equilibrium. The reaction should go in the forward direction, because then the moles of gas decrease, and the pressure (which is proportional to moles of gas) decreases. In this way, the initial pressure increase is partially reduced.

You find the same result by looking at the reaction quotient Q_c. Let [CO], [H_2], [CH_4], and [H_2O] be the molar concentrations at equilibrium for the methanation reaction. When the volume of an equilibrium mixture is halved, the partial pressures and

Figure 15.5
Effect on chemical equilibrium of changing the pressure. The effect on the methanation reaction is shown; the approximate composition is represented by the proportion of different "molecules" (see key at right). *(A)* The original equilibrium mixture of CO, H_2, CH_4, and H_2O molecules. *(B)* The gases are compressed to one-half their initial volume, increasing their partial pressures, so that the mixture is no longer at equilibrium. *(C)* Equilibrium is reestablished when the reaction goes in the forward direction: $CO + 3H_2 \longrightarrow CH_4 + H_2O$. In this way, the total number of molecules is reduced, which reduces the initial pressure increase.

Gases at equilibrium | After compression, but before equilibrium | Compressed gases at equilibrium

A   **B**   **C**

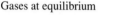

Key:
CO
H_2
CH_4
H_2O

therefore the concentrations are doubled. You obtain the reaction quotient at that moment by replacing each equilibrium concentration by double its value.

$$Q_c = \frac{(2[CH_4])(2[H_2O])}{(2[CO])(2[H_2])^3} = \frac{K_c}{4}$$

Because $Q_c < K_c$, the reaction proceeds in the forward direction.

To decide the direction of reaction when the pressure of the reaction mixture is increased (say, by decreasing the volume), you ignore liquids and solids. Consider the reaction

$$C(s) + CO_2(g) \rightleftharpoons 2CO(g)$$

The moles *of gas* increase when the reaction goes in the forward direction (1 mol CO_2 goes to 2 mol CO). Therefore, when you increase the pressure of the reaction mixture by decreasing its volume, the reaction goes in the reverse direction, as Le Chatelier's principle leads you to expect.

It is important to note that an increase or a decrease in pressure of a gaseous reaction must result in changes of partial pressures of substances in the chemical equation if it is to have an effect on the equilibrium composition. Consider the effect of increasing the pressure in the methanation reaction by adding helium gas. Although the total pressure increases, the partial pressures of CO, H_2, CH_4, and H_2O do not change. Thus, the equilibrium composition is not affected.

EXAMPLE 15.10

Applying Le Chatelier's Principle When the Pressure Is Altered

Look at each of the following equations and decide whether an increase of pressure obtained by decreasing the volume will increase, decrease, or have no effect on the amounts of products.

Table 15.2
Equilibrium Constant for Methanation at Different Temperatures

Temperature (K)	K_c
800	1.38×10^5
1000	2.54×10^2
1200	3.92

a. $CO(g) + Cl_2(g) \rightleftharpoons COCl_2(g)$

b. $2H_2S(g) \rightleftharpoons 2H_2(g) + S_2(g)$

c. $C(graphite) + S_2(g) \rightleftharpoons CS_2(g)$

SOLUTION

a. Reaction decreases the number of molecules of gas (from two to one). According to Le Chatelier's principle, an increase of pressure **increases** the amount of product.

b. Reaction increases the number of molecules of gas (from two to three); hence, an increase of pressure **decreases** the amounts of products.

c. Reaction does not change the number of molecules of gas. (You ignore the change in volume due to consumption of solid carbon, because the change in volume of a solid is insignificant. Look only at gas volumes when deciding the effect of pressure change on equilibrium composition.) Pressure change has **no effect**.

See Problems 15.47 and 15.48.

Figure 15.6
Effect on chemical equilibrium of changing the temperature.
The reaction shown is endothermic:
$Co(H_2O)_6{}^{2+}(aq) + 4Cl^-(aq)$
$\rightleftharpoons CoCl_4{}^{2-}(aq) + 6H_2O(l)$.
At room temperature, the equilibrium mixture is blue, from $CoCl_4{}^{2-}$. When cooled by the ice bath, the equilibrium mixture turns pink, from $Co(H_2O)_6{}^{2+}$.

▶ The quantitative effect of temperature on the equilibrium constant is discussed in Chapter 19.

Effect of Temperature Change

Temperature has a profound effect on most reactions. In the first place, reaction rates usually increase with an increase in temperature, meaning that equilibrium is reached sooner. Many gaseous reactions have imperceptible rates at room temperature but speed up enough at higher temperature to become commercially feasible processes.

Second, equilibrium constants vary with temperature. Table 15.2 gives values of K_c for methanation at various temperatures.

Whether you should raise or lower the temperature of a reaction mixture to increase the equilibrium amount of product can be shown by Le Chatelier's principle. Again, consider the methanation reaction,

$$CO(g) + 3H_2(g) \rightleftharpoons CH_4(g) + H_2O(g); \Delta H° = -206.2 \text{ kJ}$$

As products are formed, considerable heat is released. According to Le Chatelier's principle, as the temperature is raised, the reaction shifts to form more reactants, thereby absorbing heat and attempting to counter the increase in temperature.

$$CO(g) + 3H_2(g) \longleftarrow CH_4(g) + H_2O(g) + \text{heat}$$

You predict that the equilibrium constant will be smaller for higher temperatures, in agreement with the values of K_c given in Table 15.2. Figure 15.6 illustrates the effect of temperature on another chemical equilibrium. ◀

We conclude from Le Chatelier's principle that:

For an endothermic reaction ($\Delta H°$ positive), the amounts of products are increased at equilibrium by an increase in temperature (K_c is larger at higher T). For an exothermic reaction ($\Delta H°$ negative), the amounts of products are increased at equilibrium by a decrease in temperature (K_c is larger at lower T).

EXAMPLE 15.11

Applying Le Chatelier's Principle When the Temperature Is Altered

Carbon monoxide is formed when carbon dioxide reacts with carbon.

$$CO_2(g) + C(\text{graphite}) \rightleftharpoons 2CO(g); \Delta H° = 172.5 \text{ kJ}$$

Is a high or low temperature more favorable to the formation of carbon monoxide?

SOLUTION

The reaction absorbs heat in the forward direction.

$$\text{Heat} + CO_2(g) + C(\text{graphite}) \longrightarrow 2CO(g)$$

As the temperature is raised, reaction occurs in the forward direction, using heat and thereby tending to lower the temperature. Thus, **high temperature** is more favorable to the formation of carbon monoxide. This is one reason why combustions of carbon and organic materials can produce significant amounts of carbon monoxide.

See Problems 15.49 and 15.50.

Choosing the Optimum Conditions for Reaction

You are now in a position to understand the optimum conditions for the methanation reaction. Because the reaction is exothermic, low temperatures should favor high yields of methane; that is, the equilibrium constant is large for low temperature. But gaseous reactions are often very slow at room temperature. In practice, the methanation reaction is run at moderately elevated temperatures (230°C–450°C) in the presence of a nickel catalyst. Under these conditions, the rate of reaction is sufficiently fast but the equilibrium constant is not too small. Because the methanation reaction involves a decrease in moles of gas, the yield of methane should increase as the pressure increases. However, the equilibrium constant is large at the usual operating temperatures, so very high pressures are not needed to obtain economical yields of methane. Pressures of 1 atm to 100 atm are usual for this reaction.

As another example, consider the Haber process for the synthesis of ammonia:

$$N_2(g) + 3H_2(g) \xrightleftharpoons{\text{Fe catalyst}} 2NH_3(g); \Delta H° = -91.8 \text{ kJ}$$

▶ This technological problem has stimulated a great deal of basic research into understanding how certain bacteria "fix" nitrogen at atmospheric pressure to make NH_3. An enzyme called nitrogenase is responsible for N_2 fixation in these bacteria. This enzyme contains Fe and Mo, which may play a catalytic role.

Because the reaction is exothermic, the equilibrium constant is larger for lower temperatures. But the reaction proceeds too slowly at low temperatures to be practical, even in the presence of the best available catalysts. ◀ The optimum choice of temperature, found experimentally to be about 450°C, is a compromise between an increased rate of reaction at higher temperature and an increased yield of ammonia at lower temperature. Because the formation of ammonia decreases the moles of gases, the yield of product is improved by high pressures. Ammonia from the Haber reactor is removed from the reaction mixture by cooling the compressed gases until NH_3 liquefies. Unreacted N_2 and H_2 circulate back to the reactor.

15.9 Effect of a Catalyst

A *catalyst* is a substance that increases the rate of a reaction but is not consumed by it. The significance of a catalyst can be seen in the reaction of sulfur dioxide with oxygen to give sulfur trioxide.

$$2SO_2(g) + O_2(g) \rightleftharpoons 2SO_3(g)$$

The equilibrium constant K_c for this reaction is 1.7×10^{26}, which indicates that the reaction should go almost completely to products. Yet when sulfur is burned, it forms predominantly SO_2 and very little SO_3. Oxidation of SO_2 to SO_3 is simply too slow. However, the rate of the reaction is appreciable in the presence of a platinum or divanadium pentoxide catalyst. The oxidation of SO_2 in the presence of a catalyst is the main step in the *contact process* for the industrial production of sulfuric acid. Sulfur trioxide reacts with water to form sulfuric acid. ◀

▶ Sulfur dioxide from the combustion of coal and from other sources appears to be a major cause of the marked increase in acidity of rain in the eastern United States in the past few decades. This *acid rain* has been shown to contain sulfuric and nitric acids. The SO_2 is oxidized in moist, polluted air to H_2SO_4. Acid rain is discussed in the essay on acid rain at the end of Section 17.2.

It is important to understand that *a catalyst has no effect on the equilibrium composition of a reaction mixture. A catalyst merely speeds up the attainment of equilibrium.* For example, suppose you mix 2.00 mol SO_2 and 1.00 mol O_2 in a 100.0-L vessel. In the absence of a catalyst, these substances appear unreactive. Much later, if you analyze the mixture, you find essentially the same amounts of SO_2 and O_2. But when a catalyst is added, the rates of both forward and reverse reactions are very much increased. As a result, the reaction mixture comes to equilibrium in a short time. The amounts of SO_2, O_2, and SO_3 can be calculated from the equilibrium constant. You find that the mixture is mostly SO_3 (2.00 mol), with only 1.7×10^{-8} mol SO_2 and 8.4×10^{-9} mol O_2.

A catalyst is useful for a reaction, such as $2SO_2 + O_2 \rightleftharpoons 2SO_3$, that is normally slow but has a large equilibrium constant. However, if the reaction has an exceedingly small equilibrium constant, a catalyst is of little help. The reaction

$$N_2(g) + O_2(g) \rightleftharpoons 2NO(g)$$

has been considered for the industrial production of nitric acid. (NO reacts with O_2 and H_2O to give nitric acid.) At 25°C, however, the equilibrium constant K_c equals 4.6×10^{-31}. An equilibrium mixture would contain an extremely small concentration of NO. A catalyst would not provide a significant yield at this temperature; a catalyst merely speeds up the attainment of equilibrium. The equilibrium constant increases as the temperature is raised, so that at 2000°C, air (a mixture of N_2 and O_2) forms about 0.4% NO at equilibrium.

The Ostwald process for making nitric acid presents an interesting example of the effect of a catalyst in determining a product when several possibilities exist. Two reactions of ammonia with oxygen are possible. The reaction used in the Ostwald process is

$$4NH_3(g) + 5O_2(g) \rightleftharpoons 4NO(g) + 6H_2O(g)$$

Ammonia can also be made to burn in oxygen, and N_2 and H_2O are the products.

$$4NH_3(g) + 3O_2(g) \rightleftharpoons 2N_2(g) + 6H_2O(g)$$

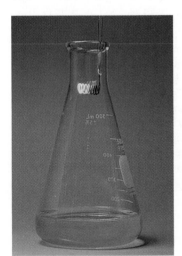

Figure 15.7
Oxidation of ammonia using a copper catalyst.
The products (formed on the glowing copper coil) are N_2 and H_2O, whereas a platinum catalyst results in NO and H_2O as products.

The reaction occurs most readily in the presence of a copper catalyst; see Figure 15.7. However, what Ostwald discovered was that the first reaction, to form NO from NH_3 and O_2, is catalyzed by platinum. Therefore, by using this catalyst at moderate temperatures, nitric oxide can be selectively formed. Many other examples could be cited in which the products of given reactants depend on the catalyst.

A Checklist for Review

Important Terms

chemical equilibrium (15.1)
equilibrium-constant
    expression (15.2)
equilibrium constant (15.2)

law of mass action (15.2)
homogeneous
    equilibrium (15.3)

heterogeneous
    equilibrium (15.3)
reaction quotient (15.5)

Le Chatelier's
    principle (15.7)

Key Equations

$$K_c = \frac{[C]^c[D]^d}{[A]^a[B]^b}$$

Summary of Facts and Concepts

Chemical equilibrium can be characterized by the *equilibrium constant K_c*. In the expression for K_c, the concentration of products is in the numerator and the concentration of reactants is in the denominator. Pure liquids and solids are ignored in writing the equilibrium-constant expression. When K_c is very large, the equilibrium mixture is mostly products, and when K_c is very small, the equilibrium mixture is mostly reactants. The reaction quotient Q_c takes the form of the equilibrium-constant expression. If you substitute the concentrations of substances in a reaction mixture into Q_c, you can predict the direction the reaction must go to attain equilibrium. You can use K_c to calculate the composition of the reaction mixture at equilibrium, starting from various initial compositions.

The *choice of conditions,* including *catalysts,* can be very important to the success of a reaction. Removing a product from the reaction mixture, for example, shifts the equilibrium composition to give more product. Changing the pressure and temperature can also affect the product yield. *Le Chatelier's principle* is useful in predicting the effect of such changes.

Operational Skills

1. **Applying stoichiometry to an equilibrium mixture** Given the starting amounts of reactants and the amount of one substance at equilibrium, find the equilibrium composition. **(EXAMPLE 15.1)**
2. **Writing equilibrium-constant expressions** Given the chemical equation, write the equilibrium-constant expression. **(EXAMPLES 15.2, 15.4)**
3. **Obtaining an equilibrium constant from reaction composition** Given the equilibrium composition, find K_c. **(EXAMPLE 15.3)**
4. **Using the reaction quotient** Given the concentrations of substances in a reaction mixture, predict the direction of reaction. **(EXAMPLE 15.5)**
5. **Obtaining one equilibrium concentration given the others** Given K_c and all concentrations of substances but one in an equilibrium mixture, calculate the concentration of this one substance. **(EXAMPLE 15.6)**
6. **Solving equilibrium problems** Given the starting composition and K_c of a reaction mixture, calculate the equilibrium composition. **(EXAMPLES 15.7, 15.8)**
7. **Applying Le Chatelier's principle** Given a reaction, use Le Chatelier's principle to decide the effect of adding or removing a substance **(EXAMPLE 15.9)**, changing the pressure **(EXAMPLE 15.10)**, or changing the temperature **(EXAMPLE 15.11)**.

Review Questions

15.1 Consider the reaction $N_2O_4(g) \rightleftharpoons 2NO_2(g)$. Draw a graph illustrating the changes of concentrations of N_2O_4 and NO_2 as equilibrium is approached. Describe how the rates of the forward and reverse reactions change as the mixture approaches dynamic equilibrium. Why is this called a *dynamic* equilibrium?

15.2 When 1.0 mol each of $H_2(g)$ and $I_2(g)$ are mixed at a certain high temperature, they react to give a final mixture consisting of 0.5 mol each of $H_2(g)$ and $I_2(g)$ and 1.0 mol $HI(g)$. Why do you obtain the same final mixture when you bring 2.0 mol $HI(g)$ to the same temperature?

15.3 Obtain the equilibrium constant for the reaction

$$HCN(aq) \rightleftharpoons H^+(aq) + CN^-(aq)$$

from the following:

$$HCN(aq) + OH^-(aq) \rightleftharpoons CN^-(aq) + H_2O(l);$$
$$K_1 = 4.9 \times 10^4$$
$$H_2O(l) \rightleftharpoons H^+(aq) + OH^-(aq); K_2 = 1.0 \times 10^{-14}$$

15.4 Which of the following reactions involve homogeneous equilibria and which involve heterogeneous equilibria? Explain the difference.
a. $2NO(g) + O_2(g) \rightleftharpoons 2NO_2(g)$
b. $2Cu(NO_3)_2(s) \rightleftharpoons 2CuO(s) + 4NO_2(g) + O_2(g)$
c. $2N_2O(g) \rightleftharpoons 2N_2(g) + O_2(g)$
d. $2NH_3(g) + 3CuO(s) \rightleftharpoons 3H_2O(g) + N_2(g) + 3Cu(s)$

Conceptual Problems

15.9 During an experiment with the Haber process, a researcher put 1 mol N_2 and 1 mol H_2 into a reaction vessel to observe the equilibrium formation of ammonia, NH_3.

$$N_2(g) + 3H_2(g) \rightleftharpoons 2NH_3(g)$$

When these reactants come to equilibrium, assume that x mol H_2 react. How many moles of ammonia form?

15.10 Suppose liquid water and water vapor exist in equilibrium in a closed container. If you add a small amount of liquid water to the container, how does this affect the amount of water vapor in the container? If, instead, you add a small amount of water vapor to the container, how does this affect the amount of liquid water in the container?

15.11 A mixture initially consisting of 2 mol CO and 2 mol H_2 comes to equilibrium with methanol, CH_3OH, as the product:

$$CO(g) + 2H_2(g) \rightleftharpoons CH_3OH(g)$$

At equilibrium, the mixture will contain which of the following?
a. less than 1 mol CH_3OH
b. 1 mol CH_3OH
c. more than 1 mol CH_3OH but less than 2 mol
d. 2 mol CH_3OH
e. more than 2 mol CH_3OH

15.5 Explain why pure liquids and solids can be ignored when writing the equilibrium-constant expression.

15.6 What qualitative information can you get from the magnitude of the equilibrium constant?

15.7 How is it possible for a catalyst to give products from a reaction mixture that are different from those obtained when no catalyst or a different catalyst is used? Give an example.

15.8 List four ways in which the yield of ammonia in the reaction

$$N_2(g) + 3H_2(g) \rightleftharpoons 2NH_3(g); \Delta H° < 0$$

can be improved for a given amount of H_2. Explain the principle behind each way.

15.12 The following reaction is carried out at 500 K in a container equipped with a movable piston.

$$A(g) + B(g) \rightleftharpoons C(g); \quad K_c = 10 \text{ (at 500 K)}$$

After the reaction has reached equilibrium, the container has the composition depicted here.

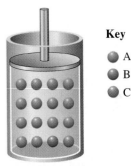

Key
- A
- B
- C

Suppose the container volume is doubled.
a. How does the equilibrium composition shift?
b. How does the concentration of each of the reactants and the product change? (That is, does the concentration increase, decrease, or stay the same?)

15.13   An experimenter introduces 4.0 mol of gas A into a 1.0-L container at 200 K to form product B according to the reaction

$$2A(g) \rightleftharpoons B(g)$$

Using the experimenter's data (one curve is for A, the other is for B), calculate the equilibrium constant at 200 K.

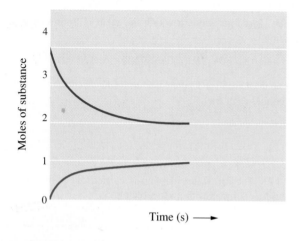

15.14   At some temperature, a 100-L reaction vessel contains a mixture that is initially 1.00 mol CO and 2.00 mol H_2. The vessel also contains a catalyst so that the following equilibrium is attained:

$$CO(g) + 2H_2(g) \rightleftharpoons CH_3OH(g)$$

At equilibrium, the mixture contains 0.100 mol CH_3OH. In a later experiment in the same vessel, you start with 1.00 mol CH_3OH. How much methanol is there at equilibrium? Explain.

Practice Problems

Reaction Stoichiometry

15.15   You place 1.50 mol of dinitrogen trioxide, N_2O_3, into a flask, where it decomposes at 25.0°C and 1.00 atm:

$$N_2O_3(g) \rightleftharpoons NO_2(g) + NO(g)$$

What is the composition of the reaction mixture at equilibrium if it contains 0.45 mol of nitrogen dioxide, NO_2?

15.16   A 1.250-mol sample of phosphorus pentachloride, PCl_5, dissociates at 160°C and 1.00 atm to give 0.169 mol of phosphorus trichloride, PCl_3, at equilibrium.

$$PCl_5(g) \rightleftharpoons PCl_3(g) + Cl_2(g)$$

What is the composition of the final reaction mixture?

15.17   Nitric oxide, NO, reacts with bromine, Br_2, to give nitrosyl bromide, NOBr.

$$2NO(g) + Br_2(g) \rightleftharpoons 2NOBr(g)$$

A sample of 0.0655 mol NO with 0.0328 mol Br_2 gives an equilibrium mixture containing 0.0389 mol NOBr. What is the composition of the equilibrium mixture?

15.18   You place 0.750 mol of nitrogen, N_2, and 2.250 mol of hydrogen, H_2, into a reaction vessel at 450°C and 10.0 atm. The reaction is

$$N_2(g) + 3H_2(g) \rightleftharpoons 2NH_3(g)$$

What is the composition of the equilibrium mixture if you obtain 0.060 mol of ammonia, NH_3, from it?

The Equilibrium Constant and its Evaluation

15.19   Write equilibrium-constant expressions K_c for each of the following reactions.
a. $N_2O_3(g) \rightleftharpoons NO_2(g) + NO(g)$
b. $2H_2S(g) \rightleftharpoons 2H_2(g) + S_2(g)$
c. $2NO(g) + O_2(g) \rightleftharpoons 2NO_2(g)$

15.20   Write equilibrium-constant expressions K_c for each of the following reactions.
a. $N_2H_4(g) \rightleftharpoons N_2(g) + 2H_2(g)$
b. $2NOCl(g) \rightleftharpoons 2NO(g) + Cl_2(g)$
c. $2NO(g) + 2H_2(g) \rightleftharpoons N_2(g) + 2H_2O(g)$

15.21   The equilibrium-constant expression for a reaction is

$$K_c = \frac{[NO_2]^4[O_2]}{[N_2O_5]^2}$$

What is the equilibrium-constant expression when the equation for this reaction is halved and then reversed?

15.22 The equilibrium-constant expression for a reaction is

$$K_c = \frac{[NH_3]^4[O_2]^5}{[NO]^4[H_2O]^6}$$

What is the equilibrium-constant expression when the equation for this reaction is halved and then reversed?

15.23 A 5.00-L vessel contained 0.0185 mol of phosphorus trichloride, 0.0158 mol of phosphorus pentachloride, and 0.0870 mol of chlorine at 230°C in an equilibrium mixture. Calculate the value of K_c for the reaction

$$PCl_3(g) + Cl_2(g) \rightleftharpoons PCl_5(g)$$

15.24 An 8.00-L reaction vessel at 491°C contained 0.650 mol H_2, 0.275 mol I_2, and 3.000 mol HI. Assuming that the substances are at equilibrium, find the value of K_c at 491°C for the reaction of hydrogen and iodine to give hydrogen iodide. The equation is

$$H_2(g) + I_2(g) \rightleftharpoons 2HI(g)$$

15.25 If a 1.50-L vessel contains 0.119 mol CO, 0.237 mol H_2 and 0.0313 mol CH_3OH, what is the value of K_c for the following reaction at 500 K:

$$CO(g) + 2H_2(g) \rightleftharpoons CH_3OH(g)$$

15.26 If a 2.00-L vessel contains 0.0104 mol SO_2, 0.0052 mol O_2 and 0.0296 mol SO_3, what is the value of K_c for the following reaction at 900 K:

$$2SO_2(g) + O_2(g) \rightleftharpoons 2SO_3(g)$$

z15.27 At 77°C, 2.00 mol of nitrosyl bromide, NOBr, placed in a 1.00-L flask dissociates to the extent of 9.4%; that is, for each mole of NOBr before reaction, $(1.000 - 0.094)$ mol NOBr remains after dissociation. Calculate the value of K_c for the dissociation reaction

$$2NOBr(g) \rightleftharpoons 2NO(g) + Br_2(g)$$

15.28 A 2.00-mol sample of nitrogen dioxide was placed in an 80.0-L vessel. At 200°C, the nitrogen dioxide was 6.0% decomposed according to the equation

$$2NO_2(g) \rightleftharpoons 2NO(g) + O_2(g)$$

Calculate the value of K_c for this reaction at 200°C. (See Problem 15.27.)

15.29 Write equilibrium-constant expressions K_p for each of the following reactions:
a. $H_2(g) + Br_2(g) \rightleftharpoons 2HBr(g)$
b. $CS_2(g) + 4H_2(g) \rightleftharpoons CH_4(g) + 2H_2S(g)$
c. $4HCl(g) + O_2(g) \rightleftharpoons 2H_2O(g) + 2Cl_2(g)$

15.30 Write equilibrium-constant expressions K_p for each of the following reactions:
a. $N_2O_4(g) \rightleftharpoons 2NO_2(g)$
b. $2NO(g) + Br_2(g) \rightleftharpoons 2NOBr(g)$
c. $2SO_2(g) + O_2(g) \rightleftharpoons 2SO_3(g)$

15.31 The reaction

$$SO_2(g) + \tfrac{1}{2}O_2(g) \rightleftharpoons SO_3(g)$$

has K_p equal to 6.55 at 627°C. What is the value of K_c at this temperature?

15.32 Fluorine, F_2, dissociates into atoms on heating.

$$\tfrac{1}{2}F_2(g) \rightleftharpoons F(g)$$

The value of K_p at 842°C is 7.55×10^{-2}. What is the value of K_c at this temperature?

15.33 Write the expression for the equilibrium constant K_c for each of the following equations:
a. $C(s) + CO_2(g) \rightleftharpoons 2CO(g)$
b. $FeO(s) + CO(g) \rightleftharpoons Fe(s) + CO_2(g)$
c. $Na_2CO_3(s) + SO_2(g) + \tfrac{1}{2}O_2(g) \rightleftharpoons Na_2SO_4(s) + CO_2(g)$

15.34 For each of the following equations, give the expression for the equilibrium constant K_c:
a. $NH_4Cl(s) \rightleftharpoons NH_3(g) + HCl(g)$
b. $C(s) + 2N_2O(g) \rightleftharpoons CO_2(g) + 2N_2(g)$
c. $2NaHCO_3(s) \rightleftharpoons Na_2CO_3(s) + H_2O(g) + CO_2(g)$

Using the Equilibrium Constant

15.35 On the basis of the value of K_c, decide whether or not you expect nearly complete reaction at equilibrium for each of the following:
a. $N_2(g) + O_2(g) \rightleftharpoons 2NO(g)$; $K_c = 4.6 \times 10^{-31}$
b. $C_2H_4(g) + H_2(g) \rightleftharpoons C_2H_6(g)$; $K_c = 1.3 \times 10^{21}$

15.36 Would either of the following reactions go almost completely to product at equilibrium?
a. $2NO(g) + 2H_2(g) \rightleftharpoons N_2(g) + 2H_2O(g)$; $K_c = 6.5 \times 10^{113}$
b. $COCl_2(g) \rightleftharpoons CO(g) + Cl_2(g)$; $K_c = 3.6 \times 10^{-16}$

15.37 The following reaction has an equilibrium constant K_c equal to 3.07×10^{-4} at 24°C.

$$2NOBr(g) \rightleftharpoons 2NO(g) + Br_2(g)$$

For each of the following compositions, decide whether the reaction mixture is at equilibrium. If it is not, decide which direction the reaction should go.
a. $[NOBr] = 0.0720\ M$, $[NO] = 0.0162\ M$, $[Br_2] = 0.0123\ M$
b. $[NOBr] = 0.121\ M$, $[NO] = 0.0159\ M$, $[Br_2] = 0.0139\ M$
c. $[NOBr] = 0.103\ M$, $[NO] = 0.0134\ M$, $[Br_2] = 0.0181\ M$

15.38   The following reaction has an equilibrium constant K_c equal to 3.59 at 900°C.

$$CH_4(g) + 2H_2S(g) \rightleftharpoons CS_2(g) + 4H_2(g)$$

For each of the following compositions, decide whether the reaction mixture is at equilibrium. If it is not, decide which direction the reaction should go.
a. $[CH_4] = 1.26\ M$, $[H_2S] = 1.32\ M$,
   $[CS_2] = 1.43\ M$, $[H_2] = 1.12\ M$
b. $[CH_4] = 1.25\ M$, $[H_2S] = 1.52\ M$,
   $[CS_2] = 1.15\ M$, $[H_2] = 1.73\ M$
c. $[CH_4] = 1.20\ M$, $[H_2S] = 1.31\ M$,
   $[CS_2] = 1.15\ M$, $[H_2] = 1.85\ M$
d. $[CH_4] = 1.56\ M$, $[H_2S] = 1.43\ M$,
   $[CS_2] = 1.23\ M$, $[H_2] = 1.91\ M$

15.39   Methanol, CH_3OH, is manufactured industrially by the reaction

$$CO(g) + 2H_2(g) \rightleftharpoons CH_3OH(g)$$

A gaseous mixture at 500 K is 0.020 M CH_3OH, 0.10 M CO, and 0.10 M H_2. What will be the direction of reaction if this mixture goes to equilibrium? The equilibrium constant K_c equals 10.5 at 500 K.

15.40   Sulfur trioxide, used to manufacture sulfuric acid, is obtained commercially from sulfur dioxide.

$$2SO_2(g) + O_2(g) \rightleftharpoons 2SO_3(g)$$

The equilibrium constant K_c for this reaction is 4.17×10^{-2} at 727°C. What is the direction of reaction when a mixture that is 0.20 M SO_2, 0.10 M O_2, and 0.40 M SO_3 approaches equilibrium?

15.41   Iodine and bromine react to give iodine monobromide, IBr.

$$I_2(g) + Br_2(g) \rightleftharpoons 2IBr(g)$$

What is the equilibrium composition of a mixture at 150°C that initially contained 0.0015 mol each of iodine and bromine in a 5.0-L vessel? The equilibrium constant K_c for this reaction at 150°C is 1.2×10^2.

15.42   Initially a mixture contains 0.850 mol each of N_2 and O_2 in an 8.00-L vessel. Find the composition of the mixture when equilibrium is reached at 3900°C. The reaction is

$$N_2(g) + O_2(g) \rightleftharpoons 2NO(g)$$

and $K_c = 0.0123$ at 3900°C.

15.43   The equilibrium constant K_c for the reaction

$$PCl_3(g) + Cl_2(g) \rightleftharpoons PCl_5(g)$$

equals 49 at 230°C. If 0.500 mol each of phosphorus trichloride and chlorine are added to a 5.0-L reaction vessel, what is the equilibrium composition of the mixture at 230°C?

15.44   Calculate the composition of the gaseous mixture obtained when 1.00 mol of carbon dioxide is exposed to hot carbon at 800°C in a 1.00-L vessel. The equilibrium constant K_c at 800°C is 14.0 for the reaction

$$CO_2(g) + C(s) \rightleftharpoons 2CO(g)$$

Le Chatelier's Principle

15.45   Consider the equilibrium

$$FeO(s) + CO(g) \rightleftharpoons Fe(s) + CO_2(g)$$

When carbon dioxide is removed from the equilibrium mixture (say, by passing the gases through water to absorb CO_2), what is the direction of net reaction as the new equilibrium is achieved?

15.46   (a) Predict the direction of reaction when chlorine gas is added to an equilibrium mixture of PCl_3, PCl_5, and Cl_2. The reaction is

$$PCl_3(g) + Cl_2(g) \rightleftharpoons PCl_5(g)$$

(b) What is the direction of reaction when chlorine gas is removed from an equilibrium mixture of these gases?

15.47   What would you expect to be the effect of an increase of pressure on each of the following reactions? Would the pressure change cause reaction to go to the right or left?
a. $CH_4(g) + 2S_2(g) \rightleftharpoons CS_2(g) + 2H_2S(g)$
b. $H_2(g) + Br_2(g) \rightleftharpoons 2HBr(g)$
c. $CO_2(g) + C(s) \rightleftharpoons 2CO(g)$

15.48   Indicate whether either an increase or a decrease of pressure obtained by changing the volume would increase the amount of product in each of the following reactions.
a. $CO(g) + 2H_2(g) \rightleftharpoons CH_3OH(g)$
b. $2SO_2(g) + O_2(g) \rightleftharpoons 2SO_3(g)$
c. $N_2O_4(g) \rightleftharpoons 2NO_2(g)$

15.49   Methanol is prepared industrially from synthesis gas (CO and H_2).

$$CO(g) + 2H_2(g) \rightleftharpoons CH_3OH(g); \Delta H° = -21.7\ kcal$$

Would the fraction of methanol obtained at equilibrium be increased by raising the temperature? Explain.

15.50   One way of preparing hydrogen is by the decomposition of water.

$$2H_2O(g) \rightleftharpoons 2H_2(g) + O_2(g); \Delta H° = 484\ kJ$$

Would you expect the decomposition to be favorable at high or low temperature? Explain.

15.51 Use thermochemical data (Appendix B) to decide whether the equilibrium constant for the following reaction will increase or decrease with temperature.

$$2NO_2(g) + 7H_2(g) \rightleftharpoons 2NH_3(g) + 4H_2O(g)$$

General Problems

15.53 A mixture of carbon monoxide, hydrogen, and methanol, CH_3OH, is at equilibrium according to the equation

$$CO(g) + 2H_2(g) \rightleftharpoons CH_3OH(g)$$

At 250°C, the mixture is 0.096 M CO, 0.191 M H_2, and 0.015 M CH_3OH. What is K_c for this reaction at 250°C?

15.54 An equilibrium mixture of SO_3, SO_2, and O_2 at 727°C is 0.0160 M SO_3, 0.0056 M SO_2, and 0.0021 M O_2. What is the value of K_c for the following reaction?

$$SO_2(g) + \tfrac{1}{2}O_2(g) \rightleftharpoons SO_3(g)$$

15.55 A 2.00-L vessel contains 1.00 mol N_2, 1.00 mol H_2, and 2.00 mol NH_3. What is the direction of reaction (forward or reverse) needed to attain equilibrium at 400°C? The equilibrium constant K_c for the reaction

$$N_2(g) + 3H_2(g) \rightleftharpoons 2NH_3(g)$$

is 0.51 at 400°C.

15.56 A vessel originally contained 0.200 mol iodine monobromide (IBr), 0.0010 mol I_2, and 0.0010 mol Br_2. The equilibrium constant K_c for the reaction

$$I_2(g) + Br_2(g) \rightleftharpoons 2IBr(g)$$

is 1.2×10^2 at 150°C. What is the direction (forward or reverse) needed to attain equilibrium at 150°C?

15.57 Hydrogen bromide dissociates when heated according to the equation

$$2HBr(g) \rightleftharpoons H_2(g) + Br_2(g)$$

The equilibrium constant K_c equals 1.6×10^{-2} at 200°C. What are the moles of substances in the equilibrium mixture at 200°C if we start with 0.010 mol HBr in a 1.0-L vessel?

15.58 Iodine monobromide, IBr, occurs as brownish-black crystals that vaporize with decomposition:

$$2IBr(g) \rightleftharpoons I_2(g) + Br_2(g)$$

The equilibrium constant K_c at 100°C is 0.026. If 0.010 mol IBr is placed in a 1.0-L vessel at 100°C, what are the moles of substances at equilibrium in the vapor?

15.52 Use thermochemical data (Appendix B) to decide whether the equilibrium constant for the following reaction will increase or decrease with temperature.

$$CH_4(g) + 2H_2S(g) \rightleftharpoons CS_2(g) + 4H_2(g)$$

15.59 Phosgene, $COCl_2$, is a toxic gas used in the manufacture of urethane plastics. The gas dissociates at high temperature.

$$COCl_2(g) \rightleftharpoons CO(g) + Cl_2(g)$$

At 400°C, the equilibrium constant K_c is 8.05×10^{-4}. Find the percentage of phosgene that dissociates at this temperature when 1.00 mol of phosgene is placed in a 25.0-L vessel.

15.60 Dinitrogen tetroxide, N_2O_4, is a colorless gas (boiling point, 21°C), which dissociates to give nitrogen dioxide, NO_2, a reddish-brown gas.

$$N_2O_4(g) \rightleftharpoons 2NO_2(g)$$

The equilibrium constant K_c at 25°C is 0.125. What percentage of dinitrogen tetroxide is dissociated when 0.0300 mol N_2O_4 is placed in a 1.00-L flask at 25°C?

15.61 The amount of nitrogen dioxide formed by dissociation of dinitrogen tetroxide,

$$N_2O_4(g) \rightleftharpoons 2NO_2(g)$$

increases as the temperature rises. Is the dissociation of N_2O_4 endothermic or exothermic?

15.62 The equilibrium constant K_c for the synthesis of methanol, CH_3OH,

$$CO(g) + 2H_2(g) \rightleftharpoons CH_3OH(g)$$

is 4.3 at 250°C and 1.8 at 275°C. Is this reaction endothermic or exothermic?

15.63 For the reaction

$$N_2(g) + 3H_2(g) \rightleftharpoons 2NH_3(g)$$

show that $\qquad$ $K_c = K_p(RT)^2$

Do not use the formula $K_p = K_c(RT)^{\Delta n}$ given in the text. Start from the fact that $P_i = [i]RT$, where P_i is the partial pressure of substance i and $[i]$ is its molar concentration. Substitute into K_c.

15.64 For the reaction

$$COCl_2(g) \rightleftharpoons CO(g) + Cl_2(g)$$

show that $\qquad$ $K_c = K_p/(RT)$

Do not use the formula $K_p = K_c(RT)^{\Delta n}$ given in the text. See Problem 15.63.

15.65 The equilibrium constant K_c for the reaction

$$PCl_3(g) + Cl_2(g) \rightleftharpoons PCl_5(g)$$

equals 4.1 at 300°C.
a. A sample of 35.8 g of PCl_5 is placed in a 5.0-L reaction vessel and heated to 300°C. What are the equilibrium concentrations of all of the species?
b. What fraction of PCl_5 has decomposed?
c. If 35.8 g of PCl_5 were placed in a 1.0-L vessel, what qualitative effect would this have on the fraction of PCl_5 that has decomposed (give a qualitative answer only; do not do the calculation)? Why?

15.66 At 25°C in a closed system, ammonium hydrogen sulfide exists as the following equilibrium:

$$NH_4HS(s) \rightleftharpoons NH_3(g) + H_2S(g)$$

a. When a sample of pure $NH_4HS(s)$ is placed in an evacuated reaction vessel and allowed to come to equilibrium at 25°C, total pressure is 0.660 atm. What is the value of K_p?
b. To this system, sufficient $H_2S(g)$ is injected until the pressure of H_2S is three times that of the ammonia at equilibrium. What are the partial pressures of NH_3 and H_2S?
c. In a different experiment, 0.750 atm of NH_3 and 0.500 atm of H_2S are introduced into a 1.00-L vessel at 25°C. How many moles of NH_4HS are present when equilibrium is established?

15.67 At moderately high temperatures, $SbCl_5$ decomposes into $SbCl_3$ and Cl_2 as follows:

$$SbCl_5(g) \rightleftharpoons SbCl_3(g) + Cl_2(g)$$

A 65.4-g sample of $SbCl_5$ is placed in an evacuated 5.00-L vessel and it is raised to 195°C and the system comes to equilibrium. If at this temperature 35.8% of the $SbCl_5$ is decomposed, what is the value of K_p?

15.68 The following reaction is important in the manufacture of sulfuric acid.

$$SO_2(g) + \tfrac{1}{2}O_2(g) \rightleftharpoons SO_3(g)$$

At 900 K, 0.0216 mol of SO_2 and 0.0148 mol of O_2 are sealed in a 1.00-L reaction vessel. When equilibrium is reached, the concentration of SO_3 is determined to be 0.0175 M. Calculate K_c for this reaction.

15.69 Sulfuryl chloride is used in organic chemistry as a chlorinating agent. At moderately high temperatures it decomposes as follows:

$$SO_2Cl_2(g) \rightleftharpoons SO_2(g) + Cl_2(g)$$

with $K_c = 0.045$ at 650 K. *(continued in next column.)*

Problem 15.69 continued.
a. A sample of 8.25 g of SO_2Cl_2 is placed in a 1.00-L reaction vessel and heated to 650 K. What are the equilibrium concentrations of all of the species?
b. What fraction of SO_2Cl_2 has decomposed?
c. If 5 g of chlorine is inserted into the reaction vessel, what qualitative effect would this have on the fraction of SO_2Cl_2 that has decomposed?

15.70 Phosgene was used as a poisonous gas in World War I. At high temperatures it decomposes as follows:

$$COCl_2(g) \rightleftharpoons CO(g) + Cl_2(g)$$

with $K_c = 4.6 \times 10^{-3}$ at 800 K.
a. A sample of 6.55 g of $COCl_2$ is placed in a 1.00-L reaction vessel and heated to 800 K. What are the equilibrium concentrations of all of the species?
b. What fraction of $COCl_2$ has decomposed?
c. If 3 g of carbon monoxide is inserted into the reaction vessel, what qualitative effect would this have on the fraction of $COCl_2$ that has decomposed?

15.71 Gaseous acetic acid molecules have a certain tendency to form dimers. (A *dimer* is a molecule formed by the association of two identical, simpler molecules.) The equilibrium constant K_c at 25°C for this reaction is 3.2×10^4.
a. If the initial concentration of CH_3COOH monomer (the simpler molecule) is 4.0×10^{-4} M, what are the concentrations of monomer and dimer when the system comes to equilibrium? (The simpler quadratic equation is obtained by assuming that all of the acid molecules have dimerized and then some of them dissociate to monomer.)
b. Why do acetic acid molecules dimerize? What type of structure would you draw for the dimer?
c. As the temperature increases, would you expect the percentage of dimer to increase or decrease? Why?

15.72 Gaseous acetic acid molecules have a certain tendency to form dimers. (A *dimer* is a molecule formed by the association of two identical, simpler molecules.) The equilibrium constant K_p at 25°C for this reaction is 1.3×10^3.
a. If the initial pressure of CH_3COOH monomer (the simpler molecule) is 7.5×10^{-3} atm, what are the pressures of monomer and dimer when the system comes to equilibrium? (The simpler quadratic equation is obtained by assuming that all of the acid molecules have dimerized and then some of them dissociate to monomer.)
b. Why do acetic acid molecules dimerize? What type of structure would you draw for the dimer?
c. As the temperature decreases, would you expect the percentage of dimer to increase or decrease? Why?

15.73   When 0.112 mol of NO and 18.22 g of bromine are placed in a 1.00-L reaction vessel and sealed, the mixture is heated to 350 K and the following equilibrium is established:

$$2NO(g) + Br_2(g) \rightleftharpoons 2NOBr(g)$$

If the equilibrium concentration of nitrosyl bromide is 0.0824 M, what is K_c?

15.74   When 0.0322 mol of NO and 1.52 g of bromine are placed in a 1.00-L reaction vessel and sealed, the mixture reacts and the following equilibrium is established:

$$2NO(g) + Br_2(g) \rightleftharpoons 2NOBr(g)$$

At 25°C the equilibrium pressure of nitrosyl bromide is 0.438 atm. What is K_p?

Cumulative-Skills Problems

15.75   The following equilibrium was studied by analyzing the equilibrium mixture for the amount of H_2S produced.

$$Sb_2S_3(s) + 3H_2(g) \rightleftharpoons 2Sb(s) + 3H_2S(g)$$

A vessel whose volume was 2.50 L was filled with 0.0100 mol of antimony(III) sulfide, Sb_2S_3, and 0.0100 mol H_2. After the mixture came to equilibrium in the closed vessel at 440°C, the gaseous mixture was removed, and the hydrogen sulfide was dissolved in water. Sufficient lead(II) ion was added to react completely with the H_2S to precipitate lead(II) sulfide, PbS. If 1.029 g PbS was obtained, what is the value of K_c at 440°C?

15.76   The following equilibrium was studied by analyzing the equilibrium mixture for the amount of HCl produced.

$$LaCl_3(s) + H_2O(g) \rightleftharpoons LaOCl(s) + 2HCl(g)$$

A vessel whose volume was 1.25 L was filled with 0.0125 mol of lanthanum(III) chloride, $LaCl_3$, and 0.0250 mol H_2O. After the mixture came to equilibrium in the closed vessel at 619°C, the gaseous mixture was removed and dissolved in more water. Sufficient silver(I) ion was added to precipitate the chloride ion completely as silver chloride. If 3.59 g AgCl was obtained, what is the value of K_c at 619°C?

15.77   Phosphorus(V) chloride, PCl_5, dissociates on heating to give phosphorus(III) chloride, PCl_3, and chlorine.

$$PCl_5(g) \rightleftharpoons PCl_3(g) + Cl_2(g)$$

A closed 2.00-L vessel initially contains 0.0100 mol PCl_5. What is the total pressure at 250°C when equilibrium is achieved? The value of K_c at 250°C is 4.15×10^{-2}.

15.78   Antimony(V) chloride, $SbCl_5$, dissociates on heating to give antimony(III) chloride, $SbCl_3$, and chlorine.

$$SbCl_5(g) \rightleftharpoons SbCl_3(g) + Cl_2(g)$$

A closed 3.50-L vessel initially contains 0.0125 mol $SbCl_5$. What is the total pressure at 248°C when equilibrium is achieved? The value of K_c at 248°C is 2.50×10^{-2}.

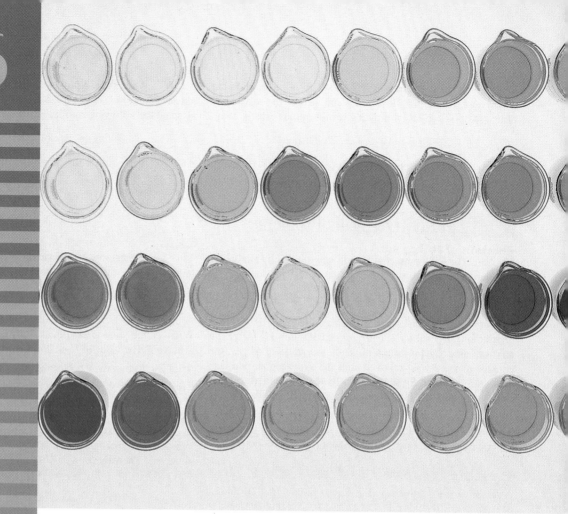

Acid–base indicator dyes in solutions of different hydronium-ion concentration.

16

Acids and Bases

Acids and bases were first recognized by simple properties, such as taste. Acids have a sour taste; bases are bitter. Also, acids and bases change the color of certain dyes called indicators, such as litmus and phenolphthalein. Acids change litmus from blue to red and basic phenolphthalein from red to colorless. Bases change litmus from red to blue and phenolphthalein from colorless to pink. Acids and bases react with each other to produce ionic substances called salts. Acids react with active metals, such as magnesium and zinc, to release hydrogen.

The Swedish chemist Svante Arrhenius framed the first successful concept of acids and bases. He defined acids and bases in terms of the effect these substances have on water. Acids are substances that increase the concentration of H^+ ion in aqueous solution, and bases increase the concentration of OH^- ion in aqueous solution. But many reactions that have characteristics of acid–base reactions in aqueous

Figure 16.1
Reaction of HCl(*g*) and NH₃(*g*) to form NH₄Cl(*s*).
Gases from the concentrated solutions diffuse from their watch glasses and react to give a smoke of ammonium chloride.

solution occur in other solvents or without a solvent. Hydrochloric acid reacts with aqueous ammonia and the reaction can be written

$$HCl(aq) + NH_3(aq) \longrightarrow NH_4Cl(aq)$$

The product is a solution of NH_4^+ and Cl^- ions. A very similar reaction occurs between hydrogen chloride and ammonia dissolved in benzene, C_6H_6. The product is again NH_4Cl, which in this case precipitates from the solution.

$$HCl(benzene) + NH_3(benzene) \longrightarrow NH_4Cl(s)$$

Hydrogen chloride and ammonia also react in the gas phase, forming dense white fumes of NH_4Cl above adjacent watch glasses of concentrated hydrochloric acid and concentrated ammonia (Figure 16.1).

$$HCl(g) + NH_3(g) \longrightarrow NH_4Cl(s)$$

These reactions of HCl and NH_3, in benzene and in the gas phase, are similar to the reaction in aqueous solution but cannot be explained by the Arrhenius concept. Broader acid–base concepts are needed.

Acid–Base Concepts

Why is a substance acidic? What are some of the characteristics of acids and bases? These questions will be explored in Sections 16.1–16.3.

16.1 Arrhenius Concept of Acids and Bases

▶ This section summarizes the discussion in Section 4.4.

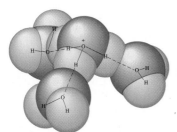

Figure 16.2
The hydronium ion, H₃O⁺.
The species is shown here hydrogen-bonded to three water molecules. The positive charge shown is actually distributed over the ion.

▶ We introduced strong and weak acids and bases in Section 4.4.

We can state the Arrhenius concept of an acid as follows: *An acid is a substance that, when dissolved in water, increases the concentration of hydronium ion, $H_3O^+(aq)$.* Chemists frequently use the notation $H^+(aq)$ for the $H_3O^+(aq)$ ion, and call it the hydrogen ion. However, the aqueous hydrogen ion is not a bare proton in water, but a proton chemically bonded to water. The $H_3O^+(aq)$ ion is itself associated through hydrogen bonding with a variable number of water molecules (Figure 16.2). *A base is a substance that, when dissolved in water, increases the concentration of hydroxide ion, $OH^-(aq)$.* ◀

The special role of the hydronium ion (or hydrogen ion) and the hydroxide ion in aqueous solutions arises from the following reaction:

$$H_2O(l) + H_2O(l) \rightleftharpoons H_3O^+(aq) + OH^-(aq)$$

The addition of acids and bases alters the concentrations of these ions in water.

In Arrhenius's theory, a *strong acid* is a substance that completely ionizes in aqueous solution to give $H_3O^+(aq)$ and an anion. An example is perchloric acid.

$$HClO_4(aq) + H_2O(l) \longrightarrow H_3O^+(aq) + ClO_4^-(aq)$$

Other examples of strong acids are H_2SO_4, HI, HBr, HCl, and HNO_3. A *strong base* (such as NaOH) completely ionizes in aqueous solution to give OH^- and a cation.

$$NaOH(s) \xrightarrow{H_2O} Na^+(aq) + OH^-(aq)$$

The principal strong bases are the hydroxides of Group IA elements and Group IIA elements (except Be). See Table 16.1. ◀

**Table 16.1
Common Strong Acids
and Bases**

Strong Acids	Strong Bases*
$HClO_4$	$LiOH$
H_2SO_4	$NaOH$
HI	KOH
HBr	$Ca(OH)_2$
HCl	$Sr(OH)_2$
HNO_3	$Ba(OH)_2$

*In general, the Group IA and Group IIA hydroxides (except beryllium hydroxide) are strong bases.

Most of the other acids and bases you encounter are *weak*. They are not completely ionized in solution. Consider acetic acid, $HC_2H_3O_2$:

$$HC_2H_3O_2(aq) + H_2O(l) \rightleftharpoons H_3O^+(aq) + C_2H_3O_2^-(aq)$$

The neutralization of a strong acid by a strong base is essentially the reaction of $H_3O^+(aq)$ with $OH^-(aq)$ and should give the same $\Delta H°$ per mole of water formed. The neutralization of $HClO_4$ with $NaOH$ in ionic form is

$$H_3O^+(aq) + \cancel{ClO_4^-(aq)} + \cancel{Na^+(aq)} + OH^-(aq) \longrightarrow \cancel{Na^+(aq)} + \cancel{ClO_4^-(aq)} + 2H_2O(l)$$

After canceling, you get the net ionic equation

$$H_3O^+(aq) + OH^-(aq) \longrightarrow 2H_2O(l)$$

All reactions between strong acids and bases have the same $\Delta H°$: -55.90 kJ per mole of H^+. The same reaction occurs in each case, as Arrhenius's theory predicts.

Despite its successes, the Arrhenius concept is limited. In addition to looking at acid–base reactions only in aqueous solutions, it singles out the OH^- ion as the source of base character, when other species can play a similar role.

16.2 Brønsted–Lowry Concept of Acids and Bases

In 1923 the Danish chemist Johannes N. Brønsted (1879–1947) and, independently, the British chemist Thomas M. Lowry (1874–1936) pointed out that acid–base reactions are proton-transfer reactions. According to the Brønsted–Lowry concept, an **acid** is *the species donating a proton in a proton-transfer reaction*. A **base** is *the species accepting the proton.* ◀

▶ To be precise, we should say *hydrogen nucleus* instead of proton (because natural hydrogen contains some ^2H as well as ^1H). The term *proton* is conventional in this context, however.

Consider the reaction of hydrochloric acid with ammonia.

$$H_3O^+(aq) + \cancel{Cl^-(aq)} + NH_3(aq) \longrightarrow H_2O(l) + NH_4^+(aq) + \cancel{Cl^-(aq)}$$

After canceling Cl^-, you obtain the net ionic equation.

$$H_3O^+(aq) + NH_3(aq) \longrightarrow H_2O(l) + NH_4^+(aq)$$

In this reaction in aqueous solution, a proton, H^+, is transferred from the H_3O^+ ion to the NH_3 molecule, giving H_2O and NH_4^+ (Figure 16.3). Here H_3O^+ is the proton donor, and NH_3 is the proton acceptor.

You can also apply the Brønsted–Lowry concept to the reaction of HCl and NH_3 dissolved in benzene. In benzene, HCl and NH_3 are not ionized. The equation is

$$HCl(benzene) + NH_3(benzene) \longrightarrow NH_4Cl(s)$$
$$\text{acid} \qquad\qquad \text{base}$$

**Figure 16.3
Representation of the reaction $H_3O^+ + NH_3 \rightleftharpoons H_2O + NH_4^+$.** Note the transfer of a proton, H^+, from H_3O^+ to NH_3. The charges indicated for ions are *overall* charges. They are not to be associated with specific locations on the ions.

Here the HCl molecule is the proton donor, and the NH_3 molecule is the proton acceptor.

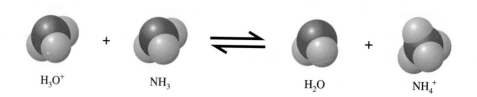

H_3O^+ + NH_3 ⇌ H_2O + NH_4^+

In any reversible acid–base reaction, both forward and reverse reactions involve proton transfers. Consider the reaction of NH_3 with H_2O.

$$NH_3(aq) + H_2O(l) \rightleftharpoons NH_4^+(aq) + OH^-(aq)$$

base   acid   acid   base

In the forward reaction, NH_3 accepts a proton from H_2O. In the reverse reaction, NH_4^+ donates a proton to OH^-.

Note that NH_3 and NH_4^+ differ by a proton. The species NH_4^+ and NH_3 are a conjugate acid–base pair. A **conjugate acid–base pair** consists of *two species in an acid–base reaction that differ by the loss or gain of a proton*. The acid in such a pair is called the *conjugate acid* of the base, whereas the base is the *conjugate base* of the acid. Here NH_4^+ is the conjugate acid of NH_3, and NH_3 is the conjugate base of NH_4^+.

EXAMPLE 16.1

Identifying Acid and Base Species

In the following equations, label each species as an acid or a base. Show the conjugate acid–base pairs.

a. $HCO_3^-(aq) + HF(aq) \rightleftharpoons H_2CO_3(aq) + F^-(aq)$

b. $HCO_3^-(aq) + OH^-(aq) \rightleftharpoons CO_3^{2-}(aq) + H_2O(l)$

SOLUTION

a. On the left, HF is the proton donor; on the right, H_2CO_3 is the proton donor. The proton acceptors are HCO_3^- and F^-.

$$HCO_3^-(aq) + HF(aq) \rightleftharpoons H_2CO_3(aq) + F^-(aq)$$

base   acid   acid   base

Thus, H_2CO_3 and HCO_3^- are a conjugate acid–base pair, as are HF and F^-.

b. You have

$$HCO_3^-(aq) + OH^-(aq) \rightleftharpoons CO_3^{2-}(aq) + H_2O(l)$$

acid   base   base   acid

Here HCO_3^- and CO_3^{2-} are a conjugate acid–base pair, as are H_2O and OH^-. Note that although HCO_3^- functions as an acid in this reaction, it functions as a base in part a.

See Problems 16.19 and 16.20.

▶ *Amphoteric* is a general term referring to a species that can act as an acid or a base. The species need not be amphiprotic, however. For example, aluminum oxide is an amphoteric oxide, because it reacts with acids and bases. It is not amphiprotic, because it has no protons.

The Brønsted–Lowry concept defines a species as an acid or a base according to its function in the proton-transfer reaction. Some species (Example 16.1) can act as either an acid or a base. An **amphiprotic species** is a *species that can act as either an acid or a base* depending on the other reactant. ◀ For example, HCO_3^- acts as an acid in the presence of OH^- but as a base in the presence of HF. Anions with ionizable hydrogens, such as HCO_3^-, and certain solvents, such as water, are amphiprotic.

The amphiprotic characteristic of water is important. Consider, for example, the reactions of water with the base NH_3 and with the acid $HC_2H_3O_2$ (acetic acid).

$$NH_3(aq) + H_2O(l) \rightleftharpoons NH_4^+(aq) + OH^-(aq)$$

base   acid   acid   base

$$HC_2H_3O_2(aq) + H_2O(aq) \rightleftharpoons C_2H_3O_2^-(aq) + H_3O^+(aq)$$

acid   base   base   acid

Water reacts as an acid with the base NH_3 and as a base with the acid $HC_2H_3O_2$.

The Brønsted–Lowry concept of acids and bases has greater scope than the Arrhenius concept. In the Brønsted–Lowry concept:

1. A base is a species that accepts protons; OH^- is only one example of a base.

2. Acids and bases can be ions as well as molecular substances.

3. Acid–base reactions are not restricted to aqueous solution.

4. Some species can act as either acids or bases, depending on what the other reactant is.

CONCEPT CHECK 16.1

Chemists in the seventeenth century discovered that the substance that gives red ants their irritating bite is an acid. They called this substance formic acid after the ant, whose Latin name is *formica rufus*. Formic acid has the following structural formula and molecular model:

$$\begin{array}{c} O \\ \parallel \\ H-C-O-H \end{array}$$

Write the acid–base equilibria connecting all components in the aqueous solution. Now list all of the species present.

16.3 Lewis Concept of Acids and Bases

Certain reactions have the characteristics of acid–base reactions but do not fit the Brønsted–Lowry concept. An example is the reaction of the basic oxide Na_2O with the acidic oxide SO_3 to give the salt Na_2SO_4. ◄

▶ Acidic and basic oxides were discussed in Section 8.7.

$$Na_2O(s) + SO_3(g) \longrightarrow Na_2SO_4(s)$$

G. N. Lewis, who proposed the electron-pair theory of covalent bonding, realized that it could be extended to the concept of acids and bases. According to this concept, a **Lewis acid** is *a species that can form a covalent bond by accepting an electron pair from another species;* a **Lewis base** is *a species that can form a covalent bond by donating an electron pair to another species.*

Consider again the neutralization of NH_3 by HCl in aqueous solution.

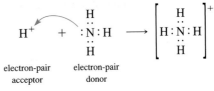

electron-pair   electron-pair
acceptor   donor

Here the red arrow shows the proton accepting an electron pair from NH_3 and an H—N bond being formed. The proton is an electron-pair acceptor, so it is a Lewis acid. Ammonia, NH_3, which has a lone pair of electrons, is an electron-pair donor and therefore a Lewis base.

Now let us look at the reaction of Na_2O with SO_3. It involves the reaction of the oxide ion, O^{2-}, from the ionic solid, Na_2O, with SO_3.

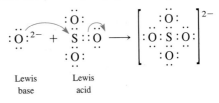

Lewis   Lewis
base   acid

Here SO_3 accepts the electron pair from the O^{2-} ion. Thus, O^{2-} is the Lewis base and SO_3 is the Lewis acid.

The Lewis concept embraces many reactions, such as the reaction of boron trifluoride with ammonia:

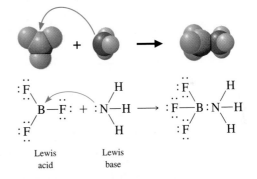

Lewis   Lewis
acid   base

The electron pair originally on the nitrogen atom is now shared between the nitrogen and boron atoms, forming a B—N bond.

The formation of *complex ions* can also be looked at as Lewis acid–base reactions. Complex ions are formed when a metal ion bonds to electron pairs from molecules such as H_2O or NH_3 or from anions such as $:C{\equiv}N:^-$. The formation of a hydrated metal ion, such as $Al(H_2O)_6^{3+}$, involves a Lewis acid–base reaction.

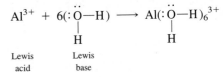

Lewis   Lewis
acid   base

$Al(OH_2)_6^{3+}$

EXAMPLE 16.2

Identifying Lewis Acid and Base Species

In the following reactions, identify the Lewis acid and the Lewis base.

a. $Ag^+ + 2NH_3 \rightleftharpoons Ag(NH_3)_2^+$

b. $B(OH)_3 + H_2O \rightleftharpoons B(OH)_4^- + H^+$

PROBLEM STRATEGY

Write the equations using Lewis electron-dot formulas. Then identify the electron-pair acceptor, or Lewis acid, and the electron-pair donor, or Lewis base.

SOLUTION

a. The silver ion, Ag^+, forms a complex ion with two NH_3 molecules.

$$Ag^+ + 2:NH_3 \rightleftharpoons Ag(:NH_3)_2^+$$

$$\underset{\text{Lewis acid}}{Ag^+} \quad \underset{\text{Lewis base}}{2:NH_3}$$

b. The reaction is

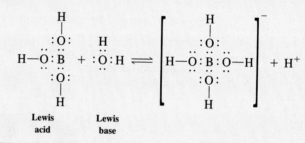

See Problems 16.23 and 16.24.

Acid and Base Strengths

The Brønsted–Lowry concept considers an acid–base reaction as a proton-transfer reaction. It is useful to consider such acid–base reactions as a competition between species for protons. From this point of view, you can order acids by their relative strengths as proton donors.

16.4 Relative Strengths of Acids and Bases

By comparing various acid–base reactions, you can construct a table of relative strengths of acids and bases (see Table 16.2). To see how you might do this, recall that an acid is strong if it completely ionizes in water. In the reaction of hydrogen chloride with water, water acts as a base, accepting the proton from HCl.

$$\underset{\text{acid}}{HCl(aq)} + \underset{\text{base}}{H_2O(l)} \longrightarrow \underset{\text{base}}{Cl^-(aq)} + \underset{\text{acid}}{H_3O^+(aq)}$$

Taking Your Medicine

The expressions "take your medicine" and "it's a bitter pill" are metaphors that refer to doing what's necessary to solve a difficult personal problem. To take your medicine might be unpleasant if it is bitter—and medicines do frequently taste bitter. Why?

A bitter taste appears to be a common feature of a base. It is a fact that many medicinal substances are nitrogen bases, substances that organic chemists call *amines*. Such substances are considered derivatives of ammonia, in which one or more hydrogen atoms have been substituted by carbon-containing groups. Here, for example, are the structures of some amines:

$$H-\overset{\overset{\displaystyle CH_3}{|}}{\underset{\cdot\cdot}{N}}-H \qquad H_3C-\overset{\overset{\displaystyle CH_3}{|}}{\underset{\cdot\cdot}{N}}-H \qquad H_3C-\overset{\overset{\displaystyle CH_3}{|}}{\underset{\cdot\cdot}{N}}-CH_3$$

The nitrogen atom in an amine has a lone pair of electrons that can be donated to form a covalent bond. So, an amine is a Lewis base. But, like ammonia, an amine accepts a hydrogen ion from an acid to form an amine ion, so it is also a Brønsted–Lowry base.

Some of our most important medicinal drugs have originated from plants. Lewis and Clark took Peruvian bark, or cinchona bark, with them as a medicine on their 1804 expedition from the eastern United States to the Pacific Coast and back. The bitter essence of cinchona bark is quinine, an amine drug that has been used to combat malaria. Quinine is responsible for the bitter taste of tonic water, a carbonated beverage (Figure 16.4). Some other amines that come from plants are caffeine (from coffee, a stimulant), atropine (from the deadly nightshade, used to dilate the pupil of the eye for eye exams), and codeine (from the opium poppy, used as a painkiller).

Figure 16.4
Tonic water. *Left:* This carbonated beverage contains quinine, which accounts for its bitter taste. *Right:* Structural formula of quinine.

Quinine (antimalarial)

Because the reaction goes almost completely to the right, you say that HCl is a strong acid. You can also consider the reverse reaction where the Cl^- ion acts as the base, accepting a proton from the acid H_3O^+.

Look at this reaction in terms of the relative strengths of the two acids, HCl and H_3O^+. Because HCl is a strong acid, it must lose its proton readily, more readily than H_3O^+ does. You would say that HCl is a stronger acid than H_3O^+:

$$\underset{\substack{\text{stronger} \\ \text{acid}}}{HCl(aq)} + H_2O(l) \rightleftharpoons Cl^-(aq) + \underset{\substack{\text{weaker} \\ \text{acid}}}{H_3O^+(aq)}$$

An acid–base reaction normally goes in the direction of the weaker acid. As another example, look at the ionization of acetic acid, $HC_2H_3O_2$, in water.

Table 16.2
Relative Strengths of Acids and Bases

	Acid	Base	
Strongest acids	$HClO_4$	ClO_4^-	Weakest bases
	H_2SO_4	HSO_4^-	
	HI	I^-	
	HBr	Br^-	
	HCl	Cl^-	
	HNO_3	NO_3^-	
	H_3O^+	H_2O	
	HSO_4^-	SO_4^{2-}	
	H_2SO_3	HSO_3^-	
	H_3PO_4	$H_2PO_4^-$	
	HNO_2	NO_2^-	
	HF	F^-	
	$HC_2H_3O_2$	$C_2H_3O_2^-$	
	$Al(H_2O)_6^{3+}$	$Al(H_2O)_5OH^{2+}$	
	H_2CO_3	HCO_3^-	
	H_2S	HS^-	
	$HClO$	ClO^-	
	$HBrO$	BrO^-	
	NH_4^+	NH_3	
	HCN	CN^-	
	HCO_3^-	CO_3^{2-}	
	H_2O_2	HO_2^-	
	HS^-	S^{2-}	
Weakest acids	H_2O	OH^-	Strongest bases

$$HC_2H_3O_2(aq) + H_2O(l) \rightleftharpoons C_2H_3O_2^-(aq) + H_3O^+(aq)$$

Experiment reveals that in a 0.1 M acetic acid solution, only about 1% of the acetic acid molecules have ionized. This implies that $HC_2H_3O_2$ is a weaker acid than H_3O^+. With 0.1 M HF, you find that about 3% of the HF molecules have dissociated. Thus, HF is a weaker acid than H_3O^+ but a stronger acid than $HC_2H_3O_2$. You have already seen that HCl is stronger than H_3O^+. Thus, the acid strengths for these four acids are in the order HCl > H_3O^+ > HF > $HC_2H_3O_2$.

This procedure of determining the relative order of acid strengths by comparing their relative ionizations in water cannot be used to obtain the relative strengths of two strong acids such as HCl and HI. In water, they are essentially 100% ionized. In another solvent that is less basic than water—say, pure acetic acid—you do see a difference. Neither acid is completely ionized, but a greater fraction of HI molecules is found to be ionized. Thus, HI is a stronger acid than HCl. In water, the acid strengths of the strong acids appear to be the same; that is, they are "leveled out." We say that water exhibits a *leveling effect* on the strengths of the strong acids.

The first column of Table 16.2 lists acids by their strength; the strongest is at the top of the table. Note that the arrow down the left side of the table points toward the weaker acid and is in the direction the reaction goes. For example, in the reaction involving HCl and H_3O^+, the arrow points from HCl toward H_3O^+.

You can also view this same reaction in terms of the bases, H_2O and Cl^-. A stronger base picks up a proton more readily than does a weaker one. Water has greater base strength than the Cl^- ion; that is, H_2O picks up protons more readily than Cl^- does. Because H_2O molecules compete more successfully for protons than Cl^- ions do, the reaction goes almost completely to the right. (Note that the arrow on the right in Table 16.2 points from H_2O to Cl^-.)

$$HCl(aq) + H_2O(l) \longrightarrow Cl^-(aq) + H_3O^+(aq)$$
$$\underset{\substack{\text{stronger} \\ \text{base}}}{} \qquad \underset{\substack{\text{weaker} \\ \text{base}}}{}$$

The reaction goes in the direction of the weaker base.

By comparing reactions between different pairs of bases, you can arrive at a relative order for base strengths, just as for acids. *The strongest acids have the weakest conjugate bases, and the strongest bases have the weakest conjugate acids.* This means that a list of conjugate bases of the acids in Table 16.2 will be in order of increasing base strength.

You can use Table 16.2 to predict the direction of an acid–base reaction. The normal direction of reaction is from the stronger acid and base to the weaker acid and base.

EXAMPLE 16.3

Deciding Whether Reactants or Products Are Favored in an Acid–Base Reaction

For the following reaction, decide which species (reactants or products) are favored at the completion of the reaction.

$$SO_4^{2-}(aq) + HCN(aq) \rightleftharpoons HSO_4^-(aq) + CN^-(aq)$$

SOLUTION

If you compare the relative strengths of the two acids HCN and HSO_4^- (Table 16.2), you see that HCN is weaker. Or, comparing the bases SO_4^{2-} and CN^-, SO_4^{2-} is weaker. Hence, the reaction would go from right to left.

$$SO_4^{2-}(aq) + HCN(aq) \longleftarrow HSO_4^-(aq) + CN^-(aq)$$
$$\underset{\substack{\text{weaker} \\ \text{base}}}{} \quad \underset{\substack{\text{weaker} \\ \text{acid}}}{} \quad \underset{\substack{\text{stronger} \\ \text{acid}}}{} \quad \underset{\substack{\text{stronger} \\ \text{base}}}{}$$

The reactants are favored.

See Problems 16.27 and 16.28.

CONCEPT CHECK 16.2

Formic acid, $HCHO_2$, is a stronger acid than acetic acid, $HC_2H_3O_2$. Which is the stronger base, formate ion, CHO_2^-, or acetate ion, $C_2H_3O_2^-$?

16.5 Molecular Structure and Acid Strength

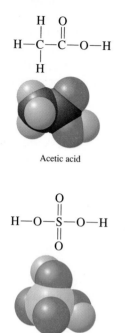

Acetic acid

The strength of an acid depends on how easily the proton, H^+, is lost or removed from an H—X bond in the acid species. By understanding the factors that determine the ease of proton loss, you will be able to predict the relative strengths of similar acids.

Two factors are important in determining relative acid strengths. One is the polarity of the bond to which the H atom is attached. The H atom should have a positive partial charge:

$$\overset{\delta+ \quad \delta-}{H—X}$$

The more polarized the bond is in this direction, the more easily the proton is removed. A hydrogen atom bonded to an oxygen atom generally has a partial positive charge, but in comparing such an atom in acetic acid and sulfuric acid, the hydrogen atom in sulfuric acid is much more positive. Thus sulfuric acid is more acidic.

The second factor determining acid strength is the strength of the bond—that is, how tightly the proton is held. This, in turn, depends on the size of atom X. The larger atom X, the weaker is the bond and the greater the acid strength.

Consider the binary acids of the Group VIIA elements: HF, HCl, HBr, and HI. As you go down the column of elements, each time adding a shell of electrons to the atom, the radius increases markedly. *As the size of atom X increases, the H—X bond strength decreases, and the strength of the binary acid increases.* You predict the following order of acid strength:

$$HF < HCl < HBr < HI$$

As you go across a row of elements of the periodic table, the atomic radius decreases slowly. For this reason, the relative strengths of the binary acids of these elements are less dependent on the size of atom X. Now the polarity of the H—X bond becomes the dominant factor in determining acid strength. *Going across a row of elements of the periodic table, the electronegativity increases, the H—X bond polarity increases, and the acid strength increases.* For example, the binary acids of the last two elements of the second period are H_2O and HF. The acid strengths are

$$H_2O < HF$$

This again is the order shown in Table 16.2.

Now consider the oxoacids. An oxoacid has the structure

$$H—O—Y—$$

The acidic H atom is always attached to an O atom, which, in turn, is attached to an atom Y. Other groups, such as O atoms or O—H groups, may also be attached to Y. Bond polarity appears to be the dominant factor determining relative strengths of the oxoacids. This, in turn, depends on the electronegativity of atom Y. If the electronegativity of atom Y is large, the H—O bond is relatively polar and the acid strength large. *For a series of oxoacids of the same structure, differing only in the atom Y, the acid strength increases with the electronegativity of Y.* Consider, for example, the acids HClO, HBrO, and HIO. ◄ The structures are

$$H—\overset{..}{\underset{..}{O}}—\overset{..}{\underset{..}{Cl}}: \qquad H—\overset{..}{\underset{..}{O}}—\overset{..}{\underset{..}{Br}}: \qquad H—\overset{..}{\underset{..}{O}}—\overset{..}{\underset{..}{I}}:$$

H—O—S—O—H structure (sulfuric acid)

Sulfuric acid

▶ The formulas of these acids may be written HXO or HOX, depending on the convention used. Formulas of oxoacids are generally written with the acidic H atoms first, followed by the characteristic element (X), then O atoms. However, the formulas of molecules composed of three atoms are often written in the order in which the atoms are bonded, which in this case is HOX.

HClO

HClO₂

HClO₃

HClO₄

The electronegativity decreases from Cl to Br to I and the order of acid strengths is

$$HIO < HBrO < HClO$$

For a series of oxoacids, (HO)$_m$YO$_n$, the acid strength increases with n, the number of O atoms bonded to Y (excluding O atoms in OH groups). The oxoacids of chlorine provide an example:

H—O—Cl: H—O—Cl—O: H—O—Cl—O: H—O—Cl—O:

With each additional O atom, the Cl atom becomes effectively more electronegative. As a result, the H atom becomes more acidic. The acid strengths increase in the following order:

$$HClO < HClO_2 < HClO_3 < HClO_4$$

Let us compare the relative acid strengths of a polyprotic acid and its corresponding acid anions. For example, H_2SO_4 ionizes by losing a proton to give HSO_4^-, which, in turn, ionizes to give SO_4^{2-}. Because of the negative charge on the HSO_4^- ion, which tends to attract protons, its acid strength is reduced from that of the uncharged species. That is, the acid strengths are in the order

$$HSO_4^- < H_2SO_4$$

We conclude that *the acid strength of a polyprotic acid and its anions decreases with increasing negative charge* (see Table 16.2).

Self-Ionization of Water and pH

In aqueous solutions, two ions have dominant roles. These ions, the hydronium ion, H_3O^+ (or hydrogen ion, H^+), and the hydroxide ion, OH^-, are available in any aqueous solution as a result of the *self-ionization* of water, a reaction of water with itself, which we will describe in the next section. This will also give us some background to acid–base equilibrium calculations, which we will discuss in Chapter 17.

16.6 Self-Ionization of Water

Although pure water is often considered a nonelectrolyte, precise measurements do show a very small conduction. This conduction results from **self-ionization**, *a reaction in which two like molecules react to give ions.* In the case of water, a proton from one H_2O molecule is transferred to another H_2O molecule.

$$H_2O(l) + H_2O(l) \rightleftharpoons H_3O^+(aq) + OH^-(aq)$$

The equilibrium expression is

$$K_c = \frac{[H_3O^+][OH^-]}{[H_2O]^2}$$

Because the concentration of ions formed is very small, the concentration of H_2O remains essentially constant, about 56 M at 25°C. If you rearrange this equation, placing $[H_2O]^2$ with K_c, the ion product $[H_3O^+][OH^-]$ equals a constant.

$$\underbrace{[H_2O]^2 K_c}_{\text{constant}} = [H_3O^+][OH^-]$$

▶ The thermodynamic equilibrium constant is defined in terms of activities. The activity of water is essentially 1, so we do not include water explicitly in the equilibrium expression.

We call *the equilibrium value of the ion product* $[H_3O^+][OH^-]$ the **ion-product constant for water,** which is written K_w. ◀ At 25°C, the value of K_w is 1.0×10^{-14}. Like any equilibrium constant, K_w varies with temperature.

$$K_w = [H_3O^+][OH^-] = 1.0 \times 10^{-14} \text{ at 25°C}$$

Using K_w, you can calculate the concentrations of H_3O^+ and OH^- ions in pure water. These ions are produced in equal numbers in pure water, so $x = [H_3O^+] = [OH^-]$. Then, substituting into the equation for the ion-product constant, you get, at 25°C,

$$1.0 \times 10^{-14} = x^2$$

Hence, x equals 1.0×10^{-7}. Thus, the concentrations of H_3O^+ and OH^- are both 1.0×10^{-7} M in pure water.

If you add an acid or a base to water, the concentrations of H_3O^+ and OH^- will no longer be equal.

16.7 Solutions of a Strong Acid or Base

Consider an aqueous solution of a strong acid or base. Suppose you dissolve 0.10 mol HCl in 1.0 L of aqueous solution, giving 0.10 M HCl. Because you started with 0.10 mol HCl, the reaction will produce 0.10 mol H_3O^+.

Now consider the concentration of H_3O^+ ion produced by the self-ionization of water. In pure water, the concentration of H_3O^+ produced is 1.0×10^{-7} M; in an acid solution, the contribution of H_3O^+ from water will be even smaller. You can see this by applying Le Chatelier's principle to the self-ionization reaction.

$$H_2O(l) + H_2O(l) \longleftarrow H_3O^+(aq) + OH^-(aq)$$

Consequently, the concentration of H_3O^+ produced by the self-ionization of water ($< 1 \times 10^{-7}$ M) is negligible in comparison with that produced from HCl (0.10 M).

In a solution of a strong acid, you can normally ignore the self-ionization of water as a source of H_3O^+. The H_3O^+ concentration is usually determined by the strong acid concentration. (This is not true when the acid solution is extremely dilute, however.)

Although you normally ignore the self-ionization of water in calculating the H_3O^+ concentration in a solution of a strong acid, the self-ionization equilibrium still exists and is responsible for a small concentration of OH^- ion. You can use the ion-product constant for water to calculate this concentration. As an example, calculate the concentration of OH^- ion in 0.10 M HCl. You substitute $[H_3O^+] = 0.10$ M into the equilibrium equation for K_w (for 25°C).

$$K_w = [H_3O^+][OH^-]$$

$$1.0 \times 10^{-14} = 0.10 \times [OH^-]$$

Solving for [OH$^-$],

$$[OH^-] = \frac{1.0 \times 10^{-14}}{0.10} = 1.0 \times 10^{-13}$$

The OH$^-$ concentration is 1.0×10^{-13} M.

Now consider a solution of a strong base, such as 0.010 M NaOH. What are the OH$^-$ and H$_3$O$^+$ concentrations in this solution? Because NaOH is a strong base, all of the NaOH is present in the solution as ions. Therefore, the concentration of OH$^-$ obtained from NaOH in 0.010 M NaOH solution is 0.010 M. The concentration of OH$^-$ produced from the self-ionization of water in this solution ($< 1 \times 10^{-7}$ M) is negligible and can be ignored. Therefore, the concentration of OH$^-$ ion in the solution is 0.010 M. To obtain the hydronium ion concentration, you substitute into the equilibrium equation for K_w (for 25°C).

$$K_w = [H_3O^+][OH^-]$$

$$1.0 \times 10^{-14} = [H_3O^+] \times 0.010$$

Solving for H$_3$O$^+$ concentration, you get $[H_3O^+] = \mathbf{1.0 \times 10^{-12}}$ M.

The following example further illustrates the calculation of H$_3$O$^+$ and OH$^-$ concentrations in solutions of a strong acid or base.

EXAMPLE 16.4

Calculating Concentrations of H$_3$O$^+$ and OH$^-$ in Solutions of a Strong Acid or Base

Calculate the concentrations of hydronium ion and hydroxide ion at 25°C in:
a. 0.15 M HNO$_3$, b. 0.010 M Ca(OH)$_2$.

SOLUTION

a. Every mole of HNO$_3$ contributes one mole of H$_3$O$^+$ ion, so the H$_3$O$^+$ concentration is **0.15 M**. The OH$^-$ concentration is obtained from the equation for K_w.

$$K_w = [H_3O^+][OH^-]$$

$$1.0 \times 10^{-14} = 0.15 \times [OH^-]$$

so

$$[OH^-] = \frac{1.0 \times 10^{-14}}{0.15} = \mathbf{6.7 \times 10^{-14}}\,M$$

b.

$$Ca(OH)_2(s) \xrightarrow{H_2O} Ca^{2+}(aq) + 2OH^-(aq)$$

Every mole of Ca(OH)$_2$ that dissolves yields two moles of OH$^-$. Therefore, 0.010 M Ca(OH)$_2$ contains 2×0.010 M OH$^-$ = **0.020 M OH$^-$**. The H$_3$O$^+$ concentration is obtained from

$$K_w = [H_3O^+][OH^-]$$

$$1.0 \times 10^{-14} = [H_3O^+] \times 0.020$$

so

$$[H^+] = \frac{1.0 \times 10^{-14}}{0.020} = \mathbf{5.0 \times 10^{-13}\,M}$$

See Problems 16.33 and 16.34.

By dissolving substances in water, you can alter the concentrations of H_3O^+ and OH^- ions. At 25°C, you can observe the following conditions:

> In an acidic solution, $[H_3O^+] > 1.0 \times 10^{-7}\,M$.
>
> In a neutral solution, $[H_3O^+] = 1.0 \times 10^{-7}\,M$.
>
> In a basic solution, $[H_3O^+] < 1.0 \times 10^{-7}\,M$.

16.8 The pH of a Solution

You can quantitatively describe the acidity of an aqueous solution by giving the hydronium-ion concentration. But because these concentration values may be very small, it is often more convenient to give the acidity in terms of **pH,** which is defined as *the negative of the logarithm of the molar hydronium-ion concentration:* ◄

▶ The Danish biochemist S. P. L. Sørensen devised the pH scale while working on the brewing of beer.

$$pH = -\log [H_3O^+]$$

When the hydronium-ion concentration is $1.0 \times 10^{-3}\,M$, the pH is

$$pH = -\log(1.0 \times 10^{-3}) = \mathbf{3.00}$$

Note that the number of places after the decimal point in the pH equals the number of significant figures reported in the hydronium-ion concentration.

Figure 16.5
The pH scale. Solutions having pH less than 7 are acidic; those having pH more than 7 are basic.

A neutral solution, has a pH of 7.00. For an acidic solution, the pH is less than 7.00. Similarly, a basic solution has a pH greater than 7.00. Figure 16.5 lists pH values of some common solutions.

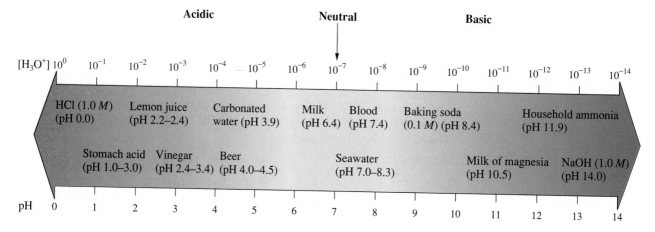

Figure 16.8
Some acid-
indicators i
different H
concentrat

1   2

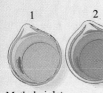

Methyl violet

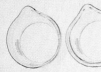

Phenolphthalein

Bromthymol blue

Bromcresol green

Universal indicator

Methyl orange
10^{-1} 10

EXAMPLE 16.5
Calculating the pH from the Hydronium-Ion Concentration

A sample of orange juice has a hydronium-ion concentration of $2.9 \times 10^{-4}\, M$. What is the pH? Is the solution acidic?

SOLUTION

$$pH = -\log[H_3O^+] = -\log(2.9 \times 10^{-4}) = \mathbf{3.54}$$

The pH is less than 7.00, so the solution is **acidic** (as you expect for orange juice).

See Problems 16.41 and 16.42.

You can find the pH of a solution of known hydroxide-ion concentration by first solving for the hydronium-ion concentration, as shown in Example 16.4 b. However, you can also find the pH simply from the *pOH*, a measure of hydroxide-ion concentration similar to the pH:

$$pOH = -\log[OH^-]$$

▶ **Taking the logarithm of both sides of the equation**

$$[H_3O^+][OH^-] = 1.0 \times 10^{-14}$$

you get

$$\log[H_3O^+] + \log[OH^-] = -14.00$$

or

$$(-\log[H_3O^+]) + (-\log[OH^-]) = 14.00$$

Hence, pH + pOH = 14.00.

Then, because $K_w = [H_3O^+][OH^-] = 1.0 \times 10^{-14}$ at 25°C, you can show that ◀

$$pH + pOH = 14.00$$

To find the pH of an ammonia solution whose hydroxide-ion concentration is $1.9 \times 10^{-3}\, M$, you first calculate the pOH.

$$pOH = -\log(1.9 \times 10^{-3}) = 2.72$$

Then the pH is

$$pH = 14.00 - pOH = 14.00 - 2.72 = \mathbf{11.28}$$

EXAMPLE 16.6
Calculating the Hydronium-Ion Concentration from the pH

The pH of human arterial blood is 7.40. What is the hydronium-ion concentration?

PROBLEM STRATEGY

The hydronium-ion concentration is

$$[H_3O^+] = \text{antilog}(-pH) = 10^{-pH}$$

On some calculators, you obtain the antilogarithm by taking the *inverse* of the logarithm. On others, you use a 10^x key.

SOLUTION

The result is

$$[H_3O^+] = \text{antilog}(-7.40) = 10^{-7.40} = \mathbf{4.0 \times 10^{-8}\, M}$$

Unclogging the Sink and Other Chores

EVERYDAY LIFE

Drain cleaner solutions, available in grocery stores, are simply solutions of sodium hydroxide, a strong base. Such solutions work by chemically reacting with fat and with hair, the usual ingredients of a stopped drain. When fat reacts with a strong base, it forms the salt of a fatty acid, a product otherwise known as soap. Hair is a protein material, and in the presence of a strong base, the protein breaks up into the salts of its constituent amino acids. That solution washes easily down the drain. Oven cleaner is a paste made of sodium hydroxide. It cleans by reacting with the fatty deposits and other residues in the oven.

Sodium hydroxide is produced commercially by sending a direct current through an aqueous solution of sodium chloride (Figure 16.9). Hydrogen is released at the negative pole and chlorine is released at the positive pole, and the resulting solution contains sodium hydroxide. The overall reaction is

$$2NaCl(aq) + 2H_2O(l) \xrightarrow{\text{electrolysis}} H_2(g) + Cl_2(g) + 2NaOH(aq)$$

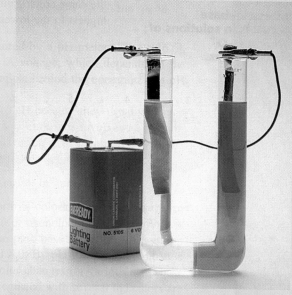

Figure 16.9
Preparation of sodium hydroxide by hydrolysis.
A solution of sodium chloride is electrolyzed. Note the pink color at the negative electrode. Phenolphthalein indicator turns pink from the hydroxide ion formed at the electrode. Later, if we evaporate the solution, sodium hydroxide crystallizes out.

Figure 16.6
A digital pH [...]
experimenter [...]
containing ele[...]
solution and re[...]
the meter.

▶ Litmus contai[...]
stances and c[...]
a broader ran[...]
pH 5 to pH 8.

Figure 16.7
Color change[...]
acid–base in[...]
Acid–base indi[...]
whose acid for[...]
and whose bas[...]
another color. [...]
diprotic acid an[...]
different color [...]
here as the acid[...]
the basic range[...]

A Checklist for Review

Important Terms

acid (Brønsted–Lowry) (16.2)
base (Brønsted–Lowry) (16.2)

conjugate acid–base pair (16.2)
amphiprotic species (16.2)

Lewis acid (16.3)
Lewis base (16.3)
self-ionization (16.6)

ion-product constant for water (K_w) (16.6)
pH (16.8)

Key Equations

$$K_w = [H_3O^+][OH^-]$$
$$= 1.0 \times 10^{-14} \text{ at } 25°C$$

$$pH = -\log[H_3O^+]$$
$$pH + pOH = 14.00$$

Summary of Facts and Concepts

The *Arrhenius concept* was the first successful theory of acids and bases. Then in 1923, Brønsted and Lowry characterized acid–base reactions as proton-transfer reactions. According to the *Brønsted–Lowry concept,* an acid is a proton donor and a base is a proton acceptor. The *Lewis concept* is even more general than the Brønsted–Lowry concept. A Lewis acid is an

electron-pair acceptor and a Lewis base is an electron-pair donor. Reactions of acidic and basic oxides and the formation of complex ions, as well as proton-transfer reactions, can be described in terms of the Lewis concept.

Acid–base reactions can be viewed as a competition for protons. Using this idea, you can construct a table of *relative strengths of acids and bases.* Reaction favors the weaker acid and the weaker base. Acid strength depends on the polarity and the strength of the bond involving the acidic hydrogen atom. Therefore, you can *relate molecular structure and acid strength.* The text gives a number of rules that allow you to predict the relative strengths of similar acids based on molecular structure.

Water ionizes to give hydronium ion and hydroxide ion. The concentrations of these ions in aqueous solution are related by the *ion-product constant for water (K_w).* Thus, you can describe the acidity or basicity of a solution by the hydronium-ion concentration. You often use the pH, which is the negative logarithm of the hydronium-ion concentration, as a measure of acidity.

Operational Skills

1. **Identifying acid and base species** Given a proton-transfer reaction, label the acids and bases, and name the conjugate acid–base pairs. **(EXAMPLE 16.1)**
2. **Identifying Lewis acid and base species** Given a reaction involving the donation of an electron pair, identify the Lewis acid and the Lewis base. **(EXAMPLE 16.2)**
3. **Deciding whether reactants or products are favored in an acid–base reaction** Given an acid–base reaction and the relative strengths of acids (or bases), decide whether reactants or products are favored. **(EXAMPLE 16.3)**
4. **Calculating concentrations of H_3O^+ and OH^- in solutions of a strong acid or base** Given the concentration of a strong acid or base, calculate the hydronium-ion and hydroxide-ion concentrations. **(EXAMPLE 16.4)**
5. **Calculating the pH from the hydronium-ion concentration, or vice versa** Given the hydronium-ion concentration, calculate the pH **(EXAMPLE 16.5)**; or given the pH, calculate the hydronium-ion concentration **(EXAMPLE 16.6)**.

Review Questions

16.1 Which of the following are strong acids? Which are weak acids? (a) $HC_2H_3O_2$; (b) $HClO$; (c) HCl; (d) HNO_3; (e) HNO_2; (f) HCN

16.2 Define an acid and a base according to the Brønsted–Lowry concept. Give an acid–base equation and identify each species as an acid or a base.

16.3 What is meant by the conjugate acid of a base?

16.4 Write an equation in which $H_2PO_3^-$ acts as an acid and another in which it acts as a base.

16.5 Describe four ways in which the Brønsted–Lowry concept enlarges on the Arrhenius concept.

16.6 Define an acid and a base according to the Lewis concept. Give a chemical equation to illustrate.

16.7 Give two important factors that determine the strength of an acid. How does an increase in each factor affect the acid strength?

16.8 What is meant by the self-ionization of water? Write the expression for K_w. What is its value at 25°C?

16.9 What is meant by the pH of a solution? Describe two ways of measuring pH.

16.10 What is the pH of a neutral solution at 37°C, where K_w equals 2.5×10^{-14}?

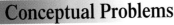 Conceptual Problems

16.11 Aqueous solutions of ammonia, NH_3, were once thought to be solutions of an ionic compound ammonium hydroxide, NH_4OH, in order to explain how the solutions could contain hydroxide ion. Using the Brønsted–Lowry concept, show how NH_3 yields hydroxide ion in aqueous solution without involving the species NH_4OH.

16.12 Blood contains several substances that minimize changes in its acidity by reacting with either an acid or a base. One of these is the hydrogen phosphate ion, HPO_4^{2-}. Write one equation showing this species acting as a Brønsted–Lowry acid and another in which the species acts as a Brønsted–Lowry base.

16.13 Self-contained environments, such as that of a space station, require that the carbon dioxide exhaled by people be continuously removed. This can be done by passing the air over solid alkali hydroxide, in which carbon dioxide reacts with hydroxide ion. What ion is produced by the addition of OH^- ion to CO_2? Use the Lewis concept to explain this.

16.14 Compare the structures of HNO_2 and H_2CO_3. Which would you expect to be the stronger acid? Explain your choice.

Practice Problems

Brønsted–Lowry Concept

16.15 Give the conjugate base to each of the following species regarded as acids.
a. $H_2PO_4^-$ b. H_2Se
c. HNO_2 d. $HAsO_4^{2-}$

16.16 Give the conjugate base to each of the following species regarded as acids.
a. HSO_4^- b. PH_4^+
c. HS^- d. $HOCl$

16.17 Give the conjugate acid to each of the following species regarded as bases.
a. ClO^- b. AsH_3
c. $H_2PO_4^-$ d. TeO_3^{2-}

16.18 Give the conjugate acid to each of the following species regarded as bases.
a. HS^- b. NH_2^-
c. BrO_2^- d. N_2H_4

16.19 For the following reactions, label each species as an acid or a base. Indicate the species that are conjugates of one another.
a. $HSO_4^- + NH_3 \rightleftharpoons SO_4^{2-} + NH_4^+$
b. $HPO_4^{2-} + NH_4^+ \rightleftharpoons H_2PO_4^- + NH_3$
c. $Al(H_2O)_6^{3+} + H_2O \rightleftharpoons Al(H_2O)_5OH^{2+} + H_3O^+$

16.20 For the following reactions, label each species as an acid or a base. Indicate the species that are conjugates of one another.
a. $H_2PO_4^- + HCO_3^- \rightleftharpoons HPO_4^{2-} + H_2CO_3$
b. $F^- + HSO_4^- \rightleftharpoons HF + SO_4^{2-}$
c. $HSO_4^- + H_2O \rightleftharpoons SO_4^{2-} + H_3O^+$

Lewis Acid–Base Concept

16.21 The following shows ball-and-stick models of the reactants in a Lewis acid–base reaction.

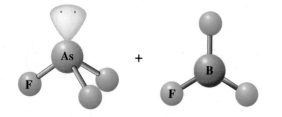

Write the complete equation for the reaction, including the product. Identify each reactant as a Lewis acid or a Lewis base.

16.22 The following shows ball-and-stick models of the reactants in a Lewis acid–base reaction.

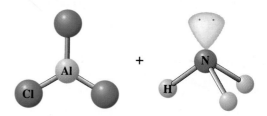

Write the complete equation for the reaction, including the product. Identify each reactant as a Lewis acid or a Lewis base.

16.23 In the following reactions, identify each reactant as a Lewis acid or a Lewis base.
a. $Cr^{3+} + 6H_2O \longrightarrow Cr(H_2O)_6^{3+}$
b. $BF_3 + (C_2H_5)_2\overset{..}{O}: \longrightarrow F_3B:\overset{..}{O}(C_2H_5)_2$

16.24 In the following reactions, label each reactant as a Lewis acid or a Lewis base.
a. $BeF_2 + 2F^- \longrightarrow BeF_4^{2-}$
b. $SnCl_4 + 2Cl^- \longrightarrow SnCl_6^{2-}$

16.25 Natural gas frequently contains hydrogen sulfide, H_2S. H_2S is removed from natural gas by passing it through aqueous ethanolamine, $HOCH_2CH_2NH_2$ (an ammonia derivative), which reacts with the hydrogen sulfide. Write the equation for the reaction. Identify each reactant as either a Lewis acid or a Lewis base. Explain how you arrived at your answer.

16.26 Coal and other fossil fuels usually contain sulfur compounds that produce sulfur dioxide, SO_2, when burned. One possible way to remove the sulfur dioxide is to pass the combustion gases into a tower packed with calcium oxide, CaO. Write the equation for the reaction. Identify each reactant as either a Lewis acid or a Lewis base. Explain how you arrived at your answer.

Acid and Base Strengths

16.27 Use Table 16.2 to decide whether the species on the left or those on the right are favored by the reaction.
a. $NH_4^+ + H_2PO_4^- \rightleftharpoons NH_3 + H_3PO_4$
b. $HCN + HS^- \rightleftharpoons CN^- + H_2S$
c. $HCO_3^- + OH^- \rightleftharpoons CO_3^{2-} + H_2O$

16.28 Use Table 16.2 to decide whether the species on the left or those on the right are favored by the reaction.
a. $NH_4^+ + CO_3^{2-} \rightleftharpoons NH_3 + HCO_3^-$
b. $HCO_3^- + H_2S \rightleftharpoons H_2CO_3 + HS^-$
c. $CN^- + H_2O \rightleftharpoons HCN + OH^-$

16.29 In the following reaction of trichloroacetic acid, $HC_2Cl_3O_2$, with formate ion, CHO_2^-, the formation of trichloroacetate ion, $C_2Cl_3O_2^-$, and formic acid, $HCHO_2$, is favored.

$$HC_2Cl_3O_2 + CHO_2^- \longrightarrow C_2Cl_3O_2^- + HCHO_2$$

Which is the stronger acid, trichloroacetic acid or formic acid? Explain.

16.30 In the following reaction of tetrafluoroboric acid, HBF_4, with the acetate ion, $C_2H_3O_2^-$, the formation of tetra-fluoroborate ion, BF_4^-, and acetic acid, $HC_2H_3O_2$, is favored.

$$HBF_4 + C_2H_3O_2^- \longrightarrow BF_4^- + HC_2H_3O_2$$

Which is the weaker base, BF_4^- or acetate ion?

16.31 For each of the following pairs, give the stronger acid. Explain your answer.
a. H_2S, HS^- b. H_2SO_3, H_2SeO_3 c. HBr, H_2Se

16.32 Order each of the following pairs by acid strength, giving the weaker acid first. Explain your answer.
a. HNO_3, HNO_2 b. HCO_3^-, H_2CO_3 c. H_2S, H_2Te

Solutions of a Strong Acid or Base

16.33 What are the concentrations of H_3O^+ and OH^- in each of the following?
a. 2.5 M HCl   b. 0.065 M $Ca(OH)_2$

16.34 What are the concentrations of H_3O^+ and OH^- in each of the following?
a. 1.50 M KOH   b. 0.25 M $Ba(OH)_2$

16.35 What are the hydronium-ion and the hydroxide-ion concentrations of a solution at 25°C that is 0.0050 M strontium hydroxide, $Sr(OH)_2$?

16.36 A saturated solution of magnesium hydroxide is 3.2×10^{-4} M $Mg(OH)_2$. What are the hydronium-ion and hydroxide-ion concentrations in the solution at 25°C?

16.37 The following are solution concentrations. Indicate whether each solution is acidic, basic, or neutral.
a. 5×10^{-6} M H_3O^+ b. 5×10^{-9} M OH^-

16.38 The following are solution concentrations. Indicate whether each solution is acidic, basic, or neutral.
a. 2×10^{-11} M OH^- b. 2×10^{-6} M H_3O^+

Calculations Involving pH

16.39 Which of the following pH values indicate an acidic solution at 25°C? Which are basic and which are neutral?
a. 10.7   b. 1.9   c. 7.0   d. 4.3

16.40 Which of the following pH values indicate an acidic solution at 25°C? Which are basic and which are neutral?
a. 1.9   b. 4.3   c. 8.7   d. 13.8

16.41 Obtain the pH corresponding to the following hydronium-ion concentrations.
a. 1.0×10^{-8} M   b. 5.0×10^{-12} M

16.42 Obtain the pH corresponding to the following hydronium-ion concentrations.
a. 1.0×10^{-4} M   b. 3.2×10^{-10} M

16.43 Obtain the pH corresponding to the following hydroxide-ion concentrations.
a. 5.25×10^{-9} M   b. 8.3×10^{-3} M

16.44 Obtain the pH corresponding to the following hydroxide-ion concentrations.
a. 4.83×10^{-11} M   b. 3.2×10^{-5} M

16.45 A solution of washing soda (sodium carbonate, Na_2CO_3) has a hydroxide-ion concentration of 0.0040 M. What is the pH at 25°C?

16.46 A solution of lye (sodium hydroxide, NaOH) has a hydroxide-ion concentration of 0.050 M. What is the pH at 25°C?

16.47 A detergent solution has a pH of 11.63 at 25°C. What is the hydroxide-ion concentration?

16.48 Morphine is a narcotic that is used to relieve pain. A solution of morphine has a pH of 9.61 at 25°C. What is the hydroxide-ion concentration?

16.49 A 1.00-L aqueous solution contained 5.80 g of sodium hydroxide, NaOH. What was the pH of the solution at 25°C?

16.50 A 1.00-L aqueous solution contained 6.78 g of barium hydroxide, $Ba(OH)_2$. What was the pH of the solution at 25°C?

16.51 A certain sample of rainwater gives a yellow color with methyl red and a yellow color with bromthymol blue. What is the approximate pH of the water? Is the rainwater acidic, neutral, or basic? (See Figure 16.7.)

16.52 A drop of thymol blue gave a yellow color with a solution of aspirin. A sample of the same aspirin solution gave a yellow color with bromphenol blue. What was the pH of the solution? Was the solution acidic, neutral, or basic? (See Figure 16.7.)

General Problems

16.53 Identify each of the following as an acid or a base in terms of the Arrhenius concept. Give the chemical equation for the reaction of the substance with water, showing the origin of the acidity or basicity.
a. BaO   b. H_2S c. CH_3NH_2 d. SO_2

16.54 Which of the following substances are acids in terms of the Arrhenius concept? Which are bases? Show the acid or base character by using chemical equations.
a. P_4O_{10} b. K_2O c. N_2H_4 d. H_2Se

16.55 Write a reaction for each of the following in which the species acts as a Brønsted acid. The equilibrium should favor the product side.
a. H_2O_2 b. HCO_3^- c. NH_4^+ d. $H_2PO_4^-$

16.56 Write a reaction for each of the following in which the species acts as a Brønsted base. The equilibrium should favor the product side.
a. H_2O b. HCO_3^- c. NH_3 d. $H_2PO_4^-$

16.57 For each of the following, write the complete chemical equation for the acid–base reaction that occurs. Describe each using Brønsted language (if appropriate) and then using Lewis language (show electron-dot formulas).
a. The ClO^- ion reacts with water.
b. The reaction of NH_4^+ and NH_2^- in liquid ammonia to produce NH_3.

16.58 For each of the following, write the complete chemical equation for the acid–base reaction that occurs. Describe each using Brønsted language (if appropriate) and then using Lewis language (show electron-dot formulas).
a. The HS^- ion reacts in water to produce H_2S.
b. Cyanide ion, CN^-, reacts with Fe^{3+}.

16.59 List the following compounds in order of increasing acid strength: HBr, H_2Se, H_2S.

16.60 List the following compounds in order of increasing acid strength: $HBrO_2$, $HClO_2$, HBrO.

16.61 A wine has a hydronium-ion concentration equal to 1.5×10^{-3} M. What is the pH of this wine?

16.62 A sample of lemon juice has a hydronium-ion concentration equal to 2.5×10^{-2} M. What is the pH of this sample?

16.63 A sample of apple cider has a pH of 3.15. What is the hydroxide-ion concentration of this solution?

16.64 A sample of grape juice has a pH of 4.05. What is the hydroxide-ion concentration of this solution?

16.65 A 2.500-g sample of a mixture of sodium hydrogen carbonate and potassium chloride is dissolved in 25.00 mL of 0.437 M H_2SO_4. Some acid remains after treatment of the sample.
a. Write both the net ionic and the molecular equations for the complete reaction of sodium hydrogen carbonate with sulfuric acid.
b. If 35.4 mL of 0.108 M NaOH were required to titrate the excess sulfuric acid, how many moles of sodium hydrogen carbonate were present in the original sample?
c. What is the percent composition of the original sample?

16.66 A 2.500-g sample of a mixture of sodium carbonate and sodium chloride is dissolved in 25.00 mL of 0.798 M HCl. Some acid remains after the treatment of the sample.
a. Write the net ionic equation for the complete reaction of sodium carbonate with hydrochloric acid.
b. If 28.7 mL of 0.108 M NaOH were required to titrate the excess hydrochloric acid, how many moles of sodium carbonate were present in the original sample?
c. What is the percent composition of the original sample?

16.67 The bicarbonate ion has the ability to act as an acid in the presence of a base and as a base in the presence of an acid, so it is said to be amphiprotic. Illustrate this behavior with water by writing Brønsted–Lowry acid–base reactions. Also illustrate this property by selecting a common strong acid and base to react with the bicarbonate ion.

16.68 The dihydrogen phosphate ion has the ability to act as an acid in the presence of a base and as a base in the presence of an acid. What is this property called? Illustrate this behavior with water by writing Brønsted–Lowry acid–base reactions. Also illustrate this property by selecting a common acid and base to react with the dihydrogen phosphate ion.

16.69 A solution contains 4.25 g of ammonia per 250.0 mL of solution. Electrical conductivity measurements at 25°C show that 0.42% of the ammonia has reacted with water. Write the equation for this reaction and calculate the pH of the solution.

16.70 A solution contains 0.675 g of ethylamine, $C_2H_5NH_2$, per 100.0 mL of solution. Electrical conductivity measurements at 20°C show that 0.98% of the ethylamine has reacted with water. Write the equation for this reaction. Calculate the pH of the solution.

Cumulative-Skills Problems

16.71 Boron trifluoride, BF_3, and ammonia, NH_3, react to produce $BF_3 \cdot NH_3$. A coordinate covalent bond is formed between the boron atom on BF_3 and the nitrogen atom on NH_3. Write the equation for this reaction, using Lewis electron-dot formulas. Label the Lewis acid and the Lewis base. Determine how many grams of $BF_3 \cdot NH_3$ are formed when 10.0 g BF_3 and 10.0 g NH_3 are placed in a reaction vessel, assuming that the reaction goes to completion.

16.72 Boron trifluoride, BF_3, and diethyl ether, $(C_2H_5)_2O$, react to produce a compound with the formula $BF_3 \cdot (C_2H_5)_2O$. A coordinate covalent bond is formed between the boron atom on BF_3 and the oxygen atom on $(C_2H_5)_2O$. Write the equation for this reaction, using Lewis electron-dot formulas. Label the Lewis acid and the Lewis base. Determine how many grams of $BF_3 \cdot (C_2H_5)_2O$ are formed when 10.0 g BF_3 and 20.0 g $(C_2H_5)_2O$ are placed in a reaction vessel, assuming that the reaction goes to completion.

Calculations with K_a

Using the value of K_a for an acid HA, you can calculate the equilibrium concentrations of species HA, A^-, and H_3O^+ for solutions of different molarities. Here we illustrate the use of a simplifying approximation that can often be used for weak acids.

EXAMPLE 17.2

Calculating Concentrations of Species in a Weak Acid Solution Using K_a (Approximation Method)

What are the concentrations of nicotinic acid, hydrogen ion, and nicotinate ion and the pH in a solution of 0.10 M nicotinic acid at 25°C? What is the degree of ionization of nicotinic acid? The acid-ionization constant, K_a, is 1.4×10^{-5}.

SOLUTION

STEP 1 Consider that initially the concentration of HNic is 0.10 M and that of Nic^- is 0. The concentration of H_3O^+ is essentially zero (~ 0). In 1 L of solution, the nicotinic acid ionizes to give x mol H_3O^+ and x mol Nic^-, leaving $(0.10 - x)$ mol of nicotinic acid. These data are summarized in the following table:

Concentration (M)	$HNic(aq) + H_2O(l) \rightleftharpoons$	$H_3O^+(aq) +$	$Nic^-(aq)$
Starting	0.10	~ 0	0
Change	$-x$	$+x$	$+x$
Equilibrium	$0.10 - x$	x	x

STEP 2 Now substitute these concentrations and the value of K_a into the equilibrium-constant equation for acid ionization:

$$\frac{[H_3O^+][Nic^-]}{[HNic]} = K_a$$

You get

$$\frac{x^2}{(0.10 - x)} = 1.4 \times 10^{-5}$$

STEP 3 Now solve this equation for x. This is actually a quadratic equation, but it can be simplified by assuming that x is much smaller than 0.10, so that $0.10 - x \simeq 0.10$. The equilibrium-constant equation becomes

$$\frac{x^2}{0.10} \simeq 1.4 \times 10^{-5}$$

or

$$x^2 \simeq 1.4 \times 10^{-5} \times 0.10 = 1.4 \times 10^{-6}$$

Hence,

$$x \simeq 1.2 \times 10^{-3} = 0.0012$$

At this point, you should check to make sure that the assumption that $0.10 - x \simeq 0.10$ is valid. You substitute the value obtained for x into $0.10 - x$.

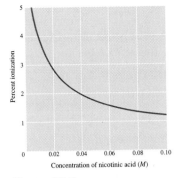

Figure 17.3
Variation of percent ionization of a weak acid with concentration.
On this curve for nicotinic acid, note that the percent ionization is greatest for the most dilute solutions.

$$0.10 - x = 0.10 - 0.0012 = 0.10 \quad \text{(to two significant figures)}$$

The assumption is indeed valid.

The concentrations of nicotinic acid, hydronium ion, and nicotinate ion are **0.10 M, 0.0012 M, and 0.0012 M,** respectively.

The pH of the solution is

$$pH = -\log[H_3O^+] = -\log(0.0012) = \textbf{2.92}$$

The degree of ionization equals the amount per liter of nicotinic acid that ionizes ($x = 0.0012$) divided by the total amount per liter of nicotinic acid initially present (0.10). Thus, the degree of ionization is 0.0012/0.10 = **0.012.**

See Problems 17.23 and 17.24.

In Example 17.2, the degree of ionization of nicotinic acid is relatively small. Only 1.2% of the molecules ionize. It is the small value of the degree of ionization that allows you to neglect x in the term $0.10 - x$ and thereby simplify the calculation.

The degree of ionization of a weak acid depends on both K_a and the concentration of the acid solution. For a given concentration, the larger the K_a, the greater is the degree of ionization. For a given value of K_a, however, the more dilute the solution, the greater is the degree of ionization. Figure 17.3 shows how the percent ionization (degree of ionization $\times$ 100) varies with the concentration of solution.

How do you know when you can use the simplifying assumption of Example 17.2, where you neglected x in the denominator of the equilibrium equation? *It can be shown that this simplifying assumption gives an error of less than 5% if the concentration of acid, C_a, divided by K_a equals 100 or more.* For example, for an acid concentration of 10^{-2} M and K_a of 10^{-5}, $C_a/K_a = 10^{-2}/10^{-5} = 10^3$, so the assumption is valid.

If the simplifying assumption of Example 17.2 is not valid, you can solve the equilibrium equation exactly by using the *quadratic formula.* We illustrate this in the next example with a solution of aspirin (Figure 17.4). ◄

▶ You might find it helpful to compare the results of some equilibrium calculations using the simplifying assumption with exact results from the "Equilibrium Calculator (Equilib)," by Robert Allendorfer, JCE: Software, Madison, WI.

EXAMPLE 17.3

Calculating Concentrations of Species in a Weak Acid Solution Using K_a (Quadratic Formula)

What is the pH at 25°C of the solution obtained by dissolving a 0.325-g tablet of aspirin (acetylsalicylic acid, $HC_9H_7O_4$) in 0.500 L of water? The acid is monoprotic, and K_a equals 3.3×10^{-4} at 25°C.

PROBLEM STRATEGY

If C_a/K_a is less than 100, solve the equation exactly using the quadratic formula.

SOLUTION

The molar mass of $HC_9H_7O_4$ is 180.2 g. From this you find that an aspirin tablet contains 0.00180 mol of the acid. Hence, the concentration of the aspirin solution is 0.00180 mol/0.500 L, or 0.0036 M. $C_a/K_a = 0.0036/3.3 \times 10^{-4} = 11$, which is less than 100, so we must solve the equilibrium equation exactly.

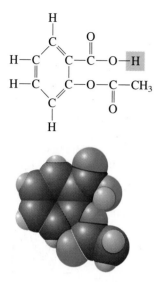

Figure 17.4
The structure of aspirin.
Top: Structural formula of aspirin (acetylsalicylic acid), a well-known headache remedy. Note the acidic hydrogen atom (in color). *Bottom:* Molecular model of aspirin.

STEP 1 Abbreviate the formula for acetylsalicylic acid as HAcs and let x be the amount of H_3O^+ ion and Acs^- ion formed in 1 L of solution. The HAcs is $(0.0036 - x)$ mol. These data are summarized in the following table:

Concentration (M)	$HAcs(aq) + H_2O(l) \rightleftharpoons$	$H_3O^+(aq) +$	$Acs^-(aq)$
Starting	0.0036	~0	0
Change	$-x$	$+x$	$+x$
Equilibrium	$0.0036 - x$	x	x

STEP 2 Substitute into the equilibrium-constant equation and you get

$$\frac{[H_3O^+][Acs^-]}{[HAcs]} = K_a$$

$$\frac{x^2}{0.0036 - x} = 3.3 \times 10^{-4}$$

STEP 3 Rearrange the preceding equation to put it in the form $ax^2 + bx + c = 0$. You get

$$x^2 = (0.0036 - x) \times 3.3 \times 10^{-4} = 1.2 \times 10^{-6} - 3.3 \times 10^{-4} x$$

or

$$x^2 + 3.3 \times 10^{-4} x - 1.2 \times 10^{-6} = 0$$

Now substitute into the quadratic formula.

$$x = \frac{-b \pm \sqrt{b^2 - 4ac}}{2a}$$

$$= \frac{-3.3 \times 10^{-4} \pm \sqrt{(3.3 \times 10^{-4})^2 + 4(1.2 \times 10^{-6})}}{2}$$

$$= \frac{-3.3 \times 10^{-4} \pm 2.2 \times 10^{-3}}{2}$$

If you take the lower sign in $\pm$, you will get a negative value for x. Therefore, taking the upper sign, you get

$$x = [H_3O^+] = 9.4 \times 10^{-4}$$

Now you can calculate the pH.

$$pH = -\log [H_3O^+] = -\log(9.4 \times 10^{-4}) = \mathbf{3.03}$$

See Problems 17.27 and 17.28.

CONCEPT CHECK 17.1

You have prepared dilute solutions of equal molar concentrations of $HC_2H_3O_2$ (acetic acid), HNO_2, HF, and HCN. Rank the solutions from the highest pH to the lowest pH. (Refer to Table 17.1.)

17.2 Polyprotic Acids

In the preceding section, we dealt only with acids releasing one H_3O^+ ion or proton. Some acids, however, have two or more such protons; these acids are called *polyprotic acids*. Sulfuric acid, for example, can lose two protons in aqueous solution. One proton is lost completely to form H_3O^+ (sulfuric acid is a strong acid).

$$H_2SO_4(aq) + H_2O(l) \longrightarrow H_3O^+(aq) + HSO_4^-(aq)$$

The hydrogen sulfate ion, HSO_4^-, may then lose the other proton.

$$HSO_4^-(aq) + H_2O(l) \rightleftharpoons H_3O^+(aq) + SO_4^{2-}(aq)$$

For a weak diprotic acid like carbonic acid, H_2CO_3, there are two simultaneous equilibria to consider.

$$H_2CO_3(aq) + H_2O(l) \rightleftharpoons H_3O^+(aq) + HCO_3^-(aq)$$

$$HCO_3^-(aq) + H_2O(l) \rightleftharpoons H_3O^+(aq) + CO_3^{2-}(aq)$$

Each equilibrium has an associated acid-ionization constant. For the loss of the first proton,

$$K_{a1} = \frac{[H_3O^+][HCO_3^-]}{[H_2CO_3]} = 4.3 \times 10^{-7}$$

and for the loss of the second proton,

$$K_{a2} = \frac{[H_3O^+][CO_3^{2-}]}{[HCO_3^-]} = 4.8 \times 10^{-11}$$

Note that K_{a1} for carbonic acid is much larger than K_{a2} (by a factor of about 1×10^4). This indicates that carbonic acid loses the first proton more easily than the second one.

> In general, the second ionization constant, K_{a2}, of a polyprotic acid is much smaller than the first ionization constant, K_{a1}. In the case of a triprotic acid, the third ionization constant, K_{a3}, is much smaller than the second one, K_{a2}.

See the values for phosphoric acid, H_3PO_4, in Table 17.1.

Calculating the concentrations of various species in a solution of a polyprotic acid might appear complicated, because several equilibria occur at once. However, reasonable assumptions can be made that simplify the calculation, as we show in the next example.

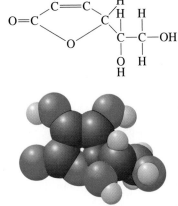

Figure 17.5
The structure of ascorbic acid. *Top:* Structural formula of ascorbic acid (vitamin C); note the acidic hydrogen atoms (in color). *Bottom:* Molecular model of ascorbic acid.

EXAMPLE 17.4

Calculating Concentrations of Species in a Solution of a Diprotic Acid

Ascorbic acid (vitamin C) is a diprotic acid, $H_2C_6H_6O_6$ (Figure 17.5). What is the pH of a $0.10\,M$ solution? What is the concentration of ascorbate ion, $C_6H_6O_6^{2-}$? The acid ionization constants are $K_{a1} = 7.9 \times 10^{-5}$ and $K_{a2} = 1.6 \times 10^{-12}$.

542 CHAPTER 17 Acid–Base Equilibria

PROBLEM STRATEGY

By considering only the first ionization, set up a table of concentrations (starting, change, and equilibrium). Substitute these values into the equilibrium equation for K_{a1}, solve for $x = [H_3O^+]$, and then solve for pH. Now calculate the ascorbate ion concentration.

SOLUTION

CALCULATION OF pH Abbreviate the formula for ascorbic acid as H_2Asc. Hydronium ions are produced by two successive acid ionizations.

$$H_2Asc(aq) + H_2O(l) \rightleftharpoons H_3O^+(aq) + HAsc^-(aq); K_{a1} = 7.9 \times 10^{-5}$$

$$HAsc^-(aq) + H_2O(l) \rightleftharpoons H_3O^+(aq) + Asc^{2-}(aq); K_{a2} = 1.6 \times 10^{-12}$$

Let x be the amount of H_3O^+ formed in the first ionization.

Concentration (M)	$H_2Asc(aq) + H_2O(l) \rightleftharpoons$	$H_3O^+(aq) +$	$HAsc^-(aq)$
Starting	0.10	~0	0
Change	$-x$	$+x$	$+x$
Equilibrium	$0.10 - x$	x	x

Now substitute into the equilibrium-constant equation for the first ionization.

$$\frac{[H_3O^+][HAsc^-]}{[H_2Asc]} = K_{a1}$$

$$\frac{x^2}{0.10 - x} = 7.9 \times 10^{-5}$$

Assuming x to be much smaller than 0.10, you get

$$\frac{x^2}{0.10} \simeq 7.9 \times 10^{-5}$$

or

$$x^2 \simeq 7.9 \times 10^{-5} \times 0.10$$

$$x \simeq 2.8 \times 10^{-3} = 0.0028$$

The hydronium-ion concentration is 0.0028 M, so

$$pH = -\log [H_3O^+] = -\log(0.0028) = \mathbf{2.55}$$

ASCORBATE-ION CONCENTRATION Ascorbate ion, Asc^{2-}, which we will call y, is produced only in the second reaction. Assume the starting concentrations of H_3O^+ and $HAsc^-$ for this reaction to be those from the first equilibrium. The amounts of each species in 1 L of solution are as follows:

Concentration (M)	$HAsc^-(aq) + H_2O(l) \rightleftharpoons$	$H_3O^+(aq) +$	$Asc^{2-}(aq)$
Starting	0.0028	0.0028	0
Change	$-y$	$+y$	$+y$
Equilibrium	$0.0028 - y$	$0.0028 + y$	y

Now substitute into the equilibrium-constant equation for the second ionization.

Acid Rain

ENVIRONMENT

Acid rain means rain having a pH lower than that of natural rain (pH 5.6). Natural rain dissolves carbon dioxide from the atmosphere to give a slightly acidic solution. The pH of rain in eastern North America and western Europe, however, is approximately 4 and sometimes lower. This acidity is primarily the result of the dissolving in rainwater of sulfur oxides and nitrogen oxides from human activities. In the northeastern United States, the strong acid components in acid rain are about 62% sulfuric acid, 32% nitric acid, and 6% hydrochloric acid.

The sulfuric acid in acid rain has been traced to the burning of fossil fuels and to the burning of sulfide ores in the production of metals, such as zinc and copper. Coal, for example, contains some sulfur mainly as pyrite, or iron(II) disulfide, FeS_2. When this burns in air, it produces sulfur dioxide.

$$4FeS_2(s) + 11O_2(g) \longrightarrow 2Fe_2O_3(s) + 8SO_2(g)$$

In the presence of dust particles and other substances in polluted air, the sulfur dioxide oxidizes further to give sulfur trioxide, which reacts with water to form sulfuric acid.

$$2SO_2(g) + O_2(g) \longrightarrow 2SO_3(g)$$

$$SO_3(g) + H_2O(l) \longrightarrow H_2SO_4(aq)$$

Acid rain can be harmful to some plants, to fish (by changing the pH of lake water), and to structural materials and monuments (Figure 17.6). Marble, for example, is composed of calcium carbonate, $CaCO_3$, which dissolves in water of low pH.

$$H_3O^+(aq) + CaCO_3(s) \longrightarrow Ca^{2+}(aq) + HCO_3^-(aq) + H_2O(l)$$

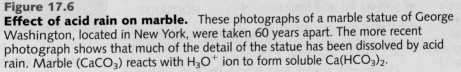

Figure 17.6
Effect of acid rain on marble. These photographs of a marble statue of George Washington, located in New York, were taken 60 years apart. The more recent photograph shows that much of the detail of the statue has been dissolved by acid rain. Marble ($CaCO_3$) reacts with H_3O^+ ion to form soluble $Ca(HCO_3)_2$.

A Checklist for Review

Important Terms

acid-ionization constant (17.1)
degree of ionization (17.1)

base-ionization constant (17.3)
hydrolysis (17.4)

common-ion effect (17.5)
buffer (17.6)
Henderson–Hasselbalch equation (17.6)

acid–base titration curve (17.7)
equivalence point (17.7)

Key Equations

$$K_a = \frac{[H_3O^+][A^-]}{[HA]} \qquad K_a K_b = K_w$$

$$K_b = \frac{[HB^+][OH^-]}{[B]} \qquad pH = pK_a + \log \frac{[base]}{[acid]}$$

Summary of Facts and Concepts

When a weak acid dissolves in water, it ionizes to give hydronium ions and the conjugate base ions. The equilibrium constant for this *acid ionization* is $K_a = [H_3O^+][A^-]/[HA]$, where HA is the general formula for the acid. The constant K_a can be determined from the pH of an acid solution of known concentration. Once obtained, the acid-ionization constant can be used to find the concentrations of species in any solution of the acid. In the case of a *diprotic* acid, H_2A, the concentration of H_3O^+ and HA^- are calculated from K_{a1}, and the concentration of A^{2-} equals K_{a2}. Similar considerations apply to *base ionizations*.

Solutions of salts may be acidic or basic because of *hydrolysis* of the ions. The equilibrium constant for hydrolysis equals K_a for a cation or K_b for an anion. Calculation of the pH of a solution of a salt in which one ion hydrolyzes is fundamentally the same as calculation of the pH of a solution of an acid or base. However, K_a or K_b for an ion is usually obtained from the conjugate base or acid by applying the equation $K_a K_b = K_w$.

A *buffer* is a solution that can resist changes in pH when small amounts of acid or base are added to it. A buffer contains either a weak acid and its conjugate base or a weak base and its conjugate acid. The concentrations of acid and base conjugates are approximately equal. Different pH ranges are possible for acid–base conjugates, depending on their ionization constants.

An *acid–base titration curve* is a plot of the pH of the solution against the volume of reactant added. During the titration of a strong acid by a strong base, the pH changes slowly at first. Then, as the amount of base nears the *equivalence point,* the pH rises abruptly, changing by several units. The pH at the equivalence point is 7.0. A similar curve is obtained when a weak acid is titrated by a strong base. The pH at the equivalence point is greater than 7.0 because of hydrolysis of the salt produced. An indicator must be chosen that changes color within a pH range near the equivalence point.

Operational Skills

1. **Determining K_a (or K_b) from the solution pH** Given the molarity and pH of a solution of a weak acid, calculate K_a for the acid (EXAMPLE 17.1).
2. **Calculating concentrations of species in a weak acid solution using K_a** Given K_a, calculate the hydronium-ion concentration and pH of a solution of a weak acid of known molarity (EXAMPLES 17.2, 17.3). Given K_{a1}, K_{a2}, and the molarity of a diprotic acid solution, calculate the pH and the concentrations of H_3O^+, HA^-, and A^{2-} (EXAMPLE 17.4).
3. **Calculating concentrations of species in a weak base solution using K_b** Given K_b, calculate the hydronium-ion concentration and pH of a solution of a weak base of known molarity. (EXAMPLE 17.5)

4. **Predicting whether a salt solution is acidic, basic, or neutral** Decide whether an aqueous solution of a given salt is acidic, basic, or neutral. (EXAMPLE 17.6)
5. **Obtaining K_a from K_b or K_b from K_a** Calculate K_a for a cation or K_b for an anion from the ionization constant of the conjugate base or acid. (EXAMPLE 17.7)
6. **Calculating concentrations of species in a salt solution** Given the concentration of a solution of a salt in which one ion hydrolyzes, and given the ionization constant of the conjugate acid or base of this ion, calculate the H_3O^+ concentration. (EXAMPLE 17.8)
7. **Calculating the common-ion effect on acid ionization** Given K_a and the concentrations of weak acid and strong acid in a solution, calculate the degree of ionization of the

weak acid **(EXAMPLE 17.9)**. Given K_a and the concentrations of weak acid and its salt in a solution, calculate the pH **(EXAMPLE 17.10)**.

8. **Calculating the pH of a buffer when a strong acid or strong base is added** Calculate the pH of a given volume of buffer solution (given the concentrations of conjugate acid and base in the buffer) to which a specified amount of strong acid or strong base is added. **(EXAMPLE 17.11)**

9. **Calculating the pH of a solution of a strong acid and a strong base** Calculate the pH during the titration of a strong acid by a strong base, given the volumes and concentrations of the acid and base. **(EXAMPLE 17.12)**

10. **Calculating the pH at the equivalence point in the titration of a weak acid by a strong base** Calculate the pH at the equivalence point for the titration of a weak acid by a strong base **(EXAMPLE 17.13)**. Be able to do the same type of calculation for the titration of a weak base by a strong acid.

Review Questions

17.1 Write an equation for the ionization of hydrogen cyanide, HCN, in aqueous solution. What is the equilibrium expression K_a for this acid ionization?

17.2 Which of the following is the weakest acid: $HClO_4$, HCN, or $HC_2H_3O_2$? See Table 16.1 and Table 17.1.

17.3 Describe how the degree of ionization of a weak acid changes as the concentration increases.

17.4 Consider a solution of $0.0010 \, M$ HF ($K_a = 6.8 \times 10^{-4}$). In solving for the concentrations of species in this solution, could you use the simplifying assumption in which you neglect x in the denominator of the equilibrium equation? Explain.

17.5 Phosphorous acid, H_2PHO_3, is a diprotic acid. Write equations for the acid ionizations. Write the expressions for K_{a1} and K_{a2}.

17.6 Write the equation for the ionization of aniline, $C_6H_5NH_2$, in aqueous solution. Write the expression for K_b.

17.7 Which of the following is the strongest base: NH_3, $C_6H_5NH_2$, or CH_3NH_2? See Table 17.2.

17.8 What is meant by the common-ion effect? Give an example.

17.9 The pH of $0.10 \, M$ CH_3NH_2 (methylamine) is 11.8. When the chloride salt of methylamine, CH_3NH_3Cl, is added to this solution, does the pH increase or decrease? Explain, using Le Chatelier's principle and the common-ion effect.

17.10 Define the term *buffer*. Give an example.

17.11 What is meant by the capacity of a buffer? Describe a buffer with low capacity and the same buffer with greater capacity.

17.12 Describe the pH changes that occur during the titration of a weak base by a strong acid. What is meant by the term *equivalence point*?

Conceptual Problems

17.13 You have 0.10-mol samples of three acids identified simply as HX, HY, and HZ. For each acid, you make up $0.10 \, M$ solutions by adding sufficient water to each of the acid samples. When you measure the pH of these samples, you find that the pH of HX is greater than the pH of HY, which in turn is greater than the pH of HZ.
a. Which of the acids is the least ionized in its solution?
b. Which acid has the largest K_a?

17.14 What reaction occurs when each of the following is dissolved in water?
a. HF   b. NaF   c. $C_6H_5NH_2$ d. $C_6H_5NH_3Cl$

17.15 You have the following solutions, all of the same molar concentrations: KBr, HBr, CH_3NH_2, and NH_4Cl. Rank them from the lowest to the highest hydroxide-ion concentrations.

17.16 Rantidine is a nitrogen base that is used to control stomach acidity by suppressing the stomach's production of hydrochloric acid. The compound is present in Zantac® as the chloride salt (rantidinium chloride; also called rantidine hydrochloride). Do you expect a solution of rantidine hydrochloride to be acidic, basic, or neutral? Explain by means of a general chemical equation.

17.17 A chemist prepares dilute solutions of equal molar concentrations of NH_3, NH_4Br, NaF, and NaCl. Rank these solutions from highest pH to lowest pH.

17.18 You are given the following acid–base titration data, where each point on the graph represents the pH after adding a given volume of titrant (substance being added during the titration).

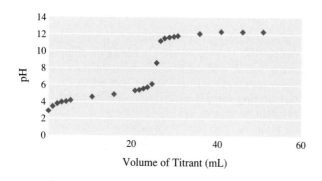

a. What substance is being titrated, a strong acid, strong base, weak acid, or weak base?
b. What is the pH at the equivalence point of the titration?
c. What indicator might you use to perform this titration? Explain.

Practice Problems

Note: For values of K_a and K_b that are not given in the following problems, see Tables 17.1 and 17.2.

Acid Ionization

17.19 Write chemical equations for the acid ionizations of each of the following weak acids (express these in terms of H_3O^+).
a. HOCN (cyanic acid) b. HF (hydrofluoric acid)
c. HIO (hypoiodous acid) d. HCO_2H (formic acid)

17.20 Write chemical equations for the acid ionizations of each of the following weak acids (express these in terms of H_3O^+).
a. HN_3 (hydrazoic acid) b. $HClO_2$ (chlorous acid)
c. HNO_2 (nitrous acid) d. HCN (hydrocyanic acid)

17.21 Acrylic acid, whose formula is $HC_3H_3O_2$ or $HO_2CCH{=}CH_2$, is used in the manufacture of plastics. A 0.10 M aqueous solution of acrylic acid has a pH of 2.63. What is K_a for acrylic acid?

17.22 Heavy metal azides, which are salts of hydrazoic acid, HN_3, are used as explosive detonators. A solution of 0.20 M hydrazoic acid has a pH of 3.21. What is the K_a for hydrazoic acid?

17.23 Boric acid, $B(OH)_3$, is used as a mild antiseptic. What is the pH of a 0.025 M aqueous solution of boric acid? What is the degree of ionization of boric acid in this solution? The hydronium ion arises principally from the reaction

$$B(OH)_3(aq) + 2H_2O(l) \rightleftharpoons B(OH)_4^-(aq) + H_3O^+(aq)$$

The equilibrium constant for this reaction is 5.9×10^{-10}.

17.24 Formic acid, $HCHO_2$, is used to make methyl formate (a fumigant for dried fruit) and ethyl formate (an artificial rum flavor). What is the pH of a 0.12 M solution of formic acid? What is the degree of ionization of $HCHO_2$ in this solution? See Table 17.1 for K_a.

17.25 A solution of acetic acid, $HC_2H_3O_2$, on a laboratory shelf was of undetermined concentration. If the pH of the solution was found to be 2.68, what was the concentration of the acetic acid?

17.26 A chemist wanted to determine the concentration of a solution of lactic acid, $HC_3H_5O_3$. She found that the pH of the solution was 2.51. What was the concentration of the solution? The K_a of lactic acid is 1.4×10^{-4}.

17.27 Hydrofluoric acid, HF, unlike hydrochloric acid, is a weak electrolyte. What is the hydronium-ion concentration and the pH of a 0.040 M aqueous solution of HF?

17.28 Chloroacetic acid, $HC_2H_2ClO_2$, has a greater acid strength than acetic acid, because the electronegative chlorine atom pulls electrons away from the O—H bond and thus weakens it. Calculate the hydronium-ion concentration and the pH of a 0.015 M solution of chloroacetic acid. K_a is 1.3×10^{-3}.

17.29 Phthalic acid, $H_2C_8H_4O_4$, is a diprotic acid used in the synthesis of phenolphthalein indicator. $K_{a1} = 1.2 \times 10^{-3}$, and $K_{a2} = 3.9 \times 10^{-6}$. (a) Calculate the hydronium-ion concentration of a 0.015 M solution. (b) What is the concentration of the $C_8H_4O_4^{2-}$ ion in the solution?

17.30 Carbonic acid, H_2CO_3, can be found in a wide variety of body fluids (from dissolved CO_2). (a) Calculate the hydronium-ion concentration of a 5.45×10^{-4} M H_2CO_3 solution. (b) What is the concentration of CO_3^{2-}?

Base Ionization

17.31 Write the chemical equation for the base ionization of methylamine, CH_3NH_2. Write the K_b expression for methylamine.

17.32 Write the chemical equation for the base ionization of aniline, $C_6H_5NH_2$. Write the K_b expression for aniline.

17.33 Ethanolamine, $HOC_2H_4NH_2$, is a viscous liquid with an ammonialike odor; it is used to remove hydrogen sulfide from natural gas. A 0.15 M aqueous solution of ethanolamine has a pH of 11.34. What is K_b for ethanolamine?

17.34 Trimethylamine, $(CH_3)_3N$, is a gas with a fishy, ammonialike odor. An aqueous solution that is 0.25 M trimethylamine has a pH of 11.63. What is K_b for trimethylamine?

17.35 What is the concentration of hydroxide ion in a 0.065 M aqueous solution of methylamine, CH_3NH_2? What is the pH?

17.36 What is the concentration of hydroxide ion in a 0.20 M aqueous solution of hydroxylamine, NH_2OH? What is the pH?

Acid–Base Properties of Salt Solutions; Hydrolysis

17.37 Note whether hydrolysis occurs for each of the following ions. If hydrolysis does occur, write the chemical equation for it. Then write the equilibrium expression for the acid or base ionization (whichever occurs).
a. I^- b. CHO_2^- c. $CH_3NH_3^+$ d. IO^-

17.38 Note whether hydrolysis occurs for each of the following ions. If hydrolysis does occur, write the chemical equation for it. Then write the equilibrium expression for the acid or base ionization (whichever occurs).
a. NO_2^- b. Br^- c. NO_3^- d. $NH_2NH_3^+$

17.39 Write the equation for the acid ionization of the $Zn(H_2O)_6^{2+}$ ion.

17.40 Write the equation for the acid ionization of the $Cu(H_2O)_6^{2+}$ ion.

17.41 For each of the following salts, indicate whether the aqueous solution will be acidic, basic, or neutral.
a. $Fe(NO_3)_3$ b. Na_2CO_3 c. $Ca(CN)_2$ d. NH_4ClO_4

17.42 Note whether aqueous solutions of each of the following salts will be acidic, basic, or neutral.
a. Na_2S b. $Cu(NO_3)_2$ c. $KClO_4$ d. CH_3NH_3Cl

17.43 Obtain (a) the K_b value for NO_2^-; (b) the K_a value for $C_5H_5NH^+$ (pyridinium ion).

17.44 Obtain (a) the K_b value for ClO^-; (b) the K_a value for NH_3OH^+ (hydroxylammonium ion).

17.45 What is the pH of a 0.025 M aqueous solution of sodium propionate, $NaC_3H_5O_2$? What is the concentration of propionic acid in the solution?

17.46 Calculate the OH^- concentration and pH of a 0.010 M aqueous solution of sodium cyanide, $NaCN$. Finally, obtain the hydronium-ion, CN^-, and HCN concentrations.

Common-Ion Effect

17.47 Calculate the degree of ionization of (a) 0.80 M HF (hydrofluoric acid); (b) the same solution that is also 0.10 M HCl.

17.48 Calculate the degree of ionization of (a) 0.20 M $HCHO_2$ (formic acid); (b) the same solution that is also 0.10 M HCl.

17.49 What is the pH of a solution that is 0.10 M KNO_2 and 0.15 M HNO_2 (nitrous acid)?

17.50 What is the pH of a solution that is 0.20 M KOCN and 0.10 M HOCN (cyanic acid)?

Buffers

17.51 A buffer is prepared by adding 45.0 mL of 0.15 M NaF to 35.0 mL of 0.10 M HF. What is the pH of the final solution?

17.52 A buffer is prepared by adding 115 mL of 0.30 M NH_3 to 145 mL of 0.15 M NH_4NO_3. What is the pH of the final solution?

17.53 What is the pH of a buffer solution that is 0.10 M NH_3 and 0.10 M NH_4^+? What is the pH if 12 mL of 0.20 M hydrochloric acid is added to 125 mL of buffer?

17.54 A buffer is prepared by mixing 525 mL of 0.50 M formic acid, $HCHO_2$, and 475 mL of 0.50 M sodium formate, $NaCHO_2$. Calculate the pH. What would be the pH of 85 mL of the buffer to which 8.5 mL of 0.15 M hydrochloric acid has been added?

17.55 How many moles of sodium acetate must be added to 2.0 L of 0.10 M acetic acid to give a solution that has a pH equal to 5.00? Ignore the volume change due to the addition of sodium acetate.

17.56 How many moles of hydrofluoric acid, HF, must be added to 500.0 mL of 0.30 M sodium fluoride to give a buffer of pH 3.50? Ignore the volume change due to the addition of hydrofluoric acid.

Titration Curves

17.57 What is the pH of a solution in which 15 mL of 0.10 M NaOH is added to 25 mL of 0.10 M HCl?

17.58 What is the pH of a solution in which 40 mL of 0.10 M NaOH is added to 25 mL of 0.10 M HCl?

17.59 Find the pH of the solution obtained when 32 mL of 0.087 M ethylamine is titrated to the equivalence point with 0.15 M HCl.

17.60 What is the pH at the equivalence point when 22 mL of 0.20 M hydroxylamine is titrated with 0.15 M HCl?

17.61 Calculate the pH of a solution obtained by mixing 500.0 mL of 0.10 M NH_3 with 200.0 mL of 0.15 M HCl.

17.62 Calculate the pH of a solution obtained by mixing 35.0 mL of 0.15 M acetic acid with 25.0 mL of 0.10 M sodium acetate.

General Problems

17.63 Calculate the base-ionization constants for CN^- and CO_3^{2-}. Which ion is the stronger base?

17.64 Calculate the base-ionization constants for PO_4^{3-} and SO_4^{2-}. Which ion is the stronger base?

17.65 Calculate the pH of a 0.15 M aqueous solution of aluminum chloride, $AlCl_3$. The acid ionization of hydrated aluminum ion is

$$Al(H_2O)_6^{3+}(aq) + H_2O(l) \rightleftharpoons$$
$$Al(H_2O)_5OH^{2+}(aq) + H_3O^+(aq)$$

and K_a is 1.4×10^{-5}.

17.66 Calculate the pH of a 0.15 M aqueous solution of zinc chloride, $ZnCl_2$. The acid ionization of hydrated zinc ion is

$$Zn(H_2O)_6^{2+}(aq) + H_2O(l) \rightleftharpoons$$
$$Zn(H_2O)_5OH^-(aq) + H_3O^+(aq)$$

and K_a is 2.5×10^{-10}.

17.67 Blood contains several acid–base systems that tend to keep its pH constant at about 7.4. One of the most important buffer systems involves carbonic acid and hydrogen carbonate ion. What must be the ratio of $[HCO_3^-]$ to $[H_2CO_3]$ in the blood if the pH is 7.40?

17.68 Codeine, $C_{18}H_{21}NO_3$, is an alkaloid ($K_b = 6.2 \times 10^{-9}$) used as a painkiller and cough suppressant. A solution of codeine is acidified with hydrochloric acid to pH 4.50. What is the ratio of the concentration of the conjugate acid of codeine to that of the base codeine?

17.69 Calculate the pH of a solution obtained by mixing 456 mL of 0.10 M hydrochloric acid with 285 mL of 0.15 M sodium hydroxide. Assume the combined volume is the sum of the two original volumes.

17.70 Calculate the pH of a solution made up from 2.0 g of potassium hydroxide dissolved in 115 mL of 0.19 M perchloric acid. Assume the change in volume due to adding potassium hydroxide is negligible.

17.71 Find the pH of the solution obtained when 25 mL of 0.065 M benzylamine, $C_7H_7NH_2$, is titrated to the equivalence point with 0.050 M hydrochloric acid. K_b for benzylamine is 4.7×10^{-10}.

17.72 What is the pH of the solution obtained by titrating 1.24 g of sodium hydrogen sulfate, $NaHSO_4$, dissolved in 50.0 mL of water with 0.180 M sodium hydroxide until the equivalence point is reached? Assume that any volume change

due to adding the sodium hydrogen sulfate or to mixing the solutions is negligible.

17.73 Methylammonium chloride is a salt of methylamine, CH_3NH_2. A 0.10 M solution of this salt has a pH of 5.82.
a. Calculate the value of the equilibrium constant for the reaction

$$CH_3NH_3^+ + H_2O \rightleftharpoons CH_3NH_2 + H_3O^+$$

b. What is the K_b value for methylamine?
c. What is the pH of a solution in which 0.450 mole of solid methylammonium chloride is added to 1.00 L of a 0.250 M solution of methylamine? Assume no volume change.

17.74 Sodium benzoate is a salt of benzoic acid, C_6H_5COOH. A 0.15 M solution of this salt has a pOH of 5.31 at room temperature.
a. Calculate the value of the equilibrium constant for the reaction

$$C_6H_5COO^- + H_2O \rightleftharpoons C_6H_5COOH + OH^-$$

b. What is the K_a value for benzoic acid?
c. Benzoic acid has a low solubility in water. What is its molar solubility if a saturated solution has a pH of 2.83 at room temperature?

17.75 Each of the following statements concerns a 0.010 M solution of a weak acid, HA. Briefly describe why each statement is either true or false.
a. [HA] is approximately equal to 0.010 M.
b. [HA] is much greater than $[A^-]$.
c. $[OH^-]$ is approximately equal to $[H_3O^+]$.
d. The pH is 2.
e. The H_3O^+ concentration is 0.010 M.
f. $[H_3O^+]$ is approximately equal to $[A^-]$.

17.76 Each of the following statements concerns a 0.10 M solution of a weak organic base, B. Briefly describe why each statement is either true or false.
a. [B] is approximately equal to 0.10 M.
b. [B] is much greater than $[HB^+]$.
c. $[H_3O^+]$ is greater than $[HB^+]$.
d. The pH is 13.
e. $[HB^+]$ is approximately equal to $[OH^-]$.
f. $[OH^-]$ equals 0.10 M.

17.77 (a) Draw a pH titration curve that represents the titration of 50.0 mL of 0.10 M NH_3 by the addition of 0.10 M HCl from a buret. Label the axes and put a scale on each axis. Show where the equivalence point and the buffer region are on

the titration curve. You should do calculations for the 0%, 30%, 50%, and 100% titration points.
(b) Is the solution neutral, acidic, or basic at the equivalence point? Why?

17.78 (a) Draw a pH titration curve that represents the titration of 25.0 mL of 0.15 M propionic acid, CH_3CH_2COOH, by the addition of 0.15 M KOH from a buret. Label the axes and put a scale on each axis. Show where the equivalence point and the buffer region are on the titration curve. You should do calculations for the 0%, 50%, 60%, and 100% titration points.
(b) Is the solution neutral, acidic, or basic at the equivalence point? Why?

17.79 The equilibrium equations and K_a values for three reaction systems are given below.

$$H_2C_2O_4(aq) + H_2O \rightleftharpoons$$
$$H_3O^+(aq) + HC_2O_4^-(aq); K_a = 5.6 \times 10^{-2}$$

$$H_3PO_4(aq) + H_2O \rightleftharpoons$$
$$H_3O^+(aq) + H_2PO_4^-(aq); K_a = 6.9 \times 10^{-3}$$

$$HCOOH(aq) + H_2O \rightleftharpoons$$
$$H_3O^+(aq) + HCOO^-(aq); K_a = 1.7 \times 10^{-4}$$

a. Which conjugate pair would be best for preparing a buffer with a pH of 2.88?
b. How would you prepare 50 mL of a buffer with a pH of 2.88 assuming that you had available 0.10 M solutions of each pair?

17.80 The equilibrium equations and K_a values for three reaction systems are given below.

$$NH_4^+(aq) + H_2O \rightleftharpoons$$
$$H_3O^+ + NH_3(aq); K_a = 5.6 \times 10^{-10}$$

$$H_2CO_3(aq) + H_2O \rightleftharpoons$$
$$H_3O^+(aq) + HCO_3^-(aq); K_a = 4.3 \times 10^{-7}$$

$$H_2PO_4^-(aq) + H_2O \rightleftharpoons$$
$$H_3O^+(aq) + HPO_4^{2-}(aq); K_a = 6.2 \times 10^{-8}$$

a. Which conjugate pair would be best for preparing a buffer with a pH of 6.96? Why?

b. How would you prepare 100 mL of a buffer with a pH of 6.96 assuming that you had available 0.10 M solutions of each pair?

17.81 A 25.0-mL sample of hydroxylamine is titrated to the equivalence point with 35.8 mL of 0.150 M HCl.
a. What was the concentration of the original hydroxylamine solution?
b. What is the pH at the equivalence point?
c. Which indicators, bromphenol blue, methyl red, or phenolphthalein, should be used to detect the end point of the titration? Why?

17.82 A 25.00-mL sample contains 0.562 g of $NaHCO_3$. This sample is used to standardize an NaOH solution. At the equivalence point, 42.36 mL of NaOH has been added.
a. What was the concentration of the NaOH?
b. What is the pH at the equivalence point?
c. Which indicator, bromthymol blue, methyl violet, or alizarin yellow R, should be used in the titration? Why?

17.83 Tartaric acid is a weak diprotic fruit acid with $K_1 = 1.0 \times 10^{-3}$ and $K_2 = 4.6 \times 10^{-5}$.
a. Letting the symbol H_2A represent tartaric acid, write the chemical equations that represent K_1 and K_2. Write the chemical equation that represents $K_1 \times K_2$.
b. Qualitatively describe the relative concentrations of H_2A, HA^-, A^{2-}, and H_3O^+ in a solution that is about 0.5 M in tartaric acid.
c. Calculate the pH of a 0.0250 M tartaric acid solution and the equilibrium concentration of $[H_2A]$.
d. What is the A^{2-} concentration?

17.84 Malic acid is a weak diprotic organic acid with $K_1 = 4.0 \times 10^{-4}$ and $K_2 = 9.0 \times 10^{-6}$.
a. Letting the symbol H_2A represent malic acid, write the chemical equations that represent K_1 and K_2. Write the chemical equation that represents $K_1 \times K_2$.
b. Qualitatively describe the relative concentrations of H_2A, HA^-, A^{2-}, and H_3O^+ in a solution that is about one molar in malic acid.
c. Calculate the pH of a 0.0100 M malic acid solution and the equilibrium concentration of $[H_2A]$.
d. What is the A^{2-} concentration?

Cumulative-Skills Problems

17.85 The pH of a white vinegar solution is 2.45. This vinegar is an aqueous solution of acetic acid with a density of 1.09 g/mL. What is the mass percentage of acetic acid in the solution?

17.86 The pH of a household cleaning solution is 11.87. This cleanser is an aqueous solution of ammonia with a density of 1.00 g/mL. What is the mass percentage of ammonia in the solution?

17.87 A chemist needs a buffer with pH 4.35. How many milliliters of pure acetic acid (density = 1.049 g/mL) must be added to 465 mL of 0.0941 M NaOH solution to obtain such a buffer?

17.88 A chemist needs a buffer with pH 3.50. How many milliliters of pure formic acid (density = 1.220 g/mL) must be added to 325 mL of 0.0857 M NaOH solution to obtain such a buffer?

Formation of the violet copper–ammonia complex ion.

Solubility and Complex-Ion Equilibria

Many natural processes depend on the precipitation or the dissolving of a slightly soluble salt. For example, caves form in limestone (calcium carbonate) over thousands of years as groundwater seeps through cracks, dissolving out cavities in the rock. Kidney stones form when salts such as calcium phosphate or calcium oxalate slowly precipitate in the kidney (Figure 18.1). Poisoning by oxalic acid is also explained by the precipitation of the calcium salt. If oxalic acid is accidentally ingested, the oxalate-ion concentration in the blood may increase sufficiently to precipitate calcium oxalate. Calcium ion, which is needed for proper muscle control, is then removed from the blood, and muscle tissues go into spasm.

To understand such phenomena quantitatively, you must be able to solve problems in solubility equilibria. Calcium oxalate kidney stones form when the concentrations of calcium ion and oxalate ion are sufficiently great. What is the relationship between the concentrations of ions and the solubility of a salt? What is the minimum concentration of oxalate ion that gives a precipitate of the calcium salt from a 0.0025 M solution of Ca^{2+} (the approximate concentration of calcium ion in blood plasma)? What is the effect of pH on the solubility of this salt? We will look at questions such as these in this chapter.

Solubility Equilibria

We will now consider equilibria of slightly soluble, or nearly insoluble, ionic compounds and their equilibrium constants. We will show some applications of the *solubility product constant*.

18.1 The Solubility Product Constant

Figure 18.1
A kidney stone. Kidney stones are usually calcium phosphate or calcium oxalate that precipitates as a crystalline mass.

▶ An ionic equilibrium is affected to a small extent by the presence of ions not directly involved in the equilibrium. We will ignore this effect here.

When an ionic compound is dissolved in water, it usually goes into solution as the ions. When an excess of a slightly soluble ionic compound is mixed with water, an equilibrium occurs between the solid compound and the ions in the saturated solution. For the salt calcium oxalate, CaC_2O_4, you have the following equilibrium:

$$CaC_2O_4(s) \xrightleftharpoons{H_2O} Ca^{2+}(aq) + C_2O_4^{2-}(aq)$$

The equilibrium constant for this solubility process is called the *solubility product constant* of CaC_2O_4. It is written

$$K_{sp} = [Ca^{2+}][C_2O_4^{2-}]$$

In general, the **solubility product constant (K_{sp})** is *the equilibrium constant for the solubility equilibrium of a slightly soluble (or nearly insoluble) ionic compound.* It equals the product of the equilibrium concentrations of the ions in the compound, each concentration raised to a power equal to the number of such ions in the formula of the compound. Like any equilibrium constant, K_{sp} depends on the temperature, but at a given temperature it has a constant value for various concentrations of the ions. ◀

Lead(II) iodide, PbI_2, is another slightly soluble salt. The equilibrium in water is

$$PbI_2(s) \rightleftharpoons Pb^{2+}(aq) + 2I^-(aq)$$

and the expression for the equilibrium constant, or solubility product constant, is

$$K_{sp} = [Pb^{2+}][I^-]^2$$

EXAMPLE 18.1

Writing Solubility Product Expressions

Write the solubility product expressions for the following salts: a. AgCl; b. Hg_2Cl_2; c. $Pb_3(AsO_4)_2$.

SOLUTION

The equilibria and solubility product expressions are

a. $AgCl(s) \rightleftharpoons Ag^+(aq) + Cl^-(aq);$ $K_{sp} = [Ag^+][Cl^-]$

b. $Hg_2Cl_2(s) \rightleftharpoons Hg_2^{2+}(aq) + 2Cl^-(aq)$; $K_{sp} = [Hg_2^{2+}][Cl^-]^2$

Note that the mercury(I) ion is Hg_2^{2+}. (Mercury also has salts with the mercury(II) ion, Hg^{2+}.)

c. $Pb_3(AsO_4)_2(s) \rightleftharpoons 3Pb^{2+}(aq) + 2AsO_4^{3-}(aq)$; $K_{sp} = [Pb^{2+}]^3[AsO_4^{3-}]^2$

See Problems 18.13 and 18.14.

The solubility product constant, K_{sp}, of a slightly soluble ionic compound is expressed in terms of the molar concentrations of ions in the saturated solution. These ion concentrations are in turn related to the *molar solubility* of the ionic compound, which is the moles of compound that dissolve to give a liter of saturated solution. The next two examples show how to determine the solubility product constant from the solubility of a slightly soluble ionic compound.

EXAMPLE 18.2

Calculating K_{sp} from the Solubility (Simple Example)

A liter of a solution saturated at 25°C with calcium oxalate, CaC_2O_4, is evaporated to dryness, giving a 0.0061-g residue of CaC_2O_4. Calculate the solubility product constant for this salt at 25°C.

PROBLEM STRATEGY

The grams of residue from a liter of saturated solution equals the solubility in grams per liter. Convert this to molar solubility (solubility in moles per liter). Then follow the three steps similar to those for an equilibrium calculation: set up a table of ion concentrations (starting, change, and equilibrium) and calculate K_{sp} from the concentrations. The starting concentrations are those before any solid has dissolved (= 0), and the change values are obtained from the molar solubility.

SOLUTION

The solubility of calcium oxalate is 0.0061 g/L of *solution*, and the formula weight of CaC_2O_4 is 128 amu. Convert grams per liter to moles per liter:

$$\text{Molar solubility of } CaC_2O_4 = 0.0061 \text{ g } CaC_2O_4/L \times \frac{1 \text{ mol } CaC_2O_4}{128 \text{ g } CaC_2O_4}$$
$$= 4.8 \times 10^{-5} \text{ mol } CaC_2O_4/L$$

Now look at the equilibrium problem.

STEP 1 Suppose you mix solid CaC_2O_4 in a liter of solution. Of this solid, 4.8×10^{-5} mol will dissolve to form 4.8×10^{-5} mol of each ion. The results are summarized in the following table.

Concentration (M)	$CaC_2O_4(s) \rightleftharpoons$	$Ca^{2+}(aq)$	$+$	$C_2O_4^{2-}(aq)$
Starting		0		0
Change		$+4.8 \times 10^{-5}$		$+4.8 \times 10^{-5}$
Equilibrium		4.8×10^{-5}		4.8×10^{-5}

STEP 2 You now substitute into the equilibrium-constant equation:

$$K_{sp} = [Ca^{2+}][C_2O_4^{2-}] = (4.8 \times 10^{-5})(4.8 \times 10^{-5}) = \mathbf{2.3 \times 10^{-9}}$$

EXAMPLE 18.3

Calculating K_{sp} from the Solubility (More Complicated Example)

It is found that 1.2×10^{-3} mol of lead(II) iodide, PbI_2, dissolves in 1 L of solution at 25°C. What is the solubility product constant at this temperature?

SOLUTION

STEP 1 Suppose the solid lead(II) iodide is mixed into 1 L of solution. You find that 1.2×10^{-3} mol dissolves to form 1.2×10^{-3} mol Pb^{2+} and $2 \times (1.2 \times 10^{-3})$ mol I^-, as summarized in the following table:

Concentration (M)	$PbI_2(s)$ $\rightleftharpoons$ $Pb^{2+}(aq)$	$+$	$2I^-(aq)$
Starting	0		0
Change	$+1.2 \times 10^{-3}$		$+2 \times (1.2 \times 10^{-3})$
Equilibrium	1.2×10^{-3}		$2 \times (1.2 \times 10^{-3})$

STEP 2 You substitute into the equilibrium-constant equation:

$$K_{sp} = [Pb^{2+}][I^-]^2 = (1.2 \times 10^{-3}) \times (2 \times 1.2 \times 10^{-3})^2 = \mathbf{6.9 \times 10^{-9}}$$

See Problems 18.15 and 18.16.

Table 18.1 lists the solubility product constants for various ionic compounds.

EXAMPLE 18.4

Calculating the Solubility from K_{sp}

The mineral fluorite is calcium fluoride. Calculate the solubility (in grams per liter) of CaF_2 in water from the solubility product constant (3.4×10^{-11}).

SOLUTION

STEP 1 Let x be the molar solubility of CaF_2. When solid CaF_2 is mixed into a liter of solution, x mol dissolves, forming x mol Ca^{2+} and $2x$ mol F^-.

Concentration (M)	$CaF_2(s)$ $\rightleftharpoons$ $Ca^{2+}(aq)$	$+$	$2F^-(aq)$
Starting	0		0
Change	$+x$		$+2x$
Equilibrium	x		$2x$

STEP 2 You substitute into the equilibrium-constant equation.

$$[Ca^{2+}][F^-]^2 = K_{sp}$$

$$(x) \times (2x)^2 = 3.4 \times 10^{-11}$$

$$4x^3 = 3.4 \times 10^{-11}$$

Table 18.1
Solubility Product Constants, K_{sp}, at 25°C

Substance	Formula	K_{sp}	Substance	Formula	K_{sp}
Aluminum hydroxide	$Al(OH)_3$	4.6×10^{-33}	Lead(II) sulfide	PbS	2.5×10^{-27}
Barium chromate	$BaCrO_4$	1.2×10^{-10}	Magnesium arsenate	$Mg_3(AsO_4)_2$	2×10^{-20}
Barium fluoride	BaF_2	1.0×10^{-6}	Magnesium carbonate	$MgCO_3$	1.0×10^{-5}
Barium sulfate	$BaSO_4$	1.1×10^{-10}	Magnesium hydroxide	$Mg(OH)_2$	1.8×10^{-11}
Cadmium oxalate	CdC_2O_4	1.5×10^{-8}	Magnesium oxalate	MgC_2O_4	8.5×10^{-5}
Cadmium sulfide	CdS	8×10^{-27}	Manganese(II) sulfide	MnS	2.5×10^{-10}
Calcium carbonate	$CaCO_3$	3.8×10^{-9}	Mercury(I) chloride	Hg_2Cl_2	1.3×10^{-18}
Calcium fluoride	CaF_2	3.4×10^{-11}	Mercury(II) sulfide	HgS	1.6×10^{-52}
Calcium oxalate	CaC_2O_4	2.3×10^{-9}	Nickel(II) hydroxide	$Ni(OH)_2$	2.0×10^{-15}
Calcium phosphate	$Ca_3(PO_4)_2$	1×10^{-26}	Nickel(II) sulfide	NiS	3×10^{-19}
Calcium sulfate	$CaSO_4$	2.4×10^{-5}	Silver acetate	$AgC_2H_3O_2$	2.0×10^{-3}
Cobalt(II) sulfide	CoS	4×10^{-21}	Silver bromide	$AgBr$	5.0×10^{-13}
Copper(II) hydroxide	$Cu(OH)_2$	2.6×10^{-19}	Silver chloride	$AgCl$	1.8×10^{-10}
Copper(II) sulfide	CuS	6×10^{-36}	Silver chromate	Ag_2CrO_4	1.1×10^{-12}
Iron(II) hydroxide	$Fe(OH)_2$	8×10^{-16}	Silver iodide	AgI	8.3×10^{-17}
Iron(II) sulfide	FeS	6×10^{-18}	Silver sulfide	Ag_2S	6×10^{-50}
Iron(III) hydroxide	$Fe(OH)_3$	2.5×10^{-39}	Strontium carbonate	$SrCO_3$	9.3×10^{-10}
Lead(II) arsenate	$Pb_3(AsO_4)_2$	4×10^{-36}	Strontium chromate	$SrCrO_4$	3.5×10^{-5}
Lead(II) chloride	$PbCl_2$	1.6×10^{-5}	Strontium sulfate	$SrSO_4$	2.5×10^{-7}
Lead(II) chromate	$PbCrO_4$	1.8×10^{-14}	Zinc hydroxide	$Zn(OH)_2$	2.1×10^{-16}
Lead(II) iodide	PbI_2	6.5×10^{-9}	Zinc sulfide	ZnS	1.1×10^{-21}
Lead(II) sulfate	$PbSO_4$	1.7×10^{-8}			

STEP 3 You now solve for x.

$$x = \sqrt[3]{\frac{3.4 \times 10^{-11}}{4}} = 2.0 \times 10^{-4}$$

The molar solubility is 2.0×10^{-4} mol CaF_2 per liter. To get the solubility in grams per liter, you convert, using the molar mass of CaF_2 (78.1 g/mol).

$$\text{Solubility} = 2.0 \times 10^{-4} \text{ mol } CaF_2/L \times \frac{78.1 \text{ g } CaF_2}{1 \text{ mol } CaF_2} = \textbf{1.6} \times \textbf{10}^{-2} \textbf{ g } CaF_2/L$$

See Problems 18.19, 18.20, 18.21, and 18.22.

In the next section, you will see how the solubility product constant can be used to calculate the solubility in the presence of other ions. K_{sp} is also useful in deciding whether to expect precipitation under given conditions.

CONCEPT CHECK 18.1

Lead compounds have been used as paint pigments, but because the lead(II) ion is toxic, the use of lead paints in homes is now prohibited. Which of the following lead(II) compounds would yield the greatest number of lead(II) ions when added to the same quantity of water (assuming that some undissolved solid always remains): $PbCrO_4$, $PbSO_4$, or PbS?

18.2 Solubility and the Common-Ion Effect

Figure 18.2
Demonstration of the common-ion effect.
When the experimenter adds lead(II) nitrate solution (colorless) from the dropper to the saturated solution of lead(II) chromate (pale yellow), a yellow precipitate of lead chromate forms.

Suppose you wish to know the solubility of calcium oxalate in a solution of calcium chloride. Each salt contributes the same cation (Ca^{2+}). The effect of the calcium ion provided by the calcium chloride is to make calcium oxalate less soluble than it would be in pure water.

You can explain this decrease in solubility in terms of Le Chatelier's principle. Suppose you first mix crystals of calcium oxalate in a quantity of pure water to establish the equilibrium

$$CaC_2O_4(s) \rightleftharpoons Ca^{2+}(aq) + C_2O_4{}^{2-}(aq)$$

Now imagine that you add some calcium chloride, which is a soluble salt. According to Le Chatelier's principle, the ions will react to remove some of the added calcium ion by precipitation.

$$CaC_2O_4(s) \longleftarrow Ca^{2+}(aq) + C_2O_4{}^{2-}(aq)$$

The solution now contains less calcium oxalate.

The decrease in solubility of calcium oxalate in a solution of calcium chloride is an example of the *common-ion effect*. In general, any ionic equilibrium is affected by a substance producing an ion involved in the equilibrium, as you would predict from Le Chatelier's principle. Figure 18.2 illustrates the principle of the common-ion effect for lead(II) chromate, $PbCrO_4$, which is only slightly soluble in water at 25°C. When the very soluble $Pb(NO_3)_2$ is added to a saturated solution of $PbCrO_4$, the concentration of the common ion, Pb^{2+}, increases and $PbCrO_4$ precipitates. The equilibrium is

$$PbCrO_4(s) \rightleftharpoons Pb^{2+}(aq) + CrO_4{}^{2-}(aq)$$

EXAMPLE 18.5

Calculating the Solubility of a Slightly Soluble Salt in a Solution of a Common Ion

What is the molar solubility of calcium oxalate in 0.15 *M* calcium chloride? Compare this molar solubility with that found earlier (Example 18.2) for CaC_2O_4 in pure water (4.8 × 10^{-5} *M*). The K_{sp} for CaC_2O_4 is 2.3 × 10^{-9}.

SOLUTION

STEP 1 Suppose solid CaC_2O_4 is mixed into 1 L of 0.15 *M* $CaCl_2$ and that the molar solubility of the CaC_2O_4 is *x M*. At the start (before CaC_2O_4 dissolves), there is 0.15 mol Ca^{2+} in the solution. Of the solid CaC_2O_4, *x* mol dissolves to give x mol of additional Ca^{2+} and x mol $C_2O_4{}^{2-}$. The following table summarizes these results:

Concentration (M)	$CaC_2O_4(s) \rightleftharpoons$	$Ca^{2+}(aq)$	$+$	$C_2O_4{}^{2-}(aq)$
Starting		0.15		0
Change		$+x$		$+x$
Equilibrium		$0.15 + x$		x

STEP 2   You substitute into the equilibrium-constant equation:

$$[Ca^{2+}][C_2O_4{}^{2-}] = K_{sp}$$

$$(0.15 + x)x = 2.3 \times 10^{-9}$$

STEP 3   Because calcium oxalate is only slightly soluble, you might expect x to be negligible compared with 0.15. In that case,

$$0.15 + x \approx 0.15$$

and the previous equation becomes

$$x \approx \frac{2.3 \times 10^{-9}}{0.15} = 1.5 \times 10^{-8}$$

The molar solubility of calcium oxalate in 0.15 M $CaCl_2$ is **1.5×10^{-8} M.** In pure water, the molar solubility is 4.8×10^{-5} M, which is over **3000 times greater.**

See Problems 18.23, 18.24, 18.25, and 18.26.

 CONCEPT CHECK 18.2

Suppose you have equal volumes of saturated solutions of $NaNO_3$, Na_2SO_4, and PbS. Which solution would dissolve the most lead(II) sulfate, $PbSO_4$?

18.3   Precipitation Calculations

In the chapter opening, we mentioned that calcium oxalate precipitates to form kidney stones. The same salt would precipitate in the body if oxalic acid (a poison) were accidentally ingested, because Ca^{2+} ion is present in the blood. Precipitation is merely another way of looking at a solubility equilibrium.

Criterion for Precipitation

Given the concentrations of substances in a reaction mixture, will the reaction go in the forward or the reverse direction? To answer this, you evaluate the *reaction quotient Q_c* and compare it with the equilibrium constant K_c. ◀ The reaction quotient has the same form as the equilibrium-constant expression. To predict the direction of reaction, you compare Q_c with K_c.

▶ The reaction quotient was discussed in Section 15.5.

If $Q_c < K_c$, the reaction should go in the forward direction.

If $Q_c = K_c$, the reaction mixture is at equilibrium.

If $Q_c > K_c$, the reaction should go in the reverse direction.

Suppose you add $Pb(NO_3)_2$ and $NaCl$ to water to give a solution that is 0.050 M Pb^{2+} and 0.10 M Cl^-. Will $PbCl_2$ precipitate? To answer this, you first write the solubility equilibrium.

$$PbCl_2(s) \overset{H_2O}{\rightleftharpoons} Pb^{2+}(aq) + 2Cl^-(aq)$$

The reaction quotient has the form of the equilibrium-constant expression, but the concentrations of the products are starting values, denoted by the subscript i.

$$Q_c = [Pb^{2+}]_i[Cl^-]_i^2$$

Here Q_c for a solubility reaction is often called the **ion product.**

To evalutae the ion product Q_c, you substitute the concentrations of Pb^{2+} and Cl^- ions in the solution at the start of the reaction.

$$Q_c = (0.050)(0.10)^2 = 5.0 \times 10^{-4}$$

K_{sp} for $PbCl_2$ is 1.6×10^{-5} (Table 18.1), so Q_c is greater than K_{sp}. Therefore, the reaction goes in the reverse direction. That is, Pb^{2+} and Cl^- react to precipitate $PbCl_2$. Precipitation ceases when the ion product equals K_{sp}.

We can summarize our conclusions in terms of the following criterion for precipitation.

▶ Precipitation may not occur even though the ion product has been exceeded. In such a case, the solution is supersaturated. Usually a small crystal forms after a time, and then precipitation occurs rapidly.

Precipitation is expected to occur if the ion product for a solubility reaction is greater than K_{sp}. If the ion product is less than K_{sp}, precipitation will not occur. If the ion product equals K_{sp}, the reaction is at equilibrium. ◀

Figure 18.3
Vats for the precipitation of magnesium hydroxide from seawater. Seawater contains magnesium ion (in addition to other ions). When base is added, magnesium hydroxide precipitates. This is the source of magnesium metal.

EXAMPLE 18.6

Predicting Whether Precipitation Will Occur

The concentration of calcium ion in blood plasma is 0.0025 M. If the concentration of oxalate ion is 1.0×10^{-7} M, do you expect calcium oxalate to precipitate? K_{sp} for calcium oxalate is 2.3×10^{-9}.

SOLUTION

The ion product for calcium oxalate is

$$\text{Ion product} = [Ca^{2+}]_i[C_2O_4^{2-}]_i = (0.0025) \times (1.0 \times 10^{-7}) = 2.5 \times 10^{-10}$$

This value is smaller than the solubility product constant, so **you do not expect precipitation to occur.**

See Problems 18.27 and 18.28.

Precipitation is an important industrial and laboratory process. Figure 18.3 shows the vats used to precipitate magnesium ion from seawater, which is a source of magnesium metal. The following example is a typical problem in precipitation.

EXAMPLE 18.7

Predicting Whether Precipitation Will Occur (Given Solution Volumes and Concentrations)

Sulfate ion, SO_4^{2-}, in solution is often determined quantitatively by precipitating it as barium sulfate, $BaSO_4$. The sulfate ion may have been formed from a sulfur compound. Analysis for the amount of sulfate ion then indicates the percentage of sulfur in the compound. Is a precipitate expected to form at equilibrium when 50.0 mL of 0.0010 M $BaCl_2$ is added to 50.0 mL of 0.00010 M Na_2SO_4? The solubility product constant for barium sulfate is 1.1×10^{-10}. Assume that the total volume of solution, after mixing, equals the sum of the volumes of the separate solutions.

SOLUTION

The molar amount of Ba^{2+} present in 50.0 mL (0.0500 L) of 0.0010 M $BaCl_2$ is

$$\text{Amount of } Ba^{2+} = \frac{0.0010 \text{ mol } Ba^{2+}}{1 \text{ L soln}} \times 0.050 \text{ L soln} = 5.0 \times 10^{-5} \text{ mol } Ba^{2+}$$

The molar concentration of Ba^{2+} in the total solution equals the molar amount of Ba^{2+} divided by the total volume (0.0500 L $BaCl_2$ + 0.0500 L Na_2SO_4).

$$[Ba^{2+}] = \frac{5.0 \times 10^{-5} \text{ mol}}{0.1000 \text{ L soln}} = 5.0 \times 10^{-4} M$$

Similarly, you find $\qquad [SO_4^{2-}] = 5.0 \times 10^{-5} M$

(Try the calculations for SO_4^{2-}.) The ion product is

$$Q_c = [Ba^{2+}]_i [SO_4^{2-}]_i = (5.0 \times 10^{-4}) \times (5.0 \times 10^{-5}) = 2.5 \times 10^{-8}$$

Because the ion product is greater than the solubility product constant (1.1×10^{-10}), **you expect barium sulfate to precipitate.**

See Problems 18.29 and 18.30.

Fractional Precipitation

Fractional precipitation is *the technique of separating two or more ions from a solution by adding a reactant that precipitates first one ion, then another, and so forth.* For example, suppose a solution is 0.10 M Ba^{2+} and 0.10 M Sr^{2+}. When you slowly add a concentrated solution of potassium chromate, K_2CrO_4, to the solution, barium chromate precipitates first. After most of the Ba^{2+} ion has precipitated, strontium chromate begins to come out of solution.

To understand why Ba^{2+} and Sr^{2+} ions can be separated in this way, calculate (1) the concentration of CrO_4^{2-} necessary to just begin the precipitation of $BaCrO_4$, and (2) the concentration of CrO_4^{2-} necessary to just begin the precipitation of $SrCrO_4$. Ignore any volume change in the solution of Ba^{2+} and Sr^{2+} ions resulting from the addition of the concentrated K_2CrO_4 solution. To calculate the CrO_4^{2-} concentration when $BaCrO_4$ begins to precipitate, you substitute the initial Ba^{2+} concentration into the solubility product equation. K_{sp} for $BaCrO_4$ is 1.2×10^{-10}.

$$[Ba^{2+}][CrO_4^{2-}] = K_{sp} \text{ (for BaCrO}_4)$$

$$(0.10)[CrO_4^{2-}] = 1.2 \times 10^{-10}$$

$$[CrO_4^{2-}] = \frac{1.2 \times 10^{-10}}{0.10} = 1.2 \times 10^{-9}\, M$$

In the same way, you can calculate the CrO_4^{2-} concentration when $SrCrO_4$ begins to precipitate. K_{sp} for $SrCrO_4$ is 3.5×10^{-5}.

$$[Sr^{2+}][CrO_4^{2-}] = K_{sp} \text{ (for SrCrO}_4)$$

$$(0.10)[CrO_4^{2-}] = 3.5 \times 10^{-5}$$

$$[CrO_4^{2-}] = \frac{3.5 \times 10^{-5}}{0.10} = 3.5 \times 10^{-4}\, M$$

Note that $BaCrO_4$ precipitates first, because the CrO_4^{2-} concentration is smaller.

These results reveal that as the solution of K_2CrO_4 is slowly added to the solution of Ba^{2+} and Sr^{2+}, barium chromate continues to precipitate. When the concentration of CrO_4^{2-} reaches $3.5 \times 10^{-4}\, M$, strontium chromate begins to precipitate.

What is the percentage of Ba^{2+} ion remaining just as $SrCrO_4$ begins to precipitate? First calculate the concentration of Ba^{2+} at this point.

$$[Ba^{2+}][CrO_4^{2-}] = K_{sp} \text{ (for BaCrO}_4)$$

$$[Ba^{2+}](3.5 \times 10^{-4}) = 1.2 \times 10^{-10}$$

$$[Ba^{2+}] = \frac{1.2 \times 10^{-10}}{3.5 \times 10^{-4}} = 3.4 \times 10^{-7}\, M$$

To calculate the percentage of Ba^{2+} ion remaining, you divide this concentration of Ba^{2+} by the initial concentration ($0.10\, M$), and multiply by 100%.

$$\frac{3.4 \times 10^{-7}}{0.10} \times 100\% = 0.00034\%$$

Most of the Ba^{2+} ion has precipitated by the time $SrCrO_4$ begins to precipitate. You conclude that Ba^{2+} and Sr^{2+} can indeed be separated by fractional precipitation. (Another application of fractional precipitation is shown in Figure 18.4.)

18.4   Effect of pH on Solubility

In discussing solubility, it is necessary to account for other reactions the ions might undergo. For example, if the anion is the conjugate base of a weak acid, the anion reacts with H_3O^+ ion. You should expect the solubility to be affected by pH.

Qualitative Effect of pH

Consider the equilibrium involving CaC_2O_4:

$$CaC_2O_4(s) \rightleftharpoons Ca^{2+}(aq) + C_2O_4^{2-}(aq)$$

Because the oxalate ion is conjugate to a weak acid (hydrogen oxalate ion, $HC_2O_4^-$), you would expect it to react with any H_3O^+ ion that is added—say, from a strong acid:

$$C_2O_4^{2-}(aq) + H_3O^+(aq) \rightleftharpoons HC_2O_4^-(aq) + H_2O(l)$$

Figure 18.4
Titration of chloride ion by silver nitrate using potassium chromate as an indicator. *Left:* A small amount of K_2CrO_4 (yellow) has been added to a solution containing an unknown amount of Cl^- ion. *Center:* The solution is titrated by $AgNO_3$ solution, giving a white precipitate of AgCl. *Right:* When nearly all of the Cl^- ion has precipitated as AgCl, silver chromate begins to precipitate. Silver chromate, Ag_2CrO_4, has a red-brown color, and the appearance of this color signals the end of the titration. An excess of Ag^+ was added to show the color of Ag_2CrO_4 more clearly.

According to Le Chatelier's principle, as $C_2O_4^{2-}$ ion is removed by reaction with H_3O^+ ion, more calcium oxalate dissolves to replenish some of the $C_2O_4^{2-}$ ion.

$$CaC_2O_4(s) \longrightarrow Ca^{2+}(aq) + C_2O_4^{2-}(aq)$$

Therefore, calcium oxalate is more soluble in acidic solution than in pure water.

In general, salts of weak acids are more soluble in acidic solutions. Consider the process of tooth decay. Bacteria on the teeth produce an acidic medium as a result of the metabolism of sugar. Teeth are normally composed of a calcium phosphate mineral, hydroxyapatite, which you can denote as either $Ca_5(PO_4)_3OH$ or $3Ca_3(PO_4)_2 \cdot Ca(OH)_2$. This mineral salt dissolves in the presence of the acid medium, producing cavities in the teeth. Fluoride toothpastes provide F^- ion, which gradually replaces the OH^- ion in the teeth to produce fluorapatite, $Ca_5(PO_4)_3F$ or $3Ca_3(PO_4)_2 \cdot CaF_2$, which is much less soluble than hydroxyapatite.

EXAMPLE 18.8

Determining the Qualitative Effect of pH on Solubility

Consider two slightly soluble salts, calcium carbonate and calcium sulfate. Which of these would have its solubility more affected by the addition of HCl, a strong acid? Would the solubility of the salt increase or decrease?

SOLUTION

Calcium carbonate gives the solubility equilibrium

$$CaCO_3(s) \rightleftharpoons Ca^{2+}(aq) + CO_3^{2-}(aq)$$

When a strong acid is added, the hydrogen ion reacts with carbonate ion.

$$H_3O^+(aq) + CO_3^{2-}(aq) \rightleftharpoons H_2O(l) + HCO_3^-(aq)$$

As carbonate ion is removed, calcium carbonate dissolves. Moreover, the hydrogen carbonate ion itself is removed in further reaction.

$$H_3O^+(aq) + HCO_3^-(aq) \longrightarrow H_2O(l) + H_2CO_3(aq) \longrightarrow 2H_2O(l) + CO_2(g)$$

Table of Atomic Numbers and Atomic Weights

Name	Symbol	Atomic number	Atomic weight
Actinium	Ac	89	(227)
Aluminum	Al	13	26.981538
Americium	Am	95	(243)
Antimony	Sb	51	121.760
Argon	Ar	18	39.948
Arsenic	As	33	74.92160
Astatine	At	85	(210)
Barium	Ba	56	137.327
Berkelium	Bk	97	(247)
Beryllium	Be	4	9.012182
Bismuth	Bi	83	208.98038
Bohrium	Bh	107	(264)
Boron	B	5	10.811
Bromine	Br	35	79.904
Cadmium	Cd	48	112.411
Calcium	Ca	20	40.078
Californium	Cf	98	(251)
Carbon	C	6	12.0107
Cerium	Ce	58	140.116
Cesium	Cs	55	132.90545
Chlorine	Cl	17	35.4527
Chromium	Cr	24	51.9961
Cobalt	Co	27	58.933200
Copper	Cu	29	63.546
Curium	Cm	96	(247)
Dubnium	Db	105	(262)
Dysprosium	Dy	66	162.50
Einsteinium	Es	99	(252)
Erbium	Er	68	167.26
Europium	Eu	63	151.964
Fermium	Fm	100	(257)
Fluorine	F	9	18.9984032
Francium	Fr	87	(223)
Gadolinium	Gd	64	157.25
Gallium	Ga	31	69.723
Germanium	Ge	32	72.61
Gold	Au	79	196.96655
Hafnium	Hf	72	178.49
Hassium	Hs	108	(265)
Helium	He	2	4.002602
Holmium	Ho	67	164.93032
Hydrogen	H	1	1.00794
Indium	In	49	114.818
Iodine	I	53	126.90447
Iridium	Ir	77	192.217
Iron	Fe	26	55.845
Krypton	Kr	36	83.80
Lanthanum	La	57	138.9055
Lawrencium	Lr	103	(262)
Lead	Pb	82	207.2
Lithium	Li	3	6.941
Lutetium	Lu	71	174.967
Magnesium	Mg	12	24.3050
Manganese	Mn	25	54.938049
Meitnerium	Mt	109	(268)
Mendelevium	Md	101	(258)
Mercury	Hg	80	200.59

Name	Symbol	Atomic number	Atomic weight
Molybdenum	Mo	42	95.94
Neodymium	Nd	60	144.24
Neon	Ne	10	20.1797
Neptunium	Np	93	(237)
Nickel	Ni	28	58.6934
Niobium	Nb	41	92.90638
Nitrogen	N	7	14.00674
Nobelium	No	102	(259)
Osmium	Os	76	190.23
Oxygen	O	8	15.9994
Palladium	Pd	46	106.42
Phosphorus	P	15	30.973762
Platinum	Pt	78	195.078
Plutonium	Pu	94	(244)
Polonium	Po	84	(209)
Potassium	K	19	39.0983
Praseodymium	Pr	59	140.90765
Promethium	Pm	61	(145)
Protactinium	Pa	91	231.03588
Radium	Ra	88	(226)
Radon	Rn	86	(222)
Rhenium	Re	75	186.207
Rhodium	Rh	45	102.90550
Rubidium	Rb	37	85.4678
Ruthenium	Ru	44	101.07
Rutherfordium	Rf	104	(261)
Samarium	Sm	62	150.36
Scandium	Sc	21	44.955910
Seaborgium	Sg	106	(263)
Selenium	Se	34	78.96
Silicon	Si	14	28.0855
Silver	Ag	47	107.8682
Sodium	Na	11	22.989770
Strontium	Sr	38	87.62
Sulfur	S	16	32.066
Tantalum	Ta	73	180.9479
Technetium	Tc	43	(98)
Tellurium	Te	52	127.60
Terbium	Tb	65	158.92534
Thallium	Tl	81	204.3833
Thorium	Th	90	232.0381
Thulium	Tm	69	168.93421
Tin	Sn	50	118.710
Titanium	Ti	22	47.867
Tungsten	W	74	183.84
Ununbium	Uub	112	(277)
Ununnilium	Uun	110	(269)
Ununquadium	Uuq	114	(289)
Unununium	Uuu	111	(272)
Uranium	U	92	238.0289
Vanadium	V	23	50.9415
Xenon	Xe	54	131.29
Ytterbium	Yb	70	173.04
Yttrium	Y	39	88.90585
Zinc	Zn	30	65.39
Zirconium	Zr	40	91.224

A value in parentheses is the mass number of the isotope of longest half-life.

Atomic weights in this table are from the IUPAC report "Atomic Weights of the Elements 1993," *Pure and Applied Chemistry*, Vol. 68, No. 12 (1996), pp. 2339–2359.

Limestone Caves

ENVIRONMENT

Limestone caves dot the landscapes of the eastern and southwestern United States and of Europe and other places as well. Though many limestone caves are inaccessible, a few, like Mammouth Cave in Kentucky and Carlsbad Caverns in New Mexico, are spectacular tourist attractions with multicolored rock draperies and icicles hanging from the ceilings of large underground rooms (Figure 18.5). How do such caves form?

Limestone itself formed millions of years ago. Seashells accumulated on ocean floors, and then layers of these shells were subsequently compacted by great pressure from overlying sediment. Later movements of the earth's crust pushed the limestone layers out of the sea, often leaving them well inland.

Seashells and limestone are primarily calcium carbonate. Calcium carbonate is fairly insoluble, but it dissolves readily in acidic solution. Water that has filtered through decomposing vegetation contains carbonic acid, as well as other acids. When such an acidic solution comes into contact with limestone, it carves out caverns. The water is now a solution of calcium hydrogen carbonate.

$$CaCO_3(s) + H_2O(l) + CO_2(aq) \longrightarrow$$
$$Ca^{2+}(aq) + 2HCO_3^-(aq)$$

This process is reversible. As the calcium hydrogen carbonate solution percolates through holes in the limestone and drips into caverns, water evaporates and carbon dioxide gas escapes; calcium carbonate precipitates onto icicle-shaped *stalactites*.

Figure 18.5
A limestone cave (Luray Caverns, Virginia).
Such caves are formed by the action of acidic groundwater on limestone deposits. The icicle-like formations (stalagmites and stalactites) in the caves are caused by the reprecipitation of calcium carbonate as carbon dioxide in the solution equilibrates with the surrounding air.

$$Ca^{2+}(aq) + 2HCO_3^-(aq) \longrightarrow$$
$$CaCO_3(s) + H_2O(l) + CO_2(g)$$

Excess solution drips onto the cavern floor, where calcium carbonate precipitates on upward-growing *stalagmites*. Iron compounds in the solution add yellow and brown colors to the columns of rock. Stalactites and stalagmites grow at the rate of about a centimeter every 300 years.

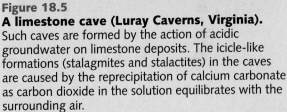

For calcium sulfate, the corresponding equilibria are

$$CaSO_4(s) \rightleftharpoons Ca^{2+}(aq) + SO_4^{2-}(aq)$$

$$H_3O^+(aq) + SO_4^{2-}(aq) \rightleftharpoons H_2O(l) + HSO_4^-(aq)$$

Again, the anion of the insoluble salt is removed by reaction with hydronium ion. You would expect calcium sulfate to become more soluble in strong acid. However, HSO_4^- is a much stronger acid than HCO_3^-, as you can see by comparing acid-ionization constants. (The values K_{a2} for H_2CO_3 and K_a for HSO_4^- in Table 17.1 are 4.8×10^{-11} and 1.1×10^{-2}, respectively.) **Thus, calcium carbonate is**

> much more soluble in acidic solution, whereas the solubility of calcium sulfate is only slightly affected.
>
> See Problems 18.35 and 18.36.

 CONCEPT CHECK 18.3

If you add a dilute acidic solution to a mixture containing magnesium oxalate and calcium oxalate, which of the two compounds is more likely to dissolve?

Separation of Metal Ions by Sulfide Precipitation

Many metal sulfides are insoluble in water but dissolve in acidic solution. The sulfide scheme discussed later in Section 18.7 uses this change in solubility of the metal sulfides with pH to separate a mixture of metal ions.

Consider a solution that is 0.10 M in zinc ion and in lead(II) ion. You add hydrogen sulfide, H_2S, to this solution. Hydrogen sulfide ionizes in water as a diprotic acid:

$$H_2S(aq) + H_2O(l) \rightleftharpoons H_3O^+(aq) + HS^-(aq)$$

$$HS^-(aq) + H_2O(l) \rightleftharpoons H_3O^+(aq) + S^{2-}(aq)$$

The ionization forms sulfide ion, S^{2-}, which can combine with the metal ions to precipitate the sulfides. Note that by increasing the hydronium-ion concentration, you can reverse both of the previous equilibria, and in that way reduce the sulfide-ion concentration. This allows you to control the sulfide-ion concentration of a hydrogen sulfide solution by varying its pH.

Whether lead(II) sulfide and zinc sulfide precipitate from the hydrogen sulfide solution depends on the solubility product constants of the sulfides and on the sulfide-ion concentration. Here are the solubility equilibria:

$$Zn^{2+}(aq) + S^{2-}(aq) \rightleftharpoons ZnS(s); K_{sp} = 1.1 \times 10^{-21}$$

$$Pb^{2+}(aq) + S^{2-}(aq) \rightleftharpoons PbS(s); K_{sp} = 2.5 \times 10^{-27}$$

By adjusting the pH of the solution, you adjust the sulfide-ion concentration in order to precipitate the least soluble metal sulfide while maintaining the other metal ion in solution. When a solution that is 0.10 M in each metal ion and 0.30 M in hydronium ion is saturated with hydrogen sulfide, lead(II) sulfide precipitates, but zinc ion remains in solution, You can now filter off the precipitate of lead(II) sulfide, leaving a solution containing the zinc ion.

Complex-Ion Equilibria

Many metal ions, especially those of the transition elements, form coordinate covalent bonds with molecules or anions having lone pairs of electrons (Lewis bases). For example, the silver ion, Ag^+, can react with NH_3 to form the $Ag(NH_3)_2^+$ ion. The lone pair of electrons on the N atom of NH_3 forms a coordinate covalent bond to the silver ion to form what is called a complex ion. (See Figure 18.6.)

$$Ag^+ + 2(:NH_3) \longrightarrow (H_3N:Ag:NH_3)^+$$

▶ The Lewis concept of acids
and bases was discussed in
Section 16.3.

A **complex ion** is *an ion formed from a metal ion with a Lewis base attached to it by a coordinate covalent bond.* A **ligand** is *a Lewis base that bonds to a metal ion to form a complex ion.* ◀

18.5 Complex-Ion Formation

The aqueous silver ion forms a complex ion with ammonia by reacting with NH_3.

$$Ag^+(aq) + 2NH_3(aq) \rightleftharpoons Ag(NH_3)_2^+(aq)$$

The **formation constant,** or **stability constant,** K_f, of a complex ion is *the equilibrium constant for the formation of the complex ion from the aqueous metal ion and the ligands.* The formation constant of $Ag(NH_3)_2^+$ is

$$K_f = \frac{[Ag(NH_3)_2^+]}{[Ag^+][NH_3]^2}$$

The value of K_f for $Ag(NH_3)_2^+$ is 1.7×10^7. Its large value means that the complex ion is quite stable. Table 18.2 lists values for the formation constants of some complex ions.

The **dissociation constant** (K_d) for a complex ion is *the reciprocal, or inverse, value of K_f.* The equation for the dissociation of $Ag(NH_3)_2^+$ is

$$Ag(NH_3)_2^+(aq) \rightleftharpoons Ag^+(aq) + 2NH_3(aq)$$

and its equilibrium constant is

$$K_d = \frac{1}{K_f} = \frac{[Ag^+][NH_3]^2}{[Ag(NH_3)_2^+]}$$

Equilibrium Calculations with K_f

The next example shows how to calculate the concentration of an aqueous metal ion in equilibrium with a complex ion.

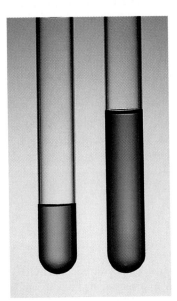

Figure 18.6
Complex ions of nickel(II).
The test tube on the left contains a solution of $Ni(H_2O)_6^{2+}$ ion (green). Adding ammonia to a similar test tube containing this ion, on the right, yields the ammonia complex ion $Ni(NH_3)_6^{2+}$ (purple).

EXAMPLE 18.9

Calculating the Concentration of a Metal Ion in Equilibrium with a Complex Ion

What is the concentration of $Ag^+(aq)$ ion in 0.010 M $AgNO_3$ that is also 1.00 M NH_3? K_f for $Ag(NH_3)_2^+$ ion is 1.7×10^7.

SOLUTION

Stoichiometry Calculation

In 1 L of solution, you initially have 0.010 mol $Ag^+(aq)$ from $AgNO_3$. This reacts to give 0.010 mol $Ag(NH_3)_2^+$, leaving $(1.00 - 2 \times 0.010)$ mol NH_3, which equals 0.98 mol NH_3 to two significant figures. You now look at the equilibrium for the dissociation of $Ag(NH_3)_2^+$.

Table 18.2
Formation Constants of Complex Ions at 25°C

Complex Ion	K_f
$Ag(CN)_2^-$	5.6×10^{18}
$Ag(NH_3)_2^+$	1.7×10^7
$Ag(S_2O_3)_2^{3-}$	2.9×10^{13}
$Cd(NH_3)_4^{2+}$	1.0×10^7
$Cu(CN)_2^-$	1.0×10^{16}
$Cu(NH_3)_4^{2+}$	4.8×10^{12}
$Fe(CN)_6^{4-}$	1.0×10^{35}
$Fe(CN)_6^{3-}$	9.1×10^{41}
$Ni(CN)_4^{2-}$	1.0×10^{31}
$Ni(NH_3)_6^{2+}$	5.6×10^8
$Zn(NH_3)_4^{2+}$	2.9×10^9
$Zn(OH)_4^{2-}$	2.8×10^{15}

Equilibrium Calculation

STEP 1   One liter of the solution contains 0.010 mol $Ag(NH_3)_2^+$ and 0.98 mol NH_3. The complex ion dissociates slightly, so that 1 L of solution contains x mol Ag^+. These data are summarized in the table:

Concentration (M)	$Ag(NH_3)_2^+ (aq) \rightleftharpoons$	$Ag^+(aq) +$	$2NH_3(aq)$
Starting	0.010	0	0.98
Change	$-x$	$+x$	$+2x$
Equilibrium	$0.010 - x$	x	$0.98 + 2x$

STEP 2   The dissociation constant corresponds to $1/K_f$.

$$\frac{[Ag^+][NH_3]^2}{[Ag(NH_3)_2^+]} = K_d = \frac{1}{K_f}$$

Substituting into this equation gives

$$\frac{x(0.98 + 2x)^2}{(0.010 - x)} = \frac{1}{1.7 \times 10^7}$$

STEP 3   The right-hand side of the equation equals 5.9×10^{-8}. If you assume x to be small compared with 0.010,

$$\frac{x(0.98)^2}{0.010} \simeq 5.9 \times 10^{-8}$$

and

$$x \simeq 5.9 \times 10^{-8} \times 0.010/(0.98)^2 = 6.1 \times 10^{-10}$$

The silver-ion concentration is **$6.1 \times 10^{-10} M$** (over 10 million times smaller than its value in $0.010 M$ $AgNO_3$ that does not contain ammonia).

See Problems 18.39 and 18.40.

Amphoteric Hydroxides

An **amphoteric hydroxide** is *a metal hydroxide that reacts with both bases and acids.* Zinc is an example of a metal that forms such a hydroxide. Zinc hydroxide, $Zn(OH)_2$, is an insoluble hydroxide. Zinc hydroxide reacts with a strong acid, as you would expect of a base, and the metal hydroxide dissolves.

$$Zn(OH)_2(s) + 2H_3O^+(aq) \longrightarrow Zn^{2+}(aq) + 4H_2O(l)$$

With a base, however, $Zn(OH)_2$ reacts to form the complex ion $Zn(OH)_4^{2-}$.

$$Zn(OH)_2(s) + 2OH^-(aq) \longrightarrow Zn(OH)_4^{2-}(aq)$$

In this way the hydroxide also dissolves in a strong base.

When a strong base is slowly added to a solution of $ZnCl_2$, a white precipitate of $Zn(OH)_2$ first forms.

$$Zn^{2+}(aq) + 2OH^-(aq) \longrightarrow Zn(OH)_2(s)$$

But as more base is added, the white precipitate dissolves, forming the complex ion $Zn(OH)_4^{2-}$. (See Figure 18.7.)

Other common amphoteric hydroxides are those of aluminum, chromium(III), lead(II), tin(II), and tin(IV). The amphoterism of $Al(OH)_3$ is commercially used to separate aluminum oxide from the aluminum ore bauxite.

18.6 Complex Ions and Solubility

Example 18.9 demonstrated that the formation of the complex ion $Ag(NH_3)_2^+$ reduces the concentration of silver ion, $Ag^+(aq)$, in solution. The solution was initially 0.010 M $AgNO_3$, or 0.010 M $Ag^+(aq)$. When this solution is made 1.00 M in NH_3, the $Ag^+(aq)$ ion concentration decreases to 6.1×10^{-10} M. You can see that the ion product for a slightly soluble silver salt could be decreased to below K_{sp}. Then the slightly soluble salt might not precipitate in a solution containing ammonia, whereas it otherwise would.

Figure 18.7
Demonstration of the amphoteric behavior of zinc hydroxide. *Left:* The beaker on the left contains $ZnCl_2$; the one on the right contains NaOH. *Center:* As some of the solution of NaOH is added to the solution of $ZnCl_2$, a white precipitate of $Zn(OH)_2$ forms. *Right:* After more NaOH is added, the precipitate dissolves by forming the hydroxo complex $Zn(OH)_4^{2-}$.

EXAMPLE 18.10
Predicting Whether a Precipitate Will Form in the Presence of the Complex Ion

a. Will silver chloride precipitate from a solution that is 0.010 M $AgNO_3$ and 0.010 M NaCl? b. Will silver chloride precipitate from this solution if it is also 1.00 M NH_3?

SOLUTION

a. To determine whether a precipitate should form, you calculate the ion product and compare it with K_{sp} for AgCl (1.8×10^{-10}).

$$\text{Ion product} = [Ag^+]_i[Cl^-]_i = (0.010)(0.010) = 1.0 \times 10^{-4}$$

This is greater than $K_{sp} = 1.8 \times 10^{-10}$, so **a precipitate should form.**

b. You first need to calculate the concentration of $Ag^+(aq)$ in a solution containing 1.00 M NH_3. We did this in Example 18.9 and found that $[Ag^+]$ equals 6.1×10^{-10}. Hence,

$$\text{Ion product} = [Ag^+]_i[Cl^-]_i = (6.1 \times 10^{-10})(0.010) = 6.1 \times 10^{-12}$$

Because the ion product is smaller than $K_{sp} = 1.8 \times 10^{-10}$, **no precipitate should form.**

In Example 18.10, we determined whether a precipitate of AgCl is expected to form from a solution made up with given concentrations of Ag^+, Cl^-, and NH_3. The problem involved the $Ag(NH_3)_2^+$ complex in equilibrium. From it, we determined the concentration of free $Ag^+(aq)$ and then calculated the ion product of AgCl.

Suppose you want to find the solubility of AgCl in a solution of aqueous ammonia of given concentration. This problem involves the solubility equilibrium for AgCl, in addition to the complex-ion equilibrium. As silver chloride dissolves to give ions, the Ag^+ ion reacts with NH_3 to give the complex ion $Ag(NH_3)_2^+$. The equilibria are

$$AgCl(s) \rightleftharpoons Ag^+(aq) + Cl^-(aq)$$

$$Ag^+(aq) + 2NH_3(aq) \rightleftharpoons Ag(NH_3)_2^+(aq)$$

When $Ag^+(aq)$ reacts to give the complex ion, more AgCl dissolves to partially replenish the $Ag^+(aq)$ ion, according to Le Chatelier's principle. Therefore, silver chloride is more soluble in aqueous ammonia than in pure water.

An Application of Solubility Equilibria

Qualitative analysis involves *the determination of the identity of the substances present in a mixture.* One scheme for the qualitative analysis of a solution of metal ions is based on the relative solubilities of the metal sulfides. This scheme is no longer widely used in practical analysis, having been replaced largely by modern instrumental methods. But it is still taught in general chemistry laboratory as a way of imparting some knowledge of inorganic chemistry while illustrating the concepts of solubility, buffers, and so forth. We will look at the main outline of the sulfide scheme.

18.7   Qualitative Analysis of Metal Ions

In the qualitative analysis scheme for metal ions, a cation is usually detected by the presence of a characteristic precipitate. For example, silver ion gives a white precipitate with chloride ion. But other ions also give a white precipitate with chloride ion. Therefore, you must first subject a mixture to a procedure that separates individual ions before applying a precipitation test to any particular ion.

Figure 18.8 shows a flowchart illustrating how the metal ions in an aqueous solution are first separated into five analytical groups. The Analytical Group I ions, Ag^+, Hg_2^{2+}, and Pb^{2+}, are separated from a solution of other ions by adding dilute hydrochloric acid. The ions are precipitated as the chlorides AgCl, Hg_2Cl_2, and $PbCl_2$, which are removed by filtering.

The other metal ions remain in the filtrate, the solution that passes through the filter. Many of these ions can be separated by precipitating them as metal sulfides with H_2S. This separation takes advantage of the fact that only the least soluble metal sulfides precipitate in acidic solution. After these are removed and the solution is made basic, other metal sulfides precipitate. ◄

▶ The solubility of the metal sulfides at different pH was discussed in Section 18.4.

Figure 18.8

Flowchart of the qualitative analysis scheme for the separation of metal ions. The first group of ions consists of those precipitated by HCl. The separation of the next two groups is based on the differences in solubilities of the metal sulfides. The fourth group consists of ions of the alkaline earth elements, which are precipitated as carbonates or phosphates. The final group of ions is in the filtrate after these precipitations.

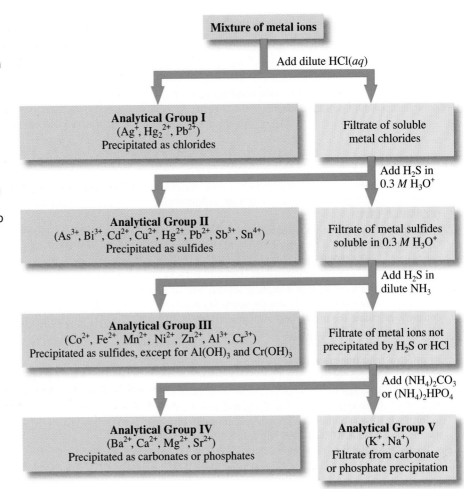

A Checklist for Review

Important Terms

solubility product constant (K_{sp}) (18.1)
fractional precipitation (18.3)

complex ion (18.5)
ligand (18.5)
formation (stability) constant (K_f) (18.5)

dissociation constant (of a complex ion) (K_d) (18.5)
amphoteric hydroxide (18.5)

qualitative analysis (18.7)

Summary of Facts and Concepts

The equilibrium constant for the equilibrium between a slightly soluble ionic solid and its ions in solution is called the *solubility product constant, K_{sp}*. Its value can be determined from the solubility of the solid. Conversely, when the solubility product constant is known, the solubility of the solid can be calculated. The solubility is decreased by the addition of a soluble salt that supplies a common ion. Qualitatively, this can be seen to follow from Le Chatelier's principle. Quantitatively, the *common-ion effect* on solubility can be obtained from the solubility product constant.

Rather than look at the solubility process as the dissolving of a solid in a solution, you can look at it as the *precipitation* of the solid from the solution. You can decide whether precipitation will occur by computing the *ion product*. Precipitation occurs when the ion product is greater than K_{sp}.

Solubility is affected by the pH if the compound supplies an anion conjugate to a weak acid. As the pH decreases (H_3O^+ ion concentration increases), the anion concentration decreases because the anion forms the weak acid. Therefore, the ion product decreases and the solubility increases.

The concentration of a metal ion in solution is decreased by *complex-ion formation*. The equilibrium constant for the formation of the complex ion from the aqueous metal ion and the ligands is called the *formation constant* (or *stability constant*), K_f. Because complex-ion formation reduces the concentration of aqueous metal ion, an ionic compound of the metal is more soluble in a solution of the ligand.

The sulfide scheme of *qualitative analysis* separates the metal ions in a mixture by using precipitation reactions. Variation of solubility with pH and with complex-ion formation is used to aid in the separation.

Operational Skills

1. **Writing solubility product expressions** Write the solubility product expression for a given ionic compound. **(EXAMPLE 18.1)**
2. **Calculating K_{sp} from the solubility, or vice versa** Given the solubility of a slightly soluble ionic compound, calculate K_{sp} **(EXAMPLES 18.2, 18.3)**. Given K_{sp}, calculate the solubility of an ionic compound **(EXAMPLE 18.4)**.
3. **Calculating the solubility of a slightly soluble salt in a solution of a common ion** Given the solubility product constant, calculate the molar solubility of a slightly soluble ionic compound in a solution that contains a common ion. **(EXAMPLE 18.5)**
4. **Predicting whether precipitation will occur** Given the concentrations of ions originally in solution, determine whether a precipitate is expected to form **(EXAMPLE 18.6)**. Determine whether a precipitate is expected to form when two solutions of known volume and molarity are mixed **(EXAMPLE 18.7)**. For both problems, you will need the solubility product constant.
5. **Determining the qualitative effect of pH on solubility** Decide whether the solubility of a salt will be greatly increased by decreasing the pH. **(EXAMPLE 18.8)**
6. **Calculating the concentration of a metal ion in equilibrium with a complex ion** Calculate the concentration of aqueous metal ion in equilibrium with the complex ion, given the original metal-ion and ligand concentrations **(EXAMPLE 18.9)**. The formation constant K_f of the complex ion is required.
7. **Predicting whether a precipitate will form in the presence of the complex ion** Predict whether an ionic compound will precipitate from a solution of known concentrations of cation, anion, and ligand that complexes with the cation **(EXAMPLE 18.10)**. K_f and K_{sp} are required.

Review Questions

18.1 Suppose the molar solubility of nickel hydroxide, $Ni(OH)_2$, is x M. Show that K_{sp} for nickel hydroxide equals $4x^3$.

18.2 Explain why calcium sulfate is less soluble in sodium sulfate solution than in pure water.

18.3 Discuss briefly how you could predict whether a precipitate will form when solutions of lead nitrate and potassium iodide are mixed. What information do you need to have?

18.4 Explain why barium fluoride dissolves in dilute hydrochloric acid but is insoluble in water.

18.5 Explain how metal ions such as Pb^{2+} and Zn^{2+} are separated by precipitation with hydrogen sulfide.

18.6 A precipitate forms when a small amount of sodium hydroxide is added to a solution of aluminum sulfate. This precipitate dissolves when more sodium hydroxide is added. Explain what is happening.

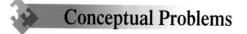

Conceptual Problems

18.7 Which compound in each of the following pairs of compounds is the more soluble one?
a. silver chloride or silver iodide
b. magnesium hydroxide or copper(II) hydroxide

18.8 You are given two mineral samples: halite, which is NaCl, and fluorite, which is CaF_2. Describe a simple test you could use to discover which mineral is fluorite.

18.9 Which of the following pictures best represents a solution made by adding 10 g of silver chloride, AgCl, to a liter of water? In these pictures, the dark spheres represent Ag^+ ions and the light spheres represent chloride ions. For clarity, water molecules are not shown.

18.10 Which of the following pictures best represents an unsaturated solution of sodium chloride, NaCl? In these pictures, the dark spheres represent Na^+ ions and the light spheres represent chloride ions. For clarity, water molecules are not shown.

18.11 You are given a solution of the ions Mg^{2+}, Ca^{2+}, and Ba^{2+}. Devise a scheme to separate these ions using sodium sulfate. Note that magnesium sulfate is soluble.

18.12 When ammonia is first added to a solution of copper(II) nitrate, a pale blue precipitate of copper(II) hydroxide forms. As more ammonia is added, however, this precipitate dissolves. Describe what is happening.

Practice Problems

Solubility and K_{sp}

18.13 Write solubility product expressions for the following compounds.
a. $SrSO_4$ b. $Mg(OH)_2$ c. $Ca_3(PO_4)_2$ d. Ag_2S

18.14 Write solubility product expressions for the following compounds.
a. MgC_2O_4 b. $Al(OH)_3$
c. $Zn(OH)_2$ d. $Ca_3(AsO_4)_2$

18.15 Calculate the solubility product constant for copper(II) iodate, $Cu(IO_3)_2$. The solubility of copper(II) iodate in water is 0.13 g/100 mL.

18.16 The solubility of silver chromate, Ag_2CrO_4, in water is 0.022 g/L. Calculate K_{sp}.

18.17 The pH of a saturated solution of magnesium hydroxide (milk of magnesia) was found to be 10.52. From this, find K_{sp} for magnesium hydroxide.

18.18 A solution saturated in calcium hydroxide (limewater) has a pH of 12.35. What is K_{sp} for calcium hydroxide?

18.19 Strontianite (strontium carbonate) is an important mineral of strontium. Calculate the solubility of strontium carbonate, $SrCO_3$, from the solubility product constant (see Table 18.1).

18.20 Magnesite (magnesium carbonate, $MgCO_3$) is a common magnesium mineral. From the solubility product constant (Table 18.1), find the solubility of magnesium carbonate in grams per liter of water.

18.21 What is the solubility of PbF_2 in water? The K_{sp} for PbF_2 is 2.7×10^{-8}.

18.22 What is the solubility of MgF_2 in water? The K_{sp} for MgF_2 is 7.4×10^{-11}.

Common-Ion Effect

18.23 What is the solubility (in grams per liter) of strontium sulfate, $SrSO_4$, in 0.20 M sodium sulfate, Na_2SO_4?

18.24 What is the solubility (in grams per liter) of lead(II) chromate, $PbCrO_4$, in 0.15 M potassium chromate, K_2CrO_4?

18.25 The solubility of magnesium fluoride, MgF_2, in water is 0.016 g/L. What is the solubility (in grams per liter) of magnesium fluoride in 0.015 M sodium fluoride, NaF?

18.26 The solubility of silver sulfate, Ag_2SO_4, in water has been determined to be 8.0 g/L. What is the solubility in 0.65 M sodium sulfate, Na_2SO_4?

Precipitation

18.27 From each of the following ion concentrations in a solution, predict whether a precipitate will form in the solution.
a. $[Ba^{2+}] = 0.020\ M$, $[F^-] = 0.015\ M$
b. $[Mg^{2+}] = 0.0012\ M$, $[CO_3^{2-}] = 0.041\ M$

18.28   From each of the following ion concentrations in a solution, predict whether a precipitate will form in the solution.
a. $[Sr^{2+}] = 0.012\ M$, $[CrO_4^{2-}] = 0.0015\ M$
b. $[Pb^{2+}] = 0.0035\ M$, $[Cl^-] = 0.15\ M$

18.29   The following solutions are mixed: 1.0 L of 0.00010 M NaOH and 1.0 L of 0.0020 M $MgSO_4$. Is a precipitate expected? Explain.

18.30   A 45-mL sample of 0.015 M calcium chloride, $CaCl_2$, is added to 55 mL of 0.010 M sodium sulfate, Na_2SO_4. Is a precipitate expected? Explain.

18.31   How many moles of calcium chloride, $CaCl_2$, can be added to 1.5 L of 0.020 M potassium sulfate, K_2SO_4, before a precipitate is expected? Assume that the volume of the solution is not changed significantly by the addition of calcium chloride.

18.32   Magnesium sulfate, $MgSO_4$, is added to 456 mL of 0.040 M sodium hydroxide, NaOH, until a precipitate just forms. How many grams of magnesium sulfate were added? Assume that the volume of the solution is not changed significantly by the addition of magnesium sulfate.

Effect of pH on Solubility

18.33   Write the net ionic equation in which the slightly soluble salt barium fluoride, BaF_2, dissolves in dilute hydrochloric acid.

18.34   Write the net ionic equation in which the slightly soluble salt lead(II) carbonate, $PbCO_3$, dissolves in dilute hydrochloric acid.

18.35   Which salt would you expect to dissolve readily in acidic solution, barium sulfate or barium fluoride? Explain.

18.36   Which salt would you expect to dissolve readily in acidic solution, calcium phosphate, $Ca_3(PO_4)_2$, or calcium sulfate, $CaSO_4$? Explain.

Complex Ions

18.37   Write the chemical equation for the formation of the $Cu(CN)_2^-$ ion. Write the K_f expression.

18.38   Write the chemical equation for the formation of the $Ni(NH_3)_6^{2+}$ ion. Write the K_f expression.

18.39   Sufficient sodium cyanide, NaCN, was added to 0.015 M silver nitrate, $AgNO_3$, to give a solution that was initially 0.100 M in cyanide ion, CN^-. What is the concentration of silver ion, Ag^+, in this solution after $Ag(CN)_2^-$ forms? The formation constant K_f for the complex ion $Ag(CN)_2^-$ is 5.6×10^{18}.

18.40   The formation constant K_f for the complex ion $Zn(OH)_4^{2-}$ is 2.8×10^{15}. What is the concentration of zinc ion, Zn^{2+}, in a solution that is initially 0.20 M in $Zn(OH)_4^{2-}$?

Qualitative Analysis

18.41   Describe how you could separate the following mixture of metal ions: Cd^{2+}, Pb^{2+}, and Sr^{2+}.

18.42   Describe how you could separate the following mixture of metal ions: Na^+, Hg^{2+}, and Ca^{2+}.

General Problems

18.43   Lead(II) sulfate is often used as a test for lead(II) ion in qualitative analysis. Calculate the molar solubility of lead(II) sulfate in water.

18.44   Mercury(II) ion is often precipitated as mercury(II) sulfide in qualitative analysis. Calculate the molar solubility of mercury(II) sulfide in water, assuming that no other reactions occur.

18.45   What is the solubility of magnesium hydroxide in a solution buffered at pH 8.80?

18.46   What is the solubility of silver oxide, Ag_2O, in a solution buffered at pH 10.50? The equilibrium is $Ag_2O(s) + H_2O(l) \rightleftharpoons 2Ag^+(aq) + 2OH^-(aq)$; $K_c = 2.0 \times 10^{-8}$.

18.47   What is the molar solubility of $Mg(OH)_2$ in a solution containing $1.0 \times 10^{-1}\ M$ NaOH? See Table 18.1 for K_{sp}.

18.48   What is the molar solubility of $Al(OH)_3$ in a solution containing $1.0 \times 10^{-3}\ M$ NaOH? See Table 18.1 for K_{sp}.

18.49   What must be the concentration of sulfate ion in order to precipitate calcium sulfate, $CaSO_4$, from a solution that is 0.0030 M Ca^{2+}?

18.50   What must be the concentration of chromate ion in order to precipitate strontium chromate, $SrCrO_4$, from a solution that is 0.0025 M Sr^{2+}? K_{sp} for strontium chromate is 3.5×10^{-5}.

18.51   How many grams of sodium chloride can be added to 785 mL of 0.0015 M silver nitrate before a precipitate forms?

18.52   How many grams of sodium sulfate can be added to 435 mL of 0.0028 M barium chloride before a precipitate forms?

18.53   Calculate the molar solubility of silver bromide, AgBr, in 5.0 M NH_3.

18.54   Calculate the molar solubility of silver iodide, AgI, in 2.0 M NH_3.

18.55 Ammonia, NH_3, is a base that ionizes to give NH_4^+ and OH^- ($K_b = 1.8 \times 10^{-5}$). You add magnesium sulfate to an ammonia solution. Calculate the concentration of Mg^{2+} ion when magnesium hydroxide, $Mg(OH)_2$, just begins to precipitate from 0.10 M NH_3. K_{sp} for $Mg(OH)_2$ is 1.8×10^{-11}.

18.56 Hydrazine, N_2H_4, is a base that ionizes to give $N_2H_5^+$ and OH^- ($K_b = 1.7 \times 10^{-6}$). You add magnesium sulfate to a hydrazine solution. Calculate the concentration of Mg^{2+} ion when magnesium hydroxide, $Mg(OH)_2$, just begins to precipitate from 0.20 M N_2H_4. K_{sp} for $Mg(OH)_2$ is 1.8×10^{-11}.

18.57 A saturated solution of copper(II) iodate in pure water has a copper-ion concentration of 2.7×10^{-3} M.
a. What is the molar solubility of copper iodate in a 0.35 M potassium iodate solution?
b. What is the molar solubility of copper iodate in a 0.35 M copper nitrate solution?
c. Should there be a difference in the answers to parts a and b? Why?

18.58 A saturated solution of lead iodate in pure water has an iodate-ion concentration of 8.0×10^{-5} M.
a. What is the molar solubility of lead iodate in a 0.15 M lead nitrate solution at the same temperature?
b. Should the molar solubility of lead iodate in part a be the same as, greater than, or less than that of lead iodate in pure water? Why?

18.59 A solution contains 0.0150 M lead(II) ion. A concentrated sodium iodide solution is added to precipitate lead iodide (assume no volume change).
a. At what concentration of I^- does precipitate start to form?
b. When $[I^-] = 2.0 \times 10^{-3}$ M, what is the lead-ion concentration? What percentage of the lead(II) originally present remains in solution?

18.60 A solution contains 0.00750 M calcium ion. A concentrated sodium fluoride solution is added to precipitate calcium fluoride (assume no volume change).
a. At what concentration of F^- does precipitate start to form?
b. When $[F^-] = 9.5 \times 10^{-4}$ M, what is the calcium-ion concentration? What percentage of the calcium ion has precipitated?

18.61 a. If the molar solubility of cobalt(II) hydroxide is 5.4×10^{-6} mol/L in pure water, what is its K_{sp} value?
b. What is the molar solubility of $Co(OH)_2$ in a buffered solution that has a pH of 10.43?
c. Account for the differences in molar solubility in parts a and b.

18.62 a. If the molar solubility of beryllium(II) hydroxide is 8.6×10^{-7} M in pure water, what is its K_{sp} value?
b. What is the molar solubility of beryllium(II) hydroxide in a solution that is 1.50 M in NH_3 and 0.25 M in NH_4Cl?
c. Account for the differences in molar solubility in parts a and b.

18.63 A 1.0-L solution that is 4.2 M in ammonia is mixed with 26.7 g of ammonium chloride.
a. What is the hydroxide-ion concentration of this solution?
b. 0.075 mol of $MgCl_2$ is added to the above solution. Assume that there is no volume change. After $Mg(OH)_2$ has precipitated, what is the molar concentration of magnesium ion? What percent of the Mg^{2+} is removed from solution?

18.64 A 1.0-L solution that is 1.6 M in ammonia is mixed with 75.8 g of ammonium sulfate.
a. What is the hydroxide-ion concentration of this solution?
b. 0.058 mol of $MnCl_2$ is added to the above solution. Assume that there is no volume change. After $Mn(OH)_2$ has precipitated, what is the molar concentration of manganese ion? What percent of the Mn^{2+} is removed from solution (K_{sp} of $Mn(OH)_2$ is 4.6×10^{-14})?

Cumulative-Skills Problems

18.65 What is the solubility of calcium fluoride in a buffer solution containing 0.45 M $HCHO_2$ (formic acid) and 0.20 M $NaCHO_2$? (*Hint:* Consider the equation $CaF_2(s) + 2H^+(aq) \rightleftharpoons Ca^{2+}(aq) + 2HF(aq)$, and solve the equilibrium problem.)

18.66 What is the solubility of magnesium fluoride in a buffer solution containing 0.45 M $HC_2H_3O_2$ (acetic acid) and 0.20 M $NaC_2H_3O_2$? The K_{sp} for magnesium fluoride is 6.5×10^{-9}. (See the hint for Problem 18.65.)

18.67 A 67.0-mL sample of 0.350 M $MgSO_4$ is added to 45.0 mL of 0.250 M $Ba(OH)_2$. What is the net ionic equation for the reaction that occurs? What are the concentrations of ions in the mixture at equilibrium?

18.68 A 50.0-mL sample of 0.0150 M Ag_2SO_4 is added to 25.0 mL of 0.0100 M $PbCl_2$. What is the net ionic equation for the reaction that occurs? What are the concentrations of ions in the mixture at equilibrium?

Rust forms in a spontaneous reaction of iron with oxygen in air.

Thermodynamics and Equilibrium

U rea, NH_2CONH_2, is an important industrial chemical. It is used to make synthetic resins for adhesives and melamine plastics (Figure 19.1). Its major use, however, is as a nitrogen fertilizer for plants. Urea is produced by reacting ammonia with carbon dioxide.

$$2NH_3(g) + CO_2(g) \longrightarrow NH_2CONH_2(aq) + H_2O(l)$$

You might ask the following questions: Does the reaction naturally go in the direction it is written? Will the reaction mixture contain a sufficient amount of the product at equilibrium? We want to discuss these questions from the point of view of thermodynamics.

Thermodynamics is the study of the relationship between heat and other forms of energy involved in a chemical or physical process. With only heat measurements, you can predict the natural direction of a chemical reaction, and you can also determine the composition of a reaction mixture at equilibrium.

19.1   First Law of Thermodynamics; Enthalpy

We introduced the thermodynamic property of enthalpy, *H*, in Chapter 6. Now we want to look at this property again, but define it more precisely, in terms of the energy of the system. We begin by discussing the first law of thermodynamics.

First Law of Thermodynamics

▶ According to the law of conservation of energy, the total energy remains constant; energy is neither created nor destroyed.

The first law of thermodynamics is essentially the law of conservation of energy applied to thermodynamic systems. ◀ To state the law, you need to understand what the internal energy of a system is and how you can change the internal energy.

The **internal energy, *U*,** is *the sum of the kinetic and potential energies of the particles making up the system.* The kinetic energy includes the energy of motion of electrons, nuclei, and molecules. The potential energy results from the chemical bonding of atoms and from the attractions between molecules.

Internal energy is a **state function.** That is, it is *a property of a system that depends only on its present state, which is completely determined by variables such as temperature and pressure.* Thus, 1 mol of water at 0°C and 1 atm pressure has a definite quantity of energy. When a system changes from one state to another, its internal energy changes. You can calculate the change in internal energy, ΔU, from the initial value of the internal energy, U_i, and the final value of the internal energy, U_f.

$$\Delta U = U_f - U_i$$

Normally you are interested in changes in internal energy. You measure these changes for a thermodynamic system by noting the exchanges of energy between the system and its surroundings. These exchanges of energy are of two kinds: heat and work. **Work** is *the energy exchange that results when a force F moves an object through a distance d; work (w) equals F × d.*

Figure 19.1
Urea. *Top:* Urea is used as a plant fertilizer because it slowly decomposes in the soil to provide ammonia. *Bottom:* A molecular model of urea.

The distinction between heat and work is illustrated in the system shown in Figure 19.2. This system consists of a gas in a vessel equipped with a movable piston. On top of the piston is a weight, which you can consider as part of the surroundings. Fix the position of the piston so it does not move. Suppose the temperature of the surroundings is raised, and as a result, heat passes from the surroundings to the vessel. If you find that the energy of the surroundings decreases by 165 J in this way, you know from the law of conservation of energy that the internal energy of the system must have increased by just this quantity. You write $q = +165$ J, using the sign convention that heat absorbed by the system is positive and that heat evolved by the system is negative.

As the temperature of the gas in the vessel increases, the gas pressure increases. You now allow the piston to move, so the gas expands and lifts the piston and the weight on top of it. In lifting the weight, the system does work. The energy gained by the weight equals the force of gravity on the weight times the height to which the weight was raised. Suppose this energy is 92 J. Because the surroundings, which include the weight, have gained 92 J of energy, the system must have lost 92 J of energy. You write $w = -92$ J, adhering to the sign convention that work done *on* the system is positive and work done *by* the system is negative. ◀

▶ The convention given here is that adopted by IUPAC (International Union of Pure and Applied Chemistry).

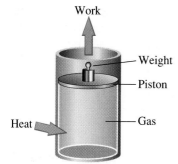

Figure 19.2
Exchanges of heat and work with the surroundings. A gas is enclosed in a vessel with a piston. Heat flows into the vessel from the surroundings, which are at a higher temperature. As the temperature of the gas increases, the gas expands, lifting the weight (doing work).

The system in Figure 19.2 gains internal energy from the heat absorbed and loses internal energy via the work done. In general, the net change of internal energy equals heat plus work.

$$\Delta U = q + w$$

The **first law of thermodynamics** states that *the change in internal energy of a system, ΔU, equals $q + w$.* For the system shown in Figure 19.2,

$$\Delta U = (+165\ \text{J}) + (-92\ \text{J}) = +73\ \text{J}$$

Heat of Reaction and Internal Energy

Now consider the reaction of zinc with hydrochloric acid.

$$Zn(s) + 2H_3O^+(aq) \longrightarrow Zn^{2+}(aq) + 2H_2O(l) + H_2(g)$$

When this reaction is carried out in a beaker open to the atmosphere, the reaction is exothermic, evolving 152.4 kJ of heat per mole of zinc. You write $q_p = -152.4$ kJ, where the subscript p indicates that the process occurs at constant pressure (the pressure of the atmosphere).

The hydrogen that is produced increases the volume of the system. As hydrogen is evolved, work must be done by the system to push back the atmosphere. How can you calculate this work? Imagine for the moment that the atmosphere is replaced by a piston and weights, whose downward force from gravity F creates a pressure on the gas equivalent to that of the atmosphere. The pressure P equals F divided by the cross-sectional area of the piston, A. (See Figure 19.3.)

Because the volume of a cylinder equals its height h times its area A, the change in volume is $\Delta V = A \times h$, and $h = \Delta V/A$. The work done by the system in expanding equals the force of gravity times the distance the piston moves.

$$w = -F \times h = -F \times \frac{\Delta V}{A} = -\frac{F}{A} \times \Delta V$$

The negative sign is given because w is work done by the system. Note that F/A is the pressure, P, which equals that of the atmosphere. Therefore,

$$w = -P\Delta V$$

Figure 19.3
Reaction of zinc metal with hydrochloric acid at constant pressure. *(A)* The beginning of the reaction. The constant pressure of the atmosphere has been replaced by a piston and weight to give an equivalent pressure. *(B)* The reaction produces hydrogen gas. This increases the volume of the system, so that the piston and weight are lifted upward. Work is done by the system on the piston and weight.

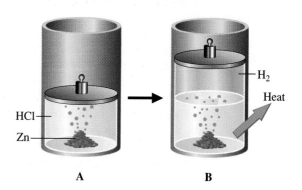

This formula tells you that you can calculate the work done by a chemical reaction carried out in an open vessel by multiplying the atmospheric pressure P by the change in volume of the chemical system, ΔV. For example, when 1.00 mol Zn reacts with excess hydrochloric acid, 1.00 mol H_2 is produced. At 25°C and 1.00 atm (= 1.01×10^5 Pa), this amount of H_2 occupies 24.5 L (= 24.5×10^{-3} m^3). The work done by the chemical system in pushing back the atmosphere is

$$w = -P\Delta V = -(1.01 \times 10^5 \text{ Pa}) \times (24.5 \times 10^{-3} \text{ m}^3)$$
$$= -2.47 \times 10^3 \text{ J, or } -2.47 \text{ kJ}$$

If you apply the first law to this chemical system, you can relate the change in internal energy of the system to the heat of reaction. You have

$$\Delta U = q_p + w = q_p - P\Delta V$$

For the reaction of Zn with HCl, $q_p = -152.4$ kJ and $w = -P\Delta V = -2.47$ kJ, so

$$\Delta U = -152.4 \text{ kJ} - 2.47 \text{ kJ} = -154.9 \text{ kJ}$$

We can now summarize what happens when 1.00 mol Zn reacts with excess acid. This energy change, ΔU, equals -154.9 kJ. Energy leaves the system mostly as heat ($q_p = -152.4$ kJ) but partly as expansion work ($w = -2.47$ kJ).

Enthalpy and Enthalpy Change

In Chapter 6, we defined enthalpy in terms of the heat of reaction at constant pressure, q_p. We now define **enthalpy, H,** precisely as *the quantity U + PV.* Because U, P, and V are state functions, H is also a state function. This means that for a given temperature and pressure, a given amount of a substance has a definite enthalpy.

Let us now show the relationship of ΔH to the heat of reaction q_p. Note that ΔH is the final enthalpy H_f minus the initial enthalpy H_i.

$$\Delta H = H_f - H_i$$

Now, substitute defining expressions for H_f and H_i, using the subscripts f and i to indicate *final* and *initial,* respectively. Note, however, that the pressure is constant.

$$\Delta H = (U_f + PV_f) - (U_i + PV_i) = (U_f - U_i) + P(V_f - V_i) = \Delta U + P\Delta V$$

Earlier we found that $\Delta U = q_p - P\Delta V$. Hence,

$$\Delta H = (q_p - P\Delta V) + P\Delta V = q_p$$

In practice, what you do is measure certain heats of reaction and use them to tabulate enthalpies of formation, ΔH_f°. The standard enthalpy change for a reaction is

$$\Delta H^\circ = \Sigma \, n\Delta H_f^\circ(\text{products}) - \Sigma \, m\Delta H_f^\circ(\text{reactants})$$

Standard enthalpies of formation are listed in Table 6.2 and in Appendix B.

How much heat is absorbed or evolved in the reaction between NH_3 and CO_2 to produce urea and water? The following are standard enthalpies of formation at 25°C for the substances indicated (in kJ/mol): $NH_3(g)$, -45.9; $CO_2(g)$, -393.5; $NH_2CONH_2(aq)$, -319.2; $H_2O(l)$, -285.8. Substituting into the previous equation yields

$$\Delta H^\circ = [(-319.2 - 285.8) - (-2 \times 45.9 - 393.5)] \text{ kJ} = -119.7 \text{ kJ}$$

From the minus sign in the value of ΔH°, you conclude that heat is evolved.

Spontaneous Processes and Entropy

Why does a chemical reaction go naturally in a particular direction? To answer this question, you need to look at spontaneous processes. A **spontaneous process** is *a physical or chemical change that occurs by itself*. It requires no continuing outside agency to make it happen. A rock at the top of a hill rolls down (Figure 19.4, top). Heat flows from a hot object to a cold one. An iron object rusts in moist air (Figure 19.5). These processes occur spontaneously without requiring an outside force. If these processes were to go in the opposite direction, they would be *nonspontaneous*. The rolling of a rock uphill by itself is nonspontaneous (see Figure 19.4). The rock could be moved to the top of the hill, but work would have to be expended. Heat can be made to flow from a cold to a hot object, but a heat pump or refrigerator is needed. Rust can be converted to iron, but the process requires a reducing agent.

19.2 Entropy and the Second Law of Thermodynamics

When you ask whether a chemical reaction goes in the direction in which it is written, you are asking whether the reaction is spontaneous. The first law of thermodynamics cannot help you answer such a question. It does help you keep track of the various forms of energy in a chemical change using the conservation of energy. At one time it was thought that spontaneous reactions must be exothermic, but many spontaneous reactions are now known to be endothermic (see Figure 6.1).

The second law of thermodynamics, which we will discuss in this section, provides a way to answer questions about the spontaneity of a reaction. The second law is expressed in terms of a quantity called *entropy*.

Entropy

Entropy, S, is *a thermodynamic quantity that is a measure of the randomness or disorder in a system*. Entropy, like enthalpy, is a state function. That is, the quantity of entropy in a given amount of substance depends only on variables, such as temperature and pressure, that determine the state of the substance.

You would expect the entropy (disorder) to increase when ice melts to the liquid (see Figure 11.39). In ice, the H_2O molecules occupy regular, fixed positions in a crystal lattice. But in the liquid, the molecules move about freely, giving a disordered structure—one having greater entropy.

You calculate the entropy change, ΔS, for a process similarly to the way you calculate ΔH. If S_i is the initial entropy and S_f is the final entropy, the change in entropy is

$$\Delta S = S_f - S_i$$

For the melting of ice to liquid water,

$$H_2O(s) \longrightarrow H_2O(l)$$
$$\Delta S = (63 - 41)\,J/K = 22\,J/K$$

When 1 mol of ice melts at 0°C, the water increases in entropy by 22 J/K.

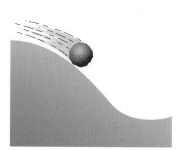

Spontaneous process

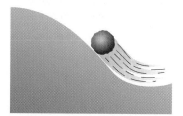

Nonspontaneous process

Figure 19.4
Examples of a spontaneous and a nonspontaneous process.
Top: The rolling of a rock downhill is a spontaneous process. The rock eventually comes to equilibrium at the bottom of the hill. *Bottom:* The rolling of a rock uphill is a nonspontaneous process.

CONCEPT CHECK 19.1

> You have a sample of 1.0 mg of solid iodine at room temperature. Later, you notice that the iodine has sublimed (passed into the vapor state). What can you say about the change of entropy of the iodine?

Second Law of Thermodynamics

A process occurs naturally as a result of an *overall* increase in disorder of a system plus its surroundings. There is a natural tendency for things to mix and to break down, events that represent increasing disorder. Where you see order or structure being built, it results from the use of greater order elsewhere. Order in one place is used to build order in another.

▶ Consider a simple analogy. A baker makes (creates) cookies. She also buys cookies from a wholesaler and sells cookies at her store. The change in number of cookies in her store in any time interval equals the number created plus the flow of cookies in and out of her store (number bought minus number sold). The analogy breaks down if someone eats cookies in the store. Whereas entropy can only be created, cookies can be created and destroyed (eaten).

The **second law of thermodynamics** states that *the total entropy of a system and its surroundings always increases for a spontaneous process.* Note that entropy is quite different from energy. Energy can be neither created nor destroyed during chemical change. But entropy is created during a spontaneous, or natural, process.

For a spontaneous process carried out at a given temperature, the second law can be restated in a form that refers only to the system. As the process takes place, entropy is created. At the same time, heat flows into or out of the system, and entropy accompanies that heat flow. Thus, when heat flows into the system, entropy flows into the system. In general, the entropy change associated with heat q at absolute temperature T can be shown to equal q/T. The net change in entropy of the system, ΔS, equals the sum of the entropy created during the spontaneous process and the change in entropy associated with the heat flow (entropy flow). ◀

$$\Delta S = \text{entropy created} + \frac{q}{T} \qquad \text{(spontaneous process)}$$

Normally the quantity of entropy created during a spontaneous process cannot be directly measured. Because it is a positive quantity, however, if you delete it from the right side of the equation, you can conclude that ΔS must be greater than q/T for a spontaneous process.

$$\Delta S > \frac{q}{T} \qquad \text{(spontaneous process)}$$

The restatement of the second law is as follows: *for a spontaneous process at a given temperature, the change in entropy of the system is greater than the heat divided by the absolute temperature.*

Entropy Change for a Phase Transition

Certain processes occur at equilibrium. For example, ice at 0°C is in equilibrium with liquid water at 0°C. If heat is slowly absorbed by the system, it remains very near equilibrium, but the ice melts. Under these conditions, no entropy is created. The entropy change results from the absorption of heat. Therefore,

$$\Delta S = \frac{q}{T} \qquad \text{(equilibrium process)}$$

Figure 19.5
The rusting of iron in moist air is a spontaneous reaction. This sculpture by Picasso in Chicago was allowed to rust to give a pleasing effect.

Free-Energy Concept

At the end of Section 19.2, you saw that the quantity $\Delta H - T\Delta S$ can serve as a criterion for spontaneity of a reaction at constant temperature and pressure. If the value of this quantity is negative, the reaction is spontaneous. If it is positive, the reaction is nonspontaneous. If it equals zero, the reaction is at equilibrium.

As an application of this criterion, consider the preparation of urea from NH_3 and CO_2. The heat of reaction, $\Delta H°$, was calculated (Section 19.1) as -119.7 kJ. Then, in Example 19.3, you calculated the entropy of reaction $\Delta S°$ and found a value of -356 J/K, or -0.356 kJ/K. Substitute these values and $T = 298$ K ($25°C$) into the expression $\Delta H° - T\Delta S°$.

$$\Delta H° - T\Delta S° = (-119.7 \text{ kJ}) - (298 \text{ K})(-0.356 \text{ kJ/K}) = -13.6 \text{ kJ}$$

You see that $\Delta H° - T\Delta S°$ is a negative quantity, from which you conclude that the reaction is spontaneous under standard conditions.

19.4 Free Energy and Spontaneity

It is very convenient to define a new thermodynamic quantity in terms of H and S that will be directly useful as a criterion of spontaneity. For this purpose, the American physicist J. Willard Gibbs (1839–1903) introduced the concept of **free energy, G,** which is *a thermodynamic quantity defined by the equation $G = H - TS$.*

The change in free energy, ΔG, is given by the equation

$$\Delta G = \Delta H - T\Delta S$$

If you can show that ΔG for a reaction at a given temperature and pressure is negative, you can predict that the reaction will be spontaneous.

Standard Free-Energy Change

The standard states are as follows: for pure liquids and solids, 1 atm pressure; for gases, 1 atm partial pressure; for solutions, 1 M concentration. The temperature is the temperature of interest, usually $25°C$ (298 K).

The standard free-energy change, $\Delta G°$, is the free-energy change that occurs when reactants in their standard states are converted to products in their standard states. Example 19.4 illustrates the calculation of the standard free-energy change, $\Delta G°$.

EXAMPLE 19.4

Calculating $\Delta G°$ from $\Delta H°$ and $\Delta S°$

What is the standard free-energy change, $\Delta G°$, for the following reaction at $25°C$?

$$N_2(g) + 3H_2(g) \longrightarrow 2NH_3(g)$$

Use values of $\Delta H_f°$ and $S°$ from Tables 6.2 and 19.1.

SOLUTION

Write the balanced equation and place below each formula the values of ΔH_f° and S° multiplied by stoichiometric coefficients.

$$N_2(g) + 3H_2(g) \longrightarrow 2NH_3(g)$$

ΔH_f°: 0   0 $2 \times (-45.9)$ kJ

S°: 191.5 3×130.6 2×193 J/K

Now you can calculate ΔH° and ΔS°.

$$\Delta H^\circ = \Sigma \, n\Delta H_f^\circ(\text{products}) - \Sigma \, m\Delta H_f^\circ(\text{reactants})$$
$$= [2 \times (-45.9) - 0] \text{ kJ} = -91.8 \text{ kJ}$$

$$\Delta S^\circ = \Sigma \, nS^\circ(\text{products}) - \Sigma \, mS^\circ(\text{reactants})$$
$$= [2 \times 193 - (191.5 + 3 \times 130.6)] \text{ J/K} = -197 \text{ J/K}$$

You now substitute into the equation for ΔG° in terms of ΔH° and ΔS°. Note that you substitute ΔS° in units of kJ/K.

$$\Delta G^\circ = \Delta H^\circ - T\Delta S^\circ = -91.8 \text{ kJ} - (298 \text{ K})(-0.197 \text{ kJ/K}) = \mathbf{-33.1 \text{ kJ}}$$

Standard Free Energies of Formation

The **standard free energy of formation, ΔG_f°,** is *the free-energy change that occurs when 1 mol of substance is formed from its elements in their stablest states at 1 atm and at a specified temperature* (usually 25°C). For example, the standard free energy of formation of $NH_3(g)$ is the free-energy change for the reaction

$$\tfrac{1}{2}N_2(g) + \tfrac{3}{2}H_2(g) \longrightarrow NH_3(g)$$

In Example 19.4, you found ΔG° for the formation of 2 mol NH_3 from its elements to be -33.1 kJ. Hence, $\Delta G_f^\circ(NH_3) = -33.1$ kJ/2 mol $= -16.6$ kJ/mol.

As in the case of standard enthalpies of formation, the standard free energies of formation of elements in their stablest states are assigned the value zero. By tabulating ΔG_f° for substances, as in Table 19.2, you can calculate ΔG° for any reaction involving those substances.

$$\Delta G^\circ = \Sigma \, n\Delta G_f^\circ(\text{products}) - \Sigma \, m\Delta G_f^\circ(\text{reactants})$$

EXAMPLE 19.5

Calculating ΔG° from Standard Free Energies of Formation

Calculate ΔG° for the combustion of 1 mol of ethanol, C_2H_5OH, at 25°C.

$$C_2H_5OH(l) + 3O_2(g) \longrightarrow 2CO_2(g) + 3H_2O(g)$$

SOLUTION

Write the balanced equation with values of ΔG_f° multiplied by stoichiometric coefficients below each formula.

$$C_2H_5OH(l) + 3O_2(g) \longrightarrow 2CO_2(g) + 3H_2O(g)$$

ΔG_f°: -174.8 0 $2(-394.4)$ $3(-228.6)$ kJ

Review Questions

19.1   What is a spontaneous process? Give three examples of spontaneous processes. Give three examples of nonspontaneous processes.

19.2   Which contains greater entropy, a quantity of frozen benzene or the same quantity of liquid benzene at the same temperature? Explain in terms of the degree of order of the substance.

19.3   State the second law of thermodynamics.

19.4   Describe how the standard entropy of hydrogen gas at 25°C can be obtained from heat measurements.

19.5   Describe what you would look for in a reaction involving gases in order to predict the sign of $\Delta S°$. Explain.

19.6   Define the free energy G. How is ΔG related to ΔH and ΔS?

19.7   What is meant by the standard free-energy change $\Delta G°$ for a reaction? What is meant by the standard free energy of formation $\Delta G_f°$ of a substance?

19.8   Explain how $\Delta G°$ can be used to decide whether a chemical equation is spontaneous in the direction written.

19.9   Give an example of a chemical reaction used to obtain useful work.

19.10   Explain how the free energy changes as a spontaneous reaction occurs. Show by means of a diagram how G changes with the extent of reaction.

19.11   Discuss the different sign combinations of $\Delta H°$ and $\Delta S°$ that are possible for a process carried out at constant temperature and pressure. For each combination, state whether the process must be spontaneous or not, or whether both situations are possible. Explain.

19.12   Consider a reaction in which $\Delta H°$ and $\Delta S°$ are positive. Suppose the reaction is nonspontaneous at room temperature. How would you estimate the temperature at which the reaction becomes spontaneous?

Conceptual Problems

19.13   For each of the following statements, indicate whether it is true or false.
a. A spontaneous reaction always releases heat.
b. A spontaneous reaction is always a fast reaction.
c. The entropy of a system always increases for a spontaneous change.
d. The entropy of a system and its surroundings always increases for a spontaneous change.
e. The energy of a system always increases for a spontaneous change.

19.14   Which of the following are spontaneous processes?
a. A cube of sugar dissolves in a cup of hot tea.
b. A rusty crowbar turns shiny.
c. Butane from a lighter burns in air.
d. A clock pendulum, initially stopped, begins swinging.
e. Hydrogen and oxygen gases bubble out from a glass of pure water.

19.15   For each of the following series of pairs, indicate which one of each pair has the greater quantity of entropy.
a. 1.0 mol of carbon dioxide gas at 20°C, 1 atm, or 2.0 mol of carbon dioxide gas at 20°C, 1 atm
b. 1.0 mol of butane liquid at 20°C, 10 atm, or 1.0 mol of butane gas at 20°C, 10 atm
c. 1.0 mol of solid carbon dioxide at −80°C, 1 atm, or 1.0 mol of solid carbon dioxide at −90°C, 1 atm
d. 25 g of solid bromine at −7°C, 1 atm, or 25 g of bromine vapor at −7°C, 1 atm

19.16   Predict the sign of the entropy change for each of the following processes.
a. A drop of food coloring diffuses throughout a glass of water.
b. A tree leafs out in the spring.
c. Flowers wilt and stems decompose in the fall.
d. A lake freezes over in the winter.
e. Rainwater on the pavement evaporates.

Practice Problems

First Law of Thermodynamics

19.17   A gas is cooled and loses 82 J of heat. The gas contracts as it cools, and work done on the system equal to 29 J is exchanged with the surroundings. What are q, w, and ΔU?

19.18   An ideal gas is cooled isothermally (at constant temperature). The internal energy of an ideal gas remains constant during an isothermal change. If q is −76 J, what are ΔU and w?

Entropy Changes

19.19   Chloroform, $CHCl_3$, is a solvent and has been used as an anesthetic. The heat of vaporization of chloroform at its boiling point (61.2°C) is 29.6 kJ/mol. What is the entropy change when 1.20 mol $CHCl_3$ vaporizes at its boiling point?

19.20   Diethyl ether (known simply as ether), $(C_2H_5)_2O$, is a solvent and an anesthetic. The heat of vaporization of diethyl

ether at its boiling point (35.6°C) is 26.7 kJ/mol. What is the entropy change when 1.34 mol $(C_2H_5)_2O$ vaporizes at its boiling point?

19.21 The enthalpy change when liquid methanol, CH_3OH, vaporizes at 25°C is 37.4 kJ/mol. What is the entropy change when 1.50 mol of vapor in equilibrium with liquid condenses to liquid at 25°C? The entropy of the vapor at 25°C is 252 J/(mol · K). What is the entropy of the liquid at this temperature?

19.22 The heat of vaporization of carbon disulfide, CS_2, at 25°C is 29 kJ/mol. What is the entropy change when 1.85 mol of vapor in equilibrium with liquid condenses to liquid at 25°C? The entropy of the vapor at 25°C is 248 J/(mol · K). What is the entropy of the liquid at this temperature?

19.23 Predict the sign of $\Delta S°$, if possible, for each of the following reactions. If you cannot predict the sign for any reaction, state why.
a. $C_2H_2(g) + 2H_2(g) \longrightarrow C_2H_6(g)$
b. $N_2(g) + O_2(g) \longrightarrow 2NO(g)$
c. $2C_2H_2(g) + 3O_2(g) \longrightarrow 4CO(g) + 2H_2O(g)$

19.24 Predict the sign of $\Delta S°$, if possible, for each of the following reactions. If you cannot predict the sign for any reaction, state why.
a. $2SO_3(g) \longrightarrow 2SO_2(g) + O_2(g)$
b. $C_2H_5OH(l) + 3O_2(g) \longrightarrow 2CO_2(g) + 3H_2O(l)$
c. $P_4(g) \longrightarrow P_4(s)$

19.25 Calculate $\Delta S°$ for the following reactions, using standard entropy values.
a. $2Na(s) + Cl_2(g) \longrightarrow 2NaCl(s)$
b. $NaCl(s) \longrightarrow NaCl(aq)$
c. $CS_2(l) + 3O_2(g) \longrightarrow CO_2(g) + 2SO_2(g)$

19.26 Calculate $\Delta S°$ for the following reactions, using standard entropy values.
a. $2Ca(s) + O_2(g) \longrightarrow 2CaO(s)$
b. $CaCl_2(s) \longrightarrow CaCl_2(aq)$
c. $CS_2(g) + 4H_2(g) \longrightarrow CH_4(g) + 2H_2S(g)$

Free-Energy Change and Spontaneity

19.27 The free energy of formation of one mole of compound refers to a particular chemical equation. For each of the following, write that equation.
a. $KBr(s)$ b. $CH_3Cl(l)$ c. $H_2S(g)$ d. $AsH_3(g)$

19.28 The free energy of formation of one mole of compound refers to a particular chemical equation. For each of the following, write that equation.
a. $MgO(s)$ b. $COCl_2(g)$ c. $CF_4(g)$ d. $PCl_5(g)$

19.29 Calculate the standard free energy of the following reactions at 25°C, using standard free energies of formation.
a. $CH_4(g) + 2O_2(g) \longrightarrow CO_2(g) + 2H_2O(g)$
b. $CaCO_3(s) + 2H^+(aq) \longrightarrow Ca^{2+}(aq) + H_2O(l) + CO_2(g)$

19.30 Calculate the standard free energy of the following reactions at 25°C, using standard free energies of formation.
a. $C_2H_4(g) + 3O_2(g) \longrightarrow 2CO_2(g) + 2H_2O(g)$
b. $Na_2CO_3(s) + H^+(aq) \longrightarrow 2Na^+(aq) + HCO_3{}^-(aq)$

19.31 On the basis of $\Delta G°$ for each of the following reactions, decide whether the reaction is spontaneous or nonspontaneous as written. Or, if you expect an equilibrium mixture with significant amounts of both reactants and products, say so.
a. $2H_2O_2(aq) \longrightarrow O_2(g) + 2H_2O(l)$; $\Delta G° = -211$ kJ
b. $HCOOH(l) \longrightarrow CO_2(g) + H_2(g)$; $\Delta G° = 119$ kJ
c. $I_2(s) + Br_2(l) \longrightarrow 2IBr(g)$; $\Delta G° = 7.5$ kJ

19.32 For each of the following reactions, state whether the reaction is spontaneous or nonspontaneous as written or is easily reversible (that is, is a mixture with significant amounts of reactants and products).
a. $HCN(g) + 2H_2(g) \longrightarrow CH_3NH_2(g)$; $\Delta G° = -92$ kJ
b. $N_2(g) + O_2(g) \longrightarrow 2NO(g)$; $\Delta G° = 173$ kJ
c. $H_2(g) + I_2(s) \longrightarrow 2HI(g)$; $\Delta G° = 2.6$ kJ

Maximum Work

19.33 Consider the reaction of 2 mol $H_2(g)$ at 25°C and 1 atm with 1 mol $O_2(g)$ at the same temperature and pressure to produce liquid water at these conditions. If this reaction is run in a controlled way to generate work, what is the maximum useful work that can be obtained? How much entropy is produced in this case?

19.34 Consider the reaction of 1 mol $H_2(g)$ at 25°C and 1 atm with 1 mol $Cl_2(g)$ at the same temperature and pressure to produce gaseous HCl at these conditions. If this reaction is run in a controlled way to generate work, what is the maximum useful work that can be obtained? How much entropy is produced in this case?

Calculation of Equilibrium Constants

19.35 Give the expression for the thermodynamic equilibrium constant for each of the following reactions.
a. $CO(g) + H_2O(g) \rightleftharpoons CO_2(g) + H_2(g)$
b. $2Li(s) + 2H_2O(l) \rightleftharpoons 2Li^+(aq) + 2OH^-(aq) + H_2(g)$

19.36 Write the expression for the thermodynamic equilibrium constant for each of the following reactions.
a. $CO(g) + 2H_2(g) \rightleftharpoons CH_3OH(g)$
b. $CaCO_3(s) + 2H^+(aq) \rightleftharpoons Ca^{2+}(aq) + H_2O(l) + CO_2(g)$

19.37 Calculate the standard free-energy change and the equilibrium constant K_p for the following reaction at 25°C. See Table 19.2 for data.

$$CO(g) + 3H_2(g) \rightleftharpoons CH_4(g) + H_2O(g)$$

19.38 Calculate the standard free-energy change and the equilibrium constant K_p for the following reaction at 25°C. See Appendix B for data.

$$CO(g) + 2H_2(g) \rightleftharpoons CH_3OH(g)$$

19.39 Obtain the equilibrium constant K_c at 25°C from the free-energy change for the reaction

$$Fe(s) + Cu^{2+}(aq) \rightleftharpoons Fe^{2+}(aq) + Cu(s)$$

See Appendix B for data.

19.40 Calculate the equilibrium constant K_c at 25°C from the free-energy change for the following reaction:

$$Zn(s) + 2Ag^+(aq) \rightleftharpoons Zn^{2+}(aq) + Ag(s)$$

See Appendix B for data.

Free Energy and Temperature Change

19.41 Use data given in Tables 6.2 and 19.1 to obtain the value of K_p at 1000°C for the reaction

$$C(graphite) + CO_2(g) \rightleftharpoons 2CO(g)$$

Carbon monoxide is known to form during combustion of carbon at high temperatures. Do the data agree with this? Explain.

19.42 Use data given in Tables 6.2 and 19.1 to obtain the value of K_p at 2000°C for the reaction

$$N_2(g) + O_2(g) \rightleftharpoons 2NO(g)$$

Nitric oxide is known to form in hot flames in air, which is a mixture of N_2 and O_2. It is present in auto exhaust from this reaction. Are the data in agreement with this result? Explain.

19.43 Sodium carbonate, Na_2CO_3, can be prepared by heating sodium hydrogen carbonate, $NaHCO_3$.

$$2NaHCO_3(s) \longrightarrow Na_2CO_3(s) + H_2O(g) + CO_2(g)$$

Estimate the temperature at which $NaHCO_3$ decomposes to products at 1 atm. See Appendix B for data.

19.44 Oxygen was first prepared by heating mercury(II) oxide, HgO.

$$2HgO(s) \longrightarrow 2Hg(g) + O_2(g)$$

Estimate the temperature at which HgO decomposes to O_2 at 1 atm. See Appendix B for data.

General Problems

19.45 Find the sign of $\Delta S°$ for the reaction

$$2N_2O_5(s) \longrightarrow 4NO_2(g) + O_2(g)$$

The reaction is endothermic and spontaneous at 25°C. Explain the spontaneity of the reaction in terms of enthalpy and entropy changes.

19.46 The combustion of acetylene, C_2H_2, is a spontaneous reaction given by the equation

$$2C_2H_2(g) + 5O_2(g) \longrightarrow 4CO_2(g) + 2H_2O(l)$$

As expected for a combustion, the reaction is exothermic. What is the sign of $\Delta H°$? What do you expect for the sign of $\Delta S°$? Explain the spontaneity of the reaction in terms of the enthalpy and entropy changes.

19.47 Acetic acid, CH_3COOH, freezes at 16.6°C. The heat of fusion, ΔH_{fus}, is 69.0 J/g. What is the change of entropy, ΔS, when 1 mol of liquid acetic acid freezes to the solid?

19.48 Acetone, CH_3COCH_3, boils at 56°C. The heat of vaporization of acetone at this temperature is 29.1 kJ/mol. What is the entropy change when 1 mol of liquid acetone vaporizes at 56°C?

19.49 Without doing any calculations, decide what the sign of $\Delta S°$ will be for each of the following reactions.
a. $2LiOH(aq) + CO_2(g) \longrightarrow Li_2CO_3(aq) + H_2O(l)$
b. $(NH_4)_2Cr_2O_7(s) \longrightarrow N_2(g) + 4H_2O(g) + Cr_2O_3(s)$
c. $2N_2O_5(g) \longrightarrow 4NO_2(g) + O_2(g)$

19.50 For each of the following reactions, decide whether there is an increase or a decrease in entropy. Why do you think so? (No calculations are needed.)
a. $N_2(g) + 3H_2(g) \longrightarrow 2NH_3(g)$
b. $NH_4Cl(s) \longrightarrow NH_3(g) + HCl(g)$
c. $CO(g) + 2H_2(g) \longrightarrow CH_3OH(l)$

19.51 Acetic acid in vinegar results from the bacterial oxidation of ethanol.

$$C_2H_5OH(l) + O_2(g) \longrightarrow CH_3COOH(l) + H_2O(l)$$

What is $\Delta S°$ for this reaction? Use standard entropy values. (See Appendix B for data.)

19.52 Methanol is produced commercially from carbon monoxide and hydrogen.

$$CO(g) + 2H_2(g) \longrightarrow CH_3OH(l)$$

What is $\Delta S°$ for this reaction? Use standard entropy values.

19.53 Is the following reaction spontaneous as written? Explain. Do whatever calculation is needed to answer the question.

$$SO_2(g) + H_2(g) \longrightarrow H_2S(g) + O_2(g)$$

19.54 Is the following reaction spontaneous as written? Explain. Do whatever calculation is needed to answer the question.

$$CH_4(g) + N_2(g) \longrightarrow HCN(g) + NH_3(g)$$

19.55 The reaction

$$CO_2(g) + H_2(g) \longrightarrow CO(g) + H_2O(g)$$

is nonspontaneous at room temperature but becomes spontaneous at a much higher temperature. What can you conclude from this about the signs of $\Delta H°$ and $\Delta S°$, assuming that the enthalpy and entropy changes are not greatly affected by the temperature change? Explain your reasoning.

19.56 The reaction

$$N_2(g) + 3H_2(g) \longrightarrow 2NH_3(g)$$

is spontaneous at room temperature but becomes nonspontaneous at a much higher temperature. From this fact alone, obtain the signs of $\Delta H°$ and $\Delta S°$, assuming that $\Delta H°$ and $\Delta S°$ do not change much with temperature. Explain your reasoning.

19.57 Calculate $\Delta G°$ at 25°C for the reaction

$$CaF_2(s) \rightleftharpoons Ca^{2+}(aq) + 2F^-(aq)$$

The value of $\Delta G_f°$ at 25°C for $CaF_2(s)$ is -1162 kJ/mol. See Table 19.2 for other values. What is the value of the solubility product constant, K_{sp}, for this reaction at 25°C?

19.58 Calculate $\Delta G°$ at 25°C for the reaction

$$BaSO_4(s) \rightleftharpoons Ba^{2+}(aq) + SO_4^{2-}(aq)$$

The values of $\Delta G_f°$ at 25°C are (in kJ/mol): $BaSO_4(s)$, -1353; $Ba^{2+}(aq)$, -561; $SO_4^{2-}(aq)$, -742. What is the value of the solubility product constant, K_{sp}, for this reaction at 25°C?

19.59 Consider the decomposition of phosgene, $COCl_2$.

$$COCl_2(g) \longrightarrow CO(g) + Cl_2(g)$$

Calculate $\Delta H°$ and $\Delta S°$ at 25°C for this reaction. See Appendix B for data. What is $\Delta G°$ at 25°C? Assume that $\Delta H°$ and $\Delta S°$ are constant with respect to a change of temperature. Now calculate $\Delta G°$ at 800°C. Compare the two values of $\Delta G°$. Briefly discuss the spontaneity of the reaction at 25°C and at 800°C.

19.60 Consider the reaction

$$CS_2(g) + 4H_2(g) \rightleftharpoons CH_4(g) + 2H_2S(g)$$

Calculate $\Delta H°$, $\Delta S°$, and $\Delta G°$ at 25°C for this reaction. Assume $\Delta H°$ and $\Delta S°$ are constant with respect to a change of temperature. Now calculate $\Delta G°$ at 650°C. Compare the two values of $\Delta G°$. Briefly discuss the spontaneity of the reaction at 25°C and at 650°C.

19.61 For the reaction

$$CH_3OH(l) + \tfrac{3}{2}O_2(g) \longrightarrow 2H_2O(l) + CO_2(g)$$

the value of $\Delta G°$ is -702.6 kJ at 25°C. Other data are as follows:

	$\Delta H_f°$ (kJ/mol) at 25°C	$S°$ (J/mol · K) at 25°C
$CH_3OH(l)$	-238.6	127
$H_2O(l)$	-285.8	70
$CO_2(g)$	-393.5	214

Calculate the absolute entropy, $S°$, per mole of $O_2(g)$.

19.62 For the reaction

$$CH_2O(g) + \tfrac{2}{3}O_3(g) \longrightarrow CO_2(g) + H_2O(g)$$

the value of $\Delta G°$ is -621.7 kJ/mol at 25°C. Other data are as follows:

	$\Delta H_f°$ (kJ/mol) at 25°C	$S°$ (J/mol · K) at 25°C
$CH_2O(g)$	-116.0	219
$H_2O(g)$	-241.8	189
$CO_2(g)$	-393.5	214
$O_3(g)$	143	?

Calculate the absolute entropy, $S°$, per mole of $O_3(g)$.

19.63 Tin(IV) oxide can be reacted with either hydrogen or carbon to form tin and water vapor or carbon dioxide, respectively. Data not available in the text are $\Delta H_f°$ $(SnO_2(s)) = -580.7$ kJ/mol and $S°$ $(SnO_2(s)) = 52.3$ J/mol · K.

a. Calculate $\Delta H°$ and $\Delta S°$ at 25°C for the reaction of SnO_2 with H_2 and for the reaction of SnO_2 with C.
b. At what temperature will each of these processes become spontaneous?
c. Industrially, which process is preferred? Why?

19.64 Tungsten is usually produced by the reduction of WO_3 with hydrogen

$$WO_3(s) + 3H_2(g) \longrightarrow W(s) + 3H_2O(g)$$

Consider the following data:

	$WO_3(s)$	$H_2O(g)$
$\Delta H_f°$ (kJ/mol)	-839.9	-241.8
$\Delta G_f°$ (kJ/mol)	-763.1	-228.6

a. Is $K > 1$ or < 1 at 25°C? Explain your answer.
b. What is the value of $\Delta S°$ at 25°C?
c. What is the temperature at which $\Delta G°$ equals zero for this reaction at 1 atm pressure?
d. What is the driving force of this reaction?

19.65 For the reaction

$$2Cu(s) + S(s) \longrightarrow Cu_2S(s)$$

$\Delta H°$ and $\Delta G°$ are negative and $\Delta S°$ is positive.

a. At equilibrium, will reactants or products predominate? Why?

b. Why must the reaction system be heated in order to produce copper(I) sulfide?

19.66 For the reaction

$$C_6H_6(l) + Br_2(l) \longrightarrow C_6H_5Br(l) + HBr(g)$$

the products have a lower energy than the reactants.

a. Predict the sign of the entropy change for this reaction. Explain.

b. Predict whether reactants or products would predominate in an equilibrium mixture. Why?

c. How can you account for the fact that a mixture of the reactants must be heated in the presence of a catalyst in order to produce products?

19.67 a. Calculate K_1 at 25°C for phosphoric acid:

$$H_3PO_4(aq) \rightleftharpoons H^+(aq) + H_2PO_4^-(aq)$$

	$H_3PO_4(aq)$	$H^+(aq)$	$H_2PO_4^-(aq)$
$\Delta H_f°$ (kJ/mol)	−1289	0	−1302
$S°$ (J/mol · K)	176	0	89

b. Which thermodynamic factor is the most significant in accounting for the fact that phosphoric acid is a weak acid? Why?

19.68 a. Calculate K_1 at 25°C for sulfurous acid:

$$H_2SO_3(aq) \rightleftharpoons H^+(aq) + HSO_3^-(aq)$$

	$H_2SO_3(aq)$	$H^+(aq)$	$HSO_3^-(aq)$
$\Delta H_f°$ (kJ/mol)	−608	0	−635
$S°$ (J/mol · K)	234	0	109

b. Which thermodynamic factor is the most significant in accounting for the fact that sulfurous acid is a weak acid? Why?

Cumulative-Skills Problems

19.69 Hydrogen bromide dissociates into its gaseous elements, H_2 and Br_2, at elevated temperatures. Calculate the percent dissociation at 375°C and 1.00 atm. What would be the percent dissociation at 375°C and 10.0 atm? Use data from Appendix B and any reasonable approximation to obtain K.

19.70 Hydrogen gas and iodine gas react to form hydrogen iodide. If 0.500 mol H_2 and 1.00 mol I_2 are placed in a closed 10.0-L vessel, what is the mole fraction of HI in the mixture when equilibrium is reached at 205°C? Use data from Appendix B and any reasonable approximations to obtain K.

19.71 K_a for acetic acid at 25.0°C is 1.754×10^{-5}. At 50.0°C, K_a is 1.633×10^{-5}. What are $\Delta H°$ and $\Delta S°$ for the ionization of acetic acid?

19.72 K_{sp} for silver chloride at 25.0°C is 1.782×10^{-10}. At 35.0°C, K_{sp} is 4.159×10^{-10}. What are $\Delta H°$ and $\Delta S°$ for the reaction?

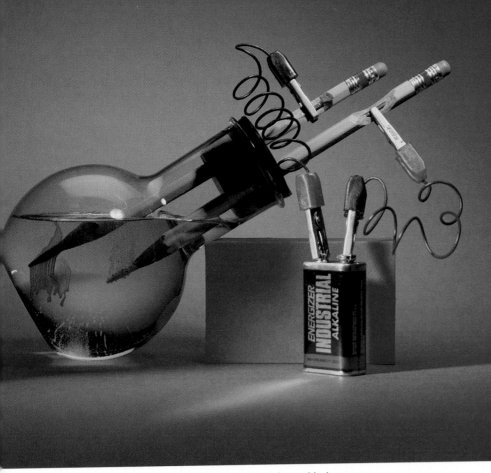

Electrolysis of a potassium iodide solution producing iodine and hydrogen gas.

Electrochemistry

The first battery was invented by Alessandro Volta about 1800. He assembled a pile consisting of pairs of zinc and silver disks separated by paper disks soaked in salt water. With a tall pile, he could detect a weak electric shock when he touched the two ends of the pile. Later Volta showed that any two different metals could be used to make such a voltaic pile (Figure 20.1).

Today many different kinds of batteries are in use. They include miniature button batteries for pocket calculators and watches, as well as larger batteries to power electric cars or to store the energy from the solar collectors of communications satellites. But all such batteries operate on the same general principles of Volta's pile.

Half-Reactions

▶ **You may wish to review Sections 4.5 and 4.6.**

A voltaic cell employs a spontaneous oxidation–reduction reaction as a source of energy. It is constructed to separate the reaction physically into two half-reactions, one involving oxidation and the other reduction, with electrons moving from the oxidation half-reaction through the external circuit to the reduction half-reaction. ◀

20.1 Balancing Oxidation–Reduction Reactions in Acidic and Basic Solutions

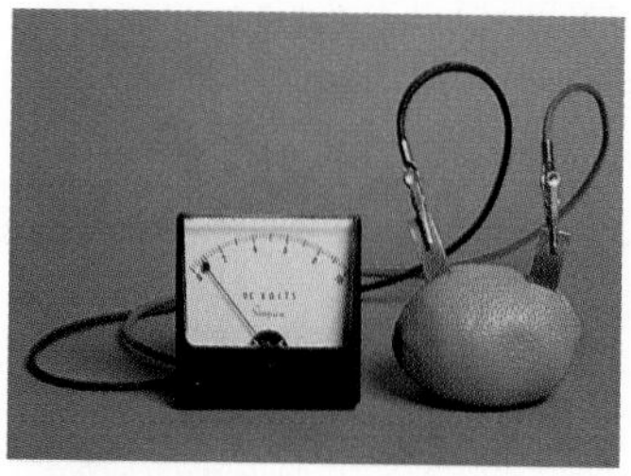

In Chapter 4 (Section 4.6) we introduced the *half-reaction method* of balancing simple oxidation–reduction reactions. We now extend this method to reactions that occur in acidic or basic solution. The steps used to balance these equations successfully are the same as those presented earlier. Keep in mind that oxidation–reduction reactions involve a transfer of electrons from one species to another.

In this chapter, we will represent the hydronium ion, $H_3O^+(aq)$, by its simpler notation $H^+(aq)$, which chemists call the hydrogen ion. In the chapters that focused on acid–base reactions, we stressed proton transfer in solution, so we adopted the hydronium-ion notation. Here, we shift our focus to the electron-transfer aspect of certain reactions, so we adopt the simpler hydrogen-ion notation.

Skeleton Oxidation–Reduction Equations

We need to discuss the essential information required to describe an oxidation–reduction reaction, which is called a *skeleton equation*. To set up the skeleton equation and then balance it, you need answers to the following questions:

1. What species is being oxidized (or, what is the reducing agent)? What species is being reduced (or, what is the oxidizing agent)?

2. What species result from the oxidation and reduction?

3. Does the reaction occur in acidic or basic solution?

For example, iron(II) ion may be oxidized by permanganate ion in acidic, aqueous solution to Fe^{3+}. In the process of oxidizing Fe^{2+}, the MnO_4^- is reduced to Mn^{2+}. We can express these facts in the following skeleton equation:

$$\overset{+2}{Fe^{2+}}(aq) + \overset{+7}{MnO_4^-}(aq) \longrightarrow \overset{+3}{Fe^{3+}}(aq) + \overset{+2}{Mn^{2+}}(aq) \qquad \text{(acidic solution)}$$

We have written oxidation numbers over the appropriate atoms. Note that the skeleton equation is not balanced, nor is it complete.

Steps in Balancing Oxidation–Reduction Equations in Acidic Solution

1. Assign oxidation numbers to each atom so you know what is oxidized and what is reduced.

2. Split the skeleton equation into two half-reactions, proceeding as follows. Note the species containing the element that increases in oxidation number and write those species to give the oxidation half-reaction. Similarly, note the species

**Figure 20.1
Lemon battery.** Zinc and copper strips in a lemon generate a voltage.

containing the element that decreases in oxidation number and write the reduction half-reaction.

3. Complete and balance each half-reaction.

   a. Balance all atoms except O and H.

   b. Balance O atoms by adding H_2O's to one side of the equation.

   c. Balance H atoms by adding H^+ ions to one side of the equation.

   d. Balance electric charge by adding electrons (e^-) to the more positive side.

4. Combine the two half-reactions to obtain the final balanced oxidation–reduction equation.

   a. Multiply each half-reaction by a factor such that when the half-reactions are added, the electrons cancel. (Electrons cannot appear in the final equation.)

   b. Simplify the balanced equation by canceling species that occur on both sides and reduce the coefficients to smallest whole numbers.

Let us apply this method to the skeleton equation we wrote for the oxidation of iron(II) ion by permanganate ion in acidic solution. We have already determined the elements that change oxidation number (Step 1). So, following Step 2, we write

$$\overset{+2}{Fe^{2+}}(aq) \longrightarrow \overset{+3}{Fe^{3+}}(aq)$$

Note that the equation is balanced in Fe as it stands (Step 3a). Moving to Step 3d, balance the half-reaction in electric charge. You do that by adding one electron to the more positive side.

$$Fe^{2+}(aq) \longrightarrow Fe^{3+}(aq) + e^- \qquad \text{(oxidation half-reaction)}$$

Now let us look at the reduction half-reaction.

$$\overset{+7}{MnO_4^-}(aq) \longrightarrow \overset{+2}{Mn^{2+}}(aq)$$

The equation is balanced in Mn as it stands (Step 3a). For Step 3b, you add four H_2O's to the right side of the equation to balance the O atoms.

$$MnO_4^-(aq) \longrightarrow Mn^{2+}(aq) + 4H_2O(l)$$

Balance the H atoms by adding eight H^+ ions to the left side (Step 3c).

$$MnO_4^-(aq) + 8H^+(aq) \longrightarrow Mn^{2+}(aq) + 4H_2O(l)$$

Finally, you balance the half-reaction in electric charge by adding electrons to the more positive side of the equation (Step 3d). Note that the left side has a charge of $(-1 + 8) = +7$, whereas the right side has a charge of $+2$. So you add five electrons to the left side:

$$MnO_4^-(aq) + 8H^+(aq) + 5e^- \longrightarrow Mn^{2+}(aq) + 4H_2O(l) \qquad \text{(reduction half-reaction)}$$

In Step 4a, you need to multiply each half-reaction by a factor such that when the two half-reactions are added, the electrons cancel. You multiply the oxidation half-reaction by 5 (the number of electrons in the reduction half-reaction) and multiply the reduction half-reaction by 1 (the number of electrons in the oxidation half-reaction). Then you add the two half-reactions:

$$5 \times (Fe^{2+} \longrightarrow Fe^{3+} + e^-)$$
$$\underline{1 \times (MnO_4^- + 8H^+ + 5e^- \longrightarrow Mn^{2+} + 4H_2O)}$$
$$5Fe^{2+} + MnO_4^- + 8H^+ + \cancel{5e^-} \longrightarrow 5Fe^{3+} + Mn^{2+} + 4H_2O + \cancel{5e^-}$$

After canceling electrons, the final balanced equation is

$$5Fe^{2+}(aq) + MnO_4^-(aq) + 8H^+(aq) \longrightarrow 5Fe^{3+}(aq) + Mn^{2+}(aq) + 4H_2O(l)$$

EXAMPLE 20.1

Balancing Equations by the Half-Reaction Method (Acidic Solution)

Zinc metal reacts with nitric acid, HNO_3, to produce a number of products depending on how dilute the acid solution is. In a concentrated solution, zinc reduces nitrate ion to ammonium ion; zinc is oxidized to zinc ion, Zn^{2+}. Write the net ionic equation for this reaction.

SOLUTION

Nitric acid is a strong acid, so it exists in solution as H^+ and NO_3^- ions. For the skeleton equation, write just the NO_3^- ion. The skeleton equation, with oxidation numbers for those atoms that change values (Step 1), is

$$\overset{0}{Zn}(s) + \overset{+5}{NO_3^-}(aq) \longrightarrow \overset{+2}{Zn^{2+}}(aq) + \overset{-3}{NH_4^+}(aq)$$

Separate this equation into two incomplete half-reactions (Step 2). Note that zinc is oxidized (increases in oxidation number), and nitrogen is reduced (decreases in oxidation number). The two incomplete half-reactions are

$$Zn(s) \longrightarrow Zn^{2+}(aq)$$
$$NO_3^-(aq) \longrightarrow NH_4^+(aq)$$

The oxidation half-reaction is balanced except for the charge. Following Step 3d, add electrons to the more positive side to balance the charge.

$$Zn(s) \longrightarrow Zn^{2+}(aq) + 2e^- \qquad \text{(oxidation half-reaction)}$$

The reduction half-reaction is balanced in N. Add three H_2O's to the right side to balance O atoms (Step 3b), and add ten H^+ ions to the left side to balance the four H atoms in the NH_4^+ ion and the six H atoms in the H_2O's you just added (Step 3c).

$$10H^+(aq) + NO_3^-(aq) \longrightarrow NH_4^+(aq) + 3H_2O(l)$$

Now, you add electrons to the more positive side to balance the charge.

$$10H^+(aq) + NO_3^-(aq) + 8e^- \longrightarrow NH_4^+(aq) + 3H_2O(l)$$
$$\text{(reduction half-reaction)}$$

Finally (Step 4a), you multiply the two half-reactions by factors so that when added, the electrons cancel. You multiply the oxidation half-reaction by 4 and the reduction half-reaction by 1:

$$4 \times (Zn \longrightarrow Zn^{2+} + 2e^-)$$
$$\underline{10H^+ + NO_3^- + 8e^- \longrightarrow NH_4^+ + 3H_2O}$$
$$4Zn + 10H^+ + NO_3^- + \cancel{8e^-} \longrightarrow 4Zn^{2+} + NH_4^+ + 3H_2O + \cancel{8e^-}$$

Cancel electrons (Step 4b) and the net ionic equation is

$$\mathbf{4Zn(s) + 10H^+(aq) + NO_3^-(aq) \longrightarrow 4Zn^{2+}(aq) + NH_4^+(aq) + 3H_2O(l)}$$

See Problems 20.19 and 20.20.

To balance an oxidation–reduction equation in basic solution, you begin by balancing the equation *as if it were a reaction in acidic solution*. Then, you add the following steps. ◄

► You can apply Steps 5 and 6 to the individual half-reactions, rather than to the overall equation. Although this is more work, it does yield the complete half-reactions, which can be useful in discussing batteries (later in this chapter).

Additional Steps for Balancing Oxidation-Reduction Equations in Basic Solution

5. Note the number of H^+ ions in the equation. Add this number of OH^- ions to both sides of the equation.

6. Simplify the equation by noting that H^+ reacts with OH^- to give H_2O. Cancel any H_2O's that occur on both sides of the equation and reduce the equation to simplest terms.

EXAMPLE 20.2

Balancing Equations by the Half-Reaction Method (Basic Solution)

Permanganate ion oxidizes sulfite ion in basic solution according to the following skeleton equation:

$$MnO_4^-(aq) + SO_3^{2-}(aq) \longrightarrow MnO_2(s) + SO_4^{2-}(aq)$$

Use the half-reaction method to complete and balance this equation.

SOLUTION

After balancing the equation as if it were in acid solution, you obtain the following:

$$2MnO_4^- + 3SO_3^{2-} + 2H^+ \longrightarrow 2MnO_2 + 3SO_4^{2-} + H_2O$$

Following Step 5, you add $2OH^-$ to both sides of the equation.

$$2MnO_4^- + 3SO_3^{2-} + \underbrace{2H^+ + 2OH^-}_{2H_2O} \longrightarrow 2MnO_2 + 3SO_4^{2-} + H_2O + 2OH^-$$

You replace $2H^+ + 2OH^-$ by $2H_2O$ on the left side, then cancel one of these with the H_2O on the right side. The balanced equation is

$$2MnO_4^-(aq) + 3SO_3^{2-}(aq) + H_2O(l) \longrightarrow 2MnO_2(s) + 3SO_4^{2-}(aq) + 2OH^-(aq)$$

See Problems 20.21 and 20.22.

Voltaic Cells

Battery cells, or voltaic cells (also called galvanic cells), are a kind of electrochemical cell. An **electrochemical cell** is *a system consisting of electrodes that dip into an electrolyte and in which a chemical reaction either uses or generates an electric current*. A **voltaic**, or **galvanic**, **cell** is *an electrochemical cell in which a spontaneous reaction generates an electric current*. An **electrolytic cell** is *an electrochemical cell in which an electric current drives an otherwise nonspontaneous reaction*.

Nernst Equation

Recall that the free-energy change, ΔG, is related to the standard free-energy change, $\Delta G°$, by the following equation (Section 19.6):

$$\Delta G = \Delta G° + RT \ln Q$$

You can apply this equation to a voltaic cell. In that case, the concentrations and gas pressures are those that exist in the cell at a particular instant. If you substitute $\Delta G = -nFE_{cell}$ and $\Delta G° = -nFE°_{cell}$ into this equation, you obtain

$$-nFE_{cell} = -nFE°_{cell} + RT \ln Q$$

This result rearranges to give the **Nernst equation.**

$$E_{cell} = E°_{cell} - \frac{RT}{nF} \ln Q \quad \text{or} \quad E_{cell} = E°_{cell} - \frac{2.303\, RT}{nF} \log Q$$

If you substitute 298 K (25°C) for the temperature in the Nernst equation and put in values for R and F, you get (using common logarithms)

$$E_{cell} = E°_{cell} - \frac{0.0592}{n} \log Q \quad \text{(values in volts at 25°C)}$$

You can show from the Nernst equation that the cell emf, E_{cell}, decreases as the cell reaction proceeds. As the reaction occurs in the voltaic cell, the concentrations of products increase and the concentrations of reactants decrease. Eventually the cell emf goes to zero, and the cell reaction comes to equilibrium.

The next example illustrates a complete calculation of emf from the ion concentrations in a voltaic cell.

EXAMPLE 20.10

Calculating the Cell emf for Nonstandard Conditions

What is the emf of the following voltaic cell at 25°C?

$$Zn(s)\,|\,Zn^{2+}(1.00 \times 10^{-5}M)\,\|\,Cu^{2+}(0.100\,M)\,|\,Cu(s)$$

The standard emf of this cell is 1.10 V.

SOLUTION

The cell reaction is

$$Zn(s) + Cu^{2+}(aq) \rightleftharpoons Zn^{2+}(aq) + Cu(s)$$

The number of electrons transferred is two; hence, $n = 2$. The reaction quotient is

$$Q = \frac{[Zn^{2+}]}{[Cu^{2+}]} = \frac{1.00 \times 10^{-5}}{0.100} = 1.00 \times 10^{-4}$$

The standard emf is 1.10 V, so the Nernst equation becomes

$$E_{cell} = E°_{cell} - \frac{0.0592}{n} \log Q$$

$$= 1.10 - \frac{0.0592}{2} \log (1.0 \times 10^{-4})$$

$$= 1.10 - (-0.12) = 1.22 \text{ V}$$

Figure 20.5

A glass electrode. *Left:* A small, commercial glass electrode. *Right:* A sketch showing the construction of a glass electrode for measuring hydrogen-ion concentrations.

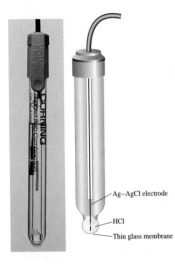

Ag–AgCl electrode

HCl

Thin glass membrane

Figure 20.6

Leclanché dry cell. The cell has a zinc anode and a graphite rod with a paste of MnO_2, NH_4Cl, $ZnCl_2$, and C as the cathode.

Positive terminal

Air space

Asphalt seal

Porous paper

Paste of MnO_2, $ZnCl_2$, NH_4Cl, and C

Graphite rod (cathode)

Zinc can (anode)

Metal jacket

Insulator

Negative terminal

The cell emf is **1.22 V.** This result is qualitatively what you would expect. Because the concentration of product (Zn^{2+}) is much less than the standard value (1 *M*), whereas the concentration of reactant (Cu^{2+}) is 0.100 *M*, the spontaneity of the reaction as measured by E_{cell} is greater than the standard value.

See Problems 20.53 and 20.54.

Determination of pH

The pH of a solution can be obtained very accurately from emf measurements, using the Nernst equation to relate emf to pH. To see how this is done, suppose you have a test solution whose pH you would like to determine. You set up a voltaic cell as follows: You use the test solution as the electrolyte for a hydrogen electrode and bubble in hydrogen gas at 1 atm. (The hydrogen electrode is shown in Figure 20.3.) Now connect this hydrogen electrode to a standard zinc electrode to give the following cell:

$$Zn\,|\,Zn^{2+}(1\ M)\,\|\,H^+(\text{test solution})\,|\,H_2(1\ \text{atm})\,|\,Pt$$

The cell reaction is

$$Zn(s) + 2H^+(\text{test solution}) \longrightarrow Zn^{2+}(1\ M) + H_2(1\ \text{atm})$$

The emf of this cell depends on the hydrogen-ion concentration of the test solution, according to the Nernst equation. In this way, measurement of the cell emf gives you the H^+ concentration.

The hydrogen electrode is seldom employed in routine laboratory work, because it is awkward to use. It is often replaced by a *glass electrode*. This compact electrode (see Figure 20.5) consists of a silver wire coated with silver chloride immersed in a solution of dilute hydrochloric acid. The electrode solution is separated from the test solution by a thin glass membrane, which develops a potential across it depending on the hydrogen-ion concentrations on its inner and outer surfaces. A mercury–mercury(I) chloride (calomel) electrode is often used as the other electrode. The emf of the cell depends linearly on the pH. In a common arrangement, the emf is measured with a voltmeter that reads pH directly (see Figure 16.6).

CONCEPT CHECK 20.3

Consider a voltaic cell, $Fe(s)\,|\,Fe^{2+}(aq)\,\|\,Cu^{2+}(aq)\,|\,Cu(s)$, being run under standard conditions.

a. Is $\Delta G°$ positive or negative for this process?

b. Change the concentrations from their standard values in such a way that E_{cell} is reduced. Write your answer using the shorthand notation of Section 20.3.

20.8 Some Commercial Voltaic Cells

We commonly use voltaic cells as convenient, portable sources of energy. Flashlights and radios are examples of devices that are often powered by the **zinc–carbon,** or **Leclanché, dry cell** (Figure 20.6). *This voltaic cell has a zinc can as the anode; a graphite rod in the center, surrounded by a paste of manganese dioxide, ammonium*

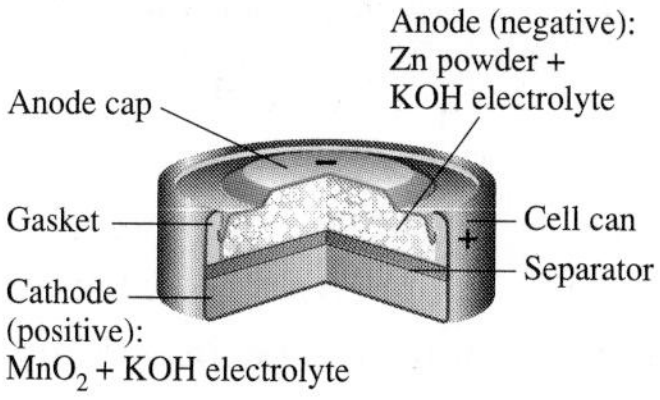

Anode (negative):
Zn powder +
KOH electrolyte

Anode cap

Gasket — Cell can

Cathode — Separator
(positive):
MnO_2 + KOH electrolyte

and zinc chlorides, and carbon black, is the cathode. The electrode reactions are complicated but are approximately these:

$$Zn(s) \longrightarrow Zn^{2+}(aq) + 2e^- \qquad \text{(anode)}$$

$$2NH_4^+(aq) + 2MnO_2(s) + 2e^- \longrightarrow Mn_2O_3(s) + H_2O(l) + 2NH_3(aq) \quad \text{(cathode)}$$

The voltage of this dry cell is initially about 1.5 V, but it decreases as current is drawn off. The voltage also deteriorates rapidly in cold weather.

An **alkaline dry cell** (Figure 20.7) is *similar to the Leclanché cell, but it has potassium hydroxide in place of ammonium chloride.* This cell performs better under current drain and in cold weather. The half-reactions are

$$Zn(s) + 2OH^-(aq) \longrightarrow Zn(OH)_2(s) + 2e^- \qquad \text{(anode)}$$

$$2MnO_2(s) + H_2O(l) + 2e^- \longrightarrow Mn_2O_3(s) + 2OH^-(aq) \qquad \text{(cathode)}$$

Once a dry cell is completely discharged (has come to equilibrium), the cell is not easily reversed, or recharged, and is normally discarded. Some types of cells are rechargeable after use, however. An important example is the **lead storage cell.** *This voltaic cell consists of electrodes of lead alloy grids; one electrode is packed with a spongy lead to form the anode, and the other electrode is packed with lead dioxide to form the cathode* (see Figure 20.8). Both are bathed in an aqueous solution of sulfuric acid, H_2SO_4. The half-cell reactions during discharge are

$$Pb(s) + HSO_4^-(aq) \longrightarrow PbSO_4(s) + H^+(aq) + 2e^- \qquad \text{(anode)}$$

$$PbO_2(s) + 3H^+(aq) + HSO_4^-(aq) + 2e^- \longrightarrow PbSO_4(s) + 2H_2O(l) \qquad \text{(cathode)}$$

White lead(II) sulfate coats each electrode during discharge, and sulfuric acid is consumed. Each cell delivers about 2 V, and a battery consisting of six cells in series gives about 12 V.

After the lead storage battery is discharged, it is recharged from an external electric current. The previous half-reactions are reversed. Some water is decomposed into hydrogen and oxygen gas during this recharging, so more water may have to be added at intervals. However, newer batteries use lead electrodes containing some calcium metal; the calcium–lead alloy resists the decomposition of water. These *maintenance-free* batteries are sealed (Figure 20.9).

The **nickel–cadmium cell** (nicad cell) is a common storage battery. It is *a voltaic cell consisting of an anode of cadmium and a cathode of hydrated nickel oxide*

Figure 20.8 *(left)*
A lead storage battery.
Each cell delivers about 2 V, and a battery consisting of six cells in series gives about 12 V.

Figure 20.9 *(right)*
A maintenance-free battery. This lead storage battery contains calcium–lead alloy grids for electrodes. This alloy resists the decomposition of water, so the battery can be sealed.

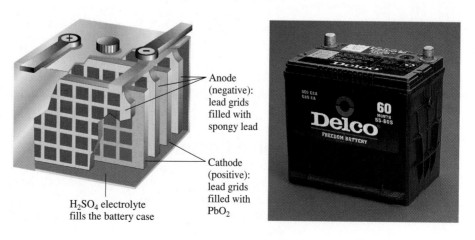

Anode
(negative):
lead grids
filled with
spongy lead

Cathode
(positive):
lead grids
filled with
PbO_2

H_2SO_4 electrolyte
fills the battery case

(approximately NiOOH) on nickel; the electrolyte is potassium hydroxide. Nicad batteries are used in calculators, portable power tools, shavers, and toothbrushes. The half-cell reactions during discharge are

$$Cd(s) + 2OH^-(aq) \longrightarrow Cd(OH)_2(s) + 2e^- \qquad \text{(anode)}$$

$$NiOOH(s) + H_2O(l) + e^- \longrightarrow Ni(OH)_2(s) + OH^-(aq) \qquad \text{(cathode)}$$

These half-reactions are reversed when the cell is recharged. Nicad batteries can be recharged and discharged many times (see Figure 20.10).

A **fuel cell** is *essentially a battery, but it differs by operating with a continuous supply of energetic reactants, or fuel.* Figure 20.11 shows a fuel cell that uses hydrogen and oxygen. ◄ At one electrode, oxygen passes through a porous material that catalyzes the following reaction:

$$O_2(g) + 2H_2O(l) + 4e^- \longrightarrow 4OH^-(aq) \qquad \text{(cathode)}$$

At the other electrode, hydrogen reacts.

$$2H_2(g) + 4OH^-(aq) \longrightarrow 4H_2O(l) + 4e^- \qquad \text{(anode)}$$

The sum of these half-cell reactions is

$$2H_2(g) + O_2(g) \longrightarrow 2H_2O(l)$$

which is the net reaction in the fuel cell. Such cells are used in the space shuttle orbiters to supply electric energy. Other types of cells employing hydrocarbon fuels have been constructed.

Another use of voltaic cells is to control the corrosion of underground pipelines and tanks. Such pipelines and tanks are usually made of steel, an alloy of iron, and their corrosion or rusting is an electrochemical process.

Consider the rusting that occurs when a drop of water is in contact with iron. The edge of the water drop exposed to the air becomes one pole of a voltaic cell (see Figure 20.12). At this edge, molecular oxygen from air is reduced to hydroxide ion in solution.

▶ The first fuel cell was invented in 1839 by William Groves. He obtained the cell by electrolyzing water and collecting the hydrogen and oxygen in small, inverted test tubes. After removing the battery current from the electrolysis cell, he found that a current would flow from the cell. In effect, the electrolysis charged the cell, after which it operated as a fuel cell.

Figure 20.10
Nicad storage batteries.
These rechargeable cells have a cadmium anode and a hydrated nickel oxide cathode.

Figure 20.11
A hydrogen–oxygen fuel cell. Hydrogen gas passes into a chamber where the gas is in contact with a porous material in contact with a hot aqueous solution of potassium hydroxide. This forms the anode at which hydrogen is oxidized to water. Oxygen gas enters a similar electrode and is reduced to hydroxide ion. The net chemical change is the reaction of hydrogen with oxygen to form water.

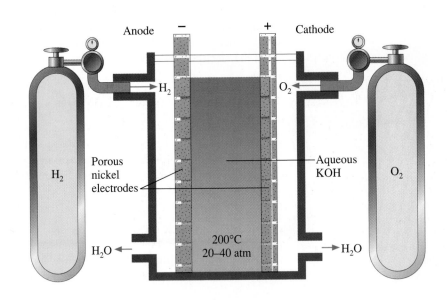

Figure 20.12
The electrochemical process involved in the rusting of iron. Here a single drop of water containing ions forms a voltaic cell in which iron is oxidized to iron(II) ion at the center of the drop (this is the anode). Oxygen gas from air is reduced to hydroxide ion at the periphery of the drop (the cathode). Hydroxide ions and iron(II) ions migrate together and react to form iron(II) hydroxide. This is oxidized to iron(III) hydroxide by more O_2 that dissolves at the surface of the drop. Iron(III) hydroxide precipitates, and this settles to form rust on the surface of the iron.

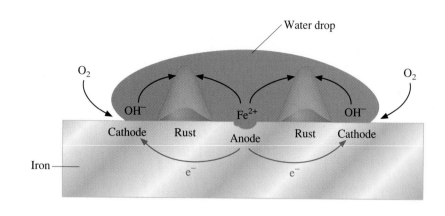

$$O_2(g) + 2H_2O(l) + 4e^- \longrightarrow 4OH^-(aq)$$

The electrons for this reduction are supplied by the oxidation of metallic iron at the center of the drop, which acts as the other pole of the voltaic cell.

$$Fe(s) \longrightarrow Fe^{2+}(aq) + 2e^-$$

These electrons flow from the center of the drop through the metallic iron to the edge of the drop. The metallic iron functions as the external circuit between the cell poles.

Ions move within the water drop, completing the electric circuit. Iron(II) ions move outward from the center of the drop, and hydroxide ions move inward from the edge. The two ions meet in a doughnut-shaped region, where they react to precipitate iron(II) hydroxide.

$$Fe^{2+}(aq) + 2OH^-(aq) \longrightarrow Fe(OH)_2(s)$$

This precipitate is quickly oxidized by oxygen to rust (approximated by the formula $Fe_2O_3 \cdot H_2O$).

$$4Fe(OH)_2(s) + O_2(g) \longrightarrow 2Fe_2O_3 \cdot H_2O(s) + 2H_2O(l)$$

Figure 20.13
Cathodic protection of a buried steel pipe. Iron in the steel becomes the cathode in an iron–magnesium voltaic cell. Magnesium is then oxidized in preference to iron.

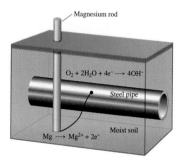

If a buried steel pipeline (Figure 20.13) is connected to an active metal (that is, a highly electropositive substance) such as magnesium, a voltaic cell is formed; the active metal is the anode and iron becomes the cathode. Wet soil forms the electrolyte, and the electrode reactions are

$$Mg(s) \longrightarrow Mg^{2+}(aq) + 2e^- \qquad \text{(anode)}$$
$$O_2(g) + 2H_2O(l) + 4e^- \longrightarrow 4OH^-(aq) \qquad \text{(cathode)}$$

As the cathode, the iron-containing steel pipe is protected from oxidation. Of course, the magnesium rod is eventually consumed and must be replaced, but this is cheaper than digging up the pipeline. This use of an active metal to protect iron from corrosion is called *cathodic protection*. See Figure 20.14 for a laboratory demonstration of cathodic protection.

CONCEPT CHECK 20.4

Keeping in mind that seawater contains a number of ions, explain why seawater corrodes iron much faster than freshwater.

Figure 20.14
A demonstration of cathodic protection. The nails are in a gel containing phenolphthalein indicator and potassium ferricyanide. Iron corrosion yields Fe^{2+}, which reacts with ferricyanide ion to give a dark blue precipitate. Where OH^- forms, phenolphthalein appears pink. *Left:* The unprotected nail. *Right:* The nail has magnesium wrapped around the center. Note that no corrosion (dark blue or black region) appears on it.

Electrolytic Cells

An *electrolytic cell* is an electrochemical cell in which an electric current drives an otherwise nonspontaneous reaction. *The process of producing a chemical change in an electrolytic cell* is called **electrolysis.** Many important substances, including aluminum and chlorine, are produced commercially by electrolysis.

20.9 Electrolysis of Molten Salts

Figure 20.15 shows a simple electrolytic cell. Wires from a battery are connected to electrodes that dip into molten sodium chloride. At the electrode connected to the negative pole of the battery, globules of sodium metal form; chlorine gas evolves from the other electrode.

As noted earlier (Section 20.2), the *anode* is the electrode at which oxidation occurs, and the *cathode* is the electrode at which reduction occurs (these definitions hold for electrolytic cells as well as for voltaic cells).

This electrolysis of molten NaCl is used commercially to obtain sodium metal from sodium chloride. A **Downs cell** is *a commercial electrochemical cell used to obtain sodium metal by the electrolysis of molten sodium chloride* (Figure 20.16).

Figure 20.15
Electrolysis of molten sodium chloride. Sodium metal forms at the cathode from the reduction of Na^+ ion; chlorine gas forms at the anode from the oxidation of Cl^- ion. Sodium metal is produced commercially this way, although the commercial cell must be designed to collect the products and to keep them away from one another.

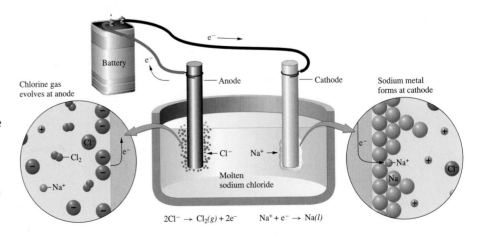

$$2Cl^- \rightarrow Cl_2(g) + 2e^- \qquad Na^+ + e^- \rightarrow Na(l)$$

Figure 20.16
A Downs cell for the preparation of sodium metal. In this commercial cell, sodium is produced by the electrolysis of molten sodium chloride. The salt contains calcium chloride, which is added to lower the melting point of the mixture. Liquid sodium forms at the cathode, where it rises to the top of the molten salt and collects in a tank. Chlorine gas is a by-product.

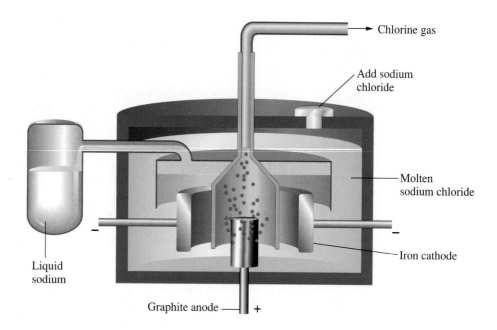

Chlorine gas

Add sodium chloride

Molten sodium chloride

Iron cathode

Liquid sodium

Graphite anode +

The cell is constructed to keep the products of the electrolysis separate, because they would otherwise react. Calcium chloride is added to the sodium chloride to lower the melting point from 801°C for NaCl to about 580°C for the mixture. You obtain the cell reaction by adding the half-reactions.

$$Na^+(l) + e^- \longrightarrow Na(l)$$
$$Cl^-(l) \longrightarrow \tfrac{1}{2}Cl_2(g) + e^-$$
$$\overline{Na^+(l) + Cl^-(l) \longrightarrow Na(l) + \tfrac{1}{2}Cl_2(g)}$$

A number of other reactive metals are obtained by the electrolysis of a molten salt or ionic compound. Lithium, magnesium, and calcium metals are all obtained by the electrolysis of the chlorides.

20.10 Aqueous Electrolysis

In the electrolysis of a molten salt, reactions are usually limited to those involving ions from the salt. When you electrolyze an aqueous solution of an ionic compound, however, you must consider the possibility that water is involved. Water can be reduced to H_2 or oxidized to O_2.

Consider the reduction half-reaction. The only species in which there are no changes in oxidation states from those in H_2O is OH^-. The balanced half-reaction is

$$2H_2O(l) + 2e^- \longrightarrow H_2(g) + 2OH^-(aq)$$

You can obtain the oxidation half-reaction for water in a similar way. The only species in which there is no change in oxidation state is H^+. The balanced half-reaction is

$$2H_2O(l) \longrightarrow O_2(g) + 4H^+(aq) + 4e^-$$

Now let us consider the electrolysis of different aqueous solutions and try to decide what the half-reactions might be.

Electrolysis of Sulfuric Acid Solutions

To decide what is likely to happen during the electrolysis of a solution of sulfuric acid, H_2SO_4, you must consider the half-reactions involving ionic species from H_2SO_4, in addition to those involving water. Sulfuric acid is a strong acid and ionizes completely into H^+ and HSO_4^-. The HSO_4^- ion partially ionizes into H^+ and SO_4^{2-}. Therefore, the species you have to look at are H^+, SO_4^{2-}, and H_2O.

At the cathode, the possible reduction half-reactions are

$$2H^+(aq) + 2e^- \longrightarrow H_2(g)$$
$$2H_2O(l) + 2e^- \longrightarrow H_2(g) + 2OH^-(aq)$$

At the anode, the possible oxidation half-reactions are

$$2SO_4^{2-}(aq) \longrightarrow S_2O_8^{2-}(aq) + 2e^-$$
$$2H_2O(l) \longrightarrow O_2(g) + 4H^+(aq) + 4e^-$$

In the first half-reaction, the sulfate ion, SO_4^{2-}, is oxidized to the peroxydisulfate ion, $S_2O_8^{2-}$. Oxygen in this half-reaction is oxidized from oxidation state -2 to -1.

Consider the cathode reaction. The two reduction half-reactions (reduction of H^+ and reduction of H_2O) in acidic solution are essentially equivalent. We will consider the cathode reaction in acid solution to be the reduction of H^+.

For the anode, consider the oxidation half-reactions and their oxidation potentials.

$$2SO_4^{2-}(aq) \longrightarrow S_2O_8^{2-}(aq) + 2e^-; \; -E° = -2.01 \text{ V}$$
$$2H_2O(l) \longrightarrow O_2(g) + 4H^+(aq) + 4e^-; \; -E° = -1.23 \text{ V}$$

The species whose oxidation half-reaction has the larger (less negative) oxidation potential is the more easily oxidized. Therefore, under standard conditions, you expect H_2O to be oxidized in preference to SO_4^{2-}.

A note of caution must be added here. Electrode potentials are measured under conditions in which the half-reactions are at or very near equilibrium. In electrolysis, in which the half-reactions can be far from equilibrium, you may have to supply a larger voltage than predicted from electrode potentials. This additional voltage, called *overvoltage,* can be fairly large (several tenths of a volt), particularly for forming a gas. This means that in trying to predict which of two half-reactions is the one that actually occurs at an electrode, you must be especially careful if the electrode potentials are within several tenths of a volt of one another. For example, the standard oxidation potential for O_2 is -1.23 V, but because of overvoltage, the effective oxidation potential may be several tenths of a volt lower (more negative). The sulfate ion is so difficult to oxidize ($-E° = -2.01$ V), however, that the previous conclusion about the anode half-reaction in electrolyzing aqueous sulfuric acid is unaltered.

You obtain the cell reaction by adding the half-reactions that occur at the electrodes.

$$2 \times [2H^+(aq) + 2e^- \longrightarrow H_2(g)] \qquad \text{(cathode)}$$
$$\underline{2H_2O(l) \longrightarrow O_2(g) + 4H^+(aq) + 4e^- \qquad \text{(anode)}}$$
$$2H_2O(l) \longrightarrow 2H_2(g) + O_2(g)$$

The net cell reaction is simply the *electrolysis of water.*

Electrolysis of Sodium Chloride Solutions

When you electrolyze an aqueous solution of sodium chloride, NaCl, the possible species involved in half-reactions are Na^+, Cl^-, and H_2O. The possible cathode half-reactions are

$$Na^+(aq) + e^- \longrightarrow Na(s); E° = -2.71 \text{ V}$$
$$2H_2O(l) + 2e^- \longrightarrow H_2(g) + 2OH^-(aq); E° = -0.83 \text{ V}$$

Under standard conditions, you expect H_2O to be reduced in preference to Na^+, which agrees with what you observe. Hydrogen gas evolves at the cathode. ◀

The possible anode half-reactions are

$$2Cl^-(aq) \longrightarrow Cl_2(g) + 2e^-; -E° = -1.36 \text{ V}$$
$$2H_2O(l) \longrightarrow O_2(g) + 4H^+(aq) + 4e^-; -E° = -1.23 \text{ V}$$

▶ The OH^- concentration is actually 1×10^{-7} *M*. From Nernst's equation, *E* for the reduction of H_2O is −0.41 V; so you still conclude that H_2O is reduced in preference to Na^+.

Under standard-state conditions, you might expect H_2O to be oxidized in preference to Cl^-. However, the potentials are close and overvoltages at the electrodes could alter this conclusion.

It is possible, nevertheless, to give a general statement about the product expected at the anode. Electrode potentials, as you have seen, depend on concentrations. It turns out that when the solution is concentrated enough in Cl^-, Cl_2 is the product; but in dilute solution, O_2 is the product.

The half-reactions and cell reaction for the electrolysis of aqueous sodium chloride to chlorine and hydroxide ion are as follows:

$$2H_2O(l) + 2e^- \longrightarrow H_2(g) + 2OH^-(aq) \qquad \text{(cathode)}$$
$$\underline{2Cl^-(aq) \longrightarrow Cl_2(g) + 2e^- \qquad \text{(anode)}}$$
$$2H_2O(l) + 2Cl^-(aq) \longrightarrow H_2(g) + Cl_2(g) + 2OH^-(aq)$$

Because the electrolysis started with sodium chloride, the cation in the electrolyte solution is Na^+. When you evaporate the electrolyte solution at the cathode, you obtain sodium hydroxide, NaOH. Figure 20.17 shows the similar electrolysis of aqueous potassium iodide, KI.

The electrolysis of aqueous sodium chloride is the basis of the *chlor-alkali industry,* the major commercial source of chlorine and sodium hydroxide. Commercial cells are of several types, but in each the main problem is to keep the products separate, because chlorine reacts with aqueous sodium hydroxide.

Figure 20.17
Electrolysis of aqueous potassium iodide. Iodine forms at the anode by the oxidation of I^- ion. Iodine reacts with iodide ion to form I_3^- ion, which is red-brown. Hydrogen bubbles and hydroxide ion form at the cathode by the reduction of water.

Electroplating of Metals

Many metals are protected from corrosion by plating them with other metals. Iron coatings are often used to protect steel, because the coat protects the steel by cathodic protection even when the iron coat is scratched (see end of Section 20.8). A thin iron coating can be applied to steel by *electrogalvanizing,* or iron electroplating. (Galvanized steel has a thick iron coating obtained by dipping the object in molten iron.)

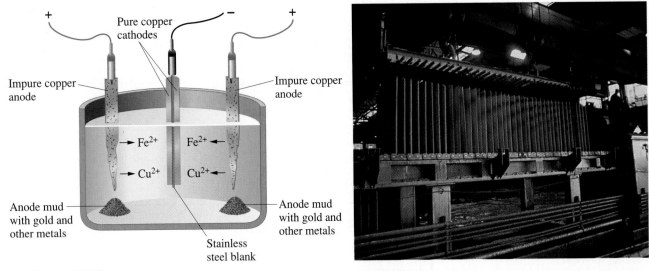

Figure 20.18
Purification of copper by electrolysis. *Left:* Copper(II) ions leave the anodes and plate out on the cathode. Reactive ions, such as Fe²⁺, remain in solution, and unreactive substances (including gold) collect as a mud under the anode. *Right:* Stainless steel cathode blanks that produce pure copper sheets alternate with impure copper slabs in this electrolytic tank. The pure copper sheets grow in size and are removed in about twenty days.

The steel object is placed in a bath of iron salts and made the cathode in an electrolytic cell. The cathode half-reaction is

$$Fe^{2+}(aq) + 2e^- \longrightarrow Fe(s)$$

Electrolysis is also used to purify some metals. For example, copper for electrical use, which must be very pure, is purified by electrolysis. Slabs of impure copper serve as anodes, and stainless steel sheets serve as cathodes; the electrolyte bath is copper(II) sulfate, $CuSO_4$ (Figure 20.18). During the electrolysis, copper(II) ions leave the anode slabs and plate out on stainless steel cathode sheets. Less reactive metals, such as gold, silver, and platinum, that were present in the impure copper form a valuable mud that collects on the bottom of the electrolytic cell. Metals more reactive than copper remain as ions in the electrolytic bath. After about twenty days in the electrolytic cell, the pure copper cathodes are removed from the cell bath.

EXAMPLE 20.11

Predicting the Half-Reactions in an Aqueous Electrolysis

What do you expect to be the half-reactions in the electrolysis of aqueous copper(II) sulfate?

SOLUTION

The species you should consider for half-reactions are $Cu^{2+}(aq)$, $SO_4^{2-}(aq)$, and H_2O.
  Possible cathode half-reactions are

$$Cu^{2+}(aq) + 2e^- \longrightarrow Cu(s); E° = 0.34 \text{ V}$$
$$2H_2O(l) + 2e^- \longrightarrow H_2(g) + 2OH^-(aq); E° = -0.83 \text{ V}$$

Because the copper electrode potential is much larger than the reduction potential of water, you expect Cu^{2+} to be reduced.

Possible anode half-reactions are

$$2SO_4{}^{2-}(aq) \longrightarrow S_2O_8{}^{2-}(aq) + 2e^-; \ -E° = -2.01 \text{ V}$$
$$2H_2O(l) \longrightarrow O_2(g) + 4H^+(aq) + 4e^-; \ -E° = -1.23 \text{ V}$$

You expect H_2O to be oxidized.

The expected half-reactions are

$$Cu^{2+}(aq) + 2e^- \longrightarrow Cu(s)$$
$$2H_2O(l) \longrightarrow O_2(g) + 4H^+(aq) + 4e^-$$

See Problems 20.57 and 20.58.

20.11 Stoichiometry of Electrolysis

In 1831 and 1832, the British chemist and physicist Michael Faraday showed that the amounts of substances released at the electrodes during electrolysis are related to the total charge that has flowed in the electric circuit. ◄ If you look at the electrode reactions, you see that the relationship is stoichiometric.

When molten sodium chloride is electrolyzed, sodium ions migrate to the cathode, where they react with electrons to produce sodium. Similarly, chloride ions migrate to the anode and release electrons by producing chlorine. Therefore, when one mole of electrons reacts with sodium ions, one faraday of charge passes through the circuit. One mole of sodium metal is deposited at one electrode, and one-half mole of chlorine gas evolves at the other electrode.

What is new in this type of stoichiometric problem is the measurement of numbers of electrons. You measure the quantity of electric charge that has passed through the circuit. You use the following fact:

▶ Faraday's results were summarized in two laws: (1) The quantity of a substance liberated at an electrode is directly proportional to the quantity of electric charge that has flowed in the circuit. (2) For a given quantity of electric charge, the amount of any metal deposited is proportional to its equivalent weight (atomic weight divided by the charge on the metal ion). These laws follow directly from the stoichiometry of electrolysis.

One faraday (9.65×10^4 C) is equivalent to the charge on one mole of electrons.

If you know the current in a circuit and the length of time it has been flowing, you can calculate the electric charge.

Electric charge = electric current × time lapse

Corresponding units are

Coulombs = amperes × seconds

The **ampere (A)** is *the base unit of current in the International System (SI)*. The coulomb (C), the SI unit of electric charge, is equivalent to an ampere-second. So a current of 0.50 amperes flowing for 84 seconds gives a charge of 0.50 A × 84 s = 42 A · s, or 42 C.

If you are given the amount of substance produced at an electrode and the time of electrolysis, you can determine the current. If you are given the current and the time of electrolysis, you can calculate the amount of substance produced at an electrode. The next two examples illustrate these calculations.

EXAMPLE 20.12

Calculating the Amount of Charge from the Amount of Product in an Electrolysis

When an aqueous solution of copper(II) sulfate, $CuSO_4$, is electrolyzed, copper metal is deposited.

$$Cu^{2+}(aq) + 2e^- \longrightarrow Cu(s)$$

(The other electrode reaction gives oxygen: $2H_2O \longrightarrow O_2 + 4H^+ + 4e^-$.) If a constant current was passed for 5.00 h and 404 mg of copper metal was deposited, what was the current?

SOLUTION

The conversion of grams of Cu to coulombs required to deposit 404 mg Cu is

$$0.404 \text{ g Cu} \times \frac{1 \text{ mol Cu}}{63.6 \text{ g Cu}} \times \frac{2 \text{ mol } e^-}{1 \text{ mol Cu}} \times \frac{9.65 \times 10^4 \text{ C}}{1 \text{ mol } e^-} = 1.2\underline{2}6 \times 10^3 \text{ C}$$

The time lapse, 5.00 h, equals 1.80×10^4 s. Thus,

$$\text{Current} = \frac{\text{charge}}{\text{time}} = \frac{1.226 \times 10^3 \text{ C}}{1.80 \times 10^4 \text{ s}} = \mathbf{6.81 \times 10^{-2} \text{ A}}$$

See Problems 20.59 and 20.60.

EXAMPLE 20.13

Calculating the Amount of Product from the Amount of Charge in an Electrolysis

When an aqueous solution of potassium iodide is electrolyzed using platinum electrodes, the half-reactions are

$$2I^-(aq) \longrightarrow I_2(aq) + 2e^-$$
$$2H_2O(l) + 2e^- \longrightarrow H_2(g) + 2OH^-(aq)$$

How many grams of iodine are produced when a current of 8.52 mA flows through the cell for 10.0 min?

SOLUTION

When the current flows for 6.00×10^2 s (10.0 min), the amount of charge is

$$8.52 \times 10^{-3} \text{ A} \times 6.00 \times 10^2 \text{ s} = 5.11 \text{ C}$$

Note that two moles of electrons are equivalent to one mole of I_2. Hence,

$$5.11 \text{ C} \times \frac{1 \text{ mol } e^-}{9.65 \times 10^4 \text{ C}} \times \frac{1 \text{ mol } I_2}{2 \text{ mol } e^-} \times \frac{254 \text{ g } I_2}{1 \text{ mol } I_2} = \mathbf{6.73 \times 10^{-3} \text{ g } I_2}$$

See Problems 20.61 and 20.62.

A Checklist for Review

Important Terms

electrochemical cell (20.1)
voltaic (galvanic)
  cell (20.1)
electrolytic cell (20.1)
half-cell (20.2)
salt bridge (20.2)
anode (20.2)

cathode (20.2)
cell reaction (20.2)
potential difference (20.4)
volt (V) (20.4)
Faraday constant (F) (20.4)
electromotive force
  (emf) (20.4)

standard emf (20.5)
Nernst equation (20.7)
zinc–carbon (Leclanché)
  dry cell (20.8)
alkaline dry cell (20.8)
lead storage cell (20.8)
nickel–cadmium cell (20.8)

fuel cell (20.8)
electrolysis (20.9)
Downs cell (20.9)
ampere (A) (20.11)

Key Equations

$$w_{max} = -nFE_{cell}$$

$$E°_{cell} = E°_{cathode} - E°_{anode}$$

$$\Delta G° = -nFE°_{cell}$$

$$E°_{cell} = \frac{0.0592}{n} \log K \quad \text{(values in volts at 25°C)}$$

$$E_{cell} = E°_{cell} - \frac{0.0592}{n} \log Q \quad \text{(values in volts at 25°C)}$$

Summary of Facts and Concepts

Oxidation–reduction reactions involve a transfer of electrons from one species to another. The *half-reaction method* can be applied to balancing oxidation–reduction reactions in acidic and basic solutions. Many of the principles required for balancing these reactions were presented in Chapter 4.

Electrochemical cells are of two types: voltaic and electrolytic. A *voltaic cell* uses a spontaneous chemical reaction to generate an electric current. It does this by physically separating the reaction into its oxidation and reduction half-reactions. These half-reactions take place in *half-cells*. The half-cell in which reduction occurs is called the *cathode;* the half-cell in which oxidation occurs is called the *anode.* Electrons flow in the external circuit from the anode to the cathode.

The *electromotive force (emf)* is the maximum voltage of a voltaic cell. It can be directly related to the maximum work that can be done by the cell. A table of standard electrode potentials is useful for establishing the direction of spontaneity of an oxidation–reduction reaction and for calculating the *standard emf* of a cell.

The standard free-energy change, standard emf, and equilibrium constant are all related. Knowing one, you can calculate the others. Electrochemical measurements can therefore provide equilibrium or thermodynamic information.

An electrode potential depends on concentrations of the electrode substances, according to the *Nernst equation.* Because of this relationship, cell emf's can be used to measure ion concentrations. This is the basic principle of a pH meter, a device that measures the hydrogen-ion concentration.

Voltaic cells are used commercially as portable energy sources (batteries). In addition, the basic principle of the voltaic cell is employed in the *cathodic protection* of buried pipelines and tanks.

Electrolytic cells represent another type of electrochemical cell. They use an external voltage source to push a reaction in a nonspontaneous direction. The electrolysis of an aqueous solution often involves the oxidation or reduction of water at the electrodes. Electrolysis of concentrated sodium chloride solution, for example, gives hydrogen at the cathode. The amounts of substances released at an electrode are related to the amount of charge passed through the cell. This relationship is stoichiometric and follows from the electrode reactions.

Operational Skills

1. **Balancing equations in acidic and basic solutions by the half-reaction method** Given the skeleton equation for an oxidation–reduction equation, complete and balance it. **(EXAMPLES 20.1, 20.2)**

2. **Writing the cell reaction from the cell notation** Given the notation for a voltaic cell, write the overall cell reaction **(EXAMPLE 20.3)**. Alternatively, given the cell reaction, write the cell notation.

3. **Calculating the quantity of work from a given amount of cell reactant** Given the emf and overall reaction for a voltaic cell, calculate the maximum work that can be obtained from a given amount of reactant. **(EXAMPLE 20.4)**

4. **Determining the relative strengths of oxidizing and reducing agents** Given a table of standard electrode potentials, list oxidizing or reducing agents by increasing strength. **(EXAMPLE 20.5)**

5. **Determining the direction of spontaneity from electrode potentials** Given standard electrode potentials, decide the direction of spontaneity for a reaction under standard conditions. **(EXAMPLE 20.6)**

6. **Calculating the emf from standard potentials** Given standard electrode potentials, calculate the standard emf of a voltaic cell. **(EXAMPLE 20.7)**

7. **Calculating the free-energy change from electrode potentials** Given standard electrode potentials, calculate the standard free-energy change. **(EXAMPLE 20.8)**

8. **Calculating the equilibrium constant from cell emf** Given standard potentials (or standard emf), calculate the equilibrium constant. **(EXAMPLE 20.9)**

9. **Calculating the cell emf for nonstandard conditions** Given standard electrode potentials and the concentrations of substances in a voltaic cell, calculate the cell emf. **(EXAMPLE 20.10)**

10. **Predicting the half-reactions in an aqueous electrolysis** Using values of electrode potentials, decide which electrode reactions actually occur in the electrolysis of an aqueous solution. **(EXAMPLE 20.11)**

11. **Relating the amounts of charge and product in an electrolysis** Given the amount of product obtained by electrolysis, calculate the amount of charge that flowed **(EXAMPLE 20.12)**. Given the amount of charge that flowed, calculate the amount of product obtained by electrolysis **(EXAMPLE 20.13)**.

Review Questions

20.1 Describe the difference between a voltaic cell and an electrolytic cell.

20.2 What is the SI unit of electrical potential?

20.3 Why is it necessary to measure the voltage of a voltaic cell when no current is flowing to obtain the cell emf?

20.4 How are standard electrode potentials defined?

20.5 Give the mathematical relationships between the members of each possible pair of the three quantities $\Delta G°$, $E°_{cell}$, and K.

20.6 Describe the zinc–carbon, or Leclanché, dry cell and the lead storage battery.

20.7 Explain the electrochemistry of rusting.

20.8 Iron may be protected by coating with tin (tin cans) or with zinc (galvanized iron). Galvanized iron does not corrode as long as zinc is present. By contrast, when a tin can is scratched, the exposed iron underneath corrodes rapidly. Explain the difference between zinc and tin as protective coatings against iron corrosion.

20.9 The electrolysis of water is often done by passing a current through a dilute solution of sulfuric acid. What is the function of the sulfuric acid?

20.10 Describe a method for the preparation of sodium metal from sodium chloride.

20.11 Potassium was discovered by the British chemist Humphry Davy when he electrolyzed molten potassium hydroxide. What would be the anode reaction?

20.12 Briefly explain why different products are obtained from the electrolysis of molten NaCl and the electrolysis of a dilute aqueous solution of NaCl.

Conceptual Problems

20.13 Keeping in mind that aqueous Cu^{2+} is blue and aqueous Zn^{2+} is colorless, predict what you would observe over a several-day period if you performed the following experiments.

a. A strip of Zn is placed into a beaker containing aqueous Zn^{2+}.

b. A strip of Cu is placed into a beaker containing aqueous Cu^{2+}.

c. A strip of Zn is placed into a beaker containing aqueous Cu^{2+}.

d. A strip of Cu is placed into a beaker containing aqueous Zn^{2+}.

20.14 You place a battery in a flashlight in which all of the electrochemical reactions have reached equilibrium. What do you expect to observe when you turn on the flashlight? Explain your answer.

20.15 The difference between a "heavy-duty" and a regular zinc–carbon battery is that the zinc can in the heavy-duty battery is thicker walled. What makes this battery heavy-duty in terms of output?

20.16 From an electrochemical standpoint, what metal, other than zinc, would be a reasonable candidate to coat a piece of iron to prevent corrosion (oxidation)?

20.17 Household bleach is a solution containing NaClO, which is a very strong oxidizing agent. Provide a brief explanation as to how bleach is able to "whiten" stains in clothing.

20.18 The development of lightweight batteries is an ongoing research effort combining many of the physical sciences. You are a member of an engineering team trying to develop a lightweight battery that will effectively react with $O_2(g)$ from the atmosphere as an oxidizing agent. A reducing agent must be chosen for this battery that will be lightweight, have nontoxic products, and react spontaneously with oxygen. Using data from Table 20.1, suggest a likely reducing agent, being sure the above conditions are met. Are there any drawbacks to your selection?

Practice Problems

Balancing Oxidation–Reduction Equations

20.19 Balance the following oxidation–reduction equations. The reactions occur in acidic solution.
a. $Cr_2O_7{}^{2-} + C_2O_4{}^{2-} \longrightarrow Cr^{3+} + CO_2$
b. $Cu + NO_3{}^- \longrightarrow Cu^{2+} + NO$
c. $MnO_2 + HNO_2 \longrightarrow Mn^{2+} + NO_3{}^-$
d. $PbO_2 + Mn^{2+} + SO_4{}^{2-} \longrightarrow PbSO_4 + MnO_4{}^-$

20.20 Balance the following oxidation–reduction equations. The reactions occur in acidic solution.
a. $Mn^{2+} + BiO_3{}^- \longrightarrow MnO_4{}^- + Bi^{3+}$
b. $Cr_2O_7{}^{2-} + I^- \longrightarrow Cr^{3+} + IO_3{}^-$
c. $MnO_4{}^- + H_2SO_3 \longrightarrow Mn^{2+} + SO_4{}^{2-}$
d. $Cr_2O_7{}^{2-} + Fe^{2+} \longrightarrow Cr^{3+} + Fe^{3+}$

20.21 Balance the following oxidation–reduction equations. The reactions occur in basic solution.
a. $Mn^{2+} + H_2O_2 \longrightarrow MnO_2 + H_2O$
b. $MnO_4{}^- + NO_2{}^- \longrightarrow MnO_2 + NO_3{}^-$
c. $Mn^{2+} + ClO_3{}^- \longrightarrow MnO_2 + ClO_2$
d. $MnO_4{}^- + NO_2 \longrightarrow MnO_2 + NO_3{}^-$

20.22 Balance the following oxidation–reduction equations. The reactions occur in basic solution.
a. $Cr(OH)_4{}^- + H_2O_2 \longrightarrow CrO_4{}^{2-} + H_2O$
b. $MnO_4{}^- + Br^- \longrightarrow MnO_2 + BrO_3{}^-$
c. $Co^{2+} + H_2O_2 \longrightarrow Co(OH)_3 + H_2O$
d. $Pb(OH)_4{}^{2-} + ClO^- \longrightarrow PbO_2 + Cl^-$

20.23 Balance the following oxidation–reduction equations. The reactions occur in acidic or basic aqueous solution, as indicated.
a. $H_2S + NO_3{}^- \longrightarrow NO_2 + S_8$ (acidic)
b. $NO_3{}^- + Cu \longrightarrow NO + Cu^{2+}$ (acidic)
c. $MnO_4{}^- + SO_2 \longrightarrow SO_4{}^{2-} + Mn^{2+}$ (acidic)
d. $Bi(OH)_3 + Sn(OH)_3{}^- \longrightarrow Sn(OH)_6{}^{2-} + Bi$ (basic)

20.24 Balance the following oxidation–reduction equations. The reactions occur in acidic or basic aqueous solution, as indicated.
a. $Hg_2{}^{2+} + H_2S \longrightarrow Hg + S_8$ (acidic)
b. $S^{2-} + I_2 \longrightarrow SO_4{}^{2-} + I^-$ (basic)
c. $Al + NO_3{}^- \longrightarrow Al(OH)_4{}^- + NH_3$ (basic)
d. $MnO_4{}^- + C_2O_4{}^{2-} \longrightarrow MnO_2 + CO_2$ (basic)

Electrochemical Cells

20.25 A voltaic cell is constructed from the following half-cells: a magnesium electrode in magnesium sulfate solution and a nickel electrode in nickel sulfate solution. The half-reactions are

$$Mg(s) \longrightarrow Mg^{2+}(aq) + 2e^-$$
$$Ni^{2+}(aq) + 2e^- \longrightarrow Ni(s)$$

Sketch the cell, labeling the anode and cathode (and the electrode reactions), and show the direction of electron flow and the movement of cations.

20.26 Half-cells were made from a nickel rod dipping in a nickel sulfate solution and a copper rod dipping in a copper sulfate solution. The half-reactions in a voltaic cell using these half-cells were

$$Cu^{2+}(aq) + 2e^- \longrightarrow Cu(s)$$
$$Ni(s) \longrightarrow Ni^{2+}(aq) + 2e^-$$

Sketch the cell and label the anode and cathode, showing the corresponding electrode reactions. Give the direction of electron flow and the movement of cations.

20.27 A silver oxide–zinc cell maintains a fairly constant voltage during discharge (1.60 V). The button form of this cell is used in watches, hearing aids, and other electronic devices. The half-reactions are

$$Zn(s) + 2OH^-(aq) \longrightarrow Zn(OH)_2(s) + 2e^-$$
$$Ag_2O(s) + H_2O(l) + 2e^- \longrightarrow 2Ag(s) + 2OH^-(aq)$$

Identify the anode and the cathode reactions. What is the overall reaction in the voltaic cell?

20.28 A mercury battery, used for hearing aids and electric watches, delivers a constant voltage (1.35 V) for long periods. The half-reactions are

$$HgO(s) + H_2O(l) + 2e^- \longrightarrow Hg(l) + 2OH^-(aq)$$
$$Zn(s) + 2OH^-(aq) \longrightarrow Zn(OH)_2(s) + 2e^-$$

Which half-reaction occurs at the anode and which occurs at the cathode? What is the overall cell reaction?

Voltaic Cell Notation

20.29 Write the cell notation for a voltaic cell with the following half-reactions.

$$Ni(s) \longrightarrow Ni^{2+}(aq) + 2e^-$$
$$Pb^{2+}(aq) + 2e^- \longrightarrow Pb(s)$$

20.30 Write the cell notation for a voltaic cell with the following half-reactions.

$$Al(s) \longrightarrow Al^{3+}(aq) + 3e^-$$
$$2H^+(aq) + 2e^- \longrightarrow H_2(g)$$

20.31 Give the notation for a voltaic cell constructed from a hydrogen electrode (cathode) in 1.0 M HCl and a nickel electrode (anode) in 1.0 M $NiSO_4$ solution. The electrodes are connected by a salt bridge.

20.32 A voltaic cell has an iron rod in 0.30 M iron(III) chloride solution for the cathode and a zinc rod in 0.40 M zinc sulfate solution for the anode. The half-cells are connected by a salt bridge. Write the notation for this cell.

20.33 Consider the voltaic cell

$$Cd(s)\,|\,Cd^{2+}(aq)\,\|\,Ni^{2+}(aq)\,|\,Ni(s)$$

Write the half-cell reactions and the overall cell reaction. Make a sketch of this cell and label it. Include labels showing the anode, cathode, and direction of electron flow.

20.34 Consider the voltaic cell

$$Zn(s)\,|\,Zn^{2+}(aq)\,\|\,Cr^{3+}(aq)\,|\,Cr(s)$$

Write the half-cell reactions and the overall cell reaction. Make a sketch of this cell and label it. Include labels showing the anode, cathode, and direction of electron flow.

Electrode Potentials and Cell emf's

20.35 A voltaic cell whose cell reaction is

$$2Fe^{3+}(aq) + Zn(s) \longrightarrow 2Fe^{2+}(aq) + Zn^{2+}(aq)$$

has an emf of 0.72 V. What is the maximum electrical work that can be obtained from this cell per mole of iron(III) ion?

20.36 A particular voltaic cell operates on the reaction

$$Zn(s) + Cl_2(g) \longrightarrow Zn^{2+}(aq) + 2Cl^-(aq)$$

giving an emf of 0.853 V. Calculate the maximum electrical work generated when 20.0 g of zinc metal is consumed.

20.37 Order the following oxidizing agents by increasing strength under standard-state conditions: $O_2(g)$; $MnO_4^-(aq)$; $NO_3^-(aq)$ (in acidic solution).

20.38 Order the following oxidizing agents by increasing strength under standard-state conditions: $Ag^+(aq)$; $I_2(aq)$; $MnO_4^-(aq)$ (in acidic solution).

20.39 Consider the reducing agents $Cu^+(aq)$, $Zn(s)$, and $Fe(s)$. Which is strongest? Which is weakest?

20.40 Consider the reducing agents $Sn^{2+}(aq)$, $Cu(s)$, and $I^-(aq)$. Which is strongest? Which is weakest?

20.41 Consider the following reactions. Are they spontaneous in the direction written, under standard conditions at 25°C?
a. $Sn^{4+}(aq) + 2Fe^{2+}(aq) \longrightarrow Sn^{2+}(aq) + 2Fe^{3+}(aq)$
b. $4MnO_4^-(aq) + 12H^+(aq) \longrightarrow$
$$4Mn^{2+}(aq) + 5O_2(g) + 6H_2O(l)$$

20.42 Answer the following questions by referring to standard electrode potentials at 25°C.
a. Will dichromate ion, $Cr_2O_7^{2-}$, oxidize iron(II) ion in acidic solution under standard conditions?
b. Will copper metal reduce 1.0 M $Ni^{2+}(aq)$ to metallic nickel?

20.43 What would you expect to happen when chlorine gas, Cl_2, at 1 atm pressure is bubbled into a solution containing 1.0 M F^- and 1.0 M Br^- at 25°C? Write a balanced equation for the reaction that occurs.

20.44 Dichromate ion, $Cr_2O_7^{2-}$, is added to an acidic solution containing Br^- and Mn^{2+}. Write a balanced equation for any reaction that occurs. Assume standard conditions at 25°C.

20.45 Calculate the standard emf of the following cell at 25°C.

$$Cr(s)\,|\,Cr^{3+}(aq)\,\|\,Hg_2^{2+}(aq)\,|\,Hg(l)$$

20.46 Calculate the standard emf of the following cell at 25°C.

$$Sn(s)\,|\,Sn^{2+}(aq)\,\|\,Cu^{2+}(aq)\,|\,Cu(s)$$

Relationships Among $E°_{cell}$, $\Delta G°$, and K

20.47 Calculate the standard free-energy change at 25°C for the following reaction.

$$3Cu(s) + 2NO_3^-(aq) + 8H^+(aq) \longrightarrow$$
$$3Cu^{2+}(aq) + 2NO(g) + 4H_2O(l)$$

Use standard electrode potentials.

20.48 Calculate the standard free-energy change at 25°C for the following reaction.

$$4Al(s) + 3O_2(g) + 12H^+(aq) \longrightarrow 4Al^{3+}(aq) + 6H_2O(l)$$

Use standard electrode potentials.

20.49 Calculate the standard emf at 25°C for the following cell reaction from standard free energies of formation (Appendix B).

$$Mg(s) + 2Ag^+(aq) \longrightarrow Mg^{2+}(aq) + 2Ag(s)$$

20.50 Calculate the standard emf at 25°C for the following cell reaction from standard free energies of formation (Appendix B).

$$2Al(s) + 3Cu^{2+}(aq) \longrightarrow 2Al^{3+}(aq) + 3Cu(s)$$

20.51 Calculate the equilibrium constant K for the following reaction at 25°C from standard electrode potentials.

$$Fe^{3+}(aq) + Cu(s) \longrightarrow Fe^{2+}(aq) + Cu^{2+}(aq)$$

The equation is not balanced.

20.52 Calculate the equilibrium constant K for the following reaction at 25°C from standard electrode potentials.

$$Sn^{4+}(aq) + Hg(l) \longrightarrow Sn^{2+}(aq) + Hg_2^{2+}(aq)$$

The equation is not balanced.

Nernst Equation

20.53 Calculate the emf of the following cell at 25°C.

$$Cr(s)\,|\,Cr^{3+}(1.0 \times 10^{-3}\,M)\,\|\,Ni^{2+}(1.5\,M)\,|\,Ni(s)$$

20.54 What is the emf of the following cell at 25°C?

$$Ni(s)\,|\,Ni^{2+}(1.0\,M)\,\|\,Sn^{2+}(1.0 \times 10^{-4}\,M)\,|\,Sn(s)$$

20.55 The voltaic cell

$$Cd(s)\,|\,Cd^{2+}(aq)\,\|\,Ni^{2+}(1.0\,M)\,|\,Ni(s)$$

has an electromotive force of 0.240 V at 25°C. What is the concentration of cadmium ion? ($E_{cell}^\circ = 0.170$ V.)

20.56 The emf of the following cell at 25°C is 0.475 V.

$$Zn\,|\,Zn^{2+}(1\,M)\,\|\,H^+(\text{test solution})\,|\,H_2(1\text{ atm})\,|\,Pt$$

What is the pH of the test solution?

Electrolysis

20.57 Describe what you expect to happen when the following solutions are electrolyzed: (a) aqueous Na_2SO_4; (b) aqueous KBr. That is, what are the electrode reactions? What is the overall reaction?

20.58 Describe what you expect to happen when the following solutions are electrolyzed: (a) aqueous $CuCl_2$; (b) aqueous $Cu(NO_3)_2$. That is, what are the electrode reactions? What is the overall reaction? Nitrate ion is not oxidized.

20.59 In the commercial preparation of aluminum, aluminum oxide, Al_2O_3, is electrolyzed at 1000°C. (The mineral cryolite is added as a solvent.) Assume that the cathode reaction is

$$Al^{3+} + 3e^- \longrightarrow Al$$

How many coulombs of electricity are required to give 3.61 kg of aluminum?

20.60 Chlorine, Cl_2, is produced commercially by the electrolysis of aqueous sodium chloride. The anode reaction is

$$2Cl^-(aq) \longrightarrow Cl_2(g) + 2e^-$$

How long will it take to produce 1.18 kg of chlorine if the current is 5.00×10^2 A?

20.61 When molten lithium chloride, LiCl, is electrolyzed, lithium metal is liberated at the cathode. How many grams of lithium are liberated when 5.00×10^3 C of charge passes through the cell?

20.62 How many grams of cadmium are deposited from an aqueous solution of cadmium sulfate, $CdSO_4$, when an electric current of 1.51 A flows through the solution for 221 min?

General Problems

20.63 Balance the following skeleton equations. The reactions occur in acidic or basic aqueous solution, as indicated.
a. $MnO_4^- + S^{2-} \longrightarrow MnO_2 + S_8$ (basic)
b. $IO_3^- + HSO_3^- \longrightarrow I^- + SO_4^{2-}$ (acidic)
c. $Fe(OH)_2 + CrO_4^{2-} \longrightarrow Fe(OH)_3 + Cr(OH)_4^-$ (basic)

20.64 Balance the following skeleton equations. The reactions occur in acidic or basic aqueous solution, as indicated.
a. $MnO_4^- + H_2S \longrightarrow Mn^{2+} + S_8$ (acidic)
b. $Zn + NO_3^- \longrightarrow Zn^{2+} + N_2O$ (acidic)
c. $MnO_4^{2-} \longrightarrow MnO_4^- + MnO_2$ (basic)

20.65 Iron(II) hydroxide is a greenish precipitate that is formed from iron(II) ion by the addition of a base. This precipitate gradually turns to the yellowish-brown iron(III) hydroxide from oxidation by O_2 in the air. Write a balanced equation for this oxidation by O_2.

20.66 A sensitive test for bismuth(III) ion consists of shaking a solution suspected of containing the ion with a basic solution of sodium stannite, Na_2SnO_2. A positive test consists of the formation of a black precipitate of bismuth metal. Stannite ion is oxidized by bismuth(III) ion to stannate ion, SnO_3^{2-}. Write a balanced equation for the reaction.

20.67 Give the notation for a voltaic cell that uses the reaction

$$Ca(s) + Cl_2(g) \longrightarrow Ca^{2+}(aq) + 2Cl^-(aq)$$

What is the half-cell reaction for the anode? for the cathode? What is the standard emf of the cell?

20.68 Give the notation for a voltaic cell whose overall cell reaction is

$$Mg(s) + 2Ag^+(aq) \longrightarrow Mg^{2+}(aq) + 2Ag(s)$$

What are the half-cell reactions? Label them as anode or cathode reactions. What is the standard emf of this cell?

20.69 Use electrode potentials to answer the following questions. (a) Is the oxidation of nickel by iron(III) ion a spontaneous reaction under standard conditions?

$$Ni(s) + 2Fe^{3+}(aq) \longrightarrow Ni^{2+}(aq) + 2Fe^{2+}(aq)$$

(b) Will iron(III) ion oxidize tin(II) ion to tin(IV) ion under standard conditions?

$$2Fe^{3+}(aq) + Sn^{2+}(aq) \longrightarrow 2Fe^{2+}(aq) + Sn^{4+}(aq)$$

20.70 Use electrode potentials to answer the following questions, assuming standard conditions. (a) Do you expect permanganate ion (MnO_4^-) to oxidize chloride ion to chlorine gas in acidic solution? (b) Will dichromate ion ($Cr_2O_7^{2-}$) oxidize chloride ion to chlorine gas in acidic solution?

20.71 a. Calculate the equilibrium constant for the following reaction at 25°C.

$$Sn(s) + Pb^{2+}(aq) \rightleftharpoons Sn^{2+}(aq) + Pb(s)$$

The standard emf of the corresponding voltaic cell is 0.010 V.
b. If an excess of tin metal is added to 1.0 M Pb^{2+}, what is the concentration of Pb^{2+} at equilibrium?

20.72 a. Calculate the equilibrium constant for the following reaction at 25°C.

$$Ag^+(aq) + Fe^{2+}(aq) \rightleftharpoons Ag(s) + Fe^{3+}(aq)$$

The standard emf of the corresponding voltaic cell is 0.030 V.
b. When equal volumes of 1.0 M solutions of Ag^+ and Fe^{2+} are mixed, what is the equilibrium concentration of Fe^{2+}?

20.73 How many faradays are required for each of the following processes? How many coulombs are required?
a. Reduction of 1.0 mol Na^+ to Na
b. Reduction of 1.0 mol Cu^{2+} to Cu
c. Oxidation of 1.0 g H_2O to O_2

20.74 How many faradays are required for each of the following processes? How many coulombs are required?
a. Reduction of 1.0 mol Fe^{3+} to Fe^{2+}
b. Reduction of 1.0 mol Fe^{3+} to Fe
c. Oxidation of 1.0 g Sn^{2+} to Sn^{4+}

20.75 A standard electrochemical cell is made by dipping a silver electrode into a 1.0 M Ag^+ solution and a cadmium electrode into a 1.0 M Cd^{2+} solution.
a. What is the spontaneous chemical reaction, and what is the maximum potential produced by this cell?
b. What would be the effect on the potential of this cell if sodium sulfide were added to the Cd^{2+} half-cell and CdS were precipitated? Why?
c. What would be the effect on the potential of the cell if the size of the silver electrode were doubled?

20.76 A standard electrochemical cell is made by dipping an iron electrode into a 1.0 M Fe^{2+} solution and a copper electrode into a 1.0 M Cu^{2+} solution.
a. What is the spontaneous chemical reaction, and what is the maximum potential produced by this cell?
b. What would be the effect on the potential of this cell if sodium carbonate were added to the Cu^{2+} half-cell and $CuCO_3$ were precipitated? Why?
c. What would be the effect on the potential of the cell if the size of the iron electrode were halved?

20.77 A solution of copper(II) sulfate is electrolyzed by passing a current through the solution using inert electrodes. Consequently, there is a decrease in the Cu^{2+} concentration and an increase in the hydronium ion concentration. Also, one electrode increases in mass and a gas evolves at the other electrode. Write half-reactions that occur at the anode and at the cathode.

20.78 A potassium chloride solution is electrolyzed by passing a current through the solution using inert electrodes. A gas evolves at each electrode, and there is a large increase in pH of the solution. Write the half-reactions that occur at the anode and at the cathode.

20.79 A constant current of 1.50 amp is passed through an electrolytic cell containing a 0.10 M solution of $AgNO_3$ and a silver anode and a platinum cathode until 2.48 g of silver is deposited.
a. How long does the current flow to obtain this deposit?
b. What mass of chromium would be deposited in a similar cell containing 0.10 M Cr^{3+} if the same amount of current were used?

20.80 A constant current of 1.25 amp is passed through an electrolytic cell containing a 0.050 M solution of $CuSO_4$ and a copper anode and a platinum cathode until 2.10 g of copper is deposited.
a. How long does the current flow to obtain this deposit?
b. What mass of silver would be deposited in a similar cell containing 0.10 M Ag^+ if the same amount of current were used?

20.81 An electrochemical cell is made by placing a zinc electrode in 1.00 L of 0.200 M $ZnSO_4$ solution and a copper electrode in 1.00 L of 0.0100 M $CuCl_2$ solution.

a. What is the initial voltage of this cell when it is properly constructed?

b. Calculate the final concentration of Cu^{2+} in this cell if it is allowed to produce an average current of 1.00 amp for 225 s.

20.82 An electrochemical cell is made by placing an iron electrode in 1.00 L of 0.15 M $FeSO_4$ solution and a copper electrode in 1.00 L of 0.036 M $CuSO_4$ solution.

a. What is the initial voltage of this cell when it is properly constructed?

b. Calculate the final concentration of Cu^{2+} in this cell if it is allowed to produce an average current of 1.25 amp for 335 s.

20.83 a. Calculate ΔG° for the following cell reaction:

$$Cd(s)\,|\,Cd^{2+}(aq)\,\|\,Co^{2+}(aq)\,|\,Co(s)$$

b. From ΔG°, calculate the standard emf for the cell reaction and from this, determine the standard potential for $Co^{2+}(aq) + 2e^- \longrightarrow Co(s)$.

20.84 a. Calculate ΔG° for the following cell reaction:

$$Tl(s)\,|\,Tl^+(aq)\,\|\,Pb^{2+}(aq)\,|\,Pb(s)$$

The ΔG_f° for $Tl^+(aq)$ is -32.4 kJ/mol.

b. From ΔG°, calculate the standard emf for the cell reaction and from this, determine the standard potential for $Tl^+(aq) + e^- \longrightarrow Tl(s)$.

Cumulative-Skills Problems

20.85 Consider the following cell reaction at 25°C.

$$2Cr^{3+}(aq) + 3Zn(s) \longrightarrow 3Zn^{2+}(aq) + 2Cr(s)$$

Calculate the standard emf of this cell from the standard electrode potentials, and from this obtain ΔG° for the cell reaction. Use data in Appendix B to calculate ΔH°; note that $Cr(H_2O)_6^{3+}(aq)$ equals $Cr^{3+}(aq)$. Use these values of ΔH° and ΔG° to obtain ΔS° for the cell reaction.

20.86 Consider the following cell reaction at 25°C.

$$2Cr(s) + 3Fe^{2+}(aq) \longrightarrow 2Cr^{3+}(aq) + 3Fe(s)$$

Calculate the standard emf of this cell from the standard electrode potentials, and from this obtain ΔG° for the cell reaction. Use data in Appendix B to calculate ΔH°; note that $Cr(H_2O)_6^{3+}(aq)$ equals $Cr^{3+}(aq)$. Use these values of ΔH° and ΔG° to obtain ΔS° for the cell reaction.

20.87 Under standard conditions for all concentrations, the following reaction is spontaneous at 25°C.

$$O_2(g) + 4H^+(aq) + 4Br^-(aq) \longrightarrow 2H_2O(l) + 2Br_2(l)$$

If $[H^+]$ is decreased so that the pH = 3.60, what value will E_{cell} have, and will the reaction be spontaneous at this $[H^+]$?

20.88 Under standard conditions for all concentrations, the following reaction is spontaneous at 25°C.

$$O_3(g) + 2H^+(aq) + 2Co^{2+}(aq) \longrightarrow$$
$$O_2(g) + H_2O(l) + 2Co^{3+}(l)$$

If $[H^+]$ is decreased so that the pH = 9.10, what value will E_{cell} have, and will the reaction be spontaneous at this $[H^+]$?

20.89 Under standard conditions for all concentrations, the following reaction is spontaneous at 25°C.

$$O_2(g) + 4H^+(aq) + 4Br^-(aq) \longrightarrow 2H_2O(l) + 2Br_2(l)$$

If $[H^+]$ is adjusted by adding a buffer of 0.10 M NaOCN and 0.10 M HOCN ($K_a = 3.5 \times 10^{-4}$), what value will E_{cell} have, and will the reaction be spontaneous at this $[H^+]$?

20.90 Under standard conditions for all concentrations, the following reaction is spontaneous at 25°C.

$$O_3(g) + 2H^+(aq) + 2Co^{2+}(aq) \longrightarrow$$
$$O_2(g) + H_2O(l) + 2Co^{3+}(aq)$$

If $[H^+]$ is adjusted by adding a buffer of 0.10 M NaClO and 0.10 M HClO ($K_a = 3.5 \times 10^{-8}$), what value will E_{cell} have, and will the reaction be spontaneous at this $[H^+]$?

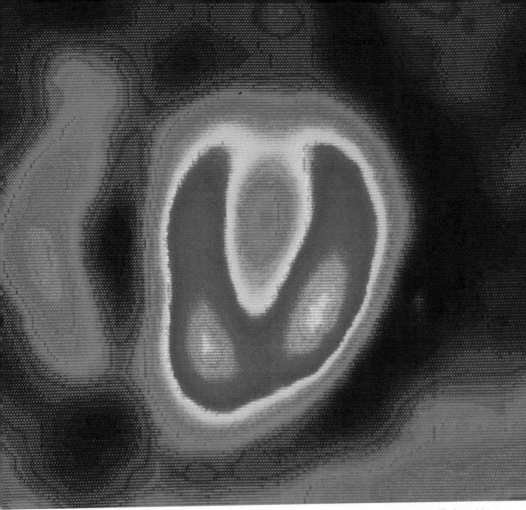

Image of healthy human heart produced by recording gamma emissions of the radionuclide, thallium-201. In order to produce this image, the radioactive thallium-210 was injected into the patient.

Nuclear Chemistry

Technetium is an unusual element. Although a *d*-transition element (under manganese in Group VIIB) with a small atomic number ($Z = 43$), it has no stable isotopes. The nucleus of every technetium isotope is radioactive and decays, or disintegrates, to give an isotope of another element. Many of the technetium isotopes decay by emitting an electron from the nucleus.

Because of its nuclear instability, technetium is not found naturally on earth. Nevertheless, it is produced commercially in kilogram quantities from other elements by nuclear reactions. Technetium was the first new element produced in the laboratory from another element. It was discovered in 1937 by Carlo Perrier and Emilio Segrè when the element molybdenum was bombarded with deuterons (nuclei of hydrogen, each having one proton and one neutron). Later, technetium was found to be a product of the fission, or splitting, of uranium nuclei. Technetium is produced in nuclear fission reactors used to generate electricity.

Technetium-99 is used in medical diagnostics. A compound of technetium injected into a vein concentrates in certain body organs. The energy emitted by technetium nuclei is detected and gives an image of these body organs (Figure 21.1). The technetium is eliminated by the body after several hours.

In this chapter, we will look at nuclear processes such as those we have described for technetium. We will answer such questions as the following: How do you describe the radioactive decay of technetium? How do you describe the transformation of a molybdenum nucleus into technetium? How is technetium produced from uranium by nuclear fission, or splitting? What are some practical applications of nuclear processes?

Radioactivity and Nuclear Bombardment Reactions

In chemical reactions, only the outer electrons of the atoms are disturbed. In nuclear reactions, however, the nuclear changes that occur are independent of the chemical environment of the atom. For example, the nuclear changes in a radioactive ^3_1H atom are the same if the atom is part of an H_2 molecule or incorporated into H_2O.

We will look at two types of nuclear reactions. One type is **radioactive decay,** *the process in which a nucleus spontaneously disintegrates, giving off radiation.* The radiation usually consists of nuclear particles and electromagnetic radiation.

The second type of nuclear reaction is a **nuclear bombardment reaction,** *a nuclear reaction in which a nucleus is bombarded, or struck, by another nucleus or by a nuclear particle.* If there is sufficient energy in this collision, the nuclear particles of the reactants rearrange to give a product nucleus or nuclei.

21.1 Radioactivity

The phenomenon of radioactivity was discovered by Antoine Henri Becquerel in 1896. He discovered that photographic plates develop bright spots when exposed to uranium minerals, and he concluded that the minerals give off some sort of radiation.

The radiation from uranium minerals was later shown to be separable by electric (and magnetic) fields into three types, alpha (α), beta (β), and gamma (γ) rays (Figure 21.2). *Alpha rays* bend away from a positive plate and toward a negative plate, indicating that they have a positive charge; they are now known to consist of helium-4 nuclei. *Beta rays* bend in the opposite direction, indicating that they have a negative charge; they are now known to consist of high-speed electrons. *Gamma rays* are unaffected by electric and magnetic fields: they have been shown to be a form of electromagnetic radiation that is similar to x rays, except they are higher in energy with shorter wavelengths. Uranium minerals contain a number of radioactive elements, each emitting one or more of these radiations.

Nuclear Equations

▶ Nuclide symbols were introduced in Section 2.3. For uranium-238, you write

mass number ⟶ 238
atomic number ⟶ 92 U

You can write an equation for the nuclear reaction corresponding to the decay of uranium-238 much as you would write an equation for a chemical reaction. You represent the uranium-238 nucleus by the *nuclide symbol* $^{238}_{92}U$. ◀ The radioactive decay of $^{238}_{92}U$ by alpha-particle emission yields thorium-234.

$$^{238}_{92}U \longrightarrow\ ^{234}_{90}Th + {}^4_2He$$

This is an example of a **nuclear equation,** which is *a symbolic representation of a nuclear reaction.* Normally, only the nuclei are represented. It is not necessary to indicate the chemical compound or the electron charges for any ions involved, because the chemical environment has no effect on nuclear processes.

Reactant and product nuclei are represented in nuclear equations by their nuclide symbols. Other particles are given the following symbols, in which the subscript equals the charge and the superscript equals the total number of protons and neutrons in the particle (mass number):

Proton	$_{1}^{1}\text{H}$	or	$_{1}^{1}\text{p}$
Neutron	$_{0}^{1}\text{n}$		
Electron	$_{-1}^{0}\text{e}$	or	$_{-1}^{0}\beta$
Positron	$_{1}^{0}\text{e}$	or	$_{1}^{0}\beta$
Gamma photon	$_{0}^{0}\gamma$		

EXAMPLE 21.1

Writing a Nuclear Equation

Write the nuclear equation for the radioactive decay of radium-226 by alpha decay to give radon-222.

SOLUTION

The nuclear equation is

$$_{88}^{226}\text{Ra} \longrightarrow {}_{86}^{222}\text{Rn} + {}_{2}^{4}\text{He}$$

See Problems 21.19, 21.20, 21.21, and 21.22.

The total charge is conserved during a nuclear reaction. This means that the sum of the subscripts (number of protons, or positive charges, in the nuclei) for the products must equal the sum of the subscripts for the reactants.

Similarly, the total number of *nucleons* (protons and neutrons) is conserved during a nuclear reaction. This means that the sum of the superscripts (the mass numbers) for the reactants equals the sum of the superscripts for the products.

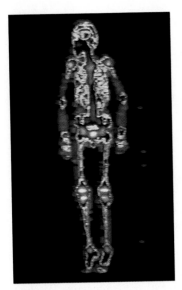

Figure 21.1
Image of a person's skeleton obtained using an excited form of technetium-99. A technetium compound was injected into the body, where it concentrated in bone tissue. Gamma rays (similar to x rays) emitted by technetium were detected by special equipment to produce this image.

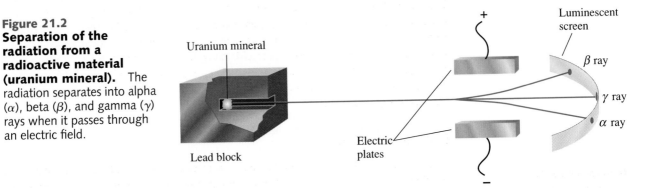

Figure 21.2
Separation of the radiation from a radioactive material (uranium mineral). The radiation separates into alpha (α), beta (β), and gamma (γ) rays when it passes through an electric field.

EXAMPLE 21.2

Deducing a Product or Reactant in a Nuclear Equation

Technetium-99 is a long-lived radioactive isotope of technetium. Each nucleus decays by emitting one beta particle. What is the product nucleus?

SOLUTION

Technetium-99 has the nuclide symbol $^{99}_{43}\text{Tc}$. A beta particle is an electron; the symbol is $^{0}_{-1}\text{e}$. The nuclear equation is

$$^{99}_{43}\text{Tc} \longrightarrow {}^{A}_{Z}\text{X} + {}^{0}_{-1}\text{e}$$

From the superscripts, you can write

$$99 = A + 0, \text{ or } A = 99$$

Similarly, from the subscripts, you get

$$43 = Z - 1, \text{ or } Z = 43 + 1 = 44$$

Hence $A = 99$ and $Z = 44$, so the product is $^{99}_{44}\text{X}$. Because element 44 is ruthenium, symbol Ru, you write the product nucleus as $^{99}_{44}\textbf{Ru}$.

See Problems 21.23 and 21.24.

Nuclear Stability

At first glance, the existence of several protons in the small space of a nucleus is puzzling. Why wouldn't the protons be strongly repelled by their like electric charges? The existence of stable nuclei with more than one proton is due to the nuclear force. The **nuclear force** is *a strong force of attraction between nucleons that acts only at very short distances (about 10^{-15} m)*. Beyond nuclear distances, these nuclear forces become negligible. This force in a nucleus can more than compensate for the repulsion of electric charges and thereby give a stable nucleus.

Evidence points to the special stability of pairs of protons and pairs of neutrons, analogous to the stability of pairs of electrons in molecules. Table 21.1 lists the number of stable isotopes that have an even number of protons and an even number of neutrons (157). By comparison, only 5 stable isotopes have an odd number of protons and an odd number of neutrons.

When you plot each stable nuclide on a graph with the number of protons (Z) on the horizontal axis and the number of neutrons (N) on the vertical axis, these sta-

Table 21.1
Number of Stable Isotopes with Even and Odd Numbers of Protons and Neutrons

	Number of Stable Isotopes			
	157	52	50	5
Number of protons	Even	Even	Odd	Odd
Number of neutrons	Even	Odd	Even	Odd

Figure 21.3
Band of stability. The stable nuclides, indicated by black dots, cluster in a band. Nuclides to the left of the band of stability usually decay by beta emission, whereas those to the right usually decay by positron emission or electron capture. Nuclides of Z > 83 often decay by alpha emission.

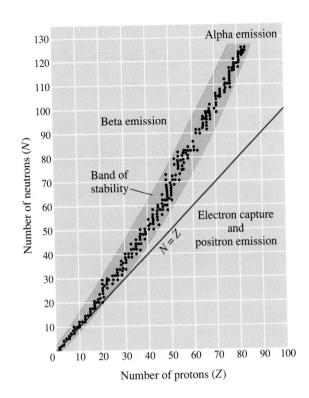

ble nuclides fall in a certain region, or band, of the graph. The **band of stability** is *the region in which stable nuclides lie in a plot of number of protons against number of neutrons.* Figure 21.3 shows the band of stability; the rest of the figure is explained later in this section. For nuclides up to Z = 20, the ratio of neutrons to protons is about 1.0 to 1.1. As Z increases, however, the neutron-to-proton ratio increases to about 1.5. This increase in neutron-to-proton ratio with increasing Z is believed to result from the increasing repulsions of protons from their electric charges. More neutrons are required to give attractive nuclear forces to offset these repulsions.

When the number of protons becomes very large, stable nuclides are impossible. No stable nuclides are known with atomic numbers greater than 83. All elements with Z equal to 83 or less have one or more stable nuclides, with the exception of technetium (Z = 43) and promethium (Z = 61).

EXAMPLE 21.3

Predicting the Relative Stabilities of Nuclides

One of the nuclides in each of the following pairs is radioactive; the other is stable. Which is radioactive and which is stable? Explain.

a. $^{208}_{84}$Po, $^{209}_{83}$Bi   b. $^{39}_{19}$K, $^{40}_{19}$K   c. $^{71}_{31}$Ga, $^{76}_{31}$Ga

SOLUTION

a. Polonium has an atomic number greater than 83, so $^{208}_{84}$Po is radioactive. Bismuth-209 has 126 neutrons (an even number of neutrons), and $^{209}_{83}$Bi is **the only stable isotope of bismuth.**

b. Of these two isotopes, $^{39}_{19}$K has an even number of neutrons (20), so $^{39}_{19}$K is **expected to be stable.** The isotope $^{40}_{19}$K has an odd number of protons (19) and an odd number of neutrons (21). Because stable odd–odd nuclei are rare, **you might expect $^{40}_{19}$K to be radioactive.**

c. Of the two isotopes, $^{76}_{31}$Ga lies farther from the center of the band of stability, so it is more likely to be radioactive. For this reason, **you expect $^{76}_{31}$Ga to be radioactive. $^{71}_{31}$Ga has an even number of neutrons and its mass number is close to the atomic mass of gallium, so it would be expected to be stable.**

See Problems 21.25 and 21.26.

Types of Radioactive Decay

There are six common types of radioactive decay; the first five are listed in Table 21.2.

1. **Alpha emission:** *emission of a $^4_2$He nucleus, or alpha particle, from an unstable nucleus.* An example is the radioactive decay of radium-226.

$$^{226}_{88}\text{Ra} \longrightarrow {}^{222}_{86}\text{Rn} + {}^4_2\text{He}$$

2. **Beta emission:** *emission of a high-speed electron from an unstable nucleus.* Beta emission is equivalent to the conversion of a neutron to a proton.

$$^1_0\text{n} \longrightarrow {}^1_1\text{p} + {}^{\ 0}_{-1}\text{e}$$

An example of beta emission is the radioactive decay of carbon-14.

$$^{14}_6\text{C} \longrightarrow {}^{14}_7\text{N} + {}^{\ 0}_{-1}\text{e}$$

► Positrons are annihilated as soon as they encounter electrons. When a positron and an electron collide, both particles vanish with the emission of two gamma photons that carry away the energy.

$$^{\ 0}_{1}\text{e} + {}^{\ 0}_{-1}\text{e} \longrightarrow 2^0_0\gamma$$

3. **Positron emission:** *emission of a positron from an unstable nucleus.* A positron, denoted in nuclear equations as $^0_1$e, is a particle identical to an electron in mass but having a positive instead of a negative charge. Positron emission is equivalent to the conversion of a proton to a neutron. ◄

$$^1_1\text{p} \longrightarrow {}^1_0\text{n} + {}^0_1\text{e}$$

The radioactive decay of technetium-95 is an example of positron emission.

$$^{95}_{43}\text{Tc} \longrightarrow {}^{95}_{42}\text{Mo} + {}^0_1\text{e}$$

Table 21.2
Types of Radioactive Decay

Type of Decay	Radiation	Equivalent Process	Resulting Nuclear Change		Usual Nuclear Condition
			Atomic Number	**Mass Number**	
Alpha emission (α)	$^4_2$He	—	−2	−4	$Z > 83$
Beta emission (β)	$^{\ 0}_{-1}$e	$^1_0\text{n} \longrightarrow {}^1_1\text{p} + {}^{\ 0}_{-1}\text{e}$	+1	0	N/Z too large
Positron emission (β^+)	$^0_1$e	$^1_1\text{p} \longrightarrow {}^1_0\text{n} + {}^0_1\text{e}$	−1	0	N/Z too small
Electron capture (EC)	x rays	$^1_1\text{p} + {}^{\ 0}_{-1}\text{e} \longrightarrow {}^1_0\text{n}$	−1	0	N/Z too small
Gamma emission (γ)	$^0_0\gamma$	—	0	0	Excited

4. **Electron capture:** *the decay of an unstable nucleus by capturing, or picking up, an electron from an inner orbital of an atom.* In effect, a proton is changed to a neutron, as in positron emission.

$$_1^1p + _{-1}^0e \longrightarrow _0^1n$$

▶ Most of the argon in the atmosphere is believed to have resulted from the radioactive decay of $_{19}^{40}K$.

An example is given by potassium-40, which has a natural abundance of 0.012%. ◀ Potassium-40 can decay by electron capture, as well as by beta and positron emissions. The equation for electron capture is

$$_{19}^{40}K + _{-1}^0e \longrightarrow _{18}^{40}Ar$$

When another orbital electron fills the vacancy in the inner-shell orbital created by electron capture, an x-ray photon is emitted.

5. **Gamma emission:** *emission from an excited nucleus of a gamma photon.* In many cases, radioactive decay results in a product nucleus that is in an excited state. The excited state is unstable and goes to a lower-energy state with the emission of electromagnetic radiation.

Often gamma emission occurs very quickly after radioactive decay. In some cases, however, an excited state has significant lifetime before it emits a gamma photon. A **metastable nucleus** is *a nucleus in an excited state with a lifetime of at least one nanosecond.* An example is metastable technetium-99, denoted $_{43}^{99m}Tc$, which is used in medical diagnosis (Section 21.5).

$$_{43}^{99m}Tc \longrightarrow _{43}^{99}Tc + _0^0\gamma$$

6. **Spontaneous fission:** *the spontaneous decay of an unstable nucleus in which a heavy nucleus of mass number greater than 89 splits into lighter nuclei, and energy is released.* For example, a uranium-236 atom can spontaneously undergo the following nuclear reaction:

$$_{92}^{236}U \longrightarrow _{39}^{96}Y + _{53}^{136}I + 4_0^1n$$

Nuclides outside the band of stability (Figure 21.3) are generally radioactive. Nuclides to the left of the band of stability have a neutron-to-proton ratio (N/Z) larger than that needed for stability. These nuclides tend to decay by beta emission. Beta emission reduces the neutron-to-proton ratio, because in this process a neutron is changed to a proton. In contrast, nuclides to the right of the band of stability have a neutron-to-proton ratio smaller than that needed for stability. These nuclides tend to decay by either positron emission or electron capture. Both processes convert a proton to a neutron, increasing the neutron-to-proton ratio.

Consider a series of isotopes of a given element, such as carbon. Carbon-12 and carbon-13 are stable isotopes, whereas the other isotopes of carbon are radioactive. The isotopes of mass number smaller than 12 decay by positron emission. The isotopes of carbon with mass nunber greater than 13 decay by beta emission.

You can predict the expected type of radioactive decay of an isotope by noting whether the mass number is less than or greater than the mass number of stable isotopes. Generally, the mass numbers of stable isotopes will be close to the numerical value of the atomic weight of the element. However, in the very heavy elements, especially those with Z greater than 83, radioactive decay is often by alpha emission. $_{92}^{238}U$, $_{88}^{226}Ra$, and $_{90}^{232}Th$ are examples of alpha emitters.

EXAMPLE 21.4

Predicting the Type of Radioactive Decay

Predict the expected type of radioactive decay for each of the following radioactive nuclides: a. $_{20}^{47}$Ca; b. $_{13}^{25}$Al.

SOLUTION

a. The atomic weight of calcium is 40.1 amu, so you expect calcium-40 to be a stable isotope. Calcium-47 has a mass number greater than that of the stable isotope, so you expect it to decay by **beta emission.**

b. The atomic weight of aluminum is 27.0 amu, so you expect aluminum-27 to be a stable isotope. The mass number of aluminum-25 is less than 27, so you expect it to decay by either **positron emission or electron capture.**

See Problems 21.27 and 21.28.

Radioactive Decay Series

All nuclides with atomic number greater than $Z = 83$ are radioactive. Many of these nuclides decay by alpha emission. By emitting an alpha particle, the nucleus reduces its atomic number, becoming more stable. However, if the nucleus has a very large atomic number, the product nucleus is also radioactive. Natural radioactive elements, such as uranium-238, give a **radioactive decay series,** *a sequence in which one radioactive nucleus decays to a second, which then decays to a third, and so forth.* Eventually, a stable nucleus, which is an isotope of lead, is reached.

Three radioactive decay series are found naturally. One of these series begins with uranium-238. Figure 21.4 shows the sequence of nuclear decay processes.

CONCEPT CHECK 21.1

You have two samples of water, each made up of different isotopes of hydrogen: one contains $_{1}^{1}$H$_2$O and the other, $_{1}^{3}$H$_2$O.

a. Would you expect these two water samples to be chemically similar?

b. Would you expect these two water samples to be physically the same?

c. Which one of these water samples would you expect to be radioactive?

21.2 Nuclear Bombardment Reactions

In 1919, Ernest Rutherford discovered that it is possible to change the nucleus of one element into the nucleus of another element. **Transmutation** is *the change of one element to another by bombarding the nucleus of the element with nuclear particles or nuclei.*

Figure 21.4
Uranium-238 radioactive decay series. Each nuclide occupies a position on the graph determined by its atomic number and mass number. Alpha decay is shown by a red diagonal line. Beta decay is shown by a short blue horizontal line.

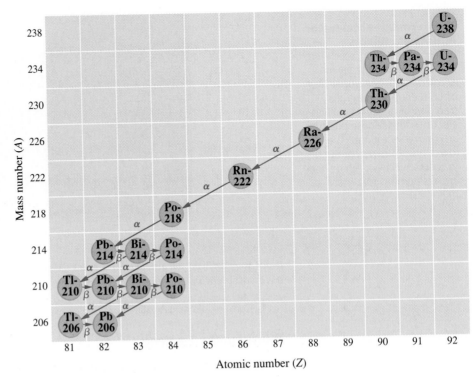

Transmutation

Rutherford used a radioactive element as a source of alpha particles and allowed these particles to collide with nitrogen nuclei. He discovered that protons are ejected in the process. The equation for the nuclear reaction is

$$^{14}_{7}N + ^{4}_{2}He \longrightarrow ^{17}_{8}O + ^{1}_{1}H$$

When beryllium is bombarded with alpha particles, a penetrating radiation is given off that is not deflected by electric or magnetic fields. The British physicist James Chadwick (1891–1974) suggested in 1932 that the radiation from beryllium consists of neutral particles, each with a mass approximately that of a proton. The particles are called neutrons. The reaction that resulted in the discovery of the neutron is

$$^{9}_{4}Be + ^{4}_{2}He \longrightarrow ^{12}_{6}C + ^{1}_{0}n$$

In 1933, a nuclear bombardment reaction was used to produce the first artificial radioactive isotope. Irène and Frédéric Joliot-Curie found that aluminum bombarded with alpha particles produces phosphorus-30.

$$^{27}_{13}Al + ^{4}_{2}He \longrightarrow ^{30}_{15}P + ^{1}_{0}n$$

Phosphorus-30 was the first radioactive nucleus produced in the laboratory. Since then over a thousand radioactive isotopes have been made. ◄

▶ Some of the uses of radioactive isotopes are discussed in Section 21.5.

Elements of large atomic number merely scatter, or deflect, alpha particles from natural sources, rather than giving a transmutation reaction. These elements have nuclei of large positive charge, and the alpha particle must be traveling very fast in

Figure 21.5
The Fermilab accelerator in Batavia, Illinois. The tunnel of the main accelerator is shown here. Protons are accelerated in the upper ring of conventional magnets (red and blue) to 400 billion electron volts. These protons are then injected into the lower ring of superconducting magnets (yellow), where they are accelerated to almost a trillion electron volts.

order to penetrate the nucleus and react. Alpha particles from natural sources do not have sufficient kinetic energy. To shoot charged particles into heavy nuclei, it is necessary to accelerate the charged particles to very high speeds (Figure 21.5).

Transuranium Elements

The **transuranium elements** are *elements with atomic numbers greater than that of uranium (Z = 92), the naturally occurring element of greatest Z.* In 1940, E. M. McMillan and P. H. Abelson, at the University of California at Berkeley, discovered the first transuranium element. They produced an isotope of element 93, which they named neptunium, by bombarding uranium-238 with neutrons. This gave uranium-239, by the capture of a neutron, and this nucleus decayed in a few days by beta emission to neptunium-239.

$$^{238}_{92}\text{U} + ^{1}_{0}\text{n} \longrightarrow ^{239}_{92}\text{U}$$
$$^{239}_{92}\text{U} \longrightarrow ^{239}_{93}\text{Np} + ^{0}_{-1}\text{e}$$

The next transuranium element to be discovered was plutonium (Z = 94). Deuterons, the positively charged nuclei of hydrogen-2, were accelerated by a cyclotron and directed at a uranium target to give neptunium-238, which decayed to plutonium-238.

$$^{238}_{92}\text{U} + ^{2}_{1}\text{H} \longrightarrow ^{238}_{93}\text{Np} + 2^{1}_{0}\text{n}$$
$$^{238}_{93}\text{Np} \longrightarrow ^{238}_{94}\text{Pu} + ^{0}_{-1}\text{e}$$

Another isotope of plutonium, plutonium-239, is now produced in large quantity in nuclear reactors. Plutonium-239 is used for nuclear weapons.

The transuranium elements have a number of commercial uses. Plutonium-238 emits only alpha radiation, which is easily stopped by shielding. The isotope has been used as a power source for space satellites, navigation buoys, and heart pacemakers. Americium-241 is both an alpha-ray and a gamma-ray emitter. The gamma rays are used in devices that measure the thickness of materials such as metal sheets. Americium-241 is also used in home smoke detectors, in which the alpha radiation ionizes the air in a chamber within the detector and renders it electrically conducting. Smoke reduces the conductivity of the air, and this reduced conductivity is detected by an alarm circuit.

Figure 21.6
The High Flux Isotope Reactor at Oak Ridge National Laboratory.   The reactor in which transuranium isotopes are produced lies beneath a protecting pool of water. Visible light is emitted by high-energy particles moving through the water.

Figure 21.6 shows the High Flux Isotope Reactor at Oak Ridge National Laboratory, where technicians produce transuranium elements by bombarding plutonium-239 with neutrons.

EXAMPLE 21.5

Determining the Product Nucleus in a Nuclear Bombardment Reaction

Plutonium-239 was bombarded by alpha particles. Each $^{239}_{94}\text{Pu}$ nucleus was struck by one alpha particle and emitted one neutron. What was the product nucleus?

SOLUTION

This is similar to Example 21.2. You can write the nuclear equation as follows:

$$^{239}_{94}\text{Pu} + {}^{4}_{2}\text{He} \longrightarrow {}^{A}_{Z}\text{X} + {}^{1}_{0}\text{n}$$

To balance this equation, write

$$239 + 4 = A + 1 \quad \text{(from superscripts)}$$
$$94 + 2 = Z + 0 \quad \text{(from subscripts)}$$

Hence,

$$A = 239 + 4 - 1 = 242$$
$$Z = 94 + 2 = 96$$

The product is $^{242}_{96}\text{Cm}$.

See Problems 21.31, 21.32, 21.33, and 21.34.

21.3   Radiations and Matter: Detection and Biological Effects

Radiations from nuclear processes affect matter in part by dissipating energy in it. An alpha, beta, or gamma particle traveling through matter dissipates energy by ionizing atoms or molecules, producing positive ions and electrons. In some cases, these radiations may also excite electrons in matter. When these electrons undergo transitions back to their ground states, light is emitted. The ions, free electrons, and light produced in matter can be used to detect nuclear radiations. Two types of devices—*ionization counters* and *scintillation counters*—are used to count particles emitted from radioactive nuclei and other nuclear processes. Ionization counters detect the production of ions in matter. Scintillation counters detect the production of scintillations, or flashes of light.

   A radiation counter can be used to measure the rate of nuclear disintegrations in a radioactive material. The **activity of a radioactive source** is *the number of nuclear disintegrations per unit time occurring in a radioactive material.* A **curie (Ci)** is *a unit of activity equal to 3.700×10^{10} disintegrations per second.* ◀ For example, a sample of technetium having an activity of 1.0×10^{-2} Ci is decaying at the rate of $(1.0 \times 10^{-2}) \times (3.7 \times 10^{10}) = 3.7 \times 10^{8}$ nuclei per second.

▶ The curie was originally defined as the number of disintegrations per second from 1.0 g of radium-226.

Biological Effects and Radiation Dosage

Although the quantity of energy dissipated in a biological organism from a radiation dosage might be small, the effects can be quite adverse because important chemical bonds may be broken. DNA in the chromosomes of the cell is especially affected, which interferes with cell division. Cells that divide the fastest, such as those in the blood-forming tissue in bone marrow, are most affected by nuclear radiations.

To monitor the effect of nuclear radiations on biological tissue, it is necessary to have a measure of radiation dosage. The **rad** (from *r*adiation *a*bsorbed *d*ose) is *the dosage of radiation that deposits 1×10^{-2} J of energy per kilogram of tissue.* However, the biological effect of radiation depends not only on the energy deposited in the tissue but also on the type of radiation. For example, neutrons are more destructive than gamma rays of the same radiation dosage measured in rads. A **rem** is *a unit of radiation dosage used to relate various kinds of radiation in terms of biological destruction. It equals the rad times a factor for the type of radiation, called the relative biological effectiveness (RBE).*

$$\text{rems} = \text{rads} \times \text{RBE}$$

▶ Sources of alpha radiation outside the body are relatively harmless, because the radiation is absorbed by the skin. Internal sources, however, are very destructive.

Beta and gamma radiations have an RBE of about 1, whereas neutron radiation has an RBE of about 5 and alpha radiation an RBE of about 10. ◀

The effects of radiation on a person depend not only on the dosage but also on the length of time in which the dose was received. A series of small doses has less overall effect than these dosages given all at once. A single dose of about 500 rems is fatal to most people, and survival from a much smaller dose can be uncertain or leave the person chronically ill. Detectable effects are seen with doses as low as 30 rems. Continuous exposure to such low levels of radiation may result in cancer or leukemia. At even lower levels, the answer to whether the radiation dose is safe depends on the possible genetic effects of the radiation. Because radiation can cause chromosome damage, heritable defects are possible.

CONCEPT CHECK 21.2

If you are internally exposed to 10 rads of α, β, and γ radiation, which form of radiation will cause the greatest biological damage?

21.4 Rate of Radioactive Decay

Although technetium-99 is radioactive and decays by emitting beta particles, it is impossible to say when a particular nucleus will disintegrate. A sample of technetium-99 continues to give off beta rays for millions of years. Thus, a particular nucleus might disintegrate the next instant or several million years later. The rate of this radioactive decay cannot be changed by varying the temperature, the pressure, or the chemical environment of the technetium nucleus.

Rate of Radioactive Decay and Half-Life

The rate of radioactive decay—that is, the number of nuclei disintegrating per unit time—is found to be proportional to the number of radioactive nuclei in the sample.

You can express this rate mathematically as

$$\text{Rate} = kN_t$$

Here N_t is the number of radioactive nuclei at time t, and k is the **radioactive decay constant**, or *rate constant for radioactive decay*. This rate constant is a characteristic of the radioactive nuclide, each nuclide having a different value. ◄

You can obtain the decay constant for a radioactive nucleus by counting the nuclear disintegrations. The original definition of the curie (now 3.7×10^{10} disintegrations per second) was the activity or decay rate of 1.0 g of radium-226. You can use this with the rate equation to obtain the decay constant of radium-226.

▶ The rate equation for radioactive decay has the same form as the rate law for a first-order chemical reaction. Indeed, radioactive decay is a first-order rate process, and the mathematical relationships used in Chapter 14 for first-order reactions apply here also.

$$1.0 \text{ g Ra-226} \times \frac{1 \text{ mol Ra-226}}{226 \text{ g Ra-226}} \times \frac{6.02 \times 10^{23} \text{ Ra-226 nuclei}}{1 \text{ mol Ra-226}} = 2.7 \times 10^{21} \text{ Ra-226 nuclei}$$

This equals the value of N_t. When you solve Rate $= kN_t$ for k, you get

$$k = \frac{\text{rate}}{N_t}$$

Substituting into this gives

▶ Recall from Chapter 7 that /s is equivalent to the unit s^{-1}.

$$k = \frac{3.7 \times 10^{10} \text{ nuclei/s}}{2.7 \times 10^{21} \text{ nuclei}} = 1.4 \times 10^{-11}/\text{s} \quad ◄$$

EXAMPLE 21.6

Calculating the Decay Constant from the Activity

A 1.0-mg sample of technetium-99 has an activity of 1.7×10^{-5} Ci (Ci = curies), decaying by beta emission. What is the decay constant for $^{99}_{43}\text{Tc}$?

◆ PROBLEM STRATEGY

Solve the equation Rate $= kN_t$ for k; obtain Rate and N_t from values in the problem statement. The rate of decay (Rate) of Tc-99 equals the activity in curies expressed as nuclei decaying per second, noting that 1 Ci $= 3.7 \times 10^{10}$ nuclei/s. Obtain the number of Tc-99 nuclei, N_t.

SOLUTION

The rate of decay in this sample is

$$\text{Rate} = 1.7 \times 10^{-5} \text{ Ci} \times \frac{3.7 \times 10^{10} \text{ nuclei/s}}{1.0 \text{ Ci}} = 6.3 \times 10^5 \text{ nuclei/s}$$

The number of nuclei in this sample of 1.0×10^{-3} g $^{99}_{43}\text{Tc}$ is

$$1.0 \times 10^{-3} \text{ g Tc-99} \times \frac{1 \text{ mol Tc-99}}{99 \text{ g Tc-99}} \times \frac{6.02 \times 10^{23} \text{ Tc-99 nuclei}}{1 \text{ mol Tc-99}}$$

$$= 6.1 \times 10^{18} \text{ Tc-99 nuclei}$$

The decay constant is

$$k = \frac{\text{rate}}{N_t} = \frac{6.3 \times 10^5 \text{ nuclei/s}}{6.1 \times 10^{18} \text{ nuclei}} = \mathbf{1.0 \times 10^{-13}/s}$$

See Problems 21.35 and 21.36.

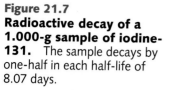

Figure 21.7
Radioactive decay of a 1.000-g sample of iodine-131. The sample decays by one-half in each half-life of 8.07 days.

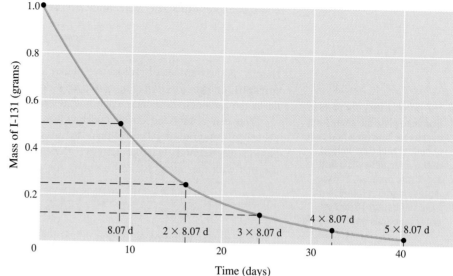

▶ The half-life of a reaction was discussed in Section 14.4.

▶ Iodine is attracted to the thyroid gland, which incorporates the element into the growth hormone thyroxine. Radiation from iodine-131 kills cancer cells in the thyroid gland.

We define the **half-life** of a radioactive nucleus as *the time it takes for one-half of the nuclei in a sample to decay.* ◀ The half-life is independent of the amount of sample. For example, you find that 1.000 g of iodine-131, an isotope used in treating thyroid cancer, decays to 0.500 g in 8.07 days. ◀ Thus, its half-life is 8.07 days. Figure 21.7 shows this decay pattern.

Although you might be able to obtain the half-life of a radioactive nucleus by direct observation in some cases, this is impossible for many nuclei because they decay too quickly or too slowly. Uranium-238, for example, has a half-life of 4.51 billion years. The usual method of determining the half-life is by measuring decay rates.

You relate the half-life for radioactive decay, $t_{1/2}$, to the decay constant k by the equation

$$t_{1/2} = \frac{0.693}{k}$$

EXAMPLE 21.7

Calculating the Half-Life from the Decay Constant

The decay constant for the beta decay of $^{99}_{43}$Tc was obtained in Example 21.6. We found that k equals 1.0×10^{-13}/s. What is the half-life of this isotope in years?

SOLUTION

Substitute the value of k into the preceding equation.

$$t_{1/2} = \frac{0.693}{k} = \frac{0.693}{1.0 \times 10^{-13}/s} = 6.9 \times 10^{12} \text{ s}$$

Then you convert this half-life in seconds to years.

$$6.9 \times 10^{12} \text{ s} \times \frac{1 \text{ min}}{60 \text{ s}} \times \frac{1 \text{ h}}{60 \text{ min}} \times \frac{1 \text{ d}}{24 \text{ h}} \times \frac{1 \text{ y}}{365 \text{ d}} = \mathbf{2.2 \times 10^5 \text{ y}}$$

See Problems 21.37 and 21.38.

Tables of radioactive nuclei often list the half-life. When you want the decay constant or the activity of a sample, you can calculate them from the half-life.

EXAMPLE 21.8

Calculating the Decay Constant and Activity from the Half-Life

Tritium, $_1^3H$, is a radioactive nucleus of hydrogen. It is used in luminous watch dials. Tritium decays by beta emission with a half-life of 12.3 years. What is the decay constant (in /s)? What is the activity (in Ci) of a sample containing 2.5 μg of tritium? The atomic mass of $_1^3H$ is 3.02 amu.

SOLUTION

The conversion of the half-life to seconds gives

$$12.3 \text{ y} \times \frac{365 \text{ d}}{1 \text{ y}} \times \frac{24 \text{ h}}{1 \text{ d}} \times \frac{60 \text{ min}}{1 \text{ h}} \times \frac{60 \text{ s}}{1 \text{ min}} = 3.88 \times 10^8 \text{ s}$$

Because $t_{1/2} = 0.693/k$, you solve this for k and substitute the half-life in seconds.

$$k = \frac{0.693}{t_{1/2}} = \frac{0.693}{3.88 \times 10^8 \text{ s}} = \mathbf{1.79 \times 10^{-9}/s}$$

Before substituting into the rate equation, you need to know the number of tritium nuclei in a sample containing 2.5×10^{-6} g of tritium. You get

$$2.5 \times 10^{-6} \text{ g H-3} \times \frac{1 \text{ mol H-3}}{3.02 \text{ g H-3}} \times \frac{6.02 \times 10^{23} \text{ H-3 nuclei}}{1 \text{ mol H-3}} = 5.0 \times 10^{17} \text{ H-3 nuclei}$$

Now you substitute into the rate equation.

$$\text{Rate} = kN_t = 1.79 \times 10^{-9}/s \times 5.0 \times 10^{17} \text{ nuclei} = 9.0 \times 10^8 \text{ nuclei/s}$$

You obtain the activity of the sample by converting the rate in disintegrations of nuclei per second to curies.

$$\text{Activity} = 9.0 \times 10^8 \text{ nuclei/s} \times \frac{1.0 \text{ Ci}}{3.7 \times 10^{10} \text{ nuclei/s}} = \mathbf{0.024 \text{ Ci}}$$

See Problems 21.39 and 21.40.

Once you know the decay constant for a radioactive isotope, you can calculate the fraction of the radioactive nuclei that remains after a given period of time by the following equation. ◀

▶ A similar equation is given in Chapter 14 for the reactant concentration at time t, $[A]_t$:

$$\ln \frac{[A]_t}{[A]_0} = -kt$$

where $[A]_0$ is the concentration of A at $t = 0$.

$$\ln \frac{N_t}{N_0} = -kt$$

Here N_0 is the number of nuclei in the original sample ($t = 0$). After a period of time t, the number of nuclei decreases by decay to the number N_t. The fraction of nuclei remaining after time t is N_t/N_0. The next example illustrates the use of this equation.

EXAMPLE 21.9

Determining the Fraction of Nuclei Remaining after a Specified Time

Phosphorus-32 is a radioactive isotope with a half-life of 14.3 d (d = days). A biochemist has a vial containing a compound of phosphorus-32. If the compound is used in an experiment 5.5 d after the compound was prepared, what fraction of the radioactive isotope originally present remains? Suppose the sample in the vial originally contained 0.28 g of phosphorus-32. How many grams remain after 5.5 d?

SOLUTION

If N_0 is the original number of P-32 nuclei in the vial and N_t is the number after 5.5 d, the fraction remaining is N_t/N_0. You can obtain this fraction from the equation

$$\ln \frac{N_t}{N_0} = -kt$$

You substitute $k = 0.693/t_{1/2}$.

$$\ln \frac{N_t}{N_0} = \frac{-0.693t}{t_{1/2}}$$

Because $t = 5.5$ d and $t_{1/2} = 14.3$ d, you obtain

$$\ln \frac{N_t}{N_0} = \frac{-0.693 \times 5.5 \text{ d}}{14.3 \text{ d}} = -0.267$$

$$\text{Fraction nuclei remaining} = \frac{N_t}{N_0} = e^{-0.267} = \mathbf{0.77}$$

Thus, 77% of the nuclei remain. The mass of $_{15}^{32}\text{P}$ in the vial after 5.5 d is

$$0.28 \text{ g} \times 0.766 = \mathbf{0.21 \text{ g}}$$

See Problems 21.43 and 21.44.

Radioactive Dating

Fixing the dates of relics and stone implements or pieces of charcoal from ancient campsites is an application based on radioactive decay rates. Because the rate of radioactive decay of a nuclide is constant, this rate can serve as a clock for dating. Dating wood and similar carbon-containing objects that are several thousand to fifty thousand years old can be done with carbon-14, which has a half-life of 5730 y.

Carbon-14 is present in the atmosphere as a result of collisions with neutrons that arise from cosmic radiations from space. The collision of a neutron with a nitrogen-14 nucleus can produce a carbon-14 nucleus.

$$_{7}^{14}\text{N} + _{0}^{1}\text{n} \longrightarrow _{6}^{14}\text{C} + _{1}^{1}\text{H}$$

Carbon dioxide containing carbon-14 mixes with the lower atmosphere. Because of the constant production of $_{6}^{14}\text{C}$ and its radioactive decay, a small, constant fractional abundance of carbon-14 is maintained in the atmosphere.

Living plants, which continuously use atmospheric carbon dioxide, also maintain a constant abundance of carbon-14. Similarly, living animals, by feeding on plants, have a constant fractional abundance of carbon-14. But once an organism dies, it is no longer in chemical equilibrium with atmospheric CO_2. The ratio of carbon-14 to carbon-12 begins to decrease by radioactive decay of carbon-14. In this way, this ratio of carbon isotopes becomes a clock measuring the time since the death of the organism.

You can deduce the age of any dead organic object by measuring the level of beta emissions that arise from the radioactive decay of carbon-14. ◄

▶ Analyses of tree rings have shown that this assumption is not quite valid. Before 1000 B.C., the levels of carbon-14 were somewhat higher than they are today. Moreover, recent human activities (burning of fossil fuels and atmospheric nuclear testing) have changed the fraction of carbon-14 in atmospheric CO_2.

$$^{14}_{6}\text{C} \longrightarrow {}^{14}_{7}\text{N} + {}^{0}_{-1}\text{e}$$

EXAMPLE 21.10

Applying the Carbon-14 Dating Method

A piece of charcoal from a tree killed by the volcanic eruption that formed the crater in Crater Lake (in Oregon) gave 7.0 disintegrations of carbon-14 nuclei per minute per gram of total carbon. Present-day carbon (in living matter) gives 15.3 disintegrations per minute per gram of total carbon. Determine the date of the volcanic eruption. Recall that the half-life of carbon-14 is 5730 y.

 PROBLEM STRATEGY

You substitute the ratio of rates of disintegration (N_0/N_t) and $k = 0.693/t_{1/2}$ into the equation for the number of nuclei in a sample after time t.

$$\ln \frac{N_t}{N_0} = -kt$$

SOLUTION

Substituting the ratio of N_0/N_t and $t_{1/2} = 5730$ y and solving for t gives

$$t = \frac{t_{1/2}}{0.693} \ln \frac{N_0}{N_t} = \frac{5730 \text{ y}}{0.693} \ln 2.2 = 6.5 \times 10^3 \text{ y}$$

Thus, the date of the eruption was about **4500 B.C.**, 6500 years ago.

See Problems 21.47 and 21.48.

CONCEPT CHECK 21.3

Why do you think that carbon-14 dating is limited to less than 50,000 years?

21.5 Applications of Radioactive Isotopes

Chemical Analysis

A **radioactive tracer** is *a very small amount of radioactive isotope added to a chemical, biological, or physical system to study the system.* The advantage of a radioactive tracer is that it behaves chemically just as a nonradioactive isotope does, but it can be detected in exceedingly small amounts by measuring the radiations emitted.

A series of experiments using tracers was carried out in the late 1950s by Melvin Calvin at the University of California, Berkeley, to discover the mechanism of photosynthesis in plants. ◄ The overall process of photosynthesis involves the reaction of CO_2 and H_2O to give glucose, $C_6H_{12}O_6$, and O_2. Energy for photosynthesis comes from the sun.

► Melvin Calvin received the Nobel Prize in chemistry in 1961 for his work on photosynthesis.

$$6CO_2(g) + 6H_2O(l) \xrightarrow{\text{sunlight}} C_6H_{12}O_6(aq) + 6O_2(g)$$

This equation represents only the net result of photosynthesis. As Calvin was able to show, the actual process consists of many separate steps. In several experiments, algae (single-celled plants) were exposed to carbon dioxide containing much more radioactive carbon-14 than occurs naturally. Then the algae were extracted with a solution of alcohol and water. The various compounds in this solution were separated by chromatography and identified. ◄ Compounds that contained radioactive carbon were produced in the different steps of photosynthesis. Eventually, Calvin was able to use tracers to show the main steps in photosynthesis.

► Chromatography was discussed in the essay at the end of Section 1.4.

Another example of the use of radioactive tracers in chemistry is **isotope dilution,** *a technique to determine the quantity of a substance in a mixture or of the total volume of solution by adding a known amount of an isotope to it.* After removal of a portion of the mixture, the fraction by which the isotope has been diluted provides a way of determining the quantity of substance or volume of solution.

A typical chemical example of isotope dilution is the determination of the amount of vitamin B_{12}, a cobalt-containing substance, in a sample of food. Although part of the vitamin in food can be obtained in pure form, not all of the vitamin can be separated. Therefore, you cannot precisely determine the quantity of vitamin B_{12} in a sample of food by separating the pure vitamin and weighing it. But you can determine the amount of vitamin B_{12} by isotope dilution. Suppose you add 2.0×10^{-7} g of vitamin B_{12} containing radioactive cobalt-60 to 125 g of food and mix well. You then separate 5.4×10^{-7} g of pure vitamin B_{12} from the food and find that the activity in curies of this quantity of the vitamin contains 5.6% of the activity added from the radioactive cobalt. This indicates that you have recovered 5.6% of the total amount of vitamin B_{12} in the sample. Therefore, the mass of vitamin B_{12} in the food, including the amount added (2.0×10^{-7} g), is

$$5.4 \times 10^{-7} \text{ g} \times \frac{100}{5.6} = 9.6 \times 10^{-6} \text{ g}$$

Subtracting the amount added in the analysis gives

$$9.6 \times 10^{-6} \text{ g} - 2.0 \times 10^{-7} \text{ g} = 9.4 \times 10^{-6} \text{ g}$$

Neutron activation analysis is *an analysis of elements in a sample based on the conversion of stable isotopes to radioactive isotopes by bombarding a sample with neutrons.* Human hair contains trace amounts of many elements. By determining the exact amounts and the position of the elements in the hair shaft, you can identify whom the hair comes from (assuming you have a sample known to be that person's hair for comparison). Consider the analysis of human hair for arsenic, for example. When the natural isotope $^{75}_{33}As$ is bombarded with neutrons, a metastable nucleus $^{76m}_{33}As$ is obtained. A metastable nucleus is in an excited state. It decays, or undergoes a transition, to a lower state by emitting gamma rays. The frequencies, or energies, of the gamma rays emitted are characteristic of the element and serve to identify it. Also, the intensities of the gamma rays emitted are proportional to the amount of the element present. The method is very sensitive; it can identify as little as 10^{-9} g of arsenic. ◄

► Neutron activation analysis has been used to authenticate oil paintings by giving an exact analysis of pigments used. Pigment compositions have changed, so it is possible to detect fraudulent paintings done with more modern pigments. The analysis can be done without affecting the painting.

Medical Therapy and Diagnosis

The use of radioactive isotopes has had a profound effect on the practice of medicine. Radioisotopes were first used in medicine in the treatment of cancer. This treatment is based on the fact that rapidly dividing cells, such as those in cancer, are more adversely affected by radiation from radioactive substances than are cells that divide more slowly.

Cancer therapy is only one of the ways in which radioactive isotopes are used in medicine. The greatest advances in the use of radioactive isotopes have been in the diagnosis of disease. Radioactive isotopes are used for diagnosis in two ways. They are used to develop images of internal body organs so that their functioning can be examined. And they are used as tracers in the analysis of minute amounts of substances, such as a growth hormone in blood, to deduce possible disease conditions.

Technetium-99m is the radioactive isotope used most often to develop images of internal body organs. It has a half-life of 6.02 h, decaying by gamma emission to technetium-99 in its nuclear ground state. The image is prepared by scanning part of the body for gamma rays with a scintillation detector. (Figure 21.1).

Thallium-201 is a radioisotope used to determine whether a person has heart disease (caused by narrowing of the arteries to the heart). This isotope decays by electron capture and emits x rays and gamma rays, which can be used to obtain images similar to those obtained from technetium-99m (Figure 21.8). Thallium-201 injected into the blood binds particularly strongly to heart muscle. Diagnosis of heart disease depends on the fact that only tissue that receives sufficient blood flow binds thallium-201. When someone exercises strenuously, some part of the person's heart tissue may not receive sufficient blood because of narrowed arteries. These areas do not bind thallium-201 and show up on an image as dark spots.

More than a hundred different radioactive isotopes have been used in medicine. Besides technetium-99m and thallium-201, other examples include iodine-131, used to measure thyroid activity; phosphorus-32, used to locate tumors; and iron-59, used to measure the rate of formation of red blood cells.

Figure 21.8
Using thallium-201 to diagnose heart disease.
Left: A patient is undergoing a heart scan using a portable thallium-201 scintillation counter (on the pivoted arm). Gamma-ray scintillations are counted and collected by a computer (in the foreground) and then presented on the screen as an image. *Right:* A series of cross-sectional images of a patient's heart after exercise (labeled "stress") and then some time afterward (labeled "rest"). By comparing the stress and rest images, a physician can see if there is impaired blood flow to an area of the heart (the area is dark in the stress image but light in the rest image) or if the heart muscle has been damaged through a heart attack (the area is dark in both stress and rest images).

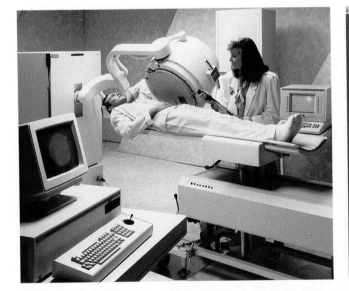

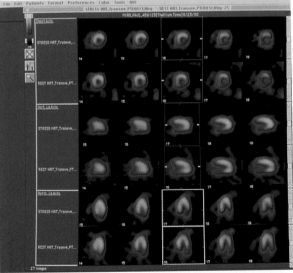

Positron Emission Tomography (PET)

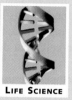

LIFE SCIENCE

Positron emission tomography (PET) is a technique for following biochemical processes within the organs (brain, heart, and so forth) of the human body. A PET scan produces an image of a two-dimensional slice through a body organ of a patient. The image shows the distribution of some positron-emitting isotope present in a compound that was administered earlier by injection. By comparing the PET scan of the patient with that of a healthy subject, a physician can diagnose the presence or absence of disease (Figure 21.9). The PET scan of the brain of an Alzheimer's patient differs markedly from that of a healthy subject.

Some isotopes used in PET scans are carbon-11, nitrogen-13, oxygen-15, and fluorine-18. All have short half-lives, so the radiation dosage to the patient is minimal. However, because of the short half-life, a chemist must prepare the diagnostic compound containing the radioactive nucleus shortly before the physician administers it. The preparation of this compound requires a cyclotron, whose cost (several million dollars) is a major deterrent to the general use of PET scans.

A PET scan instrument detects gamma radiation. When a nucleus emits a positron within the body, the positron travels only a few millimeters before it reacts with an electron. This reaction is an example of the annihilation of matter (an electron) by antimatter (a positron). Both the electron and the positron disappear and produce two gamma photons. The gamma photons easily pass through human tissue, so they can be recorded by scintillation detectors placed around the body.

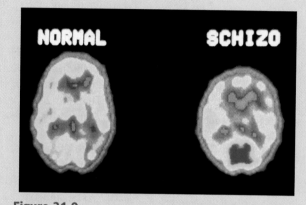

Figure 21.9
PET scans of normal and diseased patients.
The scans are of the brains of a normal patient and a schizophrenic patient. The scans show differences in brain activities measured by glucose usages—yellow-to-red colors show high values; green-to-blue are low values.

Energy of Nuclear Reactions

Nuclear reactions, like chemical reactions, involve changes of energy. However, the changes of energy in nuclear reactions are enormous by comparison with those in chemical reactions. The energy released by certain nuclear reactions is used in nuclear power reactors and to provide the energy for nuclear weapons. In the case of a power plant, the energy from the reactions is released in controlled, small amounts by the fission of uranium-235. With a nuclear weapon, the objective is to release the energy as rapidly as possible and produce a nuclear explosion. Although the source of energy for nuclear power plants and weapons can be the same, a typical nuclear power plant does not contain enough fissionable material in high enough concentration to produce a nuclear explosion.

21.6 Mass–Energy Calculations

When nuclei decay, they form products of lower energy. The change of energy is related to changes of mass, according to the mass–energy equivalence relation derived

by Albert Einstein in 1905. Using this relation, you can obtain the energies of nuclear reactions from mass changes.

Mass–Energy Equivalence

According to Einstein, energy and mass are equivalent and are related by the equation

$$E = mc^2$$

Here c is the speed of light, 3.00×10^8 m/s.

If a system loses energy, it must also lose mass. For example, when carbon burns in oxygen, it releases heat energy.

$$C(graphite) + O_2(g) \longrightarrow CO_2(g); \Delta H = -393.5 \text{ kJ}$$

► The change in mass gives the change in internal energy, ΔU. The enthalpy change, ΔH, equals $\Delta U + P\Delta V$. For the reaction given in the text, $P\Delta V$ is essentially zero.

In principle, you could obtain ΔH for the reaction by measuring the change in mass and relating this by Einstein's equation to the change in energy. ◄ However, weight measurements are of no practical value in determining heats of reaction—the changes in mass are simply too small to detect.

Calculation of the mass change for a typical chemical reaction, the burning of carbon in oxygen, will show just how small this quantity is. When the energy changes by an amount ΔE, the mass changes by an amount Δm. You can write Einstein's equation in the form

$$\Delta E = (\Delta m)c^2$$

The change in energy when 1 mol of carbon reacts is -3.935×10^5 J, or -3.935×10^5 kg · m^2/s^2. Hence,

$$\Delta m = \frac{\Delta E}{c^2} = \frac{-3.935 \times 10^5 \text{ kg} \cdot \text{m}^2/\text{s}^2}{(3.00 \times 10^8 \text{ m/s})^2} = -4.37 \times 10^{-12} \text{ kg}$$

For comparison, a good analytical balance can detect a mass as small as 1×10^{-7} kg, but this is ten thousand times greater than the mass change caused by the release of heat during the combustion of 1 mol of carbon.

By contrast, the mass changes in nuclear reactions are approximately a million times larger per mole of reactant than those in chemical reactions. Consider the alpha decay of uranium-238 to thorium-234. The nuclear equation is

$$\underset{238.05078}{^{238}_{92}\text{U}} \longrightarrow \underset{234.04359}{^{234}_{90}\text{Th}} + \underset{4.00260 \text{ amu}}{^{4}_{2}\text{He}}$$

► From Einstein's equation, one can show that 1 amu is equivalent to 931 MeV. This number can be used to obtain energies of nuclear reactions. For the reaction given in Example 21.11, the mass change is -0.01960 amu. Multiplying this by 931 MeV gives the energy change, -18.2 MeV, which agrees with the answer given to three significant figures.

The change in mass for this nuclear reaction, starting with molar amounts, is

$$\Delta m = (234.04359 + 4.00260 - 238.05078) \text{ g} = -0.00459 \text{ g}$$

The minus sign indicates a loss of mass. This loss of mass is clearly large enough to detect.

From a table of atomic masses, such as Table 21.3, you can use Einstein's equation to calculate the energy change for a nuclear reaction. This is illustrated in the next example. An **electron volt (eV)** is *the quantity of energy that would have to be imparted to an electron (whose charge is 1.602×10^{-19}) to accelerate it by one volt potential difference.* Therefore, 1 MeV equals 1.602×10^{-13} J. ◄

Table 21.3
Masses of Some Elements and Other Particles

Symbol	Z	A	Mass (amu)	Symbol	Z	A	Mass (amu)
e^-	−1	0	0.000549	Co	27	59	58.93320
n	0	1	1.008665	Ni	28	58	57.93535
p	1	1	1.00728		28	60	59.93079
H	1	1	1.00783	Pb	82	206	205.97444
	1	2	2.01400		82	207	206.97587
	1	3	3.01605		82	208	207.97663
He	2	3	3.01603	Po	84	208	207.98122
	2	4	4.00260		84	210	209.98285
Li	3	6	6.01512	Rn	86	222	222.01757
	3	7	7.01600	Ra	88	226	226.02540
Be	4	9	9.01218	Th	90	230	230.03313
B	5	10	10.01294		90	234	234.04359
	5	11	11.00931	Pa	91	234	234.04330
C	6	12	12.00000	U	92	233	233.03963
	6	13	13.00336		92	234	234.04095
O	8	16	15.99492		92	235	235.04392
Cr	24	52	51.94051		92	238	238.05078
Fe	26	56	55.93494	Pu	94	239	239.05216

EXAMPLE 21.11

Calculating the Energy Change for a Nuclear Reaction

a. Calculate the energy change in joules (four significant figures) for the following nuclear reaction per mole of $_1^2H$:

$$_1^2H + _2^3He \longrightarrow _2^4He + _1^1H$$

b. What is the energy change in MeV for one $_1^2H$ nucleus?

SOLUTION

a. You write the atomic masses below each nuclide symbol and then calculate Δm. Once you have Δm, you can obtain ΔE.

$$_1^2H + _2^3He \longrightarrow _2^4He + _1^1H$$
$$2.01400 \quad 3.01603 \quad\quad 4.00260 \quad 1.00783 \text{ amu}$$

Hence,

$$\Delta m = (4.00260 + 1.00783 - 2.01400 - 3.01603) \text{ amu}$$
$$= -0.01960 \text{ amu}$$

The mass change for molar amounts in this nuclear reaction is -0.01960 g, or -1.960×10^{-5} kg. The energy change is

$$\Delta E = (\Delta m)c^2 = (-1.960 \times 10^{-5} \text{ kg})(2.998 \times 10^8 \text{ m/s})^2$$
$$= -1.762 \times 10^{12} \text{ kg} \cdot \text{m}^2/\text{s}^2 \quad \text{or} \quad \mathbf{-1.762 \times 10^{12} \text{ J}}$$

b. The mass change for the reaction of one $_1^2\text{H}$ atom is -0.01960 amu. First change this to grams. Recall that 1 amu equals 1/12 the mass of a $_6^{12}\text{C}$ atom, whose mass is 12 g/6.022×10^{23}. Thus, 1 amu = 1 g/6.022×10^{23}. Hence, the mass change in grams is

$$\Delta m = -0.01960 \text{ amu} \times \frac{1 \text{ g}}{1 \text{ amu} \times 6.022 \times 10^{23}}$$
$$= -3.255 \times 10^{-26} \text{ g} \quad (\text{or } -3.255 \times 10^{-29} \text{ kg})$$

Then,

$$\Delta E = (\Delta m)c^2 = (-3.255 \times 10^{-29} \text{ kg})(2.998 \times 10^8 \text{ m/s})^2$$
$$= -2.926 \times 10^{-12} \text{ J}$$

Now convert this to MeV:

$$\Delta E = -2.926 \times 10^{-12} \text{ J} \times \frac{1 \text{ MeV}}{1.602 \times 10^{-13} \text{ J}} = \mathbf{-18.26 \text{ MeV}}$$

See Problems 21.49 and 21.50.

Nuclear Binding Energy

The equivalence of mass and energy explains the otherwise puzzling fact that the mass of an atom is always less than the sum of the masses of its constituent particles. For example, the helium-4 atom consists of two protons, two neutrons, and two electrons, giving the following sum:

Mass of 2 electrons = 2 × 0.000549 amu =	0.00110 amu	
Mass of 2 protons = 2 × 1.00728 amu =	2.01456 amu	
Mass of 2 neutrons = 2 × 1.00867 amu =	2.01734 amu	
Total mass of particles =	4.03300 amu	

The mass of the helium-4 atom is 4.00260 amu (see Table 21.3), so the mass difference is

$$\Delta m = (4.00260 - 4.03300) \text{ amu} = -0.03040 \text{ amu}$$

This mass difference is explained as follows. When the nucleons come together to form a nucleus, energy is released. (The nucleus has lower energy and is therefore more stable than the separate nucleons.) According to Einstein's equation, there must be an equivalent decrease in mass.

The **binding energy** of a nucleus is *the energy needed to break a nucleus into its individual protons and neutrons.* Thus, the binding energy of the helium-4 nucleus is the energy change for the reaction

$$_2^4\text{He} \longrightarrow 2_1^1\text{p} + 2_0^1\text{n}$$

The **mass defect** of a nucleus is *the total nucleon mass minus the atomic mass.* In the case of helium-4, the mass defect is 4.03300 amu − 4.00260 amu = 0.03040 amu.

Figure 21.10
Plot of binding energy per nucleon versus mass number. The binding energy of each nuclide is divided by the number of nucleons (total of protons and neutrons), then plotted against the mass number of the nuclide. Note that nuclides near mass number 50 have the largest binding energies per nucleon. Thus, heavy nuclei are expected to undergo fission to approach this mass number, whereas light nuclei are expected to undergo fusion.

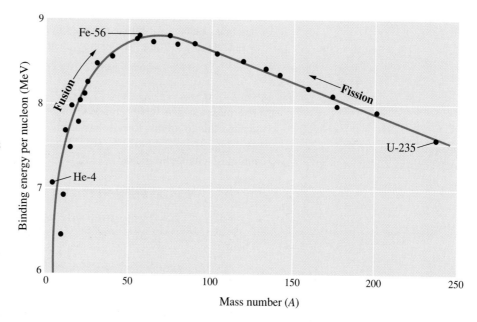

(This is the positive value of the mass difference we calculated earlier.) Both the binding energy and the corresponding mass defect are reflections of the stability of the nucleus.

To compare the stabilities of various nuclei, it is useful to compare binding energies per nucleon. Figure 21.10 shows values of this quantity (in MeV per nucleon) plotted against the mass number for various nuclides. Most of the points lie near the smooth curve drawn on the graph.

Note that nuclides near mass number 50 have the largest binding energies per nucleon. This means that a group of nucleons would tend to form those nuclides, because they would form nuclei of the lowest energy. For this reason, heavy nuclei might be expected to split to give lighter nuclei, and light nuclei might be expected to react, or combine, to form heavier nuclei.

Nuclear fission is *a nuclear reaction in which a heavy nucleus splits into lighter nuclei and energy is released.* For example, californium-252 decays both by alpha emission (97%) and by spontaneous fission (3%). There are many possible ways in which the nucleus can split. One way is represented by the following equation:

$$^{252}_{98}\text{Cf} \longrightarrow {}^{142}_{56}\text{Ba} + {}^{106}_{42}\text{Mo} + 4{}^{1}_{0}\text{n}$$

In some cases, a nucleus can be induced to undergo fission by bombarding it with neutrons. An example is the nuclear fission of uranium-235. When a neutron strikes the $^{235}_{92}\text{U}$ nucleus, the nucleus splits into roughly equal parts, giving off several neutrons. Three possible splittings are shown in the following equations:

$$^{1}_{0}\text{n} + {}^{235}_{92}\text{U} \nearrow {}^{142}_{54}\text{Xe} + {}^{90}_{38}\text{Sr} + 4{}^{1}_{0}\text{n}$$
$$\longrightarrow {}^{139}_{56}\text{Ba} + {}^{94}_{36}\text{Kr} + 3{}^{1}_{0}\text{n}$$
$$\searrow {}^{144}_{55}\text{Cs} + {}^{90}_{37}\text{Rb} + 2{}^{1}_{0}\text{n}$$

Nuclear fission is discussed further in Section 21.7.

Nuclear fusion is *a nuclear reaction in which light nuclei combine to give a stabler, heavier nucleus plus possibly several neutrons, and energy is released.* An example of nuclear fusion is

$$\ce{^2_1H + ^3_1H \longrightarrow ^4_2He + ^1_0n}$$

Even though a nuclear reaction is energetically favorable, the reaction may be imperceptibly slow unless the proper conditions are present (see Section 21.7).

21.7 Nuclear Fission and Nuclear Fusion

Nuclear Fission; Nuclear Reactors

Nuclear fission was discovered as a result of experiments to produce transuranium elements. Soon after the neutron was discovered in 1932, experimenters realized that this particle, being electrically neutral, should easily penetrate heavy nuclei. They began using neutrons in bombardment reactions, hoping to produce isotopes that would decay to new elements. In 1938, Otto Hahn, Lise Meitner, and Fritz Strassmann in Berlin identified barium in uranium samples that had been bombarded with neutrons. Soon afterward, the presence of barium was explained as a result of fission of the uranium-235 nucleus. When this nucleus is struck by a neutron, it splits into two nuclei. Fissions of uranium nuclei produce approximately 30 different elements of intermediate mass, including barium.

When the uranium-235 nucleus splits, approximately two or three neutrons are released. If the neutrons from each nuclear fission are absorbed by other uranium-235 nuclei, these nuclei split and release even more neutrons. In this way, a chain reaction can occur. A nuclear **chain reaction** is *a self-sustaining series of nuclear fissions caused by the absorption of neutrons released from previous nuclear fissions.* The number of nuclei that split quickly multiply as a result of the absorption of neutrons released from previous nuclear fissions. Figure 21.11 shows how such a chain reaction occurs. The chain reaction of nuclear fissions is the basis of nuclear power and nuclear weapons.

To sustain a chain reaction in a sample of fissionable material, a nucleus that splits must give an average of one neutron that results in the fission of another nucleus, and so on. If the sample is too small, many of the neutrons leave the sample before they have a chance to be absorbed. There is thus a **critical mass** for a particular fissionable material, which is *the smallest mass of fissionable material in which a chain reaction can be sustained.* If the mass is much larger than this (a *supercritical* mass), the number of nuclei that split multiply rapidly. An atomic bomb is detonated with a small amount of chemical explosive that pushes together two or more masses of fissionable material to get a supercritical mass (Figure 21.12). A rapid chain reaction results in the splitting of most of the fissionable nuclei—and in the explosive release of an enormous amount of energy.

A **nuclear fission reactor** is *a device that permits a controlled chain reaction of nuclear fissions.* In power plants, a nuclear reactor is used to produce heat, which in turn is used to produce steam to drive an electric generator. A nuclear reactor consists of fuel rods alternating with control rods contained within a vessel. The **fuel rods** are *the cylinders that contain fissionable material.* In the light-water (ordinary water) reactors commonly used in the United States, these fuel rods contain uranium dioxide pellets in a zirconium alloy tube. Natural uranium contains only 0.72% uranium-235, which is the isotope that undergoes fission. The uranium used for fuel in these

Figure 21.11
Representation of a chain reaction of nuclear fissions. Each nuclear fission produces two or more neutrons, which can in turn cause more nuclear fissions. At each stage, a greater number of neutrons are produced, so that the number of nuclear fissions multiplies quickly. Such a chain reaction is the basis of nuclear power and nuclear weapons. The original nucleus splits into pieces of varying mass number.

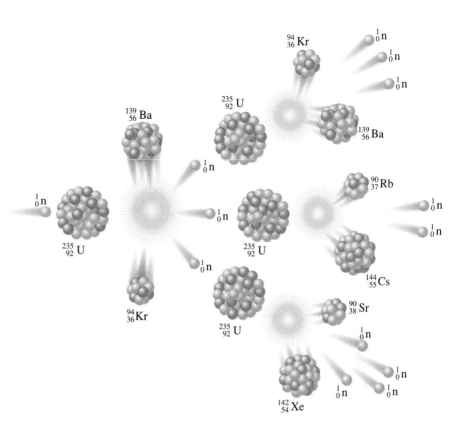

reactors is "enriched" so that it contains about 3% of the uranium-235 isotope. **Control rods** are *cylinders composed of substances that absorb neutrons, such as boron and cadmium, and can therefore slow the chain reaction.* By varying the depth of the control rods within the fuel-rod assembly (reactor core), one can increase or decrease the absorption of neutrons. If necessary, these rods can be dropped all the way into the fuel-rod assembly to stop the chain reaction.

Nuclear Fusion

As we noted in Section 21.6, energy can be obtained by combining light nuclei into a heavier nucleus by nuclear fusion. Such fusion reactions have been observed in the laboratory by means of bombardment using particle accelerators. Deuterons (^2_1H nuclei), for example, can be accelerated toward targets containing deuterium (^2_1H atoms) or tritium (^3_1H atoms). The reactions are

$$^2_1H + {}^2_1H \longrightarrow {}^3_2He + {}^1_0n$$
$$^2_1H + {}^3_1H \longrightarrow {}^4_2He + {}^1_0n$$

To get the nuclei to react, the bombarding nucleus must have enough kinetic energy to overcome the repulsion of electric charges of the nuclei.

Energy cannot be obtained in a practical way using particle accelerators. Another way to give nuclei sufficient kinetic energy to react is by heating the nuclear materials to a sufficiently high temperature. The reaction of deuterium, ^2_1H, and tritium, ^3_1H, turns out to require the lowest temperature of any fusion reaction. For this reason it is likely to be the first fusion reaction developed as an energy source. For practical purposes, the temperature will have to be about 100 million °C. At this temperature,

all atoms have been stripped of their electrons so that a plasma results. A **plasma** is *an electrically neutral gas of ions and electrons.* At 100 million °C, the plasma is essentially separate nuclei and electrons.

It is now believed that the energy of stars, including our sun, where extremely high temperatures exist, derives from nuclear fusion. The hydrogen bomb also employs nuclear fusion for its destructive power. High temperature is first attained by a fission bomb. This then ignites fusion reactions in surrounding material of deuterium and tritium.

The main problem in developing controlled nuclear fusion is how to heat a plasma to high temperatures and maintain those temperatures. When a plasma touches any material whatsoever, heat is quickly conducted away, and the plasma temperature quickly falls.

Figure 21.12

An atomic bomb. In the gun-type bomb, one piece of uranium-235 of subcritical mass is hurled into another piece by a chemical explosive. The two pieces together give a supercritical mass, and a nuclear explosion results. Bombs using plutonium-239 require an implosion technique, in which wedges of plutonium arranged on a spherical surface are pushed into the center of the sphere by a chemical explosive, where a supercritical mass of plutonium results in a nuclear explosion.

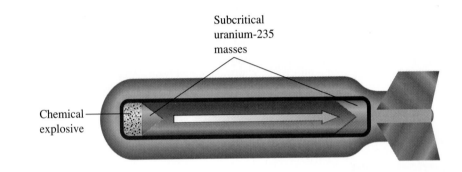

Subcritical uranium-235 masses

Chemical explosive

A CHEMIST LOOKS AT

The Chernobyl Nuclear Accident

ENVIRONMENT

The nuclear accident that occurred at the Chernobyl reactor, north of Kiev, Ukraine, on April 26, 1986, renewed fears in some about the safety of nuclear reactors. It is important to understand the nature of the accident at Chernobyl. A nuclear reactor using normal fuel elements cannot become an atomic bomb. However, without proper design and safeguards, it is possible for a malfunction of a reactor to disperse dangerous radioactivity over a populated area. This is in fact what occurred at Chernobyl.

The Chernobyl reactor had serious design problems. Unlike U.S. reactors, the Chernobyl reactor had no confinement shell surrounding it to contain radioactive spills. Another problem was the design of the cooling

system. The fuel core had a graphite moderator with water cooling in which some of the liquid water changed to steam. Liquid water is a good absorber of neutrons, whereas steam is not. This meant that as the reactor ran hotter, producing a greater percentage of steam, more neutrons became available for nuclear fission. Unless these extra neutrons were absorbed, the reactor would become even hotter and then would run out of control. Normally, as the reactor became hotter, control rods were automatically pushed in to absorb the extra neutrons. In a U.S. light-water reactor, the water coolant is always under pressure to maintain the liquid phase. As long as coolant water is present, the reactor is under control.

On the day of the Chernobyl accident, operators disabled the safety system on the reactor while they

(continued)

proceeded with an experimental test of the reactor. During the test, the reactor cooled excessively and threatened to shut down. If this happened, the reactor could not be restarted for a long period. Therefore, the operators removed most of the control rods (counter to safety instructions). Then the reactor began to overheat. With the safety system disabled, the operators were unable to work fast enough to correct the overheating, and the reactor went out of control.

The fuel rods melted and spilled their hot contents into the superheated water, which flashed into steam. The sudden pressure increase blew the top off the reactor and the roof off the building, spewing radioactive material skyward (see Figure 21.13). Hot steam reacted with the zirconium shell of the fuel rods and with the graphite moderator to produce hydrogen gas, which ignited. The graphite moderator burned for a long period, spreading more radioactivity.

The cost of the Chernobyl accident was enormous. Many people died, and several hundred were hospitalized. Thousands of people had to be evacuated and

Figure 21.13
View of the Chernobyl nuclear plant. This photograph was taken in 1987, one year after the nuclear accident.

resettled. The accident was the direct result of a faulty reactor coupled with a disregard of reactor safety procedures.

A Checklist for Review

Important Terms

radioactive decay (21.1)
nuclear bombardment
  reaction (21.1)
nuclear equation (21.1)
nuclear force (21.1)
band of stability (21.1)
alpha emission (21.1)
beta emission (21.1)
positron emission (21.1)
electron capture (21.1)
gamma emission (21.1)

metastable nucleus (21.1)
spontaneous fission (21.1)
radioactive decay
  series (21.1)
transmutation (21.2)
transuranium
  elements (21.2)
ionization counters (21.3)
scintillation counter (21.3)
activity of a radioactive
  source (21.3)

curie (Ci) (21.3)
rad (21.3)
rem (21.3)
radioactive decay
  constant (21.4)
half-life (21.4)
radioactive tracer (21.5)
isotope dilution (21.5)
neutron activation
  analysis (21.5)
binding energy (21.6)

mass defect (21.6)
nuclear fission (21.6)
nuclear fusion (21.6)
chain reaction (21.7)
critical mass (21.7)
nuclear fission
  reactor (21.7)
fuel rods (21.7)
control rods (21.7)
plasma (21.7)

Key Equations

$$\text{Rate} = kN_t$$

$$t_{1/2} = \frac{0.693}{k}$$

$$\ln \frac{N_t}{N_0} = -kt$$

$$\Delta E = (\Delta m)c^2$$

Summary of Facts and Concepts

Nuclear reactions are of two types, *radioactive decay* and *nuclear bombardment*. Such reactions are represented by nuclear equations, each nucleus being denoted by a nuclide symbol. The equations must be balanced in charge (subscripts) and in nucleons (superscripts).

When placed on a plot of N versus Z, stable nuclei fall in a *band of stability*. Those radioactive nuclides that fall to the left of the band of stability in this plot usually decay by beta emission. Those radioactive nuclides that fall to the right of the

band of stability usually decay by positron emission or electron capture. However, nuclides with $Z > 83$ often decay by alpha emission. Uranium-238 forms a *radioactive decay series*. In such series, one element decays to another, which decays to another, and so forth, until a stable isotope is reached (lead-206, in the case of the uranium-238 series).

Transmutation of elements has been carried out in the laboratory by bombarding nuclei with various atomic particles. Alpha particles from natural sources can be used as reactants

with light nuclei. For heavier nuclei, positive ions such as alpha particles must first be accelerated. Many of the *transuranium elements* have been obtained by bombardment of elements with accelerated particles. For example, plutonium was first made by bombarding uranium-238 with deuterons ($_1^2H$ nuclei).

Particles of radiation from nuclear processes can be counted by ionization counters or scintillation counters. The activity of a radioactive source, or the number of nuclear disintegrations per unit time, is measured in units of *curies* (3.700×10^{10} disintegrations per second).

Radiation affects biological organisms by breaking chemical bonds. The *rad* is the measure of radiation dosage that deposits 1×10^{-2} J of energy per kilogram of tissue. A *rem* equals the number of rads times a factor to account for the relative biological effectiveness of the radiation.

Radioactive decay is a first-order rate process. The rate is characterized by the *decay constant, k,* or by the *half-life, $t_{1/2}$.* The quantities k and $t_{1/2}$ are related. Knowing one or the other, you can calculate how long it will take for a given radioactive sample to decay by a certain fraction. Methods of *radioactive*

dating depend on determining the fraction of a radioactive isotope that has decayed and, from this, the time that has elapsed.

Radioactive isotopes are used as *radioactive tracers* in chemical analysis and medicine. *Isotope dilution* is one application of radioactive tracers in which the dilution of the tracer can be related to the original quantity of nonradioactive isotope. *Neutron activation analysis* is a method of analysis that depends on the conversion of elements to radioactive isotopes by neutron bombardment.

According to Einstein's *mass–energy equivalence,* mass is related to energy by the equation $E = mc^2$. A nucleus has less mass than the sum of the masses of the separate nucleons. The positive value of this mass difference is called the *mass defect;* it is equivalent to the *binding energy* of the nucleus. Nuclides having mass numbers near 50 have the largest binding energies per nucleon. It follows that heavy nuclei should tend to split, a process called *nuclear fission,* and that light nuclei should tend to combine, a process called *nuclear fusion.* Tremendous amounts of energy are released in both processes. Nuclear fission is used in conventional nuclear power reactors. Nuclear fusion reactors are in the experimental stage.

Operational Skills

1. **Writing a nuclear equation** Given a word description of a radioactive decay process, write the nuclear equation. **(EXAMPLE 21.1)**
2. **Deducing a product or reactant in a nuclear equation** Given all but one of the reactants and products in a nuclear reaction, find that one nuclide. **(EXAMPLES 21.2, 21.5)**
3. **Predicting the relative stabilities of nuclides** Given a number of nuclides, determine which are most likely to be radioactive and which are most likely to be stable. **(EXAMPLE 21.3)**
4. **Predicting the type of radioactive decay** Predict the type of radioactive decay that is most likely for given nuclides. **(EXAMPLE 21.4)**
5. **Calculating the decay constant from the activity** Given the activity (disintegrations per second) of a radioactive isotope, obtain the decay constant. **(EXAMPLE 21.6)**
6. **Relating the decay constant, half-life, and activity** Given the decay constant of a radioactive isotope, obtain the half-life **(EXAMPLE 21.7)**, or vice versa **(EXAMPLE 21.8)**. Given the decay constant and mass of a radioactive isotope, calculate the activity of the sample **(EXAMPLE 21.8)**.
7. **Determining the fraction of nuclei remaining after a specified time** Given the half-life of a radioactive isotope, calculate the fraction remaining after a specified time. **(EXAMPLE 21.9)**
8. **Applying the carbon-14 dating method** Given the disintegrations of carbon-14 nuclei per gram of carbon in a dead organic object, calculate the age of the object—that is, the time since its death. **(EXAMPLE 21.10)**
9. **Calculating the energy change for a nuclear reaction** Given nuclear masses, calculate the energy change for a nuclear reaction **(EXAMPLE 21.11)**. Obtain the answer in joules per mole or MeV per particle.

Review Questions

21.1 What are the two types of nuclear reactions? Give an example of a nuclear equation for each type.

21.2 List characteristics to look for in a nucleus to predict whether it is stable.

21.3 What are the six common types of radioactive decay? What condition usually leads to each type of decay?

21.4 What are the isotopes that begin each of the naturally occurring radioactive decay series?

21.5 Give equations for (a) the first transmutation of an element obtained in the laboratory by nuclear bombardment, and (b) the reaction that produced the first artificial radioactive isotope.

21.6 Define the units *curie, rad,* and *rem.*

21.7 The half-life of cesium-137 is 30.2 y. How long will it take for a sample of cesium-137 to decay to 1/8 of its original mass?

21.8 What is a radioactive tracer? Give an example of the use of such a tracer in chemistry.

21.9 Isotope dilution has been used to obtain the volume of blood supply in a living animal. Explain how this could be done.

21.10 Briefly describe neutron activation analysis.

21.11 The deuteron, $^2_1$H, has a mass that is smaller than the sum of the masses of its constituents, the proton plus the neutron. Explain why this is so.

21.12 Certain stars obtain their energy from nuclear reactions such as

$$^{12}_6\text{C} + {}^{12}_6\text{C} \longrightarrow {}^{23}_{11}\text{Na} + {}^1_1\text{H}$$

Explain in a sentence or two why this reaction might be expected to release energy.

Conceptual Problems

21.13 When considering the lifetime of a radioactive species, a general rule of thumb is that after 10 half-lives have passed, the amount of radioactive material left in the sample is negligible. The disposal of some radioactive materials is based on this rule.
a. What percentage of the original material is left after 10 half-lives?
b. When would it be a bad idea to apply this rule?

21.14 Identify the following reactions as fission, fusion, a transmutation, or radioactive decay.
a. $4^1_1\text{H} \longrightarrow {}^4_2\text{He} + 2^0_1\text{e}$
b. $^{14}_6\text{C} \longrightarrow {}^{14}_7\text{N} + {}^0_{-1}\text{e}$
c. $^1_0\text{n} + {}^{235}_{92}\text{U} \longrightarrow {}^{140}_{56}\text{Ba} + {}^{93}_{36}\text{Kr} + 3^0_1\text{e}$
d. $^{14}_7\text{N} + {}^4_2\text{He} \longrightarrow {}^{17}_8\text{O} + {}^1_1\text{H}$

21.15 Sodium has only one naturally occurring isotope, sodium-23. Using the data presented in Table 21.3, explain how the molecular weight of sodium is 22.98976 amu and not the sum of the masses of the protons, neutrons, and electrons.

21.16 You have a mixture that contains 10 g of Pu-239 with a half-life of 2.4×10^4 years and 10 g of Np-239 with a half-life of 2.4 days. Estimate how much time must elapse before the quantity of radioactive material is reduced by 50%.

21.17 Come up with an explanation as to why α radiation is easily blocked by materials such as a piece of wood, whereas γ radiation easily passes through those same materials.

21.18 You have an acquaintance who tells you that he is going to reduce his radiation exposure to zero. What examples could you present that would illustrate this to be an impossible goal?

Practice Problems

Radioactivity

21.19 Rubidium-87, which forms about 28% of natural rubidium, is radioactive, decaying by the emission of a single beta particle to strontium-87. Write the nuclear equation for this decay of rubidium-87.

21.20 Write the nuclear equation for the decay of phosphorus-32 to sulfur-32 by beta emission. A phosphorus-32 nucleus emits a beta particle and gives a sulfur-32 nucleus.

21.21 Thorium is a naturally occurring radioactive element. Thorium-232 decays by emitting a single alpha particle to produce radium-228. Write the nuclear equation for this decay of thorium-232.

21.22 Radon is a radioactive noble gas formed in soil containing radium. Radium-226 decays by emitting a single alpha particle to produce radon-222. Write the nuclear equation for this decay of radium-226.

21.23 Fluorine-18 is an artificially produced radioactive isotope. It decays by emitting a single positron. Write the nuclear equation for this decay.

21.24 Scandium-41 is an artificially produced radioactive isotope. It decays by emitting a single positron. Write the nuclear equation for this decay.

21.25 From each of the following pairs, choose the nuclide that is radioactive. (One is known to be radioactive, the other stable.) Explain your choice.
a. $^{122}_{51}\text{Sb}$, $^{136}_{54}\text{Xe}$ b. $^{204}_{82}\text{Pb}$, $^{204}_{85}\text{At}$ c. $^{87}_{37}\text{Rb}$, $^{80}_{37}\text{Rb}$

21.26 From each of the following pairs, choose the nuclide that is radioactive. (One is known to be radioactive, the other stable.) Explain your choice.
a. $^{102}_{47}\text{Ag}$, $^{109}_{47}\text{Ag}$ b. $^{25}_{12}\text{Mg}$, $^{24}_{10}\text{Ne}$ c. $^{203}_{81}\text{Tl}$, $^{223}_{90}\text{Th}$

21.27 Predict the type of radioactive decay process that is likely for each of the following nuclides.
a. $^{228}_{92}\text{U}$ b. ^8_5B c. $^{68}_{29}\text{Cu}$

21.28 Predict the type of radioactive decay process that is likely for each of the following nuclides.
a. $^{60}_{30}\text{Zn}$ b. $^{10}_6\text{C}$ c. $^{241}_{93}\text{Np}$

Voltaic Cell Notation

20.29 Write the cell notation for a voltaic cell with the following half-reactions.

$$Ni(s) \longrightarrow Ni^{2+}(aq) + 2e^-$$
$$Pb^{2+}(aq) + 2e^- \longrightarrow Pb(s)$$

20.30 Write the cell notation for a voltaic cell with the following half-reactions.

$$Al(s) \longrightarrow Al^{3+}(aq) + 3e^-$$
$$2H^+(aq) + 2e^- \longrightarrow H_2(g)$$

20.31 Give the notation for a voltaic cell constructed from a hydrogen electrode (cathode) in 1.0 M HCl and a nickel electrode (anode) in 1.0 M NiSO$_4$ solution. The electrodes are connected by a salt bridge.

20.32 A voltaic cell has an iron rod in 0.30 M iron(III) chloride solution for the cathode and a zinc rod in 0.40 M zinc sulfate solution for the anode. The half-cells are connected by a salt bridge. Write the notation for this cell.

20.33 Consider the voltaic cell

$$Cd(s)|Cd^{2+}(aq)\|Ni^{2+}(aq)|Ni(s)$$

Write the half-cell reactions and the overall cell reaction. Make a sketch of this cell and label it. Include labels showing the anode, cathode, and direction of electron flow.

20.34 Consider the voltaic cell

$$Zn(s)|Zn^{2+}(aq)\|Cr^{3+}(aq)|Cr(s)$$

Write the half-cell reactions and the overall cell reaction. Make a sketch of this cell and label it. Include labels showing the anode, cathode, and direction of electron flow.

Electrode Potentials and Cell emf's

20.35 A voltaic cell whose cell reaction is

$$2Fe^{3+}(aq) + Zn(s) \longrightarrow 2Fe^{2+}(aq) + Zn^{2+}(aq)$$

has an emf of 0.72 V. What is the maximum electrical work that can be obtained from this cell per mole of iron(III) ion?

20.36 A particular voltaic cell operates on the reaction

$$Zn(s) + Cl_2(g) \longrightarrow Zn^{2+}(aq) + 2Cl^-(aq)$$

giving an emf of 0.853 V. Calculate the maximum electrical work generated when 20.0 g of zinc metal is consumed.

20.37 Order the following oxidizing agents by increasing strength under standard-state conditions: O$_2(g)$; MnO$_4^-$(aq); NO$_3^-$(aq) (in acidic solution).

20.38 Order the following oxidizing agents by increasing strength under standard-state conditions: Ag$^+$(aq); I$_2$(aq); MnO$_4^-$(aq) (in acidic solution).

20.39 Consider the reducing agents Cu$^+$(aq), Zn(s), and Fe(s). Which is strongest? Which is weakest?

20.40 Consider the reducing agents Sn^{2+}(aq), Cu(s), and I$^-$(aq). Which is strongest? Which is weakest?

20.41 Consider the following reactions. Are they spontaneous in the direction written, under standard conditions at 25°C?
a. $Sn^{4+}(aq) + 2Fe^{2+}(aq) \longrightarrow Sn^{2+}(aq) + 2Fe^{3+}(aq)$
b. $4MnO_4^-(aq) + 12H^+(aq) \longrightarrow$
$$4Mn^{2+}(aq) + 5O_2(g) + 6H_2O(l)$$

20.42 Answer the following questions by referring to standard electrode potentials at 25°C.
a. Will dichromate ion, Cr$_2$O$_7^{2-}$, oxidize iron(II) ion in acidic solution under standard conditions?
b. Will copper metal reduce 1.0 M Ni^{2+}(aq) to metallic nickel?

20.43 What would you expect to happen when chlorine gas, Cl$_2$, at 1 atm pressure is bubbled into a solution containing 1.0 M F$^-$ and 1.0 M Br$^-$ at 25°C? Write a balanced equation for the reaction that occurs.

20.44 Dichromate ion, Cr$_2$O$_7^{2-}$, is added to an acidic solution containing Br$^-$ and Mn^{2+}. Write a balanced equation for any reaction that occurs. Assume standard conditions at 25°C.

20.45 Calculate the standard emf of the following cell at 25°C.

$$Cr(s)|Cr^{3+}(aq)\|Hg_2^{2+}(aq)|Hg(l)$$

20.46 Calculate the standard emf of the following cell at 25°C.

$$Sn(s)|Sn^{2+}(aq)\|Cu^{2+}(aq)|Cu(s)$$

Relationships Among $E°_{cell}$, $\Delta G°$, and K

20.47 Calculate the standard free-energy change at 25°C for the following reaction.

$$3Cu(s) + 2NO_3^-(aq) + 8H^+(aq) \longrightarrow$$
$$3Cu^{2+}(aq) + 2NO(g) + 4H_2O(l)$$

Use standard electrode potentials.

20.48 Calculate the standard free-energy change at 25°C for the following reaction.

$$4Al(s) + 3O_2(g) + 12H^+(aq) \longrightarrow 4Al^{3+}(aq) + 6H_2O(l)$$

Use standard electrode potentials.

21.6   Define the units *curie, rad,* and *rem.*

21.7   The half-life of cesium-137 is 30.2 y. How long will it take for a sample of cesium-137 to decay to 1/8 of its original mass?

21.8   What is a radioactive tracer? Give an example of the use of such a tracer in chemistry.

21.9   Isotope dilution has been used to obtain the volume of blood supply in a living animal. Explain how this could be done.

21.10   Briefly describe neutron activation analysis.

21.11   The deuteron, ^2_1H, has a mass that is smaller than the sum of the masses of its constituents, the proton plus the neutron. Explain why this is so.

21.12   Certain stars obtain their energy from nuclear reactions such as

$$^{12}_6C + ^{12}_6C \longrightarrow ^{23}_{11}Na + ^1_1H$$

Explain in a sentence or two why this reaction might be expected to release energy.

Conceptual Problems

21.13   When considering the lifetime of a radioactive species, a general rule of thumb is that after 10 half-lives have passed, the amount of radioactive material left in the sample is negligible. The disposal of some radioactive materials is based on this rule.
a. What percentage of the original material is left after 10 half-lives?
b. When would it be a bad idea to apply this rule?

21.14   Identify the following reactions as fission, fusion, a transmutation, or radioactive decay.
a. $4^1_1H \longrightarrow ^4_2He + 2^0_1e$
b. $^{14}_6C \longrightarrow ^{14}_7N + ^0_{-1}e$
c. $^1_0n + ^{235}_{92}U \longrightarrow ^{140}_{56}Ba + ^{93}_{36}Kr + 3^0_1e$
d. $^{14}_7N + ^4_2He \longrightarrow ^{17}_8O + ^1_1H$

21.15   Sodium has only one naturally occurring isotope, sodium-23. Using the data presented in Table 21.3, explain how the molecular weight of sodium is 22.98976 amu and not the sum of the masses of the protons, neutrons, and electrons.

21.16   You have a mixture that contains 10 g of Pu-239 with a half-life of 2.4×10^4 years and 10 g of Np-239 with a half-life of 2.4 days. Estimate how much time must elapse before the quantity of radioactive material is reduced by 50%.

21.17   Come up with an explanation as to why α radiation is easily blocked by materials such as a piece of wood, whereas γ radiation easily passes through those same materials.

21.18   You have an acquaintance who tells you that he is going to reduce his radiation exposure to zero. What examples could you present that would illustrate this to be an impossible goal?

Practice Problems

Radioactivity

21.19   Rubidium-87, which forms about 28% of natural rubidium, is radioactive, decaying by the emission of a single beta particle to strontium-87. Write the nuclear equation for this decay of rubidium-87.

21.20   Write the nuclear equation for the decay of phosphorus-32 to sulfur-32 by beta emission. A phosphorus-32 nucleus emits a beta particle and gives a sulfur-32 nucleus.

21.21   Thorium is a naturally occurring radioactive element. Thorium-232 decays by emitting a single alpha particle to produce radium-228. Write the nuclear equation for this decay of thorium-232.

21.22   Radon is a radioactive noble gas formed in soil containing radium. Radium-226 decays by emitting a single alpha particle to produce radon-222. Write the nuclear equation for this decay of radium-226.

21.23   Fluorine-18 is an artificially produced radioactive isotope. It decays by emitting a single positron. Write the nuclear equation for this decay.

21.24   Scandium-41 is an artificially produced radioactive isotope. It decays by emitting a single positron. Write the nuclear equation for this decay.

21.25   From each of the following pairs, choose the nuclide that is radioactive. (One is known to be radioactive, the other stable.) Explain your choice.
a. $^{122}_{51}Sb, ^{136}_{54}Xe$ b. $^{204}_{82}Pb, ^{204}_{85}At$ c. $^{87}_{37}Rb, ^{80}_{37}Rb$

21.26   From each of the following pairs, choose the nuclide that is radioactive. (One is known to be radioactive, the other stable.) Explain your choice.
a. $^{102}_{47}Ag, ^{109}_{47}Ag$ b. $^{25}_{12}Mg, ^{24}_{10}Ne$ c. $^{203}_{81}Tl, ^{223}_{90}Th$

21.27   Predict the type of radioactive decay process that is likely for each of the following nuclides.
a. $^{228}_{92}U$ b. ^8_5B c. $^{68}_{29}Cu$

21.28   Predict the type of radioactive decay process that is likely for each of the following nuclides.
a. $^{60}_{30}Zn$ b. $^{10}_6C$ c. $^{241}_{93}Np$

Nuclear Bombardment Reactions

21.29 A proton is accelerated to 14.3 MeV per particle. What is this energy in kJ/mol?

21.30 An alpha particle is accelerated to 23.1 MeV per particle. What is this energy in kJ/mol?

21.31 Fill in the missing part of the following reaction.

$$^6_3\text{Li} + {}^1_0\text{n} \longrightarrow ? + {}^3_1\text{H}$$

21.32 Fill in the missing part of the following reaction.

$$^{27}_{13}\text{Al} + {}^3_1\text{H} \longrightarrow {}^{27}_{12}\text{Mg} + ?$$

21.33 Curium was first synthesized by bombarding an element with alpha particles. The products were curium-242 and a neutron. What was the target element?

21.34 Californium was first synthesized by bombarding an element with alpha particles. The products were californium-245 and a neutron. What was the target element?

Rate of Radioactive Decay

21.35 Tritium, or hydrogen-3, is prepared by bombarding lithium-6 with neutrons. A 0.250-mg sample of tritium decays at the rate of 8.94×10^{10} disintegrations per second. What is the decay constant (in /s) of tritium, whose atomic mass is 3.02 amu?

21.36 The first isotope of plutonium discovered was plutonium-238. It is used to power batteries for heart pacemakers. A sample of plutonium-238 weighing 2.8×10^{-6} g decays at the rate of 1.8×10^6 disintegrations per second. What is the decay constant of plutonium-238 in reciprocal seconds (/s)?

21.37 Tellurium-123 is a radioactive isotope occurring in natural tellurium. The decay constant is 1.7×10^{-21}/s. What is the half-life in years?

21.38 Neptunium-237 was the first isotope of a transuranium element to be discovered. The decay constant is 1.03×10^{-14}/s. What is the half-life in years?

21.39 Gold-198 has a half-life of 2.69 d. What is the activity (in curies) of a 0.86-mg sample?

21.40 Cesium-134 has a half-life of 2.05 y. What is the activity (in curies) of a 0.25-mg sample?

21.41 A sample of a phosphorus compound contains phosphorus-32. This sample of radioactive isotope is decaying at the rate of 6.0×10^{12} disintegrations per second. How many grams of ^{32}P are in the sample? The half-life of ^{32}P is 14.3 d.

21.42 A sample of sodium thiosulfate, $Na_2S_2O_3$, contains sulfur-35. Determine the mass of ^{35}S in the sample from the decay rate, which was determined to be 7.7×10^{11} disintegrations per second. The half-life of ^{35}S is 88 d.

21.43 A sample of sodium-24 was administered to a patient to test for faulty blood circulation by comparing the radioactivity reaching various parts of the body. What fraction of the sodium-24 nuclei would remain undecayed after 12.0 h? The half-life is 15.0 h. If a sample contains 5.0 μg of ^{24}Na, how many micrograms remain after 12.0 h?

21.44 A solution of sodium iodide containing iodine-131 was given to a patient to test for malfunctioning of the thyroid gland. What fraction of the iodine-131 nuclei would remain undecayed after 7.0 d? If a sample contains 2.5 μg of ^{131}I, how many micrograms remain after 7.0 d? The half-life of I-131 is 8.07 d.

21.45 If 28.0% of a sample of nitrogen-17 decays in 1.97 s, what is the half-life of this isotope (in seconds)?

21.46 If 18.0% of a sample of zinc-65 decays in 69.9 d, what is the half-life of this isotope (in days)?

21.47 Carbon from a cypress beam obtained from the tomb of Sneferu, a king of ancient Egypt, gave 8.1 disintegrations of ^{14}C per minute per gram of carbon. How old is the cypress beam? Carbon from living material gives 15.3 disintegrations of ^{14}C per minute per gram of carbon.

21.48 Carbon from the Dead Sea Scrolls, very old manuscripts found in Israel, gave 12.1 disintegrations of ^{14}C per minute per gram of carbon. How old are the manuscripts? Carbon from living material gives 15.3 disintegrations of ^{14}C per minute per gram of carbon.

Mass–Energy Equivalence

21.49 Calculate the energy change for the following nuclear reaction (in joules per mole of ^2_1H).

$$^2_1\text{H} + {}^3_1\text{H} \longrightarrow {}^4_2\text{He} + {}^1_0\text{n}$$

Give the energy change in MeV per ^2_1H nucleus. See Table 21.3.

21.50 Calculate the change in energy, in joules per mole of ^1_1H, for the following nuclear reaction.

$$^1_1\text{H} + {}^1_1\text{H} \longrightarrow {}^2_1\text{H} + {}^0_1\text{e}$$

Give the energy change in MeV per ^1_1H nucleus. See Table 21.3.

21.51 Obtain the mass defect (in amu) and binding energy (in MeV) for the ^6_3Li nucleus. What is the binding energy (in MeV) per nucleon? See Table 21.3.

21.52 Obtain the mass defect (in amu) and binding energy (in MeV) for the $^{58}_{28}\text{Ni}$ nucleus. What is the binding energy (in MeV) per nucleon? See Table 21.3.

General Problems

21.53   Sodium-23 is the only stable isotope of sodium. Predict how sodium-20 will decay and how sodium-26 will decay.

21.54   Aluminum-27 is the only stable isotope of aluminum. Predict how aluminum-24 will decay and how aluminum-30 will decay.

21.55   A bismuth-209 nucleus reacts with an alpha particle to produce an astatine nucleus and two neutrons. Write the complete nuclear equation for this reaction.

21.56   A bismuth-209 nucleus reacts with a deuteron to produce a polonium nucleus and a neutron. Write the complete nuclear equation for this reaction.

21.57   Complete the following equation by filling in the blank.

$$^{238}_{92}U + ^{12}_{6}C \longrightarrow \underline{\hspace{1cm}} + 4^{1}_{0}n$$

21.58   Complete the following equation by filling in the blank.

$$^{246}_{96}Cm + ^{12}_{6}C \longrightarrow \underline{\hspace{1cm}} + 4^{1}_{0}n$$

21.59   Tritium, or hydrogen-3, is formed in the upper atmosphere by cosmic rays, similar to the formation of carbon-14. Tritium has been used to determine the age of wines. A certain wine that has been aged in a bottle has a tritium content only 72% of that in a similar wine of the same mass that has just been bottled. How long has the aged wine been in the bottle? The half-life of tritium is 12.3 y.

21.60   The naturally occurring isotope rubidium-87 decays by beta emission to strontium-87. This decay is the basis of a method for determining the ages of rocks. A sample of rock contains 102.1 μg ^{87}Rb and 5.3 μg ^{87}Sr. What is the age of the rock? The half-life of rubidium-87 is 4.8×10^{10} y.

21.61   Calculate the energy released when 3.00 kg of uranium-235 undergoes the following fission process.

$$^{1}_{0}n + ^{235}_{92}U \longrightarrow ^{136}_{53}I + ^{96}_{39}Y + 4^{1}_{0}n$$

The masses of $^{136}_{53}I$ and $^{96}_{39}Y$ nuclei are 135.8401 amu and 95.8629 amu, respectively. Other masses are given in Table 21.3. Compare this energy with the heat released when 3.00 kg C(graphite) burns to $CO_2(g)$.

21.62   Calculate the energy released when 1.00 kg of hydrogen-1 undergoes fusion to helium-4, according to the following reaction.

$$4^{1}_{1}H \longrightarrow ^{4}_{2}He + 2^{0}_{1}e$$

This reaction is one of the principal sources of energy from the sun. Compare the energy released by 1.00 kg of $^{1}_{1}H$ in this reaction to the heat released when 1.00 kg of C(graphite) burns to $CO_2(g)$. See Table 21.3 for data.

21.63   The half-life of calcium-47 is 4.536 days and it decays by the emission of a beta particle.
a. Write a balanced equation for the decay of Ca-47.
b. If 10.0 μg of Ca-47 is needed for an experiment, what mass of $^{47}CaSO_4$ must be ordered if it takes 48 h for it to arrive from the supplier?

21.64   The radioactive isotope phosphorus-32 is often used in biochemical research. Its half-life is 14.28 days and it decays by beta emission.
a. Write a balanced equation for the decomposition of P-32.
b. If the original sample were 275 mg of $K_3{}^{32}PO_4$, what amount remains after 45.0 d? What percent of the sample has undergone decay?

Cumulative-Skills Problems

21.65   A sample of sodium phosphate, Na_3PO_4, weighing 54.5 mg contains radioactive phosphorus-32 (with mass 32.0 amu). If 15.6% of the phosphorus atoms in the compound is phosphorus-32 (the remainder is naturally occurring phosphorus), how many disintegrations of this nucleus occur per second in this sample? Phosphorus-32 has a half-life of 14.3 d.

21.66   A sample of sodium thiosulfate, $Na_2S_2O_3$, weighing 38.1 mg contains radioactive sulfur-35 (with mass 35.0 amu). If 22.3% of the sulfur atoms in the compound is sulfur-35 (the remainder is naturally occurring sulfur), how many disintegrations of this nucleus occur per second in this sample? Sulfur-35 has a half-life of 87.9 d.

21.67   What is the energy (in joules) evolved when 1 mol of helium-4 nuclei is produced from protons and neutrons? How many liters of ethane, $C_2H_6(g)$, at 25°C and 725 mmHg are needed to evolve the same quantity of energy when the ethane is burned in oxygen to $CO_2(g)$ and $H_2O(g)$? See Table 21.3 and Appendix B for data.

21.68   Plutonium-239 has been used as a power source for heart pacemakers. What is the energy obtained from the following decay of 215 mg of plutonium-239?

$$^{239}_{94}Pu \longrightarrow ^{4}_{2}He + ^{235}_{92}U$$

Suppose the electric energy produced from this amount of plutonium-239 is 25.0% of this value. What is the minimum grams of zinc that would be needed for the standard voltaic cell $Zn|Zn^{2+}\|Cu^{2+}|Cu$ to obtain the same electric energy?

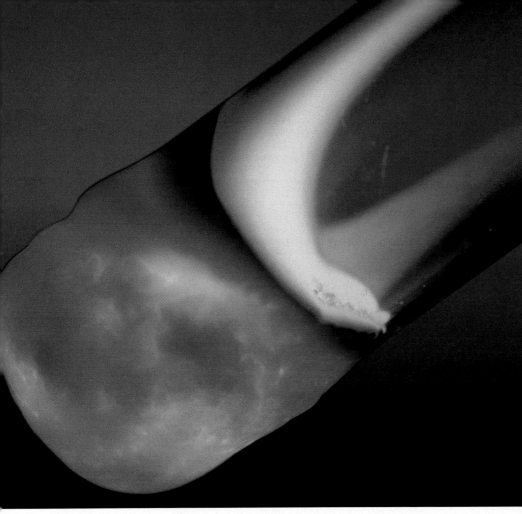

The exothermic reaction of magnesium and silicon dioxide forming magnesium oxide and silicon.

Chemistry of the Main-Group Elements

Figure 22.1 is an abridged periodic table showing the main-group elements but omitting the transition elements. Many of these elements are of interest because they are naturally abundant and, when combined with each other, form many of the necessary ingredients for life and living on our planet. In this chapter you will have the opportunity to explore the properties, chemistry, and commercial applications of several of these elements.

22

Figure 22.1
An abridged periodic table showing the main-group elements. Elements to the left of the heavy staircase line are largely metallic in character; those to the right are largely nonmetallic.

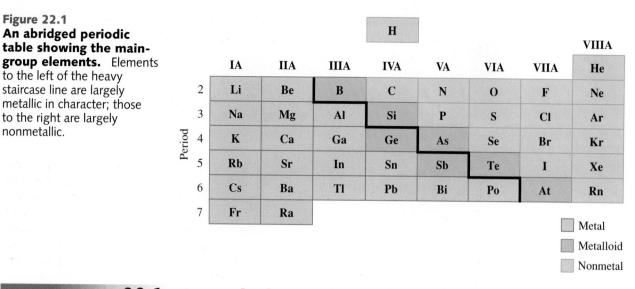

22.1 General Observations About the Main-Group Elements

In this section, we want to consider some periodic trends that you will find useful as you study the chemistry of specific elements. The elements on the left side of the periodic table in Figure 22.1 are largely metallic in character; those on the right side are largely nonmetallic.

The metallic elements generally have low ionization energies and low electronegativities compared with the nonmetallic elements. As a result, the metals tend to lose their valence electrons to form cations (Na^+, Ca^{2+}, Al^{3+}) in compounds or in aqueous solution. Nonmetals, on the other hand, form monatomic anions (O^{2-}, Cl^-) and oxoanions (NO_3^-, SO_4^{2-}).

Also, as noted earlier (Section 8.7), the oxides of the metals are usually basic. The oxides of the most reactive metals react with water to give basic solutions. For example,

$$CaO(s) + H_2O(l) \longrightarrow Ca^{2+}(aq) + 2OH^-(aq)$$

The oxides of the nonmetals are acidic. Sulfur dioxide, an oxide of a nonmetal, dissolves in water to form an acidic solution.

$$SO_2(g) + 2H_2O(l) \rightleftharpoons H_3O^+(aq) + HSO_3^-(aq)$$

Table 22.1 shows the oxidation states displayed by compounds of the main-group elements. The principal conclusions are as follows: the metallic elements generally have oxidation states equal to the group number, representing a loss of the valence electrons in forming compounds. Some of the metallic elements in the fifth and sixth periods also have oxidation states equal to the group number minus two (for example, Pb^{2+}). However, the nonmetals (except for the most electronegative elements fluorine and oxygen) have a variety of oxidation states extending from the group number (the most positive value) to the group number minus eight (the most negative value).

The metallic characteristics of the main-group elements in the periodic table generally decrease in going across a period from left to right (Figure 22.1). The trend of decreasing metallic character can be seen clearly in the third-period elements. Sodium,

magnesium, and aluminum are metals. Silicon is a metalloid, whereas phosphorus, sulfur, and chlorine are nonmetals. The oxides of these elements show the expected trend from basic to acidic.

The metallic characteristics of the main-group elements in the periodic table become more important going down any column (group). This trend is more pronounced in the middle groups of the periodic table (Groups IIIA to VA). For example, in Group IVA, carbon is a nonmetal, silicon and germanium are metalloids, and tin and lead are metals. The metallic elements tend to become more reactive as you progress down a column.

A second-period element is often rather different from the remaining elements in its group. The second-period element generally has a small atom that tends to hold electrons strongly, giving rise to a relatively high electronegativity. Another reason for the difference in behavior between a second-period element and the other elements of the same group has to do with the fact that bonding in the second-period elements involves only s and p orbitals, whereas the other elements may use d orbitals. For example, although nitrogen forms only the trihalides (such as NCl_3), phosphorus has both trihalides (PCl_3) and pentahalides (PCl_5).

Table 22.1
Oxidation States in Compounds of the Main-Group Elements*

| Period | Group | | | | | | |
	IA	IIA	IIIA	IVA	VA	VIA	VIIA
2	Li +1	Be +2	B +3	C +4 +2 −4	N +5 +4 +3 +2 +1 −3	O −1 −2	F −1
3	Na +1	Mg +2	Al +3	Si +4 −4	P +5 +3 −3	S +6 +4 +2 −2	Cl +7 +5 +3 +1 −1
4	K +1	Ca +2	Ga +3	Ge +4 +2	As +5 +3 −3	Se +6 +4 −2	Br +7 +5 +1 −1
5	Rb +1	Sr +2	In +3 +1	Sn +4 +2	Sb +5 +3 −3	Te +6 +4 −2	I +7 +5 +1 −1
6	Cs +1	Ba +2	Tl +3 +1	Pb +4 +2	Bi +5 +3	Po +4 +2	At +5 −1

*The most common oxidation state is shown in color. Some uncommon oxidation states are not shown.

Chemistry of the Main-Group Metals

In the first half of this chapter, we will focus on the chemistry of the Group IA, IIA, and IIIA metals. The sources and preparation of several of these elements were covered in Sections 13.1 and 13.2 of Chapter 13.

22.2 Group IA: The Alkali Metals

IA								VIIIA
H	IIA		IIIA IVA VA VIA VIIA					He
Li	Be			B	C	N	O	F	Ne
Na	Mg	IIIB	IIB	Al	Si	P	S	Cl	Ar
K	Ca	Sc	Zn	Ga	Ge	As	Se	Br	Kr
Rb	Sr	Y	Cd	In	Sn	Sb	Te	I	Xe
Cs	Ba	La	Hg	Tl	Pb	Bi	Po	At	Rn
Fr	Ra	Ac						

The alkali metals are all soft, silvery metals. Figure 22.2 shows sodium metal being cut with a knife. The alkali metals are the most reactive of all metals, reacting readily with air and water.

The Group IA elements occur extensively in silicate minerals, which weather to form soluble compounds of the elements. These soluble compounds eventually find their way to landlocked lakes and oceans, where they concentrate. Enormous underground beds of sodium and potassium compounds formed when lakes and seas became isolated by geologic events.

Lithium

In recent years, the commercial importance of lithium metal has risen markedly. Among the alkali metals, only sodium metal is more important. Lithium metal is obtained by electrolysis of the chloride, which was described in Section 13.2.

The use of lithium metal has greatly expanded in recent years. Lithium is used in the production of lightweight batteries wherever a reliable current is required for a lengthy period (Figure 22.3). A major use is in the production of low-density aluminum alloy for aircraft construction.

Reactions of Lithium Metal   Lithium, like the other alkali metals, reacts readily with water and with moisture in the air to produce lithium hydroxide and hydrogen gas:

$$2Li(s) + 2H_2O(l) \longrightarrow 2LiOH(aq) + H_2(g)$$

Lithium burns in air to produce lithium oxide, Li_2O, a white powder.

$$4Li(s) + O_2(g) \longrightarrow 2Li_2O(s)$$

When heated with nitrogen gas, lithium reacts to form lithium nitride, Li_3N, the only stable alkali-metal nitride.

$$6Li(s) + N_2(g) \longrightarrow 2Li_3N(s)$$

Lithium Compounds   Lithium carbonate, which is obtained from lithium ore, is the primary source of other lithium compounds. Significant quantities of lithium carbonate are also used in the preparation of porcelain enamels, glazes, and special glass. When purified, the carbonate is used as a source of lithium ion for the treatment of manic-depressive disorders.

Lithium hydroxide, LiOH, is a strong base. It is used in the production of lithium soaps, which are used in making lubricating greases from oil.

Lithium hydroxide is also used to remove carbon dioxide from the air in spacecraft and submarines. For people to live in a closed space, it is necessary to remove

Figure 22.2
Cutting of sodium metal.
The metal is easily cut with a sharp knife.

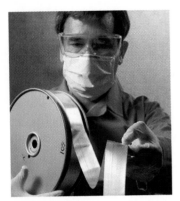

Figure 22.3
A roll of lithium metal for batteries. The metal must be handled in humidity-free rooms to prevent corrosion.

this carbon dioxide. Like the other alkali-metal hydroxides, lithium hydroxide absorbs carbon dioxide from air by forming the carbonate and hydrogen carbonate:

$$2LiOH(s) + CO_2(g) \longrightarrow Li_2CO_3(s) + H_2O(l)$$

$$Li_2CO_3(s) + H_2O(l) + CO_2(g) \longrightarrow 2LiHCO_3(s)$$

Sodium hydroxide could also be used to remove carbon dioxide from air. However, where weight is at a premium, lithium hydroxide is preferred, even though it is more expensive than sodium hydroxide.

Table 22.2 lists some uses of lithium and other alkali metal compounds.

Table 22.2
Uses of Alkali Metal Compounds

Compound	Use
Li_2CO_3	Preparation of porcelain, glazes, special glasses
	Preparation of LiOH
	Treatment of manic-depressive disorders
LiOH	Manufacture of lithium soaps for lubricating greases
	In air-regeneration systems
LiH	Reducing agent in organic syntheses
$LiNH_2$	Preparation of antihistamines and other pharmaceuticals
NaCl	Source of sodium and sodium compounds
	Condiment and food preservative
	Soap manufacture (precipitates soap from reaction mixture)
NaOH	Pulp and paper industry
	Extraction of aluminum oxide from ore
	Manufacture of viscose rayon
	Petroleum refining
	Manufacture of soap
Na_2CO_3	Manufacture of glass
	In detergents and water softeners
Na_2O_2	Textile bleach
$NaNH_2$	Preparation of indigo dye for denim (blue jeans)
KCl	Fertilizer
	Source of other potassium compounds
KOH	Manufacture of soft soap
	Manufacture of other potassium compounds
K_2CO_3	Manufacture of glass
KNO_3	Fertilizers
	Explosives and fireworks

Sodium and Potassium

Sodium compounds are of enormous economic importance. Common table salt, which is sodium chloride, has been an important article of commerce since earliest recorded history. Sodium metal and other sodium compounds are produced from sodium chloride.

The economic importance of potassium stems in large part from its role as a plant nutrient. Enormous quantities of potassium chloride from underground deposits are used as fertilizer to increase the world's food supply. In fact, plants were an early source of potassium compounds. Wood and other plant materials were burned in pots to give ashes (potash) that consist primarily of potassium carbonate.

Sodium metal is more reactive than lithium metal, and potassium metal is still more reactive.

The Group IA metals increase in softness going from the top of the column to the bottom. This increasing softness of the metals is in part a result of increasing atomic size, which results in decreasing strength of bonding by the valence-shell *s* electrons.

Sodium Metallurgy   Sodium metal is obtained by the electrolysis of molten, or fused, sodium chloride. Sodium chloride is mined from huge underground deposits. The other source of sodium chloride is seawater. We discussed electrolysis in some detail in Section 20.9. ◄

▶ Figure 20.16 depicts the commercial Downs cell for the preparation of sodium from molten sodium chloride.

Sodium metal is a strong reducing agent. For example, it is used to obtain metals such as titanium and zirconium. Titanium is a strong metal used in airplane and spacecraft manufacture. It is produced by reduction of titanium tetrachloride, $TiCl_4$.

$$TiCl_4(g) + 4Na(l) \longrightarrow Ti(s) + 4NaCl(s)$$

Sodium is also employed as a reducing agent in the production of a number of organic compounds, including dyes and pharmaceutical drugs.

Reactions of Sodium Metal   Like lithium, sodium metal reacts with water to produce hydrogen. Sodium burns in air, producing some sodium oxide, Na_2O, but mainly sodium peroxide, Na_2O_2, a compound containing the peroxide ion O_2^{2-}.

$$2Na(s) + O_2(g) \longrightarrow Na_2O_2(s)$$

The peroxide ion acts as an oxidizing agent (Figure 22.4).

Figure 22.4
Sodium peroxide as an oxidizing agent. *Left:* A couple of drops of water are added to a mixture of sodium peroxide and sulfur to initiate a reaction. *Right:* During the violent reaction, sulfur is oxidized to sulfur dioxide.

Sodium Compounds Sodium hydroxide, NaOH, is among the top ten industrial chemicals. It is produced by the electrolysis of aqueous sodium chloride. The overall electrolysis, which was described in Section 20.10, can be written as

$$2NaCl(aq) + 2H_2O(l) \xrightarrow{\text{electrolysis}} H_2(g) + Cl_2(g) + 2NaOH(aq)$$

Both chlorine gas and sodium hydroxide are major products of this electrolysis.

Sodium hydroxide is used in large quantities in aluminum production and in the refining of petroleum. It is also used to produce sodium compounds, which occur in a wide variety of commercial products (see Figure 22.5).

For example, soap is made by heating a fat with sodium hydroxide solution. The product is a mixture of sodium salts of fatty acids. Fatty acids consist of long chains of carbon atoms. Compare the structure of the acetate ion with that of the stearate ion.◄

► The action of soap in dispersing oil in water is described in Section 12.9.

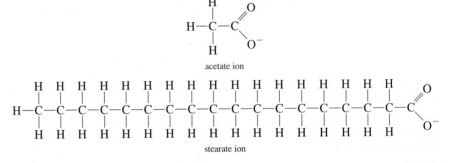

acetate ion

stearate ion

Figure 22.5
Sodium compounds available at the grocery store. *Left to right:* Salt (NaCl), baking soda (NaHCO$_3$), washing soda (Na$_2$CO$_3$ · 10H$_2$O), drain opener (NaOH), and oven cleaner (NaOH).

Sodium hydroxide is commonly known as lye or caustic soda. You can buy it in a grocery store as a drain opener and as an oven cleaner (Figure 22.5). Both usages depend on the reaction of sodium hydroxide with organic materials, such as fats and hair, to produce soluble materials. Because of these reactions, sodium hydroxide and its solutions require careful handling.

Sodium carbonate is another important compound of sodium. The anhydrous compound, Na$_2$CO$_3$, is known commercially as *soda ash*. Large quantities of soda ash are consumed with sand and lime in making glass. ◄ The decahydrate, Na$_2$CO$_3$ · 10H$_2$O, is added to many detergent preparations and is sold commercially as washing soda. In the United States, most sodium carbonate is produced from the mineral trona, whose chemical formula is approximately Na$_2$CO$_3$ · NaHCO$_3$ · 2H$_2$O. There are large deposits of trona in southwestern Wyoming.

► Soda ash decomposes to sodium oxide when heated. Sodium and calcium oxides, which are basic oxides, react with fused silicon dioxide, SiO$_2$, an acidic oxide, to produce silicate glass.

Worldwide, a large fraction of sodium carbonate is still produced by the Solvay process. The **Solvay process** is *an industrial method for obtaining sodium carbonate from sodium chloride and limestone.*

Potassium and Potassium Compounds The principal source of potassium and potassium compounds is potassium chloride, KCl, which is obtained from underground deposits. Small amounts of this salt are used to prepare the metal. Potassium is prepared by the chemical reduction of potassium chloride by sodium at 870°C.

$$Na(l) + KCl(l) \longrightarrow NaCl(l) + K(g)$$

Almost all of the potassium metal produced is used in the preparation of potassium superoxide, KO$_2$, for self-contained breathing apparatus used in situations such

as fire fighting, where toxic fumes may be present. When potassium burns in air, it produces potassium superoxide:

$$K(s) + O_2(g) \longrightarrow KO_2(s)$$

(The superoxide ion is O_2^-, and the oxidation state of oxygen in this ion is $-\frac{1}{2}$.) In a self-contained breathing apparatus, the potassium superoxide is contained in a canister through which one's breath passes. Moisture in the breath attacks the superoxide, releasing oxygen.

$$4KO_2(s) + 2H_2O(l) \longrightarrow 4KOH(s) + 3O_2(g)$$

The potassium hydroxide produced in this reaction removes the carbon dioxide from the exhaled air.

More than 90% of the potassium chloride that is mined is used directly as a plant fertilizer. The rest is used in the preparation of potassium compounds. Potassium hydroxide, used in the manufacture of liquid soaps, is obtained by the electrolysis of aqueous potassium chloride. Potassium nitrate is prepared by the reaction of potassium hydroxide and nitric acid. This compound is used in fertilizers and for explosives and fireworks (Table 22.2).

22.3 Group IIA: The Alkaline Earth Metals

The Group IIA elements are also known as the alkaline earth metals. These metals illustrate the expected periodic trends. If you compare an alkaline earth metal with the alkali metal in the same period, you see that the alkaline earth metal is less reactive and a harder metal. You also see the expected trend within the column of alkaline earth metals; the elements at the bottom of the column are more reactive and are softer metals than those at the top of the column (Figure 22.6).

Like the alkali metals, the Group IIA elements occur in nature as silicate rocks. They also occur as carbonates and sulfates, and many of these are commercial sources of alkaline earth metals and compounds.

Calcium and magnesium are the most common and commercially useful of the alkaline earth elements. Calcium is the fifth and magnesium the eighth most abundant element in the earth's crust (Table 13.1).

Magnesium

The name magnesium comes from the name of the mineral magnesite. The British chemist Humphry Davy discovered the pure element magnesium in 1808. Davy had already discovered the alkali metals potassium and sodium in late 1807. Several months later, he managed to isolate in quick succession the alkaline earth metals barium, strontium, calcium, and finally magnesium. He obtained magnesium by a procedure similar to the one he used for the other alkaline earths. He electrolyzed a moist mixture of magnesium oxide and mercury(II) oxide.

Magnesium Metallurgy Magnesium has become increasingly important as a structural metal. Its great advantages are its low density (1.74 g/cm^3, which compares with 7.87 g/cm^3 for iron and 2.70 g/cm^3 for aluminum, the other important structural metals) and the relative strength of its alloys.

Some important commercial sources of magnesium are the minerals dolomite,

$CaCO_3 \cdot MgCO_3$, and magnesite, $MgCO_3$. However, magnesium ion, Mg^{2+}, is the third most abundant dissolved ion in the oceans, after Cl^- and Na^+. The oceans are an important source of the ion. ◄

Pure magnesium metal is a relatively reactive element. Its alloys, however, contain aluminum and small quantities of other metals to impart both strength and corrosion resistance. Increasing quantities of magnesium alloy are used to make automobile and aircraft parts, as well as consumer materials such as power-tool and lawn-mower housings (Figure 22.7).

Magnesium is also used as a reducing agent in the manufacture of titanium and zirconium from their tetrachlorides.

$$ZrCl_4(g) + 2Mg(l) \longrightarrow Zr(s) + 2MgCl_2(s)$$

Reactions of Magnesium Metal   Once magnesium metal has been ignited, it burns vigorously in air to form the oxide.

$$2Mg(s) + O_2(g) \longrightarrow 2MgO(s)$$

Burning magnesium metal gives off an intense white light; this is the white light you see in the burning of some fireworks and flares.

Magnesium also burns in carbon dioxide, producing magnesium oxide and soot, or carbon (Figure 22.8).

► Magnesium metal is isolated from seawater by the Dow process, which is described in Section 13.2.

Figure 22.6
Magnesium and barium metals. Magnesium metal turnings are in the beaker (left). Barium metal (right) is much more reactive than magnesium and must be stored in a bottle of kerosene to exclude moisture and oxygen, with which barium reacts.

Figure 22.7
Magnesium in motorcycles. Magnesium alloys are used in many motorcycle parts. Shown here is a closeup of the engine area of a Harley-Davidson motorcycle. The V engine (upper right), engine block (lower right), and transmission case (lower left) are made of an aluminum–magnesium alloy.

Table 22.3
Uses of Alkaline Earth Compounds

Compound	Use
MgO	Refractory bricks (for furnaces)
	Animal feeds
$Mg(OH)_2$	Source of magnesium for the metal and compounds
	Milk of magnesia (antacid and laxative)
$MgSO_4 \cdot 7H_2O$	Fertilizer
	Medicinal uses (laxative and analgesic)
	Mordant (used in dyeing fabrics)
CaO and $Ca(OH)_2$	Manufacture of steel
	Neutralizer for chemical processing
	Water treatment
	Mortar
	Stack-gas scrubber (to remove H_2S and SO_2)
$CaCO_3$	Paper coating and filter
	Antacids, dentifrices
$CaSO_4$	Plaster, wallboard
	Portland cement
$Ca(H_2PO_4)_2$	Soluble phosphate fertilizer
$BaSO_4$	Oil-well drilling mud
	Gastrointestinal x-ray photography
	Paint pigment

IA																	VIIIA
H	IIA							IIIA	IVA	VA	VIA	VIIA					He
Li	Be									B	C	N	O	F	Ne
Na	Mg	IIIB		IIB			Al	Si	P	S	Cl	Ar
K	Ca	Sc	Zn	Ga	Ge	As	Se	Br	Kr
Rb	Sr	Y	Cd	In	Sn	Sb	Te	I	Xe
Cs	Ba	La	Hg	Tl	Pb	Bi	Po	At	Rn
Fr	Ra	Ac							

Figure 22.11
Gem-quality corundum.
This is the Logan sapphire, on exhibit at the Smithsonian Institution in Washington, D.C. It is the largest sapphire on public display in the United States.

Figure 22.12
Protective oxide coating on aluminum. In air, aluminum metal forms an adherent coating of aluminum oxide that protects the metal from further oxidation. Note the shiny surface of the metal plate on the left. When a similar plate of aluminum on the right is coated with mercury, however, needle-like crystals of aluminum oxide form at the aluminum–mercury surface. Instead of adhering to the metal, the aluminum oxide flakes away from the metal, and the aluminum continuously oxidizes in air.

Aluminum

Aluminum is the third most abundant element in the earth's crust (after oxygen and silicon). It occurs primarily in aluminosilicate minerals. Bauxite, the principal ore of aluminum, contains aluminum hydroxide, $Al(OH)_3$, and aluminum oxide hydroxide, $AlO(OH)$. Deposits of bauxite occur throughout the world but are particularly common in tropical and subtropical regions. Corundum is a hard mineral of aluminum oxide, Al_2O_3. The presence of impurities can give various colors to it. Sapphire (usually blue) and ruby (deep red) are gem-quality corundum (Figure 22.11).

Aluminum Metallurgy We discussed aluminum metallurgy in Section 13.2. Aluminum metal is obtained by electrolysis of aluminum oxide dissolved in fused cryolite, Na_3AlF_6.

Aluminum is the most important commercial metal after iron. Despite the fact that pure aluminum is soft and chemically reactive, the addition of a small quantity of other metals, such as copper and magnesium, yields hard, corrosion-resistant alloys. Aluminum is also a very good conductor of electricity.

Reactions of Aluminum Metal In air, aluminum metal reacts readily with oxygen, but the aluminum oxide that forms gives an adherent coat, which protects the underlying metal from further reaction. (The metal will burn vigorously once started, however.) Thus, unlike iron, which is chemically less reactive than aluminum but corrodes quickly in a moist environment, aluminum is corrosion resistant. The demonstration depicted in Figure 22.12 shows that aluminum does corrode quickly in air in the absence of an oxide coating. That aluminum metal does not normally corrode, or rust, makes the metal extremely useful in many practical applications.

Aluminum Compounds The most important compound of aluminum is aluminum oxide, Al_2O_3, or alumina. It is prepared by heating aluminum hydroxide, obtained from bauxite, at low temperature (550°C). Alumina is a white powder or porous solid. Although most alumina is used to make aluminum metal, large quantities are used for other purposes. For example, alumina is used as a carrier, or support, for many of the heterogeneous catalysts required in chemical processes (Figure 22.13).

When aluminum oxide is fused (melted) at high temperature (2045°C), it forms corundum, one of the hardest materials known. Corundum is used as an abrasive for grinding tools. When aluminum oxide is fused with small quantities of other metal oxides, synthetic sapphires and rubies are obtained.

Aluminum oxide is used in the manufacture of industrial ceramics. Ceramics

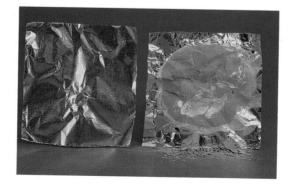

Figure 22.13
Heterogeneous catalysts.
Many heterogeneous catalysts use aluminum oxide as a carrier, or support.

made from aluminum oxide are used to line metallurgical furnaces, and the white ceramic material in automobile spark plugs is made from aluminum oxide.

Aluminum sulfate octadecahydrate, $Al_2(SO_4)_3 \cdot 18H_2O$, is the most common soluble salt of aluminum. It is prepared by dissolving bauxite in sulfuric acid. The salt is acidic in aqueous solution. Large quantities of aluminum sulfate are used in the paper industry. Printing papers require the addition of various materials such as clay to improve the capacity of the paper to hold ink without spreading. The aluminum ion does make the papers acidic, however, so paper made by this process deteriorates over time (Figure 22.14).

Aluminum sulfate is also used to treat the wastewater obtained from the process of making paper pulp, a water slurry of fibers obtained from wood. Aluminum sulfate and a base, such as calcium hydroxide, are added to the wastewater, and a gelatinous precipitate of aluminum hydroxide forms. Colloidal particles of clay and other substances adhere to the precipitate, which is then filtered from the water. The same procedure is one of the steps in the purification of municipal water supplies.

Aluminum hydroxide is amphoteric. With acids, aluminum hydroxide acts as a base. With bases, aluminum hydroxide forms a hydroxo ion (tetrahydroxoaluminate ion). The amphoteric character of aluminum hydroxide is a reflection of the partial nonmetallic character of aluminum. Table 22.4 lists the major uses of aluminum compounds.

Tin and Lead

Tin is a relatively rare element that occurs in localized deposits of the tin ore, cassiterite (SnO_2). Lead is more abundant than tin. Its most important ore is galena, a lead(II) sulfide mineral (PbS).

Tin, which is between the metalloid germanium and the metal lead in Group IVA, has two different forms, or allotropes; one is a metal and the other is a nonmetal.

The nonmetallic form of tin, called *gray tin*, is a brittle, gray powder. The metallic form of tin is called *white tin*. White tin is stable above 13°C, but at lower temperatures white tin slowly undergoes a transition to gray tin.

Metallurgy of Tin and Lead Tin metal is obtained from cassiterite. Purified tin(IV) oxide, SnO_2, from tin ore is reduced to the metal by heating with carbon in a furnace.

$$SnO_2(s) + 2C(s) \xrightarrow{\Delta} Sn(l) + 2CO(g)$$

Figure 22.14
Deterioration of paper in books. Aluminum sulfate and similar acidic compounds mixed with materials such as clay and rosin are added to paper to improve its printing characteristics. However, the acids decompose the cellulose fibers in paper, causing the paper to deteriorate over time. The Library of Congress estimates that 25% of its collection is brittle from such deteriorating paper. Calcium carbonate, a basic substance, has been suggested as an alternative in preparing paper.

Table 22.4
Uses of Aluminum Compounds

Compound	Use
Al_2O_3	Source of aluminum and its compounds
	Abrasive
	Refractory bricks and furnace linings
	Synthetic sapphires and rubies
$Al_2(SO_4)_3 \cdot 18H_2O$	Making of paper
	Water purification
$AlCl_3$	Preparation of aluminum
	Catalyst in organic reactions
$AlCl_3 \cdot 6H_2O$	Antiperspirant

Lead metal is obtained from concentrated lead(II) sulfide mineral. It is then roasted.

$$2PbS(s) + 3O_2(g) \longrightarrow 2PbO(s) + 2SO_2(g)$$

The lead(II) oxide is then reduced with carbon monoxide produced in the blast furnace.

$$PbO(s) + CO(g) \longrightarrow Pb(l) + CO_2(g)$$

Tin is used to make tin plate, which is steel (iron alloy) sheeting with a thin coating of tin. Tin plate is used for food containers ("tin cans"). The tin coating protects the iron from reaction with air and food acids. Tin is also used to make a number of alloys. Solder is a low-melting alloy of tin and lead; bronze is an alloy of copper and tin (Figure 22.15).

▶ **Lead storage batteries are discussed in Section 20.8.**

More than half of the lead produced is used to make electrodes for lead storage batteries. ◀ The manufacture of military and sporting ammunition also consumes a significant fraction of the lead produced.

Reactions of Tin and Lead Metals   Tin and lead are much less reactive than the metals of Groups IA, IIA, and IIIA. Whereas aluminum reacts vigorously with dilute hydrochloric and sulfuric acids, tin reacts only slowly with these acids. Lead metal reacts with these acids, but the products $PbCl_2$ and $PbSO_4$ are insoluble and adhere to the metal. As a result, the reaction soon stops.

Tin and Lead Compounds   Tin(II) chloride, $SnCl_2$, is used as a reducing agent in the preparation of dyes and other organic compounds. In reactions where tin(II) ion acts as a reducing agent, tin(II) ion oxidizes to tin(IV) species. Tin(II) compounds are commonly referred to as *stannous compounds,* using an older naming system.

Lead also exists in compounds in the +2 and +4 oxidation states, although lead(II) compounds are the more common. The starting compound for preparing most lead compounds is lead(II) oxide, PbO. Lead(IV) oxide, PbO_2, is a dark brown or black powder; it forms the cathode of lead storage batteries. The cathode is made by packing a paste of lead(II) oxide into a lead metal grid. When the battery is charged, the lead(II) oxide is oxidized to lead(IV) oxide. Table 22.5 gives a summary list of the uses of tin and lead compounds.

Figure 22.15
Tin alloys. Shown here are solder (and a soldering iron) and bronze.

Table 22.5
Uses of Tin and Lead Compounds

Compound	Use
SnO_2	Manufacture of tin compounds
	Glazes and enamels
$SnCl_2$	Reducing agent in preparing organic compounds
PbO	Lead storage batteries
	Lead glass
PbO_2	Cathode in lead storage batteries
Pb_3O_4	Pigment for painting structural steel

Chemistry of the Nonmetals

The two nonmetals known to the ancients, carbon and sulfur, both are associated with fire. Carbon was known in the form of charcoal and lampblack, or soot, which are products of fire. Charcoal may have been so common that it was hardly noticed at first, until its role in the reduction of metal ores was discovered. Lampblack was used by the ancient Egyptians to produce ink for writing on papyrus.

Free sulfur was less widely available than carbon, although it was probably well known because of its ready occurrence in volcanic areas. That sulfur also burned with a beautiful blue flame made it especially distinctive. Some nonmetallic elements are shown in Figure 22.16.

22.5 Hydrogen

Hydrogen is the most abundant element in the universe, comprising nearly 90% of all atoms, and is the third most abundant element on the surface of the earth. Most stars, including our sun, consist primarily of hydrogen. The hydrogen in our sun is the fuel for the fusion reactions that produce the life-sustaining energy that reaches our planet. On earth, the majority of hydrogen is found in oceans, combined with oxygen as water. Hydrogen occurs in a variety of compounds, which include the organic compounds in Chapter 24.

Properties and Preparation of Hydrogen

Hydrogen was first isolated and identified by Henry Cavendish in 1766. His experiments consisted of reacting iron, zinc, and tin with several different binary acids. For example, $H_2(g)$ can easily be produced on a small scale according to the reaction

$$2HCl(g) + Zn(s) \longrightarrow ZnCl_2(aq) + H_2(g)$$

Hydrogen is a colorless, odorless gas that is less dense than air. Even though hydrogen is often placed in Group IA of the periodic table, aside from its valence electron configuration, it has little in common with the alkali metals that make up the rest of Group IA. It is much less likely to form a cation than any of the alkali metals, with a first ionization energy of 1312 kJ/mol versus 520 kJ/mol for lithium. Because of this, hydrogen generally combines with nonmetallic elements to form covalent compounds such as CH_4, H_2S, and PH_3.

The elemental form of hydrogen is a diatomic molecule having a bond dissociation energy of 432 kJ/mol. This is a large value when compared to chlorine at 240 kJ/mol. This relatively high bond dissociation energy indicates why hydrogen is less reactive than its halogen counterparts. However, with the addition of heat or light, or in the presence of a suitable catalyst, hydrogen can be induced to react.

Hydrogen is currently produced on a massive industrial scale, with an annual U.S. production on the order of 10^{10} m^3. Approximately 40% of this production is used to manufacture ammonia. The hydrogen for this reaction is generally prepared using the **steam-reforming process** *where steam and hydrocarbons from natural gas or petroleum react at high temperature and pressure in the presence of a catalyst to form carbon monoxide and hydrogen.* For example:

$$C_3H_8(g) + 3H_2O(g) \xrightarrow[\Delta]{Ni} 3CO(g) + 7H_2(g)$$

**Figure 22.16
Some nonmetals.** *Left to right:* Sulfur, bromine, white phosphorus, and carbon.

Another viable route for hydrogen production is the *water–gas reaction,* which is no longer used commercially but may become important in the future as natural gas and petroleum become more expensive and scarce. In this reaction, steam is passed over red-hot coke or coal.

$$C(s) + H_2O(g) \longrightarrow CO(g) + H_2(g)$$

Hydrogen is also produced as a by-product of the electrolysis of concentrated aqueous NaCl solutions during the production of NaOH and chlorine. ◄

► **See Section 20.10 for a discussion of the electrolysis of sodium chloride solutions.**

Hydrogen Reactions and Compounds

In addition to the preparation of ammonia, the other major use of hydrogen is when hydrogen is added to hydrocarbon compounds containing carbon–carbon double bonds to produce compounds that contain carbon–carbon single bonds. For example, 1-butene can be reacted with hydrogen using a platinum or palladium catalyst to produce butane.

$$CH_3CH_2CH{=}CH_2 + H_2 \xrightarrow{\ Pt\ } CH_3CH_2CH_2CH_3$$

This process, called *hydrogenation,* is used in the food processing industry where oils (liquids) that contain many carbon–carbon double bonds are converted to fats (solids) that contain few or no carbon–carbon double bonds. Another important process that requires hydrogen is the cobalt-catalyzed *synthesis gas* reaction with carbon monoxide to produce methyl alcohol.

$$CO(g) + 2H_2(g) \xrightarrow{\ \text{cobalt catalyst}\ } CH_3OH(g)$$

Hydrogen is also used to reduce metal oxides to extract pure metals. For example, molybdenum and tungsten oxides can be reduced at high temperatures via the reaction

$$MO_3(s) + 3H_2(g) \longrightarrow M(s) + 3H_2O(g)$$

Hydrogen forms a **binary hydride**—that is, *a compound that contains hydrogen and one other element.* There are three categories of binary hydrides: ionic hydrides, covalent hydrides, and metallic hydrides.

Ionic hydrides, which contain the hydride ion, H^-, can be directly formed via the reaction of an active metal with hydrogen near 400°C.

$$2Li(s) + H_2(g) \longrightarrow 2LiH(s)$$
$$Ba(s) + H_2(g) \longrightarrow BaH_2(s)$$

These hydrides are white crystalline compounds in which the H atoms have an oxidation state of -1. Ionic hydrides can undergo an oxidation–reduction reaction with water to produce hydrogen and a basic solution. For example:

$$LiH(s) + H_2O(l) \longrightarrow H_2(g) + LiOH(aq)$$

Ionic hydrides are used as reducing agents during chemical reactions.

Covalent hydrides are molecular compounds in which hydrogen is covalently bonded to another element. Examples of these compounds are NH_3, H_2O, H_2O_2, and HF. Some of these compounds can be formed from the direct reaction of the elements. The reaction of hydrogen with oxygen to form water is an example of a reaction that requires the input of energy to get started; however, once it begins, the reaction is rapid and exothermic.

$$2H_2(g) + O_2(g) \longrightarrow 2H_2O(g) \qquad \Delta H = -484 \text{ kJ}$$

Because it is such an exothermic reaction and the product is a gas, it is an ideal rocket fuel (Figure 22.17). The combustion of hydrogen produces more heat per gram than any other fuel (120 kJ/g).

Metallic hydrides are compounds containing a transition metal and hydrogen. Generally, the formula of these compounds is MH_x, where x is often not an integer. These compounds contain hydrogen atoms that are spread throughout a metal crystal occupying the holes in the crystal lattice.

22.6 Group IVA: The Carbon Family

The elements of Group IVA show in more striking fashion than the previous groups the normal periodic trend of greater metallic character going down a column. Carbon, the first element in the group, is distinctly nonmetallic. Silicon and germanium are metalloids, or semimetals, although their chemical properties are primarily those of nonmetals. Tin and lead, the last two elements in the group, are metals.

We will now look at carbon.

Carbon

One of the most important features of the carbon atom is its ability to bond to other carbon atoms to form chains and rings of enormous variety. *The covalent bonding of two or more atoms of the same element to one another* is referred to as **catenation.** Although other elements display catenation, none show it to the same degree as carbon. Millions of carbon compounds are known, most classified as *organic*. In this section, we will confine ourselves to the element and its oxides and carbonates.

Carbon Black   The form of carbon known as *carbon black* is composed of extremely small crystals of carbon having an amorphous, or imperfect, graphite structure. It is produced in large quantities by burning natural gas (CH_4) or petroleum hydrocarbons in a limited supply of air.

$$CH_4(g) \xrightarrow{\Delta} \underset{\text{carbon black}}{C(s)} + 2H_2(g)$$

Figure 22.17
A liquid-hydrogen storage tank. Liquid hydrogen is used as a rocket fuel.

Carbon black is used in the manufacture of rubber tires (to increase wear) and as a pigment in black printing inks. Coke is an amorphous carbon obtained by heating coal in the absence of air; it is used in large quantities in the making of iron.

Oxides of Carbon   Carbon has two principal oxides: carbon monoxide, CO, and carbon dioxide, CO_2. Carbon and organic compounds burn in an excess of oxygen to give carbon dioxide, CO_2. However, an equilibrium exists among carbon, carbon monoxide, and carbon dioxide that favors carbon monoxide above 700°C.

$$CO_2(g) + C(s) \rightleftharpoons 2CO(g)$$

For this reason, carbon monoxide is almost always one of the products of combustion of carbon and organic compounds, unless an excess of oxygen is present.

Carbon monoxide is a colorless, odorless, toxic gas, which poisons by attaching strongly to iron atoms in the hemoglobin of red blood cells, preventing them from carrying out their normal oxygen-carrying function.

Figure 22.18
Test for carbon dioxide.
When carbon dioxide is bubbled into a solution of calcium hydroxide (limewater), a milky white precipitate of calcium carbonate forms. This is the basis of a test for carbon dioxide. The reaction is also used to manufacture pure calcium carbonate. (The needle at the right is to provide an escape for excess gas.)

Carbon monoxide is manufactured industrially from natural gas (CH_4) and petroleum hydrocarbons, either by reaction with steam or by partial oxidation. For example,

$$CH_4(g) + H_2O(g) \xrightarrow{\text{Ni}} CO(g) + 3H_2(g)$$
$$2CH_4(g) + O_2(g) \longrightarrow 2CO(g) + 4H_2(g)$$

The product in these reactions is a mixture of carbon monoxide and hydrogen, which is called *synthesis gas.* Synthesis gas can yield any of a number of organic products depending on the reaction conditions and catalyst. Methanol (CH_3OH), for example, is produced in large quantities from synthesis gas.

$$CO(g) + 2H_2(g) \xrightarrow{\text{catalyst}} CH_3OH(l)$$

Carbon dioxide is a colorless, odorless gas with a faint acid taste. Carbon dioxide does not support most combustions, which is why CO_2 fire extinguishers are used in many fire-fighting situations.

Carbon dioxide is produced whenever carbon or organic materials are burned. It is obtained commercially as a by-product in the production of ammonia (see Section 22.7) and in the calcining (strong heating) of limestone to give calcium oxide. Carbonated beverages are made by dissolving carbon dioxide gas under pressure in an aqueous solution of sugar and flavorings. Carbonated water is acidic as the result of the formation of carbonic acid, although carbonated beverages frequently also contain fruit acids and phosphoric acid.

Carbonates Carbon dioxide dissolves in water to form an aqueous solution of carbonic acid.

$$CO_2(g) + H_2O(l) \rightleftharpoons H_2CO_3(aq)$$

The acid is diprotic (has two acidic H atoms per molecule) and dissociates to form hydrogen carbonate ion and carbonate ion.

Carbonic acid has never been isolated from solution, but its salts, hydrogen carbonates and carbonates, are well known. When you bubble carbon dioxide gas into an aqueous solution of calcium hydroxide, a milky white precipitate of calcium carbonate forms (Figure 22.18).

$$CO_2(g) + Ca(OH)_2(aq) \longrightarrow CaCO_3(s) + H_2O(l)$$

This is a standard test for carbon dioxide. The reaction is also used to manufacture a pure calcium carbonate for antacids and other products.

Carbonate minerals are very common and many are of commercial importance. Limestone contains the mineral calcite, which is calcium carbonate, $CaCO_3$. Much of this was formed by marine organisms, although some limestone was also formed by direct precipitation from water solution.

22.7 Group VA: Nitrogen and the Phosphorus Family

Like the carbon family of elements, the Group VA elements show the distinct trend of increasing metallic character as you go from top to bottom of the column. The first members, nitrogen and phosphorus, are nonmetallic; arsenic and antimony are metalloids; bismuth is a metal.

As expected for a second-period element, nitrogen is in many respects different from the other elements in its group. You can see this in the formulas of the elements

and compounds. Elementary nitrogen is N_2, while white phosphorus is P_4. Similarly, the common +5 oxoacid of nitrogen is HNO_3, that of phosphorus is H_3PO_4, and that of arsenic is H_3AsO_4. [Nitrogen exists in many molecular compounds in a variety of oxidation states. Phosphorus exists in many compounds in the +5 and +3 oxidation states. Many phosphorus compounds have P—O—P bonding.]

Nitrogen

The element nitrogen is crucial to life: it is a component of all proteins, which are involved in almost every biochemical process that occurs in living organisms. Most of the available nitrogen on earth, however, is present as nitrogen gas (dinitrogen, N_2) in the atmosphere, which consists of 78.1% N_2, by mass. Most organisms cannot use dinitrogen from the atmosphere as their source of the element. However, certain soil bacteria, as well as bacteria that live in nodules on the roots of beans, clover, and similar plants, can convert dinitrogen to ammonium and nitrate compounds. Plants use these simple nitrogen compounds to make proteins and other complex nitrogen compounds. Animals eat these plants, and other animals eat those animals. Finally, bacteria in decaying organic matter convert the nitrogen compounds back to dinitrogen. In this way, nitrogen in our environment continually cycles from dinitrogen to living organisms and back. Figure 22.19 depicts this *nitrogen cycle*.

Figure 22.19
The nitrogen cycle.
Nitrogen, N_2, is fixed (converted to compounds) by bacteria, by lightning, and by the industrial synthesis of ammonia. Fixed nitrogen is used by plants and enters the food chain of animals. Later, plant and animal wastes decompose. Denitrifying bacteria complete the cycle by producing free nitrogen again.

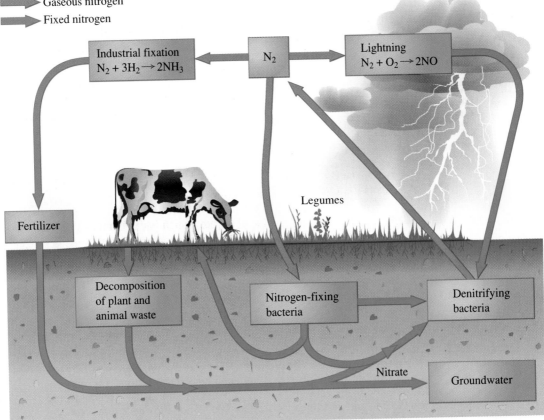

Figure 22.20
Liquid nitrogen. When liquid nitrogen is poured on a table (which, although at room temperature, is about 220°C above the boiling point of nitrogen), the liquid sizzles and boils away violently.

Properties and Uses of Nitrogen Daniel Rutherford, a chemist and physician, discovered nitrogen, N_2, in air in 1772. Nitrogen is a relatively unreactive element because of the stability of the nitrogen–nitrogen triple bond. (The $N\equiv N$ bond energy is 942 kJ/mol, compared with 167 kJ/mol for the N—N bond energy.) When substances burn in air, they generally react with oxygen, leaving the nitrogen unreacted. Some very reactive metals do react directly with nitrogen, however. For example, when magnesium metal burns in air, it forms the nitride, as well as the oxide.

$$3Mg(s) + N_2(g) \longrightarrow Mg_3N_2(s)$$

Because the nitride ion, N^{3-}, is a very strong base, ionic nitrides react with water, producing ammonia.

$$N^{3-}(aq) + 3H_2O(l) \longrightarrow NH_3(g) + 3OH^-(aq)$$

Air is the major commercial source of nitrogen. Liquid nitrogen is used as a refrigerant to freeze foods, to freeze soft or rubbery materials prior to grinding them, and to freeze biological materials (Figure 22.20). Large quantities of nitrogen are also used as a *blanketing gas*, whose purpose is to protect a material from oxygen during processing or storage. Thus, electronic components are often made under a nitrogen atmosphere.

Nitrogen Compounds Ammonia, NH_3, is the most important commercial compound of nitrogen. A colorless gas with a characteristic irritating or pungent odor, it is prepared commercially from N_2 and H_2 by the *Haber process*. Figure 22.21 shows a flowchart of the industrial preparation of ammonia from natural gas, steam, and air.

Ammonia is easily liquefied, and the liquid is used as a nitrogen fertilizer. Ammonium salts, such as the sulfate and nitrate, are also sold as fertilizers. Large quantities of ammonia are converted to urea, NH_2CONH_2, which is used as a fertilizer, as a livestock

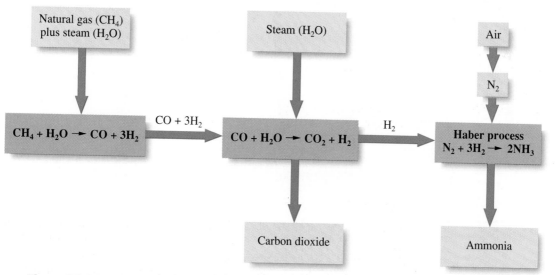

Figure 22.21
Industrial preparation of ammonia (flowchart). The raw materials are natural gas, water, and air. Hydrogen for the Haber process is obtained by reacting natural gas with steam to give carbon monoxide and hydrogen. In the next step, carbon monoxide is reacted with steam to give carbon dioxide and additional hydrogen. The carbon dioxide is removed by dissolving it in water solution.

feed supplement, and in the manufacture of urea–formaldehyde plastics. Ammonia is also the starting compound for the preparation of most other nitrogen compounds.

Nitrous oxide, N_2O, is a colorless gas with a sweet odor. It can be prepared by careful heating of molten ammonium nitrate. (If heated strongly, it explodes.)

$$NH_4NO_3(s) \xrightarrow{\Delta} N_2O(g) + 2H_2O(g)$$

Nitrous oxide, or laughing gas, is used as a dental anesthetic. It is also useful as a propellant in whipped-cream dispensers.

Nitric oxide, NO, is a colorless gas that is of great industrial and biological importance. ◄ Although it can be prepared by the direct combination of the elements at elevated temperatures, large amounts are prepared from ammonia as the first step in the commercial preparation of nitric acid.

▶ **The biological importance of NO was discussed in an essay in Chapter 5.**

Nitric acid, HNO_3, is the third most important industrial acid (after sulfuric and phosphoric acids). It is used to prepare explosives, nylon, and polyurethane plastics. Nitric acid is produced commercially by the **Ostwald process,** which is *an industrial preparation of nitric acid starting from the catalytic oxidation of ammonia to nitric oxide.* In this process, ammonia is burned in the presence of a platinum catalyst to give NO, which is then reacted with oxygen to form NO_2. The NO_2 is dissolved in water, where it reacts to form nitric acid and nitric oxide.

$$4NH_3(g) + 5O_2(g) \xrightarrow{Pt} 4NO(g) + 6H_2O(g)$$
$$2NO(g) + O_2(g) \longrightarrow 2NO_2(g)$$
$$3NO_2(g) + H_2O(l) \longrightarrow 2HNO_3(aq) + NO(g)$$

The nitric oxide produced in the last step is recycled for use in the second step.

Nitric acid is a strong oxidizing agent. Although copper metal is unreactive to most acids, it is oxidized by the nitrate ion in acid solution. In dilute acid, nitric oxide is the principal reduction product. With concentrated nitric acid, nitrogen dioxide is obtained (Figure 22.22).

$$Cu(s) + 4H_3O^+(aq) + 2NO_3{}^-(aq) \longrightarrow Cu^{2+}(aq) + 2NO_2(g) + 6H_2O(l)$$

Table 22.6 lists uses of some compounds of nitrogen (and of phosphorus, discussed in the next section).

Figure 22.22
Reaction of copper metal with concentrated nitric acid. The principal reduction product from concentrated nitric acid is nitrogen dioxide, NO_2 (reddish-brown gas).

▸ CONCEPT CHECK 22.1

Considering the fact that N_2 makes up about 80% of the atmosphere, why don't animals use the abundant N_2 instead of O_2 for biological reactions?

Phosphorus

Phosphorus, the most abundant of the Group VA elements, occurs in phosphate minerals, such as fluorapatite, whose formula is written either $Ca_5(PO_4)_3F$ or $3Ca_3(PO_4)_2 \cdot CaF_2$. Unlike nitrogen, which exists in important compounds with oxidation states between -3 and $+5$, the most important oxidation states of phosphorus are $+3$ and $+5$. Like nitrogen, however, phosphorus is an important element in living organisms. DNA (deoxyribonucleic acid), a chainlike biological molecule in which information about inheritable traits resides, contains phosphate groups along the length of its chain. Similarly, ATP (adenosine triphosphate), the energy-containing molecule of living organisms, contains phosphate groups.

Figure 22.23
Allotropes of phosphorus.
Left: White phosphorus. *Right:*
Red phosphorus.

Table 22.6
Uses of Some Compounds of Nitrogen and of Phosphorus

Compound	Use
NH_3	Nitrogen fertilizer
	Manufacture of nitrogen compounds
N_2H_4	Blowing agent for foamed plastics
	Water treatment
HNO_3	Explosives
	Polyurethane plastics
$Ca(H_2PO_4)_2 \cdot H_2O$	Phosphate fertilizer
	Baking powder
$CaHPO_4 \cdot 2H_2O$	Animal feed additive
	Toothpowder
H_3PO_4	Manufacture of phosphate fertilizers
	Soft drinks
PCl_3	Manufacture of $POCl_3$
	Manufacture of pesticides
$POCl_3$	Manufacture of plasticizers (substances that keep plastics pliable)
	Manufacture of flame retardants
$Na_5P_3O_{10}$	Detergent additive

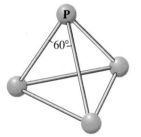

Figure 22.24
Structure of the P_4
molecule. *Top:* The reactivity
of white phosphorus results
from the small P—P—P angle
(60°). *Bottom:* Space-filling
molecular model.

Allotropes of Phosphorus Phosphorus has two common allotropes: white phosphorus and red phosphorus (Figure 22.23). If white phosphorus is left exposed to air, it bursts spontaneously into flame. Because of its reactivity with oxygen, white phosphorus is stored under water, in which it is insoluble. As you might expect from its low melting point (44°C), white phosphorus is a molecular solid, with the formula P_4. The phosphorus atoms in the P_4 molecule are arranged at the corners of a regular tetrahedron such that each atom is single-bonded to the other three (Figure 22.24).

White phosphorus is a major industrial chemical and is prepared by heating phosphate rock with coke (C) and quartz sand (SiO_2) in an electric furnace. The overall reaction is

$$2Ca_3(PO_4)_2(s) + 6SiO_2(s) + 10C(s) \xrightarrow{1500°C} 6CaSiO_3(l) + 10CO(g) + P_4(g)$$
$$\text{calcium silicate}$$

Most of the white phosphorus produced is used to manufacture phosphoric acid, H_3PO_4. Some white phosphorus is converted to red phosphorus, which has a chain structure (Figure 22.25). Red phosphorus is much less reactive than white phosphorus and can be stored in the presence of air. This phosphorus allotrope is relatively nontoxic and is used in the striking surface for safety matches. Red phosphorus is made by heating white phosphorus at about 400°C for several hours.

Phosphorus Oxides Phosphorus has two common oxides, P_4O_6 and P_4O_{10}. Their common names are phosphorus trioxide and phosphorus pentoxide, respectively. We

Figure 22.25
Chain structure of red phosphorus. The structure is obtained by linking P_4 tetrahedra together after breaking a bond in each tetrahedron.

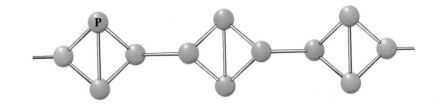

will refer to these compounds as phosphorus(III) oxide (P_4O_6) and phosphorus(V) oxide (P_4O_{10}).

These oxides have interesting structures (Figure 22.26). Phosphorus(III) oxide has a tetrahedron of phosphorus atoms, as in P_4, but with oxygen atoms between each pair of phosphorus atoms. Phosphorus(V) oxide is similar, except that each phosphorus atom has an additional oxygen atom bonded to it. These phosphorus–oxygen bonds are much shorter than the ones in the P—O—P bridges (139 pm versus 162 pm); hence, they are best represented as double bonds.

Phosphorus(III) oxide is a low-melting solid (m.p. 23°C) and the anhydride of phosphorous acid, H_3PO_3. Phosphorus(V) oxide is the most important oxide; it is a white solid that sublimes at 360°C. This oxide is the anhydride of phosphoric acid, H_3PO_4. The reaction with water is quite vigorous, making phosphorus(V) oxide useful in the laboratory as a drying agent. It is prepared by burning white phosphorus in air.

$$P_4(s) + 5O_2(g) \longrightarrow P_4O_{10}(s)$$

▶ Because phosphate rock contains CaF_2, hydrofluoric acid is a by-product in the preparation of phosphoric acid. HF is used in aluminum production.

Phosphorus(V) oxide is used to manufacture phosphoric acid, an oxoacid.

Oxoacids of Phosphorus Phosphorus has many oxoacids, but the most important of these can be thought of as derivatives of orthophosphoric acid. Orthophosphoric acid is triprotic; possible sodium salts are sodium dihydrogen phosphate (NaH_2PO_4), disodium hydrogen phosphate (Na_2HPO_4), and trisodium phosphate (Na_3PO_4). The acid has the electron-dot formula shown in Figure 22.27.

Orthophosphoric acid is produced in enormous quantity either directly from phosphate rock or from phosphorus(V) oxide. In the direct process, phosphate rock is treated with sulfuric acid, from which a solution of phosphoric acid is obtained by filtering off the calcium sulfate and other solid materials. The product is an impure phosphoric acid, which is used in the manufacture of phosphate fertilizers. ◀

$$Ca_3(PO_4)_2(s) + 3H_2SO_4(aq) \longrightarrow 3CaSO_4(s) + 2H_3PO_4(aq)$$

Figure 22.26
Structures of the phosphorus oxides.
Left: The phosphorus atoms in P_4O_6 have tetrahedral positions, as in P_4; however, the phosphorus atoms are bonded to oxygen atoms, forming P—O—P bridges between each pair of phosphorus atoms. *Right:* The P_4O_{10} molecule is similar, except that an additional oxygen atom is bonded to each phosphorus atom. Both ball-and-stick and space-filling models are shown.

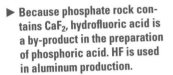

P_4O_6

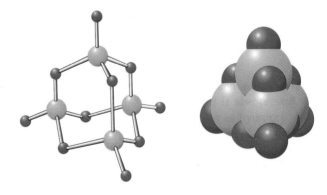

P_4O_{10}

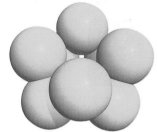

Figure 22.29
Structure of the S_8 molecule. *Top:* Each molecule consists of eight S atoms arranged in a ring (in the shape of a crown). *Bottom:* Space-filling molecular model.

Sulfur also occurs in several amino acids, which are the building blocks of the proteins in living organisms. Animals and decay bacteria derive most of their nutritional sulfur from organic sources.

Allotropes of Sulfur Sulfur has a fascinating array of allotropes, including two common crystal forms, rhombic sulfur and monoclinic sulfur (see Figure 6.12). *Rhombic sulfur* is the stablest form of the element under normal conditions; natural sulfur is rhombic sulfur. It is a yellow, crystalline solid with a lattice consisting of crown-shaped S_8 molecules (Figure 22.29). Rhombic sulfur melts at 113°C to give an orange-colored liquid.

When this liquid is cooled, it crystallizes to give *monoclinic sulfur.* Monoclinic sulfur differs from rhombic sulfur only in the way the molecules are packed to form crystals. It is unstable below 96°C, and in a few weeks at room temperature it reverts to rhombic sulfur.

When molten sulfur above 160°C (but below 200°C) is poured into water, the liquid changes to a rubbery mass, called *plastic sulfur.* Plastic sulfur is an amorphous mixture of sulfur chains. The rubberiness of this sulfur results from the ability of the spiral chains of sulfur atoms to stretch and then relax to their original length. Plastic sulfur reverts to rhombic sulfur after a period of time.

Production of Sulfur Free sulfur that occurs in deep underground deposits is mined by the **Frasch process,** *a mining procedure in which underground deposits of solid sulfur are melted in place with superheated water, and the molten sulfur is forced upward as a froth using air under pressure.* A sulfur well is similar to an oil well but consists of three concentric pipes. Superheated water flows down the outer pipe, and compressed air flows down the inner pipe. The superheated water melts the sulfur, which is then pushed up the middle pipe by the compressed air. Molten sulfur spews from the well and solidifies in large storage areas. Sulfur obtained this way is 99.6% pure.

Hydrogen sulfide, H_2S, recovered from natural gas and petroleum is also a source of free sulfur. The sulfur is obtained from the hydrogen sulfide gas by the **Claus process,** *a method of obtaining free sulfur by the partial burning of hydrogen sulfide.*

Most of the sulfur produced (almost 90%) is used to make sulfuric acid. Other uses include the vulcanization of rubber, the production of carbon disulfide (to make cellophane), and the preparation of sulfur dioxide for bleaching.

Sulfur Oxides and Oxoacids Sulfur dioxide, SO_2, is a colorless, toxic gas with a characteristic suffocating odor. Its presence in polluted air (from the burning of fossil fuels) is known to cause respiratory ailments. The SO_2 molecule has a bent geometry with a bond angle of 119.5° (Figure 22.30).

Sulfur dioxide gas is very soluble in water, producing acidic solutions. These solutions are often referred to as solutions of *sulfurous acid.* An aqueous solution of sulfur dioxide, $SO_2(aq)$, does apparently contain small amounts of the ions HSO_3^- and SO_3^{2-}, which would be expected to be produced by the ionization of H_2SO_3. When an appropriate amount of base is added to an aqueous solution of sulfur dioxide, the corresponding hydrogen sulfite salt or sulfite salt is obtained. Sodium hydrogen sulfite (also called sodium bisulfite) and sodium sulfite are produced this way commercially using sodium carbonate as the base:

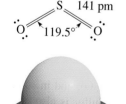

Figure 22.30
Structure of the SO_2 molecule. The molecule has a bent geometry.

$$Na_2CO_3(aq) + 2SO_2(aq) + H_2O(l) \longrightarrow 2NaHSO_3(aq) + CO_2(g)$$
$$Na_2CO_3(aq) + SO_2(aq) \longrightarrow Na_2SO_3(aq) + CO_2(g)$$

Figure 22.31
Bleaching of a rose by sulfur dioxide.   The dye in the rose is reduced by sulfur dioxide (contained in the beaker) to a colorless substance.

▶ Sulfuric acid and acid rain are discussed in an essay in Section 17.2.

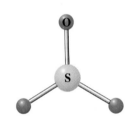

SO_3

Figure 22.32
Structure of SO_3.

Sulfur dioxide is produced on a large scale by burning sulfur, S_8. It is also obtained as a by-product of the roasting of sulfide ores (such as FeS_2, CuS, ZnS, and PbS). Most of this sulfur dioxide is used to prepare sulfuric acid. Some is used as a bleach for wood pulp and textiles (Figure 22.31) and as a disinfectant and food preservative (for example, in wine and dried fruit). Because some people are allergic to sulfur dioxide, foods containing it must be properly labeled.

Sulfur trioxide is a liquid at room temperature. The liquid actually consists of S_3O_9 molecules in equilibrium with SO_3 molecules. The vapor-phase molecule is SO_3, which has a planar triangular geometry (Figure 22.32).

Sulfur trioxide is formed in small amounts when sulfur is burned in air, although the principal product is sulfur dioxide. Sulfur dioxide reacts slowly with oxygen in air to produce sulfur trioxide. Sulfur trioxide is produced commercially by the oxidation of sulfur dioxide in the presence of vanadium(V) oxide catalyst.

$$2SO_2(g) + O_2(g) \xrightarrow{V_2O_5} 2SO_3(g)$$

Sulfur trioxide reacts vigorously and exothermically with water to produce sulfuric acid.

$$SO_3(g) + H_2O(l) \longrightarrow H_2SO_4(aq)$$

The **contact process** is *an industrial method for the manufacture of sulfuric acid. It consists of the reaction of sulfur dioxide with oxygen to form sulfur trioxide using a catalyst of vanadium(V) oxide, followed by the reaction of sulfur trioxide with water.* Because the direct reaction of sulfur trioxide with water produces mists that are unmanageable, the sulfur trioxide is actually dissolved in concentrated sulfuric acid, which is then diluted with water. ◀

Sulfuric acid is a component of acid rain and forms in air from sulfur dioxide, following reactions that are similar to those involved in the contact process. Atmospheric sulfur dioxide has both natural and human origins. Natural sources include plant and animal decomposition and volcanic emissions. However, the burning of coal, oil, and natural gas has been identified as a major source of acid rain pollution. After persisting in the atmosphere for some time, sulfur dioxide is oxidized to sulfur trioxide, which dissolves in rain to give $H_2SO_4(aq)$.

More sulfuric acid is made than any other industrial chemical. Most of this acid is used to make soluble phosphate and ammonium sulfate fertilizers. Sulfuric acid is also used in petroleum refining and in the manufacture of many chemicals. Table 22.7 lists uses of some sulfur compounds.

22.9 Group VIIA: The Halogens

The Group VIIA elements, called the halogens, have very similar properties. All are reactive nonmetals. As a second-period element, fluorine does exhibit some differences from the other elements of Group VIIA, although these are not so pronounced as those of the second-period elements in Groups IIIA to VIA.

All of the halogens form stable compounds in which the element is in the -1 oxidation state. In fluorine compounds, this is the only oxidation state. Chlorine, bromine, and iodine also have compounds in which the halogen is in one of the positive oxidation states $+1$, $+3$, $+5$, or $+7$.

Table 22.9
Some Compounds of Xenon

Compound	Formula	Description
Xenon difluoride	XeF_2	Colorless crystals
Xenon tetrafluoride	XeF_4	Colorless crystals
Xenon hexafluoride	XeF_6	Colorless crystals
Xenon trioxide	XeO_3	Colorless crystals, explosive
Xenon tetroxide	XeO_4	Colorless gas, explosive

A Checklist for Review

Important Terms

Solvay process (22.2)
steam-reforming
  process (22.5)

binary hydride (22.5)
catenation (22.6)
Ostwald process (22.7)

oxide (22.8)
peroxide (22.8)
superoxide (22.8)

Frasch process (22.8)
Claus process (22.8)
contact process (22.8)

Summary of Facts and Concepts

Several general observations can be made about the main-group elements. First, the metallic characteristics of these elements generally decrease across a period from left to right in the periodic table. Second, metallic characteristics of the main-group elements become more pronounced going down any column (group). Finally, a second-period element is usually rather different from the other elements in its group.

The Group IA metals (alkali metals) are soft, chemically reactive elements. Lithium, sodium, and potassium are important alkali metals. In recent years, the commercial uses of lithium have grown markedly. The metal is obtained by the electrolysis of molten lithium chloride and is used in the production of low-density alloys and as a battery anode. Lithium hydroxide is used to make lithium soap for lubricating greases; it is produced by the reaction of lithium carbonate and calcium hydroxide.

Sodium metal is used as a reducing agent in the preparation of other metals, such as titanium and zirconium, and in the preparation of dyes and pharmaceuticals. Sodium compounds are of enormous economic importance. Sodium chloride is the source of sodium and most of its compounds. Sodium hydroxide is prepared by the electrolysis of aqueous sodium chloride; as a strong base, it has many useful commercial applications, including aluminum production and petroleum refining. Sodium carbonate is obtained from the mineral trona, which contains sodium carbonate and sodium hydrogen carbonate, and by the *Solvay process* from salt (NaCl) and limestone ($CaCO_3$). Sodium carbonate is used to make glass. Potassium metal is produced in relatively small quantities, but potassium compounds are important. Large quantities of potassium chloride are used as a plant fertilizer.

Magnesium and calcium are the most important of the Group IIA (alkaline earth) metals. Magnesium and its alloys are important structural metals. Calcium is important primarily as its compounds, which are prepared from natural carbonates, such as limestone, and the sulfates, such as gypsum. When limestone is heated strongly, it decomposes to calcium oxide (lime). Enormous quantities of lime are used in the production of iron from its ores.

Of the Group IIIA and Group IVA metals, aluminum, tin, and lead are especially important. Aluminum is the third most abundant element in the earth's crust. It is obtained commercially from bauxite; through chemical processing bauxite yields pure aluminum oxide. Most of this aluminum oxide is used in the production of aluminum by electrolysis. Some aluminum oxide is used as a carrier for heterogeneous catalysts and in manufacturing industrial ceramic materials.

Tin is normally a metal (called white tin) but does undergo a low-temperature conversion to a nonmetallic form (called gray tin). Tin is obtained by reduction of cassiterite, a mineral form of SnO_2. Tin is used to make tin plate, bronze, and solder. Lead is obtained from galena, which is a sulfide ore, PbS. More than half of the lead produced is used to make electrodes for lead storage batteries.

Carbon and silicon are the least metallic of the Group IVA elements. *Catenation* is an important feature of carbon chemistry and is responsible for the enormous number of organic compounds. Carbon has important industrial uses, including carbon black for rubber tires. The principal oxides of carbon are CO and CO_2. Mixtures of carbon monoxide and hydrogen (synthesis gas) are used to prepare various organic compounds. Carbon dioxide is used to make carbonated beverages.

Hydrogen is the most abundant element in the universe and is the third most abundant element on the surface of the

earth. Most of the hydrogen on earth is found in water. Elemental hydrogen is produced on an industrial scale by the *steam-reforming process* in which a hydrocarbon is reacted with water in the presence of a catalyst at high temperature. The bulk of the hydrogen produced in this manner is used to make organic compounds including methanol. Hydrogen forms three classes of binary compounds called binary hydrides: ionic hydrides, covalent hydrides, and metallic hydrides. The *ionic hydrides* are reactive solids formed either by the reaction of hydrogen with an alkali metal to form compounds with the general formula MH, or with larger alkaline earth metals to form MH_2. The *covalent hydrides* are compounds in which hydrogen is covalently bonded to another element. The *metallic hydrides* contain a transition metal element and hydrogen. In these compounds, the lattice of metal atoms forms a porous structure that allows hydrogen atoms to enter and bond. Metallic hydrides are often nonstoichiometric, meaning that the ratio of hydrogen atoms to metal atoms is not a whole number.

Of the Group VA elements, nitrogen and phosphorus are particularly important. Nitrogen, N_2, is obtained from liquid air by fractional distillation; liquid nitrogen is used as a refrigerant. Ammonia, NH_3, is the most important compound of nitrogen. It is prepared from the elements and is used as a fertilizer. Ammonia is also the starting compound for the manufacture of other nitrogen compounds. For example, in the *Ostwald process* for the preparation of nitric acid, ammonia is burned in the presence of a catalyst to nitric oxide, NO. The nitric oxide reacts with oxygen to give nitrogen dioxide, which reacts with water to give nitric acid.

Phosphorus has two common allotropes, white phosphorus (P_4) and red phosphorus (chain structure). White phosphorus is obtained by heating a phosphate mineral with sand and coke in an electric furnace. When phosphorus burns in air, it forms phosphorus(V) oxide, P_4O_{10}. This oxide reacts with water to give orthophosphoric acid, H_3PO_4.

Oxygen, a Group VIA element, occurs in the atmosphere, but mostly it is present on earth as oxide and oxoanion minerals. Oxygen has two allotropes: dioxygen, O_2, and ozone, O_3. Oxygen is obtained commercially from liquid air. Oxygen reacts with almost all elements to give *oxides* or, in some cases, *peroxides* or *superoxides*.

Sulfur, another Group VIA element, occurs in sulfate and sulfide minerals. Free sulfur, S_8, occurring in deep underground deposits, is mined by the *Frasch process*. Sulfur is also produced by the *Claus process*, in which hydrogen sulfide (obtained from natural gas and petroleum) is partially burned. Most of the sulfur is used to prepare sulfuric acid by the *contact process*. In this process, sulfur is burned to sulfur dioxide, SO_2, which in the presence of a catalyst and oxygen forms sulfur trioxide, SO_3. This oxide dissolves in concentrated sulfuric acid, which when diluted with water gives additional sulfuric acid. Sulfuric acid is the most important compound of sulfur.

The Group VIIA elements, or halogens, are reactive. Chlorine (Cl_2), a pale greenish-yellow gas, is prepared commercially by the electrolysis of aqueous sodium chloride. Its principal uses are in the preparation of chlorinated hydrocarbons and as a bleaching agent and disinfectant. Hydrogen chloride, HCl, is one of the most important compounds of chlorine; aqueous solutions of HCl are known as hydrochloric acid.

The Group VIIIA elements, the noble gases, were discovered at the end of the nineteenth century. Although the noble gases were at first thought to be unreactive, compounds of xenon, krypton, radon, and argon are now known.

Operational Skills

Note: The problem-solving skills used in this chapter are discussed in previous chapters.

Review Questions

22.1 Give equations for the reactions of lithium and sodium metals with water and with oxygen in air.

22.2 Write the equation for the reaction of lithium carbonate with barium hydroxide.

22.3 Ethanol, C_2H_5OH, reacts with sodium metal because the hydrogen atom attached to the oxygen atom is slightly acidic. Write a balanced equation for the reaction of sodium with ethanol to give the salt sodium ethoxide, $NaOC_2H_5$.

22.4 How is sodium hydroxide manufactured? What is another product in this process?

22.5 Describe common uses of the following sodium compounds: sodium chloride, sodium hydroxide, and sodium carbonate.

22.6 Because magnesium reacts with oxygen, steam, and carbon dioxide, magnesium fires are extinguished only by smothering the fire with sand. Write balanced equations for the reactions of magnesium with O_2, $H_2O(g)$, and CO_2.

22.7 (a) Calcium oxide is prepared industrially from what calcium compound? Write the chemical equation for the reaction. (b) Write the equation for the preparation of calcium hydroxide.

22.8 Calcium carbonate is used in some antacid preparations to neutralize the hydrochloric acid in the stomach. Write the equation for this neutralization.

22.9 What are some major uses of aluminum oxide?

22.10 How is aluminum sulfate used to purify municipal water supplies?

22.11 Lead(IV) oxide forms the cathode of lead storage batteries. How is this substance produced for these batteries?

22.12 Why are lead pigments no longer used in house paints?

22.13 Describe the reactions that are used in the steam-reforming process for the production of hydrogen.

22.14 Explain why hydrogen has the potential to be widely used as a fuel.

22.15 Give an example of a compound of each of the binary hydrides.

22.16 What is meant by the term *catenation*? Give an example of a carbon compound that displays catenation.

22.17 Write equations for the equilibria involving carbon dioxide in water.

22.18 Describe the natural cycle of nitrogen from the atmosphere to biological organisms and back to the atmosphere.

22.19 List the different nitrogen oxides. What is the oxidation number of nitrogen in each?

22.20 Describe the steps in the Ostwald process for the manufacture of nitric acid from ammonia.

22.21 Give chemical equations for the reaction of white phosphorus in an excess of air and for the reaction of the product with water.

22.22 What is the most important commercial means of producing oxygen?

22.23 Define *oxide*, *peroxide*, and *superoxide*. Give an example of each.

22.24 Give an example of an acidic oxide and a basic oxide.

22.25 Describe the structure of the rhombic sulfur molecule.

22.26 Give equations for the steps in the contact process for the manufacture of sulfuric acid from sulfur.

22.27 Complete and balance the following equations. Write *NR* if no reaction occurs.
a. $I_2(aq) + Cl^-(aq) \longrightarrow$
b. $Cl_2(aq) + Br^-(aq) \longrightarrow$
c. $Br_2(aq) + I^-(aq) \longrightarrow$
d. $Br_2(aq) + Cl^-(aq) \longrightarrow$

22.28 What are some major commercial uses of chlorine?

22.29 How is sodium hypochlorite prepared? Give the balanced chemical equation.

22.30 What was the argument used by Bartlett that led him to the first synthesis of a noble-gas compound?

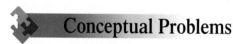

Conceptual Problems

22.31 When producing coke, why is the coal heated in the absence of air? Write the chemical reaction for what would happen if it were to be heated in air.

22.32 Even though hydrogen isn't a metal, why is it in Group IA of most periodic tables?

22.33 Aluminum hydroxide is an amphoteric substance. What does this mean? Write equations to illustrate.

22.34 Given that the reaction $Cl_2(g) + 2KBr(aq) \longrightarrow 2KCl(aq) + Br_2(aq)$ readily occurs, would you expect the reaction $I_2(s) + 2KCl(aq) \longrightarrow 2KI(aq) + Cl_2(aq)$ to occur?

Practice Problems

Group IA: The Alkali Metals

22.35 Caustic soda, NaOH, can be manufactured from sodium carbonate in a manner similar to the preparation of lithium hydroxide. Write balanced equations (in three steps) for the preparation of NaOH from slaked lime, $Ca(OH)_2$, salt (NaCl), carbon dioxide, ammonia, and water.

22.36 Sodium phosphate, Na_3PO_4, is produced by the neutralization reaction. Phosphoric acid, H_3PO_4, is obtained by burning phosphorus to P_4O_{10}, then reacting the oxide with water to give the acid. Write balanced equations (in four steps) for the preparation of Na_3PO_4 from P_4, H_2O, air, and NaCl.

22.37 Complete and balance the following equations.
a. $K(s) + Br_2(l) \longrightarrow$
b. $K(s) + H_2O(l) \longrightarrow$

c. $NaOH(s) + CO_2(g) \longrightarrow$
d. $Li_2CO_3(aq) + HNO_3(aq) \longrightarrow$
e. $K_2SO_4(aq) + Pb(NO_3)_2(aq) \longrightarrow$

22.38 Complete and balance the following equations.
a. $LiHCO_3(s) \xrightarrow{\Delta}$
b. $Na_2SO_4(aq) + BaCl_2(aq) \longrightarrow$
c. $K_2CO_3(aq) + Ca(OH)_2(aq) \longrightarrow$
d. $Li(s) + HCl(aq) \longrightarrow$
e. $Na(s) + ZrCl_4(g) \longrightarrow$

Group IIA: The Alkaline Earth Metals

22.39 Sodium hydroxide and calcium hydroxide are strong bases. What simple chemical test could you use to distinguish between solutions of these two bases?

22.40 Potassium hydroxide and barium hydroxide are strong bases. What simple chemical test could you use to distinguish between solutions of these two bases?

22.41 Complete and balance the following equations.
a. $BaCO_3(s) \xrightarrow{\Delta}$
b. $Ba(s) + H_2O(l) \longrightarrow$
c. $Mg(OH)_2(s) + HNO_3(aq) \longrightarrow$
d. $Mg(s) + NiCl_2(aq) \longrightarrow$
e. $NaOH(aq) + MgSO_4(aq) \longrightarrow$

22.42 Complete and balance the following equations.
a. $KOH(aq) + MgCl_2(aq) \longrightarrow$
b. $Mg(s) + CuSO_4(aq) \longrightarrow$
c. $Sr(s) + H_2O(l) \longrightarrow$
d. $SrCO_3(s) + HCl(aq) \longrightarrow$
e. $Ba(OH)_2(aq) + CO_2(g) \longrightarrow$

Group IIIA and Group IVA Metals

22.43 Baking powders contain sodium (or potassium) hydrogen carbonate and an acidic substance. When water is added to a baking powder, carbon dioxide is released. One kind of baking powder contains $NaHCO_3$ and sodium aluminum sulfate, $NaAl(SO_4)_2$. Write the net ionic equation for the reaction that occurs in water solution.

22.44 When aluminum sulfate is dissolved in water, it produces an acidic solution. Suppose the pH of this solution is raised by the dropwise addition of aqueous sodium hydroxide. (a) Describe what you would observe as the pH continues to rise. (b) Write balanced equations for any reactions that occur.

22.45 Lead(IV) oxide is a strong oxidizing agent. For example, lead(IV) oxide will oxidize hydrochloric acid to chlorine, Cl_2. Write the balanced equation for this reaction.

22.46 Lead(IV) oxide can be prepared by oxidizing plumbite ion, $Pb(OH)_3^-$, which exists in a basic solution of Pb^{2+}. Write the balanced equation for this oxidation by OCl^- in basic solution.

22.47 Complete and balance the following equations.
a. $Al_2O_3(s) + H_2SO_4(aq) \longrightarrow$
b. $Al(s) + AgNO_3(aq) \longrightarrow$
c. $Pb(NO_3)_2(aq) + NaI(aq) \longrightarrow$
d. $Al(s) + Mn_3O_4(s) \longrightarrow$

22.48 Complete and balance the following equations.
a. $Pb(NO_3)_2(aq) + Al(s) \longrightarrow$
b. $Pb(NO_3)_2(aq) + Na_2CrO_4(aq) \longrightarrow$
c. $Al_2(SO_4)_3(aq) + dilute\ LiOH(aq) \longrightarrow$
d. $Al(s) + HCl(aq) \longrightarrow$

Hydrogen

22.49 Calculate the amount of heat evolved when 2.5×10^4 kg of hydrogen is combusted.

$$2H_2(g) + O_2(g) \longrightarrow 2H_2O(g) \qquad \Delta H = -484 \text{ kJ}$$

22.50 How much heat will be evolved when 20.0 grams of the binary covalent hydride HF is produced via the following reaction?

$$F_2(g) + H_2(g) \longrightarrow 2HF(g) \qquad \Delta H = -546 \text{ kJ}$$

22.51 Indicate the oxidation state for the element noted in each of the following:
a. H in CaH_2 b. H in H_2O
c. C in CH_4 d. S in H_2SO_4

22.52 Indicate the oxidation state for the element noted in each of the following:
a. H in H_2 b. H in C_2H_4
c. Si in SiH_4 d. N in HNO_3

Group IVA: The Carbon Family

22.53 Describe the bonding (using valence bond theory) of carbon in each of the following:
a. C_2H_6 b. $CH_3CH{=}CH_2$

22.54 Describe the bonding (using valence bond theory) of carbon in each of the following:
a. CCl_4 b. HCN

22.55 Complete and balance the following equations.
a. $CO_2(g) + Ba(OH)_2(aq) \longrightarrow$
b. $MgCO_3(s) + HBr(aq) \longrightarrow$

22.56 Complete and balance the following equations.
a. $NaHCO_3(aq) + HC_2H_3O_2(aq) \longrightarrow$
b. $Ca(HCO_3)_2(aq) + Ca(OH)_2(aq) \longrightarrow$

Group VA: Nitrogen and the Phosphorus Family

22.57 Magnesium nitride, Mg_3N_2, reacts with water to produce magnesium hydroxide and ammonia. How many grams of ammonia can you obtain from 7.50 g of magnesium nitride?

22.58 Ammonia reacts with oxygen in the presence of a platinum catalyst to give nitric oxide, NO. How many grams of oxygen are required in this reaction to give 5.00 g NO?

22.59 Zinc metal reacts with concentrated nitric acid to give zinc ion and ammonium ion. Write the balanced equation for this reaction.

22.60 Silver metal reacts with nitric acid to give silver ion and nitric oxide. Write the balanced equation for this reaction.

22.61   Although phosphorus pentabromide exists in the vapor phase as PBr_5 molecules, in the solid phase the substance is ionic and has the structure $[PBr_4^+]Br^-$. What is the expected geometry of PBr_4^+? Describe the bonding to phosphorus in PBr_4^+.

22.62   Although phosphorus pentachloride exists in the vapor phase as PCl_5 molecules, in the solid phase the substance is ionic and has the structure $[PCl_4^+][PCl_6^-]$. What is the expected geometry of PCl_6^-? Describe the bonding to phosphorus in PCl_6^-.

Group VIA: Oxygen and the Sulfur Family

22.63   Write an equation for each of the following.
a. burning of lithium metal in oxygen
b. burning of methylamine, CH_3NH_2, in excess oxygen (N ends up as N_2)

22.64   Write an equation for each of the following.
a. burning of calcium metal in oxygen
b. burning of phosphine, PH_3, in excess oxygen

22.65   What are the oxidation numbers of sulfur in each of the following?
a. SF_6 b. SO_3 c. H_2S d. $CaSO_3$

22.66   What are the oxidation numbers of sulfur in each of the following?
a. S_8 b. CaS c. $CaSO_4$ d. SCl_4

Group VIIA: The Halogens

22.67   A solution of chloric acid may be prepared by reacting a solution of barium chlorate with sulfuric acid. Barium sulfate precipitates. Write the balanced equation for the reaction.

22.68   A solution of chlorous acid may be prepared by reacting a solution of barium chlorite with sulfuric acid. Barium sulfate precipitates. Write the balanced equation for the reaction.

22.69   Discuss the bonding in each of the following molecules or ions. What is the expected molecular geometry?
a. Cl_2O b. BrO_3^- c. BrF_3

22.70   Discuss the bonding in each of the following molecules or ions. What is the expected molecular geometry?
a. $HClO$ b. ClO_4^- c. ClF_5

22.71   Write balanced equations for each of the following.
a. Bromine reacts with aqueous sodium hydroxide to give hypobromite and bromide ions.
b. Hydrogen bromide gas forms when sodium bromide is heated with phosphoric acid.

22.72   Write balanced equations for each of the following.
a. Solid calcium fluoride is heated with sulfuric acid to give hydrogen fluoride vapor.
b. Solid potassium chlorate is carefully heated to yield potassium chloride and potassium perchlorate.

Group VIIIA: The Noble Gases

22.73   Xenon tetrafluoride, XeF_4, is a colorless solid. Give the Lewis formula for the XeF_4 molecule. What is the hybridization of the xenon atom in this compound? What geometry is predicted by the VSEPR model for this molecule?

22.74   Xenon tetroxide, XeO_4, is a colorless, unstable gas. Give the Lewis formula for the XeO_4 molecule. What is the hybridization of the xenon atom in this compound? What geometry would you expect for this molecule?

General Problems

22.75   A 50.00-mL volume of 0.4987 M HCl was added to a 5.436-g sample of milk of magnesia. This solution was then titrated with 0.2456 M NaOH. If it required 39.42 mL of NaOH to reach the endpoint, what was the mass percentage of $Mg(OH)_2$ in the milk of magnesia?

22.76   An antacid tablet consists of calcium carbonate with other ingredients. The calcium carbonate in a 0.9863-g sample of the antacid was dissolved in 50.00 mL of 0.5068 M HCl, then titrated with 41.23 mL of 0.2601 M NaOH. What was the mass percentage of $CaCO_3$ in the antacid?

22.77   A sample of limestone was dissolved in hydrochloric acid, and the carbon dioxide gas that evolved was collected. If a 0.1662-g sample of limestone gave 34.56 mL of dry carbon dioxide gas at 745 mmHg and 21°C, what was the mass percentage of $CaCO_3$ in the limestone?

22.78   A sample of rock containing magnesite, $MgCO_3$, was dissolved in hydrochloric acid, and the carbon dioxide gas that evolved was collected. If a 0.1504-g sample of the rock gave 37.71 mL of dry carbon dioxide gas at 758 mmHg and 22°C, what was the mass percentage of $MgCO_3$ in the rock?

22.79   Lithium hydroxide has been used in spaceships to absorb carbon dioxide exhaled by astronauts. Assuming that the product is lithium carbonate, determine what mass of lithium hydroxide is needed to absorb the carbon dioxide from 1.00 L of air containing 30.0 mmHg partial pressure of CO_2 at 25°C.

22.80   Potassium chlorate, $KClO_3$, is used in fireworks and explosives. It can be prepared by bubbling chlorine into hot aqueous potassium hydroxide; $KCl(aq)$ and H_2O are the other products in the reaction. How many grams of $KClO_3$ can be obtained from 156 L of Cl_2 whose pressure is 784 mmHg at 25°C?

22.81 The main ingredient in many phosphate fertilizers is $Ca(H_2PO_4)_2 \cdot H_2O$. If a fertilizer is 17.1% P (by mass), and all of this phosphorus is present as $Ca(H_2PO_4)_2 \cdot H_2O$, what is the mass percentage of this salt in the fertilizer?

22.82 A fertilizer contains phosphorus in two compounds, $Ca(H_2PO_4)_2 \cdot H_2O$ and $CaHPO_4$. The fertilizer contains 30.0% $Ca(H_2PO_4)_2 \cdot H_2O$ and 10.0% $CaHPO_4$ (by mass). What is the mass percentage of phosphorus in the fertilizer?

22.83 NaClO solution is made by electrolysis of NaCl(*aq*) by allowing the products NaOH and Cl_2 to mix. How long must a cell operate to produce 1.00×10^3 L of 5.25% NaClO solution (density = 1.00 g/mL) if the cell current is 3.00×10^3 A?

22.84 Sodium perchlorate, $NaClO_4$, is produced by electrolysis of sodium chlorate, $NaClO_3$. If a current of 2.50×10^3 A passes through an electrolytic cell, how many kilograms of sodium perchlorate are produced per hour?

Heating pink $[Co(NH_3)_5H_2O]Cl_3(s)$ drives off water molecules, forming blue $[Co(NH_3)_5Cl]Cl_2(s)$.

The Transition Elements and Coordination Compounds

In the previous chapter, we studied the main-group elements—the A groups in the periodic table. Between columns IIA and IIIA are ten columns of the transition elements—the B groups. Among these elements are metals with familiar commercial applications: iron tools, copper wire, silver jewelry and coins. Many catalysts for important industrial reactions involve transition elements.

In addition to their commercial usefulness, many transition elements have biological importance. Iron is present in hemoglobin, the molecule in red blood cells that is responsible for the transport of oxygen, O_2, from the lungs to other body tissue. Myoglobin, found in muscle tissue, is a similar molecule containing iron. Cytochromes, iron-containing compounds within each cell, are involved in the oxidation of food molecules. Hemoglobin and myoglobin are examples of *metal complexes,* or *coordination compounds,* in which the metal atom is surrounded by other atoms bonded to it by the electron pairs these atoms donate.

Properties of the Transition Elements

The **transition elements** are defined as *those metallic elements that have an incompletely filled d subshell or easily give rise to common ions that have incompletely filled d subshells.* Iron, whose electron configuration is $[Ar]3d^6 4s^2$ is an example of such an element. Copper has the configuration $[Ar]3d^{10}4s^1$, in which the $3d$ subshell is complete; copper readily forms the copper(II) ion, whose configuration $[Ar]3d^9$ has an incomplete $3d$ subshell. The Group IIB elements zinc, cadmium, and mercury have filled d subshells in the element and in the common ions, so in the strict sense these are not transition elements.

Sometimes, too, chemists include the two rows of elements at the bottom of the periodic table with the transition elements. These two rows, often referred to as the *inner-transition elements,* have partially filled f subshells in common oxidation states. The elements in the first row are called the *lanthanides,* or *rare earths,* and the elements in the second row are called the *actinides.* Figure 23.1 shows the divisions of the transition elements. The B columns of transition elements, as well as the inner-transition elements, frequently form complex ions and coordination compounds.

23.1   Periodic Trends in the Transition Elements

The transition elements have a number of characteristics that set them apart from the main-group elements:

1. All of the transition elements are metals and, except for the IIB elements, have high melting points and high boiling points and are hard solids. Of the main-group metals, only beryllium melts above 1000°C; the rest melt at appreciably lower temperatures.

2. With the exception of the IIIB and IIB elements, each transition element has several oxidation states. Vanadium, for example, exists as aqueous ions in all

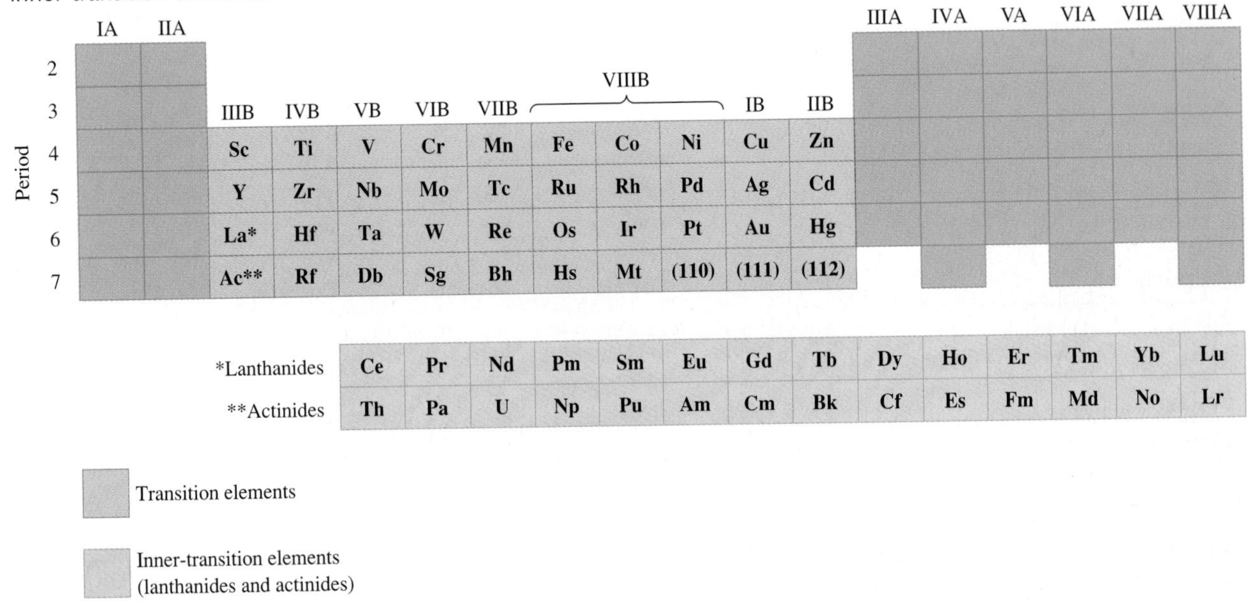

Figure 23.1
Classification of the transition elements. The classification into the B groups of transition elements and the inner-transition elements.

	IA	IIA										IIIA	IVA	VA	VIA	VIIA	VIIIA
2								VIIIB									
3			IIIB	IVB	VB	VIB	VIIB			IB	IIB						
4			Sc	Ti	V	Cr	Mn	Fe	Co	Ni	Cu	Zn					
5			Y	Zr	Nb	Mo	Tc	Ru	Rh	Pd	Ag	Cd					
6			La*	Hf	Ta	W	Re	Os	Ir	Pt	Au	Hg					
7			Ac**	Rf	Db	Sg	Bh	Hs	Mt	(110)	(111)	(112)					

Period

*Lanthanides	Ce	Pr	Nd	Pm	Sm	Eu	Gd	Tb	Dy	Ho	Er	Tm	Yb	Lu
**Actinides	Th	Pa	U	Np	Pu	Am	Cm	Bk	Cf	Es	Fm	Md	No	Lr

 Transition elements

Inner-transition elements
(lanthanides and actinides)

Figure 23.2
Oxidation states of vanadium. The test tubes contain, left to right: $VO_3^-(aq)$, pale yellow; $VO^{2+}(aq)$, bright blue; V^{3+}, gray-blue; V^{2+}, violet. The oxidation states of vanadium are, left to right, +5, +4, +3, and +2.

oxidation states from +2 to +5 (Figure 23.2). Because of their multiplicity of oxidation states, the transition elements are often involved in oxidation–reduction reactions.

3. Transition-metal compounds are often colored, and many are paramagnetic. Most compounds of the main-group metals are colorless and diamagnetic.

Electron Configurations

Electronic structure is central to any discussion of the transition elements. Table 23.1 lists the ground-state electron configurations of the fourth-period transition elements. For the most part, these configurations are predicted by the building-up principle. With chromium the configuration predicted by the building-up principle is $[Ar]3d^44s^2$, but the actual configuration is $[Ar]3d^54s^1$. With copper the predicted configuration is $[Ar]3d^94s^2$, but the actual configuration is $[Ar]3d^{10}4s^1$.

Melting Points and Boiling Points

Table 23.1 reveals that the melting points of the transition metals increase from 1541°C for scandium to 1890°C for vanadium and 1857°C for chromium, then decrease from 1535°C for iron to 1083°C for copper and 420°C for zinc. The same pattern is observed in the fifth-period and sixth-period elements. These properties depend on the strength of metal bonding, which in turn depends roughly on the number of unpaired electrons in the metal atoms. The number of unpaired d electrons increases across a period until Group VIB, after which the electrons begin to pair.

Table 23.1
Properties of the Fourth-Period Transition Elements

Property	Scandium	Titanium	Vanadium	Chromium	Manganese
Electron configuration	$[Ar]3d^14s^2$	$[Ar]3d^24s^2$	$[Ar]3d^34s^2$	$[Ar]3d^54s^1$	$[Ar]3d^54s^2$
Melting point, °C	1541	1660	1890	1857	1244
Boiling point, °C	2831	3287	3380	2672	1962
Density, g/cm^3	3.0	4.5	6.0	7.2	7.2
Electronegativity (Pauling scale)	1.3	1.5	1.6	1.6	1.5
Covalent radius, pm	144	132	122	118	117
Ionic radius (for M^{2+}), pm	—	100	93	87	81

Property	Iron	Cobalt	Nickel	Copper	Zinc
Electron configuration	$[Ar]3d^64s^2$	$[Ar]3d^74s^2$	$[Ar]3d^84s^2$	$[Ar]3d^{10}4s^1$	$[Ar]3d^{10}4s^2$
Melting point, °C	1535	1495	1453	1083	420
Boiling point, °C	2750	2870	2732	2567	907
Density, g/cm^3	7.9	8.9	8.9	8.9	7.1
Electronegativity (Pauling scale)	1.8	1.8	1.8	1.9	1.6
Covalent radius, pm	117	116	115	117	125
Ionic radius (for M^{2+}), pm	75	79	83	87	88

Figure 23.3
Covalent radii for transition elements of Periods 4 to 6. Note that elements of the fifth and sixth periods of the same column have approximately the same radii.

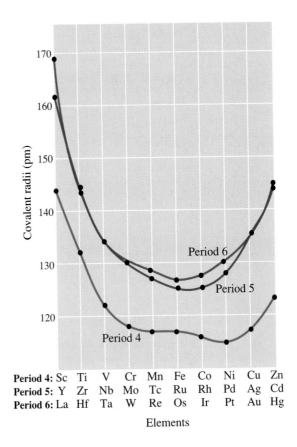

Period 4:	Sc	Ti	V	Cr	Mn	Fe	Co	Ni	Cu	Zn
Period 5:	Y	Zr	Nb	Mo	Tc	Ru	Rh	Pd	Ag	Cd
Period 6:	La	Hf	Ta	W	Re	Os	Ir	Pt	Au	Hg

Elements

Atomic Radii

Trends in atomic radii are of concern because chemical properties are determined in part by atomic size. Looking at the fourth-period covalent radii in Table 23.1, you see that they decrease in atomic size across a row. This is due to an increase in *effective nuclear charge*. The effective nuclear charge is the positive charge "felt" by an electron; it equals the nuclear charge minus the shielding or screening of the positive charge by intervening electrons. The covalent radii decrease slowly from for chromium to for nickel. Then the covalent radii increase slightly for copper and for zinc. ◀

Figure 23.3 compares the covalent radii of the transition elements. The atomic radii increase in going from a fourth-period to a fifth-period element within any column. You would expect an increase of radius because of the addition of a shell of electrons. The elements of the sixth period have nearly the same covalent radii as the corresponding elements in the fifth period.

This similarity of radii in the fifth- and sixth-period transition elements can be explained in terms of the *lanthanide contraction*. Between lanthanum and hafnium in the sixth period are 14 lanthanide elements (cerium to lutetium), in which the 4f subshell fills. The covalent radius decreases slowly from cerium to lutetium, but the total decrease is substantial. By the time the 4f subshell is complete, the covalent radii of the transition elements from hafnium on are similar to those of the elements in the preceding row of the periodic table.

The chemical properties of the transition elements parallel the pattern seen in covalent radii. That is, the fourth-period elements have substantially different properties from the elements in the same group in the fifth and sixth periods. However,

▶ The small increase in covalent radius for copper and zinc has no simple explanation.

Table 23.2
First Ionization Energies of the Transition Elements (kJ/mol)

Period	IIIB	IVB	VB	VIB	VIIB	VIIIB			IB	IIB
Fourth	Sc	Ti	V	Cr	Mn	Fe	Co	Ni	Cu	Zn
	631	658	650	652	717	759	758	737	745	906
Fifth	Y	Zr	Nb	Mo	Tc	Ru	Rh	Pd	Ag	Cd
	616	660	664	685	702	711	720	805	731	868
Sixth	La	Hf	Ta	W	Re	Os	Ir	Pt	Au	Hg
	538	680	761	770	760	840	880	870	890	1007

elements in the same group of the fifth and sixth periods are very much alike. Hafnium, for example, is so much like zirconium that it remained undiscovered until it was identified in 1923 in zirconium ores from its x-ray spectrum.

Ionization Energies

Looking at the first ionization energies of the fourth-period transition elements (Table 23.2), you see that although they vary somewhat irregularly, they tend to increase from left to right. The other rows of transition elements behave similarly. Most noteworthy, however, is that all of the sixth-period elements after lanthanum have ionization energies higher than those of the fourth-period and fifth-period transition elements in the same group.

Oxidation States

Table 23.3 gives oxidation states in compounds of the fourth-period transition elements. Scandium occurs as the 3+ ion and zinc as the 2+ ion. The other elements exhibit several oxidation states.

Table 23.3
Oxidation States of the Fourth-Period Transition Elements

IIIB	IVB	VB	VIB	VIIB	VIIIB			IB	IIB
Sc	Ti	V	Cr	Mn	Fe	Co	Ni	Cu	Zn
							+1	+1	
	+2	+2	+2	+2	+2	+2	+2	+2	+2
+3	+3	+3	+3	+3	+3	+3	+3	+3	
	+4	+4	+4	+4	+4	+4	+4		
		+5	+5	+5	+5				
			+6	+6	+6				
				+7					

Key: Common oxidation states are in boldface. Additional oxidation states, particularly zero and negative values, may be observed in complexes with CO and with organic compounds.

Most of the transition elements have a doubly filled ns orbital. Because the ns electrons ionize before the $(n-1)d$ electrons, you might expect the $+2$ oxidation state to be common. This oxidation state is in fact seen in all of the fourth-period elements except scandium.

The maximum oxidation state possible equals the number of s and d electrons in the valence shell. Titanium in Group IVB has a maximum oxidation state of $+4$. Similarly, vanadium (Group VB), chromium (Group VIB), and manganese (Group VIIB) exhibit maximum oxidation states of $+5$, $+6$, and $+7$, respectively. ◀

▶ The maximum oxidation state is generally found in compounds of the transition elements with very electronegative elements, such as F and O (in oxides and oxoanions).

The total number of oxidation states actually observed increases from scandium $(+3)$ to manganese (all states from $+2$ to $+7$). From iron on, however, the maximum oxidation state is not attained. Thereafter, the highest observed oxidation number decreases until, for zinc compounds, you find only the $+2$ oxidation state.

23.2 The Chemistry of Two Transition Elements

Chromium

The origin of the name chromium comes from the Greek word *khroma*, meaning "color." The common oxidation states of chromium in compounds are $+2$, $+3$, and $+6$.

Chromium(II), $Cr(H_2O)_6^{2+}(aq)$, has a bright blue color. The ion is obtained when an acid, such as $HCl(aq)$, reacts with chromium metal (see Figure 23.4, top).

$$Cr(s) + 2HCl(aq) + 6H_2O(l) \longrightarrow Cr(H_2O)_6^{2+}(aq) + 2Cl^-(aq) + H_2(g)$$

The chromium(II) ion is easily oxidized to chromium(III) ion by O_2 (Figure 23.4, bottom).

The most common oxidation state of chromium is $+3$. Chromium metal burns in air to give chromium(III) oxide, Cr_2O_3. This oxide is a dark green solid, which has been used as a paint pigment (chrome green). Chromium(III) oxide dissolves in acid solution to form the violet-colored ion $Cr(H_2O)_6^{3+}(aq)$. (Hydrochloric acid solution, however, yields green-colored complex ions of Cr^{3+} with Cl^-.)

The $+6$ oxidation state is represented by the chromate ion, CrO_4^{2-}, and the dichromate ion, $Cr_2O_7^{2-}$. Sodium chromate, Na_2CrO_4, which is yellow, and sodium dichromate, $Na_2Cr_2O_7$, which is orange, are the primary sources of compounds of chromium. Sodium chromate, Na_2CrO_4, is prepared from chromite, $FeCr_2O_4(s)$, the principal chromium ore, by strongly heating it with sodium carbonate in air.

$$4FeCr_2O_4(s) + 8Na_2CO_3(s) + 7O_2(g) \xrightarrow{1100°C} 8Na_2CrO_4(s) + 2Fe_2O_3(s) + 8CO_2(g)$$

Sodium chromate is soluble and can be leached from the mixture with water.

Sodium chromate can be easily converted to sodium dichromate by treating it with an acid. When the yellow solution of a chromate salt is acidified, it turns orange from the formation of dichromate ion, $Cr_2O_7^{2-}(aq)$ (Figure 23.5).

$$\underset{\text{yellow}}{2CrO_4^{2-}(aq)} + 2H^+(aq) \rightleftharpoons \underset{\text{orange}}{Cr_2O_7^{2-}(aq)} + H_2O(l)$$

The chromate and dichromate ions are in equilibrium, which is sensitive to pH changes; lower pH favors dichromate ion.

The dichromate ion is a strong oxidizing agent in acid solution.

$$Cr_2O_7^{2-}(aq) + 14H^+(aq) + 6e^- \longrightarrow 2Cr^{3+}(aq) + 7H_2O(l); \; E° = 1.33 \text{ V}$$

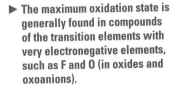

Figure 23.4
Aqueous chromium ion.
Top: Chromium metal reacts with hydrochloric acid to produce $Cr(H_2O)_6^{2+}$, which is blue. *Bottom:* $Cr(H_2O)_6^{2+}$ oxidizes with air to give a chromium(III) species, which is green.

CrO_4^{2-}

$Cr_2O_7^{2-}$

Figure 23.5
Chromate–dichromate equilibrium. The beaker on the top contains CrO_4^{2-} (yellow). When the experimenter adds sulfuric acid to a similar solution in the beaker on the bottom, CrO_4^{2-} is converted to $Cr_2O_7^{2-}$ (orange).

Pure chromium metal is prepared by the reaction of chromium(III) oxide, obtained from chromite ore, with aluminum (the *Goldschmidt process*). Once the reaction mixture is ignited, the large heat of reaction produces molten chromium.

$$Cr_2O_3(s) + 2Al(s) \longrightarrow 2Cr(l) + Al_2O_3(s); \Delta H^\circ = -488 \text{ kJ}$$

One of the chief uses of chromium is in steel making.

Copper

Copper and the other Group IB elements, silver and gold, can be found in nature as the free metals. This is a reflection of the stability of their 0 oxidation states. For example, copper is not attacked by most acids.

Copper metal does react with nitric acid. The anion of the acid acts as the oxidizing agent (rather than H^+, the usual oxidizing agent in acids). Dilute nitric acid is reduced to NO; concentrated nitric acid, to NO_2:

$$3Cu(s) + 2NO_3^-(aq) + 8H^+(aq) \longrightarrow 3Cu^{2+}(aq) + 2NO(g) + 4H_2O(l)$$
$$Cu(s) + 2NO_3^-(aq) + 4H^+(aq) \longrightarrow Cu^{2+}(aq) + 2NO_2(g) + 2H_2O(l)$$

The copper(II) ion, given in these equations as $Cu^{2+}(aq)$, is written more precisely as $Cu(H_2O)_6^{2+}$. It has a bright blue color. Hydrated copper(II) salts, such as copper(II) sulfate pentahydrate, $CuSO_4 \cdot 5H_2O$, also have a blue color. Four of the water molecules of copper(II) sulfate pentahydrate are associated with Cu^{2+}, and the fifth is hydrogen-bonded to the sulfate ion as well as to the water molecules on the copper ion. You can write the formula as $[Cu(H_2O)_4]SO_4 \cdot H_2O$.

Although most of the aqueous chemistry of copper involves the +2 oxidation state, there are a number of important compounds of copper(I). When copper is heated in oxygen below 1000°C, it forms the black copper(II) oxide, CuO. But above this temperature, it forms the brick-red copper(I) oxide, Cu_2O.

The principal commercial use of copper metal is as an electrical conductor. Most copper is obtained by open-pit mining of low-grade rock containing only a small percentage of copper as copper sulfides. The ore is concentrated in copper by flotation. In this process, a slurry of the crushed ore is agitated with air, and the copper sulfides are carried away in the froth. This concentrated ore is then treated in several steps that result in the production of copper(I) sulfide, Cu_2S. At a high temperature Cu_2S is reduced to copper by blowing air through it.

$$Cu_2S(l) + O_2(g) \longrightarrow 2Cu(l) + SO_2(g)$$

The metal produced is called *blister copper* and is about 99% pure. For electrical use, the copper must be further purified or refined by electrolysis (see Figure 20.18).

Complex Ions and Coordination Compounds

As shown in the previous section, ions of the transition elements exist in aqueous solution as *complex ions*. Iron(II) ion, for example, exists in water as $Fe(H_2O)_6^{2+}$. The water molecules in this ion are arranged about the iron atom with their oxygen atoms bonded to the metal by donating electron pairs to it. Replacing the H_2O molecules by six CN^- ions gives the $Fe(CN)_6^{4-}$ ion. Some of the transition elements have biological activity and their role in human nutrition (Table 23.4) depends in most cases on the formation of *complexes*, or *coordination compounds*, which exhibit the type of bonding that occurs in $Fe(H_2O)_6^{2+}$ and $Fe(CN)_6^{4-}$.

Table 23.4
Transition Elements Essential to Human Nutrition

Element	Some Biochemical Substances	Function
Chromium	Glucose tolerance factor	Utilization of glucose
Manganese	Isocitrate dehydrogenase	Cell energetics
Iron	Hemoglobin and myoglobin	Transport and storage of oxygen
	Cytochrome c	Cell energetics
	Catalase	Decomposition of hydrogen peroxide
Cobalt	Cobalamin (vitamin B_{12})	Development of red blood cells
Copper	Ceruloplasmin	Synthesis of hemoglobin
	Cytochrome oxidase	Cell energetics
Zinc	Carbonic anhydrase	Elimination of carbon dioxide
	Carboxypeptidase A (pancreatic juice)	Protein digestion
	Alcohol dehydrogenase	Oxidation of ethanol

23.3 Formation and Structure of Complexes

▶ Lewis acid–base reactions are discussed in Section 16.3.

A metal atom, particularly a transition-metal atom, often functions in chemical reactions as a Lewis acid, accepting electron pairs from molecules or ions. ◀ For example, Fe^{2+} and H_2O can bond to one another in a Lewis acid–base reaction.

$$Fe^{2+} \ + \ :\overset{..}{\underset{H}{O}}-H \ \longrightarrow \ \left[Fe:\overset{..}{\underset{..}{O}}-H \right]^{2+}$$

coordinate covalent bond

Lewis acid   Lewis base

A pair of electrons on the oxygen atom of H_2O forms a coordinate covalent bond to Fe^{2+}. In water, the Fe^{2+} ion forms the $Fe(H_2O)_6^{2+}$ ion.

The Fe^{2+} ion also undergoes a similar Lewis acid–base reaction with cyanide ions. The Fe^{2+} ion bonds to the electron pair on the carbon atom of CN^-.

$$Fe^{2+} + (:C\equiv N:)^- \ \longrightarrow \ (Fe:C\equiv N:)^+$$

▶ Although the ion $Fe(CN)_6^{4-}$ is a complex of Fe^{2+} and CN^- ions, a solution of $Fe(CN)_6^{4-}$ contains negligible concentration of CN^-. Therefore, a substance such as $K_4[Fe(CN)_6]$ is relatively nontoxic, even though the free cyanide ion is a potent poison.

A very stable ion, $Fe(CN)_6^{4-}$, is formed. ◀ The charge on the $Fe(CN)_6^{4-}$ ion equals the sum of the charges on the ions from which it is formed: $+2 + 6(-1) = -4$.

In some cases, a neutral species is produced from a metal ion and anions. Cisplatin, the anticancer drug discussed in the opening of Chapter 1, has the structure

$$\begin{array}{c} :\overset{..}{\underset{..}{Cl}}: \\ H_3N:\overset{..}{\underset{..}{Pt}}:\overset{..}{\underset{..}{Cl}}: \\ NH_3 \end{array}$$

It consists of Pt^{2+} with two NH_3 molecules (neutral) and two Cl^- ions, giving a neutral species. Iron pentacarbonyl, $Fe(CO)_5$, is another example of a neutral species.

Table 23.5
Examples of Complexes of Various Coordination Numbers

Complex	Coordination Number
$Ag(NH_3)_2^+$	2
HgI_3^-	3
$PtCl_4^{2-}$, $Ni(CO)_4$	4
$Fe(CO)_5$, $Co(CN)_5^{3-}$	5
$Co(NH_3)_6^{3+}$, $W(CO)_6$	6
$Mo(CN)_7^{3-}$	7
$W(CN)_8^{4-}$	8

Basic Definitions

A **complex ion** is *a metal ion with Lewis bases attached to it through coordinate covalent bonds.* A **complex** (or **coordination compound**) is *a compound consisting either of complex ions and other ions of opposite charge* (for example, the compound $K_4[Fe(CN)_6]$) *or of a neutral complex species* (such as cisplatin).

Ligands are *the Lewis bases attached to the metal atom in a complex.* They are electron-pair donors, so ligands may be neutral molecules (such as H_2O or NH_3) or anions (such as CN^- or Cl^-).

The **coordination number** of a metal atom in a complex is *the total number of bonds the metal atom forms with ligands.* In $Fe(H_2O)_6^{2+}$, the iron atom bonds to each oxygen atom in the six water molecules. Therefore, the coordination number of iron in this ion is 6, by far the most common coordination number. Coordination number 4 is also well known. Table 23.5 gives some examples of complexes for the coordination numbers 2 to 8. The coordination number for an atom depends on several factors, but size of the metal atom is important. ◄

▶ Very high coordination numbers (9 to 12) are known for some complex ions of the lanthanide elements.

Polydentate Ligands

The ligands we have discussed so far bond to the metal atom through one atom of the ligand. This type of bonding indicates a **monodentate ligand** (meaning "one-toothed" ligand). A **bidentate ligand** ("two-toothed" ligand) is *a ligand that bonds to a metal atom through two atoms of the ligand.* Ethylenediamine is an example.

ethylenediamine

Figure 23.6
Structure of tris(ethylenediamine) cobalt(III) ion, Co(en)₃³⁺.
(A) Ball-and-stick model. *(B)* Shorthand notation. Note that N⌒N represents ethylenediamine.

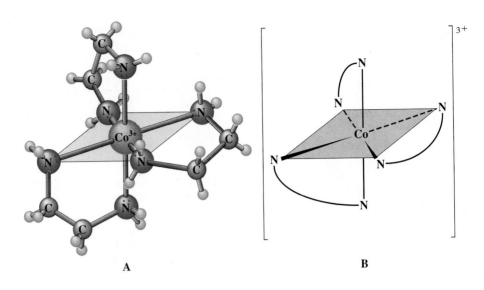

A

B

In forming a complex, the ethylenediamine molecule bends around so that both nitrogen atoms coordinate to the metal atom, M.

Because ethylenediamine is a common bidentate ligand, it is frequently abbreviated in formulas as "en." Figure 23.6 shows the structure of the stable ion $Co(en)_3^{3+}$.

The hemoglobin molecule in red blood cells is an example of a complex with a *quadridentate ligand*—one that bonds to the metal atom through four ligand atoms. Hemoglobin consists of the protein *globin* chemically bonded to *heme,* whose structure is shown in Figure 23.7 (left).

Figure 23.7
Complexes with ligands that bond with more than one atom to the metal atom (polydentate ligands). *Left:* Structure of heme. *Right:* Complex of Fe^{2+} and ethylenediaminetetra-acetate ion (EDTA). Note how the EDTA ion envelops the metal ion.

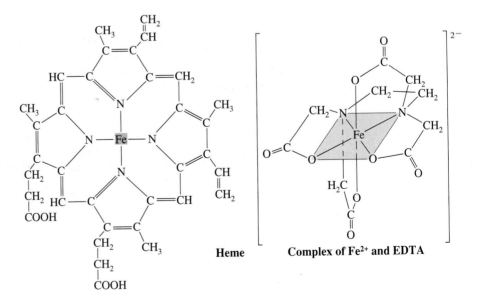

Heme   **Complex of Fe²⁺ and EDTA**

Ethylenediaminetetraacetate ion (EDTA) is a ligand that bonds through six of its atoms.

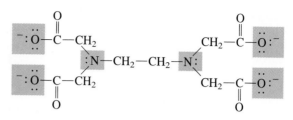

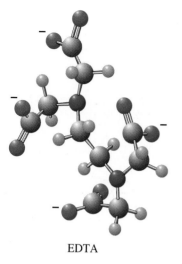

EDTA

▶ The term *chelate* is derived from the Greek *chele* for "claw," because a polydentate ligand appears to attach itself to the metal atom like crab claws to some object.

It can completely envelop a metal atom, simultaneously occupying all six positions in an octahedral geometry (Figure 23.7, right).

A **polydentate ligand** is *a ligand that can bond with two or more atoms to a metal atom. A complex formed by polydentate ligands is frequently quite stable and is called a* **chelate.** ◀ Polydentate ligands are often used to remove metal ions from a chemical system. EDTA, for example, is added to certain canned foods to remove transition-metal ions that can catalyze the deterioration of the food. The same chelating agent has been used to treat lead poisoning.

Discovery of Complexes; Formula of a Complex

Alfred Werner, a Swiss chemist, studied a series of platinum(IV) chloride and ammonia complexes (Table 23.6). In 1893 he explained that a metal atom exhibits two kinds of valences, a primary valence and a secondary valence. The primary valence is what we now call the oxidation number of the metal. The secondary valence corresponds to what we now call the coordination number, which is often 6.

In each case in Table 23.6 platinum has a coordination number of six and there is a charge balance. According to Werner's theory, these complexes should dissolve to give different numbers of ions per formula unit. Werner was able to show that the electrical conductances of solutions of these complexes were equal to what was expected for the number of ions predicted by his formulas. He also demonstrated that the chloride ions in the platinum complexes were of two kinds: those that could be precipitated from solution as AgCl using silver nitrate and those that could not. The chloride ions within the platinum complex ion are securely attached to the metal atom and only those outside the complex ion can be precipitated with silver nitrate. Werner received the Nobel Prize for his work in 1913.

Table 23.6
Some Platinum(IV) Complexes Studied by Werner

Old Formula	Modern Formula	Number of Ions	Number of Free Cl^- Ions
$PtCl_4 \cdot 6NH_3$	$[Pt(NH_3)_6]Cl_4$	5	4
$PtCl_4 \cdot 4NH_3$	$[Pt(NH_3)_4Cl_2]Cl_2$	3	2
$PtCl_4 \cdot 3NH_3$	$[Pt(NH_3)_3Cl_3]Cl$	2	1
$PtCl_4 \cdot 2NH_3$	$[Pt(NH_3)_2Cl_4]$	0	0

Salad Dressing and Chelate Stability

EVERYDAY LIFE

The list of ingredients for a particular mayonnaise reads: vegetable oil, eggs, vinegar, and calcium disodium EDTA (ethylenediaminetetraacetate). EDTA? Yes, commercial mayonnaise and salad dressings use EDTA to remove traces of metal ions. Metal ions can catalyze undesirable reactions or else provide nutrient for bacteria, resulting in off-flavors and spoilage of the product. EDTA is a polydentate ligand that forms particularly stable chelates with many metal ions, effectively removing those ions from the product. Many commercial products contain chelating agents such as EDTA (Figure 23.8).

What accounts for the special stability of chelates? Their stability stems from the additional entropy obtained when they are formed. This leads to a large negative $\Delta G°$, which is equivalent to a large equilibrium constant for the formation of the chelate. To understand how this happens, consider the formation of the chelate $Co(en)_3^{3+}$ from the complex ion $Co(NH_3)_6^{3+}$, with monodentate ligands NH_3.

$$Co(NH_3)_6^{3+} + 3en \rightleftharpoons Co(en)_3^{3+} + 6NH_3$$

Each en molecule replaces two NH_3 molecules. Therefore, the number of particles in the reaction mixture is increased when the reaction goes to the right. In most cases, an increase in number of particles increases the possibilities for randomness or disorder. Therefore, when the reaction goes to the right, there is an increase in entropy; that is, $\Delta S°$ is positive.

The reaction involves very little change in enthalpy because the bonds are similar; all consist of a nitrogen

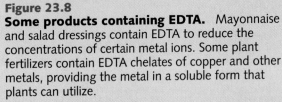

Figure 23.8
Some products containing EDTA. Mayonnaise and salad dressings contain EDTA to reduce the concentrations of certain metal ions. Some plant fertilizers contain EDTA chelates of copper and other metals, providing the metal in a soluble form that plants can utilize.

atom coordinated to a cobalt atom. Therefore, $\Delta H° \simeq 0$.

The spontaneity of a reaction depends on $\Delta G°$, which equals $\Delta H° - T\Delta S°$. But because $\Delta H° \simeq 0$,

$$\Delta G° = \Delta H° - T\Delta S°$$
$$= -T\Delta S°$$

The entropy change is positive, so $\Delta G°$ is negative.

The fact that the equilibrium for the reaction favors the chelate $Co(en)_3^{3+}$ means that the chelate has thermodynamic stability.

CONCEPT CHECK 23.1

Another complex studied by Werner had a composition corresponding to the formula $PtCl_4 \cdot 2KCl$. From electrical-conductance measurements, he determined that each formula unit contained three ions. He also found that silver nitrate did not give a precipitate of AgCl with this complex. Write a formula for this complex that agrees with this information.

23.4 Naming Coordination Compounds

Thousands of coordination compounds are now known. A systematic method of naming such compounds, or *nomenclature,* needs to provide basic information about the structure of a coordination compound. The rules of nomenclature are those of the International Union of Pure and Applied Chemistry (IUPAC).

1. In naming a salt, the name of the cation precedes the name of the anion.

    $K_4[Fe(CN)_6]$ is named <u>potassium</u> <u>hexacyanoferrate(II)</u>
                                     cation   anion

    $[Co(NH_3)_6]Cl_3$ is named <u>hexaamminecobalt(III)</u> <u>chloride</u>
                                        cation   anion

2. The name of the complex consists of two parts written together as one word. Ligands are named first, and the metal atom is named second.

    $Fe(CN)_6^{4-}$ is named <u>hexacyano</u><u>ferrate(II)</u> ion
                               ligand   metal
                               name   name

    $Co(NH_3)_6^{3+}$ is named <u>hexaammine</u><u>cobalt(III)</u> ion
                               ligand   metal
                               name   name

3. The complete ligand name consists of a Greek prefix denoting the number of ligands, followed by the specific name of the ligand. When there are two or more ligands, the ligands are written in alphabetical order (disregarding Greek prefixes).

    a. Anionic ligands end in *-o.* Some examples:

Anion Name	Ligand Name	Anion Name	Ligand Name
Bromide, Br^-	Bromo	Fluoride, F^-	Fluoro
Carbonate, CO_3^{2-}	Carbonato	Hydroxide, OH^-	Hydroxo
Chloride, Cl^-	Chloro	Oxalate, $C_2O_4^{2-}$	Oxalato
Cyanide, CN^-	Cyano	Oxide, O^{2-}	Oxo

    b. Neutral ligands are usually given the name of the molecule. There are, however, several important exceptions:

Molecule	Ligand Name
Ammonia, NH_3	Ammine
Carbon monoxide, CO	Carbonyl
Water, H_2O	Aqua

    c. The prefixes used to denote the number of ligands are *mono-* (1, usually omitted); *di-* (2); *tri-* (3); *tetra-* (4); *penta-* (5); *hexa-* (6); and so forth. To see how the ligand name is formed, consider the complex ions

    $Fe(CN)_6^{4-}$ or <u>hexa</u>cyanoferrate(II) ion
                         6 CN^- ligands

    $Co(NH_3)_6^{3+}$ or <u>hexa</u>amminecobalt(III) ion
                         6 NH_3 ligands

    d. When the name of the ligand also has a number prefix, the number of ligands is denoted with *bis* (2), *tris* (3), *tetrakis* (4), and so forth. The name of the

ligand follows in parentheses. For example, the complex [Co(en)$_3$]Cl$_3$ is named as follows:

tris(ethylenediamine)cobalt(III) chloride
  ‿‿‿‿   ‿‿‿‿‿‿‿‿
   3   ligand name

4. The complete metal name consists of the name of the metal, followed by *-ate* if the complex is an anion, followed by the oxidation number of the metal as a Roman numeral in parentheses. When there is a Latin name for the metal, it is usually used to name the anion.

English Name	Latin Name	Anion Name
Copper	Cuprum	Cuprate
Gold	Aurum	Aurate
Iron	Ferrum	Ferrate
Lead	Plumbum	Plumbate
Silver	Argentum	Argentate
Tin	Stannum	Stannate

Examples are

hexacyanoferrate(II)
           ‿‿‿‿ ‿‿‿‿‿‿‿‿
          ferrum | oxidation
          = iron | number 2
                 indicates
                 an anion

hexaamminecobalt(III)
           ‿‿‿‿ ‿‿‿‿‿‿‿‿
          metal oxidation
          name   number 3

EXAMPLE 23.1

Writing the IUPAC Name Given the Structural Formula of a Coordination Compound

Give the IUPAC name of each of the following coordination compounds: a. [Pt(NH$_3$)$_4$Cl$_2$]Cl$_2$; b. [Pt(NH$_3$)$_2$Cl$_2$]; c. K$_2$[PtCl$_6$].

SOLUTION

a. The cation is listed first in the formula.

There are two Cl$^-$ anions, so the charge on the cation is $+2$: Pt(NH$_3$)$_4$Cl$_2^{2+}$. The oxidation number of platinum plus the sum of the charges on the ligands (-2) equals the cation charge $+2$. Therefore, the oxidation number of Pt is $+4$. Hence, the name of the compound is **tetraamminedichloroplatinum(IV) chloride**. Note that the ligands are listed in alphabetical order (that is, ammine before chloro).

b. This is a neutral complex species. The oxidation number of platinum must balance that of the two chloride ions. The name of the compound is **diamminedichloroplatinum(II)**.

c. The complex anion is PtCl$_6^{2-}$. The oxidation number of platinum is 4, and the name of the compound is **potassium hexachloroplatinate(IV)**.

See Problems 23.35 and 23.36.

Figure 23.15
Covalent bond formation between atoms X and Y.
(A) In the usual case, each overlapping orbital contains one electron. *(B)* When a coordinate covalent bond forms, one orbital containing a lone pair of electrons overlaps an empty orbital.

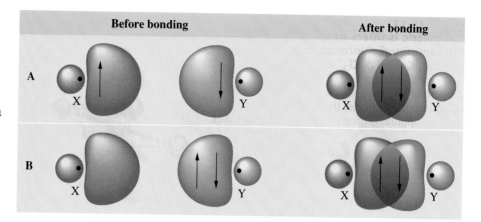

According to this view, a covalent bond is formed by the overlap of two orbitals, one from each bonding atom. In the usual covalent bond formation, each orbital originally holds one electron, and after the orbitals overlap, a bond is formed that holds two electrons. In the formation of a coordinate covalent bond in a complex, however, a ligand orbital containing two electrons overlaps an unoccupied orbital on the metal atom. Figure 23.15 diagrams these two bond formations.

Let's look at the formation of the yellow $Cr(NH_3)_6^{3+}$ ion. We can think of this ion as formed by complexing $:NH_3$ ligands with the free chromium(III) ion. First, we need the configuration of the free ion. The chromium atom configuration is $[Ar]3d^5 4s^1$, while Cr^{3+} has the configuration $[Ar]3d^3$. The orbital diagram is

Note that the $3d^3$ electrons are placed in separate orbitals with their spins in the same direction, following Hund's rule. Two empty $3d$ orbitals, in addition to orbitals of the $n = 4$ shell, can be used for bonding to ligands.

To bond electron pairs from six $:NH_3$ ligands to Cr^{3+}, forming six equivalent bonds, octahedral hybrid orbitals are required. These hybrid orbitals will use two d orbitals, the $4s$ orbital, and the three $4p$ orbitals. The d orbitals could be either $3d$ or $4d$; the two available $3d$ orbitals are used because they have lower energy. We can now write the orbital diagram for the metal atom in the complex:

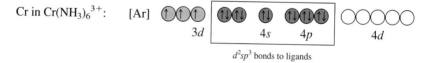

Electron pairs donated from ligands are shown in color. Note that there are three unpaired electrons in $3d$ orbitals on the chromium atom, which explains the paramagnetism of the complex ion. The bonding in other complexes can be explained in a similar fashion.

23.7 Crystal Field Theory

Although valence bond theory explains the bonding and magnetic properties of complexes, it is limited in two important ways. First, the theory cannot easily explain the color of complexes. Second, the theory is difficult to extend quantitatively. Consequently, another theory—crystal field theory—has emerged.

Crystal field theory is *a model of the electronic structure of transition-metal complexes that considers how the energies of the d orbitals of a metal ion are affected by the electric field of the ligands.* According to this theory, the ligands in a transition-metal complex are treated as point charges. So a ligand becomes simply a point of negative charge. In the electric field of these negative charges, the five *d* orbitals of the metal atom no longer have exactly the same energy.

The simplifications used in crystal field theory are drastic. Treating the ligands as point charges is essentially the same as treating the bonding as ionic. However, it turns out that the theory can be extended to include covalent character in the bonding. This simple extension is usually referred to as *ligand field theory.*

Effect of an Octahedral Field on the *d* Orbitals

All five *d* orbitals of an isolated metal atom have the same energy. But if the atom is brought into the electric field of several point charges, these *d* orbitals may be affected in different ways and therefore may have different energies.

Figure 23.16 shows the shapes of the five *d* orbitals. The orbital labeled d_{z^2} has a dumbbell shape along the *z*-axis. The other four *d* orbitals have "cloverleaf" shapes, each differing from one another only in the orientation of the orbitals in space. The orbital $d_{x^2-y^2}$ has its lobes along the *x*-axis and the *y*-axis. Orbitals d_{xy}, d_{yz}, and d_{xz} have their lobes directed between the two sets of axes designated in the orbital label.

A complex ion with six ligands will have the ligands arranged octahedrally about the metal atom to reduce mutual repulsion. Imagine that the ligands are replaced by negative charges. The six charges are placed at equal distances from the metal atom. (See Figure 23.16.)

Fundamentally, the bonding in this model of a complex is due to the attraction of the positive metal ion for the negative charges of the ligands. However, an electron in a *d* orbital of the metal atom is repelled by the negative charge of the ligands. This repulsion alters the energy of the *d* orbital depending on whether it is directed *toward* or *between* ligands. For example, consider the difference in the repulsive effect of ligands on metal-ion electrons in the d_{z^2} and the d_{xy} orbitals. Because the d_{z^2} orbital is directed at the two ligands on the *z*-axis, an electron in the orbital is rather strongly repelled by them. The energy of the d_{z^2} orbital becomes greater. Similarly, an electron in the d_{xy} orbital is repelled by the negative charge of the ligands, but because the orbital is not pointed directly at the ligands, the repulsive effect is smaller. The energy is raised, but less than the energy of the d_{z^2} orbital is raised.

If you look at the five *d* orbitals in an octahedral field (electric field of octahedrally arranged charges), you see that you can divide them into two sets. Orbitals d_{z^2} and $d_{x^2-y^2}$ are directed toward ligands, and orbitals d_{xy}, d_{yz}, and d_{xz} are directed between ligands. The orbitals in the first set (d_{z^2} and $d_{x^2-y^2}$) have higher energy than those in the second set (d_{xy}, d_{yz}, and d_{xz}). Figure 23.17 shows the energy levels of the *d* orbitals in an octahedral field. *The difference in energy between the two sets of d orbitals on a central metal ion that arises from the interaction of the orbitals with the electric field of the ligands* is called the **crystal field splitting, Δ.**

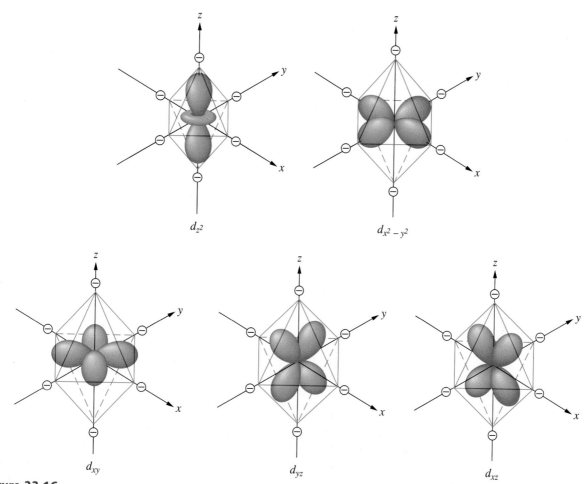

Figure 23.16
The five d orbitals. The d_{z^2} orbital has a dumbbell shape with a collar; the other orbitals have cloverleaf shapes. In an isolated atom, these orbitals have the same energy. However, in an octahedral complex, the orbitals split into two sets, with the d_{z^2} and $d_{x^2-y^2}$ orbitals having higher energy than the other three. Note that the lobes of the d_{z^2} and $d_{x^2-y^2}$ orbitals point toward the ligands (represented here by negative charges), whereas the lobes of the other orbitals point between ligands. The repulsion is greater in the case of the d_{z^2} and $d_{x^2-y^2}$ orbitals.

High-Spin and Low-Spin Complexes

Once you have the energy levels for the d orbitals in an octahedral complex, you can decide how the d electrons of the metal ion are distributed in them. Knowing this distribution, you can predict the magnetic characteristics of the complex.

Consider the complex ion $Cr(NH_3)_6^{3+}$. According to crystal field theory, it consists of the Cr^{3+} ion surrounded by NH_3 molecules treated as partial negative charges. The effect of these charges is to split the d orbitals of Cr^{3+} into two sets as shown in Figure 23.17. The question now is how the d electrons are distributed among the d orbitals of the Cr^{3+} ion. Because the electron configuration of Cr^{3+} is $[Ar]3d^3$, there are three d electrons to distribute. They are placed in the d orbitals of lower energy, following Hund's rule (Figure 23.18). You see that the complex ion $Cr(NH_3)_6^{3+}$ has three unpaired electrons and is therefore paramagnetic.

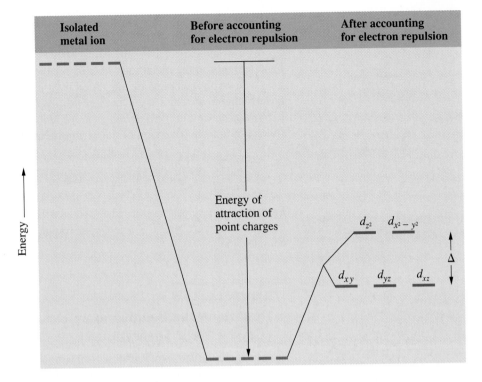

Figure 23.17
Energy levels of *d* orbitals in an octahedral field. The positive metal ion is attracted to the negative charges (ligands), but electrons in the *d* orbitals are repelled by them. Thus, although there is an overall attraction, the *d* orbitals no longer have the same energy.

If you examine any transition-metal ion that has configurations d^4, d^5, d^6, or d^7, you will find two bonding possibilities, which yield high-spin complexes in one case and low-spin complexes in the other case. A **high-spin complex** is *a complex in which there is minimum pairing of electrons in the d orbitals of the metal atom.* A **low-spin complex** is *a complex in which there is more pairing of electrons in the d orbitals of the metal atom than in a corresponding high-spin complex.*

Consider the complex ion $Fe(H_2O)_6^{2+}$. What are its magnetic characteristics? The electron configuration of the ion is $[Ar]3d^6$. You distribute six electrons among the d orbitals of the complex in such a way as to get the lowest total energy. If you place all six electrons into the lower three d orbitals, you get the distribution shown by the energy-level diagram in Figure 23.19A. All of the electrons are paired, so you would predict that this distribution gives a diamagnetic complex. The $Fe(H_2O)_6^{2+}$ ion, however, is paramagnetic. What was the mistake?

The mistake was to ignore the **pairing energy, *P***, *the energy required to put two electrons into the same orbital.* When an orbital is already occupied by an electron, it requires energy to put another electron into that orbital because of their mutual repulsion. Suppose that this pairing energy is greater than the crystal field splitting; that is, suppose $P > \Delta$. In that case, once the first three electrons have singly occupied the three lower-energy d orbitals, the fourth electron will go into one of the higher-energy d orbitals. It will take less energy to do that than to pair up with an electron in one of the lower-energy orbitals. Similarly, the fifth electron will go into the last empty d orbital. The sixth electron must pair up, so it goes into one of the lower-energy orbitals. Figure 23.19B shows this electron distribution. In this case, there are four unpaired electrons and the complex is paramagnetic.

We see that crystal field theory predicts two possibilities: a *low-spin complex* when $P < \Delta$ and a *high-spin complex* when $P > \Delta$. The value of Δ, as we will explain

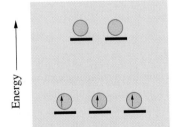

Figure 23.18
Occupation of the 3*d* orbitals in an octahedral complex of Cr^{3+}. The electrons occupy different lower-energy orbitals but have the same spin (Hund's rule).

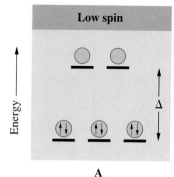

Figure 23.19
Occupation of the 3d orbitals in complexes of Fe²⁺. *(A)* Low spin. *(B)* High spin.

later, can be obtained from the spectrum of a complex, and the value of P can be calculated theoretically.

You would expect low-spin diamagnetic Fe^{2+} complexes to occur for ligands that bond strongly to the metal ion—that is, for those giving large Δ. Ligands that might give a low-spin complex are suggested by a look at the spectrochemical series. The **spectrochemical series** is *an arrangement of ligands according to the relative magnitudes of the crystal field splittings they induce in the d orbitals of a metal ion.* The following is a short version of the spectrochemical series:

weak-bonding ligands   strong-bonding ligands

$$I^- < Br^- < Cl^- < F^- < OH^- < H_2O < NH_3 < en < NO_2^- < CN^- < CO$$

increasing $\Delta \longrightarrow$

From this series, you see that the CN^- ion bonds more strongly than H_2O, which explains why $Fe(CN)_6^{4-}$ is a low-spin complex ion and $Fe(H_2O)_6^{2+}$ is a high-spin complex ion.

EXAMPLE 23.3

Describing the Bonding in an Octahedral Complex Ion (Crystal Field Theory)

Describe the distribution of d electrons in the complex ion $Co(H_2O)_6^{2+}$, using crystal field theory. The hexaaquacobalt(II) ion is a high-spin complex ion. What would be the distribution of d electrons in an octahedral cobalt(II) complex ion that is low spin? How many unpaired electrons are there in each ion?

SOLUTION

The electron configuration of Co^{2+} is $[Ar]3d^7$. The high-spin and low-spin distributions in the d orbitals are

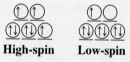

High-spin   Low-spin

$Co(H_2O)_6^{2+}$, a high-spin complex, has three unpaired electrons. A low-spin complex would have one unpaired electron.

See Problems 23.41 and 23.42.

Visible Spectra of Transition-Metal Complexes

Frequently, substances absorb light only in regions outside the visible spectrum and reflect, or pass on (transmit), all of the visible wavelengths. As a result, these substances appear white or colorless. However, some substances absorb certain wavelengths in the visible spectrum and transmit the remaining ones; they appear colored. Many transition-metal complexes, as we have noted, are colored substances. The color results from electron transitions between the two closely spaced d orbital energy levels that come from the crystal field splitting. ◄

▶ The colors of many gemstones are due to transition-metal-ion impurities in the mineral. Ruby has Cr^{3+} in alumina, Al_2O_3, and emerald has Cr^{3+} in beryl, $Be_3Al_2(SiO_3)_6$.

The spectrum of a d^1 configuration complex is particularly simple. Hexaaquatitanium(III) ion, $Ti(H_2O)_6^{3+}$, is an example. Titanium has the configuration $[Ar]3d^24s^2$, and Ti^{3+} has the configuration $[Ar]3d^1$. According to crystal field theory, the d electron occupies one of the lower-energy d orbitals of the octahedral complex.

Figure 23.20
Color and visible spectrum of Ti(H$_2$O)$_6$$^{3+}$.
Left: A test tube containing a solution of Ti(H$_2$O)$_6$$^{3+}$. *Right:* Visible spectrum of Ti(H$_2$O)$_6$$^{3+}$. Unlike atomic spectra, which show absorption lines, spectra of ions and molecules in solution yield broad bands resulting from changes in nuclear motion that accompany the electronic transitions.

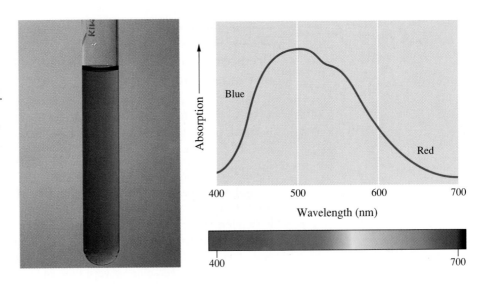

▶ **Recall from Section 7.3 that the energy change during a transition equals *hν*.**

Figure 23.20 shows the visible spectrum of Ti(H$_2$O)$_6$$^{3+}$. It results from a transition of the *d* electron from a lower-energy *d* orbital to a higher-energy *d* orbital, as shown in Figure 23.21. Note that the energy change equals the crystal field splitting, Δ. Consequently, the wavelength, λ, of light absorbed is related to Δ. ◀

$$\Delta = h\nu = \frac{hc}{\lambda} \quad \text{or} \quad \lambda = \frac{hc}{\Delta}$$

When white light, which contains all visible wavelengths (400 nm to 750 nm), falls on a solution containing Ti(H$_2$O)$_6$$^{3+}$, blue-green light is absorbed. (The maximum light absorption is observed at 500 nm, which is blue-green light. See Table 23.7.) The other wavelengths of visible light, including red and some blue light, pass through the solution, giving it a red-purple color.

When the ligands in the Ti^{3+} complex are changed, Δ changes and therefore the color of the complex changes. For example, replacing H$_2$O ligands by weaker F$^-$ ligands should give a smaller crystal field splitting and therefore an absorption at longer wavelengths. The absorption of TiF$_6$$^{3-}$ is at 590 nm, in the yellow rather than the blue-green, and the color observed is violet-blue.

From this discussion, you see that the visible spectrum can be related to the crystal field splitting, and values of Δ can be obtained by spectroscopic analysis. However, when there is more than one *d* electron, several excited states can be formed. Consequently, the spectrum generally consists of several lines, and the analysis is more complicated than for Ti^{3+}.

Figure 23.21
The electronic transition responsible for the visible absorption in Ti(H$_2$O)$_6$$^{3+}$.
An electron undergoes a transition from a lower-energy *d* orbital to a higher-energy *d* orbital. The energy change equals the crystal field splitting, Δ.

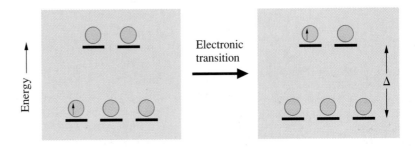

EXAMPLE 23.4

Predicting the Relative Wavelengths of Absorption of Complex Ions

When water ligands in $Ti(H_2O)_6^{3+}$ are replaced by CN^- ligands to give $Ti(CN)_6^{3-}$, the maximum absorption shifts from 500 nm to 450 nm. Is this shift in the expected direction? Explain. What color do you expect to observe for this ion?

SOLUTION

According to the spectrochemical series, CN^- is a more strongly bonding ligand than H_2O. Consequently, Δ should increase, and the wavelength of the absorption ($\lambda = hc/\Delta$) should decrease. **So the shift of the absorption is in the expected direction.** Because the absorbed light is between blue and violet-blue (see Table 23.7), the observed color is **orange-yellow** (this is the *complementary color* of the color between blue and violet-blue).

See Problems 23.43, 23.44, 23.45, and 23.46.

Table 23.7
Color Observed for Given Absorption of Light by an Object

Wavelength Absorbed (nm)	Color Absorbed	Approximate Color Observed*
410	Violet	Green-yellow
430	Violet-blue	Yellow
480	Blue	Orange
500	Blue-green	Red
530	Green	Purple
560	Green-yellow	Violet
580	Yellow	Violet-blue
610	Orange	Blue
680	Red	Blue-green
720	Purple-red	Green

*The exact color depends on the relative intensities of various wavelengths coming from the object and on the response of the eye to those wavelengths.

The Cooperative Release of Oxygen from Oxyhemoglobin

LIFE SCIENCE

Hemoglobin is an iron-containing substance in red blood cells responsible for the transport of O_2 from the lungs to various parts of the body. Myoglobin is a similar substance in muscle tissue, acting as a reservoir for the storage of O_2 and as a transporter of O_2 within muscle cells. The explanation for the different actions of these two substances involves some fascinating transition-metal chemistry.

Myoglobin consists of heme—a complex of Fe(II) bonded to a quadridentate ligand (Figure 23.7)—and globin. Globin, a protein, is attached through a nitrogen atom to one of the octahedral positions of Fe(II). The sixth position is vacant in free myoglobin but is occupied by O_2 in oxymyoglobin. Hemoglobin is essentially a four-unit structure of myoglobinlike units—that is, a *tetramer* of myoglobin.

Myoglobin and hemoglobin exist in equilibrium with the oxygenated forms oxymyoglobin and oxyhemoglobin, respectively. For example, hemoglobin (Hb) and O_2 are in equilibrium with oxyhemoglobin (HbO_2).

$$Hb + O_2 \rightleftharpoons HbO_2$$

Although hemoglobin is a tetramer of myoglobin, it does not function simply as four independent units of myoglobin. For it to function efficiently as a transporter of O_2 from the lungs and then be able to release that O_2 easily to myoglobin, hemoglobin must be less strongly attached to O_2 in the vicinity of a muscle cell than is myoglobin. In hemoglobin, the release of O_2 from one heme group triggers the release of O_2 from another heme group of the same molecule. In other words, there is a *cooperative release* of O_2 from hemoglobin that makes it possible for it to give up its O_2 to myoglobin.

The mechanism postulated for this cooperative release of O_2 depends on a change of iron(II) from a low-spin to a high-spin form, with a corresponding change in radius of the iron atom. In oxyhemoglobin, iron(II) exists in the low-spin form. When O_2 leaves, the iron atom goes to a high-spin form with two electrons in the higher-energy d orbital. These higher-energy orbitals are somewhat larger than the lower-energy d orbitals.

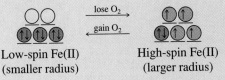

Low-spin Fe(II)　　　High-spin Fe(II)
(smaller radius)　　　(larger radius)

When an O_2 molecule leaves a heme group, the radius of the iron atom increases, and the atom pops out of the heme plane by about 70 pm. In hemoglobin, this change triggers the cooperative release of another O_2 molecule. As the iron atom moves, the attached globin group moves with it. This motion of one globin group causes an adjacent globin group in the tetramer to alter its shape, which in turn makes possible the easy release of an O_2 molecule from its heme unit.

A Checklist for Review

Important Terms

transition elements (23.1)
complex ion (23.3)
complex (coordination
　compound) (23.3)
ligands (23.3)
coordination number (23.3)

monodentate ligand (23.3)
bidentate ligand (23.3)
polydentate ligand (23.3)
chelate (23.3)
constitutional
　isomers (23.5)

stereoisomers (23.5)
geometric isomers (23.5)
crystal field theory (23.7)
crystal field splitting,
　Δ (23.7)
high-spin complex (23.7)

low-spin complex (23.7)
pairing energy, P (23.7)
spectrochemical
　series (23.7)

Summary of Facts and Concepts

The *d-block transition elements* are defined as those elements having a partially filled d subshell in any common oxidation state. They have a number of characteristics, including high melting points and a multiplicity of oxidation states. Compounds of transition elements are frequently colored and many are paramagnetic. These properties are due to the participation of d orbitals in bonding. We described the chemical properties of two transition elements: Cr and Cu. Chromium metal reacts with acids to give Cr^{2+} ion, which is readily oxidized to Cr^{3+}. The $+3$ state is the most common oxidation

state; chromium(III) oxide is a green pigment. The $+6$ oxidation state is represented by chromates and dichromates. The dichromate ion, $Cr_2O_7^{2-}$, in acid solution is a strong oxidizing agent. Copper metal reacts only with acids having strongly oxidizing anions, such as HNO_3; it gives Cu^{2+} ion. Copper(I) oxide, Cu_2O, occurs naturally as a copper mineral.

Transition-metal atoms often function as Lewis acids, reacting with groups called *ligands* by forming coordinate covalent bonds to them. The metal atom with its ligands is a *complex ion* or neutral *complex*. Ligands that bond through more than one atom are called *polydentate,* and the complex formed is called a *chelate*. The IUPAC has agreed on a nomenclature of complexes that gives basic structural information about the species. The presence of isomers in coordination compounds is evidence for particular geometries. For example, $[Pt(NH_3)_2Cl_2]$ has two isomers, which is evidence for a square planar geometry having *cis–trans isomers.*

Valence bond theory gave the earliest theoretical description of the electronic structure of a complex. According to this theory, a complex forms when doubly occupied ligand orbitals overlap unoccupied orbitals of the metal atom.

Crystal field theory treats the ligands as electric charge points that affect the energy of the d orbitals of the metal ion. In an octahedral complex, two of the d orbitals have higher energy than the other three. A high-spin complex forms when the *pairing energy* is greater than the *crystal field splitting,* so that electrons would "prefer" to occupy a higher-energy d orbital rather than pair up with an electron in a lower-energy orbital. When the pairing energy is smaller than the crystal field splitting, the d orbitals are occupied in the normal fashion, giving a low-spin complex. Color in transition-metal complexes is explained as due to the transition of an electron from the lower-energy to the higher-energy d orbitals. The crystal field splitting can be obtained experimentally from the visible spectrum of a complex.

Operational Skills

1. **Writing the IUPAC name given the structural formula of a coordination compound, and vice versa** Given the structural formulas of coordination compounds, write the IUPAC names **(EXAMPLE 23.1)**. Given the IUPAC names of complexes, write the structural formulas **(EXAMPLE 23.2)**.
2. **Deciding whether isomers are possible** Given the formula of a complex, decide whether geometric isomers are possible and, if so, draw them.
3. **Describing the bonding in a complex ion** Given a transition-metal complex ion, describe the bonding types (high-spin and low-spin, if both exist), using crystal field theory. **(EXAMPLE 23.3)**
4. **Predicting the relative wavelengths of absorption of complex ions** Given two complexes that differ only in the ligands, predict, on the basis of the spectrochemical series, which complex absorbs at higher wavelength. Given the absorption maxima, predict the colors of the complexes. **(EXAMPLE 23.4)**

Review Questions

23.1 What characteristics of the transition elements set them apart from the main-group elements?

23.2 According to the building-up principle, what is the electron configuration of the ground state of the technetium atom (atomic number 43)?

23.3 The highest melting point for metals in the fifth period occurs for molybdenum. Explain why this is expected.

23.4 Iron, cobalt, and nickel are similar in properties and are sometimes studied together as the "iron triad." For example, each is a fairly active metal that reacts with acids to give hydrogen and the $+2$ ions. In addition to the $+2$ ions, the $+3$ ions of the metals also figure prominently in the chemistries of the elements. Explain why these elements are similar.

23.5 Palladium and platinum are very similar to one another. Both are unreactive toward most acids. However, nickel, which is in the same column of the periodic table, is an active metal. Explain why this difference exists.

23.6 Write balanced equations for the reactions of Cr and Cu metals with $HCl(aq)$. If no reaction occurs, write *NR*.

23.7 Describe the structure of copper(II) sulfate pentahydrate. What color change occurs when the salt is heated? What causes the color change?

23.8 What evidence did Werner obtain to show that the platinum complex $PtCl_4 \cdot 4NH_3$ has the structural formula $[Pt(NH_3)_4Cl_2]Cl_2$?

23.9 Define the terms *complex ion, ligand,* and *coordination number.* Use an example to illustrate the use of these terms.

23.10 Define the term *bidentate ligand.* Give two examples.

23.11 Rust spots on clothes can be removed by dissolving the rust in oxalic acid. The oxalate ion forms a stable complex with Fe^{3+}. Using an electron-dot formula, indicate how an oxalate ion bonds to the metal ion.

23.12 What three properties of coordination compounds have been important in determining the details of their structure and bonding?

23.13 Describe the formation of a coordinate covalent bond between a metal-ion orbital and a ligand orbital.

23.14 (a) Describe the steps in the formation of a high-spin octahedral complex of Fe^{2+} in valence bond terms. (b) Do the same for a low-spin complex.

23.15 Explain why d orbitals of a transition-metal atom may have different energies in the octahedral field of six negative charges. Describe how each of the d orbitals is affected by the octahedral field.

23.16 (a) Use crystal field theory to describe a high-spin octahedral complex of Fe^{2+}. (b) Do the same for a low-spin complex.

23.17 What is meant by the term *crystal field splitting*? How is it determined experimentally?

23.18 What is the spectrochemical series? Use the ligands CN^-, H_2O, Cl^-, and NH_3 to illustrate the term. Then arrange them in order, describing the meaning of this order.

23.19 What is meant by the term *pairing energy*? How do the relative values of pairing energy and crystal field splitting determine whether a complex is low-spin or high-spin?

23.20 A complex absorbs red light from a single electron transition. What color is this complex?

Conceptual Problems

23.21 A cobalt complex whose composition corresponded to the formula $Co(NO_2)_2Cl \cdot 4NH_3$ gave an electrical conductance equivalent to two ions per formula unit. Excess silver nitrate solution immediately precipitates 1 mol AgCl per formula unit. Write a structural formula consistent with these results.

23.22 For the following coordination compounds, identify the geometric isomer(s) of compound X.

23.23 A complex has a composition corresponding to the formula $CoBr_2Cl \cdot 4NH_3$. What is the structural formula if conductance measurements show two ions per formula unit? Silver nitrate solution gives an immediate precipitate of AgCl but no AgBr. Write the structural formula of an isomer.

23.24 For the complexes shown here, which would have the d electron distribution shown in the diagram below: MF_6^{3-}, $M(CN)_6^{3-}$, MF_6^{4-}, or $M(CN)_6^{4-}$? Note that the neutral metal atom, M, in each complex is the same and has the ground state electron configuration $[Ar]4s^23d^6$.

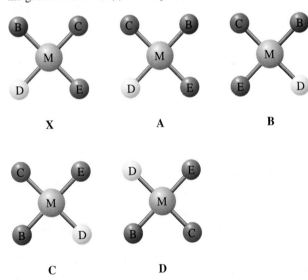

Practice Problems

Properties of the Transition Elements

23.25   Find the oxidation numbers of the transition metal in each of the following compounds:
a. $FeCO_3$ b. MnO_2 c. $CuCl$ d. CrO_2Cl_2

23.26   Find the oxidation numbers of the transition metal in each of the following compounds:
a. $CoSO_4$ b. Ta_2O_5 c. $Cu_2(OH)_3Cl$

23.27   Write the balanced equation for the reaction of iron(II) ion with nitrate ion in acidic solution. Nitrate ion is reduced to NO.

23.28   Write the balanced equation for the reaction of sulfurous acid with dichromate ion.

Structural Formulas and Naming of Complexes

23.29   Give the coordination number of the transition-metal atom in each of the following complexes.
a. $Au(CN)_4^-$ b. $[Co(NH_3)_4(H_2O)_2]Cl_3$
c. $[Au(en)_2]Cl_3$ d. $Cr(en)_2(C_2O_4)^+$

23.30   Give the coordination number of the transition element in each of the following complexes.
a. $[Ni(NH_3)_6](ClO_3)_2$ b. $[Cu(NH_3)_4]SO_4$
c. $[Cr(en)_3]Cl_3$ d. $K_2[Ni(CN)_4]$

23.31   Determine the oxidation number of the transition element in each of the following complexes.
a. $K_2[Ni(CN)_4]$ b. $Mo(en)_3^{3+}$
c. $Cr(C_2O_4)_3^{3-}$ d. $[Co(NH_3)_5(NO_2)]Cl_2$

23.32   For each of the following complexes, determine the oxidation state of the transition-metal atom.
a. $[CoCl(en)_2(NO_2)]NO_2$ b. $PtCl_4^{2-}$
c. $K_3[Cr(CN)_6]$ d. $Fe(H_2O)_5(OH)^{2+}$

23.33   Consider the complex ion $Cr(NH_3)_2Cl_2(C_2O_4)^-$.
a. What is the oxidation state of the metal atom?
b. Give the formula and name of each ligand in the ion.
c. What is the coordination number of the metal atom?

23.34   Consider the complex ion $Mn(NH_3)_2(H_2O)_3(OH)^{2+}$.
a. What is the oxidation state of the metal atom?
b. Give the formula and name of each ligand in the ion.
c. What is the coordination number of the metal atom?

23.35   Write the IUPAC name for each of the following coordination compounds.
a. $K_3[FeF_6]$ b. $Cu(NH_3)_2(H_2O)_2^{2+}$
c. $(NH_4)_2[Fe(H_2O)F_5]$ d. $Ag(CN)_2^-$

23.36   Name the following complexes, using IUPAC rules.
a. $K_4[Mo(CN)_8]$ b. CrF_6^{3-}
c. $V(C_2O_4)_3^{2-}$ d. $K_2[FeCl_4]$

23.37   Write the structural formula for each of the following compounds.
a. potassium hexacyanomanganate(III)
b. tetraamminedichlorocobalt(III) nitrate
c. hexaamminechromium(III) tetrachlorocuprate(II)

23.38   Give the structural formula for each of the following complexes.
a. diaquadicyanocopper(II)
b. potassium hexachloroplatinate(IV)
c. tetraammineplatinum(II) tetrachlorocuprate(II)

Isomerism

23.39   Draw *cis–trans* structures of any of the following square planar or octahedral complexes that exhibit geometric isomerism. Label the drawings *cis* or *trans*.
a. $[Pd(NH_3)_2Cl_2]$ b. $Pd(NH_3)_3Cl^+$
c. $Pd(NH_3)_4^{2+}$ d. $Ru(NH_3)_4Br_2^+$

23.40   If any of the following octahedral complexes display geometric isomerism, draw the structures and label them *cis* or *trans*.
a. $Co(NO_2)_4(NH_3)_2^-$ b. $Co(NH_3)_5(NO_2)^{2+}$
c. $Co(NH_3)_6^{3+}$ d. $Cr(NH_3)_5Cl^{2+}$

Crystal Field Theory

23.41   Using crystal field theory, sketch the energy-level diagram for the *d* orbitals in an octahedral field; then fill in the electrons for the metal ion in each of the following complexes. How many unpaired electrons are there in each case?
a. $V(CN)_6^{3-}$ b. $Co(C_2O_4)_3^{4-}$ (high-spin)
c. $Mn(CN)_6^{3-}$ (low-spin)

23.42   Using crystal field theory, sketch the energy-level diagram for the *d* orbitals in an octahedral field; then fill in the electrons for the metal ion in each of the following complexes. How many unpaired electrons are there in each case?
a. $ZrCl_6^{4-}$
b. $OsCl_6^{2-}$ (low-spin)
c. $MnCl_6^{4-}$ (high-spin)

Color

23.43   The $Co(SCN)_4^{2-}$ ion has a maximum absorption at 530 nm. What color do you expect for this ion?

23.44   The $Co(en)_3^{3+}$ ion has a maximum absorption at 470 nm. What color do you expect for this ion?

23.45 The $Co(NH_3)_6^{3+}$ ion has a yellow color, but when one NH_3 ligand is replaced by H_2O to give $Co(NH_3)_5(H_2O)^{3+}$, the color shifts to red. Is this shift in the expected direction? Explain.

23.46 The $Co(en)_3^{3+}$ ion has a yellow color, but the CoF_6^{3-} ion has a blue color. Is the shift from yellow to blue expected when ethylenediamine ligands are replaced by F^- ligands? Explain.

General Problems

23.47 The hexaaquascandium(III) ion, $Sc(H_2O)_6^{3+}$, is colorless. Explain why this might be expected.

23.48 The tetraaquazinc(II) ion, $Zn(H_2O)_4^{2+}$, is colorless. Explain why this might be expected.

23.49 There are only two geometric isomers of the tetraamminedichlorocobalt(III) ion, $Co(NH_3)_4Cl_2^+$. How many geometric isomers would be expected for this ion if it had a regular planar hexagonal geometry? Give drawings for them. Does this rule out a planar hexagonal geometry for $Co(NH_3)_4Cl_2^+$? Explain.

23.50 There are only two geometric isomers of the triamminetrichloroplatinum(IV) ion, $Pt(NH_3)_2Cl_4^+$. How many geometric isomers would be expected for this ion if it had a regular planar hexagonal geometry? Give drawings for them. Does this rule out a planar hexagonal geometry for $Pt(NH_3)_2Cl_4^+$? Explain.

23.51 Find the concentrations of $Cu^{2+}(aq)$, $NH_3(aq)$, and $Cu(NH_3)_4^{2+}(aq)$ at equilibrium when 0.10 mol $Cu^{2+}(aq)$ and 0.40 mol $NH_3(aq)$ are made up to 1.00 L of solution. The dissociation constant, K_d, for the complex $Cu(NH_3)_4^{2+}$ is 2.1×10^{-13}.

23.52 Find the concentrations of $Ag^+(aq)$, $NH_3(aq)$, and $Ag(NH_3)_2^+(aq)$ at equilibrium when 0.10 mol $Ag^+(aq)$ and 0.20 mol $NH_3(aq)$ are made up to 1.00 L of solution. The dissociation constant, K_d, for the complex $Ag(NH_3)_2^+$ is 5.9×10^{-8}.

Sugar crystals viewed with a microscope under polarized light.

Organic Chemistry

O rganic chemistry is the chemistry of compounds containing carbon. As discussed in Chapter 2, carbon-containing compounds make up the majority of known compounds. Organic chemistry plays a central role in most of the substances that you encounter and use every day: the food you ate this morning, the shampoo that cleans your hair, the fuel that you use to heat your house and generate electricity, and the list goes on. Also, your life and that of every living organism on earth depends on the chemical reactions of organic molecules.

In this chapter, we introduce this fascinating area of chemistry with a discussion of the structural features of organic molecules, nomenclature, and a few important chemical reactions.

Figure 24.1
Petroleum-based products contain long chains of carbon atoms.
Left: The black, tarry substance in asphalt consists of molecules with 30 or more carbon atoms bonded together. *Right:* Polyethylene contains chains of many thousands of carbon atoms bonded together. Polyethylene is used to create a variety of everyday materials.

24.1 The Bonding of Carbon

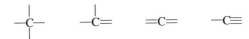

Because carbon is in Group IVA of the periodic table, it has four valence electrons. To fill its octet, it requires four additional electrons, which can be obtained through the formation of four covalent bonds. Carbon forms single, double, and triple bonds to achieve a filled octet. Therefore, the possible bonding combinations for carbon are as follows.

$$-\overset{|}{\underset{|}{C}}- \qquad -\overset{|}{C}= \qquad =C= \qquad -C\equiv$$

The molecular geometry around an atom is dictated by the number of regions of electron density. Keeping in mind that double and triple bonds count as one area of electron density, carbon can have a tetrahedral, trigonal planar, or linear geometry.

A unique feature of carbon is its ability to bond with other carbon atoms to form chains and rings of various lengths. Some petroleum-containing products, such as asphalt, consist of molecules with 30 or more carbon atoms bonded together; the molecules that make up polyethylene can have chains with thousands of carbon atoms (Figure 24.1).

Hydrocarbons

The simplest organic compounds are **hydrocarbons,** *compounds containing only carbon and hydrogen.* All other organic compounds, for example, those containing O, N, and the halogen atoms, are classified as being derived from hydrocarbons. At first glance, you might think that the hydrocarbons represent a very limited set of molecules; however, several hundred thousand molecules exist that contain only hydrogen and carbon atoms.

Hydrocarbons can be separated into three main groups:

1. **Saturated hydrocarbons** are *hydrocarbons that contain only single bonds between the carbon atoms.* Saturated hydrocarbon molecules can be *cyclic* or *acyclic.* A cyclic hydrocarbon is one in which a chain of carbon atoms has formed a ring. An acyclic hydrocarbon is one that does not contain a ring of carbon atoms.

2. **Unsaturated hydrocarbons** are *hydrocarbons that contain double or triple bonds between carbon atoms.*

3. **Aromatic hydrocarbons** are *hydrocarbons that contain benzene rings or similar features.*

24.2 Alkanes and Cycloalkanes

The Alkane Series

▶ The term *paraffin* comes from the Latin *parum affinus*, meaning "little affinity." The alkanes do not react with many reagents.

The **alkanes** are *acyclic saturated hydrocarbons,* and the **cycloalkanes** are *cyclic saturated hydrocarbons.* The alkanes, also called *paraffins,* ◀ have the general formula C_nH_{2n+2}. For $n = 1$, methane, the formula is CH_4; for $n = 2$, C_2H_6; for $n = 3$, C_3H_8; and so on. Note that the general formula conveys no information about how the atoms are connected. For now, we will assume that the carbon atoms are bonded together in a straight chain with hydrogen atoms completing the four required bonds to each carbon atom; these are called *straight-chain* or *normal* alkanes. Three-dimensional models of methane are shown in Figure 24.2. The structural formulas for the first four straight chain alkanes are shown.

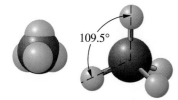

Figure 24.2
Three-dimensional models of methane, CH_4.
Left: Space-filling model of methane. *Right:* Ball-and-stick model of methane with bond angle.

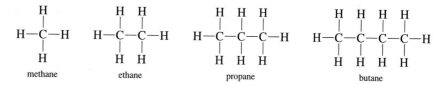

Because carbon atoms typically have four bonds, chemists often write the structures of the parts of organic compounds using **condensed structural formulas,** or **condensed formulas,** *where the bonds around each carbon atom in the compound are not explicitly written.* Condensed formulas of the first four alkanes ($n = 1$ to 4) are

$$CH_4 \qquad CH_3CH_3 \qquad CH_3CH_2CH_3 \qquad CH_3CH_2CH_2CH_3$$
methane   ethane   propane   butane

▶ Formerly, the names of the straight-chain alkanes were distinguished from branched-chain isomers by the prefix *n-* for *normal.* This designation is still common (butane is called *n*-butane) but is not used in the IUPAC name, which we will discuss in Section 24.5.

Note that the condensed formula of an alkane differs from that of the preceding alkane ($n - 1$) by a $—CH_2—$ group. The alkanes constitute a **homologous series,** which is *a series of compounds in which one compound differs from a preceding one by a fixed group of atoms.* Members of a homologous series have similar chemical properties, and their physical properties change throughout the series in a regular way. Table 24.1 lists the boiling points of the first ten straight-chain alkanes. ◀ Note that the boiling points generally increase in the series with an increase in the number of carbon atoms in the chain. This is a result of increasing intermolecular forces, which increase with molecular weight.

Table 24.1
Formulas, Names, Structures, and Boiling Points of the First Ten Straight-Chain Alkanes

Name	Elemental Formula	Condensed Structural Formula	Boiling Point (°C)
Methane	CH_4	CH_4	−162
Ethane	C_2H_6	CH_3CH_3	−89
Propane	C_3H_8	$CH_3CH_2CH_3$	−42
Butane	C_4H_{10}	$CH_3CH_2CH_2CH_3$	0
Pentane	C_5H_{12}	$CH_3CH_2CH_2CH_2CH_3$	36
Hexane	C_6H_{14}	$CH_3CH_2CH_2CH_2CH_2CH_3$	69
Heptane	C_7H_{16}	$CH_3CH_2CH_2CH_2CH_2CH_2CH_3$	98
Octane	C_8H_{18}	$CH_3CH_2CH_2CH_2CH_2CH_2CH_2CH_3$	126
Nonane	C_9H_{20}	$CH_3CH_2CH_2CH_2CH_2CH_2CH_2CH_2CH_3$	151
Decane	$C_{10}H_{22}$	$CH_3CH_2CH_2CH_2CH_2CH_2CH_2CH_2CH_2CH_3$	174

Constitutional Isomerism and Branched-Chain Alkanes

In addition to the straight-chain alkanes, *branched-chain* alkanes are possible. For example, isobutane (or 2-methylpropane) has the structure

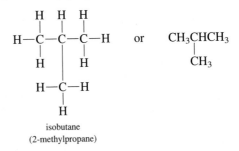

isobutane
(2-methylpropane)

Isobutane, C_4H_{10}, has the same molecular formula as butane, the straight-chain hydro-carbon. However, isobutane and butane have different molecular structures. Butane and isobutane are **constitutional** (or **structural**) **isomers,** *compounds with the same molecular formula but different structural formulas.* Figure 24.3 depicts molecular models of isobutane and butane. Because these isomers have different structures, they have different properties. For example, isobutane boils at −12°C whereas butane boils at 0°C. Here the difference in boiling point can be attributed to the fact that isobutane

Figure 24.3
Constitutional isomers of butane. Ball-and-stick models of isobutane (2-methylpropane) and butane.

Isobutane

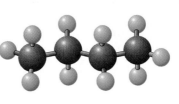

Butane

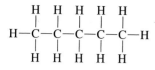

pentane, b.p. 36°C

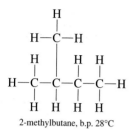

2-methylbutane, b.p. 28°C

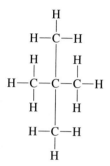

2,2-dimethylpropane, b.p. 9.5°C

Figure 24.4
Isomers of pentane. Note that each isomer is a different compound with a different boiling point.

has a more compact molecular structure than butane, which results in weaker inter-molecular interactions between isobutane molecules.

The number of constitutional isomers rapidly increases with the number of carbons in the series. For example, there are three constitutional isomers with the molecular formula C_5H_{12} (pentanes; see Figure 24.4), five of C_6H_{14} (hexanes), and 75 of $C_{10}H_{22}$ (decanes).

Branched alkanes are usually written using condensed structural formulas. The following example will help you make the transition to writing hydrocarbons in this way.

EXAMPLE 24.1
Writing Condensed Structural Formulas

Write the condensed structural formula of each of the following alkanes.

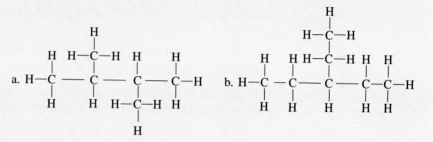

SOLUTION

a. To write the condensed structural formula, it helps to identify the carbon backbone, or the skeleton of carbon atoms in the molecule. It is good practice to write the longest chain of carbon atoms in the molecule in a straight line. For this case, the skeleton looks like:

$$
\begin{array}{c}
\quad\;\; C \\
\;\; | \\
C-C-C-C \\
\quad\;\; | \\
\quad\;\; C
\end{array}
$$

Next, write the appropriate number of H atoms next to each carbon atom.

Finally, find the straight chain(s) of carbon atoms that contain more than two carbon atoms and remove only the bonds (dashes) that connect those carbon atoms.

After a little practice you will be able to combine these steps.

b. Following the steps outlined above, the condensed structural formula for the part b molecule is

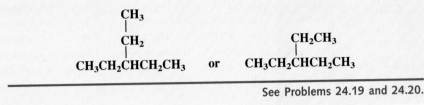

See Problems 24.19 and 24.20.

Cycloalkanes

The general formula for cyclic cycloalkanes is C_nH_{2n}. Figure 24.5 gives the names and structural formulas for the first four members of the cycloalkane series. In the condensed structural formulas, a carbon atom and its attached hydrogen atoms are assumed to be at each corner.

CONCEPT CHECK 24.1

In the model shown here, C atoms are black and H atoms are light blue:

a. Write the molecular formula.

b. Write the condensed structural formula.

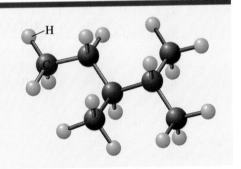

Figure 24.5
First four members of the cycloalkane series. These are saturated aliphatic hydrocarbons characterized by carbon-atom rings.

Molecular formula	C_3H_6	C_4H_8	C_5H_{10}	C_6H_{12}
Ball-and-stick model				
Full structural formula				
Condensed structural formula	△	▢	⬠	⬡
Name	Cyclopropane	Cyclobutane	Cyclopentane	Cyclohexane

Figure 24.6
Consumer products derived from petroleum.
All of these common items contain organic compounds that used petroleum as a starting material.

Sources and Uses of Alkanes and Cycloalkanes

Fossil fuels (natural gas, petroleum, and coal) are the principal sources of all types of organic chemicals. Petroleum refining involves the conversion of the relatively abundant alkanes to unsaturated hydrocarbons, aromatic hydrocarbons, and hydrocarbon derivatives. The alkanes serve as the starting point for the majority of organic compounds, including plastics and pharmaceutical drugs. It is amazing to think about the number of everyday materials we use that started out as an alkane or cycloalkane (Figure 24.6).

Reactions of Alkanes with Oxygen

We react alkanes every day through combustion with O_2; all hydrocarbons burn (combust) in an excess of O_2 at elevated temperatures to produce carbon dioxide, water, and heat. For example, a propane gas grill uses the reaction

$$C_3H_8(g) + 5O_2(g) \longrightarrow 3CO_2(g) + 4H_2O(l); \Delta H° = -2220 \text{ kJ/mol}$$

The large negative $\Delta H°$ value for this reaction and all hydrocarbon reactions with oxygen demonstrates why we rely on these molecules to meet our energy needs.

Substitution Reactions of Alkanes

Under the right conditions, alkanes can react with other molecules. An important example is the reaction of alkanes with the halogens F_2, Cl_2, and Br_2. Reaction with Cl_2, for example, requires light (indicated by hv) or heat.

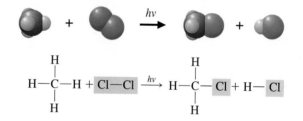

This is an example of a substitution reaction. A **substitution reaction** is *a reaction in which a part of the reacting molecule is substituted for an H atom on a hydrocarbon or hydrocarbon group.* All of the H atoms of an alkane may undergo substitution, leading to a mixture of products.

$$CH_3Cl + Cl_2 \xrightarrow{hv} CH_2Cl_2 + HCl$$

$$CH_2Cl_2 + Cl_2 \xrightarrow{hv} CHCl_3 + HCl$$

$$CHCl_3 + Cl_2 \xrightarrow{hv} CCl_4 + HCl$$

The CCl_4 product can be reacted with HF in the presence of a $SbCl_5$ catalyst to produce trichlorofluoromethane, CCl_3F, also known as CFC-11.

$$CCl_4 + HF \xrightarrow{SbCl_5} CCl_3F + HCl$$

This compound is one of a number of chlorofluorocarbons (CFCs) that have been used as a refrigerant. Data obtained in the 1970s revealed that these compounds survived long enough to travel to the stratospheric region of our atmosphere, where they facilitate the destruction of the ozone layer. (See the essay in Chapter 10 on this topic.)

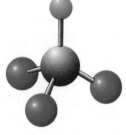

CCl_3F

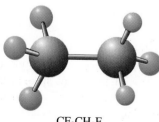

CF_3CH_2F

Recent refrigerants that do not as readily contribute to ozone destruction include hydrofluorocarbons (HFCs), such as CF_3CH_2F, which do not contain chlorine atoms.

CONCEPT CHECK 24.2

For gasoline to function properly in an engine, it should not begin to burn before it is ignited by the spark plug. If it does, it makes the noise we think of as engine "knock." The octane-number scale rates the anti-knock characteristics of a gasoline. This linear scale is based on heptane, given an octane number of 0, and on 2,2,4-trimethylpentane (an octane constitutional isomer), given an octane number of 100. The higher the octane number, the better the anti-knock characteristics. If you had a barrel of heptane and a barrel of 2,2,4-trimethylpentane, how would you blend these compounds to come up with a 90 octane mixture?

24.3   Alkenes and Alkynes

Alkenes and alkynes are unsaturated hydrocarbons. Because they contain carbon–carbon multiple bonds, they are typically much more reactive than alkanes.

Under the proper conditions, molecular hydrogen can be added to an alkane or alkyne to produce a saturated compound in a process called *catalytic hydrogenation*. For example, ethylene adds hydrogen to give ethane.

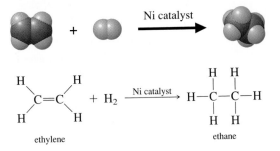

ethylene   ethane

Catalytic hydrogenation is also used in the food industry to hydrogenate carbon–carbon double bonds. For example, margarine can be manufactured by hydrogenating some of the double bonds present in corn oil to change it from oil to a solid (fat).

Alkenes and Geometric Isomerism

Alkenes are *hydrocarbons that have the general formula C_nH_{2n} and contain a carbon–carbon double bond.* (These compounds are also called *olefins.*) The simplest alkene, ethylene, has the condensed formula $CH_2{=}CH_2$ and the structural formula

$$\begin{array}{ccc} H & & H \\ & C{=}C & \\ H & & H \end{array}$$

Ethylene is a gas with a sweet odor. It is obtained from the refining of petroleum and is an important raw material in the chemical industry. For example, when ethylene molecules are linked they make polyethylene, which is commonly used to make soda bottles and milk jugs. Plants also produce small amounts of ethylene. Fruit suppliers have found that exposure of fruit to ethylene speeds ripening.

Figure 24.7
Representations of the ethylene molecule. Pi bonding forces the two CH_2 groups in ethylene to bond so as to give a planar structure.

In ethylene and other alkenes, all of the atoms connected to the two carbon atoms of the double bond lie in a single plane, as Figure 24.7 shows. This is due to the overlap of $2p$ orbitals on the carbon atoms to form a pi (π) bond. As a result, rotation about a carbon–carbon double bond cannot occur without breaking the π bond. (This is in contrast to carbon–carbon single bonds, which have very low energy requirements for rotation so they freely rotate under most conditions.) This inability to rotate gives rise to geometric isomers in certain alkenes. **Geometric isomers** are *isomers in which the atoms are joined to one another in the same way but differ because some atoms occupy different relative positions in space.* For example, 2-butene, C_4H_8, has two geometric isomers, called *cis*-2-butene and *trans*-2-butene.

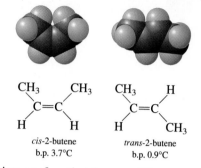

The different boiling points confirm that they are different compounds. An alkene with the general formula

exists as geometric isomers only if groups A and B are different and groups C and D are different. For instance, geometric isomers do not exist for propene, $CH_2{=}CHCH_3$.

EXAMPLE 24.2

Predicting *cis–trans* Isomers

For each of the following alkenes, decide whether *cis–trans* isomers are possible. If so draw structural formulas of the isomers.

a. $CH_3CH_2CH{=}C(CH_3)_2$ b. $CH_3CH{=}CHCH_2CH_3$

SOLUTION

a. Writing the structure, you have

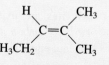

Because two methyl groups are attached to the second carbon atom of the double bond, **geometric isomers are not possible.** b. **Geometric isomers are possible.** They are

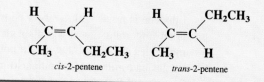

See Problems 24.21 and 24.22.

Figure 24.8
Test for unsaturation using KMnO₄(aq). *Top:* A purple KMnO₄ solution prior to the addition of an alkene. *Bottom:* KMnO₄ solution that has turned brown due to the formation of MnO₂(s) due to reaction with an alkene.

Cis and *trans* isomers of dietary fats and oils are suspected to play a role in human health. All unprocessed fats and oils contain only *cis* isomers. During processing, some of the *cis* isomers are converted to *trans* isomers, called *trans* fatty acids. *Trans* fatty acids are suspected of raising the amount of serum cholesterol in the blood stream, which can cause health problems.

Oxidation Reactions of Alkenes

Because alkenes are hydrocarbons, they undergo complete combustion reactions with oxygen at high temperatures to produce carbon dioxide and water. Unsaturated hydrocarbons can also be partially oxidized under relatively mild conditions. For example, when aqueous potassium permanganate, $KMnO_4(aq)$, is added to an alkene (or alkyne), the purple color of $KMnO_4$ fades and a brown precipitate of manganese dioxide forms (Figure 24.8).

$$3C_4H_9CH{=}CH_2 + 2MnO_4{}^-(aq) + 4H_2O \longrightarrow$$
1-hexene

$$3C_4H_9\overset{\overset{\displaystyle H}{|}}{\underset{\underset{\displaystyle H}{|}}{C}}{-}\overset{\overset{\displaystyle H}{|}}{\underset{\underset{\displaystyle H}{|}}{C}}{-}H + 2MnO_2(s) + 2OH^-(aq)$$

Addition Reactions of Alkenes

Alkenes are more reactive than alkanes because of the presence of the double bond. Many reactants *add* to the double bond. A simple example is the addition of a halogen, such as Br_2, to propene.

$$CH_3CH{=}CH_2 + Br_2 \longrightarrow CH_3\underset{\underset{\displaystyle Br}{|}}{CH}{-}\underset{\underset{\displaystyle Br}{|}}{CH_2}$$
propene

An **addition reaction** is *a reaction in which parts of a reactant are added to each carbon atom of a carbon–carbon double bond, which converts to a carbon–carbon single bond.* Bromine dissolved in carbon tetrachloride, CCl_4, is a useful reagent to test for unsaturation. When a few drops of the solution are added to an alkene, the red-brown color of the Br_2 disappears immediately.

Unsymmetrical reagents, such as HCl and HBr, add to unsymmetrical alkenes to give two products that are constitutional isomers. For example,

$$CH_3CH{=}CH_2 + HBr \rightarrow \overset{3}{C}H_3\overset{2}{\underset{\underset{\displaystyle Br}{|}}{CH}}{-}\overset{1}{\underset{\underset{\displaystyle H}{|}}{CH_2}}$$
2-bromopropane

and

$$CH_3CH{=}CH_2 + HBr \rightarrow \overset{3}{C}H_3\overset{2}{\underset{\underset{\displaystyle H}{|}}{CH}}{-}\overset{1}{\underset{\underset{\displaystyle Br}{|}}{CH_2}}$$
1-bromopropane

In one case, the hydrogen atom of HBr adds to carbon atom 1, giving 2-bromopropane; in the other case, the hydrogen atom of HBr adds to carbon atom 2, giving 1-bromopropane. However, the two products are not formed in equal amounts; one is

more likely to form. **Markownikoff's rule** is *a generalization that states that the major product formed by the addition of an unsymmetrical reagent such as H—Cl, H—Br, or H—OH is the one obtained when the H atom of the reagent adds to the carbon atom of the multiple bond that already has the greater number of hydrogen atoms attached to it.* In the preceding example, the H atom of HBr should add preferentially to carbon atom 1, which has two hydrogen atoms attached to it. The major product then is 2-bromopropane.

EXAMPLE 24.3

Predicting the Major Product of an Addition Reaction

What is the major product of the following reaction?

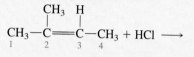

SOLUTION

One H atom is attached to carbon 3 and none to carbon 2. Therefore the major product is

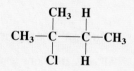

See Problems 24.25 and 24.26.

Some simple alkenes can be made to react with each other to form an addition polymer. An **addition polymer** is *a polymer formed by linking together many molecules by addition reactions.* The monomers must have multiple bonds that will undergo addition reactions. For example, when propylene is heated under pressure with a catalyst, it forms polypropylene.

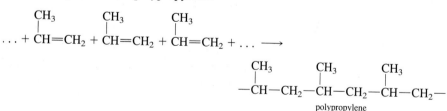

Note that the π electrons in the double bonds of the propylene molecules form new σ bonds that unite the monomers.

The preparation of an addition polymer is often induced by an *initiator,* a compound that produces free radicals (species having an unpaired electron). Organic peroxides (organic compounds with the —O—O— group) are frequently used as initiators (Figure 24.9).

Some additional polymers are listed in Table 24.2. These polymers are used extensively in our daily living and you have probably heard of them many times.

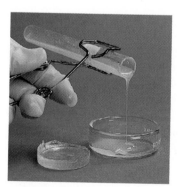

Figure 24.9
Preparation of polystyrene. When styrene, $C_6H_5CH{=}CH_2$, is heated with a small amount of benzoyl peroxide, ROOR (R = C_6H_5CO—), it yields a viscous liquid. After some time, this liquid sets to a hard plastic (sample shown at left).

Table 24.2
Some Addition Polymers

Polymer	Monomer	Uses
Polyethylene	$CH_2{=}CH_2$	Bottles, plastic tubing
Polypropylene	$CH_2{=}\overset{\displaystyle CH_3}{\overset{\displaystyle \vert}{CH}}$	Bottles, carpeting, textiles
Polytetrafluoroethylene (Teflon®)	$CF_2{=}CF_2$	Nonstick surface for frying pans
Poly(vinyl chloride) (PVC)	$CH_2{=}\overset{\displaystyle Cl}{\overset{\displaystyle \vert}{CH}}$	Plastic pipes, floor tile
Polyacrylonitrile (Orlon®, Acrilan®)	$CH_2{=}\overset{\displaystyle CN}{\overset{\displaystyle \vert}{CH}}$	Carpets, textiles
Polystyrene	$CH_2{=}\overset{\displaystyle C_6H_5}{\overset{\displaystyle \vert}{CH}}$	Plastic foam insulation, cups

Alkynes

Alkynes are *unsaturated hydrocarbons containing a carbon–carbon triple bond.* The general formula is C_nH_{2n-2}. The simplest alkyne is acetylene (ethyne).

$$H{-}C{\equiv}C{-}H$$

Acetylene is a very reactive gas that is used to produce a variety of other chemical compounds. It burns with oxygen in the oxyacetylene torch to give a very hot flame (about 3000°C). Acetylene is produced commercially from methane.

$$2CH_4 \xrightarrow{\;1600°C\;} CH{\equiv}CH + 3H_2$$

Acetylene is also prepared from calcium carbide, CaC_2. Calcium carbide is obtained by heating calcium oxide and coke (carbon) in an electric furnace. The carbide ion, C_2^{2-}, is strongly basic and reacts with water to produce acetylene (Figure 24.10).

$$CaC_2(s) + 2H_2O(l) \longrightarrow Ca(OH)_2(aq) + C_2H_2(g)$$

The alkynes, like the alkenes, undergo addition reactions, usually adding two molecules of the reagent for each $C{\equiv}C$ bond. The major product is the isomer predicted by Markownikoff's rule, as shown in the following reaction.

Figure 24.10
Preparation of acetylene gas. Here acetylene is prepared by the reaction of water with calcium carbide. The acetylene burns with a sooty flame.

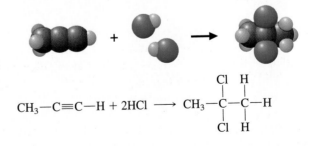

$$CH_3{-}C{\equiv}C{-}H + 2HCl \longrightarrow CH_3{-}\overset{\displaystyle Cl}{\underset{\displaystyle Cl}{\overset{\displaystyle \vert}{\underset{\displaystyle \vert}{C}}}}{-}\overset{\displaystyle H}{\underset{\displaystyle H}{\overset{\displaystyle \vert}{\underset{\displaystyle \vert}{C}}}}{-}H$$

24.4 Aromatic Hydrocarbons

Figure 24.11
Bond delocalization in benzene. The space-filling model of benzene shows that all carbon–carbon bond distances are identical.

Aromatic hydrocarbons usually contain benzene rings—six membered rings of carbon atoms with alternating carbon–carbon single and carbon–carbon double bonds. The electronic structure of benzene can be represented by resonance formulas. For benzene,

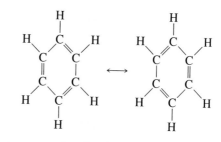

This electronic structure can also be described using molecular orbitals. In this description, π molecular orbitals encompass the entire carbon-atom ring, and the π electrons are said to be delocalized. Delocalization of π electrons means that the double bonds in benzene do not behave as isolated double bonds. Two condensed formulas for benzene are

where the circle in the formula at the right represents the delocalization of the π electrons (and therefore the double bonds). Although you will often encounter benzene represented with alternating double and single bonds as shown on the left, the better representation is the one that indicates the bond delocalization. The space-filling model of benzene is shown in Figure 24.11.

Aromatic compounds are found everywhere. The term *aromatic* implies that compounds that contain a benzene ring have aromas and, indeed, this is the case. Flavoring agents that can be synthesized in the laboratory or found in nature include the flavor and aroma of cinnamon, cinnamaldehyde, and the wintergreen flavor of candies and gum, methyl salicylate (Figure 24.12).

Figure 24.12
Examples of aromatic compounds. *Left:* A molecular model of cinnamaldehyde, the molecule responsible for the taste and smell of cinnamon. *Right:* A molecular model of methylsalicylate, the molecule that produces the flavor of wintergreen.

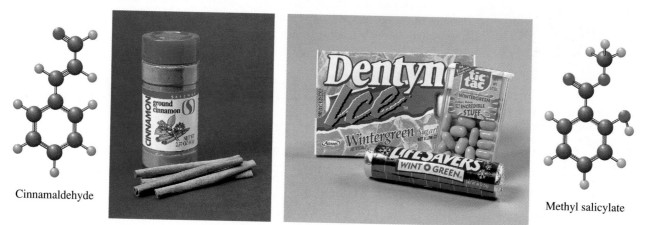

Cinnamaldehyde

Methyl salicylate

Benzene rings also exist in the pain relievers acetylsalicylic acid (aspirin), acetaminophen (Tylenol), and the illicit drug mescaline.

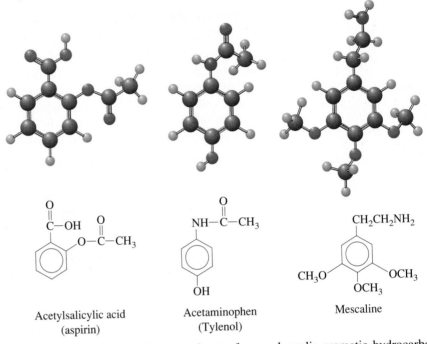

Acetylsalicylic acid
(aspirin)

Acetaminophen
(Tylenol)

Mescaline

Benzene rings can also fuse together to form polycyclic aromatic hydrocarbons in which two or more rings share carbon atoms. Figure 24.13 gives the formula for naphthalene. Naphthalene is one of the compounds that gives mothballs their characteristic odor.

Substitution Reactions of Aromatic Hydrocarbons

Although benzene, C_6H_6, is an unsaturated hydrocarbon, it does not usually undergo addition reactions because the delocalized π electrons of benzene are more stable than the localized π electrons. Because of this, benzene does not undergo an addition reaction with Br_2 in carbon tetrachloride like alkenes do. The usual reactions of benzene are substitution reactions. For example, in the presence of an iron(III) bromide catalyst, a hydrogen atom on the benzene ring is substituted for a bromine atom.

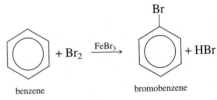

benzene   bromobenzene

Similarly, benzene undergoes substitution with nitric acid in the presence of sulfuric acid to give nitrobenzene.

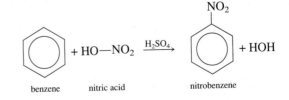

benzene   nitric acid   nitrobenzene

Figure 24.13
A polycyclic aromatic hydrocarbon. An example is naphthalene. It is a white, crystalline substance used in manufacturing plastics and plasticizers (to keep plastics pliable). Small amounts are used for mothballs.

24.5 Naming Hydrocarbons

Nomenclature of Alkanes

The first four straight-chain alkanes (methane, ethane, propane, and butane) have long-established names. Higher members of the series are named from the Greek words indicating the number of carbon atoms in the molecule, plus the suffix -ane. For example, the straight-chain alkane C_5H_{12} is named pentane. Table 24.1 gives the formulas, names, and structures of the first ten straight-chain alkanes.

The following four International Union of Pure and Applied Chemistry (IUPAC) rules are applied in naming the branched-chain alkanes:

1. Determine the longest continuous (not necessarily straight) chain of carbon atoms in the molecule. The base name of the branched-chain alkane is that of the straight-chain alkane (Table 24.1) corresponding to the number of carbon atoms in this longest chain. For example, in

$$CH_3CH_2CH_2CH_2-\overset{\overset{\displaystyle H}{|}}{\underset{\underset{\displaystyle CH_3}{|}}{\underset{\displaystyle CH_2}{\overset{\displaystyle |}{C}}}}-CH_3$$

the longest continuous carbon chain, shown in color, has seven carbon atoms, giving the base name heptane. The full name for the alkane includes the names of any branched chains. These names are placed in front of the base name, as described in the remaining rules.

2. Any chain branching off the longest chain is named as an alkyl group. An alkyl group is an alkane less one hydrogen atom. (Table 24.3 lists some alkyl groups.) When a hydrogen atom is removed from an end carbon atom of a straight-chain alkane, the resulting alkyl group is named by changing the suffix -ane of the alkane to -yl. For example, removing a hydrogen atom from methane gives the methyl group, $-CH_3$. The structure shown in Rule 1 has a methyl group as a branch on the heptane chain.

Table 24.3
Important Alkyl Groups

Original Alkane	Structure of Alkyl Group	Name of Alkyl Group
Methane, CH_4	CH_3-	Methyl
Ethane, CH_3CH_3	CH_3CH_2-	Ethyl
Propane, $CH_3CH_2CH_3$	$CH_3CH_2CH_2-$	Propyl
Propane, $CH_3CH_2CH_3$	$CH_3\underset{\underset{\displaystyle CH_3}{\overset{\displaystyle |}{}}}{CH}CH_3$	Isopropyl
Butane, $CH_3CH_2CH_2CH_3$	$CH_3CH_2CH_2CH_2-$	Butyl
Isobutane, $CH_3\underset{\underset{\displaystyle CH_3}{\overset{\displaystyle |}{}}}{CH}CH_3$	$CH_3\underset{\underset{\displaystyle CH_3}{\overset{\displaystyle |}{}}}{\overset{\overset{\displaystyle CH_3}{\overset{\displaystyle |}{}}}{C}}CH_3$	*Tertiary*-butyl (*t*-butyl)

3. The complete name of a branch requires a number that locates the branch on the longest chain. For this purpose, you number each carbon atom on the longest chain in the direction that gives the smaller numbers for the locations of all branches. The structural formula in Rule 1 is numbered as shown:

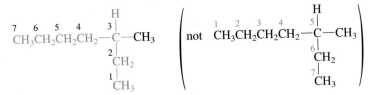

The methyl branch is located at carbon 3 of the heptane chain (not carbon 5). The complete name of the branch is 3-methyl, and the compound is named 3-methyl-heptane.

4. When there are more than one alkyl branch of the same kind (say, two methyl groups), this number is indicated by a prefix, such as *di-*, *tri-*, or *tetra-*, used with the name of the alkyl group. The position of each group on the longest chain is given by numbers. For example,

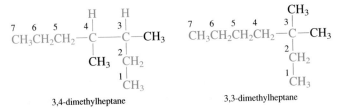

When there are two or more different alkyl branches, the name of each branch, with its position number, precedes the base name. The branch names are placed in alphabetical order. For example,

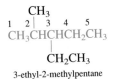

3-ethyl-2-methylpentane

EXAMPLE 24.4

Writing the IUPAC Name of an Alkane Given the Structural Formula

Give the IUPAC name for each of the following compounds.

a. CH₃CH₂CH₂
           \
            CHCH₂CH₂CH₃
           /
    CH₃CH₂CH₂CH₂

b.
        CH₃
        |
  CH₃—C—CH₂—CH₂—CH₂—CH₃
        |
        CH₃

SOLUTION

a. The longest continuous chain is numbered as follows:

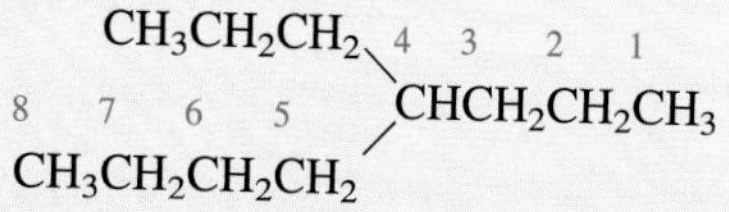

$$CH_3CH_2CH_2 \underset{8 \quad 7 \quad 6 \quad 5}{CH_3CH_2CH_2CH_2} \overset{4 \quad 3 \quad 2 \quad 1}{\diagdown} CHCH_2CH_2CH_3$$

The name of the compound is **4-propyloctane.** If the longest chain had been numbered in the opposite direction, you would have the name 5-propyloctane. But because 5 is larger than 4, this name is unacceptable. b. The numbering of the longest chain is

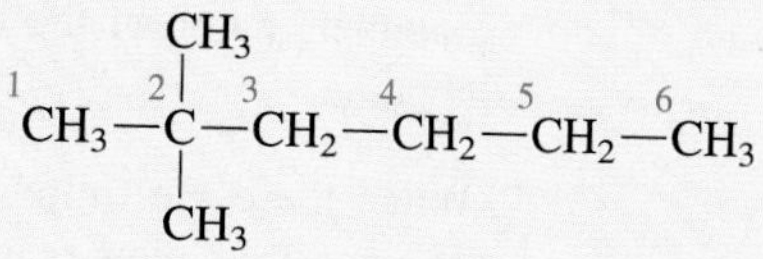

$$\overset{\displaystyle CH_3}{\underset{\displaystyle CH_3}{\overset{1}{CH_3}-\overset{2}{\underset{|}{C}}-\overset{3}{CH_2}-\overset{4}{CH_2}-\overset{5}{CH_2}-\overset{6}{CH_3}}}$$

Any of the three clustered methyl carbon atoms could be given the number 1, and the other methyl groups branch off carbon atom 2. Hence, the name is **2,2-dimethylhexane.**

See Problems 24.27 and 24.28.

EXAMPLE 24.5

Writing the Structural Formula of an Alkane Given the IUPAC Name

Write the condensed structural formula of 4-ethyl-3-methylheptane.

SOLUTION

First write the carbon skeleton for heptane.

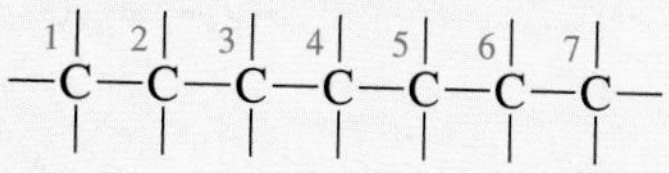

$$-\overset{1}{\underset{|}{C}}-\overset{2}{\underset{|}{C}}-\overset{3}{\underset{|}{C}}-\overset{4}{\underset{|}{C}}-\overset{5}{\underset{|}{C}}-\overset{6}{\underset{|}{C}}-\overset{7}{\underset{|}{C}}-$$

Then attach the alkyl groups.

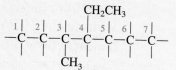

After filling out the structure with H atoms, you have

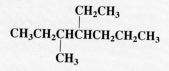

See Problems 24.29 and 24.30.

Nomenclature of Alkenes and Alkynes

The IUPAC name for an alkene is determined by first identifying the longest chain containing the double bond. As with the alkanes, the longest chain provides the stem name, but the suffix is *-ene* rather than *-ane*. The carbon atoms of the longest chain are then numbered from the end nearer the carbon–carbon double bond, and the position of the double bond is given the number of the first carbon atom of that bond (the smaller number). This number is written in front of the stem name of the alkene. Branched chains are named the same way as the alkanes. The simplest alkene, $CH_2{=}CH_2$, is called ethene, although the common name is ethylene.

EXAMPLE 24.6

Writing the IUPAC Name of an Alkene Given the Structural Formula

What is the IUPAC name of the following alkene?

$$CH_2{=}CHCH CH_2CH_3$$
$$| $$
$$CH_3$$

SOLUTION

The numbering of the carbon chain is

$$\overset{1}{CH_2}{=}\overset{2}{C}H\overset{3}{C}H\overset{4}{C}H_2\overset{5}{C}H_3$$
$$|$$
$$CH_3$$

Because the longest chain containing a double bond has five carbon atoms, this is a pentene. It is a 1-pentene, because the double bond is between carbon atoms 1 and 2. The name of the compound is **3-methyl-1-pentene.** If the numbering had been in the opposite direction, you would have named the compound as a 4-pentene. But this is unacceptable, because 4 is greater than 1.

See Problems 24.31 and 24.32.

Alkynes are named using the IUPAC rules in the same way as the alkenes, except that the stem name is determined from the longest chain that contains the carbon–carbon triple bond. The suffix for the stem name is *-yne*.

Nomenclature of Aromatic Hydrocarbons

Simple benzene-containing hydrocarbons that have one group substituted on a benzene ring use the name *benzene* as the suffix and the name of the group as the prefix. For example, the following compound is named ethylbenzene.

$$CH_2CH_3$$

ethylbenzene

When two groups are on the benzene ring, three isomers are possible. The isomers may be distinguished by using the prefixes *ortho-* (*o-*), *meta-* (*m-*), and *para-* (*p-*). For example,

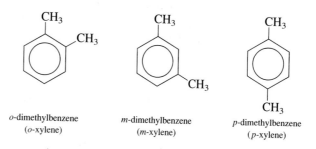

o-dimethylbenzene
(*o*-xylene)

m-dimethylbenzene
(*m*-xylene)

p-dimethylbenzene
(*p*-xylene)

A numbering system is also used to show the positions of two or more groups.

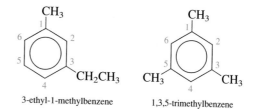

3-ethyl-1-methylbenzene

1,3,5-trimethylbenzene

It is sometimes preferable to name a compound containing a benzene ring by regarding the ring as a group in the same manner as alkyl groups. Pulling a hydrogen atom from benzene leaves the phenyl group, C_6H_5—. For example, you name the following compound diphenylmethane by using methane as the stem name.

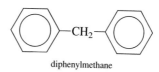

diphenylmethane

CONCEPT CHECK 24.3

In the model shown here, C atoms are black and H atoms are light blue:

a. Write the molecular formula.

b. Write the condensed structural formula.

c. Write the IUPAC name.

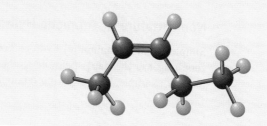

Derivatives of Hydrocarbons

Certain groups of atoms in organic molecules are part'
acteristic chemical properties. A **functional group** is
that undergoes predictable reactions. As discussed
in a compound reacts readily and predictably with the ι
tion reactions. Thus, the C=C bond is a functional group. ιν.
contain an atom other than carbon that has lone pair(s) of electrons. ᴵ
of electrons contribute to the reactivity of the functional group. Other ιᴸ
groups, such as C=O, have multiple bonds that are reactive. Table 24.4 lists soι.
common functional groups.

In the previous sections of the chapter, we discussed the hydrocarbons and their
reactions. All other organic compounds can be considered to be derivatives of hydrocarbons. In these compounds, one or more hydrogen atoms of a hydrocarbon have
been replaced by atoms other than carbon to give a functional group.

Table 24.4
Some Organic Functional Groups

Structure of General Compound* (Functional Group in Color)	Name of Functional Group
R—Cl: R—Br:	Organic halide
R—O—H	Alcohol
R—O—R′	Ether
:O: ‖ R—C—H	Aldehyde
:O: ‖ R—C—R′	Ketone
:O: ‖ R—C—O—H	Carboxylic acid
:O: ‖ R—C—O—R′	Ester
R—N—H R—N—H R—N—R″ H R′ R′	Amine
:O: ‖ R—C—N—R′ H	Amide

*R, R′, and R″ are general hydrocarbon groups.

24.6 Organic Compounds Containing Oxygen

Many of the important functional groups in organic compounds contain oxygen. Examples are alcohols, ethers, aldehydes, ketones, carboxylic acids, and esters. In this section, we will look at characteristics of these compounds.

Alcohols and Ethers

Structurally, you may think of an **alcohol** as *a compound obtained by substituting a hydroxyl group (—OH) for an —H atom on a tetrahedral (sp³ hybridized) carbon atom of a hydrocarbon group.* Some examples are

CH_3OH	CH_3CH_2OH	CH_3CHCH_3
methanol	ethanol	2-propanol
(methyl alcohol)	(ethyl alcohol)	(isopropyl alcohol)

Alcohols are named by IUPAC rules similar to those for the hydrocarbons, except that the stem name is determined from the longest chain containing the carbon atom to which the —OH group is attached. The suffix for the stem name is *-ol*. The position of the —OH group is indicated by a number preceding the stem name.

Alcohols are usually classified by the number of carbon atoms attached to the carbon atom to which the —OH group is bonded. A *primary alcohol* has one such carbon atom, a *secondary alcohol* has two, and a *tertiary alcohol* has three. The following are examples:

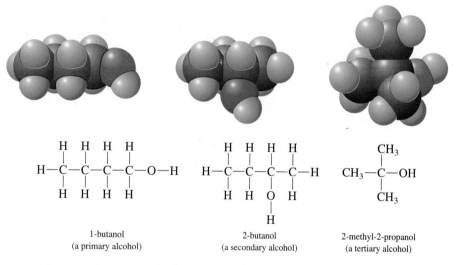

1-butanol
(a primary alcohol)

2-butanol
(a secondary alcohol)

2-methyl-2-propanol
(a tertiary alcohol)

Methanol, ethanol, ethylene glycol, and glycerol are some common alcohols. Methanol, CH_3OH, was at one time separated from the liquid distilled from sawdust—hence the common name wood alcohol. It is a toxic liquid prepared in large quantities by reacting carbon monoxide with hydrogen at high pressure in the presence of a catalyst. It is used as a solvent and as the starting material for the preparation of formaldehyde. Ethanol is manufactured by the fermentation of glucose (a

sugar) or by the addition of water to the double bond of ethylene. The latter reaction is carried out by heating ethylene with water in the presence of sulfuric acid.

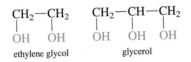

$$CH_2{=}CH_2 + HOH \xrightarrow{\ H_2SO_4\ } \underset{\substack{| \quad\ \ | \\ H \quad OH}}{CH_2{-}CH_2}$$

ethylene   ethanol

Alcoholic beverages contain ethanol. Ethanol is also a solvent and a starting material for many organic compounds. It is mixed with gasoline and sold as gasohol, an auto-motive fuel (Figure 24.14). Ethylene glycol (IUPAC name: 1,2-ethanediol) and glyc-erol (1,2,3-propanetriol) are alcohols containing more than one hydroxyl group.

Figure 24.14
**Gasoline containing
ethanol.** The gasoline in
this pump was formulated
with ethanol. Ethanol
is added to boost octane and
oxygenate the fuel.

$$\underset{\substack{| \quad\ \ | \\ OH \quad OH}}{CH_2{-}CH_2} \qquad \underset{\substack{| \qquad\ | \qquad | \\ OH \quad OH \quad OH}}{CH_2{-}CH{-}CH_2}$$

ethylene glycol   glycerol

Ethylene glycol is a liquid prepared from ethylene and is used as an antifreeze agent. It is also used in the manufacture of polyester plastics and fibers. Glycerol is a non-toxic, sweet-tasting liquid obtained from fats during the making of soap. It is used in foods and candies to keep them soft and moist.

Just as an alcohol may be thought of as a derivative of water in which one H atom of H_2O has been replaced by a hydrocarbon group, R, an **ether** is *a compound formally obtained by replacing both H atoms of H_2O by the hydrocarbon groups R and R'*.

$$H{-}O{-}H \qquad R{-}O{-}H \qquad R{-}O{-}R'$$

water   an alcohol   an ether

Common names for ethers are formed from the hydrocarbon groups plus the word *ether*. For example, $CH_3OCH_2CH_2CH_3$ is called methyl propyl ether. By IUPAC rules, the ethers are named as derivatives of the longest hydrocarbon chain. For example, $CH_3OCH_2CH_2CH_3$ is 1-methoxypropane; the methoxy group is $CH_3O{-}$. The best-known ether is diethyl ether, $CH_3CH_2OCH_2CH_3$, often called simply ether, a volatile liquid used as a solvent and as an anesthetic.

Aldehydes and Ketones

Aldehydes and ketones are compounds containing a *carbonyl group*.

carbonyl group

An **aldehyde** is *a compound containing a carbonyl group with at least one H atom attached to it*.

Ethanal

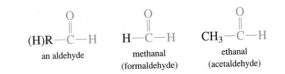

$$\underset{\text{an aldehyde}}{(H)R{-}\overset{\displaystyle O}{\overset{\|}{C}}{-}H} \qquad \underset{\substack{\text{methanal} \\ \text{(formaldehyde)}}}{H{-}\overset{\displaystyle O}{\overset{\|}{C}}{-}H} \qquad \underset{\substack{\text{ethanal} \\ \text{(acetaldehyde)}}}{CH_3{-}\overset{\displaystyle O}{\overset{\|}{C}}{-}H}$$

Figure 24.15
Some aldehydes of aromatic hydrocarbons.
The name *aromatic* in aromatic hydrocarbons derives from the aromatic odors of many derivatives of these hydrocarbons, including the aldehydes shown here.

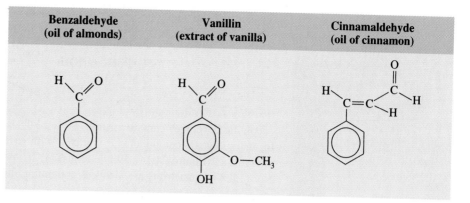

| Benzaldehyde (oil of almonds) | Vanillin (extract of vanilla) | Cinnamaldehyde (oil of cinnamon) |

Here (H)R indicates a hydrocarbon group or H atom. The aldehyde function is usually abbreviated —CHO, and the structural formula of acetaldehyde is written CH_3CHO.

A **ketone** is *a compound containing a carbonyl group with two hydrocarbon groups attached to it.*

Propanone

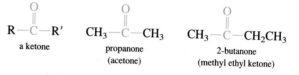

The ketone functional group is abbreviated —CO—, and the structural formula of acetone is written CH_3COCH_3.

Aldehydes and ketones are named according to IUPAC rules similar to those for naming alcohols. You first locate the longest carbon chain containing the carbonyl group to get the stem hydrocarbon name. Then you change the *-e* ending of the hydrocarbon to *-al* for aldehydes and *-one* for ketones. In the case of aldehydes, the carbon atom of the —CHO group is always the number-1 carbon. In ketones, however, the carbonyl group may occur in various nonequivalent positions on the carbon chain. For ketones, the position of the carbonyl group is indicated by a number before the stem name, just as the position of the hydroxyl group is indicated in alcohols.

The aldehydes of lower molecular weight have sharp, penetrating odors. With increasing molecular weight, the aldehydes become more fragrant. Some aldehydes of aromatic hydrocarbons have especially pleasant odors (see Figure 24.15). Formaldehyde is a gas produced by the oxidation of methanol. The gas is very soluble in water, and a 37% aqueous solution called Formalin is marketed as a disinfectant and as a preservative of biological specimens. The main use of formaldehyde is in the manufacture of plastics and resins. Acetone, CH_3COCH_3, is the simplest ketone. It is a liquid with a fragrant odor. The liquid is an important solvent for lacquers, paint removers, and nail polish removers.

Carboxylic Acids and Esters

A **carboxylic acid** is *a compound containing the carboxyl group,* —*COOH.*

Ethanoic acid

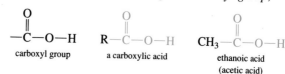

These compounds are named by IUPAC rules like those for the aldehydes, except that the ending on the stem name is *-oic* followed by the word *acid*. Many carboxylic acids have been known for a long time and are usually referred to by common names (see Table 24.5). The carboxylic acids are weak acids because of the acidity of the H atom on the carboxyl group. Acid-ionization constants are about 10^{-5}.

An **ester** is *a compound formed from a carboxylic acid, RCOOH, and an alcohol, R'OH*. The general structure is

$$\overset{:O:}{\underset{}{RC}}\text{—O—R'}$$

The reaction between ethanol and acetic acid in the presence of a catalyst is a typical reaction for the preparation of an ester.

$$CH_3CH_2\text{—OH} + HOC\text{—}CH_3 \longrightarrow CH_3CH_2\text{—O—C—}CH_3 + H_2O$$

ethanol   acetic acid   ethyl acetate

This is called a condensation reaction. A **condensation polymer** is *a polymer formed by linking together many molecules by condensation reactions*.

Polyesters   A substance with two alcohol groups reacts with a substance with two carboxylic acid groups to form a *polyester*, a polymer whose repeating units are joined by ester groups. The reactant molecules join as links of a chain to form a very long molecule. The polyester Dacron, used as a textile fiber, is prepared from ethylene glycol and terephthalic acid.

Table 24.5
Common Carboxylic Acids

Carbon Atoms	Formula	Source	Common Name	IUPAC Name
1	HCOOH	Ants (Latin, *formica*)	Formic acid	Methanoic acid
2	CH_3COOH	Vinegar (Latin, *acetum*)	Acetic acid	Ethanoic acid
3	CH_3CH_2COOH	Milk (Greek, *protos pion*, "first fat")	Propionic acid	Propanoic acid
4	$CH_3(CH_2)_2COOH$	Butter (Latin, *butyrum*)	Butyric acid	Butanoic acid
5	$CH_3(CH_2)_3COOH$	Valerian root (Latin, *valere*, "to be strong")	Valeric acid	Pentanoic acid
6	$CH_3(CH_2)_4COOH$	Goats (Latin, *caper*)	Caproic acid	Hexanoic acid
7	$CH_3(CH_2)_5COOH$	Vine blossom (Greek, *oenanthe*)	Enanthic acid	Heptanoic acid
8	$CH_3(CH_2)_6COOH$	Goats (Latin, *caper*)	Caprylic acid	Octanoic acid
9	$CH_3(CH_2)_7COOH$	Pelargonium (an herb with stork-shaped capsules; Greek, *pelargos*, "stork")	Pelargonic acid	Nonanoic acid
10	$CH_3(CH_3)_8COOH$	Goats (Latin, *caper*)	Capric acid	Decanoic acid

From Harold Hart, *Organic Chemistry: A Short Course*, Eighth Edition, p. 272. Copyright © 1991 by Houghton Mifflin Company. Used with permission.

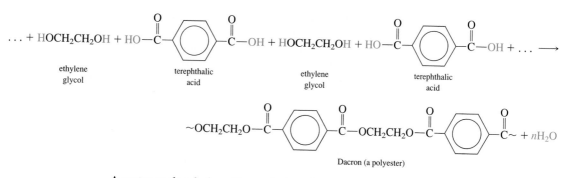

A water molecule is split out during the formation of each ester linkage.

24.7 Organic Compounds Containing Nitrogen

Alcohols and ethers, you may recall, can be considered derivatives of H_2O, where one or both H atoms are replaced by hydrocarbon groups, R. Thus, the general formula of an alcohol is ROH, and that of an ether is ROR'. Another important class of organic compounds is obtained by similarly substituting R groups for the H atoms of ammonia, NH_3.

Most organic bases are **amines,** which are *compounds that are structurally derived by replacing one or more hydrogen atoms of ammonia with hydrocarbon groups.*

Table 24.6 lists some common amines.

Amines are bases, because the nitrogen atom has an unshared electron pair that can accept a proton to form a substituted ammonium ion. For example,

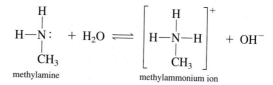

Like ammonia, amines are weak bases. Table 17.2 gives base-ionization constants of some amines.

Amides are *compounds derived from the reaction of ammonia, or of a primary or secondary amine, with a carboxylic acid.* For example, when ammonia is strongly heated with acetic acid, they react to give the amide acetamide.◄

► The condensation to give an amide occurs under milder conditions if the ammonia or amine is reacted with the acid chloride, a derivative of the carboxylic acid obtained by replacing

In this reaction, HCl is a product.

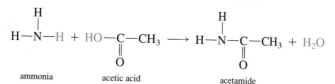

Table 24.6
Some Common Amines

Name	Formula	Boiling Point (°C)
Methylamine	$CH_3—NH_2$	−6.5
Dimethylamine	$CH_3—\underset{\underset{H}{\vert}}{N}—CH_3$	7.4
Trimethylamine	$CH_3—\underset{\underset{CH_3}{\vert}}{N}—CH_3$	3.5
Ethylamine	$CH_3CH_2—NH_2$	16.6
Piperidine	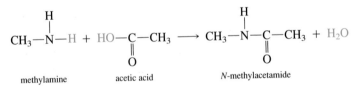 N—H	106
Aniline	—NH_2	184

From Robert D. Whitaker et al., *Concepts of General, Organic, and Biological Chemistry*, p. 343. Copyright © 1981 by Houghton Mifflin Company. Used with permission.

Figure 24.16
Nylon being pulled from a reaction mixture.
Nylon is formed by reacting hexamethylene diamine (bottom water layer) with adipic acid or, as in this experiment, with adipoyl chloride, $ClCO(CH_2)_4COCl$ (top hexane layer). Nylon, formed at the interface of the two layers, is continuously removed.

Methylamine and acetic acid react to give *N*-methylacetamide.

$$CH_3—\underset{\underset{H}{\vert}}{\overset{\overset{H}{\vert}}{N}}—H + HO—\underset{\underset{O}{\parallel}}{C}—CH_3 \longrightarrow CH_3—\underset{\underset{H}{\vert}}{\overset{\overset{H}{\vert}}{N}}—\underset{\underset{O}{\parallel}}{C}—CH_3 + H_2O$$

methylamine   acetic acid   *N*-methylacetamide

Polyamides When a compound containing two amine groups reacts with a compound containing two carboxylic acid groups, a condensation polymer called a *polyamide* is formed (Figure 24.16). Nylon-6,6 is an example. It is prepared by heating hexamethylene diamine (1,6-diaminohexane) and adipic acid (hexanedioic acid).

$$\ldots + H—\underset{\underset{H}{\vert}}{N}(CH_2)_6\underset{\underset{H}{\vert}}{N}—H + HO—\underset{\underset{O}{\parallel}}{C}(CH_2)_4\underset{\underset{O}{\parallel}}{C}—OH + H—\underset{\underset{H}{\vert}}{N}(CH_2)_6\underset{\underset{H}{\vert}}{N}—H + HO—\underset{\underset{O}{\parallel}}{C}(CH_2)_4\underset{\underset{O}{\parallel}}{C}—OH + \ldots \xrightarrow{\Delta}$$

hexamethylene diamine   adipic acid   hexamethylene diamine   adipic acid

$$\sim\underset{\underset{H}{\vert}}{N}(CH_2)_6\underset{\underset{H}{\vert}}{N}—\underset{\underset{O}{\parallel}}{C}(CH_2)_4\underset{\underset{O}{\parallel}}{C}—\underset{\underset{H}{\vert}}{N}(CH_2)_6\underset{\underset{H}{\vert}}{N}—\underset{\underset{O}{\parallel}}{C}(CH_2)_4\underset{\underset{O}{\parallel}}{C}\sim + nH_2O$$

nylon-6,6 (a polyamide)

A Checklist for Review

Important Terms

hydrocarbons (24.1)
saturated
   hydrocarbons (24.1)
unsaturated
   hydrocarbons (24.1)
aromatic
   hydrocarbons (24.1)
alkanes (24.2)
cycloalkanes (24.2)

condensed structural
   formula (24.2)
homologous series (24.2)
constitutional (structural)
   isomer (24.2)
substitution reaction (24.2)
alkenes (24.3)
geometric isomers (24.3)
addition reaction (24.3)

addition polymer (24.3)
Markownikoff's rule (24.3)
alkynes (24.3)
aromatic hydrocarbons
   (24.4)
nomenclature (24.5)
functional group (24.5)
alcohol (24.6)
ether (24.6)

aldehyde (24.6)
ketone (24.6)
carboxylic acid (24.6)
ester (24.6)
condensation polymer
   (24.6)
amines (24.7)
amides (24.7)

Summary of Facts and Concepts

Organic compounds are hydrocarbons or derivatives of hydrocarbons. Hydrocarbons contain only H and C atoms. The three main groups of hydrocarbons are *saturated hydrocarbons,* hydrocarbons with only single bonds between the carbon atoms; *unsaturated hydrocarbons,* hydrocarbons that contain double or triple bonds between carbon atoms; and *aromatic hydrocarbons,* hydrocarbons that contain a benzene ring (a six-membered ring of carbon atoms with alternating single and double carbon–carbon bonds described by resonance formulas).

The *alkanes* are *acyclic,* saturated hydrocarbons that form a *homologous series* of compounds, with the general formula C_nH_{2n+2}. The *cycloalkanes* are *cyclic,* saturated hydrocarbons that form another homologous series with the general formula C_nH_{2n} in which the carbon atoms are joined in a ring. The *alkenes* and *alkynes* are unsaturated hydrocarbons (cyclic or acyclic) that contain carbon–carbon double or triple bonds.

The alkanes and aromatic hydrocarbons usually undergo *substitution reactions.* Alkenes and alkynes undergo *addition*

reactions. *Markownikoff's rule* predicts the major product in the addition of an unsymmetrical reagent to an unsymmetrical alkene.

A *functional group* is a portion of an organic molecule that reacts readily in predictable ways. Important functional groups containing oxygen are *alcohols* (ROH), *aldehydes* (RCHO), *ketones* (RCOR′), and *carboxylic acids* (RCOOH). When a functional group is present in an organic compound, the compound is considered to be a *hydrocarbon derivative. Amines* are organic derivatives of ammonia. They react with carboxylic acids to give *amides.*

Polymers are species of very high molecular weight consisting of many units *(monomers).* They are classified as *addition polymers* or *condensation polymers,* depending on the type of reaction used in forming them. Rubber is an addition polymer. Nylon-6,6 is an example of a condensation polymer; it is produced by condensing hexamethylene diamine with adipic acid.

Operational Skills

1. **Writing a condensed structural formula** Given the structural formula of a hydrocarbon, write the condensed structural formula. (EXAMPLE 24.1)
2. **Predicting *cis–trans* isomers** Given a condensed structural formula of an alkene, decide whether *cis* and *trans* isomers are possible, and, if so, draw the structural formulas. (EXAMPLE 24.2)
3. **Predicting the major product of an addition reaction** Predict the major product in the addition of an unsymmetrical reagent to an unsymmetrical alkene. (EXAMPLE 24.3)
4. **Writing the IUPAC name of a hydrocarbon given the structural formula, and vice versa** Given the structure of a hydrocarbon, state the IUPAC name (EXAMPLES 24.4, 24.6). Given the IUPAC name of a hydrocarbon, write the structural formula (EXAMPLE 24.5).

Review Questions

24.1 Give the molecular formula of an alkane with 30 carbon atoms.

24.2 Draw structural formulas of the five isomers of C_6H_{14}.

24.3 Explain why there are two isomers of 2-butene. Draw their structural formulas and name the isomers.

24.4 Draw structural formulas for the isomers of ethylmethylbenzene.

24.5 Give condensed structural formulas of all possible substitution products of ethane and Cl_2.

24.6 Define the terms *substitution reaction* and *addition reaction*. Give examples of each.

24.7 What would you expect to be the major product when two molecules of HCl add successively to acetylene? Explain.

24.8 Write the structure of propylbenzene. Write the structure of *p*-dichlorobenzene.

24.9 What is a functional group? Give an example and explain how it fits this definition.

24.10 An aldehyde contains the carbonyl group. Ketones, carboxylic acids, and esters also contain the carbonyl group. What distinguishes these latter compounds from an aldehyde?

24.11 Identify and name the functional group in each of the following.
a. CH_3COCH_3
b. $CH_3OCH_2CH_3$
c. $CH_3CH\!=\!CH_2$
d. CH_3CH_2COOH
e. CH_3CH_2CHO
f. $CH_3CH_2CH_2OH$

24.12 What is the difference between an addition polymer and a condensation polymer? Give an example of each, writing the equation for its formation.

Conceptual Problems

24.13 Explain why you wouldn't expect to find a compound with the formula CH_5.

24.14 A classmate tells you that the following compound has the name 3-propylhexane.

$$CH_3CH_2CHCH_2CH_2CH_3$$
with a branch $CH_2CH_2CH_3$

a. Is he right? If not, what error did he make and what is the correct name?
b. How could you redraw the condensed formula to better illustrate the correct name?

24.15 In the models shown here, C atoms are black, H atoms are light blue, O atoms are red, and N atoms are dark blue:

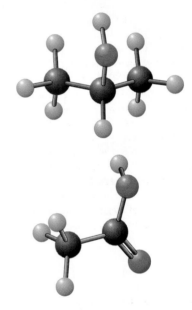

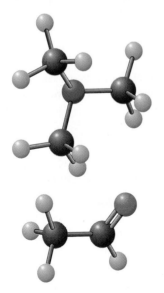

a. Write the molecular formula of each molecule.
b. Write the condensed structural formula for each molecule.
c. Identify the functional group for each molecule.

24.16   In the models shown here, C atoms are black and H atoms are light blue:

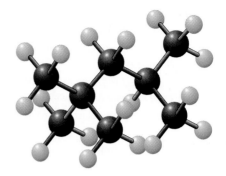

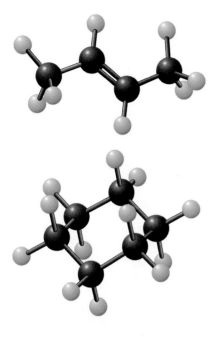

a. Write the molecular formula of each molecule.
b. Write the condensed structural formula for each molecule.
c. Give the IUPAC name of each molecule.

24.17   Why would you expect the melting points of the alkanes to increase in the series methane, ethane, propane, and so on?

24.18   Consider the following formulas of two esters:

$$CH_3CH_2-\overset{\displaystyle O}{\overset{\|}{C}}-O-CH_3 \qquad CH_3CH_2-O-\overset{\displaystyle O}{\overset{\|}{C}}-CH_3$$

One of these is ethyl ethanoate (ethyl acetate) and one is methyl propanoate (methyl propionate). Which is which?

Practice Problems

Condensed Structural Formulas

24.19   Write the condensed structural formula of the following alkane.

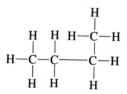

24.20   Write the condensed formula of the following alkane.

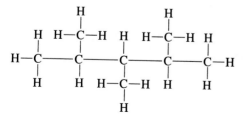

Geometric Isomers

24.21   If there are geometric isomers for the following, draw structural formulas showing the isomers.
a. $CH_3CH_2CH{=}CHCH_2CH_3$
b. $CH_3C{=}CHCH_2CH_3$
      $\quad\ \ |$
      $\quad\ CH_2CH_3$

24.22   One or both of the following have geometric isomers. Draw the structures of any geometric isomers.
a. $CH_3CHCH{=}CHCH_3$
      $\quad\ \ |$
      $\quad\ CH_3$
b. $CH_3C{=}CHCH_2CH_3$
      $\quad\ \ |$
      $\quad\ CH_3$

Reactions of Hydrocarbons

24.23   Complete and balance the following equations. Note any catalyst used.
a. $C_2H_4 + O_2 \longrightarrow$
b. $CH_2{=}CH_2 + MnO_4^- + H_2O \longrightarrow$

c. $CH_2{=}CH_2 + Br_2 \longrightarrow$

d.

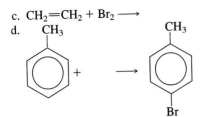

24.24 Complete and balance the following equations. Note any catalyst used.

a. $C_4H_{10} + O_2 \longrightarrow$

b.

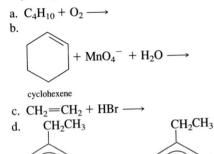

cyclohexene

$+ MnO_4^- + H_2O \longrightarrow$

c. $CH_2{=}CH_2 + HBr \longrightarrow$

d.

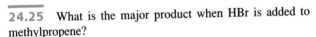

24.25 What is the major product when HBr is added to methylpropene?

24.26 Complete the following equation, giving only the main product.

$$CH_2{=}CHCH_3 + H{-}OH \xrightarrow[\text{catalyst}]{H_2SO_4}$$

Naming Hydrocarbons

24.27 Give the IUPAC name for each of the following hydrocarbons.

a.
$$\overset{\overset{\displaystyle CH_3}{|}}{CH_3CHCH_2CHCH_3} \\ \underset{\displaystyle CH_3 \quad CH_3}{}$$

b.
$$CH_3\overset{\overset{\displaystyle CH_3}{|}}{\underset{\underset{\displaystyle CH_3}{|}}{C}}CH_2CH_2CH_2\overset{\overset{\displaystyle CH_3}{|}}{\underset{\underset{\displaystyle CH_3}{|}}{C}}CH_3$$

c.
$$CH_3CH_2\overset{\overset{}{|}}{\underset{\underset{\displaystyle CH_2CH_2CH_3}{|}}{C}}HCH_2CH_2CH_2CH_3$$

24.28 What is the IUPAC name of each of the following compounds?

a.
$$CH_3CH_2\overset{\overset{\displaystyle CH_3}{|}}{C}HCHCH_3 \\ \underset{\displaystyle CH_3}{}$$

b.
$$CH_3CH_2CH_2\overset{\overset{\displaystyle CH_2CH_3}{|}}{C}HCH_2CH_3$$

c.
$$CH_3CH_2CH_2\overset{\overset{}{|}}{C}HCH_2CH_2CH_3 \\ \underset{\displaystyle CH_3CHCH_3}{}$$

24.29 Write the condensed structural formula for each of the following compounds.

a. 2,3-dimethylhexane

b. 3-ethylhexane

c. 2-methyl-4-isopropylheptane

24.30 Write the condensed structural formula for each of the following compounds.

a. 2,2-dimethylpentane

b. 3-isopropylhexane

c. 3-ethyl-4-methyloctane

24.31 Give the IUPAC name of each of the following.

a. $CH_3CH{=}CHCH_2CH_3$

b.
$$CH_3C{=}CHCH_2CHCH_3 \\ \underset{\displaystyle CH_3 \qquad\;\; CH_3}{}$$

24.32 For each of the following, write the IUPAC name.

a.
$$\begin{matrix} CH_3CH_2 \\ \\ CH_3CH_2 \end{matrix} C{=}CHCH_2CH_3$$

b.
$$CH_3CH_2\overset{\overset{}{|}}{C}CH_2CH_2CH_3 \\ \underset{\displaystyle CH_2}{\|}$$

24.33 Give the condensed structural formula for each of the following compounds.

a. 3-ethyl-2-pentene

b. 4-ethyl-2-methyl-2-hexene

24.34 Write the condensed structural formula for each of the following compounds.

a. 2,3-dimethyl-2-pentene

b. 2-methyl-4-propyl-3-heptene

24.35 Give the IUPAC names and include the *cis* or *trans* label for each of the isomers of $CH_3CH{=}CHCH_2CH_3$.

24.36 Give the IUPAC names and include the *cis* or *trans* label for each of the isomers of $CH_3CH{=}CHCH_3$.

Naming Oxygen-Containing Organic Compounds

24.37 Circle and name the functional group in each compound.

a.
$$CH_3\overset{\overset{\displaystyle O}{\|}}{-}C{-}CH_2CH_2CH_3$$

b. $CH_3—\overset{OH}{\underset{H}{C}}—CH_2CH_3$

c. $HO—\overset{O}{\overset{\|}{C}}—CH_2CH_3$

d. $H—\overset{O}{\overset{\|}{C}}—CH_2CH_3$

24.38 Circle and name the functional group in each compound.

a. $CH_3—\overset{O}{\overset{\|}{C}}—OH$

b. $O{=}C—OH$ (C bonded below to CH_3CHCH_3)

c. $HO—CH_2CH$ (CH bonded above to CH_3 and below to CH_3)

d. $O{=}C—CH_3$ (C bonded below to CH_3CHCH_3)

24.39 Give the IUPAC name for each of the following.

a. $HOCH_2CH_2CH_2CH_2CH_3$

b. $CH_3\underset{OH}{CH}CH_2CH_2CH_3$

c. $CH_3CH_2CH_2CHCH_2CH_2CH_3$ (CH bonded below to $H—\underset{H}{C}—OH$)

24.40 Write the IUPAC name for each of the following.

a. $HOCH_2\underset{CH_2CH_2CH_3}{CH}CH_2CH_3$

b. $HOCH_2CH_2CH_2\underset{CH_3}{CH}CH_3$

c. $CH_3CHCH_2CH_3$ (CH bonded below to $H—\underset{CH_3}{C}—OH$)

24.41 State whether each of the following alcohols is primary, secondary, or tertiary.

a. $CH_3\underset{OH}{CH}CH_2CH_3$

b. $\overset{CH_2CH_3}{\underset{OH}{HC}}CH_2CH_3$

c. $HOCH_2\overset{CH_3}{\underset{CH_3}{CH}}$

d. $HOCH_2\overset{CH_3}{\underset{CH_3}{C}}CH_3$

24.42 Classify each of the following as a primary, secondary, or tertiary alcohol.

a. $CH_3\overset{CH_3}{\underset{CH_3}{C}}CH_2\overset{CH_3}{CH}OH$

b. $CH_3CH_2CH_2\overset{CH_3}{\underset{CH_3}{C}}OH$

c. $CH_3CH_2\overset{CH_2OH}{\underset{CH_3}{C}}CH_3$

d. $CH_3CH_2\overset{CH_2CH_3}{\underset{CH_2CH_3}{C}}OH$

24.43 According to IUPAC rules, what is the name of each of the following compounds?

a. $CH_3COCH_2CH_3$

b. $CH_3CH_2CH_2CHO$

c. $H—\overset{O}{\overset{\|}{C}}CH_2CH_2\overset{CH_3}{\underset{CH_3}{C}}CH_3$

d. $CH_3\underset{CH_2CH_3}{CH}\overset{O}{\overset{\|}{C}}CH_3$

24.44 Write the IUPAC name of each of the following compounds.

a. $CH_3\underset{CHO}{CH}CH_3$

b. $CH_3\underset{COCH_3}{CH}CH_3$

c. CH_3CH_2 (bonded below to $CH_2\underset{O}{\underset{\|}{C}}—H$)

d. CH_3CHCH_3 (CH bonded below to $CH_2—\underset{O}{\underset{\|}{C}}—CH_2CH_3$)

Organic Compounds Containing Nitrogen

24.45 Identify each of the following compounds as a primary, secondary, or tertiary amine, or as an amide.

a. 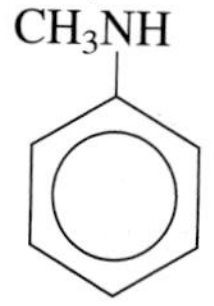

b. $CH_3CH_2NHCH_2CH_3$

24.46 Identify each of the following compounds as a primary, secondary, or tertiary amine, or as an amide.

a. 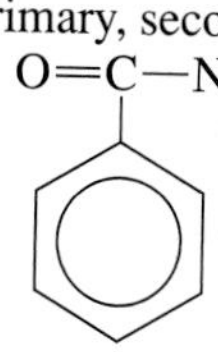

b. $CH_3CH_2CH_2NH_2$

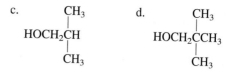

General Problems

24.47 Give the IUPAC name of each of the following compounds.

a. CH_3CHCH_2COOH
   $\quad\quad\ |$
   $\quad\quad CH_3$

b.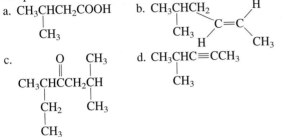

c.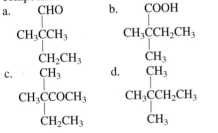

d. $CH_3CHC{\equiv}CCH_3$
   $\quad\quad |$
   $\quad\quad CH_3$

24.48 Give the IUPAC name of each of the following compounds.

a. $\quad\quad CHO$
   $\quad\quad\ |$
   CH_3CCH_3
   $\quad\quad |$
   $\quad\ CH_2CH_3$

b. $\quad\quad COOH$
   $\quad\quad\ |$
   $CH_3CCH_2CH_3$
   $\quad\quad |$
   $\quad\quad CH_3$

c. $\quad\ CH_3$
   $\quad\quad |$
   CH_3CCOCH_3
   $\quad\quad |$
   $\quad\ CH_2CH_3$

d. $\quad\ CH_3$
   $\quad\quad |$
   $CH_3CCH_2CH_3$
   $\quad\quad |$
   $\quad\ CH_3$

24.49 Write the structural formula for each of the following compounds.

a. isopropyl propionate
b. *t*-butylamine
c. 2,2-dimethylhexanoic acid
d. *cis*-3-hexene

24.50 Write the structural formula for each of the following compounds.

a. 3-ethyl-1-pentene
b. 1,1,2,2-tetraphenylethane
c. 1-phenyl-2-butanone
d. cyclopentanone

24.51 Describe chemical tests that could distinguish between:

a. propionaldehyde (propanal) and acetone (propanone)
b. $CH_2{=}CH{-}C{\equiv}C{-}CH{=}CH_2$ and benzene

24.52 Describe chemical tests that could distinguish between:

a. acetic acid and acetaldehyde (ethanal)
b. toluene (methylbenzene) and 2-methylcyclohexene

24.53 Identify each of the following compounds.

a. A gas with a sweetish odor that promotes the ripening of green fruit.
b. An unsaturated compound of the formula C_7H_8 that gives a negative test with bromine in carbon tetrachloride.

c. A compound with an ammonia-like odor that acts as a base; its molecular formula is CH_5N.
d. An alcohol used as a starting material for the manufacture of formaldehyde.

24.54 Identify each of the following compounds.

a. An acidic compound that also has properties of an aldehyde; its molecular formula is CH_2O_2.
b. A compound used as a preservative for biological specimens and as a raw material for plastics.
c. A saturated hydrocarbon boiling at 0°C that is liquefied and sold in cylinders as a fuel.
d. A saturated hydrocarbon that is the main constituent of natural gas.

24.55 Teflon is an addition polymer of 1,1,2,2-tetrafluoroethene. Write the equation for the formation of the polymer.

24.56 Poly(vinyl chloride) (PVC) is an addition polymer of vinyl chloride, $CH_2{=}CHCl$. Write the equation for the formation of the polymer.

24.57 Consider the following polymer:

$$-OCH_2CH_2O\overset{\overset{\displaystyle O}{||}}{C}CH_2CH_2\overset{\overset{\displaystyle O}{||}}{C}OCH_2CH_2O\overset{\overset{\displaystyle O}{||}}{C}CH_2CH_2\overset{\overset{\displaystyle O}{||}}{C}-$$

From which two monomers is the polymer made?

24.58 Consider the following polymer:

$$-\overset{\overset{\displaystyle O}{||}}{C}(CH_2)_{10}\overset{\overset{\displaystyle O}{||}}{C}NH(CH_2)_6\,NH\overset{\overset{\displaystyle O}{||}}{C}(CH_2)_{10}\overset{\overset{\displaystyle O}{||}}{C}NH(CH_2)_6\,NH-$$

From which two monomers is the polymer made?

24.59 A compound that is 85.6% C and 14.4% H and has a molecular weight of 56.1 amu reacts with water and sulfuric acid to produce a compound that reacts with acidic potassium dichromate solution to produce a ketone. What is the name of the original hydrocarbon?

24.60 A compound with a fragrant odor reacts with dilute acid to give two organic compounds, A and B. Compound A is identified as an alcohol with a molecular weight of 32.0 amu. Compound B is identified as an acid. It can be reduced to give a compound whose composition is 60.0% C, 13.4% H, and 26.6% O and whose molecular weight is 60.1 amu. What is the name of the original compound?

Appendixes

Appendix A | Mathematical Skills

Only a few basic mathematical skills are required for the study of general chemistry. But to concentrate your attention on the concepts of chemistry, you will find it necessary to have a firm grasp of these basic mathematical skills. In this appendix, we will review scientific (or exponential) notation, logarithms, simple algebraic operations, the solution of quadratic equations, and the plotting of straight-line graphs.

A.1 Scientific (Exponential) Notation

In chemistry, you frequently encounter very large and very small numbers. For example, the number of molecules in a liter of air at 20°C and normal barometric pressure is 25,000,000,000,000,000,000,000, and the distance between two hydrogen atoms in a hydrogen molecule is 0.000,000,000,074 meters. In these forms, such numbers are inconvenient to write and difficult to read. For this reason, you normally express them in scientific, or exponential, notation. Scientific calculators also use this notation.

In scientific notation, a number is written in the form $A \times 10^n$. A is a number greater than or equal to 1 and less than 10, and the exponent n (the nth power of ten) is a positive or negative integer. For example, 4853 is written in scientific notation as 4.853×10^3, which is 4.853 multiplied by three factors of 10:

$$4.853 \times 10^3 = 4.853 \times 10 \times 10 \times 10 = 4853$$

The number 0.0568 is written in scientific notation as 5.68×10^{-2}, which is 5.68 divided by two factors of 10:

$$5.68 \times 10^{-2} = \frac{5.68}{10 \times 10} = 0.0568$$

Any number can be conveniently transformed to scientific notation by moving the decimal point to obtain a number, A, greater than or equal to 1 and less than 10. If the decimal point is moved to the left, you multiply A by 10^n, where n equals the number of places moved. If the decimal point is moved to the right, you multiply A by 10^{-n}. Consider the number 0.00731. You must move the decimal point to the right three places. Therefore, 0.00731 equals 7.31×10^{-3}. To transform a number written in scientific notation to the usual form, the process is reversed. If the exponent is positive, the decimal point is shifted right. If the exponent is negative, the decimal point is shifted left.

EXAMPLE 1

Expressing Numbers in Scientific Notation

Express the following numbers in scientific notation:

a. 843.4   b. 0.00421   c. 1.54

SOLUTION

Shift the decimal point to get a number between 1 and 10; count the number of positions shifted.

a. $843.4 = 8.434 \times 10^2$

b. $0.00421 = 4.21 \times 10^{-3}$

c. leave as is or write 1.54×10^0

EXAMPLE 2

Converting Numbers in Scientific Notation to Usual Form

Convert the following numbers in scientific notation to usual form:

a. 6.39×10^{-4} b. 3.275×10^2

SOLUTION

a. $0.0006.39 \times 10^{-4} = 0.000639$

b. $3.275 \times 10^2 = 327.5$

Addition and Subtraction

Before adding or subtracting two numbers written in scientific notation, it is necessary to express both to the same power of ten. After adding or subtracting, it may be necessary to shift the decimal point to express the result in scientific notation.

EXAMPLE 3

Adding and Subtracting in Scientific Notation

Carry out the following arithmetic; give the result in scientific notation.

$$(9.42 \times 10^{-2}) + (7.6 \times 10^{-3})$$

SOLUTION

You can shift the decimal point in either number to obtain both to the same power of ten. For example, to get both numbers to 10^{-2}, you shift the decimal point one place to the left and add 1 to the exponent in the expression 7.6×10^{-3}.

$$7.6 \times 10^{-3} = 0.76 \times 10^{-2}$$

Now you can add the two numbers.

$$(9.42 \times 10^{-2}) + (0.76 \times 10^{-2}) = (9.42 + 0.76) \times 10^{-2} = 10.18 \times 10^{-2}$$

Since 10.18 is not between 1 and 10, you shift the decimal point to express the final result in scientific notation.

$$10.18 \times 10^{-2} = 1.018 \times 10^{-1}$$

Multiplication and Division

To multiply two numbers in scientific notation, you first multiply the two powers of ten by adding their exponents. Then you multiply the remaining factors. Division is handled similarly. You first move any power of ten in the denominator to the

numerator, changing the sign of the exponent. After multiplying the two powers of ten by adding their exponents, you carry out the indicated division.

EXAMPLE 4

Multiplying and Dividing in Scientific Notation

Do the following arithmetic, and express the answers in scientific notation.

a. $(6.3 \times 10^2) \times (2.4 \times 10^5)$

b. $\dfrac{6.4 \times 10^2}{2.0 \times 10^5}$

SOLUTION

a. $(6.3 \times 10^2) \times (2.4 \times 10^5) = (6.3 \times 2.4) \times 10^7 = 15.12 \times 10^7 =$ **1.512×10^8**

b. $\dfrac{6.4 \times 10^2}{2.0 \times 10^5} = \dfrac{6.4}{2.0} \times 10^2 \times 10^{-5} = \dfrac{6.4}{2.0} \times 10^{-3} =$ **3.2×10^{-3}**

Powers and Roots

A number $A \times 10^n$ raised to a power p is evaluated by raising A to the power p and multiplying the exponent in the power of ten by p:

$$(A \times 10^n)^p = A^p \times 10^{n \times p}$$

You extract the rth root of a number $A \times 10^n$ by first moving the decimal point in A so that the exponent in the power of ten is exactly divisible by r. Suppose this has been done, so that n in the number $A \times 10^n$ is exactly divisible by r. Then

$$\sqrt[r]{A \times 10^n} = \sqrt[r]{A} \times 10^{n/r}$$

EXAMPLE 5

Finding Powers and Roots in Scientific Notation

Evaluate the following expressions:

a. $(5.29 \times 10^2)^3$

b. $\sqrt{2.31 \times 10^7}$

SOLUTION

a. $(5.29 \times 10^2)^3 = (5.29)^3 \times 10^6 = 148 \times 10^6 =$ **1.48×10^8**

b. $\sqrt{2.31 \times 10^7} = \sqrt{23.1 \times 10^6} = \sqrt{23.1} \times 10^{6/2} =$ **4.81×10^3**

Electronic Calculators

Scientific calculators will perform all of the arithmetic operations we have just described (as well as those discussed in the next section). On most calculators, the basic operations of addition, subtraction, multiplication, and division are similar and straightforward. However, more variation exists in raising a number to a power and extracting a root. If you have the calculator instructions, by all means read them. Otherwise, the following information may help.

Squares and square roots are usually obtained with special keys, perhaps labeled x^2 and $\sqrt{x}$. Thus, to obtain $(5.15)^2$, you enter 5.15 and press x^2. To obtain $\sqrt{5.15}$, you enter 5.15 and press $\sqrt{x}$ (or perhaps INV, for inverse, and x^2). Other powers and roots require a y^x (or a^x) key. The answer to Example 5(a), which is $(5.29 \times 10^2)^3$, would be obtained by a sequence of steps such as the following: enter 5.29×10^2, press the y^x key, enter 3, and press the = key.

The same sequence can be used to extract a root. Suppose you want $\sqrt[5]{2.18 \times 10^6}$. This is equivalent to $(2.18 \times 10^6)^{1/5}$ or $(2.18 \times 10^6)^{0.2}$. If the calculator has a $1/x$ key, the sequence would be as follows: enter 2.18×10^6, press the y^x key, enter 5, press the $1/x$ key, then press the = key. Some calculators have a $\sqrt[x]{y}$ key, so this can be used to extract the xth root of y, using a sequence of steps similar to that for y^x.

A.2 Logarithms

The *logarithm* to the *base a* of a number x, denoted $\log_a x$, is the exponent of the constant a needed to equal the number x. For example, suppose a is 10, and you would like the logarithm of 1000; that is, you would like the value of $\log_{10} 1000$. This is the exponent, y, of 10 such that 10^y equals 1000. The value of y here is 3. Thus, $\log_{10} 1000 = 3$.

Common logarithms are logarithms in which the base is 10. The common logarithm of a number x is often denoted simply as $\log x$. It is easy to see how to obtain the common logarithms of 10, 100, 1000, and so forth. But logarithms are defined for all positive numbers, not just the powers of ten. In general, the exponents, or values, of the logarithm will be decimal numbers. To understand the meaning of a decimal exponent, consider $10^{0.400}$. This is equivalent to $10^{400/1000} = 10^{2/5} = \sqrt[5]{10^2} = 2.51$. Therefore, $\log 2.51 = 0.400$. Any decimal exponent is essentially a fraction, p/r, so by evaluating the expressions $10^{p/r} = \sqrt[r]{10^p}$ one could construct a table of logarithms. In practice, power series or other methods are used.

The following are fundamental properties of all logarithms:

$$\log_a 1 = 0 \tag{1}$$

$$\log_a(A \times B) = \log_a A + \log_a B \tag{2}$$

$$\log_a \frac{A}{B} = \log_a A - \log_a B \tag{3}$$

$$\log_a A^p = p \log_a A \tag{4}$$

$$\log_a \sqrt[r]{A} = \frac{1}{r} \log_a A \tag{5}$$

These properties are very useful in working with logarithms.

Electronic calculators that evaluate logarithms are now available for about $10 or so. Their simplicity of operation makes them well worth the price. To obtain the logarithm of a number, you enter the number and press the LOG key.

Antilogarithm

The antilogarithm (abbreviated antilog) is the inverse of the common logarithm. Antilog x is simply 10^x. If your electronic calculator has a 10^x key, you obtain the antilogarithm of a number by entering the number and pressing the 10^x key. (It may be necessary to press an inverse key before pressing a 10^x/LOG key.) If your calculator has a y^x (or a^x) key, you enter 10, press y^x, enter x, then press the = key.

Natural Logarithms and Natural Antilogarithms

The mathematical constant $e = 2.71828$, like π, occurs in many scientific and engineering problems. It is frequently seen in the natural exponential function $y = e^x$. The inverse function is called the *natural logarithm*, $x = \ln y$, where $\ln y$ is simplified notation for $\log_e y$.

Finding the natural logarithm with a calculator is analogous to finding the common logarithm, only in this case you enter the number and press the LN key. For example, you may obtain the natural logarithm of 6.2 by entering 6.2 and pressing the LN key ($\ln 6.2 = 1.82$). The natural antilogarithm of a number is the constant e raised to a power equal to the number. Therefore, the expression for the natural antilogarithm of 1.34 is written as

$$\text{natural antilog } 1.34 = e^{1.34}$$

In order to evaluate the expression $e^{1.34}$ on your calculator, you can use the e^x key, or you may have to use the INV key in combination with the LN key ($e^{1.34} = 3.8$).

There are instances where you may want to relate the natural logarithm to the common logarithm; the expression is

$$\ln x = 2.303 \log x$$

A.3   Algebraic Operations and Graphing

Quadratic Formula

A quadratic equation is one involving only powers of x in which the highest power is two. The general form of the equation can be written

$$ax^2 + bx + c = 0$$

where a, b, and c are constants. For given values of these constants, only certain values of x are possible. (In general, there will be two values.) These values of x are said to be the solutions of the equation.

These solutions are given by the *quadratic formula*:

$$x = \frac{-b \pm \sqrt{b^2 - 4ac}}{2a}$$

In this formula, the symbol $\pm$ means that there are two possible values of x—one obtained by taking the positive sign, the other by taking the negative sign.

EXAMPLE 6

Obtaining the Solutions of a Quadratic Equation

Obtain the solutions of the following quadratic equation:

$$2.00x^2 - 1.72x - 2.86 = 0$$

SOLUTION

Using the quadratic formula, you substitute $a = 2.00$, $b = -1.72$, and $c = -2.86$. You get

$$x = \frac{1.72 \pm \sqrt{(-1.72)^2 - 4 \times 2.00 \times (-2.86)}}{2 \times 2.00} = \frac{1.72 \pm 5.08}{4.00} = -0.840 \text{ and } +1.70$$

Mathematically there are two solutions, but in any real problem one solution may not be allowed. For example, if the solution is some physical quantity that can have only positive values, a negative solution must be rejected.

The Straight-Line Graph

A graph is a visual means of representing a mathematical relationship or physical data. Consider the following data in which values of y from some experiment are given for four values of x.

x	y
1	-1
2	1
3	3
4	5

By plotting these x, y points on a graph (Figure A.1), you see that they fall on a straight line. This suggests (but does not prove) that other points from this type of experiment might fall on the same line. It would be useful to have the mathematical equation for this line.

The general equation form of a straight line is

$$y = mx + b$$

Figure A.1
A straight-line plot of some data.

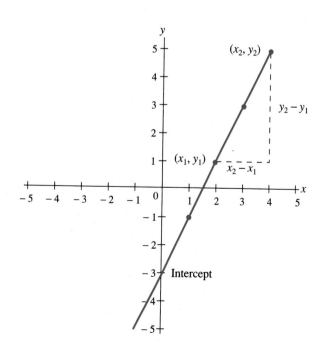

The constant m is called the *slope* of the straight line. It is obtained by dividing the vertical distance between any two points on the line by the horizontal distance. If the two points are (x_1, y_1) and (x_2, y_2), the slope is given by the following formula:

$$\text{slope} = \frac{y_2 - y_1}{x_2 - x_1}$$

Suppose you choose the points (2, 1) and (4, 5) from the data. Then

$$\text{slope} = \frac{5 - 1}{4 - 2} = 2$$

For the straight line in Figure A.1, $m = 2$.

The constant b is called the *intercept*. It is the value of y at $x = 0$. From Figure A.1, you see that the intercept is -3. Therefore, $b = -3$. Hence, the equation of the straight line is

$$y = 2x - 3$$

Appendix B

Thermodynamic Quantities for Substances and Ions at 25°C

Substance or Ion	ΔH°_f (kJ/mol)	ΔG°_f (kJ/mol)	S° (J/mol · K)	Substance or Ion	ΔH°_f (kJ/mol)	ΔG°_f (kJ/mol)	S° (J/mol · K)
$H^+(g)$	1536.3	1517.1	108.83	$H(g)$	218.0	203.30	114.60
$H^+(aq)$	0	0	0	$H_2(g)$	0	0	130.6
Group IA							
$Li^+(g)$	687.163	649.989	132.91	$Na_2CO_3(s)$	−1130.8	−1048.1	139
$Li^+(aq)$	−278.46	−293.8	14	$K^+(g)$	514.197	481.202	154.47
$Li(g)$	161	128	138.67	$K^+(aq)$	−251.2	−282.28	103
$Li(s)$	0	0	29.10	$K(g)$	89.2	60.7	160.23
$LiF(s)$	−616.9	−588.7	35.66	$K(s)$	0	0	64.672
$LiCl(s)$	−408	−384	59.30	$KF(s)$	−568.6	−538.9	66.55
$LiBr(s)$	−351	−342	74.1	$KCl(s)$	−436.68	−408.8	82.55
$LiI(s)$	−270	−270	85.8	$KBr(s)$	−394	−380	95.94
$Na^+(g)$	609.839	574.877	147.85	$KI(s)$	−328	−323	106.39
$Na^+(aq)$	−239.66	−261.87	60.2	$Cs^+(g)$	458.5	427.1	169.72
$Na(g)$	107.76	77.299	153.61	$Cs^+(aq)$	−248	−282.0	133
$Na(s)$	0	0	51.446	$Cs(g)$	76.7	49.7	175.5
$NaF(s)$	−575.4	−545.1	51.21	$Cs(s)$	0	0	85.15
$NaCl(s)$	−411.1	−384.0	72.12	$CsF(s)$	−554.7	−525.4	88
$NaBr(s)$	−361	−349	86.82	$CsCl(s)$	−442.8	−414	101.18
$NaI(s)$	−288	−285	98.5	$CsBr(s)$	−395	−383	121
$NaHCO_3(s)$	−947.7	−851.9	102	$CsI(s)$	−337	−333	130

Substance or Ion	ΔH°_f (kJ/mol)	ΔG°_f (kJ/mol)	S° (J/mol · K)	Substance or Ion	ΔH°_f (kJ/mol)	ΔG°_f (kJ/mol)	S° (J/mol · K)
Group IIA							
$Mg^{2+}(g)$	2351			$CaCl_2(s)$	−795.0	−750.2	114
$Mg^{2+}(aq)$	−461.96	−456.01	−118	$CaO(s)$	−635.1	−603.5	38.2
$Mg(g)$	150	115	148.55	$CaCO_3(s)$	−1206.9	−1128.8	92.9
$Mg(s)$	0	0	32.69	$CaSO_4(s)$	−1432.7	−1320.3	107
$MgCl_2(s)$	−641.6	−592.1	89.630	$Ba^{2+}(g)$	1649.9		
$MgO(s)$	−601.2	−569.0	26.9	$Ba^{2+}(aq)$	−538.36	−560.7	13
$MgCO_3(s)$	−1112	−1028	65.86	$Ba(g)$	175.6	144.8	170.28
$Ca^{2+}(g)$	1934.1			$Ba(s)$	0	0	62.5
$Ca^{2+}(aq)$	−542.96	−553.04	−55.2	$BaCl_2(s)$	−806.06	−810.9	126
$Ca(g)$	192.6	158.9	154.78	$BaO(s)$	−548.1	−520.4	72.07
$Ca(s)$	0	0	41.6	$BaCO_3(s)$	−1219	−1139	112
$CaF_2(s)$	−1215	−1162	68.87	$BaSO_4(s)$	−1465	−1353	132
Group IIIA							
B (β-rhombohedral)	0	0	5.87	$Al(s)$	0	0	28.3
				$Al^{3+}(aq)$	−524.7	−481.2	−313
$B_2O_3(s)$	−1272	−1193	53.8	$Al_2O_3(s)$	−1676	−1582	50.94
Group IVA							
$C(g)$	715.0	669.6	158.0	$HCN(g)$	135	125	201.7
C(graphite)	0	0	5.686	$HCN(l)$	105	121	112.8
C(diamond)	1.896	2.866	2.439	$HCN(aq)$	105	112	129
$CO(g)$	−110.5	−137.2	197.5	$CS_2(l)$	87.9	63.6	151.0
$CO_2(g)$	−393.5	−394.4	213.7	$CH_3Cl(g)$	−83.7	−60.2	234
$CO_2(aq)$	−412.9	−386.2	121	$CH_2Cl_2(l)$	−117	−63.2	179
$CO_3^{2-}(aq)$	−676.26	−528.10	−53.1	$CHCl_3(l)$	−132	−71.5	203
$HCO_3^{-}(aq)$	−691.11	−587.06	95.0	$CCl_4(g)$	−96.0	−53.7	309.7
$H_2CO_3(aq)$	−698.7	−623.42	191	$CCl_4(l)$	−139	−68.6	214.4
$CH_4(g)$	−74.87	−50.81	186.1	$COCl_2(g)$	−220	−206	283.74
$C_2H_2(g)$	227	209	200.85	$Si(s)$	0	0	18.0
$C_2H_4(g)$	52.47	68.36	219.22	$SiO_2(s)$	−910.9	−856.5	41.5
$C_2H_6(g)$	−84.667	−32.89	229.5	Sn(white)	0	0	51.5
$C_6H_6(l)$	49.0	124.5	172.8	$SnCl_4(l)$	−545.2	−474.0	259
$CH_3OH(l)$	−238.6	−166.2	127	$Pb^{2+}(aq)$	1.6	−24.3	21
$HCHO(g)$	−116	−110	219	$Pb(s)$	0	0	64.785
$HCOOH(l)$	−409	−346	129.0	PbO(s, red)	−219.0	−189.2	66.5
$C_2H_5OH(l)$	−277.63	−174.8	161	$PbO_2(s)$	−276.6	−219.0	76.6
$CH_3CHO(g)$	−166	−133.7	266	$PbCl_2(s)$	−359	−314	136
$CH_3COOH(l)$	−487.0	−392	160	$PbSO_4(s)$	−918.39	−811.24	147
$CN^{-}(aq)$	151	166	118				

Substance or Ion	ΔH_f° (kJ/mol)	ΔG_f° (kJ/mol)	S° (J/mol · K)	Substance or Ion	ΔH_f° (kJ/mol)	ΔG_f° (kJ/mol)	S° (J/mol · K)
Group VA							
$N(g)$	473	456	153.2	$NH_4NO_3(s)$	−365.6	−184.0	151.1
$N_2(g)$	0	0	191.5	$P(red)$	0	0	22.8
$NO(g)$	90.29	86.60	210.65	$P_4(white)$	69.8	48.0	164
$NO_2(g)$	33.2	51	239.9	$PCl_3(g)$	−287	−267.8	311.7
$N_2O_4(g)$	9.16	97.7	304.3	$PCl_5(g)$	−343	−278	364.5
$N_2O_5(g)$	11	118	346	$P_4O_{10}(s)$	−2940	−2675	228.9
$NH_3(g)$	−45.9	−16	193	$PO_4^{3-}(aq)$	−1277	−1018	−222
$NH_3(aq)$	−80.83	26.7	110	$HPO_4^{2-}(aq)$	−1292.1	−1089.3	−33
$NO_3^-(aq)$	−206.57	−110.5	146	$H_2PO_4^-(aq)$	−1296.3	−1130.4	90.4
$HNO_3(l)$	−173.23	−79.914	155.6	$H_3PO_4(aq)$	−1288.3	−1142.7	158
$HNO_3(aq)$	−206.57	−110.5	146				
Group VIA							
$O(g)$	249.2	231.7	160.95	$HS^-(aq)$	−17.7	12.6	61.1
$O_2(g)$	0	0	205.0	$H_2S(g)$	−20.2	−33	205.6
$O_3(g)$	143	163	238.82	$H_2S(aq)$	−39	−27.4	122
$OH^-(aq)$	−229.94	−157.30	−10.54	$SO_2(g)$	−296.8	−300.2	248.1
$H_2O(g)$	−241.826	−228.60	188.72	$SO_3(g)$	−396	−371	256.66
$H_2O(l)$	−285.840	−237.192	69.940	$SO_4^{2-}(aq)$	−907.51	−741.99	17
$H_2O_2(l)$	−187.8	−120.4	110	$HSO_4^-(aq)$	−885.75	−752.87	126.9
$H_2O_2(aq)$	−191.2	−134.1	144	$H_2SO_4(l)$	−813.989	−690.059	156.90
$S(rhombic)$	0	0	31.9	$H_2SO_4(aq)$	−907.51	−741.99	17
$S^{2-}(aq)$	41.8	83.7	22				
Group VIIA							
$F(g)$	78.9	61.8	158.64	$Br(g)$	111.9	82.40	174.90
$F^-(g)$	−255.6	−262.5	145.47	$Br^-(aq)$	−120.9	−102.82	80.71
$F^-(aq)$	−329.1	−276.5	−9.6	$Br_2(g)$	30.91	3.13	245.38
$F_2(g)$	0	0	202.7	$Br_2(l)$	0	0	152.23
$HF(g)$	−273	−275	173.67	$HBr(g)$	−36	−53.5	198.59
$Cl(g)$	121.0	105.0	165.1	$I(g)$	106.8	70.21	180.67
$Cl^-(g)$	−234	−240	153.25	$I^-(aq)$	−55.94	−51.67	109.4
$Cl^-(aq)$	−167.46	−131.17	55.10	$I_2(g)$	62.442	19.38	260.58
$Cl_2(g)$	0	0	223.0	$I_2(s)$	0	0	116.14
$HCl(g)$	−92.31	−95.30	186.79	$HI(g)$	25.9	1.3	206.33
$HCl(aq)$	−167.46	−131.17	55.06				
Group IB							
$Cu^+(aq)$	51.9	50.2	−26	$AgF(s)$	−203	−185	84
$Cu^{2+}(aq)$	64.39	64.98	−98.7	$AgCl(s)$	−127.03	−109.72	96.11
$Cu(s)$	0	0	33.1	$AgBr(s)$	−99.50	−95.939	107.1
$Ag^+(aq)$	105.9	77.111	73.93	$AgI(s)$	−62.38	−66.32	114
$Ag(s)$	0	0	42.702	$Ag_2S(s)$	−31.8	−40.3	146

Substance or Ion	ΔH°_f (kJ/mol)	ΔG°_f (kJ/mol)	S° (J/mol·K)	Substance or Ion	ΔH°_f (kJ/mol)	ΔG°_f (kJ/mol)	S° (J/mol·K)
Group IIB							
$Zn^{2+}(aq)$	−152.4	−147.21	−106.5	$CdS(s)$	−144	−141	71
$Zn(s)$	0	0	41.6	$Hg^{2+}(aq)$		164.8	
$ZnO(s)$	−348.0	−318.2	43.9	$Hg_2^{2+}(aq)$		153.9	
$ZnS(s, zinc$	−203	−198	57.7	$Hg(l)$	0	0	76.027
blende)				$HgCl_2(s)$	−230	−184	144
$Cd^{2+}(aq)$	−72.38	−77.74	−61.1	$Hg_2Cl_2(s)$	−264.9	−210.66	196
$Cd(s)$	0	0	51.5	$HgO(s)$	−90.79	−58.50	70.27
Group VIB							
$[Cr(H_2O)_6]^{3+}$	−1971			$CrO_4^{2-}(aq)$	−863.2	−706.3	38
(aq)				$Cr_2O_7^{2-}(aq)$	−1461	−1257	214
$Cr(s)$	0	0	23.8				
Group VIIB							
$Mn^{2+}(aq)$	−219	−223	−84	$MnO_2(s)$	−520.9	−466.1	53.1
$Mn(s)$	0	0	31.8	$MnO_4^-(aq)$	−518.4	−425.1	190
Group VIIIB							
$Fe^{3+}(aq)$	−47.7	−10.5	−293	$Fe_3O_4(s)$	−1121	−1018	145.3
$Fe^{2+}(aq)$	−87.9	−84.94	113	$Co^{2+}(aq)$	−67.4	−51.5	−155
$Fe(s)$	0	0	27.3	$Co(s)$	0	0	30
$FeO(s)$	−272.0	−251.4	60.75	$Ni^{2+}(aq)$	−64.0	−46.4	−159
$Fe_2O_3(s)$	−825.5	−743.6	87.400	$Ni(s)$	0	0	30.1

Appendix C   Acid-Ionization Constants at 25°C

Substance	Formula	K_a	Substance	Formula	K_a
Acetic acid	$HC_2H_3O_2$	1.7×10^{-5}	Formic acid	$HCHO_2$	1.7×10^{-4}
Arsenic acid*	H_3AsO_4	6.5×10^{-3}	Hydrocyanic acid	HCN	4.9×10^{-10}
	$H_2AsO_4^-$	1.2×10^{-7}	Hydrofluoric acid	HF	6.8×10^{-4}
	$HAsO_4^{2-}$	3.2×10^{-12}	Hydrogen peroxide	H_2O_2	1.8×10^{-12}
Ascorbic acid*	$H_2C_6H_6O_6$	6.8×10^{-5}	Hydrogen sulfate ion	HSO_4^-	1.1×10^{-2}
	$HC_6H_6O_6^-$	2.8×10^{-12}	Hydrogen sulfide*	H_2S	8.9×10^{-8}
Benzoic acid	$HC_7H_5O_2$	6.3×10^{-5}		HS^-	1.2×10^{-13}
Boric acid	H_3BO_3	5.9×10^{-10}	Hypochlorous acid	$HClO$	3.5×10^{-8}
Carbonic acid*	H_2CO_3	4.3×10^{-7}	Lactic acid	$HC_3H_5O_3$	1.3×10^{-4}
	HCO_3^-	4.8×10^{-11}	Nitrous acid	HNO_2	4.5×10^{-4}
Chromic acid*	H_2CrO_4	1.5×10^{-1}	Oxalic acid*	$H_2C_2O_4$	5.6×10^{-2}
	$HCrO_4^-$	3.2×10^{-7}		$HC_2O_4^-$	5.1×10^{-5}
Cyanic acid	$HOCN$	3.5×10^{-4}	Phenol	C_6H_5OH	1.1×10^{-10}

Substance	Formula	K_a	Substance	Formula	K_a
Phosphoric acid*	H_3PO_4	6.9×10^{-3}	Propionic acid	$HC_3H_5O_2$	1.3×10^{-5}
	$H_2PO_4^-$	6.2×10^{-8}	Pyruvic acid	$HC_3H_3O_3$	1.4×10^{-4}
	HPO_4^{2-}	4.8×10^{-13}	Sulfurous acid*	H_2SO_3	1.3×10^{-2}
Phosphorous acid*	H_2PHO_3	1.6×10^{-2}		HSO_3^-	6.3×10^{-8}
	$HPHO_3^-$	7×10^{-7}			

*The ionization constants for polyprotic acids are for successive ionizations. For example, for H_3PO_4, the equilibrium is $H_3PO_4 \rightleftharpoons H^+ + H_2PO_4^-$. For $H_2PO_4^-$, the equilibrium is $H_2PO_4^- \rightleftharpoons H^+ + HPO_4^{2-}$.

Appendix D

Base-Ionization Constants at 25°C

Substance	Formula	K_b	Substance	Formula	K_b
Ammonia	NH_3	1.8×10^{-5}	Hydroxylamine	NH_2OH	1.1×10^{-8}
Aniline	$C_6H_5NH_2$	4.2×10^{-10}	Methylamine	CH_3NH_2	4.4×10^{-4}
Dimethylamine	$(CH_3)_2NH$	5.1×10^{-4}	Pyridine	C_5H_5N	1.4×10^{-9}
Ethylamine	$C_2H_5NH_2$	4.7×10^{-4}	Trimethylamine	$(CH_3)_3N$	6.5×10^{-5}
Ethylenediamine	$NH_2CH_2CH_2NH_2$	5.2×10^{-4}	Urea	NH_2CONH_2	1.5×10^{-14}
Hydrazine	N_2H_4	1.7×10^{-6}			

Appendix E

Solubility Product Constants at 25°C

Substance	Formula	K_{sp}	Substance	Formula	K_{sp}
Aluminum hydroxide	$Al(OH)_3$	4.6×10^{-33}	Lead(II) chloride	$PbCl_2$	1.6×10^{-5}
Barium chromate	$BaCrO_4$	1.2×10^{-10}	Lead(II) chromate	$PbCrO_4$	1.8×10^{-14}
Barium fluoride	BaF_2	1.0×10^{-6}	Lead(II) iodide	PbI_2	6.5×10^{-9}
Barium sulfate	$BaSO_4$	1.1×10^{-10}	Lead(II) sulfate	$PbSO_4$	1.7×10^{-8}
Cadmium oxalate	CdC_2O_4	1.5×10^{-8}	Lead(II) sulfide	PbS	2.5×10^{-27}
Cadmium sulfide	CdS	8×10^{-27}	Magnesium arsenate	$Mg_3(AsO_4)_2$	2×10^{-20}
Calcium carbonate	$CaCO_3$	3.8×10^{-9}	Magnesium carbonate	$MgCO_3$	1.0×10^{-5}
Calcium fluoride	CaF_2	3.4×10^{-11}	Magnesium hydroxide	$Mg(OH)_2$	1.8×10^{-11}
Calcium oxalate	CaC_2O_4	2.3×10^{-9}	Magnesium oxalate	MgC_2O_4	8.5×10^{-5}
Calcium phosphate	$Ca_3(PO_4)_2$	1×10^{-26}	Manganese(II) sulfide	MnS	2.5×10^{-10}
Calcium sulfate	$CaSO_4$	2.4×10^{-5}	Mercury(I) chloride	Hg_2Cl_2	1.3×10^{-18}
Cobalt(II) sulfide	CoS	4×10^{-21}	Mercury(II) sulfide	HgS	1.6×10^{-52}
Copper(II) hydroxide	$Cu(OH)_2$	2.6×10^{-19}	Nickel(II) hydroxide	$Ni(OH)_2$	2.0×10^{-15}
Copper(II) sulfide	CuS	6×10^{-36}	Nickel(II) sulfide	NiS	3×10^{-19}
Iron(II) hydroxide	$Fe(OH)_2$	8×10^{-16}	Silver acetate	$AgC_2H_3O_2$	2.0×10^{-3}
Iron(II) sulfide	FeS	6×10^{-18}	Silver bromide	$AgBr$	5.0×10^{-13}
Iron(III) hydroxide	$Fe(OH)_3$	2.5×10^{-39}	Silver chloride	$AgCl$	1.8×10^{-10}
Lead(II) arsenate	$Pb_3(AsO_4)_2$	4×10^{-36}	Silver chromate	Ag_2CrO_4	1.1×10^{-12}

Substance	Formula	K_{sp}	Substance	Formula	K_{sp}
Silver iodide	AgI	8.3×10^{-17}	Strontium sulfate	$SrSO_4$	2.5×10^{-7}
Silver sulfide	Ag_2S	6×10^{-50}	Zinc hydroxide	$Zn(OH)_2$	2.1×10^{-16}
Strontium carbonate	$SrCO_3$	9.3×10^{-10}	Zinc sulfide	ZnS	1.1×10^{-21}
Strontium chromate	$SrCrO_4$	3.5×10^{-5}			

Appendix F

Formation Constants of Complex Ions at 25°C

Complex Ion	K_f	Complex Ion	K_f
$Ag(CN)_2^-$	5.6×10^{18}	$Fe(CN)_6^{4-}$	1.0×10^{35}
$Ag(NH_3)_2^+$	1.7×10^7	$Fe(CN)_6^{3-}$	9.1×10^{41}
$Ag(S_2O_3)_2^{3-}$	2.9×10^{13}	$Ni(CN)_4^{2-}$	1.0×10^{31}
$Cd(NH_3)_4^{2+}$	1.0×10^7	$Ni(NH_3)_6^{2+}$	5.6×10^8
$Cu(CN)_2^-$	1.0×10^{16}	$Zn(NH_3)_4^{2+}$	2.9×10^9
$Cu(NH_3)_4^{2+}$	4.8×10^{12}	$Zn(OH)_4^{2-}$	2.8×10^{15}

Answers to Concept Checks

Chapter 1

1.1 (a) two if you weigh under 100 lb, three if you weigh 100 lb or more (b) A 165-lb person would be reported as weighing 1.7 $\times 10^2$ lb. (c) A 74.8-kg or 165-lb male would weigh 50 kg or 200 lb on the truck scale. **1.2** (a) Meters, decimeters, or centimeters would be appropriate units. (b) approximately 9 m (c) 39°C would mean that you had a moderate fever. (d) You would probably be comfortable in a 23°C room. **1.3** You need to perform an experiment that will give you the density of the "fool's gold" and then compare that to the density of gold.

Chapter 2

2.1 The molecular model on the far right that contains 3 atoms with a ratio of 1:2 **2.2** One would conclude that the atom is made up primarily of a large impenetrable mass with a positive charge. **2.3** They must have similar chemical and/or physical properties. **2.4** (a) ether (b) alcohol (c) carboxylic acid (d) hydrocarbon **2.5** Q is likely to be an element in Group VIA.

Chapter 3

3.1 (a) 1.5 mol (b) 4.5 mol (c) 3.0 mol of OH$^-$ since there are 2 mol of OH$^-$ per mol of $Mg(OH)_2$ **3.2** (a) No, the empirical formula needs to be the smallest whole-number ratio of subscripts. (b) No, the subscripts are not whole numbers. (c) Yes, the empirical and molecular formulas can be the same. **3.3** (a) correct (b) incorrect (c) correct (d) incorrect (e) incorrect (f) correct **3.4** (a) $X_2(g) + 2Y(g) \longrightarrow 2XY(g)$ (b) container 1 (c) $X_2(g)$

Chapter 4

4.1 (a) $Na_2SO_4(aq) + Sr(C_2H_3O_2)_2(aq) \longrightarrow$
$$SrSO_4(s) + 2NaC_2H_3O_2(aq)$$
(b) $2Na^+(aq) + SO_4^{2-}(aq) + Sr^{2+}(aq) + 2C_2H_3O_2^-(aq) \longrightarrow$
$$SrSO_4(s) + 2Na^+(aq) + 2C_2H_3O_2^-(aq)$$
(c) $Sr^{2+}(aq) + SO_4^{2-}(aq) \longrightarrow SrSO_4(s)$
4.2 (a) MOH(s) is a base. M could be any Group I metal. (b) HA is an acid. A is an anion with a -1 charge. (c) H_2A is an acid. A is an anion with a -2 charge. (d) M $\longrightarrow Na^+$, A $\longrightarrow C_2H_3O_2^-$, A $\longrightarrow SO_3^{2-}$.
4.3 Concentration order: B > C = D > A. Leave A alone, add water to fill beaker B, add water to double the volumes of D and C. **4.4** (a) The acid in flask C has three times as many acidic protons as the acid in flask A. The acid in flask B has two times as many acidic protons as the acid in flask A. (b) Yes, if we assume that flask A contains a monoprotic acid.

Chapter 5

5.1 Since the height of the column is related to the density of the liquid, the column containing the H_2O would be higher. **5.2** (a) All contain same amount of gas. (b) Xe flask (c) He flask **5.3** (a) No change in H_2 pressure. (b) They are equal. (c) The total pressure is the sum of the pressures of the H_2 and Ar in the container.

Chapter 6

6.1 Sun's energy is changed to electric energy, to chemical energy (of the battery), and back to electric energy, then to kinetic energy of motor and water, and to potential energy of water. **6.2** c and d are exothermic; d is most exothermic **6.3** $\Delta H_{sub} = \Delta H_{fus} + \Delta H_{vap}$

Chapter 7

7.1 UV

Chapter 8

8.1 Mg and Al **8.2** IIA **8.3** As

Chapter 9

9.1 c, e **9.2** b **9.3** a

Chapter 10

10.1 The four pairs have a tetrahedral arrangement; the molecular geometry is trigonal pyramidal. **10.2** Molecule Y is likely to be trigonal planar. Molecule Z cannot have a trigonal planar geometry, but must be either trigonal pyramidal or T-shaped. **10.3** *sp*

Chapter 11

11.1 Cook it longer because the H_2O temperature would be lower. **11.2** (a) The hydrogen and oxygen would form an explosive mixture. (b) Many times greater (more positive) for the wrong reaction. (c) Apply Hess's law. **11.3** (a) AB (b) face-centered cubic

Chapter 12

12.1 At high elevations, the partial pressure of O_2 is so low that not enough dissolves in the water to sustain fish.   **12.2** A soluble, chemically similar liquid with a higher vapor pressure than water   **12.3** To prevent water from flowing into the pickle via osmosis. The pickles would be huge and would probably burst.   **12.4** Highest to lowest: $MgCl_2 > KBr > Na_2SO_4 > NaCl$.

Chapter 13

13.1 Metals in ores and minerals are in an oxidized form in which the free metal form is the neutral, elemental state.   **13.2** More energy because you would need to give the electrons enough energy to jump the gap into the unoccupied orbitals that make up the unfilled band.   **13.3** The same as in glass: curved downward from the walls of the container. It attains the shape because of hydrogen bonding interactions between the O atoms in the SiO_2 tetrahedra and the H atoms of the water molecules.

Chapter 14

14.1 (a) Point A has a faster instantaneous rate.   (b) No, it decreases with time.   **14.2** (a) Rate of aging = $(\text{diet})^w(\text{exercise})^x(\text{sex})^y(\text{occupation})^z$ (b) Gather a sample of people that have all of the factors the same except one. For example, using the equation given in part (a), you could determine the effect of diet if you had a sample of people who were the same sex, exercised the same amount, and had the same occupation. You would need to isolate each factor in this fashion to determine the exponent on each factor.   (c) 2   **14.3** Second order   **14.4** (a) the A + B reaction   (b) the E + F reaction   (c) exothermic

Chapter 15

15.1 $2A + B \longrightarrow 2C$ **15.2** Left to right.

Chapter 16

16.1 $HCHO_2(aq) + H_2O(l) \rightleftharpoons CHO_2^-(aq) + H_3O^+(aq)$; $HCHO_2(aq)$, $H_2O(l)$, $CHO_2^-(aq)$, and $H_3O^+(aq)$ **16.2** Acetate ion   **16.3** $NaOH$, NH_3, $HC_2H_3O_2$, HCl

Chapter 17

17.1 HCN, $HC_2H_3O_2$, HNO_2, HF **17.2** a, b   **17.3** d

Chapter 18

18.1 $PbSO_4$ **18.2** $NaNO_3$ **18.3** Magnesium oxalate

Chapter 19

19.1 It has increased.   **19.2** The NO concentration is low at equilibrium.   **19.3** a. $\Delta G°$ is not changed, b. ΔG increases.   **19.4** The concentration of NO_2 increases with temperature.

Chapter 20

20.1 No sustainable current would flow. The wire does not contain mobile positive and negatively charged species, which are necessary to balance the accumulation of charges in each of the half-cells.   **20.2** (a) -0.54 V   (b) No   (c) They both would be 1.10 V.   **20.3** (a) Negative   (b) Change the concentrations in a manner to increase Q, where $Q = \dfrac{[Fe^{2+}]}{[Cu^{2+}]}$. For example: $Fe(s)|Fe^{2+}(1.10\ M)\|Cu^{2+}(0.50\ M)|Cu(s)$.   **20.4** Many of the ions contained in seawater have very high reduction potentials—higher than $Fe(s)$. This means that spontaneous electrochemical reactions will occur with the $Fe(s)$, causing the iron to form ions and go into solution while at the same time, the ions in the sea are reduced and plate out on the surface of the iron.

Chapter 21

21.1 (a) Yes   (b) No, since the $_1^3H_2O$ molecule is more massive than $_1^1H_2O$.   (c) $_1^3H_2O$ **21.2** The α particle, since it has the highest RBE.   **21.3** After 50,000 years, enough half-lives have passed (about 10) that there would be almost no carbon-14 present to detect and measure.

Chapter 22

22.1 Given the high energy demands of animals to move and maintain body temperature, breaking the very strong triple bond of N_2 requires too much energy when compared to the lower-energy double bond of O_2.   **22.2** Because the only intermolecular forces in these materials are very weak van der Waals forces.

Chapter 23

23.1 $K_2[PtCl_6]$ **23.2** $[Co(NH_3)_4(H_2O)Br]Cl_2$, $[Co(NH_3)_4H_2OCl]BrCl$

Chapter 24

24.1 (a) C_7H_{14} (b)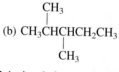

24.2 Blend of 90% 2,2,4-trimethylpentane and 10% heptane

24.3 (a) C_5H_{10} (b)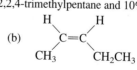

(c) *cis*-2-pentene

Answers to Odd-Numbered Problems

Chapter 1

1.13 (a) liquid and solid   (b) liquid and solid

1.15 (a) $°F = °YS \left(\dfrac{120°F}{100°YS}\right) - 100°F$ (b) $-21°F$

1.17 (a) A paper clip has the closest mass to 1 g.   (b) Your answers will vary depending on sample: grain of sand 1×10^{-5} kg, paper clip 1×10^{-3} kg, nickel 5×10^{-3} kg, 5.0-gallon bucket of water 2.0×10^1 kg, a brick 3 kg, a car 1×10^3 kg.
1.19 29.8 g   **1.21** (a) solid   (b) liquid   (c) solid
1.23 (a) physical   (b) chemical   (c) physical   (d) physical
1.25 Physical properties: lustrous; blue-black crystals that vaporize readily to a violet gas. Chemical properties: combines with many metals, including aluminum.   **1.27** (a) physical
(b) chemical   (c) physical   (d) chemical   **1.29** (a) pure substance; bromine liquid, bromine gas   (b) mixture; liquid solution and solid   (c) mixture; solid sodium hydrogen carbonate and solid potassium hydrogen tartrate   **1.31** (a) 6   (b) 3
(c) 4   (d) 5   **1.33** 4.0×10^4 km   **1.35** (a) 82.5   (b) 111
(c) 2.3×10^3 **1.37** 32 cm^3 **1.39** (a) 5.89 ps   (b) 20.1 cm
1.41 (a) 6.15×10^{-12} s   (b) 3.781×10^{-6} m   **1.43** (a) 20°C
(b) $-94°F$ **1.45** $-6.0°F$ **1.47** ethanol   **1.49** 1.3×10^2 g
1.51 4.80×10^5 mg   **1.53** 3.25×10^6 g   **1.55** 4.435×10^3 m
1.57 2.53 g   **1.59** (a) bromine   (b) phosphorus   (c) gold
(d) carbon   **1.61** no   **1.63** 6.07×10^4 cm^3 **1.65** 50.4 gal
1.67 (a) 7.6×10^{-1} (b) 1.63×10^1 (c) 4.76×10^2 (d)
1.12×10^{-1} **1.69** (a) 1.07×10^{-12} s   (b) 5.8×10^{-6} m
(c) 3.19×10^{-7} m   (d) 1.53×10^{-2} s   **1.71** $1.52 \times 10^3 °F$
1.73 907.8°C, 1180.9 K   **1.75** 1.74×10^3 kg/m^3
1.77 chloroform   **1.79** 33.24 mL   **1.81** (a) 5.91×10^6 mg
(b) 7.53×10^5 μg   (c) 9.01×10^4 kHz   **1.83** 55.0 g
1.85 2.13×10^3 g   **1.87** 2.6×10^{21} g   **1.89** 34%, 79.3 proof

Chapter 2

2.17 Equations are balanced by changing coefficients, not subscripts. A change in subscripts gives a different molecule.
2.19 (a) H_2O (b) KCl   (c) CH_3OH (d) NH_3
2.21 (a) neon   (b) zinc   (c) silver   (d) magnesium
2.23 D,C   **2.25** Cl-35: 17 protons, 18 neutrons, 17 electrons; Cl-37: 17 protons, 20 neutrons, 17 electrons   **2.27** 13.90
2.29 39.10 amu, potassium   **2.31** 24.615 amu
2.33 (a) Group IVA, Period 2, nonmetal   (b) Group VIA, Period 6, metal   (c) Group VIB, Period 4, metal   (d) Group IIA, Period 3, metal   (e) Group IIIA, Period 2, metalloid
2.35 (a) O, oxygen   (b) Na, sodium   (c) Fe, iron   (d) Ce, cerium   **2.37** 4.10×10^{22} N atoms, 2.73×10^{22} N atoms

2.39 (a) PCl_5 (b) NO_2 (c) CH_3CH_2COOH **2.41** 1 Fe to 9 O
2.43 (a) $Fe(CN)_3$ (b) K_2SO_4 (c) Li_3N (d) Ca_3P_2
2.45 (a) sodium sulfate   (b) calcium oxide   (c) copper(I) chloride   (d) chromium(III) oxide   **2.47** (a) $PbCr_2O_7$
(b) $Ba(HCO_3)_2$ (c) Cs_2S (d) $Fe(C_2H_3O_2)_2$ **2.49** (a) selenium trioxide   (b) disulfur dichloride   (c) carbon monoxide
2.51 (a) bromic acid $HBrO_3$ (b) hyponitrous acid $H_2N_2O_2$
(c) disulfurous acid $H_2S_2O_5$ (d) arsenic acid H_3AsO_4
2.53 sodium sulfate decahydrate   **2.55** $FeSO_4 \cdot 7H_2O$ **2.57** 9
2.59 (a) $Sn + 2NaOH \longrightarrow Na_2SnO_2 + H_2$
(b) $8Al + 3Fe_3O_4 \longrightarrow 4Al_2O_3 + 9Fe$
(c) $2CH_3OH + 3O_2 \longrightarrow 2CO_2 + 4H_2O$
(d) $P_4O_{10} + 6H_2O \longrightarrow 4H_3PO_4$
(e) $PCl_5 + 4H_2O \longrightarrow H_3PO_4 + 5HCl$
2.61 $Ca_3(PO_4)_2(s) + 3H_2SO_4(aq) \longrightarrow 2H_3PO_4(aq) + 3CaSO_4(s)$
2.63 The ratio of O masses is 5/4 (B to A).   **2.65** -1.6×10^{-19}
2.67 63, 60   **2.69** $^{80}_{35}Br^-$ **2.71** Ag-107: 0.518; Ag-109: 0.482
2.73 Na_2SO_4, NaCl, $NiSO_4$, $NiCl_2$ **2.75** (a) Hg_2S
(b) $Co_2(SO_3)_3$ (c) $(NH_4)_2Cr_2O_7$ (d) Al_2S_3
2.77 (a) arsenic tribromide   (b) dihydrogen selenide
(c) diphosphorus pentoxide   (d) silicon dioxide
2.79 (a) $2C_2H_6 + 7O_2 \longrightarrow 4CO_2 + 6H_2O$ (b) $P_4O_6 + 6H_2O$
$\longrightarrow 4H_3PO_3$ (c) $4KClO_3 \longrightarrow 3KClO_4 + KCl$
(d) $(NH_4)_2SO_4 + 2NaOH \longrightarrow 2NH_3 + 2H_2O + Na_2SO_4$
(e) $2NBr_3 + 3NaOH \longrightarrow N_2 + 3NaBr + 3HOBr$
2.81 26, nickel   **2.83** $NiSO_4 \cdot 7H_2O$, $NiSO_4 \cdot 6H_2O$, 4.826 g
2.85 0.8000 g, 63.54 amu, Cu

Chapter 3

3.11 (a) $3H_2(g) + N_2(g) \longrightarrow 2NH_3(g)$ (b) $H_2(g)$
(c) six moles of H_2 **3.13** (a) unreasonable   (b) unreasonable
(c) reasonable   **3.15** (a) 32.0 amu   (b) 138 amu
3.17 (a) 64.1 amu   (b) 137 amu   **3.19** (a) 3.818×10^{-23}
g/atom   (b) 8.383×10^{-23} g/molecule
3.21 (a) 3.4 g Na   (b) 236 g CH_2Cl_2 **3.23** (a) 0.238 mol
(b) 1.3 mol   **3.25** 5.81×10^{-3} mol $CaSO_4$; 1.16×10^{-2} mol
H_2O; for every mole of $CaSO_4$ there are two moles of H_2O.
3.27 (a) 2.41×10^{23} atoms   (b) 2.2×10^{24} molecules
(c) 6.59×10^{22} SO_4^{2-} ions   **3.29** 2.97×10^{19} molecules
3.31 20.2%   **3.33** 6.56×10^{-1} kg   **3.35** (a) 42.9% C,
57.1% O   (b) 19.2% Na, 1.68% H, 25.8% P, 53.3% O
3.37 91.2% C, 8.8% H   **3.39** 38.7% C, 9.79% H, 51.5% O
3.41 OsO_4 **3.43** (a) CH   (b) C_4H_4, C_6H_6 **3.45** $C_4H_{12}N_2$
3.47 $C_2H_2O_4$

3.49 C_2H_4 + $3O_2$ $\longrightarrow$

1 molecule C_2H_4 +   3 molecules O_2 $\longrightarrow$
1 mole C_2H_4 +   3 moles O_2 $\longrightarrow$
28.054 g C_2H_4 + 3×32.00 g O_2 $\longrightarrow$

     $2CO_2$ + $2H_2O$

2 molecules CO_2 +   2 molecules H_2O
2 moles CO_2 +   2 moles H_2O
2×44.01 g CO_2 + 2×18.016 g H_2O

3.51 7.82 mol O_2 **3.53** 1.46×10^5 g W   **3.55** 22.4 g CS_2
3.57 KO_2 is the limiting reactant; 0.19 moles of oxygen
3.59 40.5 g CH_3OH; H_2 remains; 5.1 g H_2 **3.61** 2.61 g is the
theoretical yield; 75.9%   **3.63** 49.5% C, 5.19% H, 28.9% N,
16.5% O   **3.65** $C_6H_4Cl_2$ **3.67** C_4H_4S, C_4H_4S **3.69** 84.2%
3.71 59.5%   **3.73** 60.3 g   **3.75** 0.21 g CuO   **3.77** 616 amu
3.79 Ag and Cl

Chapter 4

4.11 (a) $3Ca(C_2H_3O_2)_2(aq) + 2(NH_4)_3PO_4(aq) \longrightarrow$
$Ca_3(PO_4)_2(s) + 6NH_4C_2H_3O_2(aq)$ (b) $3Ca^{2+}(aq) +$
$6C_2H_3O_2^-(aq) + 6NH_4^+(aq) + 2PO_4^{3-}(aq) \longrightarrow$
$Ca_3(PO_4)_2(s) + 6NH_4^+(aq) + 6C_2H_3O_2^-(aq)$
(c) $3Ca^{2+}(aq) + 2PO_4^{3-}(aq) \longrightarrow Ca_3(PO_4)_2(s)$
4.13 Probably not, because the ionic compound that is a nonelectrolyte is not soluble.   **4.15** (a) insoluble   (b) soluble; Li^+,
SO_4^{2-} (c) insoluble   (d) soluble; Na^+, CO_3^{2-}
4.17 (a) $H^+(aq) + OH^-(aq) \longrightarrow H_2O(l)$
(b) $Ag^+(aq) + Br^-(aq) \longrightarrow AgBr(s)$
(c) $S^{2-}(aq) + 2H^+(aq) \longrightarrow H_2S(g)$
(d) $OH^-(aq) + NH_4^+(aq) \longrightarrow NH_3(g) + H_2O(l)$
4.19 (a) $FeSO_4(aq) + NaCl(aq) \longrightarrow NR$
(b) $Na_2CO_3(aq) + MgBr_2(aq) \longrightarrow MgCO_3(s) + 2NaBr(aq)$;
$CO_3^{2-}(aq) + Mg^{2+}(aq) \longrightarrow MgCO_3(s)$
(c) $MgSO_4(aq) + 2NaOH(aq) \longrightarrow$

     $Mg(OH)_2(s) + Na_2SO_4(aq)$;
$Mg^{2+}(aq) + 2OH^-(aq) \longrightarrow Mg(OH)_2(s)$
(d) $NiCl_2(aq) + NaBr(aq) \longrightarrow NR$
4.21 (a) $Ba(NO_3)_2(aq) + Li_2SO_4(aq) \longrightarrow$

     $BaSO_4(s) + 2LiNO_3(aq)$;
$Ba^{2+}(aq) + SO_4^{2-}(aq) \longrightarrow BaSO_4(s)$
(b) $NaBr(aq) + Ca(NO_3)_2(aq) \longrightarrow NR$
(c) $Al_2(SO_4)_3(aq) + 6NaOH(aq) \longrightarrow$

     $2Al(OH)_3(s) + 3Na_2SO_4(aq)$;
$Al^{3+}(aq) + 3OH^-(aq) \longrightarrow Al(OH)_3(s)$
(d) $3CaBr_2(aq) + 2Na_3PO_4(aq) \longrightarrow$

     $Ca_3(PO_4)_2(s) + 6NaBr(aq)$;
$3Ca^{2+}(aq) + 2PO_4^{3-}(aq) \longrightarrow Ca_3(PO_4)_2(s)$
4.23 (a) weak acid   (b) strong base   (c) strong acid
(d) weak acid
4.25 (a) $NaOH(aq) + HNO_3(aq) \longrightarrow NaNO_3(aq) + H_2O(l)$;
$H^+(aq) + OH^-(aq) \longrightarrow H_2O(l)$
(b) $2HCl(aq) + Ba(OH)_2(aq) \longrightarrow 2H_2O(l) + BaCl_2(aq)$;
$H^+(aq) + OH^-(aq) \longrightarrow H_2O(l)$
(c) $2HC_2H_3O_2(aq) + Ca(OH)_2(aq) \longrightarrow$

     $Ca(C_2H_3O_2)_2(aq) + 2H_2O(l)$;

$HC_2H_3O_2(aq) + OH^-(aq) \longrightarrow H_2O(l) + C_2H_3O_2^-(aq)$
(d) $NH_3(aq) + HNO_3(aq) \longrightarrow NH_4NO_3(aq)$;
$H^+(aq) + NH_3(aq) \longrightarrow NH_4^+(aq)$
4.27 (a) $2HBr(aq) + Ca(OH)_2(aq) \longrightarrow CaBr_2(aq) + 2H_2O(l)$;
$H^+(aq) + OH^-(aq) \longrightarrow H_2O(l)$
(b) $Al(OH)_3(s) + 3HNO_3(aq) \longrightarrow Al(NO_3)_3(aq) + 3H_2O(l)$;
$3H^+(aq) + Al(OH)_3(s) \longrightarrow 3H_2O(l) + Al^{3+}(aq)$
(c) $2HCN(aq) + Ca(OH)_2(aq) \longrightarrow Ca(CN)_2(aq) + 2H_2O(l)$;
$HCN(aq) + OH^-(aq) \longrightarrow H_2O(l) + CN^-(aq)$
(d) $LiOH(aq) + HCN(aq) \longrightarrow H_2O(l) + LiCN(aq)$;
$HCN(aq) + OH^-(aq) \longrightarrow H_2O(l) + CN^-(aq)$
4.29 (a) $CaS(aq) + 2HBr(aq) \longrightarrow H_2S(g) + CaBr_2(aq)$;
$2H^+(aq) + S^{2-}(aq) \longrightarrow H_2S(g)$
(b) $MgCO_3(s) + 2HNO_3(aq) \longrightarrow$

     $CO_2(aq) + H_2O(l) + Mg(NO_3)_2(aq)$;
$2H^+(aq) + MgCO_3(s) \longrightarrow CO_2(aq) + H_2O(l) + Mg^{2+}(aq)$
(c) $K_2SO_3(aq) + H_2SO_4(aq) \longrightarrow$

     $SO_2(g) + H_2O(l) + K_2SO_4(aq)$;
$2H^+(aq) + SO_3^{2-}(aq) \longrightarrow SO_2(g) + H_2O(l)$
4.31 (a) +3   (b) +4   (c) +7   (d) +6   **4.33** (a) −3
(b) +5   (c) +3   (d) +5
4.35 (a) $P_4(s)$ + $5O_2(g)$ $\longrightarrow$ $P_4O_{10}(s)$

     reducing   oxidizing
     agent   agent

(b) $Co(s)$ + $Cl_2(g)$ $\longrightarrow$ $CoCl_2(s)$

     reducing   oxidizing
     agent   agent

4.37 (a) $3Cu^{2+}(aq) + 2Al(s) \longrightarrow 2Al^{3+}(aq) + 3Cu(s)$
(b) $2Cr^{3+}(aq) + 3Zn(s) \longrightarrow 2Cr(s) + 3Zn^{2+}(aq)$
4.39 1.02 M **4.41** 1.25 L   **4.43** 1.44×10^{-4} mol
4.45 0.33 g   **4.47** 7.6 mL   **4.49** 0.302 g; 65.9%   **4.51** $FeCl_2$
4.53 50.9 mL nitric acid   **4.55** 3.19%   **4.57** $Mg(s)$ +
$2HBr(aq) \longrightarrow H_2(g) + MgBr_2(aq)$; $Mg(s) + 2H^+(aq) \longrightarrow$
$H_2(g) + Mg^{2+}(aq)$
4.59 $NiSO_4(aq) + 2LiOH(aq) \longrightarrow Ni(OH)_2(s) + Li_2SO_4(aq)$;
$Ni^{2+}(aq) + 2OH^-(aq) \longrightarrow Ni(OH)_2(s)$
4.61 (a) $Sr(OH)_2(aq) + 2HC_2H_3O_2(aq) \longrightarrow$

     $Sr(C_2H_3O_2)_2(aq) + 2H_2O(l)$;
$OH^-(aq) + HC_2H_3O_2(aq) \longrightarrow H_2O(l) + C_2H_3O_2^-(aq)$
(b) $NH_4I(aq) + CsCl(aq) \longrightarrow NR$
(c) $NaNO_3(aq) + CsCl(aq) \longrightarrow NR$
(d) $NH_4I(aq) + AgNO_3(aq) \longrightarrow AgI(s) + NH_4NO_3(aq)$
$I^-(aq) + Ag^+(aq) \longrightarrow AgI(s)$
4.63 (a) $BaCl_2(aq) + CuSO_4(aq) \longrightarrow BaSO_4(s) + CuCl_2(aq)$;
$BaSO_4$ is filtered off, and the solution evaporated to give solid
$CuCl_2$
(b) $CaCO_3(s) + 2HC_2H_3O_2(aq) \longrightarrow$

     $CO_2(g) + H_2O(l) + Ca(C_2H_3O_2)_2(aq)$;
evaporate H_2O to leave $Ca(C_2H_3O_2)_2(s)$
(c) $Na_2SO_3(s) + 2HNO_3(aq) \longrightarrow$

     $SO_2(g) + H_2O(l) + 2NaNO_3(aq)$;
evaporate H_2O to leave $NaNO_3(s)$
(d) $Mg(OH)_2(s) + 2HCl(aq) \longrightarrow 2H_2O(l) + MgCl_2(aq)$;
evaporate H_2O to leave $MgCl_2(s)$

4.65 (a) decomposition   (b) decomposition   (c) combination   (d) displacement   **4.67** 0.0405 M $CaCl_2$, 0.0405 M Ca^{2+}, 0.0811 M Cl^- **4.69** 329 mL   **4.71** 0.610 M **4.73** 0.0967 M **4.75** 21.30%   **4.77** 9.66%   **4.79** $Pb(NO_3)_2$; Cs_2SO_4;
$Pb(NO_3)_2(aq) + Cs_2SO_4(aq) \longrightarrow PbSO_4(s) + 2CsNO_3(aq)$;
$Pb^{2+}(aq) + SO_4^{2-}(aq) \longrightarrow PbSO_4(s)$;
lead(II) sulfate; cesium nitrate;
$Pb(NO_3)_2(aq) + Na_2SO_4(aq) \longrightarrow PbSO_4(s) + 2NaNO_3(aq)$
4.81 0.0121 L   **4.83** 0.930 mol/L   **4.85** 69.72 g/mol, gallium   **4.87** $P_4O_{10} + 6H_2O \longrightarrow 4H_3PO_4$; 55.7 g P_4O_{10} **4.89** 0.194 L   **4.91** 61.7%

Chapter 5

5.15 (a) At constant volume, the pressure of a gas is directly proportional to temperature ($P \propto T$).   (b) At constant volume, the pressure of a gas is directly proportional to temperature ($P \propto T$).   (c) The vapor pressure of the H_2O increases with increasing temperature (Table 5.6).   **5.17** (a) O_2 (b) H_2 (c) Both the same   (d) There will be no pressure change.   (e) 1/4   **5.19** 36 mmHg   **5.21** 2.32 L   **5.23** 3.64×10^{-4} kPa   **5.25** 2.50 mL   **5.27** 31.6 mL   **5.29** 1 volume   **5.31** $PV = nRT$; $V = nRT/P$; if temperature and moles remain constant, $V = $ constant $\times 1/P$ **5.33** 29.4 L   **5.35** 0.675 g/L   **5.37** 47.7 g/mol   **5.39** Gas density depends on molecular weight or average molecular weight of a mixture. Thus, the density of a gas of NH_4Cl would be greater than that of a mixture of NH_3 and HCl, since NH_3 and HCl have lower molecular weights than NH_4Cl.   **5.41** 1.4 L   **5.43** 160. L   **5.45** $2NH_3(g) + H_2SO_4(aq) \longrightarrow$
$(NH_4)_2SO_4(aq)$; 46.7 L
5.47 $P(CO_2) = 494$ mmHg; $P(H_2) = 190.$ mmHg; $P(HCl) = 41$ mmHg; $P(HF) = 21$ mmHg; $P(SO_2) = 13$ mmHg; $P(H_2S) = 0.8$ mmHg   **5.49** 6.34 g   **5.51** 1.53×10^2 m/s   **5.53** 6.50×10^3 K ($6.23 \times 10^{3\circ}$C)   **5.55** 146 amu   **5.57** 0.9605 atm (0.9716 atm from ideal gas law)   **5.59** 80.7 cm^3 **5.61** 6.5 dm^3 **5.63** 3.01×10^{20} atoms   **5.65** 1.2×10^3 g LiOH   **5.67** 20.4 atm   **5.69** H_2SO_4 **5.71** $-95°$C   **5.73** $P(O_2) = 0.167$ atm; $P(CO_2) = 0.333$ atm   **5.75** 0.732 g/L   **5.77** 0.194 g Na_2O_2 **5.79** 76% $CaCO_3$, 24% $MgCO_3$

Chapter 6

6.17 small car   **6.19** (a) 🔺 (b) a   **6.21** Equate ΔH for the burning to ΔH_f (products) minus ΔH_f (reactants), then solve for ΔH_f of P_4S_3. You will need ΔH_f of P_4O_{10} and SO_2.   **6.23** 7.2×10^5 J, 1.7×10^5 cal   **6.25** 5.24×10^{-21} J/molecule   **6.27** endothermic; +66.4 kJ   **6.29** $2Fe(s) + 6HCl(aq) \longrightarrow 2FeCl_3(aq) + 3H_2(g)$; $\Delta H = -175.8$ kJ   **6.31** $+2.940 \times 10^3$ kJ   **6.33** -1.90 kJ/g   **6.35** 7.96 g   **6.37** 5.8×10^4 J   **6.39** 20.5 kJ   **6.41** -1.36×10^3 kJ   **6.43** 50.6 kJ   **6.45** -137 kJ   **6.47** $+42.2$ kJ   **6.49** -23.5 kJ   **6.51** 13.9 kJ   **6.53** 1.59×10^3 J/g   **6.55** 38.0 kJ

6.57 -255 kJ/mol   **6.59** 0.128 J/g°C   **6.61** -23.6 kJ/mol   **6.63** -871.7 kJ/mol   **6.65** -20 kJ   **6.67** 178 kJ   **6.69** -39.7 kJ   **6.71** 35.5 g   **6.73** 37.4°C   **6.75** 113 kJ   **6.77** (a) 1.88 kJ   (b) 11.9%   **6.79** 1.07×10^3 g LiOH

Chapter 7

7.15 frequency-doubled beam   **7.17** x **7.19** (a)   **7.21** 219.6 m   **7.23** 6.28×10^{14}/s   **7.25** $6.05802106 \times 10^{-7}$ m, 4.959×10^{14}/s   **7.27** 9.045×10^{-28} J   **7.29** 3.72×10^{-19} J   **7.31** 2.34×10^{14}/s   **7.33** 93.9 nm   **7.35** 4.699×10^{-19} J   **7.37** 95.4 pm   **7.39** 7.28×10^7 m/s   **7.41** $l = 0, 1, 2, 3$; $m_l = -3, -2, -1, 0, 1, 2, 3$ **7.43** (a) $6d$ (b) $5g$ (c) $4f$ (d) $6p$ **7.45** (a) m_s can only be $+\frac{1}{2}$ or $-\frac{1}{2}$ (b) l can only be as large as $n - 1$ (c) m_l cannot exceed $+2$ in magnitude   (d) n cannot be 0   (e) m_s can only be $+\frac{1}{2}$ or $-\frac{1}{2}$ **7.47** 6.51×10^{14}/s, 4.31×10^{-19} J   **7.49** 485 nm, blue-green   **7.51** 6.55×10^{14}/s   **7.53** 4.35×10^{-7} m   **7.55** 1.64×10^{-7} m; near UV   **7.57** (a) five   (b) seven   (c) three   (d) one   **7.59** $6s, 6p, 6d, 6f, 6g, 6h$ **7.61** 5.01×10^{-7} m   **7.63** 5.26×10^{28} **7.65** 3.60×10^{-19} J, 2.16×10^2 kJ/mol

Chapter 8

8.21 Sr, Y   **8.23** Z would form a cation of charge 2+.   **8.25** Ge, Sn   **8.27** (a) not allowed; $2s$ electrons should have opposite spins.   (b) allowed; $1s^2 2s^2 2p^3$ (c) not allowed; second $2p$ electrons should have opposite spins.   (d) allowed; $1s^2 2s^2 2p^4$ **8.29** (a) possible   (b) possible   (c) impossible; $2p$ subshell can hold no more than six electrons.   (d) impossible; $3d$ subshell can hold no more than ten electrons.   **8.31** $1s^2 2s^2 2p^6 3s^2 3p^5$ **8.33** $1s^2 2s^2 2p^6 3s^2 3p^6 3d^3 4s^2$ **8.35** $4d^{10} 5s^2$ **8.37** [Ar] ⇅⇅⇅⇅⇅ ⇅
                                                                          $3d$ $4s$

8.39 ⇅ ⇅ ⇅⇅⇅ ⇅ ⇅↑↑ ↑
         $1s$ $2s$ $2p$ $3s$ $3p$ $4s$
Paramagnetic   **8.41** S, Se, As   **8.43** Na, Al, Cl, Ar   **8.45** (a) Br   (b) F   **8.47** $6s^2 6p^4$ **8.49** ⇅ ⇅ ⇅⇅⇅ ⇅ ⇅⇅⇅ ⇅⇅⇅⇅⇅
         $1s$ $2s$ $2p$ $3s$ $3p$ $3d$
         ⇅ ↑↑↑
         $4s$ $4p$
8.51 [Rn]$5f^{14} 6d^{10} 7s^2 7p^2$, metal, eka-PbO or eka-PbO$_2$ **8.53** [Kr] ↑↑↑↑↑ ↑
                              $4d$ $5s$
8.55 (a) Cl_2 (b) Na   (c) Sb   (d) Ar   **8.57** $1s^2 2s^2 2p^6 3s^2 3p^6 3d^3 4s^2$, Group VB, Period 4, d-block transition element   **8.59** Ba(s) + 2H_2O(l) \longrightarrow H_2(g) + Ba(OH)_2(aq)$; 447 mL   **8.61** RaO, 93.4% Ra   **8.63** 108 J   **8.65** -639 kJ/mol

Chapter 9

9.15 b **9.17** (a) Bond Cl and H to C.   (b) Bond O atoms to N, then bond H to one O.   (c) Bond O and F to N.   (d) Place N atoms in center with two O atoms bonded to each N.

9.19 (a) Ca_3X_2 (b) ionic

9.21 (a) $\cdot\overset{\cdot\cdot}{In}\cdot$ (b) In^{3+} (c) $\cdot\overset{\cdot\cdot}{P}\cdot$ (d) $\left[:\overset{\cdot\cdot}{\underset{\cdot\cdot}{P}}:\right]^{3-}$

9.23 (a) $1s^2 2s^2 2p^6 3s^2 3p^6$, Ca^{2+}

(b) $1s^2 2s^2 2p^6 3s^2 3p^6 3d^{10} 4s^2 4p^4$, $:\overset{\cdot\cdot}{Se}\cdot$

(c) $1s^2 2s^2 2p^6 3s^2 3p^6 3d^{10} 4s^2 4p^6$, $\left[:\overset{\cdot\cdot}{\underset{\cdot\cdot}{Se}}:\right]^{2-}$

(d) $1s^2 2s^2 2p^6 3s^2 3p^6 3d^{10} 4s^2 4p^6$, $\left[:\overset{\cdot\cdot}{\underset{\cdot\cdot}{Br}}:\right]^{-}$

9.25 (a) $[Xe]4f^{14}5d^{10}6s^2 6p^3$ (b) $[Xe]4f^{14}5d^{10}6s^2$
9.27 (a) $[Ar]3d^8$ (b) $[Ar]3d^7$ **9.29** (a) I, I^-; electron repulsion makes I^- larger. (b) Ca^{2+}, Ca; loss of valence electrons in Ca makes Ca^{2+} smaller. **9.31** Na^+, F^-, N^{3-}. The ions are isoelectronic, so the atomic radius increases with decreasing nuclear charge.

9.33 $3\,H\cdot\ +\ :\overset{\cdot\cdot}{P}\cdot\ \longrightarrow\ $ (see structure) — Bonding electron pairs; Lone pair

9.35 (a) P, N, O (b) Na, Mg, Al (c) Al, Si, C **9.37** C—Br, As—Br, Si—O **9.39** (a) $:\overset{\cdot\cdot}{Br}-\overset{\cdot\cdot}{Br}:$ (b) $H-\overset{\cdot\cdot}{Se}-H$

(c) $:\overset{\cdot\cdot}{\underset{\cdot\cdot}{F}}:\overset{\cdot\cdot}{\underset{:F:}{\overset{:F:}{N}}}:$ **9.41** (a) $\left[:\overset{\cdot\cdot}{\underset{\cdot\cdot}{Cl}}-\overset{\cdot\cdot}{\underset{\cdot\cdot}{O}}:\right]^-$ (b) $\left[:\overset{\cdot\cdot}{\underset{\cdot\cdot}{Cl}}-\overset{:Cl:}{\underset{\cdot\cdot}{Sn}}-\overset{\cdot\cdot}{\underset{\cdot\cdot}{Cl}}:\right]^-$

(c) $\left[:\overset{\cdot\cdot}{\underset{\cdot\cdot}{S}}-\overset{\cdot\cdot}{\underset{\cdot\cdot}{S}}:\right]^{2-}$

9.43 (a) $\left[:\overset{\cdot\cdot}{\underset{\cdot\cdot}{O}}-\overset{\cdot\cdot}{N}=\overset{\cdot\cdot}{O}:\right]^- \longleftrightarrow \left[\overset{\cdot\cdot}{O}=\overset{\cdot\cdot}{N}-\overset{\cdot\cdot}{\underset{\cdot\cdot}{O}}:\right]^-$

(b) $H-\overset{\cdot\cdot}{O}-N\overset{:O:}{\underset{O}{\diagdown}} \longleftrightarrow H-\overset{\cdot\cdot}{O}-N\overset{O}{\underset{:O:}{\diagup}}$

9.45 (a) $:\overset{\cdot\cdot}{F}-\overset{\cdot\cdot}{Xe}-\overset{\cdot\cdot}{F}:$ (b) $:\overset{\cdot\cdot}{F}-\overset{:F:}{\underset{:F:}{Se}}-\overset{\cdot\cdot}{F}:$

(c) $:\overset{\cdot\cdot}{F}-\overset{\overset{F}{|}}{\underset{\underset{F}{|}}{Te}}-\overset{\cdot\cdot}{F}:$ (d) $\left[\overset{F\quad F}{\underset{F\quad F}{Xe}}-F:\right]^+$

9.47 (a) $\overset{\cdot\cdot}{O}=\overset{\oplus}{O}-\overset{\cdot\cdot}{\underset{\ominus}{O}}:$ (b) $:C\equiv\overset{\oplus}{O}:$ (c) $H-\overset{\cdot\cdot}{O}-\overset{\oplus}{N}\overset{\overset{\displaystyle\cdot\cdot}{\underset{\ominus}{O}}}{\underset{:O:}{\diagup}}$

9.49 (a) 114 pm (b) 203 pm (c) 213 pm (d) 183 pm
9.51 −78 kJ **9.53** (a) ionic, SrO, strontium oxide (b) covalent, CBr_4, carbon tetrabromide (c) ionic, GaF_3, gallium(III) fluoride (d) covalent, NBr_3, nitrogen tribromide

9.55 $\left[:\overset{\displaystyle:O:}{\underset{\displaystyle:O:}{\underset{\cdot\cdot}{O}-As-\overset{\cdot\cdot}{O}:}}\right]^{3-}$ or $\left[:\overset{\displaystyle:O:}{\underset{\displaystyle:O:}{\underset{\cdot\cdot}{O}-As-\overset{\cdot\cdot}{O}:}}\right]^{3-}$, $Pb_3(AsO_4)_2$

9.57 $H-\overset{\cdot\cdot}{O}-\overset{\displaystyle:O:}{\underset{}{I}}-\overset{\cdot\cdot}{O}:$ or $H-\overset{\cdot\cdot}{O}-\overset{\displaystyle:O:}{\underset{}{I}}=\overset{\cdot\cdot}{O}$ **9.59** $\left[H-\overset{\displaystyle:O:}{\underset{}{N}}-H\right]^-$

9.61 (a) $:\overset{\displaystyle:O:}{\underset{}{Cl}}-\overset{}{Se}-\overset{\cdot\cdot}{Cl}:$ or $:\overset{\displaystyle:O:}{\underset{}{Cl}}-\overset{}{Se}-\overset{\cdot\cdot}{Cl}:$ (b) $:\overset{\cdot\cdot}{Se}=C=\overset{\cdot\cdot}{Se}:$

(c) $\left[:\overset{\displaystyle:Cl:}{\underset{\displaystyle:Cl:}{\underset{\cdot\cdot}{Cl}-Ga-\overset{\cdot\cdot}{Cl}:}}\right]^-$ (d) $\left[:C\equiv C:\right]^{2-}$

9.63 (a) $:\overset{\cdot\cdot}{O}=Se-\overset{\cdot\cdot}{O}: \longleftrightarrow :\overset{\cdot\cdot}{O}-Se=\overset{\cdot\cdot}{O}:$

(b)  (resonance structures)

9.65 (resonance structures) The outer N—O bonds are 118 pm, and the inner N—O bonds are 136 pm.   **9.67** -1.60×10^2 kJ

9.69 $H_3PO_3(aq) + 2NaOH(aq) \longrightarrow$
$\qquad\qquad Na_2HPO_3(aq) + 2H_2O(l)$, 0.007031 M

9.71 $Mg(ClO_4)_2$, magnesium perchlorate, Mg^{2+}, and

$\left[:\overset{\displaystyle:O:}{\underset{\displaystyle:O:}{\underset{\cdot\cdot}{O}-Cl-\overset{\cdot\cdot}{O}:}}\right]^-$ or $\left[\overset{\displaystyle:O:}{\underset{\displaystyle:O:}{\underset{\cdot\cdot}{O}=Cl=O}}\right]^-$ **9.73** $\overset{H}{\underset{:F:}{\diagdown}}C=\overset{\cdot\cdot}{O}:$

9.75 1271 kJ/mol; 860 kJ/mol compared to table value of 887 kJ/mol

Chapter 10

10.13 (a) ii (b) i (c) iv (d) iii

10.15

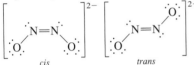

(has dipole), (no dipole)

10.17 (a) angular (b) tetrahedral (c) trigonal planar
(d) trigonal pyramidal **10.19** (a) trigonal planar (b) tetrahedral (c) linear (d) tetrahedral **10.21** (a) trigonal bipyramidal (b) T-shaped (c) square pyramidal (d) distorted
tetrahedral **10.23** (a) trigonal pyramidal and T-shaped
(b) bent **10.25** (a) uses sp^3 orbitals, (b) uses sp^3 orbitals
10.27 (a) Two single bonds and no lone pair on Hg suggests sp
hybridization. An Hg—Cl bond is formed by overlapping an Hg
hybrid orbital with a $3p$ orbital of Cl.   (b) The presence of three
single bonds and one lone pair on P suggests sp^3 hybridization.
Each hybrid orbital overlaps a $3p$ orbital of a Cl atom to form a
P—Cl bond. The fourth hybrid orbital contains a lone pair.
10.29 (a) sp^3d^2 (b) sp^3d (c) sp^3d (d) sp^3d **10.31** The P
atom in PCl_6^- has six single bonds around it and no lone pairs.
This suggests sp^3d^2 hybridization. Each bond is a σ bond formed
by overlap of an sp^3d^2 hybrid orbital on P with a $3p$ orbital on Cl.
10.33

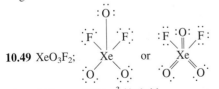

Each of the N atoms has a lone pair of electrons and is bonded to
two atoms. The N atoms are sp^2 hybridized. The two possible
arrangements of the oxygen atoms relative to one another are
shown above. Since the π bond between the nitrogen atoms must
be broken to interconvert these two forms, it is expected that the
hyponitrite ion will exhibit *cis–trans* isomerization.
10.35 (a) $KK(\sigma_{2s})^2(\sigma_{2s}^*)^2(\pi_{2p})^2$, bond order = 1, stable, paramagnetic (b) $KK(\sigma_{2s})^2(\sigma_{2s}^*)^2(\pi_{2p})^1$, bond order = 1/2, stable,
paramagnetic (c) $KK(\sigma_{2s})^2(\sigma_{2s}^*)^2(\pi_{2p})^4(\sigma_{2p})^2(\pi_{2p}^*)^3$, bond
order = 3/2, stable, paramagnetic **10.37** (a) bent (b) trigonal
planar (c) linear (d) octahedral **10.39** Left: Both carbon
atoms with the double bonds are sp^2 hybridized. The carbon atom
with the OH group is sp^3 hybridized. Right: Both carbon atoms
are sp hybridized. **10.41** The *trans* isomer is expected to have a
zero dipole moment, whereas the *cis* isomer is not.
10.43 $(\sigma_{1s})^2$, stable **10.45** $KK(\sigma_{2s})^2(\sigma_{2s}^*)^2(\pi_{2p})^4(\sigma_{2p})^2$; 3
10.47 O_2^+ has one less antibonding electron than O_2; O_2^- has a
longer bond because it has one more antibonding electron.

10.49 XeO_3F_2;   or

trigonal bipyramidal, sp^3d hybrid
10.51 (a) N_2: 110 pm, linear geometry, sp hybrids
(b) N_2F_2: 122 pm, trigonal planar geometry, sp^2 hybrids
(c) N_2H_4: 145 pm, tetrahedral geometry, sp^3 hybrids

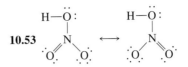

10.53

trigonal planar geometry, sp^2 hybrids, $\Delta H_f^\circ = -40$ kJ/mol,
resonance energy = 95 kJ

Chapter 11

11.15 (a) C (b) A (c) A **11.17** The heat released when
the liquid to solid phase change occurs prevents the fruit from
freezing. **11.19** (a) AB_2 (b) yes, body-centered cubic
11.21 (a) vaporization (b) freezing of eggs and sublimation of
ice (c) condensation (d) condensation (deposition)
11.23 10°C **11.25** 3.50 kJ **11.27** 27 g **11.29** 264 mmHg
11.31 53.3 kJ/mol **11.33** (a) gas (b) gas (c) liquid
(d) no **11.35** SO_2 and C_2H_2 can be liquefied at 25°C; to liquefy
CH_4, reduce temperature below -82°C, then apply pressure
greater than 46 atm; to liquefy CO, reduce temperature below
-140°C, then apply pressure greater than 35 atm.
11.37 (a) London forces (b) dipole–dipole forces, London
forces, hydrogen bonding (c) dipole–dipole forces, London
forces **11.39** CCl_4; highest molecular weight and greatest
London forces. **11.41** $HOCH_2CH_2OH$, FCH_2CH_2OH,
FCH_2CH_2F; hydrogen bonding decreases in magnitude from left
to right. **11.43** CH_4, C_2H_6, CH_3OH, CH_2OHCH_2OH
11.45 (a) metallic (b) covalent network (c) molecular
(d) molecular **11.47** $C_2H_5OC_2H_5$, C_4H_9OH, KCl, CaO
11.49 (a) LiCl (b) SiC (c) CHI_3 (d) Co **11.51** one
11.53 361 pm **11.55** 4, fcc **11.57** 19.25 g/cm^3 **11.59** 68.4%
11.61 80°C **11.63** (a) condenses to liquid (and to solid at high
pressure) (b) condenses to solid (c) remains a gas
11.65 Hydrogen bonds form between H atom of OH group on
one molecule and O atom on another molecule. **11.67** Al,
metallic; Si, covalent network; P, molecular; S, molecular.
11.69 (a) KCl; smaller charges on ions (b) CCl_4; lower molecular weight and smaller London forces (c) Zn; Group IIB
metal is lower melting (d) C_2H_5Cl; no hydrogen bonding.
11.71 3.171×10^{-22} g, 191.0 amu **11.73** 128 pm
11.75 (a) Diamond and silicon carbide are covalent network
solids with strong covalent bonds; graphite is a layered structure
with weak forces between layers. (b) Silicon dioxide is a covalent network solid; carbon dioxide is a small, discrete molecule.
11.77 (a) CO_2 consists of nonpolar molecules; SiO_2 is a covalent network solid. (b) HF has hydrogen bonding. (c) SiF_4 is
a larger molecule. **11.79** 7.6 kJ **11.81** 449 J **11.83** 0.42 g
monomer, 1.42 g dimer, 1.83 g/L

Chapter 12

12.15 There is not enough dissolved oxygen in the warm water
to support trout, which need more O_2. **12.17** The lattice energy
must be less for the XZ compound. **12.19** Because the salt
concentration outside the lettuce leaf is higher, water will pass
out of the lettuce leaf into the dressing via osmosis.

12.21 Aqueous ammonia   **12.23** H_2O, CH_2OHCH_2OH, $C_{10}H_{22}$
12.25 Al^{3+} **12.27** 0.886 g/100 mL   **12.29** Dissolve 1.81 g KI
in 70.7 g H_2O **12.31** 1.52 m **12.33** 155 g   **12.35** 0.0116
12.37 0.796 m **12.39** 41.6 mmHg, 0.631 mmHg
12.41 4.6×10^{-2} m **12.43** 163 amu   **12.45** 6.85×10^4 amu
12.47 $2.78 \approx 3$ **12.49** (a) aerosol   (b) sol   (c) foam
(d) sol   **12.51** 1.74 m, 0.0303, 1.63 M **12.53** 33 g of propane,
42 g of butane   **12.55** (a) 2.500×10^{-4} M (b) $5.000 \times$
10^{-4} M (c) 2.50×10^{-4} M **12.57** $-24°C$ **12.59** 0.30 M
12.61 0.10 m $CaCl_2$ **12.63** 1.8 g/mL; 5.0×10^2 m
12.65 $Mn_2C_{10}O_{10}$ **12.67** -783 kJ/mol, -445 kJ/mol
12.69 0.565 m **12.71** 0.701 M **12.73** $C_3H_8O_3$

Chapter 13

13.27 No. Most aluminum is available in clays and aluminosilicate minerals, which cannot be used economically as ores to obtain aluminum metal.   **13.29** Boron acts as a dopant (doping agent), which yields positive holes that become current carriers.
13.31 Examples: $MgCaSi_3O_8$ and $KNaMgSi_3O_8$ **13.33** The impure lead metal serves as the anode and pure lead serves as the cathode. Impurities either remain at the anode or go into solution.
Anode: $Pb(s) \longrightarrow Pb^{2+}(aq) + 2e^-$;
Cathode: $PbSiF_6(aq) + 2e^- \longrightarrow Pb(s) + SiF_6{}^{2-}(aq)$
13.35 3.69×10^4 g   **13.37** -417.5 kJ; exothermic   **13.39** The diagram is similar to the one for Na.   **13.41** 2.62×10^{19}
13.43 Diamond   **13.45** 0.00187 L   **13.47** The hybrid orbitals as sp^2; the other $2p$ orbital on each C atom forms pi orbitals.
13.49 (b), (d)   **13.51** Each Si atom has four tetrahedral bonds to other Si atoms, with an sp^3 orbital on one Si atom overlapping an sp^3 orbital on another Si atom.   **13.53** 10.7 kg
13.55 Spodumene consists of a long chain of SiO_4 tetrahedra linked together, with the negative charge of this chain balanced by lithium and aluminum cations. Figure 13.24 shows the linking of two SiO_4 tetrahedra. Note that in this linkage, one O atom is shared by the two tetrahedra. The repeating unit of the SiO_4 chain of spodumene consists of an Si atom with two unshared O atoms and two shared O atoms (each shared atom counting as $\frac{1}{2}$ O atom to each repeating unit). The repeating unit also has a charge of $2-$ on it. Thus, the formula of the repeating unit is $SiO_3{}^{2-}$. Two of these repeating units would have a $4-$ charge, which in spodumene is balanced by an Li^+ ion and an Al^{3+} ion, giving the formula $LiAl(SiO_3)_2$.
13.57 $2ZnS(s) + 3O_2(g) \longrightarrow 2ZnO(s)\ 2SO_2(g)$
$ZnO(s) + C(s) \longrightarrow Zn(g) + CO(g)$
13.59 Buckminsterfullerene is similar to graphite in being constructed from carbon-atom hexagons. However, buckminsterfullerene also contains carbon-atom pentagons, which allow the molecule to curl around to give a cage structure.   **13.61** 689.9 kJ

Chapter 14

14.17 (a) Rate of depletion of A $= -\dfrac{\Delta[A]}{\Delta t}$ or Rate of for-

mation of B $= +\dfrac{\Delta[B]}{\Delta t}$.   (b) No, the rate of depletion of A
would be faster than the rate of formation of B.

(c) $-\dfrac{\Delta[A]}{3\Delta t} = \dfrac{\Delta[B]}{2\Delta t}$ **14.19** (a) $x = 0$ (b) $x = 1$
(c) $x = 3$ **14.21** (a) region C   (b) region A
14.23 $-\frac{1}{2}\Delta[NO_2]/\Delta t = \Delta[O_2]/\Delta t$ **14.25** 2.3×10^{-2} M/hr
14.27 1.2×10^{-6} M/s   **14.29** Orders with respect to $MnO_4{}^-$,
$H_2C_2O_4$, and H^+ are 1, 1, 0, respectively. Overall order is 2.
14.31 rate $= k[CH_3NNCH_3]$, $k = 2.5 \times 10^{-4}$/s   **14.33** rate $=$
$k[NO]^2[H_2]$, $k = 2.9 \times 10^2/(M^2 \cdot s)$ **14.35** 5.2×10^{-4} M
14.37 1.1×10^3 s, 1.1×10^3 s, 2.2×10^3 s   **14.39** 96 hr, $1.9 \times$
10^2 hr, 2.9×10^2 hr, 3.9×10^2 hr, 4.8×10^2 hr   **14.41** First
order, $k = 0.101$/s   **14.43** $E_a = 210$ kJ   **14.45** 1.0×10^5 J/mol,
1.7×10^{-3}/s   **14.47** 1.1×10^2 kJ/mol   **14.49** $NOCl_2$; $2NO +$
$Cl_2 \longrightarrow 2NOCl$ **14.51** rate $= k[C_3H_6]^2$ **14.53** rate $=$
$(k_2k_1/k_{-1}) \times [H_2][I_2] = k[H_2][I_2]$ **14.55** $2H_2O_2 \longrightarrow 2H_2O + O_2$;
Br^-; yes (BrO^-)   **14.57** 3.5×10^{-6} M/s, 3.2×10^{-6} M/s,
2.5×10^{-6} M/s   **14.59** Average $k = 2.5 \times 10^{-4}$/s
14.61 0.22 M **14.63** 2.5×10^{-4}/s   **14.65** 1.14×10^5 J/mol,
5×10^9, 14 M/s   **14.67** rate $= k_1[NO_2Br]$ **14.69** rate $=$
$(k_2k_1/k_{-1}) \times [NH_4{}^+][OCN^-] = k[NH_4{}^+][OCN^-]$
14.71 (a) rate $= k[NO]^2[O_2]$ (b) 0.12 mol/L $\cdot$ s   **14.73** A
catalyst provides another pathway with lower activation energy.
14.75 (a) Rate is the change in concentration with time.
(b) It changes because the concentration of reactant changes.
(c) It equals k when all reactants are 1.00 M.   **14.77** 0.210 kJ/s
14.79 6.60×10^{-8} mmHg/s

Chapter 15

15.9 $\frac{2}{3}x$ **15.11** a   **15.13** $K_c = 0.25$ **15.15** 1.05 mol N_2O_3,
0.45 mol NO_2, 0.45 mol NO   **15.17** 0.0266 mol NO, 0.0134

mol Br_2, 0.0389 mol NOBr   **15.19** (a) $K_c = \dfrac{[NO_2][NO]}{[N_2O_3]}$

(b) $K_c = \dfrac{[H_2]^2[S_2]}{[H_2S]^2}$ (c) $K_c = \dfrac{[NO_2]^2}{[NO]^2[O_2]}$

15.21 $K_c = \dfrac{[N_2O_5]}{[NO_2]^2[O_2]^{1/2}}$ **15.23** 49.1

15.25 10.5   **15.27** 1.0×10^{-3} **15.29** (a) $K_p = \dfrac{P_{HBr}{}^2}{P_{H_2}P_{Br_2}}$

(b) $K_p = \dfrac{P_{CH_4}P_{H_2S}{}^2}{P_{CS_2}P_{H_2}{}^4}$ (c) $K_p = \dfrac{P_{H_2O}{}^2P_{Cl_2}{}^2}{P_{HCl}{}^4P_{O_2}}$

15.31 56.3   **15.33** (a) $K_c = \dfrac{[CO]^2}{[CO_2]}$ (b) $K_c = \dfrac{[CO_2]}{[CO]}$

(c) $K_c = \dfrac{[CO_2]}{[SO_2][O_2]^{1/2}}$ **15.35** (a) not

complete   (b) nearly complete   **15.37** (a) goes to left
(b) goes to right   (c) equilibrium   **15.39** goes to left
15.41 $[I_2] = [Br_2] = 4.7 \times 10^{-5}$ M, $[IBr] = 5.1 \times 10^{-4}$ M
15.43 0.18 mol PCl_3, 0.18 mol Cl_2, 0.32 mol PCl_5
15.45 forward   **15.47** (a) no effect   (b) no effect   (c) goes
to left   **15.49** No, the fraction of methanol would decrease.
15.51 decrease   **15.53** 4.3   **15.55** reverse   **15.57** 0.008 mol
HBr, 0.0010 mol H_2, 0.0010 mol Br_2 **15.59** 13.2%

15.61 endothermic

15.63 $K_p = \dfrac{P_{NH_3}^2}{P_{N_2}P_{H_2}^3} = \dfrac{[NH_3]^2(RT)^2}{[N_2](RT)[H_2]^3(RT)^3} = K_c/(RT)^2$, so

$K_c = K_p(RT)^2$ **15.65** (a) $[Cl_2] = [PCl_3] = 0.0306\ M$, $[PCl_5] = 0.0038\ M$ (b) 0.89 (c) Less PCl_5 would decompose.
15.67 0.335 **15.69** (a) $[SO_2] = [Cl_2] = 0.0346$ mol/L, $[SO_2Cl_2] = 0.0266$ mol/L (b) 0.565 (c) decreases
15.71 (a) 7.2×10^{-5}, 1.6×10^{-4} (b) hydrogen bonding occurs (c) decrease **15.73** 106 **15.75** 0.430
15.77 0.408 atm

Chapter 16

16.11 Hydroxide ion forms in the reaction $NH_3(aq) + H_2O(l) \rightleftharpoons NH_4^+(aq) + OH^-(aq)$ **16.13** The hydroxide ion acts as a base and donates a pair of electrons on the O atom, forming a bond with CO_2 to give the HCO_3^- ion. **16.15** (a) HPO_4^{2-} (b) HSe^- (c) NO_2^- (d) AsO_4^{3-} **16.17** (a) $HClO$ (b) AsH_4^+ (c) H_3PO_4 (d) $HTeO_3^-$
16.19 (a) $\underset{\text{acid}}{HSO_4^-} + \underset{\text{base}}{NH_3} \rightleftharpoons \underset{\text{base}}{SO_4^{2-}} + \underset{\text{acid}}{NH_4^+}$

(b) $\underset{\text{base}}{HPO_4^{2-}} + \underset{\text{acid}}{NH_4^+} \rightleftharpoons \underset{\text{acid}}{H_2PO_4^-} + \underset{\text{base}}{NH_3}$

(c) $\underset{\text{acid}}{Al(H_2O)_6^{3+}} + \underset{\text{base}}{H_2O} \rightleftharpoons \underset{\text{base}}{Al(H_2O)_5OH^{2+}} + \underset{\text{acid}}{H_3O^+}$

16.21 $BF_3 + :AsF_3 \longrightarrow BF_3:AsF_3$
BF_3, Lewis acid; AsF_3, Lewis base.

16.23 (a) $\underset{\substack{\text{Lewis}\\\text{acid}}}{Cr^{3+}} + \underset{\substack{\text{Lewis}\\\text{base}}}{6H_2O} \longrightarrow Cr(H_2O)_6^{3+}$

(b) $\underset{\substack{\text{Lewis}\\\text{acid}}}{BF_3} + \underset{\substack{\text{Lewis}\\\text{base}}}{(C_2H_5)_2\overset{\cdot\cdot}{O}:} \longrightarrow F_3B:\overset{\cdot\cdot}{O}(C_2H_5)_2$

16.25 $\underset{\substack{\text{Lewis}\\\text{acid}}}{H_2S} + \underset{\substack{\text{Lewis}\\\text{base}}}{HOCH_2CH_2NH_2} \longrightarrow HS^- + HOCH_2CH_2NH_3^+$

The H^+ ion from H_2S accepts a pair of electrons from the N atom in $HOCH_2CH_2NH_2$. **16.27** (a) left (b) left (c) right **16.29** trichloroacetic acid; an acid–base reaction normally goes in the direction of the weaker acid. **16.31** (a) H_2S; acid strength decreases with negative anion charge (b) H_2SO_3; acid strength increases with electronegativity (c) HBr; acid strength increases with electronegativity **16.33** (a) $[H_3O^+] = 2.5\ M$, $[OH^-] = 4.0 \times 10^{-15}\ M$ (b) $[H_3O^+] = 7.7 \times 10^{-14}\ M$, $[OH^-] = 0.13\ M$ **16.35** $[H_3O^+] = 1.0 \times 10^{-12}\ M$, $[OH^-] = 0.010\ M$ **16.37** (a) acidic (b) acidic **16.39** (a) basic (b) acidic (c) neutral (d) acidic **16.41** (a) 8.00 (b) 11.30 **16.43** (a) 5.72 (b) 11.92 **16.45** 11.60 **16.47** $4.3 \times 10^{-3}\ M$ **16.49** 13.16 **16.51** 5.5 to 6.5, acidic **16.53** (a) BaO is a base; $BaO + H_2O \longrightarrow Ba^{2+} + 2OH^-$ (b) H_2S is an acid; $H_2S + H_2O \longrightarrow H_3O^+ + HS^-$ (c) CH_3NH_2 is a base; $CH_3NH_2 + H_2O \longrightarrow CH_3NH_3^+ + OH^-$ (d) SO_2 is an acid; $SO_2 + 2H_2O \longrightarrow H_3O^+ + HSO_3^-$

16.55 (a) $H_2O_2(aq) + S^{2-}(aq) \longrightarrow HO_2^-(aq) + HS^-(aq)$
(b) $HCO_3^-(aq) + OH^-(aq) \longrightarrow CO_3^{2-}(aq) + H_2O(l)$
(c) $NH_4^+(aq) + CN^-(aq) \longrightarrow NH_3(aq) + HCN(aq)$
(d) $H_2PO_4^-(aq) + OH^-(aq) \longrightarrow HPO_4^{2-}(aq) + H_2O(l)$
16.57 (a) $\underset{\text{base}}{ClO^-} + \underset{\text{acid}}{H_2O} \rightleftharpoons \underset{\text{acid}}{HClO} + \underset{\text{base}}{OH^-}$

$$\left[:\overset{\cdot\cdot}{\underset{\cdot\cdot}{Cl}}-\overset{\cdot\cdot}{\underset{\cdot\cdot}{O}}:\right]^- + H-\overset{\cdot\cdot}{\underset{\cdot\cdot}{O}}-H \longrightarrow :\overset{\cdot\cdot}{\underset{\cdot\cdot}{Cl}}-\overset{\cdot\cdot}{\underset{\cdot\cdot}{O}}-H + \left[:\overset{\cdot\cdot}{\underset{\cdot\cdot}{O}}-H\right]^-$$

(b) $\underset{\text{acid}}{NH_4^+} + \underset{\text{base}}{NH_2^-} \rightleftharpoons 2NH_3$

$$\left[\overset{\displaystyle H}{\underset{\displaystyle H}{H-N-H}}\right]^+ + \left[\overset{\displaystyle H}{\underset{\displaystyle H}{:N:}}\right]^- \longrightarrow 2\ \overset{\displaystyle H}{\underset{\displaystyle H}{:N-H}}$$

16.59 H_2S, H_2Se, HBr **16.61** 2.82 **16.63** $1.4 \times 10^{-11}\ M$
16.65 (a) $H_3O^+(aq) + HCO_3^-(aq) \longrightarrow CO_2(g) + 2H_2O(l)$
$H_2SO_4(aq) + 2NaHCO_3(aq) \longrightarrow$
$\qquad\qquad Na_2SO_4(aq) + CO_2(g) + H_2O(l)$
(b) 0.0180 mol (c) 60.5% $NaHCO_3$
16.67 $HCO_3^-(aq) + H_2O(l) \rightleftharpoons H_3O^+(aq) + CO_3^{2-}(aq)$
$HCO_3^-(aq) + H_2O(l) \rightleftharpoons H_2CO_3(aq) + OH^-(aq)$
$HCO_3^-(aq) + \cancel{Na^+(aq)} + OH^-(aq) \rightleftharpoons$
$\qquad\qquad \cancel{Na^+(aq)} + CO_3^{2-}(aq) + H_2O$
$HCO_3^-(aq) + H^+(aq) + \cancel{Cl^-(aq)} \rightleftharpoons H_2O + CO_2 + \cancel{Cl^-(aq)}$
16.69 $NH_3(aq) + H_2O(l) \rightleftharpoons NH_4^+(aq) + OH^-(aq)$, 11.62
16.71 $BF_3 + :NH_3 \longrightarrow F_3B:NH_3$; BF_3 is the Lewis acid, and NH_3 is the Lewis base; 12.5 g

Chapter 17

17.13 HX, HZ **17.15** HBr, NH_4Cl, KBr, CH_3NH_2
17.17 NH_3, NaF, NaCl, NH_4Br **17.19** (a) $HOCN(aq) + H_2O(l) \rightleftharpoons OCN^-(aq) + H_3O^+(aq)$
(b) $HF(aq) + H_2O(l) \rightleftharpoons H_3O^+(aq) + F^-(aq)$
(c) $HIO(aq) + H_2O(l) \rightleftharpoons H_3O^+(aq) + IO^-(aq)$
(d) $HCO_2H(aq) + H_2O(l) \rightleftharpoons CO_2H^-(aq) + H_3O^+(aq)$
17.21 5.6×10^{-5} **17.23** 5.42, 1.5×10^{-4} **17.25** 0.26 M
17.27 $4.9 \times 10^{-3}\ M$, 2.31 **17.29** (a) $3.7 \times 10^{-3}\ M$
(b) $3.9 \times 10^{-6}\ M$ **17.31** $CH_3NH_2(aq) + H_2O(l) \rightleftharpoons$
$\qquad\qquad\qquad CH_3NH_3^+(aq) + OH^-(aq)$

$K_b = \dfrac{[CH_3NH_3^+][OH^-]}{[CH_3NH_2]}$ **17.33** 3.2×10^{-5}

17.35 $5.1 \times 10^{-3}\ M$, 11.71 **17.37** (a) No hydrolysis
(b) $CHO_2^-(aq) + H_2O(l) \rightleftharpoons HCHO_2(aq) + OH^-(aq)$

$K_b = \dfrac{[HCHO_2][OH^-]}{[CHO_2^-]}$

(c) $CH_3NH_3^+(aq) + H_2O(l) \rightleftharpoons H_3O^+(aq) + CH_3NH_2(aq)$

$K_a = \dfrac{[H_3O^+][CH_3NH_2]}{[CH_3NH_3^+]}$

(d) $IO^-(aq) + H_2O(l) \rightleftharpoons HIO(aq) + OH^-(aq)$

$K_b = \dfrac{[HIO][OH^-]}{[IO^-]}$

17.39 $Zn(H_2O)_6^{2+}(aq) + H_2O(l) \rightleftharpoons$
$$Zn(H_2O)_5(OH)^+(aq) + H_3O^+(aq)$$
17.41 (a) acidic (b) basic (c) basic (d) acidic
17.43 (a) 2.2×10^{-11} (b) 7.1×10^{-6} **17.45** 8.64, $4.4 \times 10^{-6}\ M$ **17.47** (a) 0.029 (b) 0.0065 **17.49** 3.17
17.51 3.45 **17.53** 9.26, 9.09 **17.55** 0.34 mol **17.57** 1.60
17.59 5.97 **17.61** 9.08 **17.63** $K_b(CN^-) = 2.0 \times 10^{-5}$;
$K_b(CO_3^{2-}) = 2.1 \times 10^{-4}$; CO_3^{2-} **17.65** 2.84 **17.67** 11/1
17.69 2.4 **17.71** 3.11 **17.73** (a) 2.3×10^{-11}
(b) 4.3×10^{-4} (c) 10.38 **17.75** (a) true (b) true
(c) false (d) false (e) false (f) true **17.77** (a) Initially:
pH = 11.13; 30% titration: pH = 9.62; 50% titration: pH = 9.26;
100% titration: pH = 5.28 (b) acidic **17.79** (a) H_3PO_4 and
$H_2PO_4^-$ (b) 8 mL $H_3PO_4(aq)$ and 42 mL $H_2PO_4^-(aq)$
17.81 (a) $0.215\ M$ (b) 3.55 (c) bromphenol blue
17.83 (a) $H_2A + 2H_2O \rightleftharpoons 2H_3O^+ + A^{2-}$ (b) $[H_2A] \gg$
$[H_3O^+] = [HA^-] \gg [A^{2-}]$ (c) pH = 2.34; $[H_2A] = 0.0205\ M$
(d) $4.6 \times 10^{-5}\ M$ **17.85** 4.1% **17.87** 9.1 mL

Chapter 18

18.7 From the K_{sp} values, these are the more soluble: (a) silver
chloride (b) magnesium hydroxide **18.9** beaker on the right
18.11 Add just enough Na_2SO_4 to precipitate all Ba^{2+}; filter off
$BaSO_4$; add more Na_2SO_4 to precipitate all Ca^{2+}; filter off
$CaSO_4$; Mg^{2+} remains in solution. **18.13** (a) $K_{sp} = [Sr^{2+}][SO_4^{2-}]$ (b) $K_{sp} = [Mg^{2+}][OH^-]^2$ (c) $K_{sp} = [Ca^{2+}]^3[PO_4^{3-}]^2$ (d) $K_{sp} = [Ag^+]^2[S^{2-}]$ **18.15** 1.2×10^{-7}
18.17 1.8×10^{-11} **18.19** 0.0045 g/L **18.21** $1.9 \times 10^{-3}\ M$
18.23 2.3×10^{-4} g/L **18.25** 1.9×10^{-5} g/L **18.27** (a) yes
(b) yes **18.29** no precipitate **18.31** 0.0018 mol
18.33 $BaF_2(s) + 2H_3O^+(aq) \longrightarrow Ba^{2+}(aq) + 2HF(aq) +$
$2H_2O(l)$ **18.35** BaF_2; F^- is the conjugate base of the weak acid
HF. **18.37** $Cu^+(aq) + 2CN^-(aq) \longrightarrow Cu(CN)_2^-(aq)$
$$K_f = \frac{[Cu(CN)_2^-]}{[Cu^+][CN^-]^2}$$ **18.39** $5.5 \times 10^{-19}\ M$ **18.41** Add HCl;
Pb^{2+} will be precipitated as the chloride. Filter off $PbCl_2$; add
H_2S in $0.3\ M\ H^+$. Cd^{2+} will precipitate as the sulfide; filter off
CdS. The filtrate contains Sr^{2+}. **18.43** $1.3 \times 10^{-4}\ M$
18.45 26 g/L **18.47** $1.8 \times 10^{-9}\ M$ **18.49** $8.0 \times 10^{-3}\ M$
18.51 5.5×10^{-6} g **18.53** $1.4 \times 10^{-2}\ M$ **18.55** $1.0 \times 10^{-5}\ M$
18.57 (a) $6.4 \times 10^{-7}\ M$ (b) $2.4 \times 10^{-4}\ M$ (c) yes
18.59 (a) $6.6 \times 10^{-4}\ M$ (b) $1.6 \times 10^{-3}\ M$, 11%
18.61 (a) 6.3×10^{-16} (b) $8.7 \times 10^{-9}\ M$ (c) common ion
effect **18.63** (a) $1.5 \times 10^{-4}\ M$ (b) $7.9 \times 10^{-4}\ M$, 99.0%
18.65 $2.7 \times 10^{-4}\ M$ **18.67** $Ba^{2+}(aq) + 2OH^-(aq) +$
$Mg^{2+}(aq) + SO_4^{2-}(aq) \longrightarrow BaSO_4(s) + Mg(OH)_2(s)$;
$[SO_4^{2-}] = 0.109\ M$; $[Ba^{2+}] = 1.0 \times 10^{-9}\ M$;
$[Mg^{2+}] = 0.109\ M$; $[OH^-] = 1.3 \times 10^{-5}\ M$

Chapter 19

19.13 (a) false (b) false (c) false (d) true (e) false
19.15 (a) 2.0 mol CO_2 (b) butane gas (c) CO_2 at $-80°C$
(d) bromine vapor **19.17** -82 J, 29 J, -53 J **19.19** 106 J/K

19.21 -188 J/K, 1.90×10^2 J/K for 1.50 mol **19.23** (a) negative (b) not predictable (c) positive **19.25** (a) -181.6 J/K
(b) 43.2 J/K (c) -56.1 J/K
19.27 (a) $K(s) + \frac{1}{2}Br_2(l) \longrightarrow KBr(s)$
(b) $C(graphite) + \frac{3}{2}H_2(g) + \frac{1}{2}Cl_2(g) \longrightarrow CH_3Cl(l)$
(c) $H_2(g) + \frac{1}{8}S_8$ (rhombic) $\longrightarrow H_2S(g)$
(d) $As(s) + \frac{3}{2}H_2(g) \longrightarrow AsH_3(g)$ **19.29** (a) -800.8 kJ
(b) -55.8 kJ **19.31** (a) spontaneous (b) nonspontaneous
(c) equilibrium mixture with significant amounts of reactants and
products **19.33** -474.4 kJ; 0 **19.35** (a) $K = \dfrac{P_{CO_2}P_{H_2}}{P_{CO}P_{H_2O}}$
(b) $K = [Li^+]^2[OH^-]^2 P_{H_2}$ **19.37** -142.2 kJ; 8.3×10^{24}
19.39 -149.92 kJ; 1.84×10^{26} **19.41** 1.2×10^2; yes
19.43 383 K **19.45** Reaction is spontaneous because $\Delta S°$ is
large and positive. **19.47** -14.3 J/K **19.49** (a) negative
(b) positive (c) positive **19.51** -136 J/K **19.53** no, $\Delta G°$ is
$+267$ kJ **19.55** $\Delta H°$ is positive; $\Delta S°$ is positive **19.57** 56 kJ;
2×10^{-10} **19.59** 1.10×10^2 kJ; 137 J/K; 69 kJ; -37 kJ; nonspontaneous at 25°C, but spontaneous at 800°C
19.61 205 J/mol · K **19.63** (a) 97.1 kJ; 116 J/K; 187.2 kJ;
208 J/K (b) 837 K; 900 K (c) carbon **19.65** (a) products
(b) Need to heat to increase rate of reaction.
19.67 (a) 5.4×10^{-3} (b) change in entropy
19.69 0.0040%, same; $K = 4.0 \times 10^{-10}$ **19.71** -2.29 kJ/mol;
-98.7 J/(mol · K)

Chapter 20

20.13 (a) No change (b) No change (c) The Zn strip
would dissolve, the blue color of the solution would fade, and
solid copper would precipitate out of the solution. (d) No
change **20.15** Because more zinc is present, the oxidation–
reduction reactions in the battery will run for a longer period of
time. This assumes that the zinc is the limiting reactant.
20.17 A stain is visible in clothing because it is absorbing only a
portion of the visible light striking the material (which means
some light is reflected off of the cloth, giving the stain color).
When you bleach the stain, the compounds that make up the stain
are oxidized. This oxidation often changes the electronic properties of the stain to make it absorb no colors in the visible region,
thus reflecting back all wavelengths of visible light, making the
stain "disappear." **20.19** (a) $Cr_2O_7^{2-} + 14H^+ + 3C_2O_4^{2-} \longrightarrow$
$$2Cr^{3+} + 6CO_2 + 7H_2O$$
(b) $3Cu + 2NO_3^- + 8H^+ \longrightarrow 3Cu^{2+} + 2NO + 4H_2O$
(c) $MnO_2 + H^+ + HNO_2 \longrightarrow Mn^{2+} + H_2O + NO_3^-$
(d) $5PbO_2 + 2Mn^{2+} + 5SO_4^{2-} + 4H^+ \longrightarrow$
$$5PbSO_4 + 2MnO_4^- + 2H_2O$$
20.21 (a) $Mn^{2+} + H_2O_2 + 2OH^- \longrightarrow MnO_2 + 2H_2O$
(b) $2MnO_4^- + 3NO_2^- + H_2O \longrightarrow 2MnO_2 + 3NO_3^- + 2OH^-$
(c) $Mn^{2+} + 2ClO_3^- \longrightarrow MnO_2 + 2ClO_2$
(d) $MnO_4^- + 3NO_2 + 2OH^- \longrightarrow MnO_2 + 3NO_3^- + H_2O$
20.23 (a) $8H_2S + 16NO_3^- + 16H^+ \longrightarrow 16NO_2 + S_8 + 16H_2O$
(b) $2NO_3^- + 3Cu + 8H^+ \longrightarrow 2NO + 3Cu^{2+} + 4H_2O$
(c) $2MnO_4^- + 5SO_2 + 2H_2O \longrightarrow 5SO_4^{2-} + 2Mn^{2+} + 4H^+$
(d) $2Bi(OH)_3 + 3Sn(OH)_3^- + 3OH^- \longrightarrow 3Sn(OH)_6^{2-} + 2Bi$

20.25 Anode: $Mg \longrightarrow Mg^{2+} + 2e^-$
Cathode: $Ni^{2+} + 2e^- \longrightarrow Ni$
Electron flow in circuit is from Mg to Ni.
20.27 Anode: $Zn(s) + 2OH^-(aq) \longrightarrow Zn(OH)_2(s) + 2e^-$
Cathode: $Ag_2O(s) + H_2O(l) + 2e^- \longrightarrow 2Ag(s) + 2OH^-(aq)$
Overall: $Zn(s) + Ag_2O(s) + H_2O(l) \longrightarrow 2Ag(s) + Zn(OH)_2(s)$
20.29 $Ni(s)|Ni^{2+}(aq)\|Pb^{2+}(aq)|Pb(s)$
20.31 $Ni(s)|Ni^{2+}(1\ M)\|H^+(1\ M)|H_2(g)|Pt$
20.33 Anode: $Cd(s) \longrightarrow Cd^{2+}(aq) + 2e^-$
Cathode: $Ni^{2+}(aq) + 2e^- \longrightarrow Ni(s)$
Overall: $Cd(s) + Ni^{2+}(aq) \longrightarrow Cd^{2+}(aq) + Ni(s)$
20.35 -69 kJ **20.37** $NO_3^-(aq)$, $O_2(g)$, $MnO_4^-(aq)$
20.39 strongest is $Zn(s)$; weakest is $Cu^+(aq)$ **20.41** (a) not
spontaneous (b) spontaneous **20.43** Cl_2 will oxidize Br^-;
$Cl_2(g) + 2Br^-(aq) \longrightarrow Br_2(l) + 2Cl^-(aq)$ **20.45** 1.54 V
20.47 -3.6×10^5 J **20.49** 3.56 V **20.51** 3.4×10^{14}
20.53 0.57 V **20.55** $4 \times 10^{-3}\ M$
20.57 (a) Anode: $2H_2O(l) \longrightarrow O_2(g) + 4H^+(aq) + 4e^-$;
cathode: $2H_2O(l) + 2e^- \longrightarrow H_2(g) + 2OH^-(aq)$;
overall: $2H_2O(l) \longrightarrow 2H_2(g) + O_2(g)$
(b) Anode: $2Br^-(aq) \longrightarrow Br_2(l) + 2e^-$;
cathode: $2H_2O(l) + 2e^- \longrightarrow H_2(g) + 2OH^-(aq)$;
overall: $Br^-(aq) + 2H_2O(l) \longrightarrow Br_2(l) + H_2(g) + 2OH^-(aq)$
20.59 3.87×10^7 C **20.61** 0.360 g
20.63 (a) $16MnO_4^- + 24S^{2-} + 32H_2O \longrightarrow$
$$16MnO_2 + 3S_8 + 64OH^-$$
(b) $IO_3^- + 3HSO_3^- \longrightarrow I^- + 3SO_4^{2-} + 3H^+$
(c) $3Fe(OH)_2 + CrO_4^{2-} + 4H_2O \longrightarrow$
$$3Fe(OH)_3 + Cr(OH)_4^- + OH^-$$
20.65 $4Fe(OH)_2 + O_2 + 2H_2O \longrightarrow 4Fe(OH)_3$
20.67 $Ca(s)|Ca^{2+}(aq)\|Cl^-(aq)|Cl_2(g)|Pt(s)$; anode: $Ca(s) \longrightarrow$
$Ca^{2+}(aq) + 2e^-$; cathode: $Cl_2(g) + 2e^- \longrightarrow 2Cl^-(aq)$; 4.12 V
20.69 (a) spontaneous (b) Fe^{3+} will oxidize Sn^{2+} to Sn^{4+}.
20.71 (a) 2.2 (b) $0.3\ M$ **20.73** (a) 1.0 F, 9.6×10^4 C
(b) 2.0 F, 1.9×10^5 C (c) 0.11 F, 1.1×10^{-4} C
20.75 (a) $Cd(s) + 2Ag^+(aq) \rightleftharpoons 2Ag(s) + Cd^{2+}(aq)$; 1.20 V
(b) It would increase. (c) no effect
20.77 Anode: $2H_2O(l) \longrightarrow O_2(g) + 4H^+(aq) + 4e$;
cathode: $Cu^{2+}(aq) + 2e \longrightarrow Cu(s)$ **20.79** (a) 1.48×10^3 s
(b) 0.399 g **20.81** (a) 1.06 V (b) $8.8 \times 10^{-3}\ M$ **20.83**
(a) -26.2 kJ (b) -0.26 V for $Co^{2+} + 2e \longrightarrow Co(s)$
20.85 11.7 kJ/K **20.87** -0.05 V; not spontaneous
20.89 -0.04 V; not spontaneous

Chapter 21

21.13 (a) 0.1% (b) If you had a large quantity of material,
0.1% still would be a significant quantity. Also, if the material
were particularly toxic in addition to being radioactive.
21.15 Some of the expected mass is in the form of energy: the
mass defect. **21.17** The large, positively charged He nucleus
that makes up alpha (α) radiation is unable to pass through the
atoms that make up solid materials such as wood without coming
into contact or being deflected by the nuclei. Gamma (γ) radiation, however, with its small wavelength and high energy, can

pass through large amounts of material without interaction, just
like x rays can pass through skin and other soft tissue.
21.19 $^{87}_{37}Rb \longrightarrow\ ^{87}_{38}Sr +\ ^{0}_{-1}e$ **21.21** $^{232}_{90}Th \longrightarrow\ ^{4}_{2}He +\ ^{228}_{88}Ra$
21.23 $^{18}_{9}F \longrightarrow\ ^{18}_{8}O +\ ^{0}_{1}e$ **21.25** (a) $^{122}_{51}Sb$ (b) $^{204}_{85}At$
(c) $^{80}_{37}Rb$ **21.27** (a) α emission (b) positron emission or
electron capture (c) β emission **21.29** 1.38×10^9 kJ/mol
21.31 ^4_2He **21.33** plutonium-239 **21.35** $1.79 \times 10^{-9}\ s^{-1}$
21.37 1.3×10^{13} years **21.39** 2.1×10^2 Ci
21.41 5.7×10^{-4} g **21.43** 0.574, 2.9 μg **21.45** 4.16 s
21.47 5.3×10^3 y **21.49** -1.688×10^{12} J; -17.50 MeV
21.51 0.032715 amu; 30.0489 MeV; 5.082 MeV **21.53** Na-20
is expected to decay by electron capture or positron emission.
Na-26 is expected to decay by beta emission.
21.55 $^{209}_{83}Bi +\ ^{4}_{2}He \longrightarrow\ ^{211}_{85}At + 2\ ^{1}_{0}n$ **21.57** $^{246}_{98}Cf$
21.59 5.8 y **21.61** -3.61×10^{14} J; 9.84×10^4 kJ
21.63 (a) $^{47}_{20}Ca \longrightarrow\ ^{47}_{21}Sc +\ ^{0}_{-1}\beta$ (b) 46.1 μg
21.65 1.75×10^{13} nuclei/s **21.67** 2.63×10^9 kJ; 4.73×10^7 L

Chapter 22

22.31 It would react with the oxygen by the following equation:
$C(s) + O_2(g) \longrightarrow CO_2(g)$. **22.33** This means that aluminum
hydroxide reacts with both acids and bases. For example,
$Al(OH)_3(s) + 3HCl(aq) \longrightarrow 3H_2O(l) + AlCl_3(aq)$
$Al(OH)_3(s) + NaOH(aq) \longrightarrow Na^+(aq) + Al(OH)_4^-(aq)$
22.35 $CO_2(g) + NH_3(g) + NaCl(aq) + H_2O(l) \longrightarrow$
$$NaHCO_3(s) + NH_4Cl(aq)$$
$2NaHCO_3(s) \xrightarrow{\Delta} Na_2CO_3(s) + CO_2(g) + H_2O(g)$
$Ca(OH)_2(aq) + Na_2CO_3(aq) \longrightarrow 2NaOH(aq) + CaCO_3(s)$
22.37 (a) $2K(s) + Br_2(l) \longrightarrow 2KBr(s)$
(b) $2K(s) + 2H_2O(l) \longrightarrow 2KOH(aq) + H_2(g)$
(c) $2NaOH(s) + CO_2(g) \longrightarrow Na_2CO_3(s) + H_2O(l)$
(d) $Li_2CO_3(aq) + 2HNO_3(aq) \longrightarrow$
$$2LiNO_3(aq) + H_2O(l) + CO_2(g)$$
(e) $K_2SO_4(aq) + Pb(NO_3)_2(aq) \longrightarrow PbSO_4(s) + 2KNO_3(aq)$
22.39 Add CO_3^{2-} to each solution. Calcium carbonate will
precipitate from the calcium hydroxide solution.
22.41 (a) $BaCO_3(s) \xrightarrow{\Delta} BaO(s) + CO_2(g)$
(b) $Ba(s) + 2H_2O(l) \longrightarrow Ba(OH)_2(aq) + H_2(g)$
(c) $Mg(OH)_2(s) + 2HNO_3(aq) \longrightarrow Mg(NO_3)_2(aq) + 2H_2O(l)$
(d) $Mg(s) + NiCl_2(aq) \longrightarrow MgCl_2(aq) + Ni(s)$
(e) $2NaOH(aq) + MgSO_4(aq) \longrightarrow Mg(OH)_2(s) + Na_2SO_4(aq)$
22.43 $Al(H_2O)_6^{3+}(aq) + HCO_3^-(aq) \longrightarrow$
$$Al(H_2O)_5OH^{2+}(aq) + H_2O(l) + CO_2(g)$$
22.45 $PbO_2 + 4HCl \longrightarrow PbCl_2 + Cl_2 + 2H_2O$
22.47 (a) $Al_2O_3(s) + 3H_2SO_4(aq) \longrightarrow$
$$Al_2(SO_4)_3(aq) + 3H_2O(l)$$
(b) $Al(s) + 3AgNO_3(aq) \longrightarrow Al(NO_3)_3(aq) + 3Ag(s)$
(c) $Pb(NO_3)_2(aq) + 2NaI(aq) \longrightarrow PbI_2(s) + 2NaNO_3(aq)$
(d) $8Al(s) + 3Mn_3O_4(s) \longrightarrow 4Al_2O_3(s) + 9Mn(s)$
22.49 -5.8×10^9 kJ **22.51** (a) -1 (b) $+1$ (c) -4
(d) $+6$ **22.53** (a) Each C atom has sp^3 hybridization and
forms four σ bonds, one bond to each of three H atoms, and one
bond to the other C atom. (b) The double bond between two of

the C atoms consists of a σ bond and a π bond. The σ bond is formed by the overlap of an sp^2 hybrid orbital from each of the two C atoms. The π bond is formed by the overlap of the unhybridized $2p$ orbital from each of the two C atoms. The single bond between the center C atom and the other C atom is a σ bond formed by the overlap of an sp^2 hybrid orbital from the center C atom with an sp^3 hybrid orbital from the outer C atom.

22.55 (a) $CO_2(g) + Ba(OH)_2(aq) \longrightarrow BaCO_3(s) + H_2O(l)$
(b) $MgCO_3(s) + 2HBr(aq) \longrightarrow MgBr_2(aq) + H_2O(l) + CO_2(g)$
22.57 2.53 g NH_3
22.59 $4Zn + 10H^+ + NO_3^- \longrightarrow 4Zn^{2+} + NH_4^+ + 3H_2O$
22.61 Tetrahedral. Each bond is formed by the overlap of a $4p$ orbital from Br with an sp^3 orbital from P.
22.63 (a) $4Li + O_2 \longrightarrow 2Li_2O$
(b) $4CH_3NH_2 + 9O_2 \longrightarrow 4CO_2 + 10H_2O + 2N_2$
22.65 (a) +6 (b) +6 (c) −2 (d) +4
22.67 $Ba(ClO_3)_2(aq) + H_2SO_4(aq) \longrightarrow$
$$2HClO_3(aq) + BaSO_4(s)$$
22.69 (a) sp^3; bent (b) sp^3; trigonal pyramidal (c) sp^3d; T-shaped **22.71** (a) $Br_2(l) + 2NaOH(aq) \longrightarrow$
$$BrO^-(aq) + Br^-(aq) + H_2O(l) + 2Na^+(aq)$$
(b) $NaBr(s) + H_3PO_4(aq) \xrightarrow{\Delta}$
$$HBr(g) + Na^+(aq) + H_2PO_4^-(aq)$$

22.73 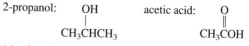 sp^3d^2, square planar

22.75 8.184% **22.77** 84.5% **22.79** 7.73×10^{-2} g
22.81 70.0% **22.83** 12.6 hr

Chapter 23

23.25 (a) +2 (b) +4 (c) +1 (d) +6
23.27 $3Fe^{2+} + NO_3^- + 4H^+ \longrightarrow 3Fe^{3+} + NO + 2H_2O$
23.29 (a) 4 (b) 6 (c) 4 (d) 6 **23.31** (a) +2 (b) +3
(c) +3 (d) +3
23.33 (a) +3 (b) *Name* *Formula* (c) 6

Name	Formula
ammine	NH_3
chloro	Cl^-
oxalato	$C_2O_4^{2-}$

23.35 (a) potassium hexafluoroferrate(III)
(b) diamminediaquacopper(II) ion
(c) ammonium aquapentafluoroferrate(III)
(d) dicyanoargentate(I) ion
23.37 (a) $K_3[Mn(CN)_6]$ (b) $[Co(NH_3)_4Cl_2]NO_3$
(c) $[Cr(NH_3)_6]_2[CuCl_4]_3$

23.39 (a)

cis   trans

(b) no geometric isomerism (c) no geometric isomerism

(d)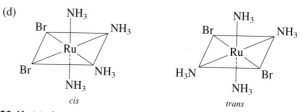

cis   trans

23.41 (a) 2 unpaired electrons (b) 3 unpaired electrons
(c) 2 unpaired electrons **23.43** purple **23.45** Yes, H_2O is a weaker bonding ligand. **23.47** no d electrons **23.49** 2
23.51 $[Cu^{2+}] = 6.1 \times 10^{-4}$ M; $[NH_3] = 2.4 \times 10^{-3}$ M; $[Cu(NH_3)_4^{2+}] = 0.10$ M

Chapter 24

24.13 This compound would violate the octet rule.
24.15 (a) trimethylamine, C_3H_9N; acetaldehyde, C_2H_4O; 2-propanol, C_3H_8O; acetic acid, $C_2H_4O_2$
(b) trimethylamine: acetaldehyde:

2-propanol: acetic acid:

(c) trimethylamine: amine (tertiary); acetaldehyde: aldehyde; 2-propanol: alcohol; acetic acid: carboxylic acid
24.17 The molecules increase regularly in molecular weight. Therefore, you expect their intermolecular forces, and thus their melting points, to increase. **24.19** $CH_3CH_2CH_2CH_3$

24.21 (a)

cis-3-hexene

trans-3-hexene

(b)

cis-3-methyl-3-hexene

trans-3-methyl-3-hexene

24.23 (a) $C_2H_4 + 3O_2 \longrightarrow 2CO_2 + 2H_2O$
(b) $3CH_2{=}CH_2 + 2MnO_4^- + 4H_2O \longrightarrow$
$$3CH_2{-}CH_2 + 2MnO_2 + 2OH^-$$
(c) $CH_2{=}CH_2 + Br_2 \longrightarrow CH_2{-}CH_2$

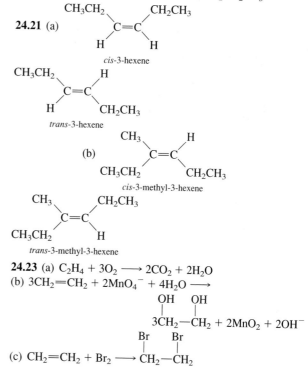

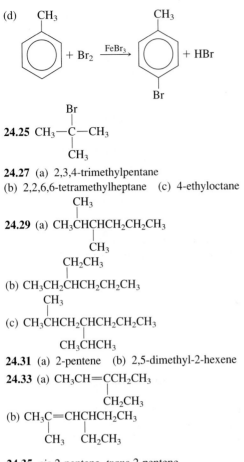

(d)

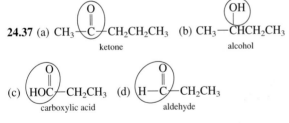

24.25 $CH_3-\overset{\overset{\displaystyle Br}{|}}{\underset{\underset{\displaystyle CH_3}{|}}{C}}-CH_3$

24.27 (a) 2,3,4-trimethylpentane
(b) 2,2,6,6-tetramethylheptane   (c) 4-ethyloctane

24.29 (a) $CH_3\overset{\overset{\displaystyle CH_3}{|}}{CH}CHCH_2CH_2CH_3$

$\overset{\overset{\displaystyle CH_3}{|}}{}$

(b) $CH_3CH_2\overset{\overset{\displaystyle CH_2CH_3}{|}}{CH}CH_2CH_2CH_3$

$\overset{\overset{\displaystyle CH_3}{|}}{}$

(c) $CH_3\overset{\overset{\displaystyle CH_3}{|}}{CH}CH_2\overset{}{CH}CH_2CH_2CH_3$

$\underset{\underset{\displaystyle CH_3CHCH_3}{|}}{}$

24.31 (a) 2-pentene   (b) 2,5-dimethyl-2-hexene

24.33 (a) $CH_3CH{=}CCH_2CH_3$

$\underset{\underset{\displaystyle CH_2CH_3}{|}}{}$

(b) $CH_3C{=}CHCHCH_2CH_3$

$\underset{\underset{\displaystyle CH_3 \quad CH_2CH_3}{| \qquad |}}{}$

24.35 *cis*-2-pentene, *trans*-2-pentene

24.37 (a) $CH_3-\overset{\overset{\displaystyle O}{\|}}{C}-CH_2CH_2CH_3$ (b) $CH_3-\overset{\overset{\displaystyle OH}{|}}{CH}CH_2CH_3$
   ketone   alcohol

(c) $HO\overset{\overset{\displaystyle O}{\|}}{C}-CH_2CH_3$ (d) $H-\overset{\overset{\displaystyle O}{\|}}{C}-CH_2CH_3$
   carboxylic acid   aldehyde

24.39 (a) 1-pentanol   (b) 2-pentanol   (c) 2-propyl-1-pentanol   24.41 (a) secondary alcohol   (b) secondary alcohol   (c) primary alcohol   (d) primary alcohol   24.43 (a) butanone   (b) butanal   (c) 4,4-dimethylpentanal   (d) 3-methyl-2-pentanone   24.45 (a) secondary amine   (b) secondary amine   24.47 (a) 3-methylbutanoic acid   (b) *trans*-5-methyl-2-hexene   (c) 2,5-dimethyl-4-heptanone   (d) 4-methyl-2-pentyne

24.49 (a) $CH_3CH_2\overset{\overset{\displaystyle O}{\|}}{C}-O-\overset{\overset{\displaystyle CH_3}{|}}{\underset{\underset{\displaystyle CH_3}{|}}{CH}}$ (b) $CH_3-\overset{\overset{\displaystyle CH_3}{|}}{\underset{\underset{\displaystyle CH_3}{|}}{C}}-NH_2$

(c) $CH_3CH_2CH_2CH_2\overset{\overset{\displaystyle CH_3}{|}}{\underset{\underset{\displaystyle CH_3}{|}}{C}}COOH$

(d) $\underset{\underset{\displaystyle H \qquad\qquad H}{|\qquad\qquad\;|}}{\overset{\overset{\displaystyle CH_3CH_2 \qquad CH_2CH_3}{\diagdown \qquad\qquad \diagup}}{C}{=}C}$

24.51 (a) Addition of dichromate ion in acidic solution to propionaldehyde will cause the reagent to change from orange to green as the aldehyde is oxidized. Under similar conditions, acetone would not react.   (b) Add Br_2 in CCl_4 to each compound. First compound reacts; the color of Br_2 fades. Benzene does not react.   24.53 (a) ethylene   (b) toluene   (c) methylamine   (d) methanol   24.55 $nCF_2{=}CF_2 \longrightarrow$ $-CF_2-CF_2-CF_2-CF_2-CF_2-CF_2-$
24.57 $HOCH_2CH_2OH$ and $HOOCCH_2CH_2COOH$
24.59 1-butene or 2-butene

Glossary

The number given in blue at the end of a definition indicates the section (or page) where the term was introduced. In a few cases, several sections are indicated.

Absolute entropy see *Standard entropy.*

Absolute temperature scale a temperature scale in which the lowest temperature that can be attained theoretically is zero. (1.6)

Accuracy the closeness of a single measurement to its true value. (1.5)

Acid (Arrhenius definition) a substance that produces hydrogen ions, H^+, (hydronium ion, H_3O^+) when it dissolves in water. (4.4 and 16.1)   (Brønsted–Lowry definition) the species (molecule or ion) that donates a proton to another species in a proton-transfer reaction. (4.4 and 16.2)

Acid–base titration curve a plot of the pH of a solution of acid (or base) against the volume of added base (or acid). (17.7)

Acid-ionization constant (K_a) the equilibrium constant for the ionization of a weak acid. (17.1)

Acid rain rain having a pH lower than that of natural rain, which has a pH of 5.6. (p. 543)

Acid salt a salt that has an acidic hydrogen atom and can undergo neutralization with bases. (4.4)

Actinides elements in the last of the two rows at the bottom of the periodic table; the 14 elements following actinium in the periodic table, in which the $5f$ subshell is filling. (p. 725)

Activated complex (transition state) an unstable grouping of atoms that can break up to form products. (14.5)

Activation energy (E_a) the minimum energy of collision required for two molecules to react. (14.5)

Activity of a radioactive source the number of nuclear disintegrations per unit time occurring in a radioactive material. (21.3)

Activity series a listing of the elements in order of their ease of losing electrons during reactions in aqueous solution. (4.5)

Addition polymer a polymer formed by linking together many molecules by addition reactions. (24.3)

Addition reaction a reaction in which parts of a reactant are added to each carbon atom of a carbon–carbon double bond, which then becomes a C—C single bond. (24.3)

Adsorption the binding or attraction of molecules to a surface. (14.9)

Aerosol a colloid consisting of liquid droplets or solid particles dispersed throughout a gas. (12.9)

Alcohol a compound obtained by substituting a hydroxyl group (—OH) for an —H atom on a tetrahedral (sp^3 hybridized) carbon atom of a hydrocarbon group. (24.6)

Aldehyde a compound containing a carbonyl group with at least one H atom attached to it. (24.6)

Alkaline dry cell a voltaic cell that is similar to the Leclanché dry cell but uses potassium hydroxide in place of ammonium chloride. (20.8)

Alkane an acyclic saturated hydrocarbon; a saturated hydrocarbon with the general formula C_nH_{2n+2}. (24.2)

Alkene a hydrocarbon that has the general formula C_nH_{2n} and contains a carbon–carbon double bond. (24.3)

Alkyl group an alkane less one hydrogen atom. (24.5)

Alkyne an unsaturated hydrocarbon containing a carbon–carbon triple bond. The general formula is C_nH_{2n-2}. (24.3)

Allotrope one of two or more distinct forms of an element in the same physical state. (6.8)

Alpha emission emission of a ^4_2He nucleus, or alpha particle, from an unstable nucleus. (21.1)

Aluminosilicate mineral a mineral consisting of silicate sheets or three-dimensional networks in which some of the SiO_4 tetrahedra of a silicate structure have been replaced by AlO_4 tetrahedra. (13.6)

Amide a compound derived from the reaction of ammonia, or a primary or secondary amine, with a carboxylic acid. (24.7)

Amine a compound that is structurally derived by replacing one or more hydrogen atoms of ammonia with hydrocarbon groups. (24.7)

Amorphous solid a solid that has a disordered structure; it lacks the well-defined arrangement of basic units (atoms, molecules, or ions) found in a crystal. (11.6)

Ampere (A) the base unit of current in the International System (SI). (20.11)

Amphiprotic species a species that can act as either an acid or a base (that is, it can lose or gain a proton). (16.2)

Amphoteric hydroxide a metal hydroxide that reacts with both bases and acids. (18.5)

Amphoteric oxide an oxide that has both acidic and basic properties. (8.7)

Angular momentum quantum number (l) also known as the azimuthal quantum number. The quantum number that distinguishes orbitals of given n having different shapes; it can have any integer value from 0 to $n - 1$. (7.5)

Anion a negatively charged ion. (2.6)

Anode the electrode at which oxidation occurs. (**20.2**)

Antibonding orbitals molecular orbitals that are concentrated in regions other than between two nuclei. (**10.5**)

Aromatic hydrocarbon a hydrocarbon that contains a benzene ring or similar structural feature. (**24.1**)

Arrhenius equation the mathematical equation $k = Ae^{-E_a/RT}$, which expresses the dependence of the rate constant on temperature. (**14.6**)

Atmosphere (atm) a unit of pressure equal to exactly 760 mmHg; 1 atm = 101.325 kPa (exact). (**5.1**)

Atom an extremely small particle of matter that retains its identity during chemical reactions. (**2.1**)

Atomic mass unit (amu) a mass unit equal to exactly one-twelfth the mass of a carbon-12 atom. (**2.4**)

Atomic number (Z) the number of protons in the nucleus of an atom. (**2.3**)

Atomic orbital a wave function for an electron in an atom; pictured qualitatively by describing the region of space where there is a high probability of finding the electron. (**7.5**)

Atomic weight the average atomic mass for the naturally occurring element, expressed in atomic mass units. (**2.4**)

Avogadro's law a law stating that equal volumes of any two gases at the same temperature and pressure contain the same number of molecules. (**5.2**)

Avogadro's number (N_A) the number of atoms in a 12-g sample of carbon-12, equal to 6.02×10^{23} to three significant figures. (**3.2**)

Band of stability the region in which stable nuclides lie in a plot of number of protons against number of neutrons. (**21.1**)

Band theory molecular orbital theory of metals. (**13.3**)

Base (Arrhenius definition) a substance that produces hydroxide ions, OH^-, when it dissolves in water. (**4.4 and 16.1**) (Brønsted–Lowry definition) the species (molecule or ion) that accepts a proton in a proton-transfer reaction. (**4.4 and 16.2**)

Base-ionization (or base-dissociation) constant (K_b) the equilibrium constant for the ionization of a weak base. (**17.3**)

Beta emission emission of a high-speed electron from an unstable nucleus. (**21.1**)

Bidentate ligand a ligand that bonds to a metal atom through two atoms of the ligand. (**23.3**)

Bimolecular reaction an elementary reaction that involves two reactant molecules. (**14.7**)

Binding energy (of a nucleus) the energy needed to break a nucleus into its individual protons and neutrons. (**21.6**)

Body-centered cubic unit cell a cubic unit cell in which there is a lattice point at the center of the unit cell as well as at the corners. (**11.6**)

Boiling point the temperature at which the vapor pressure of a liquid equals the pressure exerted on the liquid. (**11.2**)

Boiling-point elevation a colligative property of a solution equal to the boiling point of the solution minus the boiling point of the pure solvent. (**12.6**)

Bond energy the average enthalpy change for the breaking of a bond in a molecule in the gas phase. (**9.11**)

Bonding orbitals molecular orbitals that are concentrated in regions between nuclei. (**10.5**)

Bond length (bond distance) the distance between the nuclei in a bond. (**9.10**)

Bond order in a Lewis formula, the number of pairs of electrons in a bond. (**9.10**)   In molecular orbital theory, one-half the difference between the number of bonding electrons and the number of antibonding electrons. (**10.5**)

Boyle's law the volume of a sample of gas at a given temperature varies inversely with the applied pressure. (**5.2**)

Brønsted–Lowry concept a concept of acids and bases in which an acid is the species donating a proton in a proton-transfer reaction, whereas a base is the species accepting a proton in such a reaction. (**4.4 and 16.2**)

Buffer a solution characterized by the ability to resist changes in pH when limited amounts of acid or base are added to it. (**17.6**)

Building-up principle (Aufbau principle) a scheme used to reproduce the electron configurations of the ground states of atoms by successively filling subshells with electrons in a specific order (the building-up order). (**8.2**)

Calorimeter a device used to measure the heat absorbed or evolved during a physical or chemical change. (**6.6**)

Carboxylic acid a compound containing the carboxyl group, —COOH. (**24.6**)

Catalyst a substance that increases the rate of reaction without being consumed in the overall reaction. (**2.9 and p. 436**)

Cathode the electrode at which reduction occurs. (**20.2**)

Cation a positively charged ion. (**2.6**)

Charles's law the volume occupied by any sample of gas at a constant pressure is directly proportional to the absolute temperature. (**5.2**)

Chemical formula a notation that uses atomic symbols with numerical subscripts to convey the relative proportions of atoms of the different elements in a substance. (**2.6**)

Chemical kinetics the study of how reaction rates change under varying conditions and of what molecular events occur during the overall reaction. (**p. 435**)

Chemical property a characteristic of a material involving its chemical change. (**1.4**)

Chemical reaction (chemical change) a change in which one or more kinds of matter are transformed into a new kind of matter or several new kinds of matter. (**1.4 and 2.1**)

Clausius–Clapeyron equation an equation that expresses the relation between the vapor pressure P of a liquid and the absolute temperature T: $\ln P = -\Delta H_{vap}/RT + B$, where B is a constant. (11.2)

Colligative properties properties that depend on the concentration of solute molecules or ions in a solution but not on the chemical identity of the solute. (p. 384)

Collision theory a theory that assumes that, in order for reaction to occur, reactant molecules must collide with an energy greater than some minimum value and with proper orientation. (14.5)

Colloid a dispersion of particles of one substance (the dispersed phase) throughout another substance or solution (the continuous phase). (12.9)

Combination reaction a reaction in which two substances combine to form a third substance. (4.5)

Combustion reaction a reaction of a substance with oxygen, usually with the rapid release of heat to produce a flame. (4.5)

Common-ion effect the shift in an ionic equilibrium caused by the addition of a solute that provides an ion that takes part in the equilibrium. (17.5 and 18.2)

Complex ion an ion formed from a metal ion with a Lewis base attached to it by a coordinate covalent bond. (p. 580 and 23.3)

Compound a substance composed of two or more elements chemically combined. (1.4 and 2.1)

Concentration a general term referring to the quantity of solute in a standard quantity of solvent or solution. (4.7 and 12.4)

Condensation polymer a polymer formed by linking many molecules together by condensation reactions. (24.6)

Condensation reaction a reaction in which two molecules or ions are chemically joined by the elimination of a small molecule such as water. (13.6 and 24.6)

Conjugate acid in a conjugate acid–base pair, the species that can donate a proton. (16.2)

Conjugate acid–base pair two species in an acid–base reaction, one acid and one base, that differ by the loss or gain of a proton. (16.2)

Conjugate base in a conjugate acid–base pair, the species that can accept a proton. (16.2)

Constitutional (structural) isomers isomers that differ in how the atoms are joined together (23.5); compounds with the same molecular formula but different structural formulas. (24.2)

Contact process an industrial method for the manufacture of sulfuric acid. (22.8)

Continuous spectrum a spectrum containing light of all wavelengths. (7.3)

Conversion factor a factor equal to 1 that converts a quantity expressed in one unit to a quantity expressed in another unit. (1.8)

Coordinate covalent bond a bond formed when both electrons of the bond are donated by one atom. (9.4)

Coordination compound a compound consisting of complex ions. (23.3)

Coordination number in a crystal, the number of nearest-neighbor atoms of an atom. (11.7)   In a complex, the total number of bonds the metal atom forms with ligands. (23.3)

Covalent bond a chemical bond formed by the sharing of a pair of electrons between atoms. (p. 268)

Covalent network solid a solid that consists of atoms held together in large networks or chains by covalent bonds. (11.5)

Covalent radii values assigned to atoms in such a way that the sum of covalent radii of atoms A and B predicts an approximate A—B bond length. (9.10)

Critical mass the smallest mass of fissionable material in which a chain reaction can be sustained. (21.7)

Critical pressure the vapor pressure at the critical temperature; it is the minimum pressure that must be applied to a gas at the critical temperature to liquefy it. (11.3)

Critical temperature the temperature above which the liquid state of a substance no longer exists regardless of the pressure. (11.3)

Crystal a kind of solid having a regular three-dimensional arrangement of atoms, molecules, or ions. (2.6)

Crystal field splitting (Δ) the difference in energy between the two sets of d orbitals on a central metal ion that arises from the interaction of the orbitals with the electric field of the ligands. (23.7)

Crystal field theory a model of the electronic structure of transition-metal complexes that considers how the energies of the d orbitals of a metal ion are affected by the electric field of the ligands. (23.7)

Cubic close-packed structure (ccp) a crystal structure composed of close-packed atoms (or other units) with the stacking ABCABCABCA $\cdots$. (11.7)

Curie (Ci) a unit of activity, equal to 3.700×10^{10} disintegrations per second. (21.3)

Cycloalkane a cyclic saturated hydrocarbon; the general formula is C_nH_{2n}. (24.2)

Dalton's law of partial pressures the sum of the partial pressures of all the different gases in a mixture is equal to the total pressure of the mixture. (5.5)

d-Block transition elements those transition elements with an unfilled d subshell in common oxidation states. (8.2)

de Broglie relation the equation $\lambda = h/mv$. (7.4)

Decomposition reaction a reaction in which a single compound reacts to give two or more substances. (4.5)

Delocalized bonding a type of bonding in which a bonding pair of electrons is spread over a number of atoms rather than localized between two. (9.7)

Density the mass per unit volume of a substance or solution. (1.7)

Diamagnetic substance a substance that is not attracted by a magnetic field or is very slightly repelled by such a field. (8.4)

Diffusion the process whereby a gas spreads out through another gas to occupy the space uniformly. (5.7)

Dipole–dipole force an attractive intermolecular force. (11.4)

Dipole moment a quantitative measure of the degree of charge separation in a molecule. (10.2)

Dispersion forces see *London forces.*

Displacement reaction (single-replacement reaction) a reaction in which an element reacts with a compound, displacing an element from it. (4.5)

Double bond a covalent bond in which two pairs of electrons are shared by two atoms. (9.4)

Dry cell see *Alkaline dry cell* and *Zinc–carbon dry cell.*

Effective nuclear charge the positive charge that an electron experiences from the nucleus. (8.6)

Effusion the process in which a gas flows through a small hole in a container. (5.7)

Electrochemical cell a system consisting of electrodes that dip into an electrolyte and in which a chemical reaction either uses or generates an electric current. (p. 623)

Electrolysis the process of producing a chemical change in an electrolytic cell. (p. 641)

Electrolyte a substance, such as sodium chloride, that dissolves in water to give an electrically conducting solution. (4.1)

Electrolytic cell an electrochemical cell in which an electric current drives an otherwise nonspontaneous reaction. (p. 623)

Electromagnetic spectrum the range of frequencies or wavelengths of electromagnetic radiation. (7.1)

Electromotive force (emf) the maximum potential difference between the electrodes of a voltaic cell. (20.4)

Electron affinity the energy change for the process of adding an electron to a neutral atom in the gaseous state to form a negative ion. (8.6)

Electron capture the decay of an unstable nucleus by capturing an electron from an inner orbital of an atom. (21.1)

Electronegativity a measure of the ability of an atom in a molecule to draw bonding electrons to itself. (9.5)

Electron volt (eV) the quantity of energy that would have to be imparted to an electron (whose charge is 1.602×10^{-19} C) to accelerate it by one volt potential difference. (21.6)

Element a substance that cannot be decomposed by any chemical reaction into simpler substances. (1.4)   A type of matter composed of only one kind of atom, each atom of a given kind having the same properties. (2.1 and 2.3)

Elementary reaction a single molecular event, such as a collision of molecules, resulting in a reaction. (14.7)

Empirical formula (simplest formula) the formula of a substance written with the smallest integer (whole number) subscripts. (3.5)

Emulsion a colloid consisting of liquid droplets dispersed throughout another liquid. (12.9)

Endothermic process a chemical reaction or physical change in which heat is absorbed. (6.2)

Energy levels specific energy values in an atom. (7.3)

Enthalpy (*H*) an extensive property of a substance that can be used to obtain the heat absorbed or evolved in a chemical reaction at constant pressure. (6.3 and 19.1)

Enthalpy of formation see *Standard enthalpy of formation.*

Enthalpy of reaction (ΔH) the change in enthalpy for a reaction at a given temperature and pressure; it equals the heat of reaction at constant pressure. (6.3)

Entropy (*S*) a thermodynamic quantity that is a measure of the randomness or disorder in a system. (19.2)

Equilibrium constant K_c the value obtained for the equilibrium-constant expression when equilibrium concentrations are substituted. (15.2)

Equilibrium constant K_p an equilibrium constant for a gas reaction, similar to K_c, but in which concentrations of gases are replaced by partial pressures (in atm). (15.2)

Equilibrium-constant expression an expression obtained for a reaction by multiplying the concentrations of products, dividing by the concentrations of reactants, and raising each concentration term to a power equal to the coefficient in the chemical equation. (15.2)

Equivalence point the point in a titration when a stoichiometric amount of reactant has been added. (17.7)

Ester a compound formed from a carboxylic acid, RCOOH, and an alcohol, R'OH. (24.6)

Ether a compound formally obtained by replacing both H atoms of H_2O by hydrocarbon groups R and R'. (24.6)

Exchange (metathesis) reaction a reaction between compounds that, when written as a molecular equation, appears to involve the exchange of parts between the two reactants. (4.3)

Excited state a quantum-mechanical state of an atom or molecule associated with any energy level except the lowest, which is the ground state. (8.2)

Exclusion principle see *Pauli exclusion principle.*

Exothermic process a chemical reaction or physical change in which heat is evolved. (6.2)

Face-centered cubic unit cell a cubic unit cell in which there are lattice points at the centers of each face of the unit cell in addition to those at the corners. (11.6)

Faraday constant (*F*) the magnitude of charge on one mole of electrons, equal to 9.65×10^4 C. (20.4)

f-Block transition elements (inner-transition elements) the elements with a partially filled *f* subshell in common oxidation states. (8.2)

First ionization energy (first ionization potential) the energy needed to remove the highest-energy electron from a neutral atom in the gaseous state. (8.6)

First law of thermodynamics the change in internal energy of a system, ΔU, equals $q + w$. (19.1)

Formal charge (of an atom in a Lewis formula) the hypothetical charge you obtain by assuming that bonding electrons are equally shared between bonded atoms and that electrons of each lone pair belong completely to one atom. (9.9)

Formation constant (stability constant) of a complex ion (K_f) the equilibrium constant for the formation of a complex ion from the aqueous metal ion and the ligands. (18.5)

Formula weight (FW) the sum of the atomic weights of all atoms in a formula unit of a compound. (3.1)

Free energy (G) a thermodynamic quantity defined by the equation $G = H - TS$. (19.4)

Free energy of formation see *Standard free energy of formation.*

Freezing point the temperature at which a pure liquid changes to a crystalline solid, or freezes. (11.2)

Freezing-point depression a colligative property of a solution equal to the freezing point of the pure solvent minus the freezing point of the solution. (12.6)

Frequency (ν) the number of wavelengths of a wave that pass a fixed point in one unit of time (usually one second). (7.1)

Fuel cell essentially a battery, but it differs by operating with a continuous supply of energetic reactants, or fuel. (20.8)

Functional group a reactive portion of a molecule that undergoes predictable reactions. (p. 773)

Gamma emission emission from an excited nucleus of a gamma photon, corresponding to radiation with a wavelength of about 10^{-12} m. (21.1)

Gas the form of matter that is an easily compressible fluid; a given quantity of gas will fit into a container of almost any size and shape. (1.4)

Geometric isomers isomers in which the atoms are joined to one another in the same way but differ because some atoms occupy different relative positions in space. (10.4, 23.5, and 24.3)

Glass electrode a compact electrode used to determine pH by emf measurements. (20.7)

Goldschmidt process a method of preparing a metal by reduction of its oxide with powdered aluminum. (23.2)

Graham's law of effusion the rate of effusion of gas molecules from a particular hole is inversely proportional to the square root of the molecular weight of the gas. (5.7)

Gravimetric analysis a type of quantitative analysis in which the amount of a species in a material is determined by converting the species to a product that can be isolated completely and weighed. (4.9)

Ground state a quantum-mechanical state of an atom or molecule associated with the lowest energy level. (8.2)

Group (of the periodic table) the elements in any one column of the periodic table. (2.5)

Haber process an industrial process for the preparation of ammonia from nitrogen and hydrogen. (3.6 and 22.7)

Half-life $(t_{1/2})$ the time it takes for a sample to decrease to one-half of its initial value. (14.4 and 21.4)

Half-reaction one of two parts of an oxidation–reduction reaction, one part of which involves a loss of electrons (or increase of oxidation number) and the other a gain of electrons (or decrease of oxidation number). (4.5 and 20.1)

Halogens the Group VIIA elements; they are reactive nonmetals. (2.5 and 8.7)

Heat the energy that flows into or out of a system because of a difference in temperature between the thermodynamic system and its surroundings. (6.2)

Heat capacity (C) the quantity of heat needed to raise the temperature of a sample of substance one degree Celsius. (6.6)

Heat of formation see *Standard enthalpy of formation.*

Heat of fusion (enthalpy of fusion) the heat needed for the melting of a solid. (11.2)

Heat of reaction the heat absorbed (or evolved) during a chemical reaction. (6.2)

Heat of solution the heat absorbed (or evolved) when an ionic substance dissolves in water. (12.3)

Heat of vaporization (enthalpy of vaporization) the heat needed for the vaporization of a liquid. (11.2)

Hess's law of heat summation the enthalpy change for the overall equation equals the sum of the enthalpy changes for the individual steps. (6.7)

Heterogeneous catalysis the use of a catalyst that exists in a different phase from the reacting species. (14.9)

Heterogeneous equilibrium an equilibrium involving reactants and products in more than one phase. (15.3)

Heterogeneous mixture a mixture that consists of physically distinct parts, each with different properties. (1.4)

Hexagonal close-packed structure (hcp) a crystal structure composed of close-packed atoms (or other units) with the stacking ABABABA · · ·; the structure has a hexagonal unit cell. (11.7)

High-spin complex ion a complex ion in which there is minimum pairing of electrons in the orbitals of the metal atom. (23.7)

Homogeneous catalysis the use of a catalyst in the same phase as the reacting species. (14.9)

Homogeneous equilibrium an equilibrium that involves reactants and products in a single phase. (15.3)

Homogeneous mixture (solution) a mixture that is uniform in its properties throughout given samples. (1.4)

Homologous series a series of compounds in which one compound differs from a preceding one by a fixed group of atoms, for example, a —CH$_2$— group. (24.2)

Hund's rule the lowest-energy arrangement of electrons in a subshell is obtained by putting electrons into separate orbitals of the subshell with the same spin before pairing electrons. (8.4)

Hydrate a compound that contains water molecules weakly bound in its crystals. (2.8)

Hydration the attraction of ions for water molecules. (12.2)

Hydrocarbons compounds containing only carbon and hydrogen. (p. 775)

Hydrogen bonding an attractive force that exists between a hydrogen atom covalently bonded to a very electronegative atom and a lone pair of electrons on another small, electronegative atom. (11.4)

Hydrolysis the reaction of an ion with water to produce the conjugate acid and hydroxide ion or the conjugate base and hydrogen ion. (17.4)

Hydronium ion the H$_3$O$^+$ ion; also called the hydrogen ion and written H$^+$(aq). (4.4)

Hypothesis a tentative explanation of some regularity of nature. (1.2)

Ideal gas law the equation $PV = nRT$, which combines all of the gas laws. (5.3)

Immiscible fluids fluids that do not mix but form separate layers. (12.1)

Inner-transition elements the two rows of elements at the bottom of the periodic table (2.5); the elements with a partially filled f subshell in common oxidation states. (8.2 and p. 237)

Inorganic compounds compounds composed of elements other than carbon. A few simple compounds of carbon, including carbon monoxide, carbon dioxide, carbonates, and cyanides, are generally considered to be inorganic. (2.8)

Intermolecular forces the forces of interaction between molecules. (11.4)

Internal energy (U) the sum of the kinetic and the potential energies of the particles making up a system. (6.1 and 19.1)

International System of Units (SI) a particular choice of metric units that was adopted by the General Conference of Weights and Measures in 1960. (1.6)

Ion an electrically charged particle obtained from an atom or a chemically bonded group of atoms by adding or removing electrons. (2.6)

Ion exchange a process in which a water solution is passed through a column of a material that replaces one kind of ion in solution with another kind. (p. 364)

Ionic bond a chemical bond formed by the electrostatic attraction between positive and negative ions. (9.1)

Ionic radius a measure of the size of the spherical region around the nucleus of an ion within which the electrons are most likely to be found. (9.3)

Ionic solid a solid that consists of cations and anions held together by the electrical attraction of opposite charges. (11.5)

Ionization energy the energy needed to remove an electron from an atom (or molecule). (8.6)

Ion product (Q_c) the product of ion concentrations in a solution, each concentration raised to a power equal to the number of ions in the formula of the ionic compound. (18.3)

Ion-product constant for water (K_w) the equilibrium value of the ion product [H$_3$O$^+$][OH$^-$]. (16.6)

Isoelectronic refers to different species having the same number and configuration of electrons. (9.3)

Isomers compounds of the same molecular formula but with different arrangements of the atoms. (10.4)

Isotopes atoms whose nuclei have the same atomic number but different numbers of neutrons. (2.3)

Joule (J) the SI unit of energy; 1 J = 1 kg $\cdot$ m^2/s^2. (6.1)

Kelvin (K) the SI base unit of temperature; a unit on an absolute temperature scale. (1.6)

Ketone a compound containing a carbonyl group with two hydrocarbon groups attached to it. (24.6)

Kinetic energy the energy associated with an object by virtue of its motion. (6.1)

Kinetic-molecular theory (kinetic theory) the theory that a gas consists of molecules in constant random motion. (p. 155)

Lanthanides the 14 elements following lanthanum in the periodic table, in which the 4f subshell is filling. (p. 725)

Laser a source of intense, highly directed beam of monochromatic light; the word *laser* is an acronym meaning *l*ight *a*mplification by *s*timulated *e*mission of *r*adiation. (p. 216)

Lattice energy the change in energy that occurs when an ionic solid is separated into isolated ions in the gas phase. (9.1 and 12.2)

Law of combining volumes a relation stating that gases at the same temperature and pressure react with one another in volume ratios of small whole numbers. (5.2)

Law of conservation of energy energy may be converted from one form to another, but the total quantity of energy remains constant. (6.1)

Law of conservation of mass the total mass remains constant during a chemical change (chemical reaction). (1.3)

Law of definite proportions (law of constant composition) a pure compound, whatever its source, always contains definite or constant proportions of the elements by mass. (1.4)

Law of effusion see *Graham's law of effusion.*

Law of heat summation see *Hess's law of heat summation.*

Law of mass action the values of the equilibrium-constant expression K_c are constant for a particular reaction at a given temperature, whatever equilibrium concentrations are substituted. (15.2)

Law of multiple proportions when two elements form more than one compound, the masses of one element in these compounds for a fixed mass of the other element are in ratios of small whole numbers. (2.1)

Law of partial pressures see *Dalton's law of partial pressures.*

Le Chatelier's principle when a system in equilibrium is disturbed by a change of temperature, pressure, or concentration variable, the system shifts in equilibrium composition in a way that tends to counteract this change of variable. (12.3 and 15.7)

Lewis acid a species that can form a covalent bond by accepting an electron pair from another species. (16.3)

Lewis base a species that can form a covalent bond by donating an electron pair to another species. (16.3)

Lewis electron-dot formula a formula using dots to represent valence electrons. (9.4)

Ligand a Lewis base that bonds to a metal ion to form a complex ion. (p. 581 and 23.3)

Limiting reactant (limiting reagent) the reactant that is entirely consumed when a reaction goes to completion. (3.8)

Line spectrum a spectrum showing only certain colors or specific wavelengths of light. (7.3)

Liquefaction the process in which a substance that is normally a gas changes to the liquid state. (11.2)

Liquid the form of matter that is a relatively incompressible fluid; a liquid has a fixed volume but no fixed shape. (1.4)

London forces (dispersion forces) the weak attractive forces between molecules resulting from the small, instantaneous dipoles that occur because of the varying positions of the electrons during their motion about nuclei. (11.4)

Lone pair (nonbonding pair) an electron pair that remains on one atom and is not shared. (9.4)

Low-spin complex ion a complex ion in which there is more pairing of electrons in the orbitals of the metal atom than in a corresponding high-spin complex ion. (23.7)

Magnetic quantum number (m_l) the quantum number that distinguishes orbitals of given n and l—that is, of given energy and shape—but having a different orientation in space; the allowed values are the integers from $-l$ to $+l$. (7.5)

Main-group element (representative element) an element in an A column of the periodic table, in which an outer s or p subshell is filling. (2.5 and 8.2)

Manometer a device that measures the pressure of a gas or liquid in a sealed vessel. (5.1)

Markownikoff's rule a generalization stating that the major product formed by the addition of an unsymmetrical reagent such as H—Cl, H—Br, or H—OH is the one obtained when the H atom of the reagent adds to the carbon atom of the multiple bond that already has the greater number of hydrogen atoms attached to it. (24.3)

Mass defect the total nucleon mass minus the atomic mass of a nucleus. (21.6)

Mass number (A) the total number of protons and neutrons in a nucleus. (2.3)

Mass percentage parts per hundred parts of the total, by mass. (3.3)

Mass percentage of solute the percentage by mass of solute contained in a solution. (12.4)

Mass spectrometer an instrument that measures the mass-to-charge ratios of ions. (2.4)

Matter all of the objects around you (p. 2)   whatever occupies space and can be perceived by our senses. (1.3)

Maxwell's distribution of molecular speeds a theoretical relationship that predicts the relative number of molecules at various speeds for a sample of gas at a particular temperature. (5.7)

Melting point the temperature at which a crystalline solid changes to a liquid, or melts. (11.2)

Metal a substance or mixture that has a characteristic luster, or shine, is generally a good conductor of heat and electricity, and is malleable and ductile. (2.5 and p. 408)

Metallic solid a solid that consists of positive cores of atoms held together by a surrounding "sea" of electrons (metallic bonding). (11.5)

Metalloid (semimetal) an element having both metallic and nonmetallic properties. (2.5)

Metallurgy the scientific study of the production of metals from their ores and the making of alloys having various useful properties. (13.2)

Metathesis reaction see *Exchange reaction.*

Meter (m) the SI base unit of length. (1.6)

Micelle a colloidal-sized particle formed in water by the association of molecules or ions each of which has a hydrophobic end and a hydrophilic end. (12.9)

Mineral a naturally occurring inorganic solid substance or solid solution with definite crystalline form. (13.1)

Miscible fluids fluids that mix with or dissolve in each other in all proportions. (12.1)

Mixture a material that can be separated by physical means into two or more substances. (1.4)

Molality the moles of solute per kilogram of solvent. (12.4)

Molar gas constant (R) the constant of proportionality relating the molar volume of a gas to T/P. (5.3)

Molar gas volume the volume of one mole of gas. (5.2)

Molarity, *M* the moles of solute dissolved in one liter (cubic decimeter) of solution. (4.7)

Molar mass the mass of one mole of substance. In grams, it is numerically equal to the formula weight in atomic mass units. (3.2)

Mole (mol) the quantity of a given substance that contains as many molecules or formula units as the number of atoms in exactly 12 g of carbon-12. The amount of substance containing Avogadro's number of molecules or formula units. (3.2)

Mole fraction the fraction of moles of a component in the total moles of a mixture. (5.5)   The moles of a component substance divided by the total moles of solution. (12.4)

Molecular equation a chemical equation in which the reactants and products are written as if they were molecular substances, even though they may actually exist in solution as ions. (4.2)

Molecular formula a chemical formula that gives the exact number of different atoms of an element in a molecule. (2.6)

Molecular geometry the general shape of a molecule, as determined by the relative positions of the atomic nuclei. (p. 293)

Molecularity the number of molecules on the reactant side of an elementary reaction. (14.7)

Molecular orbital theory a theory of the electronic structure of molecules in terms of molecular orbitals, which may spread over several atoms or the entire molecule. (p. 315)

Molecular solid a solid that consists of atoms or molecules held together by intermolecular forces. (11.5)

Molecular weight (MW) the sum of the atomic weights of all the atoms in a molecule. (3.1)

Molecule a definite group of atoms that are chemically bonded together—that is, tightly connected by attractive forces. (2.6)

Monatomic ion an ion formed from a single atom. (2.8)

Monodentate ligand a ligand that bonds to a metal atom through one atom of the ligand. (23.3)

Monomer the small molecules that are linked together to form a polymer (2.6); a compound used to make a polymer (and from which the polymer's repeating unit arises). (p. 765)

Monoprotic acid an acid that yields one acidic hydrogen per molecule. (4.4)

Nernst equation an equation relating the cell emf, E_{cell}, to its standard emf, $E°_{cell}$, and the reaction quotient, Q. At 25°C, the equation is $E_{cell} = E°_{cell} - (0.0592/n)\log Q$. (20.7)

Net ionic equation an ionic equation from which spectator ions have been canceled. (4.2)

Neutralization reaction a reaction of an acid and a base that results in an ionic compound and possibly water. (4.4)

Neutron activation analysis an analysis of elements in a sample based on the conversion of stable isotopes to radioactive isotopes by bombarding a sample with neutrons. (21.5)

Noble-gas core an inner-shell configuration corresponding to one of the noble gases. (8.2)

Nonbonding pair an electron pair that remains on one atom and is not shared. (9.4)

Nonelectrolyte a substance that dissolves in water to give a nonconducting or very poorly conducting solution. (4.1)

Nonstoichiometric compound a compound whose composition varies from its idealized formula. (11.6)

Nuclear bombardment reaction a nuclear reaction in which a nucleus is bombarded, or struck, by another nucleus or by a nuclear particle. (p. 656)

Nuclear equation a symbolic representation of a nuclear reaction. (21.1)

Nuclear fission a nuclear reaction in which a heavy nucleus splits into lighter nuclei and energy is released. (21.6)

Nuclear fusion a nuclear reaction in which light nuclei combine to give a stabler, heavier nucleus plus possibly several neutrons, and energy is released. (21.6)

Nucleus the atom's central core; it has most of the atom's mass and one or more units of positive charge. (2.2)

Nuclide a particular atom characterized by a definite atomic number and mass number. (2.3)

Nuclide symbol a symbol for a nuclide, in which the atomic number is given as a left subscript and the mass number is given as a left superscript to the symbol of the element. (2.3)

Number of significant figures the number of digits reported for the value of a measured or calculated quantity, indicating the precision of the value. (1.5)

Octahedral geometry the geometry of a molecule in which six atoms occupy the vertices of a regular octahedron (a figure with eight faces and six vertices) with the central atom at the center of the octahedron. (10.1)

Octet rule the tendency of atoms in molecules to have eight electrons in their valence shells (two for hydrogen atoms). (9.4)

Orbital diagram a diagram to show how the orbitals of a subshell are occupied by electrons. (8.1)

Ore a rock or mineral from which a metal or nonmetal can be economically produced. (13.1)

Organic compounds compounds that contain carbon combined with other elements, such as hydrogen, oxygen, and nitrogen. (2.7)

Osmosis the phenomenon of solvent flow through a semipermeable membrane to equalize the solute concentrations on both sides of the membrane. (12.7)

Osmotic pressure a colligative property of a solution equal to the pressure that, when applied to the solution, just stops osmosis. (12.7)

Overall order of a reaction the sum of the orders of the reactant species in the rate law. (14.3)

Oxidation the part of an oxidation–reduction reaction in which there is a loss of electrons by a species (or an increase in the oxidation number of an atom). (4.5)

Oxidation number (oxidation state) either the actual charge on an atom in a substance, if the atom exists as a monatomic ion, or a hypothetical charge assigned by simple rules. (4.5)

Oxidation potential the negative of the standard electrode potential. (20.5)

Oxidation–reduction reaction (redox reaction) a reaction in which electrons are transferred between species or in which atoms change oxidation number. (4.5)

Oxidizing agent a species that oxidizes another species; it is itself reduced. (4.5)

Oxoacid an acid containing hydrogen, oxygen, and another element. (2.8)   A substance in which O atoms (and possibly other electronegative atoms) are bonded to a central atom, with one or more H atoms usually bonded to the O atoms. (9.6)

Pairing energy (P) the energy required to put two electrons into the same orbital. (23.7)

Paramagnetic substance a substance that is weakly attracted by a magnetic field; this attraction generally results from unpaired electrons. (8.4)

Partial pressure the pressure exerted by a particular gas in a mixture. (5.5)

Pascal (Pa) the SI unit of pressure; $1 \text{ Pa} = 1 \text{ kg/(m} \cdot \text{s}^2)$. (5.1)

Pauli exclusion principle no two electrons in an atom can have the same four quantum numbers. It follows from this that an orbital can hold no more than two electrons and can hold two only if they have different spin quantum numbers. (8.1)

Percentage composition the mass percentages of each element in a compound. (p. 76)

Percentage yield the actual yield (experimentally determined) expressed as a percentage of the theoretical yield (calculated). (3.8)

Period (of the periodic table) the elements in any one horizontal row of the periodic table. (2.5)

Periodic law when the elements are arranged by atomic number, their physical and chemical properties vary periodically. (8.6)

Periodic table a tabular arrangement of elements in rows and columns, highlighting the regular repetition of properties of the elements. (2.5)

Peroxide a compound with oxygen in the -1 oxidation state. (22.8)

pH the negative of the logarithm of the molar hydrogen-ion concentration. (16.8)

Phase diagram a graphical way to summarize the conditions under which the different states of a substance are stable. (11.3)

Photoelectric effect the ejection of electrons from the surface of a metal or other material when light shines on it. (7.2)

Photon particle of electromagnetic energy with energy E proportional to the observed frequency of light: $E = h\nu$. (7.2)

Physical change a change in the form of matter but not in its chemical identity. (1.4)

Physical property a characteristic that can be observed for a material without changing its chemical identity. (1.4)

Pi (π) bond a bond that has an electron distribution above and below the bond axis. (10.4)

Planck's constant (h) a physical constant with the value 6.63×10^{-34} J $\cdot$ s. It is the proportionality constant relating the frequency of light to the energy of a photon. (7.2)

Polar covalent bond a covalent bond in which the bonding electrons spend more time near one atom than near the other. (9.5)

Polyatomic ion an ion consisting of two or more atoms chemically bonded together and carrying a net electric charge. (2.8)

Polydentate ligand a ligand that can bond with two or more atoms to a metal atom. (23.3)

Polyester a polymer formed by reacting a substance containing two alcohol groups with a substance containing two carboxylic acid groups. (24.6)

Polymer a very large molecule made up of a number of smaller molecules repeatedly linked together. (2.6); a chemical species of very high molecular weight that is made up from many repeating units of low molecular weight. (p. 777)

Polyprotic acid an acid that yields two or more acidic hydrogens per molecule. (4.4 and 17.2)

Positron a particle that is similar to an electron and has the same mass but a positive charge. (21.1)

Positron emission emission of a positron from an unstable nucleus. (21.1)

Potential difference the difference in electrical potential (electrical pressure) between two points. (20.4)

Potential energy the energy an object has by virtue of its position in a field of force. (6.1)

Precipitate an insoluble solid compound formed during a chemical reaction in solution. (4.3)

Precision the closeness of the set of values obtained from identical measurements of a quantity. (1.5)

Pressure the force exerted per unit area of surface. (5.1)

Principal quantum number (n) the quantum number on which the energy of an electron in an atom principally depends; it can have any positive value: $1, 2, 3, \cdots$. (7.5)

Pseudo-noble-gas core the noble-gas core together with $(n-1)d^{10}$ electrons. (8.2)

Qualitative analysis the determination of the identity of substances present in a mixture. (p. 584)

Quantitative analysis the determination of the amount of a substance or species present in a material. (p. 126)

Quantum (wave) mechanics the branch of physics that mathematically describes the wave properties of submicroscopic particles. (7.4)

Rad the dosage of radiation that deposits 1×10^{-2} J of energy per kilogram of tissue. (21.3)

Radioactive decay the process in which a nucleus spontaneously disintegrates, giving off radiation. (p. 660)

Radioactive decay constant (k) rate constant for radioactive decay. (21.4)

Radioactive decay series a sequence in which one radioactive nucleus decays to a second, which then decays to a third, and so on. (21.1)

Radioactive tracer a very small amount of radioactive isotope added to a chemical, biological, or physical system to facilitate study of the system. (21.5)

Radioactivity spontaneous radiation from unstable elements. (21.1)

Raoult's law the partial pressure of solvent, P_A, over a solution equals the vapor pressure of the pure solvent, P_A°, times the mole fraction of the solvent, X_A, in solution: $P_A = P_A^\circ X_A$. (12.5)

Rate constant a proportionality constant in the relationship between rate and concentrations. (14.3)

Rate-determining step the slowest step in a reaction mechanism. (14.8)

Rate law an equation that relates the rate of a reaction to the concentrations of reactants (and catalyst) raised to various powers. (14.3)

Reaction intermediate a species produced during a reaction that does not appear in the net equation because it reacts in a subsequent step in the mechanism. (14.7)

Reaction mechanism the set of elementary reactions whose overall effect is given by the net chemical equation. (14.7)

Reaction order the exponent of the concentration of a given reactant species in the rate law, as determined experimentally. (14.3) See also *Overall reaction order*.

Reaction quotient (Q_c) an expression that has the same form as the equilibrium-constant expression but whose concentration values are not necessarily those at equilibrium. (15.5)

Reaction rate the increase in molar concentration of product of a reaction per unit time or the decrease in molar concentration of reactant per unit time. (14.1)

Reducing agent a species that reduces another species; it is itself oxidized. (4.5)

Reduction the part of an oxidation–reduction reaction in which there is a gain of electrons by a species (or a decrease of oxidation number of an atom). (4.5)

Reduction potential see *Standard electrode potential*.

Rem a unit of radiation dosage used to relate various kinds of radiation in terms of biological destruction. It equals the rad times a factor for the type of radiation, called the relative biological effectiveness (RBE): rems = rads $\times$ RBE. (21.3)

Representative element see *Main-group element*.

Resonance description a representation in which you describe the electron structure of a molecule having delocalized bonding by writing all possible electron-dot formulas. (9.7)

Roasting the process of heating a mineral in air to obtain the oxide. (13.2)

Root-mean-square (rms) molecular speed a type of average molecular speed, equal to the speed of a molecule having the average molecular kinetic energy. It equals $\sqrt{3RT/M_m}$, where M_m is the molar mass. (5.7)

Rounding the procedure of dropping nonsignificant digits in a calculation result and adjusting the last digit reported. (1.5)

Salt an ionic compound that is a product of a neutralization reaction. (4.4)

Salt bridge a tube of an electrolyte in a gel that is connected to the two half-cells of a voltaic cell; it allows the flow of ions but prevents the mixing of the different solutions that would allow direct reaction of the cell reactants. (20.2)

Saturated hydrocarbon a hydrocarbon that has only single bonds between carbon atoms; all carbon atoms are bonded to the maximum number of hydrogen atoms (that is, the hydrocarbon is *saturated* with hydrogen). A saturated hydrocarbon molecule can be cyclic or acyclic. (24.1)

Saturated solution a solution that is in equilibrium with respect to a given dissolved substance. (12.2)

Scientific notation the representation of a number in the form $A \times 10^n$, where A is a number with a single nonzero digit to the left of the decimal point and n is an integer, or whole number. (1.5)

Second law of thermodynamics the total entropy of a system and its surroundings always increases for a spontaneous process. Also, for a spontaneous process at a given temperature, the change in entropy of the system is greater than the heat divided by the absolute temperature. (19.2)

Self-ionization (autoionization) a reaction in which two like molecules react to give ions. (16.6)

Semiconducting element an element that exhibits very low electrical conductivity at room temperature when pure, but whose electrical conductivity rises with temperature or with the addition of certain other elements. The process of adding small quantities of these other elements to a semiconducting element to increase its conductivity is called *doping*. (13.5)

Sigma (σ) bond a bond that has a cylindrical shape about the bond axis. (10.4)

Significant figures those digits in a measured number (or result of a calculation with measured numbers) that include all certain digits plus a final one having some uncertainty. (1.5)

Silica a covalent network solid of SiO_2 in which each silicon atom is covalently bonded in tetrahedral directions to four oxygen atoms; each oxygen atom is in turn bonded to another silicon atom. (13.6)

Silicate a compound of silicon and oxygen (with one or more metals) that may be formally regarded as a derivative of silicic acid, H_4SiO_4 or $Si(OH)_4$. (13.6)

Simple cubic unit cell a cubic unit cell in which lattice points are situated only at the corners of the unit cell. (11.6)

Simplest formula see *Empirical formula*.

Single bond a covalent bond in which a single pair of electrons is shared by two atoms. (9.4)

Single-replacement reaction see *Displacement reaction*.

Solubility the amount of a substance that dissolves in a given quantity of solvent (such as water) at a given temperature to give a saturated solution. (12.2)

Solubility-product constant (K_{sp}) the equilibrium constant for the solubility equilibrium of a slightly soluble (or nearly insoluble) ionic compound. (18.1)

Solute in the case of a solution of a gas or solid dissolved in a liquid, the gas or solid; in other cases, the component in smaller amount. (12.1)

Solvent in a solution of a gas or solid dissolved in a liquid, the liquid; in other cases, the component in greater amount. (12.1)

Specific heat capacity (specific heat) the quantity of heat required to raise the temperature of one gram of a substance by one degree Celsius (or one kelvin) at constant pressure. (6.6)

Spectator ion an ion in an ionic equation that does not take part in the reaction. (4.2)

Spectrochemical series an arrangement of ligands according to the relative magnitudes of the crystal field splittings they induce in the *d* orbitals of a metal ion. (23.7)

Spin quantum number (m_s) the quantum number that refers to the two possible orientations of the spin axis of an electron; possible values are $+\frac{1}{2}$ and $-\frac{1}{2}$. (7.5)

Spontaneous fission the spontaneous decay of an unstable nucleus in which a heavy nucleus of mass number greater than 89 splits into lighter nuclei and energy is released. (21.1)

Spontaneous process a physical or chemical change that occurs by itself. (p. 594)

Square planar geometry the geometry of a molecule in which a central atom is surrounded by four other atoms arranged in a square and in a plane containing the central atom. (10.1)

Square pyramidal geometry the geometry of a molecule in which a central atom is at the apex of a pyramid and four other atoms form the square base of the pyramid. (10.1)

Standard electrode potential ($E°$) the electrode potential when the concentrations of solutes are 1 M, the gas pressures are 1 atm, and the temperature has a specified value—usually 25°C. (20.5)

Standard emf ($E°_{cell}$) the emf of a voltaic cell operating under standard-state conditions (solute concentrations are 1 M, gas pressures are 1 atm, and the temperature has a specified value—usually 25°C). (20.5)

Standard enthalpy of formation (standard heat of formation), $\Delta H_f°$ the enthalpy change for the formation of one mole of a substance in its standard state from its elements in their reference forms and in their standard states. (6.8)

Standard entropy ($S°$) the entropy value for the standard state of a species. (19.3)

Standard free energy of formation ($\Delta G_f°$) the free-energy change that occurs when one mole of substance is formed from its elements in their stablest states at 1 atm and at a specified temperature (usually 25°C). (19.4)

Standard state the standard thermodynamic conditions (1 atm and usually 25°C) chosen for substances when listing or comparing thermochemical data. (6.8)

Standard temperature and pressure (STP) the reference conditions for gases chosen by convention to be 0°C and 1 atm. (5.2)

State function a property of a system that depends only on its present state, which is determined by variables such as temperature and pressure, and is independent of any previous history of the system. (6.3 and 19.1)

States of matter the three forms that matter can assume—solid, liquid, and gas. (1.4)

Stereoisomers isomers in which the atoms are bonded to each other in the same order but that differ in the precise arrangement of the atoms in space. (23.5)

Stoichiometry the calculation of the quantities of reactants and products involved in a chemical reaction. (p. 83)

Strong acid an acid that ionizes completely in water; it is a strong electrolyte. (4.4 and 16.1)

Strong base a base that is present in aqueous solution entirely as ions, one of which is OH^-; it is a strong electrolyte. (4.4 and 16.1)

Strong electrolyte an electrolyte that exists in solution almost entirely as ions. (4.1)

Structural isomers isomers that differ in how the atoms are joined together. (24.2)

Substitution reaction a reaction in which a part of the reacting molecule is substituted for an H atom on a hydrocarbon or a hydrocarbon group. (24.2)

Superoxide a binary compound with oxygen in the $-\frac{1}{2}$ oxidation state; it contains the superoxide ion, O_2^-. (22.8)

Supersaturated solution a solution that contains more dissolved substance than does a saturated solution; the solution is not in equilibrium with the solid substance. (12.2)

Surroundings everything in the vicinity of a thermodynamic system. (6.2)

System (thermodynamic) the substance or mixture of substances under study in which a change occurs. (6.2)

Termolecular reaction an elementary reaction that involves three reactant molecules. (14.7)

Tetrahedral geometry the geometry of a molecule in which four atoms bonded to a central atom occupy the vertices of a tetrahedron with the central atom at the center of this tetrahedron. (10.1)

Theoretical yield the maximum amount of product that can be obtained by a reaction from given amounts of reactants. (3.8)

Theory a tested explanation of basic natural phenomena. (1.2)

Thermal equilibrium a state in which heat does not flow between a system and its surroundings because they are both at the same temperature. (6.2)

Thermochemical equation the chemical equation for a reaction (including phase labels) in which the equation is given a molar interpretation, and the enthalpy of reaction for these molar amounts is written directly after the equation. (6.4)

Thermochemistry the study of the quantity of heat absorbed or evolved by chemical reactions. (p. 175)

Thermodynamic equilibrium constant (K) the equilibrium constant in which the concentrations of gases are expressed in partial pressures in atmospheres, whereas the concentrations of solutes in liquid solutions are expressed in molarities. (19.6)

Thermodynamics the study of the relationship between heat and other forms of energy involved in a chemical or physical process. (p. 175 and p. 591)

Third law of thermodynamics a substance that is perfectly crystalline at 0 K has an entropy of zero. (19.3)

Titration a procedure for determining the amount of substance A by adding a carefully measured volume of a solution with known concentration of B until the reaction of A and B is just complete. (4.10)

Transition element the B columns of elements in the periodic table. (2.5); the d-block transition elements in which a d subshell is being filled. (8.2); those metallic elements that have an incompletely filled d subshell or easily give rise to common ions that have incompletely filled d subshells. (p. 724)

Transition-state theory a theory that explains the reaction resulting from the collision of two molecules in terms of an activated complex (transition state). (14.5)

Transmutation the change of one element to another by bombarding the nucleus of the element with nuclear particles or nuclei. (21.2)

Transuranium elements those elements with atomic numbers greater than that of uranium ($Z = 92$), the naturally occurring element of greatest Z. (21.2)

Trigonal bipyramidal geometry the geometry of a molecule in which five atoms bonded to a central atom occupy the vertices of a trigonal bipyramid (formed by placing two trigonal pyramids base to base) with the central atom at the center of this trigonal bipyramid. (10.1)

Trigonal planar geometry the geometry of a molecule in which a central atom is surrounded by three other atoms arranged in a triangle and in a plane containing the central atom. (10.1)

Trigonal pyramidal geometry the geometry of a molecule in which a central atom is at the apex of a pyramid and three other atoms form the triangular base of the pyramid. (10.1)

Triple bond a covalent bond in which three pairs of electrons are shared by two atoms. (9.4)

Triple point the point on a phase diagram representing the temperature and pressure at which three phases of a substance coexist in equilibrium. (11.3)

T-shaped geometry the geometry of a molecule in which three atoms are bonded to a central atom to form a T. (10.1)

Tyndall effect the scattering of light by colloidal-size particles. (12.9)

Uncertainty principle a relation stating that the product of the uncertainty in position and the uncertainty in momentum (mass times speed) of a particle can be no smaller than Planck's constant divided by 4π. (7.4)

Unimolecular reaction an elementary reaction that involves one reactant molecule. (14.7)

Unit cell the smallest boxlike unit (each box having faces that are parallelograms) from which you can imagine constructing a crystal by stacking the units in three dimensions. (11.6)

Unsaturated hydrocarbon a hydrocarbon that has at least one double or triple bond between carbon atoms; all carbon atoms are not bonded to the maximum number of hydrogen atoms (that is, the hydrocarbon is *unsaturated* with hydrogen). (24.1)

Unsaturated solution a solution that is not in equilibrium with respect to a given dissolved substance and in which more of the substance can dissolve. (12.2)

Valence bond theory an approximate theory to explain the electron pair or covalent bond in terms of quantum mechanics. (10.3)

Valence electron an electron in an atom outside the noble-gas or pseudo-noble-gas core. (8.2)

Valence-shell electron-pair repulsion (VSEPR) model predicts the shapes of molecules and ions by assuming that the valence-shell electron pairs are arranged about each atom so that electron pairs are kept as far away from one another as possible, thus minimizing electron-pair repulsions. (10.1)

van der Waals forces a general term for those intermolecular forces that include dipole–dipole and London forces. (11.4)

van der Waals equation an equation similar to the ideal gas law, but includes two constants, *a* and *b*, to account for deviations from ideal behavior. (5.8)

Vapor pressure the partial pressure of the vapor over the liquid, measured at equilibrium at a given temperature. (11.2)

Vapor-pressure lowering a colligative property equal to the vapor pressure of the pure solvent minus the vapor pressure of the solution. (12.5)

Volt (V) the SI unit of potential difference. (20.4)

Voltaic cell (galvanic cell) an electrochemical cell in which a spontaneous reaction generates an electric current. (p. 624)

Wavelength (λ) the distance between any two adjacent identical points of a wave. (7.1)

Weak acid an acid that is only partly ionized in water; it is a weak electrolyte. (4.4)

Weak base a base that is only partly ionized in water; it is a weak electrolyte. (4.4)

Weak electrolyte an electrolyte that dissolves in water to give a relatively small percentage of ions. (4.1)

Work the energy exchange that results when a force *F* moves an object through a distance *d*; it equals $F \times d$. (19.1)

Zinc–carbon (Leclanché) dry cell a voltaic cell that has a zinc can as the anode; a graphite rod in the center, surrounded by a paste of manganese dioxide, ammonium and zinc chlorides, and carbon black, is the cathode. (20.8)

Index

Photo Credits

All photographs are identified by figure number, unless otherwise noted.

Photographs by Jim Scherer: Page vii (top), xiv (bottom); Figures 1.10 (two photos), 1.15 (two photos), 1.17 (three photos), 1.23, 1.24, 2.16, 2.19, 3.2, 3.3, 4.1, 4.8 (two photos), 4.11, 4.12, 4.14, 4.19 (two photos), 5.1, 5.6 (two photos), 5.11, 6.14, 8.13, 9.6, 9.22 (three photos), 11.3 (two photos), 11.22, 12.3 (three photos), 12.23 (two photos), 13.2 (two photos), 13.5, 13.15, 14.1, 14.2, 15.7, 16.1, 16.9, 18.7 (three photos), 19.1, 20.1, 22.2, 22.4 (two photos), 22.5, 22.6, 22.8, 22.9, 22.15, 22.16, 22.18, 22.20, 22.22, 22.23, 22.35, 23.5, 23.20, 24.9.

Credits for other photos are listed below, with figure numbers (or page numbers) listed in boldface. Photos not listed specifically are by American Color and are owned by Houghton Mifflin Company.

Title page: E.R. Degginger; **vii** (bottom) Photo courtesy of Digital Instruments; **viii** (top) © 2001 PhotoDisc, Inc. (bottom) © Corbis; **ix** (top) © Corbius, **xi** (top) NASA; **xii** (top) Courtesy O.D. Lavrentovich, Liquid Crystal Institute, Kent State University; **xiii** (top) Argoone National Laboratory; **xvii** (top) Siemens Medical Systems, Inc.;

Chapter 1 opener © Yoav Levy/Phototake; **1.1** © Douglas Elbinger; **1.5** Courtesy of Corning, Inc.; **1.6** Courtesy of International Business Machines Corporation; **1.19** Courtesy of PerkinElmer Instruments, Shelton, Connecticut; **Example 1.8** © Spencer Grant/PhotoEdit; **Chapter 2 opener** IBM Almaden Research Center; **2.1** Edgar Fahs Smith Collection, University of Pennsylvania Library; **2.2** The Cavendish Laboratory; **2.14** Teflon® is a registered trademark of DuPont; **2.15** © E. R. Degginger; **4.5** © Corbis; **5.9** © Van Bucher/Photo Researchers; **5.1** © Corbis; **5.12** Courtesy of DaimlerChrysler Corporation; **Chapter 6 opener** © Joseph P. Sinnot/Fundamental Photographs, NYC; **6.2** (left) University of Missouri-Rolla Solar Car Team; **6.2** (right) © Michael Siluk/The Image Works; **6.3** Tennessee Valley Authority; **6.14** NASA; **7.1** © Robert Holmgren/Peter Arnold, Inc.; **7.5** Photo Courtesy of MDS Nordion; **7.6** AP/Wide World Photos; **7.14** © Biophoto Associates/Photo Researchers; **7.17** IBM Almaden Research Center; **7.19** IBM Almaden Research Center; **Chapter 8 opener** © Richard Megna/Fundamental Photographs, NYC; **8.1** Culver Pictures; **Chapter 9 opener** © Clyde H. Smith/Peter Arnold, Inc.; **9.1** © E. R. Degginger; **9.4** Reproduced with permission from *Chemical Engineering News,* March 30, 1998, 76(13), p. 33. Copyright 1998 American Chemical Society; **9.13** © Craig Hammell/Corbis Stock Market; **Chapter 10 opener** © Alfred Pasieka/Science Photo Library/Photo Researchers; **10.25** © Yoav Levy/Phototake, NYC; **10.32** NASA; **Chapter 11 opener** Photo courtesy of Oleg D. Lavrentovich, Liquid Crystal Institute, Kent State University; **11.2** Used with permission from the *Journal of Chemical Education,* Vol. 74, No. 7, 1997, cover; copyright © 1997, Division of Chemical Education, Inc.; **11.36** UCAR Carbon Company; **11.37** From G. D. Preston, 1926, Proceedings of the Royal Society, A, Volume 172, page 126, figure 5A. Reprinted with permission.; **11.38** Courtesy Bruker Novius Crystallographic Systems, Bruker AXS, Inc.; **Chapter 12 opener** © Richard Megna/Fundamental Photographs, NYC; **12.9** (left) © Bill Longcore/Photo Researchers; **12.9** (right) © Bill Longcore/Photo Researchers; **12.13** © Bill Stanton/Rainbow; **12.20** (left) © David Phillips/Photo Researchers; **12.20** (center)© David Phillips/Photo Researchers; **12.20** (right) © David Phillips/Photo Researchers; **Chapter 13 opener** IBM Research/Peter Arnold, Inc.; **13.1** © Raleigh America Inc./Michael Hammond, Photographer; **13.2** Alcoa Photo; **13.3** National Park Service; **13.4** © M. Claye/Jacana/Photo Researchers; **13.7** © Hank Morgan/Photo Researchers; **13.14** Argonne National Laboratory; **13.18** Dr. Richard Smalley/Rice University; **13.21** © Charles Falco/Photo Researchers; **13.22** © Dick Luria/Photo Researchers; **13.27** Stuart Gray; **13.28** World Kitchens, Inc., Elmira, NY; **13.29** Wright Medical Technology; **13.3** Textron Systems; **14.14** Engelhard Corporation; **15.1** Eastman Kodak Company; **Chapter 17 opener** © Richard Megna/Fundamental